Environmental Science
Toward a Sustainable Future

TWELFTH EDITION

Richard T. Wright
Dorothy F. Boorse
Gordon College

Boston Columbus Indianapolis New York San Francisco Upper Saddle River
Amsterdam Cape Town Dubai London Madrid Milan Munich Paris Montréal Toronto
Delhi Mexico City São Paulo Sydney Hong Kong Seoul Singapore Taipei Tokyo

VP/Editor in Chief: Beth Wilbur
Executive Director of Development: Deborah Gale
Acquisitions Editor: Alison Rodal
Project Editor/Development Editor: Leata Holloway
Executive Marketing Manager: Lauren Harp
Editorial Media Producer: Lee Ann Doctor
Managing Editor, Environmental Science: Gina M. Cheselka
Full Service/Composition: Element LLC
Full Service Project Manager: Katrina Ostler
Illustrations: Precision Graphics

Photo Manager: Maya Melenchuk
Photo Researchers: Jennifer Nonenmacher and Peter Jardim, PreMediaGlobal
Text Permissions Manager: Joseph Croscup
Text Permissions Researcher: Stephen Barker, Creative Compliance
Operations Specialist: Michael Penne
Design Manager: Marilyn Perry
Interior Design: Jerilyn Bockorick
Cover Design: Yvo Riezebos Design
Cover Image Credit: Jon Hicks/Corbis

Credits and acknowledgments for material borrowed from other sources and reproduced, with permission, in this textbook appear on the appropriate page or in the end matter.

Library of Congress Cataloging-in-Publication Data available upon request from Publisher.

1 2 3 4 5 6 7 8 9 10—CRK—16 15 14 13 12

www.pearsonhighered.com

ISBN-10: 0-321-81153-4; ISBN-13: 978-0-321-81153-0

ABOUT THE AUTHORS

Richard T. Wright is Professor Emeritus of Biology at Gordon College in Massachusetts, where he taught environmental science for 28 years. He earned a B.A. from Rutgers University and an M.A. and a Ph.D. in biology from Harvard University. For many years, Wright received grant support from the National Science Foundation for his work in marine microbiology, and in 1981, he was a founding faculty member of Au Sable Institute of Environmental Studies in Michigan, where he also served as Academic Chairman for 11 years. He is a Fellow of the American Association for the Advancement of Science, Au Sable Institute, and the American Scientific Affiliation. In 1996, Wright was appointed a Fulbright Scholar to Daystar University in Kenya, where he taught for two months. He is a member of many environmental organizations, including the Nature Conservancy, Habitat for Humanity, the Union of Concerned Scientists, and the Audubon Society, and is a supporting member of the Trustees of Reservations. He volunteers his services at the Parker River National Wildlife Refuge in Newbury, Massachusetts, and is an elder in First Presbyterian Church of the North Shore. Wright and his wife, Ann, live in Byfield, Massachusetts, and they drive a Toyota Camry hybrid vehicle as a means of reducing their environmental impact. Wright spends his spare time birding, fishing, hiking, and enjoying his three children and seven grandchildren.

Dorothy F. Boorse is a professor of biology at Gordon College in Wenham, Massachusetts. Her research interest is in drying wetlands, such as vernal pools and prairie potholes, and in salt marshes. Her research with undergraduates has included wetland and invasive species projects. She earned a B.S. in biology from Gordon College, an M.S. in entomology from Cornell University, and a Ph.D. in oceanography and limnology from the University of Wisconsin–Madison. Boorse teaches, writes, and speaks about biology, the environment, ecological justice, and care of creation. She was recently an author on a report on poverty and climate change. In 2005, Boorse provided expert testimony on wildlife corridors and environmental ethics for a congressional House subcommittee hearing. Boorse is a member of a number of ecological and environmental societies, including the Ecological Society of America, the Society of Wetland Scientists, the Nature Conservancy, the Audubon Society, the New England Wildflower Society, and the Trustees of Reservations (the oldest land conservancy group in the United States). She and her family live in Beverly, Massachusetts. They belong to Appleton Farms, a CSA (community-supported agriculture) farm. At home, Boorse has a native plant garden and has planted two disease-resistant elm trees.

ABOUT OUR SUSTAINABILITY INITIATIVES

Pearson recognizes the environmental challenges facing this planet, as well as acknowledges our responsibility in making a difference. This book is carefully crafted to minimize environmental impact. The paper is certified by the Forest Stewardship Council™ (FSC®). The binding, cover, and paper come from facilities that minimize waste, energy consumption, and the use of harmful chemicals. Pearson closes the loop by recycling every out-of-date text returned to our warehouse.

Along with developing and exploring digital solutions to our market's needs, Pearson has a strong commitment to achieving carbon-neutrality. In 2009, Pearson became the first carbon- and climate-neutral publishing company. Since then, Pearson remains strongly committed to measuring, reducing, and offsetting our carbon footprint.

The future holds great promise for reducing our impact on Earth's environment, and Pearson is proud to be leading the way. We strive to publish the best books with the most up-to-date and accurate content, and to do so in ways that minimize our impact on Earth. To learn more about our initiatives, please visit **www.pearson.com/responsibility**.

MIX
Paper from
responsible sources
FSC® C100141
www.fsc.org

PEARSON

CONTENTS

PART FOUR
Harnessing Energy for Human Societies   342

PART FIVE
Pollution and Prevention   414

ESSAY LIST

PREFACE

We are now well into the second decade of the 21st century, and it is promising to be a crucial decade. Globally, major changes are taking place in the atmosphere and climate, the human population and its well-being, and the Earth's natural resources. We are still recovering from a major global economic recession, pointing to the reach of globalization; societies are reeling from disasters such as the Japanese tsunami, the Haitian earthquake, and the huge oil leak in the Gulf of Mexico; there is no letup in the accumulating evidence of climate change, brought on by increased fossil fuel burning; terrorism and conflict continue to grip the Middle East; and crises in food and agricultural production, brought on by heat waves and extended drought in the United States, are leading to greater numbers of hungry and impoverished people.

In contrast to these and other disturbing trends, there are some changes that point to a brighter future. Renewable energy is ramping up swiftly in its share of the world's energy portfolio; many of the UN Millennium Development Goals (MDGs) are on track to be achieved by their target date of 2015; even though international accord on climate change is slow in coming, many countries are achieving major reductions in greenhouse gas emissions; death by tobacco use is being addressed in a global campaign; AIDS, tuberculosis, and malaria are on the defensive as public-health agencies expand treatment options and research; and population growth in the developing countries is continuing to decline.

The most profound change that must happen, and soon, is the transition to a sustainable civilization—one in which a stable human population recognizes the finite limits of Earth's systems to produce resources and absorb wastes, and acts accordingly. This is hard to picture at present, but it is the only future that makes any sense. If we fail to achieve it by our deliberate actions, the natural world will impose it on us in highly undesirable ways.

New to This Edition

To help implement our core values, we have made changes to the twelfth edition. The major changes are listed here, followed by a list of updates for each part and chapter. All of these changes were made in the hope of creating a book with a solid, rigorous framework that is also easily comprehensible.

- In this edition, Learning Objectives have been added to the beginning of each chapter. These objectives describe the knowledge or skills that students should have upon completing each chapter.

- The art program has been substantially updated. Each chapter features new photos and updated graphics.

- Content has been thoroughly updated: More than 45% of the research cited is from 2010 or later, and material has been carefully edited for currency.

- New Video Field Trips give students the chance to learn about wind power, invasive species, and campus sustainability in an all-new format.

- New Core Content and Data Analysis Coaching activities in MasteringEnvironmentalScience activities accompany the twelfth edition of the book.

Content Updates to the Twelfth Edition of *Environmental Science*

Part One—Framework for a Sustainable Future

Part One has a new opener that focuses on the vision of sustainability and the challenges facing us that are inconsistent with that vision. Chapter 1 (Science and the Environment) has a new opening story, the inspiring work of Rachel Carson and her seminal book, *Silent Spring*. Then, as we look at the state of the planet, we investigate the "environmentalist's paradox," where human well-being seems to be improving while the environment continues to decline. We then introduce the American environmental movement timeline, a new graphic that shows the major impacts of events, people, and policies. We continue with our three unifying themes (sustainability, sound science, and stewardship) and introduce hypothesis formation with a new essay, "When Oysters Became Canaries." The final section in Chapter 1 explores social and environmental changes in the context of globalization. Chapter 2 (Economics, Politics, and Public Policy) explores the concepts of a green versus a brown economy and incorporates results from a new World Bank study that updates the "wealth of nations" concept the bank introduced earlier. A new essay ("California's Green Economy") illustrates beneficial environmental policy impacts, and other benefits of environmental policy are updated. Because of its time-sensitive content, the section on politics and the environment has been heavily rewritten. If we only had a crystal ball. . .

Part Two—Ecology: The Science of Organisms and Their Environment

Chapters 3–5 were reorganized in the last edition, and that organization has been maintained in this one, with topics flowing from basic to more complex, small to large, and species to ecosystems and humans. Chapter 3 (Basic Needs of Living Things) has new material on the sulfur cycle—another geochemical cycle important to living things—and a new Sustainability essay ("Planetary Boundaries"). Chapter 4 (Populations and Communities) is somewhat reorganized and has additional material on Darwin's finches in a new Sound Science box ("Studying Finches: The Life of a Scientist"). An example that allows students to calculate population growth has been moved from the main text into a Sound Science essay ("The Story Behind the Numbers: It's a Wormy Project!") to make the text more readable. Chapter 5 (Ecosystems: Energy, Patterns and Disturbance) begins with a new opener—the effects of an Icelandic volcano—and includes a new Sustainability essay ("Ecological Restoration: Aldo Leopold and the Shack on the Prairie"). More-modern ecological studies are described, including the LTER and NEON programs. Chapter 6 (Wild Species and Biodiversity) begins with a new opener on the Ganges River dolphin. Two new essays describe the return of the Lake Erie Water snake (Stewardship, "Lake Erie's Island Snake Lady") and the use of DNA in forensic wildlife science (Sound Science, "Using DNA to Catch Wildlife Criminals"). The section on protecting species was reorganized to cover topics from individual to international scales, and the section on the consequences of losing biodiversity was reorganized to tie into the values of biodiversity. Citizen science is highlighted with several examples. Chapter 7 (The Value, Use, and Restoration of Ecosystems) has a new opener on the Caribbean reef. The chapter was reorganized, with restoration and conservation moved to the end. The chapter has an even greater focus on ecosystem goods and services and two new essays (Sustainability, "How Much for That Irrigation Water?", and Sound Science, "Restoration Science: Learning How to Restore").

Part Three—The Human Population and Essential Resources

Part 3 begins with a new section opener about world population hitting 7 billion in October 2011. Chapter 8 (The Human Population) includes a new essay on calculating an ecological footprint (Stewardship, "Lessening Your Ecological Footprint") and updated population profiles. The chapter returns throughout to three countries—Burkina Faso, Indonesia, and Sweden—which are introduced in the chapter opener. The section on the IPAT equation has been updated to include some newer concepts on calculating ecological impacts, and the concept of the GINI index of inequality is added to the chapter. Chapter 9 (Population and Development) has a new chapter opener on declining population growth rates in Thailand. The Stewardship essay on China ("Protecting People's Options") is expanded to include issues prior to the one-child policy and more information on human rights abuses. A new figure illustrates the concept of the speed of demographic change by contrasting different countries. The former chapter opener on Kerala, India, was updated and made into a Sustainability essay ("Looking at Change in India"). Chapter 10 (Water: Hydrologic Cycle and Human Use) has an new chapter opener on drought in China, a new essay on thermohaline circulation (Sound Science,

"What's New in the Water Cycle?"), and another on the connection between water, energy, and food (Stewardship, "The Energy/Water/Food Trade: Waste One, Waste Them All"). There is more material on impervious surfaces, extreme-distance transport of water, and the relationship of climate change and the water cycle. Chapter 11 (Soil: Foundation for Land Ecosystems) has a new chapter opener on the Dust Bowl as well as an essay on two-stage ditches in the American Midwest (Sustainability, "New Ditches Save Soil"). The effects of mining on soils, acid rain, soil and carbon storage, and nanoparticles are also added. Chapter 12 (The Production and Distribution of Food) was altered significantly, with an entirely new section on sustainable agriculture and cutting-edge agricultural techniques as well as a new chapter opener on urban agriculture. The chapter is organized around three ideas—production, environmental sustainability, and effective distribution—and with a new summary table on actions that increase or decrease food availability. The food pyramid has been updated with the new USDA food "plate." Golden rice, the subject of the former chapter opener, was moved to a new Stewardship essay ("Golden Rice"). A new Sustainabilty essay ("Preventing Food Crises") connects photographer Dorothea Lange and the Great Depression to modern food crises. Chapter 13 (Pests and Pest Control) has a new chapter opener on bedbugs, a new Sound Science essay on biofouling organisms ("Marine Fouling Organisms: Keeping One Step Ahead of the Barnacles"), and more information on stored product pests and on the decline of bees. The latter part of the chapter is reorganized to include policies regulating both pesticides and pests themselves, including more on the government agency APHIS and its role in monitoring pests. It should be mentioned that for all of these chapters, photos, graphics, and statistical data have been brought up to date.

Part Four—Harnessing Energy for Human Societies

Chapter 14 (Energy from Fossil Fuels) has a new part opener with bad news and good news and a new chapter opener on the Deepwater Horizon disaster in the Gulf of Mexico. A new figure plots the different kinds of power plants constructed in recent years, and a new energy flow scheme from Livermore Labs is employed. The discussion on oil sands is expanded, and the Keystone pipeline proposal is presented and illustrated. Natural gas and coal are given separate sections in this edition. Hydraulic fracturing (fracking) is explored, and mountaintop removal coal mining is presented under the heading "Mining Coal" (Section 14.4). Policy from the Recovery Act is added, and there is a new Sound Science essay ("Energy Returned on Energy Invested"). Chapter 15 (Nuclear Power) opens with the earthquake and tsunami that rocked northern Japan in March 2011, leading to the nuclear disaster at the Fukushima Daiichi power plant. This event is used as an illustration of the consequences of a loss-of-coolant accident. Yucca Mountain is revisited, and the President's Blue Ribbon Commission report of 2011 is presented. The relicensing situation for U.S. nuclear power plants is reviewed, and global nuclear power is summarized. Chapter 16 (Renewable

Energy) opens with the latest on Cape Wind and a look at portable solar travel bags. Then follows a new section ("Strategic Issues") that examines the issues surrounding calls for great changes in renewable energy, calls for achieving 80% of electricity by renewable sources within a few decades. The chapter highlights the utilities' trend toward large photovoltaic installations and reviews work by MIT scientists that for the first time connects solar cell power to water oxidation under standard conditions, something that could pave the way to major energy gains.

Part Five—Pollution and Prevention

Chapter 17 (Environmental Hazards and Human Health) opens with a reference to the movie *Contagion* and moves from there to some controversial research on the avian flu virus, a disease that has the potential for great harm. The obesity epidemic in the United States is highlighted, with a new graphic on body mass index. ToxCast and the new IRIS system are discussed, and new information on the causation of asthma is added. A new Sound Science essay is "The Grisly Seven," where neglected tropical diseases that affect millions are discussed. Chapter 18 (Global Climate Change) starts with a new opener on the threats of rising sea level to island populations. A new section on "The Atmosphere's Control Knob" gives a perspective on the key importance of CO_2. Another new section is "Skeptics, Deniers, and Ethics," where opposition to climate change is put into perspective. The major international climate conferences are summarized, and the gap between promised emission reduction pledges and what needs to be done is emphasized and illustrated. In Chapter 19 (Atmospheric Pollution), new material on ammonia points to the great importance of nitric acid in acid rain as sulfur dioxides are being brought under greater control. Mercury emissions are highlighted, and the new Mercury and Air Toxics Standards are presented as a highly important development in curbing pollution from older coal-burning power plants. Nitrous oxide is presented as the newly recognized key ozone-depleting substance of the 21st century. Chapter 20 (Water Pollution and Its Prevention) has new illustrations of sludge digesters and septic systems and documents changes in the Chesapeake Bay restoration efforts, ending with comments on partisan efforts in Congress to block EPA enforcement of water quality regulations. Chapter 21 (Municipal Solid Waste: Disposal and Recovery) has a new emphasis on composting, and single-stream recycling is described as the new trend in recycling. EPA's waste management hierarchy is added as an illustration. The opening story has a new focus: installing solar arrays on closed landfills. The Internet is referenced as a great way to accomplish resale and reuse, and options for getting rid of e-waste are presented. Chapter 22 (Hazardous Chemicals: Pollution and Prevention) has a new opener on the controversial issue of endocrine disruptors and bisphenol A. A new section ("Evaluating New Chemicals") illustrates the different approaches of TSCA and REACH in the United States and Europe. The many abandoned uranium mines on Navajo reservations provide a new example of environmental (in)justice.

Part Six—Stewardship for a Sustainable Future

Chapter 23 (Sustainable Communities and Lifestyles) begins with a new story on renewal in Atlanta, Georgia, exploring the Atlanta BeltLine. A new paragraph examines the Rust Belt as a way of understanding urban blight. A new paragraph on "Fixing Suburbia" presents two recent books on how to repair suburban sprawl. Finally, the essay on Tangier Island is updated with reference to a meeting between the islanders and a farming community in Pennsylvania.

Coauthors

Like the global scene, the field of environmental science evolves and continues to change, and so has this text. The most significant change has been the work of a coauthor, Dr. Dorothy Boorse. Boorse began as coauthor in the eleventh edition, and in the twelfth, she has taken responsibility for half of the book (Chapters 1 and 3–13). For Wright, this has been an especially gratifying development, as Boorse was his student at Gordon College and is now back at the college as Professor of Biology. It has been a delight to work together on this edition.

Core Values

Environmental science stands at the interface between humans and Earth and explores the interactions and relations between them. This relationship will need to be considered in virtually all future decision making. This text considers a full spectrum of views and information in an effort to establish a solid base of understanding and a sustainable formula for the future. What you have in your hands is a readable guide and up-to-date source of information that will help you to explore the issues in more depth. It will also help you to connect them to a framework of ideas and values that will equip you to become part of the solution to many of the environmental problems confronting us.

In this new edition, we hope to continue to reflect accurately the field of environmental science; in so doing, we have constantly attempted to accomplish each of the following objectives:

- To write in a style that makes learning about environmental science both interesting to read and easy to understand, without overwhelming students with details;

- To present well-established scientific principles and concepts that form the knowledge base for an understanding of our interactions with the natural environment;

- To organize the text in a way that promotes sequential learning yet allows individual chapters to stand on their own;

- To address all of the major environmental issues that confront our society and help to define the subject matter of environmental science;

- To present the latest information available by making full use of resources such as the Internet, books, and journals; every possible statistic has been brought up to date;

- To assess options or progress made in solving environmental problems; and

- To support the text with excellent supplements for instructors and students that strongly enhance the teaching and learning processes.

Because we believe that learning how to live in the environment is one of the most important subjects in every student's educational experience, we have made every effort to put in their hands a book that will help the study of environmental science come alive.

Reviewers

Eric C. Atkinson, *Northwest College*

William Brown, *State University of New York, Fredonia*

John S. Campbell, *Northwest College*

Kathy Carvalho-Knighton, *University of South Florida*

Christopher G. Coy, *Indiana Wesleyan University*

Christopher F. D'Elia, *Louisiana State University*

Lisa Doner, *Plymouth State University*

Matthew J. Eick, *Virginia Tech*

Carri Gerber, *Ohio State University Agricultural Technical Institute*

Nisse Goldberg, *Jacksonville University*

Sherri Hitz, *Florida Keys Community College*

Bryan Hopkins, *Brigham Young University*

David W. Hoferer, *Eastern University*

Aixin Hou, *Louisiana State University*

Melinda Huff, *Northeastern Oklahoma A&E College*

Ali Ishaque, *University of Maryland, Eastern Shore*

Owen Lawlor, *Northeastern University*

Charles A. McClaugherty, *University of Mount Union*

Blodwyn McIntyre, *University of the Redlands*

Karen E. McReynolds, *Hope International University*

Glynda Mercier, *Austin Community College*

Grace Ju Miller, *Indiana Wesleyan University*

Michael Moore, *Delaware State University*

Natalie Moore, *Lone Star College*

Carrie Morjan, *Aurora University*

Carole Neidich-Ryder, *Nassau Community College*

Virginia Rivers, *Truckee Meadows Community College*

Annelle Soponis, *Reading Area Community College*

Bradley Turner, *McLennan Community College*

Alison Kate Varty, *College of the Siskiyous*

Robert S. Whyte, *California University of Pennsylvania*

Todd Christian Yetter, *University of the Cumberlands*

Reviewers of Previous Editions

Clark Adams, *Texas A & M University*

Anthony Akubue, *St. Cloud State University*

Mary Allen, *Hartwick College*

Walter Arenstein, *San Jose State University*

Abbed Babaei, *Cleveland State University*

Kenneth Banks, *University of North Texas*

Raymond Beiersdorfer, *Youngstown State University*

Lisa Bonneau, *Metropolitan Community College*

Geralyn Caplan, *Owensboro Community and Technical College*

Kelly Cartwright, *College of Lake County*

Jason Cashmore, *College of Lake County*

Anne Ehrlich, *Stanford University*

Cory Etchberger, *Johnson County Community College*

Marcia Gillette, *Indiana University–Kokomo*

Sue Habeck, *Tacoma Community College*

Crista Haney, *Mississippi State University*

Les Kanat, *Johnson State College*

Barry King, *College of Santa Fe*

Cindy Klevickis, *James Madison University*

Steven Kolmes, *University of Portland*

Peter Kyem, *Central Connecticut State University*

Marcie Lehman, *Shippensburg University*

Kurt Leuschner, *College of the Desert*

Gary Li, *Cal State–East Bay Campus*

Kenneth Mantai, *SUNY Fredonia*

Heidi Marcum, *Baylor University*

Allan Matthias, *University of Arizona*

Dan McNally, *Bryant University*

Kiran Misra, *Edinboro University of Pennsylvania*

Robert Andrew Nichols, *City College–Gainesville Campus*

Tim Nuttle, *Indiana University of Pennsylvania*

Bruce Olszewski, *San Jose State University*

John Pleasants, *Iowa State University*

John Reuter, *Portland State University*

Kim Schulte, *Georgia Perimeter College*

Brian Shmaefsky, *Lone Star College–Kingwood*

Shamili Stanford, *College of DuPage*

Keith Summerville, *Drake University*

Todd Tarrant, *Michigan State University*

Dave Wartell, *Harrisburg Area Community College*

John Weishampel, *University of Central Florida*

Arlene Westhoven, *Ferris State University*

Danielle Wirth, *Des Moines Area Community College*

ACKNOWLEDGMENTS

More than 30 years ago, Bernard Nebel published the first edition of this text. He did it because he was frustrated with existing environmental science texts and was convinced he could produce a more readable and effective book—and he did! By the fourth edition, Richard Wright joined Nebel and collaborated for two more editions. From the seventh to tenth editions, Wright was responsible for the text, and now with the eleventh and twelfth editions, there are once again two authors. This is a good development, one that is already adding richness and a new voice to many of the chapters. We are deeply indebted to Nebel for his diligent work in developing the text and producing successive editions. We offer this new edition of the book as our contribution to the students who are now well into this new century, in the hope that they will join us in helping to bring about the environmental revolution—the transition to a sustainable society—that must come, we hope sooner rather than later.

Although the content and accuracy of this text are the responsibility of the authors, it would never have seen the light of day without the dedicated work of many other people. We want to express our heartfelt thanks to all those at Pearson Education who have contributed to the book in so many ways.

We salute our editor, Alison Rodal, for her encouragement in helping us reorganize and write this new edition. Our developmental and project editor, Leata Holloway, worked closely with us on every aspect of the book; thank you, Leata, for your good contributions. Katie Ostler at Element LLC was our production editor, keeping us focused on the details of transforming manuscript into a finished product. We are grateful to Gina Cheselka for her management and guidance and to Lee Ann Doctor for her work on the media. We'd also like to thank Lauren Harp for her help in marketing the text. Thanks also go to Jennifer Nonenmacher and Peter Jardim, photo researchers.

In addition, we thank Todd Tracy, Ed Zalisko, Trixi Beeker, Kayla Rihani, Thomas Pliske, and Heidi Marcum for their work on MasteringEnvironmentalScience and other text supplements.

I (Richard Wright) wish to offer some very personal thanks to my wife, Ann, who has been my companion since the beginning of my work in biology and has provided the emotional base and care without which I would be far less of a person and a biologist. Her love and patience have sustained me in immeasurable ways.

I (Dorothy Boorse) would like to thank my husband, Gary, my biggest supporter. I am also very thankful to my mentors—particularly Richard Wright, who has known me for years and helped me inestimably, and Calvin DeWitt of the University of Wisconsin–Madison, who has been one of the foremost figures in motivating young people to care about the environment and ethics.

Finally, it is our hope that this book can inspire a new generation to work toward bringing healing to a Creation suffering from human misuse.

Richard T. Wright
Dorothy F. Boorse

DEDICATION

This edition is dedicated to the memory of Wangari Muta Maathai (1940–2011) tireless advocate for the environment. Maathai became the voice of modern Africa, a voice of hope and strength protecting the environment for the next generation. We take pride in dedicating this edition to her memory.

Maathai became the first eastern African woman to get a PhD when she received a Doctorate in Anatomy from the University College of Nairobi, (now the University of Nairobi) where she also began her teaching career. There, she became an advocate for women's rights and the environment. She was convinced environmental degradation was the root of many of Kenya's problems. She served in Kenya's Parliament, became an assistant minister for Environment and Natural Resources and served international organizational roles. She is best known for founding the Green Belt Movement, an effort to help solve unemployment and deforestation while empowering women. To date, the organization has planted more than 47 million trees in Kenya.

Maathai received many international awards. She was a prolific author and frequent speaker. In 2004, she became the first African woman to receive the Nobel Peace Prize. It was the first awarded for work in protecting the natural world as a route to promoting justice and peace.

Maathai died of cancer in 2011. Her whole life, featured in the documentary film *Taking Root: The Vision of Wangari Maathai*, was dedicated to the poor and the land. In her Nobel acceptance speech, she said, "Today we are faced with a challenge that calls for a shift in our thinking, so that humanity stops threatening its life-support system. We are called to assist the Earth to heal her wounds and in the process heal our own—indeed, to embrace the whole creation in all its diversity, beauty and wonder."

MasteringEnvironmentalScience®
www.masteringenvironmentalscience.com

Mastering is the most effective and widely used online homework, tutorial, and assessment system for the sciences. It delivers self-paced tutorials that focus on your course objectives, provides individualized coaching, and responds to your individual progress. Mastering helps motivate you to learn outside of class and arrive prepared for lecture or lab.

Prescribed Fire Burn

EXPANDED! **Video Field Trips** allow you to observe everyday environmental impacts and solutions, including wind power, invasive species, a prescribed fire burn, solar energy, a coal-fired power plant, a wastewater treatment facility, a landfill, a sustainable farm, and more. Each Video Field Trip includes assessment questions that can be assigned as homework.

Solar Energy

UPDATED! **Current Events Quizzes** are regularly updated and connect recent news articles to course topics.

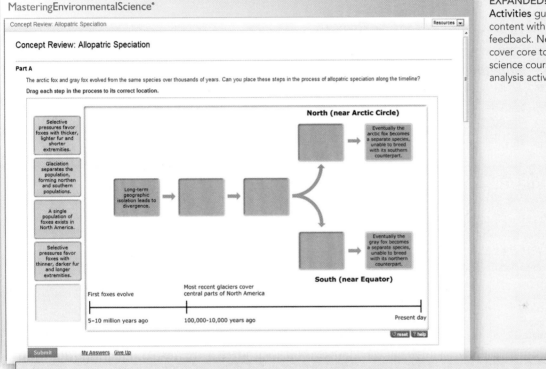

MasteringEnvironmentalScience®

Concept Review: Allopatric Speciation

Resources

Concept Review: Allopatric Speciation

Part A

The arctic fox and gray fox evolved from the same species over thousands of years. Can you place these steps in the process of allopatric speciation along the timeline?

Drag each step in the process to its correct location.

Selective pressures favor foxes with thicker, lighter fur and shorter extremities.

Glaciation separates the population, forming northern and southern populations.

A single population of foxes exists in North America.

Selective pressures favor foxes with thinner, darker fur and longer extremities.

North (near Arctic Circle)

Eventually the arctic fox becomes a separate species, unable to breed with its southern counterpart.

Long-term geographic isolation leads to divergence.

Eventually the gray fox becomes a separate species, unable to breed with its northern counterpart.

South (near Equator)

Most recent glaciers cover central parts of North America

First foxes evolve

5–10 million years ago

100,000–10,000 years ago

Present day

reset | help

Submit   My Answers   Give Up

EXPANDED! Concept Review Activities guide you through complex content with specific wrong answer feedback. New assignment options cover core topics in the environmental science course and include data analysis activities.

Try Again

You labeled 5 of 5 targets incorrectly. Recall that a single population is the precondition for allopatric speciation. Where on the timeline should this precondition be located?

Additional Study Tools in MasteringEnvironmentalScience®

- **The Pearson eText** gives you access to the text whenever and wherever you can access the Internet. The eText pages look exactly like the printed text, and include powerful interactive and customization functions.

- **Self Study Area** offers a 24/7 study resource.

- **GraphIt! Activities** help you explore the data behind environmental issues and research.

- **Reading Quizzes** help you stay on track with your reading assignments and test your understanding of the content.

New to the Twelfth Edition

A revitalized art program makes the text more visually appealing and accessible with images and figures that teach more clearly. Learning Objectives also introduce you to important concepts that you should know.

NEW! Twelfth Edition

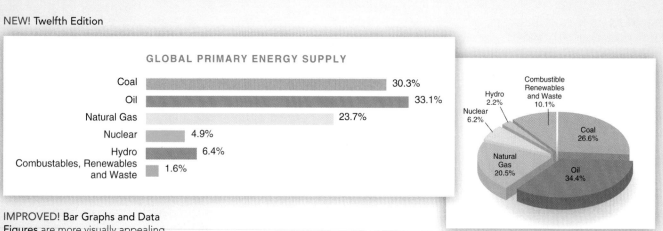

IMPROVED! **Bar Graphs and Data Figures** are more visually appealing and easier to interpret.

Corresponding Graph from Eleventh Edition

NEW! Twelfth Edition

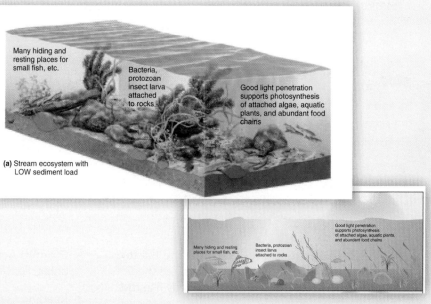

NEW! **Learning Objectives** open each chapter to introduce key concepts that you should understand at the conclusion of the chapter.

Corresponding Figure from Eleventh Edition

PART ONE

Framework for a Sustainable Future

NIGHTS SPENT POURING over blueprints, days looking at budgets, time spent deciding what parts need to be purchased and who will maintain a structure after it is built: you might imagine these activities being performed by professionals building a road or school, but they aren't. This is the work of a team of dedicated engineering undergraduates from Purdue University and their advisors. They are designing a small hydropower facility in Banyang, Cameroon, which will provide water and irrigation for a village.

THE PROJECT IS DESIGNED to promote sustainability—the ongoing thriving of people in a context of the natural world, living in such a way that doesn't use up resources, harm other creatures, or degrade our environment in the long term. Right now, there is evidence that we are not living sustainably: Our global economy relies on the use of fossil fuels and nuclear power, but continued emissions of carbon dioxide to the atmosphere and oceans are bringing major changes in the climate as Earth warms up and the oceans acidify. The global population will likely reach 9 billion by 2050, and the range and population size of many wild plant and animal species are declining. Food and clean water availability and basic health care remain low for the poorest people. Natural disasters such as the Japanese tsunami of 2011 interact with the built environment to cause radiation leaks and other problems. Human activities such as oil well drilling can interact with the natural environment to cause events such as the 2010 Gulf oil spill, which released 200 million gallons of crude oil into the Gulf, devastating sea life. We are all bound together: humanity, the other creatures, and the world around us.

We begin our framework for a sustainable future in Chapter 1, where we will introduce environmental science and what it might mean for you. In Chapter 2, we look at economics, politics, and public policy, three features of human societies that interact with science as we come to understand the environment and address the challenges we face.

CHAPTER 1

Science and the Environment

"**T**HERE WAS ONCE a town in the heart of America where all life seemed to live in harmony with its surroundings. The town lay in the midst of a checkerboard of prosperous farms, with fields of grain and hillsides of orchards where, in spring, white clouds of blossom drifted above the green fields. . . . The countryside was, in fact, famous for the abundance and variety of its bird life, and when the flood of migrants was pouring through in spring and fall people traveled from great distances to observe them. . . . So it had been from the days many years ago when the first settlers raised their houses, sank their wells, and built their barns.[1]"

These are words from the classic *Silent Spring*, written by biologist Rachel Carson **(Fig. 1–1)** to open her first chapter, titled "A Fable for Tomorrow." After painting this idyllic picture, the chapter goes on to describe "a strange blight" that began to afflict the town and its surrounding area. Fish died in streams, farm animals sickened and died, families were plagued with illnesses and occasional deaths. The birds had disappeared, their songs no longer heard—it was a "silent spring." And on the roofs and lawns and fields remnants of a white powder could still be seen, having fallen from the skies a few weeks before.

Rachel Carson explained that no such town existed, but that all of the problems she described had already happened somewhere, and that there was the very real danger that ". . . this imagined tragedy may easily become a stark reality we all shall know."[2] She published her book in 1962, during an era when pesticides and herbicides

---
[1]Rachel Carson, *Silent Spring*. Boston: Houghton Mifflin Company, pp. 1, 2.
[2]Ibid., p. 3.

Figure 1–1 **Rachel Carson.**

Figure 1–2 **Aerial spraying of pesticide on a tree farm.**

were sprayed widely on the landscape to control pests in agricultural crops, forests, and towns and cities **(Fig. 1–2)**. In *Silent Spring*, Carson was particularly critical of the widespread spraying of DDT. This pesticide was used to control Dutch elm disease, a fungus that invades trees and eventually kills them. The fungus is spread by elm bark beetles, and DDT was used to kill the beetles. In towns that employed DDT spraying, birds began dying off, until in some areas people reported their yards were empty of birds. Many thousands of songbirds were recovered dead and analyzed in laboratories for DDT content; all had toxic levels in their tissues. DDT was also employed in spraying salt marshes for mosquito control, and the result was a drastic reduction in the fish-eating bald eagle and osprey.

FALLOUT. Rachel Carson brought two important qualities to her work: she was very careful to document every finding reported in the book, and she had a high degree of personal courage. She was sure of her scientific claims, and she was willing to take on the establishment and defend her work. In spite of the fact that her work was thoroughly documented, the book ignited a firestorm of criticism from the chemical and agricultural establishment. Even respected institutions such as the American Medical Association joined in the attack against her.

Despite this criticism, Carson's book caught the public's eye, and it quickly made its way to the President's Science Advisory Committee when John F. Kennedy read a serialized version

of it in the *New Yorker*. Kennedy charged the committee with studying the pesticide problem and recommending changes in public policy. In 1963, Kennedy's committee made recommendations that fully supported Carson's thesis. Congress began holding hearings, public debate followed, and Carson's voice was joined by others who called for new policies to deal not only with pesticides, but also with air and water pollution and more protection for wild areas. Finally, in 1969, Congress passed a bill known as the Environmental Policy Act, the first legislation to recognize the interconnectedness of ecological systems and human enterprises. Shortly after that, a commission appointed by President Richard Nixon to study environmental policy recommended the creation of a new agency that would be responsible for dealing with air, water, solid waste, the use of pesticides, and radiation standards. The new agency, called the Environmental Protection Agency (EPA), was given a mandate to protect the environment against pressures from other governmental agencies and from industry, on behalf of the public. The year was 1970, the same year that 20 million Americans celebrated the first Earth Day.

In what must be seen as a triumph of Rachel Carson's work, DDT was banned in the United States and most other industrialized countries in the early 1970s. (The DDT story is more fully documented in Chapter 13.) Unfortunately, Rachel Carson did not live long after her world-shaking book was published; she died of breast cancer in 1964. Her legacy, however, is a lasting one: she is credited not only with major reforms in pesticide policy, but with initiating an environmental awareness that eventually led to the modern environmental movement and the creation of the EPA.

MOVING ON. This is a story of science and the environment, but it is more than that; it is a story of a courageous woman who changed the course of history. In this chapter, we briefly explore the current condition of our planet and then introduce three themes that provide structure to the primary goal of this text: *to promote a sustainable future.*

1.1 A Paradox: What Is the Real State of the Planet?

Paradox (n.): *A statement exhibiting contradictory or inexplicable aspects or qualities.*[3] A group of scientists from McGill University recently published a paper in which they identified a so-called **environmentalist's paradox.**[4] The paradox, they said, is this: over the past 40 years, human well-being has been steadily improving, while natural ecosystems (from which we derive many goods and services) have been declining.

To explain this paradox, the authors advanced four hypotheses:

1. The measurements of human well-being are flawed; it is actually declining.

2. Food production, a crucial ecosystem service that has been enhanced, outweighs the effects of declines in other ecosystem services.

3. Human technology makes us less dependent on ecosystem services.

4. There is a time lag between ecosystem decline and human well-being; the worst is yet to come.

We will take a brief look at four important global trends and keep in mind these hypotheses (the scientific method and hypotheses are explained later in the chapter) as we engage in our initial examination of the state of our planet: (1) human population and well-being, (2) the status of vital ecosystem services, (3) global climate change, and (4) a loss of biodiversity. Each of these is explored in greater depth in later chapters.

Population Growth and Human Well-Being

The world's human population, more than 7.1 billion in 2013, has grown by 2 billion in just the past 25 years. It is continuing to grow, at the rate of about 80 million persons per year. Even though the growth rate (now 1.2 %/year) is gradually slowing, the world population in 2050 could be 9.3 billion, according to the most recent projections from the United Nations (UN) Population Division **(Fig. 1–3)**. The 2.2 billion persons added to the human population by 2050 will all have to be fed, clothed, housed, and, hopefully, supported by gainful employment. Virtually all of the increase will be in developing countries.

Human Development Index. Each year since 1990, the United Nations Development Program (UNDP) has published a Human Development Report.[5] A key part of the report is the *Human Development Index* (HDI), a comprehensive assessment of human well-being in most countries of

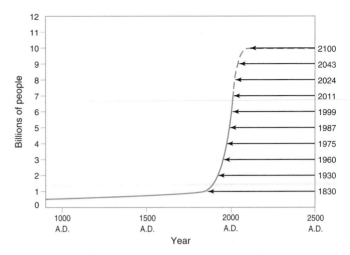

Figure 1–3 World population explosion. World population started a rapid growth phase in the early 1800s and has increased sixfold in the past 200 years. It is growing by 80 million people per year. (See Chapter 8.) Future projections are based on assumptions that birthrates will continue to decline. (Data from UN Population Division, 2010 revision, and from Population Reference Bureau reports.)

the world. With this index, well-being is measured in health, education, and basic living standards. In the 2010 report, worldwide trends in HDI were plotted over the 40 years since 1970 **(Fig. 1–4)**. Even though there are some countries showing little or no progress, most of the 135 countries analyzed show marked improvement over the past 40 years. It is this overall progress that has provided one side of the environmentalist's paradox. During that time, life expectancy rose from 59 years to 70, school enrollment climbed from 55% to 70%, and per capita **gross domestic product** (GDP) doubled to more than $10,000. There has also been significant progress in the number of countries achieving democratic rule. Studies have shown that the HDI correlates well with other indicators of well-being, such as literacy, gender equality, and overall "happiness." As a result of these facts, the McGill team concluded that its first hypothesis is not supported; there are too many indications that human well-being has indeed improved markedly.

Is It All Good? However, overall progress can, and does, mask serious inequalities. Economic growth has been extremely unequal, both between and within countries. And there are huge gaps in human development across the world, as Figure 1–4 shows. For example, in developing countries, an estimated 1.29 billion people still experience extreme poverty, existing on an income of $1.25 a day. More than 925 million people—one out of every five in developing countries—remain undernourished. Some 8.1 million children per year do not live to see their fifth birthday. Addressing these tragic outcomes of severe poverty has been a major concern of the UNDP, and in 2000, all UN member countries adopted a set of goals—the **Millennium Development Goals (MDGs)**—to reduce extreme poverty and its effects on human well-being (see Table 9–2 for a list of the eight goals). If the goals are achieved by the target date of 2015, more than 400 million people will be lifted out of extreme

---

[3]From *Webster's II New College Dictionary*. Boston: Houghton Mifflin Company, 1995.
[4]Ciara Raudsepp-Hearne et al., Untangling the Environmentalist's Paradox: Why Is Human Well-Being Increasing as Ecosystem Services Degrade? *Bioscience* 60: 576–589. September 2010.
[5]United Nations Development Program, *Human Development Report 2010: The Real Wealth of Nations*. New York: UNDP. http://hdr.undp.org. August 30, 2011.

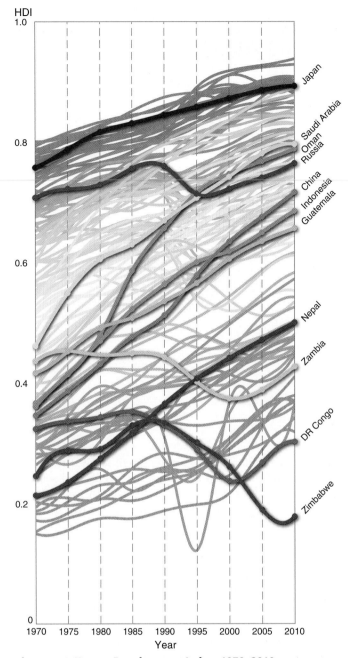

Figure 1–4 **Human Development Index, 1970–2010.** Advances in the HDI have occurred across all regions and almost all countries. The fastest progress has been in East Asia and the Pacific, followed by South Asia. All but three of the 135 countries have a higher level of human development today than in 1970.

poverty, and many millions of lives will be saved. Indeed, much progress has already been made in many developing countries, as signaled by the overall rise in the HDI values.

Virtually all of the discouraging issues outlined by the UNDP are most severe in developing countries, where population growth and fertility rates remain high. There is no doubt that these challenges will be easier to meet if population growth is brought down. When other countries have lowered their population growth, they have also made great advances in education, health care, and greater freedom and opportunities for women.

Is the prospect of 2.2 billion more people by 2050 a recipe for disaster? Can we feed that many more people? Certainly we have demonstrated our ability to continue to feed a growing human population, but at what cost to the environment? Are there enough energy and water and other ecosystem resources for 2.2 billion more of us?

Ecosystem Goods and Services

Natural and managed ecosystems support human life and economies with a range of goods and services. As crucial as they are, there is evidence that these vital resources are not being managed well. Around the world, human societies are depleting groundwater supplies, degrading agricultural soils, overfishing the oceans, and cutting forests faster than they can regrow. The world economy depends heavily on many renewable resources, as we exploit these systems for **goods**—fresh water (water with less than 0.1% salt), all of our food, much of our fuel, wood for lumber and paper, leather, furs, raw materials for fabrics, oils and alcohols, and much more. Just three sectors—agriculture, forestry, and fishing—are responsible for 50% of all jobs worldwide and 70% of all jobs in sub-Saharan Africa, eastern Asia, and the Pacific islands.

These same ecosystems also provide a flow of **services** that support human life and economic well-being, such as the breakdown of waste, regulation of the climate, erosion control, pest management, maintenance of crucial nutrient cycles, and so forth. In a very real sense, these goods and services can be thought of as capital—ecosystem capital. Human well-being and economic development are absolutely dependent on the products of this capital—its income, so to speak. As a result, the stock of ecosystem capital in a nation and its income-generating capacity represents a major form of the wealth of the nation (see Chapter 2). These goods and services are provided year after year, as long as the ecosystems producing them are protected.

Millennium Ecosystem Assessment. Launched on World Environment Day, 2001, the *Millennium Ecosystem Assessment* (ME) took four years to gather available information on the state of ecosystems across the globe. This monumental effort involved some 1,360 scientists from 95 countries gathering, analyzing, and synthesizing information from published, peer-reviewed research. The project focused especially on the linkages between ecosystem services and human well-being on global, regional, and local scales. Ecosystem goods and services (see Table 1–1) were grouped into *provisioning* services (goods like food and fuel), *regulating* services (processes like flood protection), and *cultural* services (nonmaterial benefits like recreation). The reports are now published and are also available on the Internet.

In a summary report,[6] the most prominent finding of the scientists was the widespread abuse and overexploitation of ecosystem resources. Humans have altered the

—————————
[6]Millennium Ecosystem Assessment, *Ecosystems and Human Well-Being: Synthesis.* Washington, DC: Island Press, 2005.

TABLE 1–1 The Global Status of Ecosystem Services. Human use has degraded almost two-thirds of these identified services; 20% are mixed, meaning they are degraded in some areas and enhanced in others; and 17% have been enhanced by human use.

Ecosystem Services	Degraded	Mixed	Enhanced
Provisioning (goods obtained from ecosystems)	Capture fisheries Wild foods Wood fuel Genetic resources Biochemicals Fresh water	Timber Fiber	Crops Livestock Aquaculture
Regulating (services obtained from the regulation of ecosystem processes)	Air quality regulation Regional and local climate regulation Erosion regulation Water purification Pest regulation Pollination Natural hazard regulation	Water regulation (flood protection, aquifer recharge) Disease regulation	Carbon sequestration (trapping atmospheric carbon in trees, etc.)
Cultural (nonmaterial benefits from ecosystems)	Spiritual and religious values Aesthetic values	Recreation and ecotourism	

Source: Millennium Ecosystem Assessment, Ecosystems and Human Well-Being: Synthesis. Washington, DC: Island Press, 2005.

world's ecosystems more rapidly and profoundly over the past 50 years than at any time in human history. More than 60% of the classes of ecosystem goods and services assessed by the team are being degraded or used unsustainably (Table 1–1). The scientists concluded that if this is not reversed, the next half-century could see deadly consequences for humans as the ecosystem services that sustain life are further degraded. The overall intent of the project is to build a knowledge base for sound policy decisions and management interventions; it remains for policy makers and managers to act on that knowledge.

However, one type of provisioning ecosystem services has actually been enhanced over recent years: the production of crops, livestock, and aquaculture. This production has more than kept pace with population growth, meeting the basic needs of people everywhere and improving nutrition for millions. Consequently, human well-being has benefited from the increased food provisioning services, as shown by increases in life expectancy and health improvements. These benefits apparently outweigh the costs of ecosystem decline, although the loss of certain goods and services has had serious impacts on human well-being.

Paradox Resolved? The McGill team set out to explain the environmentalist's paradox—the fact that human well-being has been improving while natural ecosystems have been declining. It rejected hypothesis 1, which stated that human well-being is actually declining. The team concluded that *hypothesis 2 was confirmed*: enhanced food production outweighs the effects of declines in other ecosystem services. Two further hypotheses remain: (3) our use of technology

to substitute for natural goods and services (e.g., irrigation, artificial fertilizers, improved sanitation), and (4) the existence of a time lag between the loss of goods and services and their impact on human well-being, with the possibility of exceeding limits and bringing on ecosystem collapse. The McGill University team concluded that these last two hypotheses also help explain the environmentalist's paradox, although not as clearly as hypothesis 2. They concluded that the paradox is not fully explained by any of the hypotheses, although hypothesis 1 was rejected. It also concluded that ecosystem conditions are indeed continuing to decline, with unknown and perhaps severe impacts on human well-being in the future. The most serious concern, one that affects all ecosystems, is global climate change.

Global Climate Change

The global economy runs on fossil fuel. Every day in 2012 we burned some 89 million barrels of oil, 340 billion cubic feet of natural gas, and 17 million tons of coal. All of this combustion generates carbon dioxide (CO_2), which is released to the atmosphere at a rate of 80 million tons a day. Because of past and present burning of fossil fuels, the CO_2 content of the atmosphere increased from 280 parts per million (PPM) in 1900 to 395 PPM in 2012. For the past decade, the level of atmospheric CO_2 has increased by 2 PPM per year, and given our dependency on fossil fuels, there is no end in sight.

Carbon dioxide is a natural component of the lower atmosphere, along with nitrogen and oxygen. It is required by plants for photosynthesis and is important to the Earth-atmosphere energy system. Carbon dioxide gas absorbs

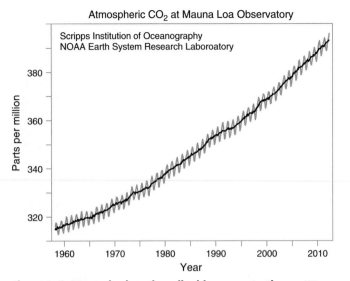

Figure 1–5 Atmospheric carbon dioxide concentrations. This record of CO_2 has been measured at the Mauna Loa Observatory since 1958. The atmospheric content of CO_2 has risen by 41% since the Industrial Revolution began.

(*Source:* Mauna Loa Observatory, Hawaii, NOAA Research Laboratory, Scripps Institution of Oceanography)

infrared (heat) energy radiated from Earth's surface, thus slowing the loss of this energy to space. The absorption of infrared energy by CO_2 and other gases warms the lower atmosphere in a phenomenon known as the **greenhouse effect.** This effect is well known, and the greenhouse gases keep Earth at hospitable temperatures as they trap heat. Although the concentration of CO_2 is a small percentage of the atmospheric gases, increases in the volume of this gas affect temperatures. **Figure 1–5** graphs changes in carbon dioxide from 1958 to the present, and **Figure 1–6** shows changes in global temperature since 1880. Both of these parameters are increasing—and rather steeply. This is not *proof* that the CO_2 caused the increase, but the argument based on the known greenhouse effect is quite convincing.

The Intergovernmental Panel on Climate Change (IPCC) was established by the UN in 1988 and given the responsibility to report its assessment of climate change at five-year intervals. The latest of these assessments, the Fourth Assessment Report (FAR), was released during 2007. The work of thousands of scientific expert reviewers and hundreds of authors, this assessment produced convincing evidence of human-induced global warming that is already severely affecting global climate. Polar ice is melting at an unprecedented rate, glaciers are retreating, storms are increasing in intensity, and sea level is rising. Because the oceans are absorbing half of the CO_2 being emitted, the seawater is becoming more acid; the Sound Science essay "When Oysters Became Canaries" (see p. 8) explores one consequence of ocean acidification. The IPCC concluded that future climate change could be catastrophic if something is not done to bring the rapidly rising emission of CO_2 and other greenhouse gases under control. With few exceptions, the scientists who have studied these phenomena have reached a clear consensus: this is a huge global problem, and it must be addressed on a global scale.

Responses. There are real solutions to this problem, but they are not easy. The most obvious need is to reduce global CO_2 emissions (called **mitigation**). At issue for many countries is the conflict between the short-term economic impacts of reducing the use of fossil fuels and the long-term consequences of climate change for the planet and all its inhabitants. Future climate changes are likely to disrupt the ecosystem goods and services essential to human well-being, and because the extreme poor depend especially on natural ecosystems, they will suffer disproportionately. International agreements to reduce greenhouse gas emissions have been forged, and some limited mitigation has been achieved.

Most observers believe that the best course forward is to aim toward an effective, binding international treaty to reduce emissions, but on a more immediate scale, for countries to act independently. The United States has backed away from any binding treaties, but the government is moving forward to regulate greenhouse gas emissions under existing air pollution laws and to encourage renewable energy development.

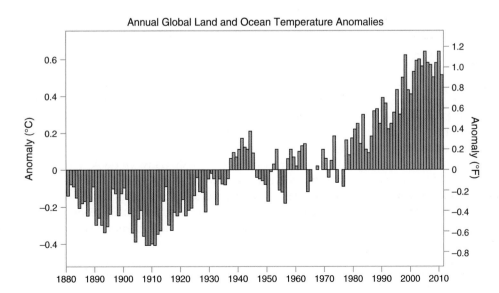

Figure 1–6 Global temperatures since 1880. This plot shows the course of global temperatures as recorded by thousands of stations around the world. The baseline, or zero point, is the 20th century average temperature.

(*Source:* National Climatic Data Center, NOAA, 2011)

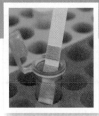

SOUND SCIENCE When Oysters Became Canaries

Oysters are big business on the U.S. West Coast, netting some $85 million a year and employing 3,000 workers. The oyster of choice on that coast is *Crassostrea gigas*, a native of Japanese waters imported many years ago and planted in many coastal waters. Because this oyster species needs warm water in order to spawn and West Coast waters are cold, oyster farms purchase so-called "seed, " young oysters 2 to 3 mm in diameter that are raised in hatcheries (see illustration). They then "plant" the seed on selected sandy banks or in special trays that are suspended in coastal embayments, and the oysters grow to maturity in a year or two as they feed on natural food in the saline waters. Sue Cudd, co-owner of Whiskey Creek Shellfish Hatchery on Netarts Bay, Oregon, ships out millions of oyster seed on a good day to oyster farmers.

At Whiskey Creek, oysters spawn under laboratory conditions. A female oyster produces millions of eggs, which are then fertilized by male oysters. The fertile eggs, now held in tanks, develop into swimming larvae, which start to form shells as they grow. The larvae settle down as they grow into "seed" oysters; all this time they have been bathed in filtered coastal seawater and provided with microscopic algae for food.

Unfortunately, the good days at Whiskey Creek came to a halt in the summer of 2007. During this time, oyster larvae and seed oysters began dying off, sometimes simply disappearing from the incubation tanks. Owners Cudd and her husband, Mark Wiegardt, immediately

started looking for the culprit; they quickly identified a naturally occurring pathogenic bacterium, *Vibrio tubiashii*, that had become established in their tanks. They installed an expensive filtration system that removed the bacteria from incoming water, but the larvae continued to die. So it wasn't a pathogen killing the baby oysters.

Then, in July 2008, all of the remaining larvae in their tanks suddenly died. At the same time, the bay water brought into the tanks had turned acidic and colder. The timing was the key; this was water that had recently been brought up to the surface as offshore winds pushed the warmer surface water away, a process called upwelling. The pH of the upwelled water was in the range of 7.5; normal surface seawater pH is around 8.5, a 10-fold difference in acidity. The reason for the acidic waters was high concentrations of CO_2, released as plankton died and sank to the deeper waters and decomposed. However, new studies by oceanographers showed that waters rising from Pacific Ocean depths hold the CO_2 from normal decomposition plus CO_2 absorbed from the atmosphere of past decades; the oceans have been acidified as they have been absorbing atmospheric CO_2. It has been well understood for some years now that the oceans have taken up at least half of the CO_2 released as fossil fuels have been burned over the past 250 years. Now it begins to appear that ocean acidification is well under way and is now beginning to affect shellfish.

The solution for the Whiskey Creek Hatchery owners was now clear; they had to avoid the acidic waters that could simply dissolve away the thin shells of the young oysters. They installed chemical monitors that measured the

CO_2 and pH of incoming waters and were able to time water withdrawals to daytime hours when natural algae were removing CO_2 from the water for photosynthesis and avoid recently upwelled and nighttime hours where CO_2 levels were elevated and pH lowered. Whiskey Creek Hatchery has recovered, and seed oysters are being shipped out once again. For a time, their oyster larvae were the "canary in the cage," a term that refers to the old practice by coal miners of bringing cages of canaries down into the mines with them. Because canaries are more sensitive to coal gas than are humans, the miners would quickly leave the mines if the canaries began dying. The oyster larvae were an early sign of a process that is now under way and looks to become increasingly serious—ocean acidification. As acidification intensifies, it is likely that more sturdy shellfish will succumb to the rising acid conditions, and whole coastal and coral ecosystems could be degraded, perhaps to the point of no return. The solution? Reduce the carbon emissions from burning fossil fuels (see Chapter 18).

Without doubt, this is one of the defining environmental issues of the 21st century. (Chapter 18 is wholly devoted to exploring the many dimensions of global climate change.)

Loss of Biodiversity

The Convention on Biological Diversity (CBD) defines biodiversity as "the variability among living organisms from all sources including terrestrial, marine, other aquatic ecosystems, and the ecological complexes of which they are part; this includes diversity within species, between species and of ecosystems."[7] The rapidly growing human population, with its growing appetite for food, water, timber, fiber, and fuel, is accelerating the conversion of forests, grasslands,

and wetlands to agriculture and urban development. The inevitable result is the loss of many of the wild plants and animals that occupy those natural habitats. Pollution also degrades habitats—particularly aquatic and marine habitats—eliminating the species they support. Further, hundreds of species of mammals, reptiles, amphibians, fish, birds, and butterflies, as well as innumerable plants, are exploited for their commercial value. Even when species are protected by law, many are hunted, killed, and marketed illegally. According to the Global Biodiversity Outlook 3 (GBO 3) assessment,[8] the majority of wild plant and animal species are declining in their range and/or population size.

As a result, Earth is rapidly losing many of its species, although no one knows exactly how many. About 2 million

---

[7]Convention on Biological Diversity. Text of Convention, Article 2. Use of Terms. http://www.cbd.int/convention/articles/?a=cbd-02

[8]Secretariat of the Convention on Biological Diversity, *Global Biodiversity Outlook 3*. Montreal. 2010. www.cbd.int/gbo3. September 5, 2011.

species have been described and classified, but scientists estimate that 5 to 30 million species may exist on Earth. Because so many species remain unidentified, the exact number of species becoming extinct can only be estimated. However, GBO 3 reported that the abundance of vertebrate species fell by nearly a third between 1970 and 2006. The most dramatic declines have occurred in the tropics, in both ground-dwelling and ocean species.

Risks of Losing Biodiversity. Why is losing biodiversity so critical? Biodiversity is the mainstay of agricultural crops and of many medicines. The loss of biodiversity can only curtail development in these areas. Biodiversity is also a critical factor in maintaining the stability of natural systems and enabling them to recover after disturbances such as fires or volcanic eruptions. Most of the essential goods and services provided by natural systems are derived directly from various living organisms, and we threaten our own well-being when we diminish the biodiversity within those natural systems. These goods and services are especially important in sustaining the poor in developing countries. There are also aesthetic and moral arguments for maintaining biodiversity. Shall we continue to erase living species from the planet, or do we have a moral responsibility to protect and preserve the amazing diversity of life on Earth? Once a species is gone, it is gone forever. Finding ways to protect the planet's biodiversity is one of the key challenges of environmental science. (Chapter 6 covers loss of biodiversity.)

1.2 Environmental Science and the Environmental Movement

As you read this book, you will encounter descriptions of the natural world and how it works. You will also encounter a great diversity of issues and problems that have arisen because human societies live in this natural world. We use materials from it (goods taken from natural ecosystems); we convert parts of it into the built environment of our towns, cities, factories, and highways; and we transform many natural ecosystems into food-producing agricultural systems. We also use the environment as a place to dump our wastes—solids, liquids, gases—all proceeding from human activities, most of them legitimate, but all of them affecting the environment. The **environment**, then, includes the natural world, human societies, and the human-built world; it is an extremely inclusive concept.

Environmental Science

Things go wrong in the environment, sometimes badly. We have already considered four trends (human population growth, ecosystem decline, global climate change, loss of biodiversity) that signal to us that we are creating problems for ourselves that we ignore at our peril. Lest you think that all we do is create problems, however, consider some of the great successes human societies have achieved. We have learned how to domesticate landscapes and ecosystems, converting them into highly productive food-producing systems that provide

sustenance for 7 billion people. We have learned how to convert natural materials into an endless number of manufactured goods and structures, all useful for the successful building of cities, roadways, vehicles, and all that makes up a 21st-century human world. We have successfully tapped into the abundant pool of carbon-based fuels laid down in the geological past, using it for energy to produce electricity, heat our homes, power our vehicles, and manufacture our products. All of these successes, however, carry with them hazards. Burning fossil fuels creates air pollutants that can make us sick and even raise the temperature of our planet; using chemicals to control diseases and pests leaves a residue of contaminants that can poison whole food chains—and ourselves; overharvesting forests and fisheries can lead to their collapse, erasing vital services that natural ecosystems provide.

Human actions have negative effects on the environment in two broad categories: cumulative impacts and unintended consequences. Sometimes we simply do too much of any one activity—too much burning, too much tree cutting, too much mining on steep slopes. Activities that would not pose a problem if a few people engaged in them are big problems if millions of people do. Sometimes it isn't the accumulation of an activity; it is that we are not paying attention to how the world works. There are unintended consequences of using chemical pesticides, as we've seen, or of putting trash dumps in wetlands. These two concepts, cumulative impacts and unintended consequences, will come up in the chapters ahead. This is where environmental science comes in. Simply put, **environmental science** is the study of how the world works. Scientists help figure out ways to lower the negative impacts of our actions, to find alternative ways to meet the same needs, and to better anticipate the likely effects of what we are doing.

All sorts of disciplines contribute to environmental science: history, engineering, geology, physics, medicine, biology, and sociology, to name a few. It is perhaps the most multidisciplinary of all sciences. As you study environmental science, you have an opportunity to engage in something that can change your life and that will certainly equip you to better understand the world you live in now and will encounter in the future.

The Environmental Movement

To understand how the world works today, we need some sense of history. Although what we now term the environmental movement began less than 60 years ago, it had its roots in the late 19th century, when some people realized that the unique, wild areas of the United States were disappearing.

The history of the environmental movement in the United States has been one in which environmental degradation, resource misuse, and disastrous events sparked scientific study and sometimes grassroots action. Scientific study yielded information on how the world works. Individuals and groups worked as stewards to make changes. Policies were put into place to better protect resources and people from environmental degradation. **Figure 1–7** (on the next page) is a timeline of some of the events and scientific findings, people, and policies of the American environmental movement that will come up throughout the book.

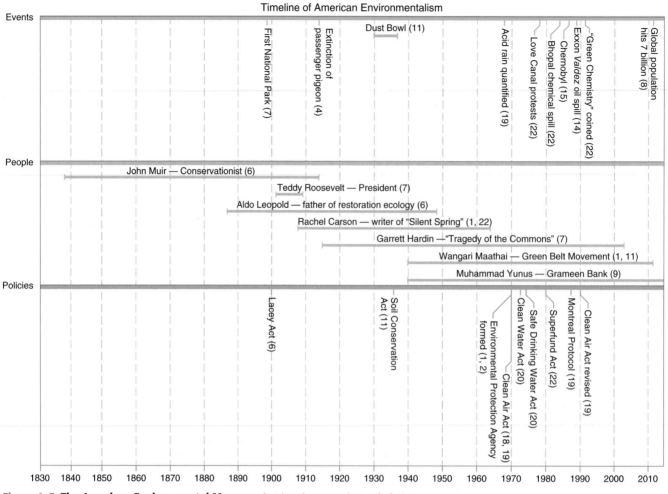

Figure 1–7 The American Environmental Movement. Selected events and scientific findings, people, and policies of the environmental movement in the United States. Numbers next to them are the chapters in which they are covered.

In the late 19th century, the indiscriminate killing of birds and animals and the closing of the frontier sparked a reaction. Around that time, several groups devoted to conservation formed: the National Audubon Society, the National Wildlife Federation, and the Sierra Club, which was founded in California by naturalist John Muir, who helped popularize the idea of wilderness. President Theodore Roosevelt promoted the conservation of public lands, and the national parks were formed.

The first half of the 20th century was marked by increasing awareness of the environment. Unwise agricultural practices following World War I eventually helped create an environmental crisis—the Dust Bowl of the 1930s (enormous soil erosion in the American and Canadian prairielands). During the Great Depression (1930–1936), conservation provided a means of both restoring the land and providing work for the unemployed, such as workers in the Civilian Conservation Corps (CCC), who built trails and did erosion control in national parks and forests.

The two decades following World War II (1945–1965) were full of technological optimism. New developments, ranging from rocket science to computers and from pesticides to antibiotics, were redirected to peacetime applications. However, although economic expansion enabled most families to have a home, a car, and other possessions, certain problems became obvious. The air in and around cities was becoming murky and irritating to people's eyes and respiratory systems. Rivers and beaches were increasingly fouled with raw sewage, garbage, and chemical wastes from industries, sewers, and dumps. Conspicuous declines occurred in many bird populations (including that of our national symbol, the bald eagle), aquatic species, and other animals. The decline of the bald eagle and other bird populations was traced to the accumulation in their bodies of DDT, the long-lasting pesticide that had been used in large amounts since the 1940s. In short, it was clear that we were seriously contaminating our environment.

The Modern Environmental Movement. In 1962, biologist Rachel Carson wrote *Silent Spring*, presenting her scenario of a future with no songbirds. Carson's voice was soon joined by others, many of whom formed organizations to focus and amplify the voices of thousands more in demanding a cleaner environment. This was the beginning of the modern **environmental movement**, in which a newly motivated citizenry demanded the curtailment of pollution, the cleanup of polluted environments, and the protection of pristine areas.

Pressured by concerned citizens, Congress created the Environmental Protection Agency (EPA) in 1970 and passed numerous laws promoting pollution control and wildlife protection, including the National Environmental Policy Act of 1969, the Clean Air Act of 1970, the Clean Water Act of 1972, the Marine Mammals Protection Act of 1972, the Endangered Species Act of 1973, and the Safe Drinking Water Act of 1974. A disastrous hazardous waste problem in Love Canal, New York, prompted the Superfund Act of 1980; scientific studies on acid rain, the decline of the ozone layer, and global climate change caused more environmental groups to form and policies to be put into place. Concerns about hazardous waste disposal triggered the rise of the Green Chemistry Movement. You will read more about these and more recent laws and their effects later in this book.

Persons and organizations with a strong focus on environmental concerns became known as **environmentalists,** and the resulting widespread development of this movement is often called **environmentalism.** As a result of these efforts, the air in our cities and the water in our lakes and rivers are far cleaner than they were in the late 1960s. Without a doubt, absent the environmental movement, our air and water would now be a toxic brew. By almost any measure, the environmental movement has been successful, at least in solving some pollution problems.

Environmentalism Acquires Critics. In the early stages of the environmental movement, sources of problems were specific and visible, and the solutions seemed relatively straightforward: install waste treatment and pollution control equipment and ban the use of very toxic pesticides, substituting safer pesticides, for example. At first, many industries conformed to the reforms, but as regulatory demands increased, a number of the more polluting industries, such as mining and utilities, pressed for a halt in regulation, and some began demanding significant deregulation.

Today, there are political battles surrounding almost every environmental issue. Bitter conflicts have emerged over issues involving access to publicly owned resources such as water, grazing land, and timber. The Endangered Species Act, which protects wild plants and animals whose existence is threatened, has been a lightning rod for controversy. When the EPA develops regulations to address air and water pollution, for example, some business special interests oppose the regulations and often environmental special interests support them. Climate change has been so controversial that it has triggered a host of antienvironmental groups. Instead of civility and persuasive discussion, discourse on environmental issues has become one of polemics and anger. As environmental public policies are developed and enforced, the controversies that accompany those policies remind us that politics always accompanies policy.

1.3 Three Unifying Themes

What will it take to move our civilization in the direction of a long-term sustainable relationship with the natural world? The answer to this question is not simple, but we would like to present three themes that provide coherence to the issues and topics covered in this text. In reality, these themes deal with how we should conceptualize our task of forging a sustainable future **(Fig. 1–8)**. Each theme is a concept that we believe is essential to environmental science. The first theme is **sustainability**—the practical goal toward which our interactions

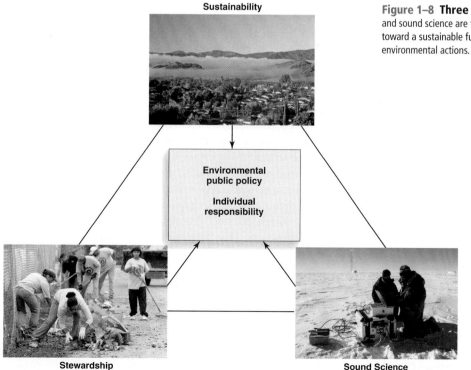

Figure 1–8 Three unifying themes. Sustainability, stewardship, and sound science are three vital concepts or ideals that can move societies toward a sustainable future if they are applied to public policies and private environmental actions.

with the natural world should be working. The second is **sound science**—the basis for our understanding of how the world works and how human systems interact with it. The third theme is **stewardship**—the actions and programs that manage natural resources and human well-being for the common good. These themes will be applied to public policy and individual responsibility throughout the text and will also be explored in brief essays within each chapter. As each chapter of this text develops, the concepts behind these themes will come into play at different points. At the end of each chapter, we will revisit the themes, summarizing their relevance to the chapter topics and often adding our own perspective. We explore each of these themes here, appropriately beginning with sustainability.

Sustainability

A system or process is sustainable if it can be continued indefinitely, without depleting any of the material or energy resources required to keep it running. The term was first applied to the idea of **sustainable yields** in human endeavors such as forestry and fisheries. Trees, fish, and other biological species normally grow and reproduce at rates faster than those required just to keep their populations stable. This built-in capacity allows every species to increase or replace a population following some natural disaster.

It is possible, then, to harvest a certain percentage of trees or fish every year without depleting the forest or reducing the fish population below a certain base number. As long as the number harvested stays within the capacity of the population to grow and replace itself, the practice can be continued indefinitely. The harvest then represents a sustainable yield. It becomes unsustainable only when trees are cut or fish are caught at a rate that exceeds the capacity of their present population to reproduce and grow. The concept of a sustainable yield can also be applied to freshwater supplies, soils, and the ability of natural systems to absorb pollutants without being damaged.

The notion of sustainability can be extended to include ecosystems. Sustainable ecosystems are entire natural systems that persist and thrive over time by recycling nutrients, maintaining a diversity of species, and using the Sun as a source of sustainable energy. As we'll see, ecosystems are enormously successful at being sustainable (Chapters 3, 4, and 5).

Sustainable Societies. Applying the concept of sustainability to human systems, we say that a **sustainable society** is a society in balance with the natural world, continuing generation after generation, neither depleting its resource base by exceeding sustainable yields nor producing pollutants in excess of nature's capacity to absorb them. Many primitive societies were sustainable in this sense for thousands of years.

When the concept of sustainability is applied to modern societies, we generally picture certain desirable, healthy characteristics of the people, their communities, and the ecosystems on which they depend. Many of our interactions with the environment are *not* sustainable, however. This is demonstrated by such global trends as the decline of biodiversity and essential ecosystems and the increased emissions of greenhouse gases. Although population growth in industrialized countries has almost halted, these countries are using energy and other resources at unsustainable rates, producing pollutants that are accumulating in the atmosphere, water, and land. In contrast, developing countries are experiencing continued population growth, yet are often unable to meet the needs of many of their people in spite of heavy exploitation of natural resources. As our modern societies pursue continued economic growth and consumption, we are apparently locked into an unsustainable mode. How do we resolve this dilemma? One answer is the concept of sustainable development.

Sustainable Development. **Sustainable development** is a term that was first brought into common use by the World Commission on Environment and Development, a group appointed by the United Nations. The commission made sustainable development the theme of its final report, *Our Common Future*, published in 1987. The report defined sustainable development as a form of development or progress that "meets the needs of the present without compromising the ability of future generations to meet their own needs." The concept arose in the context of a debate between the respective *environmental* and *developmental* concerns of groups of developed and developing countries. **Development** refers to the continued improvement of human well-being, usually in the developing countries. Both developed and developing countries have embraced the concept of sustainable development, although the industrialized countries are usually more concerned about environmental sustainability, while the developing countries are more concerned about economic development. The basic idea, however, is to maintain and improve the well-being of both humans and ecosystems.

The concept of sustainable development is now so well entrenched in international circles that it has become almost an article of faith. It sounds comforting, so people want to believe that it is possible, and it appears to incorporate some ideals that are sorely needed, such as **equity**—whereby the needs of the present are actually met and future generations are seen as equally deserving as those living now. Sustainable development means different things to different people, however, as illustrated by the viewpoints of three important disciplines traditionally concerned with the processes involved. Economists are concerned mainly with growth, efficiency, and the optimum use of resources. Sociologists mainly focus on human needs and on concepts like equity, empowerment, social cohesion, and cultural identity. Ecologists show their greatest concern for preserving the integrity of natural systems, for living within the carrying capacity of the environment, and for dealing effectively with pollution. It can be argued, however, that sustainable solutions will be found mainly where the concerns of these three groups intersect, as illustrated in **Figure 1–9**.

There are many dimensions to sustainable development—environmental, social, economic, political—and no societies today have achieved it. Nevertheless, as with justice,

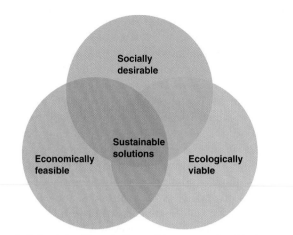

Figure 1–9 Sustainable solutions. If the concerns of sociologists, economists, and ecologists can be brought into intersection, sustainable solutions in a society are much more achievable.

equality, and freedom, it is important to uphold sustainable development as an ideal—a goal toward which all human societies need to be moving, even if we have not achieved it completely anywhere. For example, policies and actions in a society that reduce infant mortality, increase the availability of family planning, improve the air quality, provide abundant and pure water, preserve and protect natural ecosystems, reduce soil erosion, reduce the release of toxic chemicals to the environment, and restore healthy coastal fisheries are all moving that society in the right direction—toward a sustainable future. At the same time, these actions are all quite measurable, so progress in achieving sustainable development can be assessed. Communities and organizations are forging sustainable development indicators and goals to track their progress, such as the Environmental Sustainability Index (produced by Yale University, 1999–2005), which evaluated a society's natural resources and its stewardship of those resources and the people they support.

An Essential Transition. The transition to a truly sustainable civilization is hard to picture at present. It requires achieving a stable human population that recognizes the finite limits of the ability of Earth's systems to produce resources and absorb wastes and that acts accordingly. To achieve sustainability will require a special level of dedication and commitment to care for the natural world and to act with justice and equity toward one another. What will it take to make the transition to a sustainable future? There is broad agreement on the following major points:

- A population transition from a continually increasing human population to one that is stable or even declining.

- A resource transition to an economy that is not dedicated to growth and consumption, but instead relies on nature's income and protects ecosystem capital from depletion.

- A technology transition from pollution-intensive economic production to environmentally friendly processes—in particular, moving from fossil fuel energy to renewable energy sources.

- A political/sociological transition to public policies that embrace a careful and just approach to people's needs and that eliminate large-scale poverty.

- A community transition from the present car-dominated urban sprawl of developed countries to the "smart growth" concepts of smaller, functional settlements and more livable cities.

This is not an exhaustive list, but it does give a glimpse of what a sustainable future might look like. One thing is certain: we cannot continue on our present trajectory and hope that everything will turn out well. The natural world—our life support system—is being degraded, its ecosystem capital eroded. One essential tool for achieving sustainability is sound science.

Sound Science

Some environmental issues are embroiled in controversies so polarized that no middle ground seems possible. On the one side are those who argue using seemingly sound scientific facts and well-found theories. On the other are those who disagree and present opposing explanations of the same facts. Both groups may have motives for arguing their case that are not all apparent to the public. In the face of such controversy, many people are understandably left confused. It is our objective to give first a brief overview of the nature of science and the scientific method so that you can evaluate for yourself the two sides of such controversies and then to show how science forms the foundation for understanding controversial issues.

The Scientific Method. In its essence, science is simply a way of gaining knowledge; that way is called the **scientific method**. The term **science** further refers to all the knowledge gained through that method. We employ the term **sound science** to distinguish legitimate science from what can been called **junk science**, information that is presented as valid science but that does not conform to the rigors of the methods and practice of legitimate science. Sound science involves a disciplined approach to understanding how the natural world works, an approach that has become known as the scientific method.

Assumptions. To begin, the scientific method rests on four basic assumptions that most of us accept without argument. The first assumption is that what we perceive with our basic five senses represents an objective reality; that is, our perceptions are not some kind of mirage or dream. The second assumption is that this "objective reality" functions according to certain basic principles and natural laws that remain consistent through time and space.

The third assumption, which follows directly from the second, is that every result has a cause and every event in turn will cause other events. In other words, we assume that events do not occur without reason and that there is an explainable cause behind every happening. The fourth and final assumption is that, through our powers of observation, manipulation, and reason, we can discover and understand the basic principles and natural laws by which the universe functions.

Although assumptions, by definition, are premises that are simply accepted, the fact is that the assumptions underlying science have served us well and are borne out by everyday experience. For example, we suffer severe consequences if we do not accept our perception of fire as real. Similarly, our experience confirms that gravity is a predictable force acting throughout the universe and that it is not subject to unpredictable change. And the same can be said for any number of other phenomena that we observe. If our car fails to start, we know that there is a logical reason (such as lights left on), and we can get it jump-started.

Thus, whether we are conscious of the fact or not, we all accept the basic assumptions of science in the conduct and understanding of our everyday lives. Scientists and scientific investigations focused on the natural world (the **natural sciences**) extend the boundaries of everyday experience, deepen our understanding of cause-and-effect relationships, and enable us to have a greater appreciation of the principles and natural laws that seem to determine the behavior of all things, from the outcome of a chemical reaction to the functioning of the biosphere.

Observation. The foundation of all science and scientific discovery is **observation**. Indeed, many branches of science, such as natural history (the study of where and how various plants and animals live, mate, reproduce, etc.), astronomy, anthropology, and paleontology, are based almost entirely on observation because experimentation is either inappropriate or impossible. For example, it simply is impossible to conduct experiments on stars or past events. **Figure 1–10** shows how observations alone can lead to answers or explanations of some natural phenomena.

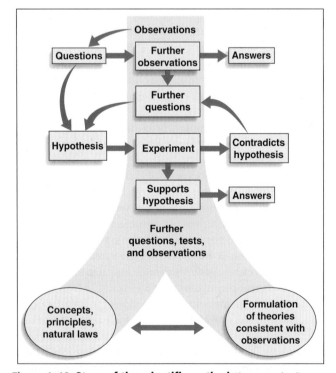

Figure 1–10 Steps of the scientific method. The process leading to the formation of theories and the postulation of natural laws and concepts is a continual interplay among observations, hypotheses, tests (experiments), explanations or theories, and further refinement.

Even experimentation, as we will discuss shortly, is conducted to gain another window of observation. Simply put, in all science, careful observation is the keystone. So how can we be sure that observations are accurate? Well, as a matter of fact, not every reported observation is accurate, for reasons ranging from honest misperceptions or instrument error to calculated mischief. Therefore, an important aspect of science, and a trait of scientists, is to be *skeptical* of any new report until it is *confirmed* or verified. As observations are confirmed by more and more investigators, they gain the status of factual data. Things or events that do not allow such confirmation—a reported visit by Martian aliens, for example—are very questionable from a scientific standpoint.

Various observations by themselves, like the pieces of a puzzle, are often put together into a larger picture—a **model** of how a system works. To give a simple example, we observe that water evaporates and leads to moist air and that water from moist air condenses on a cool surface. We also observe clouds and precipitation. Putting these observations together logically, we derive the concept of the hydrologic cycle, described later (Chapter 10). Water evaporates and then condenses as air is cooled, condensation forms clouds, and precipitation follows. Water thus makes a cycle from the surface of Earth into the atmosphere and back to Earth (Fig. 10–3). Note how this simple example incorporates the four assumptions described previously: there is an objective reality, it operates according to principles, every result has a cause, and we can discover the principles according to which reality operates. Note also how the example broadens our everyday experiences of the evaporation of water and the falling of rain into an understanding of a cycle involving both.

So, the essence of science and the scientific method may be seen as a process of making observations and logically integrating those observations into a model of how some natural system works. To be sure, many areas of science get more complex and difficult to comprehend. Still, the basic process—constructing a logically coherent picture of causes and effects from basic observations—is the same.

Where, then, does experimentation, that additional hallmark of science, fit in?

Experimentation. **Experimentation** is simply setting up situations to make more systematic observations regarding causes and effects. For example, the number of chemical reactions one can readily observe in nature is limited. However, in the laboratory it is possible to purify elements or compounds, mix them together in desired combinations and proportions, and carefully observe and measure how they react (or fail to react). From the way chemicals react, chemists have constructed a coherent cause-and-effect picture, the **atomic theory**, and they have determined the attributes of each element. Similarly, biologists put organisms into specific situations in which they can carefully observe and measure their responses to particular conditions or treatments **(Fig. 1–11)**. To solve a particular problem (e.g., What is the cause of *this* event?), a more or less systematic line of experimentation is used. This is where the sequence consisting of observation, hypothesis, test (experiment), and explanation comes in.

Figure 1–11 Experiment demonstrating the importance of mycorrhizae. Mycorrhizae is a root–fungus symbiosis that greatly enhances the ability of a plant to absorb nutrients. The pine seedlings shown were grown under identical conditions, except that the seedling on the left was grown with mycorrhizal fungi and the one on the right was grown without the fungi.

Hypotheses. Let's consider the problem of the death of oyster larvae at the Whiskey Creek hatchery (see Sound Science, p. 8). What was killing the oyster larvae in the incubation tanks? The first step in solving the problem was to make educated guesses as to the cause. Each such educated guess is a **hypothesis.** Each hypothesis is then tested by making further field observations or conducting experiments to determine whether or not the hypothesis really accounts for the observed effect. The owners found high numbers of pathogenic bacteria in the tanks (hypothesis: the bacteria were killing the larvae). They installed a new filtration system that removed bacteria before they could multiply, essentially an experiment. The larvae still died. They rejected that hypothesis. Then a crucial observation was made about the timing of the bay water brought into the tanks, water that had come from recently upwelled coastal surface water. This water had higher acidity (lower pH) than usual (hypothesis: acid waters were killing the larvae by dissolving their shells). By monitoring the pH of incoming water, the owners were able to avoid the acid waters and time water withdrawals to daytime hours when the pH was higher. The results were dramatic: the larvae were able to thrive and grow. The hypothesis was confirmed, and the larval death was satisfactorily explained. The experimental results brought what started out as a hypothesis to the status of a satisfactory explanation. Figure 1–10 shows how such a process moves from observations to hypotheses, experiments, and a satisfactory answer.

Theories. We have already noted how various specific observations (e.g., evaporation, precipitation) may fit together to give a logically coherent conceptual framework

(e.g., the hydrologic cycle). At first, the conceptual framework is termed a hypothesis because it is really a tentative explanation as to how various observations are related. It becomes a **theory** only after much testing and confirmation showing that it is logically consistent with all observations (Fig. 1–10).

Through logical reasoning, theories will generally suggest or predict certain events. If an event predicted by a theory is indeed observed, the observation provides evidence for the truth of the theory. Predictions require experiments, testing, further data gathering, and more observation. When theories reach a state of providing a logically consistent framework for all relevant observations and when they can be used to reliably predict outcomes, they deserve to be accepted. For example, we have never seen atoms directly (or, until recently, even indirectly), but innumerable observations and experiments are coherently explainable by the theory that all gases, liquids, and solids consist of various combinations of slightly over a hundred kinds of atoms. Hence, we fully accept the atomic theory of matter.

Science has successfully brought us a high degree of understanding of the natural world and has also brought a host of technologies based on that understanding. Doesn't this success mean that our theories about how the world works must be true? A philosopher of science would answer, "No, not necessarily." The best we can do is to establish a theory beyond a reasonable doubt. Theories are always less than absolutely certain, because we can never know whether some other theory exists that will do a better job of explaining the data. Generally speaking, our confidence in scientific explanations should be proportional to the evidence supporting them.

Natural Laws and Concepts. Like a theory, the second assumption underlying science—that the universe functions according to certain basic principles and natural laws that remain consistent through time and space—cannot be established with absolute certainty. Still, every observation and test that scientists have performed up to the present have borne it out. All our observations, whether direct or through experimentation, demonstrate that matter and energy behave neither randomly nor even inconsistently, but in precise and predictable ways. We refer to these principles by which we can define and precisely predict the behavior of matter and energy as **natural laws.** Examples are the *Law of Gravity*, the *Law of Conservation of Matter*, and the various *Laws of Thermodynamics.* Our technological success in space exploration and many other fields is in no small part due to our recognition of these principles and our precise calculations based on them. Conversely, trying to make something work in a manner contrary to a natural law invariably results in failure.

In many situations, the outcome of scientific work results in well-established explanations that must be expressed in the mathematical language of probability and statistics. This is especially true of biological phenomena such as predator–prey relationships and the effects of pesticides. Accordingly, here it is appropriate to speak of concepts rather than laws. **Concepts** are perfectly valid explanations of data gathered from the natural world, and they can also be predictive, but they don't reach the status of laws. They model the way we believe the natural world works and enable us to make qualified predictions of future outcomes. Thus, we might say that, on the basis of our understanding of the effects of DDT (a potent pesticide) on mosquitoes and on the basis of our observation that mosquitoes can, and do, develop resistance to DDT, continued spraying of the local salt marshes with DDT to eradicate unwelcome mosquitoes is likely to result in the development of a more resistant mosquito population in the future, an unwelcome outcome. Here the concept of pest resistance may be used to modify public policy. In general, explanations in biology do not lead to the development of statements with such high predictability that we are justified in calling them laws.

The Role of Instruments in Science. Complex instrumentation is another hallmark of science, one that often gives it an aura of mystery. Yet, regardless of complexity, all scientific **instruments** perform one of three basic functions. First, they may extend our powers of observation, as do telescopes, microscopes, X-ray machines, and CAT scans, for example. Second, instruments are used to quantify observations—that is, they enable us to measure exact quantities. For example, we may feel cold, but a thermometer enables us to measure and quantify exactly how cold it is. Comparisons, communication, and verification of different observations and events would be impossible if it were not for such measurement and quantification. Third, instruments such as growth chambers and robots help us achieve conditions and perform manipulations required to make certain observations or perform certain experiments.

All instruments used in science need to be subjected to testing and verification to be sure that they are giving as accurate a representation of reality as we can attain with their use. The need for this can be seen in the case of instruments measuring ocean temperatures (see Sound Science "Getting It Right," in Chapter 18). Failure to calibrate the instruments properly led to a serious underestimation of ocean temperature changes. It is safe to say that modern scientific instruments have made possible many important discoveries that were out of reach of the human senses.

The Scientific Community. It is important to understand that science and its outcomes take place in the context of a scientific community and a larger society. There is no single authoritative source that makes judgments on the validity of scientific explanations. Instead, it is the collective body of scientists working in a given field who, because of their competence and experience, establish what is sound science and what is not. They do so by communicating their findings to each other and to the public as they publish their work in peer-reviewed journals. The process of peer review is crucial; it requires experts in a given field to analyze the results of their colleagues' work. Careful scrutiny is given to research publications, with the objective of rooting out poor or sloppy science and affirming work that is clearly meritorious.

Controversies. Even with these checks and balances that work to establish scientific knowledge and with all the

procedures of the scientific method, controversies arise that pit one group against another, with both groups claiming that they have science on their side. For example, we will make the claim that the global climate is changing because of human activities (Chapter 18). A few scientists disagree with this claim, although for most the evidence is convincing. There are at least four reasons why such controversies occur.

1. **New information.** We are continually confronted by new observations—the hole in the ozone layer, for instance (see Chapter 19). It takes some time before all the hypotheses regarding the cause of what we have observed can be adequately tested. During this time, there may be honest disagreement as to which hypothesis is most likely. Such controversies are gradually settled by further observations and testing, but the process leads into the second reason for continuing controversy.

2. **Complex phenomena.** Certain phenomena, such as the hole in the ozone layer, do not lend themselves to simple tests or experiments. Therefore, it is difficult and time consuming to confirm the causative role of one factor or to rule out the involvement of another. Gradually, different lines of evidence come together to support one hypothesis and exclude another, enabling the issue to be resolved. When is there enough evidence to say unequivocally that one hypothesis is right and another wrong? At some point, the scientists working in the appropriate field will reach a consensus in making that judgment.

3. **Multiple perspectives.** The third reason for controversy is that multiple perspectives may be involved. This is particularly true in environmental science because many issues have social, economic, or political ramifications. For example, there is no controversy regarding nuclear power, as long as it is considered at the purely scientific level of physics. However, when it comes to deciding whether to promote the further use of nuclear energy to generate electrical power, controversy arises because different groups have different perspectives about the relative costs, risks, and benefits involved.

4. **Bias.** The fourth reason for controversy is that there are many vested interests that wish to maintain and promote disagreement because they stand to profit by doing so. For example, tobacco interests argued for years that the connection between smoking and illness had not been established and that more studies were necessary. By harping on the absence of absolute proof (a scientific impossibility anyway) and downplaying the overwhelming body of evidence supporting the connection between smoking and illness, the tobacco lobby succeeded in keeping the issue controversial and thereby delayed regulatory restrictions on smoking. In fact, some businesses hire "product-defense firms," whose strategy is to do everything possible to challenge scientific evidence damaging to their products or interests.

Some controversy is the inevitable outcome of the scientific process itself, but much of it is attributable to less-noble causes. Unfortunately, the media and the public may be unaware of the true nature of the information and will often give equal credibility to opposing views on an issue. This can be seen, for example, in the media's coverage of global climate change, where the opinions of a few industry-sponsored deniers are often given equal coverage with the overwhelming scientific consensus. This leads us to the topic of junk science.

Junk Science. Junk science traditionally refers to information that is presented as scientifically valid but that does not conform to the rigors imposed by the scientific community. In recent years, another use of this term has appeared that has created some confusion. In some cases, people with a bias have used *junk science* to refer to anything that threatens their preferred viewpoint; conversely, *sound science* becomes anything that supports their viewpoint. One example of this is the use of the term "junk science" by people who do not support the Endangered Species Act. They label as "junk science" the legitimate use of computer models to make predictions about the survival of endangered species. This limits scientific input to a much smaller number of studies and could mean we make decisions without the full range of information we might have.

In contrast, true junk science can take many forms: the presentation of selective results (picking and choosing only those results that agree with one's preconceived ideas), politically motivated distortions of scientifically sound information, and the publication of poor work in quasi-scientific (unreviewed) journals and books. Quite often, junk science is generated by special interest groups trying to influence the public debate about science-related concerns. The result is that the junk science may be given equal respectability with the sound science in the public arena. Information presented by the Tobacco Institute, for example, is clearly suspect, due to its corporate-related funding. In particular, junk science has been heavily employed in the debate surrounding climate change and what to do about it in the United States.

Evaluating Issues. Whether scientist or layperson, we can use facets of the scientific method to deal with controversies and to develop our own capacity for logical reasoning. Here are some basic questions to ask:

- What are the data underlying a particular claim? Can they be satisfactorily confirmed?
- Do the explanations and theories follow logically from the data?
- Are there reasons that a particular explanation is favored? Who profits from having that explanation accepted broadly?
- Is the conclusion supported by the community of scientists with the greatest competence to judge the work? If not, it is highly suspect.

In sum, sound science is absolutely essential to forging a sustainable relationship with the natural world. Our

planet is dominated by human beings—our activities have reached such an intensity and such a scale that we are now one of the major forces affecting nature. Our influence on the natural ecosystems that support most of the world economy and process our waste is strong and widespread, and we need to know how to manage the planet so as to maintain a sustainable relationship with it. As a result, the information gathered by scientists needs to be accurate, credible, and communicated clearly to policy makers and the public, and the policy makers need to handle that information responsibly (the interaction between science and politics is explored in Chapter 2).

Stewardship

The third of our themes is **stewardship**—the actions and programs that manage natural resources and human well-being for the common good. Stewardship is a concept that can be traced back to ancient civilizations. A steward was put in charge of his master's household, responsible for maintaining the welfare of the people and the property of the owner. Because a steward did not own the property himself, the steward's ethic involved caring for something on behalf of someone else.

Applying this concept to the world today, stewards are those who care for something—from the natural world or from human culture—that they do not own and that they will pass on to the next generation. Modern-day environmental stewardship, therefore, incorporates an ethic that guides actions taken to benefit the natural world and other people. It recognizes that even our ownership of land is temporary; the land will be there after we die, and others will own it in turn. Stewardship is compatible with the goal of sustainability, but it is different from it, too, because stewardship deals more directly with how sustainability is to be achieved—what actions are taken, and what values and ethical considerations are behind those actions. (See Stewardship, "Exploring a Stewardship Ethic." p. 19.)

Who Are the Stewards? How is stewardship achieved? Sometimes stewardship leads people to try to stop the destruction of the environment or to stop the pollution that is degrading human neighborhoods and health. For example, Rachel Carson (from our opening story) was a steward who alerted the public about the dangers of pesticides and their role in decimating bird populations. Francisco Pineda (**Fig. 1–12**, on the next page) is a farmer in El Salvador who organized rural communities to protest the damages to local irrigation streams by Canadian mining giant Pacific Rim. Their challenges led the Salvadoran government to stop the mining by Pacific Rim, and Pineda received the 2011 Goldman Environmental Prize. Stewardship is not without risks, however: Pineda is under 24-hour police protection because three of his colleagues have been murdered by supporters of the mining project. The late Wangari Maathai (the first Kenyan woman to earn a Ph.D.) founded the Green Belt Movement to help rural Kenyan women. The grassroots movement eventually

Figure 1–12 Goldman Prize winner, 2011. At the risk of his life, Francisco Pineda organized opposition to a damaging mining operation by the Canadian mining company Pacific Rim. "My work is as an environmentalist; this is the principle for my life. When I began to understand the impact of the mining exploration, I couldn't stay quiet."

(*Source:* Goldman Environmental Prize).

Figure 1–13 Stewardship at work. Wangari Maathai receiving the 2004 Nobel Peace Prize from Ole Danbolt Mjøs, chairman of the Norwegian Nobel Committee. She is an example of unusual environmental stewardship at work.

involved more than 6,000 groups of women, who planted 47 million trees in Kenya. Maathai was beaten and jailed when she went on to protest government corruption. Eventually, she was appointed deputy environment minister under a new Kenyan president and received the Nobel Peace Prize in 2004 **(Fig. 1–13)**—the first environmental activist to receive that honor.

More often, stewardship is a matter of everyday people caring enough for each other and for the natural world that they do the things that are compatible with that care. These include participating fully in recycling efforts, purchasing cars that pollute less and use less energy, turning off the lights in an empty room, refusing to engage in the conspicuous consumption constantly being urged on them by commercial advertising, supporting organizations that promote sustainable practices, staying informed on environmentally sensitive issues, and expressing their citizenship by voting for candidates who are sympathetic to environmental concerns and the need for sustainable development.

Justice and Equity. The stewardship ethic is concerned not only with caring for the natural world, but also with establishing just relationships among humans. This concern for justice is seen in the United States in the *environmental justice movement*. The major problem addressed by the movement is **environmental racism**—the placement of waste sites and other hazardous industries in towns and neighborhoods where most of the residents are nonwhite.

The flip side of this problem is seen when wealthier, more politically active, and often predominantly white communities receive a disproportionately greater share of facility improvements, such as new roads, public buildings, and water and sewer projects.

People of color are seizing the initiative to correct these wrongs, creating citizen groups and watchdog agencies to bring effective action and to monitor progress. For example, Chattanooga's Southside community is populated largely by African Americans. This part of Chattanooga was bypassed when the city underwent a huge renewal that changed its reputation from being the dirtiest city in America to the "Renaissance City of the South" (see Chapter 23). Formed in 2010 by Southside community members, Chattanooga Organized for Action (COA) has quickly become a voice for social justice. In 2011, COA organized the restoration of funding for family planning programs for low-income and minority women and forced a more thorough cleanup of a former chemical plant site in Southside owned by Velsicol Chemical Corporation, among other actions.

Justice is especially crucial for the developing world, where unjust relationships often leave people without land, with inadequate food, and in poor health. Extreme poverty is the condition of more than a billion people; this poverty is often brought on by injustices within societies where wealthy elites maintain political power and, through corruption and nepotism, steal money and give corporations preferential treatment. It takes significant government reform to turn this situation around so that the poor can achieve justice in securing property rights and gain greater access to ecosystem goods and services.

Some of the poverty of the developing countries can be attributed to unjust economic practices of the wealthy

STEWARDSHIP   Exploring a Stewardship Ethic

One of the major concerns of philosophers is the development of codes of practice and thought that define right and wrong, known as ethics. Ethics is about the good—those values and virtues we should encourage—and about the right—our moral duties as we face practical problems. Ethics, therefore, is a "normative" discipline—it tells us what we ought to do. Some things are right, and some things are wrong. How do we know what we ought to do? In other words, how does an ethic work? A fully developed ethic has four ingredients:

1. **Cases.** These refer to specific acts and ask whether a particular act is morally justified. The answer must be based on moral rules.

2. **Moral rules.** These are general guidelines that can be applied to various areas of concern, such as the rules that govern how we should treat endangered species.

3. **Moral principles.** Moral rules are based on more-general principles, which are the broadest ethical concepts. They are considered to be valid in all cases. An example is the principle of distributive justice, which states that all persons should be treated equitably.

4. **Bases.** Ethical principles are justified by reference to some philosophical or theological basis. This is the foundation for an ethical system.

A stewardship ethic is concerned with right and wrong as they apply to actions taken to care for the natural world and the people in it. Because a steward is someone who cares for the natural world on behalf of others, we might ask, "To whom is the steward responsible?" Many

would answer, "The steward is responsible to present and future generations of people who depend on the natural world as their life-support system." For people with religious convictions, stewardship stems from a belief that the world and everything in it belongs to a higher being; they may be stewards on behalf of a higher power. For others, stewardship becomes a matter of concern that stems from a deep understanding and love of the natural world and the necessary limitations on our use of that world.

Is there a well-established stewardship ethic? Insofar as human interests are concerned, ethical principles and rules are fairly well established, even if they are often violated by public and private acts. However, there is no firmly established ethic that deals with care for natural lands and creatures for their own sake. Most of our ethic concerning natural things really deals with how those things serve human purposes; that is, our current ethic is highly anthropocentric. For example, a basic ethical principle from the UN Declaration on Human Rights and the Environment is "All persons have the right to a secure, healthy, and ecologically sound environment." Here, the point of the healthy environment is for humanity. There are several underlying reasons for a stewardship ethic, some stronger than others. Ethics for protection of other species will be covered later (Chapter 6).

Many organizations have developed stewardship principles for their own guidance, and these principles can sometimes serve as broader guidelines for society, although they usually address specific areas of concern and are often anthropocentric. One excellent group is the Forest Service Employees for Environmental Ethics (www.fseee.org), with a mission of protecting national forests and bringing reform to the U.S. Forest Service by advocating environmental

ethics. The group is active in halting unauthorized logging and protecting Forest Service employees from unjust disciplinary actions, as well as lobbying in Congress to maintain stewardship of the National Forests.

Here are some of the issues that a stewardship ethic must grapple with in order to be truly helpful in accomplishing our stated goal of providing the "actions and programs that manage natural resources and human well-being for the common good":

1. How to define the common good in cases where conflicting needs emerge, such as the economic need to extract resources from natural systems versus the need to maintain those systems in a healthy state.

2. How to balance the needs of present generations against those of future generations.

3. How to preserve species when doing so clearly means limiting some of the property rights commonly enjoyed by people and organizations.

4. How to encourage people and governments to exercise compassion and care for others who suffer profoundly from a lack of access to the basic human needs of food, shelter, health, and gainful work.

5. How to promote virtues that are conducive to stewardship of the natural world, such as benevolence, accountability, frugality, and responsibility. (There are many others.)

6. How to limit excessive consumption while at the same time allowing people the freedom to choose their lifestyles. In other words, how do you balance individual environmental rights and responsibilities with those of the broader community?

industrialized countries. The current pattern of international trade is a prime example. The industrialized countries have maintained inequities that discriminate against the developing countries by taxing and restricting imports from the developing countries and by flooding the world markets with agricultural products that are subsidized (and, therefore, priced below real cost). This is the number one issue at every meeting of the World Trade Organization, where global trade agreements are negotiated (see Chapter 2). Such barriers deprive people in developing countries of jobs and money that would go far to improve their living conditions. Although international justice is a thorny issue, it is part of the mission of stewardship to address it.

1.4 Moving Toward a Sustainable Future

Without doubt, one of the dominant trends of our times is globalization—the accelerating interconnectedness of human activities, ideas, and cultures. For centuries, global connections have brought people together through migration, trade, and the exchange of ideas. What is new about the current scene is the increasing intensity, speed, extent, and impact of connectedness, to the point where its consequences are profoundly changing many human enterprises. These changes are most evident in the globalization of economies, cultural patterns, political arrangements, environmental resources, and pollution. For many in the world, globalization has brought undeniable improvements in well-being and health: as economic exchanges have connected people to global markets, information exchanges have improved public health practices, and agricultural research has improved crop yields. For others, however, globalization has brought the dilution and even destruction of cultural and religious ideals and norms and has done little to improve economic well-being. Whether helpful or harmful, globalization is taking place, and it is important not only to recognize its impacts, but also to point out where its impacts are harmful or undesirable. Moving toward a sustainable future will occur in the context of globalization.

Social Changes

The major elements of globalization may be the economic reorganization of the world and changes wrought by social media. These transformations have been facilitated by the globalization of communication, whereby people are instantly linked through the Internet, satellite, and cable. Social media play a role not only in the economy, but also in the rapid mobilization of people during times of change such as the 2011 riots in Egypt, Greece, and Libya. Other components of the global economic reorganization are the relative ease of transportation and financial transactions and the dominance of transnational corporations (e.g., Nokia, Roche, Philips Electronics, Nestlé) with unprecedented wealth and power. Moreover, economic changes bring cultural, environmental, and technological changes (to name a few) in their wake. Western diet, styles, and culture are marketed throughout the world, to the detriment of local customs and diversity.

Environmental Changes

On the one hand, the increased dissemination of information has made it possible for environmental organizations and government agencies to connect with the public and has enabled people to become politically involved with current issues. It has also enabled consumers to find more environmentally friendly consumer goods and services, such as shade-grown coffee and sustainably harvested timber. On the other hand, globalization has contributed to some notoriously harmful outcomes, such as the worldwide spread of emerging diseases like the SARS virus, H1N1 swine influenza, and AIDS; the global dispersion of exotic species; the trade in hazardous wastes; the spread of persistent organic pollutants; the radioactive fallout from nuclear accidents; the exploitation of oceanic fisheries; the destruction of the ozone layer; and global climate change. These outcomes must be addressed in the move toward a sustainable future.

A New Commitment

People in all walks of life—scientists, sociologists, workers and executives, economists, government leaders, and clergy, as well as traditional environmentalists—are recognizing

that "business as usual" is not sustainable. Many global trends are on a collision course with the fundamental systems that maintain our planet as a tolerable place to live. A finite planet cannot continue to grow by 80 million persons annually without significant detrimental effects. The current degradation of ecosystems, atmospheric changes, losses of species, and depletion of water resources inevitably lead to a point where resources are no longer adequate to support the human population without significant changes in the way we live.

The news is not all bad, however. Food production has improved the nutrition of millions in the developing world, life expectancy continues to rise, and the percentage of individuals who are desperately poor continues to decline. Population growth rates continue to fall in many countries. A rising tide of environmental awareness in the industrialized countries has led to the establishment of policies, laws, and treaties that have improved environmental protection. Professional groups such as ecologists, climatologists, and engineers are proposing scientific and technological solutions to environmental problems.

Numerous caring people are beginning to play an important role in changing society's treatment of Earth. For example, in January 2011, a group of 70 engineers produced a report proposing engineering development goals[9] and stated, ". . . the engineers of today, and the future, will need to be innovative in the application of sustainable solutions and increasingly engaged with the human factors that influence their decisions." Likewise, people in business have formed the Business Council for Sustainable Development, economists have formed the International Society for Ecological Economics, religious leaders have formed the National Religious Partnership for the Environment, and environmental philosophers are calling for a new ethic of "caring for creation."

The movement toward sustainability is growing as people, businesses, and governments slowly but surely begin to change direction. Some examples: wind energy is the fastest-growing source of electricity; cap-and-trade markets are being established to curb carbon emissions; car-sharing companies are springing up in our cities; chemists have embraced the initiatives of "green chemistry"; microfinance is opening up people's small-scale business opportunities; "green investments" are attracting increasing capital; and the nations of the world are working on agreements to bring down greenhouse gas emissions. These developments show that people's behavior toward the environment *can* be transformed from exploitative to conserving.

---

[9]Institution of Mechanical Engineers, Population: One Planet, Too Many People? 2011. http://www.imeche.org/Libraries/2011_Press_Releases/Population_report.sflb.ashx. September 21, 2011.

REVISITING THE THEMES

It will be our practice in each chapter to revisit the three themes that we have introduced in this chapter, summarizing the ways in which the chapter topics have connected to the themes and sometimes adding editorial comments.

Sustainability

The story of DDT and other chemicals widely broadcast over vast areas of land only to be revealed later as serious water and soil pollutants is a great example of unsustainable actions. Many unsustainable practices result from a combination of cumulative impacts and unintended consequences. For example, we did not expect that DDT would affect the food chain. At first, DDT seemed a wise way to protect food supplies and eliminate malaria. As we gathered more information, we discovered the unintended consequences of those actions. Likewise, as the human population expands, the effects of individuals are multiplied. Both natural and managed ecosystems are being pressed to provide increasing goods and services. We may have reached the limit in some resources, and most of the ways we use the other resources are currently unsustainable. In addition, we extract and burn fossil fuels much faster than they could ever be replenished, bringing on global climate change, and we tolerate the continued loss of biodiversity. Many suffer from hunger and other effects of poverty. The Millennium Development Goals are moving us in a better direction, but it remains to be seen if those goals will be met. A movement toward sustainability is under way, and it will occur within the context of globalization.

Sound Science

Our brief look at the global environment was based largely on the information uncovered by sound science. The Millennium Ecosystem Assessment and the Intergovernmental Panel on Climate Change demonstrate the crucial value of scientific study of the status of and trends in global ecosystems and the ways in which human systems affect them. Environmental science itself is a multidisciplinary enterprise, one that can readily lead to disputes and misuse. To understand these, it is essential to understand the scientific method as a disciplined way to investigate the natural world. Rachel Carson is a good example of a scientist who communicated sound science to popular audiences. The issue of "junk science" points to the importance of a functioning scientific community and the need for an educated public able to evaluate opposing claims of scientific support for a given issue.

Stewardship

Although current cultures and religions place some value on stewardship, it is not always a fundamental part of the way we think about caring for the natural world and for our fellow humans. If it were, there would not be the widespread degradation of essential ecosystems and the devastating impacts of poverty and injustice. However, many people work toward care for the natural world and its human inhabitants. Rachel Carson is an example of a person using her expertise to promote stewardship and advocating for policies that would better reflect sustainability. Francisco Pineda and Wangari Maathai also exemplify stewardship in this chapter. As we will see throughout the book, it is no easy matter to establish a valid stewardship ethic to guide the actions and programs that manage natural resources and human well-being for the common good.

REVIEW QUESTIONS

1. What information made Rachel Carson concerned about chemical pollution?

2. What is the paradox of human population and well-being?

3. Cite four global trends that indicate that the health of planet Earth is suffering.

4. Define *environment*, and describe the general development and successes of modern environmentalism.

5. What are the concepts behind each of the three unifying themes—sound science, sustainability, and stewardship?

6. Explain the role of assumptions, observation, experimentation, and theory formation in the operation of scientific research and thinking.

7. Cite some reasons for the existence of controversies within science, and carefully distinguish between sound science and junk science.

8. Define *sustainability* and *sustainable development*. What is a sustainable society?

9. List five transitions that are necessary for a future sustainable civilization.

10. Describe the origins of stewardship and its modern applications.

11. What is an ethic, and what are some of the difficult questions confronting the stewardship ethic?

12. What are the concerns of the environmental justice movement?

13. How does justice become an issue between the industrialized countries and the developing countries?

14. What is globalization? What are its most significant elements?

15. Cite some of the recent developments in the movement toward sustainability.

THINKING ENVIRONMENTALLY

1. Imagine a class debate between people representing the developing countries and people representing the developed countries. Characterize the arguments of the two sides in terms of the issues surrounding population growth, energy use, resource use, and sustainable development. Then describe what should be the common interests between the two.

2. Some people say that the concept of sustainable development either is an oxymoron or represents going back to some kind of primitive living. Develop an argument contradicting this opinion, and give your own opinion on sustainable development.

3. Study Table 1–1, pick one of the vital ecosystem services listed there, and investigate the reasons why it is placed where it is by the Millennium Ecosystem Assessment.

4. Set up a debate between proponents and opponents of building a new coal-fired power plant. Use the strategies of sound science and junk science to conduct the debate, and show how a perverted use of these terms can help to obscure the real issues. Other interesting debate options include the "ozone hoax" and global warming; the Internet can provide ample material for both sides of the debate.

MAKING A DIFFERENCE

1. Find out whether there is an environmental organization or club on your campus or in your community. Join it if there is, or help to create one if there isn't. Once in the organization, join efforts to maintain awareness of important environmental issues and problems.

2. Use letters to the local newspapers or visits to town meetings to raise awareness in your community about sustainable systems and sustainable societies. Examples of issues to raise include smart growth; cluster zoning, which provides a healthy mix of open space and buildings; recycling; and ways to encourage public transit systems.

3. Encourage sound science on a national level by researching and voting in favor of political candidates who support your environmental viewpoint. Share your views with your friends and relatives.

4. Investigate the placement of waste-treatment facilities and industrial plants in your city or community, especially if they are in the planning stage. If they are aimed at poorer or largely nonwhite communities, join the local population in protesting these situations.

MasteringEnvironmentalScience®

Go to **www.masteringenvironmentalscience.com** for practice quizzes, Pearson eText, videos, current events, and more.

An observer looks out over the Benxi Iron and Steel Group in Liaoning, China.

CHAPTER 2

Economics, Politics, and Public Policy

BY ALL MEASURES, China has become the economic giant of the early 21st century. Take your pick: umbrellas, toys, shoes, appliances, electronics, souvenirs, clothes—chances are that they were made in China. China has become the factory to the world. The Chinese economy has doubled in the past decade, growing at close to 10% per year. In contrast, the U.S. economy grew only 2% per year over the same period. A new Chinese middle class numbers between 100 million and 150 million. Middle-class Chinese citizens commonly own a car and an apartment and have opportunities to travel and purchase pricey consumer goods. With a population of 1.35 billion, China has a huge pool of labor. Now that the extreme control of private life is a thing of the past, people are able to choose where they live, work, and travel. Millions of rural Chinese move to urban centers each year, attracted by jobs created by factories that spring up almost overnight. Chinese schools have created a workforce with a literacy rate that is over 95%. Because of the accelerating pace of cultural life, children are now guiding their parents into the realities of modernity. Traditions are being overthrown by the Internet, by the fierce drive to learn English, and by the social mobility that accompanies the economic boom.

TOXIC AIR. There are clear signs that China's race to achieve economic superpower status carries with it some costs that go well beyond

LEARNING OBJECTIVES

1. **Economics and the Environment:** Describe how economic activity relates to environmental goods and services, and differentiate between green and brown economies.

2. **Resources in a Sustainable Economy:** Summarize the components of the wealth a nation draws on to establish and maintain an economy, and identify new efforts to measure true economic progress.

3. **Environmental Public Policy:** Explain the kinds of policies employed to regulate the use of natural resources and deal with pollution, and describe a typical policy life cycle.

4. **Benefit-Cost Analysis of Environmental Public Policy:** Discuss how benefit-cost analysis is applied to environmental policy regulations, and give examples of the impact of regulations.

5. **Politics and the Environment:** Assess the role played by partisan politics in recent environmental public policy.

Figure 2–1 Smog in Linfen, China. The city is reputed to have the worst air in China.

a stretching of the social fabric. Seven Chinese cities are among the "10 most polluted places" in the world, according to the World Health Organization. The air is loaded with sulfur dioxide from coal-burning power plants and coal stoves, and acid rain costs some $4.3 billion annually in crop damage. Up to 750,000 premature deaths a year are believed to be traceable to air pollution. In the city of Linfen in Shangxi province, the hospital treats countless numbers of patients with respiratory disease each day. The city is dotted with coke ovens, coal-processing plants, factories, and a coal-fired power plant (Fig. 2–1). It has been at the top of China's pollution charts for years. One of its hamlets, Beilu, is known as "the cancer village" because of the extremely high death rate of those 55 and older.

TOXIC WATER. Xiao Sizhu, a farmer in the village of Xiaojiadian, died at 55 of esophageal cancer. In an interview shortly before his death, Xiao spoke of swimming and fishing in the tributary of the Yellow River that runs through the village. However, he says, "Now I never go close to the water because it smells awful and has foam on top." One oncologist calls the area "the cancer capital of the world" because of esophageal and stomach cancer rates that are 25 times higher than the national average. The Yellow River itself is being sucked dry by irrigation projects stemming from some 20 dams, and the remaining water is grossly polluted with sewage and industrial wastes. Three chemical spills in 2006 turned the Yellow River red, and another spill turned it white. More than 500 million rural Chinese are forced to use water that is contaminated by human and industrial waste; only 20% of wastewater receives any treatment.

ENVIRONMENTAL POLICY. The Chinese economy has flourished in large degree because of the decentralization of power in the 1980s, when local governments and private indus-

tries were encouraged to develop their resources relatively free of the restraints of central planning. The booming economy has also put great power in the hands of the local governments and industries and has enabled them to flaunt existing environmental laws and policies. China does have its Ministry of Environmental Protection, similar to the U.S. Environmental Protection Agency (EPA), but the country is so large that the ministry has been unable to keep up with the need for enforcement. Recently, the Chinese government has been turning to international nongovernmental organizations (NGOs) and has encouraged concerned citizens to form homegrown NGOs that focus attention on pollution and other environmental issues. At least 2,000 such groups are now registered, and they are beginning to make a difference. Typically, a group—like Green Camel Bell in the western Chinese city of Lanzhou—identifies a case of blatant pollution by some factory, documents the pollution, and reports it to the central government. Green Camel Bell was successful in shutting down more than 30 factories dumping wastes into tributaries of the Yellow River.

BOTTOM LINE. China's example demonstrates that economic growth can lift millions out of poverty and establish a country's strong place in the world market. But when this growth is at the cost of natural resources and the health of many of its people, it is clearly unsustainable. The economic development happening in China has already taken place in the industrialized countries of Japan and the West, but in these countries, pollution has been addressed by environmental laws and regulations. The Chinese example shows us that economic growth must be accompanied by the development and enforcement of just and effective environmental public policies.

We look first at economics and its connections to the natural environment.

2.1 Economics and the Environment

Economics is *the social science that deals with the production, distribution, and consumption of goods and services and with the theory and management of economies or economic systems.* Because economic goods and services are usually connected to—and indeed depend on—environmental goods and services, economics needs to be grounded in an understanding of the biological and physical world. Simply put, no environmental goods and services, no economy.

Relationships Between Economic Development and the Environment

In a human society, an **economy** is *the system of exchanges of goods and services worked out by members of the society.* Goods and services are produced, distributed, and consumed as people make economic decisions about what they need and want and what they will do to become players in the system—what they might provide that others would need and want. As societies develop, economic activity assumes increasingly broader dimensions, with increasingly pervasive impacts on the whole society. In future chapters, we will see numerous instances in which pollution from economic activity has damaged the environment and human health. Furthermore, an unregulated economy can make intolerable inroads into natural resources. In particular, during the latter part of the 19th century in the United States, private enterprise had unrestrained access to forests, grazing lands, and mineral deposits. This unsustainable exploitation of resources began to be addressed at the turn of the century, as government rules and regulations imposed necessary limits.

Fortunately, economic activity in a nation can also provide the resources needed to solve the problems that this very activity creates. A strong relationship exists between the level of development of a nation and the effectiveness of its environmental public policies. **Figure 2–2** shows three patterns that describe relationships between various environmental problems and per capita (per individual in the population) income levels.

- Many problems decline (for example, sickness due to inadequate sanitation or water treatment) as income levels rise, because the society decides to apply the resources available—usually from taxes—to address the problems with effective technologies.

- Some problems increase and then decline (for example, urban air pollutants like sulfur dioxide, particulate matter, and carbon monoxide) when the serious consequences of the problem are recognized and public policies are developed to address them.

- Increased economic activity causes some problems to increase without any clear end in sight (for example, suburban sprawl, municipal wastes, and CO_2 emissions).

The key to solving all of these problems brought on by economic activity is the *development of effective public policies and institutions.* When this does not happen, environmental degradation and human disease and death are the inevitable outcomes. What is happening today in China is a sad case in point. To understand this relationship better, let us look at how economic systems work.

Economic Systems

Economic systems are *social and legal arrangements people construct in order to satisfy their needs and wants and improve their well-being.* Economic systems range from

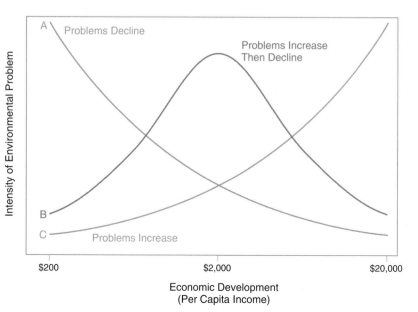

Figure 2–2 Environmental problems and per capita income. Some of the most serious environmental problems can be improved with income growth; other problems get worse and then improve, while some problems just worsen.

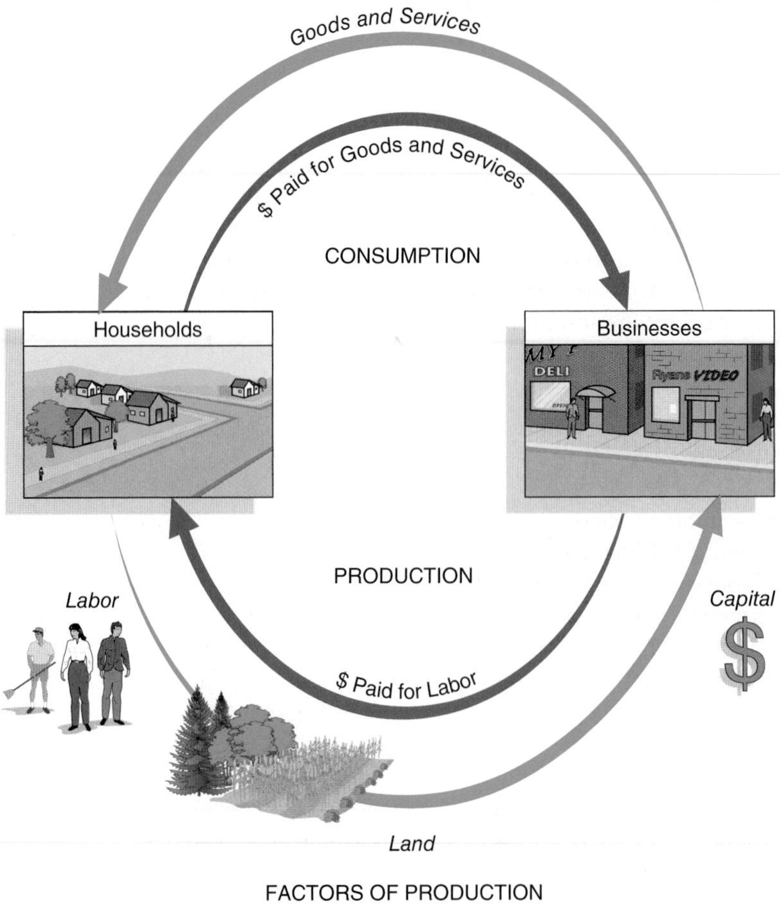

Figure 2–3 **Classical view of economic activity.** Land (natural resources), labor, and capital are the three elements constituting the "factors of production" (bottom green arrow). Economic activity involves the circular flow of money (blue arrows) and products (top green arrow).

the communal barter systems of primitive societies to the current global economy, which is so complex that it defies description. In the organized societies of today's world, two kinds of economic systems have emerged: the **centrally planned economy**, characteristic of dictatorships, and the **free-market** (or **capitalist**) **economy**, characteristic of democracies.

In any kind of economy, there are basic components that determine the economic flow of goods and services. Classical economic theory considers **land** (natural resources), **labor**, and **capital** as the three elements that constitute the "factors of production." *Economic activity* involves the circular flow of money and products, as shown in **Figure 2–3**. Money flows from households to businesses as people pay for goods and services and from businesses to households as people are paid for their labor. Labor, land, and capital are invested in the production of goods and services by businesses, and the products provided by businesses are "consumed" by households.

The Rulers Decide. The cycles shown in Figure 2–3 apply to both kinds of economic systems in existence today; the two systems differ mostly in how economic decisions are made. In a centrally planned economy, the rulers make the basic decisions regarding what and how much will be produced where and by whom. The government owns or manages most industries, and the workers often lack the freedom

to choose or change jobs. This system was most prominent during the 20th century in the communist regimes of the Soviet Union and its satellites and in Mainland China and some smaller Asian countries. It proved to be a disaster, both economically and environmentally. The failure of the centrally planned economy in the former Soviet Union led to economic chaos in the countries that it once comprised or dominated. North Korea, Cuba, and a few African countries are the last to retain this economic model.

The Market Decides. In a pure free-market economy, the market itself determines what will be exchanged. Goods and services are offered in a market that is free from government interference. The system is completely open to competition and the interplay of supply and demand. If supply is limited relative to demand, prices rise; if there is an oversupply, prices fall because people and corporations will economize and pay the lowest price possible. The whole system is in private hands and is driven by the desire of people and businesses to acquire goods, services, and wealth as they act in their self-interest. All "players" have free access to the market. Competition spurs efficiency as inferior products and services are forced out of the market. People can make informed decisions about their purchases because there is sufficient information about the benefits and harms associated with economic goods. The free-market economic system is thought to be at its best when left completely alone.

The Role of Governments. The preceding discussion describes only the very basics of the two economic systems. In reality, no country has a pure form of either economy. Recently, a hybrid economic system has emerged, called **state capitalism**, most fully developed in China, Russia, and the Gulf oil states. These countries, rich in natural resources or in abundant labor (China), have used their exports to amass capital that has enabled them to invest in building up the country's infrastructure and form state-owned corporations. Although major players in the global economy, these are countries where democracy is restricted. As they develop, their people are increasingly joining a "global middle class," and the tensions between state control of the economy and political freedom are becoming evident (as in the 2011 "Arab spring").

The developed countries and many developing countries are democracies with market economies, but even in these, government involvement occurs at many levels. Governments can, for example, own and operate different facets of the country's infrastructure, such as postal services, power plants, and public transport systems. Governments may also control interest rates, determine the amount of money in circulation, and adopt policies that stimulate economic growth in times of slowdown. Governments also maintain surveillance over financial processes such as stock markets and bank operations (although not always well, based on the U.S. 2008 market meltdown). On the downside, powerful business interests can manipulate facets of a market economy, and unscrupulous people can exploit the freedom of the market to defraud people. Only governments can provide the policing that the society needs.

Is There a Conscience? Many people believe that the conventional capitalist economic system lacks a "conscience." They argue that many workings of the market economy create hardships for people (and countries) at the bottom of the economic ladder. For example, if population growth is rapid and there are not enough jobs to go around, the result often is exploitation of workers. Furthermore, a market economy only offers people *access* to goods and services; if people lack the means to pay, access alone will not meet their needs. In this situation, a society could employ a "safety net" to provide food security for impoverished people—something the market does not do (see Chapter 12). Also, it is too easy for self-interest to lead businesses and individuals to exploit natural resources or avoid costs by polluting the environment instead of producing a "clean" product.

Presumably acting in the self-interest of their people, governments also find ways to manipulate the market in their own favor. This is especially true for international trade, where protective barriers and subsidies prevent the free flow of goods and services. The World Trade Organization (WTO) is supposed to enforce trade rules between nations, but wealthy nations have been able to dominate the WTO to the detriment of other countries.

International Trade and the World Trade Organization

Trade is a fundamental economic activity. The exchange of goods drives the engine of the free-market economy. In today's world, trade has become increasingly globalized, and in the process, world trade has become one of the most dominant forces in society. However, many question whether this is good, and their concerns become headline news every time the WTO meets. For example, in Seattle, Washington, in December 1999, tens of thousands of protesters (called "globophobes" by some) took to the streets to block the delegates from assembling—and were subjected to tear gas, pepper spray, and rubber bullets as things began to turn ugly **(Fig. 2–4)**. The

Figure 2–4 Protestors at the WTO meeting. The World Trade Organization meeting in Seattle, Washington, in 1999, drew thousands of protestors, who were repelled by police using tear gas and water cannons.

protesters were an unusual alliance of environmental activists and organized labor, groups that are often at odds over timber harvests and fishing. What were their issues?

The WTO was created in 1993 by trade ministers from many countries meeting to build on the foundation for global trade laid by the General Agreement on Tariffs and Trade (GATT). The WTO was given the responsibility for implementing numerous trade rules and, in doing so, was empowered to enforce the rules, with the option of imposing stiff trade penalties on violators. In effect, the WTO became a force for the globalization of trade, facilitating it by encouraging the reduction or elimination of trade barriers (tariffs) between countries and of subsidies to domestic industries that tended to discriminate against foreign-produced goods. Those who support the WTO claim that free trade between countries is the key to global economic growth and prosperity. This all sounds very good, but as always, the devil is in the details.

Free-Trade Rules. Since its creation, the WTO has adjudicated a number of international trade issues and, in the process, has established a reputation for elevating free trade over substantial human rights and environmental resource concerns. Furthermore, many trade agreements and the dispute rulings are carried out in closed sessions. For example, the WTO ruled against a U.S. law requiring any imports of shrimp to be certified as turtle safe (that is, the shrimp were to be certified as harvested by trawlers employing "turtle exclusion devices," or TEDs). Even though TEDs have been proven to reduce turtle mortality by 97%, the WTO ruled that the restriction was a violation of WTO rules. Other issues that have come before the WTO are a U.S. ban on the import of wooden crates from China because they harbored the Asian long-horned beetle (a serious threat to forest trees) and bans against developing countries' products that utilize child labor.

More Talk. In the Seattle meetings, the talks broke down not because of the protests, but because of a refusal of the key players—the United States, Japan, and Europe—to move from their positions. In the WTO meetings held in Cancún, Mexico, in September 2003, the talks broke down again before negotiations could proceed. This time a coalition of 20 developing nations (G-20) representing 60% of the world's people simply walked out of the meetings. Their issue was primarily the import barriers and subsidies that distort global trade in agricultural products. The developed nations are major exporters of agricultural products, and they subsidize their farmers heavily; the result is depressed world markets for these products. The developing nations want to protect their farmers from competition with these low-priced farm products, so they impose tariffs.

The future of the WTO is unclear. The **Doha Development Round** of WTO negotiations began in 2001 with the objective of lowering trade barriers around the world, but the subsequent WTO meetings in Seattle, Cancún, Hong Kong (2005), Geneva (2006), Davos (2007), and Geneva (2008) have failed to reach agreement on the issues of farming subsidies and trade barriers. Currently, there is limited hope that the Doha negotiations will come to fruition; the developed nations continue to insist on maintaining subsidies to agriculture, and the developing nations remain firm in their resolve to protect their industries and farmers from unfair competition.

The WTO supposedly retains the power already delegated to it, but the continued failure to reach agreements over tariffs and subsidies signals a serious weakening in global trade governance. The likely outcome is a proliferation of bilateral arrangements between countries, which may not be all bad, given the tendency of the developed countries to dominate WTO issues in the past.

The Need for a Sustainable Economy

The global economy has proven to be quite successful in mobilizing people to engage in economic activity. The Chinese experiment is a case in point. The global economy has also been successful in improving the well-being of people: Life expectancy has never been higher; labor-saving devices remove much of the toil in wealthy countries; and aircraft, cars, computers, and cell phones open up lifestyles and work opportunities unimaginable a few decades ago. However, the global economy has also helped to create environmental problems: continued poverty and hunger in developing countries, the decline of ecosystem goods and services, rising global temperatures and seas, and the loss of biodiversity (Chapter 1). Given the vital connections between the environment and the economy, there is a clear need for a different kind of economy; "business as usual" does not work. The path to a sustainable future will require a sustainable, *green economy*.

What will such an economy look like? Instead of promoting growth, the primary goal of the economy will be improving human well-being. Instead of drawing down natural capital, the economy will value and preserve ecosystem services and goods. Instead of plunging ahead with "business as usual" technologies, societies will embrace technologies that reduce pollutants and use energy more efficiently. A rising consumer demand for green products—products like compact fluorescent bulbs, hybrid and electric vehicles, solar and wind power systems, and organic foods—will be met by businesses eager to provide them. These technologies, and the economy they will build, are in contrast to what can be called the *brown economy*—all the current economic activities that are contributing to environmental degradation and resource depletion. In fact, this shift is revolutionary, and appropriately, the UN Department of Economic and Social Affairs has titled its 2011 World Economic and Social Survey publication "The Great Green Technological Transformation."[1]

---

[1]United Nations Department of Economic and Social Affairs. World Economic and Social Survey 2011: The Great Green Technological Transformation. New York: United Nations, 2011.

2.2 Resources in a Sustainable Economy

What are the resources on which a country draws to establish and maintain an economy? The classic economic paradigm says that *land* (which represents the environment), *labor*, and *capital* are the essential resources needed for a country to be able to mount its economy. However, a new breed of economists, often called **ecological economists**, has emerged in recent years and has taken issue with this view. These economists point out that the classic view (Fig. 2–3) sees the environment simply as one set of resources within the larger sphere of the human economy. The environment's vital role in supplying the goods and services on which human activities depend is minimized. They argue that the classical approach looks at things backward and that the natural environment actually *encompasses* the economy, which is constrained by the limits of resources in the environment **(Fig. 2–5)**. The production and consumption that define the classical economic view are just part of the system. Without the vital raw materials provided by the environment and without the capacity of the environment to absorb wastes (pollution), there is no economy. Thus, the ecosystems and natural resources found in a given country provide the context for that country's economy. Similarly, the total global biosphere is the context for the global economy.

As described by Thomas Prugh, Robert Costanza, and Herman E. Daly in *The Local Politics of Global Sustainability*, **economic production** can thus be seen as it really is: *the process of converting the natural world to the manufactured world*. Renewable and nonrenewable resources and the ecosystems containing them are turned into cars, toys, books, food, buildings, highways, computers, and so forth. Thus, to quote these authors, "Resources flow into the economy from the enfolding ecosystem and are transformed by labor and capital (using energy, also a resource), and then pass out of the economy and back into the ecosystem in the form of wastes."[2] (See Figure 2–5.)

The ecological economists' view emphasizes the role of natural ecosystems as essential life-support elements—*ecosystem capital* (Chapter 1). Their approach brings sustainability sharply into focus. If the economy continues to grow, the natural world continues to shrink. In time, economic production and consumption must come to a steady

---

[2]Thomas Prugh, Robert Costanza, and Herman E. Daly, *The Local Politics of Global Sustainability*. Washington, DC: Island Press, 2000, p. 19.

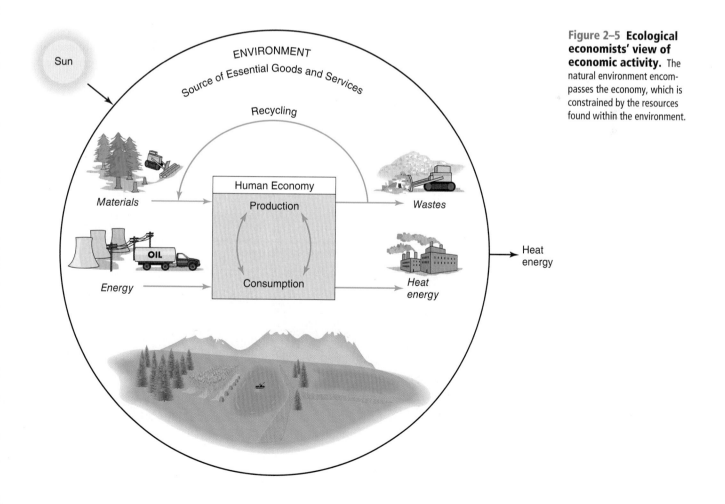

Figure 2–5 Ecological economists' view of economic activity. The natural environment encompasses the economy, which is constrained by the resources found within the environment.

state, and to be sustainable, this must happen before our natural capital is consumed beyond its ability to support our economy. If we add to ecosystem capital the nonrenewable mineral resources, such as fossil fuels and metal ores, we can use the more inclusive term **natural capital**. Thus, *natural capitalism* would be an economic system that promotes sustainability and recognizes the central role of natural ecosystems. It would replace *conventional capitalism*. It would be a green economy.

Without a doubt, the ecosystems and mineral resources of a given country—its *natural capital*—are a major element in the wealth of that country. In recent years, the World Bank's Environment Department has been working on ways of measuring the wealth of nations and has produced some insightful analyses, which we consider next.

Measuring the Wealth of Nations

Environmentally sustainable development means improving human well-being over time, a process that means everything to the world's impoverished people. It is a process that requires societies to manage all of their assets that can contribute to improving human well-being. What are the components of a nation's wealth? In a classic document titled *Where Is the Wealth of Nations? Measuring Capital for the 21st Century*,[3] the World Bank indicates three major components of capital that determine a nation's wealth: *produced capital, natural capital,* and so-called *intangible capital.*

Produced capital (Fig. 2–6a) is the human-made buildings and structures, machinery and equipment, vehicles and ships, monetary savings and stocks, highways and power lines, and so forth that are essential to the production of economic goods and services. Produced capital is often the major focus of national economic planning, and it is the most easily measured of the three components of a nation's wealth. Produced

[3]World Bank, *Where Is the Wealth of Nations? Measuring Capital for the 21st Century.* Washington, DC: World Bank, 2006.

capital assets have very clear income-earning potential, but they may also be subject to obsolescence and must be renewed continually. For example, a clothing factory produces a flow of goods destined for consumers, but the machinery wears out over time, and the building itself ages. Thus, we have income—a flow of goods—and depletion, which is referred to as capital consumption.

Natural capital (Fig. 2–6b) refers to the goods and services supplied by natural ecosystems and the mineral resources in the ground. Natural capital represents an indispensable set of resources, some of which are *renewable* and some *nonrenewable*. Renewable natural capital (*ecosystem capital*) is represented by forests, fisheries, agricultural soil, water resources, and the like, which can be employed in the production of a flow of goods (lumber, fish, grain). Often overlooked is the fact that this same natural capital also provides a flow of services in the form of waste breakdown, climate regulation, and oxygen production. (The World Bank authors offer a caution in interpreting their work: they did not assess the service components of natural capital because of difficulties in measuring them and calculated mainly the actual resource inputs into production—that is, forests as a source of timber, mineral assets, etc.).

As we will see in subsequent chapters, renewable natural capital is subject to depletion, but it also has the capacity to yield a sustained income if it is managed wisely. Nonrenewable natural capital, such as oil and mineral deposits, also is subject to depletion and provides no services unless it is extracted and converted into some useful form. It is part of the wealth of a nation, however, as long as it is in the ground. Some components of natural capital are easier to measure than others. For instance, the lumber of a forest can be reckoned in board feet, but how does one measure the impact of a forest on the local climate and on its capacity to sustain a high level of biodiversity?

Intangible capital (Fig. 2–6c) is divided into three elements. The first is **human capital**, which refers to the population and its physical, psychological, and cultural attributes

| (a) Produced capital | (b) Natural capital | (c) Intangible capital |

Figure 2–6 The wealth of nations. Three major components of a nation's capital determine its wealth: (a) produced capital, (b) natural capital, and (c) intangible capital, which largely consists of human capital and the quality of formal and informal institutions.

(innate talents, competencies, abilities, etc.). To this is added the value imparted by education, which enables members of a population to acquire skills that are useful in an economy. Education can be either formal or informal; the point is that people acquire a productive capacity through learning. Investments in health and nutrition also contribute to increases in human capital.

The second element of intangible capital is what the World Bank has called **social capital**—the social and political environment that people create for themselves in a society. Social capital includes more formal structures and relationships as defined by government, the rule of law, the court system, and civil liberties. It also includes the horizontal relationships of people as they associate into religious or ethnic affiliations or join organizations in which they have a common interest. Social capital is considered a vital element of a nation's wealth, because social relationships clearly affect, and are affected by, economic processes.

The third element of intangible capital is **knowledge assets**—the codified or written fund of knowledge that can be readily transferred to others across space and time. Although these assets constantly grow, they are useful only if people have access to them (hence, the vital importance of libraries, schools, universities, and the Internet).

New Work by the World Bank

The World Bank's Environment Department devised various means of measuring each of these components of the wealth of nations and then set about evaluating the wealth of different nations (their assets). In a recent extension of this effort,[4] a 10-year record of wealth accounting has provided some new insights into wealth measurements. Some of the results of the World Bank's analysis are shown in Table 2–1. One finding that should come as no surprise is the notable differences in per capita wealth among groups of nations ranked by income.

Interestingly, the major source of wealth for most nations is intangible wealth, but natural capital ranks high in a number of countries. Although natural capital often ranks

[4]World Bank, *The Changing Wealth of Nations: Measuring Sustainable Development in the New Millennium.* Washington, DC: World Bank, 2011.

third in many countries, this does not mean that it is unimportant. In particular, natural capital plays a more important role in low-income countries (Table 2–1), a signal that as they develop, they use these assets to build the other forms of wealth. The 10-year analysis performed by the World Bank indicates that per capita total wealth has grown by 17%, most notably in the lower-middle-income countries. This grouping is dominated by China, whose human capital doubled over the 10-year period—a consequence of China's recent heavy investment in education.

The dominance of the intangible (human resources and social assets) component emphasizes the importance of investing in health, education, and nutrition in a society that desires to move further along in development. It also emphasizes the importance of having an efficient judicial system, clear property rights, and an effective and uncorrupt government—elements of social capital. In these key works on defining the wealth of nations, the World Bank has provided the basics of a tool for measuring true economic progress. It is the *growth in wealth, and all the components of that wealth*—not the growth in gross national product (GNP), which was once commonly employed to indicate economic success, or in its now more commonly used alternative, gross domestic product (GDP).

Measuring True Economic Progress

The gross national product, or GNP, refers to *the sum of all goods and services produced* (really, consumed) *by a country in a given time frame.* The GNP was once the most commonly used indicator of the economic health and wealth of a country. Most countries now use the comparable *gross domestic product*, or GDP. The GDP is the GNP minus net income from abroad. As a *per capita* index, the GDP is often used to compare rich and poor countries and to assess economic progress in the developing countries. An important element in calculating a *net* GDP is accounting for the assets that are used in the production processes. Buildings and equipment, for example, are essential to production, but they gradually wear out or depreciate. Their depreciation is charged against the value of production (called *capital depreciation*); that is, an accounting of capital depreciation is routinely subtracted from the production of goods and

TABLE 2–1 Total Wealth and Per Capita Wealth by Type of Wealth, 2005

Income Group	Total Wealth ($ billions)	Per Capita Wealth ($ dollars)	Intangible Capital (%)	Produced Capital (%)	Natural Capital (%)
Low income	3,597	6,138	57	13	30
Lower middle income	58,038	16,903	51	24	25
Upper middle income	47,183	81,354	69	16	15
High income	551,964	588,315	81	17	2
World	673,593	120,475	77	18	5

Source: World Bank, *The Changing Wealth of Nations: Measuring Sustainable Development in the New Millennium.* Washington, DC: World Bank, 2011.

services in order to generate a net GDP. The GDP measurement is flawed, however, because it does not include goods created and services performed by a family for itself. Thus, subsistence farming in developing countries is not measured, and this omission surely produces a falsely lower GDP for these countries. The GDP also omits the natural services provided by ecosystems. For example, even though clean air is as important in treating asthma as medicine is to asthmatics, only the medicine is included in the GDP. There is a more fundamental flaw, however.

Natural Capital Depreciation. The economists who invented the GNP as a measuring device years ago never took into account the depreciation of natural capital—an omission that has been pointed out by ecological economists and others. Such concerns are regarded as external to a country's balance sheet. Following the classical view of economics, the natural resources and their associated natural services are considered a "gift from nature." Given this state of affairs, it is possible for a nation to cut down a million acres of forest and count the sale of the timber on the income side of the GDP ledger, whereas on the expense side, only the depreciation of chain saws and trucks would be included. Completely hidden from accounting is the loss of all the natural services once performed by the forests, and incredibly, the disappearance of the million acres of forest is shown as an economic asset. As long as this discrepancy remains, it is possible for nations to underestimate the value of their natural resources. They are able to deplete fisheries, lose soil by intensive farming, remove forests, and degrade rangelands by overgrazing and still *account for these activities as economic productivity*. The GDP can increase due to production by industries that deplete and pollute the environment and can increase again with cleanups of that pollution—a double benefit!

Correcting this system of accounting would require calculating the current market value of natural assets and considering it to be part of the stock of a nation's wealth, as recommended by the World Bank. To this value would be added the value of the natural services performed by the ecosystems in which the resources are found (omitted by the World Bank analysis). The natural assets and the services they perform are the nation's natural capital. When the natural resources are drawn down, the depreciation represented by the loss of natural capital would be entered into the ledger as the GDP is calculated. Similar arguments can be made for human resources as a measure of a nation's wealth. A more accurate GDP would have to take all of these depreciations into account.

Genuine Progress Indicator. If the GDP is simply unable to measure true economic progress, some other index is necessary. The nonprofit organization Redefining Progress has proposed the **Genuine Progress Indicator (GPI)** as a measure of economic progress. This index is calculated by assuming that some kinds of economic activity are positive and sustainable and others are not, even if the latter are included in the GDP. Some of the positive factors in the

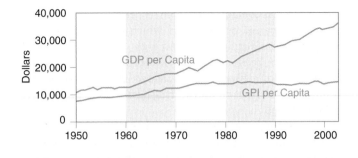

Figure 2–7 GDP and GPI per capita, 1950–2004 (in year 2000 dollars). The GDP (gross domestic product) is the conventional measure of economic progress used by economists. The GPI (Genuine Progress Indicator) is a more realistic measure of progress. Note how it is falling behind the GDP over the years.

(*Source:* John Talberth, Clifford Cobb, and Noah Slattery, *The Genuine Progress Indicator 2006: A Tool for Sustainable Development*, Redefining Progress, 2006, with permission.)

GPI—such as labor that goes into housework, parenting, and volunteering—are not calculated in the GDP. Negatives in the GPI include factors like the costs of crime, loss of leisure time, costs of pollution, depletion of nonrenewable resources, and loss of farmland. Thus, for the United States in 2004, positive contributions to the GPI totaled $11.06 trillion, while negative contributions were $6.64 trillion, yielding a net GPI of $4.42 trillion. In contrast, the 2004 GDP was $11.71 trillion. Over the years, the per capita GPI has remained fairly constant, while the GDP has risen sharply (**Fig. 2–7**), suggesting that the benefits of GDP growth are increasingly offset by the rising environmental and social costs of economic activity. The thinking behind the GPI has been carefully reviewed in the scientific literature, and several countries, the state of Maryland, and Canadian provinces have made a first attempt to measure and apply the GPI as an index of sustainability.

Environmental Accounting. One of the major documents that came out of the 1992 Earth Summit in Rio de Janeiro, *Agenda 21*, recognized the need for "a program to develop national systems of *integrated* environmental and economic accounting in all countries." The UN's Statistical Division, which coordinates the accounting procedures of member countries, has prepared a manual designed to accomplish *Agenda 21*'s objective (*Handbook of National Accounting: Integrated Environmental and Economic Accounting*). In the manual, countries are encouraged to perform environmental accounting by putting their environmental assets and services in monetary units and keeping a parallel account of their *net domestic product* (similar to the GDP) as it is affected by environmental accounting. The document includes case studies and step-by-step instructions and strongly advocates the adoption of integrated environmental and economic accounting by UN member nations.

In 2007, environment ministers from the leading global economies initiated a study intended to show how to apply economic thinking to the important ecosystem services that are so beneficial to societies. That study, *The Economics of*

Ecosystems and Biodiversity (TEEB), has produced seven reports, culminating in a 2010 final report[5] that incorporates elements of former reports and shows "how economic concepts and tools can help equip society with the means to incorporate the values of nature into decision making at all levels," to quote from the preface. This work, coordinated by the UN Environment Program (UNEP), shows how payments for ecosystem services can improve a country's stewardship of ecosystem capital. The good news is that many developing countries are working with the UNEP team to put TEEB into practice.

Resource Distribution

None of the resources that are crucial to an economy are evenly distributed among nations. Table 2–1 shows that great differences in wealth exist among the nations of the world, something we will see many times in other chapters. It is significant that the greatest differences occur in the two categories of assets that are most closely tied to the process of development: the human-resource-dominated *intangible capital* and *produced capital*. With the exception of the oil-rich Middle East, the per capita differences in natural capital are unremarkable. Thus, *economic development*, which fosters growth in human resources and produced capital, is the pathway to achieving more equity in the distribution of resources.

Essential Conditions. There are many reasons why some countries have undergone rapid development and others have not; population growth is one that is cited in this text (Chapter 9). Perhaps the most important reasons relate to the major institutions that coordinate social and political life and that make up the social capital of a society. For example, the definition of rights, the enforcement of those rights, and the facilitation of economic exchange are all considered essential to the successful development of human capital and produced capital. A well-developed body of law, an honest legal system, inclusiveness, broad civic participation, and a free press are essential for maintaining rights within a society. A well-developed market economy and free entry into and exit from markets for people and businesses are also conducive to development. Functioning communication and transportation networks and viable financial markets are needed to sustain a market economy. History tells us that if these resources are in place, a society will make progress in the development of human resources and produced assets. In societies where such resources are not in place, corruption, inefficiency, banditry, injustice, and intolerance often prevail; continuing poverty and environmental degradation are the predictable outcome. For example, one of the major components of *intangible capital* as measured by the World Bank is the "rule of law index." Sweden scores 87.5 out of 100 on this index, while Pakistan, with its corruption and

human rights abuses, scores 33.1. In fact, some countries (e.g., Kuwait, Angola) are so poorly run that they have negative intangible capital; their people are becoming more impoverished.

Intergenerational Equity. What we have just discussed is a matter of *intragenerational equity*—the "golden rule" of *making possible for others what is possible for you*. All forms of current assets need to be managed well to improve the well-being of people who already exist. Sustainable development, by contrast, is also about *intergenerational equity*—*meeting the needs of the present without compromising the ability of future generations to meet their needs*. Thus, we can conserve resources and avoid long-term pollution partly out of a concern for future generations.

Economists point to some problems with this perspective, however. Economics deals with future resources by applying what is called the **discount rate**. By this, we mean the *rate* used for finding the present value of some future benefit or cost. The discount rate often approximates current interest rates. For example, a dollar is worth more to us today than it will be five years from now, even without inflation, because we can put the dollar to use now and can earn interest on that dollar if we invest it. At a 5% rate of interest, in five years our dollar would be worth $1.28. Turning this idea around, a dollar five years from now is worth only $0.78 today.

Following this reasoning, it can be concluded that some resources or benefits are worth more now than they will be in the future. This concept can be applied to a stand of trees. The owner has several options: (1) cut all the trees and sell the timber today; (2) hold onto the trees in the hope that they will bring a better price at some time in the future; or (3) harvest the trees on a sustainable basis, spreading out profits over time. If it turns out that the trees grow more slowly than current interest rates, the economically profitable decision would be to cut them now and invest the proceeds. This is a problem for natural capital, because it would appear to argue in many cases for maximizing short-term profit over sustainable, long-term use of the resource. In other words, we should reap the economic value now and let future generations bear the ecological costs (in reality, the resource is likely to have additional value in the form of services that is not included in the economic assessment of its worth). Similarly, if we let future generations cope with global climate change, we can continue to enjoy the present benefits of the unrestrained use of fossil fuels. The same can be said about other resources, such as swordfish: we can harvest them now, and when they're gone, we can switch to some other species of fish.

This approach presents a problem for intergenerational equity. The *self-interest* of present individuals and generations is at odds with the longer term, *sustainable interests* of a community or a society. To make things a bit more challenging, some have argued that conserving resources for the future could actually be thought of as putting the interests of future generations above those of, say, the poor today. In answer, it can be said that both sets of interests are legitimate,

[5]TEEB, *The Economics of Ecosystems and Biodiversity: Mainstreaming the Economics of Nature: A synthesis of the approach, conclusions and recommendations of TEEB.* 2010.

but given a choice, many people might want to see their children and grandchildren enjoy an improved well-being and might be more inclined to seriously consider making present sacrifices in order to protect a more hopeful future.

We turn our attention now to environmental public policy, which makes possible both present and future well-being for human societies.

2.3 Environmental Public Policy

Environmental public policy includes all of a society's laws and agency-enforced regulations that deal with that society's interactions with the environment. Public policies are developed at all levels of government: local, state, federal, and international.

The Need for Environmental Public Policy

The purpose of environmental public policy is *to promote the common good.* Just what the common good consists of may be a matter of debate, but at least two goals stand out: *the improvement of human welfare* and *the protection of the natural*

world. Environmental public policy addresses two sets of environmental issues: (1) the prevention or reduction of air, water, and land pollution and (2) the use of natural resources like forests, fisheries, oil, and land. All public policy is developed in a sociopolitical context that we will simply call *politics.*

What are the consequences of not having an effective environmental public policy? Human societies and their economic activities have the potential for doing great damage to the environment, and that damage has a direct impact on present and future human welfare (Table 2–2). The effects of pollution and the misuse of resources are seen most clearly in those parts of the world where environmental public policy is often not well established and implemented—the developing world. As Table 2–2 shows, millions of deaths and widespread disease are directly traceable to degraded environments. The costs to human welfare are felt in the areas of health, economic productivity, and the ongoing ability of the natural environment to support human life needs. Therefore, laws to protect the environment are not luxuries to be tolerated only if they do not interfere with individual freedoms or economic activities; they are part of the foundation of any well-organized human society and essential for achieving sustainability.

TABLE 2–2 Principal Health and Productivity Consequences of Poor Environmental Management

Environmental Problem	Effect on Health	Effect on Productivity
Water pollution and water scarcity	More than 3 million deaths and billions of illnesses a year are attributable to water pollution and poor household hygiene caused by water scarcity.	Fisheries are declining; rural household time (time spent fetching water) and municipal costs of providing safe water are increasing; water shortages constrain economic activity.
Air pollution	Many acute and chronic health impacts exist: excessive levels of urban particulate matter and smoky indoor air are responsible for 2 million premature deaths annually.	Acid rain and ozone have harmful impacts on forests, agricultural crops, bodies of water, and human artifacts.
Solid and hazardous wastes	Diseases are spread by rotting garbage and blocked drains. Risks from hazardous wastes are typically local, but often acute.	Groundwater resources are polluted and rendered unusable for irrigation or domestic use.
Soil degradation	Effects include reduced nutrition for poor farmers on depleted soils and greater susceptibility to drought.	Some 23% of land used for crops, grazing, and forestry has been degraded by erosion. Field productivity losses in the range of 0.5–1.5% of the gross national product are common on tropical soils.
Deforestation	Localized flooding leads to death and disease.	Effects include reduced potential for sustainable logging and for prevention of erosion, increased watershed instability, and diminished carbon storage capability for forests. Nontimber forest products are also lost.
Loss of biodiversity	Effects include the potential loss of new drugs.	Ecosystem adaptability is reduced, and goods and services are lost.
Atmospheric changes	Such changes result in possible shifts in vector-borne diseases, risks from climatic natural disasters, and skin cancers attributable to depletion of ozone shield.	Coastal investments are damaged by the rise in sea level; regional changes in agricultural productivity occur.

Sources: World Bank, *World Development Report, 1992.* New York: Oxford University Press, 1992; *Millennium Ecosystem Assessment,* 2005; UNEP Global Environment Outlook GEO$_4$, 2007.

Figure 2–8 **The three branches of government in the United States.** The legislative, executive, and judicial branches provide a balance of power as laws are enacted, administered, and sometimes reviewed.

Policy in the United States

The U.S. Constitution is widely regarded as responsible for establishing a very effective form of government. The conveners of the Constitution decided on three branches of government, all of which play a role in establishing and enforcing environmental public policy **(Fig. 2–8)**. The **legislative branch,** or the Congress, consists of the Senate and the House of Representatives. All of these legislators are elected by the people from the 50 states. The Congress has the power to establish laws, which must pass both houses of the legislature and then go to the **executive branch** of the government for approval by the President. A host of administrative agencies, most headed by members of the President's cabinet, report to the President. We will encounter many of these agencies as we examine different environmental problems. The **judicial branch** of government is responsible for interpreting the law, from the Constitution and its amendments to the many pieces of legislation passed by the Congress.

Rules and Regulations. Environmental *public-policy law* is the responsibility of the Congress. Once a piece of legislation is passed by the House of Representatives and the Senate and is signed into law by the President, it is implemented by the appropriate government agencies. The EPA and most other federal agencies are required to develop *rules and regulations* for issues only broadly covered in the laws, and it is these rules that put the bite in the laws passed by Congress. In making rules, agencies are required to solicit public comments, and these most often come from major stakeholders interested in or affected by the rules—business and public interest groups. The agencies then evaluate the comments and may or may not adjust the rules, depending on the validity of the comments. Finally, agencies publish their new rules in the *Federal Register,* where they are "filed" for all interested parties to see.

One final role the Congress must play is to pass appropriations for the various programs that have already been authorized by law. The executive branch (the President) draws up a budget asking for appropriations for all of the government agencies and programs, and the Congress decides how much of what the President asks for will actually get funded. Thus, the two times the Congress can influence a given issue are when it is formulated into law and when it receives appropriations. The appropriations must be authorized every year, and new laws can always amend or abolish existing laws. Given this basic process, there are many opportunities for the political system to influence environmental public policy.

State and Local Levels. Some policies are developed at the state and community levels to solve more-local problems. States replicate much of the structure of the federal government, with legislatures, governors, a judiciary, and various agencies. Cities and towns employ conservation commissions, planning boards, health boards, water and sewer commissions, and the like. Cities and towns, for example, establish zoning regulations to protect citizens from haphazard and incompatible land uses. Some municipalities and regions go further and establish broader policies like "smart growth" to address the problems of urban sprawl. Cities and towns are adopting mandatory recycling and pay-as-you-throw regulations for municipal solid-waste management. Many states are responding to global climate change with policies aimed at reducing greenhouse gas emissions and mandating renewable-energy portfolios. Eleven states have bottle laws that facilitate recycling. Indeed, states and municipalities are often out in front of the federal government on a number of environmental issues, often initiating valuable experiments in environmental public policy that could result in national policy changes.

Policy Options: Market or Regulatory?

Suppose that a legislative or administrative decision has been made to develop a policy to deal with an environmental problem. Now what? The objective of environmental public policy is to *change the behavior of polluters or resource users so as to benefit public welfare and the environment.* The two main ways to accomplish these behavior changes are by using a direct regulatory approach, in which standards are set and technologies are prescribed—the so-called *command-and-control* strategy—and by using basic *market* approaches to set prices on pollution and the use of resources. Both approaches have their advantages and disadvantages. Many newer policies include elements of each.

Command and Control. Traditionally, most of the laws that give the EPA its authority have prescribed command-and-control solutions. The laws direct the EPA to regulate pollution problems, often specifying many of the details to be used. For example, if they know the health effects of pollutants, the legislators may choose to set standards that protect the health of the most vulnerable members of the population. This is the case with the basic criteria covering air pollutants (see Chapter 19). To meet these standards, regulations are established, and polluters are required by law to comply with the regulations. The regulatory approach also works well with land use issues in which certain values are upheld that will not necessarily be protected by a straightforward market approach. Indeed, many environmental problems are not readily amenable to market-based policies.

One of the shortcomings of the regulatory approach is that it practically guarantees a certain sustained level of pollution. If a polluter is told to use a particular technology or is given a cap on emissions, the policy gives the polluter no incentive to invest in technologies that would keep pollution at lower levels than allowed. A better approach might be to set a standard (for air quality or water quality, for example) and let the polluter decide how best to achieve the standard. Here there is a *command*, but *control* is in the hands of the polluter. Pollution *prevention* is the preferred course of action, and it is encouraging to see the EPA moving deliberately in that direction (Chapter 22).

Market Approaches. Market-based policies have the virtues of simplicity, efficiency, and, in theory, equity. All polluters are treated equally and will choose their responses on the basis of economic principles having to do with profitability. With a market-based approach, there are strong incentives to reduce the costs of using resources or of generating pollution. The cap-and-trade system for SO_2 emissions under the Clean Air Act of 1990 is a good example of the market approach to pollution control (see Chapter 19). User fees (pay as you throw) for the disposal of municipal solid waste are another (Chapter 21). Several countries have restored and sustained fishery stocks by assigning individual quotas that add up to a sustainable fishing take—the total allowable catch (see Chapter 7).

A new market-based initiative for conserving the services provided by ecosystems is **Payments for Ecosystem (or Environmental) Services (PES)**. Forest, farm, and wetlands owners hold properties that provide valuable services to societies as they sequester carbon, preserve biodiversity, and prevent soil erosion. They are seldom compensated for these services. A PES program would use some form of environmental accounting (think TEEB) to establish a monetary value for a given ecosystem service, and those who benefit from the service would pay the landowner to maintain the service over the years and not convert the land to some environmentally undesirable use. Often the government does the paying on behalf of the society, as in the Conservation Reserve Program, where farmers are paid to protect open space and/or wetlands they own (Chapter 11).

Whether regulation based or market based, public policy does not appear out of nowhere. We will now examine how public policy is developed.

Public Policy Development: The Policy Life Cycle

Environmental public policy is developed in a sociopolitical context, usually in response to a problem. As we saw (Fig. 2–2), less developed countries or regions face a group of poverty-related environmental issues, such as a lack of sanitation facilities or unsafe water supplies. As these countries develop, they become better able to deal with these basic issues through effective public policies. However, more affluence can bring new problems (such as industrial air pollution) that require further public policy development, and sometimes unexpected problems, such as stratospheric ozone depletion, arise as a result. When specific problems are addressed through public policy in democratic societies, the development of that policy often takes a predictable course, which we will call the *policy life cycle.*

The typical policy life cycle has four stages: *recognition, formulation, implementation,* and *control* (**Fig. 2–9**). Each

Figure 2–9 The policy life cycle. Most environmental issues pass through a policy life cycle, and each issue is accorded a different degree of political "weight" as it moves through each stage of the cycle. The final result is a policy that has been incorporated into the society and a problem that is under control.

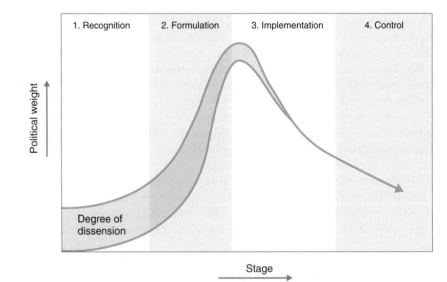

of the stages carries a certain amount of political "weight," which varies over time and is represented in the figure by the course of the life-cycle line. We will use the *discovery of ozone depletion in the stratosphere* as a case study to illustrate the policy life cycle.

Recognition Stage. In the early 1970s, chlorofluorocarbons (CFCs) were hailed as miracle chemicals: virtually inert, nontoxic, nonflammable, noncorrosive. They appeared in thousands of products and applications, from computers, telecommunications, pharmaceuticals, and refrigerators to food processing and cosmetics. They were ideal aerosol propellants, were extremely useful as refrigerants, and were effective as cleaning agents for electronics.

Physical chemist Sherwood Rowland, while working at the University of California at Irvine in 1972, read a paper by English scientist James Lovelock, who proposed using CFCs as markers to measure atmospheric winds because of their increasing presence in the atmosphere. Rowland and new postdoctoral student Mario Molina **(Fig. 2–10)** decided to investigate the question of what happens to CFCs in the atmosphere. The two scientists reasoned that CFCs would not break down unless they entered the stratosphere, where solar radiation is strong enough to break the tough molecules. They hypothesized that CFC breakdown would release chlorine and that the chlorine would in turn react with ozone molecules in the stratosphere (see Chapter 19 for details).

Molina and Rowland suddenly realized that the process they had uncovered had the potential to reduce the stratospheric ozone layer that blocked ultraviolet radiation from entering the lower atmosphere, where it could seriously damage biological tissues. They published their results in 1974.[6] The immediate response was unremarkable, so the pair began calling public attention to the dangers of CFCs. The chemical industry (especially Dow and DuPont, major CFC manufacturers) responded by attacking them wherever they spoke, ridiculing their work, and even accusing them of being agents of the Soviet KGB intent on destroying capitalism.

Illustrated by these events, the *recognition stage* of public policy is low in political weight. The stage begins with the early perceptions of an environmental problem, often coming as a result of scientific research. Scientists then publish their findings, and the media pick up the information and popularize it. When the public becomes involved, the political process is under way. During the recognition stage, dissension is high. Opposing views on the problem surface as businesses or technological industries respond to the bad news that they are at the root of some new environmental problem. Eventually, the problem gets attention from some level of the government, and the possibility of addressing it with public policy is considered.

Formulation Stage. A turning point came in 1976, when the National Academy of Sciences appointed a committee to

Figure 2–10 Mario Molina and Sherwood Rowland. These two physical chemists discovered the threat posed by chlorofluorocarbons to the ozone layer in the stratosphere. Their work led to the Montreal Protocol, an international agreement to phase out the use of ozone-destroying chemicals around the world.

look into the claims of Rowland and Molina. The committee issued a report that confirmed the dangers of CFCs and recommended banning the chemicals. The EPA considered the evidence and banned the use of CFCs in spray cans in 1978. Environmental groups applauded the decision. CFCs continued to be used in increasing amounts, however, in other applications. Then, in 1985, a British team working in Antarctica reported a huge "hole" in the ozone layer over that continent, and further research showed that the ozone hole was due to unique conditions that led to high release of chlorine from CFCs and to subsequent ozone destruction. Ozone destruction was seen to be occurring more widely, and the world scientific and environmental organizations began calling for global action. This led to a meeting convened by the United Nations in Montreal, Canada, in 1987. There, member nations agreed to scale back CFC production 50% by 2000.

Thus, the *formulation stage* is a stage of rapidly increasing political weight. The public is now aroused, and debate about policy options occurs in the corridors of power. The political battles may become fierce, as questions dealing with regulation and who will pay for the proposed changes are addressed. Media coverage is high, and politicians begin to hear from their constituencies. Lobbyists for special interests or environmental groups pressure legislators to soften or harden the policy under consideration. During the formulation stage, policy makers consider the "Three E's" of environmental public policy: **effectiveness** (whether the policy accomplishes what it intends to do in improving the environment), **efficiency** (whether the policy accomplishes its objectives at the least possible cost), and **equity** (whether the policy parcels out the financial burdens fairly among the different parties involved).

Implementation Stage. The ozone hole continued to grow, however, and the signers of the Montreal Protocol agreed to speed up the phaseout of CFCs, moving the target date for completion to 1996. The major chemical

---

[6]Mario Molina and F. S. Rowland, Stratospheric Sink for Chlorofluoromethanes: Chlorine Atom-Catalysed Destruction of Ozone. *Nature* 24(9). 810–812. June 28, 1974.

Figure 2–11 Environmental problems in the policy life cycle. Different environmental problems are in different stages of the policy life cycle in industrial societies.

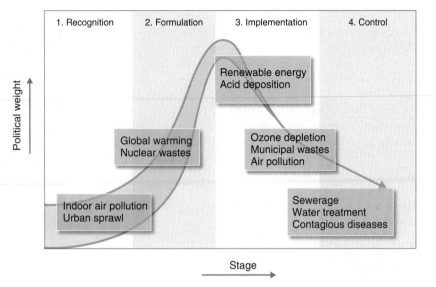

companies capitulated and were already developing more benign substitutes for CFCs. In fact, DuPont effectively opposed an ill-advised attempt in the 104th Congress to terminate U.S. participation in the Montreal Protocol. Several other chemicals in wide use that also had ozone-destroying potential were added to CFCs as substances to be regulated. It fell to regulatory agencies in the various countries to enforce the Montreal Protocol agreements.

At this point, the policy has reached the *implementation stage*, in which its real political and economic costs are exacted. The policy has been determined, and the focal point moves to regulatory agencies. During the implementation stage, public concern and political weight are declining. By now, the issue is not very interesting to the media, and the emphasis shifts to the development of specific regulations and their enforcement. Industry learns how to comply with the new regulations. Over time, greater attention may be given to efficiency and equity as all the players in the process gain experience with the policy.

Control Stage. CFCs are no longer being manufactured and released into the atmosphere. Substitutes for them are readily available. The size of the ozone hole has stopped increasing, and the amount of CFCs in the stratosphere is starting to decline. The Montreal Protocol stands as a remarkable achievement of international diplomacy and concord. Sherwood Rowland and Mario Molina were awarded the Nobel Prize in chemistry in 1995 for their discovery.

The final stage in the policy life cycle is the *control stage*. By this point, years have passed since the early days of the recognition stage. Problems are rarely completely resolved, but the environment is improving, with things moving in the right direction. Policies (and their derived regulations) are broadly supported and often become embedded in the society, although their vulnerability to political shifts continues. Regulations may become more streamlined. The policy makers must now see that the problem is kept under control, and in due time, the public often forgets that there ever was a serious problem.

The policy life cycle is a simplified view of what is frequently a highly complex and contentious political process. At any one time, different problems will be in different stages of the life cycle, as shown in **Figure 2–11** for a number of environmental problems in the industrialized countries. Also, countries in different stages of economic development will be in different stages of the policy life cycle for a given problem. For example, sickness and death due to polluted water are still serious problems in many developing countries because public policies have not yet caught up with the need for water treatment and sewage treatment. In contrast, this problem is in the control stage in the industrialized societies. The reason for the discrepancy can often be traced directly to the costs that lie behind the development and implementation of public policy.

Economic Effects of Environmental Public Policy

What are the relationships between a country's economy and its environmental public policy? Are there some policies that are too costly? How should a society parcel out its limited resources to address environmental problems? Are environmental regulations a burden on the economy? Let us begin to sort out these questions by looking at the relationship between environmental policies and costs.

Costs of Policies. As we have seen, policies do not just appear out of thin air. Instead, they are the result of some version of the policy life cycle. Some policies have relatively little or no direct *monetary cost*—they do not require major investments of administration or resources. For example, removing subsidies to special interests and denying special access to national resources can result in a more efficient and equitable operation of the economy and can protect the environment. The use of national lands for cattle grazing **(Fig. 2–12a)** and timber harvesting (Fig. 2–12b) is subsidized in the United States. As a result, the real costs—and environmental consequences—of these activities are borne not by the special interest groups that

(a)   (b)

Figure 2–12 Subsidized access to pubic resources. (a) Cattle grazing in BLM-managed land near Baker City, Oregon. (b) Timber harvesting in the Olympic National Forest, Washington. Both of these activities are subsidized by taxpayers because the special interests with access to the resources pay only a fraction of the market value of the resource.

have access to these resources, but by all taxpayers. Removing such subsidies, however, can have very real *political* costs, as powerful interests do everything they can to hold onto their privileges.

Most environmental policies impose some very real economic costs that must be paid by some segment of society. For instance, we establish policies that control pollution because their benefits are judged to outweigh their costs—that is, the public welfare is improved, and the environment is protected. Who pays the costs? The equity principle implies that those benefiting from the policies—the customers of businesses whose activities are regulated and the people whose health is protected by laws against polluting—should pay, through taxes or higher charges. (When businesses are taxed or industries are forced to add new technologies, they pass their costs along to consumers, so the costs of public policies are borne by the public in one way or another.)

Impact on the Economy. For years, special interest groups and conservative think tanks (e.g., the Competitive Enterprise Institute) have argued that environmental policies are excessive and bad for the economy—that our concern for environmental protection costs hundreds of thousands of jobs, reduces our competitiveness in the marketplace, drives up the price of products and services, and, in general, imposes costs on the economy that are nonproductive. These views are widespread and are routinely used to try to roll back the laws and regulations that represent environmental public policy.

How sound are these arguments? This concern has been addressed by many economists in recent years, and the general findings of their studies indicate that the "environment versus economy" trade-off is a myth. Even without considering the environmental or health benefits, economists found

that environmental protection has no significant adverse effect on the economy and in fact has had a pronounced positive effect.

Anecdotal Evidence. First, we must recognize that much of the evidence used to support the charges that have been made is anecdotal, and therefore suspect. For example, according to the American Petroleum Institute, "environmental restrictions on energy extraction and production have caused the loss of 400,000 jobs," and according to Florida sugar growers, "Measures to protect the Everglades will cause the loss of 15,000 jobs." Indeed, numerous cases can be cited where *some* jobs were lost because of environmental regulations.

Anecdotal evidence can also be used in *favor* of environmental policy. According to the California Planning and Conservation League Foundation, "Recycling has created 14,000 jobs in California," and according to the EPA, "The Clean Air Act of 1990 will generate 60,000 new jobs." Indeed, *many* sectors of the U.S. economy that are most subject to environmental regulations—plastics, fabrics, and so forth—have improved their efficiency and their competitiveness in the international market.

Careful Studies. Is there something more substantial to help us with this important controversy? The best evidence comes from careful studies of what is actually happening. These studies reveal the following interesting findings:

- Estimated pollution-control costs are only 1.72% of **value added.** (Value added is the increase in value from raw materials to final production.) Industries are not going to shut down, move overseas, or lay off thousands of workers for costs of this size.

- Only 0.1% of job layoffs were attributed by employers to environment-related causes, according to a study by the U.S. Bureau of Labor Statistics.

- In general, states with the strictest environmental regulations also had the highest rates of job growth and economic performance. California established new energy efficiency policies in 1978, and since then, the energy sector has added more than a million jobs, far outpacing other economic sectors. Indeed, California has set a remarkable pace in creating a "green economy" (see Sustainability, "California's Green Economy" p. 41).

In short, the results of careful studies of the relationship between environmental protection and economic growth refute the suggestion that environmental policies are bad for jobs and the economy. In fact, economic performance is highest where environmental public policy is most highly developed. Healthier workers are more productive and spend less on medical care. It may be no coincidence that in the former communist countries in Europe, where environmental public policy was deliberately suppressed in order to favor industry, both the economy and the environment have been disaster areas. The evidence indicates that concerns over energy efficiency, pollution control, and conservation of resources have encouraged businesses to modify their technologies in ways that make them *more* competitive, not less, in the national and international marketplace. The evidence also suggests that many more jobs have been created by environmental policy than have been lost.

The Environmental Protection Industry. Taken as a whole, environmental protection is a huge industry. In a landmark 1990 study,[7] the EPA calculated that the estimated cost of complying with environmental regulations by 1997 would be about 2.7% of the GDP—about $210 billion at that time. Admittedly, this estimate was very rough—some believe it was too low, while others think it was too high. To put the percentage into perspective, in 1997, the United States spent an estimated 10.6% of the GDP on health care and 4.3% for national defense. Is it worth 2.7% of the GDP to provide healthy air to breathe and clean water and ecosystems that can continue to provide essential goods and services if we are willing to spend 15% on our personal health and national security?

A study of total expenditures for the environmental protection industry for one year revealed that it was responsible for the following:

- $355 billion in total industry sales;
- $14 billion in corporate profits;
- $63 billion in federal, state, and local government revenues; and
- 4 million jobs.

In sum, we can draw the following conclusions from our examination of the impact of environmental policy on the economy:

- Environmental public policy does not *diminish* the wealth of a nation; rather, it *transfers* wealth from polluters to pollution controllers and to less polluting companies.

- The "environmental protection industry" is a major job-creating, profit-making, sales-generating industry.
- The argument that environmental protection is bad for the economy is simply unsound. Not only is environmental protection good for the economy, but environmental public policy is responsible for a safer, healthier, and more enjoyable environment.

We turn next to the details of *benefit-cost analysis*, a specific tool that is employed in developing environmental public policy. In doing so, we will address the question of how to measure the cost-effectiveness of regulating pollution.

2.4 Benefit-Cost Analysis of Environmental Public Policy

The President of the United States has the power to issue executive orders, which are usually directives to government agencies or departments designed to accomplish a presidential initiative. President Ronald Reagan, concerned that the U.S. economy and business interests might be overregulated, issued Executive Order 12291 in 1981. This order required that all *significant* regulations be supported by a *benefit-cost analysis*. A *significant* regulation is defined as one that imposes annual costs of at least $100 million. Subsequent presidents, including President Obama (in Executive Order 13563), issued similar executive orders that basically continued most of the policies of Reagan's order.

A **benefit-cost analysis** begins by examining the need for the proposed regulation and then describes a range of alternative approaches. Afterward, it compares the estimated costs of the proposed action and the main alternatives to the benefits that will be achieved. All costs and benefits are given monetary values (where possible) and compared by means of what is commonly referred to as a *benefit-cost* (or *cost-benefit*) *ratio*. A favorable ratio for an action means that the benefits outweigh the costs, and the action is said to be cost effective. There is an economic justification for proceeding with it. If costs are projected to outweigh benefits, the project may be revised, dropped, or shelved for later consideration.

Benefit-cost analysis of environmental regulations is intended to build efficiency into policy so that society does not have to pay more than is necessary for a given level of environmental quality. If the analysis is done properly, it will consider *all* of the costs and benefits associated with a regulatory option. In so doing, it must address the problem of *external costs*.

External Costs

In the language of economics, an **external cost** is some effect of a business process that is not included in the usual calculations of profit and loss. For example, when a business pollutes the air or water, the pollution imposes a cost on society in the form of poor health or the need to treat water before using it. This is an *external bad*. When workers improve their job performance as a result of experience and learning,

---

[7]*Environmental Investments: The Cost of a Clean Environment*, Report of the Administrator of the Environmental Protection Agency to the Congress of the United States. EPA-230-1 1-90-083. November 1990. National Center for Environmental Economics, USEPA.

SUSTAINABILITY   California's Green Economy

The global transition to a sustainable green economy from the current brown economy is under way. Globally, even nationally, it is a hit-and-miss proposition, but one location in which this transition has been well documented is the state of California. *Next 10*, a San Francisco nonprofit, recently published a well-documented report on the green economy in that state.[8] Relative to 1995, jobs in the green economy have grown by 56%, while the total economy grew by 18%. The state employment department reported that people in more than 263,000 jobs spend half or more of their time on the production of green products

or services, with an additional 170,000 people performing jobs that involve part-time activities in the green economy.

Just what are green jobs? *Next 10* distinguished between a "core" green economy and an "adaptive" green economy. Core green economy businesses provide the products and services that deal in alternatives to carbon-based energy sources (wind, solar, biomass, etc.), conserve energy (e.g., building efficiency products and services) and other natural resources, and reduce pollution and recycle waste (e.g., emissions monitoring, recycling machinery, etc.). Adaptive green economic activities involve the businesses that are adjusting their processes to improve resource efficiency and change consumption habits of households. The report shows that green economic activities employ people across every region of the state.

An example of a California green business is Sol Focus, located in Palo Alto. This company develops and manufactures innovative photovoltaic systems that are operating in several states and are being deployed in Spain, Greece, Italy, Portugal, and many other countries. The company holds more than 40 patents and specializes in the use of advanced optical systems that focus and convert large amounts of solar energy to electrical energy. One of its unique products is a system called concentrator photovoltaic technology (CPV). This technology uses systems of cells employing reflective mirrors that concentrate the solar energy up to 500 times, greatly reducing the number of solar cells needed to capture solar energy and also greatly reducing the cost of the collection panels. With 270 employees, Sol Focus is a great example of a green company that is making a difference in California and throughout the world.

[8]Next 10, Many Shades of Green: Regional Distribution and Trends in California's Green Economy. 2010. August 15, 2011.

---

the improvement is not credited in the company's ledgers, although it is considered an *external good*.

Amenities such as clean air, uncontaminated groundwater, and a healthy ozone layer are not privately owned. In the absence of regulatory controls, there are no direct costs to a business for degrading these amenities. In other words, they are *externalities*. Therefore, there are no economic incentives to refrain from polluting the air or water. By including *all* of the costs and benefits of a project or a regulation, benefit-cost analysis effectively brings the externalities into the economic accounting.

Environmental Regulations Impose Real Costs

The costs of pollution control include the price of purchasing, installing, operating, and maintaining pollution-control equipment and the price of implementing a control strategy. Even the banning of an offensive product costs money because jobs may be lost, new products must be developed, and machinery may have to be scrapped. In some instances, a pollution-control measure may result in the discovery of a less expensive way of doing something. In most cases, however, pollution control costs money. Thus, the effect of most regulations is to prevent an external bad by imposing economic costs that are ultimately shared by government, industry, and consumers.

Costs Go Up. Pollution-control costs tend to increase exponentially with the level of control to be achieved (**Fig. 2–13**). That is, a partial reduction in pollution may be achieved by a few relatively inexpensive measures, but further reductions generally require increasingly expensive measures, and 100%

control is likely to be impossible at any cost. Because of this exponential relationship, regulatory control often has to focus on an optimum cost-effectiveness, as Figure 2–13 indicates. In any case, the costs of pollution control constitute a powerful incentive to make substitutions, to recycle materials, or to redesign industrial processes.

Costs Go Down. In most cases, pollution-control technologies and strategies are understood and available. Thus, equipment, labor, and maintenance costs can be estimated fairly accurately. Unanticipated problems that increase costs may occur, but as technology advances and becomes more reliable, experience is gained, and lower-cost alternatives frequently emerge. The costs of pollution control are likely

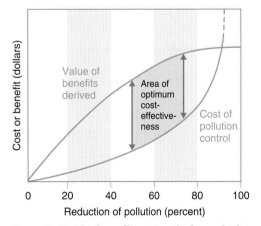

Figure 2–13 The benefit-cost ratio for reducing pollution. The cost of pollution control increases exponentially with the degree of control to be achieved. However, benefits to be derived from pollution control tend to level off and become negligible as pollutants are reduced to near or below threshold levels. The optimum cost-effectiveness is achieved at less than 100% control.

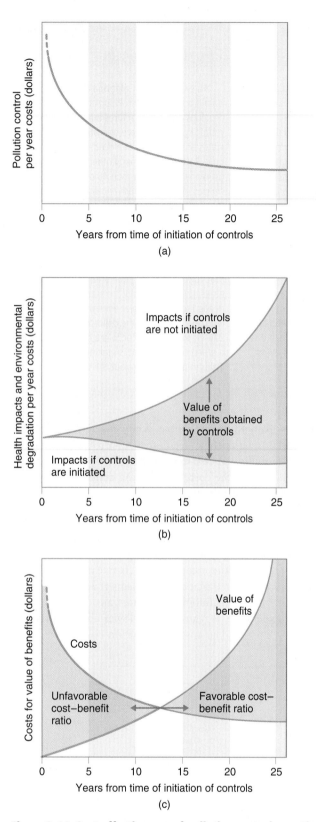

Figure 2–14 Cost-effectiveness of pollution control over time.
(a) Pollution-control strategies generally demand high initial costs. The costs then generally decline as those strategies are absorbed into the overall economy. (b) Benefits may be negligible in the short term, but they increase as environmental and human health recovers from the impacts of pollution or is spared increasing degradation. (c) When the two curves are compared, we see that what may appear as cost-ineffective expenditures in the short term (5–10 years) may in fact be highly cost-effective expenditures in the long term.

to be highest at the time they are initiated; then they decrease as time passes (**Fig. 2–14a**). The importance of this trend will become evident when we consider the time span over which costs and benefits are compared.

How accurate are the cost estimates? A look at some historical examples reveals that industry reports predict higher costs of regulations than government studies do, and surprisingly, even government analysts often overestimate the costs of regulations. For example, industry sources claimed that the costs of eliminating CFCs from automobile air conditioners would increase the price of new cars by anywhere from $650 to $1,200. In reality, the price increase turned out to be much lower—from $40 to $400. A recent study of benefit-cost analyses revealed that overestimates of costs occurred in 12 of 14 regulations. The primary reason given for the overestimates is a failure to anticipate the impact of technological innovation—admittedly a factor that is difficult to predict—although another reason might be a deliberate ploy to project costs so high that the regulators will back down.

The costs of improving the quality of the environment represent a major economic outlay. What benefits have we received in return?

The Benefits of Environmental Regulation

The Regulatory Right to Know Act directs the Office of Management and Budget (OMB) to submit an annual report to Congress containing, among other things, an estimate of the costs and benefits of federal rules. The 2011 report includes an up-to-date study of costs and benefits over the October 1, 2000–September 30, 2010 decade. The OMB found that the benefits (those that could be measured in monetary units) far exceeded the costs. The benefits were estimated at between $132 billion and $655 billion, while industries, states, and municipalities spent an estimated $44 billion to $62 billion to comply with the rules. The OMB report demonstrated that the benefits of major regulations issued between 2000 and 2010 exceeded the costs manyfold (**Fig. 2–15**). The lion's share of both the costs and the benefits for 2010 was attributed to the EPA's regulation of sulfur dioxide in the air. Fewer hospital visits and reductions in both premature deaths and lost workdays were benefits associated with reductions in this air pollutant mandated by the Clean Air Act.

Calculating Benefits. Such benefits of regulatory policies are seldom as easy to calculate as the costs. Estimating benefits is often a matter of estimating the costs of poor health and damages that *would have* occurred if the regulations had not been imposed (Fig. 2–14b). For example, the projected environmental damage that would have resulted from a given level of sulfur dioxide emissions from a coal-fired power plant (an external bad) becomes a benefit (an external good) when a regulatory action prevents half of those emissions. Benefits include such things as improved human health, reduced corrosion and deterioration of materials, reduced damage to natural resources, preservation of aesthetic and spiritual qualities of the environment, increased opportunities for outdoor recreation, and continued opportunities to use the environment in

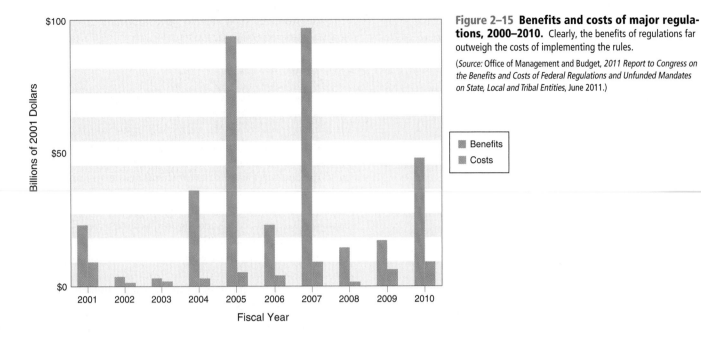

Figure 2–15 Benefits and costs of major regulations, 2000–2010. Clearly, the benefits of regulations far outweigh the costs of implementing the rules.

(*Source:* Office of Management and Budget, *2011 Report to Congress on the Benefits and Costs of Federal Regulations and Unfunded Mandates on State, Local and Tribal Entities,* June 2011.)

the future. The dollar value of these benefits is derived by estimating, for instance, the reduction in health care costs, the reduction in maintenance and replacement costs, and the economic value generated by the enhanced recreational activity. Examples of benefits are listed in Table 2–3.

The values of some benefits can be estimated fairly accurately. For example, air-pollution episodes increase the number of people seeking medical attention. Because the medical attention provided has a known dollar value, eliminating a certain number of air-pollution episodes provides a health benefit of that value.

Shadow Pricing. Many benefits, however, are difficult to estimate. Accurate benefit–cost analysis depends on assigning monetary values to every benefit, but how can a dollar value be put on maintaining the integrity of a coastal wetland or the enjoyment of breathing cleaner air? The answer is to find out how much people are willing to pay to maintain these benefits. How can this be done, though, if there is no free market for the benefits? Again, economists have an answer: *shadow pricing* involves assessing what people *might* pay for a particular benefit if it were up to them to decide. For example, to evaluate the costs of noise nuisance from aircraft (an external bad) in the vicinity of Schiphol Airport in Amsterdam, researchers evaluated the housing prices in the vicinity of many airports around the world and found that the closer the airport, and therefore the greater the noise, the lower the house value when compared with similar houses farther away. In this case, the *benefit* is seen as being free from the nuisance of low-flying aircraft, and people's willingness to pay is seen in the price of houses.

TABLE 2–3   Benefits That May Be Gained by the Reduction or Prevention of Pollution

1. Improved human health	Reduction and prevention of pollution-related illnesses and premature deaths Reduction of worker stress caused by pollution Increased worker productivity
2. Improved agricultural and forest production	Reduction of pollution-related damage More vigorous growth by removal of stress due to pollution Higher farm profits, benefiting all agriculture-related industries
3. Enhanced commercial or sport fishing	Increased value of fish and shellfish harvests Increased sales of boats, motors, tackle, and bait Enhancement of businesses serving fishermen
4. Enhanced recreational opportunities	Direct uses, such as swimming and boating Indirect uses, such as observing wildlife Enhancement of businesses serving vacationers
5. Extended lifetime of materials and less necessity for cleaning	Reduction of corrosive effects of pollution, extending the lifetime of metals, textiles, rubber, paint, and other coatings Reduction of cleaning costs
6. Enhanced real estate values	Increased property values as the environment improves

The Value of Human Life. Shadow pricing becomes difficult when the analysis has to place a value on human life. Many of the pollutants to which people are exposed are hazardous, exacting a toll on health and life expectancy. To estimate the benefits of regulating such pollutants, it is necessary to calculate how many lives will be saved or how many people will enjoy better health. The less the value that is put on a human life, the less the likelihood that a regulation will pass the benefit-cost test. Finding a value for these benefits is fraught with ethical difficulties. For example, the EPA calculated that new clean-air standards for ozone and particulates would prevent 18,700 premature deaths annually, and the monetary benefit of this prevention was valued at $110 billion (close to $6 million per life saved). Different approaches have resulted in a range of values for human life from tens of thousands of dollars to $10 million per life saved. Any benefit-cost analysis that must factor in the risk to human life requires that human lives be valued somewhere within this very broad range. The OMB valued each cancer case at $5 million, each traffic fatality at $3 million, and each lost-workday injury at $50,000.

How does shadow pricing work for the nonhuman components of natural environments—for a population of rare wildflowers, for example, or a wilderness site? The values of these things depend entirely on how willing people are to pay for their preservation. Again, monetary values must be assigned to their existence, and again, the outcome will be strongly influenced by a highly subjective element in the analysis.

Cost-Effectiveness Analysis

Cost-effectiveness analysis is an alternative option for evaluating the costs of regulations. Here the merits of the goal are accepted, and the question is: How can that goal be achieved at the least cost? To find out, alternative strategies for reaching the goal are analyzed for costs, and the least costly method is adopted.

Cost-effectiveness can be applied to the desired level of pollution control. As Figure 2–13 shows, a significant benefit may be achieved by modest degrees of cleanup. Note, however, how differently the cost and benefit curves behave with increasing reduction of pollution. At some point in the cleanup effort, the lines cross and the costs exceed the benefits. Few additional benefits are realized when cleanup begins to approach 100%, yet costs increase exponentially. This behavior follows from the fact that living organisms—including humans—can often tolerate a threshold level of pollution without ill effect. Therefore, reducing the level of a pollutant below its threshold level will not yield an observable improvement. Optimum cost-effectiveness that meets the efficiency criterion for public policy is achieved at the point where the benefit curve is the greatest distance above the cost curve.

However, the perspective of time should be considered in calculating cost-effectiveness (Fig. 2–14c). A situation that appears to be cost ineffective in the short term may prove extremely cost effective in the long term. This is particularly true for problems such as acid deposition and groundwater contamination from toxic wastes.

Progress

What has been the result of environmental regulation to date? Pollution of air and surface water reached critical levels in many areas of the United States in the late 1960s, and since that time, huge sums of money have been spent on pollution abatement. Benefit-cost analysis shows that, overall, these expenditures have paid for themselves many times over in decreased health care costs and enhanced environmental quality. Consider the phaseout of leaded gasoline as just one example. The project cost about $3.6 billion, according to an EPA benefit-cost report. Benefits were valued at more than $50 billion, $42 billion of which were for medical costs that were avoided!

The following are some of the accomplishments of environmental public policy, as reported by the EPA:

- From 1970 to 2011, total emissions of five principal air pollutants have decreased by a total of 67%.
- Average blood levels of lead for children have declined more than 90% since 1980.
- Between 1986 and 2010, releases of toxic chemicals, as reported to the Toxics Release Inventory, decreased 62%.
- In 2010, states reported that 92% of the population on community water systems had drinking water that met all standards, as opposed to 79% in 1993.
- Acid deposition has decreased by 41–69% over the past 18 years.
- More than 413,000 leaking underground gasoline storage tanks have been cleaned up since 1988.
- Between 1980 and 2011, 1,123 Superfund sites on the National Priorities List of highly contaminated locations have been completely cleaned up.
- Recovery of municipal solid wasted by recycling has increased from 7% in 1970 to 34.1% in 2010, diverting 85 million tons away from disposal in 2010.

Between 1980 and 2010, these improvements occurred even as the GDP grew by 127%, the population rose by 36%, and the number of motor vehicle miles driven increased by 96%.

A benefit-cost analysis accompanies every new regulatory rule from the EPA and other federal agencies. Even when a rule is not classified as a major regulation, the accompanying documentation must demonstrate, by benefit-cost analysis, that the rule does not qualify as significant. Like it or not, benefit-cost analysis is now part of public policy in the United States.

2.5 Politics and the Environment

Given that environmental public policies exact a cost and can seriously affect businesses and private life, establishing and defending these policies can be a contentious affair. We have described the modern environmental movement and its corollary, environmentalism (Chapter 1). Our brief look at this movement pointed out that political battles have accompanied almost every environmental issue, and that is still the situation. Politics always accompanies policy.

Political Parties and the Battle for Control

In the first decades of the modern environmental movement, pollution-control policies were enacted by broad, bipartisan coalitions and signed and implemented by Presidents of both parties. Grassroots citizen involvement played a key role in mobilizing this political accomplishment. By the 1990s, however, a very well-articulated environmental backlash emerged. The conservative mood of the country that elected Ronald Reagan embraced a large ideological contingent that called for policy reform to reduce the regulation of industry so that market forces and private enterprise could be freed up "to move the economy forward." The same contingent also defended private property rights threatened by programs like the Endangered Species Act and wetlands regulations. This movement viewed environmentalists as the enemy of free enterprise, responsible for promoting a false view of Earth as fragile and in crisis because of human mismanagement. The anti-environmental movement received the support of many new conservative legislators, and the result was gridlock in Congress.

The Republican-controlled 104th Congress (1995–1996) continued this attack on the environment with an effort to dismantle many key laws, especially the Clean Water Act, the Clean Air Act, and Superfund. When it became apparent what was happening, heavy grassroots work by many environmental organizations and extensive media coverage made it clear that the public did not support this anti-environmental thrust. As a result, most of the anti-environmental legislation of the 1990s died in the Senate or was vetoed by President Clinton. However, more recent Congresses have made it clear that environmental policy continues to be a sharply partisan matter, resulting in continued legislative gridlock. Anti-environmentalism now characterizes the Congressional Republicans, while the Democrats in the Congress are seen as pro-environment. The issues that most energize the current debate are global climate change and the EPA, although there is a broad scope of regulatory agencies and environmental legislation that have become the target of Congressional Republicans. They are aided and abetted by some powerful special interest lobbies and conservative media.

Special Interest Politics. The role and power of special interest groups became evident during these struggles. In the 1970s and 1980s, special interests (such as the coal industry and the fisheries industry) hired lobbyists who used their influence to try to convince congressional committee members to act in their favor on issues that were important to them. Environmental NGOs would then use the media to expose the issues to the public and call on their constituencies to

telephone and send mail to their representatives in Congress. The anti-environmental interests caught on to the NGO strategy and began to mimic what they were doing, but more effectively because they had deeper pockets. Soon a constant battle between special interest groups emerged in which each would use the media and telephone calls (enhanced by phone banks, which can generate thousands of calls to legislators in a short period) to pressure members of Congress. Added to this picture was the massive explosion in campaign contributions, largely from businesses and industry.

What all this means is that environmental concerns are often a battle of special interest groups—that environmental protection is seen as one among many interests in American politics. The issues are up for grabs, and protecting the environment often depends largely on the strength of environmental interest groups. James Skillen, former executive director of the Center for Public Justice, says that this is a seriously flawed development. In this setting, ecological well-being not only must have scientific proof of its validity, but also has to intrude into the political arena in competition with opposing special interests and *win*. Skillen believes that environmental justice—meaning the protection of land, air, water, and the associated biota—should not have to depend on an interest group for its advocacy. Rather, it is a matter of the *public good* and, as such, should be seen as a basic responsibility of government. To quote Skillen, "Recognition in basic law of the necessity of ecological health as the precondition of all public and market relations should become as fundamental as the recognition that certain human rights exist as the precondition of all public rules and market regulations."[9] The NGOs stepped into the gap because they found that doing so was necessary.

The Presidential Picture. Presidents have great power to influence national affairs, among them many that concern the environment. Presidents appoint agency heads who are loyal to their administration, and the regulations that emerge from federal agencies understandably reflect the ideological makeup of the party in the White House. If lobbyists have access to important members of the administration, they will have undue influence over agency rule making. Republican President George W. Bush (2000–2008) identified himself as an environmentalist during the 2000 presidential campaign. However, shortly after his election, he began a process of environmental policy redirection. With the support of a host of anti-environmental special interests—such as the energy, timber, and mining industries; the construction industry; road builders; and "free-market" think tanks—Bush used the regulatory process to make changes in environmental public policy. Many existing policies were weakened or canceled. Most prominently, the Bush administration withdrew the United States

---

[9]James Skillen, How Can We Do Justice to Both Public and Private Trusts? Paper presented at the Global Stewardship Initiative National Conference, Gloucester, MA, October 1996. Available at cesc.montreat.edu/GSI/GSI-Conf/discussion/Skillen.html.

from the Kyoto Protocol (an international agreement to curb greenhouse gas emissions), canceled support for international family planning, softened air and water quality standards, and crafted an energy policy that favored heavy exploitation of fossil fuels.

In November 2008, a majority of the American electorate voted Barack Obama, a Democrat, to the office of President. With the support of a decisive Democratic majority in the House and Senate between 2008 and 2010, Obama proceeded to reverse the Bush-era tide of antienvironmental policies. Following are some of the most prominent actions:

- Under Bush, federal scientists often had their work censored if it failed to conform to White House policy, especially in the realm of climate change; Obama issued a directive aimed at preventing any political interference in science—to ensure that "we make scientific decisions based on facts, not ideology," in his words. The key word in this directive was *integrity*; scientific integrity was to be maintained in all executive departments and agencies.

- The Bush EPA refused to act after the Supreme Court ruled that CO_2 was a pollutant under the Clean Air Act and should be regulated if it is a health threat; the Obama EPA issued a finding that CO_2 and other greenhouse gases pose a serious threat to public health and the environment because of their clear link to climate change. Subsequent regulatory rulings, like new fuel-economy standards, have implemented this finding.

- A last-minute Bush rule abolished the requirement that federal agencies consult with wildlife experts before taking actions that could harm endangered species; this requirement was reinstated by the Obama administration.

- As a major tool for fixing the stalled economy, the Stimulus Bill (the American Recovery and Reinvestment Act of 2009) included more than $40 billion for clean energy and energy efficiency programs, the largest clean energy bill in U.S. history.

Mid-Term Blues

The easy days of a Democratic Congress and a Democratic administration came to an end with the mid-term elections of 2010. When the dust cleared, the House reestablished a Republican majority, while the Senate remained in Democratic control. Once again, the House began to pass a host of antienvironmental legislative bills and riders (a rider is a provision added to a bill that has little connection to the substance of the bill and would likely not pass if considered independently). In its first year of operation, the House voted to prohibit action on climate change, to stop actions to prevent water and air pollution, to undermine protections of coastal areas and other public lands, and to dismantle Obama administration initiatives promoting clean energy and energy efficiency. Because

of Democratic Senate control, these votes fell on barren soil; Obama would have vetoed them anyway.

To conclude, we have seen a partisan struggle in recent years over the environment. It has involved special interest groups, Congress, Presidents, and their regulatory agencies. It is a battle of ideologies, and the outcome of the battle has enormous consequences for the environment and for the public.

Citizen Involvement

How is the general public involved in environmental public policy? The public is involved directly in public policy because not only are we the recipients of the outcome of the policies, but also we pay for the benefits we receive through higher taxes and higher costs of the products we consume. We should care deeply about public policies. And the evidence of polls indicates that the American public does care. A 2011 bipartisan survey[10] showed that American voters overwhelmingly reject congressional efforts to reign in the EPA and instead trust the EPA more than Congress to set clean air standards. This finding indicates a serious disconnect between many congressional representatives and the people they are supposed to represent.

What can you do to have an impact on environmental public policy? Here are some suggestions:

- Take note of the viewpoint of any political candidate on matters of environmental policy. Public concern about environmental policy can play an important role in the election or defeat of political candidates.

- Contact your legislators to inform them of your support for a particular environmental public policy under consideration by the Congress. You will very likely receive an acknowledgment of your concern, and if enough people do the same, your legislator is likely to reflect your views on the issue.

- Become involved in local environmental problems. Local jurisdictions often have the power to make crucial decisions affecting the environment, and groups of citizens working together can often make a difference in how the decisions go.

- Become a member of an environmental NGO that informs its constituencies of environmental concerns around the country and lobbies for those concerns. There is a list of these organizations in Appendix A in the back of this book.

- Stay informed on environmental affairs; only when the public is informed is it likely that grassroots support will be maintained and environmental policy will really reflect public opinion. Incidentally, taking a course in environmental science is an excellent way to start getting informed!

[10]Ayres McHenry Associates, Inc., American Voters Strongly Oppose Congressional Action Against Clean Air Standards: Voters Want EPA, Not Congress, to Set Standards. http://www.lungusa.org/healthy-air/outdoor/resources/clean-air-survey/clean-air-memo.pdf. August 16, 2011.

REVISITING THE THEMES

Sound Science

One component of the wealth of nations is *intangible capital*. Two elements of that capital are essential to maintaining science as a key to coping with environmental issues: education and the accumulation of scientific information as part of the fund of human knowledge that can be transmitted across boundaries of space and time. A truly wealthy nation will maintain a high level of scientific competence in its populace. The work of Sherwood Rowland and Mario Molina demonstrates the vital role science can play in the development of public policy—in this case, leading to protection of Earth's ozone shield. However, scientific integrity can be corrupted by politics, as seen in the workings of the George W. Bush administration, where scientists were silenced or had their work censored by politicians. The "new broom" of the Barack Obama administration has done much to restore scientific integrity in our federal offices.

Sustainability

The ecological economists have made sustainability central to their economic theories. They rightly point out that the natural environment encompasses the economy. Therefore, there are limits to global economic growth that are set by the environment, and the economy must come to a steady state at some point. A nation must manage its assets, but it can't do that unless it knows what they are and how to measure them. Depleting assets of any kind are a clear signal of unsustainable practice, and characteristic of the current business-as-usual *brown economy*. A future sustainable *green economy* must focus on human well-being, not growth. It will value and preserve ecosystem services and goods. It will develop a host of "green" products that will revolutionize our technologies. It will discard the GDP as a measure of economic progress. Instead, it may even embrace the GPI as a better measure of economic well-being. Without question, societies must develop effective environmental public policies if they are to prevent the millions of deaths and widespread disease that a lack of policies exacts. Laws to protect the environment are essential for achieving sustainability. This is well illustrated by the accomplishments of environmental public policy as documented by the EPA.

Stewardship

Table 2–2 demonstrates the consequences of poor environmental stewardship; laws to protect the environment are not a luxury, but an essential foundation for a society. Former EPA Administrator Christine Todd Whitman (a Republican) introduced the agency's *Draft Report on the Environment 2003* with these words: "We are all stewards of this shared planet, responsible for protecting and preserving a precious heritage for our children and grandchildren. As long as we work together and stay firmly focused on our goals, I am confident we will make our air cleaner, our water purer, and our land better protected for future generations." Environmental public policy over the past 40 years has accomplished much in achieving a safer and healthier environment. This represents our society's clear commitment to planetary stewardship.

REVIEW QUESTIONS

1. What is happening these days in the Chinese economy and environment?

2. Three patterns of environmental indicators are associated with differences in the level of development of a nation. What problems decline, what problems increase and then decline, and what problems increase with the level of development?

3. Name three basic kinds of economic systems and explain how they differ.

4. What is the role of the World Trade Organization, and how has this role broken down recently?

5. What are some characteristics of a sustainable economy?

6. Describe the three components of a nation's wealth. What insights emerge from the World Bank's analysis of the wealth of different groups of nations?

7. Why is the gross national product an inaccurate indicator of a nation's economic status? How can it be corrected?

8. What conditions are necessary for a country to make progress in the development of human resources and produced assets?

9. Explain the importance of considering both intragenerational equity and intergenerational equity in addressing resource allocation issues.

10. What is the overall objective of environmental public policy, and what are the objective's two most central concerns?

11. Describe the various bodies responsible for environmental public policy at the federal, state, and local levels.

12. What are the advantages and disadvantages of the regulatory approach versus a market approach to policy development?

13. List the four stages of the policy life cycle, and show how the discovery of ozone layer destruction and the subsequent responses to it illustrate the cycle.

14. What is the conclusion of careful studies regarding the relationship between environmental policies, on the one hand, and jobs and the economy, on the other?

15. Define the term *benefit–cost analysis* as it relates to environmental regulation. How does this analysis method address external costs? Distinguish it from cost-effectiveness analysis.

16. Discuss the cost-effectiveness of pollution control. How much progress have we made in pollution control in recent years?

17. How has the political arena affected environmental public policy in the last few years? What has been the role of special interest groups? Political parties? Presidents?

THINKING ENVIRONMENTALLY

1. Consider the ultimate impact that environmental law has on the economy. Do short-term drawbacks justify long-term goals? If there is a conflict, which should come first—the economy or the environment?

2. Research the Doha Round of meetings of the World Trade Organization. Analyze the positions of the developing countries and the industrial countries, and propose a solution that might meet the concerns of both groups.

3. Examine the three types of capital that make up the wealth of nations. Which do you think is the most valuable? Justify your answer.

4. What different roles do you think the federal and state governments should have in environmental policy?

5. Suppose it was discovered that the bleach that is commonly used for laundry is carcinogenic. Referring to the policy life cycle, describe a predictable course of events until the problem is brought under control.

MAKING A DIFFERENCE

1. Our government makes the decisions that have the biggest long-term effects on the environment. Find a cause that you feel strongly about, and find a way to support it. Demonstrating is easier than it sounds—you don't necessarily have to drive a hundred miles to join in some kind of march. It's easy to join an online movement (for example, visit www.StopGlobalWarming.org) or write a letter to a senator.

2. Help call attention to instances of pollution in your community. You can talk to regulators in the county health department, the EPA, or whoever is in charge of monitoring pollution in your region. It's their job to listen.

3. Instead of taking a long-distance vacation and relaxing during your break, join a volunteer group and get something done. There are many opportunities to help out.

4. Find out the costs and benefits of your own lifestyle on a purely economic level. Decide what other changes you could make to save you money and save the environment at the same time.

MasteringEnvironmentalScience®

Go to **www.masteringenvironmentalscience.com** for practice quizzes, Pearson eText, videos, current events, and more.

PART TWO

Ecology:
The Science of Organisms and Their Environment

EVEN IF YOU'VE never stepped foot in one, you probably know that tropical rain forests are humid, warm, dense, and full of unusual plants and animals. You might not know, however, that they are home to a greater diversity of living things than anywhere else on Earth; that they are the result of millions of years of adaptive evolution; and that they store more carbon than is in the entire atmosphere. Nor can you see with your eyes how energy flows through these forests or how nitrogen and phosphorus are cycled and recycled. This knowledge comes from the work of many scientists who have been asking basic questions about how such natural systems function, how they came to be, and how they relate to the rest of the world.

HOW IMPORTANT IS the knowledge that comes from these scientists? In this second part of the text, we'll see that information about how the natural world works is absolutely crucial to understanding how to manage and protect it. In Part 2, we will look at the science underlying how living things survive in the world and how matter and energy move throughout systems. We will use a hierarchical approach: we will start small and move large—from the molecules necessary for life and the solar energy that sustains it to how plants return to a habitat after a fire and why large regions of the world have different vegetation (such as forests, desert, and grassland). Understanding environmental science begins with an understanding of how living organisms are structured, survive, relate to their environment (and each other), and change over time.

CHAPTER 3

Basic Needs of Living Things

China

Quake epicenter

Historic Giant Panda range

LEARNING OBJECTIVES

1. **Organisms in Their Environment:** Explain how the science of ecology can be described as a hierarchy of questions; describe the questions different types of ecologists ask, and anticipate different experiments they might do, all within the same natural systems.

2. **Environmental Factors:** Explain how organisms need specific conditions and resources to survive; define an organism's niche.

3. **Matter in Living and Nonliving Systems:** Explain how living things need materials to build tissues and energy to carry out life processes such as photosynthesis and cellular respiration.

4. **Matter and Energy:** List the simple laws that control the transformation of energy from one type to another and its flow through Earth's systems and back into space.

5. **The Cycling of Matter in Ecosystems:** Explain the biogeochemical cycles of matter (particularly the elements of carbon, phosphorus, nitrogen, and sulfur) as they cycle through living and nonliving spheres.

O N MAY 12, 2008, a massive earthquake shook Sichuan Province, China. It closed roads, caused buildings to collapse, and triggered mudslides. The earthquake and an estimated 11,600 aftershocks over the next several weeks left more than 88,000 dead or missing and up to 5 million people homeless.

The earthquake also damaged two reserves for China's giant pandas: the Wolong Giant Panda Reserve Center in southwest Sichuan Province and the nearby Chengdu Panda Breeding and Research Center. Enclosures were destroyed, pandas were frightened and escaped, and food was difficult to obtain. Although most of the pandas were unharmed in the earthquake, one was missing, and an adult female named Maomao was killed. In the face of this bad news, two weeks after the earthquake, to the relief of keepers and the great rejoicing of panda conservationists, a pregnant female gave birth. Eventually, the pandas in the earthquake area had to be moved to other facilities.

A SAD DAY. China's sorrow and concern for the pandas made international news. Maomao's funeral was covered on the Internet and television. Why so much concern about pandas in a disaster that harmed so many humans? The giant panda is a cultural icon of China, beloved and found in books and stories. Sadly, it is endangered, with only about 1,600 pandas still alive in the wild, many in small reserves, and another approximately 270 in captivity. Despite their endangered species status, giant pandas remain under pressure from habitat loss and poaching. To many Chinese citizens, the loss of a giant panda and the fear that they might be even closer to extinction after the earthquake were sorrows added to the tremendous loss of human life. The survival of most of the protected pandas brought joy in a time of despair.

PANDA SURVIVAL. The giant panda lives on bamboo, found in broad-leaved forests in a decreasing habitat in China. Care for the declining panda population involves not only care for their individual health and the interbreeding population, but also protection of their habitat, including their food source. Often populations decline because damage is done to a habitat or food resource rather than to individuals directly. The panda and its food, bamboo, give us an opportunity to look at the fundamental science behind the survival of organisms in their natural spaces.

In this chapter, we will follow the panda and the bamboo as examples of organisms living in their environment. All organisms get nutrients and energy from the environment and use them to build their body tissues and carry out activities such as moving, getting food, protecting themselves, reproducing, and getting rid of wastes. Pandas and bamboo both need ways to deal with the conditions under which they live—the amount of light, food, and water available, for example. We will see how the basic functions of organisms fit into the scientific discipline of ecology. We will explore basic physical rules about matter and energy and how they control the lives of organisms in their environments. To do this, we will need to understand some basic chemistry and physics. At the chapter's end, we will look at the global movements of important nutrients and how those huge cycles both affect individual organisms and are affected by human activities.

3.1 Organisms in Their Environment

Bamboo is a group of about 1,000 related types in the grass family, but unlike most grasses you may be familiar with, bamboos are woody and evergreen. As a group, bamboo is the fastest-growing plant on Earth. The areas in which bamboo is found are a function of several factors—its habitat requirements, its evolutionary history (where it first developed), and its interactions with other organisms, including humans. Bamboo needs certain resources (such as sunlight and water) to survive. A panda also needs certain resources to survive, including bamboo (as a food source). The distribution and abundance of the giant panda are likewise functions of its requirements for survival, its growth, and the history of interaction between pandas and other species, especially humans. To understand why giant pandas are in decline and what conditions bamboo requires for growth, we turn to the science of ecology.

The Hierarchy of Ecology

Ecology (from *Oikos*, meaning "house or home") is *the study of all processes influencing the distribution and abundance of organisms and the interactions between living things and their environment.* A very basic understanding of ecological terms and concepts will help later as we look at how changes in the environment affect living things.

Ecologists studying the conservation of giant pandas ask an array of different questions. They investigate what type of soil bamboo requires to grow most rapidly or look for the factors that influence successful breeding. They might also study land use around Sichuan Province to see how it affects giant panda habitat. Any of these inquiries constitutes an ecological question, and several different branches of ecology exist to address them. Ecology can be described as a group of different, but related, disciplines: a hierarchy of studies. Scientists in different branches of the discipline might operate at different scales and ask different questions about the same system. The concept of ecological hierarchy is depicted in **Figure 3–1**.

As we investigate the science of ecology, we will look at these different levels in the hierarchy. We start by examining individual organisms at the species level. We will see how species are grouped into larger categories and how these groups interact. Species exist in specific habitats because those habitats provide the nutrients and other factors that a species needs to survive; we will look at how those needs are met. Lastly, we will take a broader view of matter and energy, and how energy and nutrients cycle through the various systems on Earth. All of these different processes are interlinked. Ecologists study the connections as well as the whole.

Species. A **species** is a group of individuals that share certain characteristics distinct from other such groups (robins versus redwing blackbirds, for example). Species are classified into closely related groups called *genera* (singular, *genus*), and they in turn are grouped into *families*. These are grouped into larger groups: *orders, classes, phyla* (singular, *phylum*), *kingdoms,* and *domains.* The official scientific name of any species is always Latin and has two parts: the genus name and the species epithet, or descriptive term. For example, the giant panda's name, *Ailuropoda melanoleuca* ("black-and-white cat-foot"), tells us first the genus (*Ailuropoda*) and second a descriptive word (the species epithet, *melanoleuca*). Although the giant panda is in the bear family (*Ursidae*), it is considered so different from other bears that it is the only species in this genus. In contrast, when we say "bamboo," we actually mean a group of about 1,000 related species in the same subcategory of the grass family. The bamboo eaten by the giant panda is composed of members of two of these bamboo species.

Levels of organization within an organism

Biome

Biosphere

Landscape

Ecosystem
organisms,
sun, and rain.

Organism

Population

Organ

Tissue

Cell

Community

Figure 3–1 The hierarchy of life. Organisms can be seen as a hierarchy of complex systems from cells to tissues to organs to systems to whole individuals. Likewise, individuals function in a set of increasingly complex systems that make up the disciplines of ecology. These include populations, communities, ecosystems, and landscapes. The very largest regions are called biomes. Together, all living things make up the biosphere.

Sometimes it is difficult to determine exactly what constitutes a species. The biological definition of a species is *all members that can interbreed and produce fertile offspring*. Members of different species generally do not interbreed. This definition is good for many organisms, but does not work for those that do not need to mate to produce offspring. In those cases, scientists use a variety of other classification methods. New species can arise over periods of time through evolution, and species classifications are occasionally altered to reflect this.

Populations. Each species in a biotic community is represented by a certain **population**—that is, by a certain number of individuals that make up the interbreeding, reproducing group. The distinction between population and species is that *population* refers only to those individuals of a certain species that live within a given area—for example, all of the gray wolves in Yellowstone National Park. *Species* is all inclusive, such as all of the gray wolves in the world.

Biotic Communities. The grouping of populations we observe when studying a forest, grassland, or pond is referred to as the area's **biota** (*bio* means "living") or **biotic community**. The biota includes all vegetation, from large trees through microscopic algae, and all animals, from large mammals, birds, reptiles, and amphibians through earthworms, tiny insects, and mites. It also includes microscopic creatures such as bacteria, fungi, and protozoans.

The particular kind of biotic community found in a given area is, in large part, determined by **abiotic** (nonliving, chemical, and physical) factors. Abiotic factors include the amount of water or moisture present, climate, salinity (saltiness), and soil type. Sometimes, we quickly refer to a community by its most obvious members, such as the dominant plants. For example, we might describe a "hemlock/beech forest" or an "oak/hickory forest" **(Fig. 3–2)**. This is because vegetation is easily observed (it doesn't usually run away) and is a strong indicator of the environmental conditions of a site.

The species within a community depend on one another. For example, certain animals will not be present unless certain plants that provide food and shelter are also present. In this way, the plant community supports (or limits by its absence) the animal community. In addition, every plant and animal species is adapted to cope with the abiotic factors of its region. For example, every species that lives in a temperate region is adapted in one way or another to survive the winter season, which includes a period of freezing temperatures **(Fig. 3–3)**. These interactions among organisms and their environments will be discussed later. For now, keep in mind that the populations of different species within a biotic community are constantly interacting with each other and with the abiotic environment.

Ecosystems. An ecosystem is an interactive complex of communities and the abiotic environment affecting them within a particular area. Forests, grasslands, wetlands, sand dunes, and coral reefs each contain specific species in particular spatial areas and are all distinct ecosystems. Humans are parts of ecosystems, too, and some branches of ecology, such as urban ecology, focus on human-dominated

Figure 3–2 Predictable vegetation. An oak/hickory forest is dominated by oak and hickory trees, which define the type of forest habitat other species experience.

Figure 3–3 Winter in the forest. Many trees and other plants of temperate forests are so adapted to the winter season that they actually require a period of freezing temperature in order to grow again in the spring.

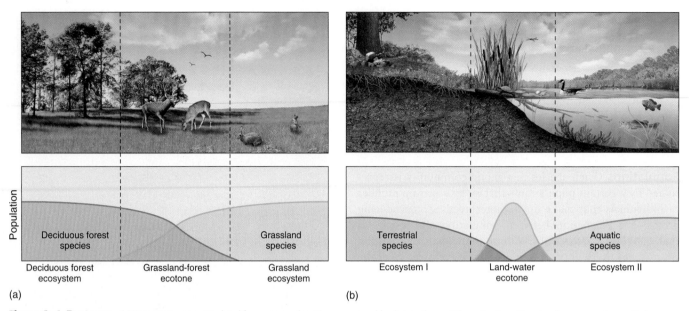

(a)   (b)

Figure 3–4 Ecotones. (a) Ecosystems are not isolated from one another. One ecosystem blends into the next through a transitional region—an ecotone—that contains many species common to both systems, though in lower numbers than in those systems. (b) An ecotone may create a unique habitat that harbors specialized species not found in either of the ecosystems bordering it. For example, cattails, reeds, and lily pads typically grow in the ecotone between land and freshwater ponds.

systems. Other ecologists study the interaction between humans and non-human ecosystems, such as salt marshes or the open ocean, which are deeply affected by humans.

While it is convenient to divide the living world into different ecosystems, you will find that there are seldom distinct boundaries between them and that they are not totally isolated from one another. Many species occupy (and are a part of) two or more ecosystems or, as in the case of migrating birds, may move from one ecosystem to another at different times.

Transitional regions between ecosystems are known as **ecotones.** An ecotone shares many of the species and characteristics of both ecosystems it passes through. Ecotones may lack unique (and contain fewer overall) species than bordering ecosystems, as in the ecotone between grassland and forest shown in **Figure 3–4a.** Alternatively, ecotones may also house and support more different species than those they pass through, as in the marshy ecotone that often occurs between the open water of a lake and dry land (Fig. 3–4b).

Landscapes and Biomes. Clusters of interacting ecosystems like forests, open meadows, and rivers together constitute **landscapes.** Landscape ecologists study issues such as how much suitable habitat will be left for fish to lay eggs in a riverbed if the surrounding forest is clear-cut or how far squirrels will be able to travel if the forest is broken into small segments. Landscapes are usually on a scale humans can envision; for example, if you look out at the horizon and see a forest merge into a field or meadow, you are seeing a landscape. However, landscapes can also be observed on a large scale: one could also pull back from Earth and look at large areas of its surface that have similar characteristics (such as dominant vegetation). Many scientists use digital technology tools such as **Geographic Information Systems** (GIS) to study landscape features. They also use **remote sensing,** including images from satellites **(Fig. 3–5).**

Tropical rain forests, grasslands, and deserts are **biomes** (discussed in Chapter 5). A biome is a large area of Earth's surface that shares climate and has similar vegetation. Biomes are much larger than landscapes. For example, grasslands (such as prairies) are geographical regions whose locations can be predicted by their rainfall and temperatures. On different continents, they may have different particular species, but they are still grasslands, described by the dominant type of vegetation. As with ecosystems, there are generally no sharp boundaries between biomes. Instead, one grades into the next through transitional regions. The term *biome* is limited to terrestrial systems. There are major categories of aquatic and wetland ecosystems that are determined primarily by the depth, salinity, and permanence of water in them (Chapter 5).

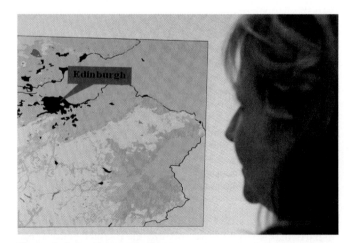

Figure 3–5 Landscape Ecology. Here a researcher demonstrates a mapping tool used in landscape ecology. This is a prototype of a Geographical Information System (GIS) called Soilfit.

Biosphere. Regardless of how we choose to categorize and name different ecosystems, they are interconnected and interdependent. Terrestrial ecosystems are connected by the flow of rivers and by migrating animals. Sediments and nutrients washing from the land may nourish or pollute the ocean. Seabirds and marine mammals connect the oceans with the land, and all biomes share a common atmosphere and water cycle. All the species on Earth live in ecosystems, and ecosystems are interconnected so that all living things form one huge system called the **biosphere**. This concept is analogous to the idea that the cells of our bodies form tissues, tissues form organs, organs are connected into organ systems, and all are interconnected to form the whole body.

3.2 Environmental Factors

Let's look back at the giant panda eating bamboo at the reserve. Why are pandas living there and not in another part of the world? The answer is both the environment and history. The bamboo and the panda have evolved to live in the environment of the reserve, with the physical, chemical, and biological factors it contains. As already discussed, we describe these factors as either biotic (living) or abiotic (nonliving). The factors can be further characterized as either conditions or resources. **Conditions** are any factors that vary in space and time, but that are not used up or made unavailable to other species. These include temperature (extremes of heat and cold as well as average temperature), wind, pH (acidity), salinity (saltiness), and fire. **Resources** are any factors—biotic or abiotic—that are consumed by organisms. These include water, chemical nutrients (like nitrogen and phosphorus), light (for plants), oxygen, food (like the bamboo for the panda), and space for resting or nesting.

Some factors can be conditions in one situation and resources in another. For example, plants in the desert compete for water as a *resource*; however, in a pond, water is a *condition* for which plants like water lilies and cattails need special adaptations and for which they are not competing.

The degree to which each factor is present or absent profoundly affects the ability of organisms to survive. However, each species may be affected differently by each factor, and its response determines whether or not it will occupy a given region.

Optimums, Zones of Stress, and Limits of Tolerance

Different species thrive with different combinations of environmental factors. This principle applies to all living things, both plants and animals. Bamboo thrives in full sun and warm temperatures, with plenty of water and well-drained, nitrogen-rich soil. In contrast, desert cacti are generally adapted for low amounts of water, poor soil fertility, and extremes of heat and cold. Aquatic systems also vary in their environmental factors and residents—for example, the amounts of freshwater and saltwater determine the varieties of fish and other organisms present.

Organismal ecologists study the adaptations of organisms to specific environmental factors. Organisms can be grown under controlled conditions in which one factor is varied while other factors are held constant. Such experiments demonstrate that, for every factor, there is a certain level at which the organisms grow or survive most; an **optimum** (plural, *optima*). At higher or lower levels than the optimum, the organisms do less well, and at further extremes, they may not be able to survive at all **(Fig. 3–6)**. This pertains to almost any abiotic factor that might be tested.

The point at which the best response occurs is the optimum, but this may be a range of several degrees (or other units), so it is common to speak of an *optimal range*. The entire span that allows any growth at all is called the **range of tolerance**. The points at the high and low ends of the range of tolerance are called the **limits of tolerance**. Between the optimal range and the high or low limit of tolerance are **zones of stress**. In other words, as the factor is raised or lowered from the optimal range, the organisms experience increasing stress, until, at either limit of tolerance, they cannot survive.

A Fundamental Biological Principle. Based on the consistency of observations and experiments, the following is considered to be a fundamental biological principle: *Every species has an optimal range, zones of stress, and limits of tolerance with respect to every abiotic factor, and these characteristics vary between species.* For instance, some plants cannot tolerate any freezing temperatures; others can tolerate slight freezing; and some actually require several weeks of freezing temperatures to complete their life cycles. Some species have a very broad range of tolerance, whereas others have a much narrower range.

The range of tolerance for any factor affects more than just the growth of individuals. Because the health and vigor of individuals affect their ability to reproduce, the survival of the next generation is also affected, which in turn influences the population as a whole. Consequently, the population density (individuals per unit area) of a species is greatest where all conditions are optimal, and it decreases as any one or more conditions depart from the optimum. You expect to see the most bamboo in the places where all conditions are best for bamboo and less where conditions are not so favorable.

Law of Limiting Factors. In 1840, Justus von Liebig studied the effects of chemical nutrients on plant growth. He observed that restricting any one of the many different nutrients at any given time had the same effect: it limited growth. A factor that limits growth is called (not surprisingly) a **limiting factor.** If any one factor is outside the optimal range, it will cause stress and limit the growth, reproduction, or even survival of a population. This observation is referred to as the **law of limiting factors** or Liebig's law of the minimum.

A limiting factor may be a problem of *too much* instead of a problem of *too little*. For example, plants may be stressed or killed not only by underwatering or underfertilizing, but also by overwatering or overfertilizing, which are common pitfalls for beginning gardeners. Note also that the limiting factor may change from one time to another and that if one limiting factor is corrected, growth will increase only until another factor comes into play. For instance, temperature may

Figure 3–6 Survival curve. For every factor influencing growth, reproduction, and survival, there is an optimum level. Above and below the optimum, stress increases until survival becomes impossible at the limits of tolerance. The total range between the high and low limits is the range of tolerance.

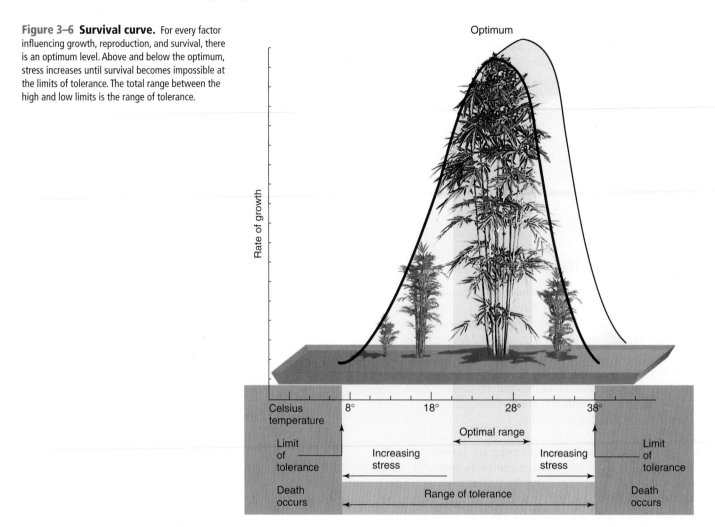

be limiting in the early spring, nutrients may be limiting in the winter, and water may be limiting if a drought occurs.

While one factor may be limiting at a given time, several factors outside the optimum may combine to cause additional stress or even death. In particular, pollutants may act in a way that causes organisms to become more vulnerable to disease or drought. Such cases are examples of **synergistic effects,** or **synergisms,** which are defined as two or more factors interacting in a way that causes an effect much greater than one would anticipate from each of the two acting separately.

Habitat and Niche. Habitat refers to the kind of place— defined by the plant community and the physical environment—where a species is biologically adapted to live. For example, deciduous forests, swamps, and grassy fields are types of habitats. Different types of forests (for instance, coniferous versus tropical) provide markedly different habitats and support different species of wildlife. Because some organisms operate on very small scales, we use the term *microhabitat* for things like puddles, spaces sheltered by rocks, and holes in tree trunks that might house their own small community **(Fig. 3–7).**

Even when different species occupy the same habitat, competition may be slight or nonexistent because each species has its own niche. An ecological **niche** is the sum of all of the conditions and resources under which a species can live: what and where it feeds, what it feeds on, where it finds shelter and nests, and how it responds to abiotic factors.

Similar species can coexist in the same habitat but have separate niches. Competition is minimized because potential competitors are using different resources. For example, bats and swallows both feed on flying insects, but they do not compete because bats feed on night-flying insects and swallows feed during the day.

In the next section, we will see how organisms take these materials from the environment and use them to grow and maintain their bodies and do activities necessary for survival.

3.3 Matter in Living and Nonliving Systems

Living organisms need to be able to take in matter and energy from their environment to grow and function. For example, the bamboo that grows in the panda reserve can survive only because it is able to take the Sun's energy and combine it with materials from the soil, air, and water around it to make the basic building blocks of its own tissues and to power its cellular activities. In turn, the giant panda will eat

Figure 3–7 Habitat and niche. For part of their lives, these damselflies live in the pool of water trapped by a type of tropical plant; the tiny pool is a microhabitat. The specific conditions required by the damselfly, such as the pool and the types of food it eats there or where it lays its eggs as an adult, are its niche.

the bamboo and get energy and matter necessary to build its own tissues and to do the work it needs to do in each of its cells. To understand this simple ecological relationship, we need to know some basic chemistry and physics. Later, we will explore what happens to energy and matter as they move through the biosphere.

Basic Units of Matter

Matter is defined as anything that occupies space and has mass. This definition covers all solids, liquids, and gases as well as all living and nonliving things. Matter is composed of **atoms**—very small pieces—that are combined to form molecules, which in turn can be combined into more complex structures.

Atoms. The basic building blocks of all matter are atoms. Only 94 different kinds of atoms occur in nature, and these are known as the naturally occurring **elements**. Atoms are made up of protons, neutrons, and electrons, which in turn are made up of still smaller particles. For example, lead (Pb) is an element. An atom of lead is the smallest amount of lead existing that has the characteristics of lead. A lead atom has a characteristic number of protons (positive particles), neutrons (neutral particles), and electrons (negative particles).

How can these relatively few building blocks make up the countless materials of our world, including the tissues of living things? Picture each kind of atom as a different-sized Lego® block. Like Legos, atoms can build a great variety of things. Also like Legos, natural materials can be taken apart into their separate constituent atoms, and the atoms can then be reassembled into different materials. All chemical reactions, whether occurring in a test tube, in the environment, or inside living things (and whether they occur very slowly or very quickly), involve rearrangements of atoms to form different kinds of matter.

Atoms do not change during the disassembly and reassembly of different materials. A carbon atom will always remain a carbon atom. In chemical reactions, atoms are neither created nor destroyed. The same number and kind of different atoms exist before and after any reaction. This constancy of atoms is regarded as a fundamental natural law, the **Law of Conservation of Matter.** Nuclear reactions differ from the chemical reactions we are discussing and can result in the splitting of an atom of one element into multiple atoms of another element. However, this is a very specific, rare instance and is not a chemical reaction. (Nuclear reactions will be discussed in Chapter 15.)

Now we turn our attention to the ways atoms are put together, which atoms make up the bulk of the bodies of living organisms, and how atoms are incorporated into organisms.

Molecules and Compounds. A molecule consists of two or more atoms (either the same kind or different kinds) bonded in a specific way. The properties of a material depend on the exact way in which atoms are bonded to form molecules as well as on the atoms themselves. A **compound** consists of two or more *different* kinds of atoms that are bonded. For example, the fundamental units of oxygen gas, which consist of two bonded oxygen atoms, are molecules (O_2) but not compounds. By contrast, when an oxygen atom binds with hydrogen atoms to create water, it is both a molecule and a compound (H_2O).

On the chemical level, the cycle of growth, reproduction, death, and decay of organisms is a continuous process of using various molecules and compounds from the environment (food), assembling them into living organisms (growth), disassembling them (decay), and repeating the process.

Four Spheres

During growth and decay, atoms move from the environment into living things and then return to the environment. To picture this process, think of the environment as three open, nonliving systems, or "spheres," interacting with the biosphere **(Fig. 3–8)**. The **atmosphere** is the thin layer of gases (including water vapor) separating Earth from outer space. The **hydrosphere** is water in all of its liquid and solid compartments: oceans, rivers, ice, and groundwater. The **lithosphere** is Earth's crust, made up of rocks and minerals. Matter is constantly being exchanged within and between these four spheres.

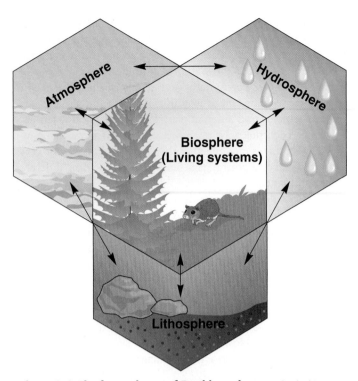

Figure 3–8 The four spheres of Earth's environment. The biosphere is all of life on Earth. It depends on, and interacts with, the atmosphere (air), the hydrosphere (water), and the lithosphere (soil and rocks).

(*Source*: Adapted from *Geosystems*, 5th ed., by Robert W. Christopherson, © 2005 Pearson Education, Inc.)

Atmosphere. The lower atmosphere is a mixture of molecules of three important gases—oxygen (O_2), nitrogen (N_2), and carbon dioxide (CO_2)—along with water vapor and trace amounts of several other gases. The gases in the atmosphere are normally stable, but under some circumstances, they react chemically to form new compounds. For example, ozone is produced from oxygen in the upper atmosphere (Chapter 18). Plants take carbon dioxide in from the atmosphere, usually through their leaves. Animals usually take oxygen in through some type of specialized organ such as a lung, but some, like earthworms, can simply absorb oxygen through their skin.

Hydrosphere. While the atmosphere is a major source of carbon and oxygen for organisms, the hydrosphere is the source of hydrogen. Each molecule of water consists of two hydrogen atoms bonded to an oxygen atom, so the chemical formula for water is H_2O. A weak attraction known as *hydrogen bonding* exists between water molecules.

Water is an important molecule for living things and usually needs to be available in liquid form. Water occurs in three different states. At temperatures below freezing, hydrogen bonding holds the molecules in position with respect to one another, and the result is a solid **crystal** structure (ice or snow). At temperatures above freezing but below vaporization, hydrogen bonding still holds the molecules close but allows them to move past one another, producing the liquid state. Vaporization (evaporation) occurs as hydrogen bonds break and water molecules move into the air independently. As temperatures are lowered again, all of these changes of state go in the reverse direction. Generally, water undergoes melting and evaporation, but sometimes water molecules leave snow or ice and go directly into the air. This is *sublimation* (**Fig. 3–9**). Moving from one state to another either releases or requires a great deal of energy. This is one reason why many animals sweat to cool off: changing from

Figure 3–9 Water and its three states. (a) Water consists of molecules, each of which is formed when two hydrogen atoms bond to an oxygen atom (H_2O). (b) In water vapor, the molecules are separate and independent. (c) In liquid water, the weak attraction between water molecules known as hydrogen bonding gives the water its liquid property. (d) At freezing temperatures, hydrogen bonding holds the molecules firmly, giving the solid state—ice.

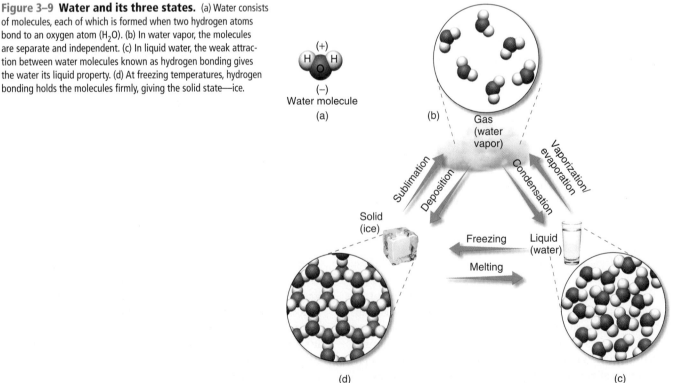

Figure 3–10 Minerals. Minerals (hard, crystalline compounds) are composed of dense clusters of atoms of two or more elements. The atoms of most elements gain or lose one or more electrons, becoming negative (−) or positive (+) ions, and form a predictable pattern held closely together. This photograph is of a mineral, gypsum in a mine in Texas.

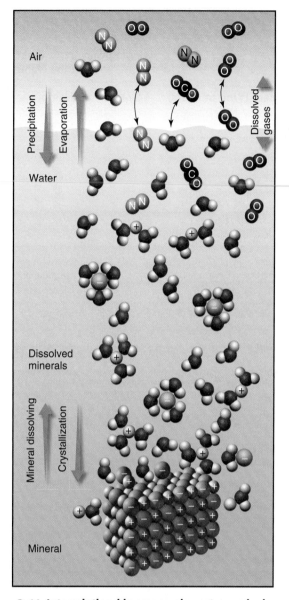

Figure 3–11 Interrelationship among air, water, and minerals. Minerals and gases dissolve in water, forming solutions. Water evaporates into air, causing humidity. These processes are all reversible: minerals in solution recrystallize, and water vapor in the air condenses to form liquid water.

the liquid to gas state requires heat and removes it from the organism. Despite the changes of state, the water molecules themselves retain their basic chemical structure of two hydrogen atoms bonded to an oxygen atom; only the *relationship* between the molecules changes.

Lithosphere. All the other elements that are required by living organisms, as well as the approximately 72 elements that they do not require, are found in the lithosphere, in the form of rock and soil minerals. A **mineral** is a naturally occurring solid, made by geologic processes; it is a hard, crystalline structure of a given chemical composition **(Fig. 3–10)**. Most rocks are made up of relatively small crystals of two or more minerals, and soil generally consists of particles of many different minerals. Each mineral is made up of dense clusters of two or more kinds of atoms bonded by an attraction between positive and negative charges on the atoms (Appendix C).

Interactions. Air, water, and minerals interact with each other in a simple, but significant, manner. Gases from the air and ions (charged atoms) from minerals may dissolve in water. Therefore, natural water is inevitably a **solution** containing variable amounts of dissolved gases and minerals. This solution is constantly subject to change because various processes may remove any dissolved substances in it or additional materials may dissolve in it. Molecules of water enter the air by evaporation and leave it again via condensation and precipitation (see the hydrologic cycle, Chapter 10). Thus, the amount of moisture in the air fluctuates constantly. Wind may carry dust or mineral particles, but the amount changes constantly because the particles gradually settle out from the air. These interactions are summarized in **Figure 3–11**. The materials in these three spheres interact with the biosphere as living organisms take materials from the spheres and use them to create complex molecules in their bodies. We will discuss the process of building these molecules next.

Organic Compounds

Your body, like that of the panda or even the bamboo it eats, is composed of relatively large chemical compounds in a number of broad categories, such as proteins, carbohydrates (sugars and starches), lipids (fatty substances), and nucleic acids (DNA and RNA). These compounds usually contain six key elements: carbon (C), hydrogen (H), oxygen (O), nitrogen (N), phosphorus (P), and sulfur (S) (Table 3–1). Living things need other elements as well, sometimes in smaller, but no less important, amounts.

Unlike the relatively simple molecules that occur in the environment (such as CO_2, H_2O, and N_2), the key chemical elements in living organisms bond to form very large, complex molecules. The chemical compounds making up

TABLE 3–1   Elements Found in Living Organisms and the Locations of Those Elements in the Environment

Element (Kind of Atom)	Biologically Important Molecule or Ion in Which the Element Occurs[a]			Location in the Environment[b]		
	Symbol	Name	Formula	Atmosphere	Hydrosphere	Lithosphere
Carbon	C	Carbon dioxide	CO_2	X	X	X (CO_3)
Hydrogen	H	Water	H_2O	X	(Water itself)	
Atomic oxygen (required in respiration)	O	Oxygen gas	O_2	X	X	
Molecular oxygen (required in photosynthesis)	O_2	Water	H_2O		(Water itself)	
Nitrogen	N	Nitrogen gas	N_2	X	X	Via fixation
		Ammonium ion	NH_4		X	X
		Nitrate ion	NH_3		X	X
Sulfur	S	Sulfate ion	SO_4^2		X	X
Phosphorus	P	Phosphate ion	PO_4^3		X	X
Potassium	K	Potassium ion	K^+		X	X
Calcium	Ca	Calcium ion	Ca^2		X	X
Magnesium	Mg	Magnesium ion	Mg^2		X	X
Trace Elements[c]						
Iron	Fe	Iron ion	Fe^2, Fe^3		X	X
Manganese	Mn	Manganese ion	Mn^2		X	X
Boron	B	Boron ion	B^3		X	X
Zinc	Zn	Zinc ion	Zn^2		X	X
Copper	Cu	Copper ion	Cu^2		X	X
Molybdenum	Mo	Molybdenum ion	Mo^2		X	X
Chlorine	Cl	Chloride ion	Cl^-		X	X

Note: These elements are found in *all* living organisms—plants, animals, and microbes. Some organisms require certain elements in addition to the ones listed. For example, humans require sodium and iodine.

[a]A molecule is a chemical unit of two or more atoms that are bonded. An ion is a single atom or group of bonded atoms that has acquired a positive or negative charge as indicated.

[b]X means that element exists in the indicated sphere.

[c]Only small or trace amounts of these elements are required.

the tissues of living organisms are referred to as **organic.** Some of these molecules may contain millions of atoms. These molecules are constructed mainly from carbon atoms bonded into chains, with hydrogen and oxygen atoms attached. Nitrogen, phosphorus, or sulfur may be present also, but the key common denominator is carbon–carbon and carbon–hydrogen or carbon–oxygen bonds. **Inorganic,** then, refers to molecules or compounds with neither carbon–carbon nor carbon–hydrogen bonds. While this is the general rule, some exceptions occur; by convention, a few compounds that contain carbon bonds, such as carbon dioxide, are still considered inorganic.

All plastics and countless other human-made compounds are also based on carbon bonding and are, chemically speaking, organic compounds. To resolve any confusion this may cause, the compounds making up living organisms are referred to as **natural organic compounds** and the human-made ones as **synthetic organic compounds**. The term *organic* can have a completely different meaning, such as in *organic farming* (Chapter 12).

To summarize: the elements essential to life (C, H, O, and so on) are present in the atmosphere, hydrosphere, or lithosphere in relatively simple molecules. In the living organisms of the biosphere, on the other hand, these elements are organized into highly complex organic compounds. In the case of the panda and bamboo, simple molecules and atoms in the soil, air, and water are combined to form the tissues of bamboo and eaten, digested, and recombined to make the tissues of the giant panda. When the bamboo or panda dies, the reverse process occurs through decomposition and decay. Each of these processes is discussed in more detail later in the chapter. Next, we discuss energy, the force that helps change chemical matter into substances that support life.

3.4 Matter and Energy

The universe is made up of matter and energy. Recall that matter is *anything that occupies space and has mass*. By contrast, light, heat, movement, and electricity do not have mass, nor do they occupy space. These are the common forms of energy that you are familiar with. What do the various forms of energy have in common? They affect matter, causing changes in its *position* or *state*. For example, the release of energy in an explosion causes things to go flying—a change in position. Heating water causes it to boil and change to steam, a change of state. On a molecular level, changes of state are actually also movements of atoms or molecules. For instance, the degree of heat energy contained in a substance is a measure of the relative vibrational motion of the atoms and molecules of the substance. Therefore, we can define **energy** as *the ability to move matter*.

Energy Basics

Energy can be categorized as either kinetic or potential **(Fig. 3–12)**. **Kinetic energy** is *energy in action or motion*. Light, heat, physical motion, and electrical current are all forms of kinetic energy. **Potential energy** is *energy in storage*. A substance or system with potential energy has the capacity, or potential, to release one or more forms of kinetic energy. A stretched rubber band has potential energy; it can send a paper clip flying. Numerous chemicals, such as gasoline and other fuels, release kinetic energy—heat, light, and movement—when ignited. The potential energy contained in such chemicals and fuels is called **chemical energy**.

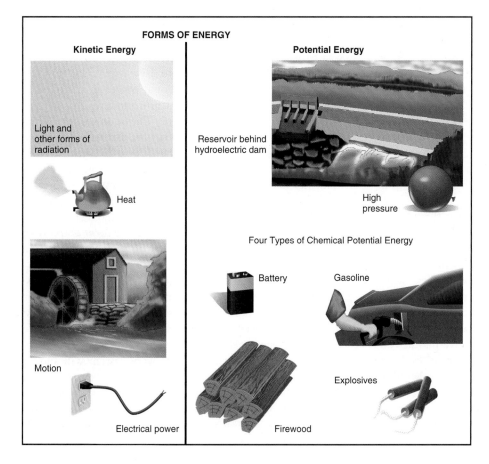

FORMS OF ENERGY

Kinetic Energy

Light and other forms of radiation

Heat

Motion

Electrical power

Potential Energy

Reservoir behind hydroelectric dam

High pressure

Four Types of Chemical Potential Energy

Battery

Gasoline

Explosives

Firewood

Figure 3–12 Forms of energy. Energy is distinct from matter in that it neither has mass nor occupies space. It has the ability to act on matter, though, changing the position or the state of the matter. Kinetic energy is energy in one of its active forms. Potential energy is the potential that systems or materials have to release kinetic energy.

Figure 3–13 Energy conversions.
Any form of energy except heat can spontaneously transform into any other form. Heat is a form of energy that flows from one system or object to another because the two are at different temperatures; therefore, heat can spontaneously transform only to something cooler.

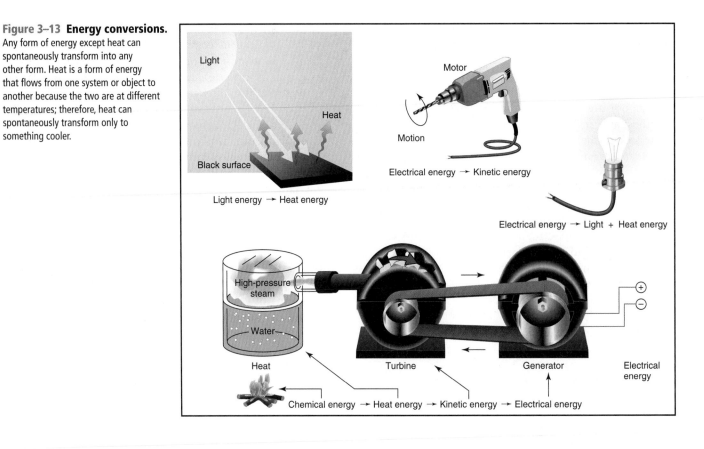

Light

Heat

Black surface

Light energy → Heat energy

Motor

Motion

Electrical energy → Kinetic energy

Electrical energy → Light + Heat energy

High-pressure steam

Water

Heat

Turbine

Generator

Electrical energy

Chemical energy → Heat energy → Kinetic energy → Electrical energy

Energy may be changed from one form to another in innumerable ways **(Fig. 3–13)**. Besides understanding that potential energy can be converted to kinetic energy, it is especially important to recognize that kinetic energy can be converted to potential energy. (Consider, for example, charging a battery or pumping water into a high-elevation reservoir.) We shall see later in this section that photosynthesis does just that.

Because energy does not have mass or occupy space, it cannot be measured in units of weight or volume. Instead, energy is measured in other kinds of units. One of the most common energy units is the **calorie,** which is defined as the amount of heat required to raise the temperature of 1 gram (1 milliliter) of water 1 degree Celsius. This is a very small unit, so it is frequently more convenient to use the kilocalorie (1 kilocalorie = 1,000 calories), the amount of thermal (heat) energy required to raise 1 liter (1,000 milliliters) of water 1 degree Celsius. Kilocalories are sometimes denoted as "Calories" with a capital "C." Food Calories, which measure the energy in given foods, are actually kilocalories. *Temperature* measures the molecular motion in a substance caused by the kinetic energy present in it.

If energy is defined as the ability to move matter, then no matter can be moved *without* the absorption or release of energy. Indeed, no change in matter—from a few atoms coming together or apart in a chemical reaction to a major volcanic eruption—can be separated from its respective change in energy.

Laws of Thermodynamics. Because energy can be converted from one form to another, numerous would-be inventors over the years have tried to build machines that produce more energy than they consume. One common idea was to use the output from a generator to drive a motor that in turn would be expected to drive a generator to keep the cycle going and yield additional power in the bargain (a perpetual-motion machine). Unfortunately, all such devices have one feature in common: They don't work. When all the inputs and outputs of energy are carefully measured, they are found to be equal. There is no net gain or loss in total energy. This observation is now accepted as a fundamental natural law, the **Law of Conservation of Energy.** It is also called the **First Law of Thermodynamics,** and it can be expressed as follows: *Energy is neither created nor destroyed, but may be converted from one form to another.* What this law really means is that you can't get something for nothing.

Imaginative "energy generators" fail for two reasons: First, in every energy conversion, a portion of the energy is converted to heat energy. Second, there is no way of trapping and recycling heat energy without expending even more energy in doing so. Consequently, in the absence of energy inputs, any and every system will sooner or later come to a stop as its energy is converted to heat and lost. This is now accepted as another natural law, the **Second Law of Thermodynamics,** and it can be expressed as follows: *In any energy conversion, some of the usable* *energy is always lost.* Thus, you can't get something for nothing (the first law), and in fact, you can't even break even (the second law)!

Entropy. The principle of increasing entropy underlies the loss of usable energy and its transformation to heat energy. **Entropy** is *a measure of the degree of disorder in a system*, so increasing entropy means increasing disorder. Without energy inputs, everything goes in one direction only—toward increasing entropy.

The conversion of energy and the loss of usable energy to heat are both aspects of increasing entropy. Heat energy is the result of the random vibrational motion of atoms and molecules. It is the lowest (most disordered) form of energy, and its spontaneous flow to cooler surroundings is a way for that disorder to spread. Therefore, the Second Law of Thermodynamics may be more generally stated as follows: *Systems will go spontaneously in one direction only—toward increasing entropy* **(Fig. 3–14)**. The second law also states that systems will go spontaneously only toward lower potential energy, a direction that releases heat from the systems.

It is possible to pump water uphill, charge a battery, stretch a rubber band, and compress air. All of these require the input of energy in order to increase the potential energy of a system. The verbs *pump, charge, stretch,* and *compress* reflect the fact that energy is being put into the system. In contrast, flow in the opposite direction (which releases energy) occurs spontaneously **(Fig. 3–15)**.

Whenever something gains potential energy, therefore, keep in mind that the energy is being obtained from somewhere else (the first law). Moreover, the amount of usable energy lost from that "somewhere else" is greater than the amount gained (the second law). We now relate these concepts of matter and energy to organic molecules, organisms, ecosystems, and the biosphere.

Energy Changes in Organisms

Picture someone building a fire in a fireplace. The wood is largely composed of complex organic molecules. These molecules required energy to form, and therefore, their chemical bonds contain high potential energy. Lighting a fire causes a series of reactions that break these molecules and release energy. When released, the energy contained in those bonds can be used to do work. This chemical reaction involves **oxidation**, a loss of electrons, usually accomplished by the addition of oxygen (which causes *burning*). The heat and light of the flame are the potential energy being released as kinetic energy. In contrast, most inorganic compounds, such as carbon dioxide, water, and rock-based minerals, are nonflammable because they have very low potential energy. Thus, the production of organic material from inorganic material represents a gain in potential energy. Conversely, the breakdown of organic matter *releases* energy. Inside the cells of the giant panda eating bamboo, the reaction that

Figure 3–14 Entropy. Systems go spontaneously only in the direction of increasing entropy. This spray can illustrates the idea. It took energy to concentrate a liquid in the can. Once it is dispersed, the energy is dissipated and entropy is increased.

takes place is similar to that of burning wood in a fireplace, except that it occurs in a set of small steps and is much more controlled.

Producers and Photosynthesis. The relationship between the formation and breakdown of organic matter, where energy is gained and released from chemical bonds, forms the basis of the energy dynamics of ecosystems. **Producers** (primarily green plants) make *high-potential-energy organic molecules* for their needs from *low-potential-energy raw materials* in the environment—namely, carbon dioxide, water, and a few dissolved compounds of nitrogen, phosphorus, and other elements **(Fig. 3–16)**. This "uphill" conversion is possible because producers use chlorophyll to absorb light energy, which "powers" the production of the complex, energy-rich organic molecules. On the other hand, **consumers** (all organisms that live on the production of others) obtain energy for movement and growth from *feeding on and breaking down organic matter made by producers* **(Fig. 3–17** on page 65).

Green plants use the process of **photosynthesis** to make sugar (glucose, which contains stored chemical energy) from carbon dioxide, water, and light energy (Fig. 3–16). This process, which also releases oxygen gas as a by-product, is described by the following chemical equation:

Photosynthesis
(An energy-demanding process)

$$6\,CO_2 + 12\,H_2O \xrightarrow{\text{light energy}} C_6H_{12}O_6 + 6\,O_2 + 6\,H_2O$$

carbon dioxide　water　light energy　glucose　oxygen　water
(gas)　　　　　input　　　　　　　　　　　　(gas)
(low potential energy)　　　　　　　　　　(high potential energy)

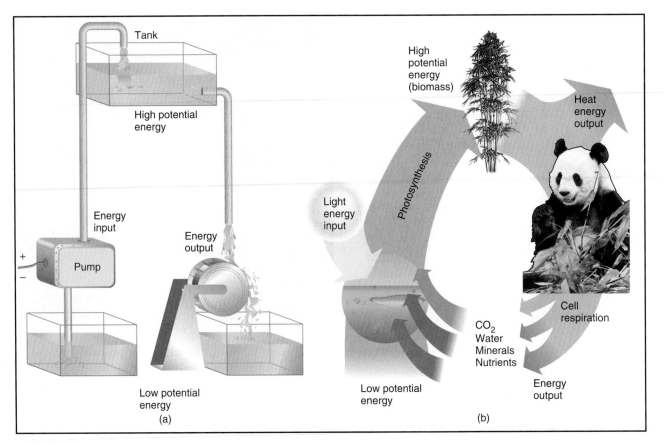

Figure 3–15 **Storage and release of potential energy.** (a) A simple physical example of the storage and release of potential energy. (b) The same principle applied to ecosystems.

Figure 3–16 Producers as chemical factories. Using light energy, producers make glucose from carbon dioxide and water, releasing oxygen as a by-product. Breaking down some of the glucose to provide additional chemical energy, they combine the remaining glucose with certain nutrients from the soil to form other complex organic molecules that the plant then uses for growth.

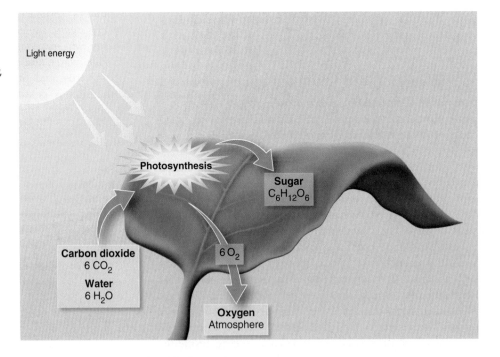

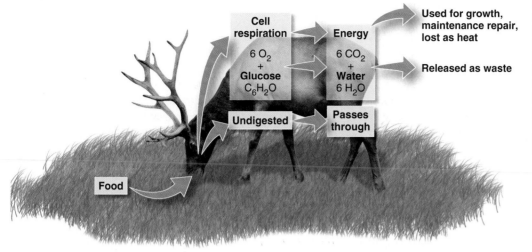

Figure 3–17 Consumers.
Only a small portion of the food ingested by a consumer is assimilated into body growth and repair. Some food provides energy in chemical bonds that are broken by the process of cell respiration: this is the reverse process of photosynthesis, and waste products are carbon dioxide, water, and various mineral nutrients. A third portion is not digested and becomes fecal waste.

Chlorophyll in the plant cells absorbs the kinetic energy of light and uses it to remove the hydrogen atoms from water (H_2O) molecules. The hydrogen atoms combine with carbon atoms from carbon dioxide to form a growing chain of carbons that eventually becomes a glucose molecule. After the hydrogen atoms are removed from water, the oxygen atoms that remain combine with each other to form oxygen gas, which is released into the air. Water appears on both sides of the equation because 12 molecules are consumed and 6 molecules are newly formed during photosynthesis.

The key energy steps in photosynthesis take the hydrogen from water molecules and join carbon atoms together to form the carbon–carbon and carbon–hydrogen bonds of glucose. These steps convert the low-potential-energy bonds in water and carbon dioxide molecules to the high-potential-energy bonds of glucose.

Within the Plant. The glucose produced in photosynthesis serves three purposes in the plant: (1) Glucose, with the addition of other molecules, can be the backbone used for making all the other organic molecules (proteins, carbohydrates, and so on) that make up the stem, roots, leaves, flowers, and fruit of the plant. (2) It takes energy to run the cell activities of plants, such as growth. This energy is obtained when the plant breaks down a portion of the glucose to release its stored energy in a process called *cell respiration*, which is discussed later. (3) A portion of the glucose produced may be stored for future use. For storage, the glucose is generally converted to starch, as in potatoes or grains, or to oils, as in many seeds. This stored energy is what is available to be eaten by consumers (Fig. 3–17).

None of these reactions, from the initial capture of light by chlorophyll to the synthesis of plant structures, takes place automatically. Each step is catalyzed by specific **enzymes,** *proteins that promote the synthesis or breaking of chemical bonds.* The same is true of cell respiration.

Cell Respiration. Inside each cell, organic molecules may be broken down through a process called **cell respiration** to release the energy required for the work done by that

cell. Most commonly, cell respiration involves the breakdown of glucose, and the overall chemical equation is basically the reverse of that for photosynthesis:

Cell Respiration
(An energy-releasing process)

$$C_6H_{12}O_6 + 6\,O_2 \;\rightarrow\; 6\,CO_2 \;+\; 6\,H_2O \;+\; \text{energy}$$
glucose   oxygen   carbon dioxide   water
(high potential energy)   (low potential energy)

The purpose of cell respiration is to release the potential energy contained in organic molecules to perform the activities of the organism. Note that oxygen is *released* in photosynthesis but *consumed* in cell respiration to break down glucose to carbon dioxide and water. All organisms need to perform cell respiration, even producers like bamboo. In order to have the oxygen they need, animals like giant pandas eating bamboo, and deer grazing on grass (Fig. 3–17), must obtain oxygen through some type of respiratory organ. This means that their ability to use food depends on the availability not only of food, but also of oxygen. Many animals, such as deer, use lungs to obtain oxygen and release carbon dioxide. Oxygen is more limited in water, and fish use gills to obtain it, but many aquatic places are severely oxygen limited.

In keeping with the Second Law of Thermodynamics, converting the potential energy of glucose to the energy required to do the body's work is not 100% efficient. The energy released from glucose is stored in molecules that work like compressed springs—that is, they store the energy so that the body can use it in cells later. But the number of "energy units" that come from respiration is not nearly as high as the amount of solar energy required for the photosynthesis of glucose. Cell respiration is only 40–60% efficient. This is a great example of the Second Law of Thermodynamics at work. The rest of the energy is released as waste heat, and this is the source of body heat. This heat output can be measured in animals (cold-blooded or warm-blooded) and in plants.

Gaining Weight. The basis of weight gain or loss becomes apparent in this context. If you consume more calories from food than your body needs, the excess may be converted to fat and stored, and you gain weight as a result. In contrast, the principle of dieting is to eat less and exercise more, to create an energy demand that exceeds the amount of energy contained in your food.

The overall reaction for cell respiration is the same as that for simply burning glucose. Thus, it is not uncommon to speak of "burning" our food for energy. There are other ways of releasing stored energy from food that do not require oxygen. Such anaerobic respiration includes fermentation, as in the process used to make wine. Yeast cells use anaerobic respiration to convert grape sugars into usable energy, releasing alcohol as a by-product. However, these methods are less efficient than oxidation, and organisms that use them do not thrive unless oxygen is severely limited.

Photosynthesis and cell respiration are necessary for life on Earth to survive. In the next section, we will see how the energy that fuels photosynthesis moves through Earth's systems after traveling from the Sun.

One-Way Flow of Energy

What happens to all the solar energy entering ecosystems? Most of it is absorbed by the atmosphere, oceans, and land, heating them in the process. The small fraction (2–5%) captured by living plants is either passed along to the consumers and **detritivores** (organisms that live on dead or decaying organisms) that eat them or degraded into the lowest and most disordered form of energy—heat—as the plants decompose. Eventually, all of the energy entering ecosystems escapes as heat. According to the laws of thermodynamics, no energy will actually be lost. So many energy conversions are taking place in ecosystem activities, however, that entropy is increased, and all the energy is degraded to a form unavailable to do further work. That heat energy may stay in the atmosphere for some period of time, but will eventually be re-radiated out into space. The ultimate result is that *energy flows in a one-way direction through ecosystems* and eventually leaves Earth; it is not lost to the universe, but it is no longer available to ecosystems, so it must be continually resupplied by sunlight.

Energy flow is one of the two fundamental processes that make ecosystems work. In contrast with the *flow* of energy, we talk about the *cycling* of nutrients and other elements, which are continually reused from those already available on Earth.

3.5 The Cycling of Matter in Ecosystems

Earlier we saw that the molecules that make up cells, which in turn make up tissues, contain certain key elements. In bamboo, for example, all organic molecules contain carbon; photosynthesis requires water (hydrogen and oxygen) and carbon dioxide (carbon and oxygen); potassium is part of the energy-holding mechanism in each cell; nitrogen is contained in every protein; and sulfur bonds help proteins stay in the right shape. There are many other necessary nutrients. Some are only necessary in very small amounts. For example, to make the protein hemoglobin, which carries oxygen in blood, small amounts of iron are needed.

According to the Law of Conservation of Matter, atoms cannot be created or destroyed, so recycling is the only possible way to maintain a dynamic system. To see how recycling takes place in the biosphere, we now focus on the pathways of four key elements heavily affected by human activities: carbon, phosphorus, nitrogen, and sulfur. Because these pathways are circular and involve biological, geological, and chemical processes, they are known as the **biogeochemical cycle**. (The water, or hydrologic, cycle is just as important, but it will be covered in Chapter 10).

The Carbon Cycle

The global carbon cycle is illustrated in **Figure 3–18**. Boxes represent major "pools" of carbon, and arrows represent the movement, or flux, of carbon from one compartment to another. For descriptive purposes, it is convenient to start the carbon cycle with the "reservoir" of carbon dioxide (CO_2 molecules present in the air). Through photosynthesis and further metabolism, carbon atoms from CO_2 become the carbon atoms of the organic molecules making up a plant's body. The carbon atoms can then be eaten by an animal, such as the panda, and become part of the tissues of all the other organisms in the ecosystem. About half of the carbon-containing molecules are respired by plants and animals, and half are deposited to the soil (a large reservoir) in the form of **detritus** (dead plant and animal matter). Respiration by organisms in the soil that eat dead matter returns more carbon to the atmosphere (as CO_2). The cycle is different in the oceans: Photosynthesis by phytoplankton and macroalgae removes CO_2 from the huge pool of inorganic carbonates in seawater, and feeding moves the organic carbon through marine food webs. Respiration by the biota returns the CO_2 to the inorganic carbon compounds in solution.

Some other significant processes exist. The figure indicates two in particular: (1) diffusion exchange between the atmosphere and the oceans; and (2) the combustion of fossil fuels, which releases CO_2 to the atmosphere. Some geological processes of the carbon cycle are not shown in Figure 3–18. For example, the fossilization of dead plants and animals produced coal deposits in many areas of Earth. This process removed vast amounts of carbon dioxide from the atmosphere and trapped it underground. Burning the coal and oil created by this process releases the CO_2 to the atmosphere. For another example, limestone (such as that formed by ancient corals) also keeps carbon out of circulation; however, the weathering of exposed limestone releases carbon into the aquatic system.

Because the total amount of carbon dioxide in the atmosphere is about 765 Gt (gigaton, or 1,000 metric tons) and photosynthesis in terrestrial ecosystems removes about

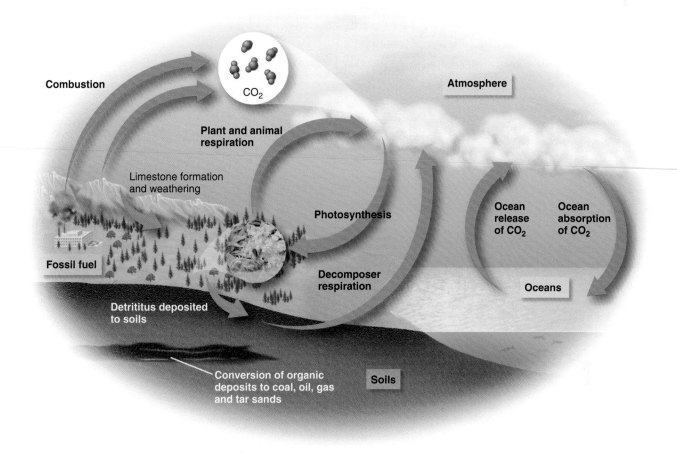

Figure 3–18 The global carbon cycle. Boxes in the figure refer to pools of carbon, and arrows refer to the movement, or flux, of carbon from one pool to another.

120 Gt/year, a carbon atom cycles from the atmosphere through one or more living things and back to the atmosphere about every six years.

Human Impacts. Human intrusion into the carbon cycle is significant. As we will see shortly, we are diverting (or removing) 40% of the photosynthetic productivity of land plants to support human enterprises. By burning fossil fuels, we have increased atmospheric carbon dioxide by 35% over preindustrial levels (Chapter 18). In addition, until the mid-20th century, deforestation and soil degradation released significant amounts of CO_2 to the atmosphere. However, more recent reforestation and changed agricultural practices have improved this somewhat.

The Phosphorus Cycle

The phosphorus cycle is similar to the cycles of other mineral nutrients—those elements that have their origin in the rock and soil minerals of the lithosphere, such as iron, calcium, and potassium. We focus on phosphorus because its shortage tends to be a limiting factor in a number of ecosystems and

its excess can stimulate unwanted algal growth in freshwater systems.

The phosphorus cycle is illustrated in **Figure 3–19**. Like the carbon cycle, it is depicted as a set of pools of phosphorus and fluxes to indicate key processes. Phosphorus exists in various rock and soil minerals as the inorganic ion phosphate (PO_4^{3-}). As rock gradually breaks down, phosphate and other ions are released in a slow process. Plants absorb PO_4^{3-} from the soil or from a water solution, and once the phosphate is incorporated into organic compounds by the plant, it is referred to as **organic phosphate**. Moving through food chains, organic phosphate is transferred from producers to the rest of the ecosystem. As with carbon, at each step it is highly likely that the organic compounds containing phosphate will be broken down in cell respiration or by decomposers, releasing PO_4^{3-} in urine or other waste material. Phosphate is then reabsorbed by plants to start the cycle again.

Phosphorus enters into complex chemical reactions with other substances that are not shown in this simplified version of the cycle. For example, PO_4^{3-} forms solid, insoluble compounds with a number of cations (positively charged ions),

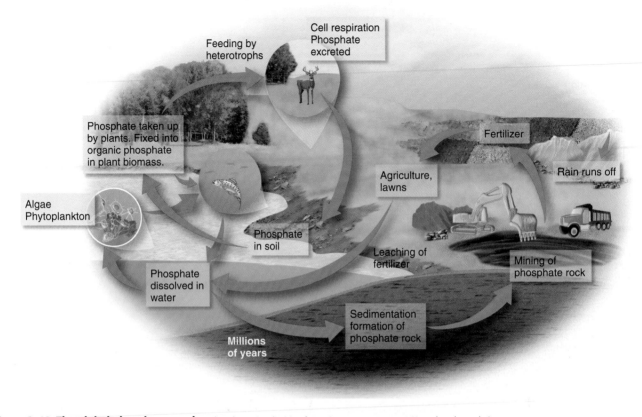

Figure 3–19 The global phosphorus cycle. Phosphorus is a limiting factor in many ecosystems. Note that the cycle is not connected to the atmosphere, which limits biosphere recycling.

such as iron (Fe^{3+}), aluminum (Al^{3+}), and calcium (Ca^{3+}). Phosphorus can bind with these ions to form chemical precipitates (solid, insoluble compounds) that are largely unavailable to plants. The precipitated phosphorus can slowly release $PO_4{}^{3-}$ as plants withdraw naturally occurring $PO_4{}^{3-}$ from soil, water, or sediments.

Human Impacts. The most serious human intrusion into the phosphorus cycle comes from the use of phosphorus-containing fertilizers. Phosphorus is mined in several locations around the world (e.g., Florida in the United States) and is then made into fertilizers, animal feeds, detergents, and other products. Phosphorus is a common limiting factor in soils and, when added to croplands, can greatly stimulate production.

Unfortunately, human applications have tripled the amount of phosphorus reaching the oceans. This increase is roughly equal to the global use of phosphorus fertilizer in agriculture. Humans are accelerating the natural phosphorus cycle as it is mined from the earth and as it subsequently moves from the soil into aquatic ecosystems, creating problems as it makes its way to the oceans. There is essentially no way to return this waterborne phosphorus to the soil, so the bodies of water end up overfertilized. This leads in turn to a severe water pollution problem known as eutrophication (Chapter 17). Eutrophication can cause the overgrowth of algae, too many bacteria, and the death of fish.

The Nitrogen Cycle

The nitrogen cycle (**Fig. 3–20**) has similarities to both the carbon cycle and the phosphorus cycle. Like carbon, nitrogen has a gas phase; like phosphorus, nitrogen acts as a limiting factor in plant growth. Like phosphorus, nitrogen is in high demand by both aquatic and terrestrial plants.

The nitrogen cycle is otherwise unique. Most notably, unlike in the other cycles, bacteria in soils, water, and sediments perform many of the steps of the nitrogen cycle.

The main reservoir of nitrogen is the air, which is about 78% nitrogen gas (N_2). This form of nitrogen is called **nonreactive nitrogen**; most organisms are not able to use it in chemical reactions. The remaining forms of nitrogen are called **reactive nitrogen** (Nr) because they are used by most organisms and can make chemical reactions more easily.

Plants in terrestrial ecosystems ("non-N-fixing producers" in Figure 3–20) take up Nr as ammonium ions ($NO_4{}^+$) or nitrate ions ($NO_3{}^-$). The plants incorporate the nitrogen into essential organic compounds such as proteins and nucleic acids. The nitrogen then moves from producers through consumers and, finally, to decomposers (organisms that live on dead plant and animal matter are referred to as "heterotrophs" in Figure 3–20). At various points, nitrogen wastes are released, primarily as ammonium compounds. A group of soil bacteria, the *nitrifying bacteria*, oxidizes the ammonium to nitrate in a process that yields energy for the

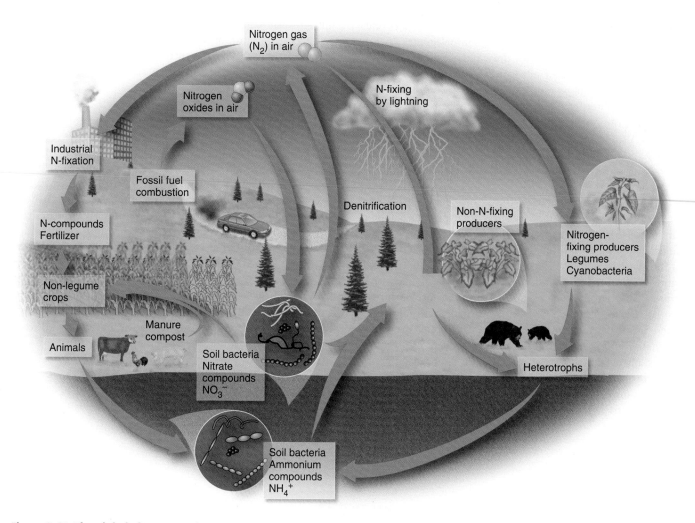

Figure 3–20 The global nitrogen cycle. Like phosphorus, nitrogen is often a limiting factor. Its cycle heavily involves different groups of bacteria.

bacteria. At this point, the nitrate is once again available for uptake by green plants—a local ecosystem cycle within the global cycle. In most ecosystems, the supply of Nr is quite limited, yet there is an abundance of nonreactive N—if it can be accessed. The nitrogen cycle in aquatic ecosystems (not shown in Figure 3–20) is similar.

Nitrogen Fixation. A number of bacteria and cyanobacteria (chlorophyll-containing bacteria, formerly referred to as blue-green algae) can use nonreactive N through a process called biological **nitrogen fixation.** In terrestrial ecosystems, the most important among these nitrogen-fixing microbes live in nodules on the roots of legumes, the plant family that includes peas and beans. The legume provides the bacteria with a place to live and with food (carbohydrates and proteins) and gains a source of nitrogen in return. From legumes, nitrogen enters the food web. The legume family includes a huge diversity of plants. Without them, plant production would be sharply impaired due to a lack of available nitrogen.

Three other important processes also "fix" nitrogen. One is the conversion of nitrogen gas to the ammonium form by discharges of lightning in a process known as *atmospheric*

nitrogen fixation; the ammonium then comes down with rainfall. The second is the *industrial fixation* of nitrogen in the manufacture of fertilizer; the Haber-Bosch process converts nitrogen gas and hydrogen to ammonia, by an industrial process using fossil fuels. This is the main source of agricultural fertilizer. The third is a consequence of the *combustion of fossil fuels*, during which nitrogen from coal and oil is oxidized; some nitrogen gas is also oxidized during high-temperature combustion. These last two processes lead to nitrogen oxides (NO_x) in the atmosphere, which are soon converted to nitric acid and then brought down to Earth as acid precipitation (Chapter 19).

Denitrification. **Denitrification** is a microbial process that occurs in soils and sediments depleted of oxygen. A number of microbes can take nitrate (which is highly oxidized) and use it as a substitute for oxygen. In so doing, the nitrogen is reduced (it gains electrons) to nitrogen gas and released back into the atmosphere. In sewage treatment systems, denitrification is a desirable process and is promoted to remove nitrogen from the wastewater before it is released in soluble form to the environment (Chapter 20).

Human Impacts. Human involvement in the nitrogen cycle is substantial. Many agricultural crops are legumes (peas, beans, soybeans, alfalfa), so they draw nitrogen from the air, thus increasing the rate of nitrogen fixation on land. Crops that are nonleguminous (corn, wheat, potatoes, cotton, and so on) are heavily fertilized with nitrogen derived from industrial fixation. Both processes benefit human welfare profoundly. Also, fossil fuel combustion fixes nitrogen from the air (Fig. 3–19). However, human actions more than double the rate at which nitrogen is moved from the atmosphere to the land.

The consequences of this global nitrogen fertilization are serious. Nitric acid (and sulfuric acid, produced when sulfur is also released by burning fossil fuels) has destroyed thousands of lakes and ponds and caused extensive damage to forests (Chapter 19). Nitrogen oxides in the atmosphere contribute to ozone pollution, global climate change, and stratospheric ozone depletion (Chapter 18). Surplus nitrogen has led to "nitrogen saturation" of many natural areas, whereby nitrogen can no longer be incorporated into living matter and is released into the soil. Washed into surface waters, nitrogen makes its way to estuaries and coastal areas of oceans, where, just like phosphorus, it triggers a series of events leading to eutrophication, resulting in dead seafood, detriments to human health, and areas of oceans that are

unfit for fish (Chapter 20). This complex of ecological effects has been called the **nitrogen cascade**, in recognition of the sequential impacts of Nr as it moves through environmental systems, creating problems as it goes.

The Sulfur Cycle

Sulfur is the last element we will investigate. Sulfur is important to living things because it is a component of many proteins, hormones, and vitamins. Sulfur is often linked in nature with oxygen atoms, as in sulfate (SO_4) **(Fig. 3–21)**. Most of Earth's sulfur is found in rocks and minerals, including deep ocean sediments. Weathering of rocks and volcanic activity sends sulfur into the atmosphere or soil. Sulfur also gets into the air when fossil fuels are burned and when mined metals are processed. In soil, sulfate can be taken up by plants and microorganisms. In air, sulfur dioxide (SO_2) contributes to acid rain when it combines with water vapor and forms sulfuric acid. Sulfates are added to water bodies as sulfur compounds fall from the atmosphere or are weathered from rocks. There they can be serious pollutants.

Human Impacts. The largest human impact of the sulfur cycle is the addition of sulfur oxides to the atmosphere and the addition of sulfates to water. Unlike phosphorus and

Figure 3–21 The global sulfur cycle. Sulfur spends little time in the atmosphere. Part of the cycle does rely on bacteria, but less than nitrogen's cycle. Like all three other cycles, the sulfur cycle is sped up by human activity.

SUSTAINABILITY

Ancient peoples thought that Earth had an edge—if you took a ship too far, perhaps you and your crew would fall right off. Today, some scientists suggest that Earth has a metaphorical edge: a boundary where use of a particular resource becomes so great that it alters natural cycles and changes ecosystems. Reaching a point where dramatic changes are likely is called the tipping point, and the levels of change at which this might occur are called planetary boundaries, a term coined by Johann Rockström and his colleagues at Stockholm University, who together published a paper in 2009 entitled "Planetary Boundaries: Exploring the Safe Operating Space for Humanity." In that paper, they estimate that a safe boundary for nitrogen would be to "limit industrial and agricultural fixation of N_2 to 35 Tg N/yr."[1] (A terragram (Tg) is one billion kilograms.)

The innovation of the Haber-Bosch process for fixing nitrogen allowed humans to produce more food on the same amount of land than was possible previously. One estimate is that without nitrogen fertilizer, we would have needed four times the agricultural land in 2000 that we did in 1900 and half of all continental land would be needed for agriculture. The downside of so much fertilizer is that it washes into oceans, where it and other pollutants produce "dead zones" where oxygen is limited and most life cannot survive. Likewise, the use of fossil fuels has brought advantages to humans, but the acid rain created by the burning of these fuels kills lake fish.

Humanity might be reaching planetary boundaries or tipping points—where we suddenly experience ecological change that is difficult to reverse and that causes great changes to our ability even to feed ourselves. Just as a ship might have a lookout, scientists have an important role in helping us understand the world so that we can recognize when we are approaching or have passed planetary boundaries. If we have a way to estimate the consequences of what we are trying to do, such as this paper suggests, then both scientists and others will be involved in measuring those consequences and finding solutions for problems that arise. One place to start is to calculate your own nitrogen footprint. Just like a carbon footprint, a nitrogen footprint is calculated from your activities and tells you something about your own effect on the globe. You can find resource footprint calculators by searching on the Web. Scientists play a role in helping to make more efficient use of fertilizer and to make technologies to remove nitrogen from smoke in burning of fossil fuels. These advancements, coupled with the will of individuals to consume less and be wise consumers, are parts of the solution to avoid taking our global ship across a planetary boundary.

[1]Rockström, J. et al., 2009. "Planetary Boundaries: Exploring the Safe Operating Space for Humanity." *Ecology and Society* *14*(2): 32.

---

nitrogen, sulfur is not usually directly added to soils in order to improve their fertility. However, soil and water sulfate levels are increased by human action. Acid rain and water pollution are the major effects. For example, in heavily developed areas of the Everglades, the percent of sulfates in water is 60 times what it would normally be expected to be (Chapter 19). Sulfate compound aerosols (small particles or drops) also play a role in the climate. They act to temporarily cool the atmosphere, although the compounds cause other problems when they fall to earth.

Comparing the Cycles

The four cycles we have looked at in depth differ in some important ways (Table 3–2). Carbon is found in large amounts in the atmosphere in a form that can be directly taken in by plants, so carbon is rarely the limiting factor in the growth of vegetation. Both nitrogen and phosphorus are often limiting factors in ecosystems. Phosphorus has no gaseous atmospheric component (though it can be found in airborne dust particles) and thus, unless added to an ecosystem artificially, enters very slowly. Nitrogen is unique because of the importance of bacteria in driving the cycle forward. Sulfur can get into the atmosphere, but only for a short time. It is also different from the others because of its concentration in mining runoff.

All four cycles have been sped up considerably by human actions. Nitrogen compounds in the atmosphere

TABLE 3–2	Major Characteristics of the Carbon, Phosphorus, and Nitrogen Cycles		
Nutrient	**Major Source**	**Interesting Feature**	**Human Impact**
Carbon	Air	Taken in directly by plant leaves	Burning fuel moves it to air from underground
Phosphorus	Rock	No atmospheric component	Fertilizer use adds it to waterways
Nitrogen	Soil/Air	Bacteria drive the cycle	Fertilizer moves it to soil, burning moves it to air
Sulfur	Rock	Spends only a short time in the atmosphere	Burning moves it to air, rain and mining move it to soil and water

contribute to acid rain, and carbon dioxide is being moved from underground storage in carbon molecules to the atmosphere, where it acts as a greenhouse gas. Both nitrogen and phosphorus are put on soil as fertilizer or get into water from sewage and runoff. Both act in water to promote overgrowth

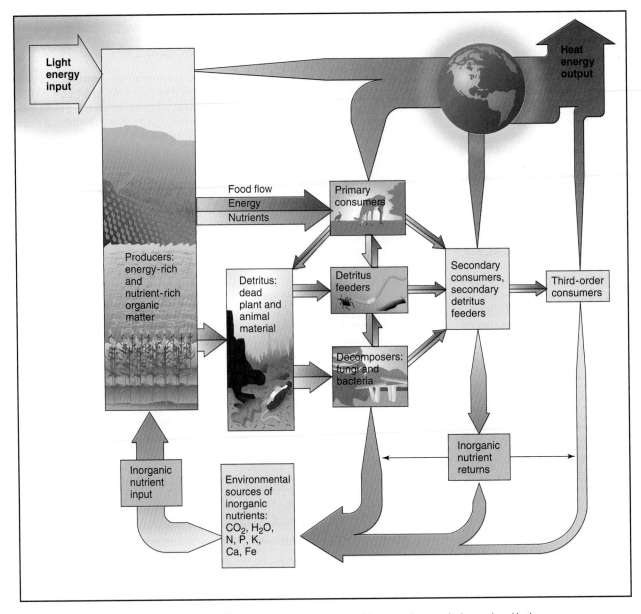

Figure 3–22 Nutrient cycles and energy flow. The movement of nutrients (blue arrows), energy (red arrows), and both (brown arrows) through the ecosystem. Nutrients follow a cycle, being used over and over. Light energy absorbed by producers is released and lost as heat energy as it is "spent."

of algae (Chapter 20). Sulfur adds to acid rain and to water pollution.

Although we have focused on carbon, phosphorus, nitrogen, and sulfur, cycles exist for oxygen, hydrogen, iron, and all the other elements that play a role in living things. While the routes taken by distinct elements may differ, all of the cycles are going on simultaneously, and all come together in the tissues of living things. As elements cycle through ecosystems, energy flows in from the Sun and through the living members of the ecosystems. The links between these two fundamental processes of ecosystem function are shown in **Figure 3–22**.

In this chapter, we introduced the various fields of ecology and showed that the science of ecology encompasses living things and their relationships with each other and the environment. We looked at two ecological levels in greater detail in this chapter—what is happening with individuals (organismal ecology) and what is happening to the large-scale movement of matter and energy through the biosphere (ecosystem ecology). To understand the flow of matter and energy, we also included some background on the basics of atoms, molecules, and the laws of energy. Later, we will investigate the other subdisciplines of ecology introduced here, especially population ecology and community ecology.

REVISITING THE THEMES

Sound Science

Ecology is the science of living things and the abiotic and biotic components interacting with them. Ecologists can study on a variety of levels. In this chapter, we looked at organismal ecology and ecosystem ecology—that is, how materials work in individual animals and how matter and energy flow throughout the world.

Much of this chapter is about how the basic processes of the natural world work, focusing on matter and energy. Both matter and energy are limited. Matter cycles in the form of molecules (sets of atoms) through living things and the abiotic environment. Energy flows in one direction—from the Sun, through activities on Earth, and back out into space. Using the giant panda and its food (the bamboo plant) as our examples, we demonstrated basic principles of organization within their bodies and the flow of energy and materials through their systems. The bamboo plant has a specific niche, or set of parameters, within which it needs to live. These resources and conditions include the amount of rainfall and nutrients it needs, for example. Because it has chlorophyll, bamboo is able to undergo photosynthesis. In this chemical process, inorganic building blocks are turned into more-complex organic molecules whose bonds contain more chemical energy than their original components do. This energy can be released when the panda eats the bamboo and its body burns those molecules through a controlled process of cellular respiration. Energy is lost at each step of this energy path. In order to build the bodies of the bamboo and its consumer, the panda, a number of nutrients need to be available. The cycles for four of these (carbon, phosphorus, nitrogen, and sulfur) were described. Humans affect these biogeochemical cycles by actions such as the burning of fossil fuels and the fertilization of agricultural land.

Sustainability

The science in this chapter that leads to specific management actions is primarily that surrounding the needs of individual species in terms of habitat and niche, the consequences of limits on nutrients and energy, and the impacts of human effects on biogeochemical cycles. These ideas will be more thoroughly explored later. For instance, the science surrounding habitat and niche underlies our understanding of why some types of organisms are particularly vulnerable to extinction (Chapter 4). The science underlying the nitrogen and phosphorus cycles is important to management decisions about water pollution (Chapter 20). Finally, the science about carbon and nitrogen cycles aids in our understanding of air pollution and global climate change (Chapters 18 and 19). The sustainability essay provides a good example of how science is used to identify boundaries, and this can be used to manage our global environment—in this case, the nitrogen cycle.

The giant panda helps us look at basic processes within organisms and the flow of matter and energy. Later, we will look at ways individuals relate to members of their own species (population ecology) and ways species relate to other species (community ecology).

Stewardship

The giant panda, focused on throughout this chapter, is in decline. It has a particular habitat, which is shrinking because of competition with human activities. As stewards, the panda conservationists in China and around the world are protecting not only the panda, but also its habitat. Human activity also speeds up biogeochemical cycles, and this can have global effects. All of us have to be stewards not only of the nutrients in our world, but also of the natural processes that move chemicals from one part of a cycle to another. Stewardship involves protecting not only resources but ecosystem functions. We will look at some of the ethical and justice implications of these scientific realities in other chapters. We will ask, "Why protect other species?" (Chapter 6). We will also look at the implications of energy loss at every transfer and apply those principles to human population growth and the need for food (Chapter 8).

REVIEW QUESTIONS

1. Distinguish between the biotic community and the abiotic environmental factors of an ecosystem.

2. Define and compare the terms *species*, *population*, and *ecosystem*.

3. Compared with an ecosystem, what are an ecotone, a landscape, a biome, and a biosphere?

4. How do the terms *organic* and *inorganic* relate to the biotic and abiotic components of an ecosystem?

5. What are the six key elements in living organisms?

6. What features distinguish between organic and inorganic molecules?

7. In one sentence, define *matter* and *energy*, and demonstrate how they are related.

8. Give four examples of potential energy. In each case, how can the potential energy be converted into kinetic energy?

9. State two energy laws. How do they relate to entropy?

10. What is the chemical equation for photosynthesis? Examine the origin and destination of each molecule referred to in the equation. Do the same for cellular respiration.

11. Describe the biogeochemical cycle of carbon as it moves into and through organisms and back to the environment. Do the same for phosphorus, nitrogen, and sulfur.

12. What are the major human impacts on the carbon, phosphorus, and nitrogen cycles?

THINKING ENVIRONMENTALLY

1. From local, national, and international news, compile a list of ways humans are altering abiotic and biotic factors on a local, regional, and global scale. Analyze ways in which local changes may affect ecosystems on larger scales and ways in which global changes may affect ecosystems locally.

2. Use the laws of conservation of matter and energy to describe the consumption of fuel by a car. That is, what are the inputs and outputs of matter and energy? (*Note:* Gasoline is a mixture of organic compounds containing carbon–hydrogen bonds.)

3. Using your knowledge of photosynthesis and cell respiration, draw a picture of the hydrogen cycle and the oxygen cycle. (*Hint:* Consult the four cycles in the book for guidance.)

4. Look up everything you can find about iron seeding of the oceans, and see if you can decide what you think we should do.

MAKING A DIFFERENCE

1. Find the closest farm that does not use artificial fertilizer. Many are listed on Web sites under the terms "community supported agriculture" or "CSA." See if you can arrange a visit. See what they do to lower their impact on the nitrogen and phosphorus cycles and still maintain fertile soil.

2. Visit the EPA's Web site to find out how you can help protect your local watershed. Its project database allows you to get involved in monitoring, cleanups, and restorations in your area.

3. Take one of the biogeochemical cycles and try to track your impact on it. This is most easily done with carbon. Search on the Web under "carbon footprint."

MasteringEnvironmentalScience®

Go to **www.masteringenvironmentalscience.com** for practice quizzes, Pearson eText, videos, current events, and more.

CHAPTER 4

Populations and Communities

The golden frog (*Atelopus zeteki*) of Panama, a highly endangered species, now lives only in captivity. The last known populations were removed from an area of Panama (inset) where a deadly fungus was known to be invading.

LEARNING OBJECTIVES

1. **Dynamics of Natural Populations:** Explain three model ways populations grow, and describe the graph that would illustrate each.

2. **Limits on Populations:** Explain factors that limit populations, including those that increase as populations become more dense (such as predation and resource limitation) and factors that are unrelated to the population density.

3. **Community Interactions:** Define the types of interactions that can occur between species in a community and the effect of those interactions on each species.

4. **Evolution as a Force for Change:** Describe the major ideas in the theory of evolution, such as inheritance and natural selection, and list examples of adaptations organisms have for survival.

5. **Implications for Human Management:** Describe at least three ways in which humans alter features in the environment to change population growth patterns.

MAGINE YOU ARE sitting outside on a summer evening listening to a chorus of frogs. In North America, you might hear bullfrogs or spring peepers through the coming dusk. Their calls mark the breeding grounds of these species and are common sounds of spring and summer. In the rain forest of Central America, you might hear a chorus of an entirely different set of amphibians—many of which are small and beautifully colored. However, one you won't hear is the golden frog (also called the golden toad) of Panama, *Atelopus zeteki*.

The golden frog, revered as a national symbol in Panama, is depicted in local mythology, sculpture, toys, lottery tickets, posters, and calendars. It is famous for its poisonous skin, which in adulthood has a bright black-and-yellow warning coloration. Unlike most other frogs and toads (of the group Anura, a subgroup of amphibians), it doesn't sing. Instead, the golden frog, a rare anuran that lacks eardrums, waves an arm to communicate with others of its species.

THREATENED. Unfortunately, devastation of Panama's tropical forests (which has caused habitat loss over the years), as well as overcollection for the pet trade have left golden frog populations at low levels. A new threat has also appeared—a chytrid fungus (*Batrachochytrium dendrobatidis*), which is decimating frog populations around the globe and threatening many with extinction. In 2007, concerned that the fungus was moving through areas of Panama and leaving dead and dying amphibians in its wake, conservationists made a bold move. They went into a remote jungle area, collected all of the amphibians (including golden frogs) that they could and removed them to Panama's El Valle Amphibian Conservation Center. This effort, unprecedented in conservation history, was documented in the film *Leap of Faith: Saving Panama's Golden Frogs*. A BBC video production crew also filmed the removal of what was then the last known population of golden frogs in the wild.

LIMITS. The golden frog is a good example of the problems a rare species can face when multiple limits affect it at once, and of the extreme measures necessary for its survival and preservation. Under changing environmental conditions, species can often adapt or find a new niche or area to exploit. However, as seen with the golden frog, environmental changes can also put species in danger of extinction. In this chapter, we will look at how populations grow, what types of limitations populations can face, how populations of different species interact with each other, how species change over time, and what changes humans impose on populations and groups of species.

4.1 Dynamics of Natural Populations

Like the golden frogs in their rain-forest habitats, each species in an ecosystem exists as a population; that is, each exists as a reproducing group. A **population** is a group of members of the same species living in an area, and the populations of different species living together in one area are known as a **community** (Chapter 3). Ecologists often want to know the pattern of change in the numbers of individuals in a population. In any population, this is easy to figure out if you have enough information. Births and immigration cause the population to grow, and deaths and emigration cause the population to shrink. This can be described by a simple equation:

(Births + Immigration) − (Deaths + Emigration)
= Change in population number

This equation shows the change in population, or **population growth**. If births plus immigration are more or less equal to deaths plus emigration over time, the population is said to be in **equilibrium**. But often, populations are either increasing or decreasing—that is, their population growth is not zero. If you take the amount a population has changed and divide it by the time it had to change, you get the **population growth rate**. Imagine there is a very small population just coming into a new ecosystem. A scientist might be interested in figuring out how fast a population could grow, how many individuals there are right now, or even what the future population size will be, using the current population and the growth rate. In order to calculate these parameters, scientists use population growth curves—graphs of how populations grow. Here we will look at several examples that represent broad types of growth patterns.

Population Growth Curves

In a natural setting, there are several patterns populations can follow as they grow or decline over time. A constant population growth rate is the simplest type of growth to model. However, constant growth is not something you

would expect to find in nature, but it serves as a good comparison to other growth patterns because the equation is so simple:

Population number at the start + (A constant × Time)
= Population number at the end

That is, the number at the end of a period of time is equal to the number at the beginning plus a constant multiplied by the number of time units. This is illustrated by the straight orange line in **Figure 4–1**.

Exponential Increase. Every species has the capacity to increase its population when conditions are favorable. For example, a beetle population growing in a bag of cereal could double every week (the first week 2, then 4, then 8, then 16 beetles). Such a series is called an **exponential increase**. This sort of growth results in a **population explosion**. In contrast to our first example of constant growth, exponential

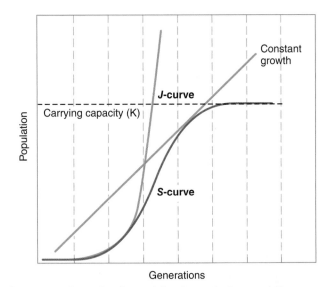

Figure 4–1 Three simple models of population growth.
Constant growth is represented by the orange line, exponential growth by the green line, and logistic growth by the red line. The carrying capacity is represented by the black line.

growth does not add a constant number of individuals to the population for each time period. Instead, the doubling time of the population remains constant. That is, if it takes two days to go from 8 worms to 16 worms, it will take two days to go from 1,000 worms to 2,000 worms. This is the reason such growth is described as an "explosion."

If we plot numbers over time during an exponential increase, the pattern produced is commonly called a *J-curve* because of its shape (Fig. 4–1). To calculate the growth rate of a population undergoing exponential growth, you need to know the number in the population right now, the time period, and the number of offspring individuals can produce in a given amount of time if resources are unlimited. This last value is often designated r_{max}, or just *r*.

The equation for exponential growth can be described as follows:

Starting population $\times$ A constant (e) multiplied
    by itself a certain number of times = Ending population

The number of times you multiply *e* by itself is equal to the value of *r* (the potential to produce offspring) multiplied by time. Under unlimited conditions, organisms with a high value of *r* will have more-rapid population growth than will those with a low *r* value. Exponential growth can only continue indefinitely if there are no limits to the size the population can reach, but in nature this is extremely rare.

There are a few times when exponential growth occurs in nature. For example, beetles getting into a big sack of cereal is a good example of a population explosion. However, exponential growth cannot continue indefinitely. There is an upper limit to the population of any particular species that an ecosystem can support. This limit is known as the **carrying capacity** (*K*) (Fig. 4–1). More precisely, the carrying capacity is the maximum population of a species that a given habitat can support without being degraded over the long term—in other words, a sustainable system.

Logistic Growth. Eventually, as a population increases, two outcomes occur: (1) The population keeps growing until it exhausts essential resources, often food, and then dies off precipitously due to starvation or other problems (this results in a crash following the *J*-curve, **Fig. 4–2**). Alternatively, (2) some processes can slow population growth so that it levels off at or near carrying capacity. The simplest version of this latter pattern is known as **logistic growth**. Plotting this type of growth on a graph results in an *S*-shaped curve and thus is called an **S-curve** (Fig. 4–1). The point at which it levels off is the carrying capacity (*K*). The equation for logistic growth can be explained this way:

Starting population + (Reproductive capacity (*r*)
    $\times$ Population $\times$ A number that represents how far
    the population is from the carrying capacity)
        = Ending population

There are two things to notice about the logistic growth curve. First, as the population nears the carrying capacity,

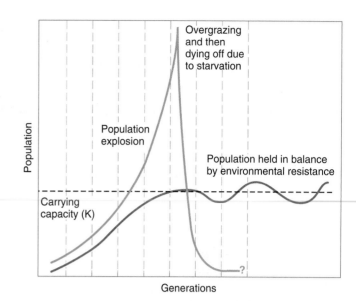

Figure 4–2 Real-life growth. In real populations, exponential growth is often followed by a crash (blue line), and logistic growth is often modified so that the population fluctuates around *K* (red line). Real-life populations often do not follow idealized equations, but such simple models help us to understand possible outcomes.

the growth of the population slows until the population size remains steady, while the population growth equals zero. Second, the maximum rate of population growth occurs halfway to carrying capacity. If you wanted, for example, to harvest the maximum number of worms, you would do better to keep your worm population at half of carrying capacity, not at a larger or smaller number (see Sound Science, "The Story Behind the Numbers: It's a Wormy Project!" and Chapter 7).

The population growth figures calculated using these simple models rarely match the exact growth figures found in real life, but many populations show *J*-shaped explosions followed by crashes (Fig. 4–2). *J*-curves come about when there are unusual disturbances, such as the introduction of a foreign species, the elimination of a predator, the sudden alteration of a habitat, or the arrival in a new section of habitat, all of which can allow rapid growth of a population. This growth is, however, followed by some type of crash. In some cases, huge overshoots of *K* by one population result in a collapse of the habitat for many species.

Other populations, often those controlled by complex relationships between species, show some type of *S*-curve, followed by cycles of lower and higher numbers roughly around *K* (Fig. 4–2). Many shoot somewhat above carrying capacity and then a little below, and eventually they cycle around the carrying capacity. This is illustrated in Figure 4–2.

Biotic Potential Versus Environmental Resistance

The rate at which members of a species reproduce under unlimited conditions (which we saw in the equations above as *r*) is a measure of its **biotic potential**. This is the number of offspring (live births, eggs laid, or seeds or spores set in plants) that members of species may produce under ideal

SOUND SCIENCE

Scientists often make models (equations or other ways of simplifying the world) so that we can understand it. One way to better understand a model such as a population growth curve is to apply it to a situation and figure out what the result would be. Suppose for a moment that you are searching the Internet and come upon a great new get-rich-quick scheme—worm farming. You order a kit, get a small number of starter worms, and plan to raise them in buckets and sell the ones you raise for composting.

In the first month, there are two worms (worms are hermaphroditic—that is, they contain both male and female organs—but they do mate). After a week, you sift the soil and find four worms, after two weeks six worms, and every week, regardless of the number of worms you started with, you have two more worms. This would be a good example of constant growth. If you started with a population of 2 worms and it displayed a constant growth rate of 2 worms per week for four weeks, the population number at the end would be 10 worms $(2 + (2 \times 4) = 10)$. At the end of 24 weeks, you would have 50 worms $(2 + (2 \times 24) = 50)$.

In contrast, with exponential growth, your two starter worms produce very different results. Suppose the two worms have an $r = 0.60$ and

a reproductive time period of one week. (Such an r value is probably higher than worms have in real life, but it works for our example. Many organisms have r values below 0.1 because most females do not reproduce in each time period.) Over four weeks, the population size is not much different than it is with constant growth:
$2 \times e^{0.60 \times 4} = 22$
worms. However, 24 weeks of exponential growth would yield an enormous ending population of more than 3.5 million $(2 \times e^{0.60 \times 24} = 3,588,150)$ worms.

Logistic growth for your worms would work slightly differently. Suppose you started the population with 2 worms where $r = 0.60$ and $K = 2,000$ worms. Four weeks after starting the worm colony, you would have 33 worms. (Each week is calculated separately—by the fourth week, the calculation is $20.8 + (0.60 \times 20.8 \times ((2,000 - 20.8)/2000)) = 33$ worms. Twenty-four weeks after the worm colony started, you

A group of earthworms

would have 2,000 worms. They would have rapidly increased and then slowed, reaching carrying capacity in week 22 and remaining there.

Scientists use models like these to describe what is happening in real-world populations. In real life, however, populations often differ from these idealized equations. For example, sometimes populations have cycles of high and low numbers, driven by resources or by other species (see Chapter 5).

conditions. The biotic potentials of different species vary tremendously, averaging from less than one birth per year in certain mammals and birds to millions per year for many plants and some invertebrates. To have any effect on the size of subsequent generations, however, the young must survive and reproduce in turn. Survival through the early growth stages to become part of the breeding population is called **recruitment**.

Environmental Resistance. Unlimited population growth is seldom seen in natural ecosystems because biotic and abiotic factors cause *mortality* (death) in populations. Among the biotic factors that cause mortality are predation, parasites, competition, and lack of food. Among the abiotic factors are unusual temperatures, moisture, light, salinity, and pH; lack of nutrients; and fire. The combination of all the biotic and abiotic factors that may limit a population's increase is referred to as **environmental resistance (Fig. 4–3)**.

Sometimes environmental resistance lowers reproduction as well as causing mortality directly. For example, the loss of suitable habitat often prevents animals from breeding, and certain pollutants affect reproduction adversely. These situations are still defined as environmental resistance, because they either block a population's growth or

cause its decline. Additionally, environmental resistance can cause migration patterns to change, as when animals leave a drought-stricken area. Migration can lower or increase local populations.

Reproductive Strategies. The interplay of environmental resistance and biotic potential drives the success of two common **reproductive strategies** in the natural world. The first is to produce massive numbers of young, but then leave survival to the whims of nature. This strategy often results in very low recruitment. Animals with this strategy have life histories with rapid reproduction, rapid movement, and often a short life span. This strategy is highly successful if a species is adapted to an environment that can suddenly change and become very favorable, like a rain-fed temporary pond. Organisms with this strategy are usually small and tend to have huge boom-and-bust populations. Because these organisms usually have a high r, they are sometimes called r-strategists or r-selected species. They are also often called "weedy" or "opportunistic" species. A familiar example of this type of species is the housefly, which multiplies quickly but also has a high mortality rate.

In contrast, the second strategy is to have a much lower reproductive rate (that is, a lower biotic potential) but then

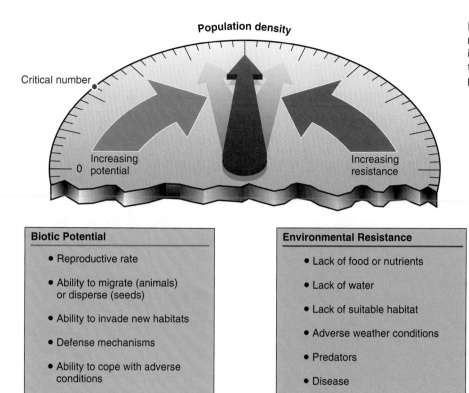

Population density

Critical number

0 Increasing potential

Increasing resistance

Biotic Potential

- Reproductive rate
- Ability to migrate (animals) or disperse (seeds)
- Ability to invade new habitats
- Defense mechanisms
- Ability to cope with adverse conditions

Environmental Resistance

- Lack of food or nutrients
- Lack of water
- Lack of suitable habitat
- Adverse weather conditions
- Predators
- Disease
- Parasites
- Competitors

Figure 4–3 Biotic potential and environmental resistance. A stable population in nature is the result of the interaction between factors tending to increase population and factors tending to decrease population.

care for and protect the young until they can compete for resources with adult members of the population. An elephant or a bird like the California condor is a good example. This strategy works best where the environment is stable and already well populated by the species. Organisms with such a strategy are larger, longer lived, and well adapted to normal environmental fluctuations. Because their populations are more likely to fluctuate around the carrying capacity, they are sometimes called *K-strategists* or *K-selected species*. They are also called *equilibrial* species (Table 4–1).

Characteristics such as the age at first reproduction and the length of life are a part of an organism's **life history**—

the progression of changes an organism undergoes in its life. Species such as the housefly and the elephant have very different life histories that reflect two extreme sets of adaptations. These life histories can be visualized in a graph called a survivorship curve, in which the number remaining from a group all born at the same time is shown decreasing over time until the maximum life span for the species is met. Some species, like humans, have relatively low mortality in early life, and most live almost the full natural life span for that species. This survivorship pattern is depicted as Type I in **Figure 4–4**. Some species have many offspring that die young, and only a few live to the end of a life span, such as oysters and dandelions. These are examples of Type III survivorship. Generally, *r*-strategists show the Type III pattern and *K*-strategists experience the Type I pattern. Some species are not easily defined as either *r*-strategists or *K*-strategists. They may also have an intermediate survivorship pattern, depicted as Type II. Squirrels and corals are likely to fall into this intermediate category. Species actually form a continuum of these survivorship curves, much as they do in terms of the characteristics of *r*- and *K*-strategists.

We have examined species in decline, such as the giant panda (Chapter 3). We will also study species experiencing rapid increase, such as pest species overtaking habitats (see Chapter 13). The reason we look at population growth equations is to show that there is a predictable connection between species characteristics, such as biotic potential (*r*), and their patterns of growth lived out in different life histories. As we will see in Section 4.4, there are foreseeable ways that environmental pressures work on species with different biotic potentials.

TABLE 4–1 General Characteristics of *r*-strategists and *K*-strategists		
	***r*-strategists**	***K*-strategists**
Environment	Advantage if less stable	Advantage if more stable
Size	Smaller	Larger
Life span	Shorter	Longer
Age at first reproduction	Younger	Older
Offspring	More	Fewer
Parental care	Little or none	Long and involved
Population stability	Wild fluctuations	Mostly stable

Figure 4–4 Survivorship curves. Some organisms, like humans (Type I), experience low mortality throughout their lives, and most live to old age. Others (Type III), represented by an oyster, have many offspring, most of which die early. Still others are intermediate (Type II), here represented by a squirrel. Type I and Type III roughly correspond to the *K*-strategists and *r*-strategists.

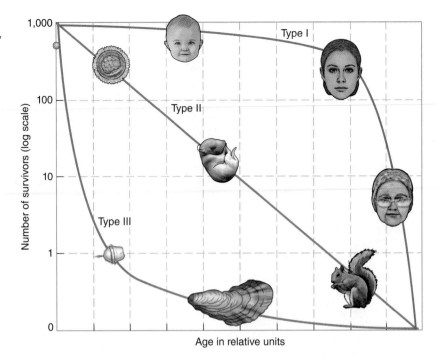

There is also a predictable pattern to the way species are affected by large-scale human activities. While there are many exceptions, it is often the case that *r*-strategists become pests when humans alter the environment and that *K*-strategists become more rare, endangered, or extinct. Houseflies, dandelions, and cockroaches increase with human activity, while eagles, bears, and oaks decline. Some exceptions to this rule include rare opportunistic species (*r*-strategists) that are so separated from new habitat that in spite of a high biotic potential, they cannot succeed.

In sum, whether a population grows, remains stable, or decreases is the result of interplay between its biotic potential and environmental resistance. In general, a population's biotic potential remains constant; it is changes in environmental resistance that allow populations to increase or cause them to decrease. Population balance is a *dynamic balance*, which means that additions (births and immigration) and subtractions (deaths and emigration) are occurring continually, and the population may fluctuate around a median. Some populations fluctuate very little, whereas others fluctuate widely. As long as decreased populations restore their numbers and the ecosystem's capacity is not exceeded, the population is considered to be at equilibrium.

4.2 Limits on Populations

The carrying capacity of a habitat reflects how large a population can be sustained there, but it may not explain what is actually limiting the population. In the case of the golden frog, habitat loss forced the frogs to compete for living space and resources, destruction of streams killed their eggs, and a fungus spread rapidly through an already weakened population. Sometimes these limiting factors, such as crowding or disease, have a greater effect when there are more individuals

in the population. The higher the **population density** (the number of individuals per unit area) is, the more that factor plays a role. In other cases, a factor like a tornado may keep a population from increasing, but the strength of the tornado has nothing to do with the density of the organisms in the population. We will see both of these types of limits in the next section.

Density Dependence and Density Independence

Rabbits thrive in a field with lush grass. As rabbits become more plentiful, foxes and coyotes may find it easier to catch and eat more of them and less of something else. This could slow the population growth of the rabbits. In this example, predation acts as a **density-dependent** limit on a population. A density-dependent limit is one that increases as population density increases—such as a disease or a food shortage. The environmental resistance caused by this limit increases mortality. Conversely, as population density decreases, the environmental resistance lessens, allowing the population to recover.

Other factors in the environment that cause mortality are **density-independent** limits—those whose effects are independent of the crowding of the population. For example, millions of a species of now-extinct Rocky Mountain locusts were once blown by a gale into glaciers in the West, where they can be found buried in ice and snow **(Fig. 4–5a,b)**. This is a good example of a density-independent cause of mortality. Similarly, a fire that sweeps through a forest may kill all small mammals in its wake. Although density-independent factors can be important sources of mortality, they are not involved in maintaining population equilibrium in the way we saw in the logistic equation. Because logistic growth occurs when populations become more crowded (and as they

(a)

(b)

Figure 4–5 Density-independent population control. (a) The Rocky Mountain locust, *Melanoplus spretus* was once quite common in Western North America. It is now extinct. (b) Millions of the members of this species died and were frozen in glaciers such as this one. Weather events such as windstorms, hurricanes, and freezes can be density-independent population controls.

on an ecosystem. Such relationships can cascade down a chain, so that the removal of one species even has effects on species that don't directly interact with it. This phenomenon is attributed to keystone species, as we will see in Section 4.3.

Critical Number

We've discussed natural mechanisms that keep populations near their carrying capacity rather than allowing them to grow indefinitely, but in nature, populations also decline. There are no guarantees that a population will recover from low numbers. Extinctions can and do occur in nature. The survival and recovery of a population depend on a certain minimum population base, which is referred to as the population's **critical number.** You can see the idea of critical number at work in a pack of wolves, a flock of birds, or a school of fish. Often, the group is necessary to provide protection and support for its members. In some cases, the critical number is larger than a single pack or flock, because interactions between groups may be necessary as well. In any case, if a population is depleted below the critical number needed to provide such supporting interactions, the surviving members become more vulnerable, breeding fails, and extinction is almost inevitable.

The loss of biodiversity is one of the most disturbing global environmental trends (Chapter 1). Human activities are clearly responsible for the decline, and even the extinction, of many plants and animals. This is happening because many human impacts are not density dependent; they can even intensify as populations decline. Examples are alteration of habitats, introduction of alien species, pollution, hunting, and other forms of exploitation. Concern for these declines eventually led to the Endangered Species Act, which calls for the recovery of two categories of species. Species whose populations are declining rapidly are classified as **threatened.** If the population is near what scientists believe to be its critical number, the species may be classified as **endangered.** These definitions, when officially assigned by the U.S. Fish and Wildlife Service, set in motion a number of actions aimed at the recovery of the species in question (see Chapter 6).

approach their carrying capacity), logistic growth is sometimes referred to as density-dependent population growth.

In the natural world, a population is subject to the sum of all the biotic and abiotic environmental factors around it. Many of these factors may cause mortality in a population, but only those that are density dependent are capable of regulating the population, or keeping it in equilibrium.

Environmental scientists distinguish between top-down and bottom-up regulation. Top-down regulation is the control of a population (or species) by predation, such as the control of rabbits by their predators, coyotes. In bottom-up regulation, the most important control of a population occurs as a result of the scarcity of some resource, as would occur if the rabbits ran out of food. Because species interact in communities, the factor that controls a population will determine what effects adding or removing a species may have

4.3 Community Interactions

Now that we have described population growth as a dynamic interplay between biotic potential and environmental resistance, we can look at the ways populations interact in a whole community. Picture all of the possible relationships between two species (A and B) as being positive (helpful), negative (harmful), or neutral for each species. These relationships are depicted in Table 4–2.

TABLE 4–2 The Major Types of Interactions Between Two Species. Interactions can be defined by whether the impact on each species is positive, negative, or neutral.

Interaction Name	Effect on Species A	Effect on Species B	Example
Predation	+	−	Wolves eat moose
Competition	−	−	Two rabbit species eat the same grass
Mutualism	+	+	Lichens (fungus/alga)
Commensalism	+	0	Water buffalo/egret
Amensalism	0	−	A plant makes a poison that accidentally harms another plant
No effect	0	0	Very little interaction

The name of each type of interaction is given in the table. For example, a relationship in which one member benefits and the other is harmed (+−) is **predation**. Parasitism is a subset of predation, as is herbivory (animals eating plants). A (− −) relationship is one in which both species are harmed. This would include competitive relationships where both species use a scarce resource. **Competition** between different species is called **interspecific** competition, and competition between members of the same species is called **intraspecific** competition. In contrast, a relationship between species in which both benefit (++) is called a **mutualism. Commensalism** occurs when one species is benefited and the other is not affected (+0).

The most important interactions are predation, competition, mutualism, and commensalism, but there are others.

One species might be unaffected and the other harmed (amensalism, 0−) through an accidental interaction such as an elephant stepping on a flower or a plant making a poison as a part of its normal metabolism that harms another organism. Similarly, it is theoretically possible to have a (00) interaction, but there is no name for this, and presumably the interaction would be a weak one. The following looks in depth at the most important species interactions.

Predation

In any relationship in which one organism feeds on another, the organism that does the feeding is called the **predator** and the organism that is fed on is called the **prey**. Predation is a conspicuous process in all ecosystems. Predation ranges from the classic predator-prey interactions between **carnivores** (meat eaters) and **herbivores** (plant eaters) to herbivores feeding on plants and **parasites** feeding on their hosts.

Parasites: A Special Kind of Predator. Parasites are organisms—either plants or animals—that become intimately associated with their "prey" and feed on it over an extended period of time, typically without killing it. However, sometimes they weaken it so that it becomes more prone to being killed by predators or adverse conditions. The plant or animal that is fed on is called the **host**. Parasites include a range of species, such as tapeworms, microscopic disease-causing bacteria, viruses, protozoans, and fungi **(Fig. 4–6)**. Parasites can feed on their hosts in different ways. Tapeworms take energy from the intestine of a host animal, whereas parasitic plants tap into the sap of other plants.

Parasitic organisms affect the populations of their host organisms in much the same way that predators affect their prey—in a density-dependent manner. As the population density of the host increases, parasites and their vectors

Figure 4–6 Several types of parasites. Nearly every biological group has at least some members that are parasitic. Shown here are (a) mistletoe, a plant parasite; (b) an intestinal parasite found in drinking water (*Giardia* sp); (c) lampreys, organisms that parasitize many types of fish; and (d) athlete's foot fungus on a human foot.

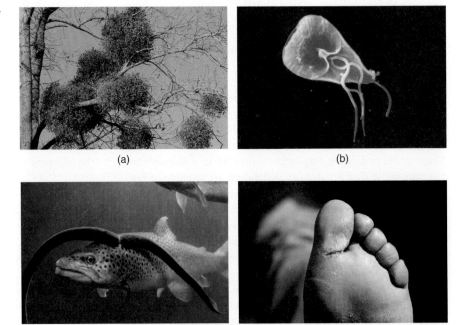

(a)

(b)

(c)

(d)

(agents that carry the parasites from one host to another, such as disease-carrying insects) have little trouble finding new hosts. Therefore, infection rates increase, causing higher mortality. Conversely, when the population density of the host is low, the transfer of infection is less efficient. This reduces the levels of infection and allows the host population to recover.

A tremendous variety of organisms may be parasitic. Various worms are well-known examples, but certain protozoans, insects, and even mammals (vampire bats) and plants (dodder) are also parasites. Many plant diseases and some animal diseases (such as athlete's foot, Figure 4–6d) are caused by parasitic fungi. Indeed, virtually every major group of organisms has at least some members that are parasitic. Parasites may live inside or outside their hosts, as the examples shown in Figure 4–6 illustrate.

In medicine, a distinction is generally made between bacteria and viruses that cause disease (known as **pathogens**) and parasites, which are usually larger organisms. Ecologically, however, there is no real distinction. Bacteria are foreign organisms, and viruses are organism-like entities feeding on, and multiplying in, their hosts over a period of time. Some fungi are parasites, and some are also referred to as disease causing. Therefore, disease-causing bacteria and viruses can be considered highly specialized parasites.

Regulation of Prey. As some populations are limited by parasites, many studies have shown that herbivores are often regulated by their predators—a type of top-down control. A well-documented example is the interaction between wolves and moose on Isle Royale, a 45-mile-long island in Lake Superior that is now a national park. During a hard winter around 1900, a small group of moose crossed lake ice to the island and stayed. Their population grew considerably in the absence of predators. Then, in 1949, a small pack of wolves also managed to reach the island. The isolation of the island provided an ideal opportunity to study a simple predator-prey system, and in 1958, wildlife biologists began carefully tracking the populations of the two species **(Fig. 4–7)**. As seen in the figure, a rise in the moose population is usually followed by a rise in the wolf population, followed by a decline in the moose population and then a decline in the wolf population. The data can be interpreted as follows: Fewer wolves represent low environmental resistance for the moose, so the moose population increases. Then the abundance of moose represents optimal conditions (low environmental resistance) for the wolves, and the wolf population increases. More wolves mean higher predation on the moose (high environmental resistance); again the moose population falls. The decline in the moose population is followed by a decline in the wolf population, because now there are fewer prey (high environmental resistance) for the wolves.

Deep snow (which makes it more difficult to get food) and an infestation of ticks in 1996 caused substantial mortality. However, the dramatic fall in the moose population in 1996 cannot be attributed entirely to predation by the small number of wolves. The sharp decline in moose is thought to be responsible for keeping the wolf population low, as there were few calves for them to catch. Wolves are generally incapable of bringing down an adult moose in good physical condition. The animals they kill are the young and those weakened by another factor, such as sickness or old age.

The observation that wolves are often incapable of killing moose that are mature and in good physical condition is extremely significant. This is what often prevents predators from eliminating their prey. As the prey population is culled down to healthy individuals who can escape attack, the predator population will necessarily decline unless it can switch to other prey. The predators are limited by the availability of their crucial food resource. Meanwhile, the survivors of the prey population are healthy and can readily reproduce the next generation. Thus, a predator-prey relationship involves both top-down (on the prey) and bottom-up (on the predator) population regulation. Furthermore, some factors such as weather can be both density-independent or dependent. For example, when a deep freeze occurs, many organisms may die, but the effect is not necessarily more dramatic if the population is higher.

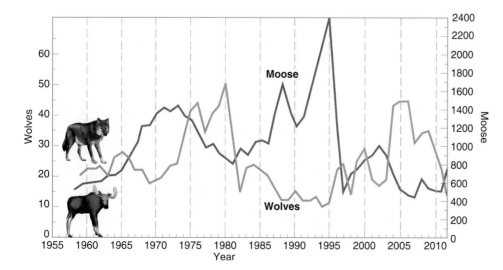

Figure 4–7 Predator-prey relationships. The fluctuations in the wolf and moose populations on Isle Royale are shown from 1955 until 2011.

A parasite can also work in conjunction with a larger predator to control a given herbivore population. When parasitic infection breaks out in a dense population of herbivores, individuals weakened by infection are more easily removed by predators, leaving a smaller, but healthier, population. Relationships between a prey population and several natural enemies are generally much more stable and less prone to wide fluctuations than are those that involve only a single predator or parasite, because different predators or parasites come into play at different population densities.

Plant-Herbivore Dynamics. Herbivores prey on plants. Just as too many wolves can bring the moose population dangerously low, too many herbivores can do the same to their plant food. In turn, whenever the predator populations are low, the prey populations encounter low environmental resistance and can overshoot their carrying capacity. These imbalances can lead to problems in the surrounding ecosystem.

Overgrazing. If herbivores eat plants faster than they can grow, the plants will eventually be depleted, and animals that eat the plants will suffer. A classic example is the case of reindeer on St. Matthew Island, a 128-square-mile island in the Bering Sea. In 1944, a herd of 29 reindeer was brought onto the island, where they had no predators; this lack of predation allowed an increase of the herd to 6,000 individuals by 1963, when most were malnourished and food had disappeared. During the winter of 1963–1964, nearly the entire herd died of starvation; there were only 42 surviving animals in 1966. Overgrazing is a common outcome in agricultural systems as well **(Fig. 4–8)**. The Food and Agriculture organization of the United Nations (FAO) estimated that the world's cattle population increased from 720 million to 1.5 billion between 1950 and 2001; domestic sheep increased from 1.04 billion to 1.75 billion during the same period. These increases have caused overgrazing of more than 20% of rangeland.

Overgrazing demonstrates that no population can escape ultimate limitation by environmental resistance, although the form of environmental resistance and the consequences may differ. If a population is not held in check, it may explode, overgraze, and then crash (exhibiting a *J*-curve and subsequent drop) as a result of starvation.

Predator Removal. Eliminating predators or other natural enemies of herbivores upsets basic plant-herbivore relationships in the same way as introducing an animal without natural enemies disrupts normal population controls on the introduced species. In much of the United States, for example, white-tailed deer populations were originally controlled by wolves, mountain lions, and bears. Many of these animals were killed because they were believed to be a threat to livestock and humans. At present, deer populations in many areas would increase to the point of overgrazing if humans didn't hunt them in place of these natural predators. Indeed, population explosions do occur where hunting is prevented. If the St. Matthew reindeer population had been kept in check by predators, for example, their population might not have increased to the point at which they overgrazed the vegetation on the island.

Keystone Species. In the case of white-tailed deer, the removal of a number of predator species allowed the deer population to explode. However, sometimes the removal of a single species can cause a cascade of effects that affect far more than just the other species with which they primarily interact. For example, in the West Coast kelp forest, the sea otter (*Enhydra lutris*) eats sea urchins, which in turn eat kelp **(Fig. 4–9)**. As a result, hundreds of other species, including the young of many fish, are able to survive in the kelp forests. Overhunting of sea otters broke this food chain and left the urchins unchecked. Sea urchins ate kelp down to the sea floor, creating areas called "urchin barrens." Subsequent protection of the otter has allowed it to return, again protecting the entire habitat. Scientists refer to the sea otter as a **keystone species**, in recognition of its crucial role in maintaining ecosystem biotic structure (in architecture, the keystone is a fundamental part of the support and structure of a building). The presence of the keystone species moderates other species that would otherwise take over the area (the sea urchin) or eat the most important species in the habitat (the kelp). Keystone species allow less-competitive species to flourish as well, thereby increasing overall diversity. Keystone species do not have to be predators, as shown in this example, but they often are predators themselves.

Figure 4–8 Plant-herbivore interaction: Overgrazing. Domestic animals such as these goats can overgraze land as well.

Figure 4–9 Keystone species. The sea otter eats sea urchins, which eat kelp in the Pacific Northwest.

Competition

Recall from Table 4–2 that there are species interactions in which both species are harmed. This result usually means that species are competing for some scarce resource. We have discussed the concept of each species having a *niche*, the conditions and resources under which a species can live (Chapter 3). Species are said to have overlapping ecological niches when they compete with each other. Competition can also occur between members of the same species. In the case of both types of competition, between and within species, you might expect that, over time, there would be a pressure to lower the overlap and have less negative interaction. Indeed, this is the case.

Intraspecific Competition. When members of the same species compete with each other, it is called **intraspecific competition**. This type of competition can occur over different types of resources. In many species, some members defend a **territory,** or limited space. The males of many species of songbirds claim a territory that they will defend vigorously at the time of nesting. Their song warns other males to keep away **(Fig. 4–10)**. The males of many carnivorous mammals, including dogs, stake out a territory by marking it with urine. The smell warns others to stay away. If other dogs do encroach, there may be a fight, but in most species, a large part of the battle is intimidation—an actual fight rarely results in serious physical harm. Sometimes, as in sea anemones, individuals can band together to defend a territory for a colony **(Fig. 4–11)**.

In territoriality, what is really being protected by the defender or sought after by the invader is the claim to an area suitable for nesting, space for establishing a harem, or adequate food resources. Hence, the territory is defended only

Figure 4–11 Competition. Organisms encrusting this coral reef compete for living space. Many have early life stages that live in the water and must settle down on a surface and attach in order to continue to grow. Sometimes sea anemone colonies such as these will fight for living space.

against others that would cause the most direct competition for those resources—members of the same species. Because territoriality is competition for space with resources, a lack of suitable territories is a density-dependent limitation on a population—the larger the population grows, the more intense the competition for space is.

Territoriality as an Advantage. Territoriality may seem like a negative factor, but it actually protects a population and ensures that some of its members will get enough resources to survive. What if the population had no territories? If so, when resources were scarce, two things might happen: First, every encounter with another member of the species could end in a potentially lethal fight. Second, instead of some members getting enough resources to raise the next generation, it might be possible for all members to get only a portion of what they need, and a large percentage—or all—of the population would die.

Whether between species or between members of the same species, competition harms both participants. It lowers their fitness and the production of offspring. Territoriality is an adaptation that lowers the direct effects of competition on members of the same species. Individuals unable to claim a territory are usually the young of the previous generations. Some may survive on the fringes and will get an opportunity to hold a territory later. Others may be driven to disperse, which may open new habitats to the species. Some may die without ever getting territories, but as a whole, territories protect the population from dying out. As you might expect, territoriality is most likely in species with long life spans, multiple years to reproduce, parental care of young, and other characteristics of *K*-strategists. Territoriality is an adaptation that helps organisms disperse and stabilizes populations.

Impact on the Species. The observation that individuals compete for scarce resources led Charles Darwin to identify the "survival of the fittest" as one of the forces in nature leading to evolutionary change in species. Those individuals in a competing group of young plants, animals, or

Figure 4–10 Territoriality. This male red-winged blackbird sings to defend his territory.

microbes that are able to survive and reproduce, while others are not, demonstrate superior **fitness** to the environment. Indeed, every factor of environmental resistance is a selective pressure resulting in the survival and reproduction of those individuals with a genetic endowment that enables them to better cope with their surroundings. This is the essence of **natural selection,** which is discussed in Section 4.4.

Intraspecific competition, therefore, has two distinct impacts on the population of a species. In the short term, it can lead to the density-dependent regulation of a species population through such factors as territoriality and self-thinning, which is a phenomenon whereby crowded organisms such as trees become less numerous as they grow bigger. Intraspecific competition can also lead, however, to long-term changes as the species adapts to its environment. Those better able to compete are the ones who survive and reproduce, and their superior traits are passed on to successive generations.

Interspecific Competition. Now we turn to the effects of competition between different species, or **interspecific competition.** Consider a natural ecosystem that contains hundreds of species of green plants, all competing for nutrients, water, and light. What prevents one plant species from driving out others? How do species survive competition? The most obvious answer is that sometimes they do not. If two species compete directly in many respects, as sometimes occurs when a species is introduced from another continent, one of the two species generally perishes in the competition. This is the competitive exclusion principle. For example, the introduction of the European rabbit to Australia (described later in Section 4.5) has led to the decline and disappearance of several small marsupial animal species due to direct competition for food and burrows.

In fact, competitive exclusion would be expected to be the norm in circumstances where a habitat is very simple and species require the same resources. However, in nature, we often see many species occupying the same habitats without becoming extinct. This has intrigued scientists. G. E. Hutchinson, a famous limnologist (a scientist who studies freshwater systems), once wrote a paper asking about this phenomenon: "A homage to Santa Rosalia or Why are there so many kinds of animals?"[1] Hutchinson suggested, and further work has supported, that the answer lies in variability over space and time. Differences in topography, type of soil, and so on mean that the landscape is far from uniform, even within a single ecosystem. Instead, it is composed of numerous microclimates or microhabitats. The specific abiotic conditions of moisture, temperature, light, and so on differ from location to location and through the course of time. This is true even in lakes, where tiny pockets of water may have different temperatures, chemistry, or light conditions. Often just when one competitor begins to get the upper hand, the seasons change and a different species benefits. Thus, the adaptation of a species to specific conditions enables it to thrive and overcome its competitors in one location or time, but not in another.

These examples simply mean that ecosystems are heterogeneous (variable) enough to support species with many different niches. Similarly, changes over time can allow multiple species to live in the same place. For example, the spring wildflowers of temperate deciduous forests (lady slippers, hepatica, trillium) sprout from perennial roots or bulbs in the early part of that season. While they would not be able to outcompete trees for light during the summer, these plants take advantage of the light that can reach the forest floor before the trees grow leaves.

Species that seem to be in competition can coexist in the same habitat, but have separate niches. Competition is minimized because potential competitors are using different resources. For example, woodpeckers, which feed on insects in deadwood, do not compete with birds that feed on seeds. Bats and swallows both feed on flying insects, but they do not compete because bats feed on night-flying insects and swallows feed during the day. Sometimes the "resource" can be the space used by different species as they forage for food. This is illustrated in **Figure 4–12,** which shows five species of warblers that coexist in the spruce forests of Maine. The birds feed at different levels in the trees and on different parts of the trees. This is called resource partitioning—the division of a resource and specialization in different parts of it. Subsequent study of these warblers has shown that when competition is more intense, they separate themselves even more.

All of these examples show that species can live in the same habitat because their adaptations, including behavior, limit competition. By limiting those competitive interactions, each species can put more energy into reproduction. However, the reason that many species can share a habitat is not simply because their niches prevent them from competing. Instead, when species live in the same place, competition can change some of their characteristics so that they use only a portion of their potential niche (which they could have used all of when living alone). For more discussion on this, see Sound Science, "Studying Finches: The Life of a Scientist" and **Figure 4–13.**

Mutualism

In contrast to competition, **mutualism** is an arrangement between two species whereby both benefit. There are many mutualistic relationships. For example, the relationship between pollinators (such as bees, which receive nutrition from flowers) and plants they pollinate (which receive pollination) is an example of mutualism (**Fig. 4–14a** on page 88). Some estimates infer that 70% of plant species have mutualistic relationships with fungi connected to their roots. The fungi benefit from the plant nutrition and in exchange make it easier for the plants to take in nitrogen and other soil nutrients. Lichens are actually a mutualistic relationship between a fungus and an alga (Fig. 4–14b). In animals, one example of a mutualism is the relationship of the anemone fish and the sea anemone. The fish protects the anemone from predation by the butterfly fish, and in turn, the anemone's stinging tentacles (to which the anemone fish is immune) protect the fish from other predators.

---

[1]*The American Naturalist,* Vol. XCIII, No. 870, May–June 1959, pp. 145–159.

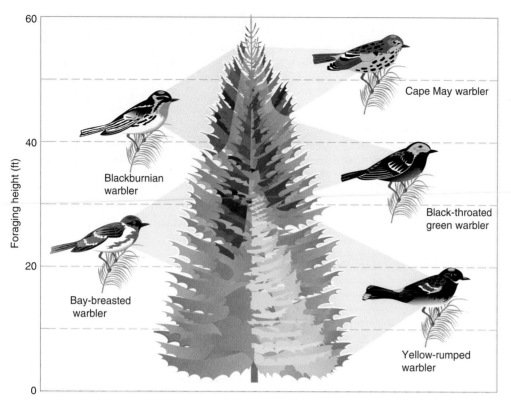

Figure 4–12 Resource partitioning. Five species of North American warblers experience less competition than it might first appear because they feed at different heights and in different parts of trees.

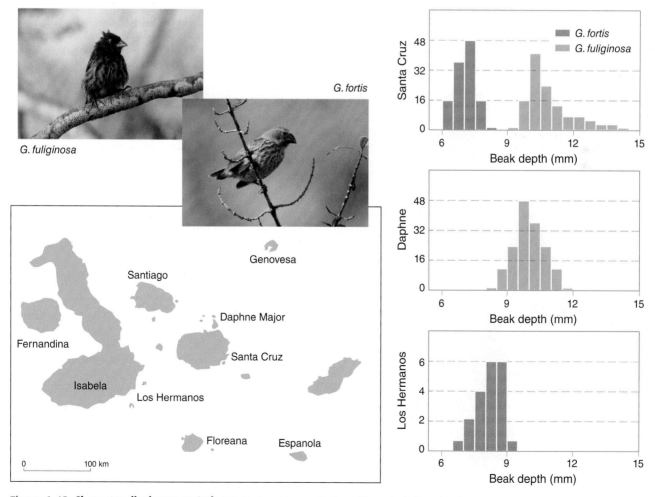

Figure 4–13 Character displacement/release. Finches of two species have different beak sizes when they occur together but do not when the species occur separately. (*Source:* Evolution of Character Displacement in Darwin's Finches" by Peter R. Grant and B. Rosemary Grant, from SCIENCE, July 2006, Volume 313(5784). Copyright © 2006 by AAAS. Reprinted with permission.)

(a) (b)

Figure 4–14 Mutualism. (a) Pollinators and their flowering plants are mutualists. (b) A lichen is a mutualism between a fungus and an alga living symbiotically.

Commensalism

Commensalism is more rare. It occurs when two species interact and one is benefited while the other is unaffected. An example would be cattle egrets and water buffalo: the grazer stirs up insect prey, and the bird follows it and eats the insects (**Fig. 4–15**). Commensalism also includes animals that hitch rides on other species without harming them or grow on them for support. Many orchids exist in this type of relationship—they grow as vines on large trees, but do not harm or feed off of them like parasitic plants.

Figure 4–15 Commensalism. A cattle egret eats insects stirred up by a grazing water buffalo, which is neither harmed nor benefited by the egret's presence.

In contrast to commensalism is **amensalism,** where one species is *harmed* while the other is unaffected. Usually, this is accomplished by the natural chemical compounds produced by an organism, which are harmless to itself but harmful to other organisms around it. An example is the black walnut tree, which secretes a chemical compound from its roots that can harm or kill other plants in the area.

In summary, the relationships between two species or between two members of one species can be described by the positive, negative, or neutral effects on the individuals. These potential outcomes are studied by ecologists. Not only do relationships such as predation, competition, mutualism, and commensalism affect individuals, but also they are part of the constant pressure that causes species to adapt to environmental change.

Some of these relationships with positive and negative effects stem from a type of **symbiosis**—a relationship in which two species live closely connected to each other. Many of the examples of parasites, predators, prey, commensalisms, and mutualisms we have studied are symbiotic. As we've seen, these relationships can be beneficial (as when mutualisms benefit both parties) or detrimental (as in the case of parasites and host).

4.4 Evolution as a Force for Change

Predation and competition are two important mechanisms that keep natural populations under control. They also support the survival of individuals with qualities that can lower the impacts of negative interactions. Predators and prey become well adapted to each other's presence. Predators are rarely able to eliminate their prey species, largely because the prey has various defenses against its predators. Intraspecific competition also represents a powerful force that can lead to improved adaptations of a species to its environment. Interspecific competition, by contrast, promotes adaptations in the competitors that allow them to specialize in exploiting

SOUND SCIENCE

How do scientists know what happened to species in the past? For the most part, they look at what happens today. They look for physical patterns and mechanisms and at DNA and other parts of cells to see patterns that suggest relationships.

Darwin himself first observed evidence of past speciation. In 1835, Darwin collected a number of finches from different islands of the Galápagos archipelago during his famous voyage on the HMS *Beagle*. Later, in consultation with ornithologist John Gould, Darwin speculated that differences in size and beak structure of the finches could have happened after subpopulations of an original parental species were isolated from one another on separate islands.

Although Darwin himself did not work out the details of how the different Galápagos finches evolved, modern scientists have found what looks to be the most likely explanation. It is probable

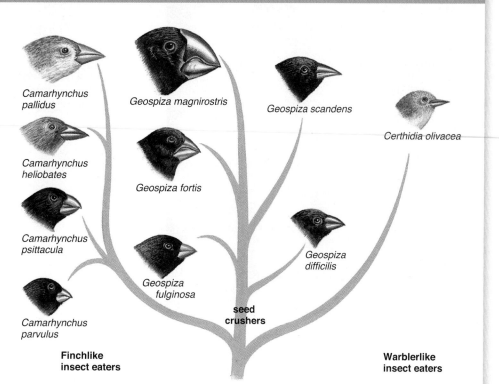

Camarhynchus pallidus

Geospiza magnirostris

Geospiza scandens

Certhidia olivacea

Camarhynchus heliobates

Geospiza fortis

Camarhynchus psittacula

Geospiza difficilis

Camarhynchus parvulus

Geospiza fulginosa

seed crushers

Finchlike insect eaters

Warblerlike insect eaters

Ancestral South American finch

Some of Darwin's finches. The similarities among these birds attest to their common ancestor. Selective pressures to feed on different foods have caused modification and speciation in adapting subpopulations. (*Source:* From *Biology: Life on Earth,* 4th ed., by Teresa Audesirk and Gerald Audesirk, copyright © 1994 by Prentice Hall, Inc., reprinted by permission of Pearson Education, Inc., Upper Saddle River, NJ 07458.)

Peter and Rosemary Grant have spent decades studying finches

that at some time in the past, a few finches from the South American mainland were blown westward by a strong storm and became the first terrestrial birds to inhabit the relatively new (10,000-year-old) volcanic islands of the Galápagos. Later, some birds dispersed to nearby islands, where they were separated from the main population. These subpopulations encountered different selective pressures and became specialized for feeding on different things (cactus fruit, insects, and seeds, to name a few). In time, when the changed populations dispersed back to their original islands, they were different enough from the parent species that interbreeding among them did not occur, and the changed populations were distinguishable as new species.

Peter and Rosemary Grant have studied the Galápagos finches for decades. They have investigated the fact that when the finch *Geospiza magnirostris* arrived on the same island as the similar species *Geospiza fortis,* the two species competed heavily for food seeds. Over 25 years of study, the scientists showed that the beak sizes of both species changed and competition was reduced as the finches began to eat slightly different-sized seeds. In populations where each species was alone, the beaks remained in a middle size. This is called character displacement, a physical change that lessens competition when two species co-occur (see Fig. 4–13).

The Grants have spent six months of every year in the Galápagos since 1973 and have studied the fates of hundreds of birds, taking blood samples and measurements. This type of careful study is useful in understanding how evolution occurs.

Figure 4–16 Resource partitioning in plants. Prairies have plants with short fibrous roots and others with long tap roots that use different areas of water in the soil.

Figure 4–17 Cryptic coloration. The ability to hide (as this chameleon does) is a defense against predation, but it also can be an adaptation that allows predators to sneak up on or wait for prey.

a resource. This specialization can lead to resource partitioning, which allows many potential competitors to share a basic resource—such as water in the soil, as illustrated by prairie grasses with long tap roots and those with short fibrous roots, shown in **Figure 4–16**. In this section, we discuss how these various adaptations have come about.

Selective Pressure

Most young plants and animals in nature do not survive, instead falling victim to various environmental resistance factors. These factors—such as predators, parasites, drought, lack of food, and temperature extremes—are known as **selective pressures**. These pressures can affect which individuals survive and reproduce and which are eliminated. If a predator is present, for example, prey animals with traits that protect them or that allow them to escape from their enemies (such as coloration that blends in with the background) tend to survive and reproduce (**Fig. 4–17**). Those without such traits tend to become the predator's dinner. Any individual with a genetic trait that slows it down or makes it conspicuous will tend to be eaten. Thus, predators function as a selective pressure, favoring the survival of traits that enhance the prey's ability to escape or protect itself and causing the elimination of any traits handicapping the organism's escape. The need for food can also be seen as a selective pressure acting on the predator. Characteristics that enhance survival, such as keen eyesight and swift speed, benefit survival and reproduction. Sometimes predators use the ability to blend in as an advantage, too. The process of

specific traits favoring the survival of certain individuals is known as **natural selection**.

Discovered independently by Charles Darwin and Alfred Russell Wallace, these concepts were first presented in detail by Darwin in his book *On the Origin of Species by Means of Natural Selection* (1859). The modification of the gene pool of a species by natural selection over the course of many generations is the sum and substance of **biological evolution**. Darwin and Wallace deserve tremendous credit for constructing their theory purely from their own observations, without any knowledge of genetics. The role of genetic information wasn't discovered until several decades later. Our modern understanding of DNA, mutations, and genetics supports the theory of evolution by natural selection.

Adaptations to the Environment

The gene pool of a population is continually tested by the selective pressures exerted by environmental resistance. Indeed, virtually all traits of an organism can be seen as features that adapt the organism for survival and reproduction, or, in Darwinian terms, **fitness**. Such characteristics of organisms include the adaptations needed for coping with climatic and other abiotic factors; obtaining nutrients, energy, and water; defending against predation and disease-causing or parasitic organisms; finding or attracting mates (or pollinating and setting seed); and migrating and dispersing. These are summarized in **Figure 4–18**.

The fundamental question about any trait is this: Does it increase survival and reproduction of the organism? If the answer is yes, the trait will be maintained through natural selection. Consequently, various organisms have evolved different traits that accomplish the same function. For example, the ability to run fast, to fly, or to burrow and protective features such as quills, thorns, or an obnoxious smell or taste all help reduce predation and can be found in various organisms.

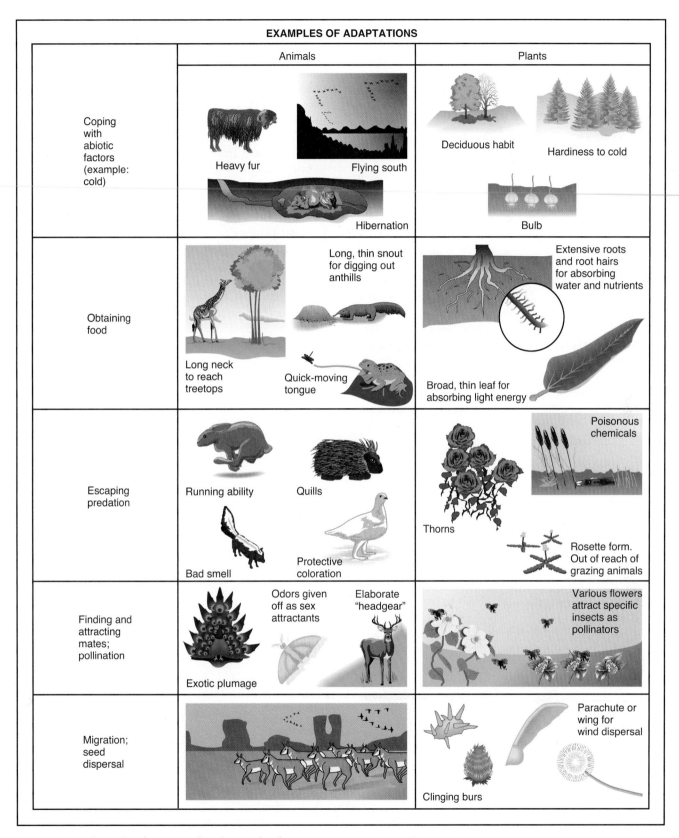

Figure 4–18 Adaptation for survival and reproduction. The five general features listed at the left are essential for the continuation of every species. Each feature is found in each species as particular adaptations that enable the species to survive or reproduce. Across the various species, a multitude of adaptations will accomplish the same function. Thus, a tremendous diversity of species exists, with each species adapted in its own special way.

The Limits of Change. As we saw in the opening essay, the Panamanian golden frog is facing a new, powerful selective pressure—a disease-causing fungus. In such a position, a species may experience three outcomes:

1. **Adaptation.** The population of survivors may gradually adapt to the new condition through natural selection.

2. **Migration.** The surviving population may migrate and find an area where conditions are suitable.

3. **Extinction.** Failing the first two possibilities, extinction is inevitable.

Migration and extinction need no further explanation. The critical question is: What factors determine whether a species will be able to adapt to new conditions instead of becoming extinct?

Recall that adaptation occurs as selective pressures and eliminates those individuals that cannot tolerate the new condition. For adaptation to occur, there must be some individuals with traits (alleles—variations of genes or new combinations of genes) that enable them to survive and reproduce under the new conditions. Even with such genetic variation, there must also be enough survivors to maintain a viable breeding population. If a breeding population is maintained, the process of natural selection should lead to increased adaptation over successive generations. If a viable population is *not* maintained at any stage, extinction results. In the case of the golden frog, the species has not yet been able to adapt to the fungus, in part because the frog was already facing pressure from habitat loss. The frog had to be removed to captivity by humans to prevent extinction. However, there are some golden frogs that have survived the fungus. If their genes can be preserved, it is possible that someday the frog will come back from the brink of extinction and be reintroduced to the wild.

In a similar example, the California condor (a very large scavenger) had declined to about 20 birds by the early 1980s because of deaths from poaching, lead poisoning, and habitat destruction. Only nine condors were left in the wild. To save the species from extinction, wildlife biologists captured the remaining wild condors and initiated a breeding program. As of March 2011, there were 369 condors, 178 of which were in captivity. This is still not a viable breeding population, but the wild condors are now forming pairs and nesting.

Keys to Survival. There are four key variables among species that will affect whether or not a viable population of individuals is likely to survive new conditions: (1) geographical distribution, (2) specialization to a given habitat or food supply, (3) genetic variation within the gene pool of the species, and (4) the reproductive rate relative to the rate of environmental change.

Consider how these factors work. It is unlikely that any change will affect all locations uniformly or equally. Therefore, a species such as the housefly—which is present over most of Earth, has no requirements for a specialized habitat or food supply, and possesses a high degree of genetic variation—is likely to survive almost any conceivable alteration in Earth's environment. The California condor, on the other hand, requires an extensive habitat of mountains, ravines, and the carcasses of large animals to eat but has little genetic variation in its population. The less vulnerable species are likely to be *r*-strategists, described in Section 4.1. The more vulnerable species are likely to be *K*-strategists, although many species fall somewhere in between these extremes, and in some cases, even very rapidly reproducing species can be vulnerable if there is enough environmental change. The factors affecting survival of these two species types are summarized in **Figure 4–19**.

Genetic Change. Assuming the survival of a viable population, how fast can that population evolve further adaptations to enable it to better cope with new or changing conditions? Over the lifetime of the individual, there is no genetic change—and hence no genetic adaptation. In the population, genetic variation is brought about over generations by

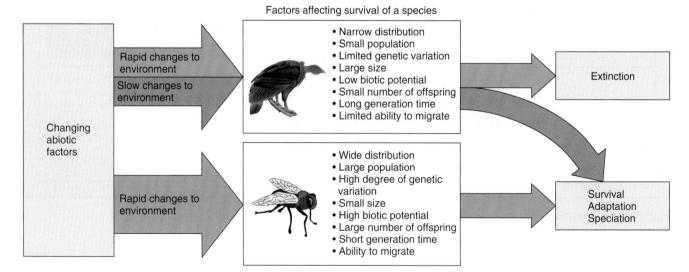

Factors affecting survival of a species

Changing abiotic factors

Rapid changes to environment

Slow changes to environment

- Narrow distribution
- Small population
- Limited genetic variation
- Large size
- Low biotic potential
- Small number of offspring
- Long generation time
- Limited ability to migrate

Rapid changes to environment

- Wide distribution
- Large population
- High degree of genetic variation
- Small size
- High biotic potential
- Large number of offspring
- Short generation time
- Ability to migrate

Extinction

Survival
Adaptation
Speciation

Figure 4–19 Vulnerable and highly adaptive species. Species differ broadly in their vulnerability to environmental changes. When confronted with rapid change in the environment, a housefly population is likely to explode and a condor population is likely to decline.

a number of mechanisms, including hybridization, mutation, and crossover (a process by which sections of chromosomes are swapped). Bacteria, for example, readily exchange small bits of genetic material with other bacteria, and recent studies have shown that microbes sometimes insert their genetic material into other organisms, including mammals, bringing new DNA variations into the genome. Genetic variation is the norm in living things.

The Evolution of Species. The newest estimates are that there are around 8.7 (+/− 1.3) million species of plants, animals, and microbes currently in existence, all living and functioning in ecosystems and all contributing to an amazing biodiversity. This estimate changes with new information because, of course, we haven't discovered most of the organisms that exist on Earth. However, biodiversity is diminishing because present species are becoming extinct faster than new species are appearing. In order to understand why, we need to understand how new species appear.

The infusion of new variations from mutations and the pressures of natural selection serve to adapt a species to the biotic community and the environment in which it exists. Over time, in this process of adaptation, the final "product"—giraffe, anteater, redwood tree—may be so different from the population that started the process that it is considered a different species. This is one aspect of the process of speciation.

Two Species from One. The same process also may result in two or more species developing from one original species. There are only two prerequisites. The first is that the original population must separate into smaller populations that do not interbreed with one another. This **reproductive isolation** is crucial because if the subpopulations continue to interbreed, all the genes will continue to mix through the entire population, keeping it as one species. The second prerequisite is that separated subpopulations must be exposed to different selective pressures. As the separated populations adapt to these pressures, they may gradually become so different that they are not considered the same species. Thus, they are unable to interbreed with one another, even if they come together again later. Consider the example of the arctic and gray foxes. Many generations ago, they may have been part of a single ancestral population, separated eventually by glaciation. In the Arctic, selective pressures favor individuals that have heavier fur (to protect them from the cold); shorter tail, legs, ears, and nose (the shortness helps conserve body heat); and white fur (which helps the animals hide in the snow). In the southern regions, selective pressures favor individuals with a thinner coat to dissipate excessive body heat and fur that blends in with darker surroundings. The geographic isolation of the two subpopulations over many generations made it possible for one ancestral fox population to develop into two separate species (**Fig. 4–20**).

As we might expect, we also see new species arise in real time today, and we have examples of species forming rapidly. The saltmarsh cordgrass, *Spartina angelica*, is one such species. It developed as a hybrid of two other related species: *Spartina alterniflora*, a native of North America, and *Spartina maritima*, a native of England. The two do not usually interbreed. *Spartina alterniflora*, accidentally introduced to England's mud flats, did not thrive there, but when it bred with *Spartina maritima*, the hybrid daughter species (*S. angelica*) spread rapidly. It has a different number of chromosomes from either parent and does not interbreed successfully with either parent species. Now, where there were two similar species, there are three, and the daughter

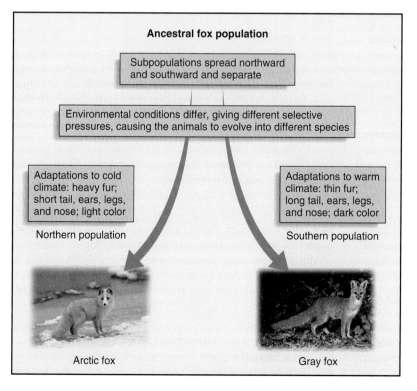

Figure 4–20 Evolution of new species. A population spread over a large area may face various selective pressures. If the population splits so that interbreeding does not occur, the different pressures may result in the subpopulations evolving into new species. Two species of foxes show the pattern we would expect if this had occurred.

species has a different set of characteristics and can live in a slightly different niche than either parent species.

In sum, new species are not formed from scratch; they are formed only by the modification of existing species. Different selective pressures act on the constant variation of genetic material that naturally occurs. Combined with isolation and other processes (such as the unusual hybridization that occurred in the genus *Spartina*), new adaptations can arise. This concept is consistent with the observation that groups of closely related species are generally found in nature, instead of distinct species without close relatives. Patterns among larger groups suggest the same thing—that organisms can be grouped into large, closely related groups differentiated by older or newer characteristics.

Although it can be difficult to imagine, overwhelming evidence from today and patterns from the past inform us that the present array of plants, animals, and microbes has developed through evolution over long periods of time and in every geographic area on Earth. This is the source of our current biodiversity.

Drifting Continents

Geographic isolation of populations is the foundation of the process of speciation. The slow movement of Earth's continents has allowed the isolation of broad groups of organisms. Using evidence from geology, German scientist Alfred Wegener proposed in 1915 that about 225 million years ago the continents were connected into a giant landmass called Pangaea. This is part of a theory that the continents have been in motion (and continue to move) since Earth formed. Wegener's theory was hotly debated for decades until, by the mid-20th century, the evidence for drifting continents became irrefutable. His theory has blossomed into a grand theory known as **plate tectonics**. This theory helps us to understand earthquakes and volcanic activity and is key to understanding the geographic distribution of present-day biota.

Much of the interior of Earth is molten rock, kept hot by the radioactive decay of unstable isotopes remaining from the time when the solar system was formed (about 5 billion years ago). Earth's crust, the lithosphere—which includes the bottoms of oceans as well as the continents—is a relatively thin layer (ranging from 10 to 250 kilometers, or 16 to 160 miles) that can be visualized as huge slabs of rock floating on an elastic layer beneath. This is much like crackers floating next to each other in a bowl of soup. These slabs of rock are called **tectonic plates**. Some 14 major plates and a few minor ones make up the lithosphere (**Fig. 4–21a**).

Within Earth's molten interior, hot material rises toward the surface and spreads out at some locations, while cooler material sinks toward the interior at other locations. Riding atop these convection currents, the plates move slowly, but inexorably, with respect to one another. The spreading process of the past 225 million years has broken up the original mass of Pangaea and brought the continents to their present positions. It also accounts for the other interactions between tectonic plates, such as volcanoes and earthquakes. The

average rate of a plate's movement is about 6 centimeters (2.4 inches) per year, but over 100 million years, this adds up to almost 8,000 kilometers (5,000 miles) in the fastest-moving segments. The plate boundaries are locked by friction and hence are regions of major disturbances.

Adjacent tectonic plates move with respect to each other by separating (as in mid-ocean ridges), sliding past each other (at fault lines like the San Andreas Fault in California), or colliding. These plate movements have large environmental effects. A catastrophic earthquake off the Pacific coast of Tōhoku, Japan, on March 11, 2011, resulted from the Pacific plate sliding under the North American plate, forming a rift under the sea floor 300 kilometers (186 miles) long, pulling eastern Japan closer to North America by 4 meters (13 feet) and even slightly changing Earth's axis. The sudden uplift triggered a powerful **tsunami** (tidal wave) that displaced an enormous quantity of seawater and sent waves up to 10 kilometers (6 miles) inland in the Sendai area of Japan. The tsunami triggered a number of nuclear accidents at the Fukushima II Nuclear Power Plant. The earthquake and tsunami resulted in more than 15,845 deaths, with thousands more missing or injured. Such movements of the crust shape our world. Over geologic time, plate collisions produce volcanic mountain chains and uplift regions into mountain ranges similar to the way the hoods of colliding cars crumple. An example of mountains caused by this type of movement is the Himalayas. The fact that volcanic eruptions and earthquakes continue to occur is evidence that tectonic plates are continuing to move today, as they have over millions of years. In fact, volcanoes and earthquakes mark the boundaries between the plates and have provided some of the best evidence for the plate tectonic theory.

In addition to the periodic catastrophic destruction that may be caused in localized regions by earthquakes and volcanic eruptions, tectonic movement may gradually lead to major shifts in climate in three ways. First, as continents gradually move to different positions on the globe, their climates change accordingly. Second, the movement of continents alters the direction and flow of ocean currents, which in turn has an effect on climate. Third, the uplifting of mountains alters the movement of air currents, which also affects climate. (For example, see the rain-shadow effect in Fig. 10–7.)

Large-scale changes in Earth's crust have been part of evolutionary pressures that have increased diversity. As continents moved to their present position, they isolated organisms into broad regions. Sometimes scientists divide the world into six zoogeographic regions—areas of Earth's surface with distinctive animal groups. Australia, for example, is part of the Australasian region and is home to almost all of the animals with pouches—the marsupials. Australia is located on the Indo-Australian tectonic plate. Its separation from other continents appears to be very ancient, which explains why so many species in the Australasian region are unique: They have had millions of years to evolve separately from species on other continental plates.

Sometimes the place where several tectonic plates meet will have an unusual mix of species. For example, the country

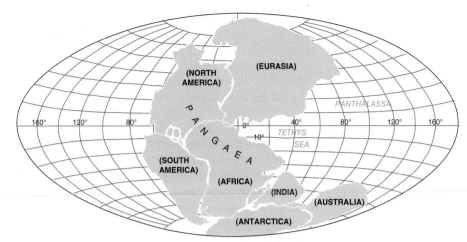

(a) 225 million years ago

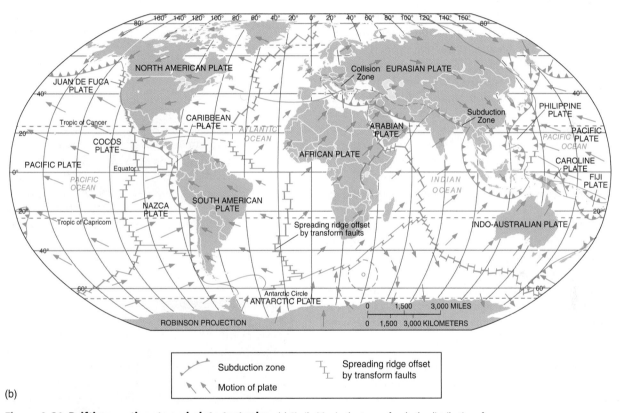

(b)

Figure 4–21 Drifting continents and plate tectonics. (a) Similarities in the types of rock, the distribution of fossil species, and other lines of evidence indicate that 225 million years ago all the present continents were formed into one huge landmass that we now call Pangaea. (b) The 14 major tectonic plates making up Earth's crust and their directions of movement. (*Source:* From *Geosystems: An Introduction to Physical Geography*, 5th ed., by Robert W. Christopherson, copyright © 2005 by Prentice Hall, Inc., reprinted by permission of Pearson Education, Inc., Upper Saddle River, NJ 07458.)

of Indonesia, which consists of about 17,000 islands, is located on the junction of three tectonic plates—the Pacific, Eurasian, and Indo-Australian plates. Papua New Guinea is part of the Indo-Australian plate and has plants and animals similar to those in Australia. Other islands have biota more like those in parts of Asia. The island of Sulawesi lies right on a line between plates and has been formed by islands being pushed together. It is home to an odd combination of animals and plants with relatives in far distant parts of the world.

Because they have been isolated on islands, many Indonesian species have changed over time, developing adaptations not seen in their Asian or Australian relatives. Both Australia, long separated from other continents, and Indonesia have a high number of species that occur nowhere else in the world. Australia and Indonesia serve as examples of how long-term, large-scale changes, such as the movement of continents and the rise of islands, can change the environment for organisms and increase their diversity.

4.5 Implications for Human Management

Ecologists study populations and communities for a number of reasons. One is simply to further our understanding of the world. Another is to better manage natural resources by protecting declining species such as the condor and the golden frog and by controlling pest species such as the chytrid fungus and the spotted knapweed. It is also important for us to study our own impacts on populations, as many of the causes of explosive growth or decline in populations stem from human actions, such as the removal of an important species or the introduction of non-native species (pests will be covered in Chapter 13 and biodiversity in Chapter 6).

The keystone species mentioned in Section 4.3 are such an integral part of their ecosystems that their removal, as we discussed, can cause the collapse of the whole ecosystem. For instance, beavers were once one of the dominant ecological forces in eastern North American, creating openings in forests and wetlands that became habitats for other species and allowed them to survive. The loss of beavers due to the fur industry and habitat destruction removed them from much of the North American landscape, to the detriment of species that rely on small wetland meadows.

Introduced Species

Humans have altered population equilibria by introducing species from foreign ecosystems. These introductions alter community and population ecology relationships such as competition and predation. Over the past 500 years, and especially now that there is vast global commerce, thousands of species of plants, animals, and microbes have been accidentally or deliberately introduced onto new continents and islands. A recent attempt to quantify the economic losses due to introduced species in the United States estimated the total cost to be in excess of $138 billion per year.

Invading Animals. A classic example of the impact an introduced species can have on a native ecosystem comes from Australia, where rabbits were introduced for sport hunting in 1859. Lacking predators, the rabbit population exploded and devastated vast areas of rangeland by overgrazing, damaging the health of both native marsupials and ranchers' sheep. The problem was temporarily brought under control by introducing a disease-causing virus in rabbits; however, the rabbits adapted to the virus and repopulated the area. Removal of rabbits on some islands resulted in restoration of the island vegetation. On the mainland, the rabbit is Australia's most destructive pest animal.

In 2008, researchers on Gough Island in the South Atlantic were shocked to discover one of the biggest reasons for the decline of nesting sea birds: predation by mice. Mice and rats have been introduced to thousands of oceanic islands by ships. Often, the descendants of those first immigrants grow to be much larger than ordinary members of the species. Researchers knew there were many rodents on sea

bird colonies; however, they did not understand the impact of the rodents until they used infrared cameras to capture night images of nesting birds. There they saw the mice directly preying on much larger baby birds. Restoration of sea bird populations requires removal of these mice; such removal has been found to be effective in trial cases (**Fig. 4–22**).

Plants. Introduced species can have a devastating effect on plants as well as animals. Prior to 1900, the dominant tree in the eastern deciduous forests of the United States was the American chestnut, which was highly valued for both its high-quality wood and its prolific production of chestnuts, eaten by wildlife and people alike. In 1904, however, a fungal disease called the chestnut blight was accidentally introduced when some Chinese chestnut trees carrying the disease were planted in New York. The fungus spread through

(a)

(b)

Figure 4–22 Invasive animals. (a) Rodents such as this house mouse often overrun oceanic islands when they are introduced by ships. Unfortunately, they have few predators and often grow larger than on mainlands. Invasive rodents are one of the chief causes of decline for sea birds. (b) This albatross chick is very vulnerable to rodent predators.

Figure 4–23 Introduced plant species. Kudzu overgrowing forests.

the forests, killing nearly every American chestnut tree by 1950. Although oaks filled in where the chestnuts died, the ecological and commercial loss was incalculable. There is hope, however, because researchers have recently crossbred the American and Chinese chestnuts, creating a hybrid that is both 94% native and resistant to the blight. However, it will be several decades before enough of the hybrids are produced to make them available for sale to nurseries and the general public.

There are also examples of introduced plant species in the United States that have caused ecological damage. Kudzu, a vigorous vine introduced in 1876 from Japan for animal fodder and erosion control, is a famous example. Kudzu invaded forests and now occupies more than 3 million hectares (7 million acres) of the Deep South (Fig. 4–23).

Pests. Most of the important insect pests we try to control—Japanese beetles, fire ants, apple snails, and gypsy moths, for example—are species introduced from other continents. The rabbits of Australia are another good example. While they

may not be seen as pests by many people, domestic cats have often proved to be effective predators and have decimated small animal populations where they have been allowed to roam free. They are responsible for greatly diminished songbird populations in urban and suburban areas, including parks (Chapter 6).

The problems of introduced species may also be exported as well as imported. In 1982, several species of jellyfish-like animals known as ctenophores were transported from the East Coast of the United States to the Black Sea and the Sea of Azov in eastern Europe. The ctenophores have cost Black Sea fisheries an estimated $250 million and have completely shut down fisheries in the Sea of Azov because they kill larval fish directly and deprive larger fish of food (Fig. 4–24b).

Lessons. There are two ecological lessons to be learned from the introduction of undesirable species. First, the regulation of populations is a matter of complex interactions among the members of the biotic community. Second, the relationships are specific to the organisms in each particular ecosystem. Therefore, when a species is transported over a physical barrier from one ecosystem to another, it is unlikely to fit into the framework of relationships in the new biotic community. In most cases, it finds the environmental resistance of the new system too severe and dies out. No harm is then done. In some instances, the introduced species simply joins the native flora or fauna and does no measurable harm. The new species has then become naturalized. In the worst cases, the transported species becomes invasive. It finds physical conditions and a food supply that are hospitable, together with an insufficient number of natural enemies to stop its population growth. Then its population explodes and it drives out native species by outcompeting them for space, food, or other resources (if not by predation). Fortunately, only a small percentage of successfully transplanted species become invasive. (Invasive species will be discussed again in Chapters 6 and 7.)

(a)   (b)

Figure 4–24 Introduced species. (a) Zebra mussels introduced from Europe are now spread throughout the Great Lakes and the Mississippi valley. The only benefit from these aliens is that they can dramatically increase water clarity where they are numerous, but that is far outweighed by the harm they do. (b) Ctenophores, originating along the Atlantic Coast in the southern United States and introduced into Europe, have destroyed fishing in the Black Sea and the Sea of Azov.

Remedies. Invasive species are such a concern for the health of ecosystems that there are state and federal agencies working to slow their entry and spread. The National Invasive Species Information Center, for example, acts as a clearinghouse of information on invasive species in the United States. The United States has a number of regulations about invasive species, including the Nonindigenous Aquatic Nuisance Prevention and Control Act of 1990 and the National Invasive Species Act of 1996.

Another solution to the takeover by an invasive species is to introduce a natural enemy. Indeed, this approach has been used in a number of cases, including in that of the spotted knapweed (see Sound Science, "The Biological Detective: The Case of Spotted Knapweed"; other approaches are discussed later, Chapter 13, in connection with the biological control of pests). Unfortunately, control

with a natural enemy is more easily said than done. Recall that control is a consequence of many factors of environmental resistance, which often include several natural enemies as well as *all* the abiotic factors. The natural enemy that is introduced could even end up as an invasive pest itself. In short, to prevent doing more harm than good, a great deal of research needs to be done before a natural enemy is introduced.

Most invasive species are not problems in their native lands. Their impact is different in a new setting, however, because ecosystems on different continents or remote islands have been isolated by physical barriers for millions of years. Consequently, the species within each ecosystem have developed adaptations to other species within their own ecosystem, and these do not prepare them to interact with species that have developed in other ecosystems.

SOUND SCIENCE — The Biological Detective: The Case of Spotted Knapweed

Millions of acres in the western United States are infested with a plant that produces more seeds than do others and is barely edible to grazers such as cattle, whose growth is harmed by the plant's toxins. Spotted knapweed (*Centaurea biebersteinii*), a member of the Aster family, was accidentally introduced into the United States as seeds from Europe in the late 1800s. It thrives here not only because its seeds can remain in the ground for five years, sprouting whenever there is an opening (high biotic potential), but also because it has few natural predators (low environmental resistance).

Many herbivores eat specific plants and have adaptations that allow them to get around the plant's defenses, but when plants are moved from one place to another, their accompanying herbivores are sometimes not moved. Without the limit of environmental resistance, the plant outcompetes other plants for light and water.

For years, scientists searched for a control for knapweed, and in the early 1970s, they thought they had the ideal solution. They introduced a gallfly that specializes in eating spotted knapweed, hoping to avoid the problems that occur when generalist herbivores are moved from one part of the globe to another and eat numerous species. It should have worked. The

gallfly lays its eggs in the knapweed flowers. Larvae hatch and trigger the plant tissues to grow a gall, which protects the insect. The larvae overwinter in the galls and emerge in the spring. Because spotted knapweed plants have to put a great deal of energy into making the woody tissue in the galls, they make fewer seeds. This should have made the population decline, but the decline was not as high as expected. Why? Recently, scientists have discovered part of the answer to the mystery.

In a study reported in 2008 in the journal *Ecological Applications*, researcher Dean Pearson and his colleagues at the U.S. Forest Service's Rocky Mountain Research Station found that part of the answer lies in the relationship between the gallfly and local deer mice.* Gallfly larvae are eaten by native deer mice, which also eat seeds and other insects. Because there are so many spotted knapweed plants, there are plenty of gallflies. Because there are plenty of gallflies, there are plenty of deer mice. But the deer mice also eat the seeds of other native plants. Because the native plants make many fewer seeds than the introduced spotted knapweed, the knapweed seeds remaining can still

*D. E. Pearson and R. M. Callaway, "Weed Biocontrol Insects Reduce Native Plant Recruitment Through Second-Order Apparent Competition," *Ecological Applications* 18 (2008): 1489–1500.

Spotted knapweed

outgrow the native plants. This unexpectedly complex set of relationships means that even a biological control agent chosen because of its specificity to one host may have unintended consequences.

REVISITING THE THEMES

Sound Science

The golden frog of Panama is in decline. The frog's population is controlled by a disease-causing fungus as well as by human impacts like habitat loss and degradation. The chytrid fungus itself is not well controlled because it is an introduced exotic species, occurring in places where organisms have not evolved adaptations to it. This example illustrates some of the ideas of the chapter, such as the limits to population growth, the relationships between parasites (or pathogens) and hosts, the problem of introduced species, and the role humans can play in managing ecosystems.

The science of population and community ecology allows people to better understand why species live where they do and what drives their abundances. Populations can be controlled by abiotic factors such as precipitation or by a complexity of biotic factors such as competition, predation, mutualism, and commensalism—all important relationships. Species typically have one of two reproductive strategies in their life history. Opportunistic or *r*-selected species make many small offspring, reproduce early, and do not care for their young, most of which die. In contrast, equilibrial or *K*-selected species have fewer offspring at older ages and care for the young. These differences correlate to differences in the ways in which populations grow: *r*-selected species often overshoot their carrying capacity and crash, while *K*-selected species are more likely to rise slowly and then fluctuate around the carrying capacity. Either way, all species are limited. When a few individuals start a new population, the growth may initially be exponential, but the population cannot maintain continual growth. Finally, sound science shows us that biological systems change because they are constantly varying genetically and because the environment changes over space and time to exert different selection pressures.

Sustainability

To follow the best management practices for sustainability can sometimes be difficult. This was shown with numerous examples of human actions resulting in negative unexpected consequences, particularly with the introduction of rabbits into Australia and with the attempts to slow the spread of spotted knapweed.

However, in this chapter, we see that best management means paying attention to the relationships in a community that already exist and to constraints such as carrying capacity. The movement of species from a place where they have coevolved in a community to other places lacking those evolutionary relationships was highlighted as a particular problem that often results in either the explosion of an unwanted pest or the collapse of some other species. We have had to develop agencies and policies to minimize problems from the transport of species. We also saw, in the case of the golden frog, that drastic action in the short term may be required to protect species until longer-term solutions can be found.

Stewardship

When questions arise about what's best for populations and communities, the answers indicate that what seems to benefit one species (importing a plant or animal, for example) may not benefit another. A species decline can result from activities that disrupt its environment on scales of time or space that it cannot adjust to. This is definitely the case with the golden frog, which has had to adapt to too many environmental changes in a short time span. The destruction of its forest habitat, for example, has benefited human enterprises but has also harmed other populations. Human impacts are not always to blame, however. Disruptive changes can occur simply through natural shifts in an ecosystem (such as a forest fire). One might ask, "If change is natural, why are we concerned about species loss?" (This will be discussed further in Chapter 6). Throughout much of this book, we will ask how we can continue to support the needs of people for agriculture and other activities, while still maintaining healthy community structures in our ecosystems.

This chapter also shows that interactions such as predation and competition can be positive parts of ecosystems even when we might have a gut feeling that they are negative. They act as checks and balances on populations. For example, we saw that natural predators and competition keep species in check in the places where they have evolved in a community. Likewise, it might seem like territoriality would be a bad thing, but these behaviors actually lower competition and ensure that some members of the species survive.

REVIEW QUESTIONS

1. What are three basic population growth curves? When might you see them in nature?

2. Define *biotic potential* and *environmental resistance*, and give factors of each. Which generally remains constant, and which controls a population's size?

3. Differentiate between the terms *critical number* and *carrying capacity*. What is density dependence?

4. Explain the difference between *r*- and *K*-strategists. Where do these terms come from, and what are the characteristics of each broad category?

5. Describe the predator-prey relationship between the moose and wolves of Isle Royale. What other factors influence these two populations?

6. Distinguish between intraspecific and interspecific competition. How do they affect species as a form of environmental resistance?

7. What is meant by territoriality, and how does it limit the effects of competition in nature?

8. What problems arise when a species is introduced from a foreign ecosystem? Why do these problems occur?

9. What are selective pressures, and how do they relate to natural selection?

10. Describe several types of adaptations a species might have that would allow it to survive in a dry environment, escape predation, or lower competition with another species.

11. What factors determine whether a species will adapt to a change or whether the change will render it extinct?

12. How may evolution lead to the development of new species (speciation)?

13. What is plate tectonics, and how does this theory explain past movements of the continents? How have past tectonic movements affected the present-day distribution of plant and animal species on Earth's surface?

THINKING ENVIRONMENTALLY

1. Describe, in terms of biotic potential and environmental resistance, how the human population is affecting natural ecosystems.

2. Explain the key differences between *r*- and *K*-strategists, their life histories, and their reactions to environmental changes. Where do the *r* and *K* come from in population equations? What do they stand for?

3. Consider the various kinds of relationships humans have with other species, both natural and domestic. Give examples of relationships that (a) benefit humans but harm other species, (b) benefit both humans and other species, and (c) benefit other species but harm humans. Give examples in which the relationship may be changing—for instance, from exploitation to protection. Discuss the ethical issues involved in changing relationships.

4. Consider the research of the Grants over decades of work on the Galápagos. Why is such research important? How does this basic science benefit us as we try to protect ecosystem functions?

5. Describe how a human action such as removing a top predator or adding a species can have an impact on many species in an ecosystem.

MAKING A DIFFERENCE

1. If you can, plant native plants instead of alien ones in your garden. Otherwise, seeds from introduced plants could escape into the surrounding ecosystems. Native grasses, flowers, shrubs, and trees are more likely to attract native birds, butterflies, and other insects.

2. Visit a nearby national park or nature reserve. Talk to the rangers to find out about any threatened species and how they are being protected. Get involved by volunteering at your local nature center or wildlife refuge. Go wildlife or bird watching in nearby parks. Wildlife-based recreation creates millions of jobs and supports local businesses.

3. If you are a hunter or fisherman, use non-lead shot and fishing gear so birds eating spent shot or lost sinkers will not suffer lead poisoning. Follow the regulations regarding seasons and catch limits, and report poachers to your local wildlife agency.

4. When choosing outdoor recreational activities, consider cross-country skiing and canoeing instead of snowmobiling or motor boating. These activities are quieter, and they increase your chances of seeing wildlife. Loud noises, especially in winter, disturb animals when they need to rest and conserve energy.

MasteringEnvironmentalScience®

Go to **www.masteringenvironmentalscience.com** for practice quizzes, Pearson eText, videos, current events, and more.

CHAPTER 5

Ecosystems:
Energy, Patterns, and Disturbance

Volcano. Iceland's Eyjafjallajökull volcano shut down airspace for days in April 2010, injecting millions of tons of ash into the air but did not kill people or spew lava as many volcanoes do.

LEARNING OBJECTIVES

1. **Characteristics of Ecosystems:** Describe how matter and energy flow through ecosystems by moving from one trophic level to another.

2. **The Flow of Energy Through the Food Web:** Explain three main ideas relating trophic pyramids with the relative numbers and biomasses of different levels in a food chain.

3. **From Ecosystems to Global Biomes:** Define and recognize characteristics of major broad regions called biomes, major aquatic regions, and factors that determine their placement on the globe.

4. **Ecosystem Reponses to Disturbance:** Explain the effects of ecological disturbances, such as a fire or volcanic eruption, which are normal in ecosystems and can even be beneficial.

5. **Human Values and Ecosystem Sustainability:** Describe ways that humans alter ecosystem services both positively and negatively, and explain why we need to manage ecosystems to protect their components from overuse.

CLOUDS OF SMOKE and ash accompanied rumblings of the earth as Iceland's volcano at Eyjafjallajökull ("EY-ya-fyat-lah-YOH-kuht") belched awake on April 14, 2010. The tremblors that had affected the area over the previous month increased, and a billowing cloud of ash erupted into a plume that rose 9 kilometers (30,000 feet) into the air. The ash spread across Western Europe, clouding the sky and closing down all air travel for days, the largest air traffic disruption since World War II. Thousands of travelers were stranded as the volcano spewed particles and gases into the atmosphere. Ash fell on the ground for miles, making the grasses unfit for consumption by cattle and horses and making driving impossible. Even so, Eyjafjallajökull was a moderate volcano, without the massive lava flows and mudslides that accompany many volcanic eruptions. People were relatively safe and, except for the cessation of air travel, life returned to normal quickly.

PREPARATION. The eruption of Mount Pinatubo on June 15, 1991, in the Philippines was larger and had more-serious global effects (Fig. 5–1). The eruption of this volcano, with no recorded history of eruptions, blew the top of the densely forested mountain into the air. This was the beginning of a process that spilled between 6 and 16 cubic kilometers of ash

(a)

(b)

Figure 5–1 Recovery from volcanic eruption. (a) The eruption of Mt. Pinatubo in 1991. (b) Changes in the ecosystem from that eruption continued even 15 years later. Sections of the hillsides remain denuded as intermittent landslides removed vegetation.

into the atmosphere, along with 20 million tons of sulfur dioxide and 10 billion metric tons of magma. Small particles known as aerosols were thrown into the air and traveled around the globe, blocking sunlight and lowering the global temperature by 0.5°C (0.9°F). More aerosols were produced by this volcanic eruption than any since Krakatoa in 1883. The sulfur found in the aerosols combined with water to form sulfuric acid, a component of acid rain. Fortunately, even though the volcano had not

erupted in years prior to the event, scientists were able to predict that it would, as seismic activity picked up in the months before the eruption, and they successfully prompted the largely successful evacuation of 60,000 people. Unfortunately, people did die in the aftermath, when typhoon rains and wet ash combined to collapse roofs and from illness due to crowded emergency housing. Today, years after the eruption, parts of Mount Pinatubo still lie denuded as periodic landslides move material down the hillside.

REBUILDING. Volcanoes, triggered by the movements of tectonic plates, relieve pressure building under the earth's crust and spread minerals into the soil. They may build new lands altogether by causing islands, such as the Hawaiian Islands, to rise out of the ocean. Often, volcanic eruptions kill many of the organisms living in their immediate vicinity, but such organisms return and recolonize in a process of succession. In this chapter, we will look at ecosystems and how they function, including the processes that occur after a disturbance like a wild fire, large flood, volcano, or retreat of a glacier. These processes offer us a chance to better understand the workings of an ecosystem.

Earlier, the focus was on the components of ecosystems (Chapters 3 and 4). We considered the environmental conditions and resources required by living things, the basic movement of energy and nutrients that sustain living things, and how populations work and relate to populations of other species. Now, we will look at how basic processes such as photosynthesis and respiration in individuals play out in complex systems and how energy moves from one part of the community to another. We will see how concepts like ranges of tolerance and limiting factors, which we applied to individuals, apply to large groups and control their distributions over large portions of Earth (Chapter 3). We will also look at how ecosystems change over time as a result of disturbances such as volcanoes. We'll see how vital ecosystems are to human well-being and look at efforts to assign value to the goods and services we derive from them. We will also see that ecosystem sustainability is being severely challenged by human use.

It is a popular belief that nature will be in balance if left alone. You will learn in this chapter, however, that an ecosystem is a dynamic system in which changes are constantly occurring. The notion of a balanced ecosystem, then, must be carefully defined. We will ask questions about balance, or equilibrium, in ecosystems: What does it mean for an ecosystem to be "in balance"? Are populations normally in a state of equilibrium? What happens when a major disturbance (like a fire) interrupts the "balance" in an ecosystem? Can ecosystems and the natural populations in them change over time yet maintain the important processes that make them sustainable?

Our objective in this chapter is to examine more of the basic mechanisms that underlie the sustainability of all ecosystems, including those heavily influenced by human use. We will look at the values, goods, and services of ecosystems and their importance to humans. The chapter concludes with some perspectives on the challenge of managing ecosystems and a summary of ecosystem sustainability issues.

5.1 Characteristics of Ecosystems

Mt. Pinatubo is located on the island of Luzon in the Philippines, an area with one of the largest forests in the country and home to some of the few remaining individuals of the national bird, the highly endangered Philippine eagle. We will look at this area to better understand the characteristics of ecosystems. The forest around Mt. Pinatubo, like all ecosystems, contains communities of interacting species and their abiotic factors. However, the species living in this ecosystem, or any other ecosystem, may function on very different scales, which makes it difficult to delineate a fixed boundary for ecosystems. For example, the lowland rain-forest ecosystem contains migratory birds that live there only seasonally. Some species inhabiting the ecosystem may also have little to no interaction with each other, such as the orchids high on branches of trees and organisms that live in leaf litter.

The rain forest around Mt. Pinatubo contrasts sharply with the treeless spaces of Iceland, where Eyjafjallajökull sent ash into the air. As we will see later in the chapter, what lives in an ecosystem is a function not only of characteristics of the regional climate, but of the landscape's history—including the actions of humans. Even so, there are predictable patterns to the distribution of ecosystems around the globe. Ecosystems that have a similar type of vegetation and similar climactic conditions are grouped into broader areas called **biomes**, such as the tropical rain-forest biome. (We will look at biomes in more depth in Section 5.4.)

Just as the study of communities raises questions about relationships that cannot be answered by looking at species separately, the study of ecosystems allows us to understand concepts that cannot be predicted by understanding community interactions alone. For example, community relationships such as predation and parasitism (discussed in Chapter 4) provide the mechanisms for the flow of energy—as well as the flow of carbon, nitrogen, and phosphorus—through the natural world (discussed in Chapter 3). But these community relationships may not tell us about ecosystem properties, such as how much photosynthesis occurs in all plants or how much water can be absorbed by all of the plants at once. In this chapter, we will pull together the concepts presented in the previous two chapters, abiotic forces and biotic interactions, to look at characteristics of ecosystems—specifically, **trophic levels**, **productivity**, and **consumption**.

Trophic Levels, Food Chains, and Food Webs

Imagine a grassy meadow containing a variety of plants. As discussed in previous chapters, plants use energy from the Sun through the process of photosynthesis to produce complex chemical chains from more-simple molecules such as carbon dioxide and water. These plants can then be eaten by their predators, such as grasshoppers, mice, or even elk (remember that herbivory is a subset of predation). In turn, the grasshopper could be eaten by its predator, a bird; the mouse eaten by an owl; and the elk fed on by a parasite or eaten by a wolf. Energy in the form of chemical bonds and nutrients in the form of molecules in the bodies of the prey are then used by the next **trophic level** (feeding level)—the predators—for energy and other physiological needs. This breakdown of chemical bonds to release energy is called cellular respiration. Recall that plants need to respire, too.

You can imagine two things from this. First, organisms can be linked into feeding chains—the grass to the mouse, to the snake, to the eagle, for example. Such **food chains** describe where the energy and nutrients go as they move from one organism to another. We usually refer to this as energy moving "up" the food chain. Second, you can imagine that the energy and nutrients are not all passed up the chain from one level to the next. Because of the energy laws we discussed, there is inefficiency every time one organism eats another (Chapter 3). While it is interesting to trace these feeding pathways, it is important to recognize that food chains seldom exist as isolated entities. Mice feed on several kinds of plants, are preyed on by several kinds of animals, and so on. Some organisms, including humans, eat at more than one level in the food chain. Consequently, virtually all food chains are interconnected and form a complex web of feeding relationships—the **food web** (Fig. 5–2).

Rather than simply asking what predators are eating snakes and how they affect the snake population or what mice are eating and how food availability affects mice, we ask different types of questions in ecosystem ecology—those that focus on trophic levels in general, in order to study bigger trends in the movement of energy and materials. Examples of these questions include the following: How much of the Sun's energy do the plants in the meadow trap? How much do they use, and how much is available to plant eaters? How many predators can the ecosystem support? These questions emphasize the broader aspects of the flow of energy and materials through ecosystems, not through individual species. In order to answer these questions, ecologists divide members in the food web into categories.

Trophic Categories

All organisms in the biosphere can be categorized as either autotrophs or heterotrophs, depending on whether they produce the organic compounds they need to survive and grow. Green plants, photosynthetic single-celled organisms, and chemosynthetic bacteria are **autotrophs** (*auto* = self; *troph* = feeding) because they produce their own organic material from inorganic constituents in their environment through the use of an external energy source. They are also referred to as producers. Organisms that must consume organic material to obtain energy are **heterotrophs** (*hetero* = other; *troph* = feeding). Heterotrophs may be divided into numerous categories, the two major ones being consumers (which eat living prey) and **decomposers** (which break down dead organic material and can include scavengers, detritus feeders, and chemical decomposers). Together, producers, consumers, and decomposers produce food, pass it along

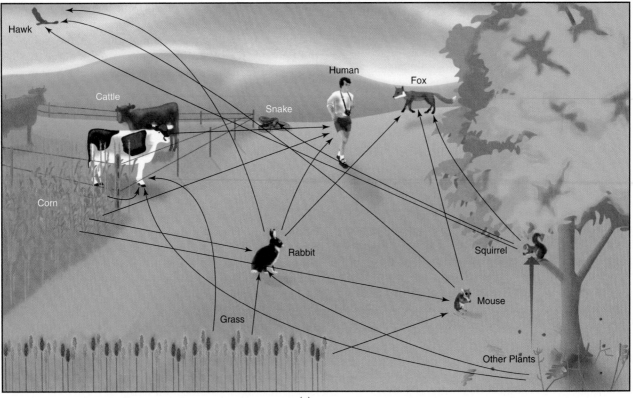

(a)

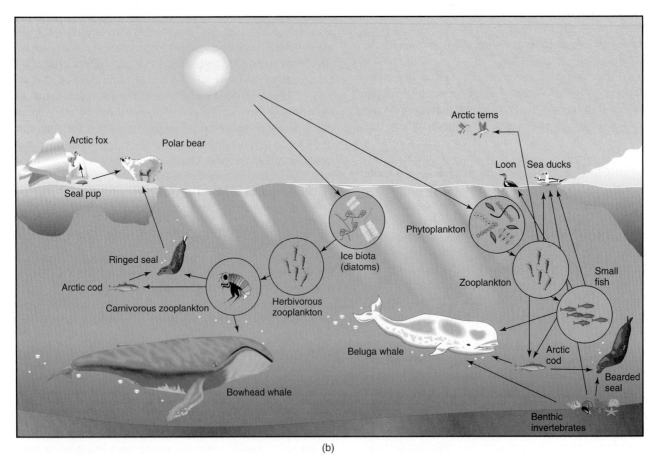

(b)

Figure 5–2 Food webs. (a) The arrows show herbivores feeding on plants and carnivores feeding on herbivores. Specific pathways, such as that from nuts to squirrels to foxes, are referred to as *food chains*. A food web
is the collection of all food chains, which are invariably interconnected. (b) The marine food web in the offshore
Beaufort Sea (north of Alaska and Canada) in spring.

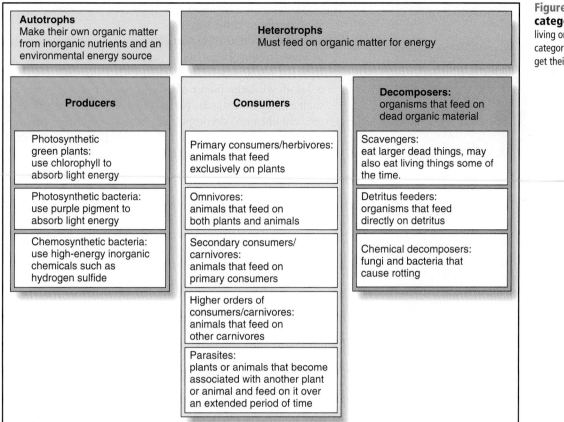

Figure 5–3 Trophic categories. A summary of how living organisms are ecologically categorized according to how they get their energy.

food chains, and return the starting materials to the abiotic parts of the environment **(Fig. 5–3)**.

Producers. As we have seen, producers are organisms that capture energy from the Sun or from chemical reactions to convert carbon dioxide (CO_2) to organic matter (Chapter 3). Most producers are green plants. Through the process of photosynthesis, these plants use light energy to convert CO_2 and water to organic compounds such as glucose (a sugar) and then release oxygen as a by-product. Plants use a variety of molecules to capture light energy in photosynthesis, but the most predominant of these is **chlorophyll**, a green pigment. Hence, plants that photosynthesize are easily identified by their green color. In some producers, additional red or brown photosynthetic pigments (in red and brown algae, for example) may mask the green. Producers range in diversity from microscopic photosynthetic bacteria and single-celled algae through medium-sized plants such as grass, daisies, and cacti to gigantic trees. Every major ecosystem, both aquatic and terrestrial, has its particular producers that are actively engaged in photosynthesis.

Interestingly, there are bacteria, including some in the hot springs at Yellowstone National Park, that are able to use the energy in some inorganic chemicals to form organic matter from CO_2 and water. This process is called **chemosynthesis**, and the organisms using it are producers, too. Producers are absolutely essential to every ecosystem. The photosynthesis and growth of producers such as green plants constitute the **primary production** of organic matter, which sustains all other organisms in the ecosystem.

Consumers. All organisms other than producers in an ecosystem feed on organic matter as their source of energy. These heterotrophs include not only all animals, but also fungi (mushrooms, molds, and similar organisms) and most bacteria. For ease of discussion, heterotrophs are often divided into consumers, which eat living prey, and decomposers, which break down dead organic material. Consumers encompass a wide variety of organisms, ranging in size from plankton to blue whales. Among consumers are such diverse groups as protozoans, worms, fish, shellfish, insects, reptiles, amphibians, birds, and mammals (including humans).

For the purpose of understanding ecosystem structure, consumers are divided into various subgroups according to their food source. Animals—as large as elephants or as small as mites—that feed directly on producers are called **primary consumers** or **herbivores** (*herb* = grass). Animals that feed on primary consumers are called **secondary consumers**. Thus, mice, which feed on vegetation, are primary consumers, whereas snakes are secondary consumers because they feed on mice **(Fig. 5–4)**. There may also be third (tertiary), fourth (quaternary), or even higher levels of consumers. Many animals occupy more than one position on the consumer scale. For instance, humans are primary consumers when they eat vegetables, secondary consumers when they eat beef, and tertiary consumers when they eat fish that feed on smaller fish that feed on algae. Secondary and higher-order consumers are also called **carnivores** (*carni* = meat). Consumers that feed on both plants and animals are called **omnivores** (*omni* = all).

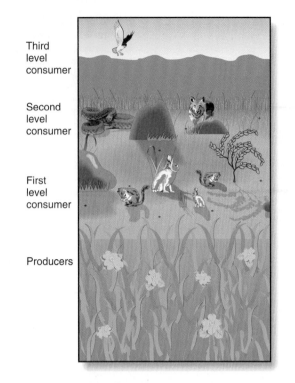

Third level consumer

Second level consumer

First level consumer

Producers

Figure 5–4 A grassland food chain. A grassland food chain, showing the producers (plants) and the first-, second-, and third-order consumers (mouse, snake, and hawk, respectively).

Decomposers. A very large proportion of the primary producer trophic level is not consumed in the grazing food web. As this material dies (leaves drop, grasses wither, and so on), it is joined by the fecal wastes and dead bodies from higher trophic levels and is referred to as *detritus*. This represents the starting point for a separate food web. In many cases, most of the energy in an ecosystem flows through the detritus food web **(Fig. 5–5)**.

Detritus is composed largely of cellulose because it consists mostly of dead leaves, the woody parts of plants, and animal fecal wastes. Nevertheless, it is still organic and high in potential energy for those organisms that can digest it—namely, decomposers. Scavengers, such as vultures, help break down large pieces of organic matter; detritus feeders, such as earthworms, eat partially decomposing organic matter; and chemical decomposers (various species of fungi and bacteria as well as a few other microbes) break down dead material on the molecular scale. Decomposers act as any other consumer does, using the organic matter as a source of both energy and nutrients. Some, such as termites, can digest woody material because they maintain decomposer microorganisms in their guts in a mutualistic, symbiotic relationship. The termite (a detritus feeder) provides a cozy home for microbes (chemical decomposers) and takes in cellulose, which the microbes digest for both their own and the termites' benefit **(Fig. 5–6)**. The decomposers, breaking down detritus, serve as the primary consumers in this food chain. The detritus is similar to the "producer" trophic level, and the decomposers have their own predators that form higher trophic levels.

Most decomposers use oxygen for cell respiration, which breaks detritus down into carbon dioxide, water, and mineral nutrients. Likewise, there is a release of waste heat, which you may observe as the "steaming" of a manure or compost pile on a cold day. The release of nutrients by decomposers is vitally important to the primary producers because it is the major source of nutrients in most ecosystems. However, some decomposers (certain bacteria and yeasts) meet their energy needs through the partial breakdown of glucose, which can occur in the absence of oxygen. This modified form of cell respiration, called **fermentation,** results in end products such as ethyl alcohol, methane gas,

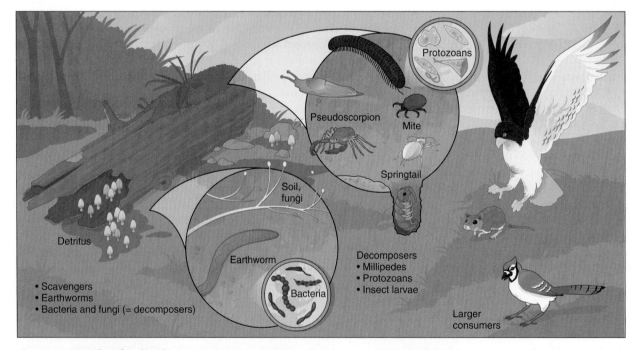

Protozoans

Pseudoscorpion

Mite

Springtail

Soil, fungi

Earthworm

Bacteria

Detritus

• Scavengers
• Earthworms
• Bacteria and fungi (= decomposers)

Decomposers
• Millipedes
• Protozoans
• Insect larvae

Larger consumers

Figure 5–5 Detritus food web. Several types of decomposers work to break up large, small, and tiny bits of dead organic matter. These organisms in turn are eaten by other, higher-level consumers. In the detritus food web, the role of the producer is played by dead plant and animal matter.

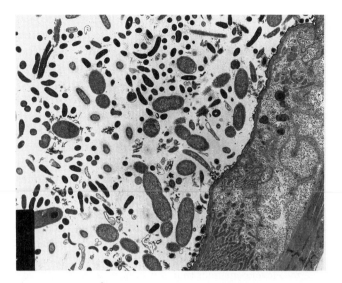

Figure 5–6 Termite gut. Termites can live on cellulose-based woody matter because their guts contain a consortium of symbiotic microbes able to digest cellulose. In this electron micrograph, the symbionts are on the left and the termite intestinal wall is on the right.

and acetic acid. In nature, **anaerobic**, or oxygen-free, environments commonly exist in the sediments of lakes, marshes, and swamps and in the guts of animals, where oxygen does not penetrate readily. Methane gas is commonly produced in these locations. A number of large grazing animals, including cattle, maintain fermenting bacteria in their digestive systems

in a mutualistic, symbiotic relationship similar to that just described for termites. As a result, both cattle and termites release methane.

Limits on Trophic Levels. How many trophic levels are there? Usually, there are no more than three or four in terrestrial ecosystems and sometimes five in marine systems (Fig. 5–2). This answer comes from straightforward observations. If you were to capture all of the organisms on each trophic level over a particular area or volume, dry them, and weigh them, you would find a pattern: The **biomass**, or total combined (net dry) weight, would be roughly 90% less at each higher trophic level for a terrestrial ecosystem **(Fig. 5–7)**. For example, if you dried and weighed all of the grass and other producers in an acre of grassland, you might have 907 kilograms (1 ton or 2,000 lbs) per acre. If you could capture, dry, and weigh all of the herbivores (everything from grasshoppers to bison) in the same acre, you would find the biomass of herbivores would be about 90.7 kilograms (200 lbs) per acre. Following that pattern, the same acre would yield about 9.7 kilograms (20 lbs) of primary carnivores (animals living on grasshoppers and bison) if you captured, dried, and weighed them. At this rate, you can't go through very many trophic levels before the biomass approaches zero. If you graphed the different levels of producer and consumer mass, you would have what is commonly called a **biomass pyramid** (Fig. 5–7). We'll see the reason for this pattern as we look at where energy and nutrients go.

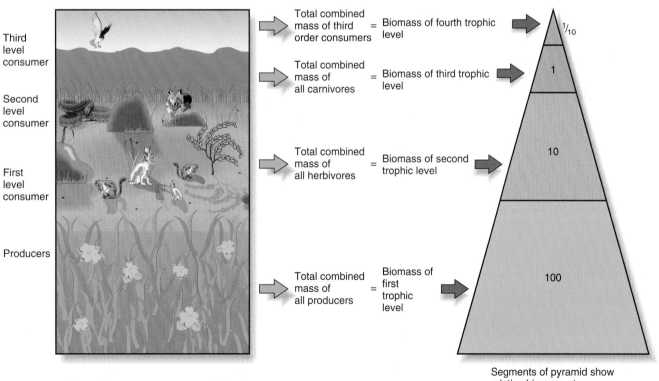

Segments of pyramid show relative biomass at each trophic level

Figure 5–7 A biomass pyramid. A graphical representation of the biomass (the total combined mass of organisms) at successive trophic levels has the form of a pyramid, with about 10% of the energy contained in one trophic level being incorporated into the bodies of the next trophic level, giving each level a smaller biomass.

5.2 The Flow of Energy in Ecosystems

In most ecosystems, sunlight (or solar energy) is the initial source of energy absorbed by producers through the process of photosynthesis. The only exceptions are rare ecosystems, such as those near the ocean floor, where the producers are chemosynthetic bacteria. Primary production—the production of organic molecules from the Sun's energy through photosynthesis—captures only about 2%, at most, of incoming solar energy. Even though this seems like a small fraction, the resulting terrestrial net production—estimated at some 120 gigatons (billion metric tons, 1.1 billion U.S. tons) of organic matter per year—is enough to fuel all of the life in a biome. In a given ecosystem, the actual biomass of primary producers at any given time is referred to as the standing-crop biomass. Both biomass and primary production vary greatly in different ecosystems. For example, a forested ecosystem maintains a very large biomass compared with tropical grassland because so much carbon is captured in the wood of trees. However, the rate of primary production could be higher in the grassland; grazing animals might simply eat the produced organic matter. Consequently, standing biomass is not always a good measure of productivity.

The Fate of Food

Whereas 60–90% of the food that consumers eat, digest, and absorb is oxidized (burned) for energy, the remaining 10–40%, which is converted to the body tissues of the consumer, is no less important. This is the fraction that enables a body to grow, maintain, and repair itself. A portion of what is ingested by consumers is not digested, but simply passes through the digestive system and out as fecal wastes. For consumers that eat plants, this waste is largely **cellulose**, the material of plant cell walls. It is often referred to as fiber, bulk, or roughage, and some of it is a necessary part of the diet. The intestines need to push some fiber through them so that they can stay clean and open. Waste products can also include compounds of nitrogen, phosphorus, and any other elements present, in addition to the usual carbon dioxide and water. These by-products are excreted in the urine (or as similar waste in other kinds of animals) and returned to the environment. This process was illustrated previously (Chapter 3, Figure 3–17).

Energy Flow and Efficiency

As the biomass pyramid suggests, when energy flows from one trophic level to the next, only a small fraction is actually passed on. This is due to three things: (1) Much of the preceding trophic level is biomass that is not consumed by herbivores, (2) much of what is consumed is used as energy to fuel the heterotroph's cells and tissues, and (3) some of what is consumed is undigested and passes through the organism as waste (e.g., feces). Thus, there is a tremendous inefficiency at each trophic level, and energy is lost to the food web. In an ecosystem, therefore, only the portion of food that becomes the body tissue of the consumer can become food for the next organism in the food chain. Incorporating matter and energy from a lower trophic level into the body of a consumer is often referred to as secondary production, and like primary production, it can be expressed as a rate (the amount of growth of the consumer, or consumer trophic level) over time.

The inefficiency of trophic levels means two things. First, individuals at higher levels in the biomass pyramid represent a greater amount of the Sun's energy for the same amount of body tissue. For example, a pound of hamburger requires a great deal more photosynthesis to create than a pound of soybeans. This may be counterintuitive, because we just saw that there is less energy at each trophic level, but it requires a great deal more energy to produce a top-order consumer than a mid-level consumer and, likewise, more energy to produce a mid-level consumer than a producer of the same body weight. It usually also takes a longer time to produce a top-order consumer than a mid-level consumer or a producer. The top-order consumer also requires more water and other resources for a unit of body weight. In later chapters, as we look at food production in a crowded world, we will see why this matters.

Second, some materials are difficult to get rid of (such as chemicals that dissolve in fat) and therefore remain throughout an organism's life. These chemicals remain in the bodies of predators at higher rates than in their prey and thus biomagnify or **bioaccumulate** (build up in the tissues) as you go up the food chain. This is the reason, as we will see later, that some toxins build up in the tissues of higher-order consumers such as eagles and tuna.

Aquatic Systems

For simplicity, the focus of this chapter has been on terrestrial ecosystems. Keep in mind, though, that exactly the same processes occur in aquatic ecosystems. As aquatic plants and algae absorb dissolved carbon dioxide and mineral nutrients from the water, they use photosynthesis to produce food and dissolved oxygen that sustain consumers and other heterotrophs. Likewise, aquatic heterotrophs return carbon dioxide and mineral nutrients to the aquatic environment.

There are two differences between aquatic and terrestrial systems in this regard, however. First, the transfer of energy is often more efficient in an aquatic system. There are also more cold-blooded animals here, and less energy is required to run the bodies of cold-blooded creatures than warm-blooded creatures. It also takes less energy to support body weight in water than on land or in the air. Consequently, when less energy is used at each level, more energy is available to the next level, and food chains can be longer.

Second, aquatic systems do not always result in the same kind of biomass pyramid as terrestrial systems. If you went out to a lake and captured all of the algae, all of the zooplankton eating the algae, all of the fish that live on

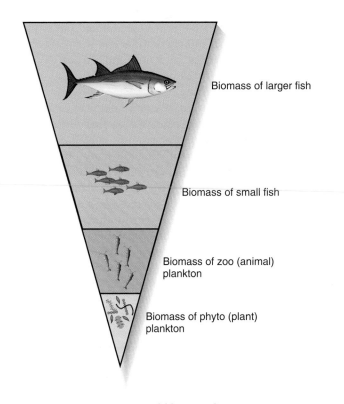

Figure 5–8 Reverse pyramid in aquatic systems. In aquatic systems such as this lake, there is a small amount of algae and a much greater biomass of large fish at any one moment. This is because the turnover is very high and life spans are so short at the lower trophic levels. The longer life spans of the upper trophic levels mean that the biomass of a fish, for example, includes all of its growth over a period of time.

zooplankton, and all of the fish that live on fish, and if you then dried and weighed them in their group, you might find a reverse biomass pyramid **(Fig. 5–8)**. The reason for this upside-down pyramid is that the fish that live on fish are the oldest and largest and the algae have very short life spans and turn over rapidly. Thus, there might be more biomass in the fish, but they had to eat many times that amount of algae to form that tissue. The algae actually in the water at any one time can be small, but they have an extremely rapid rate of new production.

5.3 From Ecosystems to Global Biomes

Just as photosynthesis processes and basic energy laws translate into broad patterns in food chain length in an ecosystem, broad patterns in ecosystems translate into a pattern we see all over the globe: There is a predictable set of organisms that live under particular conditions, such as deserts occurring in dry areas and rain forests occurring in very wet areas, for example. We can now use the concepts of optimums and limiting factors and the concepts of differences in light and productivity to gain a better understanding of why different

regions (or even localized areas) may have distinct biotic communities, creating an amazing variety of ecosystems, landscapes, and biomes.

A **biome** can be defined as a large geographical biotic community, controlled by climate. They are usually named after the dominant type of vegetation, such as deciduous forest or grassland. While the Greater Yellowstone region is an ecosystem (and possibly could be viewed as more than one), it is contained within a vast geographical region, or global biome, called the northern temperate forest. Like ecosystems, biomes have fuzzy edges. Aquatic areas such as the tidal zone and the open ocean can also be characterized by temperature and light and divided into categories. These are not typically called biomes, but function similarly.

The Role of Climate

The climate of a given region is a description of the average temperature and precipitation—the weather—that may be expected on each day throughout the entire year. Climates in different parts of the world vary widely. Equatorial regions are continuously warm, with high rainfall and no discernible seasons. Above and below the equator, temperatures become increasingly seasonal (characterized by warm or hot summers and cool or cold winters); the farther we go toward the poles, the longer and colder the winters become, until at the poles it is perpetually cold. Likewise, colder temperatures are found at higher elevations, so that there are even snowcapped mountains on or near the equator **(Fig. 5–9)**.

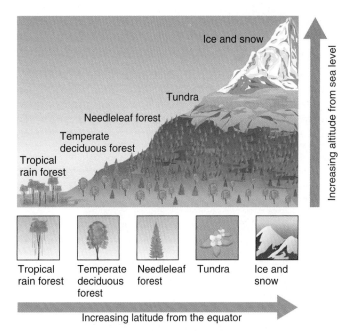

Figure 5–9 Effects of latitude and altitude. Decreasing temperatures, which result in the biome shifts depicted, occur with both increasing latitude (distance from the equator) and increasing altitude. (*Source:* Redrawn from *Geosystems*, 5th ed., by Robert W. Christopherson. Copyright © 2005 by Pearson Prentice Hall, Inc., Upper Saddle River, NJ 07458.)

Annual precipitation in an area also may vary greatly, from virtually zero to well over 250 centimeters (100 in.) per year. Precipitation may be evenly distributed throughout the year or concentrated in certain months, dividing the year into wet and dry seasons. A given climate will support only those species that find the temperature and precipitation levels within their ranges of tolerance. Population densities will be greatest where conditions are optimal and will decrease as any condition departs from the optimum (Chapter 3, Fig. 3–7). A species will be excluded from a region (or local areas) where any condition is beyond its limit of tolerance. How will this variation affect the biotic community?

Biome Examples. Individual ranges of tolerance, particularly for temperature and precipitation, determine where dominant species can live. In turn, the distribution of these species describes the placement of the biomes. To illustrate, let us consider six major types of biomes and their global distribution. Table 5–1 describes these terrestrial biomes and their major characteristics, and **Figure 5–10** shows the distribution of the biomes as they occur globally. Within the temperate zone (between 30° and 50° of latitude), the amount of rainfall is the key limiting factor. The **temperate deciduous forest** biome is found where annual precipitation

is 75–200 centimeters (30–80 in.). Where rainfall tapers off or is highly seasonal (25–150 cm or 10–60 in. per year), the **grassland** and **prairie** biomes are found, and regions receiving an average of less than 25 centimeters (10 in.) per year are occupied by a **desert biome**.

The effect of temperature, the other dominant parameter of climate, is largely superimposed on that of rainfall. That is, 75 centimeters (30 in.) or more of rainfall per year will usually support a forest, and generally temperature will determine the type of forest that grows there. For example, broad-leaved evergreen species, which are extremely vigorous and fast growing but cannot tolerate freezing temperatures, predominate in the **tropical rain forest.** By dropping their leaves and becoming dormant each autumn, deciduous trees are well adapted to freezing temperatures. Therefore, wherever rainfall is sufficient, deciduous forests predominate in temperate latitudes. Most deciduous trees, however, cannot tolerate the extremely harsh winters and short summers that occur at higher latitudes and higher elevations. These northern regions and high elevations are occupied by the **coniferous forest biome**, because conifers are better adapted to those conditions.

Temperature by itself limits forests only when it becomes low enough to cause **permafrost** (permanently frozen subsoil). Permafrost prevents the growth of trees, because

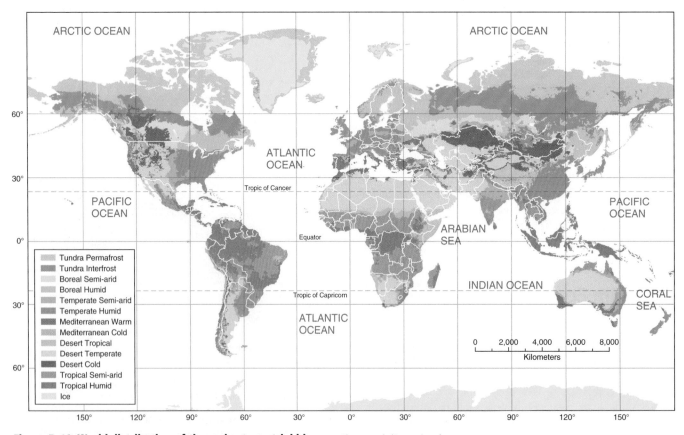

Figure 5–10 World distribution of the major terrestrial biomes. Photos and climate data for these biomes are maintained by the National Aeronautics and Space Administration's Earth Observatory.

USDA Natural Resources Conservation Service

roots cannot penetrate deeply enough in these conditions to provide adequate support. However, a number of grasses, clovers, and other small flowering plants can grow in the topsoil above permafrost. Consequently, where permafrost sets in, the coniferous forest biome gives way to the **tundra biome** (Table 5–1). At still colder temperatures, the tundra gives way to permanent snow and ice cover.

The same relationship of rainfall effects being primary and temperature effects secondary applies in deserts. Any region receiving less than about 25 centimeters (10 in.) of rain per year will be a desert, but the unique plant and animal species found in hot deserts are different from those found in cold deserts. A summary of the relationship between biomes and temperature and rainfall conditions is given in **Figure 5–11**. The average temperature for a region varies with both **latitude** (distance from equator) and **altitude** (distance from sea level), as shown in Figure 5–9.

Examples of Aquatic Systems. The biome category is reserved exclusively for terrestrial systems. However, there are major categories of aquatic and wetland ecosystems similar to biomes that are determined primarily by the depth, salinity, and permanence of water in them. Among these ecosystems are lakes, marshes, streams, rivers, estuaries, bays, and ocean systems. As units of study, these aquatic systems may be viewed as ecosystems, as parts of landscapes, or as major

TABLE 5–1 Major Terrestrial Biomes

Biome	Climate and Soils	Dominant Vegetation	Dominant Animal Life	Geographic Distribution
Deserts	Very dry; hot days, cold nights; rainfall less than 10 in. (25 cm)/yr; soils thin and porous	Widely scattered thorny bushes and shrubs, cacti	Rodents, lizards, snakes, numerous insects, owls, hawks, small birds	North and southwest Africa, parts of the Middle East and Asia, Southwestern North America
Grasslands and Prairies	Seasonal rainfall, 10 to 60 in. (25–152 cm)/yr; fires frequent; soils rich and often deep	Grass species, both tall and short; bushes and woodlands in some areas	Large grazing mammals (bison, wild horses; kangaroos); prairie dogs, coyotes, lions; termites important	Central North America, central Asia, subequatorial Africa and South America, southern India, northern Australia
Tropical Rain Forests	Nonseasonal; annual average temperature 28°C; rainfall frequent, heavy; average over 95 in. (240 cm)/yr; soils thin, poor in nutrients	High diversity of broad-leafed evergreen trees, dense canopy, vines	Enormous biodiversity; exotic, colorful insects, amphibians, birds, snakes; monkeys, tigers, jaguars	Northern South America, Central America, western central Africa, islands in Indian and Pacific Oceans, Southeast Asia
Temperate Forests	Seasonal; temperature below freezing in winter; summers warm, humid; rainfall from 30–80 in. (76–203 cm)/yr; soils well developed	Broad-leafed deciduous trees, some conifers; shrubby undergrowth, ferns, mosses	Squirrels, raccoons, opossums, skunks, deer, foxes, black bears, snakes, amphibians, many soil organisms, birds	Western and central Europe, eastern Asia, eastern North America
Coniferous Forests	Seasonal; winters long, cold; precipitation light in winter, heavier in summer; soils acidic, with leaf litter	Coniferous trees (spruce, fir, pine, hemlock), some deciduous trees (birch, maple)	Large herbivores (mule deer, moose, elk); mice, hares, squirrels; lynx, bears, foxes; nesting area for migrating birds	Northern portions of North America, Europe, Asia, extending southward at high elevations
Tundra	Bitter cold, except for short growing season with long days and moderate temperatures; precipitation low; soils thin—and below, permafrost	Low-growing sedges, dwarf shrubs, lichens, mosses, and grasses	Year round: lemmings, arctic hares, arctic foxes, lynx, caribou; summers: abundant insects, many migrant shorebirds	North of the coniferous forest in Northern Hemisphere, extending southward at elevations above the coniferous forest

Note: See http://earthobservatory.nasa.gov/Experiments/Biome/index.php and click on the links for different biomes for photos and climatic data.

Figure 5–11 Climate and major biomes. Moisture is generally the overriding factor determining the type of biome that may be supported in a region. Given adequate moisture, an area will likely support a forest. Temperature, however, determines the kind of forest. The situation is similar for grasslands and deserts. At cooler temperatures, there is a shift toward less precipitation because lower temperatures reduce evaporative water loss. Temperature becomes the overriding factor only when it is low enough to sustain permafrost. (*Source*: Redrawn from *Geosystems*, 5th ed., by Robert W. Christopherson. Copyright © 2005 by Pearson Prentice Hall, Inc., Upper Saddle River, NJ 07458.)

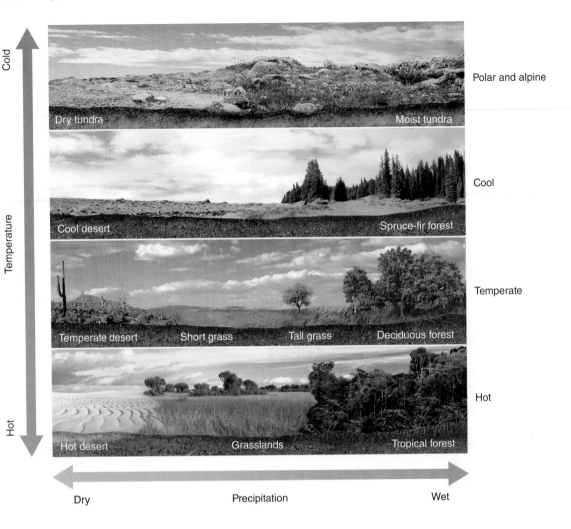

biome-like features such as seas or oceans. Table 5–2 (on page 114) lists the six major aquatic systems and their primary characteristics.

Microclimate and Other Abiotic Factors

A specific site may have temperature and moisture conditions that are significantly different from the overall climate of the region in which it is located. For example, a south-facing slope, which receives more direct sunlight in the Northern Hemisphere, will be relatively warmer and hence drier than a north-facing slope **(Fig. 5–12)**. Similarly, the temperature range in a sheltered ravine will be narrower than that in a more exposed location, and so on. The conditions found in a specific localized area are referred to as the **microclimate** of that location. In the same way that different climates determine the major biome of a region, different microclimates result in variations of ecosystems within a biome. For example, while the Greater Yellowstone region may be in the northern temperate forest biome, it includes areas of grassland and, above the tree line on mountains, areas of permafrost. This pattern of changing habitat with elevation means that the tops of mountains function ecologically as if they were more polar than their actual latitude (Fig. 5–9).

Soil type and topography also contribute to the diversity found in a biome, because these two factors affect the availability of moisture. In the eastern United States, for example, oaks and hickories generally predominate on rocky, sandy soils and on hilltops, which retain little moisture, whereas beeches and maples are found on richer soils, which hold more moisture, and red maples and cedars inhabit low, swampy areas. In the transitional region between desert and grassland, with 25–50 centimeters (10–20 in.) of rainfall per year, a soil capable of holding water will support grass, but a sandy soil with little ability to hold water will support only desert species.

Biome Productivity. All biomes have varying levels of primary productivity. Those that have high productivity may help support organisms from other biomes. For example, many seabirds migrate through salt marshes and live off the vast productivity there before traveling to other places. Highly productive biomes, like rain forests, also remove carbon dioxide from the atmosphere and trap it (Chapter 18). Biomes with high productivity aren't better than others, though. Low-productivity biomes can have other important roles, such as habitat for rare species.

Why are some biomes or types of ecosystems more productive than others? **Figure 5–13** presents (a) the average

Figure 5–12 Microclimates. Abiotic factors such as terrain, wind, and type of soil create different microclimates by influencing temperature and moisture in localized areas.

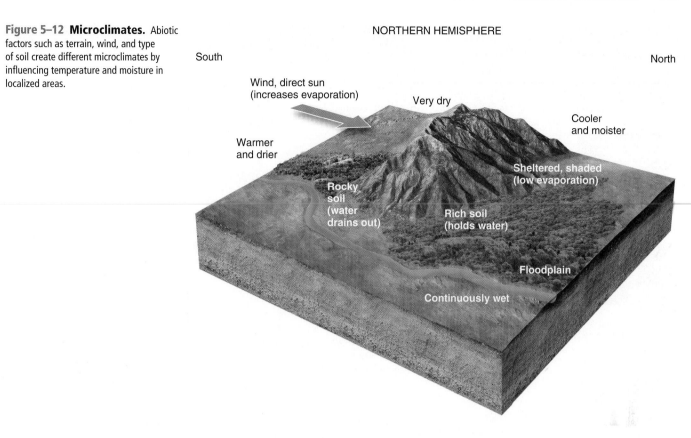

NORTHERN HEMISPHERE

South

North

Wind, direct sun (increases evaporation)

Very dry

Cooler and moister

Warmer and drier

Sheltered, shaded (low evaporation)

Rocky soil (water drains out)

Rich soil (holds water)

Floodplain

Continuously wet

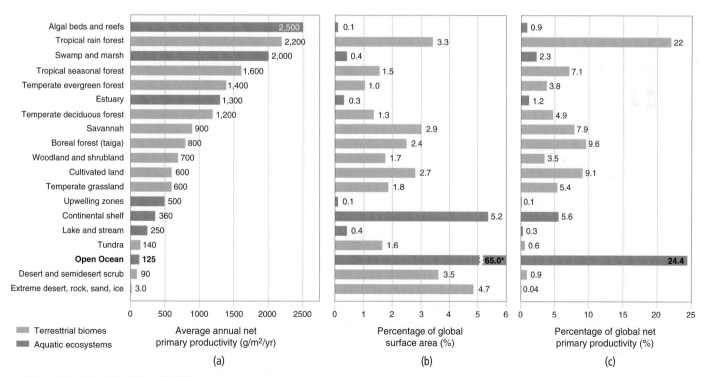

(a) Average annual net primary productivity (g/m²/yr)

(b) Percentage of global surface area (%)

(c) Percentage of global net primary productivity (%)

Terrestrial biomes

Aquatic ecosystems

Ecosystem	(a)	(b)	(c)
Algal beds and reefs	2,500	0.1	0.9
Tropical rain forest	2,200	3.3	22
Swamp and marsh	2,000	0.4	2.3
Tropical seasonal forest	1,600	1.5	7.1
Temperate evergreen forest	1,400	1.0	3.8
Estuary	1,300	0.3	1.2
Temperate deciduous forest	1,200	1.3	4.9
Savannah	900	2.9	7.9
Boreal forest (taiga)	800	2.4	9.6
Woodland and shrubland	700	1.7	3.5
Cultivated land	600	2.7	9.1
Temperate grassland	600	1.8	5.4
Upwelling zones	500	0.1	0.1
Continental shelf	360	5.2	5.6
Lake and stream	250	0.4	0.3
Tundra	140	1.6	0.6
Open Ocean	**125**	**65.0***	**24.4**
Desert and semidesert scrub	90	3.5	0.9
Extreme desert, rock, sand, ice	3.0	4.7	0.04

Figure 5–13 Productivity of different ecosystems. (a) The annual net primary productivity of different ecosystems, (b) the percentage of different ecosystems over Earth's surface area, and (c) the percentage of global net primary productivity. The outlined column for open ocean in (b) shows that the open ocean is such a high percentage of Earth's surface that it is off the scale relative to all of the other ecosystems. (*Source: Biology: Concepts and Connections,* 4th Edition, by Neil A. Campbell, Jane B. Reece, Lawrence G. Mitchell and Martha R. Taylor. Copyright © 2003 by Pearson Education. Reprinted with permission.)

TABLE 5–2 Major Aquatic Systems

Aquatic Systems	Major Environmental Parameters	Dominant Vegetation	Dominant Animal Life	Distribution
Lakes and Ponds (freshwater)	Bodies of standing water; low concentration of dissolved solids; seasonal vertical stratification of water	Rooted and floating plants, phytoplankton	Zooplankton, fish, insect larvae, turtles, ducks, geese, swans, wading birds	Physical depressions in the landscape where precipitation and groundwater accumulate
Streams and Rivers (freshwater)	Flowing water; low level of dissolved solids; high level of dissolved oxygen; often turbid from runoff	Attached algae, rooted plants	Insect larvae, fish, amphibians, otters, raccoons, wading birds, ducks, geese	Landscapes where precipitation and groundwater flow by gravity toward oceans or lakes
Inland Wetlands (freshwater)	Standing water, at times seasonally dry; thick organic sediments; high nutrients	Marshes (grasses, reeds); cattail swamps (water-tolerant trees); bogs (sphagnum moss, shrubs)	Amphibians, snakes, numerous invertebrates, wading birds, alligators, turtles	Shallow depressions, poorly drained, often occupy sites of lakes and ponds that have filled in
Estuaries (mixed)	Variable salinity; tides create two-way currents; often rich in nutrients, turbid	Phytoplankton, rooted grasses (salt-marsh grass), mangrove swamps in tropics with salt-tolerant trees, shrubs	Zooplankton, rich shellfish, worms, crustaceans, fish, wading birds, sandpipers, ducks, geese	Coastal regions where rivers meet the ocean; may form bays behind sandy barrier islands
Coastal Ocean (saltwater)	Tidal currents promote mixing; nutrients high	Phytoplankton, large benthic algae, turtle grass, symbiotic algae in corals	Zooplankton; rich bottom fauna of worms, shellfish, crustaceans; coral colonies, jellyfish, fish, turtles, gulls, terns, sea lions, seals, dolphins, penguins, whales	From coastline outward over continental shelf; coral reefs abundant in tropics
Open Ocean (saltwater)	Great depths (to 11,000 m, 36,100 ft); all but upper 200 m (656 ft) dark and cold; poor in nutrients except in upwelling regions	Phytoplankton, sargassum weed	Zooplankton and fish adapted to different depths; seabirds, whales, tuna, sharks, squid, flying fish	Covering 70% of Earth, from edge of continental shelf outward

annual net primary productivity, (b) the percentage of different ecosystems over Earth's surface, and (c) the percentage of global net primary productivity attributed to 19 of the most important ecosystems. Some key relationships between abiotic factors and specific ecosystems can be seen in the data. Tropical rain forests both are highly productive and contribute considerably to global productivity; they cover a large area of the land and are characterized by ideal climatic conditions

for photosynthesis—warm temperatures and abundant rainfall. The open oceans cover 65% of Earth's surface, so they account for a large portion of global productivity, yet their actual rate of production is low enough that they are veritable biological deserts. Although light, temperature, and water are abundant, primary production in the oceans is limited by the scarcity of nutrients—a good lesson in the significance of limiting factors. The seasonal effects of differences in latitude can also be seen by comparing productivity in tropical, temperate, and boreal (coniferous) forests.

The biosphere consists of a great variety of environments, both aquatic and terrestrial. In each environment, plant, animal, and microbial species are adapted to all the abiotic factors. In addition, they are adapted to each other, in various feeding and nonfeeding relationships. Each environment supports a group of organisms interacting with each other and with the environment in a way that perpetuates or sustains the entire group. Every ecosystem is tied to others through species that migrate from one system to another and through exchanges of air, water, and minerals common to the whole planet.

5.4 Ecosystem Responses to Disturbance

People often picture ecosystems at equilibrium, as stable environments in which species interact constantly in well-balanced predator-prey and competitive relationships. This idea has led to the popular idea of "the balance of nature": natural systems maintain a delicate balance over time that lends them great stability. A more recent understanding of ecosystems, however, shows that they operate in a dynamic, changing way. Species' abundances and ecosystem characteristics are not unchanging. Rather, the landscape is composed of a shifting mosaic of patches, much like patches of lava and unburned areas after a volcano. Volcanoes such as those in Iceland and the Philippines give us a great example of how ecosystems respond to a **disturbance**—a significant change that kills or displaces many community members.

Ecological Succession

After a disturbance like a fire or volcanic eruption, new plants grow from the soil, new trees sprout, and eventually the original forest stands are replaced by new and different stands. First, however, these communities go through a series of intermediate stages. This phenomenon of transition from one biotic community to another is called **ecological succession**. Ecological succession occurs because the physical environment may be gradually modified by the growth of a biotic community itself, such that the area becomes more favorable to another group of species and less favorable to the current occupants. Views on ecological succession have changed over time as scientists have done studies on changes near retreating glaciers and moving dune ecosystems. In the past, succession was viewed as a very predictable and orderly process through a series of stages to a stable end point. That view is useful for understanding generally how ecosystems change. However, new work, particularly through large-scale studies at Long Term Ecological Research (LTER) sites or the National Ecological Observatory Network (NEON)

project, suggests that succession is more complicated, something we will discuss shortly.

In the most basic version of succession, pioneer species—the first to colonize a newly opened area in the aftermath of a fire, flood, volcanic lava flow, or other large disturbance—start the process. As these grow, they create conditions that are favorable to longer-lived colonizers, and succession is driven forward by the changing conditions that pave the way for other species. Driving succession forward by improving conditions for subsequent species is known as **facilitation**.

The succession of species does not go on indefinitely. A stage of development is eventually reached in which there appears to be a dynamic balance between all of the species and the physical environment. In the past, this final state was called a **climax ecosystem** because the assemblage of species continues on in space and time. The term *climax ecosystem* is still used, but many ecologists are moving away from its use because they believe it implies more stability than actually occurs in nature. In the real world, even climax communities experience change. Also, within the area of a climax ecosystem there may be patches of disturbance that open spaces for new growth. Some parts of the ecosystem may be in an earlier stage of succession; sometimes there are several "final" stages, with adjoining ecosystems in the same environment at different stages. We will discuss some of these alternatives later in the chapter.

Three classic examples of succession are presented next: primary and secondary succession on land and succession in aquatic ecosystems.

Primary Succession. If an area lacks plants and soil—such as might occur after volcanic eruption or the recession of a glacier—the process of initial invasion and progression from one biotic community to the next is called **primary succession**. An example is the gradual invasion of a bare rock or gravel surface by what eventually becomes a climax forest ecosystem. Bare rock is an inhospitable environment. It has few places for seeds to lodge and germinate or for seedlings to survive. However, certain species, such as some mosses, are able to exploit this environment. Their tiny spores (specialized cells that function reproductively) can lodge and germinate in minute cracks, and moss can withstand severe drying simply by becoming dormant. With each bit of moisture, moss grows and gradually forms a mat that acts as a sieve, catching and holding soil particles as they are broken from the rock or as they blow or wash by. Thus, a layer of soil gradually accumulates, held in place by the moss (**Fig. 5–14**).

The mat of moss and soil provides a suitable place for seeds of larger plants to lodge. The larger plants in turn collect and build additional soil, and eventually there is enough soil to support shrubs and trees. In the process, the fallen leaves and other litter from the larger plants smother and eliminate the moss and most of the smaller plants that initiated the process. Thus, there is a gradual succession from

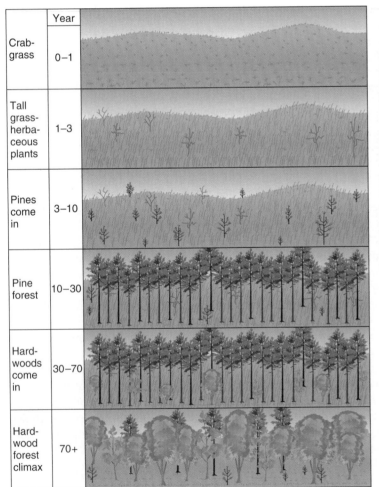

moss through small plants and, finally, to trees or whatever the regional climate supports, yielding the biomes typical of different climatic regions.

Secondary Succession. When an area has been cleared by fire, human activity, or flooding and then left alone, plants and animals from the surrounding ecosystem may gradually reinvade the area—through a series of distinct stages called **secondary succession.** The major difference between primary and secondary succession is that secondary succession starts with preexisting soil. Thus, the early, prolonged stages of soil building are bypassed. Hence, the process of reinvasion begins with different species. The steps leading from abandoned agricultural fields in the eastern United States back to deciduous forests provide a classic example of secondary succession (**Fig. 5–15**).

Aquatic Succession. Natural succession also takes place in lakes or ponds. Succession occurs because soil particles inevitably erode from the land and settle out in ponds or lakes, gradually filling them. Aquatic vegetation produces detritus that also contributes to the filling process. As the buildup occurs, terrestrial species from the surrounding

Figure 5–14 Primary succession on bare rock. Moss invades bare rock and acts as a collector, accumulating a layer of soil sufficient for additional plants to become established.

	Year	
Crab-grass	0–1	
Tall grass-herbaceous plants	1–3	
Pines come in	3–10	
Pine forest	10–30	
Hard-woods come in	30–70	
Hard-wood forest climax	70+	

(a)

(b)

Figure 5–15 Secondary succession. (a) Reinvasion of an agricultural field by a forest ecosystem occurs in the stages shown. (b) Hardwoods (in this case, species of oak) are growing up underneath and displacing pines in eastern Maryland.

Figure 5–16 Aquatic succession. In this photograph, taken in Banff National Park in the Canadian Rockies, you can visualize the lake that used to exist in the low-level area. It is now filled with sediment and covered by scrub willow. Spruce and fir forest is gradually encroaching.

ecosystem can advance in the ecotone and aquatic species must move farther out into the lake. In time, the shoreline advances toward the center of the lake until, finally, the lake disappears altogether (**Fig. 5–16**). Here the climax ecosystem may be a bog or a forest. Just as disturbances affect terrestrial ecosystems, disturbances such as droughts—or even large floods and erosion—can send **aquatic succession** back to an earlier stage.

Disturbance and Resilience

In order for natural succession to occur, the spores and seeds of various invading plants and the breeding populations of various invading animals must already be present in the vicinity. Ecological succession is not a matter of new species developing or even old species adapting to new conditions; it is a matter of populations of existing species taking advantage of a new area as conditions become favorable. Where do these early-stage species come from if their usual fate is to be replaced by late-stage or climax species? They come from surrounding ecosystems in the early stages of succession. Similarly, the late-stage species are recruited from ecosystems in the later stages of succession. In any given landscape, therefore, all stages of succession are likely to be represented in the ecosystems. This is the case because disturbances constantly create gaps or patches in the landscape. When a variety of successional stages is present in a landscape, as opposed to one single climax stage, a greater diversity of species can be expected; in other words, biodiversity is enhanced by disturbance.

If certain species have been eliminated, natural succession will be blocked or modified. For example, Iceland was colonized by the Norse in the 11th century, and, as a result, trees were cut until, in 1850, not a tree was left standing (**Fig. 5–17**). Iceland had become a barren, tundra-like habitat. Succession was unable to proceed naturally, and Iceland has remained treeless. (See Sustainability, "Ecological Restoration: Aldo Leopold and the Shack on the Prairie.")

Fire and Succession. Fire is an abiotic factor that has particular relevance to succession. It is a major form of disturbance common to terrestrial ecosystems, as we saw in the Yellowstone National Park example. About 80 years ago, forest managers interpreted the potential destructiveness of fire to mean that all fire was bad, whereupon they embarked on fire prevention programs that eliminated fires from many areas. Unexpectedly, fire prevention did not preserve all ecosystems in their existing state. In pine forests of the southeastern United States, for instance, economically worthless scrub oaks and other broad-leaved species began to displace the more valuable pines. Grasslands were gradually taken over by scrubby, woody species that hindered grazing. Pine forests of the western United States that were once clear and open became cluttered with the trunks and branches of trees that had died in the normal aging process. This deadwood became the breeding ground for wood-boring insects that

Figure 5–17 Iceland. Forests that originally covered much of this island nation were totally stripped for fuel in the 18th and 19th centuries. With natural succession impossible, Iceland has remained barren and tundra-like, as seen here.

proceeded to attack live trees. In California, the regeneration of redwood seedlings began to be blocked by the proliferation of broad-leaved species.

Scientists now recognize that fire, which is often started by lightning, is a natural and important abiotic factor. As with all abiotic factors, different species have different degrees of tolerance to fire. In particular, the growing buds of grasses and pines are located deep among the leaves or needles, where they are protected from most fires. By contrast, the buds of broad-leaved species, such as oaks, are exposed, so they are sensitive to damage from fire. Consequently, in regions where these species coexist and compete, periodic fires tip the balance in favor of pines, grasses, or redwood trees. In relatively dry ecosystems, where natural decomposition is slow, fire may also help release nutrients from dead organic matter. Some plant species even depend on fire. The cones of lodgepole pine, for example, will not release their seeds until they have been scorched by fire.

Ecosystems that depend on the recurrence of fire to maintain their existence are referred to as **fire climax ecosystems**. The category includes various grasslands and pine forests **(Fig. 5–18)**. Fire is being increasingly used as a tool in the management of such ecosystems. In pine forests, if ground fires occur every few years, relatively little deadwood accumulates. With only small amounts of fuel, fires usually just burn along the ground, harming neither pines nor wildlife significantly.

Crown fires, such as the fierce Yellowstone fires that occurred in 1980, sometimes occur naturally because not every area is burned on a regular basis. In fact, exceedingly dry conditions can make a forest vulnerable to crown fires even when no large amounts of deadwood are present. Even crown fires, however, serve to clear deadwood and sickly trees, release nutrients, and provide for a fresh ecological start. Burned areas soon become productive meadows as secondary succession starts anew. Thus, periodic crown fires create a patchwork of meadows and forests at different stages of succession that lead to a more-varied, healthier

Figure 5–18 Ground fire. Periodic ground fires are necessary to preserve the balance of pine forests. Such fires remove excessive fuel and kill competing species.

habitat that supports a greater diversity of wildlife than does a uniform, aging conifer forest.

Long-Term Ecosystem Studies. A great deal of our understanding of ecosystem changes comes from studies of changes in fields after fires, on mountain slopes after volcanoes, and near retreating glaciers. However, as human-caused changes to the environment have become global in scope, our studies have had to become longer and cover wider areas. Two ongoing research efforts of note help us to better understand ecosystems. The first is a system of ecosystems designated as Long-Term Ecological Research (LTER) sites. These 26 sites represent different types of ecosystems and different levels of human disturbance. At each site and between sites, research projects focus on five core areas:

1. Pattern and control of primary production;

2. Spatial and temporal distribution of populations selected to represent trophic structure;

3. Pattern and control of organic matter accumulation in surface layers and sediments;

4. Patterns of inorganic inputs and movements of nutrients through soils, groundwater, and surface waters;

5. Patterns and frequency of site disturbances.

Some of these sites, such as the Hubbard Brook Experimental Station, which was critical to the discovery of the problem of acid rain, have been studied for decades.

The second ecosystem study initiative, funded by the National Science Foundation, is the National Ecological Observatory Network (NEON), a network operating in the United States and Puerto Rico. It is a system for collecting data about ecosystems and climate, pulling them together, and making them available for scientists to study. The data will be available to the public as well. NEON's mission is to "enable understanding and forecasting of the impacts of climate change, land-use change and invasive species on continental-scale ecology—by providing infrastructure and consistent methodologies to support research and education in these areas."[1] Planned for years, construction on NEON infrastructure was begun in 2011.

Resilience. The ability of an ecosystem to return to normal functioning after a disturbance, such as a fire or flood, is called its **resilience**. Fire can appear to be a highly destructive disturbance to a forested landscape. Nevertheless, fire releases nutrients that nourish a new crop of plants, and in a short time, the burned area is repopulated with trees and is indistinguishable from the surrounding area. The processes of replenishment of nutrients, dispersion by surrounding plants and animals, rapid regrowth of plant cover, and succession to a forest can all be thought of as **resilience mechanisms (Fig. 5–19)**. Resilience helps maintain the sustainability of ecosystems.

[1]NEON. 2012. Mission. http://www.neoninc.org/about/mission477. May11, 2012.

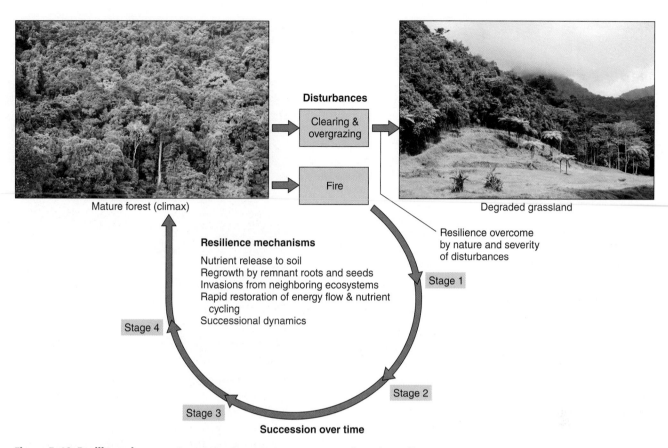

Figure 5–19 Resilience in ecosystems. Disturbances in ecosystems are usually ameliorated by a number of resilience mechanisms that restore normal ecosystem function in a short time.

Ecosystem resilience has its limits, however. If forests are removed from a landscape by human intervention and the area is prevented from undergoing reforestation by overgrazing, the soil may erode, leaving a degraded state that carries out few of the original ecosystem's functions. As in Iceland, some disturbances may be so profound that they can overcome normal resilience mechanisms and create an entirely new and far less productive ecosystem. This new ecosystem can also have its own resilience mechanisms, which may resist any restoration to the original state. We can view this as a "tipping point"—a situation in an ecosystem where a small action catalyzes a major change in the state of the system.

Tipping points act as levers that can change a system to be better able—or less able—to function as an ecosystem. They stimulate the action of **feedback loops,** where system dynamics influence a system in ways that amplify (positive loop) or dampen (negative loop) the behavior of the system. For example, predator-prey cycles function as negative feedback loops that keep both populations from exploding in terms of numbers and then crashing. Removal of predators (to favor the prey species) would be a tipping point that could allow the positive feedback of prey reproduction to bring devastation to their plant food supply.

You might wonder why it matters if an ecosystem such as the forests of Iceland moves into an alternative state, such as an Iceland with no trees, and stays there. After all, don't

treeless, cold places exist? As we will see below, ecosystems provide a number of services valuable to humans and other species. These include large-scale effects such as lowering flood surges, maintaining soil, maintaining even ground temperatures, and taking up carbon dioxide.

If just a few parts of the planet moved from one type of ecosystem to another, it really wouldn't matter. The problem is one of frequency. Large-scale human actions tend to lower ecosystem services such as flood control, wind-erosion control, and carbon dioxide storage. They tend to produce ecosystems either with very low productivity or with high primary productivity but low food chain length (such as a lake with too many nutrients). Such ecosystems tend to have less physical complexity and lower biological diversity. We will see this more when we discuss the problem of desertification (Chapter 11).

5.5 Human Values and Ecosystem Sustainability

Ecosystems have existed for millions of years, maintaining natural populations of biota and the processes that these carry out (which in turn sustain the ecosystems). One of the reasons for studying natural ecosystems is that they are models of sustainability. In this chapter, we started with trophic levels

SUSTAINABILITY

Ecological Restoration: Aldo Leopold and the Shack on the Prairie

Conservationist Aldo Leopold (1887–1948) sat in a converted chicken coop writing at a table. His essays on the value of nature and the damage inflicted by unthinking use and his scientific papers on the impacts of fire and of predator suppression were to change the course of history. Around him lay acres of degraded farmland, its sandy soil depleted by years of unrelenting use, its plants gone, sparse grasses standing as tufts across the eroded fields. In 1935, Leopold had bought this land near the Wisconsin River and begun its restoration. For years, he brought friends, neighbors, and students to the farm, encouraging them to sow seeds of prairie plants and promote the use of fire as a restoring agent. Gradually, the agricultural weed pests were outcompeted by the native grasses, shrubs, and flowers. Wild lupine, its purple flowers flecking the horizon, vied for space with purple coneflower and wild indigo. Native butterflies returned, along with the prairie voles and 13 lined ground squirrels. The work he did at the "shack," as his home away from home was called, paralleled the restoration of native prairie at the University of Wisconsin Madison Arnold Arboretum, begun by Leopold and other scientists. It was the beginning of the field of restoration ecology.

The potential for restoration of any ecosystem rests on the following three assumptions: (1) abiotic factors must have remained unaltered or, if not, can at least be returned to their original state; (2) viable populations of the species formerly inhabiting the ecosystem must still exist; and (3) the ecosystem must not have been upset by the introduction of one or more foreign species that cannot be eliminated and that may preclude the survival of reintroduced native species. If these conditions are met, revival efforts

have the potential to restore the ecosystem to some semblance of its former state.

Restoration ecologists use a set of actions described in a primer on ecological restoration published by the Society for Ecological Restoration in 2004.* The basic processes include taking an *inventory*, developing a *model* of the desired ecosystem in structural and functional terms, setting goals for the restoration efforts, designing an *implementation plan*, and carrying out the plan. For example, it may be necessary to remove all herbaceous vegetation and then plow and plant the land with native grasses. To maintain the prairie, it may be necessary to burn it regularly or, in the case of larger landholdings, to introduce bison. Clear *performance standards* should be stated, along with the means of evaluating progress. Finally, the results of the implementation plan should be monitored and strategies developed for the long-term protection and maintenance of the restored ecosystem.

Leopold practiced these strategies on his own land in Wisconsin, bringing a broken and eroded abandoned farm back to a health it had not seen for years. In his well-known *Sand County Almanac*, as well as his other essays, Leopold wrote of this goal of understanding the natural world and caring for it. Leopold died trying to control a brush fire, but his work

lives on in a whole field of endeavor. We will discuss more on his contribution to conservation biology when we look at the ethical principles that underlie efforts to protect species and their habitats (Chapter 6).

---

*Society for Ecological Restoration, International Science and Policy Working Group, *The SER International Primer on Ecological Restoration (2004)*. Available at www.ser.org.

---

and then examined the flow of energy and nutrients in an ecosystem. These discussions have focused on how natural ecosystems work in theory. In reality, however, it is Earth's specific ecosystems that we depend on for goods and services (also known as **ecosystem capital**). Is our use of natural and managed ecosystems a serious threat to their long-term sustainability? Many ecosystem scientists believe that it is.

Appropriation of Energy Flow

As we have seen, it is the Sun that energizes the processes of energy flow and nutrient cycling. Humans make heavy use of the energy that starts with sunlight and flows through natu-

ral and agricultural ecosystems. Agriculture, for example, provides most of our food. To produce food, we have converted almost 11% of Earth's land area from forest and grassland biomes to agricultural ecosystems. Grasslands sustain our domestic animals for labor, meat, wool, leather, and milk. Forest biomes provide us with 3.3 billion cubic meters (117 billion cubic feet) of wood annually for fuel, building material, and paper. Finally, some 15% of the world's energy consumption is derived directly from plant material, like firewood and peat.

Calculations of the total annual global net primary production of land ecosystems range around 120 petagrams (1 petagram = 1 billion metric tons) of dry matter, including agricultural as well as the more natural ecosystems **(Fig. 5–20)**.

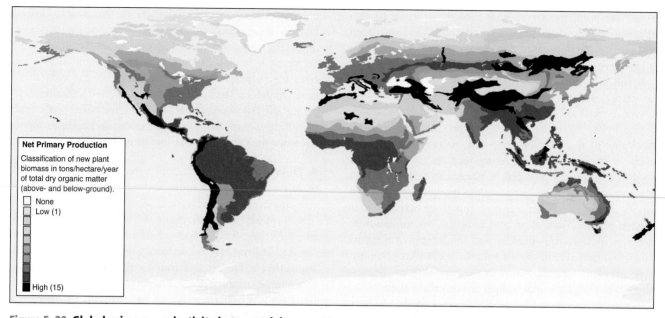

Figure 5–20 Global primary productivity in terrestrial ecosystems. The primary production values range from 1 ton/hectare/year (yellow) to 15 tons/hectare/year (black) of dry organic matter. The global annual total is 120 billion metric tons (1 billion metric tons = 1 petagram). (*Source*: Bazilevich global primary productivity map, NOAA Satellite and Information Service.)

In 1986, researchers calculated that humans appropriate around 30% of this total production for agriculture, grazing, forestry, and human-occupied lands.[2] More-recent calculations have supported this figure.[3] These estimates indicate that humans are using a large fraction of the whole. Further, because humans convert many natural and agricultural lands to urban and suburban housing, highways, dumps, factories, and the like, we cancel out an additional 8% of potential primary production. Thus, we appropriate about 40% of the land's primary production to support human needs. In using this much, we have become the dominant biological force on Earth, despite representing about 5% of the biomass. As ecologist Stuart Pimm put it, "Man eats Planet! Two-fifths already gone!"[4]

Involvement in Nutrient Cycling

Nutrients are replenished in ecosystems through the breakdown of organic compounds and the release of the chemicals that make them up—processes we have described as biogeochemical cycles. This maintains the sustainability of ecosystems indefinitely. However, human intrusion into these natural cycles is substantial, and it is bound to increase. As we have seen, our alterations of the nutrient cycles have

long-term consequences (Chapter 3). Climate change is one consequence of these alterations.

Currently, global climate change is one of the most serious environmental issues facing society. The burning of fossil fuels has increased the level of carbon dioxide in the atmosphere to the point where, by the mid–21st century, it will have doubled since the beginning of the Industrial Revolution. Methane is also released from the drying and decomposition of areas that used to be permafrost and from cattle. Because both carbon dioxide and methane are powerful greenhouse gases, a general warming of the atmosphere is occurring, leading to many other effects on Earth's climate (Chapter 20). Numerous ecosystems are being affected by the changes, and it is questionable whether they can function sustainably in the face of such rapid changes. The threat of disturbance on such an enormous scale is a major reason why many nations of the world have begun to take steps to reduce emissions of greenhouse gases.

Value of Ecosystem Capital

How much are natural ecosystems worth to us? To the poor in many developing countries, the products of natural ecosystems keep life going and may enable them to make economic gains that can lift them from deep poverty. Theirs is a direct and absolute dependence. For example, 1.6 billion people, many of them the world's poorest, depend on forests for much of their sustenance (in the form of fuelwood, construction wood, wild fruits, herbs, animal fodder, and bush meat). Their access to these ecosystem goods is often uncertain. Those of us in the developed world, however, are susceptible to a very different

[2]Peter Vitousek, Paul R. Ehrlich, Anne H. Ehrlich, and Pamela Matson, "Human Appropriation of the Products of Photosynthesis," *BioScience* 36, No. 6. (June 1986): 368–373.
[3]N. Ramankutty, A. T. Evan, C. Monfreda, and J. A. Foley, "Farming the Planet: 1. Geographic Distribution of Global Agricultural Lands in the Year 2000," *Global Biogeochemical Cycles* 22 (2008): GB1003.
[4]Pimm, Stuart. 2001. *The World According to Pimm: A scientist audits the Earth.* McGraw-Hill.

perspective: We often assign little thought or value to natural ecosystems and are unconcerned about (and insulated from) human impacts on natural ecosystems. This perspective can have dangerous consequences.

Recall that we defined the goods and services we derive from natural systems as ecosystem capital (Chapter 1). In a first-ever attempt of its kind, a team of 13 natural scientists and economists collaborated in 1997 to produce a report titled "The Value of the World's Ecosystem Services and Natural Capital."[5] Their reason for making such an effort was that the goods and services provided by natural ecosystems are not easily seen in the market (meaning the market economy that normally allows us to place value on things) or may not be in the market at all. Thus, clean air to breathe, the formation of soil, the breakdown of pollutants, and similar items never pass through and thus are not given a value by the market economy. People are often not even aware of their importance. Because of this, these things are undervalued or not valued at all.

The team identified 17 major ecosystem goods and services that provide vital functions we depend on. The team also identified the ecosystem functions that actually carry out the vital human support and gave examples of each (Table 5–3). They made the point that it is useless to consider human welfare without these ecosystem services; so, in one sense, their value as a whole is infinite. However, the **incremental value** of each type of service can be calculated. That is, changes in the quantity or quality of various types of services may influence human welfare, and an economic value can be placed on that relationship. For example, removing a forest will affect the ability of the forest to provide lumber in the future, as well as to perform other services, such as soil formation and promotion of the hydrologic cycle. The economic value of this effect can be calculated. By calculating incremental values, making many approximations, and collecting data from other researchers who have worked on individual processes, the research team tabulated the annual global value of ecosystem services performed by all types of ecosystems. According to their calculations, the total value to human welfare of a year's services amounts to $41 trillion (in 2004 dollars)—and that is considered a conservative estimate. This is close to the $55 trillion calculated for the gross world economic product in 2004.

Using Ecosystem Value in Decision Making. The real power of the team's analysis lies in its use for making local decisions. For instance, the value of wetlands cannot be represented solely by the amount of soybeans that could be grown on the land if it were drained. Instead, wetlands provide other vital ecosystem services, and these should be balanced against the value of the soybeans in calculating the costs and benefits of a proposed change in land use. The bottom line of their analysis is, in their words, "that ecosystem services provide an important portion of the total contribution to human welfare on this planet." For this reason, the ecosystem capital stock (the ecosystems and the populations in them, including the lakes and wetlands) must be given adequate weight in public

policy decisions involving changes to that stock. However, because these ecosystem services are outside the market and uncertain, they are too often ignored or undervalued, and the net result is human changes to natural systems whose social costs far outweigh their benefits. For example, coastal mangrove forests in Thailand are often converted to shrimp farms, yet an analysis showed that the economic value of the forests (for timber, charcoal, storm protection, and fisheries support) exceeded the value of the shrimp farms by 70%.

A New Look. In 2002, a new team looked at the 1997 team's calculation of value and examined the benefit-cost consequences of converting ecosystems to more-direct human uses (e.g., converting a wetland to a soybean field).[6] In every case, the net balance of value was a loss—that is, services lost outweighed services gained. The team examined five different biomes and found consistent losses in ecosystem capital, running at −1.2% per year. Based on the calculated total ecosystem value of $41 trillion, this percentage represents an annual loss of $250 billion through habitat conversion alone. If this is so, the authors asked, why is this conversion still happening? They suggested that the benefits of conversion were often exaggerated through various government subsidies, seriously distorting the market analysis. Thus, the market does not adequately measure many of the benefits from natural ecosystems, nor does it deal well with the fact that many major benefits are realized on regional or global scales, while the benefits of conversion are narrowly local. Other analyses, including a National Research Council report published in 2005[7] that investigated the value of wild nature and ecosystem services, support the idea that conservation is more achievable when the real costs of harming ecosystems are included in our thinking. Beyond conservation, our tremendous dependence on natural systems should lead, logically, to proper management of those systems, a topic discussed later in this chapter.

Can Ecosystems Be Restored?

The human capacity for destroying ecosystems is well established. To some degree, however, we also have the capacity to restore them (see Sustainability, "Ecological Restoration: Aldo Leopold and the Shack on the Prairie"). In many cases, restoration simply involves stopping the abuse. For example, after pollution is curtailed, water quality often improves, and fish and shellfish gradually return to previously polluted lakes, rivers, and bays. Similarly, forests may gradually be restored to areas that have been cleared. Humans can speed up the process by seeding and planting seedling trees and by reintroducing populations of fish and animals that have been eliminated.

In some cases, however, specific ecosystems have been destroyed or disturbed to such an extent that they require the efforts of a new breed of scientist: the restoration ecologist.

[5]Robert Costanza et al., "The Value of the World's Ecosystem Services and Natural Capital," *Nature* 387 (1997): 253–260.

[6]Andrew Balmford et al., "Economic Reasons for Conserving Wild Nature," *Science* 297 (August 9, 2002): 950–953.
[7]National Research Council, *Valuing Ecosystem Services: Toward Better Environmental Decision-Making* (Washington, D.C.: National Academies Press, 2005).

TABLE 5–3 Ecosystem Services and Function

Ecosystem Service	Ecosystem Functions	Examples
Gas regulation	Regulation of atmospheric chemical composition	CO_2–O_2 balance, O_3 for UVB protection, and SO_x levels
Climate regulation	Regulation of global temperature, precipitation, and other biologically mediated climatic processes at global or local levels	Greenhouse gas regulation, dimethylsulfoxide production affecting cloud formation
Disturbance regulation	Capacitance, damping, and integrity of ecosystem response to environmental fluctuations	Storm protection, flood control, drought recovery, other aspects of habitat response to environmental variability controlled mainly by vegetation structure
Water regulation	Regulation of hydrological flows	Provisioning of water for agricultural (such as irrigation) or industrial (such as milling) processes or transportation
Water supply	Storage and retention of water	Provisioning of water by watersheds, reservoirs, and aquifers
Erosion control and sediment retention	Retention of soil within an ecosystem	Prevention of loss of soil by wind, runoff, or other removal processes; storage of silt in lakes and wetlands
Soil formation	Soil-formation processes	Weathering of rock and the accumulation of organic material
Nutrient cycling	Storage, internal cycling, processing, and acquisition of nutrients	Nitrogen fixation, N, P, and other elemental or nutrient cycles
Waste treatment	Recovery of mobile nutrients and removal or breakdown of excess nutrients and compounds	Waste treatment, pollution control, detoxification
Pollination	Movement of floral gametes	Provisioning of pollinators for the reproduction of plant populations
Biological control	Trophic-dynamic regulations of populations	Keystone predator to control prey species, reduction of herbivory by top predators
Refuges	Habitat for resident and transient populations	Nurseries, habitat for migratory species, regional habitats for locally harvested species, or overwintering grounds
Food production	The portion of primary production extractable as food	Production of fish, game, crops, nuts, and fruits by hunting, gathering, subsistence farming, or fishing
Raw materials	That portion of primary production extractable as raw materials	The production of lumber, fuel, or fodder
Genetic resources	Sources of unique biological materials and products	Medicine, products for science, genes for resistance to plant pathogens and crop pests, ornamental species (pets and horticultural varieties of plants)
Recreation	Provision of opportunities for recreational activities	Ecotourism, sport fishing, and other outdoor recreational activities
Cultural	Provision of opportunities for noncommercial uses	Aesthetic, artistic, educational, spiritual, or scientific values of ecosystems

Source: Reprinted, by permission, from Robert Costanza, Ralph d'Arge, Rudolf de Groot, Stephen Farber, Monica Grasso, Bruce Hannon, Karin Limburg, Shahid Naeem, Robert V. O'Neill, Jose Paruelo, Robert G. Raskin, Paul Sutton, and Marjan van den Belt, "The Value of the World's Ecosystem Services and Natural Capital," *Nature* 387 (1997): 253–260.

Unfortunately, sometimes the ecosystem is too greatly disturbed for scientists to repair. In a prairie, for example, lack of grazing by native herbivores, overgrazing by cattle or sheep, and suppression of regular fires can lead to much woody vegetation. Exotic species may abound in the region and can continuously disperse seeds on the experimental prairie, and there may be no remnants of the original prairie grasses and herbs on the site. Such sites may take intensive efforts to bring back any native species or may continue in an alternative ecosystem.

Why should we restore ecosystems? Restoration ecologists Steven Apfelbaum and Kim Chapman cite several reasons.[8] First, we should do so for aesthetic reasons. Natural ecosystems are often beautiful, and the restoration of something beautiful and pleasing to the eye is a worthy project that can be uplifting to many people. Second, we should do so for the benefit of human use. The ecosystem services of a restored wetland, for example, can be enjoyed by present and future generations. The degradation of ecosystems robs them of their ecosystem capital, and restoring them returns their value. Finally, we should do so for the benefit of the species and ecosystems themselves. It can be argued that nature has value separate from human use and that people should act to preserve and restore ecosystems and species in order to preserve that value.

The Future

On a global scale, the future growth of the human population and its rising consumption levels will severely challenge ecosystem sustainability. Over the next 50 years, the world's population is expected to increase by at least 2.5 billion people (see Chapter 8). The needs and demands of this expanded human population will create unprecedented pressures on the ability of Earth's systems to provide goods and services. Estimates based on current trends indicate that impacts on the nitrogen and phosphorus cycles will raise the amounts of these nutrients being added to the land and water to two to three times current levels. Demands on agriculture will likely require at least 15% more agricultural land; irrigated land area is expected to double, greatly stressing an already strained hydrologic cycle. Such growing demands cannot be met without facing serious trade-offs between different goods and services. More agricultural land means more food, but less forest—and, therefore, less of the important services forests perform. More water for irrigation means more rivers will be diverted, so less water will be available for domestic use and for sustaining the riverine ecosystems. These dire projections are not necessarily predictions, and they don't have to be inevitable outcomes. By looking ahead, we may choose other alternatives if we clearly understand the consequences of simply continuing present practices. There is a powerful

need to understand how essential natural and managed ecosystems are for human well-being and to bring to bear the wisdom and political will necessary to promote their sustainability.

Ecosystems are resilient, living units of the landscape, often subject to natural disturbances and capable of changing over time. They also provide essential goods and services that improve the quality of human life and, in many areas, allow those in poverty to improve their lot. However, we have seen that ecosystem goods and services are routinely undervalued, with the result that, as the Millennium Ecosystem Assessment found, ecosystem capital is being degraded and overexploited in many areas of the world. To turn human efforts in a sustainable direction, we need to protect or manage the natural environment in a way that maintains the goods and services vital to the human way of life, and we need to manage ourselves. The two aspects are interrelated.

Managing Ecosystems

Virtually no ecosystem can escape human impact. We depend on ecosystems for vital goods and services—and by using ecosystems, we are in fact managing them. Forests are a good example. Management can be a matter of fire suppression (or, more recently, controlled burns), clear-cutting, or selective harvesting. The objective of management can be to maximize profit from logging or to maintain the forest as a sustainable and diverse ecosystem that yields multiple products and services. Good ecosystem management is based on understanding how ecosystems function, how they respond to disturbances, and what goods and services they can best provide to the human societies living in or around them. Since 1992, the U.S. Forest Service, the U.S. Bureau of Land Management, the U.S. Fish and Wildlife Service, the U.S. National Park Service, and the U.S. Environmental Protection Agency have all officially adopted ecosystem management as their management paradigm. What, then, is ecosystem management?

Ecosystem management is composed of several main principles. First, it looks at ecosystems on both small and large scales. For example, the impact of logging on a single river and the impact of that logging on the entire watershed are both considered in managing the ecosystem. Good management preserves the range of possible landscapes in an ecosystem instead of allowing one type to take over the area. In addition, management is not static—managers are given the opportunity to learn from experiments and to employ new knowledge from forestry and landscape science in their practice. Such "managers" can be volunteer land stewards or neighborhood groups, not just government agencies. As part of adapting to changing circumstances, management must also look at the human element of these ecosystems. The valuable goods and services of ecosystems are recognized, and through active involvement in monitoring and management, all stakeholders (all people with a stake in the ecosystem's health) are included as important elements in stewardship of the resources (see Stewardship, "Ecosystem Stakeholders").

---

[8]Steven I. Apfelbaum and Kim Alan Chapman, "Ecological Restoration: A Practical Approach," in *Ecosystem Management: Applications for Sustainable Forest and Wildlife Resources*, ed. Mark S. Boyce and Alan Haney (New Haven, CT: Yale University Press, 1997).

STEWARDSHIP

Ecosystem Stakeholders

The pressures on ecosystems illustrate why it is crucially important to understand how ecosystems work and how human societies interact with them. Although many pressures are large-scale—even global—the decisions (and decision makers) that will most directly determine ecosystem sustainability are local or regional. The Millennium Ecosystem Assessment identifies a number of "drivers of change": factors that bring about alterations in ecosystems, such as changes in land use, species introductions, and resource harvesting. Decision makers are those who are responsible for these "drivers." To understand how and why they make the decisions they do, it is useful to employ the concept of **stakeholders.** People who have an interest in or may be affected by a given approach to environmental management, including government decision makers, are all considered to be stakeholders. They have a stake in the ecosystem resource and what happens to it. Primary stakeholders are those most dependent on the resource; secondary stakeholders may be those living near the resource, but not greatly dependent on it; government officials; scientists studying the resource; and conservation organizations.

People living within and adjacent to an ecosystem are usually the primary stakeholders.

In ecosystems of the developing world, these are among the world's poorest people, and their dependence on what has been called environmental income is almost absolute. Primary stakeholders may remove vegetation, harvest wood, withdraw water, harvest game animals, and so on. They may use these products in their own households; or where markets exist, they may sell them for cash or exchange them for services. Their local decisions will have both intended and unintended consequences. If they harvest firewood from a forest (intended), soil erosion may increase (unintended). If many local decision makers make similar decisions, the impacts may have cumulative undesirable regional and, eventually, global consequences. Preventing such outcomes requires effective management of the ecosystem so as to maintain its sustainability and at the same time meet the needs of the various stakeholders.

One way to promote good stewardship of resources is to involve the stakeholders in management. One example of a case where poor rural communities were able to pursue community-based natural resource management and restore and manage crucial environmental resources is documented in the World Resources Institute report titled *The Wealth of the Poor: Managing Ecosystems to Fight Poverty.** The Shinyanga region of Tanzania was once an abundant woodland environment that supported

the Sukuma people with subsistence agriculture and forage for livestock. An indigenous natural resource management system of maintaining protected vegetation enclosures was removed by government decree in the 1970s, and the result was devastating to the region. Land and soil were overused, wood was difficult to find, and traditional wild fruit and plants became scarce. In 1986, the Tanzanian government reversed its approach and promoted the revival of the original indigenous scheme of enclosures. As a result, people exerted local ownership over the natural resources, and gradually, the landscape changed from an eroded, dry land to one where vegetation and wildlife have been restored on more than 350,000 hectares (860,000 acres). The outcome greatly improved livelihoods for the Sukuma, with dividends in fuelwood, construction timber, fodder for animals, and traditional foods. In this situation, the involvement of the primary stakeholders in decisions affecting local ecosystems was the key to restoration. One conclusion of the World Resources Institute report was that democratic reform may be the most important step toward sustainable use of ecosystem goods and services, particularly in developing countries.

---

*World Resources Institute, *World Resources 2005: The Wealth of the Poor: Managing Ecosystems to Fight Poverty* (Washington, D.C.: World Resources Institute, 2005).

Above all, the paradigm incorporates the objective of ecological sustainability.

The U.S. Forest Service is responsible for managing some 192 million acres (77.7 million hectares) of national forests and is a key player in overseeing public natural resources. A decade ago, the Forest Service convened an interdisciplinary committee of scientists to advise the service on what sound science would recommend for managing national lands. According to the committee's report, "the first priority for management is to retain and restore the ecological sustainability of watersheds, forests, and rangelands for present and future generations. The Committee believes that the policy of sustainability should be the guiding star for stewardship of the national forests and grasslands to assure the continuation of this array of benefits" (referring to the goods and services

provided by the national lands).[9] These were fine words, but have they been followed?

The governing board of the Millennium Ecosystem Assessment process examined the findings of the more than 1,350 experts who produced the many reports of the assessment and issued its own statement of the key messages to be derived from the process. The title of this work is significant: *Living Beyond Our Means: Natural Assets and Human Well-Being.*[10] As stated in the first few lines of this document: "At the heart of this assessment is a stark warning. Human activity is putting such strain on the natural functions of Earth that the ability of the planet's ecosystems to sustain future generations can no longer be taken for granted." The following are the key messages from the governing board:

- Everyone in the world depends on nature and ecosystem services to provide the conditions for a decent, healthy, and secure life.

- Humans have made unprecedented changes to ecosystems in recent decades, to meet growing demands for food, fresh water, fiber, and energy.

---

[9]Committee of Scientists, Department of Agriculture, *Sustaining the People's Lands: Recommendations for Stewardship of the National Forests and Grasslands into the Next Century* (1999). Available at www.fs.fed.us/news/science.
[10]Millennium Ecosystem Assessment, *Living Beyond Our Means: Natural Assets and Human Well-Being* (Statement of the Millennium Assessment Board, 2005). Available at www.millenniumassessment.org/en/Products.BoardStatement.aspx.

- These changes have helped to improve the lives of billions, but at the same time, they have weakened nature's ability to deliver other key services such as the purification of air and water, protection from disasters, and the provision of medicines.

- Among the outstanding problems identified by this assessment are the dire state of many of the world's fish stocks; the intense vulnerability of the 2 billion people living in dry regions to the loss of ecosystem services, including water supply; and the growing threat to ecosystems from climate change and nutrient pollution.

- Human activities have taken the planet to the edge of a massive wave of species extinctions, further threatening our own well-being.

- The loss of services derived from ecosystems is a significant barrier to the achievement of the Millennium Development Goals to reduce poverty, hunger, and disease.

- The pressures on ecosystems will increase globally in coming decades unless human attitudes and actions change.

- Measures to conserve natural resources are more likely to succeed if local communities are given ownership of them, share the benefits, and are involved in decisions.

- Even today's technology and knowledge can reduce considerably the human impact on ecosystems. They are unlikely to be deployed fully, however, until ecosystem services cease to be perceived as free and limitless and their full value is taken into account.

- Better protection of natural assets will require coordinated efforts across all sections of governments, businesses, and international institutions. The productivity of ecosystems depends on policy choices in the areas of investment, trade, subsidy, taxation, and regulation, among others.

We are living off the future, in other words. Our current well-being is apparently dependent on drawing down the ecosystem capital that provides the goods and services; as a result, agricultural soils erode, fish stocks decline, forests shrink, and pollution of land, water, and air increases. This situation is unsustainable.

REVISITING THE THEMES

Sound Science

Ecosystems have natural processes that drive their functions. They are not static entities—change is a fundamental property of ecosystems. They can stand alone or be grouped into large global areas called biomes, which are predicted largely by precipitation and temperature. Ecosystems can also include sub-ecosystems, caused by variations or microclimates. Ecosystems can be characterized by their primary productivity and how that energy and the associated nutrients move up through a food chain or food web. These structures contain several trophic levels of organisms that gain their energy from their prey (or in the case of producers, primarily from the Sun). Ecosystems pull together several concepts we saw in earlier chapters. The concepts of energy and matter (and the inefficiency of their transfer) play out in forming the biomass pyramid and determining the length of food chains. Individual limits on distribution of species are the cause of large predictable areas of Earth having similar ecosystems—the biomes. The relationships between species determine the trophic levels in ecosystems. A functioning ecosystem and its food webs are resilient to disturbance—indeed, some even depend on disturbances to reset succession and allow diversity. However, extremely large or unusual disturbances might tip an ecosystem into an altered state from which it might not recover. In some instances, this leads to much lower productivity or diversity, and in others, sound environmental management can tip a degraded ecosystem back into a restored state. Scientists and others study the processes of restoring degraded ecosystems to health, as did Aldo Leopold.

Sustainability

The message of the Millennium Ecosystem Assessment is that human impacts on the resilient natural world are so unprecedented and extensive that we crossed the line into unsustainable consumption some time ago and are now depleting ecosystem capital stock instead of living off its sustainable goods and services. Converting ecosystems into farms and other human uses has frequently depleted their ability to provide us with pivotal assets. We are just beginning to calculate the lost economic value of these altered landscapes, not to mention their lost aesthetic and natural value. Any attempt at sustainability will need to include monitoring the world's ecosystems, setting international policies to protect them, and involving all stakeholders in their management. Basic science will inform much of the policy making and other solutions for sustainability discussed in later chapters.

Stewardship

Ecosystem management has shown us that we need to monitor the impacts we have on our ecosystems, because, while they are resilient, large disturbances can do irreparable damage. Indeed, stewardship means, first of all, understanding how ecosystems, such as the Greater Yellowstone Ecosystem, do just what they do and then stepping back and restraining our activities to allow ecosystems to continue to provide us with essential goods and services. Human agency has become the dominant force on Earth, affecting living systems all over the planet. In the process, we have the ability to drastically reduce biodiversity, or we have the ability to act as responsible stewards of our environment. Ecosystem management is, above all, a stewardship approach to the natural world. As the U.S. Forest Service's Committee of Scientists put it, "the policy of sustainability should be the guiding star" for our actions, fittingly linking sustainability and stewardship.

REVIEW QUESTIONS

1. How did the described volcanoes in Iceland and the Philippines change the environment to lesser or greater extents?

2. Name and describe the attributes of the two categories into which all organisms can be divided based on how they obtain nutrition.

3. Name and describe the roles of the three main trophic categories that make up the biotic structure of every ecosystem. Give examples of organisms from each category.

4. Give four categories of consumers in an ecosystem and the role that each plays.

5. Describe different members of the decomposition food web.

6. Differentiate among the concepts of *food chain*, *food web*, and *trophic levels*.

7. Relate the concept of the biomass pyramid to the fact that all heterotrophs depend on autotrophic production.

8. Describe how differences in climate cause Earth to be partitioned into major biomes.

9. What are three situations that might cause microclimates to develop within an ecosystem?

10. Identify and describe the biotic and the abiotic components of the biome of the region in which you live.

11. Define the terms *ecological succession* and *climax ecosystem*. How do disturbances allow for ecological succession?

12. What role may fire play in ecological succession, and how may fire be used in the management of certain ecosystems?

13. What is meant by ecosystem *resilience*? What can cause it to fail? How does this relate to environmental tipping points?

14. What is meant by the term *stakeholder*? How does ecosystem management involve stakeholders?

15. Succinctly describe ecosystem management.

16. Can ecosystems be restored? What has to happen for that to work?

17. How much of Earth's primary productivity is used or preempted by humans?

18. Examine the key messages from the Millennium Ecosystem Assessment's governing board, and evaluate how these points affect you now and will affect you and your children in the future.

THINKING ENVIRONMENTALLY

1. Write a scenario describing what would happen to an ecosystem or to the human system in the event of one of the following: (a) all producers are killed through a loss of fertility of the soil or through toxic contamination, or (b) decomposers and detritus feeders are eliminated. Support all of your statements with reasons drawn from your understanding of the way ecosystems function.

2. Look at the following description of a broad global region, and describe what the biome name and main biota would be and how you know that: 40 centimeters (16 in.) of precipitation a year, seasons, frozen much of the year, high winds.

3. Consider the plants, animals, and other organisms present in a natural area near you, and then do the following: (a) Imagine how the area may have undergone ecological succession, and (b) analyze the population-balancing mechanisms that are operating among the various organisms. Choose one species, and predict what will happen to it if two or three other native species are removed from the area. Then predict what will happen to your chosen species if two or three foreign species are introduced into the area.

4. Consult the Web site www.ecotippingpoints.org, and read several of the success stories from the site. Compare the tipping points as presented, and explain the common properties of the tipping points.

5. Explore how the human system can be modified into a sustainable ecosystem in balance with (i.e., preserving) other natural ecosystems without losing the benefits of modern civilization.

MAKING A DIFFERENCE

1. Find out where the nearest wilderness area or national park is to where you live. If possible, make a plan to visit it. Find out what management issues the park has. Do they have advisories about fire risk, invasive species, or other ecosystem management issues?

2. Using an online map, find out what river drains your watershed and where that water goes. Is there a watershed stewardship group for the river? Who might be looking out for the protection of the ecosystem? If you are interested, organize a group volunteer activity.

3. In your area, is there a citizens' group for restoring any degraded ecosystems? There are many such small groups of citizens in the United States and elsewhere. See if you can link to a citizens' group. Volunteer if you can. One such activity is the annual CoastSweep, usually held in September, in which people all over the world clean up beaches. If you are coastal, see if you can help.

MasteringEnvironmentalScience®

Go to **www.masteringenvironmentalscience.com** for practice quizzes, Pearson eText, videos, current events, and more.

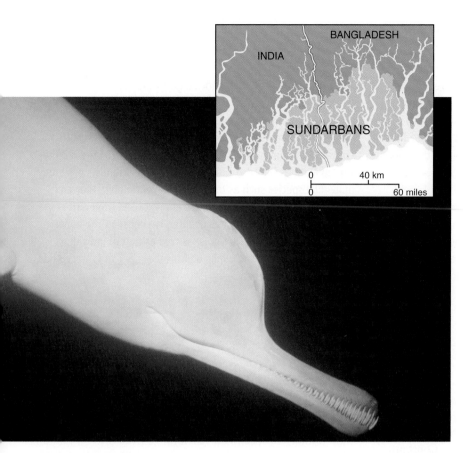

INDIA

BANGLADESH

SUNDARBANS

0   40 km

0   60 miles

CHAPTER 6

Wild Species and Biodiversity

Indus River dolphin

A BROWN RIVER WINDS slowly past mangroves filled with flocking birds, carrying with it plankton, crustaceans, and young fish that will support a vast food chain. A fisherman in a boat casts a net to the side. The place is the Sundarbans, the largest mangrove forest in the world. The long nose of a Ganges River dolphin (*Platanista gangetica gangetica*) breaks the water surface in the early morning mist. This endangered cetacean swims past mangrove roots and edges past a fishing net, its body narrowly avoiding entanglement.

THREATENED. This area of the Bay of Bengal is home to populations of two of the most endangered cetaceans in the world: the Ganges River dolphin (*Platanista gangetica gangetica*) and a related subspecies, the Indus River dolphin (*Platanista gangetica minor*). These rare beasts are nearly blind and swim sideways to get hints of sunlight through the water as they find their prey by echolocation. Freshwater dolphins, like their saltwater cousins, live on fish. Because they live in rivers instead of oceans, they are in some of the most populous areas of the world. These rivers may be highly

polluted, in habitats heavily divided by dams, and in places already crowded with fishermen.

River dolphins are among the most threatened large vertebrates in the world. In fact, another species of river dolphin, the Baiji, in China,

LEARNING OBJECTIVES

1. **The Value of Wild Species and Biodiversity:** Define and give examples of both instrumental and intrinsic value of wild species.

2. **Biodiversity and Its Decline:** Explain the causes, extent, relation to human activities, and impacts of the tremendous loss of species occurring today.

3. **Saving Wild Species:** Describe how cutting-edge science, policies for protection, and changes in the way people think can be used to protect wild species.

4. **Protecting Biodiversity Internationally:** Explain at least three international efforts to protect wild species.

was declared extinct, to the great sorrow of conservationists worldwide. The Baiji, nicknamed "the goddess of the Yangtze," declined because of loss of habitat, pollution, crowding, and injury from contact with fishing vessels, as well as overharvest. The species, once spread throughout several large rivers in China, became rarer and rarer. In 2006, an extensive 45-day search by experts with sonar and other equipment failed to find any in the last areas the Baiji was known to inhabit, and it was declared extinct.

AREAS OF IMPORTANCE. Recently, to protect the Ganges and Indus river dolphins from the same fate, Bangladesh has identified dolphin **hot spots,** areas of importance, within the sprawling Sundarbans mangrove forest. Three of these areas are being set aside as breeding and feeding areas for dolphins. Inten-

sive multicountry protections are needed to keep these unique animals from further decline. In an area already crowded with people, protecting species may seem impossible, but it is important. The effort of the Bangladeshi government is a hopeful move in the right direction.

The focus in this chapter is wild species. The world is experiencing a dramatic decline in wild organisms. This is such a concern that the United Nations has designated 2011–2020 the United Nations Decade of Biodiversity. While scientists can study individual species such as the Ganges River dolphin, they can also work on species decline in the context of biodiversity—the variety of life on Earth. In this chapter, we consider the importance of biodiversity as a whole, as well as that of individual species, and look at the public policies that seek to protect wild species and biodiversity.

6.1 The Value of Wild Species and Biodiversity

Recall that ecosystem capital is the sum of all the goods and services provided to human enterprises by natural systems (Chapter 1). This capital is enormously valuable to humankind. The 1997 mean estimate of ecosystem capital by Robert Constanza and his coauthors was $46.7 trillion, adjusted for 2009 dollars (see Chapter 5). Such a value is so difficult to pull together that there is no subsequent estimate of the same set of parameters. However, in 2008, economist Pavan Sukhdev reported that capital loss from the worldwide financial crisis stood at $1–$1.5 trillion, and capital loss from ecosystem degradation was estimated to be $2–$4.5 trillion per year.[1] The basis of much of our natural capital is ecosystems, and the basis of ecosystems is the plants, animals, and microbes—the wild species—that make them work. To maintain the sustainability of these ecosystems, their integrity must be preserved; this means maintaining their resilience, natural processes, and biodiversity, among other things. Placing a high priority on protecting wild species and the ecosystems in which they live is therefore crucial.

Even those who agree on the need to protect wild species, however, may disagree on the kind of protection that should be afforded to them. Some want wildlife protected so as to provide recreational hunting. Others feel strongly that hunting for sport should be banned. Many have a deeper concern for the ecological survival of wild species, and they view the ongoing loss of biodiversity as an impending tragedy. In the developing world, however, many collect or kill wild species in order to eat or make money, so their personal survival is at stake. All of these attitudes stem from the values we place on wild species. Can the differing values be reconciled in a way that still leads to sustainable management of these important natural resources?

Biological Wealth

The Millennium Ecosystem Assessment states that about 2 million species of plants, animals, and microbes have been examined, named, and classified, but that many more species have not yet been found or cataloged. Most estimates of the total number of species on Earth range from 5 to 30 million. The most recent estimate is around 8.7 million (+/− 1.3 million). About 15,000 new species are described each year. A 2011 survey of habitats on Luzon Island and surrounding waters in the Philippines revealed more than 300 new species, including an "inflatable" shark that swells by filling its belly with water (**Fig. 6–1**).

These natural species of living things, collectively referred to as **biota,** are responsible for the structure and maintenance of all ecosystems. They, and the ecosystems that they form, represent wealth—the **biological wealth**—that makes up most of the ecosystem capital that sustains human life

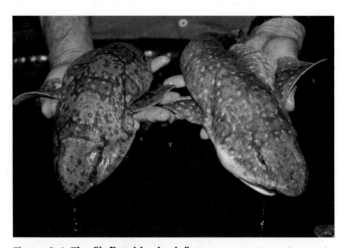

Figure 6–1 The "inflatable shark." New species are being discovered regularly. These sharks were among 300 new species discovered in the Philippines in 2011.

---

[1]Pavan Sukhdev, *The Economics of Ecosystems and Biodiversity* (Interim Report funded by the German Ministry for Environment and the European Commission on the Economics of Ecosystems and Biodiversity). Presented at the UN's Convention on Biological Diversity, November 2008.

and economic activity with goods and services. From this perspective, the biota found in each country can represent a major component of the country's total wealth. More broadly, this richness of living species constitutes Earth's biodiversity.

Humans have always depended on Earth's wild species for food and materials. Approximately 12,000 years ago, however, our ancestors began to select certain plant and animal species from the natural biota and propagate them, and the natural world has never been the same. Forests, savannas, and plains were converted to fields and pastures as the human population grew and human culture developed. In time, many living species were exploited to extinction, and others disappeared as their habitats were destroyed. The Center for Biological Diversity has reported that 631 species and subspecies went extinct in North America between 1642 and 2001. A 2008 report said that global biodiversity loss costs as much as $4.5 trillion a year.[2] We have been drawing down our biological wealth, with unknown consequences.

For many Americans who live in cities and suburbs and get all their food from supermarkets, the connection between everyday life and nature may seem remote. That is an illusion, however. Our interactions with the natural world may have changed, but we still depend on biological wealth. Moreover, millions of our neighbors in the developing world are not so insulated from the natural world; they draw sustenance and income directly from forests, grasslands, and fisheries. For them, "environmental income" is often the safety net that sustains them during lean times and can enable them in good times to generate wealth to improve their well-being. However, these people, too, are often engaged in practices that draw down their biological wealth. A root source of this problem is the way we regard and value wild nature.

Two Kinds of Value

In the 19th century, hunters on horseback could ride out to the vast herds of bison roaming the North American prairies and shoot them by the thousands, often taking only the tongues for markets back in the east. The passenger pigeons that once darkened the skies in huge flocks were killed at their roosts to fill a lively demand for their meat, until the species was gone. Plume hunters decimated the populations of egrets and other shorebirds to satisfy the demands of fashion in the late 1800s. Appalled at this wanton destruction, 19th-century naturalists called for an end to the slaughter, and the U.S. public became sensitized to the losses. People began to see natural species as worthy of preservation, and naturalists began to look for ways to justify their calls to conserve nature. Thus, there was an emerging sense that species should not be hunted to extinction. But why not? Their problem then, and ours now, is to establish the thesis

that wild species have some value that makes it essential to preserve them. If we can identify that value, then we will be able to assess what our moral duties are in relation to wild species.

Instrumental Value. Philosophers who have addressed this problem identify two kinds of value that should be considered. The first is **instrumental value.** A species or individual organism has instrumental value if its existence or use benefits some other entity (by providing food, shelter, or a source of income). This kind of value is usually anthropocentric; that is, the beneficiaries are human beings. Many species of plants and animals have instrumental value to humans and will tend to be preserved (conserved, that is) so that we can continue to enjoy the benefits derived from them.

Intrinsic Value. The second kind of value that must be considered is **intrinsic value.** Something has intrinsic value when it has value for its own sake; that is, it does not have to be useful to us to possess value. How do we know that something has intrinsic value? This is a philosophical question and comes down to a matter of moral reasoning. People often disagree about intrinsic value, as illustrated by the controversies about the rights of animals. Some people argue that animals have certain rights and should be protected, for example, from being used for food, fur and hides, or experimentation. On the other side, some people hold that animals are simply property and have no rights at all. Others, probably the majority, hold an intermediate position: They agree that cruelty to animals is wrong, but they do not believe that animals should be given rights approaching those afforded to humans **(Fig. 6–2).**

Many people believe that no species on Earth except *Homo sapiens* has any intrinsic value. Accordingly, they then find it difficult to justify preserving species that are apparently insignificant. Still, in spite of the problems of

Figure 6–2 Intrinsic value. Do species like this Blandings turtle have value separate from their value to humans? The question of intrinsic value is one of debate amongst philosophers.

[2]Ibid.

establishing intrinsic value for species, support is growing in favor of preserving species that not only may appear useless to humans, but also may never be seen by anyone except a few naturalists or taxonomists (biologists who are experts on classifying organisms).

The value of natural species can be categorized as follows:

* Value as sources for food and raw materials,
* Value as sources for medicines and pharmaceuticals,
* Recreational, aesthetic, and scientific value, and
* Value for their own sake (intrinsic value).

Sources for Food and Raw Materials

Most of our food comes from agriculture, rather than being collected directly from wild populations, so it is tempting to falsely believe that food is independent of natural biota. In nature, both plants and animals are continuously subjected to the rigors of natural selection. Only the fittest survive. Consequently, wild populations have numerous traits for competitiveness, resistance to parasites, tolerance to adverse conditions, and other aspects of vigor.

In contrast, populations grown for many generations under the "pampered" conditions of agriculture tend to lose these traits because they are selected for production, not resilience. For example, a high-producing plant that lacks resistance to drought is irrigated, and the resistance to drought is ignored. Also, in the process of breeding plants for maximum production, virtually all genetic variation is eliminated. Indeed, the cultivated population is commonly called a cultivar (for *culti*vated *vari*ety), indicating that it is a highly selected strain of the original species, with a minimum of genetic variation. When provided with optimal water and fertilizer, cultivars produce outstanding yields under the specific climatic conditions to which they are adapted. With their minimum genetic variation, however, they have virtually no capacity to adapt to any other conditions. For example, winter wheat varieties with high yields may be less winter hardy or more prone to pests and disease.

Wild Genes. To maintain vigor in cultivars and to adapt them to different climatic conditions, plant breeders comb wild populations of related species for the desired traits. When found, these traits are introduced into the cultivar through crossbreeding or, more recently, biotechnology (Chapter 12). For example, in the 1970s, the U.S. corn crop was saved from blight by genes from a wild strain of maize. The point is that this trait came from a related wild population—that is, from natural biota. Another area in which wild species have instrumental value to humans is pest control. Later, we will discuss the tremendous and invaluable opportunities there are to control pests by increasing genetic resistance (Chapter 13). Genes for increasing this resistance can come only from natural biota. If wild populations are

lost, the options for continued improvements in food plants will be greatly reduced.

New Food Plants. The potential for developing new agricultural cultivars would also be lost if wild populations were destroyed. From the hundreds of thousands of plant species existing in nature, humans have used perhaps 7,000 in all, and modern agriculture has tended to focus on only about 30. Of these, three species—wheat, maize (corn), and rice—fulfill about 50% of global food demands. This limited diversity in agriculture makes modern plants ill suited to production under many different environmental conditions. For example, arid regions are generally considered to be unproductive without irrigation, but there are wild plants, including many wild species of the bean family, that produce abundantly under dry conditions.

Scientists estimate that 30,000 plant species with edible parts could be brought into cultivation. Many of these could increase production in environments that are less than ideal. For example, every part of the winged bean, a native of New Guinea, is edible: pods, flowers, stems, roots, and leaves. Recently introduced into many developing countries, this legume has already contributed significantly to improving nutrition. Loss of biological diversity undercuts similar future opportunities.

Woods and Other Raw Materials. Because species are selected from nature for animal husbandry, forestry, and aquaculture, essentially all the same arguments apply to those important enterprises, too. For example, nearly 3 billion people rely on wood for fuel for heating and cooking. Wood products rank behind only oil and natural gas as commodities on the world market. Demand for wood is increasing so rapidly that scientists are predicting a "timber famine" or "fuelwood crisis" in the next decade. Products such as rubber, oils, nuts, fruits, spices, and gums also come from forests (**Fig. 6–3**). All of these products are part of the value of wild species for humans, and forests are only one example of ecosystems that provide wild and natural living resources.

Figure 6–3 Forest products. These are all nonwood forest products developed to increase income in African forested areas. (*Source*: FAO.)

Banking Genes. Living organisms can be thought of as a bank in which the gene pools of all the species involved are deposited. Thus, biota is frequently referred to as a **genetic bank.** Because of concerns about agriculture, many scientists suggest that wild relatives of cultivated crops be conserved. In Kew, England, the Royal Botanic Gardens house the Millennium Seed Bank (MSB). In 2007, the MSB banked the billionth seed in its vast collection of the genetic diversity of world plants—the largest wild seed collection in the world. There is also the massive Svalbard Global Seed Vault, built into the permafrost deep in a mountain on a remote arctic island in Norway. This ultimate seed bank officially opened in 2008 and holds seeds that will act as a backup for seed banks around the world. Similarly, zoos act as a type of genetic bank for animals, and many zoos are active in conservation and captive breeding. In the United Kingdom, the Frozen Ark Project is collecting cells and DNA from species likely to go extinct in the near future. This collaborative effort with organizations around the world is designed to preserve genetic diversity while we try to stabilize species and slow currently high extinction rates.

Sources for Medicine

Earth's genetic bank is also important for the field of medicine. For thousands of years, the indigenous people of the island of Madagascar used an obscure plant, the rosy periwinkle, in their folk medicine (Fig. 6–4). If this plant, which grows only on Madagascar, had become extinct before 1960, hardly anyone outside Madagascar would have cared. In the 1960s, however, scientists extracted two chemicals with medicinal properties—vincristine and vinblastine—from the plant. These chemicals have revolutionized the treatment of childhood leukemia and Hodgkin's disease. Before their discovery, leukemia was almost always fatal in

Figure 6–4 The rosy periwinkle. This plant, native to Madagascar, is a source of two anticancer agents that are highly successful in treating childhood leukemia and Hodgkin's disease.

children; today, with vincristine treatment, there is a 99% chance of remission.

The story of the rosy periwinkle is just one of hundreds. Chinese star anise, a tree in the magnolia family, yields star-shaped fruits that are harvested and processed to produce shikimic acid, the raw material for making Tamiflu®, an effective treatment against influenza viruses. Paclitaxel (trade name Taxol®), an extract from the bark of the Pacific yew, has proven valuable for treating ovarian, breast, and small-cell cancers. For a time, this use threatened to decimate the Pacific yew because six trees were required to treat one patient for a year. Now, the substance is extracted from the leaves of the English yew tree, a horticultural plant that is easy to maintain. Today, Taxol® is the most prescribed anti-tumor drug.

Stories like these have created a new appreciation for the field of *ethnobotany*, the study of the relationships between plants and people. To date, more than 3,000 plants have been identified as having anticancer properties. Drug companies have been financing field studies of the medicinal use of plants by indigenous peoples (called *bioprospecting*) for years. The search for beneficial drugs in the tropics has led to the creation of parks and reserves to promote the preservation of natural ecosystems that are home to both the indigenous people and the plants they use in traditional healing. In Surinam, for example, a consortium made up of drug companies, the U.S. government, and several conservationist and educational groups has created the Central Surinam Reserve, a 4-million-acre (1.64-million-hectare) protected area. Table 6–1 lists a number of well-established drugs that were discovered as a result of analyzing the chemical properties of plants used by traditional healers. Who knows what other medicinal cures are out there?

Recreational, Aesthetic, and Scientific Value

The species in natural ecosystems also provide the foundation for numerous recreational and aesthetic interests, ranging from sportfishing and hunting to hiking, camping, bird-watching, photography, and more (Fig. 6–5). Periodically, the U.S. government conducts a survey on outdoor recreation; the latest was carried out from 2005 to 2006. Of all people age 16 and older who were surveyed, an estimated 87.5 million were involved in some form of outdoor recreation. The largest activity was viewing natural scenery, an activity reported by 71.1 million people. Very likely, much of the public support for preserving wild species and habitats stems from the aesthetic and recreational enjoyment people derive from them. While hunting has declined in recent decades, other outdoor activities remain popular. A 2006 U.S. Fish and Wildlife Service survey on fishing, hunting, and wildlife found that 87.5 million U.S. resident adults (age 16 and older) participated in wildlife-centered recreation. This number included 30 million people who fished, 12.5 million who hunted, and 71.1 million who did some type of wildlife observation such as bird-watching. Hunting, fishing, and wildlife-watching result in employment for 2.6 million people—nearly as

Figure 6–5 Recreational, aesthetic, and scientific uses. Natural biota provides numerous values, a few of which are depicted here.

many as employed by the computer industry—and produce $108 billion in revenues in the United States.

Recreational and aesthetic values constitute a very important source of support for maintaining wild species, because activities inspired by these values also support commercial interests. **Ecotourism**—whereby tourists visit a place in order to observe wild species or unique ecological sites—represents the largest foreign-exchange-generating enterprise in many developing countries. Because some percentage of recreational dollars will be spent on activities related to the natural environment, any degradation of that environment affects commercial interests. As the national surveys indicated, recreational, aesthetic, and scientific activities involve a great number of people and represent a huge economic enterprise.

Scientific value comes from our need to have living things available so we can learn basic laws of nature, the way ecosystems function, and the way the world works generally. Societies have long held that such "pure science" is valuable. Many scientists are engaged in this endeavor by sheer curiosity about the world, including about the functioning of individual types of organisms, and biota provides the nature that we get to study. Realistically, though, most of the science done with these species aims for more-pragmatic goals—new medicines, protection from gene loss in agriculture and forestry, and other outcomes already discussed.

A Cautionary Note. The use of wild species and biodiversity for instrumental value can raise some problems of justice as we navigate the needs of people with competing claims. For example, while the rosy periwinkle was a success story because of its effectiveness in combating cancer, little of the money returned to Madagascar, a very poor country. In other cases, such as the case of turmeric (a spice commonly used in India), ancient herbal remedies have been patented by large companies. While the patent on turmeric was eventually overturned, the patent on the rosy periwinkle was not, and money from the rosy periwinkle came to Madagascar only when the patent ran out. The indigenous people in the

areas where biodiversity is highest may not be the ones to benefit from their resources. The Nagoya Protocol on Access and Benefit Sharing is a recent International Treaty designed to protect the interests of indigenous peoples as products are discovered in wild sources.

Similarly, ecotourism may bring money to poor areas of the world, but such tourism can also increase pollution, directly harm wildlife, or alter local societies in negative ways. For example, whale-watching boats on the St. Lawrence River disrupt the feeding of whales. Tourist boats on Mexican lakes frighten flocks of flamingoes and reduce their feeding time. In the Caribbean, coral reefs are damaged by divers, boat anchors, and shopkeepers who remove coral to sell. Ecotourism can marginalize indigenous people as part of the scenery, or they can remain poor while tourism operators gain wealth. Nonetheless, there is sustainable tourism, and people intending to travel in the hopes of promoting conservation can find tourism that is done well.

The Loss of Instrumental Value. Overall, a loss of biodiversity has a tremendous negative effect on the world in real, practical ways. The Economics of Ecosystems and Biodiversity (TEEB) report, released in May 2008, was the first international report to detail the economic and life-quality effects of biodiversity loss. Researchers have estimated that the loss of ecosystem services and biodiversity costs $78 billion annually and, even more importantly, that these costs are higher for the world's poorest people. Continued loss of biological diversity, the report said, could halve the income of the poorest billion and a half people. Because the people who benefit most from the rapid use of resources are not the people who are most harmed by the loss of those resources, many ethicists believe such an outcome to be morally wrong. Thus, even the instrumental values of Earth's resources have an ethical component.

Value for Their Own Sake

The usefulness (instrumental value) of many wild species is apparent. But instrumental value alone may not be a

TABLE 6–1 Modern Drugs from Traditional Medicines

Drug	Medical Use	Source	Common Name
Aspirin	Reduces pain and inflammation	*Filipendula ulmaria*	Queen of the meadow
Codeine	Eases pain; suppresses coughing	*Papaver somniferum*	Opium poppy
Ipecac	Induces vomiting	*Psychotria ipecacuanha*	Ipecac
Pilocarpine	Reduces pressure in the eye	*Pilocarpus jaborandi*	Jaborandi plant
Pseudoephedrine	Reduces nasal congestion	*Ephedra sinica*	*Ma-huang* shrub
Quinine	Combats malaria	*Cinchona pubescens*	Cinchona tree
Reserpine	Lowers blood pressure	*Rauwolfia serpentina*	Rauwolfia
Scopolamine	Eases motion sickness	*Datura stramonium*	Jimsonweed
Theophylline	Opens bronchial passages	*Camellia sinensis*	Tea
Tubocurarine	Relaxes muscles during surgery	*Chondrodendron tomentosum*	Curare vine
Vinblastine	Combats Hodgkin's disease	*Catharanthus roseus*	Rosy periwinkle

compelling reason to protect many wild species. What about the species that have no obvious value to anyone—including many plant and animal species? Another strategy for preserving all wild species is to emphasize the intrinsic value of species rather than their unknown or uncertain ecological and economic values. From this viewpoint, the extinction of any species is an irretrievable loss of something valuable.

Environmental ethicists argue that the long-established existence of species means they have a right to continued existence. Living things have ends and interests of their own, and the freedom to pursue them is often frustrated by human actions. Do humans have the right to terminate a species that has existed for thousands or millions of years? Environmental philosopher Holmes Rolston III puts it this way: "Destroying species is like tearing pages out of an unread book, written in a language humans hardly know how to read, about the place where they live."[3] Some support this view by arguing that there is value in every living thing and that one kind of living thing (e.g., humans) has no greater value than any other. This argument, however, can lead to some difficulties, such as having to defend the rights of pathogens and parasites. A more common viewpoint held by ethicists is that, because humans have the ability to make moral judgments, they also have a special responsibility toward the natural world, and that this responsibility includes concern for other species.

Religious Support. Many ethicists find their basis for intrinsic value in religious thought. For example, in the Jewish and Christian traditions, Old Testament writings express God's concern for wild species when he created them. Judeo-Christian scholars maintain that by declaring his creation good and giving it his blessing, God was saying that all wild things have intrinsic value and, therefore, deserve moral consideration and stewardly care. In the story of Noah, described in the Hebrew Bible, God makes a covenant not only with

humans, but also with all creatures, "every living thing." Similarly, in the Islamic tradition, the Quran (Koran) proclaims that the environment is the creation of Allah and should be protected because its protection praises the Creator. Many Native American religions have a strong environmental ethic (a system of moral values or beliefs); for example, the Lakota hold that humans and other forms of life should interact like members of a large, healthy family. Drawing on Hindu philosophy, the Chipko movement of northern India represents a strong grassroots environmentalism. Ethical concern for wild species underlies several religious traditions and represents a potentially powerful force for preserving biodiversity. Even if species have no demonstrable use to humans, it can still be argued that they have a value and a right to exist and that humans thus have a responsibility to protect them from harm that results from human actions.

The Land Ethic. Further, there is an ethic about the preservation not just of species, but also of their ecosystems. This view of nature was most famously described by conservationist Aldo Leopold in an essay published in 1949 entitled "The Land Ethic." Leopold is known for being one of the first scientists to understand the importance of both fire and predators in maintaining ecosystem health. He advocated for protection of wild places—*wilderness*—as well as for better care of agricultural and other human-dominated land. Leopold wrote a number of essays and is well known in the literary community as well as among conservationists. *A Sand County Almanac*, in which "The Land Ethic" appeared, sold more than 2 million copies and is printed in nine languages. "The Land Ethic" was instrumental in generating interest in the protection of wilderness. In it, Leopold describes his viewpoint: "A thing is right when it tends to preserve the integrity, stability, and beauty of the biotic community. It is wrong when it tends otherwise."[4]

---

[3]Holmes Rolston III. *Environmental Ethics: Duties to and Values in the Natural World* (Philadelphia: Temple University Press, 1988), p. 129.

[4]Aldo Leopold. "The Land Ethic," in *A Sand County Almanac and Sketches Here and There* (Oxford, England: Oxford University Press, 1949). Available at http://www.luminary.us/leopold/land_ethic.html.

6.2 Biodiversity and Its Decline

In the introduction to this chapter, biodiversity was defined simply as the variety of life on Earth. We can be more specific than that. The dimensions of biodiversity include the genetic diversity within a species and the variety of species as well as the range of communities and ecosystems. When scientists use the term *biodiversity*, however, sometimes they mean a characteristic that they can calculate and compare between habitats. In this definition, scientists use two measures to calculate biodiversity—the number of species and a measure of how "even" the species are. A habitat has low biodiversity if it is dominated by only one species, with few members of other species, because those few might easily leave or die out. Given the same number of species, diversity is considered to be higher in habitats where the dominance by any single species is low. In this chapter, our main focus is the diversity of species. Recall that over long periods of time, natural selection leads to speciation (the creation of new species) as well as extinction (the disappearance of species) (Chapter 4). Now, we are interested in how humans interact with wild species and impact biodiversity over a relatively short time.

How Many Species?

Almost 2 million species have been discovered and described, and certainly many more than that exist. Most people are completely unaware of the great diversity of species that occurs within any given taxonomic category. Groups especially rich in species are the flowering plants (270,000 species) and the insects (950,000 species), but even less-diverse groups, such as birds or ferns, are rich with species that are unknown to most people (Table 6–2). Groups that are conspicuous or commercially important—such as birds, mammals, fish, and trees—are much more fully explored and described than, say, insects and very small invertebrates (such as soil nematodes, fungi, and bacteria). Taxonomists are aware that their work in finding and describing new species is incomplete. Trained researchers who can identify major groups of organisms are few, and the task of fully exploring the diversity of life would require a major, sustained effort in systematic biology. For example, Costa Rica, less than half the size of New York State, is estimated to contain at least 5% of all living species. Estimates of unknown species keep rising as taxonomists explore the rain forests more and more. Whatever the number of species, the planet's biodiversity represents an amazing and diverse storehouse of biological wealth.

The Decline of Biodiversity

Biodiversity is declining in the United States as well as around the world. Because U.S. ecosystems are well studied, the trends that show themselves here (including the decline of many plant and animal species) may represent larger global

---

TABLE 6–2 Known and Estimated Species on Earth

Domain	Eukaryote kingdoms	No. of described species	Estimated total
Archaea		175	?
Bacteria		10,000	?
Eukarya			
	Animalia		1,320,000
	Craniata (vertebrates), total	52,500	55,000
	Mammals	4,630	
	Birds	9,750	
	Reptiles	8,002	
	Amphibians	4,950	
	Fishes	25,000	
	Mandibulata (insects and myriapods)	963,000	8,000,000
	Chelicerata (arachnids, etc.)	75,000	750,000
	Mollusca	70,000	200,000
	Crustacea	40,000	150,000
	Nematoda	25,000	400,000
	Fungi	72,000	1,500,000
	Plantae	270,000	320,000
	Protoctista	80,000	600,000
Total		1,750,000	14,000,000

Notes: This table presents estimates of the number of species of living organisms in the high-level groups recognized and in some selected groups within them. Vertebrate classes are distinguished because of the general interest in these groups. The described species column refers to species named by taxonomists. Most groups lack a formal list of species. All estimates are approximations. They are inevitably inaccurate because new species will have been described since publication of any checklist; more are continually being described, and other names turn out to be redundant synonyms. In general, the diversity of microorganisms, small-sized species, and those from habitats difficult to access are likely to be seriously underestimated. Among Archaea and Bacteria, the figures of 175 and 10,000 are very rough estimates of "species" defined on features shown in culture; there appears to be no sound estimate of the total amount of prokaryote diversity. The estimated total column includes provisional working estimates of the number of described species plus the number of unknown species; the total figure is highly imprecise. Only a small selection of animal phyla is shown, but the figure for Animalia applies to all. Figures in the total row are for all species in all domains.

Source: Groombridge B. and Jenkins M.D. (2002) *World Atlas of Biodiversity*. Prepared by the UNEP World Conservation Monitoring Centre. University of California Press, Berkeley, USA.

trends. **Endemic** species (those found in only one habitat and nowhere else) are especially at risk. Some areas of the country and of the world are particularly vulnerable to species loss. These areas are often the focus of special conservation efforts.

North America. The biota of the United States is as well known as any. Even so, not much is known about most of the 200,000-plus species of plants, animals, and microbes estimated to live in the nation. At least 500 species native to the United States, including at least 100 vertebrates, are known to have gone extinct (or have "gone missing" and are believed to have gone extinct) since the early days of colonization. An inventory of 20,897 wild plant and animal species in the 50 states, carried out in the late 1990s by a team of biologists from the Nature Conservancy and the National Heritage Network (and more recently Nature-Serve), concluded that one-third are vulnerable (threatened), imperiled (endangered), or already extinct. In the United States, mussels, crayfish, fishes, and amphibians—all species that depend on freshwater habitats—are at greatest risk. In part, this potential for loss stems from the high number of endemic aquatic species in the American Southeast. For example, there are close to 200 species of freshwater mussels and clams in North America, giving these bivalves the greatest species diversity of any freshwater bivalve group in the world. Flowering plants are also of great concern, with one-third of their numbers in decline in North America.

Species populations are a more important element of biodiversity than even the species' existence. Populations of species that occupy different habitats and ranges contribute to biological wealth, as they provide goods and services important in ecosystems. Across North America, even populations of well-studied species are in decline. Commercial landings of many species of fish are down, puzzling reductions in amphibian populations are occurring all over the world, and many North American songbird species (such as the Cerulean warbler, wood thrush, and scarlet tanager) are declining and have disappeared entirely from some local regions. These birds are neotropical migrants; that is, they winter in the neotropics of Central and South America and breed in temperate North America. According to a report from the National Audubon Society, more than one-fourth of North American birds are already declining or are in danger of doing so.

Global Outlook. Worldwide, the loss of biodiversity is even more disturbing. One of the most recent global analyses of biodiversity is the third *Global Biodiversity Outlook*, published by the UN (2010). **Figure 6–6**, from this assessment, shows groups from a comprehensive survey of more than 47,000 species, categorized by level of risk

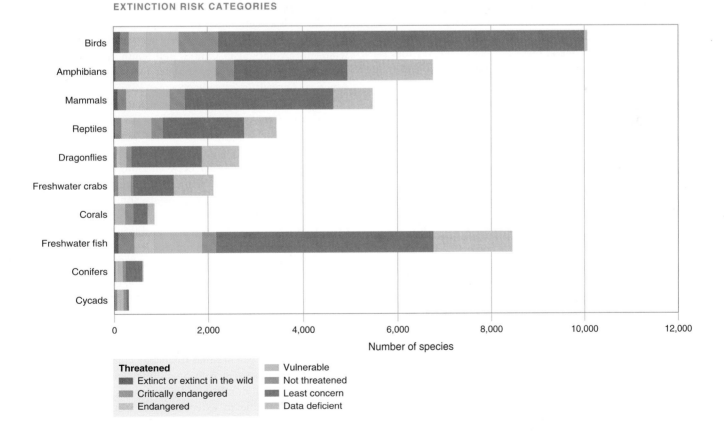

Figure 6–6 Who is going extinct? This bar graph shows groups from a comprehensive survey of more than 47,000 species, categorized by level of risk of extinction.

(*Source: Global Biodiversity Outlook 3.*)

of extinction. Estimates of extinction rates in the past show alarming increases in loss of species. For mammals, the background (past) extinction rate was less than one extinction every thousand years, although rates were higher during five great extinction events. (Scientists estimate the background extinction rate from the fossil record of marine animals and from rates of change in DNA.) This rate can be compared with known extinctions of the past several hundred years (for mammals and birds, 20 to 25 species per hundred years and, for all groups, 850 species over about 500 years), indicating that the current rate is 100 to 1,000 times higher than past rates.

The International Union for Conservation of Nature (IUCN) keeps a running list of threatened species. In 2011, approximately 23% (1,138) of mammal species and 12% (1,253) of bird species were globally threatened. These two groups have been thoroughly studied.[5] The status of all other groups of organisms is much less well known. The majority of threatened species are concentrated in the tropics, where biodiversity is so rich as to be almost unimaginable. Biologist Edward O. Wilson identified 43 species of ants on a single tree in a Peruvian rain forest, a level of diversity equal to the entire ant fauna of the British Isles.[6] Other scientists found 300 species of trees in a 1-hectare (2.5-acre) plot and almost 1,000 species of beetles on 19 specimens of a species of tree in Panama.[7] Unfortunately, the same tropical forests that hold such high biodiversity are also experiencing the highest rate of deforestation. Because the inventory of species in these ecosystems is so incomplete, it is virtually impossible to assess extinction rates in such forests. Regardless of the exact rates, numerous species are in decline, and some are becoming extinct. No one knows how many, but the loss is real, and species loss represents a continuing depletion of the biodiversity of our planet.

Reasons for the Decline

Extinctions of the distant past were caused largely by processes of climate change, plate tectonics, and even asteroid impacts. Current threats to biodiversity are sometimes described with the acronym **HIPPO**, which refers to five factors: habitat destruction, **invasive species,** pollution, population, and overexploitation. Many declining species experience combinations of several or all of these factors. Indeed, global climate change in particular increases the effects of some of these factors. We will discuss each factor in more detail below. Future losses will be greatest in the developing world, where biodiversity is greatest and human population growth is highest. Africa and Asia have lost almost two-thirds of their original natural habitat. People's desire for a better way of life, the desperate poverty of rural populations, and the global market for timber and other natural resources are powerful forces that will continue to draw down biological wealth on those continents.

One key to slowing the loss of biodiversity lies in bringing human population growth down (Chapter 8). If the human population increases to 10 billion, as some demographers believe that it will, the consequences for the natural world may be frightening. Another key is to pull individual consumption down to sustainable levels in parts of the world with high consumption of energy and materials. These two concepts will be vital to solving several environmental issues and will be further discussed throughout the rest of the book.

Habitat Change. By far the greatest source of biodiversity loss is the physical alteration of habitats through the processes of conversion, fragmentation, simplification, and intrusion. Habitat destruction has already been responsible for 36% of the known extinctions and is the key factor in the currently observed population declines. Natural species are adapted to specific habitats, so if the habitat changes or is eliminated, the species go with it.

Conversion. Natural areas are commonly converted to farms, housing subdivisions, shopping malls, marinas, and industrial centers. When a forest is cleared, for example, destruction of the trees is not the only result; every other plant or animal that occupied the destroyed ecosystem, either permanently or temporarily (e.g., migrating birds), also suffers. Global forest cover has been reduced by 40% already, and the decline continues. For example, the decline in North American songbird populations has been traced to the loss of winter forest habitat in Central and South America and the increasing fragmentation of summer forest habitat in North America. Agriculture already occupies 38% of land on Earth; habitats like croplands that replace natural habitats such as grasslands are quite inhospitable to all but a few species that tend to be well adapted to the new managed landscapes. Haiti provides one extreme example of habitat loss. **Figure 6–7** shows the border between Haiti and the Dominican Republic, with Haiti on the left. Land use in Haiti has resulted in a tremendous loss of forests, something important not only in the context of loss of species, but also as a driver of soil erosion and water quality problems (as we will see in later chapters). In the Dominican Republic, social trends such as greater stability and democratization have helped the country avoid unrest and environmental destruction.

Fragmentation. Natural landscapes generally have large patches of habitat that are well connected to other similar patches. Human-dominated landscapes, however, consist of a mosaic of different land uses, resulting in small, often geometrically configured patches that frequently contrast highly with neighboring patches **(Fig. 6–8)**. Small fragments of habitat can support only small numbers and populations of species, making them vulnerable to extermination. Species that require large areas are the first to go; those that grow slowly or have naturally unstable populations are also vulnerable.

Reducing the size of a habitat creates a greater proportion of **edges**—breaks between habitats that expose species

[5]Red List Category summary for all animal classes and orders, IUCN Red List version 2011.2: Table 4a. http://www.iucnredlist.org/documents/summarystatistics/2011_2_RL_Stats_Table4a.pdf. March 3, 2012.

[6]Edward O. Wilson. "The Arboreal Ant Fauna of Peruvian Amazon Forests: A First Assessment," *Biotropica* 19 (1987): 245–251.

[7]T. L. Erwin. "The Tropical Forest Canopy: The Heart of Biotic Diversity." In *Biodiversity,* edited by E. O. Wilson, National Academy Press, Washington, D.C., pp. 123–129.

Figure 6–7 The border of Haiti and the Dominican Republic. Severe deforestation in Haiti, on the left, leads to habitat loss for many species. It also causes a number of problems for humans. The forested area on the right is the Dominican Republic.

Figure 6–8 Fragmentation. Development usually leads to the breaking up of natural areas, resulting in a mosaic of habitats that may not support local populations of species.

to predators. Increased edge favors some species but may be detrimental to others. For example, the Kirtland's warbler is an endangered species that is highly dependent on large patches of second-growth jack pines in Michigan. The species is endangered because its habitat has been greatly fragmented, creating edges that favor the brown-headed cowbird, a nest parasite that can invade the forest and lay its eggs in the nest of the rare warbler. Edges also favor species that prey on nests, such as crows, magpies, and jays. These predators are thought to be partly responsible for declines in many neotropical bird populations.

Roadways offer particular dangers to wildlife. The number of animals killed on roadways now far exceeds the number killed by hunters. As rural areas are developed, the increasing numbers of animals found on roadways are a serious hazard to motorists. With 5.7 million miles (9.2 million km) of paved roadways in the United States, more than a million animals a day become roadkill. Overpasses and tunnels are increasingly being built to provide wildlife with safe corridors. A 2008 study at Purdue University identified amphibians as especially affected by road mortality.[8] More than 95% of the roadkill that researchers identified over 11 months along a stretch of road in the Midwest were amphibians.

Simplification. Human use of habitats often simplifies them. Removing fallen logs and dead trees from woodlands for firewood, for example, diminishes an important microhabitat on which species ranging from woodpeckers to germinating forest trees depend. When a forest is managed for the production of one (or a few) species of tree, tree diversity declines, and as a result, so does the diversity of the plant and animal species that depend on the less favored trees. Streams are sometimes "channelized"—their beds are cleared of fallen trees and stony riffle areas (places with more rapidly moving water), and sometimes the stream is straightened out by dredging the bottom. Such alterations inevitably reduce the diversity of fish and invertebrates that live in the stream.

Intrusion. Birds use the air as a highway, especially as they migrate north and south in the spring and fall. Recently, telecommunications towers have presented a new hazard to birds. Although television towers have been around for decades, new towers are sprouting up on hilltops and in countrysides in profusion. Cell phone signal towers are also common. The lights often placed on these towers can attract birds, which usually migrate at night, and the birds simply collide with the towers and supporting wires. In 2000, a study estimated that the towers kill somewhere between 5 and 50 million birds a year.[9] Three organizations petitioned the Federal Communications Commission (FCC) to incorporate bird mortality in its decisions about new towers in 2002, but nothing was done. The scientific research from the study gave environmentalists the power to go to court to

force the FCC to consider risk to birds when making decisions. Finally, a federal appeals court ruled in 2008 that the FCC did in fact have to come up with a plan to protect birds. This ruling resulted in a meeting of five agencies with the FCC in June of 2008. A draft plan followed in 2011, with a time for public comment and hopes of an agreement to make more-bird-friendly towers. The clear surfaces of windows are one of the greatest dangers for birds. Ornithologist Daniel Klem estimated that between 100 million and 1 billion birds die every year from crashing into glass windows. Birds are just one example of organisms harmed by the intrusion of structures into habitats. Any solution to wild species' decline must include creative ways to lower the impact of human structures on other organisms.

Invasive Species. An exotic, or alien, species is one that is introduced into an area from somewhere else, often a different continent (see Chapter 4, where they were covered extensively). Because the species is not native to the new area, it is frequently unsuccessful in establishing a viable population and quietly disappears. Such is the fate of many pet birds, reptiles, and fish that escape or are deliberately released. Some exotic species establish populations that remain at low levels without becoming pests. Occasionally, however, an alien species finds the new environment to be very favorable, and it becomes an **invasive species** thriving, spreading out, and perhaps eliminating native species by predation or competition for space or food. You will learn that most of the insect pests and plant parasites that plague agricultural production were accidentally introduced (Chapter 13). Invasive species are major agents in driving native species to extinction and are responsible for an estimated 39% of all animal extinctions since 1600.

Accidental Introductions. As humans travelled around the globe in the days of sailing ships and exploration, they brought with them more than curiosity and dreams of exotic lands. They brought rats and mice to ports all over the world. As a result, rats are the most invasive mammal worldwide. The two most common invasive rat species are the black rat (*Rattus rattus*, called the ship rat) and the brown rat (*Rattus norvegicus*, called the Norway rat). Fleas infesting black rats carried the bubonic plague bacterium to Europe, eventually killing a third of Europe's human population in the Middle Ages. In terms of species loss, the arrival of rats is especially deadly to birds. One report found that rats have been found preying on nearly one-quarter of seabird species. Thirty percent of seabirds are threatened with extinction, and rats are the biggest single cause. Rats also eat crops, destroy property, and cause other harm directly to humans, but their effect on other species is profound. Rats are only one example of accidental introduction of non-native species. **Figure 6–9** shows another: the red imported fire ant (*Solenopsis invicta*), which arrived in Mobile, Alabama, from South America, possibly in shipments of wood. The fire ant is a major pest across the southern United States, inflicting damage to crops and domestic animals. It also contributes to the decline of wild species. In Texas, it is estimated that fire ants kill a fifth of songbird babies before they leave the nest.

[8]D. J. Glista, T. L. DeVault, and J. A. DeWoody. "Vertebrate Road Mortality and Its Impact on Amphibians," *Herpetological Conservation and Biology* 3 (2008): 77–87.
[9]G. Shire. *New Report Documents Hundreds of Thousands of Birds Killed at Communications Towers* (2000). Available from the American Bird Conservancy at www.abcbirds.org/newsandreports/releases/000629.html.

Figure 6–9 The red imported fire ant. This ant, *Solenopsis invicta*, has invaded the southern United States, where it kills other animals and sometimes starves them by eating their prey.

Figure 6–10 Japanese knotweed. Japanese knotweed can break foundations and walkways, shade other plants, and grow without seeds from a tiny bit of root or underground stem.

May I Introduce . . . Deliberate introductions sanctioned by the U.S. Natural Resources Conservation Service (in the interest of "reclaiming" eroded or degraded lands) have brought us kudzu and other runaway plants, such as the autumn olive and multiflora rose (Chapter 4). Salt cedar was introduced in the American Southwest for erosion control and has taken over riverbanks there. Many other exotic plants are introduced as horticultural desirables **(Fig. 6–10)**. *Fallopia japonica*, Japanese knotweed, is a good example. Tall, with attractive flowers, it originated in Asia and is now found in North America and Europe. It forms dense patches and outcompetes native plants. Unfortunately, the strong underground stem (rhizome) and root systems cause the plant to spread even when it does not produce seeds. Whole plants can grow from tiny fragments of root. In the United Kingdom, it is illegal to dispose of Japanese knotweed except by burning or placement in a landfill. Japanese knotweed is on the World Conservation Union's list of 100 worst invasive species. The site of the 2012 Olympics in London had a large infestation, which was removed over a period of years in a process estimated to cost 70 million pounds (more than $110 million).

There are specific instances of both deliberate and accidental introductions in **aquaculture**—the farming of shellfish, seaweed, and fish—which produces one-third of all seafood consumed worldwide. More than 100 species of aquatic plants and animals are raised in aquaculture in the United States, most of which are not native to the farming locations. Parasites, seaweeds, invertebrates, and pathogens have been introduced together with the desired aquaculture species, often escaping from the pens or ponds and entering the sea or nearby river system. For example, Asian black carp are used in raising catfish because the carp eat snails that harbor parasites capable of ruining the catfish meat. So far, no carp are known to have escaped in the United States, but everywhere else that they have been used in farming, they have escaped and spread. Sometimes the aquaculture species itself is a problem. Atlantic salmon,

for example, are now raised along the Pacific Coast, and many have escaped from the pens and have been breeding in some West Coast rivers, where they compete with native species and spread diseases.

Over Time. The transplantation of species by humans has occurred throughout history, to the point where most people are unable to distinguish between the native and exotic species living in their lands. European colonists brought hundreds of weeds and plants to the Americas, and now most of the common field, lawn, and roadside plant species in eastern North America are exotics. For example, almost one-third of the plants in Massachusetts are alien or introduced—about 725 species, of which 66 have been specifically identified as invasive or likely to become so. In Europe, there are about 11,000 alien species, 15% of which harm biodiversity. Among the animals introduced to North America, the most notorious have been the house mouse and the Norway rat; others include the wild boar, the starling, the horse, and the nutria. One of the most destructive exotics is the house cat: Studies in the 1990s showed that the millions of domestic and feral cats in the United States are highly efficient at catching small mammals and birds **(Fig. 6–11)**. They kill an estimated 1 billion small mammals (chipmunks, deer mice, and ground squirrels, as well as rats) and hundreds of millions of birds annually. One report calculates the annual cost of invasive species in the United States to be $137-billion.[10]

Species originating in the Americas can cause havoc in other places, too. For example, the North American gray squirrel is invasive in Europe, where it outcompetes the native red squirrels and may be infecting them with a fatal virus. The gray squirrel is one of the 100 most invasive species worldwide.

[10]D. Pimental, L. Lach, R. Zuniga, and D. Morrison. "Environmental and Economic Costs of Nonindigenous Species in the United States," *BioScience* 50 (2000): 53–65.

Figure 6–11 The house cat. House cats allowed to roam freely kill more than a billion small animals every year.

Invasive Species and Trophic Levels. Plants provide food for herbivores that in turn provide food for carnivores. Non-native plants may be difficult for herbivores to eat and thus may keep energy and materials from passing up the food chain, even if a species has been in a new ecosystem for a long time. For example, the Norway maple was introduced to North America in 1756, but in spite of being in the United States for hundreds of years, these maples provide much less food up the food chain for herbivores (like caterpillars) and their predators (like songbirds) than native trees do.

Pollution. Another factor that decreases biodiversity is pollution, which can directly kill many kinds of plants and animals, seriously reducing their populations. For example, nutrients (such as phosphorus and nitrogen) that travel down the Mississippi River from the agricultural heartland of the United States have created a fluctuating "dead zone" in the Gulf of Mexico, an area of more than 6,765 square miles (in 2011) where oxygen completely disappears from depths below 20 meters every summer[11] (see Chapter 20). Shrimp, fish, crabs, and other commercially valuable sea life are either killed or forced to migrate away from this huge area along the Mississippi and Louisiana coastline.

Pollution destroys or alters habitats, with consequences just as severe as those caused by deliberate conversions. Every oil spill kills seabirds and often sea mammals, sometimes by the thousands (water pollution is discussed further in Chapter 20). Some pollutants, such as the pesticide DDT, can travel up food chains and become more concentrated in higher consumers (Chapter 5). Acid deposition and air pollution kill forest trees; sediments and large amounts of nutrients kill species in lakes, rivers, and bays; and global climate change, brought on by greenhouse gas emissions, is

already having an impact on many species. Climate change is likely to accelerate in the next 50 years (discussed in depth in Chapter 18).

Pollution can spread disease. Human wastes can spread pathogenic microorganisms to wild species, a threat called pathogen pollution. For example, the manatee (an endangered aquatic mammal) has been infected by human papillomavirus, cryptosporidium, and microsporidium, with fatal outcomes. Pollution can also disrupt hormones and other body functions. Fish exposed to polluted river water may develop tumors or deformed organs. Sometimes they develop both male and female tissues, a condition called "intersex" **(Fig. 6–12)**. The recent and rapid rise in the incidence of these deformities has been traced to habitats that have been altered by human use, specifically run-off, industry, and sewage in populated areas.

Population. Human population growth puts pressure on wild species through several mechanisms—through direct use of wild species, through conversion of habitat into agriculture or other use, through pollution, and through having to compete for resources with growing human populations (human population trends will be described in depth in Chapters 8 and 9). Overconsumption is tied to overpopulation and to the concept of overexploitation—the last letter in the HIPPO acronym—which will be discussed next. Even if each person using a resource used a modest amount, large numbers of people might still use too much, leading to a total overconsumption. Thus, overpopulation is a problem even if people use few resources individually. However, a small group of people can also overuse resources. Indirectly, overconsumption can drive pollution, conversion of habitat, and other effects that harm wildlife. In a world in which some individuals own literally billions of times the money and resources that

Figure 6–12 Intersex fish. Up to 80% of male fish in some rivers are found to have a condition where they also have female tissues, such as eggs developing in their testes. This appears to result from pollution.

[11]Conners, D. (2011). 2011 Gulf of Mexico dead zone smaller than scientists predicted. EarthSky. http://earthsky.org/earth/2011-gulf-of-mexico-dead-zone-smaller-than-scientists-predicted.

others own, people with the most consumptive lifestyles have a disproportionate effect on the environment. Differences in level of consumption and in sheer numbers of people drive some of the tensions among the high-, middle-, and low-income countries (see Chapter 8).

Overexploitation. The overharvest of a particular species is known as overexploitation. Removing whales, fish, or trees faster than they can reproduce will lead to their ultimate extinction. In addition, overuse of individual species harms ecosystems. Overexploitation is estimated to have caused 23% of recent extinctions (past 500 years) and is often driven by a combination of greed, ignorance, and desperation. Poor resource management often leads to a loss of biodiversity. Forests and woodlands are overcut for firewood, grasslands are overgrazed, game species are overhunted, fisheries are overexploited, and croplands are overcultivated. These practices not only deplete the resource in question, but also often set into motion a cycle of erosion and desertification, with effects far beyond the exploited area (this problem is explored in depth in Chapter 7).

Trade in Exotics. One prominent form of overuse is the trafficking in wildlife and in products derived from wild species. Much of this "trade" is illegal. Worldwide, the International Criminal Police Organization (Interpol) estimates that it generates at least $12 billion a year, making it the third-largest illicit income source, after drugs and guns. This illegal activity flourishes because some consumers are willing to pay exorbitant prices for furniture made from tropical hardwoods (like teak), exotic pets, furs from wild animals, traditional medicines from animal parts, and innumerable other "luxuries," including polar-bear rugs, ivory-handled knives, and reptile-skin handbags. For example, some Indonesian and South American parrots sell for up to $10,000 in the United States, and a panda-skin rug can bring in $25,000.

One example of illegal wildlife trade that still continues is that of rhinoceros horn. In 2011, the West African rhino was declared extinct. In October 2011, the Javan rhino was declared extinct in mainland Asia, only surviving in a small population in Indonesia. Rhino species are all very rare because of the high value placed on their horn. Rhino horn is valued as a traditional medicine in Asia and used for ornamental knife handles in the Middle East.

Rhinos have suffered, as have all large African mammals, from loss of habitat and increasing human population. However, the biggest danger to rhinos is direct illegal killing **(Fig. 6–13)**. In November 2011, customs agents in Hong Kong seized $2.2 million worth of rhino horns along with ivory from elephant tusks. The horns represented the deaths of 17 rhinos. The news is not all bad, however. In 2011, a survey of one-horned rhinos in Nepal found that conservation efforts and protection was working and that a small population was increasing. That small glimmer of hope shows us what conservation can do if poaching can be stopped.

Greed. The long-term prospect of extinction does not curtail the activities of exploiters because, to them, the prospect of a huge immediate profit outweighs it. Even when the species is protected, the economic incentive is such that poaching and black-market trade continue. Particularly severe is the growing fad for exotic pets—fish, reptiles, birds, and houseplants. In many cases, these plants and animals are taken from the wild; often, only a small fraction even survives the transition. When animals and plants are removed from their natural breeding populations, these individuals are no better than dead when it comes to maintaining the species. At least 40 of the world's 330 known species of parrots are threatened because of this fad. Many pet stores carry imported species. Not purchasing wild-caught species is one way individuals can help preserve biodiversity.

In 1992, the United States, the largest importer of exotic birds, adopted the Wild Bird Conservation Act, a law designed to stop wild capture of declining birds, uphold international treaties in wildlife trade, and support sustainable breeding programs. In 2004, more than 200 nongovernmental organizations (NGOs) signed the European Union Wild Bird Declaration, a call for wild bird protection in the European Union (EU). In a huge step forward in bird conservation, in 2007 the EU decided unanimously to prohibit the import of wild birds.

(a)

(b)

Figure 6–13 Black rhino and rhino horn products. (a) Rhino horn is prized in traditional Asian medicine and as ornamentation. (b) The black rhino is often killed by poachers for its horn.

Consequences of Losing Biodiversity

Why is biodiversity loss a concern? The answer goes back to the values of wild species. Intrinsic value is self-explanatory: the loss of species matters simply because that species is valuable, regardless of human benefit. However, much of the discussion about the consequences of losing species is focused on instrumental value—on the fact that we humans need biodiversity. We need wild species for all the reasons we value them: we use them directly for food, fiber, fuel; we value them aesthetically, want medicines from them, want recreation in natural habitats, or need them to maintain healthy ecosystems. It is this last reason that most people understand the least. The complexity of ecosystems means that the loss of diversity causes them to break down.

The Millennium Ecosystem Assessment team agreed: biodiversity is essential for the ecosystem services and goods that human societies derive from the natural world. For example, richly diverse mangrove forests and coral reefs are buffers against coastal storms and floods. The well-being of millions of rural poor in the developing world is tied to the provision of products of natural ecosystems, especially during economic hard times. In fact, biodiversity loss threatens the achievement of virtually all of the Millennium Development Goals (see Chapter 8). Nature-based tourism would not happen without the rich diversity of wildlife and flora that attract and delight people; for many countries of sub-Saharan Africa, nature-based tourism represents the largest earner of foreign currency.

Can ecosystems lose some species and still retain their functional integrity? Research evidence indicates that the dominant plants and animals determine major ecosystem processes such as energy flow and nutrient cycling. Thus, simplifying ecosystems by driving rare species to extinction will not necessarily lead to ecosystem decline. This has risks, however: It is possible to lose keystone species—species whose role is absolutely vital to the survival of many other species in an ecosystem (see Chapter 4). For example, through felling trees and building dams, beavers make local habitats that undergo succession as beavers move throughout the landscape, increasing ecosystem diversity in a region. Alternatively, the keystone species in an ecosystem may be a predator at the top of a food chain that keeps herbivore populations under control. Sometimes keystone species are the largest animals in the ecosystem, and because of their size, they have the greatest demands for unspoiled habitat. Thus, wolves, elephants, tigers, moose, and other large animals are considered "umbrella" species for whole ecosystems.

As we have seen, one thing many declining species have in common is that they are often long-lived, large, and older at first reproduction and have high parental care of offspring (as *K*-strategists are) (Chapter 4). Thus, they are vulnerable to rapid changes in the environment. Sometimes species go

---

STEWARDSHIP
Lake Erie's Island Snake Lady

A three-foot-long grayish-tan snake basks on a rock at the shoreline of an island in Lake Erie, Pennsylvania. As is warms itself, the sun glints on its scales. Slowly it moves, one coil sliding over another, and quietly glides into the water. Head still, it watches a small round goby fish (*Neogobius melanostomus*). Suddenly, it strikes. The goby caught, the snake swallows slowly and slithers off to resume basking in the sun. The snake is the Lake Erie water snake *Nerodia sipedon insularum*, and it is endemic to the area. It was once endangered and limited to only 1,000 individuals and was subject to conservation measures.

Out of the vegetation steps a woman in field gear. Kristen Stanford, known as the "Island Snake Lady," is the water snake recovery plan coordinator for this rare reptile, working out of the Ohio State University Stone Laboratory. Over the past several years, she and her colleagues caught, measured, tagged, and protected water snakes; as a result, the small snake recovered in numbers.

The hard work of scientists, policy makers, and enthusiasts paid off. The snake, returning to a population of 12,000 individuals, was taken off of the threatened species list in 2011. Species are placed on the threatened and endangered species lists when they are already so rare that, for many, recovery is a long, drawn-out process. The rapid recovery of the snake is great news for conservationists.

What makes populations of one species increase rapidly and causes populations of another to decline? How do species relate to the other species around them? In the case of the water snake, intensive human killing and loss of habitats were the main causes of decline. Ironically, the presence of large numbers of invasive goby fish helped the snake come back from the brink of extinction. This unexpected benefit is not the norm for invasive species, however.

Many people find it hard to feel concerned about a declining snake. They may feel that snakes are threatening or are unattractive. Many people need a reason to care for a species in decline and find it much easier to relate to rabbits, whales, a beautiful bird like the whooping crane, or perhaps the rare and majestic Bengal tiger. These species are considered "charismatic" because of their wide appeal. But many of the world's creatures that are disappearing are the small and seemingly unimportant parts of the

Lake Erie Water Snake

ecosystem, like a rare water snake or small beetles. Why care about them? Conservationists suggest several reasons for concern about the loss of species. For people like Kristen Stanford, protection of species, even the small ones, is worth the cold, the bitten hands, and the wet feet that come with being "the snake lady."

Sources: Water Snake, Lake Erie. (2001). Ohio Public Library Information Network. http://www.oplin.org/snake/fact%20 pages/water_snake_lake_erie/water_snake_lake_erie.html. March 6, 2012.

Michael Scott, The Plain Dealer. "Lake Erie water snakes are making a comeback." Published Sunday, June 08, 2008, http://blog.cleveland.com/metro/2008/06/snakes.html.

into decline even though they are quite common. Species such as the passenger pigeon, once the most common bird in North America, can leave a huge ecological gap when they decline or disappear. Those species least likely to be harmed by human actions include widely distributed species with small size, rapid reproduction, short generation time, low parental care of offspring, and an ability to migrate (characteristics of *r*-strategists). Such species (roaches, houseflies, and others) are most likely to become pest species. Thus, the overarching effect of human disturbance of the environment is not only to decrease species that we may like, but also to increase a number that will become major pests.

Moving Forward

What will happen as rare or unknown species go extinct because of human activities? Some people will mourn their passing. We do not really know, however, what we are losing when species become extinct. One thing is certain: The natural world is less beautiful and less interesting as a result. We might be inclined to feel overwhelmed by these numbers and the significant decline in species and in abundances of organisms. But there are glimmers of hope. Sometimes species we thought were extinct are found not to be. Two of these are the Beck's petrel, not seen for 80 years before being found by an ornithologist in 2008, and the pygmy tarsier, not seen since 1921 and thought extinct but found on the Indonesian island of Sulawesi in 2008. Sometimes new populations of very rare species are discovered, as happened in the case of the highly threatened greater bamboo lemur. A new population of this rare species was found in 2007, around 400 kilometers (240 mi) from the only previously known population. As we saw in Stewardship, "Lake Erie's Island Snake Lady," sometimes the hard work of scientists and conservation volunteers pays off in the recovery of a rare and declining species such as the modest water snake. While success in protecting species begins with the courage and determination of individuals, conservation requires the coordinated efforts over regional, national, and international scales, involving the work of both scientists and policy makers.

Sometimes new protections emerge from a change in policy. Ironically, the conservation of birds was enhanced in the EU when concern over avian flu made it possible to pass regulations limiting the import of wild birds. Sometimes scientific accomplishments make conservation doable. For example, 2008 saw the first live rhinoceros birth from frozen semen, and black-footed ferrets, an extremely rare weasel-like animal, have been born from sperm stored from fathers who had died eight years before **(Fig. 6–14)**. These breakthroughs in captive breeding may save species. Once in a while, conservationists catch a break, as they did in 2008, when a rare Vietnamese turtle, one of the last four of its species in the world, was swept from a lake by a flood and caught by a fisherman who was going to sell the reptile to a restaurant for soup. Conservationists were able to purchase the 68-kilogram (150-lb) turtle and return it to the lake. In the next section, we will see more of the efforts scientists and others are putting into saving wild species.

Figure 6–14 Captive breeding with new techniques. Captive breeders are now able to use frozen semen to produce black-footed ferrets, an extremely rare animal.

6.3 Saving Wild Species

If wild species and biodiversity as a whole have so much value and if they are in such decline, there must be efforts in the works to solve the problem. Indeed, there are. Those efforts involve basic science, individuals, corporations, public policy, and decisions at the local, national, and international levels (Table 6–3). Scientists are on the front lines of biodiversity protection, as they often have a better knowledge of what is out there in the world and whether or not it is declining. However, scientists cannot stop biodiversity loss by themselves. That requires laws, and it demands enforcement of those laws. Protection of biodiversity also requires stepping back to look at the big picture, seeing how people relate to the rest of the natural world, and examining how we can do so in more sustainable ways.

The Science of Conservation

The scientific projects that have been used to save the Lake Erie water snake, rhinos, and black-footed ferrets are just a few of many research activities in conservation biology, the branch of science that is most focused on the protection of populations and species. Conservation biologists use techniques like captive breeding, telemetry (transmitting data through remote communication), and tracking devices in order to learn more about the natural history of particular species and protect those that are vanishing. Scientists monitor species to see which are threatened with extinction and which are invasive and need control. They study the relationships between species and between biodiversity as a whole and ecosystem functions.

There is another critical element in the science of protecting species—**taxonomy**. Taxonomy is the cataloging of species and the naming of new ones. It allows us to understand what species are out there and to identify those in the most trouble. Right now, the field of taxonomy is suffering from a lack of experts, and this lack is harming biodiversity conservation. By 2018, there may be no specialists in fungi

TABLE 6-3 Actions Taken to Protect Species by Individuals and Different Groups

Group	Type of action	Example
Individual	Support policy	Vote to support biodiversity
	Smart consumerism	Purchase sustainable materials, use fewer resources
	Citizen science	Great Sunflower Project
Business	Sustainable practices	Coffee plantations with plants for birds
	Smart investment	Go into green technologies and businesses
Nonprofit	Land conservation	Protection of marine reserves/wildlife
	Captive breeding	Zoo programs
	Education	Environmental education
	Sustainable development	Working with local groups for conservation goals
Science	Basic research in natural history, taxonomy, captive breeding techniques	Rhino breeding
	Develop wildlife forensics	Scottish Wildlife Forensics labs
	Monitor species	IUCN Red List
	Control invasive species, and better understand other causes of species decline	Study effects of house cats or cell phone towers on birds
Government	Make sound policies for local and national needs	Lacy Act, Marine Mammal Act, Endangered Species Act
	Develop international treaties	CITES, Convention on Biodiversity

in the United Kingdom, despite the fact that fungi provide medicines, can be major crop pests, and are important parts of soil health.[12] Because most people cannot even identify the species we are dealing with, it is difficult to promote solutions. The loss of expertise as taxonomists retire and the pressure from international treaties to protect species may require increased funding to train new taxonomists.

Individuals and Corporations

Individuals can directly support conservation by being careful in their consumption of resources, use of chemicals, driving habits, and other personal choices. However, many of the actions individuals take are through support of nonprofits, pushing for policy actions, and pushing corporations to work in more-biodiversity-friendly ways. For example, consumers are insisting on more organic foods and certification of wild-caught fish or forest products as sustainable (some of these approaches are discussed in Chapter 7). Individuals can act as "citizen scientists" and take part in massive information-gathering efforts such as the Great Sunflower project, which monitors pollinators (**Fig. 6–15**), the Firefly Project (which tracks firefly abundance), the annual Audubon Society Christmas Bird count (in which volunteers count birds), Project Budburst (a project that tracks when plants change seasonally), and others.

Consumers also encourage businesses to care about species. Many businesses are beginning to realize that biodiversity loss could harm their profits. Biodiversity-friendly investing is

on the rise as well. A survey in 2009 of global CEOs found that 27% expressed concern about the impacts of biodiversity loss on their business prospects.

Nonprofit Efforts

One way individuals contribute to conservation of biodiversity is through the funding of nonprofit organizations such as the World Wildlife Fund, Audubon Society, and many others. With some fundraising campaigns, people are able to "sponsor" individual whales, tigers, polar bears, and other creatures considered "charismatic"—that is, widely appealing. The efforts to protect these species, which often require

Figure 6–15 Citizen science. The Great Sunflower Project is one of many that links a network of amateur scientists to collect data on species. In this case, they study pollinators, which are in decline.

---

[12]S. Connor. "Fungi scientists are endangered species: Budget cuts threaten research into basics that propagate life on Earth," *The Independent* (November 2008).

large areas of habitat, protect other species as well. Land conservation is used to protect multiple species at the same time (discussed in more detail in Chapter 7). The oldest regional land conservancy in the United States is the Trustees of Reservations in Massachusetts.

Zoos have a special role in biodiversity conservation. Not everyone believes it is a good thing for animals to be kept in zoos, away from their natural habitat. However, in many cases, zoos have a two-fold role in protecting species. First, they actively educate the public so people care about species protection. Second, many are involved in research and in captive breeding programs. For some species, the breeding done at zoos and other captive breeding facilities is the difference between survival and extinction. Botanical gardens can have the same role for plants. Of course, breeding many species solely in captivity is not a sustainable long-term solution for species loss, but it may bridge the gap while humans figure out how to live more sustainably.

Governments: Local, State, and National Policies

Public policies, and the agencies that make and support them, are necessary for the protection of species. Governments from local to international are involved in monitoring, conserving, and protecting biodiversity. If we want to save wild species, we first need to ask who owns wildlife. In the United States, property owners do not own the wildlife living on their lands; wildlife resources are public resources, protected under the Public Trust Doctrine. The government holds these resources in trust for all its people and is obliged to provide protection for these resources. In the United States, this protective role is usually exercised by state fisheries and wildlife agencies. Sometimes the law mandates federal jurisdiction, as in the case of endangered species. Game animals were some of the first animals people recognized as having the potential to be overharvested; consequently, they are a group that receives special oversight, even if they are not currently rare.

Game Animals in the United States. Game animals are those animals traditionally hunted for sport, meat, or pelts. In the early days of the United States, there were no restrictions on hunting, and a number of species were hunted to extinction (the great auk, heath hen, and passenger pigeon) or near extinction (the bison and wild turkey). As game animals became scarce in the face of unrestrained hunting, regulations were enacted. State governments, backed up by the federal government, enacted laws establishing hunting seasons and bag limits and hired wardens to enforce the laws. Some species were given complete protection in order to allow their populations to build up to numbers that would once again allow hunting. The wild turkey, for example, was hunted to the brink of extinction. It made a slow comeback by the 1930s as a result of hunting restrictions and returned strongly after reintroduction into areas it had once inhabited. The turkey is now found in 49 states. Unfortunately, the turkey was also introduced to western states, where it

was not native, and it is now a pest in those areas. Ideally, in the future it would be protected in states where it is native and removed from states where it is invasive.

Hunting and Conservation. Using hunting and trapping fees as a source of revenue, state wildlife managers enhance the habitats that support important game species. In spite of what seems on the surface to be destruction of wildlife, hunting has many positive aspects. Besides the fees hunters pay, many hunters belong to organizations dedicated to the game they are interested in hunting. Organizations such as Ducks Unlimited raise funds that are used for the restoration and maintenance of natural ecosystems vital to the game they are interested in hunting.

Defenders of hunting and trapping argue that their prey are often animals that lack natural predators and would increase to the point of destroying their own habitat, something that frequently happens in the United States with larger animals such as deer and elk. For example, there are so many deer in Pennsylvania that, in 2004, they were estimated to be at three times the natural carrying capacity. The deer eat crops, damage cars in collisions, and seriously harm forest understories through extensive browsing. It might seem like hunting is the answer, but sometimes it is not that simple. Despite high deer numbers, some hunters have actually lobbied the state to take action to *increase* the deer herd. Why? These hunters are hunting in the northern and western part of the state, where deer populations were once heavy. Unfortunately, the deer have left these forests and have moved to the southern and eastern parts of the state, where they eat agricultural crops and browse in suburban developments. They wreak economic damage but are very difficult to hunt. Thus, the perception of hunters is that deer populations are declining, while the perception of many other people (including wildlife managers) is that there are too many deer for ecosystems to support.

Bring Back the Predators? Many members of the non-hunting public object to the killing of wildlife, and some groups, such as the Humane Society of the United States and People for the Ethical Treatment of Animals (PETA), actively campaign to limit or end hunting and trapping. Some practices that are regarded as especially cruel, such as the use of leghold steel traps, have been banned in several states as a result of ballot initiatives. Even people who are not opposed to hunting may feel that encouraging hunting because of a lack of other predators may not be getting at the fundamental problems with species loss. They argue for the return of predators such as grizzly bears, mountain lions, and wolves in order to restore more natural checks and balances to ecosystems. Because humans do not always get along well with top predators, such reintroductions are unlikely to occur on a large scale (**Fig. 6–16**).

Too Many Animals. Because wild animals are in the public trust, their numbers are managed by government agencies. Now that many populations are increasing, the government sometimes has to limit them. Many nuisance animals are thriving in highly urbanized areas, creating various health

Figure 6–16 Wild cougar. As humans encroach on wild habitat, animals come into human areas. This cougar was on a suburban roof in California. It was shot with a tranquilizer. Returning big predators to ecosystems is not likely to happen in human-inhabited areas.

hazards. Opossums, skunks, and deer are attracted to urban areas by opportunities for food; unsecured garbage cans and pet food left outside will ensure visits from raccoons. In 2010, the Centers for Disease Control and Prevention reported 6,153 cases of rabid animals in the United States, 60% of which were raccoons and skunks.[13]

One highly controversial effort to control unwanted animals is carried out by an agency of the U.S. Department of Agriculture. **Wildlife Services** responds to requests from livestock owners, farmers, homeowners, and others to remove nuisance animals and birds. "Removal" virtually always means killing, and Wildlife Services routinely uses poisons, traps, and other devices to kill around 2.5 million animals yearly. Of these, most are native species such as raccoons or skunks, but others, such as wild hogs and starlings, are exotic invasive species. Wildlife Services plays the important role of keeping both invasive species and native species from overtaking human areas and limiting negative human–wildlife interactions.

Protecting Endangered Species

In colonial days, huge flocks of snowy egrets inhabited the coastal wetlands and marshes of the southeastern United States. In the 1800s, when fashion dictated fancy hats adorned with feathers, egrets and other birds were hunted for their plumage. By the late 1800s, egrets were almost extinct. In 1886, the newly formed National Audubon Society began a press campaign to shame "feather wearers" and end the practice. The campaign caught on, and gradually attitudes changed; new laws followed. Government policies that protect animals from overharvesting are essential to keep species from the brink of extinction. Even when cultural standards change due to the efforts of individual groups (such as the National Audubon Society), laws and policy measures must

follow to ensure that endangered populations remain protected. Since the 1800s, several important laws have been passed to protect a wide variety of species.

Lacey Act. Florida and Texas were the first states to pass laws protecting plumed birds. Then, in 1900, Congress passed the **Lacey Act,** forbidding interstate commerce in illegally killed wildlife—and therefore making it more difficult for hunters to sell their kill. Since then, numerous wildlife refuges have been established to protect the egrets' breeding and migratory habitats. With millions of people visiting these refuges and seeing the birds in their natural locales, attitudes have changed significantly. Today, the thought of hunting these birds would be abhorrent to most of us, even if official protection were removed. Thus protected, egret populations were able to recover substantially. In the meantime, the Lacey Act and its amendments have become the most important piece of legislation protecting any wildlife from illegal killing or smuggling. Under the act, the U.S. Fish and Wildlife Service (USFWS) can bring federal charges against anyone violating a number of wildlife laws. In 2007, the Combat Illegal Logging Act of 2007 (S. 1930) was proposed to expand the Lacey Act to cover rare timber and timber products. It is still waiting to move forward.[14]

For examples of the work of the U.S. Fish and Wildlife Service: In 2007, special agents worked on more than 12,177 violations of the Lacey Act and other wildlife-related laws. Their efforts resulted in fines of more than $14 million and a total of 31 years of jail time for those convicted. In one case, owners of an animal park in South Dakota were caught trafficking bear gall bladders and buying and selling threatened grizzly bears; in another, a Utah man was fined for having 106 illegally caught bobcat skins that he had intended to sell

[13]J. D. Blanton, et al. *Rabies Surveillance in the United States During 2010.* Available at http://avmajournals.avma.org/doi/pdf/10.2460/javma.239.6.773.

[14]USFWS Actions. *US FWS Office of Law Enforcement.* Annual Report FY 2007 (2008). Available at http://www.fws.gov/le/aboutle/annual.htm.

in Europe (where their value was $38,000). In 2009, agents helped catch a man bringing endangered fish into the United States from Indonesia in a suitcase. In 2011, a Washington State couple was prosecuted for sale of bald and golden eagle body parts from at least 45 birds. As wildlife smuggling increases and budgets tighten, USFWS agents work harder to catch people harming the rarest species.

Endangered Species Act. Congress took another major step when it passed a series of acts to protect endangered species. The most comprehensive and recent of these acts is the **Endangered Species Act** of 1973 (reauthorized in 1988). Recall that an endangered species is a species that has been reduced to the point where it is in imminent danger of becoming extinct if protection is not provided (**Fig. 6–17**).

(a) Male Pine Barrens Tree Frog

(b) Silversword

(c) Red-cockaded Woodpecker

(d) Karner Blue Butterfly

(e) Swamp Pink

(f) Whooping Cranes

(g) Devil's Hole Pupfish

(h) White Oryx

(i) Manatee

Figure 6–17 Endangered species. Shown are some examples of endangered species—species whose populations in nature have dropped so low that they are in imminent danger of becoming extinct unless protection is provided.

Genetically distinct populations (subspecies) such as the Florida panther and the peninsular bighorn sheep may also be protected. Further, the act provides for the protection of threatened species, which are judged to be in jeopardy but not on the brink of extinction. When a species is officially recognized as being either endangered or threatened, the law specifies substantial fines for killing, trapping, uprooting (in the case of plants), modifying significant habitat of, or engaging in commerce in the species or its parts. The legislation forbidding commerce includes wildlife threatened with extinction anywhere in the world.

The Endangered Species Act is administered by the U.S. Fish and Wildlife Service for terrestrial and freshwater species (and a few marine mammals) and by the National Oceanic and Atmospheric Administration (NOAA) Fisheries Service for marine and anadromous species (those that migrate into freshwater). There are three crucial elements in the process of designating a species as endangered or threatened:

1. **Listing.** Species may be listed by the appropriate agency or by petition from individuals, groups, or state agencies.

Listing must be based on the best available information and must not take into consideration any economic impact the listing might have. Current listed species are summarized in Table 6–4.

2. **Critical Habitat.** When a species is listed, the agency must also designate as *critical habitat* the areas where the species is currently found or where it could likely spread as it recovers. A 1995 Supreme Court decision made it clear that federal authority to conserve critical habitats extended to privately held lands.

3. **Recovery Plans.** The agency is required to develop recovery plans that are designed to allow listed species to survive and thrive.

As Table 6–4 indicates, as of November 2011, 1,990 U.S. species were currently listed for protection under the act; recovery plans are in place for 1,138 of them. In addition, 251 candidate species are waiting to be listed.

Alternatives and Roadblocks. The Endangered Species Act was scheduled for reauthorization in 1992, but because of political battles, it has been operating on year-to-year

TABLE 6-4 Federal Listings of Threatened and Endangered U.S. Plant and Animal Species, 2011

Summary of Listed Species, Listed Populations,[a] and Recovery Plans[b] as of November 19, 2011

	United States			Foreign			Total Listings (U.S. and Foreign)	U.S. Listings with Active Recovery Plans[c]
	Endangered	Threatened	Total Listings	Endangered	Threatened	Total Listings		
Mammals	70	14	84	256	20	276	360	60
Birds	76	16	92	204	14	218	310	85
Reptiles	13	24	37	66	16	82	119	37
Amphibians	15	10	25	8	1	9	34	17
Fishes	77	68	145	11	1	12	157	101
Clams	64	8	72	2	0	2	74	70
Snails	25	12	37	1	0	1	38	29
Insects	51	10	61	4	0	4	65	40
Arachnids	12	0	12	0	0	0	12	12
Crustaceans	19	3	22	0	0	0	22	18
Corals	0	2	2	0	0	0	2	0
Animal Subtotal	422	167	589	552	44	604	1,193	469
Flowering Plants	613	147	760	1	0	1	761	638
Conifers and Cycads	2	1	3	0	2	2	5	3
Ferns and Allies	27	2	29	0	0	0	29	26
Lichens	2	0	2	0	0	0	2	2
Plant Subtotal	644	150	794	1	2	3	797	669
Grand Total	1,066	317	1,383	553	54	607	1990	1,138

Source: Department of the Interior, U.S. Fish and Wildlife Service, Division of Endangered Species, November 19, 2011.

[a]A listing has an E or a T in the "status" column of the tables in 50 C.F.R. §17.11(h) or 50 C.F.R. §17.12(h) (the "List of Endangered and Threatened Wildlife and Plants").

[b]Sixteen animal species (11 in the United States and 5 in foreign countries) are counted more than once in the above table, primarily because these animals have distinct population segments (each with its own individual listing status).

[c]Recovery plans pertain only to U.S. species.

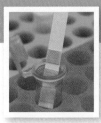

SOUND SCIENCE   Using DNA to Catch Wildlife Criminals

Scotland: A group of men stand around, holding shovels, nets, and the leashes of dogs wearing radio collars. A local police officer stops them, wondering what they are up to. They claim to be hunting rabbits, but the police think otherwise. In other years, there might not be anything for the police to do, but now, it is easier to gather evidence. The police swab the mouths of the dogs, and their suspicions are confirmed when the gathered specimens are found to have traces of fox DNA.

Sound like an odd episode of *CSI: Wildlife*? You are right! But in this real-life drama, Scotland's own Wildlife DNA Forensic Unit came to the rescue to give the police the data they needed to apprehend criminals. The illegal fox hunters pleaded guilty, and the work of the first dedicated wildlife DNA forensic testing facility in Europe was showcased.

New technologies can often help in crimes against wildlife, much as it is widely used to solve crimes against humans. In Italy, scientists were able to show that the blood on a knife owned by a suspected poacher had come from a specific poached wild boar. Scientists at the Center for Conservation Biology at the University of Washington have worked to find the origin of contraband ivory shipped from Africa to Singapore. Were the elephants slaughtered for their tusks all from one place or scattered throughout Africa? They compared DNA from the ivory with DNA patterns found in elephant dung collected all over the African continent. By doing so, they were better able to identify where the poaching had occurred. In this instance, they found that the origin of the tusks was a small area centered in Zambia. The group's efforts could be increased to stop poaching. They were able to do the same with a second shipment of contraband ivory, detected in a shipping container in Malaysia. In this case, the elephants came from another region. Knowing how wildlife traffickers are attacking rare species is important to their protection.

Wildlife forensics is increasing as wildlife trafficking increases. Some people believe that the solution is not to ban traffic in rare species outright, but to "farm" them so that limited legal amounts could be harvested. This is a controversial approach, however. Regardless of how we try to protect them, crimes against wild species populations occur, and when they do, it is helpful to have cutting-edge science and technology to unravel the secrets.

Source: http://conservationbiology.net/research-programs/tracking-poached-ivory-2/

budget extensions. It has been a political lightning rod since its enactment. Opposition to the act comes from development, timber, recreational, and mineral interests as well as other lobby groups. Organizations such as the American Land Rights Association and the American Forest Resource Council continually lobby for weakening or abolishing the Endangered Species Act because they believe it limits their property rights to log, build buildings, and carry out other activities. In 2005, then-congressman Richard Pombo, who once chaired the House Resources Committee, sponsored H.R. 3824, the Threatened and Endangered Species Recovery Act (TESRA). TESRA had significant limitations from a scientific standpoint. The objections of scientists to TESRA were so strong that six scientific societies signed a joint statement outlining their concerns in 2006.[15] TESRA passed the House in the 109th Congress but failed to make it through the Senate and was not reintroduced subsequently.

Conflicting Values. Some critics claim that the Endangered Species Act has been a failure because only 23 of the 1,900 listed species have recovered and been taken off the list and 25 others downlisted from endangered to threatened (2011). In response, proponents state that the two main causes of extinction—habitat loss and invasive species—are on the increase. Furthermore, species are listed only when they have already reached dangerously low populations, which means that they almost certainly would have gone extinct without the protection. If success were measured by the number of species that have stabilized or increased their populations, then the act might be judged a success, as some 41% of listed species have responded this way after being listed.

Other critics of the Endangered Species Act believe that it does not go far enough. A major shortcoming is that protection is not provided until a species is officially listed as endangered or threatened and a recovery plan is established. For half of the protected vertebrate species, fewer than 1,000 individuals remained in each population at the time of listing; for half of the listed plant species, fewer than 120 individuals remained. Currently, the list's backlog includes some 253 "candidate species," all of which are acknowledged to be in need of listing. In the first 20 years following the passage of the Endangered Species Act, 114 U.S. species went extinct (or "missing," which means that, despite at least 10 years of survey, they haven't been seen). The majority of these never made it to the list and became extinct during a delay in listing procedures.

Another contentious issue is the establishment of critical habitat. Opponents believe that the critical habitat designation puts unwanted burdens on property owners and assert that the critical habitat findings have little merit in conserving species. TESRA, for example, would have completely eliminated critical habitat protections and replaced them with a statement that recovery plans must identify areas of "special value" to the protection of the species. However, nothing in this designation is intended to be mandatory. Critics of TESRA answered that critical habitat works; species with critical habitat have been twice as likely to be on the road to recovery as those without it. In fact, current efforts are

---

[15]*Scientific Societies' Statement on the Endangered Species Act* (February 27, 2006). Available at http://www.conbio.org/Sections/NAmerica/ScientificSocieties OnUSESA.pdf.

to protect the critical habitat of multiple species at once, an ecosystem approach.

Improving Legislation. In March 2006, a letter signed by almost 6,000 scientists from around the country was delivered to the U.S. Senate, urging them to pass legislation that maintains and strengthens the provisions and funding of the Endangered Species Act and to reject proposals that would weaken the act.

In the final analysis, the Endangered Species Act is a formal recognition of the importance of preserving wild species, regardless of whether they are economically important. Species listed as endangered or threatened have legal rights to protection under the law. The act is something of a last resort for wild species, but it embodies an encouraging attitude toward nature that has now become public policy. Some changes may make the act function better. These might include tax breaks as incentives to landowners caring for endangered species or other measures to lift the burden of critical habitat designations. In 2007, a bipartisan group of 17 senators introduced another bill—the Endangered Species Recovery Act (ESRA) of 2007 (S.700)—to suggest just that. The reauthorization of a strong Endangered Species Act, possibly with changes to better protect whole ecosystems, will be a major signal of our continuing commitment to consider the value of wild species on grounds other than economic or political.

Seeing Success

Efforts to protect species—such as the Lacey Act, the Marine Mammals Protection Act, the Endangered Species Act, and many other laws designed to keep our biotic heritage from going extinct—have resulted in some successes. The Lake Erie water snake is one such example. Another example of a species restored to many former areas of its range because of these protection and recovery efforts is the gray wolf. Other species have also benefited significantly from these laws.

Birds of Prey. One of the species recently removed from the Endangered Species List because its numbers have recovered significantly is the American bald eagle, the U.S. national symbol **(Fig. 6–18)**. The peregrine falcon, another previously endangered species, has also been removed for the same reason. Both are large raptors—predatory birds at the top of the food chain. Both the bald eagle and the peregrine falcon were driven to extremely low numbers because of the use of DDT as a pesticide from the 1940s through the 1960s. Carried up to these predators through the food chain, DDT caused a serious thinning of the birds' eggshells that led to nesting failures in the two species and in numerous other predatory birds (see Chapter 1). DDT use was banned in both the United States and Canada in the early 1970s, and the stage was set for the birds to recover. On August 20, 1999, the peregrine falcon was delisted, and on June 28, 2007, the Interior Department took the American bald eagle off the Endangered Species List. Both raptors will still be protected by the Migratory Bird Treaty Act, and the eagle will be protected by the Bald and Golden Eagle Protection Act.

Figure 6–18 The American bald eagle. The Endangered Species Act worked well in bringing this magnificent national symbol to the point where it was dropped from the list.

Fly Away Home. A few species have gained exceptional public attention, and heroic efforts have been mounted to save them. Efforts to save the whooping crane, for example, include virtually full-time monitoring and protection of the single remaining flock, which for years numbered only in the teens. The historic migratory wild flock is increasing steadily and reached a population of 279 individuals in the spring of 2011 from a low of 14 cranes in 1939. This flock migrates from their wintering grounds on the Texas Gulf Coast to Wood Buffalo National Park in Canada. Nonmigratory flocks have been reintroduced in central Florida and Louisiana. Some 30 birds now make up these flocks. A spectacular effort has established a new migratory flock of 105 cranes that summer in Wisconsin and migrate to Florida. This flock was first "taught" its migratory path by following researchers in ultralight aircraft **(Fig. 6–19)**. Although each of these flocks is extremely vulnerable to environmental changes, whooping cranes have suffered setbacks before, and conservationists remain hopeful.

Whooping cranes are one of the examples of successful conservation, and, although they still need help, their numbers are clearly increasing. Conservationists treasure these examples in a time when so many species are in decline. It is important to know that the work you do can make a difference. Another source of encouragement for people interested in saving wild species are the so-called "Lazarus species"—species that were thought extinct and later found to still have some members alive. The Bermuda petrel is one such species. Not seen since 1620, it was thought extinct for 330 years until 1951, when 18 nesting pairs were found on a remote island. The Takahe, a flightless bird of New Zealand, is another example. The last four known specimens were taken in 1898. In 1948, it was rediscovered. Both birds are endangered today, but not extinct. These stories of hope help motivate people in the hard work of protecting species.

Figure 6–19 Whooping cranes and pilot. A pilot in an ultralight aircraft leads a flock of whooping cranes in their migration from Wisconsin to Florida.

6.4 Protecting Biodiversity Internationally

Serious efforts are being made to preserve biodiversity around the world, especially in the tropics, where so much of the world's biodiversity exists. These protective efforts require an immense amount of coordination among local, state, and federal authorities. The National Biological Information Infrastructure is a collaborative program that makes protecting species easier for agencies in the United States, but the United States also has to coordinate with the rest of the world. These partnerships not only create treaties, but also monitor species, share scientific advances, and find solutions to the needs of people who are clashing with wild species.

International Developments

There are numerous international developments for the protection of species. Some, like the International Union for Conservation of Nature (IUCN, formerly called the World Conservation Union), are organizations that monitor the successes and failures of our conservation efforts. Others are organizations that help coordinate scientists or policy makers around the globe. For example, the Invasive Species Specialist Group (ISSG) is a part of the Species Survival Commission of IUCN. The ISSG is made up of invited world experts who act as advisors and maintain a global invasive species database. The policy and treaty makers have to hammer out documents like the Convention on Biological Diversity. Finally, there has to be funding for species protection, some of which is provided by the Critical Ecosystem Partnership Fund and other foundations.

The Red List. The IUCN maintains a "Red List" of threatened species.[16] Similar to the U.S. Endangered Species List,

the Red List uses a set of criteria to evaluate the risk of extinction for thousands of species throughout the world. The list is updated frequently, and there were 18,678 species of plants and animals on it in 2011. Once published in book form, the Red List is now available in searchable electronic format on the Internet. Each species on the list is given its proper classification, and its distribution, documentation, habitat, ecology, conservation measures, and data sources are described.

The IUCN is not actively engaged in preserving endangered species in the field, but its findings are often the basis of conservation activities throughout the world, and it provides crucial leadership to the world community on issues involving biodiversity. One of its efforts was to help establish the Trade Records Analysis for Flora and Fauna in International Commerce (TRAFFIC) network, which monitors wildlife trade and helps to implement the decisions and provisions of the Convention on International Trade in Endangered Species of Wild Fauna and Flora (CITES).

CITES. Endangered species outside the United States are only tangentially addressed by the Endangered Species Act. However, under the leadership of the United States, CITES was established in the early 1970s. CITES is not specifically a device to protect rare species—instead, it is an international agreement (signed by 169 nation-states) that focuses on trade in wildlife and wildlife parts. The treaty recognizes three levels of vulnerability of species—the highest being species threatened with extinction—and covers 5,000 animals and 28,000 plants, fungi, and other species. Restrictive trade permits and agreements between exporting and importing countries are applied to those species, sometimes resulting in a complete ban on trade if the nations agree. Every two or three years, the signatory countries meet at a Conference of the Parties. The latest such conference was held in 2011 in Bergen, Norway.

Perhaps the best-known act of CITES was to ban the international trade in ivory in 1989 in order to stop the rapid decline of the African elephant (from 2.5 million animals in 1950 to about 609,000 in 1989 and only around 470,000 in 2008). Recently, several countries have applied to CITES to resume ivory sales and conduct a limited harvest of elephants; permission was given for sales of stocks of ivory at several conferences, and each time poaching of elephants resumed. Unfortunately, legal sales of ivory from the few stable populations may make poaching in other populations worse. In 2006, 3.9 tons of ivory shipped from Cameroon and seized in Hong Kong were shown (via DNA testing) to have come from elephants in Gabon (see Sound Science, "Using DNA to Catch Wildlife Criminals"). Any plan to protect African elephants needs to enable people in Africa to manage wildlife without overexploitation, persuade others to stop using ivory, and use science to stop ivory crime. These changes require a world outcry against ivory collection and an expanded effort to stop ivory crime to save the African elephant from extinction.

---
[16]http://www.iucnredlist.org/

Convention on Biological Diversity. Although CITES provides some protection for species that might be involved in international trade, it is inadequate to address broader issues pertaining to the loss of biodiversity. With the support of the United Nations Environment Programme (UNEP), an ad hoc (formed for this purpose) working group proposed an international treaty that would move to conserve biological diversity worldwide. After several years of negotiation, the Convention on Biological Diversity was drafted and became one of the pillars of the 1992 Earth Summit in Rio de Janeiro. The Biodiversity Treaty, as the convention is called, was ratified in December 1993 and is now in force. One hundred ninety-two states and the European Union are treaty members. As originally organized, the treaty addresses three complementary objectives: (1) the conservation of biodiversity, (2) the sustainable use of biodiversity services, and (3) the equitable sharing of the use of genetic resources found in a country. The treaty establishes a secretariat and designates the Conference of the Parties (the treaty members) as the governing body. Ten conferences have been held, the latest in Nagoya, Aichi Prefecture, Japan, in October 2010. At that meeting, delegates adopted a Revised and Updated Strategic Plan, including Aichi Biodiversity Targets for the period 2011–2020, designated as the UN Decade of Biodiversity. After the joint international meeting, individual countries agreed to form national biodiversity strategies and action plans within two years. National reports on those strategies and plans are due by 2014.

Some examples of the Aichi Biodiversity Targets are:

- At least halve and, where feasible, bring close to zero the rate of loss of natural habitats, including forests

- Establish a conservation target of 17% of terrestrial and inland water areas and 10% of marine and coastal areas

- Restore at least 15% of degraded areas through conservation and restoration activities

- Make special efforts to reduce the pressures faced by coral reefs

The secretariat of the Convention on Biological Diversity and the UNEP jointly published the report *Global Biodiversity Outlook 1, 2,* and *3* in 2001, 2006, and 2010, respectively. *Outlook 3* lays out issues for biodiversity in the 21st century and analyzes why some of the goals for 2010 were not met. The bottom line: Although the targets laid out in the *Strategic Plan* are achievable and the tools for achieving them are largely in place, it will take an unprecedented and costly effort to reach success, and more people need to become convinced of the immediacy of the problems associated with biodiversity loss. The loss of biodiversity is continuing at all levels. Deforestation continues; population trends of key land, freshwater, and marine species all show declines; and more species are threatened with extinction. One positive note is that the area of protected spaces is increasing.

Even though President Clinton signed the Convention on Biological Diversity in 1993, lobbying by a number of key organizations—especially those concerned about property rights—prevented ratification of the treaty by the U.S. Senate, leaving U.S. involvement in a state of limbo. However, many of the provisions of the treaty are being carried out by different U.S. agencies. The National Biological Service, for example, is working on several of the treaty's provisions. The United States continues to send delegations to the Conference of the Parties and participates in other related meetings, such as those dealing with genetically modified organisms.

Critical Ecosystem Partnership Fund. The Critical Ecosystem Partnership Fund emerged in August 2000. Jointly sponsored by the World Bank, Conservation International, the government of Japan, the MacArthur Foundation, and the Global Environment Facility, the fund provides grants to NGOs and community-based groups for conservation activities in biodiversity "hot spots"—34 regions, making up just 2.3% of Earth's land surface **(Fig. 6–20)**, in which 75% of the most threatened mammals, birds, and amphibians are located. By 2011, the fund had already provided $137 million in grants to some 1,627 partners to work on preserving biodiversity in the hot spots.

Stewardship Concerns

What we do about wild species and biodiversity reflects on our wisdom and our values. Wisdom dictates that we take steps to protect the biological wealth that sustains so many of our needs and economic activities. Our values come from our view of species and their instrumental and intrinsic worth. Our values also reflect what we think about the importance of justice—whether we believe, for example, that the people who most benefit economically from the use of an ecosystem should be the same people who bear the risks of its decline, or whether we value maintaining choices and options for future generations.

Wisdom. The scientific team that put together the UN Global Biodiversity Assessment focused on four themes in its recommendations:

1. **Reforming policies that often lead to declines in biodiversity.** Many governments (including that of the United States) subsidize the exploitation of natural resources and agricultural activities that consume natural habitats.

2. **Addressing the needs of people who live adjacent to or in high-biodiversity areas or whose livelihood is derived from exploiting wild species.** Although protecting biodiversity benefits entire societies, the people who are closest to the protected areas are key to the success of efforts to maintain sustainability of natural resources. These people must be given ownership rights or at least communal-use rights to natural resources, and they must be involved in the protection and management of wildlife resources.

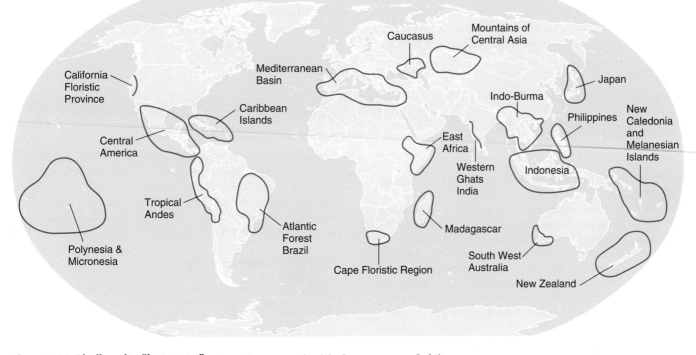

Figure 6–20 Biodiversity "hot spots." These regions are considered the "emergency rooms" of planetary biodiversity and represent most of the hot spots being funded by the Critical Ecosystem Partnership Fund. (*Source:* Redrawn from various sources.)

3. **Practicing conservation at the landscape level.** Too often, we tend to believe that a given refuge or park will protect the biodiversity of a region, when, in reality, many wild species require larger contiguous habitats or a variety of habitats. For example, to address the needs of migratory birds, summer and winter ranges and flyways between them must be considered.

4. **Promoting more research on biodiversity.** Far too little is known about how to manage diverse ecosystems and landscapes, and much remains to be learned about the species within the boundaries of different countries. Toward that end, a new field called biodiversity informatics has been established. Its goal is to put taxonomic data and other information on species online.[17] Species 2000, a program with the "objective of enumerating all known species of organisms on Earth," is one such effort.[18] Another is the Global Biodiversity Information Facility, whose express purpose "is to make the world's biodiversity data freely and universally available."[19]

Values. Our values also are demonstrated by our approach to wild species. Is the natural world simply grist for

our increasing consumption, or is nature something to be managed sustainably and in a stewardly fashion and then passed on to our children and their children? Because we live together in society, we must be prepared to persuade others in our belief that wild species should be conserved. Our own consumption patterns—and our actions as citizens—can help, but coordination with nonprofit, scientific, business, and government agencies is necessary to slow species loss.

At the start of the chapter, we learned about breeding reserves for river dolphins in Bangladesh, developed through the efforts of conservation biologists. The success of other conservation efforts shows us that we can indeed stem the tide of species declines with hard work. Further strategies to address the loss of biodiversity must focus on preserving the natural ecosystems that sustain wild species. Ecosystems are discussed in detail later, including how we depend on them, what we are doing to them, and what we must do to resist the forces that seem to be leading to a future in which all that is left of wild nature is what we have managed to protect in parks, preserves, and zoos (Chapter 7).

Thomas Jefferson once said, "For if one link in nature's chain might be lost, another might be lost, until the whole of things will vanish by piecemeal."[20] As this quote

[17]F. A. Bisby. "The Quiet Revolution: Biodiversity Informatics and the Internet," *Science*, 89, 5488 (2000): pp. 2309–2312.

[18]Goddard Space Flight Center, An Online Catalogue of the World's Known Species. http://gcmd.nasa.gov/records/Species2000-99.html.

[19]Global Biodiversity Information Facility. http://www.gbif.org/informatics/infrastructure/.

[20] Jefferson, Thomas. Memoir on the Discovery of Certain Bones of a Quadruped of the Clawed King in the Western Parts of Virginia. *Transactions of the American Philosophical Society.* 1799. Vol. A, pp. 246–260.

suggests, loss of wild species and their associated genetic and ecosystem diversity could lead to a breakdown in the natural functioning of ecosystems. Indeed, as we look at the world around us, we see that Jefferson was correct.

Currently, many animal and plant populations are in decline. However, efforts made in science and policy and the political will of all of us can result in successes such as the comeback of the puffin.

REVISITING THE THEMES

Sound Science

Sometimes we do not know enough science to manage species well. Fortunately, many universities provide excellent programs in wildlife management. The Millennium Ecosystem Assessment and the Heinz State of the Nation's Ecosystems, as well as annual reports by the IUCN, pull together the solid work of thousands of scientists. Conservation biology, like restoration ecology, is a science that is focused on the protection of living things. The work of conservation biologists helps us to understand why some species are going extinct, where they travel, when they are recuperating, and sometimes when wildlife-related crimes have occurred. Sometimes scientists work together with "citizen scientists," networks of observers, to get more information. Other times they use cutting-edge technology such as DNA analysis to catch criminals. Furthermore, the science of taxonomy is foundational to the protection of species, and making sure we have scientists ready to study has to be part of the solution to biodiversity decline.

Sustainability

Because ecosystem capital is of such high value to human enterprises, the pathway to a sustainable future depends on protecting the wild species that make ecosystems work. Such protection is afforded by local and federal policies and international treaties. Sustainability is enhanced when we take steps to encourage healthy populations of organisms; if they thrive, so do their natural ecosystems. Sometimes, however, it is necessary to hunt or otherwise remove animals (such as deer) from populations when their numbers increase too much. Long-term plans for sustainability have to address the root issues that cause biodiversity declines, including those caused by overpopulation and overconsumption.

Stewardship

The reestablishment of the Lake Erie water snake is good stewardship at work. Stewardship places a high value on wild species. Although instrumental value is sufficient reason for preserving many species, perceiving intrinsic value might be the nearest to a pure stewardship ethic because it works out primarily for the sake of the species. Although some people dislike snakes, their protection is important for ecosystems. Other species, such as river dolphins, are easy to like, and people rally behind them for their protection, although they live in heavily populated areas.

The protection of biodiversity is tied to the ways people value wild species. These values can include instrumental or utilitarian values as well as intrinsic values. Wild species and their genes represent biological wealth, providing the basis for the goods and services coming from natural ecosystems and sustaining human life and economies. Species can also be thought of as a genetic bank, capable of being drawn on when new traits are needed. Losing biodiversity eventually leads to a biologically impoverished future.

REVIEW QUESTIONS

1. Define *biological wealth* and apply the concept to the human use of that wealth.

2. Compare instrumental value and intrinsic value as they relate to determining the worth of natural species. Where does Leopold's idea of the land ethic fit into these two categories?

3. What are the four categories into which the human value of natural species can be divided? Give examples of each one.

4. What means are used to protect game species, and what are some problems emerging from the adaptations of many game species to the humanized environment?

5. What is the Lacey Act, why was it needed, and how is it used to protect wild species?

6. How does the Endangered Species Act preserve threatened and endangered species in the United States? Give some examples of how the act has been implemented.

7. What is one example of an endangered species that has benefited from protection under the Endangered Species Act?

8. What is biodiversity? What do scientists need to know to calculate it for a habitat?

9. What do the letters in the acronym HIPPO stand for? What is an example of each?

10. Give several examples of ways habitat change can either harm species or aid them.

11. What are IUCN, CITES, and the Convention on Biological Diversity? How do their roles differ?

12. What are the Aichi Biodiversity Targets?

THINKING ENVIRONMENTALLY

1. Some people argue that each individual animal has an intrinsic right to survival. Should this right extend to plants and microorganisms? What about the *Anopheles* mosquito, which transmits malaria, or tigers that sometimes kill people in India? What about the bacteria that cause typhoid fever? Defend your position.

2. Choose an endangered or threatened animal or plant species, and use the Internet to research what is currently being done to preserve it. What dangers is your chosen species subject to?

3. The Endangered Species Act is the most important legislation in the United States for the protection of species. However, there are critics who think it is too strict and those who think it is not strict enough. Investigate one example of the application of the law—for example, to gray wolves—and decide what you think ought to be done.

4. Log onto Conservation International's biodiversity hot spots Web page and research one of the 34 hot spots. What is being done to protect biodiversity in this location?

MAKING A DIFFERENCE

1. Instead of devoting your whole yard to grass, convert some of your yard to native plants. The National Wildlife Federation has a great deal of information on turning your backyard into wildlife habitat and even a certification program. They can be found on the Web.

2. Support groups that conserve and protect wildlife through the political process. Groups like the Center for Biological Diversity and others will alert members when there are political actions that they can support.

3. Consider volunteering or interning for a program that conserves wildlife. The Audubon Society has volunteer opportunities, as do many other wildlife conservation groups. These opportunities are found on the Internet.

4. Be careful about purchasing products that might come from a wild population. For example, it is better to buy pets bred in captivity rather than wild-caught pets.

MasteringEnvironmentalScience®

Go to **www.masteringenvironmentalscience.com** for practice quizzes, Pearson eText, videos, current events, and more.

CHAPTER 7

The Value, Use, and Restoration of Ecosystems

Caribbean Reef. Diver in the Caribbean passes a large Nassau grouper. Such fish are in decline.

LEARNING OBJECTIVES

1. **Ecosystem Capital and Services:** List several ways natural ecosystems have great economic value (as they provide goods and services vital to human well-being).

2. **Types of Uses:** Describe the differences between the consumptive and productive uses to which ecosystems may be put.

3. **Biomes and Ecosystems Under Pressure:** Describe how forests and oceans are examples of ecosystems under pressure, and describe sustainable ways to fill demands for their products.

4. **Protection and Restoration:** Explain how the public and private management of lands is key to keeping habitats both protected and productive, and describe an example of an ecosystem that needed to be restored to a more healthy state.

BRIGHT SUN FILTERS through shallow water to reveal the colors of a tropical reef. Fish flit by branching arms of a coral. A parrotfish grazes a piece of seaweed, and a lobster hides at the base of a craggy rock. As the sun warms the clear water, divers swim by. An endangered Nassau grouper floats in an eddy, eyeing a smaller fish. A non-native lionfish snaps out, catching prey. It's another day in a tropical paradise. Suddenly, a SCUBA diver snatches the venomous lionfish and heads for the surface with the catch.

Today is the Lionfish Derby, held in the Florida Keys and run by the Reef Environmental Education Foundation (REEF). In the Derby, teams compete for cash and prizes to capture these invasive fish and remove them from this critical habitat. Lionfish lack predators in the Caribbean and eat the young of other fish. Researchers will use data from the captured fish to better understand their biology and ecology, and the capture removes some of the invaders from a habitat that is important to many other species. In 2011, the Florida Keys Lionfish Derby teams netted more than 1,500 fish.

CITIZEN SCIENTISTS. On a similar day elsewhere in the Caribbean, SCUBA divers move methodically through the water, recording the fish they see. They note the condition of coral reefs and record the presence of fish species, both herbivorous and carnivorous. This is also a REEF initiative: the Volunteer Survey Project. The SCUBA divers, trained volunteers in a network of "citizen scientists," will submit the

information they collect, to be used by scientists who track what is happening with coral reef fish. Using these data, scientists have already been able to figure out a few things: Fish are declining, which affects the health of the coral reef. As herbivorous fish decline, algae and water plants can grow over the corals. Some of these produce toxic chemicals and kill the coral. As carnivorous fish decline, coral-eating sea stars are free to take over. The invasive lionfish can oust other native reef fish. Researcher Christopher Stallings of the Florida State University Coastal and Marine Lab is one of the scientists working with REEF. His work with the large database on fish in the Caribbean showed that some large-bodied fish like the Nassau grouper have completely disappeared in some highly populated areas.[1]

CRITICAL ECOSYSTEMS. Around the world, coral reefs are critical ecosystems. They act as nurseries for marine fish and as an early warning system for habitat degradation. Because they need clean, clear water and grow relatively slowly, corals show us when marine systems are in decline. Physical breakage, pollution,

heat, and invasive species combine to kill off these colonial animals and their algal symbionts. A coral reef is a good example of an ecosystem that provides humans with goods and services and that needs monitoring, protection, and sometimes restoration. The dive teams competing in the Lionfish Derby and the SCUBA divers recording species while they dive are ways that ordinary people can be involved in the care of an ecosystem.

One key to saving wild species is to preserve their habitats—the ecosystems in which they are found (Chapter 6). Populations of wild species continue to decline, however, because ecosystems are in decline throughout the world. In addition to the loss of species, the natural goods and services that ecosystems provide are also being degraded or lost. It is not only the natural world, but also human well-being that is threatened in the face of this unsustainable decline of ecosystems.

The focus in this chapter is particularly on the natural ecosystems that directly sustain human life and economy. We will pay special attention to those ecosystems that are in the deepest trouble. These ecosystems, the species in them, and the goods and services they generate are identified as **ecosystem capital**, a major component of the natural capital wealth of nations (Chapters 1, 2).

[1]C. Stallings. "Fishery-Independent Data Reveal Negative Effect of Human Population Density on Caribbean Predatory Fish Communities," *PLoS ONE* 4(5) (2009): [e5333].

7.1 Ecosystem Capital

Earth consists of ecosystems that vary greatly in their species makeup but that exhibit common functions, such as energy flow and the cycling of matter. The major types of terrestrial and aquatic ecosystems—the biomes—reflect the response of plants, animals, and microbes to different climatic conditions. Earth's land area can be divided into five broad categories for the purposes of this chapter (Table 7–1): forests and woodlands, grasslands and savannas, croplands, **wetlands**,

TABLE 7–1 Ecosystems on Earth's Surface

Ecosystem	Area (million km^2)	Area (million mi^2)	Percentage of Total Land Area
Forests and woodlands	47.3	18.2	31.7
Grasslands and savannas	30.2	11.6	20.2
Croplands	16.1	6.2	10.7
Wetlands	5.2	2.0	3.5
Desert lands and tundra	51.0	19.6	34.0
Total land area	**149.8**	**57.6**	**100.0**
Coastal ocean and bays	21.8	8.4	
Coral reefs	0.52	0.2	
Open ocean	426.4	164.0	
Total ocean area	**448.7**	**172.6**	

and **deserts** and **tundra**. The oceanic ecosystems can be categorized as coastal ocean and bays, coral reefs, and open ocean. These eight terrestrial and aquatic systems provide all of our food; much of our fuel; wood for lumber and paper; fibers—leather, furs, and raw materials for fabrics; oils and alcohols; and much more. The world economy and human well-being directly depend on the exploitation of the **natural goods** (referred to in Table 1–1 by the Millennium Ecosystem Assessment team as *provisioning services*) that can be extracted from ecosystems.

Ecosystems also perform a number of other valuable natural services, including *regulating* and *cultural services*, as they process energy and circulate matter in the normal course of their functioning (Table 1–1). Regulating services are benefits derived from regulation of ecosystem processes, such as flood control or climate moderation. Cultural services comprise nonmaterial benefits people gain from ecosystems, including spiritual enrichment, knowledge, and aesthetic enjoyment. Some of the services are general and pertain to essentially all ecosystems, whereas others are more specific. Finally, ecosystems provide services that maintain themselves: photosynthesis, for example. These are called *supporting services*.

Table 7–2 shows combinations of services provided to societies by a spectrum of ecosystems (a listing of ecosystem services can be found in Table 5–3). In their normal functioning, all natural and altered ecosystems perform some or all of these natural services, year after year. A team of natural scientists and economists calculated the total value of a year's global ecosystem goods and services to be $41 trillion (Chapter 5), and we receive them free of charge.

TABLE 7–2 Services from Various Types of Ecosystems

	Mountain and Polar	Forest and Woodlands	Inland Rivers and Other Wetlands	Drylands	Cultivated	Urban Parks and Gardens	Coastal	Island	Marine
Air quality regulation						√			
Biofuels					√				
Carbon sequestration		√							
Climate regulation (local and wide spread)	√	√		√		√	√		√
Cultural heritage				√	√	√			
Disease regulation and medicines		√	√		√				
Dyes					√				
Education						√			
Erosion control									
Fiber	√			√	√		√		
Flood regulation		√	√				√		
Food	√	√	√	√	√		√	√	√
Fresh and drinking water	√	√	√		√			√	
Fuelwood and timber		√		√	√		√		
Nutrient cycling			√		√		√		√
Pest regulation					√				
Pollution regulation			√						
Recreation, ecotourism, and aesthetic values	√	√	√	√	√	√	√	√	√
Sediment retention and transport			√						
Waste processing							√		
Water regulation						√			

Source: Millennium Ecosystem Assessment. "Living Beyond Our Means: Natural Assets and Human Well-Being" (Statement of the Millennium Ecosystem Assessment Board, 2005).

Figure 7–1 Shrimp aquaculture farms in Thailand. Carved out of mangrove forest, these shrimp farms provide only a fraction of the value of the services performed by the intact mangroves.

To illustrate the value of an ecosystem, consider a mangrove wetland in South Asia that is converted to a shrimp farm **(Fig. 7–1)**, a common conversion. Studies of shrimp aquaculture (cultivating aquatic organisms like fish and shellfish under controlled conditions) in Thailand show that the economic value of the shrimp was $3,375 per acre over five years, following which the site is abandoned as yields decline sharply. Left intact, the mangroves provide services not fully captured in the marketplace: storm protection for coastline communities, forest products from mangrove harvest, and spawning ground for local fisheries. These services have a sustained estimated value of $14,450 per acre. In addition, there are net social costs of the shrimp farm in the form of pollution and land degradation that require expensive restoration, calculated at $2,200 per acre. Thus, the net income from the shrimp farming, when social costs are subtracted, is lowered to $1,175, which is only 8% of the value of services provided by the intact mangrove wetland.[2]

The benefit is local, short term, and specific to the shrimp farmer, who is likely to be from outside the community, while the loss of services is more regional, long term, and diffuse to all those living near the coast. Fortunately, shrimp farming in mangrove areas has been largely stopped because the farms do not do well long term and construction costs are high. But mangrove loss from conversion to rice agriculture, fuel, wood pulp, and other uses has continued. Many areas of Southeast Asia hit by the 2004 tsunami had been deprived of their protective mangrove forests and were more heavily damaged than those with intact mangroves. Mangroves are also efficient at taking up carbon dioxide and are home to globally threatened species. Removal of mangroves for aquaculture often ignores other benefits and costs.

[2]Elizabeth McLeod and Rodney V. Salm. *Managing Mangroves for Resilience to Climate Change*. IUCN Science Group Working Paper Series No. 2. 2006.

Ecosystems as Natural Resources

If the natural services performed by ecosystems are so valuable, why are we still draining wetlands and removing forests? The answer is clear: a natural area will receive protection only if the value a society assigns to services provided in its natural state is higher than the value the society assigns to converting it to a more direct human use.

Natural ecosystems and the biota in them are commonly referred to as **natural resources**. As resources, they are expected to produce something of value, and the most commonly understood value is economic value. By referring to natural systems as resources, we are deliberately placing them in an economic setting, and it becomes easy to lose sight of their ecological value. Instead of natural resources, we use the concept of *ecosystem capital* in order to avoid making this mistake.

Valuing. Markets provide the mechanism for assessing economic value, but do poorly at placing monetary values on ecosystem services. People and institutions may express their preferences for different ecosystem services via the market, an instrumental value (Chapter 6). Instrumental valuation works fairly well for the provisioning services provided by ecosystems (goods such as timber, fish, and crops). However, many of the regulating and cultural services (and supporting services) provided by natural ecosystems—the more ecological values—are far more difficult to put prices on. These services are **public goods** (something that is not used up as people use it, and which can't be effectively limited, in order to charge money; for example, breathing air) not usually provided by markets, but contributing greatly to people's well-being. In fact, they are absolutely essential.

Regulating and cultural services provide benefits that are experienced over a large area, and their losses are also experienced regionally, rather than just locally. Large forested areas can modify the climate in a region and absorb and hold water in the hydrologic cycle. A small forest can be cut and the timber sold, with no noticeable impact on the weather, but with a definite economic value in a market setting, providing income to the owners. If the larger area is also cut, it will have both an immediate economic benefit and a long-term loss in regulating and supporting services. Thus, it will have an impact on people's well-being in the formerly forested area and well beyond, as short-term profits are made at the long-term environmental expense of many. When the benefits and losses in such conversions are reasonably evaluated economically, losses generally outweigh benefits (Chapter 4).

Private Versus Public Lands. Some natural ecosystems are maintained in a natural or seminatural state because that is how they provide the greatest economic (direct-use) value for their owners. For example, most of the state of Maine is owned by private corporations that exploit the forests for lumber and paper manufacturing. However, if Maine experienced a population explosion and corporations could sell their land to developers for more money than could be gained from harvesting timber, forested lands would quickly become house lots. Similar conversions can be pictured: tropical forests to grasslands for grazing, agricultural fields to suburbs, and wetlands to vacation developments.

Some natural ecosystems either are publicly owned (state and federal lands) or cannot be owned at all (ocean ecosystems). These ecosystems are still considered natural resources and may be subject to economically motivated exploitation. Sustainable exploitation of such systems will maintain the natural services they perform. Fortunately, ecosystems can not only sustain continued exploitation, but can also be restored from destructive uses that have degraded them and removed their natural biota.

Domesticated Nature. Whether private or public, human domination of the landscape is pervasive, and ecosystems have been converted or domesticated in vast areas of the planet. What we call natural may be only an illusion, because humankind has so converted and domesticated natural systems that it is difficult to find areas of the world unaffected by human impact. Even where systems appear natural, it is likely that humans have killed larger animals, suppressed wildfires, controlled rivers, introduced exotic species, and harvested wood. Elsewhere, we have converted grasslands and forests to grazing lands and crop cultivation, so that almost 50% of the land surface is now used in the production of crops and animal products and over half of the forests have been lost in the process. To be sure, domesticated ecosystems still provide provisioning and regulating and cultural services, and may do so sustainably, but they do so at our consent and largely for our benefit.

Future Pressures. As the human population continues to rise, pressures on ecosystems to provide resources will increase: more food, more wood, more water, and more fisheries products will be needed. Maintaining sustainable exploitation of ecosystem capital will become more difficult, in view of the already intense pressure being put on those ecosystems. The main objective of the thorough work of the Millennium Ecosystem Assessment was to evaluate the health of the services that people derive from the natural world. The assessment team found that 15 of the 24 services they evaluated were in decline worldwide, five were in a stable state, and four were increasing at the expense of other services (Chapter 1). The governing board of the assessment was so concerned about these findings that it published a separate document for a "nonspecialist" readership—*Living Beyond Our Means: Natural Assets and Human Well-Being.* This brief report begins: "At the heart of this assessment is a stark warning. Human activity is putting such strain on the natural functions of Earth that the ability of the planet's ecosystems to sustain future generations can no longer be taken for granted. . . . This assessment shows that healthy ecosystems are central to the aspirations of humankind."

7.2 Consumption and Production

An ecosystem's remarkable natural capacity to regenerate makes it (and its biota) a **renewable resource.** In other words, an ecosystem has the capacity to replenish itself despite certain quantities of organisms being taken from it, and this renewal can go on indefinitely—it is sustainable. Recall that every species has the biotic potential to increase its numbers and that, in a balanced population, the excess numbers fall prey to parasites, predators, and other factors of environmental resistance (Chapter 5). It is difficult to find fault with activities that effectively put some of this excess population to human use. The trouble occurs when users (hunters, fishers, loggers, whalers) take more than the excess and deplete the breeding population.

Conservation Versus Preservation

Conservation of natural biotas and ecosystems does not imply *no* use by humans whatsoever, although this may sometimes be temporarily expedient in a management program to allow a certain species to recover its numbers. Rather, the aim of conservation is to *manage or regulate use* so that it does not exceed the capacity of the species or system to renew itself. Conservation is capable of being carried out sustainably, and when sustainability is adopted as a principle, conservation has a well-defined goal.

Preservation is often confused with conservation. The objective of the preservation of species and ecosystems is to ensure their continuity, regardless of their potential utility. Effective preservation often precludes making use of the species or ecosystems in question. For example, it is impossible to maintain old-growth (virgin) forests and at the same time harvest the trees. Thus, a second-growth forest can be conserved (trees can be cut, but at a rate that allows the forest to recover), but an old-growth forest must be preserved (it must not be cut down at all).

There are times when conservation and preservation come into conflict. The Muriqui monkey of Brazil was once thought to require virgin forests, leading to a concern for protecting such forests for the sake of the species **(Fig. 7–2)**. Recent research has shown, however, that the monkeys actually do better in second-growth forests, which support

Figure 7–2 Muriqui monkey. Protection of this species of monkey in Brazil involves allowing some cutting of the forest in order to provide second-growth forests.

Figure 7–3 Consumptive use. Hunters in Papua, New Guinea bring home a captured wild pig. Wild game is an important food source for millions in many parts of the world.

a greater range of vegetation on which the monkeys feed. Thus, *conservation* of the forests is essential for *preservation* of this New World monkey, a seriously endangered species.

Patterns of Human Use of Natural Ecosystems

Before examining the specific biomes and ecosystems that are under the highest pressure from human use, we need to study a few general patterns of such use to understand what is happening in these natural systems and why it is happening.

Consumptive Use. People in more-rural areas make use of natural lands or aquatic systems that are in close proximity to them. When people harvest natural resources in order to provide for their needs for food, shelter, tools, fuel, and clothing, they are engaged in **consumptive use.** This kind of exploitation usually does not appear in the calculations of the market economy of a country. People may barter or sell such goods in exchange for other goods or services in local markets, but most consumptive use involves family members engaged in hunting and gathering to meet their own needs. Thus, people hunt for game (**Fig. 7–3**) and fish and gather fruits and nuts in order to meet their food needs; or they collect natural products like firewood, as well as wood or palm leaves to construct shelters or to use as traditional medicines. This "wild income," as it has been called, is highly important to the poor; some 90% of the 1.2 billion poorest people depend on forests for some of their income.

Wild game, or "bush meat," is harvested in many parts of Africa and provides a large proportion of the protein needs of people. In Congo, for example, bush meat may account for as much as 75% of a person's protein intake. The estimated take in the Congo basin alone is somewhere around 1 million metric tons annually. Without access to

these resources, people's living standard would decline and their very existence could be threatened. Unfortunately, this is a largely unregulated practice and often involves poaching in wildlife parks. According to the Convention on International Trade in Endangered Species (CITES), this practice has contributed to the decline of 30 endangered species. Making matters worse, commercial hunters have recently turned what was once a matter of local people using local wildlife for food into an unsustainable industry that supplies cities with tons of "fashionable" meat. For example, primate meat (e.g., chimpanzee) has been found in markets in New York, London, Paris, Brussels, and others.

Dependence on consumptive use is most commonly associated with the developing world but also exists in the more rural areas of the developed world, where people depend on wood for fuel or wild game for food. Their well-being would be diminished if they were denied access to these "free" resources.

Productive Use. In contrast to consumptive use, **productive use** is the exploitation of ecosystem resources for economic gain, as when products like timber and fish are harvested and sold for national or international markets (**Fig. 7–4**). Productive use is an enormously important source of revenue and employment for people in every country. More than 1.2 billion hectares of forests are managed for wood production. Nearly half of the wood removed, however, is used for fuel, so a great deal of the value of forests occurs on the consumptive, noneconomic side. Planted forests are on the increase, shifting timber harvest from natural to planted forests and making conservation of some natural forests easier. The global fisheries catch in 2008, which reached 90 million metric tons (83 million tons), was valued at $93.9 billion. Another productive

Figure 7–4 Productive use. Loggers load poplar logs to be transported to a lumber mill, where they will be made into waferboard for the building industry.

SUSTAINABILITY

How Much for That Irrigation Water?

The dry bed of the Santa Cruz River in Arizona lies 20 feet below the bank sides, filled with water only in large floods, but it wasn't always so. In the late 1880s, a local farmer dug deep ditches and tapped into the underground water, moving below the shallow river bed. Fierce storms sent raging torrents through the channels, eroding the river bed, stranding farm fields above the precious water. The dry deep river of today and the increased cost of irrigation result from human actions of the past.

For years, humans have used the natural world as a place to obtain goods and services, often without paying for them. Humans are checking out at the global grocery store, with a basket of things that come from ecosystems: wood products, fish, fibers, chemicals, metals, water. How much is that flood control or clean air worth? How much is the irrigation from a river worth? To estimate the "value" of goods and services—things like photosynthesis, water cycling, and absorption of carbon dioxide—we need to know how ecosystems work and how our activities affect the rest of the system. That is part of what scientists do. We also have to determine how much the price of our activities and things we produce reflects the effect of our actions on ecosystems. This is done primarily by economists. Policy makers help figure out how we can act together to change the way things are done (Chapter 2), in order to incorporate the real-world cost of ecosystem capital into our human decisions.

How do natural and social scientists determine the value of ecosystems? Some valuation is dependent on what stakeholders think. Researchers doing a study on the Santa Cruz River asked participants, "Why do people think rivers are important? What do they value about rivers?" Respondents valued water quality, presence (in a river where stretches sometimes dried up altogether), lush vegetation, and habitat for birds and fish. They also valued human connections to the river, use by others downstream, and recharge of underground water and other ecosystem services. The people around the Santa Cruz River want the river to have a role in their daily lives and desire its health. Of course, putting economic value only on the ecosystem services people recognize, value, and are willing to pay for is a faulty approach, as we often do not know all the important features of ecosystems. However, it is a beginning and often provides useful information to policy makers.

How Much Is the Flood Control Worth? This picture of the Santa Cruz River shows the pressures rivers face—drying, erosion and invasive species.

Sources: B. Fisher et al. "Ecosystem Services and Economic Theory: Integration for Policy-Relevant Research." *Ecological Applications,* 18(8) (2008): 2050–2067.

P. L. Ringold, J. Boyd, D. Landers, and M. Weber. Report from the Workshop on Indicators of Final Ecosystem Services for Streams. 56p. EPA/600/R-09/137. July 2009.

M. Weber et al. Why are rivers Important? Ecosystem Service Definition via Qualitative Research: A Southwestern Pilot Study. Oral Presentation. December 9, 2010. http://conference.ifas.ufl.edu/aces10/Presentations/Thursday/C/AM/Yes/0845%20MattWeber.pdf.

use is the collection of wild species of plants and animals for cultivation or domestication. The wild species may provide the initial breeding stock for commercial plantations or ranches, or they may be used as sources of genes to be introduced into existing crop plants or animals to improve their resistance or to add other desirable genetic traits. Finally, wild species continue to provide sources of new medicines (Chapter 6).

Tenure. Consumptive and productive uses of natural ecosystem resources are the consequence of the rights of **tenure,** or property rights, over the land and water holding those resources. Four kinds of tenure are recognized: (1) *private ownership*, which restricts access to natural resources; (2) *communal ownership*, which permits use of natural resources by members of the community; (3) *state ownership* of natural resources, which implies regulated use; and (4) *open access*, where natural resources can be used by anyone. In all of these arrangements—private, communal, state, and open access—there is the potential either for sustainable use of the resources or for unsustainable use leading to the loss of the resources and the services provided by the natural ecosystem being exploited. As a result, the perspective of stewardship is vital. Do the resource managers and harvesters see themselves as caretakers, as well as exploiters, of the natural resource? Are public policies that encourage stewardship and sustainability in place? Is the local or national government presence weak or corrupt? Are sound scientific principles employed in management strategies? Keep these questions in mind as we next examine two patterns that are commonly seen when people exploit natural resources.

Maximum Sustainable Yield. The central question in managing a renewable natural resource is this: How much continual use can be sustained without undercutting the capacity of the species or system to renew itself? A scientific model employed to describe this amount of use is the maximum sustainable yield (MSY): the highest possible rate of use that the system can match with its own rate of replacement or maintenance. Besides pertaining to the harvesting of natural biota, the concept of MSY can apply to the maintenance of parks, air quality, water quality and quantity, and soils—indeed, the entire biosphere. *Use* can refer to the cutting of timber, hunting, fishing, park visitations, the discharge of pollutants into air or water, and so on. Natural systems can withstand a certain amount of use (or abuse, in terms of pollution) and still remain viable. However, a point exists at which increasing use begins to destroy the system's regenerative capacity. Just short of that point is the MSY.

Optimal Population. An important consideration in understanding how the MSY works is the **carrying capacity** of the ecosystem—the maximum population the ecosystem can support on a sustainable basis. If a population is well below the carrying capacity of the ecosystem (**Fig. 7–5a**), then allowing that population to grow will increase the number of reproductive individuals and thus the yield that can be harvested. However, as the population approaches the carrying capacity of the ecosystem, new individuals must compete with older individuals for food and living space. As a result, recruitment (survival of new individuals in the population) may fall drastically (Fig. 7–5b). When a population is at or near the ecosystem's carrying capacity, production—and hence sustainable yield—can be increased by thinning the population so that competition is reduced and optimal growth and reproductive rates are achieved. Thus, the MSY cannot be obtained with a population that is at the carrying capacity. Theoretically, the **optimal population** for harvesting the MSY is just halfway to the carrying capacity.

The practical use of this model is complicated by the fact that the carrying capacity—and, therefore, the optimal population—is variable. Carrying capacity may vary from

year to year as the weather fluctuates. Replacement of harvested individuals may also vary from year to year, because some years are particularly favorable to reproduction and recruitment, while others are not. Human impacts, such as pollution and other alterations of habitats, adversely affect reproductive rates, recruitment, carrying capacity, and, consequently, sustainable yields. For these reasons alone, managing natural populations to achieve the MSY is fraught with difficulties. Furthermore, accurate estimates of the size of the population and the recruitment rate must be made continually, and these data are hard to come by for many species.

Precautionary Principle. The conventional approach has been to use the estimated MSY to set a fixed quota—in fishery management, the **total allowable catch (TAC)**. If the data on population and recruitment are inaccurate, it is easy to overestimate the TAC, especially when there are pressures from fishers (and the politicians who represent them) to maintain a high harvest quota. In the face of such uncertainties, and also in response to repeated cases of overuse and depletion of resources, resource managers have been turning to the **precautionary principle**. That is, where there is uncertainty, resource managers must favor the protection of the living resource. Thus, the exploitation limits must be set well enough below the MSY to allow for uncertainties. Those using the resource ordinarily want to push the limits higher, however, so the result is frequent conflicts between users and managers. These conflicts are addressed in greater detail when the problems of fisheries are discussed later in this section.

Using the Commons. Where a resource is owned by many people in common or by no one (open access), it is known as a common-pool resource, or **commons**. Strictly speaking, a commons refers to a system with open access where use by one does not subtract from use by others; for example, knowledge is a commons. Examples of natural common-pool

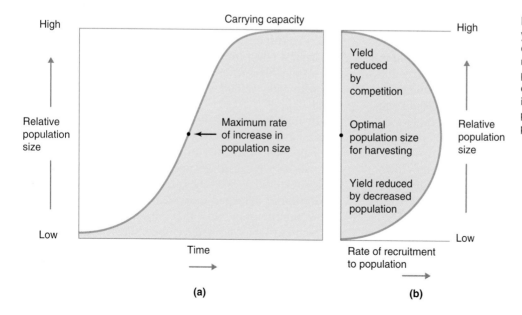

Figure 7–5 Maximum sustainable yield. The maximum sustainable yield occurs when the population is at the optimal level (meaning the rate of increase in population is at a maximum). (a) The logistic curve of population size in relation to carrying capacity. (b) Recruitment plotted against population size, showing the effects of competition and decreased population levels.

resources include federal grasslands, on which private ranchers graze their livestock; coastal and open-ocean fisheries used by commercial fishers; groundwater drawn for private estates and farms; nationally owned woodlands and forests harvested for fuel in the developing countries; and the atmosphere, which is polluted by private industry and everyone who drives motor vehicles.

The exploitation of such common-pool resources presents some serious problems and can lead to the eventual ruin of the resource—a phenomenon called the "tragedy of the commons," taken from the title of a 1968 essay by the late biologist Garrett Hardin.[3] Sustainability implies that common-pool resources will be maintained so as to continue to yield benefits not just for the present, but also for future users.

Tragedy of the Commons. As described by Hardin, the original "commons" were pastures around villages in England. These pastures were provided free by the king to anyone who wished to graze cattle. Herders were quick to realize that whoever grazed the most cattle stood to benefit the most. Even if they realized that a particular commons was being overgrazed, those who withdrew their cattle simply sacrificed personal profits, while others went on using that commons. One herder's loss became another's gain, and the commons was overgrazed in any case. Consequently, herders would add to their herds until the commons was totally destroyed. The lesson of this story is that where there is no management of the commons, the pursuit of individuals' own interests will lead to a tragedy for all.

Hardin's essay applies to a limited, but significant, set of problems that arise where there is open access to a common-pool resource, but where there is no regulating authority (or there is one, but it is ineffective) and no functioning community. Exploitation of the resource then becomes a free-for-all in which profit is the only motive. Coastal and offshore fisheries have consistently demonstrated the reality of the tragedy of the commons, with stocks of desirable fish declining all over the world. The tragedy can be avoided only by limiting freedom of access.

Limiting Freedom. One arrangement that can mitigate the tragedy is private ownership. When a renewable natural resource is privately owned, access to it is restricted and, in theory, it will be exploited in a manner that guarantees a continuing harvest for its owner(s). This theory does not hold, however, when an owner maximizes immediate profit and then moves on. Modern corporations lend themselves to this strategy: For example, corporate raiders can take over a natural resource—say, a timber company and its holdings—and clear-cut the forest in order to pay off debts incurred in the process of the takeover. Many owners, though, will be in for the long run and manage the resources more responsibly because it is in their best interests to do so.

Where private ownership is unworkable or lacking, the alternative is to regulate access to the commons. Regulation should allow for (1) protection, so that the benefits derived from the commons can be sustained; (2) fairness in access rights; and (3) mutual consent of the regulated. Such regulation can be the responsibility of the state, but it does not have to be. In fact, the most sustainable approach to maintaining the commons may be local community control, wherein the power to manage the commons resides with those who directly benefit most from its use and there are strong social ties and customs that can function well over time to protect the commons. By contrast, in many situations, state control of a commons has accelerated its ruin and has led to an associated social breakdown and the impoverishment of people.

Maine Lobsters. The lobster is a tasty crustacean found in the colder marine waters of the Northern Hemisphere, especially where there are shallow-water rocky subtidal seascapes interspersed with sandy patches. The state of Maine is famous for its lobsters, and management of the lobster fishery—a common-pool resource—has evolved into a workable combination of state and local regulation. The state has established conservation laws that protect juvenile and breeding female lobsters and limit the number of lobster traps (**Fig. 7–6**). The lobster fishers, however, have developed a second set of local rules that determine who can fish a given harbor and coastline area. It is an informal system, but it works because the fishers are part of a functioning community where they know each other and are monitoring each other's harvesting. The result is a successful outcome where the state's lobster fishery is thriving even though the number of fishers has more than doubled since 1973.

Public Policies. In sum, to achieve the objectives of conservation when harvesting living resources, it is important to consider both the concept and limitations of the MSY and the social and economic factors causing overuse and degradation that diminish the sustainable yield. After reviewing these factors, public policies can be established and enforced that protect natural resources effectively. Some principles that should be embodied in these policies are presented in Table 7–3. When policies based on principles like these are in place, it is possible for people to continue to put natural resources to sustainable consumptive and productive uses. The vast harvest taken from the natural world each year is proof that sustainable use not only is possible, but also actually occurs with many resources and in many locations around the world.

In some situations, however, exploitation and degradation have gone too far, and the resilience of an ecosystem has been overcome. In those cases, natural services and uses can be restored only if the habitats are deliberately restored as well. In recent years, the need for the restoration of damaged ecosystems has become clear, and a new subdiscipline with that objective—called **restoration ecology**—has been developed. This is discussed in Section 7.4.

The next section discusses a number of still-functioning biomes and ecosystems that are in trouble because of their great economic importance.

[3]Garrett Hardin. "The Tragedy of the Commons," *Science* 162 (1968): 1243–1247.

Figure 7–6 **Exploiting a commons.** Lobster boat unloading a lobster pot in Maine waters.

TABLE 7–3 Principles That Should Be Incorporated into Public Policies to Protect Natural Resources

1. Natural resources cannot be treated as open-access commons. Habitats and species should be held by private, communal, or state entities that are responsible for their exploitation and sustainability.

2. Scientific research should be employed to assess the health of the resource and to set sustainable limits on its use.

3. To accommodate uncertainty, the precautionary principle should be used in setting limits for exploitation.

4. Regulations should be enforced; the resource must be protected from poaching.

5. Economic incentives that encourage the violation of regulations should be eliminated.

6. Subsidies that encourage exploitation of the resource should be removed.

7. All stakeholders should be involved in resource-related decisions; those living close to the resource should share in profits from its use.

8. Security of tenure is crucial for poor people who depend on the resource for their well-being.

7.3 Biomes and Ecosystems Under Pressure

Although human activities affect virtually all biomes and ecosystems, some are under more pressure than others. This section begins with a look at the forest biomes, the most important of all biomes in terms of their economic significance and potential for active human management.

Forest Ecosystems

Forests are the normal ecosystems in regions with year-round rainfall that is adequate to sustain tree growth. Forests are also the most productive systems the land can support, and they are self-sustaining. Forests and woodlands (ecosystems with mixed trees and grasses) perform a number of vital natural services. Among other things, they conserve biodiversity,

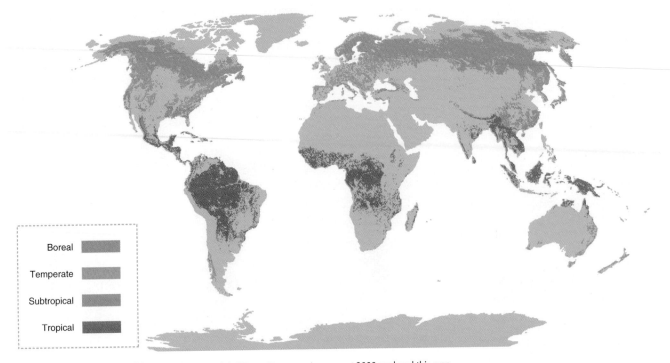

Figure 7–7 World forest biomes. The FAO *Global Forest Resources Assessment 2000* produced this map showing the locations of the major world forests.

moderate regional climates, prevent erosion, store carbon and nutrients, and provide recreational opportunities. They provide a number of vital goods, too, such as lumber (the raw material for making paper), fodder for domestic animals, fibers, gums, latex, fruit, berries, nuts, and fuel for cooking and heating. In fact, forests provide fuel, food, medicine, and income for more than 1.6 billion people, mostly in the developing countries. In spite of this value, the major threat to the world's forests is not simply their exploitation, but rather their total destruction.

Forest Resources Assessments. The UN's Food and Agricultural Organization (FAO) conducts periodic assessments of forest resources and compares the data with previous information in order to evaluate trends in forested areas. **Figure 7–7** shows the locations of the four major world forest biomes. The FAO produced the *Global Forest Resources Assessment 2010*. The most recent FAO forestry report was titled *State of the World Forests 2011*. Forest ecosystems are tracked by other groups as well, such as NASA's earth observatory. Major findings by those assessing forests are as follows:

1. In 2005, the world's forest cover was 3.95 billion hectares (9.8 billion acres), or 30% of total land area. The FAO defines "forests" as lands where trees are the dominant life form and produce a canopy greater than 10% cover. An additional 1.4 billion hectares (3.5 billion acres) are classified as "other wooded land," sometimes referred to as "woodlands," where the tree cover is less than 10% or the trees and/or shrubs are species that do not grow to a height above 5 meters. Forests account for 50% of plant productivity.

2. Deforestation continues to occur, primarily in the developing countries. Forest cover in the developed countries has remained stable. The FAO defines **deforestation** as the removal of forest and replacement by another land use. Thus, logging per se does not count as deforestation if the forests are allowed to regenerate. The current global rate of net deforestation is estimated at 7.3 million hectares (18 million acres) per year, some 18% better than in the 1990s. The annual net changes in forest area in major world regions are shown in **Figure 7–8**.

3. Throughout the world, the most important forest product is wood for industrial use; half of forest lands are designated for "production," where the wood is harvested for pulp (paper source), lumber, and fuelwood. Close to 1.2 billion hectares (2.9 billion acres), about 30% of the world's forests, are managed for production of forest products, including wood.

4. Worldwide, 13.5% of forests are in legally established protected areas. This proportion represents an increase, a recognition by many countries of the important non-timber services of forests. China in particular has increased forests in order to slow the spread of deserts and soil loss (Chapter 14).

5. The role of forests in climate change was formally acknowledged in November 2001 at a meeting of the signers of the Kyoto Protocol in Marrakech, Morocco. The outcome of this accord is a much more thorough inventory of the role of forests as carbon stores, sources of carbon emissions, and carbon sinks. Some estimates are that 45% of the carbon stored on land is in forests—

CHANGE IN FOREST AREA BY REGION, 1990–2010

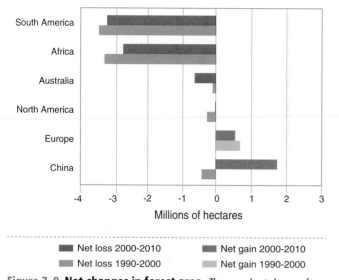

Millions of hectares

■ Net loss 2000-2010   ■ Net gain 2000-2010
■ Net loss 1990-2000   ■ Net gain 1990-2000

Figure 7–8 Net changes in forest area. The annual net change of forests from 2000 to 2010 in the different regions of the world. Virtually all of the losses occurred in tropical forests of South America and Africa.

(*Source:* Food and Agriculture Organization [FAO] of the United Nations, Global Forest Resources Assessment 2010, p. xvi.)

as much as 289 gigatonnes (GT). However, reduction in forest area caused this number to drop about 0.5 GT per year between 2005 and 2010.

6. Insect pests, disease, invasive species, and fires affect global forests significantly. In the United States and Canada, the mountain pine beetle has killed more than 11 million acres of forest since the 1990s.

7. Government spending on forestry is not replaced. On average, governments spend about $7.50 (managing timber and building and maintaining logging roads) and collect about $4.50 per hectare. Those that spend less also receive less. Overall, governments "generally spend more on forestry than they collect in revenue."[4] About 80% of the world's forests are publically owned, some managed locally and others by central national governments.

Forests as Obstacles. Even though the current rate of deforestation is lower than in the 1990s, it is still a major cause for concern. Why are forests being cleared when they could be managed for wood production on a sustainable basis? Even though forests are highly productive systems, it has always been difficult for humans to exploit them for food. Most deforestation is used to convert forests into pastures and agricultural land. Unlike forests, grasslands have short food chains in which herbaceous growth supports large herbivores that can yield meat and other animal products. Alternatively, forests can be replaced by cultivated plants that are used directly for food. Thus, forests have always been an

obstacle to conventional animal husbandry and agriculture. Indeed, the first task the early European colonists faced when they came to the Western Hemisphere was to clear the forests so they could raise crops. Once the forests were cleared, continual grazing or plowing effectively prevented the trees from regrowing.

Consequences. Clearing a forest has the following immediate consequences for the land and its people:

- Overall productivity of the area is reduced.
- Standing stock of nutrients and biomass, once stored in the trees and leaf litter, is enormously reduced.
- Biodiversity is greatly diminished.
- Soil is more prone to erosion and drying.
- The hydrologic cycle is changed as water drains off the land instead of being released by transpiration through the leaves of trees or percolating into groundwater.
- A major carbon dioxide sink (removal of CO_2 from the air) is lost.
- Land no longer yields forest products.
- People who depend on harvesting forest products lose their livelihood.

The trade-off, of course, if you cut down a forest, is that now the land will yield crops or support livestock, more people may be fed, and (usually) more profit will accrue to the owners, but sometimes those effects do not last. If forests are not cleared and converted to other uses, they could be managed to yield a harvest of wood for fuel, paper, and building materials indefinitely. Approximately 3 billion cubic meters (3.7 billion yd^3) of wood are harvested annually from the world's forests for fuel, wood, and paper (**Fig. 7–9**). It is unreasonable to expect the countries of the world—especially the developing countries—to forgo making use of their forests. However, forest-management practices vary greatly in their impact on the forest itself and on surrounding ecosystems.

Figure 7–9 Logging operations. A forest in New Guinea is logged for its tropical wood.

[4]FAO. Global Forest Resources Assessments, Key Findings 2010, p. 9. http://www.pefc.org/images/stories/documents/external/KeyFindings-en.pdf.

Types of Forest Management. The practice of forest management, usually with the objective of producing a specific crop (hardwood, pulp, softwood, wood chips, etc.), is called **silviculture.** Several practices are employed, with quite different impacts on the total forest ecosystem involved. Because trees take from 25 to 100 years to mature to harvestable size, the normal process of management is a rotation, which is a cycle of decisions about a particular stand of trees from its early growth to the point of harvest. Two options are open to the forester: even-aged management and uneven-aged management. With even-aged management, trees of a fairly uniform age are managed until the point of harvest, cut down, and then replanted, with the objective of continuing the cycle in a dependable sequence.

Clear-Cutting. Typically, fast-growing, but economically valuable, trees are favored for even-aged management. (Others are culled from the stand.) In the Pacific Northwest, Douglas fir is frequently managed this way. Harvesting is often accomplished by **clear-cutting**—removing an entire stand at one time. This management strategy has been severely criticized because it creates a fragmented landscape with serious impacts on biodiversity and adjacent ecosystems; also, a clear-cut leaves an ugly, unnatural-looking site for many years. Large timber companies often employ the practice because it is efficient at the time of harvest and it does not involve much active management and planning.

Other Methods. Uneven-aged management of a stand of trees can result in a more diverse forest and lends itself to different harvesting strategies. For example, in **selective cutting**, some mature trees are removed in small groups or singly, leaving behind a forest that continues to maintain a diversity of biota and normal ecosystem functions. Replanting is usually unnecessary, as the remaining trees provide seeds. Another strategy is **shelter-wood cutting**, which involves cutting the mature trees in groups over a period of, say, 10 or 20 years, such that at any time there are enough trees both to provide seeds and to give shelter to growing seedlings. Like selective cutting, this method takes more active management and skill, but unlike clear-cutting, it leaves a functional ecosystem standing. In effect, it should lead to a sustainable forest, an objective that the entire wood and paper industry is currently focusing on—at least in principle.

Sustainable Forestry. In forestry, sustainability must be carefully defined. A common objective of forestry management is **sustained yield,** which means that the production of wood is the primary goal and the forest is managed to harvest wood continuously without being destroyed. It is similar in concept to MSY in that its objective is to maximize harvest rates, but it ignores the ecosystem properties of forests. **Sustainable forest management,** by contrast, was defined at the United Nations Conference on Environment and Development (UNCED) of 1992: *Forests are to be managed as ecosystems, with the objectives of maintaining the biodiversity and integrity of the ecosystem, but also to meet the social, economic, cultural and spiritual needs of present and future generations.* This sentiment has been expanded into a list of principles for sustainable forestry.

Tropical Forests. Because of their continued deforestation, tropical forests are of great concern. They are the habitat for millions of plant and animal species and a storehouse of biodiversity. These forests are also crucial in maintaining Earth's climate, serving as a major sink for carbon and restraining the buildup of global carbon dioxide. Nevertheless, tropical forests continue to be removed at a rapid rate. Between 1960 and 1990, 20% of the tropical forests (over 445 million hectares, or 1.1 billion acres)—equivalent to two-fifths of the land area of the United States—were converted to other uses. According to the FAO, in 2010 the annual global deforestation rate was about 13 million hectares (32 million acres), down from about 16 million hectares per year in the 1990s. Most of this loss occurred in the tropical forests **(Fig. 7–10).**

Reasons for Removal. Deforestation is caused by a number of factors, all of which come down to the fact that the countries involved are in need of greater economic development and have rapid population growth. The FAO concluded that the current major cause of deforestation is conversion to pastures and agriculture. Many governments in the developing world are promoting deforestation by encouraging the colonization of forested lands. Indonesia, for example, has embarked on a program of resettlement and intends to convert 20% of its remaining forests to agricultural production. The clearing of forests for subsistence agriculture is especially intense in Africa, where population growth is unrelenting. In Brazil's Amazon basin, commercial ranches, home to some 74 million cattle, are replacing rain forests, enabling Brazil to dominate the beef export market. The Brazilian government has introduced antideforestation measures, which had some success in reducing the Amazon deforestation for several years (2004–2007), but recent satellite-based reports indicate that the rate is rising again.

New Trends. Left to grow, the forests of the developing world are an important resource that can generate much-needed revenue. Developing countries account for 26% of the international trade in forest products. Often,

DESIGNATED FUNCTIONS OF FORESTS, 2010

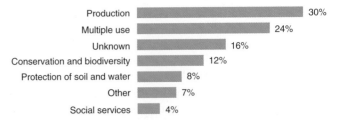

Figure 7–10 Designated uses of forests. Worldwide, forests have different ecosystem services for which they are managed.

(*Source:* Food and Agriculture Organization (FAO) of the United Nations, Global Forest Resources Assessment 2010, pg. xxvi.)

concessions are sold to multinational logging companies, which reap high profits by exploiting the economic desperation of the developing countries and harvest the timber with little regard for regenerating the forest. Chinese and other Asian companies are logging millions of acres of tropical forest in smaller countries like Belize, Suriname, and Papua New Guinea, where regulation is weak and corruption is strong. More important is the consumptive use of forests by millions of people living in the forests or on their edges. Owning few private possessions and often little land, they absolutely depend on foraging in and extracting goods from the forests.

There are encouraging trends in forest management in the developing world—trends that indicate that those in power are paying more attention to the needs of indigenous people and the importance of forest goods and services (other than just wood). These actions include practicing sustainable forest management, designating forest areas for conservation, establishing more-sustainable plantations of trees for wood or other products (cacao, rubber, etc.) **(Fig. 7–11)**, and setting aside extractive reserves (areas for nonwood products) that yield nontimber goods. Some 3.6 million hectares (9 million acres) of the Brazilian Amazon have been set aside as extractive reserves.

Another trend is the preservation of forests as part of a national heritage and putting them to use as tourist attractions. This practice can often generate much more income than logging can. Putting forests under the control of indigenous villagers is an important part of forest protection in many parts of the world. The villagers can then collectively use the forest products in traditional ways. Given tenure over the land, the villagers tend to exercise stewardship over their forests in a way that is sustainable. Where this practice has been implemented, the forests have fared better than where they have been placed under state control.

Certification. The **Forest Stewardship Council** was the result of a recent initiative directed toward the certification of wood products. This alliance of nongovernmental organizations (including the World Wide Fund for Nature and National Wildlife Federation, industry representatives, and forest scientists) has developed into a major international organization with the mission of promoting sustainable—or, as they call it, "responsible"—forestry by certifying forest products for the consumer market. By 2009, more than 110 million hectares (242 million acres) in 81 countries had been certified. Certification makes it possible for consumers to choose wood products that have been harvested sustainably.

Ocean Ecosystems

Covering 75% of Earth's surface, the oceans and coastal ecosystems provide priceless goods and services that enable commerce and enhance human well-being. As vast as the oceans are, human activities have impacted them, often severely. A team of British, Canadian, and American scientists recently constructed a global map that pictures the severity of the effects of such human impacts as commercial shipping, resource extraction, nutrient runoff from land, fishing, species invasion, and greenhouse gas emissions **(Fig. 7–12)**. The team's work provides a first attempt at documenting the human impacts throughout the oceans and is expected to stimulate strategies for minimizing the ecological impact of human activities in locations where that impact is especially severe. One of those activities is commercial fishing.

Marine Fisheries. Fisheries provide employment for at least 200 million people and account for more than 15% of the total human consumption of protein. The term *fishery* refers either to a limited aquatic area or to a group of fish or shellfish species being exploited. For years, the oceans beyond a 12-mile limit were considered international commons. By the end of the 1960s, however, numerous regions of the sea were being seriously depleted of many species by overfishing on the part of international fleets equipped with factory ships and modern fish-finding technology. In the mid-1970s, as a result of agreements forged at a series of UN Conferences on the Law of the Sea, nations extended their limits of jurisdiction to 200 miles offshore. The United States accomplished this with the Magnuson Act of 1976. Since many prime fishing grounds are located between 12 and 200 miles from shore, this action effectively removed most fisheries from the international commons and placed them under the authority of particular nations. As a result,

Figure 7–11 Plantation forest. A rubber tree plantation in Vietnam.

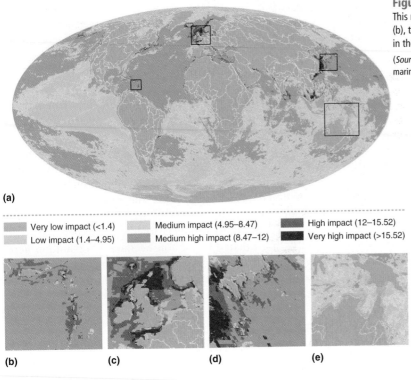

(a)

Figure 7–12 **Human impact on ocean ecosystems.**
This map indicates the highest-impact areas in the eastern Caribbean (b), the North Sea (c), and Japanese waters (d) and a least-impact area in the vicinity of northern Australia (e).

(*Source*: Halpern, Benjamin S. et al.(2008). A global map of human impact on marine ecosystems. Science, vol. 319, no. 5865, pp. 948–952)

▨ Very low impact (<1.4)	▨ Medium impact (4.95–8.47)	▨ High impact (12–15.52)
▨ Low impact (1.4–4.95)	▨ Medium high impact (8.47–12)	▨ Very high impact (>15.52)

(b)   (c)   (d)   (e)

some fishing areas recovered, while nationally based fishing fleets expanded to exploit the fisheries.

The Catch. The total recorded harvest from marine fisheries and fish farming is reported annually by the FAO. The harvest has increased remarkably since 1950, when it was just 20 million metric tons **(Fig. 7–13)**. By 2009, it had reached almost 145 million metric tons. Aquaculture accounted for 55.1 million metric tons, or 38% of world fish supplies that year. The trends in the figure show that the

"capture" fisheries leveled off in the 1990s and the continued rise in fish production is due to aquaculture. Of the total of 145 million metric tons, 90 million metric tons were the freshwater and marine fisheries catch, to which should be added approximately 7 million tons of bycatch (fish caught and discarded).

One of the effects of this increase in fishing effort is the removal of large fish. Because large and slow-growing fish take longer to replace, ocean fish are changing—to smaller individuals and those that reproduce more rapidly. As fishers capture too many of one species, they move on to the next. First, they overfish the top predators—such as tuna, marlin, and swordfish—then those smaller and lower on the food chain. This process of moving from species to species as each type is depleted is known as "fishing down the food chain." It means that commercial fishing has made long-lasting changes to basic ecological processes in marine ecosystems.

Aquaculture. Modern aquaculture, the farming of aquatic organisms, includes species ranging from seaweed to bivalve mollusks to freshwater and marine fish and has increased dramatically in the past 20 years (see Fig. 7–13). This production is offsetting the plateau in the wild capture fisheries, but coastal aquaculture is not without its problems and critics. For example, shrimp farming in many tropical areas has been criticized for destroying mangrove habitats, accelerating the decline of fisheries for some species that depend on the mangroves for a part of their life cycle. Another concern is that farming carnivorous species such as shrimp and salmon requires food—often in the form of fishmeal from marine organisms caught in the wild. More than 37% of the capture fisheries are species like whiting, herring, and anchovy that are reduced to fish products to

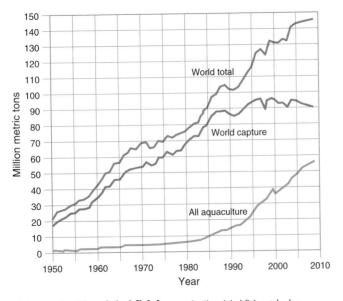

Figure 7–13 **The global fish harvest.** The global fish catch plus aquaculture equals the world total. Note how aquaculture is providing all of the increase since 1995. Shown are data for 1950–2008.

(*Source*: Data from FAO. 2010. The State of the World's Fisheries and Aquaculture.)

SOUND SCIENCE

A group of people walk across the flat landscape, bent over, moving slowly from side to side. From a distance their motion looks odd. Up close, you can see they are throwing something on the ground from containers in their hands. They look like ancient farmers, in a planting ritual going back millennia, but they are not. They are ecologists and volunteers restoring an abandoned farm to a native prairie. The seeds they cast are from native grasses like big blue stem and plants like wild lupine, purple coneflower, and Black-eyed Susan.

Volunteers have painstakingly picked the seeds over a period of months, and now, the soil plowed and ready, the seeds are being planted. Later, when both native plants and non-native agricultural weeds have grown, the team will reconvene to burn the field, giving the native, fire-adapted plants an edge and clearing out the weeds. Right now it is planting time. For this moment to happen, a lot of other things had to be done first. Seeds were collected some months ago. Some species needed to be chilled for weeks to stimulate growth. Some were scratched (or "scarified"), others soaked and exposed to fungi that would help them grow.

Humans are doing the work that a healthy soil would do in a natural setting.

In order to find the most effective ways to restore an ecosystem such as an abandoned farm, old mining area, or former construction site, scientists first must do extensive research. Much of this research owes a debt of gratitude to the restoration pioneers at the Curtis prairie (Madison, Wisconsin), the oldest ecologically restored prairie in the world. Scientists at the University of Wisconsin-Madison have been experimenting with restorations for 75 years, coming up with many of the techniques and scientific advances that are used in other places.

Even so, scientists will agree that restoration will not bring back the system that was degraded. Usually, what is restored differs significantly from the original. Prevention is a better approach, and restoration should be used when other approaches have failed.

Restoration Ecology. Prairie seeds collected for restoring prairie habitat.

Sources: R. H. Hilderbrand, A. C. Watts, and A. M. Randle. "The Myths of Restoration Ecology." *Ecology and Society*, 10(1) (2005): 19. http://www.ecologyandsociety.org/vol10/iss1/art19/.

M. P. Zedler, B. Herrick, and J. Zedler, 2008. Curtis Prairie: 75-Year-Old Restoration Research Site. http://botany.wisc.edu/zedler/images/Leaflet_16.pdf.

---

feed farm-raised fish and animals. Salmon and shrimp aquaculture farms consume several times as much fish protein as they actually produce.

Spread of diseases from farmed to wild populations and pollution of waters by fish farms' nutrients are other concerns that accompany aquaculture. The result in many cases is conflict among the various users, and the long-term outlook for coastal aquaculture is that nearshore waters are becoming less and less available for farming.

One approach is open-ocean aquaculture, which is the farming of coastal species in ocean waters several kilometers from the shoreline. This is a new practice in North America, and even in Europe and other areas where it has been long practiced, total production levels are far lower than they could be. Open-ocean waters are relatively clean. High water quality is important for aquaculture, but it is especially critical for species such as bivalve mollusks, which concentrate pathogens and other pollutants that can be passed on to human consumers. Also, new satellite-based remote-sensing efforts are revealing previously unknown nutrient-rich upwelling zones. In some areas in the Gulf of Maine, for example, phytoplankton concentrations can exceed the levels thought to occur only in enriched estuarine waters. This means that open-ocean waters may be

able to support a high productivity of farmed species such as the blue mussel. Mussels and other bivalve mollusks are herbivorous animals, feeding lower on the food web. A long-term sustainable approach to aquaculture would see more use of herbivorous species, reducing the impact on the wild fish stocks.

The Limits. The world fish catch may appear stable, but many species and areas are overfished. The FAO has concluded that 52% of fish stocks are fully exploited (at their maximum sustainable yield), 16% are overexploited and will likely decline further, and 8% are depleted and are much less productive than they once were. Much of the catch (one-third) is of such low commercial value that it is used only for fishmeal and oil production. Aquaculture is growing in developing countries, helping to alleviate poverty and improve the well-being of rural people.

The FAO has also concluded that the capture fisheries are at their upper limit, as indicated by the long-term catch data. Fisheries scientists reason that if overexploited and depleted fisheries were restored, the catch could increase by at least 10 million metric tons. That could happen, however, only if fishing for many species was temporarily eliminated or greatly reduced and the fisheries were managed on a more sustainable basis. At the 2002 Johannesburg World Summit,

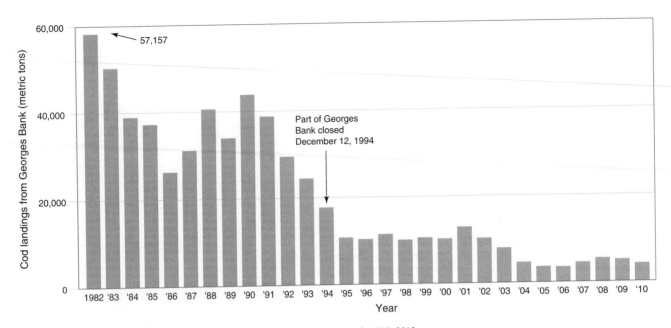

Figure 7–14 Fishery in distress. Data showing cod landings from Georges Bank, 1982–2010.
(*Source:* NOAA, 2011.)

delegates agreed (one of the few agreements to come out of the meeting) to restore depleted global fisheries to achieve an MSY "on an urgent basis" by 2015 and to establish a network of protected marine areas by 2012. Recent studies have shown that marine reserves provide sheltered nurseries for surrounding areas and can supply world-record-size fish to adjacent fisheries.

Georges Bank. Events on Georges Bank, New England's richest fishing ground, illustrate well the fisheries' problems. Cod, haddock, and flounder (called groundfish because they feed on or near the bottom and are caught by bottom trawling) were the mainstay of the fishing industry for centuries. In the early 1960s, these species accounted for two-thirds of the fish population on the bank. After 1976, fishing on Georges Bank doubled in intensity, resulting in a decline in the desirable species and a rise in the so-called rough species—dogfish (sharks) and skates. As the rough species increased their numbers, fishers focused their attention on them, even though they were less valuable. Predictably, dogfish and skate populations have also been overfished and are now subject to regulation.

Management Councils. The prized species declined on Georges Bank because the fishery was poorly managed. The Magnuson Act established eight regional management councils made up of government officials and industry representatives. The councils are responsible for creating management plans for their regions, while the National Marine Fisheries Service (NMFS) provides advice and scientifically based information on assessing stocks. The NMFS also has the authority to reject elements of the plans submitted to it by the councils. The New England Fishery Management Council (NEFMC) began its task by setting TAC quotas, but fishers claimed that the TACs were set too low and successfully argued for an indirect approach that employed mesh

with openings of a size that allowed smaller fish to escape. In less than 10 years, the number of boats fishing Georges Bank doubled, with disastrous results, as indicated by the cod-landing data **(Fig. 7–14)**.

The NEFMC discarded the mesh-size approach and initiated a number of regulations designed to take some pressure off the Georges Bank cod. Vessels were restricted in the days they could spend at sea, areas of the bank were closed to fishing, and target TACs were set at levels designed to reduce the harvest. In addition, the NMFS has been buying back fishing permits and paying fishers to scrap their boats. In spite of these measures, cod landing consistently exceeded the TAC for years, by 150–200%, until 2004–2005, when the stock had plummeted to its lowest level ever and restrictions finally led to a much-reduced harvest. Recently, the New England ground fishery (with the blessing of the NEFMC) began to operate a community-based quota system; this has been so successful that 17 additional groups of eastern fishers have petitioned to start similar systems.

Other Cod Fisheries. The Grand Banks off the coast of Newfoundland were once the richest fishing grounds in the world. The cod, however, were decimated by Canadian fishers, who were encouraged by their government to exploit the fishery and were given little warning of its imminent collapse. By 1991, the cod population had dropped to one-hundredth of its former size. In 1992, the fishery was closed, costing the jobs of 35,000 fishers. So far, the cod population has not rebounded, and cod are at their lowest levels in a century. Across the water, Europe's cod fishery is experiencing the same decline: The North Sea's cod stocks are now only 15% of what they were in the 1970s. Proposals to reduce harvests have met with fierce resistance from the European Union fishing nations, who argue that the reductions would destroy them.

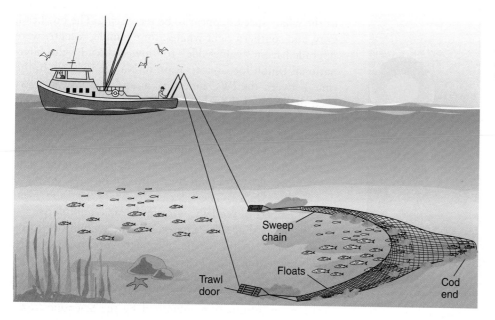

Figure 7–15 Bottom trawling. Because of the way the ocean bottom is degraded, this method, used to harvest groundfish, has been compared to clear-cutting forests. The steel trawl doors weigh up to 5 tons, and the distance between the doors is as much as 200 feet.

The basic problem of the fisheries is that there are too many boats—rigged with high technology that gives the fish little chance to escape—chasing too few fish. Equipped with sophisticated electronic equipment that can find fish in the depths and with trawlers with 200-foot-wide openings that scoop up everything on and close to the bottom, the fishing industry has become too efficient (**Fig. 7–15**). A better system of management is needed and the size of the fishing fleet must be reduced, so that the remaining fishers are better able to make a living without overfishing. One system, called **catch shares**, effectively gives fishers the equivalent of property rights in the fishery. The rights are transferable and allow the fishers to determine how and when they will harvest their catch. Coupled with accurate population assessments by fishery scientists, the catch shares system has produced an economically stable and environmentally sustainable Pacific halibut fishery, and there is no reason that it should not be applied to fisheries everywhere.

Fisheries Law Reauthorized. In December 2006, Congress passed the second reauthorization of the Magnuson Act, with strong bipartisan support: the **Magnuson-Stevens Fishery Conservation and Management Reauthorization Act.** In the new legislation, the structure of the regional councils is retained, but they are required to set catch limits based on sound scientific advice from the council's Scientific and Statistical Committee (SSC). The annual catch limits are to be set below the maximum sustainable yield, and the councils are to employ precautionary principles in setting those limits. The law also requires councils to end overfishing within two years after a species is judged to be overfished (it used to be 10 years!). As with the previous version of the act, depleted fish stocks must be rebuilt and maintained at biologically sustainable levels. The law also requires steps to be taken to assess and minimize bycatch. Limited access programs may be developed along the lines of the catch shares system. Clear accountability measures are required to ensure that the councils

and their scientists follow the mandates of the new law. Some fisheries scientists were disappointed that the new law does not require an ecosystem-based approach to fisheries management, instead maintaining the existing species-by-species approach. It does, however, strengthen the role of the SSCs, which improves the science-based approach to fisheries management.

Marine Reserves. Marine Protected Areas (MPAs) are areas of the coasts and sometimes open oceans that have been closed to all commercial fishing and mineral mining. They often allow less-harmful use like small-scale fishing, tourism, and recreational boating, although some MPAs prohibit all such activities. The number of MPAs is growing, now numbering more than 4,000 of varying sizes. More than 18% of coral reefs are covered by MPAs, but not all of these are rigorously managed. The benefits of MPAs are well known and include basic protection of these vital ecosystems, increased tourism, and the spillover of fish and larvae from the MPA to surrounding fisheries. The most direct path to the restoration of a fishery or a damaged habitat is likely to be the creation of a marine reserve. In 2009, President George W. Bush established three huge MPAs in waters surrounding several South Pacific islands, totaling 505,000 square kilometers (195,000 mi^2) in size.

International Whaling. Another fishery of great concern to the international community is the great whale. Whales, found in the open ocean and in coastal waters, were heavily depleted by overexploitation until the late 1980s. Whales once were harvested for their oil but are now prized for their meat, considered a delicacy in Japan and a few other countries. In 1974, the International Whaling Commission (IWC), an organization of nations with whaling interests, decided to regulate whaling according to the principle of maximum sustainable yield. Whenever a species of whale dropped below the optimal population for such a yield, the

Figure 7–16 Current whaling operation. Butchers slice into a Baird's beaked whale at a pier-side slaughterhouse in Wada, Japan.

IWC instituted a ban on hunting that species in order to allow the population to recover. At that time, three species (the right whale, bowhead whale, and blue whale) were at very low levels and were immediately protected. Because of difficulties in obtaining reliable data on and enforcing catch limits, the IWC took more-drastic action and placed a moratorium on the harvesting of all whales beginning in 1986. The moratorium has never been lifted; however, some limited whaling by Japan, Iceland, and Norway continues, as does harvesting by indigenous people in Alaska, the Russian Federation, and Greenland (Fig. 7–16).

Table 7–4 lists the status of 13 whale species or populations that are of commercial interest. Estimates of the numbers remaining range from 100 Northwest Pacific gray whales to up to 2 million sperm whales. Their main threat today is entanglement with fishing gear and collisions with ships. Although reliable data are still hard to acquire, many of the whale species appear to be recovering. The bowhead, gray, and southern right whales, for example, recently were upgraded from "vulnerable" to "lower risk" by the International Union for Conservation of Nature (IUCN). In the time between the IWC moratorium and the present, however, the basis of the ethical controversy over whaling has shifted from conservation to animal rights. Many people worldwide simply believe that it is wrong to kill and eat such large and unique mammals. People from whaling nations counter with the argument that their culture eats whales, just as other cultures eat cows or turkeys.

Whale Stakes. There has been heavy pressure from three members of the IWC—Japan, Iceland, and Norway—to reopen whaling. The interests of these countries focus on the small minke whale, although the Japanese have begun taking the much larger fin whales. In 1993, Norway resumed whaling for minkes, citing the country's right to refuse to accept specific IWC rulings. Norway based its decision on information submitted by the Scientific Committee of the IWC, which judged the minke population to be numerous enough to absorb a sustainable exploitation. Currently, Norway takes up to 650 and Japan more than 1,000 minke whales per year, the Norwegians for human consumption and the Japanese for "scientific purposes." It is no secret that the Japanese "scientific" harvest ends up in commercial markets; the Japanese people, however, are losing their taste for whale meat, and much of it ends up as pet food or in school dinners.

The IWC is right in the middle of this controversy. At its annual meetings, prowhaling members push for a resumption of whaling with quotas, and antiwhaling countries oppose any such resumption. The 2007 meeting of the IWC reaffirmed the moratorium on commercial whaling, and subsequent meetings have not changed that.

One contentious issue, however, is the claim that Japan and Iceland kill whales under permits given for scientific purposes but really use the law as a loophole to get around the moratorium on commercial whaling.

TABLE 7–4 Remaining Numbers of Large Whales		
Endangered	**Vulnerable**	**Lower Risk or Insufficient Data**
Blue whale (3,000–5,000)	Sperm whale (1 million–2 million)	Minke whale (935,000)
North Atlantic right whale (300–350)		Bryde's whale (40,000–80,000)
North Pacific right whale (under 1,000)		Bowhead whale (11,750)
Fin whale (50,000–90,000)		Southern right whale (7,500)
Sei whale (50,000)		Humpback whale (63,600)
Northwest Pacific gray whale (100)		Gray whale (26,300)

Sources: WWF, IUCN *Red Data List* (2008), and International Whaling Commission.

Figure 7–17 Whale watching. A group of humpback whales entertains boatloads of whale watchers on Stellwagen Bank just north of Provincetown, Cape Cod. Whale watching has replaced whaling in New England waters that were once famous as whaling centers.

Whale Watching. Whale watching has become an important tourist enterprise in coastal areas (**Fig. 7-17**). For example, Stellwagen Bank, within easy reach of boats from Boston, Cape Ann, and Cape Cod, Massachusetts, has become the center of a whale-watching industry estimated to be worth more than $30 million annually in New England alone. Whale watching now flourishes in 87 countries and generates more than $1 billion in business annually.

Besides having aesthetic and entertainment value, whale watching is of scientific value. Whale-watching tour boats usually carry along a biologist who identifies the whale species and interprets the experience for the visitors. Unfortunately, whales can be spooked by boats following too closely. In 2011, a U.S. researcher was charged with bringing boats too close and feeding whales, an activity that could teach whales to come too close to boats and lead to their injury.

Coral Reefs. In a band from 30° north to 30° south of the equator, coral reefs (such as the Caribbean reefs discussed at the chapter's start) occupy shallow coastal areas and form atolls, or islands, in the ocean. Coral reefs are among the most diverse and biologically productive ecosystems in the world. The reef-building coral animals live in a symbiotic relationship with photosynthetic algae called **zooxanthellae** and, therefore, are found only in water shallower than 75 meters (245 ft). The corals build and protect the land shoreward of the reefs and attract tourists, who enjoy the warm, shallow waters. In addition, because the reefs attract a great variety of fish and shellfish, they are important sources of food and trade for local people, generating an estimated $30 billion per year in tourism and fishing revenues. In recent years, coral reefs have been the subject of concern because they have shown signs of deterioration in areas close to human population centers. The year 1998 was a watershed: Before then, an estimated 11% of coral reefs had been destroyed by human activities. Since then, the percentage of destroyed reefs has increased to 20%. The threats are both global and local.

Extensive **coral bleaching** (wherein the coral animals expel their symbiotic algae and become white in appearance) severely damaged an estimated 16% of the world's reefs (**Fig. 7-18**) during the record-high sea surface temperatures accompanying the 1997–1998 El Niño. Some coral bleaching is an annual phenomenon related to the normal high temperatures and light intensities of summer, but the events of 1998 were unprecedented in scope. Studies have shown that increased sea surface temperatures can bring on 100% bleaching; indeed, 2010 brought the highest global water temperatures on record, and the second worst coral bleaching event since we have kept records. Corals are resilient, recovering from bleaching if sea surface temperatures decline in a short time, and some species are more tolerant of bleaching than others. Repeated bleaching, however, weakens the corals and eventually kills them. Many had been recovering from 1998 temperatures more than a decade later, when new record temperatures set them back.

The other global threat—one that is more long-term—is **ocean acidification.** As the surface oceans take up increasing amounts of the CO_2 pumped into the atmosphere by humans, seawater becomes more acid and the carbonate ion concentration is reduced. This in turn makes it more difficult for coral animals to build their calcium carbonate skeletons. Scientists point out that persistent high summer temperatures and the gradual acidification of the oceans could permanently wipe out coral reefs over vast areas of the shallow tropical oceans—an enormous loss of biodiversity and economic value.

Islanders and coastal people in the tropics often scour the reefs for fish, shellfish, and other edible sea life. Some exploitation of the reefs, however, is driven by the lucrative trade in tropical fish for food and the pet trade. Local residents sometimes use dynamite and cyanide to flush the fish out of hiding in the coral, a practice that is not only destructive to the coral, but also dangerous to the perpetrators.

Figure 7–18 Bleached coral. This coral in the Red Sea shows the effects of excessively warm temperature.

"Fishing" with cyanide is a growing (illegal) practice that has been encouraged by exporters in the Indo-Pacific region.

One solution to this destructive trade is to foster effective, sustainable management of the coral reef resources. Ideally, such management would be community based so that local people benefit from the exploitation. Another solution is to incorporate coral reefs in Marine Protected Areas (MPAs). A recent study found that 980 MPAs are already providing 19% of the world's coral reefs with some level of protection. Again, effective management and enforcement are essential, and even this is not always effective.

Mangroves. Often, just inland of coral reefs is a fringe of mangrove trees. These trees have the unique ability to take root and grow in shallow marine sediments. There they protect the coasts from damage due to storms and erosion and form a rich refuge and nursery for many marine fish. Despite these benefits, mangroves are under assault from coastal development, logging, and shrimp aquaculture. Between 1983 and the present, half of the world's 45 million acres (18 million ha) of mangroves were cut down, with percentages ranging from 40% (in Cameroon and Indonesia, for example) to nearly 80% (in Bangladesh and the Philippines, for instance). The massive removal of mangroves has brought on the destabilization of entire coastal areas, causing erosion, the siltation of sea grasses and coral reefs, and the ruin of local fisheries. Developing nations that once considered the mangroves to be useless swampland are now beginning to recognize the natural services they perform. Local and international pressure to stop the destruction of mangroves is growing. (Mangroves are highlighted in the section in this chapter on forests as well.)

We have investigated how the management (or lack thereof) of natural resources has affected several ecosystems around the world, focusing on forests and ocean systems. We could, however, have looked at freshwater ecosystems such as lakes and rivers, grasslands, or mountain areas. In each case, we could have looked at the use of natural capital from those ecosystems and whether humans are using the ecosystem goods and services in a sustainable way. In the next section, we will focus on the connection between public policy and natural resource management as it is reflected in public and private lands in the United States and at ways to restore degraded ecosystems.

7.4 Protection and Restoration

The time has passed when the loss of ecosystems could be excused on the grounds that there are suitable substitute habitats just over the hill. With the rising human population, industrial expansion, and pressure to convert natural resources to economic gain, there will always be reasons to exploit natural ecosystems. The last resort for many species and ecosystems is protection by law in the form of national parks, wildlife refuges, and reserves. The World Database on Protected Areas (WDPA) recognizes 106,900 "protected areas,"

sites whose management is aimed at "achieving specific conservation objectives," according to the Convention on Biological Diversity. These sites, representing 13% of Earth's land area, cover a spectrum of protection from strict nature preserves to managed resource use. However, establishment of a protected area is not always followed by effective management and control over misuse.

Public and Private Lands in the United States

In March 2009, the Omnibus Public Lands Management Act was signed into law by President Obama. This sweeping legislation protected an additional 2.1 million acres of wilderness, increased the wild and scenic river system by 50%, established new national parks and monuments, and provided for a new federal land system, the National Landscape Conservation System. The United States is unique among the countries of the world in having set aside a major proportion of its landmass for public ownership. Nearly 40% of the country's land is publicly owned and is managed by state and federal agencies for a variety of purposes, excluding development. The distribution of federal lands, shown in **Figure 7–19**, is greatly skewed toward the western states, a consequence of historical settlement and land distribution policies. Nonetheless, although most of the East and Midwest are in private hands, forests and other natural ecosystems still function all across these regions.

Wilderness. Land given the greatest protection (preservation) is designated as **wilderness**. Authorized by the Wilderness Act of 1964, this classification includes 109 million acres in more than 700 locations—more than 4% of the land area of the United States. The act provides for the permanent protection of these undeveloped and unexploited areas so that natural ecological processes can operate freely. Permanent structures, roads, motor vehicles and other mechanized transport are prohibited. Timber harvesting is excluded. Some livestock grazing and mineral development are allowed where such use existed previously; hiking and other similar activities are also allowed. Areas in any of the federally owned lands can be designated as wilderness; the Bureau of Land Management, National Park Service, Forest Service, and Fish and Wildlife Service all manage wilderness lands.

National Parks and National Wildlife Refuges. The **national parks**, administered by the National Park Service, and the **national wildlife refuges**, administered by the Fish and Wildlife Service, provide the next level of protection to 84 million and 96 million acres, respectively. Here the intent is to protect areas of great scenic or unique ecological significance, protect important wildlife species, and provide public access for recreation and other uses. The dual goals of protecting and providing public access often conflict with each other, because the parks and refuges are extremely popular, drawing so many visitors (more than 310 million visits a year) that protection can be threatened

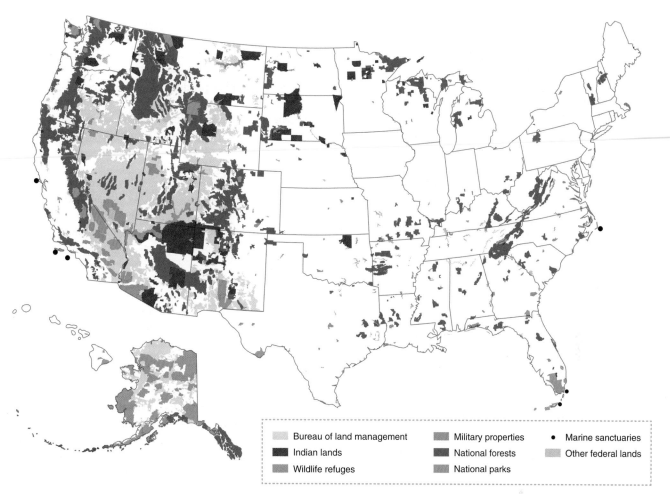

Figure 7–19 Distribution of federal lands in the United States. Because the East and Midwest were settled first, federally owned lands are concentrated in the West and Alaska.

(*Source:* Council on Environmental Quality, *Environmental Trends,* 1989.)

by those who want to see and experience the natural sites. At a number of the most popular parks, auto traffic has been replaced by shuttle vehicles because of the problems of parking and habitat destruction accompanying the use of motor vehicles within the parks. All of the federal agencies involved are trying to cope with the rising tide of recreational vehicles like snowmobiles and off-road vehicles used on most federal lands. The Blue Ribbon Coalition, a lobbying group representing owners and manufacturers of off-road vehicles, fiercely defends keeping federal lands accessible to these vehicles.

Increasingly, agencies, environmental groups, and private individuals are working together to manage natural sites as part of larger ecosystems. For example, the Greater Yellowstone Coalition has been formed to conserve the larger ecosystem that surrounds Yellowstone National Park **(Fig. 7–20)**. Yellowstone and Grand Teton National Parks form the core of the ecosystem, to which seven national forests, three wildlife refuges, and private lands are added to form the greater ecosystem of 18 million acres. The coalition acts to restrain forces threatening the ecosystem. Thus, it has challenged the Forest Service's logging activities, addressed residential sprawl and road building on the private lands, acted to protect crucial grizzly bear habitat, and curtailed

the senseless shooting of bison that wander off the national parklands in search of winter grazing.

This cooperative approach is important for the continued maintenance of biodiversity because so much of the nation's natural lands remain outside of protected areas. Cooperation may also help restrict development up to the borders of the parks and refuges, keeping them from becoming fairly small natural islands in a sea of developed landscape.

National Forests. Forests in the United States represent an enormously important natural resource, providing habitat for countless wild species as well as supplying natural services and products. These forests range over 740 million acres, of which about two-thirds are managed for commercial timber harvest. Almost three-fourths of the managed commercial forestland is in the East and privately owned; the remainder is mainly in the West and is administered by a number of government agencies, primarily the Forest Service and the Bureau of Land Management.

In the United States, although we have cut all but 5% of the forests that were here when the colonists first arrived, second-growth forests have regenerated wherever forestlands have been protected from conversion to croplands or house

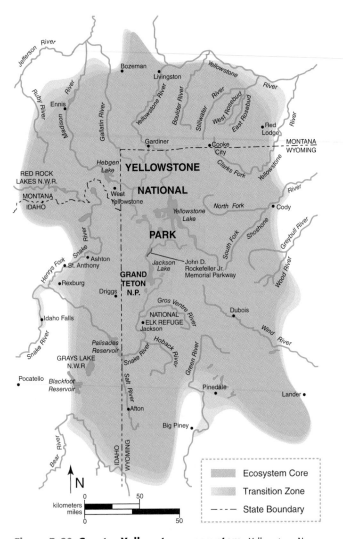

Figure 7–20 Greater Yellowstone ecosystem. Yellowstone National Park is the center of a much larger ecosystem, which has received attention from the Greater Yellowstone Coalition.

The Forest Service is responsible for managing 193 million acres of national forests. The Bureau of Land Management oversees 258 million acres that are a great mixture of prairies, deserts, forests, mountains, wetlands, and tundra. Most federal timber production comes from the national forest lands.

Multiple Use. The Forest Service's management principle in the 1950s and 1960s was multiple use, which allowed for a combination of extracting resources (grazing, logging, and mining), using the forest for recreation, and protecting watersheds and wildlife. Although the intent was to achieve a balance among these uses, multiple use actually emphasized the extractive uses; that is, it was output oriented and served to justify the ongoing exploitation of public lands by private interest groups (ranchers, miners, and especially the timber industry).

New Forestry. A forest-management strategy called **new forestry** was introduced in the late 1980s. This practice is directed more toward protecting the ecological health and diversity of forests than toward producing a maximum harvest of logs. New forestry involves cutting trees less frequently (at 350-year intervals instead of every 60 to 80 years), leaving wider buffer zones along streams to reduce erosion and protect fish habitats, leaving dead logs and debris in the forests to replenish the soil, and protecting broad landscapes—up to 1 million acres—across private and public boundaries, while involving private landowners in management decisions. The Forest Service began adopting some of these management principles in the early 1990s, and they formed the core of what is now the official management paradigm of the Forest Service: ecosystem management.

To harvest timber, government foresters select tracts of forest they judge ready for harvesting and then lease the tracts to private companies, which log and sell the timber. The timber harvest from national forests has been a controversial issue. Historic data on timber sales (**Fig. 7–21**) reflect major shifts in logging. Note the rapid rise following World War II to satisfy the housing boom that was under way at the time, the drop reflecting rising environmental concern in the 1970s, increased logging during the Reagan-Bush years, and a steep drop with the Clinton administration as a result of renewed environmental concern. According to a recent Forest Service analysis, growth of forests has exceeded harvests for

lots. There are more trees in the United States today than there were in 1920. In the East, some second-growth forests have aged to the point where they look almost as they did before their first cutting, except for the absence of the American chestnut and elm. Now scientists are trying to breed disease-resistant American chestnuts and elms.

Figure 7–21 National forest timber sales. Timber sales over the years reflect changes in resource needs and political philosophy.

(*Source:* US Forest Service FY 1905–2011 National Summary Cut and Sold Data and Graph. Available from http://www.fs.fed.us/forestmanagement/documents/sold-harvest/documents/1905-2011_Natl_Summary_Graph.pdf. May 1, 2012.)

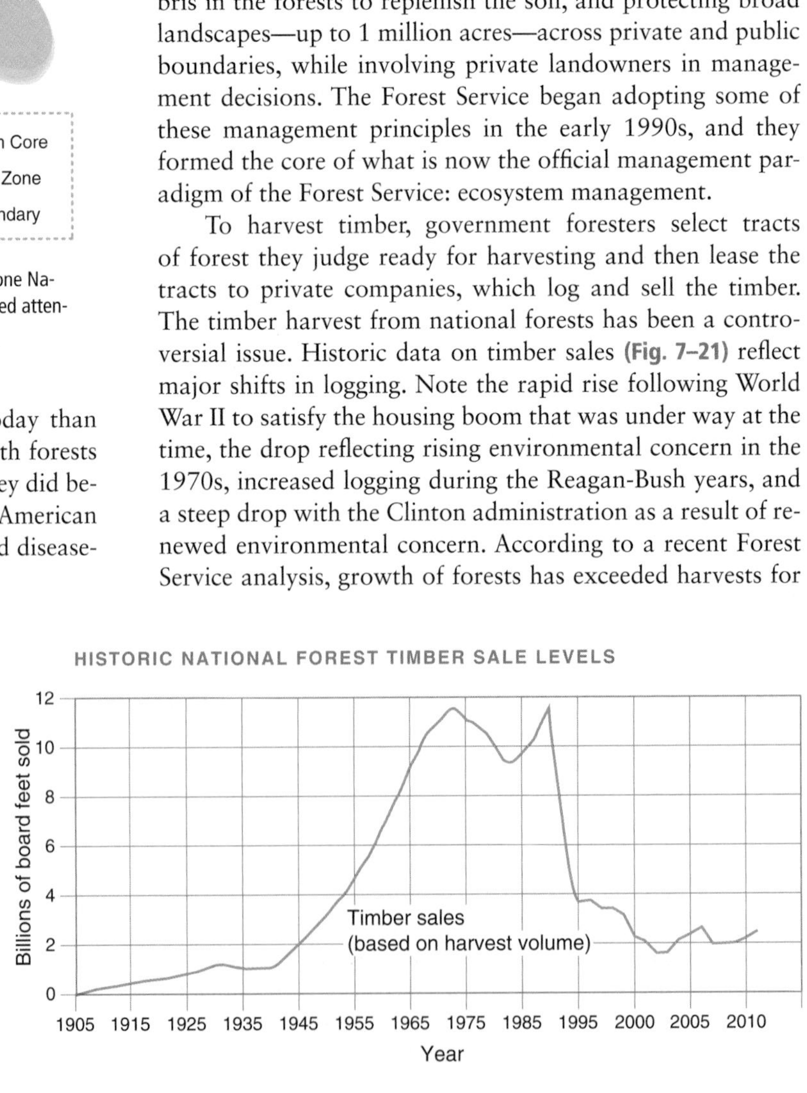

HISTORIC NATIONAL FOREST TIMBER SALE LEVELS

Timber sales (based on harvest volume)

several decades in both national and private forests, indicating that the forests are being managed sustainably.

The Roadless Controversy. The Clinton administration's Forest Service chief, Mike Dombeck, embraced the ecosystem management paradigm, and one of his first acts was to initiate a moratorium on building new logging roads, which became the Roadless Area Conservation Rule. The Forest Service had been criticized for its road building because the existing roads had contributed to the destruction of forests and streams. The timber industry and many members of Congress from states with national forests objected to the moratorium. They believed that it would halt timber harvests and prevent access by off-road vehicles in roadless areas (they were right!). Some 58 million acres of forestland were affected by the moratorium. The Bush administration, unhappy with the roadless rule, replaced the road ban with a ruling in 2005 that allows governors to ask the secretary of agriculture to open lands to roads and development. In 2006, this ruling was canceled by a federal district court order to reinstate the roadless rule. Another district court tossed out the roadless rule in 2008, creating conflicting court decisions and initiating a lengthy appeal process. Under President Barack Obama, a 2011 decision made the roadless rule the law of the land, although there have already been court challenges. Currently, this rule continues to protect 58 million acres of forests.

Fires. The years 2006 and 2007 set new records for forest fires, with 9.9 million acres burning the former year (a record) and 9.3 million acres the latter. With a decade of drought in the western states (2011 was one of the worst on record for Texas) and serious drought in the South, it is no surprise that forest fires have become more extensive. Many scientists believe that this drought pattern is a consequence of the buildup of greenhouse gases and their impacts on climate. A hot debate has focused on why so much forest land bears what is called "fuel loads"—meaning dead trees and dense understories of trees and other vegetation—and what to do about it. Most observers believe that a combination of fire suppression and common logging practices are responsible for the high fuel loads. Ecologists remind us that wildfires can also be seen as natural disturbances, rather than ecological disasters—a confirmation of the dynamic nonequilibrium nature of ecosystems (Chapter 5).

Salvage logging, one response to fires, involves harvesting timber from recent burns (fires often consume only the outer branches of trees). The Forest Service authorizes sales of timber on recently burned national forest lands but is required to follow conservation safeguards mandated by the National Environmental Policy Act and the Endangered Species Act. This takes time, but in recent years, this kind of logging has been profitable to private operators in national forests. A 2006 study of postwildfire logging in an Oregon forest showed, however, that salvage logging not only added to the continued fire risk (loggers left branches and brush behind), but also actually killed up to three-fourths of the seedlings that had regenerated on the burned land.[5] Many scientists believe that the best approach to regenerating a forest is to do nothing; removing the dead trees kills

seedlings, disturbs the soil, removes nesting sites for wildlife, and reduces nutrients and shade needed for new trees to grow.

Protecting Nonfederal Lands. In 2008, voters across the nation approved $7.3 billion in spending for the preservation of open space and for parks. California and Florida, in particular, spent $700 million together. Pennsylvania commissioned a report on public lands and parks in 2008 and discovered that in Philadelphia alone the parks provided millions of dollars in revenue, cost savings, and taxes—a total (estimated $1.1 billion) about ten times the amount the city spends on parks each year.

Land Trusts. Often, landowners and townspeople want to protect natural areas from development but are wary of turning the land over to a government authority. One creative option is the **private land trust**, a nonprofit organization that will accept either outright gifts of land or **easements**—arrangements in which the landowner gives up development rights into the future but retains ownership of the parcel. The land trust may also purchase land to protect it from development. The land trust movement is growing considerably: There were 429 land trusts in the United States in 1980 and more than 1,723 in 2010, as reported by the Land Trust Alliance, an umbrella organization in Washington, DC, serving local and regional trusts. These trusts protect almost 47 million acres of land. The Nature Conservancy, a national and international land trust, also protects 102 million acres in other countries.

Land trusts are proving to be a vital link in the preservation of ecosystems. In 1999, alarmed about signs of development in their region, 30 residents of Grand Lake Stream in northeastern Maine (pop. 150) met to share their concerns. Out of that meeting grew the Downeast Lakes Land Trust, which, over time, has carried out a $30 million conservation effort that has protected 342,000 acres of woodlands, 60 lakes, and 1,500 miles of riverfront. Members are currently in a campaign to raise funds for another 21,000 acres. These lands connected several other conservation holdings in eastern Maine and New Brunswick, and the result (to the surprise of the original 30 residents) is that more than 1 million contiguous acres are protected against development. The Downeast Land Trust is busy now constructing hiking trails and canoe-access campsites so that the public will be able to enjoy hiking, canoeing, and fishing in the conserved landscape.

Ecosystem Restoration

A great increase in restoration activity has occurred during the past 30 years, spurred on by federal and state programs and the growing science of restoration ecology. The intent of ecosystem restoration is to repair the damage to specific lands and waters so that normal ecosystem integrity,

[5]D. C. Donato, J. B. Fontaine, J. L. Campbell, W. D. Robinson, J. B. Kauffman, and B. E. Law. "Post-Wildfire Logging Hinders Regeneration and Increases Fire Risk," *Science* 311 (January 20, 2006): 352.

resilience, and productivity return. It has become a worldwide, $70 billion industry. Because of the complexity of natural ecosystems, however, the practical task of restoration is often difficult. Frequently, soils have been disturbed, pollutants have accumulated, important species have disappeared, and other—often exotic—species have achieved dominance.

For these reasons, a thorough knowledge of ecosystem and species ecology is essential to the success of restoration efforts. The ecological problems that can be mended by restoration include those resulting from soil erosion, surface strip mining, draining of wetlands, hurricane damage, agricultural use, deforestation, overgrazing, desertification, and the eutrophication of lakes. Every case is different. Some cases involve working with several different ecological problems at the same time, such as the huge restoration project under way in the (formerly) vast Everglades of southern Florida.

Everglades Restoration. A much-publicized plan to restore the Everglades received congressional approval in 2000. The plan, the **Comprehensive Everglades Restoration Plan (CERP)**, is expected to take 36 years and almost $11 billion to complete **(Fig. 7–22)**. The federal government promised to provide half of the funds, and state, tribal, and local agencies the other half. CERP is managed by the U.S. Army Corps of Engineers, the federal agency that was actually responsible for creating much of the system of water control that made the restoration necessary.

The Everglades is a network of wetland landscapes that once occupied the southern half of Florida. Now reduced to half its original size through development and wetlands draining, it is still an amazing and unique ecosystem, harboring a great diversity of species and many endangered and threatened species. It is home to 16 wildlife refuges and four national parks, the crown jewel being Everglades National Park.

Because Florida is so flat, the general topography promoted a slow movement of water south from Lake Okeechobee by way of a 40-mile-wide, 100-mile-long shallow "river of grass" ending in Florida Bay. Floridians brought the water flow under human control with a system of 1,000 miles of canals, 720 miles of levees, and hundreds of water control structures (locks, dams, and spillways), bounded on the east by a string of cities and their suburbs. In its center, 550,000 acres of croplands were created, which form a lucrative sugarcane industry. Now, water shortages in the winter leave too little for the natural systems, and in the summer rainy season, too much water is diverted to the Everglades. The water quality, once pristine, is now degraded because of nutrients (especially phosphorus) from agricultural runoff.

Water Release. A broader restoration program was originally begun in 1996, coordinated by the South Florida Ecosystem Restoration Task Force. The total task force project is expected to cost close to $15 billion. The plan calls for removing 240 miles of levees and canals and creating a system of reservoirs and underground wells to capture water for release during the dry season. The new flowage is designed to restore the river of grass, thereby restoring the 2.4 million acres of Everglades—not to the original state, but at least to a healthy system that promotes the second goal of the task force. If CERP does its job, close to 2 billion gallons of water that flow out of the Everglades each day will be redirected to flow southward. The plan calls for 80% of this water to feed the Everglades and 20% to feed the thirst of South Florida's cities and farms.

Funding. One of the most difficult parts of any restoration is funding. A major step in the restoration of the Florida Everglades project was announced in 2008, when U.S. Sugar agreed to sell its land and facilities to the state of Florida for $1.7 billion. This sale involved 187,000 acres of land to the south and east of Lake Okeechobee and could recreate wetlands to store up to 1.2 cubic kilometers (0.28 cubic miles) of water that otherwise would need to be pumped into reservoirs and underground wells during the summer. Currently, the excess water is pumped eastward and westward out to sea, disturbing the salinity and nutrient regimes in the receiving estuaries. The water stored in the new wetlands would then be free to flow south into the Everglades during the dry season, once the path to the Everglades is made clear by other planned projects. Unfortunately, this planned sale has been scaled way back because of Florida's financial uncertainties, a result of the recent national recession. The new proposal would invest $530 million for 72,500 acres of land south of Lake Okeechobee and leave open future options for the state to buy more land.

There is no blueprint for a project like the Everglades restoration. The long time frame, the need for a continued infusion of federal and state money, and the great difficulties that will arise in carrying out the restoration guarantee that the project will attract skepticism as well as effusive praise. It will be in the news for a long time.

Pending Restorations. As the values of natural ecosystems become more recognized, efforts to restore damaged or lost ones will become increasingly important. Many other large systems, such as the California Bay Delta, Chesapeake Bay, Platte River basin, and Upper Mississippi River system, have suffered from human abuse and ignorance and are now the focus of restoration projects. The Mississippi delta, battered by Hurricanes Katrina and Rita in 2005, is in need of a huge effort to restore the damaged wetlands and barrier islands. Restoration is under way at many sites that are ecological gems, including the Galápagos Islands, under siege from many exotic species; the Illinois River, confined by levees; the Brazilian Atlantic forest, severely deforested; and Tampa Bay, affected by its urban surroundings.

Final Thoughts

Ecosystems everywhere are being exploited for human needs and profit. In addition to all the examples discussed in this chapter, other areas that are in trouble include overgrazed rangelands, and rivers that are overdrawn for irrigation water. The purpose of this chapter has been to highlight some of the most critical problems and to indicate steps being taken to correct them. It is certain that greater pressures will be put on natural ecosystems as the human population

◆ Surface Water Storage Reservoirs
181,300 acres of above and in-ground reservoirs are planned to store millions of gallons of water.

● Aquifer Storage and Recovery
More than 300 underground water storage wells are proposed to store up to 1.6 billion gallons of treated water a day in confined aquifers.

◗ Stormwater Treatment Areas
35,600 acres of manmade wetlands will be constructed to remove pollutants and other harmful contaminants from water before it is discharged to the Everglades.

■ Wastewater Reuse
Two advanced treatment plants are proposed to recycle more than 220 million gallons of wastewater a day, adding a new source of high quality water for the southern Everglades.

★ Seepage Management
Barriers are proposed to be built to stop the rapid underground seepage of water out of the Everglades, which today results in the loss of millions of gallons of water each year.

▲ Removing Barriers to Sheetflow
More than 240 miles of canals and levees may be removed to restore the historic overland sheetflow through the Everglades wetlands. Sections of Tamiami Trail will be elevated to handle increased water flows contributed by CERP project features.

✦ Operational Changes
Changes will be made in the regional water management system to benefit Lake Okeechobee, the Everglades, and the coastal estuaries.

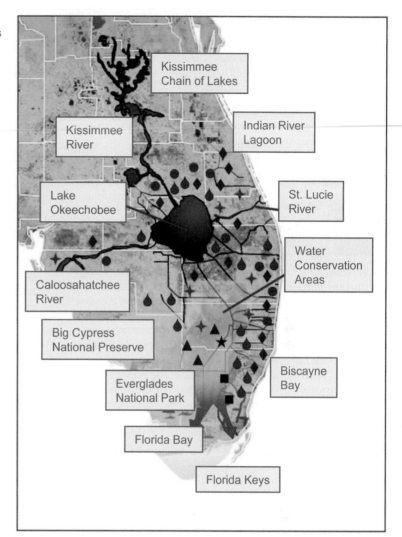

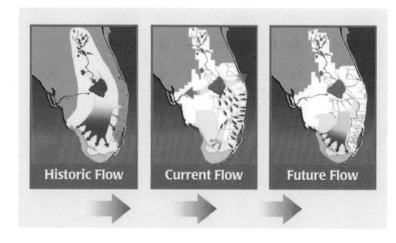

Figure 7–22 The Everglades restoration plan. The goal of this plan is to restore historical water flow patterns so that ecological restoration can occur. The seven major features of the plan are described at the left of the map. The lower three maps illustrate the historical, current, and future water flow through southern Florida.

(*Source:* Comprehensive Everglades Restoration Plan, www.evergladesplan.org.)

continues to rise. These pressures must be met with increasingly effective protective measures if we want to continue to enjoy the goods and services provided by ecosystems. The problems are most difficult in the developing world, where poverty forces people to take from nature in order to survive. If natural areas are to be preserved, the needs of people must be met in ways that do not involve destroying ecosystems. People must be provided with alternatives to overexploita-

tion, a situation requiring both wise leadership and effective international aid.

In April 2000, when he was UN Secretary-General, Kofi Annan included in his millennium report to the General Assembly a stern warning that far too little concern was being given to the sustainability of our planet: "If I could sum it up in one sentence, I should say that we are plundering our children's heritage to pay for our present unsustainable practices. . . . We must preserve our forests, fisheries, and the diversity of living species, all of which are close to collapsing under the pressure of human consumption and destruction. . . . in short," Annan stated, "we need a new ethic of stewardship." Indeed, under UN sponsorship, the Billion Tree Campaign, begun in 2006, overran its goal in 18 months and has now become the Seven Billion Tree Campaign. From individuals to heads of states, people in 155 countries are planting trees, acting in stewardship concern for the climate and for peoples' needs.

REVISITING THE THEMES

Sound Science

Calculating a total allowable catch (TAC) requires solid scientific data on fish populations and an understanding of the MSY model. These limits are difficult, if not impossible, to accomplish for mobile creatures such as fish and whales. Fisheries scientists are constantly challenged to assess fish stocks accurately, and their recommendations can often be trumped by political and economic decisions, as is well illustrated in the plight of the New England fisheries. Ensuring sustainable harvests in forests requires thorough scientific assessments of tree growth and a balancing of the needs of ecosystems with those of the harvesters. The restoration of ecosystems requires a great deal of sound science in the form of species and ecosystem ecology. It also requires the ability to make midcourse corrections, which projects like the Everglades restoration will require in the future. Scientists at the Curtic prairie set the stage for the discipline of restoration ecology. Scientists today use not only experiments, but also data collected by citizen scientists to test hypotheses, as the example of the REEF data collection showed.

Sustainability

The protection of forests, reefs, and other ecosystems requires intelligent policies and practices. Often this requires coordination, as an unregulated commons almost always tends to overuse. This is why we have the International Whaling Commission, for example. Unfortunately, there are many cases of unsustainable use: the taking of "bush meat" by commercial hunters, the cyanide poisoning of coral reefs, the exploitation of a commons where there is open access to all, and the deforestation in tropical rain forests, to name a few. The principle of sustainability is built into the model of the maximum sustainable yield, but for fisheries it turns out to be harder to accomplish in practice than in theory. For this reason, it is wise to use the precautionary principle in attempting to set such a threshold for a resource. Sustainable forestry, however, is not so difficult, as long as it is focused on managing the forests as functioning ecosystems and as long as biodiversity is respected. Where resources are being exploited in much of the developing world, sustainability is unworkable without taking into account the needs of local people and involving them in management decisions. Incorporating ecosystem services into our calculations will make protection of ecosystems more economically viable. Such an approach requires a knowledge of the value of ecosystem capital.

Stewardship

Stewardship care of ecosystems means consciously managing them so as to benefit both present and future generations. Sustainable consumptive and productive use of ecosystem resources involves tenurial rights over them; exercising these rights should be a matter of stewardship embodied in rules and policies. Table 7–3 presents a set of principles that promote such stewardship. When forests and fisheries are under the control of the people who most directly depend on their use, stewardship can flourish. Management of the fisheries exhibits both good and poor stewardship in different regions: Compare the Pacific Northwest and Georges Bank fisheries. Individuals can show stewardship by acting as citizen scientists or informed consumers or helping nonprofits such as land trusts.

Finally, it is heartening to see the former UN secretary-general endorsing a stewardship ethic as an important step in preserving the natural resources that are under tremendous pressure from present generations. The Seven Billion Tree Campaign is a tangible demonstration of that ethic.

REVIEW QUESTIONS

1. How did individuals act to help scientists in the Caribbean Sea?

2. What are some goods and services provided by natural ecosystems?

3. Compare the concept of ecosystem capital with that of natural resources. What do the two reveal about values?

4. Compare and contrast the terms *conservation* and *preservation*.

5. Differentiate between consumptive use and productive use. Give examples of each.

6. What does maximum sustainable yield mean? What factors complicate its application?

7. What is the tragedy of the commons? Give an example of a common-pool resource, and describe ways of protecting such resources.

8. When are restoration efforts needed? Describe efforts under way to restore the Everglades.

9. Describe some of the findings of the most recent FAO Global Forest Resources Assessments. What are the key elements of sustainable forest management?

10. What is deforestation, and what factors are primarily responsible for deforestation of the tropics?

11. What is the global pattern of exploitation of fisheries? Compare the yield of the capture fisheries with that of aquaculture.

12. Compare the objectives of the original Magnuson Act with those of the 2006 Magnuson-Stevens Fishery Conservation and Management Reauthorization Act.

13. What is the current status of the large whales? Discuss the controversy over continued whaling by some countries.

14. How are coral reefs and mangroves being threatened, and how is this destruction linked to other environmental problems?

15. Compare the different levels of protection versus use for the different categories of federal lands in the United States.

16. Describe the progression of the management of our national forests during the past half century. What are current issues, and how are they being resolved?

17. How do land trusts work, and what roles do they play in preserving natural lands?

THINKING ENVIRONMENTALLY

1. It is an accepted fact that both consumptive use and productive use of natural ecosystems are necessary for high-level human development. To what degree should consumptive use hold priority over productive use? Think about more than one resource (lumber, bush meat, etc.).

2. Consider the problem presented by Hardin of open access to the commons without regulation. To what degree should the freedom of the use of these areas be limited by the authorities? Make use of Table 7–3 when you defend your position.

3. Consider the benefits and problems associated with coastal and open-ocean aquaculture. Is aquaculture a useful practice overall? Justify your answer.

4. Kofi Annan stated that we are in need of a "new ethic of stewardship." What principles should this new ethic be built on?

MAKING A DIFFERENCE

1. To determine the impact you have on the ecosystems around you, calculate your ecological footprint (several sites on the Internet can help with this). What can you do to lower your ecological footprint?

2. Consider Michael Pollan's work *The Omnivore's Dilemma*. The book deals with America's eating habits, government regulations regarding food production, farming techniques, and similar topics. Decide which eating habits are ecologically sustainable and which are not. Go to the Internet and search for sites that offer ecosystem-friendly recipes.

3. By transporting firewood from its native area, you may run the risk of spreading diseases and alien insect species that can damage trees and overrun whole areas of forest. You can avoid this by using firewood harvested locally or, if you need to carry firewood a distance, by finding wood marked with a U.S. Department of Agriculture (USDA) tag confirming that the logs are safe to move.

4. Organize a volunteer tree-planting group. You can purchase young trees at a nursery, college agricultural department, or city government. To raise money for a large project, try to find individual or corporate sponsors.

5. When you visit parks and other natural areas, stay on the designated trails. Avoid making new trails that might lead to erosion, and never disturb nesting birds and wild animals raising their young.

MasteringEnvironmentalScience®

Go to **www.masteringenvironmentalscience.com** for practice quizzes, Pearson eText, videos, current events, and more.

PART THREE

The Human Population and Essential Resources

Two thousand eleven was a newsworthy year. The Occupy Wall Street protests in America, the Arab Spring uprisings in the Middle East, and the financial difficulties in the Eurozone were events that captured the world's attention and will affect us for years to come. The world also celebrated another milestone in 2011 when the United Nations Population Division estimated that on October 31 global human population reached 7 billion. On that day, in various places around the globe, babies were celebrated as the "seven billionth." Celebrations began in the Philippines, where tiny Danica May Camacho was born in Manila's Jose Fabella Memorial Hospital. Of course, naming one particular infant the 7 billionth is purely symbolic. Estimating the day on which human population will reach a specific number is impossible. We cannot count people accurately enough or frequently enough to know such a thing precisely. But we do know that human populations are continuing to rise. Like Danica May, each new person needs shelter, food, education, and, eventually, a job. We need to find ways to accommodate the population at the same time as it is increasing rapidly. Global population is projected to reach more than 10 billion by the year 2100. Virtually all of the increase will be in middle- and low-income countries, which are already densely populated and straining to meet the needs of their people for food, water, health care, shelter, and employment.

Yet in recent decades, a population decline has begun to occur in some industrialized countries. Populations are aging in these countries, bringing serious social, economic, and political consequences such as a shrinking workforce, a rising retiree population, escalating health care costs, and increasing "replacement migration" of workers from developing countries.

You will learn how human populations are similar to and different from populations of other species (Chapter 8). We will see how historic changes in technology have affected our population growth and how that growth can affect the rest of the world. We will also look at the demographic transition, which has brought an end to population growth in much of the industrialized world.

Next, we will discuss what needs to be done to bring the developing countries through the demographic transition and look at efforts being made to change consumption and population patterns (Chapter 9). Finally, we will look at the critical resources humans need, how we manage and use these resources, and their relationship to a growing population (Chapters 10–13). It is necessary to look at population and consumption concerns in tandem with resource use and pollution because population growth and its twin, consumption, affect every environmental issue.

CHAPTER 8

The Human Population

LEARNING OBJECTIVES

1. **Humans and Population Ecology:** Explain how humans, like other organisms, are subject to natural laws and ecological processes. Describe some significant differences between humans and other creatures in their ability to change their own world.

2. **Population and Consumption—Different Worlds:** Explain the relationship between income and fertility in countries around the world.

3. **Consequences of Population Growth and Affluence:** Describe the likely outcome of unlimited population growth or unlimited use of natural resources. Explain ways in which both population growth and consumption patterns must be addressed for stewardship of resources to occur.

4. **Projecting Future Populations:** Explain how age structure, population momentum, and the demographic transition help social scientists understand populations and predict future population trends in developing and developed countries.

SWEDEN. The cold winter wind cuts through the streets of Stockholm, Sweden, as Anje, a 19-year-old university student, hurries home to the warmth of her apartment, where she will settle in for an evening at her computer and books, studying for a degree in banking. She will stop for a fast-food meal on the way. Like many Swedes, Anje plans to travel extensively when her studies are over and then work for an international firm. While she is dating, Anje has no plans to marry in the near future. She is likely to marry late, if she does, and have children later than many people in the world. Like many in the European Union (EU), Anje gets a free university education. She works hard because jobs will be scarce and she wants, someday, to be established. When she is ready to have children, Anje is likely to have good benefits, with medical care and maternity leave. She is likely to care for her elderly parents in a society in which the number of elderly people is increasing, but good social and medical services will keep them active long into their seventies, maybe their eighties.

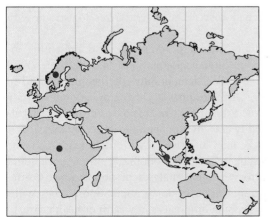

INDONESIA. A world away, in the heat of bustling Jakarta, Indonesia, 19-year-old

Atin rides a motorcycle home with her brother. They will stop at a market for chicken and noodles and come home to the small house they share with their parents and three siblings. Atin and her family have a small television and a few other luxuries. A medical emergency could set them back, though, and Atin is worried about her father's coughing. Like millions of Indonesian men, he smokes local clove and tobacco cigarettes. Their family, like many others, was hurt by the Asian financial crisis in the 1990s. Atin's father lost a stable job and took a series of temporary jobs, and their mother began to work part-time. Atin dropped out of high school to help her family. Now Atin and her brother are working to help support the family. Atin hopes to marry a young man her family knows, if they can get enough money together. Like many Indonesians, she will most likely have a smaller family than her parents did.

BURKINA FASO. In the dry season, Burkina Faso (in West Africa) is a place of dust and sun. Awa, 19-year-old married mother of two, walks with her children along a dusty road toward her home. She is carrying a large container of water on her head. Awa's husband works in agriculture in neighboring Côte d'Ivoire (Ivory Coast), where he makes a small amount of money and sends it home. Neither Awa nor her husband can read, and there is no nearby medical facility. She is worried because she is pregnant and has been feeling very ill. Awa and her husband had another child who died shortly after birth, and Awa is afraid that this coming baby will not survive either because she is sick and still nursing another child. Her feet ache and she is very tired, but there is little to be done about it. She sometimes cares for her mother and for the children of her sister, who died of AIDS. Her four surviving siblings have similar lives and live nearby. They all farm during the rainy season—growing cassava, yams, peanuts, and maize—and a few members of the extended family work on a cotton farm or have other relatives working in the Ivory Coast.

Anje, Atin, and Awa are three of millions of young people living around the world. Their life circumstances give them different health concerns, access to education, and medical care. They are also likely to have different family sizes and to use natural resources differently. In the 20th century, the global human population experienced an unprecedented explosion, more than tripling its numbers. Now, the rate of growth is slowing down, but the increase in absolute numbers continues to be substantial—and in some parts of the world, there isn't much slowing.

Remarkable changes in technology and substantial improvements in human well-being have accompanied this growth. In just the past 50 years (1960–2010), average income per capita has more than tripled, global economic output has risen more than sevenfold (from $10 trillion to $76 trillion), life expectancy has risen from 53 to 69 years, and infant mortality has been cut to one-tenth (from 40 to 4 deaths per thousand live births). Even so, Anje, Atin, and Awa do not live in a world where people experience average conditions. The remarkable improvements in living conditions are not experienced equally. Extreme poverty is still widespread. An estimated 1.2 billion people live on less than $1 per day, and the income gap between the richest and the poorest countries is enormous.

Humans are animals, and population ecologists can study human populations and ask many of the same questions they would if they were studying wolves or elk. But humans are also exceptional. They can think about and make decisions about their own fertility, they have ethical systems that inform their decisions, and they can use technology to prolong life and increase food production. Even more unusual, humans have cultural constructs, such as money. Money and its economic institutions mean that survival is driven not only by evolutionary fitness, but also by the economic circumstances into which people are born.

In this chapter, we will put the stories of people like Anje, Atin, and Awa into the context of the dynamics of population growth and the related context of consumption and its social and environmental consequences. A continually growing population is unsustainable, as is a continually increasing consumption of natural resources, so the focus in this and the next chapter is population and consumption stability—and what is required to get there.

8.1 Humans and Population Ecology

Humans are part of the natural world, and human populations are subject to processes such as birth and death. **Figure 8–1** shows human population growth over the past 2,000 years. The long, slow incline is followed by a rapid rise, giving what looks like a *J*-shaped curve (Chapter 4). However, on closer inspection, the past few decades have shown decreasing population growth rates, and projections (possible future scenarios determined by assumptions about changes we will see) suggest that the global human population may level off, so that the pattern in 2100 will be more like the *S*-shaped curve that characterizes logistic growth (Chapter 4).

The field of collecting, compiling, and presenting information about human populations like that presented in Figure 8–1 is called **demography**; the people engaged in this work are **demographers**. Demographers may study population processes such as migration or changes in fertility and mortality. In environmental science, biological science quickly becomes intertwined with other fields, so some social demographers include economic, cultural, social, and biological factors in their analyses of populations. Consequently, we cannot discuss human population ecology without quickly discussing differences in wealth or health care.

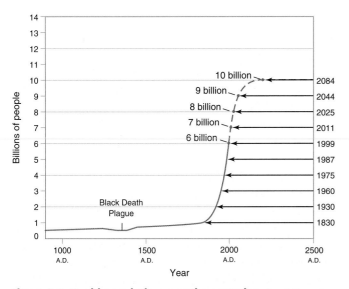

Figure 8–1 World population over the centuries. Population grew slowly for most of human history, but in modern times it has expanded greatly.

(*Sources:* Basic plot from Joseph A. McFalls, Jr. "Population: A Lively Introduction," *Population Bulletin* 46, no. 2 (1991): 4; updated from UN Population Division, medium projection, 2010.)

Demographers use a set of specialized terms, described in Table 8–1, which we will use throughout the chapter.

r- or *K*-Strategists

We have contrasted organisms like cockroaches, which breed rapidly and have boom-and-bust populations, with others like elephants, which maintain smaller, but more stable population sizes (Chapter 4). Cockroaches are described as "*r*-selected species," or "*r*-strategists," because they have a high *r*, or reproductive potential. They are contrasted with "*K*-selected species," or "*K*-strategists," which remain close to *K* (carrying capacity) and are characterized by long life spans, older age at first reproduction, high parental care of offspring, and fewer offspring. These characteristics form a continuum, with some organisms not easily placed in either group.

When we ask "Are humans *r*- or *K*-strategists?" it becomes clear that humans are like other organisms, but also significantly different. Just looking at the population growth curve of the past thousand years (Fig. 8–1), we might be tempted to conclude that we are looking at a J-shaped curve representing exponential growth typical of weedy species that move from place to place quickly and cannot outcompete other organisms over the long term (*r*-strategists). However, humans have characteristics such as high parental care and late reproduction that we would expect of the more stable *K*-strategists.

How can humans have the rapid population growth curve of a weedy, explosive species but also the life history of a slower-growing **equilibrial** species (those with population numbers remaining near the carrying capacity)? The answer lies in the fact that in some ways humans are critically different from other organisms. Even though the Millennium Ecosystem Assessment recognizes that "humans are an integral part of ecosystems," there are some differences between

TABLE 8–1 Demographic Terms

Term	Definition
Growth rate (annual rate of increase or decrease)	The rate of growth of a population, as a percentage. Multiplied by the existing population, this rate gives the net yearly increase (or decrease, if negative) for the population.
Total fertility rate	The average number of children each woman has over her lifetime, expressed as a rate based on fertility occurring during a particular year.
Replacement-level fertility	A fertility rate that will only replace a woman and her partner, theoretically 2.0, but adjusted slightly higher because of mortality and failure to reproduce.
Infant mortality	The number of infant deaths per thousand live births.
Population profile (age structure)	A bar graph plotting numbers of males and females for successive ages in the population, starting with the youngest at the bottom.
Population momentum	The effect of current age structure on future population growth. Young populations will continue growing even after replacement-level fertility has been reached, due to reproduction by already-existing age groups.
Crude birthrate	The number of live births per thousand in a population in a given year.
Crude death rate	The number of deaths per thousand in a population in a given year.
Epidemiologic transition	The shift from high death rates to low death rates in a population as a result of modern medical and sanitary developments.
Fertility transition	The decline of birthrates from high levels to low levels in a population.
Demographic transition	The tendency of a population to shift from high birth and death rates to low birth and death rates as a result of the epidemiologic and fertility transitions. The result is a population that grows very slowly, if at all.

humans and other organisms that make our role in ecosystems different from that of others.

Remember that a *population* is an interbreeding group of a particular species in the same area. We do not typically talk about a widespread species as having a "global population," because local populations rise and decline according to local conditions. Drought in one area might be offset by good conditions in another. In contrast, we do talk about "human population" and mean the whole global population. Prior to the modern era, human populations acted more separately—disease, plenty, and scarcity were local effects, and populations rose and fell with surrounding conditions. Several changes, due to society and culture, have altered the way human populations interact and have also changed the way human population ecology works. In the modern age, what happens in one part of the world affects humans in other parts of the world. Humans act both as local populations, with population characteristics such as family size and life expectancy, and as part of a larger global whole. Humans are unique in the biosphere in our ability to regulate our reproduction, use fire, store food for longer times, and adapt our environment with technology so we can live in more places. Consequently, population ecology can be applied to humans, but not in exactly the same way it is applied to other organisms.

Revolutions

The changes that have allowed humans to become such a dominant part of the landscape are so extreme that they can be considered revolutionary. Here we will look at a number of the revolutions that have brought humans from a number of isolated populations—ruled (like other species) primarily by local conditions—to the global human population today. In the past, there have been several large-scale upheavals in the way humans do things. These include the development of agriculture (which allowed settling into cities), the Industrial Revolution, modern medicine, the green agricultural revolution of the 1970s, and the current environmental revolution. In each case, we will see how new ways of doing things altered the effect of high populations.

Neolithic Revolution. Natural ecosystems have existed and perpetuated themselves on Earth for hundreds of millions of years, while humans are relative newcomers to the scene. Anatomically modern humans most likely arose in Africa around 200,000 years ago. The earliest fully modern humans in Europe appeared some 40,000 years ago, during the end of a period known as the Upper *Paleolithic* (around 50,000 to 10,000 years ago). Archeological evidence suggests that Paleolithic humans survived in small tribes as hunter-gatherers, catching wildlife and collecting seeds, nuts, roots, berries, and other plant foods **(Fig. 8–2)**. Settlements were never large and were of relatively short duration because, as one area was picked over, the tribe was forced to move on. As hunter-gatherers, these people were much like other omnivorous consumers in natural ecosystems. Popula-

tions could not expand beyond the sizes that natural food sources supported, and deaths from predators, disease, and famine were common.

About 12,000 years ago, a highly significant change in human culture occurred when humans in the Middle East began to develop animal husbandry and agriculture. The development of agriculture provided a more abundant and reliable food supply, but it was a turning point in human history for other reasons as well. Because of its profound effect, it is referred to as the **Neolithic Revolution**. Agriculture allows permanent (or at least long-term) settlements and the specialization of labor. Some members of these settlements specialize in tending crops and producing food, freeing others to specialize in other endeavors. With this specialization of labor, there is more incentive and potential for technological development, such as better tools and better means of transporting water and other materials. Trade with other settlements begins, and commerce is born. In addition, living in settlements allows a greater storage of

Figure 8–2 The Neolithic Revolution. Before the advent of agriculture, all human societies had to forage for their food; agriculture allowed people to settle in cities and to divide labor.

food and the development of preservation techniques, and food storage means that food stored from one year might be able to help people in years with less food production, evening out the risk from year to year. A consequent reduced mortality rate, coupled with more reliable food production, supports population growth, which in turn supports (and is supported by) expanding agriculture. In short, modern civilization and population growth had their origins in the invention of agriculture about 12,000 years ago.

Industrial Revolution. For about another 11,000 years, the human population increased and spread out over the planet, but these increases were not as dramatic as those seen today. Agriculture and natural ecosystems supported the growth of civilizations and cultures, which increased their knowledge and mastery over the natural world. With the birth of modern science and technology in the 17th and 18th centuries, the human population—by 1800 almost a billion strong—was on the threshold of another revolution: the **Industrial Revolution (Fig. 8–3)**. The Industrial Revolution and its technological marvels were energized by fossil fuels—first coal, then oil and gas. In the past, people had used other fuels, particularly peat (compacted plant material in wetlands) and wood. These fuels do the same thing that fossil fuels do: they allow people to use energy that was trapped by plants from the sun a long time ago. The use of coal, in particular, opened up vast resources of trapped energy from other eras. This energy was used by humans to do work that we would never have been able to do if we had relied on human power—or even on the power of domesticated animals. The burning of fossilized plants and animals, or of wood, is similar to the process of cellular respiration (Chapter 3). It breaks chemical bonds to release energy and releases water and carbon dioxide as waste products. Fossil fuels contain other chemicals as well—such as sulfur and nitrogen compounds, mercury, lead, and arsenic—

all of which can contribute to the pollution of land, air, and water and will be discussed in other chapters.

The harnessing of this extra energy allowed people to produce more food. However, the Industrial Revolution came with some unexpected costs. Pollution and exploitation took on new dimensions as the industrial powers turned to the extraction of raw materials from all over the world. In time, every part of Earth was affected by this revolution and continues to be affected today. As a result, we now live in a time of population growth and economic expansion, but unfortunately with all of the environmental problems (Chapter 1).

Medical Revolution. One of the main reasons for the slow and fluctuating population growth prior to the early 1800s was the prevalence of diseases that were often fatal, such as smallpox, diphtheria, measles, and scarlet fever. These diseases hit infants and children particularly hard, and it was not uncommon for a woman who had seven or eight live births to have only two or three children reach adulthood. In addition, epidemics of diseases such as the black plague of the 14th century, typhus, and cholera eliminated large numbers of adults. Prior to the 1800s, therefore, the human population was essentially in a dynamic balance with natural enemies—mainly diseases—and other aspects of environmental resistance. High reproductive rates were largely balanced by high mortality, especially among infants and children. With high birth and death rates, the population growth rate was low in preindustrial societies.

In the late 1800s, Louis Pasteur and others demonstrated that many diseases were caused by infectious agents (now identified as various bacteria, viruses, and parasites) and that often these organisms were transmitted via water, food, insects, and rodents. Soon, vaccinations against different diseases were developed, and whole populations were immunized against such scourges as smallpox, diphtheria, and typhoid fever **(Fig. 8–4)**. At the same time, cities and towns began treating their sewage and drinking water. Later, in the 1930s, the discovery of penicillin (the first in a long line of antibiotics) resulted in cures for often-fatal diseases such as pneumonia and blood poisoning. Improvements in nutrition began to be significant as well. In short, better sanitation, medicine, and nutrition brought about spectacular reductions in mortality—especially among infants and children—while birthrates remained high. From a biological point of view, the human population began growing almost exponentially, as does any natural population once it is freed from natural enemies and other environmental restraints.

The Green Revolution. After the Industrial and Medical Revolutions, human populations increased dramatically. Concerns about whether we would be able to feed the rapidly burgeoning global population led to intense efforts to find new ways to increase agricultural efficiency. The development of chemical pesticides in World War II, along with an increase in irrigation and fertilizer use, dramatically increased crop yields. These trends were enhanced

Figure 8–3 Industrial Revolution. The Industrial Revolution began in England in the 1700s. Coal was the energy source, and economic growth and pollution were the consequences.

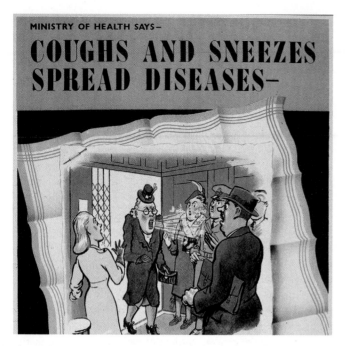

Figure 8–4 This cartoon is from a British Ministry of Health public health campaign

(*Source:* The National Archives/SSPL/Getty Images)

native (often hardier) plant varieties were some of the costs of industrial agriculture.

Fossil fuels allowed people to use energy over short time frames that had been trapped over millions of years; in the same way, industrialized agriculture allowed people to use resources, particularly soil and underground water, more rapidly than they were replaced. (This problem will be discussed in some detail when we talk about soil in Chapter 11, water in Chapter 10, and food production in Chapter 12). In addition, the use of pesticides initially saved many human lives and protected crops. In the long term, however, the overuse of pesticides resulted in **pesticide resistance**, the ineffectiveness of a pesticide when the target organisms are no longer affected by it. A current worldwide interest in sustainable agriculture is an attempt to rectify some of these problems. However, as we will see in Section 8.3, population growth is pressuring us to increase, rather than decrease, high-yield agriculture.

The Newest Revolution. In the past, humans experienced revolutions with new technological breakthroughs. Now, we have the Internet, computers, nanotechnology, robotics, and solar and other technologies. Many of these can be put to use in what some hope will be a real "green" revolution—an **Environmental Revolution.** More-efficient technologies, better urban and regional planning, policy and industrial changes, and the personal decisions of billions of people will be required to drive this revolution.

Humans are part of natural ecosystems, but they are different enough from other organisms to make their population ecology global and to allow them to change their world more extensively. They have characteristics of both *r*- and *K*-strategists. We see the *J*-shaped curve for human population growth because, at different times in history, we have increased the

by the development of high-yield crops such as specialized rice, which yielded more per acre than other crops. These advances in agriculture did increase crop yields dramatically and allowed many countries to provide food for their rapidly growing populations **(Fig. 8–5)**. However, the industrialized agriculture that allowed for such high yields came at significant costs, much as the Industrial Revolution did. Increased erosion, soil and water pollution, and the loss of

Figure 8–5 Green Revolution. Indian farmers harvest high-yield wheat. Different varieties of crops, fertilizer, irrigation, and pesticides are all part of industrialized agriculture, which helped feed growing populations but led to the pollution of soil and water.

number of people the world can support by changing the way we do things. Our past growth might seem to suggest that natural laws do not apply to humans. In the next section, we will look at carrying capacity and limits and ask whether they apply to human populations.

Do Humans Have a Carrying Capacity?

It is challenging to determine a carrying capacity for humans, a fact that makes predicting human impacts on the environment or likely future human poverty outcomes difficult. If people were close to starving many years ago and yet today there are many more people, people live longer, and proportionately fewer are extremely poor, how can we talk about humans reaching a carrying capacity? In fact, some people claim that the whole idea of a carrying capacity does not apply to humans. Economist Julian Simon, the author of *The Ultimate Resource*,[1] claimed that human ingenuity is the ultimate resource and that there is no such thing as a limit for humans. He believed we would colonize space and that human populations really have no upper limit. There are plenty of people who would agree with him, in part because every time in the past that it was thought humans had hit a carrying capacity, a problem has been solved to overcome it.

Most ecologists and demographers, however, would disagree. In their view, an escape into space is unlikely to solve the world's problems for the vast majority of humans, and humans are subject to limits and natural laws of population growth. How then do we reconcile natural laws with the pattern of human population growth in the past 2,000 years? Throughout history, humans have improved their survival rate, increased their populations, and increased their life span by doing some things other organisms cannot do (or cannot do on the scale we can). These improvements have essentially increased the carrying capacity for our species. For example, we were once limited by what we could capture or hunt in one season, but the development of agriculture and trade removed that limit. We were limited by our energy requirements, but the use of fossil fuels gave us a new resource; we were limited by child mortality, but medical breakthroughs lowered child mortality and populations exploded.

This pattern of population growth is actually exactly what you would expect to see under the law of limiting factors—each time a limit is removed, population increases until there is another limiting factor. Simon represents a group that believes that we will always be able to eliminate limiting factors. However, in the past, increases have come at the expense of using up nonrenewable resources (like fossil fuels). Increases in population have also come at the expense of creating another set of limits—this time, pollutants. The limiting factors that humans face in the 21st century include pollution from our own activities, a limit on agricul-

tural land, a limit on ocean fish, and trade-offs between ecosystems we need for their services (such as photosynthesis, flood control, and habitat) versus land and water for human activities. The argument that Simon makes also includes no value for nonhuman parts of nature except as resources for humans. That is, the rest of nature has no *intrinsic* value (Chapter 6).

Because humans seem able to increase their carrying capacity in the short term—while potentially making trade-offs that will be paid for in the long term—determining a carrying capacity for the global human population is difficult. Demographer Joel Cohen explained the reason in his book *How Many People Can the Earth Support?*[2] Cohen said that there isn't a clear-cut answer to the question of carrying capacity because, with humans, some of the answer depends on what living standard we are willing to accept. For example, if we never ate meat, we did not rely on wild organisms, and we were willing to never travel, heat our homes, or use paper, Earth could support more people. On the other hand, if everyone on Earth wanted a 21st-century American way of life, we would need multiple Earths to support the global population we already have. Because so much of the answer depends on standard of living expectations, Cohen argued, it is very difficult to determine an actual carrying capacity. Some people have used Cohen's argument to say that there is no carrying capacity at all. That was not Cohen's meaning, as he has made clear in other writings. Cohen simply meant that we can't figure out a carrying capacity for humans in the same way we do for fish, ducks, or flies, because we have (and want) more choices.

While it may be difficult to find a single number for human carrying capacity because we may not all agree on what living standard we are willing to accept, the world's scientific societies disagree with Simon and believe that there is some type of limit to human population growth. In 1993, representatives of the National Academies of Sciences from 58 countries met in New Delhi, India, to agree on and sign a statement about world population. They said, among other things, that "the earth is finite and that natural systems are being pushed ever closer to their limits."[3]

Scientists believe that, even if a concrete number may be hard to agree on, it is important to try to estimate what carrying capacity limits might be from an ecosystem perspective. In 2004, two professors in the Department of Spatial Economics at the Free University in Amsterdam, the Netherlands, reported on their meta-analysis (an analysis of multiple other analyses) of 69 papers on world population and carrying capacity, looking to see if there were any central trends. They identified a number of factors that could limit human populations, such as "water availability, energy, carbon, forest products, nonrenewable resources, heat removal, photosynthetic capacity, and the availability of land for

[1]Julian Simon. *The Ultimate Resource* (Princeton, NJ: Princeton University Press, 1983).

[2]Joel Cohen. *How Many People Can the Earth Support?* (New York: Norton, 1995).
[3]Population Summit of the World's Scientific Academies, 1993. National Academy of Sciences, National Academy of Engineering, Institute of Medicine (SEM), p. 7. Available at http://books.nap.edu/openbook.php?record_id=9148&page=7.

food production."[4] They concluded that the best estimates for global human carrying capacity centered on 7.7 billion people. One important result the analysis showed is that the best predictions of how large our human populations will get by 2050 are greater than the best estimates of our carrying capacity. Such analyses are difficult to do. While regional carrying capacities have been calculated, a similar global calculation has not been made recently. However, other types of analyses have reached similar conclusions.

Planetary Boundaries. In 2009, a team of 28 leading academics headed by Johan Rockström, executive director of the Stockholm Environment Institute, proposed a way of thinking about what parameters would need to be in place for humans to survive what they called "planetary boundaries" for a "safe operating space for humanity." The scientists claimed that there are a series of limits, which act as "tipping points" that, if passed, would keep humans from surviving (Chapter 5). Within them is a "safe space."[5] Both population growth and overconsumption could push us past our boundaries. Estimates suggest that of the nine parameters they outlined, three of the boundaries (climate change, biodiversity loss, and flow of nutrient cycles) may have already been crossed. To survive, we will have to find ways to meet human needs without pushing those parameters. (For more discussion, see Stewardship, "Lessening Your Ecological Footprint").

Picking Up the Population Pace. From the dawn of human history until the beginning of the 1800s, the population increased slowly and variably, with periodic setbacks. It was roughly 1830 before world population reached the 1 billion mark. By 1930, however, the population had doubled to 2 billion. In 1960 it reached 3 billion, and by 1975 it had climbed to 4 billion. Thus, the population doubled in just 45 years, from 1930 to 1975. Then, just 12 years later, in 1987, it crossed the 5 billion mark. In 1999 world population passed 6 billion and hit 7 billion in late 2011.

On the basis of current trends (which assume a continued decline in fertility rates), the UN Population Division (UNPD) medium projection predicts that world population will pass the 8 billion mark in 2025, the 9 billion mark in 2045, about 9.3 billion around 2050, and 10 billion around 2085 (Fig. 8–1). Future growth and projections will be described in Section 8.4.

8.2 Population and Consumption: Different Worlds

To understand what is likely to happen in the next decades, we need to understand the lives of people like Anje, Atin, and Awa. In spite of the fact that globalization is part of the modern world, people in wealthy, middle-income, and poor countries live in radically different economic and demographic conditions. In fact, even within a single country there can be tremendous disparities in wealth and in quality of life.

Rich Nations, Middle-Income Nations, Poor Nations

The World Bank, an arm of the United Nations, divides the countries of the world into three main economic categories, according to average per capita gross national income (Fig. 8–6):

1. **High-income, highly developed, industrialized countries.** This group (population 1.13 billion in 2010) includes the United States, Canada, Japan, Republic of Korea, Australia, New Zealand, the countries of Western Europe and Scandinavia, Singapore, Taiwan, Israel, and several Arab states (2010 gross national income per capita $12,276 and above; average of $38,658). Sometimes the high-income nations are further divided into those belonging to the Organisation for Economic Co-operation and Development (OECD), an economic collaboration between primarily European countries, and non-OECD countries. Anje in Sweden is in a high-income country in the OECD.

2. **Middle-income, moderately developed countries.** This group (4.92 billion) includes mainly the countries of Latin America (Mexico, Central America, and South America), northern and southern Africa, China, Indonesia and other southeastern Asian countries, many Arab states, Eastern Europe, and countries of the former U.S.S.R. The average income for this group in 2010 was $3,764. The category is further divided by the World Bank into lower middle-income and upper middle-income countries (2010 gross national income per capita: $1,005–$3,975 for the former category, $3,975–$12,276 for the latter). In 2010, China, Ecuador, Jordan, Thailand, Maldives, and Tunisia were reclassified from the lower-middle income to the upper-middle income group, indicating that their economies had improved, while Latvia was moved down from high to upper-middle income and Fiji fell from upper-middle to lower-middle income. Indonesia, home to Atin and her family, is in the lower-middle-income group.

3. **Low-income, developing countries.** This group (population 0.8 billion in 2010) is composed of the countries of eastern, western, and central Africa, India and other countries of southern Asia, and a few former Soviet republics (2010 gross national income per capita: less than $1005, average of $510). In 2010, Zambia, Solomon Islands, Mauritania, Ghana, and Lao PDR (Laos) moved out of this low category, to lower middle, showing that their economies are improving. Awa and her family in Burkina Faso are members of the low-income group.

[4]J. Van Den Bergh and P. Rietveld. "Reconsidering the Limits to World Population: Meta-Analysis and Meta-Prediction," *BioScience* 54, no. 3 (2004): 195.
[5]J. Rockström et al. "Planetary Boundaries: Exploring the Safe Operating Space for Humanity." *Ecology and Society* 14, no. 2 (2009): 32. http://www.ecologyandsociety.org/vol14/iss2/art32/.

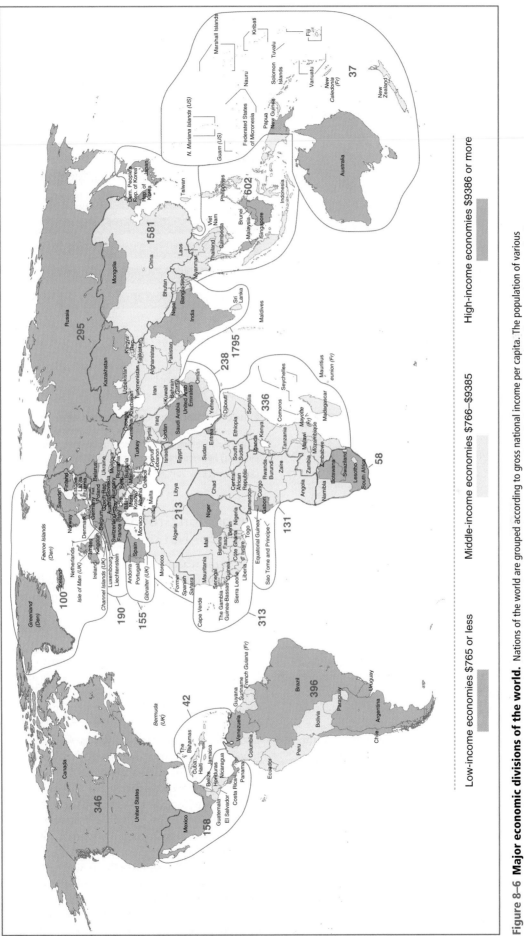

Figure 8–6 Major economic divisions of the world. Nations of the world are grouped according to gross national income per capita. The population of various regions (in millions) is also shown by magenta lines and numbers. (*Sources: World Development Report, 2005* (New York: Oxford University Press, Inc.). Copyright © 2005 by the International Bank for Reconstruction and Development/The World Bank. Populations from *World Population Data Sheet* (Washington, DC, Population Reference Bureau, 2010).)

Low-income economies $765 or less   Middle-income economies $766–$9385   High-income economies $9386 or more

The high-income nations are commonly referred to as **developed countries**, whereas the middle- and low-income countries (a much larger number) are often grouped together and referred to as **developing countries**. Of course, all countries are in the process of developing in some sense, but here the term refers to economic changes. The terms *more developed countries* (MDCs), *less developed countries* (LDCs), and *third-world countries* are being phased out, although you may still hear them used.

Moving Up: Good News

Some of the news about population is good. Worldwide, the low-income percent of the world's population has declined dramatically since 1998, falling from about 60% to less than 12%. This is a reduction of about 80% in only a few years. It used to be that living in a low-income country was the norm for humans, and now it is not. It is part of the reason people talk about the "environmentalist's paradox" (Chapter 1). This paradox refers to the fact that some measures of human development are increasing at the same time that serious problems are arising in the environment. One of the improved measures is the movement of many countries from the lowest income status even as populations are rapidly rising.

The World Bank explains that, "Developing countries have increased their share of the global economy by growing faster than rich countries, on average 6.8% per year compared to only 1.8% for high income economies over the 2000 to 2010 period."[6] While this is good news, in spite of the movement of millions of people from the extremely low-income to the middle-income categories, the disparity in distribution of wealth among the countries of the world is still mind-boggling (**Fig. 8–7**).

The disparity of wealth is difficult to understand just by looking at general income figures (Fig. 8–7). The UN Development Program (UNDP) uses the *Human Development Index* as a measure of general well-being, based on life expectancy, education, and per capita income. The UNDP also uses the *Human Poverty Index*, based on additional information about literacy and living standards, to make a more direct measurement of poverty in both low- and high-income countries. Between 10% and 15% of the people in developed countries are unable to afford adequate food, shelter, or clothing, compared with about 45% of those in developing countries. Anje, in Sweden, lives in a country with one of the lowest human poverty indexes, while Awa, in Burkina Faso, lives in a country with one of the highest. These data, available yearly in the UNDP Human Development Report, have been used to focus attention on the most deprived people in a country, to help countries develop appropriate policies, and to mark progress toward sustainable development.

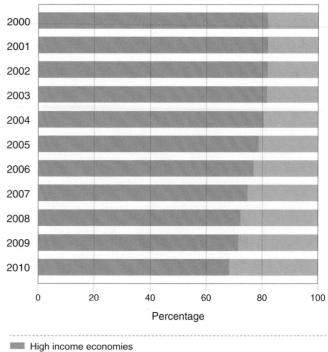

SHARE OF WORLD GROSS NATIONAL INCOME, 2000—2010

Figure 8–7 Between-country inequality. Low- and middle-income countries are gaining some in the share of world gross national income, while their populations are increasing more rapidly than those of high-income countries. Inequality between countries remains stark.

(*Source:* World Bank. 2011. http://data.worldbank.org/news/2010-GNI-income-classifications.)

Population Growth in Rich and Poor Nations

One of the reasons we spend so much time looking at wealth distributions in a study of environmental science is because population growth is related to poverty, environmental degradation, and other development measures. If you look at population growth in developed versus developing countries, you find a discrepancy that parallels the great difference in wealth between these groups of countries. According to the UNPD, the high-income countries such as Anje's Sweden, with a combined population of 1.24 billion in 2010, are growing at a rate of 0.1% annually. These countries will add less than 1 million to the world's population in a year. The middle-income countries, such as Atin's Indonesia, and the low-income countries, such as Awa's Burkina Faso, together had a 2010 total population of 5.73 billion. The populations of these countries are increasing at a rate of 1.5% annually, adding 75 million in a year. Consequently, more than 98% of world population growth is occurring in the developing (middle- and low-income) countries. In fact, population growth is highest in the least developed countries (**Figs. 8–8, 8–9**). What lies behind this discrepancy?

Population growth occurs when births outnumber deaths. In the absence of high mortality, the major determining factor for population growth is births, conventionally measured using the **total fertility rate**—the average number

---

[6]"Changes in Country Classifications." 2011. http://data.worldbank.org/news/2010-GNI-income-classifications.

Land area

Technical notes
- Data are from the United Nations Development Programme's 2004 Human Development Report.
- The United Nations Development Programme uses one Human Poverty Index for poorer territories and another for richer territories. The scores of the latter can be divided by 10 so the indices are comparable.

HUMAN POVERTY

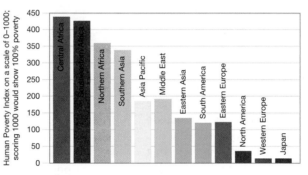

Figure 8–8 Another way to picture poverty. The territory size shows the number of poor people in the country. On the human poverty index, Burkina Faso ranks highest and Sweden ranks lowest.

(*Source:* SASI Group (University of Sheffield) and Mark Newman (University of Michigan).)

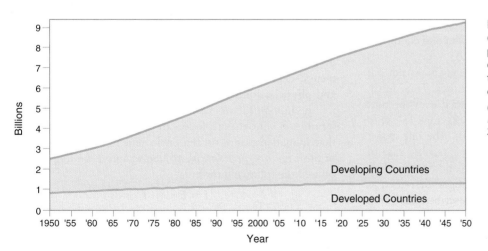

Figure 8–9 Population growth in different world regions. Because of larger populations and higher birthrates, developing countries represent an ever-growing share of the world's population. This trend is expected to continue.

(*Source:* UN Population Division, *World Population Prospects: The 2010 Revision* (New York: United Nations, 2010).)

Estimates and Projections of Total Fertility
2005 – 2010 Estimate

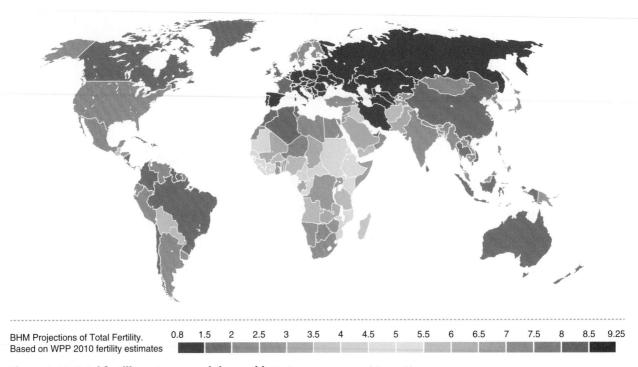

| BHM Projections of Total Fertility. Based on WPP 2010 fertility estimates | 0.8 | 1.5 | 2 | 2.5 | 3 | 3.5 | 4 | 4.5 | 5 | 5.5 | 6 | 6.5 | 7 | 7.5 | 8 | 8.5 | 9.25 |

Figure 8–10 Total fertility rates around the world. Fertility rates vary around the world.

(*Source:* UN Population Division, *World Population Prospects: The 2010 Revision* (New York: United Nations, 2008).)

of children each woman in a population has over her lifetime (**Fig. 8–10**). Theoretically, a total fertility rate of 2.0 will give a stable population, because two children per woman, on average, will replace two parents when they eventually die. A total fertility rate greater than 2.0 will result in a growing population, because each generation is replaced by a larger one; and, barring immigration, a total fertility rate less than 2.0 will lead to a declining population, because each generation will eventually be replaced by a smaller one. Given that infant and childhood mortality is not, in fact, zero and that some women do not reproduce, **replacement-level fertility**— the fertility rate that will just replace the population of parents—is 2.1 for developed countries and slightly higher for developing countries, which have higher infant and childhood mortality.

Total fertility rates have dropped all over the world, but most dramatically in the high-income countries. Total fertility rates in developed countries have declined over the past two centuries to the point where they now average 1.7, with some as low as 1.1 (Table 8–2, Fig. 8–10). The one major exception is the United States, with a total fertility rate of 2.06 in 2012. When countries industrialize and have modern medicine, they typically undergo these changes, described as the **demographic transition**. This will be described in Section 8.4 when we discuss how demographers project future population trends.

In developing countries, fertility rates have come down considerably in recent years, but they still average 2.7. Some have rates as high as 7 or more, however (Table 8–2). Thus, the populations of developing countries,

half of which are poor (low-income) countries, will continue growing, while the populations of developed countries will stabilize or decline. (An exception to the rule, because of immigration and births to immigrants, is the U.S. population.) As a consequence, the percentage of the world's population living in developing countries—already 82%—is expected to climb steadily to more than 90% by 2075 (Fig. 8–9). Nevertheless, it is not just the developing countries that have problems.

Different Populations, Different Problems

Just as we saw in the discussion of carrying capacity, both the number of people and their standards of living play a role in whether we achieve sustainability or experience environmental degradation. Some time ago, ecologist Paul Ehrlich and physicist John Holdren proposed a formula to account for the human factors that contribute to environmental deterioration and the depletion of resources. They reasoned that human pressure on the environment was the outcome of three factors: *population, affluence,* and *technology.* They offered the following formula:

$$I = P \cdot A \cdot T$$

According to this equation, called the **IPAT formula**, environmental impact (*I*) equals population (*P*), multiplied by affluence and consumption patterns (*A*), multiplied by the level of technology of the society (*T*). The IPAT formula fits well with the concept of carrying capacity. Given the high level of technology in the industrialized countries and the

TABLE 8–2 Population Data for Selected Countries

Country	Total Fertility Rate	Population 2011 (millions)	Population 2050— Projected (millions)
World	2.5	6,987	9,587
Least Developed	4.5	861	1,826
Egypt	2.9	82.6	123.5
China	1.5	1,345.90	1,312.60
Afghanistan	6.3	32.4	76.3
India	2.6	1,241.30	1691.7
Iraq	4.7	32.7	83.4
Burkina Faso	5.8	17	46.7
Haiti	3.4	10.1	14.2
Indonesia	2.3	238.2	309.4
More Developed:			
Average	1.7		
United States	2.0	311.7	422.6
Canada	1.7	34.5	47.5
Japan	1.4	128.1	95.2
Sweden	2.0	9.4	10.7
Italy	1.4	60.8	62
Spain	1.4	46.2	49.1

Source: Data from *2011 World Population Data Sheet* (Washington, DC: Population Reference Bureau, 2011).

affluent lifestyle that accompanies it, a fairly small population can have a very large impact on the environment. More recently, there have been attempts to improve the equation. The technology part of the equation has always been difficult to quantify. Some technologies improve environmental conditions and some harm it. Some economists have rewritten the IPAT formula to separate the effects of technology (*T*) into two components: consumption per unit of GDP (How much do you use as your economy increases?) and impact per unit of consumption (How inefficient, or how negative, is what you are doing?). They call this formula **ImPACT**.[7]

Effect of Wealth. Sometimes people think that wealth will solve environmental problems—that is, the rich can afford the luxury of caring for the environment. Wealthy countries, the thinking goes, can more easily afford the technologies to lower air pollution, clean water, and take care of sewage, for example (Chapter 2). Give the poor quick ways to improve their economies, and eventually they will also take care of the environment. It is true that the environment has

improved in the United States and western Europe with the passage of many environmental laws and that some wealth is required to build technologies to deal with the needs of large groups of people with minimal environmental degradation. But the relationship between economic wealth and environmental health is not clear-cut for two reasons. First, while issues like infectious disease in water might improve, other problems like municipal waste increase with wealth. The second reason is that wealth allows people to care for the environment in their area, but doing so often pushes environmental problems to other, poorer places. For example, the high-income countries may be able to clean up their own sewage because they have the money for treatment plants. However, the extraction of resources and the manufacture of goods in other countries for use in these high-income countries have impacts in many other parts of the world.

Ecological Footprint. Some people argue that because the highest-density places—like the Netherlands, Singapore, Hong Kong, and Japan—are places with healthy, long-lived, wealthy human populations, overpopulation is not real. If they can be so crowded, why can't we have a whole world like that? But high-density, wealthy areas such as Singapore use resources and affect ecosystems over a much wider area.

[7]Richard York, Eugene A. Rosa, and Thomas Dietz. "STIRPAT, IPAT, and Im-PACT: Analytic Tools for Unpacking the Driving Forces of Environmental Impacts." *Ecological Economics* 46 (2003): 351–365.

People living in high-density areas have impacts in the far reaches of the seas and land. This is why we talk about an **ecological** (or environmental) **footprint,** or an effect of something we might be doing. An environmental footprint is an estimate of the amount of land and ocean required to provide the resources you need or absorb the wastes you produce. The environmental footprint of Hong Kong, for example, is many times greater than the size of the metropolitan region (see Stewardship, "Lessening your Ecological Footprint").

Inequalities Within Countries. The developing countries, especially the low-income group of countries, do have a population problem, and it is making their progress toward sustainable development that much more difficult. Their needs are great: economic growth, more employment, wise leaders, effective public policies, fair treatment by other nations, and, especially, technological and financial help from the wealthy nations. Meeting these needs is a matter of justice, one of the key components of stewardship. Even this description is too simplistic because within each group of nations and within individual nations there is a tremendous difference in wealth and poverty levels and individual experiences. There is a rising middle class in some middle-income countries, with consumption patterns similar to those in the United States, Japan, and Europe. This group is called "the consumer class." The World Bank measures **inequality** in each country with the GINI index, a way of quantifying differences in wealth within countries. Sweden has one of the lowest values (23) and South Africa one of the highest (65). In fact, Burkina Faso (39) and Indonesia (37) each have a GINI index in the middle; they have few ultra-rich citizens.

India has hundreds of millions of extremely wealthy people and an emerging middle class (as wealthy as Europeans, Japanese, or Americans), while hundreds of millions

more live in extreme poverty. In China, there are more than 300 million people in this rising middle- and upper-class group. Even in low-income countries, the top leaders may be extremely wealthy, while the rest of the citizens are poor. The United States (GINI index 45), an extremely wealthy country, also has large disparities between groups in terms of health care and access to services. Because of these disparities, it ranks lowest among the developed countries in life expectancy, in spite of spending more per person on health care than any other country. It has more children in poverty and more people in prison than any other high-income country. Income inequality can cause social unrest. In the United States, the percent of wealth owned by the top 1% went from 28% to 47% between 1968 and 2007. This change contributed to the 2011 protests across the country called the "Occupy Wall Street" movement.

We have seen that simple statements about the life experiences of high-, middle-, and low-income countries may not apply to every person. The disparities between rich and poor within a country also matter. Differences between countries are not the only differences that justice requires us to address.

Enter Stewardship. Despite having a fairly stable population, the developed countries have their own problems with consumption, affluence, damaging technologies, and burgeoning wastes. These issues must be addressed to achieve sustainability, and this, too, requires wise leaders and effective public policies. It also requires willpower, both individually and as a society, and an ethical basis for caring for others and for the rest of the natural world.

Fortunately, the environmental impacts of affluent lifestyles may be moderated somewhat by practicing environmental stewardship. For example, suitable attention to

---

STEWARDSHIP   Lessening Your Ecological Footprint

Some estimates conclude that the average American places at least 20 times the demand on Earth's resources—including its ability to absorb pollutants—as does the average person in Bangladesh, a poor Asian country. Major world pollution problems—such as the depletion of the ozone layer, the impacts of global climate change, and the accumulation of toxic wastes in the environment—are largely the consequence of the high consumption associated with affluent lifestyles in the developed countries. How might Atin, Awa, and Anje affect the environment? Anje in Sweden probably uses fewer resources and has a smaller footprint than her cousin in the United States, because

the EU has more regulations on products and because Europe is so tightly packed and distances between cities are much smaller. Public transportation, biking, smaller vehicles, and other lower-impact services are the norm. However, Europeans still have very high impacts on the world in comparison to citizens of low-income countries. Awa in Burkina Faso likely has a lower footprint from consumer impacts, but is likely to have more children and might cause more water pollution if sanitary facilities are not available. Overall, her use of resources will be much lower than the other two. Atin in Indonesia is likely to have greater consumption levels than Awa and lower than Anje.

If you search the Internet for "ecological footprint," you can calculate your own, one of the first steps to figuring out how to be more

sustainable. A footprint calculator is likely to ask you to answer questions about your housing, use of energy, purchasing habits, recycling, eating habits, and other questions. Then it will use this information to calculate how many Earths we would need if everyone lived at the standard of living you have. The calculator will suggest steps to reduce your footprint based on your habits. You can also calculate specific types of footprints such as nitrogen, carbon, and water. If we calculated a footprint for Anje, for example, it might be less than a college student in the United States, because she would be less likely to have a car and more likely to recycle and to take public transportation. Awa would require fewer resources because she uses much less energy, does not travel much, and grows her own food. Figuring out our contribution to the overuse of resources is the first step to change!

wildlife conservation, pollution control, energy conservation and efficiency, and recycling may lower (to some extent) the negative impact of a consumer lifestyle. These changes would alter the technology part of the ImPACT equation, lessening the effect of our use of resources.

Some people argue that population growth is the main problem, others claim that our highly consumption-oriented lifestyle is chiefly to blame, and still others maintain that it is our inattention to stewardship that is the prime shortcoming. In order to make the transition to a sustainable future, however, all three areas (and more) must be addressed. That is, the population must stabilize (a demographic transition), consumption must decrease (a resource and technology transition), and stewardship action must increase (a political/ sociological transition).

8.3 Consequences of Population Growth and Affluence

Expanding populations and increasing affluence—it sounds like trouble for the environment, and it is. It also means trouble for people in high-, middle-, and low-income countries. In countries experiencing rapid population growth, people overuse land, and many flee to cities in the hopes that there will be new opportunities there. These hopes are not necessarily fulfilled. In high-income countries, increases in industrialization have resulted in huge amounts of wealth, but have caused problems as well, including the transfer of environmental degradation to places where those most responsible for generating it do not have to experience its price.

Countries with Rapid Growth

Prior to the Industrial Revolution, many humans survived through subsistence agriculture. Families lived on the land, raised livestock, and produced enough crops for their own consumption and perhaps some extra to barter for other essentials. Forests provided firewood and structural materials for housing. With small, isolated, and relatively stable populations, this system was basically sustainable. As the older generation passed away, the land and natural systems could still support the next generation. Indeed, many cultures sustained themselves in this way over thousands of years.

After World War II, modern medicines—chiefly vaccines and antibiotics—were introduced into developing nations, whereupon death rates plummeted and populations grew rapidly. Today 75% of the world's poor in developing countries live in rural areas, and most are engaged in small-scale agriculture **(Fig. 8–11)**, but only 3.8% of official development assistance goes to agriculture. Rapid growth has negative consequences on a population that is largely engaged in farming a few acres or less of land. When farms become too small to support the next generation, only a few basic options are possible, all of which are being played out to various degrees by people in these societies. They can reform the system of land ownership, intensify the cultivation of existing land to increase the production per unit area, and open

Figure 8–11 Small-scale agriculture. Seventy percent of the world's poor live in rural areas and are heavily dependent on small-scale agriculture.

up new land to farm. They may also engage in illicit activities for income, emigrate to other countries (either legally or illegally), or move to the cities and seek employment.

As we will see, each of these options has consequences.

Land Ownership Reform. Rising population growth in rural developing countries has increased the need to reform the system of land ownership. Collectivization (the gathering of farmers into group farms) and ownership by the wealthy few are two patterns of agricultural land ownership that have historically kept rural peoples in poverty. Collectivization originated within 20th-century communism and was one of the great failures of the former Soviet Union. Today, state-owned land is being privatized, with the result that agricultural production is on the rise. When China abandoned collective agriculture in 1978 and assigned most agricultural land to small-scale farmers, farm output grew more than 6% a year for the next 15 years.

Ownership by the wealthy few was a common result of colonialism in the 19th and 20th centuries. For example, under the apartheid system in South Africa, 87% of South Africa's land was available only to whites or was owned by the government, leaving a much smaller amount of poor-quality land for use by people of color. Since apartheid ended, the South African government has been slowly trying to solve the problems brought about by such land inequality. However, resolving these kinds of inequities can disrupt the social order, as events in Zimbabwe have shown. In that country, a chaotic land reform program, which was supposed to return land to landless people, has virtually destroyed the country's agriculture and helped to plunge the country into economic decline.

Intensifying Cultivation. The introduction of more highly productive varieties of basic food grains in the Green Revolution has had a dramatic beneficial effect in supporting the growing population, but is not without some concerns (Chapter 9). For example, intensifying cultivation means working the land harder. Traditional subsistence farming in Africa involved rotating cultivation among three plots and allowing some of the land to lie fallow. With pressures to increase productivity, plots have been put into continuous production with no time off. The results have been a deterioration of the soil, decreased productivity (ironically), and erosion. Given the countertrends of rapidly increasing population and the deterioration of land from overcultivation, food production per capita in Africa is currently decreasing and may decrease in China in the near future. Burkina Faso, Awa's country, is characterized by agriculture but is limited by drought and poor soil. Intensifying agriculture is done to increase cash crops such as cotton and sugar cane—changes that may or may not help local farmers.

Opening Up New Lands. Opening up new lands for agriculture may sound like a good idea, but there is really no such thing as "new land." Most good agricultural land is already in production. Opening up new land means converting natural ecosystems to agricultural production, which means losing the goods and services those ecosystems contribute. Even then, converted land is often not well suited for agriculture. **Figure 8–12** depicts an areal view of small agricultural plots around the world. These landscape views enable us to see how intensive agriculture already is and how much land has been converted to its use. Unfortunately, much of the land that is useful to agriculture is already used. The use of steep hillsides, marginal lands, or cut forests for planting is not a long-term solution.

Illicit Activities. Anyone who doesn't have a way to grow sufficient food must gain enough income to buy it—and sometimes desperate people break the law to do this. Although it is difficult to draw the line between the need and the greed that also draws people into illicit activities, it is undeniable that the shortage of adequate employment exacerbates the problem. In many developing countries, income is obtained from illegal activities such as raising drug-related crops and poaching wildlife. In Indonesia, Atin's home, there is a great deal of poaching of wildlife, including sea turtles, parrots, and many other species. The temptation is high for people who are poor in an area with rare and exotic species.

Migration Between Countries. The gap between high- and low-income countries is reflected in the perception of many in poorer countries who believe they can improve their well-being by migrating to a wealthier country. In fact, 3.4 million people move to more-developed regions annually, including about 1.4 million to North America. The fact that populations in the wealthy countries are aging suggests a strategy that, on the surface, looks appealing. That is, the wealthy countries need more and younger workers, so they should welcome the migrants from the low-income countries who are looking desperately for work. Such migration does happen. Each year many millions migrate to the United States and Europe to find a better life or to escape civil wars and ethnic persecution **(Fig. 8–13)**. Immigration, however, has its problems, too. Prejudice against foreigners is common, especially in countries with strong ethnic and cultural homogeneity. In some countries with declining birthrates, such as Taiwan and Japan, the government is trying to encourage higher birthrates in order to have enough workers to support aging populations, rather than increase immigration.

Current U.S. immigration laws officially admit around 1.2 million new immigrants per year, a number larger than any time since the 1920s and larger than the number accepted by all other countries combined. At present, immigration accounts for about 35% of U.S. population growth. A report by the National Research Council examined the economic impacts of our immigration policies and concluded that legal immigration basically benefits the U.S. economy and has little negative impact on native-born Americans, with the exception of a few areas of the country where immigrants

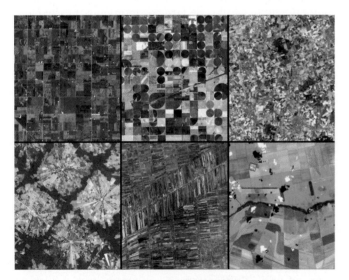

Figure 8–12 Conversion of land for agriculture. This is a set of six satellite and computer generated images of fields in Minnesota (upper left), Kansas (upper middle)—where center pivot irrigation is responsible for the round field patterns—and northwest Germany (upper right). Near Santa Cruz, Bolivia (lower left), the fields radiate from small communities. Outside of Bangkok, Thailand (lower middle), rice paddies are skinny rectangular fields. In the Cerrado in southern Brazil (lower right), inexpensive flat land has resulted in large field sizes. Scale: covers an area of 10.5 × 12 km.

(*Source:* NASA.)

Figure 8–13 Migration. Many people migrate from one country to another to better their lives.

are especially numerous. However, the United States has a high number of illegal immigrants, something that is extremely sensitive politically.

Refugees. Much migration is not from a poor country to a wealthy one, but from one low- or middle-income country to another, especially in areas with civil strife. Refugee immigration leads to temporary refugee camps, where diseases and hunger often take a terrible toll on human life. Some "migrants" are easy targets for exploitation. In some cases, recruiters from plantations in the Ivory Coast pay parents in neighboring Burkina Faso (families like Awa's) to send their children to work on the plantations, where they are used virtually as slaves.

Migration to Cities. Faced with the poverty and hardship of the countryside, many hundreds of millions of people in developing nations continue to migrate to cities in search of employment and a better life. This urbanization is one of the biggest trends in population today. In 2008, the world passed a landmark—more than half of the population lived in cities. This trend is expected to continue, so that by 2050 urban dwellers will account for 70% of the global population. In 2010, there were 21 metroregions that qualified as "megacities" of 10 million or more—the top ten of which are shown in **Figure 8–14a**—but most urbanites live in cities of less than half a million. Most of the net growth of the next 50 years in the developing countries will be absorbed in urban areas. About one-third of that urban growth is expected to occur in China and India.

Challenges to Governments. Population growth and migration to urban areas are outpacing economic growth and the provision of basic services in many developing countries. The most rapidly expanding cities, especially in sub-Saharan Africa, have fallen so far behind in providing basic services that they are getting worse, not better. Streets are potholed, sanitation and drinking water are poor, electricity and telephone service are erratic, and crime is rampant. Many are forced to live in sprawling, wretched shantytowns and slums that do not even provide adequate water and sewers, much less other services (Fig. 8–14b). Diseases like malaria and malnutrition are rampant, and the incidence of HIV/AIDS is much higher in the cities than in the countryside, a consequence of much higher numbers of single men.

(a)

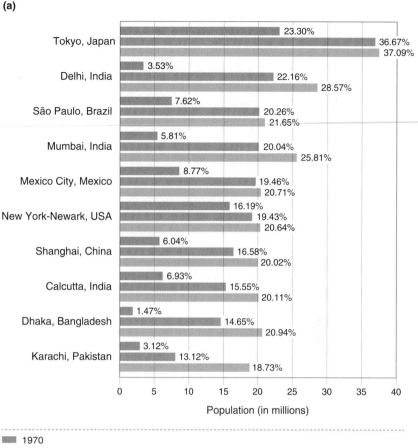

- 1970
- 2009
- 2025 (expected)

(b)

Figure 8–14 Growing cities. (a) The top 10 world metropolitan areas in 2011. Since 1970, cities in the developing world have grown phenomenally, and a number of them are now among the world's largest. (b) Slums on the outskirts of São Paulo, Brazil. Thirty-two percent of the city's population lives in these blighted areas.

(*Source:* Data for part (a) from UN Population Division, *World Urbanization Prospects: The 2010 Revision* (New York: United Nations, 2011).)

People need basic services like education, housing, health care, and transportation, and many developing countries are simply unable to provide these fast enough to keep up with population growth and migration to urban areas. Pakistan's population increased almost fivefold between 1960 and 2010, as a result of decades of 3.2% annual growth. According to Pervez Musharraf, former president of Pakistan, that country's population growth was "the main factor retarding economic growth, poverty alleviation, and action on joblessness."[8]

Although there are connections between rapid population growth in the developing world and environmental decline, the most pressing issue is the massive poverty that afflicts the low-income countries where population growth is still very high. The 2011 Human Development Report makes the argument that equitability and social progress are connected to environmental sustainability. A different set of issues connects population and environmental decline in the affluent countries.

Countries with Affluence

The United States has the dubious distinction of leading the world in the consumption of many resources. We consume the largest share of 11 of 20 major commodities: aluminum, coffee, copper, corn, lead, oil, oilseeds, natural gas, rubber, tin, and zinc. We lead in per capita consumption of many other items, such as meat. The average American eats about twice the global average of meat. We lead the world in paper consumption, at 725 pounds (330 kg) per person per year. All of these factors (and many more like them) contribute to the unusually high environmental impact each of us makes on the world. Unlike many other developed countries, however, the United States continues to have high population increase, much of which is from immigration. During the 20th century, the United States tripled its population and increased per capita use of raw materials 17 times.[9]

Despite the adverse effects of affluence, increasing the average wealth of a population can affect the environment positively. An affluent country such as ours provides amenities like safe drinking water, sanitary sewage systems and sewage treatment, and the collection and disposal of refuse. Thus, many forms of pollution are held in check, and the environment improves with increasing affluence. In addition, if we can afford gas and electricity, we are not destroying our parks and woodlands to obtain firewood. In short, we can afford conservation and management, better agricultural practices, and pollution control, thereby improving our environment.

Still, because the United States consumes so many resources, we also lead the world in the production of many pollutants. For example, by using such large quantities of fossil fuel (coal, oil, and natural gas) to drive our cars, heat and cool our homes, and generate electricity, the United States is responsible for a large share of the carbon dioxide produced. As mentioned, with about 4.5% of the world's population, the United States generates 18% of the carbon dioxide emissions that may be changing global climate.[10] Similarly, emissions of chlorofluorocarbons (CFCs) that have degraded the ozone layer, emissions of chemicals that cause acid rain, emissions of hazardous chemicals, and the production of nuclear wastes are all largely the by-products of affluent societies.

High individual consumption places enormous demands on the environment. The world's wealthiest 20% of people (some located in low- and middle-income countries) are responsible for 76% of all private consumption (the poorest fifth, only 1.5%). As a consequence, by 2005, 75% of major world fisheries were either fully exploited or overexploited, and old-growth forests in southern South America were being clear-cut and turned into chips to make paper. Oil spills are a "by-product" of our appetite for energy. Tropical forests are harvested to satisfy the desires of the affluent for exotic wood furnishings. Metals are mined, timber harvested, commodities grown, and oil extracted, all far from the industrialized countries where these goods are used. Every one of these activities has a significant environmental impact. As increasing numbers of people strive for and achieve greater affluence, it seems more than likely that such pressures, and other ones like them, will mount.

One way of generalizing the effect of affluence is to say that it enables the wealthy to clean up their immediate environment by transferring their wastes to more distant locations. It also allows them to obtain resources from more distant locations, so they neither see nor feel the impacts of getting those resources. However, affluence also provides people with opportunities to exercise lifestyle choices that are consistent with the concerns for stewardship and sustainability.

With this picture of population growth and its impacts in view, the next section discusses future population trends, how demographers identify them, and the biggest population trend—the demographic transition.

8.4 Projecting Future Populations

During the 1960s, the world population growth rate peaked at 2.1% per year, after having risen steadily for decades. Following this peak, it began a steady decline (**Fig. 8–15**). Twenty years later, the number of persons added per year peaked at 87 million. The reason for the difference is that even though the population growth rate was in decline, there were increasing numbers of people who could reproduce. The total

[8]"The President of Pakistan on the Need to Slow Population Growth in the Muslim World," *Population and Development Review* 31, no. 2 (2005): 399.
[9]U.S. Census Bureau. (April 2004). "Statistical Abstract of the United States." And L. A. Wagner. (2002). *Materials in the Economy: Material Flows, Scarcity and the Environment.* U.S. Geological Survey Circular 1221. http://pubs.usgs.gov/circ/2002/c1221/c1221-508.pdf.

[10]International Energy Agency (IEA). (2011). CO₂ *Emissions from Fuel Combustion: Highlights (2011 edition).* Paris, France: IEA.

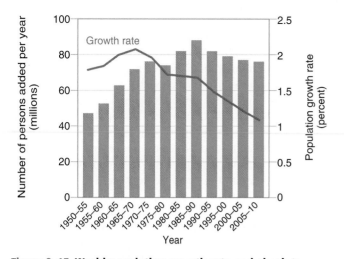

Figure 8–15 World population growth rate and absolute growth. Declining fertility rates in the past three decades have resulted in a decreasing rate of population growth. Absolute numbers, however, are still adding 78 million persons per year.

(*Source:* Data from Shiro Horiuchi, "World Population Growth Rate," *Population Today* (June 1993): 7 and (June/July 1996): 1; updated from UN Population Division, *World Population Prospects: The 2010 Revision* (New York: United Nations, 2010).)

number of persons added every year did not go down until the growth rate fell low enough to offset the increase in reproducing people.

The declines in both the population growth rate and the number of persons added per year are primarily a consequence of the decline in the total fertility rate—that is, the average number of babies born to a woman over her lifetime. In the 1960s, the total fertility rate was an average of 5.0 children per woman; it has since steadily declined to its present value of 2.6 children per woman. In addition to the total fertility rate, the age at which a woman first reproduces also controls how rapidly a population grows. A population in which women have children at a young age will grow more quickly than one in which women are slightly older when they reproduce, even if women have the same number of children. That is, there are more generations alive at the

same time when women reproduce young. Thus, total fertility rate is the most important, but not the only, factor determining population growth.

Extrapolating the trend of lower fertility rates leads to the UN Population Division's projection that the global human population will reach 9.3 billion by 2050; it will not have leveled off at that time, but is likely to level off sometime in the 22nd century. This projection is the UN's *medium* scenario; other scenarios are based on different fertility assumptions. **Figure 8–16** depicts the UN Population Division's projections until 2050 under four different sets of assumptions. The assumption of declining fertility rates is crucial; if current fertility rates remain unchanged (*constant* in the figure), the 2050 population will be 10.9 billion.[11] To achieve the medium scenario, people will need to continue to purposefully lower their fertility in the high fertility areas of the world, at a rate we are not seeing yet. The UN doesn't usually project past 50 years out. The projected leveling off over 10 billion raises serious concerns about whether Earth can sustain such numbers.

Population Profiles

In studying population growth, we must consider more than just the increase in numbers, which is simply births minus deaths. We must also consider how the number of births ultimately affects the entire population over the **longevity**, or lifetimes, of the individuals (see Sound Science, "Are We Living Longer?"). A **population profile** is a bar graph showing the number or proportion of people (males and females separately) at each age for a given population. **Figure 8–17** shows examples of such graphs. The data are collected through a census of the entire population using a combination of household questionnaires and estimates for groups such as

[11]United Nations, Department of Economic and Social Affairs, Population Division. (2011). *World Population Prospects: The 2010 Revision, Volume I: Comprehensive Tables.* ST/ESA/SER.A/313.

Figure 8–16 United Nations population projections. The most recent population projections demonstrate the vital role played by different fertility assumptions. The *Constant* projection assumes that fertility rates remain at their present level (2.6 children per woman). The *Medium* projection assumes a gradual decline in fertility in the developing countries, where 85% drop below replacement-level fertility by 2050 and the world total fertility rate is 2.02 children per woman. This projection is given the highest probability by the UN Population Division. The *High* projection assumes fertility rates a 1/2 child greater than the *Medium* projection, and the *Low* projection assumes fertility rates a 1/2 child lower than the *Medium* projection. All of these projections include the impact of the AIDS epidemic on increasing mortality.

(*Source:* UN Population Division, *World Population Prospects: The 2010 Revision* (New York: United Nations, 2010).)

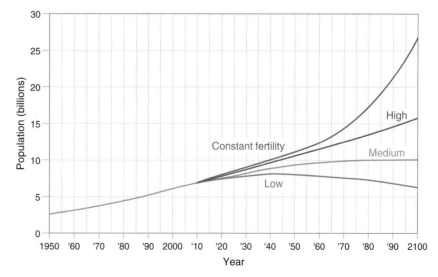

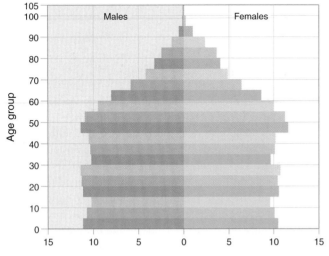

(a) United States of America: 2010

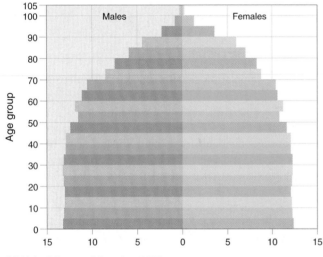

(b) United States of America: 2050

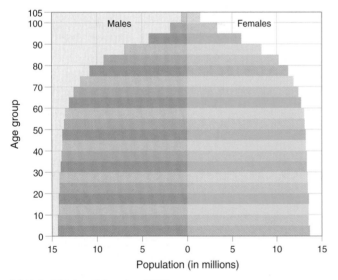

(c) United States of America: 2100

Figure 8–17 Population profiles of the United States. The age structure of the U.S. population (a) in 1985, (b) in 2005, and (c) projected to 2050. www.masteringenvironmentalscience.com.

(*Source:* U.S. Census Bureau, International Data Base.)

the homeless. In the United States and most other countries, a detailed census is taken every 10 years. Between censuses, the population profile may be adjusted using data regarding births, deaths, immigration, and the aging of the population.

A population profile shows the **age structure** of the population—that is, the number of people in each age group at a given date. It is a snapshot of the population at a given time. For example, population profiles of the United States for 1950 and 2000 and projected future profiles for 2050 and 2100 are shown in Figure 8–17. Leaving out the complication of migration for the moment, each bar in the profile represents one *cohort* (group of the same age) of the population. As you look from one graph to the next, you can see a particular group as it ages. In high-income countries such as the United States, the proportion of people who die before age 60 is relatively small. Therefore, the population profile below age 60 is an "echo" of past events that affected birthrates. Figure 8–17a shows, for example, that smaller numbers of people were born between 1931 and 1935 (ages 15–18 in 1950). This drop was a reflection of lower birthrates during the Great Depression. The dramatic increase in people born in 1946 and for 14 years thereafter (ages 50–64 in 2010, Fig. 8–17b) is a reflection of returning veterans and others starting families and choosing to have relatively large numbers of children following World War II—the "baby boom." A number of industries expanded and then contracted as the baby-boom generation moved through a particular age range, and this phenomenon is not yet past. In sequence, schools, then colleges and universities, and then the job market were affected by the large influx of baby boomers. As the large baby-boom generation, now in middle age and retirement (and projected to be extremely elderly in 2050), moves up the population profile (Fig. 8–17a–c), any business or profession that provides goods or services to seniors is looking forward to a period of growth. Planners also use population profiles to understand the probable needs for social security programs in the future.

The baby boom is not the only phenomenon we see in U.S. population profiles. The general drop in numbers of people born from 1961 to 1976 (ages 34–49 in 2010) resulted from declining fertility rates—the "baby bust." The rise in numbers of people born in more recent years (ages 20–29 in 2010) is termed the "baby boom echo" and is due to the large baby-boom generation producing children, even though the actual total fertility rate remained near 2.0. These changes in fertility are shown in **Figure 8–18**.

Predicting Populations

Current population growth in a country is calculated from three vital statistics: births, deaths, and migration. The population of the United States grew by more than 2.8 million people during the 15-month period from April 1, 2010, to July 1, 2011. This growth rate (0.92%) was the lowest since the mid-1940s. Only three years prior, the population growth of 2007–2008 had been much higher—there were 4.32 million births (15,000 more than the highest previous record in 1957). In the United Sates in January 2012,

U.S. TOTAL FERTILITY RATE, 1911—2010

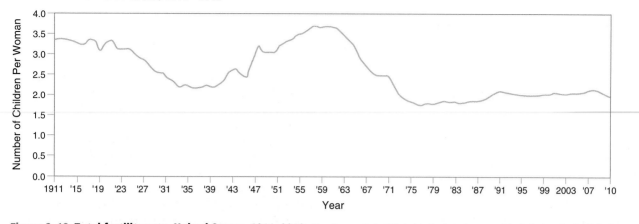

Figure 8–18 **Total fertility rate, United States, 1911–2010.** The changes in fertility led to the baby boom and the baby bust in the United States. The rate now hovers about the replacement level.

(*Sources: 2010 World Population Data Sheet* (Washington, DC: Population Reference Bureau, 2010).)

immigration was projected to add one person every 46 seconds to the population, with one birth every 8 seconds and one death every 12. Thus, the total population was increasing by one person approximately every 17 seconds (U.S. Census Bureau News),[12] although these numbers are imprecise because of illegal immigration.

Demographers used to make population *forecasts*. Most often, they took current birth and death rates, factored in expected immigration, and then extrapolated the data into the future. They were virtually always wrong. For example, in 1964, the total fertility rate (TFR) in the United States was around 3.0. The U.S. Census Bureau forecast that the TFR would range between 2.5 and 3.5 up to 2000 and projected a 2000 population of 362 million under the higher fertility assumption—81 million more than the actual number. The agency totally missed the baby bust of the late sixties and seventies, although the trend in 1964 was definitely downward.

Nowadays, demographers only make *projections*, and they cover their bases by making their assumptions about fertility, mortality, and migration very clear. As noted earlier, the projection that the world population will reach 9.3 billion in 2050 is based on the assumption that fertility rates will continue their decline. The United Nations gives three different projections of future world population (Fig. 8–16). All three projections assume fertility rates will fall. In fact, if we maintained a current rate of fertility, the human population would hit 27 billion by 2100. Instead, fertility rates in high-fertility countries are likely to fall, and those in extremely low-fertility countries are likely to rise slightly. The medium scenario assumes that world fertility will drop from 2.55 children per woman to slightly above 2 children per woman well before 2050, the high-fertility scenario assumes a total fertility rate half a child above, and the low-fertility scenario assumes a total fertility rate half a child below that of the medium variant. Recall that the constant fertility line in Figure 8–16

(orange) shows what will happen if the fertility rates around the world do not change. Note how each fertility assumption generates profoundly different world populations. Notice also that none of these projections, even the one based on a drop in total fertility to 1.5 children per woman, will lead to a 2050 population smaller than today's population.

Population Projections for Developed Countries. The population profiles of Sweden, a developed country in southern Europe, shown in **Figure 8–19**, reflect the fact that Swedish women have had a stable fertility rate for some time. Sweden has a boom group that will be getting older by 2050, but also has a stable production of new youth. In some countries, such as Italy (with a fertility rate of 1.3) or Japan, the number of workers for each dependent (child, elderly, disabled, or otherwise unable to work) will decline. In the case of these countries, it is because an aging population is not replaced by new babies.

Graying of the Population. For the next 20 years, Italy's and Japan's populations will be **graying**, a term used to indicate that the proportion of elderly people is increasing. In 2005, one-fourth of Italians were aged 60 and above; by 2050, 41% are expected to be 60 or older. Overall, a net population decrease is expected to occur. What opportunities and risks does the changing population profile imply for Italy? If you were an advisor to the Italian or Japanese governments, what would your advice be for the short term? For the longer term? Unless a smaller population is the goal, it might be wise to encourage, and to provide incentives for, Italian couples to bear more children. In fact, many European governments have policies to encourage women to have more children, with mixed results so far. France, with one of the highest fertility rates in Europe (TFR = 1.9) provides child allowances, tax credits for in-home child care, and discounts on many goods and services for larger families.

The very low fertility rate and the expected declining population seen in Italy and Japan are typical of an increasing number of highly developed nations. Europe as a whole is

---

[12]U.S. Census Bureau. "Census Bureau Projects U.S. Population of 312.8 Million on New Year's Day." http://www.census.gov/newsroom/releases/archives/population/cb11-219.html. June3, 2012.

POPULATION BY AGE GROUPS AND SEX (absolute numbers)

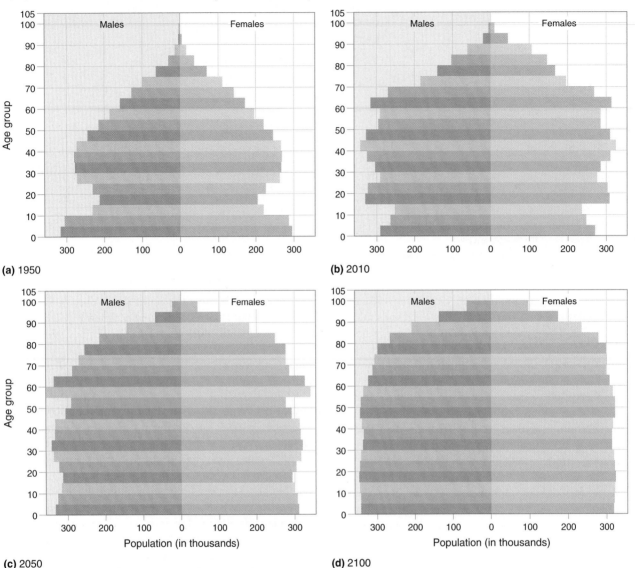

(a) 1950

(b) 2010

(c) 2050

(d) 2100

Population (in thousands)

Figure 8–19 Projecting future populations: A developed country. A population profile of Sweden, representative of a highly developed country, in four times. Note how stable the population becomes.

(*Source:* U.S. Census Bureau, International Data Base.)

on a trajectory of population decline if only **natural increase** (rather than immigration) is considered. Who will produce the goods and services needed by the European countries' aging populations? Will their economies remain competitive on the world markets? The surge of older people retiring may overwhelm government pension systems; life expectancies near 80 and retirement around 60 mean decades of retirement income. A number of countries in the Eurozone experienced financial difficulties in 2011 when it became clear that the burden of their social programs, including pensions, was unsupportable. Japan is also struggling to figure out how to care for the elderly.

One obvious solution is to allow more immigration. However, allowing the declining numbers of Italian or Japanese people to be replaced by immigrant populations has implications for art, culture, and religion. Such changes

are sometimes difficult for the host society. In fact, Europe is in the midst of a migration wave. More than 20 million immigrants now live and work in the countries of western Europe, many of whom do not have resident status. In some of these countries, fear and mistrust of these "guest workers" have led to violent attacks on them. Many Europeans do not want more immigration, even to offset declines in the labor force and population. Yet just to maintain their current populations, most of these countries will have to triple their current immigration levels in the near future.

Less Graying Here. In contrast to other developed countries, the fertility rate in the United States reversed directions in the late 1980s and started back up. On the basis of the lower fertility rate, the U.S. population had been projected to stabilize at between 290 million and 300 million

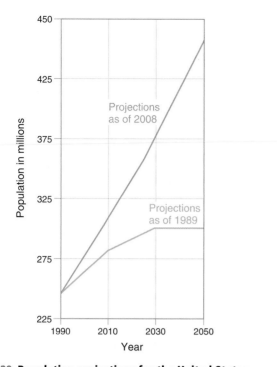

Figure 8–20 Population projections for the United States.
Projections shift drastically with changes in fertility. Contrast the 1989 projection, based on a fertility rate of 1.8, with the 2008 projection, based on an increased fertility rate of 2.1 and current immigration.

(*Source:* U.S. Census Bureau, U.S. Population Projections.)

These projections for the United States show how differences in the total fertility rate and high immigration profoundly affect population estimates when they are extrapolated forward. In light of the concerns for sustainable development, the population policy of the United States (and scenarios that could lead to lower future growth) is under heavy debate.

Population Projections for Developing Countries.

Developing countries are in a situation vastly different from that of developed countries. Fertility rates in developing countries are generally declining, but they are still well above replacement level. The average TFR (excluding China, where it is 1.6) is currently 3.2, which is similar to that of the United States at the peak of the baby boom. Because of decades of high fertility rates, the population profiles of developing countries have a pyramidal shape.

Burkina Faso and Indonesia.

A series of population profiles for Awa's home country of Burkina Faso, which has a total fertility rate of 6.3, are shown in **Figure 8–21a–d**. Burkina Faso is a good example of the lowest-income countries and has demographics that are typical of that group. Even assuming that this high fertility rate will gradually decline by 2025, the population is likely to increase from 15.3 million (2010) to 25.4 million, yielding the profiles shown in Figure 8–21c and d. Even the high infant mortality rate (currently 91 per thousand) comes nowhere close to offsetting the high fertility rate. The impact of AIDS is also included in the UN projections. AIDS has increased child and young adult mortality and lowered average life spans in a number of countries. The pyramidal form of the profile remains the same in both 2010 and 2025, because, for many years, the rising generation of young adults produces an even

toward the middle of the next century. With the higher fertility rate of 2.1 and with an influx of immigrants, the U.S. population is now projected to be 439 million by 2050 (Fig. 8–17c) and to continue growing indefinitely (**Fig. 8–20**). For this projection, immigration is assumed to remain constant at current levels.

SOUND SCIENCE   Are We Living Longer?

With the advent of modern medicine and disease control, a long life has become possible for more and more people. Average **life expectancy** (the number of years a newborn can expect to live under current mortality rates) has been increasing dramatically. Over the past two centuries, world life expectancy increased from about 25 years to 65 for men and 69 for women. Life expectancy and mortality rates are linked, and as infant mortality in particular has declined in the developing countries, their average life expectancy has climbed sharply. Life expectancies in those countries are now within 10 years of life expectancies in the developed countries (73 for men and 80 for

women). What is not well known is that life expectancy has also continued to climb about 2 1/2 years every decade for the low-mortality developed countries. Mortality rates have declined for every age.

Can we hope that this trend will continue until, perhaps a century from now, human life expectancies will be approaching or going beyond 100? There is intense debate over this question. First, however, we must differentiate between life expectancy and longevity. Longevity refers to the *maximum life span for a species.* The verifiable record for humans is held by Jeanne Calumet of France, who lived for 122 years. The record for the United States is that of a woman who died at 119 years of age in 1999. While these individuals were very old, many workers in the field of gerontology (the

study of aging) are convinced that, in humans (as in all species), there is an aging process that cannot be prevented—that built into our cells are the biochemical seeds of destruction. Cells, tissues, and organs wear out at variable rates in different people. Thus, we will not likely find life expectancies rising beyond about 85 years anywhere or anytime in the future.

Others argue that even though absolute longevity seems to be a limiting process, we can expect life expectancies to continue the increase they have shown in recent years. They ask, "Why should life expectancies stop at 85?" The trend of a 2 1/2-year increase per decade is based on reducing many of the factors that cause mortality at different ages. This trend should continue, given the medical prowess of our 21st-century science.

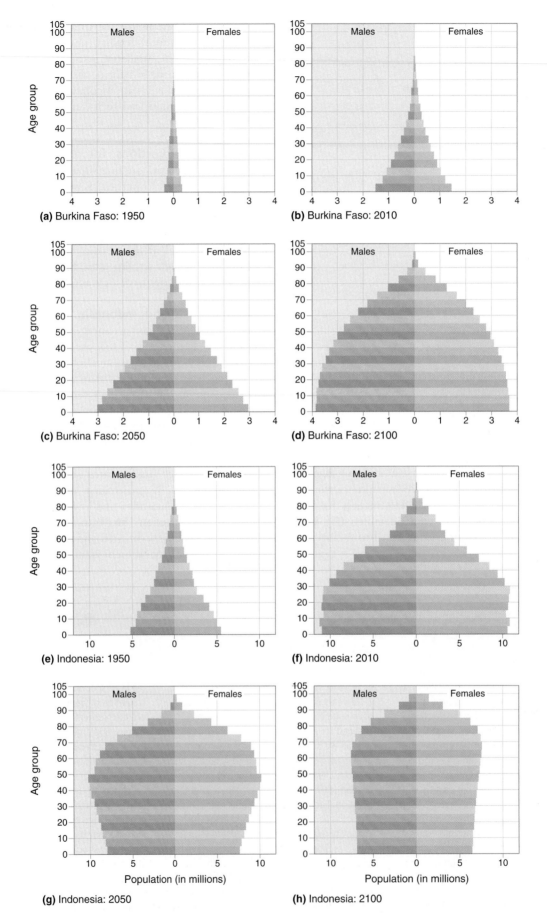

(a) Burkina Faso: 1950

(b) Burkina Faso: 2010

(c) Burkina Faso: 2050

(d) Burkina Faso: 2100

(e) Indonesia: 1950

(f) Indonesia: 2010

(g) Indonesia: 2050

(h) Indonesia: 2100

Figure 8–21 **Projecting future populations: Indonesia and Burkina Faso.** (a–d) Projections for Burkina Faso. (e–h) Projections for Indonesia.

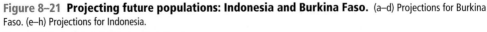

(*Source*: U.S. Census Bureau, International Data Base.)

TABLE 8–3 Populations by Age Group

Region or Country	Percent of Population in Specific Age Groups			
	<15	15 to 65	>65	Dependency Ratio*
Sub-Saharan Africa	43	54	3	85
Latin America	28	65	7	56
Asia	26	67	7	54
Indonesia	28	66	6	54
Europe	16	68	16	47
Sweden	17	65	18	54
China	17	74	9	37
United States	20	67	13	49

Source: Data from *2011 World Population Data Sheet* (Washington, DC: Population Reference Bureau, 2011).

*Number of individuals below 15 and above 65, divided by the number between 15 and 65 and expressed as a percentage.

larger generation of children. Thus, the pyramid gets wider and wider until the projected declining fertility rate begins to take effect. Figure 8-21e–h depicts the population profiles of Indonesia, Atin's home. While Indonesia is not well off, its population growth rate is lower than Burkina Faso's. Its large population may stabilize earlier, but it is so large that it has a lot of population momentum.

While highly developed countries are facing the problems of a graying population, the high fertility rates in developing countries maintain an exceedingly young population. An "ideal" population structure, with equal numbers of persons in each age group and a life expectancy of 75 years, would have one-fifth (20%) of the population in each 15-year age group. By comparison, 40% to 50% of the population is below 15 years of age in many developing countries, whereas less than 20% of the population is below the age of 15 in most developed countries (Table 8–3).

Growth Impacts. What do these differing population structures mean in terms of the need for new schools, housing units, hospitals, roads, sewage facilities, telephones, and so forth? One of the things they mean is that if a country such as Burkina Faso is simply to maintain its current standard of living, the amount of housing and all other facilities (not to mention food production) must be almost doubled in as little as 25 years. In particular, jobs must be available for a large number of young people. Unfortunately, the population growth of a developing country can easily cancel out its efforts to get ahead economically.

Enormous growth is in store for the developing world, even assuming that fertility rates in the developing world continue their current downward trend. Even bringing the fertility rates of developing countries down to 2.0 right now will not immediately stop population growth in those countries, because of a phenomenon known as population momentum.

Population Momentum

Population momentum refers to the effect of current age structures on future populations. In a young population, such as Burkina Faso's, momentum is *positive* because such a small portion of the population is in the upper age groups (where most deaths occur) and many young people are entering their reproductive years. Even if the rising generations have only two children per woman (replacement-level fertility), the number of births will far exceed the number of deaths. This imbalance will continue until the current children reach the Burkina Faso limits of life expectancy—50 to 60 years. In other words, only a population that has been at or below replacement-level fertility for decades will stop growing (achieve a stable population).

Despite population momentum, efforts to stabilize population are not fruitless. It just means that population growth cannot be halted quickly. Like a speeding train, there is a long time between applying the brakes and stopping completely. The earlier the fertility rates are reduced, the greater the likelihood is of ending population growth and achieving a sustainable society in the near future.

Europe's population, on the other hand, will soon begin to experience *negative* population momentum as a consequence of the low fertility of the past three decades. It has been calculated that if the current TFR of 1.5 births continues for 15 more years, negative momentum will shrink the EU population by some 88 million by 2100 (of course, immigration would offset this number).

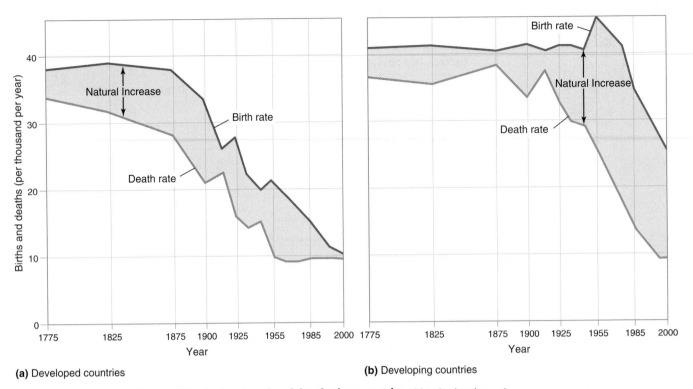

(a) Developed countries

(b) Developing countries

Figure 8–22 Demographic transition in developed and developing countries. (a) In developed countries, the decrease in birthrates proceeded soon after, and along with, the decrease in death rates, so very rapid population growth never occurred. (b) In developing countries, both birth and death rates remained high until the mid-1900s. Then the sudden introduction of modern medicine caused a precipitous decline in death rates. Birthrates remained high, however, resulting in very rapid population growth.

(*Source*: Redrawn with permission of Population Reference Bureau.)

The Demographic Transition

The concept of a stable, nongrowing global human population based on people freely choosing to have smaller families can become a reality, because it is already happening in developed countries. Early demographers observed that the modernization of a nation brings about more than just a lower death rate resulting from better health care. A decline in the fertility rate also occurs as people choose to limit the size of their families. Thus, as economic development occurs, human societies move from a primitive population stability, in which high birthrates are offset by high infant and childhood mortality, to a modern population stability, in which low infant and childhood mortality are balanced by low birthrates. A gradual shift in birth and death rates from the primitive to the modern condition in the industrialized societies is called the demographic transition (**Figs. 8–22, 8–23**). The basic premise of the demographic transition is that there is a causal link between modernization and a decline in birth and death rates.

Birth and Death Rates. To understand the demographic transition, you need to understand the **crude birthrate (CBR)** and the **crude death rate (CDR)**. The CBR and the CDR are the number of births and deaths, respectively, per thousand of the population per year. By giving the data per thousand

of the population, populations of different countries can be compared, regardless of their total size. The term *crude* is used because no consideration is given to what proportion of the population is old or young, male or female. Subtracting the CDR from the CBR gives the increase (or decrease) per thousand per year. Dividing this result by 10 then yields the *percent* increase (or decrease) of the population.

Number of births per 1,000 per year (CBR)		Number of deaths per 1,000 per year (CDR)		Natural increase (or decrease) in population per 1,000 per year		Percent increase (or decrease) in population per year
	−		=		÷10 =	

A zero-growth population is achieved if, and only if, the CBR and CDR are equal and there is no net migration.

Epidemiologic Transition. Throughout most of human history, crude death rates were high—40 or more per thousand for most societies. By the middle of the 19th century, however, the epidemics and other social conditions responsible for high death rates began to recede and death rates in Europe and North America declined. The decline was gradual in the now-developed countries, lasting for many decades and finally stabilizing at a CDR of about 10 per thousand (Fig. 8–21). At present, cancer, cardiovascular disease, and other degenerative diseases account for most mortality, and many people survive to old age. This

THE CLASSIC PHASES OF THE DEMOGRAPHIC TRANSITION

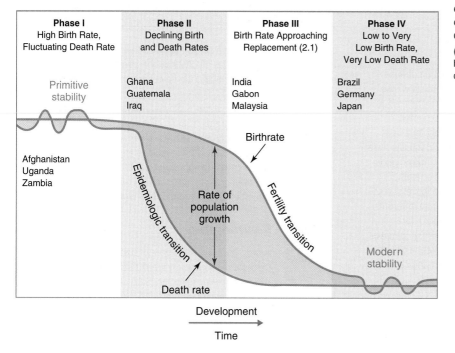

Figure 8–23 The demographic transition. The epidemiologic transition and the fertility transition combined to produce the demographic transition in the developed countries over many decades.

(*Source*: Population Bulletin, Population Reference Bureau, http://www.prb.org/Publications/Datasheets/2011/world-population-data-sheet/population-bulletin.aspx, 2011.)

pattern of change in mortality factors has been called the *epidemiologic* transition and represents one element of the demographic transition. (Epidemiology is the study of diseases in human societies.)

Fertility Transition. Another pattern of change over time can be seen in crude birthrates. In the now-developed countries, birthrates have declined from a high of 40 to 50 per thousand to 9 to 12 per thousand—a *fertility* transition. As Figure 8–22 shows, this fertility transition did not happen at the same time as the epidemiologic transition; instead, it was delayed by decades or more. Because net growth is the difference between the CBR and the CDR, the time during which these two patterns are out of phase is a time of rapid population growth. The developed countries underwent such growth during the 19th and early 20th centuries and during the baby-boom years of the mid-20th century.

Phases of the Demographic Transition. The demographic transition is typically presented as occurring in the four phases shown in Figure 8–23. Phase I is the primitive stability resulting from a high CBR being offset by an equally high CDR. Phase II is marked by a declining CDR—the epidemiologic transition. Because fertility and, hence, the CBR remain high, population growth accelerates during Phase II. The CBR declines during Phase III due to a declining fertility rate, but population growth is still significant. Finally, Phase IV is reached, in which modern stability is achieved by a continuing low CDR and an equally low CBR. Figure 8–23 also shows countries that typify each of the stages of the demographic transition.

Developed countries have generally completed the demographic transition, so they are in Phase IV. Developing countries, by contrast, are still in Phases II and III. In most of these countries, death rates declined markedly as modern medicine and sanitation were introduced in the mid-20th century. Fertility and birthrates are declining but remain considerably above replacement levels (Fig. 8–21). Therefore, populations in developing countries are still growing rapidly.

In light of these circumstances, the key question is: What must happen in the developing countries so that they complete their demographic transition and reach Phase IV? Is economic development essential to undergoing the transition, or may other factors, such as family planning programs, come into play?

In this chapter, we looked at three young people—Anje in Sweden, Atin in Indonesia, and Awa in Burkina Faso. We saw that their life experiences were dramatically different from each other—in part because of broad differences between countries in wealth, health care, and other goods and services. Even within countries, great variability exists. Those differences in life experiences drive differences in consumption patterns and fertility—two parameters that need to be brought into line to have a sustainable world.

Next, we will investigate the factors that influence birth and death rates, and you will explore the relative contributions of economic development and family planning in bringing about the fertility transition (Chapter 9). In the process, we can begin to picture what sustainable development might look like for the developing world, in particular, and what it will take to get there.

REVISITING THE THEMES

Sound Science

Demography, the science of human population phenomena, reflects the diligent work of many scientists whose concern is to document data and interpret it. Their work forms much of the data and information presented in this chapter. Demography can be viewed in the context of population ecology as a whole. Human populations can be studied like other populations and are subject to natural laws such as the law of limiting factors and carrying capacity. Despite some opinions, there is consensus in the scientific community that population growth and increased demand for resources cannot continue indefinitely. However, humans are also quite different from other organisms. It is difficult to classify them as either *r*- or *K*-strategists, because their life history of slow growth and high parental care does not easily match their current rapid global population growth and opportunistic use of new habitats. In the past, the carrying capacity of humans has been increased by using technology to trap energy from other eras and open up new habitats. However, many of the things we did to achieve modern life have prices, as we will see in later chapters.

Demographers have clarified the nature of the demographic transition, and their population projections enable us to see what the future will look like if current trends continue.

Sustainability

A sustainable world is a world whose population is generally stable—it cannot increase indefinitely, as the world's population currently is doing. World population may level off at 10 billion or even 9 billion, but we may be exceeding Earth's carrying capacity. Two or 3 billion more people will need to be fed, clothed, housed, and employed, all in an environment that continues to provide essential goods and services. Many agree that sustainability will be possible only if the additional people maintain a modest, low-consumption lifestyle, at level lower than that currently experienced in the developed countries. People already in developed countries will need to make changes to lower their own consumption so that others may have access to resources. These changes involve not only myriad personal decisions, but also coordinated action and the development of more efficient technologies.

In a sense, the United States is playing a small role in the process by accepting so many immigrants from developing countries—about 1% of the annual population growth in these countries (hardly a significant percentage). More significant is the impact of immigration on the United States itself, which is currently on a trajectory of indefinite population growth because of immigration. Sustainability requires that we somehow stabilize population both nationally and globally.

Stewardship

In order to achieve the goal of sustainability, humans need to steward resources by stabilizing both population growth and resource consumption. Such stewardship requires us to promote justice and may require difficult decisions about what is right. Calling on the developing countries to keep down their future demands on the environment could be considered unjust, as so many people live in abject poverty while others consume resources at high rates. Still, there is no way for these countries to meet their future needs by mimicking what the developed world is currently doing with resources and the environment. Justice demands (and sustainability requires) that both the rich and the poor countries reduce their environmental impact.

Stewardship requires both individual efforts and corporate action through policy. If enough people in the high- and middle-income countries practice environmental stewardship, the pathway to such a future will be easier. Stewardship can do much to alleviate the negative impacts of affluence and technology, prominent items in the *IPAT* formula. Population growth has to slow as well. Achieving that end without trampling on the rights of others is part of the job ahead of us. Immigration also offers ethical dilemmas as immigrants move to places that may or may not be prepared to receive them.

REVIEW QUESTIONS

1. In what ways is human population ecology similar to and different from that of other organisms? Why is it difficult to determine a carrying capacity for humans?

2. How has the global human population changed from prehistoric times to 1800? From 1800 to the present? What is projected over the next 50 years?

3. How does the World Bank classify countries in terms of economic categories?

4. What three factors are multiplied to give total environmental impact? Are developed nations exempt from environmental impact? Why or why not?

5. What are the environmental and social consequences of rapid population growth in rural developing countries? In urban areas?

6. Describe the negative and positive impacts of affluence (high individual consumption) on the environment.

7. What information is given by a population profile?

8. How do the population profiles and fertility rates of developed countries differ from those of developing countries?

9. Compare future population projections, and their possible consequences, for developed and developing countries.

10. Discuss the immigration issues pertaining to developed and developing countries. Which countries are sending the most immigrants and which are receiving the most?

11. What is meant by population momentum, and what is its cause?

12. Define the crude birthrate (CBR) and the crude death rate (CDR). Describe how these rates are used to calculate the percent rate of growth and the doubling time of a population.

13. What is meant by the demographic transition? Relate the epidemiologic transition and the fertility transition (two elements of the demographic transition) to its four phases.

14. What are examples of countries in different stages of the demographic transition?

THINKING ENVIRONMENTALLY

1. Consider how humans are different from and similar to other species. In what ways does it seem obvious that population ecology principles apply to humans? Why do some people see humans as not subject to the law of limiting factors?

2. List the parameters you think you would need to keep track of to make accurate predictions about future human population growth. For example, what conditions might drive mortality and fertility trends?

3. Consider the population situation in Italy or Japan: very low birthrates, an aging population, and eventual decline in overall numbers. What policies would you recommend to these countries, assuming their desire to achieve a sustainable society?

4. Consider the populations of Sweden, Indonesia, and Burkina Faso. How could people in each of these countries help those in the other countries to steward their resources better or to promote a sustainable future?

MAKING A DIFFERENCE

1. Calculate your ecological (environmental) footprint using a site such as http://www.myfootprint.org.

2. Come up with two things you can do to lower your resource consumption. They might include buying less of something you don't need or buying a product with less packaging. Agree to do it for one month, and decide whether you could continue longer.

3. Educate yourself on immigration policy. Decide whether you support immigration reform and what policies you would like to see in place. Then write to an elected official about it.

MasteringEnvironmentalScience®

Go to **www.masteringenvironmentalscience.com** for practice quizzes, Pearson eText, videos, current events, and more.

CHAPTER 9

Population and Development

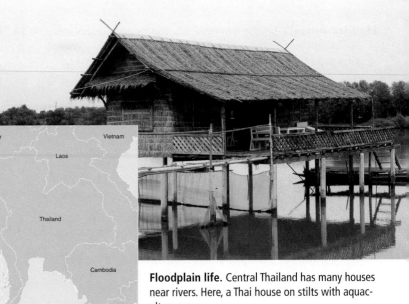

Floodplain life. Central Thailand has many houses near rivers. Here, a Thai house on stilts with aquaculture pens.

THE SUN SHINES ON THE SURFACE OF THE CHAO PHRAYA RIVER IN THAILAND. The water glistens as it runs past the rich fields in the central Thai region. In the villages by the river, extended families live in wooden structures on stilts, water buffalos and chickens safely housed beneath. Anong and her husband, Somchai, are part of such a family. They live with Anong's parents, now elderly, and their own two children, a boy and a girl. While Somchai is the head of the family in most matters, Anong is in charge of finances and the household. Both work outside the home and earn a modest income. Both are literate. Although they live in only one room, the family has access to clean drinking water and sanitation. Both parents were raised in large families; Anong had seven siblings and Somchai five. Today, their family of two children is more typical.

FERTILITY DROP. Their experience is typical of many in Thailand. In a single generation, the country experienced one of the largest drops in fertility in world history. Between 1960 and 2005, the total fertility rate dropped from 6.4 children per woman to 1.8. At the same time, GNP per capita and life expectancy rose dramatically ($1,059 to $7,274; 53 years to 69.9 years). Anong and Somchai are composites of the experience of modern Thais. Like the rest of their generation, they

have experienced rapid change in their home country. Probably the biggest change is family size. This remarkable drop in population growth rate astonished the world.

In 1960, when the Texas-sized country held 28 million people, demographers predicted the country would fall into a poverty trap: any increase in the economy would be eaten up by the needs of the growing population. Today, with population growth slowed and a population of about 70 million, the country has 25 million fewer people than demographers predicted in the 1970s.

SOCIAL CHANGES. How did Thailand accomplish this? Much was done through direct efforts to lower fertility: Voluntary contraception was increased, public health and awareness campaigns were undertaken, and health care clinics were opened, even in rural areas. For example, the public health campaign of one non-profit group made contraception a familiar and humorous topic of everyday life.

But there were other factors as well. Modern Thai women now have higher freedom, equality, and literacy rates than Thai women of the past. In Thai culture, sons are not favored over daughters, so women are not under pressure to abort female fetuses or to have more children than they can support in order to produce sons. A nutrition program implemented by the government dramatically slashed child malnutrition. Contrary to some prevailing thought, the fertility transition Thailand experienced occurred even though the country was relatively poor, and even though most people remained rural. While the drop in population growth prevented disaster, some of the economic growth and the support for millions more Thai citizens came at the expense of ecosystems. Thailand now has to contend with water and air pollution and loss of biodiversity brought on by its activities. On the plus side, the country's ability to solve ecological problems is more likely with lower population growth. This chapter follows up on the demographic information presented earlier (Chapter 8). We begin with a reassessment of the demographic transition.

9.1 Predicting the Demographic Transition

The *demographic transition* describes the shift from high birth and death rates to low birth and death rates over time in industrialized countries (Chapter 8). Using the concept of the demographic transition, current birth and death rates of the major regions of the world can be plotted to visualize where they currently fall in the transition **(Fig. 9–1)**. The vertical dividing line on this plot separates the regions (to the right of the line) that are on a fast track to completing the demographic transition (or have already done so) from the regions (to the left of the line) that are in the middle of the demographic transition. These regions (half of the world) are

mostly in their fifth decade of rapid population increase. It is as if they are trapped there, with serious consequences. One of the key questions scientists who study population ask is, "What causes some countries to move quickly through the demographic transition and others to remain in the process indefinitely?"

Different Ways Forward

Countries trapped in long-term rapid growth would all like to experience the economic development that has brought Thailand, South Korea, Indonesia, Malaysia, Brazil, and other low-income nations into the middle- and even high-income nation groups. If they did so, it is likely that they

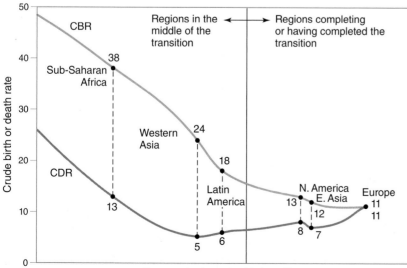

Figure 9–1 World regions in the process of demographic transition. Crude birthrates (CBRs, orange) and crude death rates (CDRs, green) are shown for major regions of the world. A dividing line separates countries at or well along in the demographic transition from those still in the middle of the transition. (*Source:* Data from *2011 World Population Data Sheet* (Washington, D.C.: Population Reference Bureau, 2011).)

Figure 9–2 Technological progress, income growth, and poverty reduction. Average annual per capita income and technology growth, 1990–2005, in selected regions. (*Source: Technology & Development (Findings from a World Bank Report). Global Economic Prospects 2008: Technology Diffusion in the Developing World* (Washington, D.C.: World Bank, 2008).)

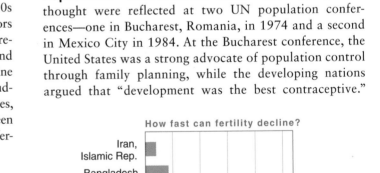

Technological progress, income growth and poverty reduction
Average annual per capita income and technology growth, 1990–2005

would then move through the demographic transition. There are great disparities in economic progress among developing countries, however. A World Bank study of developing economies found that growth in per capita income was strongly correlated with technological progress; both of these measures were at their lowest in three regions with continued high population growth and poverty: sub-Saharan Africa, the Middle East, and Latin America (**Fig. 9–2**).[1]

How did the now-developed nations come through the demographic transition without getting caught in the poverty-population trap? In these countries, the improvements in disease control that lowered death rates (the epidemiological transition) occurred gradually through the 1800s and early 1900s. Industrialization, which introduced factors that lowered fertility, occurred over the same period. Therefore, there was never a huge discrepancy between birth and death rates (see Figure 8–22). In contrast, modern medicine was introduced into the developing world relatively suddenly, bringing about a precipitous decline in death rates, while the fertility-lowering effects of development have been slow in coming. **Figure 9–3** shows how extreme the differences in speed of the demographic transition can be.

Looking at the Data

Demographers and economists try to determine which factors most predict changes in populations and why. Why is it that educated women have fewer children? If the GNP of a whole nation goes up, but inequality is high, will population growth rates still fall? Should governments simply try to increase the family economy for the poorest and assume fertility rates will fall, or do people need specific programs for family planning? Demographers seek answers to these questions, which may vary by country.

Competing Ideas

What do the developing countries need to do to fast-forward their way through the demographic transition? From early on, two basic conflicting theories offered answers to this question:

1. Speed up economic development in the high-growth countries and population growth will slow down "automatically," as it did in the developed countries.

2. Concentrate on population policies and family-planning technologies to bring down birthrates.

Population Conferences. These two schools of thought were reflected at two UN population conferences—one in Bucharest, Romania, in 1974 and a second in Mexico City in 1984. At the Bucharest conference, the United States was a strong advocate of population control through family planning, while the developing nations argued that "development was the best contraceptive."

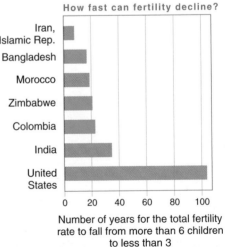

How fast can fertility decline?

Number of years for the total fertility rate to fall from more than 6 children to less than 3

Figure 9–3 Fertility transition times. Countries experience very different rates of change in fertility. (*Source:* www.gapminder.org.)

---

[1]Kavita Watsa, *Technology & Development: Findings from a World Bank Report: Global Economic Prospects 2008: Technology Diffusion in the Developing World* (Washington, D.C.: World Bank, 2008).

SUSTAINABILITY

Looking at Change in India

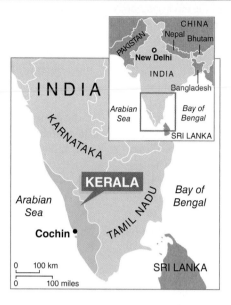

If you were to call any one of several companies for a credit card application, chances are that you might be talking to a young Indian person, one of 13,000 working for Genpact, in a call center in Gurgaon, a small city on the outskirts of Delhi, India. Genpact recruits some 1,100 new employees a month from a pool of well-educated young people who aspire to live in apartments like those of Heritage City. Outside this high-tech enclave, however, is the old India, with families living in shantytowns, sacred cows roaming the streets, garbage and trash everywhere, and shoeless children begging for rupees. This is the stark contrast found in many countries with a history of rapid population growth.

In the nearly 65 years since India gained its independence, its population has more than tripled. With 1.22 billion people already (2012), India is now growing at a rate of 18 million per year and is expected to surpass China's population by 2030. In spite of many decades of family-planning programs, India's population has continued to grow, canceling out much of the country's gains in food production, health care, and literacy. Almost half of India's children under the age of five are malnourished, and 75% of its population lives on less than $2 a day. India has the largest illiterate population in the world. With a per capita gross national income of $1,370 per year, India has a very long way to go to achieve the kind of development that characterizes the industrialized countries. It might be tempting to consider this a hopeless situation were it not for Kerala, one region of India where things are quite different.

Kerala, a state in the south of India, has a population of 33.4 million and an area of 39,900 square miles (about the size of Kentucky). This makes Kerala the second most densely populated state in India and close to the most densely populated region in the world. Situated only 10° north of the equator, Kerala is tropical, with lush plantations of coffee, tea, rubber, and spices in the highlands and rice, coconut, sugarcane, tapioca, and bananas in the lowlands. With the Arabian Sea lapping at Kerala's shore, fishing is an important industry. Kerala is very much like the rest of India in that it is crowded, per capita income is small, and food intake is low. Here, however, the similarities end.

The people of Kerala have a life expectancy of 73 years, compared with 65 for all of India. Infant mortality in the state is 12 per thousand versus 57 per thousand for India. The fertility rate is 1.8 (below replacement level) compared with 2.8 for India. Literacy is near 90%, almost all villages in the state have access to schools and modern health services, and women have achieved high political offices and are as well educated as men. Even though Kerala must be considered a poor region by every economic measure, its people are well on the way to achieving a stable population.

New and old India. Heritage City, an apartment complex in Gurgaon, on the southwest fringe of Delhi, India. Gurgaon has become a corporate and industrial center.

Their resistance to family planning was also bolstered by feelings that the developed world's promotion of population control was another form of economic imperialism or even genocide.

At the conference in Mexico City, the sides were somewhat reversed. Developing countries facing real problems of excessive population growth were asking for more assistance with family planning, whereas the United States, under pressure from anti-abortion advocates, took the position that development was the answer and terminated all contributions to international family-planning efforts, a policy that remained in effect until 1993. The other developed countries, however, remained convinced that family planning was essential and continued to support international efforts to help the developing countries implement policies designed to bring fertility rates down.

Most Recent Thinking. The most recent population conference, the International Conference on Population and Development (ICPD), was held in Cairo in 1994. Poverty, population growth, and development were clearly linked, and resources and environmental degradation were an important focus. The ideological split between developed and developing countries was a thing of the past—all agreed that population growth must be dealt with in order to make progress in reducing poverty and promoting economic development. The responsibility for bringing fertility rates down was placed firmly on the developing countries themselves, although the developed countries pledged to help with technology and other forms of aid.

There was also broad agreement on several main ideas. One of these was that women's rights to health care, education, and employment were foundational to achieving slower population growth. Development must be linked to a reduction in poverty, the participants said. Existing poverty in the developing countries is an affront to human dignity that should not be tolerated. Everyone realized that continued population growth, poverty, and development were a threat to the health of the environment and that only sustainable development would prevent a future of unprecedented biological and human impoverishment.

Five-year anniversary reviews affirmed the universal commitment to the agreements forged at the 1994 conference. Because the Cairo conference produced a 20-year plan, the next such conference is anticipated sometime around 2014.

Demographic Dividend. An interesting thing happens as a country goes through the demographic transition. As birthrates decline, the working-age population increases relative to the younger and older members of the population. This relationship is known as the **dependency ratio** and is defined as the ratio of the nonworking population (under 15 and over 65) to the working-age population (see Table 8–3). For a time, therefore, the society can spend less on new schools and old-age medical expenses and more on factors that will alleviate poverty and generate economic growth—a **demographic dividend.** This demographic window is open only once, however, as the numbers of younger children decrease, and stays open only for a few decades, until the numbers of older people increase. Graying populations then increase the dependency ratio again, though if countries have succeeded in increasing their economy, they will then be in a better position to care for dependents (Chapter 8). For countries well below replacement level, such as some in Europe and Japan, the fertility level may rebound slightly or immigration may increase to level the population out in the long term.

Figure 9–4 shows how dependency ratios have changed over time—and will likely change in the future—for several world regions. A number of East Asian and Latin American countries, such as South Korea, Brazil, and Mexico, have taken advantage of this window. They have put in place population and economic policies that are consistent with poverty reduction and economic expansion, and as a result, they have experienced rapid economic development. The countries of South Asia (for example, Pakistan, India, and Bangladesh) are currently approaching this window of opportunity. Will they make the necessary investments in health, education, and economic opportunities to take full advantage of the demographic window? To do so, they must directly help the poorest people in their societies.

Large and Small Families

The fertility transition is the most vital element in the demographic transition. Many studies have shown that high fertility and poverty are linked, including the correlation between fertility rate and gross national income per capita **(Fig. 9–5)**. Countries on the far left side of this plot are all in the early or middle stages of the demographic transition; most are low-income economies. Those in the middle and to the right are all middle-income economies and are moving through the transition.

To the observer in a developed country, it may seem strange that so many poor women in developing countries have large families. It is obvious from our perspective that more children could spread a family's income more thinly and handicap efforts to get ahead economically. In the United States, an extremely large family is so rare that it may even lead to endorsements of a reality show like *19 Kids and Counting*. In other countries, large families are more common, though in countries with poor health outcomes, members may not live to adulthood. The decision-making process that leads parents to produce large families is different in low-income countries than in the United States, however, because they live under very different sociocultural conditions.

Let's look at some examples. In western Uganda, most families are large and polygamous and make their living in agriculture. Children are an asset in the work force, and most do not receive education **(Fig. 9–6)**. Because girl children bring a bride price to their parents, both boys and girls are valued. An average bride is 16 years old and will have more than six children. In fact, about a third of women will have 10 or more children. Some women have access to family planning but do not use it, and many people see the immediate need for help in the fields as more important than their children's long-term options.

Afghanistan is another good example. In 1950, the population was 8 million; in 2009, 28 million; and estimates for 2030 range between 43 million and 50 million. Half of all Afghani adults are between the ages of 15 and 29. In comparison, that age group makes up 26% of the population in the United States. Just as American college students enter the work

Figure 9–4 The demographic window. As countries experience gradually decreasing fertility, the dependency ratio declines and opportunities for development increase. The plot compares several developing regions with some developed countries. (*Source: World Data Sheet 2011* and the World Economic Outlook Database, September 2011.)

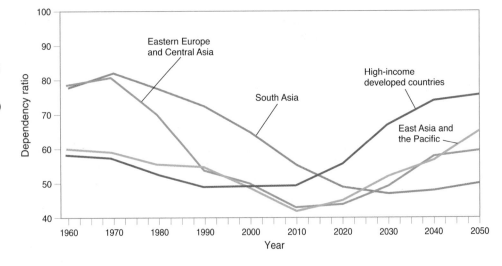

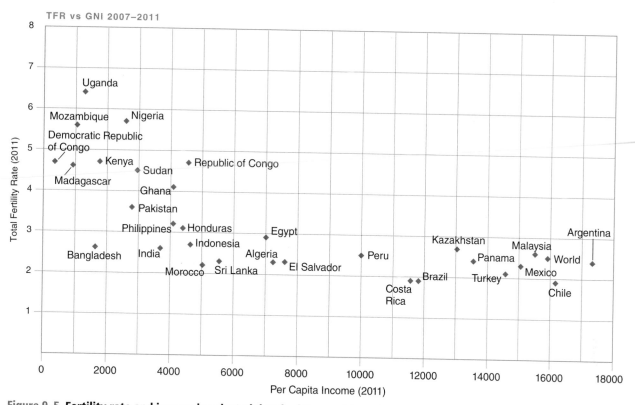

Figure 9–5 Fertility rate and income in selected developing countries. There is a correlation between income (gross national income in purchasing power parity per capita) and lower total fertility. Factors that affect fertility more directly are health care, education for women, and the availability of information and services related to contraception. (*Source: World Data Sheet 2011* and the World Economic Outlook Database, September 2011)

force uncertain if there will be jobs open to them, young Afghans feel that pressure many times over. A number of authors have described the dilemmas faced by Afghan families: War, poverty, lack of human rights, uncertainty about the future, and a lack of education combine to make life difficult. A typical Afghan woman **(Fig. 9–7)** might marry a man a dozen years older than herself and bear children as soon as she is able to get pregnant. She will have few rights to her own decision making, little schooling, and heavy social pressure to bear sons. She may not even know about contraception or may live in a culture that

doesn't accept it. Because of harsh conditions, both mothers and infants die in the birth process at higher rates than in much of the world. One in eight Afghan women is likely to die of pregnancy-related diseases. In spite of all of these difficulties, fertility in Afghanistan is currently declining and contraceptive use rose from 5% to 22% in some rural areas between 2003 and 2011.

These stories of Uganda and Afghanistan put a human face on high fertility. Numerous studies and surveys reveal the following as primary reasons that the poor in developing countries have large families.

Figure 9–6 Children as an economic asset. Children working fields in Myanmar. In many developing countries, children perform adult work and thus contribute significantly to the income of the family.

Figure 9–7 Young parenthood. Marriage by young girls (often to older husbands) is common in war-torn Afghanistan, where this mother and child live. A young age at first reproduction and lack of medical care cause high maternal and infant mortality. They also contribute to high growth rates.

1. **Old age security.** A primary reason given by poor women in developing nations for having many children is to ensure that they will be cared for in old age.

2. **Infant and childhood mortality.** The common experience of children dying leads people to try to make sure that some of their children will survive to adulthood.

3. **Helping hands.** In subsistence-agriculture societies like rural Uganda, women do most of the work relating to the direct care and support of the family. Very young children help with chores and fieldwork. In short, children are seen as a productive asset.

4. **Status of women.** The traditional social structure in many developing countries still discourages women from obtaining higher education, owning businesses or land, and pursuing many careers. As is often the case in Afghanistan, respect for a woman increases as she bears more children.

5. **Availability of contraceptives.** Poor women often lack access to reproductive health information, services, and facilities. Providing contraceptives to women is a major facet of family planning. Studies show a strong correlation between lower fertility rates and the percentage of couples using contraception **(Fig. 9–8).** In Thailand, for example, the availability of contraception was a large part of the fertility transition. Worldwide about 215 million women would like to delay or prevent pregnancy but are unable to do so. This is an unmet need for contraception.

Conclusions. The five factors supporting large families are common to preindustrialized, agrarian societies. With industrialization, urbanization, and development, however, come conditions conducive to having small families. These include: the relatively high cost of raising children, the existence of pensions and a Social Security system, opportunities for women to make money, access to affordable contraceptives, adequate health care, educational opportunities, and an older age at marriage.

Fertility rates in developing countries remain high not because people in those countries are behaving irrationally or irresponsibly, but because the sociocultural climate in which they live has favored high fertility for years and because contraceptives are often unavailable. Sometimes the rise in population triggers a vicious cycle: poverty, environmental degradation, and high fertility drive one another forward **(Fig. 9–9).** Increasing population density leads to a greater depletion of rural community resources like firewood, water, and land, which encourages couples to have more children to help gather resources, and so on. They are in a poverty trap. In addition, rapid population growth can increase inequality in a society because the abundance of labor lowers the price of labor. This feeds the same cycle, as inequality also promotes poverty.

Thus, it is not just economic development that leads to declining fertility rates. Rather, fertility rates decline insofar as development provides (1) security in one's old age apart from the help of children, (2) lower infant and childhood mortality, (3) universal education for children, (4) opportunities for higher education and careers for women, and (5) unrestricted

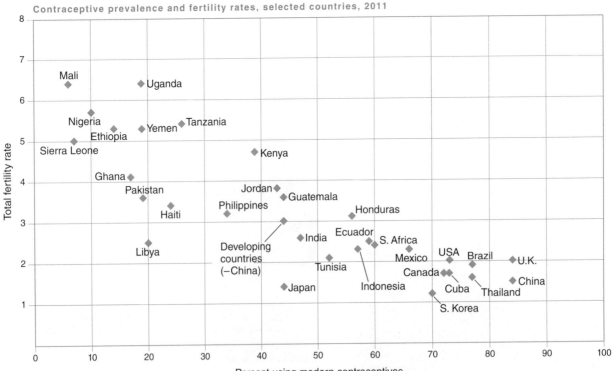

Contraceptive prevalence and fertility rates, selected countries, 2011

Figure 9–8 Prevalence of contraception and fertility rates. More than any other single factor, lower fertility rates are correlated with the percentage of the population using contraceptives. (*Source:* Data from *2011 World Population Data Sheets* (Washington, D.C.: Population Reference Bureau, 2011).)

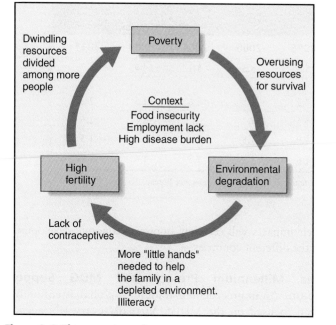

Figure 9–9 The poverty cycle. Poverty, environmental degradation, and high fertility rates become locked in a self-perpetuating vicious cycle.

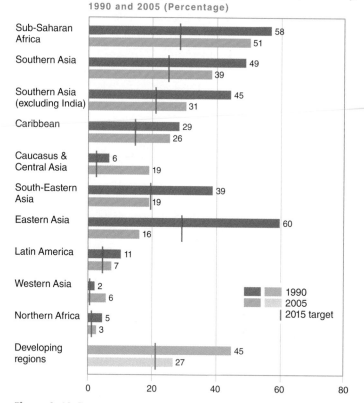

Proportion of people living on less than $1.25 a day, 1990 and 2005 (Percentage)

Figure 9–10 Extreme poverty. Extreme poverty rates around the world are falling, though numbers are still high. (*Source: Millennium Development Goals Report 2011* (New York: United Nations, 2011).

access to contraceptives and reproductive health services. These are all outcomes of public policies, and countries and regions (like Kerala and Thailand) that have made them a high priority have moved into and through the demographic transition.

9.2 Promoting Development

The good news is that many developing countries have made remarkable *economic* progress. The gross national products of some countries have increased as much as fivefold, bringing them from low- to medium-income status, and some medium-income nations have achieved high-income status. India, for example, has dramatically lowered its poverty rate—from 52% to almost 22%—despite having huge numbers of people in poverty. Although the world economy is still strongly dominated by the developed countries, the developing countries have become more and more involved in what is now an integrated global economy. At the heart of this economy is technology, and as Figure 9–2 indicates, those developing regions showing the highest growth in income also show the greatest technological progress. There is still a great gap between rich and poor countries in income, as well as technology, but the percentage of impoverished people in the world—those living on less than $1.25 per day (the new *extreme poverty* criterion)—has dropped from 42% to 21.3% of the developing world's population in the two decades since 1990. This is further expected to drop to 14.4 % by 2015 **(Fig. 9–10)**.

Tremendous *social* progress has been made in many developing countries, too. Efforts from many branches of the United Nations have aided the work of various government programs. Private charitable organizations and

nongovernmental organizations (NGOs) have also played a large role. Literacy rates, the percentage of the population with access to clean drinking water and sanitary sewers, and other social indicators of development have generally improved. Furthermore, the fertility rates of most developing countries have declined, although they are still far from the replacement level (Table 9–1).

The successes just described, however, are dulled by some sobering facts. A sixth of the world's population (1.2 billion people) still lives on less than $1.25 a day, nearly 1 billion lack access to clean water, more than 1.2 billion are living in urban slums under deplorable conditions, and more than 1 billion are malnourished. These are the world's poorest people. As Nelson Mandela, former leader of South Africa, put it, "Massive poverty and obscene inequality are such terrible scourges of our times—times in which the world boasts breathtaking advances in science, technology, industry and wealth accumulation—that they have to rank alongside slavery and apartheid as social evils."[2]

Is the solution to these problems simply to facilitate economic growth and the alleviation of poverty will follow? The United Nations Fund for Population Activities (UNFPA) puts it the following way in *State of World Population 2002*: "Development has often bypassed the poorest people, and has even increased their disadvantages. The poor need direct

[2]Speech in London's Trafalgar Square for the Campaign to End Poverty in the Developing World. BBC News, February 3, 2005.

TABLE 9–1 Decline in Total Fertility Rates

	1985	1990	1995	2000	2005	2008	2011
Africa	6.3	6.2	5.8	5.3	5.1	4.9	4.7
Latin America and the Caribbean	4.2	3.5	3.1	2.8	2.6	2.5	2.2
Asia (excluding China)	4.6	4.1	3.5	3.3	3	2.8	2.6
China	2.1	2.3	1.9	1.8	1.6	1.6	1.5
Developed countries	2	2	1.6	1.5	1.5	1.6	1.7

Source: Data from various *World Population Data Sheets* (Washington, D.C.: Population Reference Bureau).

action to bring them into the development process and create the conditions for them to escape from poverty."[3] In fact, as we have already seen, a country can have increases in wealth, but many people remain in poverty if inequality is high. To alleviate poverty is the explicit objective of the Millennium Development Goals.

Millennium Development Goals

In 1997, representatives from the United Nations, the World Bank, and the Organisation for Economic Co-Operation and Development (OECD) met to formulate a set of goals for international development that would reduce the extreme poverty in many countries and its various impacts on human well-being. The goals were sharpened and then adopted as a global compact by all UN members at the UN Millennium Summit in 2000. Table 9–2 lists the eight *Millennium Development Goals (MDGs)* and 17 targets. As well as targets, each goal has definite indicators for monitoring progress. The indicators are measurable and provide feedback to all who are concerned about whether progress is being made.

The goals reinforce each other and should all be worked on together. For example, the indicators of progress for Goal 4 (*reduce child mortality*) are the infant mortality rate, the under-five mortality rate, and the proportion of one-year-old children immunized against measles. Achieving Goals 5 and 6 (*improve maternal health* and *combat HIV/AIDS, malaria, and other diseases*) should go far toward accomplishing Goal 4.

Most importantly, the MDGs are a clear set of targets for the developing countries themselves. Each country is expected to work on its own needs, but it is expected that all of the countries will do so in partnership with developed countries and development agencies, consistent with Goal 8 (*forge a global partnership for development*). The low-income countries simply lack the resources to go it alone. It is important to note that the MDGs are not focused on the economic development normally tracked by measures like gross national income. They specifically address extreme poverty wherever it is found. As the goals are achieved, however,

their impacts will certainly improve economic development in the different countries.

The Millennium Project and MDG Support Team. An unprecedented amount of global attention has been focused on the MDGs. Virtually every UN agency, scores of NGOs, and government agencies of many countries have produced strategies and directives that demonstrate how they are mobilizing to achieve the MDGs. In the midst of this activity, the **Millennium Project** was given the mandate to develop a coordinated action plan that other agencies and organizations can consult as they address the MDGs. The Millennium Project, commissioned by the UN Secretary-General in 2002, was an independent advisory body headed by economist Jeffrey Sachs, a world expert on poverty, and involved some 250 development experts from around the world. Several years later, it appears that some of the MDGs will be met by 2015, but others will not. The Millennium Project was succeeded by the **MDG Support Team**, operating under the mandate of the United Nations Development Program (UNDP). The MDG Support Team works with countries by invitation, helping them to prepare and implement national strategies for achieving the MDGs, developing tools and methods specific for that country, coordinating technical expertise across the UN family of agencies, and obtaining the financing needed for support of MDG work in the recipient countries.

Progress Toward the MDGs. A recent summary of global and regional progress toward the MDGs is found in *The Millennium Development Goals Report 2011*, produced by statisticians from a host of UN agencies. Now that we are close to the target date of 2015, how is it going?

Extreme Poverty. The first goal (*eradicate extreme poverty and hunger*) aims to reduce by half the proportion of people whose income is less than $1 (now $1.25) a day. This goal has almost been met at the global level, and by 2015, the number in this poorest group should fall to below 900 million. Unfortunately, the worldwide recession that began in 2008 slowed economic growth. Furthermore, there are major regions—in particular, sub-Saharan Africa and Southern Asia—that are far behind on this goal.

[3]UNFPA *State of World Population 2002. People, poverty, and possibilities: making development work for the poor.* Ed Thoyaya Ahmed Obaid. UN Population Fund.

STEWARDSHIP

China's Population Policies

When a country has un-controlled population growth, it isn't always obvious how to respond. Increasing social modernization may help but isn't guaranteed to work; a country may be headed into a poverty trap. How countries respond to un-checked population growth may have a great deal to do with the level of centralization and the culture. If ignoring the problem isn't reasonable, and simply letting people die of starvation and disease is unethical (as most people would argue), then governments have to decide how strongly they will act to lower population growth.

The most acceptable option—the one that has been the theme of this chapter—is to create an economic and social climate in which people, of their own volition, will desire to have fewer children. Along with such a climate, society will provide the means (family planning) to enable people to exercise that choice. This option has been realized "unconsciously" in developed countries, but it raises certain ethical dilemmas when you attempt to apply it to developing countries.

China is a good example of a country whose government has pursued policies with which many people are uncomfortable. With its current population of 1.3 billion (a fifth of the world's people), China offers extensive economic incentives and disincentives aimed at reducing population growth. This is in part because China was in an intolerable position. Under the Communist leader Mao Tse-tung, the government encouraged people to have large families. During the early Communist years, the country went through an epidemiological transition, but not a fertility transition. While Chairman Mao ruled, China's population increased from 550 million to 900 million people.

However, during the '40s and '50s, the government pursued a series of unfortunate plans for industry and agriculture that, along with mismanagement and natural disasters, led to famine. Between 1958 and 1960, crop production dropped by 70%. Between 1958 and 1961, the greatest level of starvation in human history occurred in China; between 20 million and 43 million people died. By the early '70s, China's leaders recognized that, unless population growth was stemmed, the country would be unable to live within the limits of its resources. They began a voluntary family planning program of clinics and education, with some success. In

1979, they introduced the one-child policy for mostly urban families. Exceptions were made for rural families, those who had daughters first, families in which both parents are only children, and ethic minorities. Only about 40% of married couples actually are obligated to this policy; others are allowed two and sometimes more children. To achieve population goals, the government instituted an elaborate array of incentives and deterrents. The prime *incentives* were as follows:

- Paid leave to women who have fertility-related operations—namely, sterilization or abortion procedures;

- A monthly subsidy to one-child families;

- Job priority for only children;

- Additional food rations for only children;

- Housing preferences for single-child families; and

- Preferential medical care to parents whose only child is a girl. (There is a strong preference for sons in China, and parents generally wish to have children until at least one son is born.)

Penalties for an excessive number of children in China included the following:

- Bonuses received for the first child must be repaid to the government if a second is born.

- A tax (now called a "social compensation fee") must be paid for having a second child.

- Higher prices must be paid for food for a second child.

- Maternity leave and paid medical expenses apply only to the first child.

Besides improving economic opportunities, these incentives and deterrents have helped China achieve a precipitous drop in its total fertility rate, from a rate of about 4.5 in the mid-1970s to a rate of 1.5 in 2011. The policy is unpopular, but without it, China's population would be 400 million larger. In practice, the one-child policy has been rigorously pursued only in urban areas, home to some 36% of the population. People in cities, already feeling crowded, have experienced a drop in the desired size of their families as well. People in rural areas and minorities (who are exempted from the policy) have a higher fertility rate. The policy has been extremely controversial. Local governments in China, tasked with keeping population growth down, carried out egregious human rights violations. These included practices

such as razing the homes of people with multiple children and forced sterilizations and abortions that caused outrage in China and in the world community. In the past decade, rapid economic growth and more personal freedoms have made the Chinese people harder to control. Many are willing to pay the penalties, more people know about human rights violations through social media, and there have been activists fighting for reproductive rights for Chinese citizens.

Despite such controls, the population of China is still growing due to momentum and will continue to do so. One disturbing consequence of China's policy is the skewed ratio of males to females. Males are preferred in Chinese families, so some parents try to make sure that their one child is a male, through such extreme measures as selective abortion and even infanticide of female children. Loss of females is common in India as well and in some other countries where sons are strongly preferred. Unfortunately, this means that there are now millions more young men than women. There are concerns that this will cause social unrest.

You could look at the ethical dilemmas posed by China's one-child policy as competing claims on stewardship. In order to protect resources for current and future people, China chose not to preserve individual freedom. That is, to have the option of supporting everyone, they chose to remove the option of making individual decisions about family size and, in some cases, intimate rights to protection from sterilization and abortion. In most of the world, such a trade-off is viewed as abhorrent. In China, people have a mixed view of the policy. China's population policy has been a case study for environmental ethicists and a topic of passion for human rights workers since its inception. However, the Cairo population conference made it clear that on a worldwide basis, most countries believe people should be able to make individual choices about family size.

TABLE 9–2 Millennium Development Goals and Targets (Conditions in 1990 Are Compared with Those in 2015, the Year for Reaching the Targets)

Goal 1. Eradicate extreme poverty and hunger.
- **Target 1:** Reduce by half the proportion of people whose income is less than $1 (now $1.25) a day.
- **Target 2:** Reduce by half the proportion of people who suffer from hunger.

Goal 2. Achieve universal primary education.
- **Target 3:** Ensure that all children, boys and girls alike, are enrolled in primary school.

Goal 3. Promote gender equality and empower women.
- **Target 4:** Eliminate gender disparities in primary and secondary education by 2005 and in all levels of education by 2015.

Goal 4. Reduce child mortality.
- **Target 5:** Reduce mortality rates by two-thirds for all infants and children under five.

Goal 5. Improve maternal health.
- **Target 6:** Reduce maternal mortality rates by three-quarters.
- **Target 7:** Achieve, by 2015, universal access to reproductive health.

Goal 6. Combat HIV/AIDS, malaria, and other diseases.
- **Target 8:** Have halted, by 2015, and begun to reverse the spread of HIV/AIDS.
- **Target 9:** Achieve, by 2010, universal access to treatment for HIV/AIDS for all those who need it.
- **Target 10:** Have halted, by 2015, and begun to reverse the incidence of malaria and other major diseases.

Goal 7. Ensure environmental sustainability.
- **Target 11:** Integrate the principles of sustainable development into country policies and programs, and reverse the loss of environmental resources.
- **Target 12:** Reduce by half the proportion of people without sustainable access to safe drinking water and basic sanitation.
- **Target 13:** Achieve a significant improvement in the lives of at least 100 million slum dwellers by 2020.

Goal 8. Forge a global partnership for development.
- **Target 14:** Develop a trading system that is open, rule based, and nondiscriminatory—one that is committed to sustainable development, good governance, and a reduction in poverty.
- **Target 15:** With Official Development Assistance (ODA), address the special needs of the least developed countries (including the elimination of tariffs and quotas for their exports and an increase in programs for debt relief).
- **Target 16:** Deal with the debt problems of all developing countries by taking measures that will make debt sustainable in the long run.
- **Target 17:** In cooperation with the private sector, make available the benefits of new technologies, especially information and communications technologies.

Sources: United Nations, World Bank, Organisation for Economic Co-Operation and Development, Millennium Ecosystem Assessment.

Goals Affecting Children. Another goal that will likely be met (because of progress already made) is Goal 2 (*achieve universal primary education*). In fact, sub-Saharan Africa, which consistently ranks in the lowest on many quality of life measures, increased primary school enrollment from 58% to 76% in only a decade. The goal that seems least likely to be achieved is Goal 4 (*reduce child mortality*), with the target of reducing under-five mortality rates by two-thirds (**Fig. 9–11**). More than one-third of these deaths are associated with malnutrition, a clear link to Goal 1, Target 2 (*reduce by half the people who suffer from hunger*). In sub-Saharan Africa, though child mortality has fallen substantially, one in eight children still die before the age of five. All regions of the world except for Oceania, sub-Saharan Africa, and Southern Asia have seen drops in child mortality of 50% or more. As of 2011, researchers thought it was unlikely but still possible to reach the MDG by 2015, with aggressive action.

Some Other Successes. The *Millennium Development Goals Report 2012* lists a number of additional successes. For example, in all but two regions, primary school enrollment is 90% or higher. The number of deaths from measles has fallen dramatically; about 80% of children in developing countries are now immunized (**Fig. 9–12**). Many more women have skilled attendants during childbirth; 81% of pregnant women have some prenatal care. In fact, the number of maternal deaths from childbirth has been halved in less than 20 years. Water availability is also a success. At least 1.6 billion people have gained access to safe drinking water since 1990, making Target 7c of the MDGs one of the first attained.

Unmet Goals. The report also lists areas where greater effort is required if goals are to be met, particularly nourishment of children, ensuring environmental sustainability, lowering maternal mortality, and access to sanitation. The goal that has the greatest number of targets worsening instead of improving is Goal 7 (*ensure environmental sustainability*).

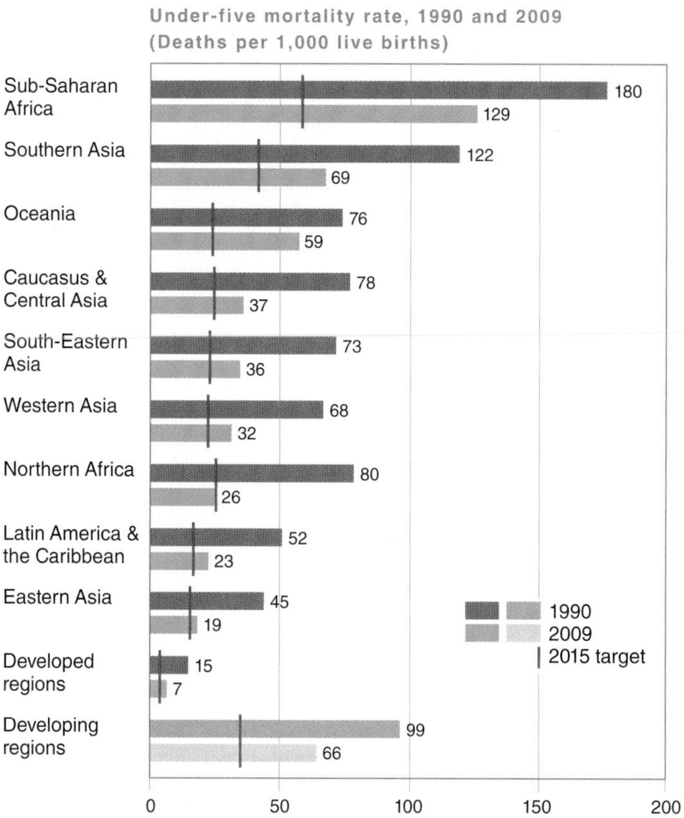

Under-five mortality rate, 1990 and 2009
(Deaths per 1,000 live births)

Sub-Saharan Africa — 180 (1990), 129 (2009)
Southern Asia — 122 (1990), 69 (2009)
Oceania — 76 (1990), 59 (2009)
Caucasus & Central Asia — 78 (1990), 37 (2009)
South-Eastern Asia — 73 (1990), 36 (2009)
Western Asia — 68 (1990), 32 (2009)
Northern Africa — 80 (1990), 26 (2009)
Latin America & the Caribbean — 52 (1990), 23 (2009)
Eastern Asia — 45 (1990), 19 (2009)
Developed regions — 15 (1990), 7 (2009)
Developing regions — 99 (1990), 66 (2009)

Legend: 1990, 2009, 2015 target

Scale: 0, 50, 100, 150, 200

Figure 9–11 Millennium Development Goal 4, Target 5. As a result of efforts toward this goal—*reduce child mortality*—deaths among children under five finally dropped below 10 million. About half of these deaths occur in sub-Saharan Africa. Death from measles has dropped dramatically; 80% of children in developing countries now receive a measles vaccine. (*Source: Millennium Development Goals Report 2011* (New York: United Nations, 2011).)

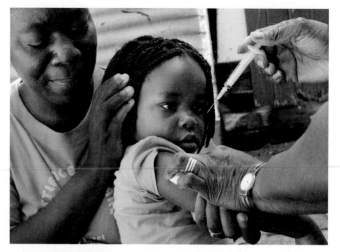

Figure 9–12 Measles vaccination.
One of the successes of the Millennium Development Goals has been the dramatic drop in measles. More than 80% of children are vaccinated worldwide, just as this child is in a primary school in Uganda.

Freshwater resources, forests, and fish stocks are all declining, and species extinctions and greenhouse gasses are increasing. However, on the good side, ozone-depleting CFCs have declined, and many more people have access to improved water.

Individual countries. Global and regional reports are important in providing a general picture of progress being made toward the MDGs, but the real action is in the individual countries. Toward that end, some 140 countries now produce reports, coordinated and often funded by the UNDP. Some, such as Albania's, are available on the Internet. It is the responsibility of the developing countries themselves to manage their own economic and social development and to create a climate that will facilitate the partnership envisioned in Goal 8 of the MDGs. They must also devote significant financial resources toward achieving the goals. The best estimates of the Millennium Project suggest that for the lowest-income countries, the costs will have to be split evenly between domestic sources and Official Development Aid (ODA) from wealthy donor countries. At a September 2010 Millennium Development Goals Summit at UN Headquarters, a coalition of governments, foundations, and businesses pledged an estimated $40 billion in new commitments over five years to meet the MDGs. This foreign investment and participation in global markets is important because it directly stimulates and helps grow the economies of the developing countries. More important yet to the poorest people, however, is the work of various organizations that directly provide assistance and funnel aid to the developing countries. These agencies are discussed next, followed by the issues of development aid and the debt crisis.

World Agencies at Work

In 1944, during World War II, delegates from around the world met in New Hampshire and conceived a vision of development for the countries ravaged by World War II. They established the International Bank for Reconstruction and Development (now called the **World Bank**). The World Bank functions as a special agency under the UN umbrella, owned by the countries that provide its funds. As the more developed countries of Europe and Asia got back on their feet, the World Bank directed its attention toward the developing countries. With deposits from governments and commercial banks in the developed world, the World Bank now lends money to governments (and only governments) of developing nations for a variety of projects at interest rates somewhat below the going market rates. In effect, the World Bank helps governments borrow large sums of money for projects they otherwise could not afford.

Annual loans from the World Bank have climbed steadily, from $1.3 billion in 1949 to $43 billion in 2011. Through its power to approve or disapprove loans and the amount of money it lends, the World Bank has become the major single agency providing aid to developing countries for the past 60 years. The World Bank is actually five closely associated agencies. They have different roles, as you can see in **Figure 9–13**. One typical action of World Bank agencies is providing loans for a wide range of development activities including supporting the Millennium Development Goals.

The **International Bank for Reconstruction and Development (IBRD)** lends to governements of middle-income and creditworthy low-income countries.

The **International Development Association (IDA)** provides interest-free loans—called credits—and grants to governments of the poorest countries.

The **International Finance Corporation (IFC)** provides loans, equity and technical assistance to stimulate private sector investment in developing countries.

![MIGA]

The **Multilateral Investment Guarantee Agency (MIGA)** provides guarantees against losses caused by non-commercial risks to investors in developing countries.

![globe]

The **International Centre for Settlement of Investment Disputes (ICSID)** provides international facilities for conciliation and arbitration of investment disputes.

Figure 9–13 Roles of the World Bank. The World Bank consists of five different agencies with different roles.

Bad Bank, Good Bank. It is inaccurate either to credit the World Bank for all the progress made in development or to blame it for all areas in which progress has been lacking. However, critics cite many examples wherein the bank's projects have actually exacerbated the cycle of poverty and environmental decline.

Bad Bank. The World Bank was criticized for a 2010 loan of $3.75 billion to South Africa to built the world's fourth largest coal-fired power plant. The country had been in an electricity crisis, and major investments in electricity generation had not been made in decades. The plant will increase electricity production but greatly increase harmful environmental effects from coal use (Chapter 14). In the 1990s, the World Bank pressured Bolivia to open its water sources to multinational purchasers. The thinking was that local governments are often too corrupt or lack management expertise to provide water, and privatization would bring infrastructure and skills. The World Bank threatened to not renew a large loan unless water rights were privatized. This led to civil unrest when prices rose and poor people had limited access to water (**Fig. 9–14**). In other situations such as hydroelectric dams in a number of countries, projects funded by the World Bank have displaced people and exacerbated their poverty.

These and many past projects funded by the World Bank have been destructive to the environment and often to the poorer segments of society. For example, the World Bank funneled $1.5 billion into Latin America from 1963 to 1985 to clear millions of acres of tropical forests. Most of the cleared land was given over to large cattle operations that produced beef for export. In other countries, projects have emphasized growing cash crops

for export and fostering huge mechanized plantations, while leaving the poor marginalized. At that time, the World Bank procedures considered environmental concerns as an add-on rather than as an integral component of its policies.

Good Bank. There has been a sea change in recent years in how the World Bank does business with the developing world. In the words of former World Bank President James D. Wolfensohn, "Poverty amidst plenty is the world's greatest challenge, and we at the Bank have made it our mission to fight poverty with passion and professionalism. This objective is at the center of all of the work we do." Toward that end, the bank helped initiate the Millennium Development Goals. It now requires *Poverty Reduction Strategy Papers* for all countries receiving IDA loans. The World Bank's *Country Assistance Strategy* identifies how its assistance reduces poverty and establishes and maintains sustainable development. Unfortunately, the World Bank is willing to suspend some of its requirements about sustainability and justice for projects in the poorest nations, a strategy that has provoked backlash.

Environmental Strategy. As evidence of a change in policies, the World Bank adopted a new environmental strategy, *Making Sustainable Commitments: An Environmental Strategy for the World Bank*, in 2002. The stated goal of the strategy "is to promote environmental improvements as a fundamental element of development and poverty reduction strategies and actions." The strategy "recognizes that sustainable development, which balances economic development, social cohesion, and environmental protection, is fundamental to the World Bank's core objective of lasting poverty alleviation."

An analysis of more than 11,000 projects funded by the World Bank to address poverty reduction or biodiversity found that these goals were often mutually reinforcing. The analysis showed that, depending on the different objectives, some 60% to

Figure 9–14 Bolivian water wars. This 2000 scene in Cochabamba Bolivia, features people protesting privatization of water, including a man holding a sign that says "what's ours is ours and it cannot be taken away."

85% of all projects were rated as satisfactory or highly satisfactory. The South Africa coal plant would be an exception.

Other UN Agencies. To quote its mission statement, the "UNDP is the UN's global development network, advocating for change and connecting countries to knowledge, experience and resources to help people build a better life."[4] The agency has offices in 131 countries and works in 166. It helps developing countries attract and use development aid effectively. As we have seen, its network also coordinates global and national efforts to reach the MDGs. The UN Food and Agricultural Organization (FAO) and the World Health Organization (WHO) are other important agencies whose work will be discussed in future chapters.

How Useful Is the UN? The United Nations and its various agencies frequently come under attack in the United States, where critics charge that the organization is useless to us. These critics would like to reduce or eliminate foreign aid. The fact is that, without the work of the World Bank and UN agencies like the UNDP and the FAO, the world would be in much worse shape than it is. We are fortunate that we do not need development aid, but billions of others around the world depend on such aid for continued economic development, especially as it is reflected in the MDGs. This is not a luxury but a necessity in this globally connected world where so many suffer from the effects of poverty.

We turn now to the debt crisis and to the role of conflict in poverty. To some extent, this crisis is an outcome of World Bank policies, but it is also a consequence of development aid in general.

The Debt Crisis

The World Bank, many private lenders, and other nations lend money to developing countries. In fact, China alone lends as much as the World Bank. Theoretically, development projects are intended to generate additional revenues that would be sufficient for the recipients to pay back their development loans with interest. Unfortunately, however, a number of things have gone wrong with this theory, such as corruption, mismanagement, and honest miscalculations.

Over time, developing countries as a group have become increasingly indebted. Their total debt reached $4.08 trillion in 2010. Interest obligations climb accordingly, and any failure to pay interest gets added to the debt, increasing the interest owed—the typical credit-debt trap. For example, Pakistan had an external debt of $52 billion in 2009, and it is expected to reach $73 billion by 2016. Pakistan has taken on heavy debt since about 1958. Since then, it has paid and repaid loans many times over in the cost of interest. Servicing the debt is so difficult that the government has sold off assets, including state-owned services, which has left thousands jobless.

A Disaster. The debt situation continues to be an economic, social, and ecological disaster for many developing countries. In order to keep up even partial interest payments, poor countries often do one or more of the following:

1. **Focus agriculture on growing cash crops for export.** Transitioning from growing food for local families to large scale operations growing food for sale abroad clearly affects the country's ability to feed its people. In fact, indigenous people bear the brunt of this change as sometimes their land is taken and used for crops for export.

2. **Adopt austerity measures.** Government expenditures are reduced so that income can go to pay interest. But what is cut? Usually, it is funds for schools, health clinics, police protection in poor areas, building and maintenance of roads in rural areas, and other goods and services that benefit not only the poor, but also the country as a whole (Fig. 9–15).

3. **Invite the rapid exploitation of natural resources (for example, logging of forests and extraction of minerals) for quick cash.** With the emphasis on quick cash, few, if any, environmental restrictions are imposed. Thus, the debt crisis has meant disaster for the environment.

In essence, these measures are all examples of liquidating ecosystem capital to raise cash for short-term needs. They do not represent sustainability. As is typical with the credit trap, many countries have paid back in interest many times what they originally borrowed, yet the debt remains. It has become obvious that many countries would never be able to repay their debts.

Debt Cancellation. The World Bank is addressing the problems of debt and poverty directly through two initiatives, one to provide small amounts of money to the poorest people and the other to relieve debt in the poorest counties. The *Consultative Group to Assist the Poorest* (CGAP) is designed to increase access to financial services for very poor households through what is called **microlending**—lending of small amounts of money to people with very little (to be discussed shortly). The *Heavily Indebted Poor Country* (HIPC) initiative and its related

Figure 9–15 Unemployment.
Romania, the second poorest European nation, has unemployement of 80% in some villages. Unemployment is particularly high among ethnic minorities such as these Sinti, camped in Alunis, Romania. Austerity measures can worsen unemployment, keeping national debt down but making the lives of people hard.

[4]UNDP Washington Representation Office. http://www.us.undp.org/Washington Office/Mission.shtml. July 5, 2012.

program, *Multilateral Debt Relief Initiative* (MDRI), address the debt problem of the low-income developing countries (mostly from sub-Saharan Africa). To qualify, countries have to demonstrate a track record of carrying out economic and social reforms that lead to greater stability and alleviate poverty. By 2011, 32 countries had gone through the application process and, proving themselves stable, had received more than $100 billion in debt relief. In most countries that qualified for debt relief, spending on education and health has risen substantially.

Zambia. To demonstrate how HIPC has worked, consider the case of Zambia. Before HIPC, Zambia's debt was $7.1 billion. Through HIPC, that country's total debt was reduced to $500 million. Although this sounds like a tremendous break for Zambia, the country routinely experiences food shortages, it has a gross national income (GNI) of close to $1,070 (2010) per capita, and servicing this debt was costing the country twice what it spends on education, more than 7% of its GNI. Thus, the news of debt cancellation was received with joy in Zambia, and the government responded by introducing free health care for people in rural areas of the country.

Development Aid

College students and their parents know all about the differences among scholarships, grants, and loans as they apply for financial aid. Paying back college loans is no picnic for the thousands of students who accept them in order to afford their education. It is all called "aid," but the best kind of aid is the grant or scholarship, which does not have to be repaid. The developing countries are well aware of the differences, too, and have been receiving grant-type aid from private and public donors that is essential to their continued development. How much aid have they been given?

Official Development Assistance. The best records are for official development assistance (ODA), coming from donor countries. The total amount in 2011 was $129 billion, part of an encouraging trend following a decade of declining aid in the 1990s. Aid rises and falls with not only the need of low-income countries but also other world events, conflicts, and donor interests.

Aid as a percentage of the GDP of donor countries declined to 0.22% around 1997 and then began to rise to its present rate of 28%. At the Rio Earth Summit of 1992, the aid target for donor countries was 0.7% of GNI. Only four European countries have met or exceeded the 0.7% target. At the 2005 Paris Declaration on Aid Effectiveness, representatives from other donor countries pledged to raise their ODA to 0.35% by 2010, well short of the 0.7% target, but a significant increase. Norway, Denmark, Luxembourg, the Netherlands, and the United States exceeded their goals, the United States achieving its donor goal a year early. Switzerland, South Korea, and Canada met their goals; the other donor countries did not.

Some of the shortfall in aid is explained as **donor fatigue,** when high-income countries have tired of handing out money. They see themselves asked for multimillion-dollar rescue packages, only to have a large percentage go to fill the pockets of corrupt leaders or to cover administrative costs. Steps have recently been taken to minimize this misdirection of funds. Allocation of aid is increasingly being tied to policy reform in the recipient countries.

Migration and Remittances. Some 215 million people live outside the country of their birth, and many more second-generation immigrants maintain ties to their native countries. Most of these migrants left their home country looking for employment; wage levels in developed countries are five times those of the countries of origin of most migrants. The migrants typically send much of their money home—**remittances,** as they are called. This is a surprisingly large flow of funds—estimated at $325 billion in 2010 (just to developing countries) and more than double the size of international aid flows. There is no doubt that these funds alleviate the severity of poverty in the developing countries and can be thought of as a "do it yourself" kind of development aid. **Figure 9–16** shows how important remittances are to the economy of Haiti, for example.

Aid from the United States. How much aid does the United States give? Our ODA donations were $28.8 billion in 2009 ($30.2 billion in 2010), the largest of any donor, but only 0.21% of our GNI. Norway, by contrast, gives the equivalent of 1.06% of its GNI in development aid. To put it in its proper perspective, though, contributions from religious organizations, foundations, corporations, nongovernmental organizations, and individuals in the United States provided some $37.5 billion in aid in 2009, outstripping our ODA. When remittances and private capital investments are added to these, the total economic contributions from the United States to developing countries amount (for 2009) to $226.7 billion (Table 9–3); however, the global financial crisis dramatically lowered some sources of philanthropic giving, so this total is actually lower than the 2007 total ($235 billion).

Global Financial Crisis. Developing world economies experienced record economic growth during 2002–2007, a result of increased export earnings and substantial foreign direct investment. The collective GNI of developing countries grew more than 5% per year, peaking at 8% in 2006 and well outpacing the developed countries. However, in late 2007, the U.S. housing bubble burst, leading to a collapse of financial markets that spread to other developed countries. The United States fell into a severe recession, and the recession spread globally. In 2011, the financial crisis deepened in the Eurozone, the European Union. This global crisis will likely perturb economies in developing countries in a number of ways, even though they are not as vulnerable as the developed countries to the meltdown of financial markets. There will be a reduction in their exports, there will be much less foreign investment in their markets, and the GNI growth they had recently experienced will likely decline. The developed

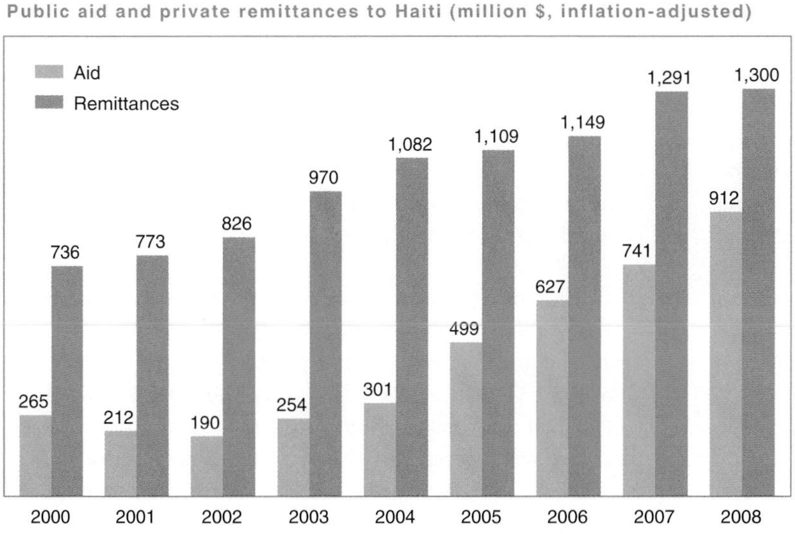

Public aid and private remittances to Haiti (million $, inflation-adjusted)

Figure 9–16 Haitian aid and remittances. Haiti has many citizens who live abroad and send money home. Their impact is greater than the aid Haiti receives from other countries. (*Source: World Bank DAC2, Center for Global Development 2010.*)

countries may be tempted to reduce development aid and raise barriers to trade from developing countries. Also, as unemployment grows, foreign-born workers are likely to be affected and will be less able to send remittances to their countries of origin. Unfortunately, worldwide food prices have also risen since 2008.

Even though foreign assistance is vital, the developing countries themselves must also address the root causes of poverty, and this usually means public policy changes at more local levels. To illustrate, the discussion in the next section

assesses the relative roles played by development and family planning.

9.3 A New Direction: Social Modernization

As the cases of Thailand and Kerala, India, demonstrate, demographic experts are recognizing that the shift from high to low fertility rates in the poorer developing countries does not require the economic trappings of a developed country. Instead, what is needed are efforts within the country, made on behalf of the poor, with particular emphasis on improving education—especially literacy and the education of girls and women; lowering infant mortality and improving life expectancy; making family planning both available and affordable; enhancing employment opportunities; and resource management (maintaining ecosystem sustainability). It might be obvious that most of these track well with the Millennium Development Goals. That is, development targeted at alleviating the problems of the poorest people will bring down population growth rates even as they alleviate dire poverty (Section 9.2). Similarly, the solutions to problems that contribute to population growth are the solutions to the problems that cause people to have large families (Section 9.2).

A great deal of focus is on women because they not only bear children, but they also are the primary providers of nutrition, child care, hygiene, and early education. When women have more education and some income of their own, they are more likely to spend money on education and health care for children and less likely to have large numbers of children. Here we will look at each part of social modernization briefly.

Improving Education

Investing in the education of children (and adults who lack schooling) is a key element of the public-policy options

TABLE 9–3 U.S. Economic Contributions to Developing Countries, 2011

Type of Funding	Billions of $
Official Development Assistance (ODA)*	28.3
Private philanthropy	37.5
Foundations	34.6
Corporations	8.9
Private and voluntary organizations	12
Volunteerism	3
Universities and colleges	1.8
Religious organizations	7.2
Remittances	90.7
Private capital investments	69.2
U.S. total economic engagement	226.2**

*Some ways to calculate U.S. government foreign aid report much larger numbers. For the most part, this is due to the inclusion of military aid. In 2011, the United States donated more than $15 billion in military assistance. Israel, Iraq, Afghanistan, and Pakistan are the largest recipients.

**Down from 235 in 2007.

Source: The Index of Global Philanthropy 2011, Executive Summary. The Center for Global Prosperity. Hudson Institute, 2009 (www.global-prosperity.org).

of a developing country, one that returns great dividends (**Fig. 9–17**). In 1960, for example, Pakistan and South Korea both had similar incomes and population growth rates (2.6%), but very different school enrollments—94% in South Korea versus 30% in Pakistan. Within 25 years, South Korea's economic growth was three times that of Pakistan's, and now the rate of population growth in South Korea has declined to 0.5% per year, while Pakistan's is still 2.2% per year. Kerala, where literacy is greater than 90%, is another case in point. The lowest net enrollment of children in primary schools (about 71%) is in sub-Saharan Africa, where 38 million children are still out of school. An educated populace is an important component of the wealth of a nation. Remember, Millennium Development Goal 2 is to ensure that children everywhere receive a primary-school education. This is a worthwhile goal, but in some ways, it is insufficient. Evidence shows that achieving at least 50% secondary schooling will help large segments of a poor country's population out of poverty and will have a strong impact in decreasing fertility and improving the health of children.

Improving Health

The health care needed most by poor communities in the developing world is the basics of good nutrition and hygiene—steps such as boiling water to avoid the spread of disease and properly treating infections and common ailments such as diarrhea. (In developing countries, diarrhea is a major killer of young children. It is easily treated by giving suitable liquids, a technique called *oral rehydration therapy.*) The discrepancy in infant mortality rates (infant deaths per 1,000 live births) between the developed countries (6/1,000) and developing countries (46/1,000) speaks for itself.

Life Expectancy. One universal indicator of health is human life expectancy. In 1955, average life expectancy globally was 48 years. Today, it around 68 years and rising; it is expected to reach 73 years by 2025. This progress is the result of social, medical, and economic advances in the latter half of the

20th century that have substantially increased the well-being of large segments of the human population. Recall, the *epidemiologic transition*: the trend of decreasing death rates that is seen in countries as they modernize (Chapter 8). Recall also that, as this transition occurred in the now-developed countries, it involved a shift from high mortality rates (due to infectious diseases) toward the present low mortality rates (due primarily to diseases of aging, such as cancer and cardiovascular diseases).

Despite the progress indicated by the general rise in life expectancy, there is a large difference in life expectancy between developed and developing countries. Monaco, for example, has an average life expectancy of 90 years and Japan's is 84 years, while Chad, Afghanistan, South Africa, and several other countries have life expectancies as low as 49 years. The discrepancy between the most developed and least developed countries can be seen quite clearly in the distribution of deaths by main causes, shown in **Figure 9–18**. Infectious diseases were responsible for 47% of mortality in the least developed countries, compared with only 6% in the developed countries.

The solution is very evident: Primary health services need to be centered on people's needs, they must provide health security in the context of people's communities, and there must be universal access to primary health care. Disease prevention and the promotion of health are critical components of this primary health care. Clear focus on these issues has come from the MDGs, and the attention given to achieving the MDGs is improving the delivery of health care to people in low-income countries. Health care in the developing world must especially emphasize maternal pre- and postnatal care for mothers and children. Many government, charitable, and religious organizations are involved in providing basic health care, and when this is extended to rural areas in the form of clinics, it is one of the most effective ways of delivering primary health care to

Figure 9–17 Education in developing countries. Providing education in developing countries can be quite cost effective because any open space can suffice as a classroom and few materials are required. These girls are Afghan refugees in Pakistan, attending an NGO-funded school.

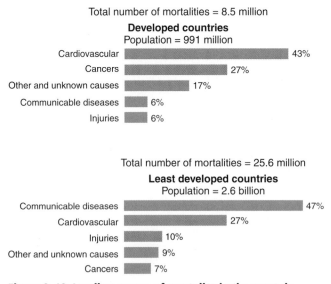

Total number of mortalities = 8.5 million
Developed countries
Population = 991 million

Cause	%
Cardiovascular	43%
Cancers	27%
Other and unknown causes	17%
Communicable diseases	6%
Injuries	6%

Total number of mortalities = 25.6 million
Least developed countries
Population = 2.6 billion

Cause	%
Communicable diseases	47%
Cardiovascular	27%
Injuries	10%
Other and unknown causes	9%
Cancers	7%

Figure 9–18 Leading causes of mortality in the most developed and least developed countries. Infectious diseases are predominant in the least developed countries, while cancer and diseases of the cardiovascular system predominate in the highly developed countries. (*Source:* Data from World Health Organization, *Projections of Mortality and Burden of Disease, 2002–2030* (2008).)

families. When women are better off, whole populations are improved. This is the focus of reproductive health.

Reproductive Health. A fairly new concept in population matters, **reproductive health** was strongly emphasized at the International Conference on Population Development held in Cairo, Egypt, in 1994. Reproductive health is a subset of general health improvement that is focused on women and infants and, in some cases, on reproductive education and health for men. Several key components are involved, mostly focused on medical care and education such as prenatal care, safe childbirth, postnatal care, information about contraception, prevention and treatment of sexually transmitted diseases (STDs), abortion services (where legal) and care afterward, and prevention and treatment of infertility. In addition, promotion of reproductive health also involves protecting women from violence and coercion. For example, reproductive health requires elimination of violence against women (coercive sex, rape), sexual trafficking, and female circumcision and infibulation (traditional practices in some African societies).

Reproductive health care underpins virtually all of the MDGs, especially Goals 4, 5, and 6. Goal 5 (*improve maternal health*) was expanded in 2005 to include reproductive health services, and the new target is to achieve universal access to reproductive health by 2015. It must be addressed particularly to the needs of young people. This is often a controversial issue, but sex education for adolescents is a very real need, and not just in the developing world. Premarital sex is on the rise in most regions of the world, resulting in unplanned pregnancies and out-of-wedlock births. This represents a cultural shift for many societies and results in many young people having sex without the information they need. Of course, marriage does not protect against AIDS, other diseases, or unplanned pregnancies. However, many agencies study the impact of premarital sex, in part because unmarried members of society may not have access to health care and sex education and because some parts of the world have extremely large adolescent populations—a group likely to affect future fertility the most.

AIDS. One of the greatest challenges to health care in developing countries is the sexually transmitted disease known as **acquired immunodeficiency syndrome (AIDS)**. Unfortunately, the global epidemic of AIDS is most severe in many of the poorest developing countries, which are least able to cope with the consequences. This communicable disease devastates the economies of communities by killing people in the prime of their working lives, increasing the dependency ratio as children are left in the care of the elderly or other children. Thailand, in spite of high literacy, life expectancy, and clean water availability, had a weak response to HIV/AIDS when the epidemic first began. Eventually, it had the highest HIV infection rate in South Asia. Recent aggressive public education campaigns have begun to pull the rate down.

These positive results are the outcome of a remarkable increase in funding from the developed countries. Recall that Target 8 (part of Goal 6) of the MDGs is to have halted and begun to reverse the spread of HIV/AIDS by 2015. Major sources of funding are the World Bank; the Global Fund to Fight AIDS,

Tuberculosis and Malaria; the U.S. President's Emergency Plan for AIDS Relief (PEPFAR); and local country budgets. PEPFAR has provided $32 billion over nine years, with a plan for increased funding in the next several years. The funds are used for a combination of testing, treatment, and prevention. By 2014, the goal is to get needed medications to 85% of those currently suffering from HIV/AIDS to prevent 12 million new infections.

Most prevention approaches involve getting people to change their sexual behavior—to practice safe sex, limit their sexual partners, and avoid prostitution. That these things can be accomplished is proven by the case of Uganda, where a 70% decline in HIV prevalence has been registered in recent years, linked to a 60% reduction in casual sex. The Ugandan government has mounted a national AIDS control program that includes a major education campaign. Recent evidence has revealed another effective preventive measure—namely, male circumcision, which causes a 60% reduction in HIV risk in males involved in heterosexual relationships and protects more women than most other preventive strategies. The WHO strongly recommends that countries offer male circumcision as one of their anti-AIDS services.

Family Planning

Information on contraceptives, related materials, and treatments are the most direct ways to bring down high population growth and, for poor nations, are supported by agencies in turn supported by governments, international groups, and NGOs. The stated policy of family-planning agencies (or similar private services) is, as the name implies, to *enable people to plan their own family size—that is, to have children only if and when they want them.* In addition to helping people avoid unwanted pregnancies, family planning often involves determining and overcoming fertility problems for those couples who are having reproductive difficulties. Family planning is a critical component of reproductive health care (**Fig. 9–19**).

The vigorous promotion and provision of contraceptives has been proven all by itself to lower fertility rates, as seen in

Figure 9–19 Family planning. Thailand brought down its high fertility in part because of a humorous public education campaign. *Cabbages and Condoms,* shown here, is a restaurant chain whose profits fund the Population and Community Development Association (PDA), a Thai non-profit.

Figure 9–8. Those countries that have implemented effective family-planning programs have experienced the most rapid decline in fertility. For example, as we saw in the example of Thailand, a vigorous family-planning program was initiated as part of a national population policy in 1971. Population growth subsequently declined from 3.1% per year to the present 0.5% per year, and the Thai economy has posted one of the most rapid rates of increase over the same years.

Women who are not currently using contraception but who want to postpone or prevent childbearing and lack access to contraception are said to have an *unmet need*. The unmet need for family planning ranges from 10% to 50% of women in the different developing countries. Because of the lack of other family planning services, some women resort to abortions to terminate unwanted pregnancies. Nearly everyone agrees that an abortion is the least desirable way to avoid having an unwanted child—especially in view of the other alternatives that are available.

Negotiating Disagreements. The major document resulting from the 1994 Cairo population conference (ICPD Program of Action) explicitly states that abortions should not be used as a means of family planning. That is, it states, family planning focuses on education and the provision of services directed at avoiding unwanted or high-risk pregnancies. This statement was not aimed at either eliminating abortion or supporting a right to abortion as an option for women worldwide. The statement was in the context of many other facets of reproductive health and was focused on maintaining support for family planning among stakeholders who disagree about abortion. Research has supported that abortions are higher when family planning services such as contraception and education are unavailable than when they are available. It is hard to make policy on an international level because of social and individual disagreement about important issues. The ICPD Program of Action helped forge a middle ground between people with very disparate views and emphasized the importance of family planning services to women worldwide.

Planned Parenthood, which operates clinics throughout the world, is probably the best-known family planning agency. Another significant player is UNFPA, which provides financial and technical assistance to developing countries at their request. The emphasis of UNFPA is on combining family planning services with maternal and child health care and expanding the delivery of such services in rural and marginal urban areas. Support for this UN agency and other family planning agencies has become a difficult subject in the United States because of the strong opinions surrounding abortion.

Fragile States. Some of the poorest countries are designated **fragile states**. These conflict-prone countries lack a stable government. It is very difficult for aid agencies to work with them because of danger to aid workers. Much of the aid such countries receive is humanitarian and disaster relief, but what they most need is to get to a point where they could build infrastructure and social capital. Such volatile situations are devastating for women and children. There are often floods of refugees, and rape and violence may be rife.

In such a situation, it may be nearly impossible to implement family planning or other long-term assistance programs. The needs people have are for a basic, stable governance and security. When that is achieved other poverty issues may be tractable. Somalia, which has been at war and lacked a central government for 20 years, is a good example of a fragile state.

U.S. Contributions to International Family Planning. In 1984 and again in 2001, the U.S. government was prohibited by a presidential order from giving aid to foreign family planning agencies if they provide abortions, counsel women about abortion if they are dealing with an unwanted pregnancy, or advocate for abortion law reforms in their own country. This was a blow to efforts to increase the availability of contraceptive services in the developing countries because the United States had been the largest international donor to family-planning programs. At one point, the United States withdrew its support for the entire 1994 ICPD Program of Action because it uses terms such as *reproductive services* and *reproductive health care*. Critics of this move pointed out (unsuccessfully) that the conference documents made clear that, "in no case should abortion be promoted as a method of family planning" and that the Cairo agreement was critical to the fight against HIV/AIDS and prevented the very unwanted pregnancies that could lead to abortions. The last word in this controversy comes from President Obama, who has taken steps to restore funding for UNFPA.

Employment and Income

The bottom line of any economic system is the exchange of goods and services. At its simplest level, this entails a barter economy in which people agree on direct exchanges of certain goods or services. Barter economies are still widespread in the developing world.

The introduction of a cash economy facilitates the exchange of a wider variety of goods and services, and everyone may prosper, as they have a wider market for what they can provide and a wider choice of what they can get in return. In a poor community, everyone may have the potential to provide certain goods or services and may want other things in return, but there may be no money to get the system off and running. In a growing economy, people who wish to start a new business venture generally begin by obtaining a bank loan to set up shop. The poor, however, are considered high credit risks. Furthermore, they may want a smaller loan than a commercial bank wants to deal with, and many of the poor may be women, who are denied credit because of gender discrimination alone. For these three reasons, poor communities have trouble getting start-up capital. Fortunately, however, the situation is changing.

Grameen Bank. In 1976, Muhammad Yunus **(Fig. 9–20)**, an economics professor in Bangladesh, conceived of and created a new kind of bank (now known as the **Grameen Bank**) that would engage in **microlending** to the poor. As the name implies, microloans are small—they averaged just $370

Figure 9–20 Muhammad Yunus. An economics professor in Bangladesh, Yunus created the Grameen Bank, which initiated the microlending movement. The Grameen Bank has loaned more than $6.6 billion over the past 35 years, and the microlending model has been duplicated in more than 40 countries. Yunus received the 2006 Nobel Peace Prize in recognition of the global impact of his work.

in 2011—and are short-term loans, usually just four to six months. Nevertheless, they provide such basic things as seed and fertilizer for a peasant farmer to start growing tomatoes, pans for a baker to start baking bread, a supply of yarn for a weaver, tools for an auto mechanic, or an inventory of goods to start a grocery store. By 2011, 3,590,923 people had taken these small loans.

Yunus secured his loans by having the recipients form **credit associations**—groups of several people who agreed to be responsible for each other's loans. With this arrangement, the Grameen Bank has experienced an exceptional rate of payback—greater than 97%. Small-scale agriculture loans from the Grameen Bank have had outstanding results. In a rural area of Bangladesh, small loans, along with horticultural advice, are now enabling peasant farmers to raise tomatoes and other vegetables for sale to the cities. These people doubled their incomes in three years. Microlending has been found to have the greatest social benefits when it is focused on women. The credit associations also create another level of cooperation and mutual support within the community.

More Microlending. The unqualified success of microlending in stimulating economic activity and enhancing the incomes of people within poor communities has been so remarkable that the Grameen concept has been adopted in more than 40 countries around the world (including the United States) with various modifications. Recently, the

World Bank dramatically increased its support of microfinancing through the CGAP, which coordinates the work of some 54 different funding agencies around the world. A major development in this arena has been the work of the Microcredit Summit Campaign, an outgrowth of the 1995 UN Fourth Conference on Women in Beijing. An update from this campaign indicates that, in 2010, some 137.5 million microloans were in service helping more than 687 million family members. These efforts may be doing more than anything else to meet the MDGs, as they empower the poor to improve their own well-being.

Resource Management

The world's poor depend on local ecosystem capital resources—particularly water, soil for growing food, and forests for firewood. Many lack access to enough land to provide an income and often depend on foraging in woodlands, forests, grasslands, and coastal ecosystems. This activity generates income and is vital to those in extreme poverty; 90% of the world's poorest people depend on forests for some of their sustenance (by extracting fuelwood, construction wood, wild fruits and herbs, fodder, and "bush meat" for both subsistence and cash). Forests, fisheries, reefs, grasslands, and waterways all can be resources for the poor—so-called common-pool resources (see Chapter 7). These can be a safety net, as well as an employment source, but if not managed in some way, common-pool resources are liable to be overused, especially when populations are increasing. **Figure 9–21** depicts a woman in Eritrea planting mangroves. In this group project, the women work together to plant trees that will, in turn, protect small fish and provide leaves to feed goats. This is a good example of protecting common resources.

Empowering the poor to manage community- or state-owned lands is one approach that has worked in Nepal. There forestry user groups are given the right to own trees,

Figure 9–21 Improving resource management. A major step in enhancing incomes and protecting the environment is to encourage better resource management. Here a Nigerian woman plants mangrove seedlings in a conservation area in order to increase fish and grow wood for harvest.

but not the land, and there are now some 6,000 user groups managing 450,000 hectares (1.1 million acres). The groups develop forest-management plans, set timber sale prices, and manage the surplus income from the operations. Protecting forest resources and tackling land degradation in drylands are two sustainable steps that can be addressed by country policies.

Putting It All Together

Each of the five components of social modernization—education, health, family planning, employment and income, and resource management—depends on and supports the other components. For example, better health and nutrition support better economic productivity, better economic productivity supports obtaining a better education, and a better education leads to a delay in marriage and the desire to have fewer children. The availability of family-planning services is essential to realizing the desire of parents to have fewer children. All the components work together to improve lives. Conversely, the lack of any component undercuts the ability to achieve all the other components.

Sustainable development. The elements of social modernization are entirely compatible with the MDGs (Table 9–2). But as we move past the MDGs, we continue to face real hurdles trying to protect valuable ecosystem services even as we promote development. Doing so needs to be compatible with **sustainable development,** development that meets the needs of the present without compromising the ability of future generations to meet their own needs.[5] In 2012, world leaders NGOs, and representatives of many stakeholder groups met in Rio De Janeiro, Brazil for a UN conference 20 years after the 1992 Earth Summit at which a global agenda for sustainable development was first described. At the 2012 **Rio +20** meeting, world leaders hashed out goals described in a document, *The Future We Want.*[6] While world leaders were not able to come to as many concrete decisions as observers hoped, particularly on climate change, they did reaffirm the need for sustainable development and agreed to set a series of "sustainable development goals."

One reality that the developing countries must confront in their progression toward economic and social development is globalization. What role can globalization play in bringing about social modernization, the demographic transition, and sustainable development in the developing countries?

Globalization. Defined as the accelerating interconnectedness of human activities, ideas, and cultures, **globalization** is profoundly changing many human enterprises, especially in its economic impacts. A small number of developing countries (China, Malaysia, and Indonesia, for example) have successfully entered the global market, and more strive to do so. Their exports have paved the way for economic progress, and these countries are becoming more connected to the economic- and information-rich high-tech world.

The poorest countries, however, are mostly populated by self-employed peasants who have not made the transition to becoming wage earners in a market economy. This transition is not easy, and they can't make it by themselves. The jobs have to be there, and that often requires investment from outside agencies—businesses and NGOs with the knowledge and capital needed. Also, there has to be some hope of marketing the products, which means entering the global market on a level playing field. At present, however, the playing field is far from level because developed countries protect their exports with subsidies, quotas, and tariffs and all of these trade barriers work against the developing countries (Chapter 2). Toward that end, Goal 8 of the MDGs (*forge a global partnership for development*) calls for eliminating these barriers and encouraging the poor countries themselves to develop a market-based trading system committed to sustainable development and attractive to foreign investment.

On the other hand, globalization can be harmful to the poor if it opens markets for agricultural products from the developed countries, which can ruin subsistence farming. Globalization can also widen the gap between the rich and poor within a society, as the wealthier members capitalize on the information flow and quick transfer of goods and services, leaving the poor farther behind. Aid can have a similar effect, undermining local economies. This is one reason that the 2005 Paris Declaration (an agreement on aid) emphasized that donor countries should buy local products where possible, rather than donating from abroad.

Critics of globalization claim that it doesn't seem capable of promoting some other goods, such as financial stability and social justice. It also has the potential to dilute and even destroy cultural and religious ideals and norms when it is uncritically welcomed into developing countries. Whether helpful or harmful, globalization is taking place, making it essential that the developing countries confront it head on and adapt its connections to their needs and cultures.

---

[5]World Commission on Environment and Development (1987). *Our Common Future.* Oxford: Oxford University Press. p. 27.
[6]UN. *The Future We Want* Outcomes Document. United Nations Conference on Sustainable Development, Rio De Janiero, Brazil June 20-22, 2012https://rio20.un.org/sites/rio20.un.org/files/a-conf.216l-1_english.pdf.pdf

REVISITING THE THEMES

Sound Science

The sciences of demography, epidemiology, and sociology have brought us the insights into what causes the success of the demographic transition and the causes of the demographic dividend, Medical advances and health care have helped bring many countries through the epidemiologic transition—however, there is still great need for basic health care in the poorer countries. Sound science has brought us improved AIDS treatment through anti-HIV medicines but so far has not come up with an effective vaccine. Demographers test hypotheses about what factors most predict, and in some cases cause, a drop in high fertility. In some parts of the world, aging populations require a different set of supports, medical and social.

Sustainability

The demographic transition is essential for a sustainable future. This chapter is about what has to be done economically and socially to bring about development that is sustainable. Economic progress is being made in many developing countries, and it is accompanied by most of the elements of social modernization. However, economic progress is still lacking in many others. For these, social modernization is the most feasible approach, as it will bring fertility rates down and enable countries to open the demographic window that may make economic progress possible.

An element of development activities that is unsustainable is the debt built up over the years by many poorer developing countries. Although this has been helped by much debt cancellation, the overall debt of developing countries continues to build up, and the repayment of these debts forces many developing countries to liquidate ecosystem capital in order just to make their payments.

Sustainable development requires facilitation by integrating principles of sustainability into country policies and programs and by reversing the loss of environmental resources. Improving the management of these resources is an important component of social modernization. The 2012 Rio +20 UN conference reaffirmed the need to protect the environment even as we improve people's lives.

Social sustainability is another aspect of sustainability. Large-scale economic efforts by the World Bank and other organizations have sometimes actually harmed the poor and increased inequality. Any solution to either economic or environmental issues needs to be just in order to be socially sustainable as well. This is one of the reasons the Grameen Bank, owned by the poor, has been successful.

Stewardship

Justice for the developing world was one of the key ethical issues identified in the discussion of stewardship in Chapter 1. *Distributive justice* is an ethical ideal that considers the gross inequality of the human condition as unjust—the kind of inequality represented by the billion-plus people who live on less than $1.25 a day versus typical people in the wealthy countries. Although equal distribution of the world's goods and services is neither attainable nor necessary, the existing poverty in the poor countries is an affront to human dignity. In order to address this injustice, the alleviation of poverty must be made a priority of public policy.

The case study of China contrasts with that of Thailand. Both have seen drops in high fertility, but Thailand did it primarily though education, clinics, free contraception, and other parts of social modernization. While China had some of these elements, it also had an involuntary policy associated with some serious human rights abuses. Because of the authoritarian approach taken by China, many people worldwide see Chinese population control as symbolic of the problems of government control.

REVIEW QUESTIONS

1. What are the two basic schools of thought regarding the demographic transition? How were these reflected in the three most recent global population conferences?

2. Discuss five specific factors that influence the number of children a poor couple may desire.

3. What are the MDGs? Where did they come from? Cite an example of one goal and a target used to measure progress in attaining it.

4. What are two major agencies that promote development in poor nations, and how do they carry out their work?

5. What is meant by the debt crisis of the developing world? What is being done to help resolve this crisis?

6. What is development aid, and how does it measure up against the need for such aid?

7. What are the five interdependent components that must be addressed to bring about social modernization?

8. Define *family planning*, and explain why it is critically important to all other aspects of development.

9. What is meant by an unmet need?

10. What is microlending? How does it work?

11. What is the significance of globalization for economic and social development in the developing countries?

THINKING ENVIRONMENTALLY

1. Is the world population below, at, or above the optimum? Defend your answer by pointing out things that may improve and things that may worsen as the population increases.

2. Suppose you are the head of an island nation with a poor, growing population and natural resources that are being degraded. What kinds of policies would you initiate, and what help would you ask for to try to provide a better, sustainable future for your nation's people?

3. List and discuss the benefits and harms of writing off debts owed by developing nations.

4. Look up the document *The Future We Want*. Describe what future you would want on a personal, regional, national and global level for 30 years from now. How much depends on environmental health? How does the future you want match what other people around the globe hope for?

5. Do you believe contraceptives should be made available free of charge by governments, as this would have the potential to curb population growth and/or abortion rates? Defend your answer.

MAKING A DIFFERENCE

1. Think carefully about your own reproduction. What concerns will you and your mate weigh as you plan your family?

2. Become involved in, and support, programs promoting effective sex education and responsible sexual behavior. Consider the advantages of abstinence and monogamy as ways to avoid needing an abortion or contracting a sexually transmitted disease.

3. If you are not convinced of the profound impacts of poverty in the developing world, log on to the Millennium Project Web site and read about the Millennium Village movement (www.unmillenniumproject.org/mv/index.htm).

4. Encourage sustainability by buying products that originate from the developing world.

MasteringEnvironmentalScience®

Go to **www.masteringenvironmentalscience.com** for practice quizzes, Pearson eText, videos, current events, and more.

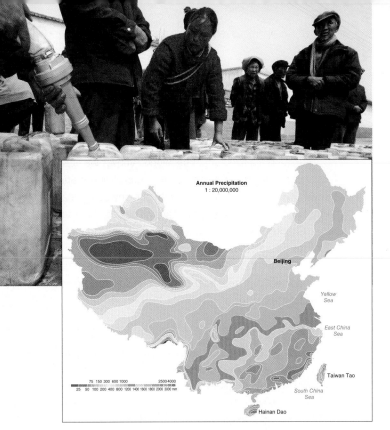

Annual Precipitation
1 : 20,000,000

Beijing

Yellow
Sea

East China
Sea

Taiwan Tao

South China
Sea

Hainan Dao

75 150 300 600 1000   2500 4000
25 50 100 200 400 800 1200 1400 1600 1800 2000 3000 mm

Running out of water. These villagers in the dry southwest of China are lined up to get water. Parts of China are in the midst of the worst drought in 60 years, an event that is making China's water problems worse.

CHAPTER 10

Water: Hydrologic Cycle and Human Use

C HINA HAS A WATER PROBLEM. The growing population requires more and more to grow food and sustain cities, and yet much of the country is dry. The wetter southeast is crowded with people, while the drier southwest is experiencing the worst drought in 60 years. In the center of the country lies the central plateau, an area of intense agricultural production; in the north and northwest is the extremely dry Gobi desert, which is creeping south and east yearly. Ten provinces of China are so water-poor that they fall beneath the World Bank's definition of water poverty: 1,000 m³ of water per year per person. The karst, or limestone regions, of the south offer a unique problem. Precipitation, while not scarce, rapidly percolates through porous rock to travel in vast underground rivers, inaccessible to most rural residents of the area. Such underground river water is undrinkable because of all of the surface pollutants that percolate down it to the water table.

DAM BUILDING. China's explosive economic growth has fueled a rising middle class that wants a more diverse diet, but such foods are energy and water intensive. China's need for power has triggered a dam-building spree. The central plateau, home of the richest agriculture, is experiencing crop loss from too much groundwater removal. Unfortunately, pumping the 100 km³ per

LEARNING OBJECTIVES

1. **Water, a Vital Resource: Explain the unique properties that make water so vital, the differences in water availability in different societies, and conflicts over availability of clean water.**

2. **Hydrologic Cycle, and Human Impacts: Explain the movement of water through the hydrologic cycle and human impacts on the cycle.**

3. **Human Controlled Flooding and Drought: Describe the ways humans try to provide clean freshwater and some of their outcomes.**

4. **Water Stewardship, Economics, and Policy: Describe options for meeting rising demands for water, new innovations in water science and technology, and public policies for water in a water-scarce world.**

239

year that China uses to irrigate requires a great deal of energy as well. Vice President Xi Jinping visited the United States in 2012 to discuss, among other things, dropping trade barriers to some U.S. agricultural products. While China is already a big importer of U.S. products, growing more water-intensive crops elsewhere will save water in China, making the trade represent not only needed food but saved water.

LONG-DISTANCE TRANSPORT. In addition to building dams, the Chinese government is working on moving water long distances. One such undertaking is the South-to-North Water Diversion Project (SNWT), a 50-year venture (two parts are scheduled to be delivering water by 2014) with an estimated cost of $64 billion that will bring water to the arid north. An eastern route will carry water from the Yangtze River 1,152 km (716 mi) to Tianjin, through a series of canals, reservoirs tunnels, and natural rivers, and a central route will carry water 1,264 km (785 mi) from the south of Beijing. Together they should carry 28 km³ (6.7 mi³) of water per year.

China is a microcosm of the world. All over, we are trying to balance needs for energy, agriculture, drinking water, and biodiversity. As we welcome new people into our crowded global community, finding clean, safe water for all will be one of our top priorities.

Chapter 10, the first of four chapters on essential resources, focuses on water. The concepts here will be strongly intertwined with those we have already seen in chapters on population and development and with later chapters on soil, food production, and climate change.

10.1 Water: A Vital Resource

Water is absolutely fundamental to life as we know it. Happily, Earth is virtually flooded with water. A total volume of some 1.4 billion cubic kilometers (325 million mi³) of water covers 75% of Earth's surface. About 97.5% of this volume is the saltwater of the oceans and seas. The remaining 2.5% is **freshwater**—water with a salt content of less than 0.1% (1,000 ppm). This is the water on which most terrestrial biota, ecosystems, and humans depend. Of the 2.5%, though, two-thirds is bound up in the polar ice caps and glaciers. Thus, only 0.77% of all water is found in lakes, wetlands, rivers, groundwater, biota, soil, and the atmosphere **(Fig. 10–1)**. Nevertheless, evaporation from the oceans combines with precipitation to resupply that small percentage continually through the solar-powered hydrologic cycle, described in detail in Section 10.2. Thus, freshwater is a continually renewable resource.

Streams, rivers, ponds, lakes, swamps, estuaries, groundwater, bays, oceans, and the atmosphere all contain water, and they all represent ecosystem capital—goods and services vital to human interests (see Table 5–2 for the array of aquatic ecosystems and their characteristics). They provide drinking water, water for industries, and water to irrigate crops. Bodies of water furnish energy through hydroelectric power and control flooding by absorbing excess water. They provide transportation, recreation, waste processing, and habitats for aquatic plants and animals. Freshwater is a vital resource for all land ecosystems, modulating the climate through evaporation and essential global warming (when water is in the atmosphere as water vapor). During the past two centuries, many of these uses (and some threats to them) have led us to construct a huge infrastructure designed to bring water under control. We have built dams, canals, reservoirs, aqueducts, sewer systems, treatment plants, water towers, elaborate pipelines, irrigation systems, and desalination plants. As a result, waterborne diseases have been brought under control in the developed countries, vast cities thrive in deserts, irrigation makes it possible to grow 40% of the world's food, and one-fifth of all electricity is generated through hydropower. These great benefits are especially available to people in the developed countries **(Fig. 10–2a)**.

In the developing world (Fig. 10–2b), by contrast, 1.1 billion people still lack access to safe drinking water, 2.6 billion do not have access to adequate sanitation services, and more than 1.8 million deaths each year are traced to waterborne diseases (mostly in children under five). All too often in these countries, water is costly or inaccessible to the poorest in society, while the wealthy have it piped into their homes. In addition, because of the infrastructure that is used to control water, whole seas are being lost, rivers are running dry, millions of people have been displaced to make room for reservoirs, groundwater aquifers are being pumped down, and disputes over water have raised tensions from local to international levels. Freshwater is a limiting resource in many parts of the world and is certain to become even more so as the 21st century unfolds.

Feeding a growing world population will challenge our water management infrastructure, especially as cities and industries compete with agriculture for scarce water resources. The growing impact of climate change on the hydrologic cycle is in turn worsening the problems brought on by droughts and floods. Rainfall variability can reduce the economic growth of nations as floods devastate some croplands and droughts dry them up in other parts of the world. Even in the United States, effects from climate change will vary. A 2011 report from the Department of the Interior suggested that one of the biggest impacts will be to lower the summer availability of water in already dry areas of the Southwest.[1]

There are two ways to consider water issues. The focus in this chapter is on *quantity*—that is, on the global water cycle and how it works, on the technologies we use to control water and manage its use, and on public policies we have put in place to govern our different uses of water. Later we

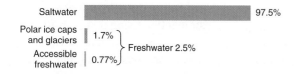

Figure 10–1 Earth's water. The Earth has an abundance of water, but terrestrial ecosystems, humans, and agriculture depend on accessible freshwater, which constitutes only 0.77% of the total.

[1]Alexander, P. et al., Reclamation, SECURE Water Act Section 9503(c), Reclamation Climate Change and Water, Report to Congress 2011, U.S. Department of the Interior, Denver, CO.

(a)

(b)

Figure 10–2 Differences in availability. Human conditions range from (a) being able to luxuriate in water to (b) having to walk long distances and pull from deep pumps to obtain enough simply to survive. In many cases, however, the abundance of water is illusory because water supplies are being overdrawn.

will focus on water *quality*—that is, on water pollution and its consequences, on sewage-treatment technologies, and on public policies for dealing with water pollution issues (Chapter 20). Water—so vital to life—moves through the land, oceans, and atmosphere in a cycle.

10.2 Hydrologic Cycle: Natural Cycle, Human Impacts

Earth's water cycle, also called the **hydrologic cycle**, is represented in **Figure 10–3**. The basic cycle consists of water rising to the atmosphere through **evaporation** (movement of water molecules from the surface of the liquid to a gas state) and **transpiration** (the loss of water vapor as it moves from the soil through green plants and exits through leaf pores) and returning to the land and oceans through **condensation** (formation of liquid water from a gas state) and **precipitation** (release of water from clouds in form of rain, sleet, snow, or hail). Modern references to the hydrologic cycle distinguish between **green water** and **blue water**. Green water is water in vapor form, while blue water is liquid water wherever it occurs. Green water and blue water are vitally linked in the hydrologic cycle. Terms commonly used to describe water are listed and defined in Table 10–1 on page 243.

Evaporation, Condensation, and Purification

Recall that a weak attraction known as hydrogen bonding tends to hold water molecules (H_2O) together (Chapter 3). Below 32°F (0°C), the kinetic energy of the molecules is so low that the hydrogen bonding is strong enough to hold the molecules in place with respect to one another, and the result is

ice. At temperatures above freezing but below boiling (212°F or 100°C), the kinetic energy of the molecules is such that hydrogen bonds keep breaking and re-forming with different molecules. The result is *liquid water*. As the water molecules absorb energy from sunlight or an artificial source, the kinetic energy they gain may be enough to allow them to break away from other water molecules entirely and enter the atmosphere. This process is known as evaporation, and the result is **water vapor**—water molecules in the gaseous state.

You will learn later that water vapor is a powerful greenhouse gas (Chapter 18). It contributes about two-thirds of the total warming from all greenhouse gases. The amount of water vapor in the air is the **humidity**. Humidity is generally measured as **relative humidity**, the amount of water vapor as a percentage of what the air can hold *at a particular temperature*. For example, a relative humidity of 60% means that the air contains 60% of the maximum amount of water vapor it could hold at that particular temperature. The amount of water vapor air can hold changes with the temperature. When warm, moist air is cooled, its relative humidity rises until it reaches 100%, and further cooling causes the excess water vapor to condense back to liquid water because the air can no longer hold as much water vapor (**Fig. 10–4**).

Condensation is the opposite of evaporation. It occurs when water molecules rejoin by hydrogen bonding to form liquid water. If the droplets form in the atmosphere, the result is fog and clouds (fog is just a very low cloud). If the droplets form on the cool surfaces of vegetation, the result is dew. Condensation is greatly facilitated by the presence of **aerosols** in the atmosphere. Aerosols are microscopic liquid or solid particles (like hair spray) originating from land and water surfaces. They provide sites that attract water vapor

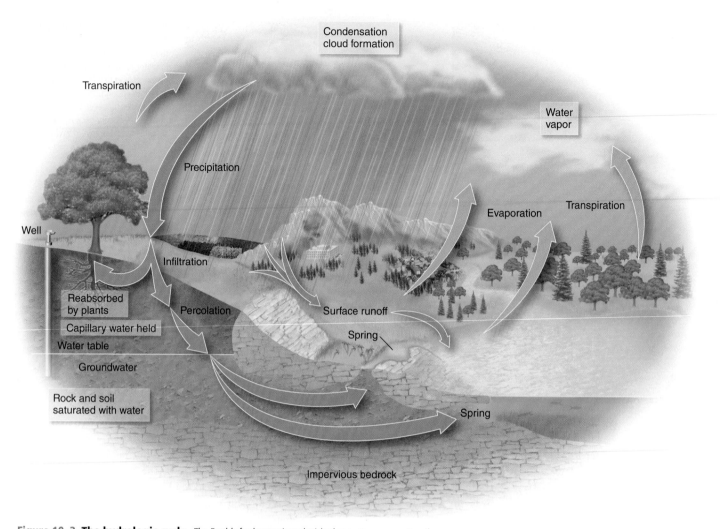

Figure 10–3 The hydrologic cycle. The Earth's freshwater is replenished as water vapor enters the atmosphere by evaporation and transpiration from vegetation, leaving salts and other impurities behind. As precipitation hits the ground, three additional pathways are possible: surface runoff, infiltration, and reabsorption by plants. Green arrows depict green water, present as vapor in the atmosphere; blue arrows show liquid water.

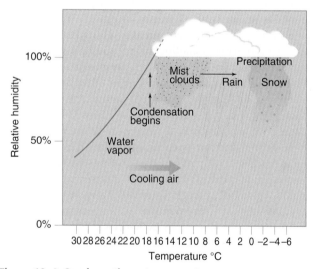

Figure 10–4 Condensation. The amount of water vapor that air can hold changes with the temperature. The red line follows an air mass that starts with 40% relative humidity (RH) at 30°C and is cooled. When the mass of air cools to 17°C, it has reached 100% RH, and condensation begins, forming clouds. If we started with a higher RH, clouds would form sooner when the air is cooled. Further cooling and condensation result in precipitation.

and promote the formation of droplets of moisture. Aerosols may originate naturally, from sources such as volcanoes, wind-stirred dust and soil, and sea salts. Anthropogenic sources—sulfates, carbon, and dust—contribute almost as much as natural sources (more in some locations) and can have a significant impact on regional and global climates.

Purification occurs when water is separated from the solutes and particles it contains. The processes of evaporation and condensation purify water naturally. When water in an ocean or a lake evaporates, only the water molecules leave the surface; the dissolved salts and other solids remain behind in solution. When the water vapor condenses again, it is thus purified water—except for the pollutants and other aerosols it may pick up from the air. The water in the atmosphere turns over every 10 days, so water is constantly being purified. (The most chemically pure water for use in laboratories is obtained by distillation, a process of boiling water and condensing the vapor.)

Thus, evaporation and condensation are the source of all natural freshwater on Earth. Freshwater that falls on the land as precipitation gradually makes its way through aquifers, streams, rivers, and lakes to the oceans (blue water flow). In the process, it carries along salts from the land,

TABLE 10–1 Terms Commonly Used to Describe Water

Term	Definition
Water quantity	The amount of water available to meet demands.
Water quality	The degree to which water is pure enough to fulfill the requirements of various uses.
Freshwater	Water having a salt concentration below 0.1%. As a result of purification by evaporation, all forms of precipitation are freshwater, as are lakes, rivers, groundwater, and other bodies of water that have a throughflow of water from precipitation.
Saltwater	Water that contains at least 3% salt (30 parts salt per 1,000 parts water), typical of oceans and seas.
Brackish water	A mixture of freshwater and saltwater, typically found where rivers enter the ocean.
Hard water	Water that contains minerals, especially calcium or magnesium, that cause soap to precipitate, producing a scum, curd, or scale in boilers.
Soft water	Water that is relatively free of minerals.
Polluted water	Water that contains one or more impurities, making it unsuitable for a desired use.
Purified water	Water that has had pollutants removed or is rendered harmless.
Storm water	Water from precipitation that runs off of land surfaces in surges.
Green water	Water in vapor form originating from water bodies, the soil, and organisms—the source of water for precipitation.
Blue water	Precipitation, renewable surface water runoff, and groundwater recharge—the focus of management and the main source of water for human withdrawals and natural ecosystems.

which eventually accumulate in the oceans. Salts also accumulate in inland seas or lakes, such as the Great Salt Lake in Utah. The salinization of irrigated croplands is a noteworthy human-made example of this process (Chapter 11). Most of the freshwater from precipitation, however, is returned directly to the atmosphere by evaporation and transpiration.

Precipitation

Warm air rises from Earth's surface because it is less dense than the cooler air above. As the warm air encounters the lower atmospheric pressure at increasing altitudes, it gradually cools as it expands—a process known as **adiabatic cooling**. When the relative humidity reaches 100% and cooling continues, condensation occurs and clouds form. As condensation intensifies, water droplets become large enough to fall as precipitation. **Adiabatic warming**, by contrast, occurs as cold air descends and is compressed by the higher air pressure in the lower atmosphere.

Precipitation on Earth ranges from near zero in some areas to more than 2.5 meters (100 in.) per year in others (**Figure 10–5** shows the pattern over land). The distribution

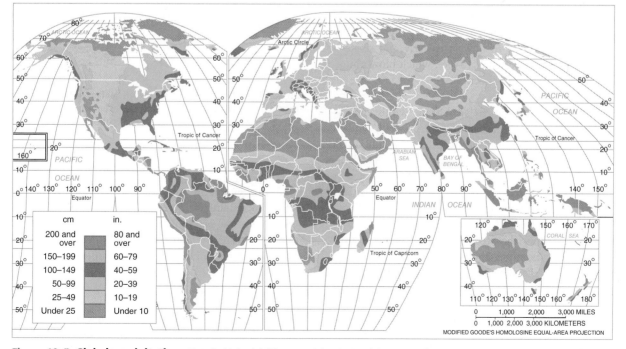

Figure 10–5 Global precipitation. Note the high rainfall in equatorial regions and the regions of low rainfall to the north and south. (*Source:* Robert W. Christopherson, *Elemental Geosystems,* 7th ed., Pearson Education, Upper Saddle River, NJ.)

depends on patterns of rising and falling air currents. As air rises, it cools, condensation occurs, and precipitation results. As air descends, it tends to become warmer, causing evaporation to increase and dryness to result. A rain-causing event that you see in almost every television weather report is the movement of a cold front. As the cold front moves into an area, the warm, moist air already there is forced upward because the cold air of the advancing front is denser. The rising warm air cools, causing condensation and precipitation along the leading edge of the cold front.

Two factors—global convection currents and rain shadow—may cause more or less continuously rising or falling air currents over particular regions, with major effects on precipitation.

Convection Currents. At the equator, the rays of the sun are nearly perpendicular to the Earth's surface; at the poles, light rays come at an angle through the atmosphere. That difference causes the surface of the Earth to be hotter in some places than others and drives great air movements, called "global convection currents." As the air at the equator is heated, it expands, rises, and cools; condensation and precipitation occur. The constant intense heat in these equatorial areas ensures that this process is repeated often, thus causing high rainfall, which, along with continuous warmth, supports tropical rain forests.

Rising air over the equator is just half of the convection current, however. The air, now dry, must come down again. Pushed from beneath by more rising air, it literally "spills over" to the north and south of the equator and descends over subtropical regions (25° to 35° north and south of the equator), resulting in subtropical deserts. The Sahara of northern Africa and the Kalahari of southern Africa are prime examples. The two halves of the system composed of the rising and falling air make up a **Hadley cell (Fig. 10–6a and b).** Because of Earth's rotation, winds are deflected from the strictly vertical and horizontal paths indicated by a Hadley cell and tend to flow in easterly and westerly directions—the **trade winds** (Fig. 10–6c), which blow almost continuously from the same direction.

Rain Shadow. When moisture-laden trade winds encounter mountain ranges, the air is deflected upward, causing cooling and high precipitation on the windward side of the range. As the air crosses the range and descends on the other side, it becomes warmer and increases its capacity to pick up moisture. Hence, deserts occur on the leeward (downwind) sides of mountain ranges. The dry region downwind of a mountain range is referred to as a **rain shadow (Fig. 10–7).** The severest deserts in the world are caused by the rain-shadow effect. For example, the westerly trade winds, full of moisture from the Pacific Ocean, strike the Sierra Nevada Mountains in California. As the winds rise over the mountains, large amounts of water precipitate out, supporting the lush forests on the western slopes. Immediately east of the southern Sierra Nevada, however, lies Death Valley, one of the driest regions of North America. Likewise, the Gobi desert in China is in the rain shadow of the Himalayas.

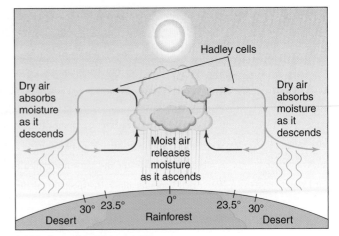

(a) Hadley cells at the equator

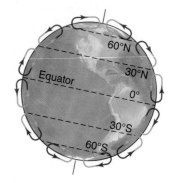

(b) Global air flow patterns

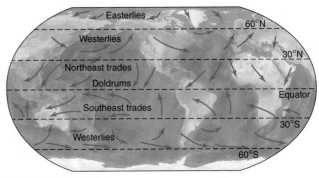

(c) Global trade winds

Figure 10–6 Global air circulation. (a) The two Hadley cells at the equator. (b) The six Hadley cells on either side of the equator, indicating general vertical airflow patterns. (c) Global trade-wind patterns, formed as a result of Earth's rotation.

Figure 10–5 shows general precipitation patterns for the land; the general atmospheric circulation (Fig. 10–6) and the effects of mountain ranges (Fig. 10–7) combine to give great variation in the precipitation reaching the land. The distribution of precipitation in turn largely determines the biomes and ecosystems found in a given region (Chapter 5).

Groundwater

As precipitation hits the ground, it may either soak into the ground (**infiltration**) or run off the surface. The amount that soaks in compared with the amount that runs off is called the **infiltration-runoff ratio**.

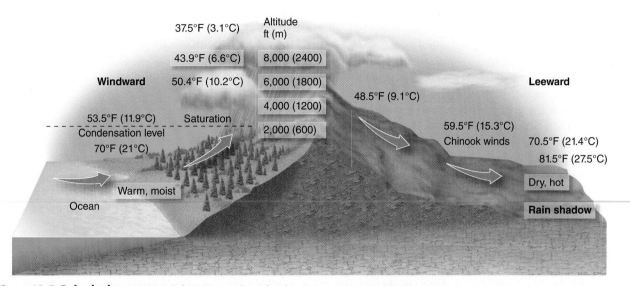

Figure 10–7　Rain shadow. Moisture-laden air in a trade wind cools as it rises over a mountain range, resulting in high precipitation on the windward slopes. Desert conditions result on the leeward side as the descending air warms and tends to evaporate water from the soil.

Runoff flows over the surface of the ground into streams and rivers, which make their way to the ocean or to inland seas. All the land area that contributes water to a particular stream or river is referred to as the **watershed** for that stream or river. All ponds, lakes, streams, rivers, and other waters on the surface of Earth are called **surface waters**.

Water that infiltrates has two alternatives (Fig. 10–3). The water may be held in the soil in an amount that depends on the water-holding capacity of the soil. This water, called **capillary water**, returns to the atmosphere either by evaporation from the soil or by transpiration through plants (a green water flow). The combination of evaporation and transpiration is referred to as **evapotranspiration**.

The second alternative is **percolation** (a blue water flow). Infiltrating water that is not held in the soil is called **gravitational water** because it trickles, or percolates, down through pores or cracks under the pull of gravity. Sooner or later, however, gravitational water encounters an impervious layer of rock or dense clay. It accumulates there, completely filling all the spaces above the impervious layer. This accumulated water is called **groundwater**, and its upper surface is the **water table** (Fig. 10–3). Gravitational water becomes groundwater when it reaches the water table in the same way that rainwater is called lake water once it hits the surface of a lake. Wells must be dug to depths below the water table. Groundwater, which is free to move, then seeps into the well and fills it to the level of the water table.

Recharge. Groundwater will seep laterally as it seeks its lowest level. Where a highway has been cut through rock layers, you can frequently observe groundwater seeping out. Layers of porous material through which groundwater moves are called **aquifers**. It is often difficult to determine the location of an aquifer. Many times these layers of porous rock are found between layers of impervious material, and the entire formation may be folded or fractured in various ways. Thus,

groundwater in aquifers may be found at various depths between layers of impervious rock. Also, the **recharge area**—the area where water enters an aquifer—may be many miles away from where the water leaves the aquifer. Underground aquifers hold some 99% of all *liquid* freshwater; the rest is in lakes, wetlands, and rivers.

Underground Purification. As water percolates through the soil, debris and bacteria from the surface are generally filtered out. However, water may dissolve and leach out certain minerals. Underground caverns, for example, are the result of the gradual leaching away of limestone (calcium carbonate). In most natural situations, the minerals that leach into groundwater are harmless. Indeed, calcium from limestone is considered beneficial to health. Thus, groundwater is generally high-quality freshwater that is safe for drinking. A few exceptions occur when the groundwater leaches minerals containing sulfide, arsenic, or other poisonous elements that make the water unsafe to drink.

Drawn by gravity, groundwater may move through aquifers until it finds some opening to the surface. These natural exits may be **seeps** or **springs**. In a seep, water flows out over a relatively wide area; in a spring, water exits the ground as a significant flow from a relatively small opening. As seeps and springs feed streams, lakes, and rivers, groundwater joins and becomes part of surface water. A spring will flow, however, only if it is *lower* than the water table. Whenever the water table drops below the level of the spring, the spring will dry up. Groundwater movement actually constitutes one of the loops in the hydrologic cycle.

Loops, Pools, and Fluxes in the Cycle

The hydrologic cycle consists of four physical processes: evaporation, condensation, precipitation, and gravitational flow. There are three principal loops in the cycle: (1) In the evapotranspiration loop (consisting of green water), the water evaporates and is returned by precipitation. On land, this water—the main source for natural ecosystems and rain-fed agriculture—is held as capillary water and then returns to the atmosphere by way of evapotranspiration. (2) In the surface runoff loop (containing blue water), the water runs across the ground surface and becomes part of the surface water system. (3) In the groundwater loop (also containing blue water), the water infiltrates, percolates down to join the groundwater, and then moves through aquifers, finally exiting through seeps, springs, or wells, where it rejoins the surface water. The surface runoff and groundwater loops are the usual focus for human water resource management.

In the hydrologic cycle, there are substantial exchanges of water among the land, the atmosphere, and the oceans; such exchanges are the cycle's fluxes. **Figure 10–8** shows estimates of these fluxes and of the sinks that hold water globally. About 97% of the Earth's water is in the oceans. Each year a volume of water evaporates off of the ocean surface that would cover it to 100 cm deep, about 425,000 cubic

kilometers. Additional water evaporates from land and ice. All together the hydrologic cycle annually moves the equivalent of 25 times the water in the Great Lakes. Absent from Figure 10–8, however, are the human impacts on the cycle, which are large and growing.

Human Impacts on the Hydrologic Cycle

A large share of the environmental problems we face today stem from direct or indirect impacts on the water cycle. These impacts can be classified into four categories: (1) changes to Earth's surface, (2) changes to Earth's climate, (3) atmospheric pollution, and (4) withdrawals (or diversions) for human use.

Changes to the Surface of Earth. Imagine a temperate forest, its trees rooted in thick soil, drawing water from below, sending water up and out leaves, each tree stopping raindrops in a storm, letting them run down the trunk. The soil and leaf litter slow the water, letting it soak in. Then imagine cutting down the trees. What would happen? Put up a

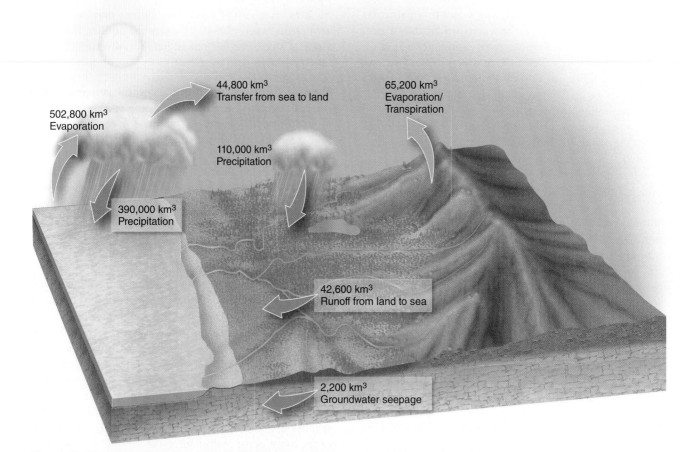

Figure 10–8 Yearly water balance in the hydrologic cycle. The data show (1) the contribution of water from the oceans to the land via evaporation and then precipitation, (2) the movement of water from the land to the oceans via runoff and groundwater seepage, and (3) the net balance of water movement between terrestrial and oceanic regions of Earth. Values are in cubic kilometers per year. (*Source:* Figure 4.2 from *Earth Science*, 8th ed., by Edward Tarbuck and Frederick K. Lutgens, copyright © 1997 by the authors, reprinted by permission of Pearson Education, Inc., Upper Saddle River, NJ; revised with data from *GEO Yearbook 2003*, United Nations Environment Programme.)

TABLE 10–2 U.S. Demands on Freshwater	
Use	**Gallons (Liters) Used per Person per Day**
Consumptive	
Irrigation and other agricultural use	485 (1,835)
Nonconsumptive	
Electric power production	477 (1,805)
Industrial use	72 (273)
Public supply and self-supply	164 (621)

Source: S. S. Hutson, N. L. Barber, J. F. Kenny, K. S. Linsey, D. S. Lumia, and M. A. Maupin, *Estimated Use of Water in the United States in 2000* (Circular 1268) (Reston, Va.: U.S. Geological Survey, 2004).

parking lot, a strip mall, a highway. Soon, the very processes that make up the water cycle will be altered. Alterations to the land surface stem primarily from removing vegetation, putting down impervious surfaces, and degrading wetlands.

Loss of Vegetation. In most natural ecosystems, precipitation is intercepted by vegetation and infiltrates into porous topsoil. From there, water provides the lifeblood of natural and human-created ecosystems. As forests are cleared or land is overgrazed and plants are not there to intercept rainfall, the pathway of the water cycle is shifted from infiltration and groundwater recharge to runoff, and the water runs into streams or rivers almost immediately. This causes floods, surface erosion, sediment, and other pollutants. Increased runoff means less infiltration and, therefore, less evapotranspiration and groundwater recharge. Lowered evapotranspiration means less moisture for local rainfall. Groundwater may be insufficient to keep springs flowing during dry periods. Dry, barren, and lifeless streambeds are typical of deforested regions—a tragedy for both the ecosystems and the humans who are dependent on the flow.

Taming Rivers and Wetlands. Wetlands function to store and release water in a manner similar to the way the groundwater reservoir does, only on a shorter time scale. Therefore, the destruction of wetlands has the same impact as deforestation: Flooding is increased, droughts are also increased, and waterways are polluted. In the Midwest, many fields have been provided with underground drainage pipes (tiles) to move water more quickly so the fields can be planted. This action lowers the water table, and the result is predictable: Rainwater is moved more quickly from the land to the rivers to the ocean. Channelizing rivers (by straightening them) has much the same effect (Chapter 7).

Building Impervious Surface. Roads, buildings, parking lots, and other structures actually cover the soil, so even a slow steady rain cannot infiltrate.[2] In some places,

impervious surface can cover 90% of the area. The lower 48 states have more than 43,000 square miles of impervious surface, about the size of Ohio **(Figure 10–9)**.

Dams. One of the chief ways people try to lower flood risk is by building dams. Dams store water and have the opposite effect from many of the other land changes: They slow the movement of water to the ocean. However, because they are not wetlands, they often do not result in increased groundwater recharge. In dry areas, the pooling of water in reservoirs can leave water exposed to evaporation. Ironically, evaporative loss of water can cause regional water loss even when the object of the reservoir building was water provision during drought. Dams also cause some serious ecological problems. The Three Gorges Dam on the Yangtze River is a good example.

Flooding on the Yangtze River took more than 300,000 lives in the 20th century. This fact, combined with a need for power, prompted the building of one of the most spectacular dams in the world, the Three Gorges Dam. Completed in 2006, it is the largest hydroelectric project in the world (in terms of its maximum capacity), holding back 1.4 trillion cubic feet of water and generating as much as 84.7 billion kilowatt-hours of electricity annually, as much as the energy produced by burning 50 million tons of coal. The power generated raised the proportion of China's electricity generated from renewable sources from 7% to 15%. It will, for a time, bring much of the major flood potential of the Yangtze under control.

However, the dam has come with a high price. More than 1.3 million people have been displaced and relocated, and another 3 million to 4 million people will be relocated in coming years to make way for the 600-kilometer-long (410-mile long) reservoir. As the reservoir fills, lowering the Yangtze River flow by 50%, saltwater has crept upriver at the mouth of the river, ships have been stranded, pollutants have accumulated, and many fish populations have been reduced due to interference with breeding cycles. In spite

Figure 10–9 Impervious surface.
Impervious surface such as seen on roofs, roads and parking lots, decreases the amount of water that can go into the soil and recharge groundwater and increases the amount of water and the speed with which water runs off surfaces.

---

[2]Schueler, Thomas R. "The Importance of Imperviousness." Reprinted in *The Practice of Watershed Protection.* 2000. Center for Watershed Protection, Ellicott City, MD.

of the many drawbacks, Chinese officials continue with the project.

Moving Water to the Ocean. Overall, almost all human actions on land surface move water more quickly from rain to the ocean and lessen the amount going into groundwater. These actions contribute to both floods and droughts and increase erosion (Chapter 11). Floods are one of the most costly problems with the water cycle and have always been common. In many parts of the world, however, the frequency and severity of flooding are increasing—not because precipitation is greater, but because land use has changed, as we have described. Deforestation, cultivation, impervious surfaces, wetland loss, and other changes have increased, causing erosion and reducing infiltration. A recent study of floods in 56 developing countries has confirmed the "sponge" effect of forests, concluding that continued loss of forests will lead to greater loss of lives and economies for the developing countries where floods are frequent. For example, extreme flooding in Bangladesh (a very flat country only a few feet above sea level) is now more common because the Himalayan foothills in India and Nepal have been deforested. Due to sediment deposited from upriver erosion, the Ganges River basin has risen 5–7 meters (15–22 ft) in recent years. Floods are also increasing as a result of climate change, discussed shortly. **Figure 10–10** shows the impacts of flood waters in flooding that occurred in 2010 in Pakistan.

Solutions to water problems caused by land use changes include changes in requirements for new construction. One change in land use planning has been the requirement in many places to build flood retention ponds near new construction. These areas do not function as wetland habitat very well and often have pollutants and generalist weedy species. However, they do stop floods and allow infiltration. The protection of wetlands and river vegetation has been important to slowing run-off. Another solution is to grow more vegetation in cities. Urban gardening, rooftop gardens, and the planning of parks can increase soil infiltration in some cases and can also lower heat. There is also a move to go to

new types of surfaces, especially for driveways, that are more environmentally friendly. These solutions will be discussed under sustainable cities as well (Chapter 23).

Climate Change. There is now unmistakable evidence that Earth's climate is warming because of the rise in greenhouse gases, and as this occurs, the water cycle is being altered (Chapter 18). A warmer climate means more evaporation from land surfaces, plants, and water bodies because evaporation increases with temperature. A wetter atmosphere means more and, frequently, heavier precipitation and more flood events. In effect, global warming is believed to be speeding up the hydrologic cycle. Already, the United States and Canada have experienced a 5–10% increase in precipitation over the past century. Regional and local changes are difficult to project; different computer models predict different outcomes for a given region. However, global climate models project a radically changing hydrological world as the 21st century unfolds (Chapter 18). The climatologists calculate that a warmer climate will likely generate more hurricanes and more droughts; extended droughts in the western United States are likely to occur more frequently. They also calculate that many of the currently water-stressed areas like East Africa and the Sahel region will get less water, which will bring large agricultural losses (Chapter 12).

Climate change has two other key impacts on water bodies. First, warm water has a greater volume. Warming the oceans contributes to sea level rise. Second is the movement of water from ice (the cryosphere) to liquid water. Glaciers and polar ice melt, and that water makes its way to the ocean. Both thermal expansion and melting ice contribute to the sea level rise that is already apparent. Melting ice adds freshwater, which is lighter than saltwater and does not mix well. The influx of great volumes of freshwater may affect ocean currents (see Sound Science, "What's New in the Water Cycle?"). Models suggest that oceans will rise about 2 to 3 feet (0.8 to 1 m) by 2100. While that does not sound like much, more than 630 million people live in low elevation areas, less than 10 meters above sea level. You can imagine that many will have to move and others will experience flooding.

Atmospheric Particles. Tiny atmospheric particles (aerosols) form nuclei that enable water to condense into droplets. The more such particles there are, the greater is the tendency for clouds to form. Anthropogenic aerosols are on the increase, primarily in the form of sulfates (from sulfur dioxide in coal), carbon (as soot), and dust. They form a brownish haze that is associated with industrial areas, tropical burning, and dust storms. Their impact on cloud formation is substantial, and where these aerosols occur, solar radiation to Earth's surface is reduced (so they have a cooling effect).

Their most significant impact, however, is on the hydrologic cycle. Because aerosols promote the formation of smaller-than-normal droplets in the clouds, aerosols actually *suppress* rainfall where they occur in abundance, even though they encourage cloud formation. As they do so, the

Figure 10–10 Flooding in Pakistan. The 2010 floods in Pakistan broke records and walls. Twenty million people were displaced.

SOUND SCIENCE

The water cycle is pretty straightforward on a first approximation. Every fourth grader has studied at least the basics. But scientists still find more to study. The cutting edge in the study of hydrologic cycle science is in trying to figure out what will cause changes in the future, especially how the global water cycle will be affected by human actions.

Peter Wadhams, a professor of oceanographic physics at the University of Cambridge, England, wanted to know more about the ocean and climate change. One big question he was interested in is how the ocean responds to changes in the water cycle as ocean temperatures rise. There are large currents (thermohaline currents) in the water that move vast quantities of water around, in a slow-moving "ocean conveyor." Wind and tides provide the energy for these currents. The speed, volume, and direction of currents can be affected by climate change because of two principles: cold water is denser than warm, and freshwater is less dense than saltwater. When deep water rises, it brings nutrients toward the surface; as surface water sinks, it brings oxygen down. Waters warmed in the tropics redistribute heat to the cold poles.

These currents help warm Europe and the east coast of North America. But there may be changes in the future. As glacial melt water and rain run-off from Siberia increase in the Arctic, cold freshwater forms a layer on top of the saltwater there. Enough freshwater might change the ocean conveyor, some scientists have hypothesized. In 2005, Wadhams and his colleagues investigated the cold dense waters of the Greenland Sea where columns of very dense

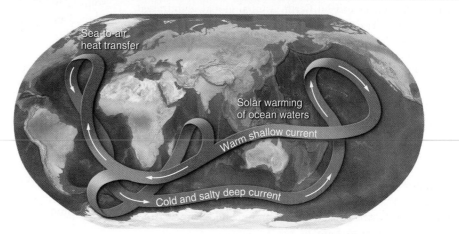

water form. These "chimneys" (which can be 100 km in diameter and which last for months) carry water down more than 2400 m towards the sea floor, contributing to the movement of masses of cold water in the deep ocean. But instead of the dozen gigantic rivers of water flowing downward that they had found other times, they found only two. Does this mean the deep ocean currents will slow? Is the change long term? Right now, Wadhams and others aren't sure what will happen. Some scientists say any shift in the thermohaline circulation is likely to occur slowly; others think a slowing of the conveyor could occur more abruptly.

The Greenland Sea isn't the only place marine currents are studied. In 2012, scientists from the University of Washington, including graduate student Sarah Purkey, explored the deepest, coldest waters on Earth—off of Antarctica. They found that those waters,

dubbed the Antarctic Bottom Water, have declined in volume by 60% (losing an equivalent of 50 times the flow of the Mississippi River) over the past several decades. The deep ocean is becoming less cold and dense, and scientists expect this to change the way the oceans work. Even though the basic hydrologic cycle is well known, the details of how it works and what changes we might expect in a changing environment are still part of today's science. Wadhams, Purkey and all of their colleagues are on the cutting edge of sound science.

Sources: Wunsch, C. (2002). "What is the thermohaline circulation?" *Science* 298(5596): 1179–1181. doi:10.1126/science.1079329. PMID 12424356.
Wadhams, P. (2004). "Convective chimneys in the Greenland Sea: A review of recent observations." *Oceanography and Marine Biology* 42, 1–27
Purkey, S. G., & Johnson, G. C. "Global contraction of Antarctic Bottom Water between the 1980s and 2000s.". *Journal of Climate*, 2012

atmospheric cleansing that would normally clear the aerosols is suppressed, and they remain in the atmosphere longer than usual. With suppressed rainfall comes drier conditions, so more dust and smoke (and more aerosols) are the result. Ironically, this type of air pollution works just opposite to the impact of the greenhouse gases on climate and the hydrologic cycle. Again, there are significant differences; the aerosol impact is more local, whereas the impact of greenhouse gases on climate and the hydrologic cycle is more global. In addition, the greenhouse gases are gradually accumulating in the atmosphere, while the aerosols have a lifetime measured in days.

Withdrawal for Human Use. Finally, there are many problems related to the fourth impact, diversion and

withdrawals of water for human use (as we make use of surface waters and groundwater). We will discuss these impacts next, as we look at how humans try to have enough clean freshwater—and not too much.

10.3 Water: Getting Enough, Controlling Excess

Meeting water needs is a primary goal in our rapidly crowding world. We are seeing both stresses and successes. Between 1990 and 2012, 2 billion people were given access to clean drinking water, an achievement that met the Millennium Development Goal for water. However, water needs are very serious in many countries (**Fig. 10–11**).

According to the Millennium Ecosystem Assessment, human use now appropriates about 3,600 cubic kilometers of water per year, which represents 27% of the accessible freshwater runoff on Earth. Annual global withdrawal is expected to rise by about 10% every decade. While the human population tripled in the 20th century, use of water increased sixfold, much of that increase to enable water intensive higher-yielding agriculture.

The U.S. Geological Service periodically prepares estimates of major uses of freshwater in the United States (Table 10–2). It is significant that Americans use less water now than they did in 1980, even with population increases (Fig. 10–12). Most of the water used in homes and industries is for washing and flushing away unwanted materials, and the water used in electric power production is for taking away waste heat. These are **nonconsumptive water uses** because the water, though now contaminated with wastes, remains available to humans for the same or other uses if its quality is adequate or if it can be treated to remove undesirable materials. In contrast, irrigation is a **consumptive water use** because the applied water does not return to the local water resource. It can only percolate into the ground or return to the atmosphere through evapotranspiration. In either case, the water does reenter the overall water cycle but is gone from human control.

Worldwide, the largest use of water is for irrigation (70%), second is for industry (20%), and third is for direct human use (10%) (Fig. 10–13). These percentages vary greatly from one region to another, depending on natural precipitation and the degree to which the region is developed (Fig. 10–10). Most increases in withdrawal are due to increases in irrigated lands. In the United States, irrigation accounts for 65% of freshwater consumption.

Sources

About 20% of domestic water comes from groundwater sources and 80% from surface waters (rivers, lakes, reservoirs) in the United States. Before municipal water supplies were developed, people drew their water from whatever source they could: wells, rivers, lakes, or rainwater. This approach is still used in rural areas in the developing countries. Women in many of these countries walk long distances each day to fetch water. Because surface waters and shallow wells often receive runoff, they frequently are polluted with various wastes, including animal excrement and human sewage likely to contain pathogens (disease-causing organisms). In fact, an estimated 90% of wastewater in developing countries is released to surface waters without any treatment. Yet, unsafe as it is, this polluted water is the only water available for an estimated 1.1 billion people in less developed countries (Fig. 10–14 on page 252). It is commonly consumed without purification, but not without consequences: According to the UN World Health Organization, contaminated water is responsible for the deaths of more than 1.6 million people in developing countries each year, 90% of whom are children. Increasing water sanitation is part of the Millennium Development Goals.

Technologies. In the industrialized countries, the collection, treatment, and distribution of water are highly developed. Larger municipalities rely primarily on surface-water sources, while smaller water systems tend to use groundwater. In the former, dams are built across rivers to create reservoirs, which hold water in times of excess flow and can be drawn down at times of lower flow. In addition, dams and reservoirs may offer power generation, recreation, and flood control. Reservoirs created by dams on rivers are also major sources of water for irrigation. In this case, no treatment is required. However, water for

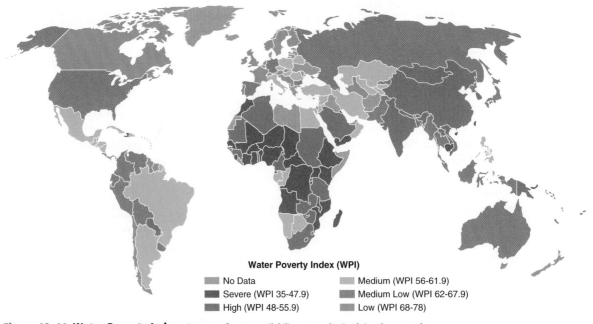

Water Poverty Index (WPI)

No Data · Severe (WPI 35-47.9) · High (WPI 48-55.9) · Medium (WPI 56-61.9) · Medium Low (WPI 62-67.9) · Low (WPI 68-78)

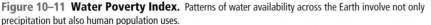

Figure 10–11 Water Poverty Index. Patterns of water availability across the Earth involve not only precipitation but also human population uses.

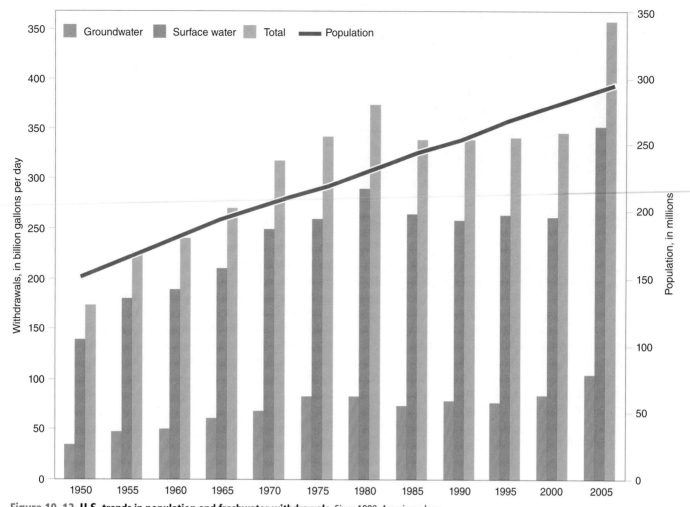

Figure 10–12 U.S. trends in population and freshwater withdrawals. Since 1980, Americans have
been using less water, even though the population has increased substantially. (*Source:* S. S. Hutson, N. L. Barber,
J. F. Kenny, K. S. Linsey, D. S. Lumia, and M. A. Maupin, *Estimated Use of Water in the United States in 2000*
(Circular 1268) (Reston, VA: U.S. Geological Survey, 2004).)

municipal use is piped from the reservoir to a treatment plant,
where it is treated to kill pathogens and remove undesirable
materials, as shown in **Figure 10–15.** After treatment, the water
is distributed through the water system to homes, schools, and
industries. Wastewater, collected by the sewage system, is carried
to a sewage-treatment plant, where it is treated before being

discharged into a natural waterway (Chapter 20). Often, wastewater is discharged into the same river from which it was withdrawn, but farther downstream.

Whenever possible, both water and sewage systems
are laid out so that gravity maintains the flow through the
system. This arrangement minimizes pumping costs and

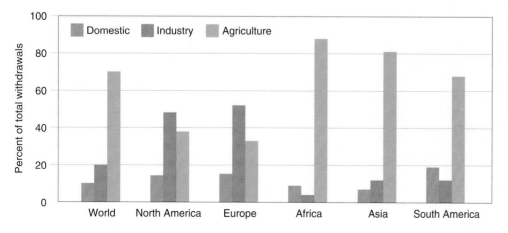

**Figure 10–13 Regional usage of
water.** The percentage used in each
category varies with the climate and relative
development of the region. A less developed
region with a dry climate (e.g., Africa) uses
most of its water for irrigation, whereas an
industrialized region (e.g., Europe) requires
the largest percentage for industry.

Figure 10–14 Water in the developing world. In many villages and cities of the developing world, people withdraw water from rivers, streams, and ponds. These sources are often contaminated with pathogens and other pollutants. This Kenyan man is collecting water from a river used by people and animals.

increases reliability. On major rivers, such as the Mississippi, water is reused many times. Each city along the river takes water, treats it, uses it, treats it again, and then returns it to the river. In developing nations, the wastewater frequently is discharged with minimal or no treatment. Thus, as the water moves downstream, each city has a higher load of pollutants to contend with than the previous city had, and ecosystems at the end of the line may be severely affected by the

pollution. Pollutants include industrial wastes as well as pollutants from households because industries and residences generally utilize the same water and sewer systems.

Smaller public drinking-water systems frequently rely on groundwater, which often may be used with minimal treatment because it has already been purified by percolation through soil. More than 1.5 billion people around the world rely on groundwater for their domestic use. Both surface water and groundwater are replenished through the water cycle (Fig. 10–3). Therefore, in theory at least, these two sources of water represent a sustainable or renewable (self-replenishing) resource, but they are not inexhaustible. Over the past 50 years, groundwater use has outpaced the rate of aquifer recharge in many areas of the world.

Is Bottled Water the Answer? One-half of Americans buy bottled water, which is sometimes seen as a solution to limited water availability, particularly after disasters. Sales of bottled water were $34 billion (about 50 billion bottles) in 2008 in the United States alone. Unfortunately, use of bottled water has a number of negative environmental effects. First, the bottling can occur on a large scale in areas where locals also want to use the water. Does a company have the right to pump out millions of gallons of water from an aquifer? This has been disputed. For example, after many years in the courts, in 2009, Michigan Citizens for Water Conservation won a lawsuit to keep Nestlé Waters North America, Inc., from pumping 400 gallons of water a minute from a

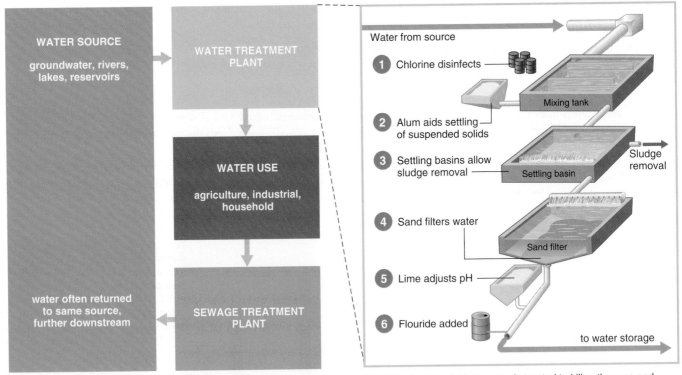

(a) Water is often taken from a river or reservoir, piped to the treatment plant, treated, used, then returned to the source.

(b) At the treatment plant, water is treated to kill pathogens and remove undesirable material.

Figure 10–15 Municipal water use and treatment. The stages of water treatment are outlined in the figure. Water is taken from a river or reservoir, piped to a treatment plant, treated with chemicals, filtered, adjusted for pH, and stored before being used.

groundwater supply in Michigan. The proposed pumping was drying a local stream. Nestlé was allowed to pump, but at a reduced rate. Bottled water also requires a great deal of energy, for the making of plastic bottles and the transportation of the product. Overall, bottled water may be good for emergencies but is not a good solution for a lack of drinking water generally.

Surface Waters

Surface waters are diverted for use, slowed and held in dams, and moved along swiftly by levees, impervious surfaces, channelized rivers, and devegetation. Of these effects, the use of dams is the most prominent for keeping and using surface water.

Dams and the Environment. To trap and control flowing rivers, more than 48,000 large dams (more than 50 feet high) have been built around the world. Dams together have a storage capacity around 6,000 km³, serving 30–40% of the world's irrigated land and providing 19% of the world's total electricity supply. Of the 42,600 cubic kilometers of annual runoff in the hydrologic cycle, only 31% is accessible for withdrawal because many rivers are in remote locations and much of the remaining water is required for navigation, flood control, and the generation of hydroelectric power. Sixty-five percent of ocean bound freshwater is stopped by dams at some point. The large dams have an enormous direct social impact, leading to the displacement of an estimated 40 million to 80 million people worldwide (primarily in China and India) and preventing access by local people to the goods and services of the now-buried ecosystems.

Dam Impacts. The United States is home to some 75,000 dams at least 6 feet in height and an estimated 2 million smaller structures (**Fig. 10–16**). These have been built to run mills (now an obsolete usage), control floods, generate electric power, and provide water for municipal and agricultural use. These dams have enormous ecological impacts. When a river is dammed, valuable freshwater habitats, such as waterfalls, rapids, and prime fish runs, are lost. When the river's flow is diverted to cities or croplands, the waterway below the diversion is deprived of that much water.

The impact on fish and other aquatic organisms is obvious, but the ecological ramifications go far beyond the river. Wildlife that depends on the water or on food chains involving aquatic organisms is also adversely affected. Wetlands dry up, resulting in frequent die-offs of waterfowl and other wildlife that depended on those habitats. Fish such as salmon, which swim from the ocean far upriver to spawn, are seriously affected by the reduced water level and have problems getting around the dam, even one equipped with fish ladders (a stepwise series of pools on the side of the dam, where the water flows in small falls that fish can negotiate). If the fish do get upriver, the hatchlings have similar problems getting back to the ocean. On the Columbia and Snake rivers, juvenile salmon suffer 95% mortality in their journey to the sea as a result of negotiating the dams and reservoirs that block their way.

Glen Canyon Dam is an example of a large dam with significant environmental effects. On the Columbia River, the dam provides hydropower and created the second largest artificial lake in America: Lake Powell. An environmental impact study in the late 1980s and early 1990s concluded that the operation of the Glen Canyon Dam on the Colorado River had seriously damaged the downstream ecology of the Colorado River and its recreational resources (hiking, river rafting). In response, Congress passed the Grand Canyon Protection Act in 1992, which required a process of "adaptive management" that would adjust

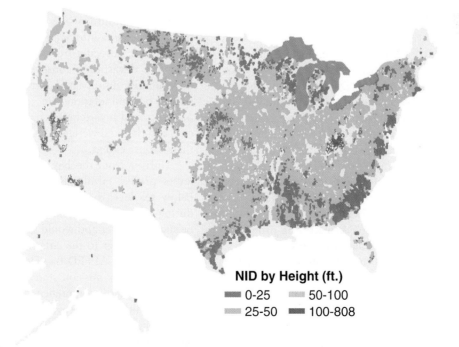

NID by Height (ft.)
- 0-25
- 25-50
- 50-100
- 100-808

Figure 10–16 Number of dams in the United States. Each point represents a dam; the colors tell the height. (Source National Inventory of Dams Map)

Figure 10–17 Removal of a dam. The 2011 removal of the Conditt Dam on the White Salmon River in Washington state cleared the way for salmon to spawn upstream.

water-release rates in a manner designed to enhance the ecological and recreational values of the river. This rule has helped balance the needs of different stakeholders.

Dams Gone. Increasingly, people are recognizing the costs of dams and finding them to be unacceptable. Removing a dam is not easy, however. Legal complexities abound where long-existing uses of a dam (to control floods or to provide irrigation water or lake-like still water for recreation) conflict with the expected advantages of removal (to reestablish historic fisheries such as salmon and steelhead runs or to restore the river for recreational and aesthetic use). Practical problems also exist. However, more than 714 dams have already been removed. **Figure 10–17** shows the successful removal of a dam on the White Salmon River.

Diversion of Surface Water. In addition to dams, humans move freshwater through diversion projects to take water to places where it is scarce. One example is the diversion project in China mentioned at the beginning of the chapter. When water is moved from one watershed to another, it is called an interbasin water transfer. The largest diversion project ever proposed for North America was the North American Water and Power Alliance (NAWAPA) XXI project. Designed in the 1950s, the ambitious project was proposed to bring water from Alaska, down through Canada, and into the dry Midwest and western states through 369 separate construction projects. As proposed, it would double the water available to the lower 48 states. Public sentiment in Canada has run against the project, and it has been shelved. China has discovered already, though, that large water diversion projects can bring unexpected environmental costs including water pollution or the effects of water loss in the original rivers and lakes.

Impacts on Estuaries. When river waters are diverted on their way to the sea, serious impacts are seen in **estuaries**—bays and rivers in which freshwater mixes with seawater. Estuaries are among the most productive ecosystems on Earth; they are rich breeding grounds for many species of fish, shellfish, and waterfowl. As a river's flow is diverted (say, to irrigate fields), less freshwater enters and flushes the estuary. Consequently, the salt concentration increases, profoundly affecting the estuary's ecology. This is exactly what is happening in San Francisco Bay. More than 60% of the freshwater that once flowed from rivers into the bay has been diverted for irrigation in the Central Valley (2.5 million acres irrigated) and for municipal use in southern California (25 million people served). In dry years (which are becoming more common), the farms and municipalities remove a third of the water that would normally flow through the delta into San Francisco Bay. Without the freshwater flows, saltwater from the Pacific has intruded into the bay, with devastating consequences. Chinook salmon runs have almost disappeared. Sturgeon, Dungeness crab, and striped bass populations are greatly reduced. Exotic species of plants and invertebrates have replaced many native species, and the tidal wetlands have been reduced to only 8% of their former extent.

Upriver from the bay, more than a thousand miles of levees protect farmlands that have sunk 20 feet or more below sea level due to compacting. If the levees collapsed during an earthquake, massive amounts of saltwater would be sucked upriver and would shut down the huge pumps that divert freshwater to the cities and farms. Given these major problems, the CALFED Bay-Delta Program was established, according to its mission statement, to "develop and implement a long-term comprehensive plan that will restore ecological health and improve water management for beneficial uses of the Bay-Delta System." However, this is not working, as major stakeholders (farmers, government agencies, environmental groups) have not been able to agree on changes to the delta recommended by scientists.

(a)

(b)

(c)

Figure 10–18 Exploitation of an aquifer. (a) Pumping up water from the Ogallala aquifer has made this arid region of the United States into some of the most productive farmland in the country. (b) Water is applied by means of center-pivot irrigation, in which the water is pumped from a central well to a self-powered boom that rotates around the well, spraying water as it goes. (c) An aerial photograph shows the extent of center-pivot irrigation throughout the region. Groundwater depletion will bring an end to this kind of farming.

The problem is not limited to the United States. The southeastern end of the Mediterranean Sea was formerly flushed by water from the Nile River. Because this water is now held back and diverted for irrigation by the Aswan High Dam in Egypt, that part of the Mediterranean is suffering severe ecological consequences (such as severe coastal erosion) as sediments normally flowing out to the coast are kept behind the dam.

Groundwater

Groundwater is the large unseen storage of water that accumulated slowly in porous rock over millennia. As mentioned before, some 99% of all liquid freshwater is in underground aquifers. Of this groundwater, more than three-fourths has a recharge rate of centuries or more. Such groundwater is considered "nonrenewable." Because groundwater supplies are vital for domestic use and especially for agriculture, they are a resource that requires management.

Groundwater Levels. The agriculture of the midwestern United States makes the country the breadbasket for the world. Water to support that agriculture is mined from deep underground. Within the seven-state High Plains region of the

United States, the Ogallala aquifer supplies irrigation water to 4.2 million hectares (10.4 million acres), one-fifth of the irrigated land in the nation **(Fig. 10–18)**. This aquifer is probably the largest in the world, but it is mostly "fossil water," recharged during the last ice age. The recent withdrawal rate has been about 28 billion cubic meters (23 million acre-ft) per year, two orders of magnitude higher than the recharge rate. Water tables have dropped 30–60 meters (100–200 ft) and continue to drop at the rate of 6 feet per year. Irrigated farming has already come to a halt in some sections, and it is predicted that over the next 20 years another 1.2 million hectares (3 million acres) in this region will be abandoned or converted to dryland farming (ranching and the production of forage crops) because of water depletion.

Renewable groundwater is replenished by the percolation of precipitation water, so it is vulnerable to variations in precipitation. In tapping groundwater, we are tapping large, but not unlimited, natural reservoirs, contained within specific aquifer systems. Its sustainability ultimately depends on balancing withdrawals with rates of recharge. For most of the groundwater in arid regions, there is essentially no recharge. The resource must be considered nonrenewable. Like oil, it can be removed, but any water removed today will be unavailable for the future. Groundwater depletion has several undesirable consequences.

Falling Water Tables. The simplest indication that groundwater withdrawals are exceeding recharge is a falling water table, a situation that is common throughout the world. Because irrigation consumes far and away the largest amount of freshwater, depleting water resources will ultimately have its most significant impact on crop production. Although running out of water is the obvious eventual conclusion of overdrawing groundwater, falling water tables can bring on other problems.

Diminishing Surface Water. Surface waters are affected by falling water tables. In various wetlands, for instance, the water table is essentially at or slightly above the ground surface. When water tables drop, these wetlands dry up, with the ecological results described earlier. Further, as water tables drop, springs and seeps dry up as well, diminishing even streams and rivers to the point of dryness. Thus, excessive groundwater removal creates the same results as the diversion of surface water.

Land Subsidence. Over the ages, groundwater has leached cavities in the ground. Where these spaces are filled with water, the water helps support the overlying rock and soil. As the water table drops, however, this support is lost. Then there may be a gradual settling of the land, a phenomenon known as **land subsidence**. The rate of sinking may be 10–15 centimeters (6–12 in.) per year. In some areas of the San Joaquin Valley in California, land has settled as much as 9 meters (29 ft) because of groundwater removal. Land subsidence causes building foundations, roadways, and water and sewer lines to crack. In coastal areas, subsidence causes flooding unless levees are built for protection. For example, where a 10,000-square-kilometer (4,000-mi²) area in the Houston–Galveston Bay region of Texas is gradually sinking because of groundwater removal, coastal properties are being abandoned as they are gradually being inundated by the sea. Land subsidence is also a serious problem in many other places throughout the world. When it happens in cities, underground pipes break, causing leaks and wastage of domestic water. Sewage pipes can also fracture, contaminating the groundwater aquifer. Subsidence is a serious problem in many cities in developing countries, where the public water supply is not dependable.

Another kind of land subsidence, a **sinkhole**, may develop suddenly and dramatically (**Fig. 10–19**). A sinkhole results when an underground cavern, drained of its supporting groundwater, suddenly collapses. Sinkholes may be 300 feet (91 m) across and as deep as 150 feet. The problem of sinkholes is particularly severe in the southeastern United States, where groundwater has leached numerous passageways and caverns through ancient beds of underlying limestone. An estimated 4,000 sinkholes have formed in Alabama alone, some of which have "consumed" buildings, livestock, and sections of highways.

Saltwater Intrusion. Another problem resulting from dropping water tables is **saltwater intrusion**. In coastal

Figure 10–19 Sinkhole. The removal of groundwater may drain an underground cavern until the roof, no longer supported by water pressure, collapses. The result is the sudden development of a sinkhole, such as this one, which appeared in Plant City, Florida, on May 29, 2000.

regions, springs of outflowing groundwater may lie under the ocean. As long as a high water table maintains a sufficient head of pressure in the aquifer, freshwater will flow into the ocean. Thus, wells near the ocean yield freshwater (**Fig. 10–20a**). However, lowering the water table or removing groundwater at a rapid rate may reduce the pressure in the aquifer, permitting saltwater to flow back into the aquifer and hence into wells (Fig. 10–20b). Saltwater intrusion is a serious problem in many European countries along the Mediterranean coast.

10.4 Water Stewardship: Supply and Public Policy

The hydrologic cycle is entirely adequate to meet human needs for freshwater because it processes several times as much water as we require today. However, the water is often not distributed where it is most needed, and the result is a persistent scarcity of water in many parts of the world. In the developing countries, there is still a deficit of infrastructure, such as wells, water-treatment systems, and (sometimes) large dams, for capturing and distributing safe drinking water. Despite the growing negative impacts of overdrawing water resources, expanding populations create an ever-increasing demand for additional water for agriculture, industry, and municipal use. Water for food dominates the current water demand, estimated at 6,800 cubic kilometers per year. Of this, some 5,000 cubic kilometers are used as green water flow in the world's rain-fed agriculture and 1,800 cubic kilometers as blue water flow in the form of irrigation water withdrawn from rivers, lakes, and groundwater. It is estimated that 4,500 cubic kilometers (of the total 6,800 km³) are attributed to the developing countries to fulfill current food needs.

If we assume that virtually all of the future population increase will be in the developing countries, then so also will be the increased need for food.

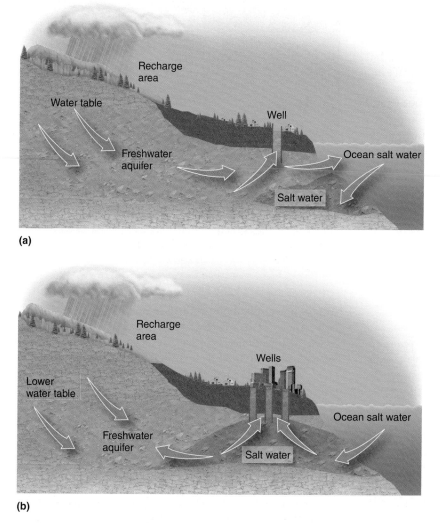

Figure 10–20 Saltwater intrusion. (a) Where aquifers open into the ocean, freshwater is maintained in the aquifer by the pressure of freshwater inland. (b) Excessive removal of water may reduce the pressure so that saltwater moves into the aquifer.

What are the possibilities for meeting existing needs and growing demands in a sustainable way? Planners call for a **Blue Revolution**, a radical change in the way we use water. There are only about five options: (1) capture more of the runoff water, (2) gain better access to existing groundwater aquifers, (3) desalt seawater, (4) conserve present supplies by using less water, and (5) make food production more efficient. Making matters more difficult, there is no single U.S. agency responsible for forming public policy on such water supply issues.

Capturing Runoff

While individuals may put out rain barrels to capture runoff from roofs and villages may use cisterns to catch rainwater, dams are the best example of surface water capture and are widely used. China is the best example of both the pros and cons of dam building, which we have already discussed. In an effort to provide hydropower while cutting greenhouse gas emissions, China has committed to an aggressive program of dam building. New dam projects threaten China's remaining biodiversity. In the United States, protection has been accorded to some rivers with the passage of the Wild and Scenic Rivers Act of 1968, which keeps rivers designated as "wild

and scenic" from being dammed or affected by other harmful operations. There are 252 designated wild and scenic rivers in the United States. Some 12,600 miles of rivers have been protected. The numerous large and small dams across the country have affected some 600,000 miles of our rivers, by comparison. Those rivers designated as wild and scenic are in many ways equivalent to national parks.[3]

Tapping More Groundwater

Already, more than 2 billion people depend on groundwater supplies. In many areas, groundwater use exceeds aquifer recharge, leading to shortages as the water table drops below pump levels. Groundwater depletion is considered the single greatest threat to irrigated agriculture, and it is happening in many parts of the world. India, China, West Asia, countries of the former Soviet Union, the western United States, and the Arabian Peninsula are all experiencing declining water tables. Calculations by the International Water Management Institute suggest that an expansion of

---

[3]Wild and Scenic Rivers: Leading the Charge to Protect the Nation's Last, Best Rivers. American Rivers. http://www.americanrivers.org/our-work/protecting-rivers/wild-and-scenic/ Sept 2012.

irrigation on the order of 20% might be possible, but this will not meet the food needs in the developing countries by 2030.[4]

Exploiting renewable groundwater will continue to be an option, but it is unlikely to provide great increases in water supply because these same sustainable supplies, recharged by annual precipitation, are being increasingly polluted. Agricultural chemicals such as fertilizers and pesticides, animal wastes, and industrial chemicals readily enter groundwater, making groundwater pollution as much of a threat to domestic water use as depletion. For example, some 60,000 square kilometers of aquifers in western and central Europe will likely be contaminated with pesticides and fertilizers within the next 50 years. Also, India and Bangladesh are plagued with arsenic-contaminated groundwater, which has extracted a heavy toll in sickness and death for millions in those countries.

Figure 10–21 Desalination plant. The Israel Ashkelon Seawater desalination plant is the largest reverse osmosis desalinization plant in the world, producing 330,000 m3/d of drinking water.

Desalting Seawater

The world's oceans are an inexhaustible source of water, not only because they are vast, but also because any water removed from them will ultimately flow back in. Not many plants can be grown in full seawater, however, and although researchers are developing salt-tolerant plants via genetic modification, we are not there yet. Thus, with increasing water shortages and most of the world's population living near coasts, there is a growing trend toward **desalination** (desalting) of seawater for domestic use. More than 15,000 desalination plants have been installed in some 120 countries, especially in the Middle East. Many Australian cities, such as Perth, also rely on desalination. Perth relies on water from the Indian Ocean, desalinized in a plant powered by wind turbines.

Two technologies—microfiltration (reverse osmosis, or RO) and distillation—are commonly used for desalination. Smaller desalination plants generally employ microfiltration, in which great pressure forces seawater through a reverse osmosis membrane filter fine enough to remove the salt. Larger facilities, particularly those that can draw from a source of waste heat (for example, from electrical power plants), generally use distillation (evaporation and condensation of vapor). Efficiency is gained by using the heat given off by condensing water to heat the incoming water. Even where waste heat is used, however, the costs of building and maintaining the plant, which is subject to corrosion from seawater, are considerable. Costs can be two to four times what most city dwellers in the United States currently pay, but it is still not a high price to pay for drinking water in areas of the world that are prone to drought and is much cheaper than bottled water.

In January 2008, the Tampa Bay seawater desalination plant was completed, signaling a new day for desalination **(Fig. 10–21)**. The plant is the country's first ever built to serve as a primary water source. The plant uses salty cooling water

from a power plant and reverse osmosis to produce 25 million gallons of drinking water per day (MGD), providing Pasco County, Pinellas County, St. Petersburg, and other cities with some 10% of their drinking water. The Tampa Bay region has experienced chronic water shortages and is exploiting an opportunity that has attracted worldwide attention. The water is expected to cost Tampa Bay Water $2.49 per 1,000 gallons wholesale over 30 years—more than the $1 per 1,000 gallons for existing groundwater sources, but still a reasonable cost.

Although the higher cost might cause some people to cut back on watering lawns and to implement other conservation measures, most people in the United States could afford desalinized water without unduly altering their lifestyles. For irrigating croplands, however, the higher cost of desalinized seawater would probably be prohibitive, except possibly for high-value cash crops like greenhouse flowers or vegetables.

The world's largest RO desalination plant employs a unique energy recovery device (a rotary displacement pump) that is helping greatly to make desalination affordable. It was built at Hadara, between Tel Aviv and Haifa in Israel, and will produce 72 MGD, with a potential expansion to 94 MGD. The largest distillation plant is the Jebel Ali desalination plant in the United Arab Emirates, producing 215 MGD, or 2,500 gallons per second.

Using Less Water

A developing-nation family living where water must be carried several miles from a well finds that one gallon per person per day is sufficient to provide for all of its essential needs, including cooking and washing. Yet, a typical household in the United States consumes an average of 380 liters (100 gal) per person per day. If all indirect uses are added (especially irrigation), this figure increases to 4,900 liters (1,300 gal) per person per day. Similarly, a peasant farmer may irrigate by carefully ladling water onto each plant with a dipper, while typical modern irrigation floods the whole field. The way water resource planners think is beginning

[4]M. Falkenmark and J. Rockström, "The New Blue and Green Water Paradigm: Breaking Ground for Water Resources Planning and Management," *Journal of Water Resources Planning and Management* 132 (May/June 2006): pg 131

to change, however: Instead of asking how much water we "need" and where can we get it, people are now asking how much water is available and how can we best use it. The good news is that the U.S. rate of water use per capita has actually begun to drop, which can be attributed to some well-needed water conservation strategies already put in place. Some specific measures that are being implemented to reduce water withdrawals are described in the next several subsections.

Agriculture. Agriculture is far and away the largest consumer of freshwater—some 40% of the world's food is grown in irrigated soils, which produce 100–400% higher yields than traditional farming. Most present-day irrigation wastes huge amounts of water, and in fact, half of it never yields any food. Where irrigation water is applied by traditional flood or center-pivot systems, 30–50% is lost to evaporation, percolation, or runoff. Several strategies have been employed recently to cut down on this waste. One is the surge flow method, in which computers control the periodic release of water, in contrast to the continuous-flood method. Surge flow can cut water use in half.

Drip, Drip. Another water-saving method is the drip irrigation system, a network of plastic pipes with pinholes that literally drip water at the base of each plant (**Fig. 10–22**). Although such systems are costly, they waste much less water. Also, they have the added benefit of retarding salinization (see Chapter 11). Studies have shown that drip irrigation can reduce water use 30–70%, while actually increasing crop yields, compared with traditional flooding methods. Although drip irrigation is spreading worldwide, especially in arid lands such as Israel and Australia, 97% of the irrigation in the United States, and 99% throughout the world, is still done by traditional flood or center-pivot methods.

The reason that so few farms have changed over to drip irrigation systems is that it costs about $1,000 per acre to install them. In comparison, water for irrigation is heavily subsidized by the government, so farmers pay next to nothing for it. Therefore, it makes financial sense to use the cheapest system for distributing water, even if it is wasteful. Calculations of construction costs and energy subsidies to provide irrigation water to farmers indicate an annual subsidy of $4.4 billion for the 11 million acres of irrigated land in the western United States, an average of $400 an acre. Governments collect an average of 10% or less of the actual cost in water fees charged to farmers. Reducing this subsidy would greatly encourage water conservation through the use of more efficient irrigation technologies.

Treadle Pumps. In the developing countries, irrigation often bypasses the rural poor, who are at the greatest risk for hunger and malnutrition. Affordable irrigation technologies can be remarkably successful, as in Bangladesh, where low-cost treadle pumps (**Fig. 10–23**) enable farmers to irrigate their rice paddies and vegetable fields at a cost of less than $35 a system. The pump is worked with the feet like a step exercise machine, is locally manufactured, and has been adopted by millions. The irrigation it affords makes it possible for farmers to irrigate small plots during

Figure 10–23 Treadle pump. This treadle pump in Zambia is operated like a step exercise machine. Millions of these pumps now allow rural farmers to raise crops during dry seasons.

Figure 10–22 Drip irrigation. Irrigation consumes the most water. Drip irrigation offers a conservative method of applying water.

STEWARDSHIP

The Energy/Water/Food Trade: Waste One, Waste Them All

Ever stand in a public restroom and see someone run the sink water and not turn it off? What about seeing someone leave their car idling? Or throwing food in the trash? Drive you crazy? It should! Because the way the world is intertwined, if you waste one (energy, water, or food), you waste the others or at least waste the opportunity to obtain them. It works this way:

It takes energy to get clean freshwater, especially if you live in a dry area or a place with water pollution. The water might have to be cleaned or moved long distances. Bottled water is even more energy intensive. For example, in California, the California State Water Project is the largest single user of energy in the state. One of its activities is moving water over the Tehachapi Mountains, which requires as much energy as a third of the household energy use in the region. Wasting water means you waste the energy it took to get it, clean it, and heat it, too. Because we need to be efficient with energy so we can avoid greater harm to the environment, water waste isn't just about the water—it hurts the whole system.

It takes water to get energy. This is true when water is used in cooling (nuclear plants), in coal mining, and in coal power plants. The newest way of getting natural gas, called hydraulic fracturing, uses enormous amounts of water. There are more than 35,000 wells fractured in the United States each year, using an estimated 70 billion to 140 billion gallons of water. When the water is used, it is mixed with toxic chemicals and injected deep into the Earth, rather than returned to the regional water cycle. The use of water for fracking, as this process is called, directly competes with other uses of water such as agriculture and has been the source of lawsuits in dry states such as Wyoming and Montana.

Another example is biofuels. Crops used for biofuels such as ethanol are usually raised using

land and water that could have been used for food crops. Their use to fuel vehicles drove up the cost of basic food staples in 2007, contributing to the world food crisis (Chapter 12). Wasting energy might mean we have to obtain more energy and use more water, making it harder to produce adequate food.

We use water and energy to grow, store, and move food. Wasting food means that water and energy cannot be used for something else. It takes 1,300 liters of water to produce a kilogram of wheat, 3,000 liters for a kilogram of eggs, and 15,500 liters for a kilogram of beef. It also takes energy to store food and a great deal to transport it. The average American foodstuff travels 1,500 miles. As much as 40% of the energy used in food production is in the manufacture of pesticides and fertilizers (Chapters 12 and 13). But much of the recent increase in food energy cost is due to increased processing and packaging, such as for single-serving foods.

Food is often wasted. A Food and Agriculture Organization (FAO)–commissioned study found that 1.3 billion tons, roughly a third, of the food we produce for human consumption is wasted each year. This is the equivalent of half of the world cereals crop. Some food waste is lost at harvest or in storage or processing. Some is in stores and at consumers' tables. Any way you work it, if we can save food, we can feed more people and use less water and energy.

Sources: NRDC. (2004). "Energy Down the Drain: The Hidden Costs of California's Water Supply." http://www.nrdc.org/water/conservation/edrain/edrain.pdf; EPA. (2011). Draft Plan to Study the Potential Impacts of Hydraulic Fracturing on Drinking Water Resources. Report Number: EPA/600/D-11/001

Chapagain, A. K., Hoekstra, A. Y. (2004). "Water Footprints of Nations." *Value of Water Research Report Series* (UNESCO-IHE) 6; Jenny Gustavsson, Christel Cederberg, Ulf Sonesson, Robert van Otterdijk, and Alexandre Meybeck. (2011). *Global Food Losses and Food Waste,* FAO.

dry seasons from groundwater lying just a few feet below the surface. More than 1.5 million treadle pumps have been sold in Bangladesh alone, increasing the productivity of more than 600,000 acres of farmland there. The pumps are now also in use in India, Nepal, Myanmar, Cambodia, Zambia, Kenya, and South Africa.

Enhance Soil Water Retention. We have already addressed the need and options for improvement in irrigation technology. However, most of the agriculture

in developing countries is rain-fed, depending on green water flow from soil and resulting in evapotranspiration. Further, most of the soils are dryland soils, receiving only 10 to 30 inches of rainfall per year (Chapter 11). Crops raised on these soils are characteristically low in yield. Storing the seasonal rainfall with "check dams" (small rock or raised soil dams across seasonal water courses) increases yields. These dams recharge aquifers, restrain soil erosion, and provide for some irrigation water. Another strategy is to enhance rainfall retention, maintain soil fertility, and restrain soil erosion through conservation tillage, where crop residues are left in place and new crops are planted with little or no tilling.

Municipal Systems. The 100 gallons of water each person consumes per day in modern homes are used mostly for washing and removing wastes—that is, for flushing toilets (3–5 gallons per flush), taking showers (2–3 gallons per minute), doing laundry (20–30 gallons per wash), and so on. Watering lawns, filling swimming pools, and other indirect consumption only adds to this use.

Water conservation has long been promoted as a "save the environment" measure, though without much effect. Now, numerous cities are facing the stark reality that it will be extremely expensive—and in many cases, impossible—to increase supplies by the traditional means of building more reservoirs or drilling more wells. The only practical alternative, they are discovering, is to take real steps toward reducing water consumption and waste. A considerable number of cities have programs whereby leaky faucets will be repaired and low-flow showerheads and water-displacement devices in toilets will be installed free of charge.

Lawns. Grass is the most heavily irrigated crop. Cities in Arizona are paying homeowners to replace their lawns with **xeriscaping**—landscaping with desert species that require no additional watering—and the business is thriving. Many cities and towns ban certain uses of water when droughts reduce the available supply.

A Flush World. In 1997, the last phase of a regulation authorized by the 1992 National Energy Act took effect, and it became illegal to sell 6-gallon commodes. In their place is the wonder of the flush world: the 1.6-gallon toilet. Newer versions use even less water than the first, less effective low-flow toilets; the newest work perfectly well and save 10 gallons or more a day per person. In fact, the newest models are dual flush, with options for a 1.28-gallon (solids) or 0.8-gallon (liquids) flush. The 50 million low-flow toilets now in place in the United States save an estimated 600 million gallons of water a day. Los Angeles is offering the low-flow toilets free as part of its efforts to restore Mono Lake, which had lost a great deal of water. In 2011, the City of San Francisco found the downside of low-flow toilets: While the city had saved 20 million gallons of water a year (just from toilets!), the low-flow toilets weren't pushing the sewage through the pipes as quickly as before. To solve the problem, city engineers are planning to kill bacteria with chemicals, though other solutions have been pointed out.

Gray Water. It seems wasteful to raise all domestic water to drinking-water standards and then use most of it to water lawns and flush toilets. Increasingly, gray water recycling systems are being adopted in some water-short areas. **Gray water**, the slightly dirtied water from sinks, showers, bathtubs, and laundry tubs, is collected in a holding tank and used for such things as flushing toilets, watering lawns, and washing cars. Going further,

a number of cities are using treated wastewater (sewage water) for irrigating golf courses and landscapes, both to conserve water and to abate the pollution of receiving waters (see Chapter 20). Residents of California use more than 160 billion gallons of treated wastewater for such purposes, and in Israel, 70% of treated wastewater is reused for irrigation of nonfood crops. If the idea of reusing sewage water turns you off, recall that *all* water is recycled by nature. There is hardly a molecule of water you drink that has not moved through organisms—including humans—numerous times. A number of communities are already treating their wastewater so thoroughly that its quality exceeds what many cities take in from lakes and rivers.

Virtual Water. Another option for reducing water loss in water-stressed regions is to import food or goods that require a great deal of water, to ship *virtual water*, as it is called. Many arid countries avoid using their own water supplies by importing farm products from regions that have excess freshwater. Indeed, an estimated 800 cubic kilometers of virtual water are traded yearly in the form of foodstuffs. However, many of the arid countries are also economically poor and have a limited ability to purchase great quantities of food. Furthermore, shipping takes energy, and as we have seen, wasting energy is eventually bad for water, too.

One thing consumers can do is purchase products that have a low "water footprint." Production of meat, for example, almost always requires a great deal more water, but locally grown vegetables, whole foods you cook yourself, and whole grains require less. Packaging and things transported across the world represent greater water inputs. Individuals can figure out their own water footprint by looking on the Web for one of several ways to calculate it.

Public-Policy Challenges

The hydrologic cycle provides a finite flow of water through each region of Earth. When humans come on the scene, this flow of water is inevitably divided between the needs of the existing natural ecosystems and the agricultural, industrial, and domestic needs of humans, with the latter usually being met first. In fact, recent calculations indicate that humans now use 26% of total terrestrial evapotranspiration and 54% of accessible precipitation runoff. We are major players in the water cycle, and as we have seen, many facets of our water use are unsustainable. It takes public policy to strike a sustainable balance between competing water needs, and this is often sadly lacking.

Conflict. Maintaining a supply of safe drinking water for people is a high-priority issue. In addition, water for irrigation is vital to food production across much of the world. Often, these two demands come into conflict, generating what have been referred to as "water wars." One example was in Cochabamba, Bolivia, when the World Bank refused to renew loans unless the country opened

Figure 10–24 Water wars. Tensions mount over water rights in a world with increasing numbers of people trying to use the same resource. Here farmers protest lack of water in Gustine, California in 2009. National and international policies can ease these tensions.

water resources to foreign privatization. Investors were supposed to put money into providing clean drinking water, but some locals found they lost access to water and rates skyrocketed. This led to protests and the government eventually had to change plans. **Figure 10–24** depicts protesters in California after a federal judge ordered that less water could be pumped from the Sacramento Delta, where endangered fish live. This ruling pitted environmentalists against farmers and urban residents in a state with continual water stress.

National Water Policy. Many countries have addressed their water resources and water needs with a national policy; not so the United States. The Clean Water Act and subsequent amendments authorize the U.S. Environmental Protection Agency (EPA) to develop programs and rules to carry out its mandate for oversight of the nation's water quality (considered in Chapter 20). The EPA, however, does not deal with water quantity. There is no federal agency to provide similar oversight for the water quantity issues discussed in this chapter.

Peter Gleick, a leading expert on global freshwater resources, addresses this public-policy need,[5] pointing out that most water policy decisions in the United States are made at local and regional levels. National policies, however, are needed to guide these lower-level decisions and also to deal with interstate issues and federally managed resources. The last time a water policy report was issued was in 1950, when President Harry S Truman established a water resources policy commission and the commission gave its report. This commission needs to be revived or a new one needs to be established, and in either case,

it must be given a mandate to collect data on water resources and problems and issue recommendations on how the federal government could facilitate water stewardship in the 21st century.

Key Issues. Any water resources policy commission needs to address several key ideas. Two of these are increasing efficiency and stopping subsidies. In the United States, scarcity and rising costs have led to significant gains in efficiency. As a result, water withdrawals peaked in the early 1980s and have leveled off since then (Fig. 10–9). This change has been attributed to steps taken in all three sectors (domestic, industry, and irrigation) to reduce water losses. The United States still has a high per capita water use compared with the rest of the world. Also, water subsidies need to be reduced or eliminated. Policy makers have persistently subsidized water resources, especially for agricultural use, and the result has been enormous waste and inefficiencies. This undervaluing of the resource harms ecosystem services. In addition, polluters must be charged according to their effluents. When municipalities, farmers, and industries are allowed to pollute waterways without any accountability, the result is that everyone downstream is forced to pay more to clean up the water and aquatic ecosystems are degraded accordingly. Various approaches are possible, such as permit fees, "green taxes," direct charges, and market-based trading (see Chapter 20). In the 2012 Water for Food conference in Lincoln, Nebraska, one of the major messages was that there is no one-size-fits-all policy for protecting water supplies.

Public policy needs to focus on the watershed level as well. Watershed management must be integrated into the pricing of water. Watersheds hold the key to water purity, aquifer recharge, and water storage, as well as the maintenance of natural wetland and riparian habitats that are vital to many plants and animals. The City of New York

[5]Peter H. Gleick, "Global Water: Threats and Challenges Facing the United States," *Environment* 43 (March 2001): 18–26.

found out that every dollar it invested in watershed management paid off handsomely, saving as much as $200 for new treatment facilities. One part of managing watersheds is to regulate dam operations so that river flow is maintained in a way that simulates natural flow regimes. When this happens, rivers below dams can sustain recreational fisheries and other uses, and the natural biota flourishes.

Finally, the United States must respond to the global water crisis with adequate levels of international development aid. Achieving the Millennium Development Goal of reducing by half the 1.1 billion people without access to safe drinking water will save millions of lives and billions of dollars annually.

International Action. On the international front, the World Commission on Water for the 21st Century sponsors the World Water Forum, which convened for the sixth time in March 2012. The findings of the forum relate directly to solving the great water-related needs of Earth, especially in the developing countries. With a large proportion of the developing world already experiencing shortages of clean water or water for agriculture, the population increases and pressure to develop these countries' resources that are certain to come in the next few decades will undoubtedly subject hundreds of millions to increased water stress. The problem is not that the Earth contains too little freshwater; rather, it is that we have not yet learned to manage the water that our planet provides. To quote the World Water Council's *World Water Vision* report, "There is a water crisis, but it is a crisis of management. We have threatened our water resources with bad institutions, bad governance, bad incentives, and bad allocations of resources. In all this, we have a choice. We can continue with business as usual, and widen and deepen the crisis tomorrow. Or we can launch a movement to move from vision to action—by making water everybody's business."[6]

---

[6]*World Water Vision: Making Water Everybody's Business.* World Water Council, 2000.

REVISITING THE THEMES

Sound Science

What is known about the hydrologic cycle is the result of sound scientific research—research that is needed to make progress in managing water resources. Work performed under the Millennium Ecosystem Assessment has given us a better understanding of such matters as groundwater recharge, essential river flows, and the interactions between different land uses and water resources. The techniques of water treatment have been fine-tuned by scientific work as well as the development of reverse osmosis for desalination. Sound science is being used to figure out current questions. One we do not yet have an answer to is what is likely to happen with oceanic thermohaline circulation.

Sustainability

The opening story showed how unsustainable China's water use is. Still, the hydrologic cycle is a remarkable global system that renews water and does so sustainably. Groundwater—especially the fraction that is nonrenewable because it was formed centuries ago—is being used in an unsustainable manner. The consequences are immediate: Water tables disappear into the depths of the Earth, farms dry up because there is no water, and land subsidence ruins properties and groundwater storage itself. Conserving water through xeriscaping, low-flush toilets, gray water use, and more efficient food production and consumption is a sustainable path.

Stewardship

Exchanges among land, water, and atmosphere represent a cyclical system that has worked well for eons and that can continue to supply all our needs into the future—if we manage water wisely. However, we are changing the cycle in ways that reflect a lack of wisdom and a lack of resolve to meet people's needs in a just way. One way to remedy this is to look at water, energy, and food as interrelated. Increased inefficiency and lower waste are necessary in all three areas if we are to provide water and food for more people. Stewardship of water resources means that we consider the needs of the natural ecosystems—and especially endangered species—when we make policy decisions. It means that we address the wastefulness that characterizes so much of our water use. It means in particular that we show concern and help the millions who lack adequate clean water and are suffering because of it. Some examples of good stewardship are the Glen Canyon Dam operation, the removal of outdated dams to restore river flow, the use of treadle pumps in many developing countries, and xeriscaping. In the United States, developing a national water policy would be a huge step in stewardship management.

REVIEW QUESTIONS

1. What are the lessons to be learned from China's water hardships?

2. Give examples of the infrastructure that has been fashioned to manage water resources. What are the challenges related to developing countries?

3. What are the two processes that result in natural water purification? State the difference between them. Distinguish between green water and blue water.

4. Describe how a Hadley cell works, and explain how Earth's rotation creates the trade winds.

5. Why do different regions receive different amounts of precipitation?

6. Define *precipitation, infiltration, runoff, capillary water, transpiration, evapotranspiration, percolation, gravitational water, groundwater, water table, aquifer, recharge area, seep,* and *spring.*

7. Use the terms defined in Question 6 to give a full description of the hydrologic cycle, including each of its three loops—namely, the evapotranspiration, surface runoff, and groundwater loops. What is the water quality (purity) at different points in the cycle? Explain the reasons for the differences.

8. How does changing Earth's surface (for example, by deforestation) change the pathway of water? How does it affect streams and rivers? Humans? Natural ecology?

9. Explain how climate change and atmospheric pollution can affect the hydrologic cycle.

10. What are the three major uses of water? What are the major sources of water to match these uses?

11. How do dams facilitate the control of surface waters? What kinds of impacts do they have?

12. Distinguish between renewable and nonrenewable groundwater resources. What are the consequences of overdrawing these two kinds of groundwater?

13. What are the five options for meeting existing water scarcity needs and growing demands?

14. Describe how water demands might be reduced in agriculture, industry, and households.

15. What is the status of water policy in the United States? Cite some key issues that should be addressed by any new initiatives to establish a national water policy.

THINKING ENVIRONMENTALLY

1. Imagine that you are a water molecule, and describe your travels through the many places you have been and might go in the future as you make your way around the hydrologic cycle time after time. Include travels through organisms.

2. Suppose some commercial interests want to create a large new development on what are presently wetlands. Describe a debate between those representing the commercial interests and those representing environmentalists who are concerned about the environmental and economic costs of development. Work toward negotiating a consensus.

3. Describe how many of your everyday activities, including your demands for food and other materials, add pollution to the water cycle or alter it in other ways. How can you be more stewardly with your water consumption?

4. An increasing number of people are moving to the arid southwestern United States, even though water supplies are already being overdrawn. If you were the governor of one of the states in this region, what policies would you advocate to address this situation?

5. Suppose some commercial interests want to develop a golf course on what is presently forested land next to a reservoir used for city water. Describe the impacts this development might have on water quality in the reservoir.

MAKING A DIFFERENCE

1. Since 1992, showerheads and faucets have been required to use no more than 2.5 gallons of water per minute. If you have older plumbing, it's possible that you may not meet these standards. Equipment such as aerating showerheads can help you reduce your water use (and your water bill) without decreasing effectiveness too much.

2. Consider taking a "navy" shower (so-called because ships at sea must conserve water): Wet down, then turn the shower off, then soap down, and then rinse down. You will save both water and energy this way.

3. Americans drink billions of single-use water bottles per year. In addition to the carbon miles it takes to ship these products, most Americans simply throw the recyclable containers away. A much more sustainable method of drinking is simply to fill reusable bottles with filtered tap water.

4. Old water heaters are most likely running at less than 50% efficiency. If this is the case for you, you can replace your water heater with a newer model or simply use a tankless heater, which heats water only when you need it.

5. A dishwasher can save more than 5,000 gallons of water per year as well as hundreds of hours of your time. Scraping your dishes first, running the washer only for full loads, and using the energy-saver settings can reduce water and energy usage even more.

MasteringEnvironmentalScience®

Go to **www.masteringenvironmentalscience.com** for practice quizzes, Pearson eText, videos, current events, and more.

Dust Storm Damage, 1930-1940

- Dust Bowl States
- Area with most severe dust storm damage
- Other areas damaged by dust storms

CHAPTER 11

Soil: The Foundation for Land Ecosystems

The Dust Bowl. The Dust Bowl era in the 1930s was a time of drought and wind erosion that devastated farming in the American Midwest. This wall of dust approached Rolla, Kansas on April 14, 1935.

"The nation that destroys its soil, destroys itself."

—*Franklin D. Roosevelt*[1]

A CHURNING WALL OF SOIL, two miles high, bore across the plains. People caught outside fled for buildings, cloths held to their faces. Some faced with the blinding swirl lost their way and died, suffocated by dust. Dust was everywhere, driven by the howling winds: blinding, choking, filling cracks and windows. People wept, prayed, cried, despaired; the Apocalypse had come. The day was Black Sunday, April 14, 1935. The dust storm was the worst of the "black blizzards" in the tragic period we call the American Dust Bowl. The almost-decade-long period of drought and dry storms was arguably the longest and worst environmental disaster in US history. The storms carried tons of soil across seven states and left devastation in their wake. Some people fled the dust states or died, while others stayed and survived, only to find themselves coughing and sickly when storm after storm filled the air with dust, victims of a disorder called "dust pneumonia."

THE DUST BOWL BEGINS. A drought beginning in 1931 precipitated the disaster. By 1934, it was profound: More than 75% of the country was in drought, and 27 states were severely affected. Although drought was the trigger, human policies and practices made the Dust Bowl worse. Rising wheat and cotton prices encouraged people to plant more and more land, even in areas more suited to ranging animals. During wet years, crops were

LEARNING OBJECTIVES

1. **Soil and Plants:** Describe the basic parts of soil, soil communities, and soil profiles.

2. **Soil Degradation:** Explain how soils are degraded by several effects, including overcultivation, overgrazing, deforestation, mining, and killing soil biota.

3. **Soil Conservation:** Describe soil conservation requires action at two levels, from individual landholders and through public policy.

[1]Letter, February 26, 1937, to state governors. Available from http://www.presidency.ucsb.edu/ws/?pid=15373

bountiful, but plowing broke through the tough roots of prairie plants that held soil together, broke open the ground, and laid the dirt open to be borne about by the winds. Lack of trees or other windbreaks allowed soil to move freely. By 1935, estimates were that 850 million tons of topsoil had been blown off of the southern plains in that year alone. By the end of the Dust Bowl, many areas had lost 75% of their topsoil, once some of the richest soil in the world.

The Homestead Act of 1862, which enabled settlers to claim land and settle it, played a role in the disaster. By encouraging hundreds of thousands of small farms (160 to 320 acres) to be established between 1880 and 1920, the act ensured that too much land was cultivated in some areas of the country. Small farm owners were unlikely to let land lie fallow, diversify into pasture, plant crops with strips of vegetation in between, or farm less intensively. These homesteads were more vulnerable to financial ruin during drought. Homesteaders fled the southern plains in the largest migration in American history, described by John Steinbeck in his novel *The Grapes of Wrath*.

A dust storm on June 5, 2011 that hit Phoenix, Arizona, gave only the tiniest glimpse of such storms to modern Americans. The storm hit a peak height of at least 5000 to 6000 ft (1500 to 1800 m). The front edge was almost 100 mi (160 km) long (**Fig. 11–1**). As was the case in the '30s, an ongoing drought in Arizona in 2011 was a contributing factor to the severity of the storm. But such storms are rare now, in part because of better soil protection practices and the movement from more intensive cultivation of marginal lands. Around the world however, loss of soil continues to plague farmland and contribute to poor crops, water pollution, and the spread of deserts.

PAST NEGLECT. Ninety percent of the world's food comes from land-based agricultural systems, and the percentage is growing as the ocean's fish and natural ecosystems are increasingly being depleted. Protecting and nurturing agricultural soils, which are the cornerstone of food production, must be a central feature of sustainability. Yet it is a feature that has been overlooked repeatedly in the past. The Dust Bowl is not the only time such land degradation has occurred. The fall of the ancient Greek, Roman, and Mayan empires was more the result of a decline in agricultural productivity due to soil erosion than of outside forces.

Figure 11–1 A dust storm in Phoenix, Arizona. A 2011 dust storm in Phoenix showed just a tiny picture of what the 1930s dust bowl was like.

As the world's population grows past 7 billion, croplands and grazing lands are being increasingly pressed to yield more crops. A landmark 1991 Global Assessment of Soil Degradation (GLASOD) study reported that 15% of Earth's land was degraded, mostly due to soil erosion. Yet a new study by the Land Degradation Assessment in Drylands (LADA) found that **land degradation** is worse than that.[2] More than 20% of cultivated areas, 30% of forests, and 10% of grasslands were degraded between 1981 and 2003. Some 2 billion people depend directly on the land undergoing degradation. Throughout the world, soils have been (and continue to be) degraded by erosion, the buildup of salts, and other problems that can only undermine future productivity. Farm soils are not the only concern. Mining affects soils, as does deforestation. Both contribute to water and soil pollution as well as loss of ecosystem services from the ecosystems that were displaced.

Our objective in this chapter is to develop an understanding of the attributes of soil required to support good plant growth, of how these attributes may deteriorate under various practices, and of what is necessary to maintain a productive soil. We will then look at what is being done nationally and internationally to rescue this essential resource and to establish practices that are sustainable.

---

[2]Bai, Z. G. et al. (2008). Global Assessment of Land Degradation and Improvement 1: Identification by Remote Sensing. Report 2008/01, FAO/ISRIC—Rome/Wageningen.

11.1 Soil and Plants

Soil can be defined as solid material of geological and biological origin that is changed by chemical, biological, and physical processes, giving it the ability to support plant growth.

A rich soil is much more than the dirt you might get out of any hole in the ground. Indeed, agriculturists cringe when anyone refers to soil as "dirt." The quality of soil can make the difference between harvesting an abundant crop and abandoning a field to weeds. You have already learned that various detritus feeders and decomposers feed on organic debris in an ecosystem (Chapter 4). Nutrients from the detritus are released into the soil and absorbed by producers, thus recycling the nutrients. In short, productive topsoil involves dynamic interactions among organisms, detritus, and mineral particles of soil **(Fig. 11–2)**.

Soil Characteristics

Most soils are hundreds or thousands of years old and change very slowly. Soil science is cross-disciplinary and is at the heart of agricultural and forestry practice. Many years of the study of soils have resulted in a system of classification of soil profiles and soil structure and a taxonomy of soil types from all over the world—as well as a large body of scientific literature investigating relevant topics. This section presents just enough of this information to enable you to open the door to soil science; you may even be motivated to follow up on your own in this fascinating field.

Soil Texture. The mineral material of soil, or **parent material**, has its origin in the geological history of an area. Parent material can be rock or sediments deposited by wind, water, or ice. Sooner or later parent material is broken down by natural **weathering** (gradual physical and chemical breakdown) to the point where often it is impossible to tell just what kind of rock the soil came from. As rock weathers, it breaks down into smaller and smaller fragments—stones and smaller "soil separates." These are classified as sand, silt, and clay. **Sand** is made up of particles from 2.0 to 0.063 millimeters in size, **silt** particles range from 0.063 to 0.004 millimeters, and **clay** is anything finer than 0.004 millimeters **(Fig. 11–3)**. Particles larger than sand are called gravel, cobbles, or boulders, depending on their size. You can see the individual rock particles in sand, and you may be familiar with the finer particles called silt, but you would need a good microscope to see clay particles.

Proportions. The sand, silt, and clay particles constitute the mineral portion of soil. **Soil texture** refers to the relative proportions of each type of particle in a given soil. If one predominates, soil is said to be sandy, silty, or clayey. A proportion that is commonly found in soil consists of roughly 40% sand, 40% silt, and 20% clay. A soil with these proportions is called a **loam**. You can determine the texture of a given soil by shaking a small amount of it with water in a large test tube to separate the particles and then allowing them to settle. Because particles settle according to their

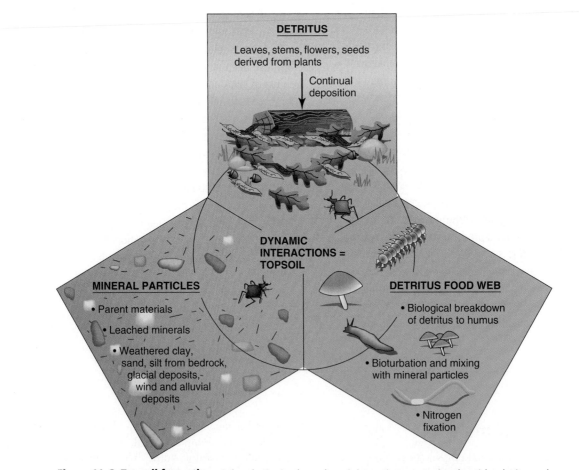

Figure 11–2 Topsoil formation. Soil production involves a dynamic interaction among mineral particles, detritus, and members of the detritus food web.

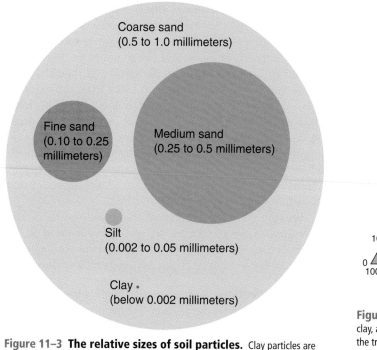

Figure 11–3 The relative sizes of soil particles. Clay particles are too small to see with the naked eye, while silt particles are slightly larger. Sand particles vary from small (but still larger than silt) to large. Gravel and boulders are larger still.

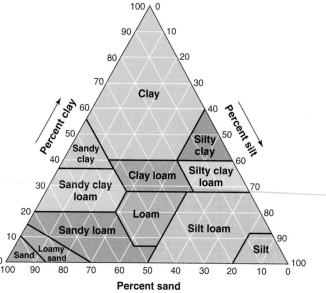

Figure 11–4 The soil texture triangle. Relative proportions of sand, clay, and silt are represented on each axis. Major classes of soil are indicated on the triangle. For clay, read across horizontally; for silt, read diagonally downward toward the left from the right axis; for sand, read diagonally upward to the left from the bottom axis. The texture content of any soil should total 100% if the triangle is read properly.

weight, sand particles settle first, silt second, and clay last. (More precise measurements require a laboratory analysis.) Soil scientists classify soil texture with the aid of a triangle that shows the relative proportions of sand, silt, and clay in a given soil **(Fig. 11–4).**

Properties. Soil scientists often ask questions about how the components of soil determine its properties. Three basic considerations determine how several important properties of soil are influenced by texture:

1. Larger particles have larger spaces separating them than smaller particles have. (Visualize the difference between packing softballs and packing golf balls in containers of the same size.)

2. Smaller particles have more surface area relative to their volume than larger particles have. (Visualize cutting

a block in half again and again. Each time you cut it, you create two new surfaces, but the total volume of the block remains the same.)

3. Nutrient ions and water molecules tend to cling to surfaces. (When you drain a nongreasy surface, it remains wet.)

These properties of matter profoundly affect such soil properties as infiltration, nutrient- and water-holding capacities, and aeration (Table 11–1). Note how soil properties correspond to particle size (sand, silt, and clay) in the table.

Soil texture also affects **workability**—the ease with which a soil can be cultivated. Workability has a tremendous impact on agriculture. Clayey soils are very difficult to work with because even modest changes in moisture content can turn them from too sticky and muddy to too hard and even bricklike.

TABLE 11–1	Relationship Between Soil Texture and Soil Properties				
Soil Texture	Water Infiltration	Water-Holding Capacity	Nutrient-Holding Capacity	Aeration	Workability
Sand	Good	Poor	Poor	Good	Good
Silt	Medium	Medium	Medium	Medium	Medium
Clay	Poor	Good	Good	Poor	Poor
Loam	Medium	Medium	Medium	Medium	Medium

Sandy soils are very easy to work because they become neither muddy when wet nor hard and bricklike when dry.

Soil Profiles.

The processes of soil formation create a vertical gradient of layers that are often (but not always) quite distinct. These horizontal layers are known as **soil horizons**, and a vertical slice through the different horizons is called the **soil profile (Fig. 11–5)**. The profile reveals a great deal about the factors that interact in the formation of a soil. The topmost layer, the O horizon, consists of dead organic matter (detritus) deposited by plants: leaves, stems, fruits, and seeds. Thus, the O horizon is high in organic content and is the primary source of energy for the soil community. Toward the bottom of the O horizon, the processes of decomposition are well advanced, and the original materials may be unrecognizable. At this point, the material is dark and is called **humus**.

Subsurface Layers.

The next layer in the profile is the A horizon, a mixture of mineral soil from below and humus from above. The A horizon is also called **topsoil**. Fine roots from the overlying vegetation permeate this layer. The A horizon is usually dark because of the humus that is present and may be shallow or thick, depending on the overlying ecosystem. Vital to plant growth, topsoil grows at the rate of an inch or two every hundred years. In many soils, the next layer is the E horizon, where *E* stands for **eluviation**—the process of **leaching** (dissolving away) many minerals due to the downward movement of water. This layer is usually paler in color than the two layers above it.

O Horizon: Humus. (surface litter, decomposing plant matter)

A Horizon: Topsoil. (mixed humus and leached mineral soil)

E Horizon: Zone of leaching. (less humus, minerals resistant to leaching)

B Horizon: Subsoil. (accumulation of leached minerals like iron and aluminum oxides)

C Horizon: Weathered parent material. (partly broken-down minerals)

Figure 11–5 Soil profile. Major horizons from the surface to the parent material in an idealized soil profile.

Below the E horizon is the B horizon. This layer is characterized by the deposition of minerals that have leached from the A and E horizons, so it is often high in iron, aluminum, calcium, and other minerals. Frequently referred to as the **subsoil**, the B horizon is typically high in clay and is reddish or yellow in color from the oxidized metals present. Below the B horizon is the C horizon, which is the parent mineral material that originally occupied the site. This layer represents weathered rock, glacial deposits, or volcanic ash and usually reveals the geological process that created the landscape. The C horizon is affected little by the biological and chemical processes that go on in the upper layers.

Because there are innumerable soils across the diverse landscapes of the continents, soil profiles will differ in the thickness and the content of the layers. However, all soils exist in layers and can be characterized by their texture.

Soil Classification.

Soils come in a wide variety of vertical structures and textures. To give some order to this diversity, soil scientists have created a taxonomy, or classification system, of soils. It works much like the biological taxonomy with which you may be familiar (see Chapter 6). The most inclusive group in the taxonomy is the **soil order**. If you are classifying a soil, you find the order first and then work your way downward through the taxonomic categories (*suborders, groups, subgroups, families*) until you come to the soil *class* that best corresponds to the soil in question. There are literally hundreds of soil classes. **Figure 11–6** displays the worldwide distribution of the 12 major orders. Following are the characteristics of the four major soil orders that are most important for agriculture, animal husbandry, and forestry and a fifth soil type that is particularly important to environmental health.

Mollisols. Mollisols are fertile, dark soils found in temperate grassland biomes. They are the world's best agricultural soils and are encountered in the Midwestern United States; across temperate Ukraine, Russia, and Mongolia; and in the pampas (a broad region of grassy plains) in Argentina. They have a deep A horizon and are rich in humus and minerals; precipitation is insufficient to leach the minerals downward into the E horizon.

Oxisols. These are soils of the tropical and subtropical rain forests. They have a layer of iron and aluminum oxides in the B horizon and have little O horizon, due to the rapid decomposition of plant matter. Most of the minerals are in the living plant matter, so oxisols provide limited fertility for agriculture. If the forests are cut (as in "slash-and-burn" agriculture, a common practice in the tropics), a few years of crop growth can be obtained, but in time, the intense rainfall leaches the minerals downward, forming an impervious hardpan that resists further cultivation.

Alfisols. Alfisols are widespread, moderately weathered forest soils. Although not deep, they have well-developed O, A, E, and B horizons. Alfisols are typical of the moist, temperate forest biome and are suitable for agriculture if they are supplemented with organic matter or mineral fertilizers to maintain soil fertility.

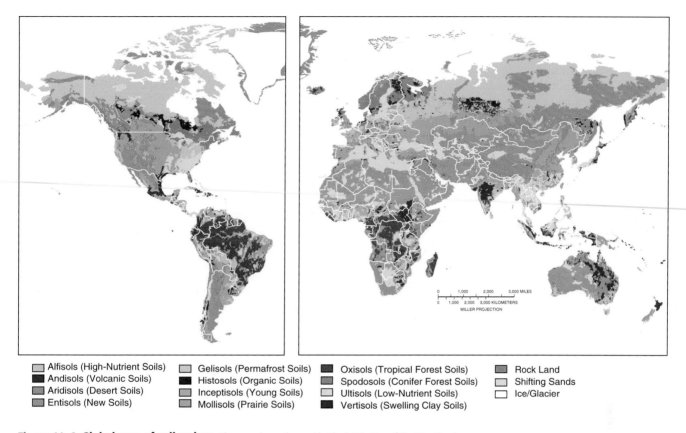

Alfisols (High-Nutrient Soils)
Andisols (Volcanic Soils)
Aridisols (Desert Soils)
Entisols (New Soils)

Gelisols (Permafrost Soils)
Histosols (Organic Soils)
Inceptisols (Young Soils)
Mollisols (Prairie Soils)

Oxisols (Tropical Forest Soils)
Spodosols (Conifer Forest Soils)
Ultisols (Low-Nutrient Soils)
Vertisols (Swelling Clay Soils)

Rock Land
Shifting Sands
Ice/Glacier

Figure 11–6 Global map of soil orders. The map shows the worldwide distribution of the 12 soil orders.
(*Source:* After the U.S. Department of Agriculture's Natural Resources Conservation Services, World Soil Resources staff.)

Aridisols. These are the very widespread soils of **dry-lands** and deserts (drylands are arid, semiarid, and seasonally dry areas). The paucity of vegetation and precipitation leaves aridisols relatively unstructured vertically. They are thin and light colored and, in some regions, may support enough vegetation for rangeland animal husbandry. Irrigation used on these soils usually leads to salinization, as high evaporation rates draw salts to surface horizons, where they accumulate to toxic levels.

The descriptions of these four classes of soil demonstrate that different soils vary.

Another Type of Soil: Hydric. Another soil type that is particularly important, although not necessarily good for agriculture, is **hydric soils.** These soils indicate a wetland or aquatic site. Hydric soils are not a soil class on their own, but are members of a number of other classes. Hydric soils are used in delineating the boundaries of wetlands, in order to get permits for construction, and to protect wetlands and water bodies from the negative impacts of development. The soil order histosols includes several types of hydric soils. They contain a buildup of undecayed vegetation called **peat** and hence are called peatlands. Peat contains a great deal of energy and can be dried and burned. Vast areas of northern zones are hydric soils, including soil order gelisols, that are permanently frozen at some depth, a condition called permafrost. One of the concerns about climate change is that in a warmer world, permafrost areas will thaw and greater

rates of decay will occur. This would act as positive feedback, putting more greenhouse gases released from decaying peat into the atmosphere (Chapter 18).

Soil Constraints. Soil scientists study more than soil classification. They often research the relationship of soil to plants it supports. Such scientists speak of "soil constraints," meaning the various characteristics of soils that limit primary productivity, especially in agriculture and forestry. These constraints include poor soil drainage (hydromorphy), low ability to retain nutrients, high levels of aluminum, low levels of phosphorous, soil cracking (from certain types of clays), the presence of soluble salts (salinity and sodicity), shallowness, and high erosion likelihood. Let's look at the characteristics that enable soils to support plant growth, especially agricultural crops.

Soil and Plant Growth

For their best growth, plants need a root environment that supplies optimal amounts of mineral nutrients, water, and air (oxygen). The pH (relative acidity) and salinity (salt concentration) of soil are also critically important. **Soil fertility**—soil's ability to support plant growth—often refers to the presence of proper amounts of nutrients, but it also includes soil's ability to meet all the other needs of plants. Farmers speak of a given soil's ability to support plant growth as the **tilth** of soil.

Mineral Nutrients and Nutrient-Holding Capacity.
Mineral nutrients—phosphate (PO_4^{3-}) potassium (K^+), calcium (Ca^{2+}), and other ions—are present in various rocks, along with non-nutrient elements. Minerals initially become available to roots through the weathering of rock. Weathering, however, is much too slow to support normal plant growth. The nutrients that support plant growth in natural ecosystems are supplied mostly through the breakdown and release (recycling) of nutrients from detritus.

Leaching. Nutrients may literally be washed from soil as water moves through it, a process called **leaching**. Leaching not only lessens soil fertility but also contributes to pollution when materials removed from soil enter waterways. Consequently, soil's capacity to bind and hold nutrient ions until they are absorbed by roots is important. This property is referred to as the soil's **nutrient-holding capacity**.

Fertilizer. In agricultural systems, there is an unavoidable removal of nutrients from soil with each crop because nutrients absorbed by plants are contained in the harvested material. Therefore, agricultural systems require inputs of nutrients to replace those removed with the harvest. Nutrients are replenished with applications of **fertilizer**—material that contains one or more necessary nutrients. Fertilizer may be organic or inorganic. Organic fertilizer includes plant or animal wastes or both; manure and compost (rotted organic material) are two examples. Organic fertilizer also includes leguminous fallow crops (alfalfa, clover) or food crops (lentils, peas), which fix atmospheric nitrogen. Inorganic fertilizers are chemical formulations of required nutrients, without any organic matter included. Inorganic fertilizers are much more prone to leaching than organic fertilizers.

Water and Water-Holding Capacity. As with nutrients, the supply of water in soil is crucial to plant growth. Water is constantly being absorbed by the roots of plants, passing up through the plant and exiting as water vapor through microscopic pores in the leaves—a process called **transpiration**. The pores, called *stomata* (singular, *stoma*), are essential to permit the entry of carbon dioxide and the exit of oxygen in photosynthesis; however, the plant's loss of water via transpiration through the stomata is dramatic. A field of corn, for example, transpires an equivalent of a field-sized layer of water 43 centimeters (17 in.) deep in a single growing season. Plants with inadequate water may wilt (close their stomata to save water and fold up) and eventually die.

Water is resupplied to soil naturally by rainfall or artificially by irrigation. Three attributes of soil are significant in this respect. First is soil's ability to allow water to infiltrate or soak in. If water runs off the ground surface, it won't be useful to the plants. Worse, it may cause erosion, which is discussed shortly.

Second is soil's ability to hold water after it infiltrates, a property called **water-holding capacity**. Poor water-holding capacity implies that most of the infiltrating water percolates on down below the reach of roots—not very far in the case of seedlings and small plants—again becoming useless. What is desired is a good water-holding capacity—the ability to hold a large amount of water like a sponge—providing a reservoir from which plants can draw between rains. Sandy soils are notorious for their poor water-holding capacity, whereas clayey and silty soils are good at holding water.

The third critical attribute of soil is its ability to minimize **evaporative water loss**. This kind of evaporation depletes soil's water reservoir without serving the needs of plants. The O horizon functions well to reduce evaporative water loss by covering soil. These aspects of soil-water relationship are summarized in **Figure 11–7**.

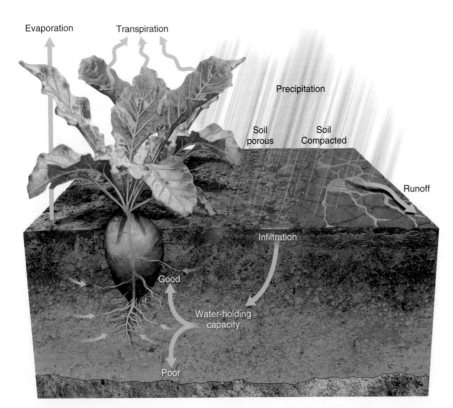

Figure 11–7 Plant-soil-water relationships. When water evaporates from the leaves of a plant, a vacuum is created that pulls the water up through the plant tissues. The roots draw water from soil to replenish the evaporated water. The sun's energy drives transpiration via the initial evaporation process. Water lost from the plant by transpiration must be replaced from a reservoir of water held in soil. In addition to the amount and frequency of precipitation, the size of this reservoir depends on soil's ability to allow water to infiltrate, to hold water, and to minimize direct evaporation.

Aeration. Novice gardeners commonly kill plants by overwatering, or "drowning," them. Roots need to "breathe." Basically, they need a constant supply of oxygen for energy via metabolism. Land plants depend on soil being loose and porous enough to allow the diffusion of oxygen into, and carbon dioxide out of, the soil, a property called soil **aeration.** Overwatering fills the air spaces in soil, preventing adequate aeration. Plants that live in very wet soil have special adaptations to get oxygen. **Compaction,** or packing of soil, which occurs with excessive foot or vehicular traffic, also lowers soil air content. Compaction also reduces infiltration and increases runoff. Again, soil texture strongly influences this property, as indicated in Table 11–1.

Relative Acidity (pH). The term **pH** refers to the acidity or alkalinity of any solution. A solution that is neither acidic nor alkaline is said to be neutral and has a pH of 7. The pH scale, which runs from 1 to 14, is discussed more fully later (Chapter 19). For now, it is important to know that different plants are adapted to different pH ranges. Most plants do best with a pH near neutral, but many plant species are specialized for distinctly acidic or alkaline soils (blueberries, for example, are notorious acid lovers).

Salt and Water Uptake. A buildup of salts in soil makes it impossible for the roots of a plant to take in water. Indeed, if salt levels in soil get high enough, water can be drawn *out* of the plant (by osmosis), resulting in dehydration and death. Only plants with special adaptations can survive saline soils, and to date, none of those are crop plants. The importance of the problem is explained in Section 11.2, when we discuss how irrigation may lead to the accumulation of salts in soil (salinization).

Soil and Carbon Storage. Soils hold carbon from dead organisms. Sometimes this matter decays rapidly and sometimes slowly, but scientists estimate that soils hold as much as three times the amount of carbon held in the atmosphere and living plants. Right now, scientists are keenly interested in knowing what drives the rate of organic matter decay in soils. It used to be thought that temperature was the main driving factor, but a study in 2011 suggested that soil factors such as moisture, physical separation of particles, and placement of roots may play a role.[3] Microbial communities also play a role. This information may be used to make more accurate predictions of the carbon cycle. Charcoal (burned wood) is an organic material that decays very slowly. Some researchers have suggested burying charcoal ("biochar") as a way of sequestering carbon in soil, after discovering that Amazonian indigenous peoples had used charcoal in fields. The newest research, however, suggests that biochar may be less effective than first hoped. Either way, soils with organic matter that decays slowly have higher fertility.

[3]Schmidt, Michael W. I. et al. (2011). Persistence of soil organic matter as an ecosystem property. *Nature*, 478(7367), 49.

The Soil Community

As we've seen, to support desired plant growth, soil must (1) have a good supply of nutrients and good nutrient-holding capacity; (2) allow infiltration, have good water-holding capacity, and resist evaporative water loss; (3) have a porous structure that permits good aeration; (4) have a pH near neutral; and (5) have low salt content. Moreover, these attributes must be sustained. How does a soil provide and sustain such attributes? Which is the best soil?

Recall the principle of limiting factors: The poorest attribute is the limiting factor (Chapter 3). The poor water-holding capacity of sandy soil, for example, may preclude agriculture altogether because soil dries out so quickly. The inability of clay soils to allow water infiltration or aeration prevents these from being good soils. As Table 11–1 indicates, the best soil textures are silts and loams because limiting factors are moderated in these two types of soil. Even though silts and loams are rated only "medium" for use in agriculture, their soil texture limitations are greatly improved by the organic parts of the soil ecosystem—the detritus and soil organisms.

Organisms and Organic Matter in Soil. The dead leaves, roots, and other detritus accumulated on and in soil support a complex food web, including numerous species of bacteria, fungi, protozoans, mites, insects, millipedes, spiders, centipedes, earthworms, snails, slugs, moles, and other burrowing animals **(Fig. 11–8)**. The most numerous and important organisms are the smallest—the bacteria; literally millions of bacteria can be seen and counted in a gram of soil.

Humus. As all these organisms feed, the bulk of the detritus is consumed through cell respiration. Carbon dioxide, water, and mineral nutrients are released as by-products (see Chapter 4). However, each organism leaves a certain portion undigested; that is, a portion resists breakdown by the organism's digestive enzymes. This residue of partly decomposed organic matter is humus, found in high concentrations at the bottom of the O horizon. A familiar example is the black or dark brown spongy material remaining in a dead log after the center has rotted out. **Composting** is the process of fostering the decay of organic wastes under more or less controlled conditions, and the resulting compost is essentially humus. Humus is good for plant growth because it has an extraordinary capacity for holding both water and nutrients—as much as 100 times greater than the capacity of clay, on the basis of weight.

Soil Structure and Topsoil. As animals feed on detritus on or in soil, they often ingest mineral soil particles as well. For example, it is estimated that as much as 37 tons per hectare (15 tons per acre) of soil pass through earthworms each year in the course of their feeding. As the mineral particles go through the worm's gut, they actually become "glued" together by the indigestible humus compounds. Thus, earthworm excrements—or castings, as they are called—are relatively stable clumps of inorganic particles plus humus. The burrowing activity of organisms keeps the clumps loose. This loose, clumpy characteristic is a desirable **soil structure**

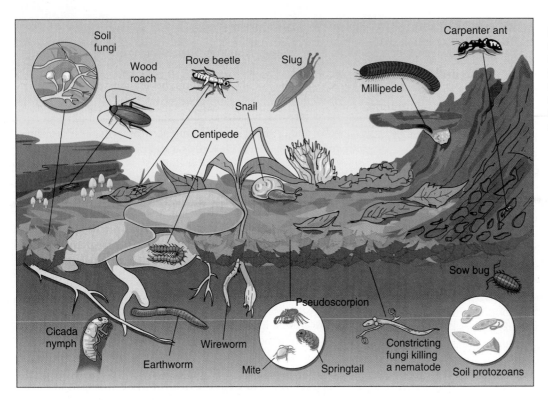

Figure 11–8 Soil as a detritus-based ecosystem. A host of organisms, major examples of which are shown here, feed on detritus and burrow through soil, forming a humus-rich topsoil with a loose, clumpy structure.

(Fig. 11–9). Whereas soil texture describes the size of soil particles, soil structure refers to their arrangement. A loose soil structure is ideal for infiltration, aeration, and workability. The clumpy, loose, humus-rich soil that is best for supporting plant growth is topsoil, represented in a soil profile as the A horizon (see Figure 11–5).

If plants are grown on adjacent plots, one of which has had all its topsoil removed, the results are striking: The yield from plants grown on subsoil is only 10% to 15% of that from plants grown on topsoil. In other words, a loss of all the topsoil would result in an 85% to 90% decline in productivity. The development of topsoil from subsoil or parent material is a process that takes hundreds of years or more.

Figure 11–10 shows a site where the topsoil was removed 50 years ago to provide fill for a highway. Although a lichen cover and some stunted trees have developed on the exposed gravel, there is still no topsoil there.

Interactions. There are some important interactions between plants and soil biota. A highly significant one is the symbiotic relationship between the roots of some plants and certain fungi called **mycorrhizae**. Drawing some nourishment from the roots, mycorrhizae penetrate the detritus, absorb nutrients, and transfer them directly to the plant (**Fig. 11–11**). Thus, there is no loss of nutrients to leaching. Another important relationship is that of certain bacteria that add nitrogen to soil (Chapter 3).

Figure 11–9 Humus and the development of soil structure. On the left is a humus-poor sample of loam. Note that it is a relatively uniform, dense "clod." On the right is a sample of the same loam, but rich in humus. Note that it has a very loose structure, composed of numerous aggregates of various sizes.

Figure 11–10 The results of removing topsoil. This "soil" is nothing but gravel with topsoil removed. After 50 years, only lichens and a few stunted trees have developed. The white object is a ballpoint pen.

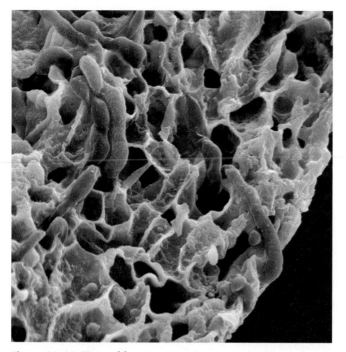

Figure 11–11 Mycorrhizae. Symbiotic fungi connect to the roots of plants; they help get nutrients and increase root absorption, while getting protection and food. Here you can see a canning electron micrograph of the small fungal hyphae entering the plant root hair. Mycorrizae can be purchased as a soil additive and increase crop yields for gardeners.

Not all soil organisms are beneficial to plants, however. For example, nematodes, small worms that feed on living roots, are highly destructive to some agricultural crops. In a flourishing soil ecosystem, however, nematode populations may be controlled by other soil organisms.

Soil Enrichment or Mineralization. You can now see how the aboveground and soil portions of an ecosystem support each other. The bulk of the detritus, which supports soil organisms, is from green-plant producers, so green plants support soil organisms. In turn, by feeding on detritus, soil organisms create the chemical and physical soil environment that is most beneficial to plant growth.

Green plants protect soil and, consequently, themselves in two other important ways: The cover of living plants and detritus (1) protects soil from erosion and (2) reduces evaporative water loss. Thus, it is desirable to maintain an organic mulch around any garden vegetables that do not maintain a complete cover themselves.

Unfortunately, the mutually supportive relationship between plants and soil can be broken all too easily. The maintenance of topsoil depends on additions of detritus in sufficient quantity to balance losses. Without continual additions of detritus, soil organisms will starve, and their benefit in keeping soil loose and nutrient rich will be lost. Additional consequences can occur as well. Although resistant to digestion by soil organisms, humus does decompose at the rate of about 2% to 5% of its volume per year. As soil's humus content declines, the clumpy aggregate structure created by soil particles glued together with the humus breaks down. Water- and nutrient-holding capacities, infiltration, and aeration decline as well. This loss of humus and the consequent collapse of topsoil are referred to as the **mineralization** of soil because what is left is just the gritty mineral content—sand, silt, and clay.

Thus, topsoil is the result of a dynamic balance between detritus additions and humus-forming processes, on the one hand, and the breakdown and loss of detritus and humus **(Fig. 11–12)** on the other. Soil deteriorates or is revitalized through loss or additions of organic matter.

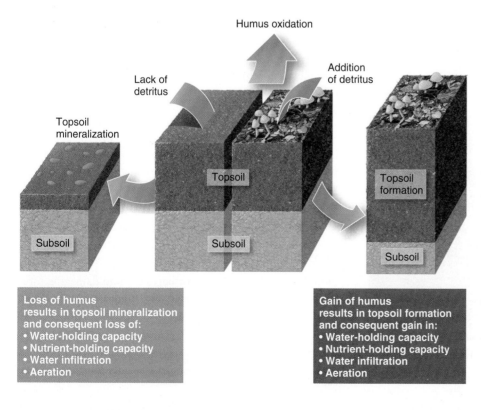

Figure 11–12 The importance of humus to topsoil. Topsoil is the result of a balance between detritus additions and humus-forming processes as well as their breakdown and loss. If additions of detritus are insufficient, soil will gradually deteriorate.

Loss of humus results in topsoil mineralization and consequent loss of:
• Water-holding capacity
• Nutrient-holding capacity
• Water infiltration
• Aeration

Gain of humus results in topsoil formation and consequent gain in:
• Water-holding capacity
• Nutrient-holding capacity
• Water infiltration
• Aeration

11.2 Soil Degradation

In natural ecosystems, there is always a turnover of plant material, so new detritus is continuously supplied. However, when humans come on the scene and cut forests, graze livestock, or grow crops, the soil is at the mercy of our management or mismanagement. When key soil attributes required for plant growth or for other ecosystem services deteriorate over time, the soil is considered degraded. Before we were able to ask how serious this problem was, we needed some way to track and assess soil changes **(Figure 11.13)**. This has now been done on a global scale, though it was very difficult.

The GLASOD report on degraded land quoted earlier (15% degraded) was for a long time the only truly global assessment. The data for the report came from questionnaires sent to soil experts around the world. Very little of the information was validated by collecting data on actual soil conditions or information on crop productivity. The authors actually pointed this out in the report. Long before GLASOD, however, scientists had been trying to gather soil data. The ISRIC World Soil Information (once the International Soil Research Information Centre) was started in 1966 as an offshoot of a World Soil Museum created in 1961 and housed in the Netherlands. All over the world, regional governments were collecting data on land, land use, agriculture, and soils, but it was often in different formats and scales or with different vocabulary. Standardizing information from around the world and providing data to researchers has been part of the job of the ISRIC World Soil Information organization.

The GLASOD map was a huge undertaking, but it was not as accurate as we needed. For example, 75% of the soils in Burkina Faso, one of the poorest and most densely populated countries in the dryland region of West Africa, were considered degraded. However, agricultural yields for virtually all crops have increased in the past 40 years, dur-

ing which time the population more than doubled. Eventually, the GLASOD data were considered suspect and out of date.

GLASOD was followed by a project launched by the Food and Agricultural Organization of the United Nations (FAO), called the Land Degradation Assessment in Drylands (LADA). Land degradation is defined as a reduction in the capacity of land to perform ecosystem functions and services that support society and development. This project used satellite remote-sensing techniques to produce an overall view of soil quality, called the Global Assessment of Land Degradation and Improvement (GLADA).[4] The GLADA assessment employed data from a 23-year time series by National Oceanic and Atmospheric Administration (NOAA) satellites and precipitation data from more than 9,300 land stations. One analysis combined plant productivity with rain-use efficiency to produce a map of regions that are worsening, calling them "hot spots," and regions that are showing improving trends, calling them "bright spots."

Updating Soil Science

GLADA was updated in 2010. The update used new techniques to estimate aboveground plant biomass from satellite images. Satellites can measure light variations, which can be used to calculate measures of greenness. Greenness is then used as an index of aboveground biomass, which in turn is used to assess land degradation because degraded land produces less vegetation. But even undergraded lands in some biomes do not produce a lot of vegetation, so the program actually measures changes—is the ability of the land to produce vegetation declining over time? In times of drought or at certain times of year, one would expect vegetation to decline anyway, so the program must account for seasons and the effect of droughts to measure whether the land is degraded for any length of time. This is an area where scientific ability to accurately model what is happening is both rapidly improving and crucial. One of the findings of the updated GLADA study was that land degradation is changing in China differently than people expected. Land degradation is most conspicuous in the humid south and east, where urban development is more intense, rather than the dry north and west, because of aggressive land protection and reforestation efforts in the north by the Chinese government. Other researchers have tried to estimate the effect of soil degradation on China's crop yields **(Fig. 11–14)**. One of the most critical land degradations is erosion, discussed next.

Erosion

How is topsoil lost? The most pervasive and damaging force is **erosion**, the process of soil and humus particles being picked up and carried away by water or wind. Erosion

Figure 11–13 Soil research. Researchers such as the one shown here measure soil radiation. They study soil in the hopes of protecting and enhancing productivity.

[4]Bai, Z. G., Jong, R. de, & van Lynden, G. W. J. (2010). An Update of GLADA—Global Assessment of Land Degradation and Improvement. ISRIC report 2010/08, ISRIC—World Soil Information, Wageningen.

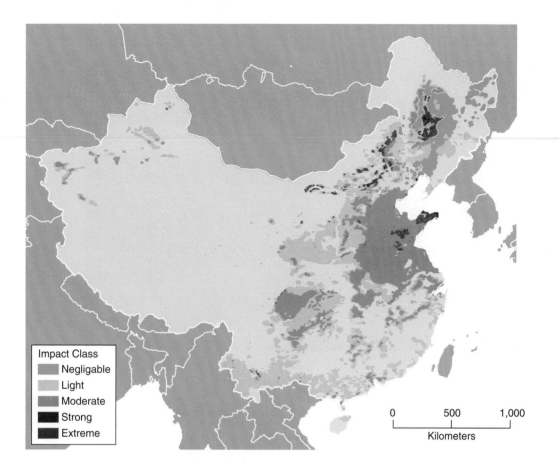

Figure 11–14 Soil degradation and crops in China. Across China, soil degradation is affecting croplands, especially in the south and east, even though the north and west are where desertification occurs. (Source: Liming Ye and Eric Van Ranst. (2009). Production scenarios and the effect of soil degradation on long-term food security in China. *Global Environmental Change, 19*(4), 464–481.)

follows any time soil is bared and exposed to the elements. The removal may be slow and subtle, as when soil is gradually blown away by wind, or it may be dramatic, as when gullies are washed out in a single storm.

In most natural terrestrial ecosystems, a vegetative cover protects against erosion. Falling raindrops are intercepted by the vegetation, and the water infiltrates gently into the loose topsoil without disturbing its structure. With good infiltration, runoff is minimal. Runoff that does occur is slowed as the water moves through the vegetative or litter mat. Similarly, vegetation slows the velocity of wind and holds soil particles.

Splash, Sheet, and Gully Erosion. When soil is left bare and unprotected, however, it is easily eroded. **Splash erosion** occurs in storms as the impact of falling raindrops breaks up the clumpy structure of the topsoil. The dislodged particles wash into spaces between other aggregates, clogging pores and thereby decreasing infiltration and aeration. The decreased infiltration results in more water running off, carrying away the fine particles from the surface, a phenomenon called **sheet erosion**. As further runoff occurs, the water converges into rivulets and streams, which have greater volume, velocity, and energy and hence greater capacity to pick up and remove soil. The result is the erosion into gullies, or **gully erosion (Fig. 11–15)**. This process of erosion can become a cycle that perpetuates.

Desert Pavement and Cryptogamic Crusts. Wind and water erosion both involve the differential removal of soil particles. The lighter particles of humus and clay are the first to be carried away, while rocks, stones, and coarse sand remain behind. Consequently, as erosion removes the finer materials, the remaining soil becomes progressively coarser—sandy, stony, and finally, rocky. Such coarse soils frequently reflect past or ongoing erosion. Did you ever wonder why deserts are full of sand? The sand is what remains after the finer, lighter clay and silt particles have blown away. However, in some deserts, the removal of fine material by wind has left a thin surface layer of stones and gravel called a **desert pavement**, which protects the underlying soil against further erosion (**Fig. 11–16**). Damaging this surface layer with vehicular or pedestrian traffic allows another episode of erosion to commence.

Figure 11–15 Erosion.
Severe erosion due to poor
farming practices near
Bosencheve, Mexico.

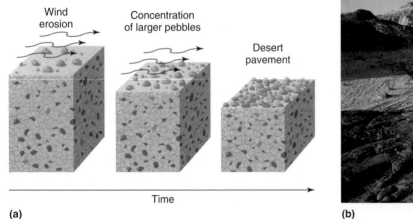

Wind
erosion

Concentration
of larger pebbles

Desert
pavement

Time

(a)

(b)

Figure 11–16 Formation of desert pavement. (a) As wind erosion removes the finer particles, the larger grains and stones are concentrated on the surface. (b) The result is desert pavement, which protects the underlying soil from further erosion. The Sinai Desert has a mixture of desert pavement, rock and tracks from vehicles, which break up the desert pavement and initiate further erosion.

Water erosion can cause other changes in soil composition. Often, when rainfall clogs soils, the soils harden when they dry. Following this, the soil may be colonized by different kinds of primitive plants (algae, lichens, and mosses) called cryptogams. Their growth and colonization create a crust on the soil, called a **cryptogamic crust**. Such crusts have a positive ecological impact; they can stabilize soil, slow erosion, and add nutrients through nitrogen fixation. However, they can also inhibit water infiltration and seed generation. They are readily broken up by livestock trampling or human

intrusion, and if the soil below is loosened, it can again be subject to wind and water erosion. Accordingly, let us now take a closer look at the practices that lead to erosion, one of the major elements of land degradation.

Causing and Correcting Erosion

Three major practices that expose soil to erosion and lead to soil degradation are overcultivation, overgrazing, and deforestation, all of which are a consequence of unsustainable management practices. All of these are practices with impacts that can be reversed, as we will see.

Overcultivation. Traditionally, the first step in growing crops has been plowing to control weeds. The drawback is that the soil is then exposed to wind and water erosion. Further, soil may remain bare for a considerable time after planting and again after harvest. Plowing is frequently considered necessary to loosen soil to improve aeration and infiltration through it, yet all too often the effect is just the reverse. Splash erosion destroys soil's aggregate structure and seals the surface, so that aeration and infiltration are decreased. The weight of tractors used in plowing adds to the compaction of soil. In addition, plowing accelerates evaporative water loss.

Despite the harmful impacts inherent in cultivation, systems of crop rotation—a cash crop such as corn every third year, with hay and clover (which fixes nitrogen as well as adding organic matter) in between—have proved sustainable. If farmers abandon rotation, degradation and erosion exceed regenerative processes, and the result is a decline in the quality of soil. This is the essence of overcultivation.

No-Till. A technique that permits continuous cropping, yet minimizes soil erosion, is **no-till agriculture**, which is now routinely practiced in areas of the United States. According to this technique, the field is first sprayed with herbicide to kill weeds, and then a planting apparatus is pulled behind a tractor to accomplish several operations at once. A steel disk cuts a furrow through the mulch of dead weeds, drops seed and fertilizer into the furrow, and then closes it **(Fig. 11–17)**. At harvest, the process is repeated, and the waste from the previous crop becomes the detritus and mulch cover for the next. Thus, the soil is never left exposed, erosion and evaporative water loss are reduced, and there is enough detritus, including roots from the previous crop, to maintain the topsoil. Of course, this requires the use of herbicides, which have their own problems and may increase the effects of pests that remain in crop residues from one season to another (Chapter 13).

Figure 11–17 Apparatus for no-till planting. This no till planter is planting corn over stubble with minimal disturbance to the protective vegetation layer on the surface of the soil. The apparatus creates furrows with cutting discs, applies herbicide and fertilizer, drops in seeds, and closes the furrows.

A variation on this theme is **low-till farming**, which is catching on in Asia. For example, to plant wheat after a rice crop has been harvested, only one pass over a field is made—to plant seed and spread fertilizer. Previously, farmers would plow 6 to 12 times to break up the muddy soil and dry it out for the wheat, and then they would irrigate the land.

Fertilizer. Another aspect of overcultivation involves the use of inorganic versus organic fertilizer. There is no question that optimal amounts of required nutrients can be efficiently provided by the suitable application of inorganic chemical fertilizer. The failing of chemical fertilizer is its lack of organic matter to support soil organisms and build soil structure. Under intensive cultivation—one or more cash crops every year—nutrient content may be kept high with inorganic fertilizer, but mineralization, and thus soil degradation, proceeds in any case. Then, with the soil's loss of nutrient-holding capacity, applied inorganic fertilizer is prone to simply leach into waterways, causing pollution.

This is not to say that chemical fertilizers do not have a valuable place in agriculture—after all, the exclusive use of organic material may not provide enough of certain nutrients to support optimal plant growth. What is required is for growers to understand the different roles played by organic material and inorganic nutrients and then to use each as necessary.

Natural Resources Conservation Service. Regardless of cropping procedures and the use of fertilizer, a number of other techniques are widely employed to reduce erosion. Among the most conspicuous are **contour farming** (plowing and cultivating at right angles to contour slopes) and **shelterbelts** (protective belts of trees and shrubs alongside plowed fields) **(Fig. 11–18)**. The U.S. **Natural Resources Conservation Service (NRCS)**, formerly the Soil Conservation Service, was established in response to the Dust Bowl tragedy of the early 1930s. Through a nationwide network of regional offices, the NRCS provides information to farmers or other interested persons regarding soil and water conservation practices. (Federal extension service offices will test soil samples, provide an analysis thereof, and make recommendations.) The NRCS performs an inventory of erosion losses in the United States (the Natural Resources Inventory, last conducted in 2007). Estimates indicate that soil erosion on croplands has lessened in recent years, from 2.1 billion tons in 1992 to 1.7 billion tons in 2007. This decrease is likely a consequence of improved conservation practices, such as windbreaks, grassed waterways, and field border strips of perennial vegetation to filter runoff.

Overgrazing. Grasslands that receive too little rainfall to support cultivated crops or are too steep for cropping have traditionally been used for grazing livestock. In fact, some 65% of dryland areas are rangelands. Unfortunately, such lands are too often overgrazed. As grass production fails to keep up with consumption, erosion follows and the land becomes barren.

Overgrazing is not a new problem. In the United States in the 1800s, the American buffalo (bison) was slaughtered to starve out Native Americans and allow for stocking the

(a)

(b)

Figure 11–18 Contour farming and shelterbelts. (a) Cultivation up and down a slope encourages water to run down furrows and may lead to severe erosion. The problem is reduced by plowing and cultivating along the contours at a right angle to the slope. In this photograph, strip cropping is also being utilized: The light green bands are one type of crop, the dark bands another. (b) Shelterbelts—belts of trees around farm fields—break the wind and protect soil from erosion, seen in New Zealand.

rangelands with cattle. Cattle overgrazing became rampant, leading to erosion and encroachment by hardy desert plants such as sagebrush, mesquite, and juniper, which are not palatable to cattle. Western rangelands now produce less than 50% of the livestock forage they produced before the advent of commercial grazing. Yet overstocking and the federal subsidizing of ranching means that in 2005 the grazing program on federal public lands cost $144 million and brought in only $21 million to the federal government. The Obama administration has proposed a grazing fee increase.

The broader ecological impact of overgrazing should not go unnoticed, either. The National Public Lands Grazing Campaign documents the many harmful effects of livestock grazing on public lands. Livestock compete with native animals for forage, reducing the latter's numbers substantially (about one-third of the endangered species in the United States are in jeopardy because of overgrazing or other practices associated with raising cattle, such as predator-control programs and the suppression of fire). Particularly hard hit

are wooded zones along streams and rivers, zones that are trampled by cattle seeking water. The resulting water pollution by sediments and cattle waste is high on the list of factors making fish species the fastest-disappearing wildlife group in the United States. A study of rangeland management concluded that (1) desertification affects some 85% of North America's drylands and (2) the most widespread cause of rangeland desertification is livestock grazing.[5]

Public Lands. In many cases, overgrazing occurs because rangelands are public lands not owned by the people who own the animals. Where this is the case, herders who choose to have fewer livestock on the range sacrifice income, while others continue to overgraze the range. This problem, known as the "tragedy of the commons," was discussed earlier (Chapter 7).

In a variation of the tragedy of the commons, the U.S. Bureau of Land Management (BLM) and the Forest Service lease grazing rights for 2 million square kilometers of federal (i.e., taxpayer-owned) lands. The income to the government is less than private land owners would be paid for grazing rights, though ranchers have to do more on federal lands, such as provide water and fences and get permits. Altogether it pays the government far less than the costs of administration: The grazing program loses money, and the 2012 grazing fee ($1.35 per cow and calf per month) is less than a tenth of what grazing fees are on private lands. A worse problem is the ecological cost exacted by livestock to watersheds, streams, wildlife, and endangered species, considered to be as high as $500 million. Attempts in Congress and the Department of the Interior (which houses the BLM) to raise grazing fees are consistently met with opposition from western congressmen, who threaten to cut the BLM budget.

One solution lies in better management. Restoration of rangelands could benefit both wildlife and cattle production. The NRCS has a program (the **Conservation Stewardship Program**) that provides information and support to enable ranchers who own their lands to burn unwanted woody plants, reseed the land with perennial grass varieties that hold water, and manage cattle so that herds are moved to a new location before overgrazing occurs. One solution gaining favor among ranchers and environmental groups is for the government to buy up some of the 26,000 existing grazing permits and "retire" rangelands. The ranchers would be offered a generous payment for the permits they hold, and the land would be used for wildlife, recreation, and watershed protection.

Deforestation. Forest ecosystems are extremely efficient systems both for holding and recycling nutrients and for absorbing and holding water because they maintain and protect a very porous, humus-rich topsoil. Investigators at Hubbard Brook Forest in New Hampshire found that converting a hillside from forest to grass doubled the amount of runoff

───────────

[5]Government Accountability Office. Livestock Grazing: Federal Expenditures and Receipts Vary, Depending on the Agency and the Purpose of the Fee Charged (September 30, 2005).

and increased the leaching of nutrients many-fold. Much worse is what occurs if the forest is simply cut and soil is left exposed. The topsoil becomes saturated with water and slides off the slope in a muddy mass into waterways, leaving barren subsoil that continues to erode. Forests continue to be cleared at an alarming rate (Chapter 7). The problem is particularly acute when tropical rain forests are cut. Tropical soils (oxisols) are notoriously lacking in nutrients from leaching by rainwater. When the forests are cleared, the thin layer of humus with nutrients readily washes away. Only the nutrient-poor subsoil is left.

The Other End of the Erosion Problem. Water that is unable to infiltrate flows over the surface immediately into streams and rivers, overfilling them and causing flooding. Eroding soil, called **sediment**, is carried into streams and rivers, clogging channels and intensifying flooding, filling reservoirs, killing fish, and generally damaging the ecosystems of streams, rivers, bays, and estuaries. Excess sediments and nutrients resulting from erosion are recognized as the greatest pollution problem of surface waters in many regions of the world. Meanwhile, groundwater resources are also depleted because the rainfall runs off rather than recharging groundwater. This is another example of the tight connection between water availability and quality and the health of soils.

Drylands and Desertification

Recall that clay and humus are the most important components of soil for both nutrient- and water-holding capacity. As clay and humus are removed, nutrients are removed as well because they are bound to those particles. The loss of water-holding capacity is even more serious, however. Regions that have sparse rainfall or long dry seasons support grasses, scrub trees, or crops only insofar as soils have good water-holding and nutrient-holding capacity. As these soil properties are diminished by the erosion of topsoil, such areas become deserts, both ecologically and from the standpoint of agricultural production. Indeed, the term **desertification** is used to denote this process. Note that desertification doesn't mean advancing deserts; rather, the term refers to a permanent reduction in the productivity of arid, semiarid, and seasonally dry areas (traditionally called drylands). It is a process of land degradation with a number of possible causes, including extended droughts, overgrazing, erosion, deforestation, and overcultivation.

Dryland ecosystems cover 41% of Earth's ice-free land area. They are defined by precipitation, not temperature. Beyond the relatively uninhabited deserts, much of the land receives only 25 to 75 centimeters (10 to 30 in.) of rainfall a year, a minimal amount to support rangeland or non-irrigated cropland (**Fig. 11–19**). Droughts are common features of the climate in drylands. When the rainfall returns, however, vegetation often returns. Despite this, according to the Millennium Ecosystem Assessment, 10–20% of these lands suffer from some form of land degradation. These drylands, occupying some 6 billion hectares, are home to more than 2 billion people and are found on every continent except

Figure 11–19 Drylands cultivation. Farmers till soil in Jinan, China, planting a crop in the semi-desert soil in anticipation of some rainfall.

Antarctica. Since these dryland areas are commonly inhabited by some of the poorest people on Earth, promoting sustainable land management is essential for improving standards of living and meeting the Millennium Development Goals (MDGs) successfully.

Recognizing the severity of this problem, the United Nations established the Convention to Combat Desertification (UNCCD), which was signed and officially ratified by more than 100 nations in 1996. Regular conferences of the parties and other UNCCD meetings have been concerned with such issues as the funding of projects to reverse land degradation, "bottom-up" programs that enable local communities to help themselves, the gathering and dissemination of traditional knowledge on effective drylands agricultural practices, and the development of a 10-year strategic plan. Affected countries have also been developing National Action Programs. One alliance, TerrAfrica, was formed in 2005 by several UN agencies and African countries to fight land degradation in sub-Saharan Africa. The alliance is designed to coordinate efforts toward arresting land degradation and promoting sustainable land management. For example, the government of Madagascar has been working on sustainable land management with several TerrAfrica partners, combating land degradation in its upland watersheds and southern arid lands. Even though variations in climate often play a role in the processes of desertification, it is human impacts that pose the greatest threat to the health of dryland ecosystems. This highlights yet again the connection between soil, water, and human actions.

Water and Soil: Irrigation, Salinization, and Leaching

How we use water has a huge effect on soil health, especially in agricultural systems. Nowhere is this more clear than with irrigation—supplying water to croplands by artificial means.

Figure 11–20 Flood irrigation. The traditional method of irrigation is to flood furrows between rows with water from an irrigation canal. This method is extremely wasteful because most water evaporates or percolates beyond the root zone or because the water table rises to waterlog soil and prevent the crop from growing.

Figure 11–21 Salinization. Millions of acres of irrigated land are now worthless because of the accumulation of salts left behind as water evaporates.

Irrigation has dramatically increased crop production in regions that typically receive low rainfall and is a major part of the Green Revolution (Chapters 8, 12). However, irrigation is also a major contributor to land degradation. Traditionally, water has been diverted from rivers through canals and flooded through furrows in fields, a technique known as **flood irrigation (Fig. 11–20). Center-pivot irrigation,** in which water is pumped from a central well through a gigantic sprinkler that slowly pivots itself around the well, has gained popularity as an irrigation method (see Figures 10–16b and c). In more recent years, the much more efficient **drip irrigation,** in which water is taken directly to plants through small tubes, is used at least on some crops.

In the United States, the Bureau of Reclamation is the major government agency involved with ongoing supplies of irrigation water to the western regions. More than 800 dams and reservoirs have been constructed by the bureau since its inception in 1902, and it now provides irrigation water for 4 million hectares (10 million acres) of farmland. Total irrigated land acreage in the United States is 22 million hectares (55 million acres), one-sixth of all cultivated cropland, according to the U.S. Department of Agriculture.

Worldwide, irrigated acreage has risen dramatically in the past few decades. The FAO estimates that total irrigated land now amounts to just more than 300 million hectares (741 million acres). New irrigation projects continue to be built, but the expansion in acreage is now being offset in large part by a negative trend: *salinization.*

Salinization. Salinization is the accumulation of salts in and on soil to the point where plant growth is suppressed **(Fig. 11–21).** Salinization occurs because even the freshest irrigation water contains at least 200 to 500 ppm (0.02% to 0.05%) dissolved salts. Adding water to dryland soils also dissolves the high concentrations of soluble minerals that are often present in those soils as salts when water evaporates. Because it happens in drylands and renders the land less productive or even useless, salinization is considered a form of desertification.

Worldwide, an estimated 1.5 million hectares (3.7 million acres) of agricultural land—or the equivalent of Connecticut and Rhode Island—is lost yearly to salinization and waterlogging. Pakistan alone has lost an estimated 6.3 million hectares (15.5 million acres) of farmland to salinization. In Mozambique, a promising irrigation project in the Limpopo Valley brought water to more than 33,000 hectares (81,500). Unfortunately by 2012, more than 10,000 of those hectares (25,000) were unusable because of salinization. Adding to the problems, water supplies in many places are fast being depleted by withdrawals for irrigation (Chapter 10).

Salinization can be avoided, and even reversed, if sufficient water is applied to leach the salts down through soil. Attention must be paid to where the salt-laden water goes. For example, the Kesterson National Wildlife Refuge in California was destroyed by pollution from irrigation drainage from selenium-enriched soils of the San Joaquin Valley. After receiving the salty drainage water for only three years, birds, fish, insects, and plants began to die off. The site was declared a toxic waste dump and has been drained and capped with soil. At least 14 other locations in the western United States have been identified as having the potential to accumulate toxic irrigation water, now known as the Kesterson Effect.

Leaching. In places that are heavily irrigated or have heavy rainfall, soils can lose nutrients to leaching. This is particularly true in tropical soils. Sometimes this is a problem because nutrients are leached out of soil, and sometimes it is a problem because toxins are released from being bound to soil particles. One good example of the latter results from a change in soil pH. Soil pH can change with acid rain, a byproduct of burning fossil fuels. As acid rain percolates through soil, it can change the amount of attraction between some chemicals and soil particles. This change can have two important effects. Nutrients such as calcium, magnesium, and potassium that are bound to clay and humus can be separated from soil particles more easily and can be leached out of soil by rain. Another effect is the mobilization of toxins.

Aluminum is usually held in soil in a nontoxic form: aluminum hydroxide. Acid rain breaks chemical bonds, releasing toxic aluminum ions from the insoluble aluminum hydroxide, a process that stunts the roots of plants, keeping them from absorbing nutrients.

Loss of Soil Biota

Soil health is largely regulated by microorganisms and detritus. Organic material holds water, binds mineral particles together, and slowly releases nutrients. Soil microbes help plants germinate, obtain nutrients, and perform other activities. Erosion, salinization, acid rain, and waterlogging can kill soil biota. Some other human activities we have not discussed can also lower soil fertility. One new effect on soil is that of nanoparticles. Nanoparticles are tiny and are used for a variety of purposes. For example, extremely small beads of silver are added to many products as an antimicrobial agent. You can buy "antimicrobial" socks and other products. Unfortunately, nanoparticles can have negative effects on water and soil. A study in the *Journal of Hazardous Materials* in 2011 showed that nanoparticles can kill beneficial soil bacteria and can keep plants from being able to take up nitrogen. However, the study looked at high concentrations of nanoparticles. It is unclear what their effects will be in soils generally. One of the problems with nanoparticles is that they became widely used before sufficient prior study on their environmental effects was completed.

Land Use Changes: Impacts of Mining

Agriculture is the human activity that probably affects soils the most. However, other activities damage soil as well. Just as in the water cycle, soils are affected by land use changes. Construction projects, for example, remove vegetation and topsoil and often lead to erosion. One of the most significant non-agricultural impacts on soil due to land use is the effect of mining.

The scene after a strip mining operation is completed is often similar to a moonscape—great piles of rocks with topsoil removed. In some parts of the United States, these areas are required to be restored, though restoration success varies. In the removed mountaintops of Appalachia, the plateaus left after coal mining are green but relatively unproductive (Chapter 14). The topsoil is gone, too poor to support crops or most trees, and the vegetation that is planted is usually the non-native *Lespedeza cuneata*.

Mining is big business around the world. In addition to coal, people mine gypsum, phosphates, salts, metals, and many other materials, even materials called rare earths that are used in making electronics. Mining removes vegetation and has huge impacts on water quality, human health, and soil erosion. Because of these connections, mining is of importance in our discussions on fossil fuels, water pollution, environmental health, and hazardous waste (Chapters 14, 20, 17, and 22). In 2008, the first International Conference on Mining Impacts on the Human and Natural Environments was held in Florida. In 2012, conferences were held in Manila, Philippines (March), and Nevada City, Nevada (May), on the impacts of mining on the environment and the need for effective land restoration. Hopefully, this is a direction scientists will pursue.

11.3 Soil Conservation

There is broad consensus that soils are vital to human societies and that they should be preserved. This is encouraging, and so long as this consensus translates into action, we will be moving in the direction of sustainability. Because topsoils grow so slowly (an inch every hundred years at best), sustainability means doing all we can to reduce soil erosion and prevent degradation. However, we will also briefly discuss soil restoration, which can be done after degradation has occurred.

Soil conservation must be practiced at two levels. The most important is the level of the individual landholder. As the story of Gabino López demonstrates (see Sound Science, "Sustainable Agriculture Research with Farmer Experimenters"), those working on the land are best situated to put into practice the conservation strategies that not only enhance their soils, but also bring better harvests and improve their way of life. A great deal of traditional knowledge exists with landholders, who routinely practice soil conservation techniques. The increased production you will read about in Burkina Faso is a prime example. This is stewardship in action.

The second level is public policy. In particular, farm policies play a crucial role in determining how soils are conserved. Policies governing the grazing of public lands and forestry practices also lead to either good or poor stewardship of soils. The two levels must work in harmony to bring about effective, sustainable stewardship of soil resources. Let us examine briefly what is being done in these two areas, beginning with individual landholders.

Helping Individual Landholders

In the developing world as well as the developed world, it is the individual landholders, farmers, and herders who hold the key to sustainable soil stewardship. They must be convinced that what they do will work, that it is affordable, and that, in the long run, their own well-being will improve. It may require help in the way of microlending, sound advice (often best coming from their peers) or simply encouragement to experiment (Chapter 9).

Toward those ends, the UN Convention to Combat Desertification is a large step in the right direction. Other efforts include the new TerrAfrica Alliance and the Sustainable Agriculture and Rural Development Initiative (SARD) facilitated by the FAO. The latter is a kind of umbrella organization that coordinates efforts to reach small farmers with sound agricultural practices; SARD puts the efforts in the context of encouraging sustainable development and achieving the MDGs. An example is found in the Bobo Diaulasso region of

SOUND SCIENCE

Sustainable Agriculture Research with Farmer Experimenters

Thirty-five years ago, Gabino López attended classes on sustainable agriculture in San Martín Jilotepeque, Guatemala, and then returned to his village to apply what he had learned to his own small farm. He planted vegetation barriers along the contours of his land to stop erosion and applied cow manure to enrich the soil. His maize harvest increased from 0.8 to 1.3 tons/hectare (1,050 lb/acre). López's harvest jumped to 3.4 tons/hectare when he plowed under his crop residues, as opposed to when he burned them or rotated crops. Aware of his success, his friends asked him for advice, and before long, López became an international consultant in *ecological agriculture*, traveling from Mexico to Ecuador and even India. He and his fellow farmers employ a number of practices that follow the "five golden rules of the humid tropics": (1) keeping soil covered, (2) using minimal or even no tillage, (3) using mulch to provide nutrients for crops, (4) maximizing biomass production, and (5) maximizing biodiversity. As a result, the maize harvest has averaged 4.3 tons/hectare in a number of local villages.

López and his fellow farmers had to deal with both erratic rainfall and the possibility of mudslides. López worked with COSECHA, a Honduran nongovernmental organization (NGO) to connect farmers to research by involving them from the start as "farmer experimenters." To overcome the problem of irregular rainfall, the

farmers tested a number of different contraptions for trapping and holding the water for small-scale irrigation. The farmers decided that 1- to 2-cubic-meter catchments were ideal and found many additional uses for the water. Water harvesting is joining green manures and natural pest control in the growing reach of agricultural development that involves villagers experimenting and teaching others what they learn.

Many Central American farmers have to work fields on hillsides, called steep lands. They are difficult to till with machinery, and mudslides are common. While much of the soil conservation work in the world is focused on drylands (lands where soil moisture is very low), steep lands offer their own problems. American researchers have worked with farmer experimenters in the Philippines, another area with steep land farming, to solve some of the unique problems such agriculture causes. In one study, experimenters used high-value crops that grow thickly as a type of hedge between

plots of other crops. They also changed when crops were planted in order to have plants that lowered erosion in the field during times when more erosion occurs. For example, corn and cabbage were grown during rainy months rather than tomato because soil loss was lessened. The participation of farmers in the research provided real-world data and made it easier to adopt better soil conservation practices.

Burkina Faso, a rural community in Africa's drylands. Here intensive farming, driven by population growth, has led to a loss of soil fertility and declining crop yields. An FAO project has stepped in to enable a group of farmers to organize and adopt a range of sustainable practices, such as mixing legume and cereal grain plots to preserve soil fertility, cultivating soybeans to diversify income sources, and producing forage to feed livestock during the extended dry season. FAO brought the landholders and other stakeholders together to consolidate these findings into a set of established practices that could be implemented in other similar situations.

The Keita Project. Preventing and reversing the degradation of dryland soils remains a key need for helping the more than 2 billion people who live in drylands, who are often among the poorest on Earth and subject to climatic variations that can destroy their livelihood and well-being. Niger is one of the hottest countries of the world, with much of its land in the Sahara Desert itself. Below the Sahara, Niger is part of the Sahel, an ecoclimatic belt that

runs across Africa from the Atlantic Ocean to the Red Sea. It is a semiarid grassland, receiving 6–20 inches of rainfall a year, but often less in drought years. In 1982, the Italian government initiated a project in the Keita district of Niger, more than 4,000 square miles of plateaus, rocky slopes, and grasslands inhabited by 230,000 people. The project targeted the serious degradation brought on by overpopulation and the effects of soil and wind erosion. Some 41 dams were established to catch water from the monsoonal summer rains, more than 18 million trees were planted, and countless check dams were constructed to stop rainwater sheet erosion and provide soil and moisture for crops (see Chapter 10, p. 260). Most of the work was carried out by local people, especially women, who were paid with food. The antierosion land management scheme took many years, but it worked. The Keita district is now a flourishing environment for crops (**Fig. 11–22**), livestock, and people. Desertification has been halted. The people now have control over their plots of land, an important step toward long-term stewardship of local resources.

SUSTAINABILITY

New Ditches Save Soil

A hard rain pounds on a dusty farm field. Water races down a slope, running in rivulets through the rows, flowing over weeds at the field's edge, swelling the ditch that runs between it and the next field. In minutes the ditch is full, its banks covered with the rushing water. Thunder and lightning fill the sky overhead. The ditch's walls begin to crack, slumping into the torrent. Eddies of water carry brown soil away. The soil leaves the field to head downstream—to a river that does not need this sediment, to clog an estuary, to cover organisms, and to remove a valued resource from the farm that needs it.

Farm ditches are a valuable way to control the flow of water. They keep fields from lying sodden after long rains, but they can also be channels of erosion and can become eroded themselves in a thundering summer downpour. Researchers in Iowa are trying to design a new kind of ditch—one that is effective under normal conditions but can also keep down erosion in extreme storm events. They call this approach "two-stage ditch design." At the bottom of the ditch is a typical narrow channel. Partway up the ditch are two broad vegetated areas (like benches) that lie on either side of the channel, enclosed by banks that go up to the level of the field. The vegetated flat areas are designed to slow water and stop sediment in the case of a storm event. The ditches do more than save farmer's soil. In rivers, they protect freshwater mussels from the effects of silt. Sediment can overload waterways and kill vulnerable organisms. Far away, less sediment and fewer nutrients will arrive to pollute the distant Gulf of Mexico, an indication of just how connected ecosystems are.

The two-stage ditch design was created by the Nature Conservancy and collaborating partners. General Mills, Cargill, and the Natural Resource Conservation Service provided funding. Other partners helped with testing, monitoring, and working with landowners. It has been used extensively in Ohio and other states. The two-stage drainage ditch is a good example of a low-tech solution to a common problem—one that will help limit erosion of valuable farm soil.

Additional filters next to fields

Two-stage ditch design

Original water table

Grass benches

Main channel

Two Chains of Events. Desertification has been happening in many places and, when made worse by climate change or extended drought, can lead to permanent degradation and to greater poverty and misery. The human factors responsible include population growth and immigration, unjust land tenure policies, and economic pressure to produce cash crops, all of which intensify the pressure on the land to increase its ecosystem services. The productivity of the land declines as overgrazing and overcultivation lead to increased erosion. These results are only made worse in the face of extended drought and some impacts of climate change.

This series of events is not inevitable, however. In contrast, in places such as such as Burkina Faso and Niger, dryland dwellers take corrective measures, including better managing livestock and grazing, adopting soil and water conservation measures, and diversifying their income in ways that relieve pressure on the land. The result is a sustainable relationship with dryland ecosystems that results in improving human well-being. Some of the processes of desertification, although distressing, may be signs of the cyclical nature of dryland ecosystems; they simply mean that in time things are likely to change and that what seemed irreversible was in fact not.

Soil Restoration. Although prevention of soil degradation is the best policy, in some cases, soil restoration can be done. Addition of organic material and use of rooted plants to trap soils as well as introduction of soil biota are all components. Sometimes contaminants need to be removed as well. When this is done using organisms, it is called bioremediation. For example, in the case of petroleum contamination, soil microbes are used to break down the contaminant. In other cases, compost, manure, other plant products, and earthworms are added to enhance degraded soils.

Figure 11–22 Land in the Keita district of Niger. This district was the recipient of a remarkable land restoration effort headed by the government of Italy.

Public Policy and Soils

Despite the crucial importance of saving agricultural land, forests, and rangeland, soil degradation in the form of erosion, desertification, and salinization continues to occur because of human activities. The losses are on a collision course with sustainability. Who is responsible for making changes? At the first level, it is the individual landowner, but landowners are subject to public policies, which is the second level of decision making for soil conservation.

Subsidies. Farm policy in the United States was originally focused solely on increasing production. This goal has been achieved, as is demonstrated by continuing high yields and surpluses. The federal government supports U.S. agriculture through a number of programs; the farm lobby is a powerful one that has successfully maintained its support through the years. In more recent years, farm policy has emphasized maintaining farm income and support of farm commodities—in other words, subsidies. Farmers are guaranteed price levels for many products, including grains, cotton, sugar, peanuts, dairy products, and soybeans. This helps keep food prices from fluctuating year to year. In their book *Perverse Subsidies*, Norman Myers and Jennifer Kent[6] point out how subsidies are bad news for the economy and the environment. Myers and Kent argue that subsidies encourage excessive use of pesticides and fertilizers, reduce crop rotation by locking farmers into annual crop support subsidies, and promote the continued drawdown of groundwater aquifers through irrigation. Recall from Section 11.2 how similar subsidies to ranchers for grazing on federal lands have undercut the objectives of soil conservation for those lands.

Sustainable Agriculture. In comparison, the goals of sustainable agriculture are to (1) maintain a productive topsoil, (2) keep food safe and wholesome, (3) reduce the use of chemical fertilizers and pesticides, and last, but far from least, (4) keep farms economically viable. Professional agriculturists are becoming increasingly aware of the shortfalls of modern farming and are at work developing alternatives. Many options actually mimic practices of the past, such as contouring, crop rotation, terracing, the use of smaller equipment, and the application of a reduced amount of chemicals or no chemicals at all. In 1988, the U.S. Department of Agriculture started the Sustainable Agriculture Research and Education (SARE) program, which provides funding for investigating ways to accomplish all of the goals of sustainable agriculture just listed. The program has continued to receive funding of $5 million to $12 million annually, a small but encouraging amount earmarked for building and disseminating knowledge about sustainable agricultural systems.

Farm Legislation. The Federal Agricultural Improvement and Reform Act (FAIR) of 1996 attempted to bring needed reforms. Subsidies and controls over many farm commodities were reduced or eliminated, giving farmers greater flexibility in deciding what to plant, but also forcing them to rely more heavily on the market to guide their decisions. Unfortunately, declining commodity prices prompted yearly emergency farm aid packages, essentially maintaining the subsidies and controls targeted by FAIR.

The 2002 Farm Security and Rural Investment Act succeeded FAIR. It continued to subsidize a host of farm products, maintaining price supports and farm income for American farmers. The latest farm bill is the Farm Bill of 2012, which as of this writing is still under discussion. Before that, the Farm Bill of 2008 was passed by Congress over President George W. Bush's veto. The president objected to the continued high level of support to big farms, arguing that payment limitations and subsidies for crops were too high. Unfortunately, because of subsidies for corn ethanol production, the bill encouraged farmers to plow under remaining native grasslands. It does, however, continue a number of programs aimed at soil and wetland conservation.

Conservation Programs. In 1985, Congress passed the **Conservation Reserve Program (CRP)**, under whose authority highly erodible cropland could be established as a "conservation reserve" of forest and grass. Farmers are paid each year for land placed in the reserve.

Farm bills continue to set out conservation funding. Funding has been provided for two newer programs, the Wildlife Habitat Incentives Program (WHIP) and the Environmental Quality Incentives Program (EQIP), both of which encourage conservation-minded landowners to set aside portions of their land or address pollution problems. The 2002 farm bill initiated the Conservation Security Program (CSP), a voluntary program that encourages stewardship of private farmlands, forests, and watersheds by providing landowners incentives to conserve soil and water resources on their lands. This was renamed the Conservation Stewardship Program in the 2008 farm bill; more than $1 billion was provided for the new CSP. More than 12 million acres of farmlands may be enrolled each year. The 2012 farm bill is still in process at the time of this writing.

Public policies aside, farmers should have incentives to nurture the soil on lands under cultivation. The more farmers incorporate conservation practices in their operations, the more productive their farms will become in the long run. The same applies to those who care for livestock that graze open rangelands. Most observers agree that the most significant obstacle to establishing sustainable soil conservation is a lack of knowledge about what such conservation can do. In particular, this applies to the developing countries, where agricultural policies often are poorly developed and the majority of landholders have little access to agricultural extension services like those provided in the United States by the NRCS. In cases where they are provided with such services, as with Gabino López, they often accomplish surprising gains in agricultural production.

---

[6]Norman Myers and Jennifer Kent, *Perverse Subsidies: How Tax Dollars Can Undercut the Environment and the Economy* (Washington, D.C.: Island Press, 2001).

REVISITING THE THEMES

Sound Science

Soil science uses the principles of sound science to understand how soils are formed, how they are structured, and how they can be sustained. As with any science, however, it is possible to draw erroneous conclusions from the data, as seen with the GLASOD data on world soils and soil degradation and with the data on soil erosion in the United States. The truth is that more field measurements are desperately needed, and the UNCCD's Land Degradation Assessment in Drylands (LADA) is a big step in that direction. In the United States, soil erosion is measured with map reading and equations, and when some scientists actually measured erosion (see p. XXX), they found that the equation approach grossly overestimated erosion rates. Science can include farmers as scientists, giving data on soil use that might be hard to obtain otherwise.

Sustainability

Topsoils take hundreds of years to form and represent the most vital component of land ecosystems. Under most conditions, they are highly sustainable. Because of heavy human use for agriculture, grazing, and forestry and activities like urban expansion and mining, however, lands are being degraded, causing the productivity of those ecosystems to decline. In particular, the drylands have been recognized as extremely vulnerable, and the UNCCD has been helping farmers to adopt new approaches

to farming. Programs like COSECHA and TerrAfrica are spreading the word on these new approaches as well as on many sound traditional approaches that are not widely known. The sustainable agriculture practices in Guatemala and around the world are a way to improve soil health. On the other hand, the fees charged for grazing livestock on public lands in the United States are woefully inadequate and promote poor and unsustainable use of public lands.

Stewardship

Those who work the land or graze their livestock are stewards of the soil. So are miners or anyone else using land. They are temporary tenants whose lives will be far shorter than the time it took to form the soil. Their stewardship can lead to sustainable soil health, or it can lead to further erosion and degradation of soil. A few short years of poor management can lead to the loss of hundreds of years' worth of topsoil. Soil is cared for at the local level. Soil stewardship takes place one field at a time, by the people who are most familiar with the land. In many countries, soil erosion is being reversed with practices like no-till agriculture and contour strip cropping. In the United States, the Conservation Stewardship Program encourages farmers to care for soils on their lands. This and other conservation programs under the U.S. Department of Agriculture have been adequately supported by the Food, Conservation, and Energy Act of 2008.

REVIEW QUESTIONS

1. Describe the events that led to the devastation of the Dust Bowl and what role human decision making played.

2. Name and describe the properties of the three main components of soil texture.

3. What are the different layers, or horizons, that compose a soil profile? What makes each one distinct from the others?

4. Name and describe four major soil orders listed in this chapter. Where are they mainly found? How do hydric soils relate to soil orders?

5. What are the main things that plant roots must obtain from soil? Name and describe a process (natural or unnatural) that can keep plants from obtaining the amounts of each of these things they need for survival.

6. Describe, in terms of both content and physical structure, the soil environment that will best fulfill the needs of plants.

7. State and define the relationships among humus, detritus, and soil organisms. Describe the role each of these factors plays in creating a soil favorable to plants.

8. What is meant by mineralization? What causes it? What are its consequences for soil?

9. What are some problems with the sweeping claims made about soil erosion and soil degradation? How have these claims been challenged?

10. What is the impact of water and wind on bare soil? Define and describe the process of erosion in detail.

11. What is meant by desertification? Describe how the process of erosion leads to a loss of water-holding capacity and, hence, to desertification.

12. What are drylands? Where are they found, and how important are they for human habitation and agriculture?

13. What are three major cultural practices that expose soil to the weather?

14. What is meant by salinization, and what are its consequences? How does salinization result from irrigation?

15. Explain how soil biota can be killed by human actions and some of the effects.

16. What are the two levels at which soil conservation must be practiced? Give an example of what is being done at each level.

THINKING ENVIRONMENTALLY

1. Why is soil considered to be a detritus-based ecosystem? Describe how the aboveground portion and the belowground portion of an ecosystem act as two interrelated and interdependent communities.

2. Suppose you have uncovered an excellent new technique for keeping dryland soils moist and fertile. How would you get your message to the subsistence farmers?

3. Critique the following argument: Erosion is always with us; mountains are formed and then erode, and rivers erode canyons. You can't ever eliminate soil erosion, so it is foolish to ask people to do so.

4. The current system of management of grazing lands by the BLM is clearly unsustainable. What would you suggest in the way of public policy to make it sustainable?

MAKING A DIFFERENCE

1. To prevent soil erosion in your gardens, mulching twice a year can hold moisture in during the summertime and keep plants warmer during the winter. It also keeps rain from washing away your precious topsoil.

2. Don't use a leaf blower. It can erode topsoil more than even some of the most violent winds. It also burns fossil fuels.

3. Try mowing your fallen leaves to make shreds. Leave the shredded leaves in place and cover some with bark chips in garden beds to generate soil that is aerated, is rich with nutrients, and holds moisture better. Both of these practices combat erosion.

4. Composting is extremely sustainable and a very good method to practice. Two-thirds of solid waste is made up of purely organic material such as table scraps, trimmings from the yard, and paper products. Information on how to compost is available on the Internet. If you don't have enough space for traditional composting, you could try vermicomposting, which utilizes the surprising sustainable powers of worms. Many municipalities have yard waste composting as well.

5. When you use fertilizers, use organic products and ones labeled "slow-release" because "quick-release" fertilizer can be easily washed out and pollutes groundwater.

MasteringEnvironmentalScience®

Go to www.masteringenvironmentalscience.com for practice quizzes, Pearson eText, videos, current events, and more.

CHAPTER 12

The Production and Distribution of Food

William Allen, at the urban farm Growing Power in Milwaukee, Wisconsin.

WHAT DOES IT TAKE to get a MacArthur "Genius Grant" (a grant given by the John D. and Catherine T. MacArthur Foundation to people who show exceptional merit and promise for creative work)? For one famous athlete it took a vision for a farm in the city. Will Allen, a former pro basketball player, had an idea for a way to provide fresh local food for the people of Milwaukee, Wisconsin, so he purchased a defunct plant nursery and began his foray into urban farming. The nonprofit he now directs, Growing Power, is a sustainable agriculture organization. It runs the last functioning farm within the city limits of Milwaukee, operates a 40-acre farm outside of the city, and has branched out as far as Chicago. Of the whole operation, the piece that most captures the imagination is the original farm carved into the north side of Milwaukee. On a 3-acre plot lies an intensive urban agricultural operation with greenhouses, fish, bees, chickens, and goats. The integrated system brings food to an estimated 10,000 people annually. Every piece of space is carefully used.

A WHOLE LIFE. Growing Power's Milwaukee farm meets needs other than simply providing food. More than 5,000 visitors come to the facility a year, and the farm uses 400 tons of the city's food waste. It produces and sells worms, worm cast-

ings, compost, fruits and vegetables, goat milk and cheese, fish, and honey. It also provides job training for urban youth. Growing Power's Chicago location also provides job skills training and fresh food.

These examples of urban agriculture highlight the themes of this chapter: The future of agriculture will be different from agriculture of the past. It must produce enough food to support people and to be accessible, while

LEARNING OBJECTIVES

1. **Major Patterns of Food Production:** Explain the ways the Industrial Revolution and the Green Revolution radically transformed the practice of farming; compare to subsistence agriculture and animal farming.

2. **From Green Revolution to Gene Revolution:** Explain genetically modified crops and their pros and cons.

3. **Food Distribution and Trade:** Describe the broad movements of world trade in foodstuffs and relate to food security for people.

4. **Hunger, Malnutrition, and Famine:** Understand the differences between hunger and malnutrition and the causes of shortages.

5. **The Future: Increasing Production and Distribution of Food:** Describe changes needed to feed the world as we move toward 2030 and 2050. Explain new trends in agriculture, sustainable agriculture, organic, and urban agriculture. Be able to describe the key elements of sustainable agricultural systems.

lessening negative effects on the environment caused by modern agriculture. The need for change is real. Throughout the world, three decades of rapid population growth have left hundreds of millions of people dependent on imported food or food aid, while one in five in the developing world remains undernourished. As the world population continues its rise, no resource is more vital than food.

This chapter examines food-production systems, problems surrounding the lack of food, and the global effort being made to reduce hunger and malnutrition. Food production connects to topics we have already covered, or will cover, such as politics and economics, nitrogen and phosphorus cycles, population growth, poverty, water availability, soil, pests and their control, environmental health, climate change, water pollution, and sustainable cities. Food is central to our lives as humans and central to topics in environmental science. Wild foods, forest products, fisheries, and aquaculture are covered elsewhere. Here, we will look at three things necessary to provide nourishment for a burgeoning population and protect the environment for the long term: the production of more food, the elimination of the harmful effects of food production on the environment, and the distribution of the right kinds of food to everyone on the planet.

12.1 Crops and Animals: Major Patterns of Food Production

By many measures, human societies have done very well at putting food on the table. More people are being fed, and with more nutritional food, than ever before. In the past 30 years, world food production has more than doubled, rising even more rapidly than population. In addition, the daily amount of food available in the developing countries has increased by 25%. A lively world trade in foodstuffs forms the bulk of economic production for many nations. How did we get to a place where we produce this much food?

Some 12,000 years ago, the Neolithic Revolution saw the introduction of agriculture and animal husbandry, which probably did more than anything else to foster the development of human civilization. Virtually all of the major crop plants and domestic animals were established in the first thousand years of agriculture. In most of the developing world, plants and animals continue to be raised for food by *subsistence farmers*, using traditional agricultural methods that go back millennia. These farmers represent the great majority of rural populations.

Subsistence Agriculture in the Developing World

Subsistence farmers live on small parcels of land that provide them with enough food for their households and, it is hoped, a small cash crop. From the point of view of the modern world, such farmers are very poor, although many do not consider themselves to be so. The food they produce and consume themselves is not counted in the global economy and is sometimes undervalued by others (Chapter 2). Like past agricultural practice in the United States, subsistence farming is labor intensive and lacks practically all of the inputs of industrialized agriculture. Also, it is often practiced on marginally productive land (**Fig. 12–1**).

Figure 12–1 Subsistence farming. Subsistence farming feeds more than 1.4 billion people in the developing world. Here we see a farmer planting rice near Yangshuo in Guangxi Province, China.

Figure 12–2 Slash-and-burn agriculture. Countryside showing slash-and-burn agriculture in Para State, Brazil. Such agriculture can be a sustainable mix of forest and fields.

Typically, a family owns a small parcel of land for growing food and maintains a few goats, chickens, or cows. The system may be sustainable if crop residues are fed to livestock, the livestock manure is used as a fertilizer, and the family's nutrition is adequate. Such a family is making the best use of very limited resources. Keep in mind, however, that subsistence agriculture is primarily practiced in regions experiencing the most rapid population growth, even though that kind of agriculture is best suited for low population densities. An estimated 1.4 billion people in Latin America, Asia, and Africa—more than one-third of the people there—depend on subsistence agriculture. It has been called the "silent giant" that feeds most of the world's rural poor.

Successes. Because subsistence agricultural practice varies with the local climate and with local knowledge, there are regions of the world where it works well. In some areas,

subsistence agriculture involves shifting cultivation within tropical forests—often called slash-and-burn agriculture **(Fig. 12–2)**. Research has shown that this practice can be sustainable. The cultivators create highly diverse ecosystems in which the cleared land supports a few years of crops and gradually shifts into agroforestry—a system of tree plantations with different ground crops employed as the trees grow. The reason this works, though, is because the local human population is low.

Africa. Parts of Africa serve as good examples of the difficulties with subsistence agriculture. Low yields, rapid population growth, poverty, and hunger are the rule in sub-Saharan Africa. Two-thirds of the region's people live in rural lands and are dependent on agriculture for their livelihood. Agricultural production in sub-Saharan Africa has lagged behind that of the rest of the developing world **(Fig. 12–3)**. Because of the failures of agriculture, many in the region are dependent on imported food, and world food prices have been soaring. With the help of foreign investment, African crop scientists are working with conventional breeding techniques to improve traditional African food crops. The alliance will also work throughout the subcontinent to help improve irrigation, soil health, and local and regional agricultural markets. The basic goal of this work is to enable the millions of subsistence farmers in sub-Saharan Africa to grow enough food to sustain their families and to produce enough to encourage the economic development of their communities and their countries. Many people would like Africa to see the changes in agriculture that much of the rest of the world has already seen.

The Development of Modern Industrialized Agriculture

In 2012, the U.S. Department of Agriculture celebrated its 150th anniversary. At the time of its founding, the majority

Figure 12–3 The yield gap for cereals. Subsistence agriculture often has lower crop yields. The difference in yield in terms of tons per acre between sub-Saharan Africa and other regions continues to widen. (*Source:* FAOSTAT.)

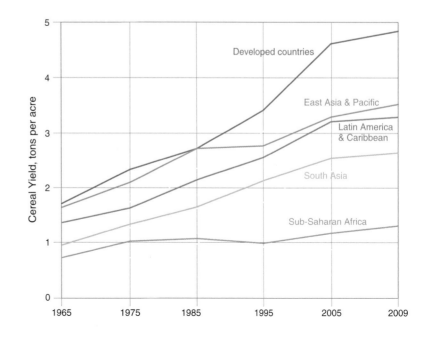

(a)   **(b)**

Figure 12–4 Traditional versus modern farming. (a) Traditional farming practiced in Arkansas.
For hundreds of years, traditional practices on American farms supported the growing U.S. population.
(b) Modern agricultural practice, illustrated by a combine loading corn into a tractor trailer in America's Midwest.

of people in the United States lived and worked on small farms. Human and animal labor turned former forests and grasslands into systems that produced enough food to supply a robust and growing nation **(Fig. 12–4a)**. Farmers used traditional approaches to combat pests and soil erosion: Crops were rotated regularly, many different crops were grown, and animal wastes were returned to the soil. Farming, although difficult, was efficient enough to allow an increasing number of people to leave the farm and join the growing ranks of merchants and workers living in cities and towns. Then, in the mid-1800s, the Industrial Revolution came to the United States, and it had a major impact on farming.

The Transformation of Traditional Agriculture. The Industrial Revolution transformed agriculture so profoundly that today, in the United States, some 2 million farmers produce enough food for all the nation's needs, plus a substantial amount for trade on world markets (Fig. 12–4b). This revolution increased the efficiency of farming remarkably. Since the mid-1930s, the number of farms has decreased by two-thirds (from 6.8 million to 2.02 million), while the size of farms has grown fourfold. (They now average 418 acres, or 170 hectares). Farming is a large and important business in the United States. Indeed, the revolution has achieved such gains in production that the United States frequently has had to cope with surpluses of many crops.

Virtually every industrialized nation has experienced this agricultural revolution. The pattern of developments in U.S. agriculture could just as well describe that in France, Canada, or Japan. Crop production has been raised to new heights, doubling or tripling yields per acre **(Fig. 12–5)**. To understand how this tremendous increase in production was made possible, you need to understand the components of the agricultural revolution. These in turn reveal some of the problems brought on by the revolution.

Infrastructure. It is not enough to simply allow farmers to produce crops on their land. The transformation of agriculture was greatly facilitated by developing a supporting infrastructure. This involved rural electrification and road building, agricultural programs in state colleges and universities, agricultural and soil extension services for farmers, the establishment of markets and efficient transportation of farm goods, banks and credit unions to handle farm loans, irrigation facilities, the formation of farm cooperatives, price and income support programs, and many other forms of government subsidies of farming. Most of these developments were essential in achieving the agricultural revolution, but many believe that the long-term maintenance of farm subsidies simply favors large corporate farms and works against the interests of farmers in developing countries.

Machinery. An incredible array of farm machinery handles virtually every imaginable need for working the soil: seeding, irrigating, weeding, and harvesting. This machinery

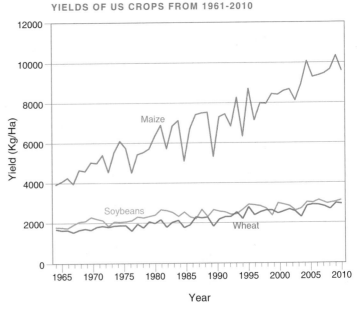

Figure 12–5 U.S. crop yields. Corn, wheat, and soybean yields (measured as kilograms per hectare) have increased over time. (*Source:* FAOSTAT.)

has enabled farmers to cultivate far more land than ever. However, the shift from animal labor to machinery has created a dependency on fossil fuel energy. When the price of oil rises, so does the cost of food production.

Land Under Cultivation. The United States has the largest area of arable land in the world. Cropland and pastureland comprise more than one billion acres (400 million ha). This is about 45% of the total land area of the United States. Agriculture, therefore, has an enormous impact on the landscape. Before 1960, much of the increased production in the United States came from bringing new land into production. Since then, increases in crop yields and consistent crop surpluses have lessened pressure to convert additional land to cropland, enabling farmers to be selective in the land they cultivate. The surpluses have also given farmers the opportunity to take erosion-prone land out of production. Under current farm policy, the **Conservation Reserve Program** reimburses farmers for "retiring" erosion-prone land and planting it with trees or grasses (Chapter 11). It is likely that some of this land will soon be brought back into production to meet a rising demand for corn used in ethanol production.

Essentially, all of the good cropland in the United States is now under cultivation or held in short-term reserve. Globally, agriculture occupies about 38% of the land, a number that has remained relatively steady since 1994, although yields have increased. Any expansion of cropland anticipated in the future will come at the expense of forests and wetlands, however, which are both economically important and ecologically fragile.

Fertilizers and Pesticides. Farmers long ago learned that animal manure and other organic supplements to soil could increase their crop yields. When chemical fertilizers, made with fossil fuels, first became available for use, however, farmers discovered that they could achieve even greater yields. Moreover, the fertilizers were more convenient to use and more available than manure. When fertilizers were first employed, 15 to 20 additional tons of grain were gained from each ton of fertilizer used (**Fig. 12–6**). Between 1950 and 1990, worldwide fertilizer use rose tenfold. Then, following a decline traced to hard times in the countries of the former Soviet Union after it broke up in the 1990s, the worldwide use of fertilizer resumed its rise and is now higher than ever (176 million tons in 2011). Most of the current increase in fertilizer use is in "underfertilized" countries such as China, India, and Brazil. However, fertilizer prices vary with the cost of fuel and other factors. World fertilizer prices doubled in 2007, for example, affecting the cost of food.

Chemical pesticides provided significant control over insect and plant pests, at least initially. However, due to natural selection, pests have become resistant to many pesticides. This leads to a cycle. More pesticides are used, and pests recover and increase. The amount of money spent on pesticides has increased more than tenfold between 1970 and now, and today 37% of food and fiber crops are still lost to pests. Currently, more than 113 weed species, 150 plant diseases, and 520 insects and mites are resistant to pests. Efforts to reduce the use of pesticides have begun because of side effects to human and environmental health. As with the use of farm

Figure 12–6 The effects of fertilizer. This 1942 photo shows the difference between fertilized and unfertilized crops and was part of a campaign to help farmers increase yields. (*Source:* Franklin D. Roosevelt Presidential Library and Museum.)

machinery and fertilizer use, pesticide manufacture depends on fossil fuel energy. Concerns about pesticides are discussed extensively elsewhere (Chapter 13).

Irrigation. Worldwide, irrigated acreage has doubled since 1950, while water withdrawals have tripled. By 2011, irrigated acreage represented 18% of all cropland—some 683 million acres (277 million ha)—and produced 40% of the world's food. Irrigation is still expanding, but at a much slower pace because of limits on water resources (irrigation represents 70% of all water use). Recall that much current irrigation is unsustainable because groundwater resources are being depleted (Chapter 10). In addition, production is being adversely affected on as much as one-third of the world's irrigated land because of waterlogging and salinization—consequences of irrigating where there is poor drainage. Unfortunately, irrigation ditches are also associated with the spread of as many as 30 diseases, because open water breeds mosquito larvae, houses snails that carry parasites, and supports other vectors of disease.

High-Yielding Varieties of Plants. Several decades ago, plant geneticists developed new varieties of wheat, corn, and rice that gave yields double to triple those of traditional varieties. This feat was accomplished by selecting strains that diverted more of the plant's photosynthate (photosynthetic product) to the seeds and away from the stems, leaves, and roots. The seeds of modern wheat, rice, and corn now receive more than 50% of the plant's photosynthate, close to the calculated physiological limit of 60%. As these new varieties were grown in the developed countries and then introduced throughout the world, production soared, and the Green Revolution was born.

The Green Revolution

The same technologies that gave rise to the agricultural revolution in the industrialized countries were eventually

Figure 12–7 Traditional versus high-yielding wheat. A comparison of an old variety of wheat, with a new, high-yielding variety of dwarf wheat growing in Mexico.

introduced into the developing world. There, they gave birth to the remarkable increases in crop production called the Green Revolution.

In 1943, the Rockefeller Foundation sent agricultural expert Norman Borlaug and three other U.S. agricultural scientists to Mexico, with the objective of exporting U.S. agricultural technology to a less developed nation that had serious food problems. Their aim was to improve the traditional crops grown in Mexico, especially wheat. Mexican wheat was well adapted to the subtropical climate, but it gave low yields and responded to fertilization by growing very tall stalks that were easily blown over. Using wheat from other areas of the world, Borlaug and his coworkers bred a dwarf hybrid with a large head and a thick stalk. The hybrid did well in warm weather when provided with fertilizer and sufficient water (Fig. 12–7). The program was highly successful: By the 1960s, Mexico had closed the gap between food production and food needs, wheat production had tripled, and Mexican wheat appeared on the export market.

Nobel Effort. Research workers with the Consultative Group on International Agricultural Research (CGIAR) extended the work done in Mexico, introducing modern varieties of high-yielding wheat and rice to other developing countries. In one success story, Borlaug induced India to import hybrid Mexican wheat seed in the mid-1960s, and in six years, India's wheat production tripled. Within a few years, many of the world's most populous countries turned the corner from being grain importers to achieving stability and, in some cases, even becoming grain exporters. Thus, while

world population was increasing at its highest rate (2% per year), rice and wheat production increased 4% or more per year. In 35 years, global food production doubled. The Green Revolution has probably done more than any other single scientific or other achievement to prevent hunger and malnutrition. Borlaug was awarded the Nobel Peace Prize in 1970 in recognition of his contribution.

Impacts. The CGIAR sponsored a study of the impact of the Green Revolution in the developing world between 1960 and 2000.[1] The study's findings have shed light on that important agricultural development. The early years of the Green Revolution (1960–1980) were only the beginning; research on high-yielding crops has continued, and more varieties continue to be developed, released, and adopted by farmers. Recent research has focused on resistance to diseases, pests, and climatic stresses. The early Green Revolution contributed greatly to expanded food production in Asia and Latin America; the later years of the Green Revolution benefited Africa and the Middle East as a greater variety of crops was developed.

Panacea? High-yielding modern crop varieties are now cultivated throughout the world and have become the basis of food production in China, Latin America, the Middle East, southern Asia, and the western nations. Because the technology raises yields without requiring new agricultural lands, the Green Revolution has also held back a significant amount of

---

[1]R. E. Evenson and D. Gollin, "Assessing the Impact of the Green Revolution, 1960 to 2000," *Science* 300 (May 2, 2003): 758–762.

deforestation in the developing world. As remarkable as it was, however, the Green Revolution is not a panacea for all of the world's food-population difficulties. Each of the Green Revolution technologies included externalities—costs not associated with the price of the food (Chapter 2). These costs mean that ecosystems have been harmed and ecosystem capital has been lost. By 2004, the hidden costs of U.S. industrial agriculture were estimated at $5.7 billion to $16.9 billion a year.[2]

The high-yielding varieties of crops can be hard to grow. Grains do best on irrigated land, and water shortages have begun to occur as a result of this dependence (Chapter 10). Also, modern varieties of grains require constant inputs of fertilizer, pesticides, and energy-using mechanized labor, all of which can be in short supply in developing countries. Further, Green Revolution agriculture did not eradicate hunger or poverty. Although India has been self-sufficient in terms of food since 1990, one-fifth of the population still suffers from malnutrition because they can't afford to purchase the food they need and there is no effective safety net.

Animal Farming and Its Consequences

Another world change has been the development of concentrated livestock and poultry production, as well as the problem of overgrazing in many areas with high human populations and traditional herds. Raising livestock—sheep, goats, cattle, buffalo, and poultry—has many parallels to raising crops, and there are many connections between the two. One-third of the world's croplands are used to feed domestic animals. In the United States alone, 70% of the grain crop—one-half the cultivated acreage—goes to animals. The care, feeding, and "harvesting" of some 4.7 billion four-footed livestock and almost 21.5 billion birds constitutes one of the most important activities on the planet. The primary force driving this livestock economy is the large number of the world's people who enjoy eating meat and dairy products—primarily, most of the developed world and growing numbers of people in less developed nations. As with crop farming, however, there are multiple patterns. In the developed world, and increasingly in the developing world, animals are raised in large numbers in confinement (the so-called concentrated animal feeding operations, or CAFOs) **(Fig. 12–8)**, and others are raised on extremely large ranches. In still other places, such as rural societies in the developing world, livestock and poultry are raised on family farms or by pastoralists who are subsistence farmers.

CAFOs. Industrial-style animal farming can damage the environment and human health in a host of ways. Because so much of the plant crop is fed to animals, all of the problems of industrialized agriculture apply to animal farming. Another serious problem is the management of animal manure. In developing countries, manure is a precious resource that is used to renew the fertility of the soil, build shelters, and provide fuel. In developed countries, it is a wasted resource.

Figure 12–8 CAFOs. Concentrated animal feeding operations (CAFOs), like this cattle lot in Arizona, are common. These cattle are held in very tight quarters while they are fed grains to fatten them. This results in high concentrations of animal wastes, sometimes leading to water pollution.

Close to 314 million tons of animal waste is produced by CAFOs each year in the United States, some of which leaks into surface waters and contributes to die-offs of fish, contamination with pathogens, and a proliferation of algae. CAFOs now produce more than 50% of the animals raised for meat in the United States. The Environmental Protection Agency (EPA) asserts that animal-based agriculture is the most widespread source of pollution in the nation's rivers.

The crowded factory farms are perfect conditions for diseases to incubate and spread among animals as well as from animals to humans. Some very serious public-health concerns have emerged, such as the avian flu. Further, many bacteria associated with animal farming are human pathogens, causing food-borne human diseases. The bacterium *Salmonella* alone is estimated to cause losses of $2.5 billion per year in the United States. Because antibiotics are used heavily in animal feed, many of the bacterial contaminants have developed resistance to antibiotics, and these are far more difficult to treat when humans are infected.

Large Ranches. Australia is home to 28 million cattle, making it the world's top beef exporter. These cattle are ranch raised, on farms called "cattle stations," grazing over vast areas of the Australian rangelands. In the wetter coastal regions, some support high numbers of cattle. While Australian cattle farms are very large, in drier regions of the continent, the number of cattle they can support rises and falls with drought cycles. Anna Creek Station, the world's largest cattle farm, is larger than the country of Israel. Because of a drought in the 2000s, the farm now supports only about 3,000 cattle and 15 workers. At its high point, it could support 16,000 cattle.

Ranches account for a great deal of the loss of rain forests. Half of the world's rain forests have been cut or degraded. In Latin America, more than 58 million acres (23 million ha) of tropical rain forests have been converted to cattle pasture, often after the land was briefly farmed for crops. Even though most of this land is best suited for growing rain-forest trees, some of it does support a rural population of subsistence farmers producing a diversity of crops.

[2]E. Tegtmeier and M. Duffy. (2012). "External Costs of Agricultural Production in the United States." *International Journal of Agricultural Sustainability, 2*(1), 1–20.

Much of the land, however, is held by relatively few ranchers who own huge spreads. Today, cattle production in the Amazon basin has expanded and is export driven; for example, Brazilian beef exports bring in more than $5 billion annually.

Climate Change. Deforestation and other changes in land use in the tropics release an estimated 980 million tons of carbon to the atmosphere annually, contributing a significant amount of carbon dioxide to the greenhouse effect. Also, because their digestive process is anaerobic, cows and other ruminant animals annually emit some 80 million tons of methane, another greenhouse gas, through belching and flatulence. The anaerobic decomposition of manure leads to an additional 30 million tons of methane per year. In the United States, methane released by livestock makes up about 20% of methane produced (Chapter 18). However, better and more efficient diets for animals can lower methane production.

Pastoralists. Even though their animals also contribute to the methane problem, it is callous to fault the subsistence farmers whose domestic animals enhance their diet and improve their quality of life. In fact, one of the most important kinds of sustainable development aid brought to rural families is the gift of a cow or a few goats or rabbits, as carried out by Heifer Project International **(Fig. 12–9)**. Working in 125 countries, the project has distributed large animals, beehives, fowl, and fish fingerlings to families. The organization works with grassroots groups of local people who oversee a given project. Those receiving the animals must agree to raise them properly and to pass on the gift in the form of donating offspring of their livestock to other needy people.

Animals that are well managed can enhance the soil and enable rural farmers to maintain a balanced farm ecosystem, unless herds overgraze local resources. Farmers in developing countries now own two-thirds of the world's domestic animals, a trend that has made an important contribution to nutrition, especially for women and small children. In sum,

Figure 12–10 Biofuels. Although they show promise as an alternative fuel, biofuels raise the price of corn by diverting it from food use.

animal farming is far more likely to be sustainable in the rural farms and pastoral herds of the developing world than in the CAFOs of the developed world, where there is a pressing need to address problems of pollution and disease.

Biofuels and Food Production

Because of the accumulation of greenhouse gases in the atmosphere, global climate change is affecting virtually every country (Chapter 18). The most basic cause of climate change is the burning of fossil fuels, which has driven atmospheric CO_2 (a greenhouse gas) more than 39% above preindustrial levels. One of the important ways to mitigate climate change is to burn **biofuels** (ethanol and oils derived from agricultural crops) instead of fossil fuels **(Fig. 12–10)**. Because there is no new CO_2

Figure 12–9 Heifer Project International. A vet inspects a dairy heifer for a Rwandan man. Heifer Project International provides animals for families in need. Participants agree to pass on offspring of their livestock to others.

being released to the atmosphere with this use, biofuels are a form of renewable energy. But there is another, more basic reason for using biofuels. In the past, fossil fuel oil was less expensive than biofuels, but in recent years, oil prices have fluctuated wildly, and biofuels have become very attractive.

The primary biofuel in production today is ethanol, derived from the fermentation of cornstarch (in the United States) or cane sugar (in Brazil). Ethanol is a suitable substitute for gasoline, and with inflated gasoline prices, ethanol manufacture is quite competitive. Corn farmers are happy to sell their corn to the highest bidder, so a rising fraction (35% in 2012) of U.S. corn is devoted to meeting the growing demand from ethanol distilleries. At the same time, food prices have risen worldwide to the point of crisis.

Consequences. The United States produces more than 40% of the world's corn, 80% of which is used for livestock, poultry, and fish feeds. Corn is also a staple food around the world. The diversion of corn to biofuel production lowered its availability for other uses. A nutritionist with the UN World Food Program stated that "global price rises mean that food is literally being taken out of the mouths of hungry children whose parents can no longer feed them."[3] How accurate is this accusation?

While it is certainly true that biofuel production has created a new demand for corn, it must be noted that the prices of wheat, rice, and soybeans also rose dramatically—in fact, more than the price of corn. So far, the new demand for corn is being met with larger corn plantings; the increase was accomplished by taking land out of soybean production. Ethanol production uses field corn, which is usually fed to livestock. Only the cornstarch is used for ethanol production; the proteins, vitamins, and fiber are converted to high-energy animal feed, sweeteners, and corn oil.

Other factors contribute to high food prices. These include (1) the price of oil (used by farm machinery and in fertilizer manufacture), which reflects a growing competition from developing countries like China and India for a resource that is limited; (2) bad weather and poor harvests in many parts of the world, and especially in Australia, which is normally a major wheat exporter; and (3) the rising demand for meat and animal products, again from emerging economies like China.

Research by the International Food Policy Research Institute (IFPRI) suggested that perhaps 30% of the grain price increases between 2000 and 2007 was linked to biofuel production. Many observers suggest that some limit be put on corn ethanol production, while making a greater investment in a new generation of biofuels based on grasses and timber waste (see Chapter 16). The surge in biofuel production in the United States is not likely to subside anytime soon. In 2012, more than 5 billion bushels of corn, more than one-third of the crop, were diverted to ethanol production, a demand that is expected to rise; consequently, less corn will be available for animal feed and export.

Part of feeding the modern world's high population has been the irrigation, fertilizers, pesticides, plant breeding, and concentrated animal operations just described, but emerging technologies promise to change the face of agriculture even more.

12.2 From Green Revolution to Gene Revolution

Genetic engineering makes it possible to combine characteristics from genetically different organisms and to incorporate desired traits into crop lines and animals, producing so-called transgenic, or genetically modified (GM), varieties. This is a radically different technology from the genetic research that produced the Green Revolution: Researchers are no longer limited to genes that already exist within a species or that could arise via mutation; it is now possible to exchange genes among bacteria, animals, and plants.

This kind of technology can help the developing world to produce more food. Genetically modified organisms are under intense research and development, with many new GM organisms coming out of the lab. Still, in spite of the obvious potential of these organisms, there are concerns about their development and use. We need to look at both sides of this controversy.

The Promise

The earliest and still most widely adopted genetically altered products to be marketed are of two types. First are cotton plants with built-in resistance to insects that comes from genes taken from a bacterium (so-called Bt, standing for *Bacillus thuringiensis,* the name of the bacterium) **(Fig. 12–11)**. Second are corn and soybeans resistant to herbicides, including the chemicals glufosinate (several brands) and glyphosate (usually the brand Roundup®), allowing farmers to spray herbicide without harming their crops and to employ no-till techniques. More recently, biotechnology has developed other genetically altered products, including sorghum (an important African crop) resistant to a parasitic plant known as witchweed, which infests many crops in Africa; corn, potatoes, and cotton resistant to insects; rice resistant to bacterial blight disease; and trees and salmon that grow very rapidly.

In the years since bioengineered seeds became commercially available, farmers have overwhelmingly turned to the transgenic breeds of soybeans, cotton, and corn. Worldwide, 370 million acres (134 million ha) in 29 countries were planted with bioengineered crops in 2011 **(Fig. 12–12)**. Most of this area is accounted for by just four crops (corn, soybeans, canola, and cotton). Bioengineered sugar beets are also increasing in use. In the United States, bioengineered crops are planted on about half of all acreage planted with crops. A rapidly increasing trend is toward the so-called "stacked" products, containing two or more biotech genes in a crop (e.g., traits directed toward different insects or insect resistance plus herbicide tolerance).

[3]World Food Program News. "Price Rises Eating Away at Children's Futures: An Interview with WFP Nutritionist Andrew Thorne-Lyman." April 28, 2008.

Figure 12–11 Bt cotton. Bioengineered cotton has a gene from a bacterium that helps it resist insect pests.

Biotech crop research that can benefit developing countries is proceeding at a rapid pace, often in those very countries. China, in particular, has emerged as a leader in plant biotechnology, having recently sequenced the rice genome. Among other aims, the objectives of agricultural biotechnology are (1) to incorporate resistance to diseases and pests that attack important tropical plants; (2) to increase tolerance to environmental conditions, such as drought and a high salt level, which stress most plants; (3) to improve the nutritional value of commonly eaten crops (like the work with golden rice described later); and (4) to produce pharmaceutical products in ordinary crop plants ("pharma" crops). Thus, products under development include virus-resistant cassavas, sweet potatoes, melons, squash, tomatoes, and papayas; protein-enhanced corn, soybeans, and cassavas (**Fig. 12–13**); fungus-resistant bananas; bananas and tomatoes that contain an antidiarrheal vaccine; and drought-tolerant sorghum and corn. The potential for transgenic crops and animals seems almost unlimited.

Among the important possible environmental benefits of bioengineered crops are reductions in the use of pesticides because the crops are already resistant to pests, less erosion because no-till cropping is facilitated by the use of herbicide-resistant crops, and less environmental damage associated with bringing more land into production because existing agricultural lands will produce more food. While the jury is still out on GM crops, countries everywhere are cautiously examining the technology, and the trend is in favor of its adoption; in 2011, 29 countries, including several in the EU, were growing biotech crops.

Marker-Assisted Breeding. Some new technologies also involve crop improvements without resorting to transgenic traits. For example, a technique called **marker-assisted breeding** essentially speeds up traditional plant breeding. Scientists identify genes for desirable traits in crop plants or

Figure 12–12 Global area of genetically modified crops. The plot shows the rapid increase in global areas planted with genetically modified crops (primarily cotton, soybeans, canola and corn) for industrial (high income) and developing countries as well as total global use. (*Source:* Clive James. "Global Status of Commercialized Biotech/GM Crops." ISAAA, 2011, p. 3.)

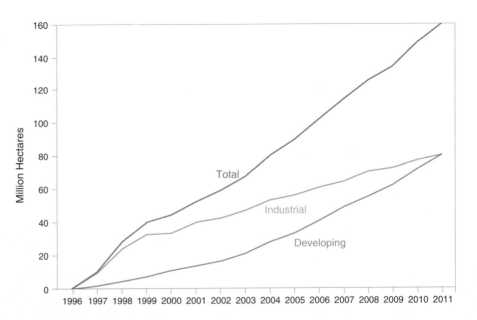

Figure 12–13 Genetically modified cassava. Cassava is a staple crop for millions, but one low in protein, vitamins, and minerals and vulnerable to viruses. GM varieties are more nutritious and resist disease.

their wild ancestors, using DNA sequencing. Next, plants with the desired gene are crossbred with a modern crop breeding line. Then DNA screening of seedlings for the desired gene makes the process move more quickly. Using this technique, a gene from a wild rice ancestor was incorporated into a modern high-yield rice variety, raising the yield by 18%. Other crops being developed with this approach are carrots (a high-calcium "super-carrot"), potatoes (more starch), and pigeon peas (higher yields). This approach has the distinct advantage of not requiring the special testing and permits needed with transgenic crops, greatly reducing costs and time. It also bypasses the objections raised to transgenic crops in many countries.

Pharma Crops. The biotech industry has recognized the advantages of producing pharmaceuticals by engineering genes for desirable products into common crop plants. Between 1991 and 2011, there were 101 field trials for crops genetically modified to produce hormones, enzymes, diagnostic drugs, and other substances, but none has yet come to market. The Union of Concerned Scientists (UCS) brought together a panel of experts to examine the risks of this technology, and its report concluded that the primary risk is that these "pharma" crops could contaminate ordinary food crops—some of the compounds being developed could be harmful if ingested by people or animals. In fact, corn altered to produce a pig vaccine spread in fields in Iowa and Nebraska in 2002; intervention by the USDA kept the altered corn out of food. Because it seems unlikely that a foolproof system could be established, UCS is urging an alternate strategy: produce the pharmaceuticals with noncrop plants.

The Problems

Concerns about genetic engineering technology involve three main considerations: environmental problems, food safety, and access to the new techniques.

Environmental Concerns. A major environmental concern focuses on the pest-resistant properties of the transgenic crops. If pests have a broad exposure to the toxin or some other resistance incorporated into the plant, it is possible that they will develop resistance to the toxin and thus render it ineffective as an independent pesticide. The transgenic crop then loses its advantage. Indeed, glyphosate resistance has been found in a number of weeds that infest soybean, corn, and cotton fields. Because so much acreage worldwide is now planted with glyphosate-resistant crops, resistant weeds could spread widely and cause economic disaster for farmers. Insect pests could also become resistant to transgenic plants that are themselves supposed to be insect resistant. For example, Bt cotton is designed to resist the cotton boll weevil by producing a toxin once made by a bacterium. However, boll weevils that can survive on the engineered cotton now abound. When Chinese cotton farmers began turning to Bt cotton in the early 2000s, they found that they used much less pesticide, saving money, time, and their own health in the process. These benefits did not last, however. In 2010, Chinese farmers suffered outbreaks of **secondary pests** (pests that were minor but become major in the absence of a competing pest), mirid bugs, which increased by twelvefold over their occurrence before use of the Bt cotton (**Fig. 12–14**).

Another concern is the ecological impact of the crops on nontarget organisms. For example, pollen from Bt corn (resistant to the corn borer) can be spread by the wind to adjacent natural areas where beneficial insects may pick it up and be killed by the toxin. This effect was shown to occur in laboratory tests with monarch butterflies. Later studies indicated, however, that the Bt corn poses a low risk to the butterflies, and the EPA cleared the corn for continued planting, although some scientists are still concerned.

A third environmental concern arises because genes for herbicide resistance or for tolerance to drought and other environmental conditions can also spread to ordinary crop plants or their wild relatives, possibly creating new or "super" weeds. Field studies indicate that this concern is

Figure 12–14 Mirid bugs. Mirid bugs, such as this one, became a major pest on cotton in China only when Bt cotton resisted other competing insects and their natural predators were killed.

valid. In fact, one study on genetically modified canola plants (used in animal feeds and vegetable oils) found that the vast majority of escaped canola plants growing in North Dakota along roads had genes from GM canola crops.

Safety Issues. Food safety issues arise because transgenic crops contain proteins from different organisms and might trigger an unexpected allergic response in people who consume the food. Tests have shown that a Brazil nut gene incorporated into soybeans was able to express (or manufacture) a protein in the soybeans that induced an allergic response in individuals already allergic to Brazil nuts. Also, antibiotic-resistance genes are often incorporated into transgenic organisms. If antibiotic resistance got into human systems by a separate process of gene transfer or a food product prevented an antibiotic from being used effectively, that would obviously pose a health risk. Another concern relates to the possibility that plants could produce substances in their tissues or that new combinations of genes could have as yet unexpected health effects on consumers. However, the World Health Organization, summarizing the finding of many researchers, has stated, "GM foods currently available on the international market have passed risk assessments and are not likely to present risks for human health. In addition, no effects on human health have been shown as a result of the consumption of such foods by the general population in the countries where they have been approved."[4]

Access in the Developing World. The problems concerning access to the new technologies relate to the developing world. For the first few years, almost all genetically modified organisms were developed by large agricultural-industrial firms, with profit as the primary motive. Accordingly, farmers were forbidden by contract from simply propagating the seeds themselves and had to purchase seeds annually. Farmers in the developing countries are less able to afford the higher costs of the new seeds, which must be paid up front each year. Fortunately, some noncommercial and donor-funded laboratories are taking aim at this problem. Further, biotech research is booming in China, India, and the Philippines, where most of the fruits of the research are being made available to developing-world farmers. In addition, the genetically modified seeds are spreading rapidly through seed "piracy."

The experience of GM cotton growers in India illustrates a number of the concerns associated with GM crops. In 2002, India began planting Bt cotton and doubled its harvest, but 10 years later, farmers were frustrated. In six years, the crop had not increased, though the amount of GM crops planted increased fourfold. Bt cotton required more nutrients, fertilizer, and water and was vulnerable to bacterial disease. Local cottonseed varieties disappeared, and farmers could not afford the GM varieties. In 2012, the harvest was half of what it was in 2011 in most areas. The "white revolution" had not panned out the way farmers had hoped. Thus,

GM crops, like the techniques of the Green Revolution, are not a magic bullet.

All of these concerns, combined with the general fear of what unknown technologies can bring, have stimulated a rather heated controversy over the spreading use of genetically modified organisms. Activists around the world have rallied to protest the development of such organisms and the "Frankenfoods" derived from them (**Fig. 12–15**). The protests have been strongest in Europe, and some countries, like Ireland, have banned the crops. Far less concern has arisen so far in the United States. Because corn and soybeans are used in so many processed foods, an estimated 70% of all processed foods in the United States are thought to contain some genetically modified substances. Some consumer groups would like to see mandatory labeling of all such foods, but U.S. policy does not require such labeling.

In 2004, the UN FAO gave its qualified approval to agricultural biotechnology in *The State of Food and Agriculture*[5] but concluded that the promise of the technology to alleviate global hunger and improve the well-being of farmers in developing countries was still more theory than reality. Developing countries have potential to produce their own GM crop applications. Iran began in 2006 to deploy a Bt rice developed in that country and hopes within a few years to eliminate its need to import rice (currently 1 million tons a year). South African researchers have developed corn resistant to the maize streak virus, which devastates corn crops throughout sub-Saharan Africa.

Clearly, the future almost certainly will have to include major advances in food production from biotechnology, and many of these advances are likely to benefit the developing countries. If food production is to keep pace with population growth, biotechnological advances will be

Figure 12–15 GM crop protests. Protesters in France rip up genetically modified corn in protest.

[4]World Health Organization. "20 Questions on Genetically Modified Foods/ Question 8." http://www.who.int/foodsafety/publications/biotech/20questions/en/ June 25, 2012.

[5]United Nations, Food and Agricultural Organization, *The State of Food and Agriculture 2003–2004: Agricultural Biotechnology: Meeting the Needs of the Poor?* (Rome: UN FAO, 2004).

STEWARDSHIP

Golden Rice

Half the people in the world—most of them in the developing countries—eat rice daily. Although an excellent food for energy, rice is a poor source of vitamins and other nutrients; hence, it must be supplemented with other foods for a balanced diet. Unfortunately, millions depend heavily on rice for their nutrition, resulting in two of the most common nutritional deficiencies: vitamin A deficiency, which leads to blindness and immune-system failures, and iron deficiency, which ultimately causes anemia and also immune-system problems. More than 195 million children suffer from vitamin A deficiency. Iron deficiency causes more than 100,000 maternal deaths a year during pregnancy and childbirth.

Biochemist Ingo Potrykus, working at the Swiss Federal Institute of Technology in Zurich, believed that rice could be supplemented with iron and vitamin A in order to conquer these two serious nutrient disorders. In the early 1990s, Potrykus, colleague Peter Beyer of the University of Freiberg, Germany, and others began research that would lead to a remarkable accomplishment of plant biotechnology: golden rice. The group incorporated several genes into a rice genome that would enable the plant to synthesize beta carotene (a precursor to vitamin A) and iron in the body of the grain, rather than just in the leaves, where they normally appear in rice. The new rice was named golden rice because of the color of the beta carotene in the grains, though it still did not have enough beta carotene to help humans.

Since its development, golden rice has been improved to the point where a new variety contains enough beta carotene to satisfy much of the vitamin A needed from foods in a typical developing country diet. Some genetic lines of golden rice are well along in the regulatory approval process. The golden rice concept has met with a less-than-universal welcome because it is a genetically modified (GM) crop, containing genes from corn and beans. The international debate is far from settled. However, golden rice lacks some of the traits that make other GM crops problematic. It is controlled by a humanitarian organization, not a for-profit company, and it does not contain genes that would benefit a super weed or to which pests would become resistant. Finally, while the genes come from two other crop plants, rice already has genes to make beta carotene in the grain; these genes have simply been turned off.

Detractors say that all GM crops are cause for concern and that the money invested in golden rice would be better spent supporting

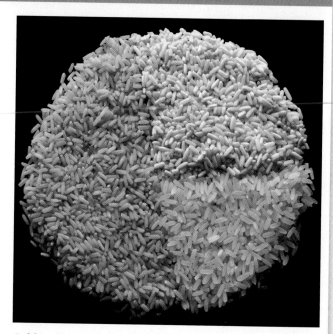

Golden rice. The yellow rice kernels contain beta carotene, inserted into them with biotechnology to enable the rice to help meet nutritional needs for vitamin A. The early golden rice (GR1) is on the upper right, GR2 is on the left, and normal rice on the lower right. (*Source:* Martin Enserink, "Tough Lessons from Golden Rice," *Science* 320 (April 25, 2008): 470.)

biodiverse diets containing green vegetables. Supporters argue that if GM crops are ever reasonable, this is a good example of a worthwhile crop that provides an opportunity to help millions. They might see it as a type of stewardship—of food capacity. GM crops offer a good example of the tensions between risk of the unknown and meeting current needs.

essential, in the view of most observers. Consequently, the world community has been developing national and international policies.

The Policies

In the United States, the EPA, the USDA, and the Food and Drug Administration (FDA) all have regulatory oversight of different elements of the application of biotechnology to food crops. Concern over their oversight led to a number of reports from the National Academy of Sciences' National Research Council. One, *Impact of Genetically Engineered Crops on Farm Sustainability in the United States* (2010),[6] concluded that herbicide resistance in weeds was a real problem but that, overall, GM crops have been a benefit. In sum,

several reports endorsed the science, technology, and regulation of GM crops, one stating, "The committee is not aware of any evidence that foods on the market are unsafe to eat as a result of genetic modification."[7] At the same time, the reports have called for more research on environmental and safety issues.

Cartagena Protocol. On the international level, the UN Convention on Biodiversity sponsored a key conference in Montreal, Canada, in January 2000 to deal with trade in genetically modified organisms. After heated negotiations, the conference reached an agreement, called the **Cartagena Protocol on Biosafety.** Surprisingly, the agreement was welcomed by governments, the private sector, and environmental groups alike.

---

[6]National Research Council, Committee on Genetically Modified Pest-Protected Plants, *Impact of Genetically Engineered Crops on Farm Sustainability in the United States* (Washington, DC: National Academy Press, 2000), 8.

[7]National Research Council, Board on Agriculture and Natural Resources, *Impact of Genetically Engineered Crops on Farm Sustainability in the United States* (Washington, DC: National Academy Press, 2010), 215.

TABLE 12–1 Drivers of Food Availability. Some factors of modern agriculture increase food availability, and some result in effects that decrease food availability.

Increase food availability	Limit or decrease food availability
Green Revolution high-yield crops	need for water, fertilizer
pesticide use	pest resistance, secondary pest outbreaks
irrigation	salinization erosion overuse of ground water
fertilizers	disruption of nitrogen cycle, nitrous oxide
conversion to farmland	loss of cropland to development use of poor quality land erosion, desertification
no-till farming with herbicides	herbicide resistance in weeds
GM crops with tolerance, high yield	resistance of pests to crops, super weeds, superpests loss of genetic diversity
monocultures	vulnerability to pathogens and pests
use of fossil fuels, equipment	cost of fuel climate change effects—plant stress from drought, insects, damage from air pollution
shipping	energy and economic costs of shipping
refrigeration	energy and economic costs of refrigeration food loss in storage and transport
large farms	loss of small farmers
subsidies in wealthy countries	subsidies in some countries reduce food availability in other countries
emphasis on cash crops	loss of mixed crops and consumptive use poverty for some—lack of money to purchase food
concentrated animal feeding operations	pollution use of grains to feed animals use of foods as biofuels
fishing efficiency	overfishing

At issue was a difference in fundamental philosophy about how technologies, particularly new ones, should be regulated. Some felt that no one should be able to limit trade in a new product unless they had sufficient scientific evidence to know that it was harmful. Others believed that, because it takes a long time to do research, governments should be able to regulate trade in goods where there was good reason to be concerned about a product before all of the research was completed. This question is particularly important in an era of globalization.

The Precautionary Principle. The Cartagena Protocol states that the "lack of scientific certainty due to insufficient relevant scientific information and knowledge . . . shall not prevent [a country] from making a decision" on the import of genetically modified organisms. The protocol gives the right to deny entry of any of these organisms to the importing country. However, countries cannot ban imports on a whim or simply to protect their own producers. The decision must be based on sound science (involving an assessment of the risks involved) and the broad sharing of information about the products. Thus, the agreement makes operational—for the first time in international environmental affairs—the so-called **precautionary principle**, laid out in the 1992 Rio Declaration, which states that "Where there are threats of serious or irreversible damage, lack of full scientific certainty shall not be used as a reason for postponing cost-effective measures to prevent environmental degradation."[8] There were other stipulations in the protocol, including one about labeling shipments of GM products. The protocol was ratified by 147 countries (but not the United States) and became legally binding in 2003.

Table 12–1 shows a list of actions we take to increase food production as well as a list of effects and actions that eventually lower our ability to produce food. Solutions to the problems modern agriculture causes are going to require getting more human health from the food we produce and modifying food production in ways that may seem hard to imagine. We will look at these topics in the latter half of the chapter. First, let's look at food distribution. We already produce a lot of food, so we could ask, "Why doesn't everyone have healthy food?"

12.3 Food Distribution and Trade

In the past, the general rule for basic foodstuffs—grains, vegetables, meat, and dairy products—was regional *self-sufficiency*. Whenever climate, blight, or war interrupted the agricultural production of a nation or region, the inevitable result was famine and death, sometimes on the scale of millions. Of course, refugees might flee a famine, and some cultures were naturally nomadic, so there was some moving around to escape famine. However, for the most part, natural disasters including crop failures drastically affected the people locally but not people far away. This is one of the reasons human population was relatively stable for thousands of years before the modern era.

Trade on ancient routes moved goods from place to place in a limited way. This was expanded when colonies were established in the New World. Timber, furs, tobacco, fish, sugar, coffee, cotton, and other raw materials began to flow back to the Old World. In turn, the Old World exported manufactured goods, which in turn altered the culture of the colonies. With the Industrial Revolution, and especially with refrigeration and faster shipping, trade between nations

---

[8]Rio Declaration on Environment and Development 1992, Article 15, upheld in Cartagena Protocol, Articles 1, 10.6, and 11.8.

intensified, and soon it became economically feasible to ship basic foodstuffs around the world. A lively and important world trade in foodstuffs arose, and, as it did, the need for self-sufficiency in food diminished. Like other sectors of the economy, food has become globalized. World trade in agricultural products was worth $1.362 trillion in 2010.

Patterns in Food Trade

Given this global pattern, agricultural production systems supply more than a country's internal food needs. The capacity to produce more basic foodstuffs than the home population needs represents an extremely important economic enterprise. For many other countries (especially those of the developing world), special commodities such as coffee, fruit, sugar, spices, palm oil, cocoa, and nuts provide the only significant export products. This trade clearly helps the exporter, and it allows importing nations to enjoy foods that they are unable to raise themselves or that are out of season. In a market economy, the exchange works well only as long as the importing nation can pay cash for the food. Cash is earned by exporting raw materials, fuel, manufactured goods, or special commodities.

Grain on the Move. The most important foodstuffs on the world market are grain (wheat, rice, corn, barley, rye, and sorghum) and oilseeds, such as soybeans. Some grain is imported by high- and middle-income countries to satisfy the rising demand for animal protein. It is instructive to examine the pattern of global trade in grain over the past seven decades (Table 12–2). In 1935, only western Europe was importing grain; Asia, Africa, and Latin America were self-sufficient. By 1950, new patterns were emerging, and today the trade in grains—as well as other basic foodstuffs—has changed again. As the table shows, North America has become the major source of exportable grains—the world's "breadbasket" or, in another sense, the world's "meat market." The trade in oilseeds is similarly lopsided. In this case,

the United States, Argentina, and Brazil have a corner on the export trade, and Europe and China are major importers. Most of this is used as a protein supplement in animal feeds.

Over the past 50 years, Asia and Africa have shown an increasing dependence on imported grain (and China, imported soybeans). These regions have also experienced 50 years of continued population growth. Although much of the food needs of these regions are met by internal production, the persistent dependency is an ominous signal. If, as is projected, the developing countries import twice as much grain by 2020, the grain-exporting countries will have to either greatly increase domestic production or use less grain (not likely). And the rising use of corn for biofuel production will make this harder.

Keeping a Reserve. At no time in recent history has the world grain supply run out. The global amount of grain produced in 2010 was 2.2 billion tons. Production was down slightly because, while rice and maize had record yields, wheat dropped significantly. Worldwide governments try to maintain 70 days' worth of grain stored in reserves; however, such reserves are often much higher, above 120 days during some years of the 80s. In 2007, world reserves fell to 64 days, rising shortly there after. In 2010, they fell again to 72. With some use of carryover stocks in weaker years, enough grain is available to satisfy the world's food needs—and enough to feed millions of animals and continue to keep some in storage. However, many people lack food. This is primarily because they cannot afford it. Even if we succeed in increasing food production as population grows, some people will still be poor. This problem is the chief cause of hunger in the world today and is not easily solved. Recently, it was exacerbated by several factors in the global food crisis of 2006–2008.

The Global Food Crisis

Between 2006 and 2008, the prices of food commodities on the world market rose precipitously—on average, 100%

TABLE 12–2 World Grain Trade Since 1935

Region	Amount Exported or Imported (Million Metric Tons)*									
	1935	1950	1960	1970	1980	1990	1995	2000	2004	2007
North America	5	23	39	56	131	123	108	98	101	120
Latin America	9	1	0	4	−10	−12	−4	−16	−8	−7
Western Europe	−19	−22	−25	−30	−62	−10	21	13	10	0.2
Africa	1	0	−2	−5	−15	−31	−31	−44	−42	−51
Asia	2	−6	−17	−37	−63	−82	−78	−74	−79	−72
Australia and New Zealand	3	3	6	12	19	15	13	21	25	9

Source: FAO FAOSTAT, FAO Food Outlook, June 2008.

*A minus sign in front of a figure indicates a net import; no sign indicates a net export.

SUSTAINABILITY   Preventing Food Crises

On a cold day in February 1936, photographer Dorothea Lange was returning from a month-long assignment where she was photographing and documenting the plight of the poor in the Great Depression. As she passed a camp of ragged tents in Nipomo, California, she was only interested in returning home. But miles down the road, something called her to return to an enclave of pea pickers in tents, starving after the collapse of a pea crop. There she met and photographed the woman who became the iconic image of the effects of the Great Depression. A widow of 32, Florence Owens Thompson and her small children were living on vegetables scrounged from the frozen ground and birds her children had caught. She had sold her tires to buy food and had nowhere to go and no plan to escape their deprivation. Lange took only five photos, but they became the images that spoke of the suffering of farmworkers across the country. One known as "Migrant Mother" is probably the most famous picture of the time. Those photos woke the nation, food was brought to those pickers isolated in a barren field, and the images Lange took gripped the imagination and care of others.

Sometimes images of children starving wring our hearts and cause us to pour out kindness and wealth. But sometimes, when the problems are not easily resolved, such images fill people with a sense of despair. Instead of doing something, no matter how small, they do nothing. Relief and development agencies also know that to solve food shortages and prevent famine, food needs to be given when we know a drought or disaster is looming, not when children are already standing, blank eyed, with protruding bellies. In recent years, there have been several years of poor crops in the Sahel region of Africa. In 2010, 2011, and 2012, people who track and predict crop outcomes predicted severe food shortages months before they were realized. They reported these findings in the news, but international aid from governments and charities was slow to come. Millions of people face starvation, when their plight could have be lessened, if not avoided, by timely aid. One worker for Save the Children said, "The humanitarian system is like an ambulance—it is focused on disaster response, not prevention."[9]

In a sustainable world, prevention has to play a larger role. Dorothea Lange could not prevent the plight of the pea pickers in California. But she could help get them aid. Today's aid organizations are trying to prevent large-scale famines before they happen, though this can be difficult to do.

"Migrant Mother." This iconic photo of a migrant pea picker by photographer Dorothea Lange captured the despair of the Great Depression. (*Source:* The Library of Congress)

[9]Andrew Wardell. "Extreme Hunger in East Africa and the Sahel: Forewarned but Not Forearmed. Poverty MattersBlog. *The Guardian.* May 9, 2012. http://www.guardian.co.uk/global-development/poverty-matters/2012/may/09/extreme-hunger-east-africa-sahel.

(Table 12–3)—triggering food riots and emergency measures in many countries. In response, the United Nations addressed the issue at a scheduled World Food Summit in Rome in June 2008, and the World Bank began a new effort to provide immediate grants for food and agriculture in the poorest countries. If nothing else, this crisis reflected the interconnected global trade in food.

A number of factors conspired to create this situation, including the high cost of oil, which increased the costs of fertilizer and fuel and thus the cost of food production. Corn was diverted to produce biofuels rather than used for food. Demand increased, especially in Asia, for more expensive foods (e.g., meat, dairy products, vegetable oils). Weather-related disasters slashed harvests in some exporting countries (Australia, Canada). Finally, world reserves of many commodities were declining.

Several of these factors were transitory, causing some relief to the food crisis, but some of the problems are here to stay. Better weather improved the exporting countries' harvest shortfalls, farmers responded to higher prices by

TABLE 12–3 Rise in World Market Prices of Selected Commodities

Commodity	Cost Basis	Average 2002–2006 Price	Average 2007–2008 Price
Wheat	$/ton	$168	$319
Coarse grains	$/ton	113	181
Rice	$/ton	262	361
Oilseeds	$/ton	293	486
Vegetable oils	$/ton	587	1,015
Butter	$/100 kg	162	294
Whole milk powder	$/100 kg	192	417

Source: OECD/FAO Agricultural Outlook 2008–2017.

investing in available technologies to increase yields and croplands, and the current severe downturn in the global economy brought on a decline in petroleum prices. However, the new biofuel economy and the demand for better diets, especially in Asian emerging economies, remain. Since the food crisis, food prices have been volatile, spiking again in 2010–2011 and remaining high in some parts of the world while declining in most.

One consequence of high commodity prices is that farmers get much higher prices for their crops; this is especially true for the farmers in higher-income countries that dominate the food export business. However, farmers in developing countries are not connected to the world markets and benefit little from high market commodity prices. The developing countries whose food deficits force them to import commodities are faced with prices they can't afford. For most people in these countries, food expenditures are more than 50% of their income, and any increase in food price can result in an increase in undernourishment and malnourishment.

RIO+20. One basic function of world conferences is to draw the attention of national leaders to international issues. For example, the World Food Summit, held in Rome in June 2008, was the third World Food Summit; earlier summits in 1996 and 2002 led to verbal commitments to eradicate hunger and malnutrition. Although climate change and bioenergy were addressed in the 2008 meeting, the conference shifted much of its focus to the existing world food crisis. The conference did indeed get the attention of national leaders as details of food shortages, high food prices, and shortfalls in aid were presented.

More recently, however, the 2012 UN Conference on Sustainable Development in Rio de Janeiro (called Rio+20) brought world leaders together to reassess progress 20 years after the 1992 Rio Earth Summit. The document produced at Rio+20, *The Future We Want*, highlighted seven themes: jobs, energy, cities, food, water, oceans, and disasters. Food and agriculture were central topics of the meeting and the outcomes document. At Rio+20, countries were given a challenge—to lower hunger to zero, a plan discussed shortly.

Food Security

Bread for the World's Hunger Institute defines **food security** as "assured access for every person to enough nutritious food to sustain an active and healthy life." Who is responsible for food security? **Figure 12–16** displays *three major levels of responsibility for food security*: the family, the nation, and the global community. At each level, the players are part of a market economy as well as a sociopolitical system. In a market economy, food flows in the direction of *economic* demand. Need is not taken into consideration. A cash economy, following the rules of the market, provides the *opportunity* to purchase food, but not the food itself. Where the economic status of the player is very low (a jobless parent, a poor country), another group or individual may be willing and able to provide the needed purchasing power or the food (but maybe not).

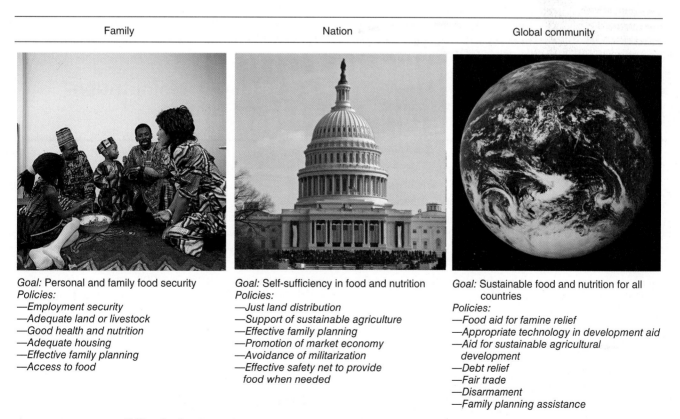

Family	Nation	Global community

Goal: Personal and family food security
Policies:
—*Employment security*
—*Adequate land or livestock*
—*Good health and nutrition*
—*Adequate housing*
—*Effective family planning*
—*Access to food*

Goal: Self-sufficiency in food and nutrition
Policies:
—*Just land distribution*
—*Support of sustainable agriculture*
—*Effective family planning*
—*Promotion of market economy*
—*Avoidance of militarization*
—*Effective safety net to provide food when needed*

Goal: Sustainable food and nutrition for all countries
Policies:
—*Food aid for famine relief*
—*Appropriate technology in development aid*
—*Aid for sustainable agricultural development*
—*Debt relief*
—*Fair trade*
—*Disarmament*
—*Family planning assistance*

Figure 12–16 Responsibility for food security. Goals and strategies for meeting food needs at three levels of responsibility: the family, the nation, and the global community.

The most important level of responsibility is the family. Family members purchase, raise, and gather food from natural ecosystems or have the food provided by someone, usually in the context of the family. When families cannot do so, they need a *safety net*—policies or programs at some national, state, or local level whose objective is to meet the food security needs of all individuals in the society. Locally, food banks and community groups such as shelters and faith communities may provide food. More than 50,000 local food charities help Americans who are hungry.

On a state and national level, safety-net policies and programs can be of two types: (1) official policies, represented by a variety of welfare measures, such as the Food Stamp Program and the Supplemental Security Income program; and (2) voluntary aid through hunger-relief organizations. For example, Feeding America, a network of food banks throughout the country, forms part of the voluntary safety net that provides emergency food help for more than 25 million Americans. Feeding America retrieves food from food processors and suppliers and distributes it to the organization's network of food banks. One law—the Charity, Aid, Recovery, and Empowerment Act of 2003 (CARE Act)—allows family farmers, ranchers, and restaurant owners to deduct the costs of food donated to agencies such as Feeding America as a charitable contribution.

National *self-sufficiency in food*—enough food to satisfy the nutritional needs of all of a country's people—can be achieved either by producing all the food people need or buying the food on the world market. This goal implies that countries have policies that allow the poorest members into the market, such as a just land-distribution policy and a functioning market economy. Many nations, though, are not self-sufficient in food and turn to the global community for *food aid* and other forms of technical development assistance.

Globally, a substantial amount of food aid flows from rich countries to food-poor countries. However, much more food is imported and paid for by the developing countries. The World Trade Organization (WTO) is the body that governs international trade. Meeting every few years, the WTO brings together developed and developing member countries for negotiations and policy decisions (see Chapter 2). Protective policies in developed countries result in tariffs—taxes the developed countries set on imported agricultural products—and in the *subsidies* given to the agricultural sector. U.S. farm bills have continued the subsidies of most agricultural products. The EU is notorious for its high tariff barriers and large subsidies for its agricultural sector, even though it has enacted agricultural reforms that change the rules for subsidies.

Trade Policies. Haiti is a good example of the problem of uneven trade policies. It cut its tariff on rice from 50% to 3% in 1995, opening the country's markets to subsidized rice from the United States (called "dumping"). By 2000, rice production in Haiti was cut in half, leaving thousands of small rice farmers without a means of livelihood. Haiti began to import rice. Haiti's problems were exacerbated by the 2010 earthquake, with the additional loss of livelihood for

thousands. After the earthquake, several NGOs and scientists collaborated to try a new rice-growing initiative, which produced yields that were 64% higher, with fewer seeds and less water and fertilizer. However, uneven trading policies still make it difficult for Haitian farmers to survive in the world market.

Other Needs. There are other initiatives that can help the poorer countries become self-sufficient in food. One of the most important is relieving the debt crisis in developing countries, an issue addressed earlier (Chapter 9). Another factor in meeting global nutrition needs is the *trade imbalance* between the industrial and the developing countries. The developing countries typically export raw goods such as cash crops, mineral ores, and petroleum and import more-sophisticated manufactured products such as aircraft, computers, and machinery. Prices for the latter have risen, while those for the commodities from developing countries have only recently begun to rise. To some extent, the increasing use of labor in the developing countries in electronic assembly and in the manufacture of clothing has offset this imbalance. Indeed, the globalization of markets is now working to the advantage of the developing countries, which have a surplus of labor and low labor costs. As a result, many industrial corporations have outsourced their manufacturing to developing countries, which helps workers in developing countries improve their well-being.

Appropriate global food security goals for the wealthy nations are listed in Figure 12–16. These are policies that promote self-sufficiency in food production and sustainable relationships between the poorer nations and their environments as well as address the trade imbalance and human exploitation problems.

12.4 Hunger, Malnutrition, and Famine

At a UN World Food Conference in 1974, delegates from all nations subscribed to the objective "that *within a decade* no child will go to bed hungry, that no family will fear for its next day's bread, and that no human being's future and capacities will be stunted by malnutrition." That was an unattained dream, though great strides have been made. With the MDGs, a new, more modest target has been proposed: *to cut world hunger in half by 2015* in the developing countries. This means a reduction in the proportion of hungry people from 20% to 10%. In 2008, 34 years after the initial goals, the United Nations held a World Food Summit and addressed the food crisis that was responsible for adding 75 million to the 848 million people still suffering from hunger and malnutrition in the world. Let's look at hunger and malnutrition in the world.

Malnutrition Versus Hunger

Hunger is the general term referring to a lack of basic food required to provide energy and to meet nutritional needs

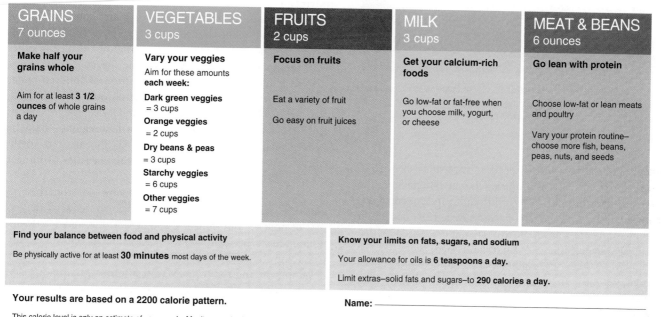

Figure 12–17 The new food guide. The new food guide produced by the U.S. Department of Agriculture to help people evaluate their food intake. Go to https://www.supertracker.usda.gov/default.aspx to get a personalized recommendation.

such that the individual is unable to lead a normal, healthy life. **Undernourishment** is the lack of adequate food energy (usually measured in Calories). **Malnutrition** is the lack of essential nutrients, such as specific amino acids, vitamins, and minerals, and can occur in people who eat nutritionally bad diets even if they have other food available. Another nutritional disorder that is becoming epidemic, even in many developing countries, is **overnourishment,** a condition of eating too much. Worldwide, more than 1.5 billion adults are overweight. More than one-third of the U.S. population is clinically obese (more than 30 pounds overweight). As the World Health Organization describes the situation, "65% of the world's population lives in countries where overweight and obesity kill more people than underweight."[10]

Food Advice in the United States

To address this and other nutritional problems in the United States, the USDA established six food groups and arranged them in a pyramid. The original pyramid, with its one-size-fits-all approach that identified six food groups and a range of servings, has been replaced more than once: first with a different pyramid that differentiated by age and physical activity, and most recently with a graphic of a food plate **(Fig. 12–17).** Bottom line? The best way to combat obesity is to increase the amount of exercise you get and to lower your total calorie intake and emphasize whole grains, vegetables, and unsaturated fats.

Although overnourishment does occur in developing countries, undernourishment is even more prevalent, especially in sub-Saharan Africa and South Asia. There the major nutritional problems are a lack of proteins and some vitamins (malnutrition), especially in children, and a lack of food for energy (undernourishment) for all ages.

Extent and Consequences of Hunger

Accurate, reliable figures on the worldwide extent of hunger are unavailable, mainly because few governments document such figures. On the basis of household surveys, the FAO estimated in 2010 that close to 925 million people are undernourished.

[10]World Health Organization. 2012 Obesity and Overweight Factsheet No. 311. http://www.who.int/mediacentre/factsheets/fs311/en/.

Figure 12–18 Malnourished children. These Guatemalan children are not suffering from a famine but from chronic malnourishment, which will leave their growth stunted.

Where? Almost two-thirds of the undernourished (642 million) are in just seven countries: India, Bangladesh, the Democratic Republic of Congo, Ethiopia, Indonesia, Pakistan, and China. Asia and the Pacific saw a 12% decline in undernourishment from 2009 to 2010. Sub-Saharan Africa has the highest percentage of undernourished, with 30% of the population, some 239 million people, afflicted. It is easy to lose track, when faced with so many numbers, of the fact that that each person is a unique individual whose potential, hopes, and dreams are limited by this lack of food.

Consequences. The effects of malnutrition and undernutrition are greatest in preschool children and next greatest in women. Hunger can prevent normal growth in children, leaving them thin, stunted, and often mentally and physically impaired (**Fig. 12–18**), a condition known as protein-energy malnutrition. Research has shown that undernutrition in early childhood can seriously limit growth and intellectual development throughout life. In Guatemala, for example, one out of two children under five are chronically undernourished, with rates much higher among indigenous people. Guatemala has the highest number of chronically malnourished children in the western hemisphere, sixth highest in the world. This state of malnourishment irreversibly stunts children and inhibits them physically and mentally. Estimates are that, as adults, their productivity is reduced by a third. UNICEF estimated that chronic malnutrition costs Guatemala $8.4 million a day because of sickness, repeated schooling, hospitalization, and lowered productivity.

Sickness and death are companions of hunger. Because poor nutrition lowers a person's resistance to disease, measles, malaria, and diarrheal diseases are common and are major causes of death in the malnourished and undernourished. Hunger is often a seasonal phenomenon in rural areas that are supported by subsistence agriculture, as people are forced to ration their stored food in order to survive until the beginning of the next harvest. Anyone who travels in the developing world cannot help but notice that few people in the rural areas look well fed. Most are thin and spare from a lifetime of difficult physical labor and limited access to food.

Root Cause of Hunger

As we already saw, *the root cause of hunger is poverty*. Hungry and malnourished people lack either the money to buy food or enough land to raise their own. One FAO report on food insecurity states that "Most of the widespread hunger in a world of plenty results from grinding, deeply rooted poverty."[11]

Inadequate food is only one of many consequences of poverty. Addressing hunger and the poverty that causes it will be essential to making progress toward all the other Millennium Development Goals. One measure of progress toward reducing hunger is the proportion of children under five who are underweight because of chronic or acute hunger. **Figure 12–19** compares data on underweight proportions from 1990 and 2009, showing that a few regions—namely, eastern Asia, Latin America and the Caribbean, and northern Africa—are on target to reduce this measure of hunger 50% by 2015.

Hunger and poverty do not always go from bad to worse. A number of countries significantly reduced the extent of poverty and hunger during the 1980s and 1990s. Indonesia has benefited from oil exports and Green Revolution technology, and it continues to put a major emphasis on rural development and the country's social infrastructure. Malawi has been able to double its agricultural productivity through a voucher program for fertilizer and seed, helping poor farmers to feed their families and produce crops for the market. It is possible, therefore, for societies to make progress in reducing the extent of absolute poverty and hunger. In contrast, in any given year, there are countries and sometimes whole regions that face exceptional food emergencies. It is here that international responsibility comes most sharply into focus.

Famine

A **famine** is a severe shortage of food accompanied by a significant increase in the death rate. Famine is a clear signal that a society is either unable or unwilling to distribute food to all segments of its population. Two factors—*drought* and *conflict*—have been immediate causes of famines in recent years.

Drought. Drought is blamed for famines that occurred in 1968–1974 and again in 1984–1985 and 2010–2012 in the Sahel region of Africa. The Sahel, a broad belt of drylands south of the Sahara Desert, is occupied by more than 70 million people who practice subsistence agriculture or tend cattle, sheep, and goats. Although the region normally has enough rainfall to support dry grassland or savanna ecosystems,

---

[11]FAO. "State of Food Insecurity in the World 2002." Rome, Italy.

Proportion of children under age five who are underweight, 1990 and 2006 (Percentage)

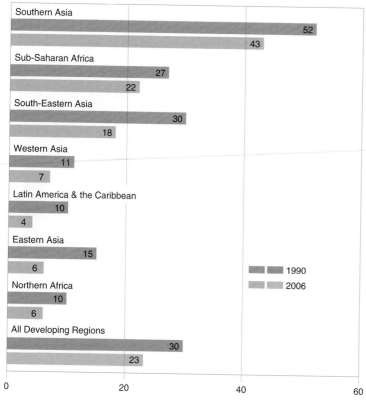

Figure 12–19 Underweight children. The figure shows the percentages of children under age five who were underweight in 1990 and 2009. Millennium Development Goal 1 aims at reducing the 1990 proportion 50% by 2015. (*Source: Millennium Development Goals Report 2011* (New York: United Nations, 2011), 10.)

the rainfall is seasonal, undependable, and prone to failure. Beginning in 1965, the region experienced 20 years of subnormal rainfall, with tragic results. Crops withered, forage for livestock declined, watering places dried up, and livestock died. Both farmers and pastoralists began abandoning their land and migrating toward urban centers, where they ended up in refugee camps. Unsanitary conditions in the camps led to the spread of infectious diseases such as dysentery and cholera, and many thousands died before effective aid could be organized. The 1984–1985 Sahelian famine is thought to have been responsible for almost a million deaths in Ethiopia alone. The number would have been higher if not for aid extended by Africans and numerous international agencies.

Since 1999, periods of drought and normal rainfall have alternated in the western Sahel. Sometimes, as in 2003–2004, favorable weather led to desert locust infestations, which devastated crops in many regions. Drought and deforestation have brought on catastrophic desertification, which in turn has triggered massive dust storms that spread clouds of Sahara dust across the Atlantic Ocean and into the Northern Hemisphere. Although rainfall is normal in the western Sahel at present, the eastern Sahel—the "Horn of Africa"—remains in the grip of an extended drought that has put more than 23 million people at risk. The years 2011 and 2012 saw a continuation of drought. While drought is frequent in that region, an aquifer of ancient water discovered in 2012 deep under Namibia may help Africa's driest sub-Saharan country.[12]

---

[12]McGrath, M. Vast Aquifer found in Namibia could last for centuries. BBC News 2012. http://www.bbc.co.uk/news/science-environment-18875385.

Warning Systems. In order to prevent droughts from leading to famines, two major efforts provide crucial warnings about food insecurity in different regions of the world. One, the FAO's *Global Information and Early Warning System (GIEWS)*, monitors food supply and demand all over the world, watching especially for those places where food supply difficulties are expected or are happening. The GIEWS issues frequent reports and alerts that are usually the result of its field missions or on-the-ground reports from other UN agencies and NGOs. The agency also relies on satellite imagery for crop-monitoring activities.

The second major effort is the *Famine Early Warning System Network (FEWS NET)*, funded by the U.S. Agency for International Development. FEWS NET is a partnership of agencies involved in satellite operations focusing on sub-Saharan Africa. It continually measures rainfall and agricultural conditions in the region and, like the GIEWS, issues regular bulletins and special reports on the Internet, giving governments and relief agencies accurate and timely assessments of the status of food security in African countries.

Conflict. Devastating and prolonged civil warfare put millions of people at risk of famine. Famines in which the causative factor was not drought, but war, threatened Ethiopia, Eritrea, Somalia, Rwanda, Sudan, Mozambique, Angola, and Congo in the 1990s. The civil wars disrupted the farmers' normal planting and harvesting and displaced millions from their homes and food sources. In Mozambique alone, 900,000 people died from direct military action or from indirect effects of the war there. At least 15 countries

World food aid (millions metric tons)

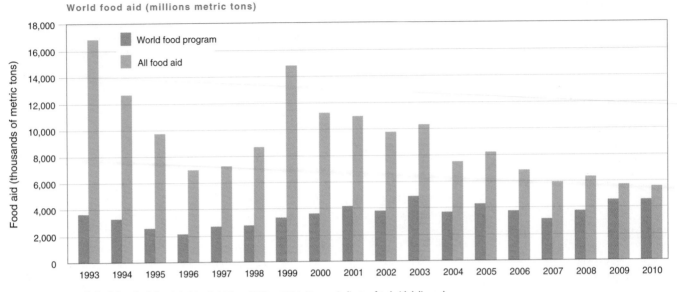

Figure 12–20 Global food aid. Global food aid from 1993 to 2010. Orange indicates food aid delivered by the World Food Program; green indicates food aid from all sources. (*Source:* World Food Program annual reports, FAIS, UNFAO.)

in sub-Saharan Africa have experienced food emergencies in recent years because of internal conflicts and their effects on neighboring countries. Continuing conflicts in Afghanistan and Sri Lanka interfere with access to food, and a 30-year civil war in Guatemala is part of what set the stage for chronic malnutrition there, though it is not an actual famine.

Famines from drought and war are preventable. India, Brazil, Kenya, and southern Africa have coped with droughts in recent years by mobilizing effective relief in the form of food, clothing, and medical assistance. Indeed, the drought of the early 1990s accelerated the peace process in Mozambique and helped change the political landscape in southern Africa. Cooperation between South Africa and the 10 nations of the Southern Africa Development Community to prevent famine lowered barriers between South Africa and its neighbors before democracy came to South Africa.

Hunger Hot Spots

Much of Africa has experienced long-term and severe droughts as well as widespread civil conflict. Erratic weather patterns and a general warming trend thought by some scientists to be the outcome of global climate change have caused crop harvests to oscillate between average and poor. And with the recent precipitous rise in food prices, many millions in the region are experiencing food insecurity. Climate change is expected to affect global food production worldwide, lowering south and southeastern Asian food production, for example, by as much as 50% over the next three decades.

Other countries with recurrent hunger and food insecurity include North Korea, Haiti, and Bangladesh. In early 2012, the FAO identified 34 countries, including 27 African countries in need, as being "in crisis and requiring external

assistance."[13] The "external assistance" here is food aid; the reasons for the need vary from conflict to drought to floods, hurricanes, and cyclones to localized crop failures. Widespread, severe hunger and famine are being averted today only because of food aid, though it does not always work as well as we would like.

Food Aid

The World Food Program (WFP) of the United Nations coordinates global food aid, receiving donations from the United States, Japan, and the countries of the European Union. In addition, many nations conduct their own programs of bilateral food aid to needy countries. Although international food aid was cut in half (to 6.2 million metric tons) between 1993 and 1996, it reached 15 million metric tons in 1999; it has since dropped as low as 5.7 million metric tons (2010) **(Fig. 12–20)**. Much of the increase in 1999 was the result of a major shortfall in grain production in the Russian Federation and a consequent extension of food aid primarily from the United States (admittedly, to boost U.S. farm exports, which had been falling).

Food aid is distributed to countries all over the world, not just where famines are threatened. Food aid is closely negotiated on the world stage. In April 2012, a new Food Assistance Convention (FAC) was agreed on (replacing one from 1999). It is the only international treaty that requires members to provide a minimum amount of food assistance. The 2012 FAC was a vast improvement over earlier treaties in some ways, in part because it addressed concerns about the problems of food aid, including fostering dependence and problems caused by aid that isn't useful to the recipients.

---

[13]FAO. Crop Prospects and Food Situation. No. 1. February 2009.

When Aid Doesn't Help. Numerous humanitarian campaigns to end world hunger have been mounted in the past 50 years. The United States and Canada have been world leaders in giving away food (which is first purchased from farmers and, therefore, represents a subsidy). A number of serious famines have been moderated or averted by these efforts, and certainly the need for ongoing emergency food aid is undeniable. However, routinely supplying food aid in an attempt to alleviate chronic hunger in developing countries may be the worst thing to do.

The problem is that people will not pay more than they have to for food. Therefore, free or very cheap foreign food undercuts the local market. In effect, local farmers must compete economically with free or low-cost imported food. When they cannot earn a profit, they stop producing and eventually enter the ranks of the poor. In the long run, the entire local economy deteriorates. Hence, the donation of food, while well intended, often aggravates the very conditions that it is meant to alleviate.

When It Does Help. The WFP is aware of this problem and targets most of its food aid to emergency situations. Some aid does go to "development projects," benefiting poor people in countries where food security is at risk and regions where malnutrition is severe. The intention of this aid is to free poor people of the need to provide food for their families and allow them to devote their efforts to other development activities. As an important component of this strategy, the WFP now makes 80% of its food purchases in the developing countries themselves. This is a "win-win" solution, where small-scale farmers receive income for their crops and needy people are fed. In fact, WFP has special programs designed to prevent hunger by better connecting poor farmers to the market.

Unfortunately, the United States has a law requiring food aid from the United States to be American grown and processed and transported by American companies. Many of our aid dollars go to pay shipping costs, but shipping food delays its arrival and uses fossil fuels. In 2008, Congress authorized a four-year pilot program to purchase local food in crises. It was successful, saved money, and got food to needy people more quickly. In addition, the 2012 FAC treaty stressed assistance in the form of grants that support local economies, rather than food grown in donor countries. It is hoped that these two changes will alter the way the United States approaches food aid.

The Goal Is Zero, the Way There Is Complex

At the 2012 Rio+20 summit, UN Secretary-General Ban Ki-moon offered the Zero Hunger Challenge to the world. Ki-moon said that achieving zero hunger would lower conflict and increase productivity and prosperity. The challenge was focused on five objectives: 100% access to adequate food all year round; zero stunted children under two years old and no more malnutrition in pregnancy and early childhood; all food systems sustainable; 100% growth in smallholder productivity and income, particularly for women; and zero loss or waste of food, including responsible consumption. It is a tall order, but to get anywhere, we must begin by setting high goals. International agreements have held that the right to food must be considered a basic human right. It follows, then, that we as a nation have a moral obligation to respond to world hunger. Has that moral obligation made its way into public policy?

Although food is our most vital resource, we do not treat it as a commons (free to all that need it). Indeed, the production and distribution of food constitute one of the most important economic enterprises on Earth. Alleviating hunger is primarily a matter of addressing the absolute poverty that afflicts one of every five people on the planet. To treat food as a commons would be to treat wealth as a commons. While this is one of the tenets of socialist systems, this notion has proved to be unworkable. We are part of the world economy now, and it is a market economy—for the perfectly good reason that nothing else seems to work, given the realities of human nature.

What has not been done, however, is to bring this market economy under the discipline of sustainability. Short-term profit crowds out long-term sustainable management of natural resources. Rich nations have been content to maintain the current high subsidies to agriculture that allow them to flood markets with underpriced products. Many developing countries, especially in sub-Saharan Africa, have been plagued with corrupt and inept elites who are more interested in staying in power than in reducing poverty in their countries. In some areas, tribal conflicts have taken a terrible toll in lives and hunger.

We understand the situation well, but the solutions are not simple. The solutions lie in the realm of political and social action at all levels of responsibility. Given the current groundswell of concern about the environment and the global attention to the Millennium Development Goals and, we hope, the Sustainable Development Goals, there may never be a better time to turn things around and take more seriously our responsibilities as stewards of the planet and as our brothers' (and sisters') keepers.

12.5 The Future: Feeding the World as We Approach 2030–2050

Looking ahead to 2020, we need to be able to feed an additional 700 million people and also make significant progress on meeting the Millennium Development Goal of reducing by half existing hunger and malnutrition in the world—and do so without the unsustainable practices agriculture currently employs. Is this possible? What are the prospects as we look even further ahead, to 2030–2050?

Future Prospects

Given the limits on suitable land for agriculture, there are several prospects for increasing food production. We could continue to increase crop yields with new technologies. We could grow food crops on land that is now used for feedstock crops, biofuels, or cash crops. We could use new and unexpected agricultural opportunities such as urban agriculture or mixed permacultures. We could also attempt to feed more people with the food we already produce by changing social structures so more people have a chance to enter the market—in turn lowering crop loss and food waste—and by eating foods that require less fuel, water, land, and pesticides to produce.

More Green Revolution

As mentioned earlier, a dramatic rise in crop yields in the developing countries was the major accomplishment of the Green Revolution: Wheat yields tripled, rice and corn yields more than doubled, potato yields rose 78%, and cassava yields rose 36%, according to the UN Food and Agricultural Organization (FAO). Interestingly, grain yields have continued to rise in some of the *developed* countries, too. In France, for example, wheat yields have quadrupled since 1950, reaching more than 6 tons per hectare. Rice yields in Japan have risen 67% in the same period. Actually, the genetic potential exists for wheat yields of 14 tons per hectare or higher, and rice yields can be as high as 13 tons per hectare under the right conditions. Can we expect yields to continue to increase up to their genetic potential?

It turns out that the great differences in grain yields among regions have less to do with the genetic strains used and more to do with the weather. Egypt and Mexico, for example, irrigate their wheat and get higher yields, while U.S. wheat is rain-fed. Australian wheat yields are even lower, the result of the country's sparse rainfall. Once the agricultural land is planted with high-yielding strains and fertilized to the maximum, other factors—soil, rainfall, and available sunlight—limit productivity. These environmental limits are a reminder that agricultural sustainability is highly dependent on soil and water conservation and on the weather (Chapters 10 and 11).

But some people still see the Green Revolution as the way forward. Gordon Conway, president of the Rockefeller Foundation, has called for a new Green Revolution. He has dubbed it a Doubly Green Revolution: "a revolution that is even more productive than the first Green Revolution and even more 'green' in terms of conserving natural resources and the environment. During the next three decades, it must aim to repeat the successes of the Green Revolution on a global scale in many diverse localities and be equitable, sustainable, and environmentally friendly."[14]

Less Meat and Biofuels. One way to produce more food is to eat less meat and divert fewer crops to biofuels. We can switch from the production of feed grain and cash crops to food for people. Almost 70% of domestic grain in the United States is used to feed livestock. The percentages drop over other regions of the world in proportion to the economic level of the region. Sub-Saharan Africa and India, for example, use only 2% of their grain to feed livestock. The trend, however, is in exactly the opposite direction. More people want to eat higher on the food chain. Finding a medium in which people have protein in their diets but are not heavily meat dependent is part of a more sustainable approach to food.

Reprising the green revolution may be a part of what we need to do, but it can't be all of it. That is because modern farming is damaging to the environment. We cannot set up our most important systems (such as food production) to constantly damage the very ecosystems we depend on and expect this situation to last long. Modern agriculture has too many externalities to be sustainable; one example is the use of fossil fuels, which increases climate change. As the predicted warming of the 21st century occurs, it is impossible to project how rainfall patterns will change. However, scientists at NASA's Goddard Institute of Space Studies estimate that developing countries could lose 334 million acres (135 million ha) of essential farmland in the next 50 years due to temperature increases in the tropics. Many farming practices are heavily dependent on fossil fuels, and farming releases other greenhouse gases. In fact, one greenhouse gas, nitrous oxide, is released from fertilized farm fields and from manure; agricultural releases are at least 10 times higher than any other source. Methane production by livestock was already discussed. Nitrous oxide has 298 times the greenhouse gas effect of carbon dioxide, and methane is 20 times more effective at trapping thermal energy than CO_2, so their production by agriculture is important, and more of the same agriculture won't solve the problem.

Feeding many people sustainably is going to take something different than more of the same. It will require sustainable agriculture, that is, agriculture that can be carried on indefinitely without environmental harm. In 2012, the Commission on Sustainable Agriculture and Climate Change put out a report, *Achieving Food Security in the Face of Climate Change*, at a conference in London. In an interview, Commission chair Professor Sir John Beddington said, "If you're going to generate enough food both to address the poverty of a billion people not getting enough food, with another billion [in the global population] in 13 years' time . . . you can't do it using the same agricultural techniques we've used before, because that would seriously increase greenhouse gas emissions for the whole world . . ."[15]

Sustainable Agriculture

What would sustainable agriculture look like? It might include some of the large farms we are used to, though with fewer chemicals, fewer fossil fuels, and fewer of the most wasteful types of irrigation (as we have seen in other chapters). We will need to eat lower on the food chain, eating less meat in the developed world and allowing others to eat more **(Fig. 12–21)**. But it will also require wild creative approaches very different from industrial agriculture, such as the intensive, mixed

[14]Gordon Conway. "*The Doubly Green Revolution: Food for All in the 21st Century.*" 1999. Cornell University Press. Ithaca, NY.

[15]Richard Black. "Farming Needs "Climate-Smart" Revolution, Says Report." *BBC News Science and Environment.* March 28, 2012. http://www.bbc.co.uk /news/science-environment-17495031 June 25, 2012.

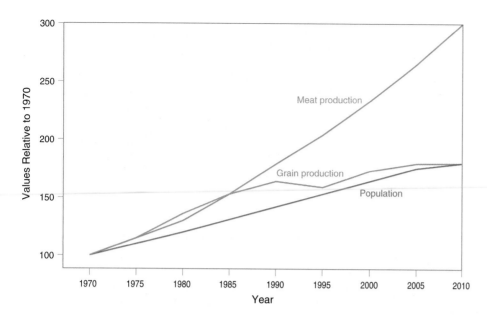

Figure 12–21 Global population and grain and meat production. Grain production has kept up with the population increase, and meat production has greatly outpaced it over the past 35 years. High meat consumption in the West makes it difficult to produce food sustainably. (*Source:* FAOSTAT.)

product (polyculture) urban farm described in the story of Growing Power. Here are some other trends in the future of food production.

Locavore, Anyone? While trade is important in order to provide jobs and thus access to food markets for all, one way to lower the environmental cost of food is by eating locally. **Locavores** are people who try to purchase or grow most of their food within their own region. Of course, this means foods have to be preserved in some way, and many foods are not available in all seasons. There are many locavore groups in the United States and elsewhere. **Community-supported agriculture** farms (CSAs) are on the rise in American urban and suburban areas. These farms are supported by people who purchase shares and who often volunteer at the farms. Because shares are purchased in advance, the risks of a poor crop are shared, and farmers borrow less money and are likely to weather difficult seasons more easily. CSA farms are popular, as they provide local fresh produce, access to farming, and support for farms that might not be economically viable if they had to compete with cheap food trucked long distances from industrial farms.

Urban Agriculture. Urban farming occurs across the world and is part of the future of food production. Cuba is the poster country for urban farming and the only country in the world with extensive state-supported urban agriculture. Food shortages of the 1990s drove a grassroots effort to produce food in Cuban city neighborhoods. To meet the needs of small-scale urban farmers, the Cuban government created an Urban Agriculture Department. It provides expertise and places to buy seeds and equipment and supports infrastructure. In Havana alone, there are more than 30,000 people growing food in more than 8,000 farms and gardens (**Fig. 12–22a**). Cubans even began to grow rice in small plots. Small-scale rice production now rivals the large-scale production of large farms in Cuba.

Urban Foraging. Another model of urban food production is foraging. **Urban foragers** find fruits and berries and other foods available in the urban landscape. Sometimes they cre-

ate urban food maps, which are available on the Web. There is a growing community of urban foragers. In Seattle, community organizers designed a 7-acre mixed foraging and garden plot area called an **urban food forest**. The Beacon Food Forest, as it is called, was designed as a project for a class in permaculture—the growing of food with perennial plants. From a small class project, the forest grew with grants to be planned as the largest urban food forest on U.S. public land. The project will be realized over a period of years.

Algae to the Rescue. Imagine a world in which nutritious algae were grown in fish tanks on nearly everyone's windowsills, algae farms in the ocean grew sea lettuce and other edible seaweeds, algae were grown in ponds and tanks for fish food, or vast bags of wastewater were treated with algae, which were then fed to fish. Algae would be used as a biofuel because of their high oil content (Fig. 12–22b), and algae ditches at the edges of farm fields would take up nutrients and in turn provide algal fertilizer for the fields. Algae as a food are very nutritious, high in omega-3 fatty acids, vitamins, and antioxidants. As a biofuel, it is competitive because of its high oil content, as a nutrient extractor, it is efficient because of their rapid growth, and as a fertilizer, it is rich.

That imagined future is now. Work is proceeding rapidly on all of these fronts to produce a world in which algae can be used in a wide array of ways. Dr. Aaron Baum is a good example of an algal entrepreneur. He is working with NASA, which has a large algae project (the OMEGA project) designed to use wastewater to raise algae for biofuel production (Fig. 12–22c). But Baum's major effort is with small food algae production. He designs kits for people to raise *Spirulina*, an extremely nutritious type of algae found in food supplements. Instead of taking a small-scale idea and trying to scale up to industrial levels, Baum is trying to make household-sized algae-growing apparatuses that can be used by any apartment dweller to grow healthy food.

Organics. Organic farming has been around a long time and overlaps with the other sustainable practices outlined here.

(a)

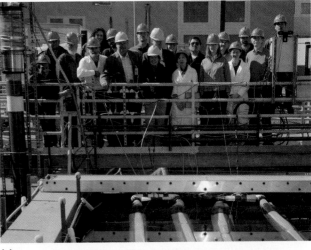

(c)

(b)

(d)

Figure 12–22 Types of sustainable agriculture. (a) An urban garden in Havana, Cuba. (b) Scientists working on algae production for biofuels. (c) NASA has a project for algal treatment of wastewater. The algae are then fed to fish. (d) The land institute works to make agricultural systems similar to prairies.

In the United States, farms can be certified as organic by the USDA. Certified organic farms have to meet strict criteria. They are limited to natural pesticide products and cannot use the artificial fertilizers typical of modern agriculture. Often, organic farm yields are somewhat lower and organic food prices higher. In one study of more than 322 research papers, scientists found that organic yields were 80% of conventional yields on average (with a great deal of variability). However, in intensive mixed cultures (such as the Growing Power farm), with practices where different types of crops are grown together (companion planting) and the natural enemies of pests are supported, yields can be high (Chapter 13).

In the United States, certified organic farms cannot use GM crops. This is a controversial issue. Many people who support organic farming also see GM crops (especially those modified for drought or salt tolerance or enhanced nutrition) as one of the tools we will need to use to feed the world. It is also difficult to keep organic farms GM free; farmers often find their crops contaminated with GM crop plants. As a result of some of these limitations, many farms, especially

CSAs, are not certified as organic by the USDA although they may use less polluting methods.

Permanent Polycultures. In Salina, Kansas, Wes Jackson (another MacArthur "Genius" award recipient) has spent a career working toward another model of sustainable agriculture—the permanent polyculture. Instead of the conventional monoculture that is susceptible to insects and diseases and drought, Jackson and his colleagues are trying to create an agricultural system modeled on a native prairie. The characteristics of a prairie include high biodiversity and permanent ground cover. If these could be duplicated even in an agricultural field, they figure, the growing of crops could be sustainable (Fig. 12–22d). What they describe is a **permanent polyculture**, a group of perennial plants of different species grown together that produce food or fuel. This is hard to achieve, in part because perennial plants put more energy into roots and leaves and less into the parts of the plant we eat. This type of research is important to a radically new approach to food production.

Policy changes. The future also needs to include policy and individual changes. U.S. farm policies currently encourage large farms, support fewer workers, discourage sustainable agriculture, and subsidize industrial agriculture to the detriment of the environment. Incorporating externalities such as the true cost of erosion, water pollution, or overuse of water resources into the actual price of food would go a long way to changing the way we view both food and the way we produce it. Finding ways to support farmers in the United States without subsidizing large industrial agriculture and CAFOs at the expense of small or organic farmers and small

farmers abroad would go a long way toward solving job and food insecurity. Providing means to get crop insurance for small farmers, including those practicing subsistence, would help lessen the vulnerability of farmers to environmental changes we anticipate with climate change.

Individuals need to feel connected to the rest of the world and empowered to do something about the big issues of our time. Food production is central to every topic in environmental science. In order to solve environmental problems, we will need to have new ways of approaching them.

REVISITING THE THEMES

Sound Science

The new high-yield grains that led to the Green Revolution were developed using science. So is agricultural research that continues to seek new ways to improve crop yields and research that shows the environmental costs of conventional agriculture. Sound science is also needed to provide an accurate evaluation of all genetically modified crops and foods before they are implemented. Most importantly, science, using all of its tools, is necessary to find long-term solutions to food production. New science techniques are used to grow algae for food and biofuels and to accomplish other sustainable agriculture goals.

Sustainability

Global land and water used for agricultural crops and animal husbandry represent two of the most crucial and irreplaceable components of ecosystem capital. However, the land must be nourished; it is a renewable resource, but it can be eroded and worn down by unsustainable practices. Water can be overused. As populations increase, pressure will mount to turn more forests and wetlands into croplands, a transformation with trade-offs that are not usually beneficial due to the loss of the goods and services of the natural lands.

Many of the techniques of modern agriculture are not sustainable, including heavy use of fossil fuels, ancient groundwater, pesticides, artificial fertilizer, and monocultures. Many consider the diversion of food crops to biofuel production unsustainable. Likewise, some food

aid is harmful to sustainability in developing countries, where self-sufficiency in food is essential to establishing a sustainable society. A number of technologies and approaches make up sustainable agriculture, the approach most likely to both feed populations and maintain ecosystem health.

Stewardship

The remarkable progress made in keeping the world's people fed is certainly good stewardship in action. This is especially true where efforts are being made, such as those by the national and international agencies and the many NGOs involved in hunger and development aid, to identify those suffering from food insecurity and to ensure that they have sufficient food for their needs. Thousands of organizations and people show that they care as this work goes on. The work of Borlaug and his team, of the CGIAR, and of the IRRI is particularly noteworthy because these groups have sought to improve the yields of the more important grains and make the new high-yielding varieties available to the developing countries. Stewardship requires us to protect access of all people to crops and seeds, rather than amassing power in a few companies. The Feeding America organization plays an important role in the United States in keeping people from going hungry in a land of plenty. We still need to do a better job of stewardship of ecosystem services affected by unsustainable food production. In part, this will involve changing ways people relate to the market—a focus on access rather than production, on lowering waste and crop loss, and on diverting less food to livestock feed or biofuels.

REVIEW QUESTIONS

1. Describe the differences between subsistence and modern agriculture. What are major components of the agricultural revolution. What are the environmental costs of each?

2. What is the Green Revolution? What have been its limitations and its gains?

3. How do sustainable animal farming and industrial-style animal farming affect the environment on different scales?

4. What are biofuels, and what are their impacts on food production?

5. What are the major advantages and problems associated with using biotechnology in food production?

6. Describe the patterns in grain trade between different world regions over the past 70 years. How do the food crisis of 2006–2008 and food shortages of 2010–2011 relate to these patterns?

7. Describe the levels of responsibility for food security. At each level, list several ways food security can be improved.

8. Define *hunger*, *malnutrition*, and *undernourishment*. What is the extent of these problems in the world? What is their root cause?

9. Discuss the causes of famine, severe hunger, and chronic malnutrition, and name the geographical areas most threatened by them.

10. How is relief from famine and severe hunger accomplished? Why does food aid sometimes aggravate poverty and hunger?

11. Define sustainability in an agricultural context. In what ways is modern industrial agriculture unsustainable?

12. What are at least four cutting-edge concepts in sustainable agriculture?

THINKING ENVIRONMENTALLY

1. Imagine that you have been sent as a Peace Corps volunteer to a poor African nation experiencing widespread hunger. Design a strategy for assessing the needs of the people and for contacting appropriate sources for help.

2. Calculate your "food footprint" by looking for one of several Web sites where this is calculated. Usually it gives a value of either the land or ocean area required to sustain you or, in some cases, your carbon dioxide production. How many Earths would it take to feed everyone if we all ate like you do?

3. Use the Web to evaluate the work of FEWS NET (www.fews.net) and the GIEWS (www.fao.org/GIEWS). How do these organizations function in bringing food aid to people with the greatest need?

MAKING A DIFFERENCE

1. Try to buy cloth bags for food shopping instead of using paper or plastic. These are reusable, as opposed to the disposable varieties, and are very sustainable. Even though paper bags come from trees (and are both renewable and recyclable), they generate air and water pollutants in their manufacture, more so than plastic.

2. If you are cautious about genetically altered foods and you want to know whether a product is organic or not, check the barcode number on the package: If the first digit is 9, the food is organic. If that number is 4, the item was a result of conventional farming. If the number is 8, then it has been genetically engineered.

3. Plant your own garden; you control all of the chemical inputs (if any), and no carbon dioxide is emitted in getting the goods onto your plate.

4. Bring your lunch to work; you'll save your own money as well as the environment.

5. Demonstrate your concern for the hungry and homeless in your city by giving your time (or donating foodstuffs) to a local soup kitchen or food pantry.

MasteringEnvironmentalScience®

Go to **www.masteringenvironmentalscience.com** for practice quizzes, Pearson eText, videos, current events, and more.

CHAPTER 13

Pests and Pest Control

W HAT DO NEW YORK City's Abercrombie and Fitch stores, the University of Nebraska, Lincoln's housing system, and the Fort Worth, Texas, housing authority have in common? In the past few years, they have all dealt with outbreaks of bedbugs (*Cimex lectularius*). Bedbugs, once common in the United States, had been well controlled since after WWII but have resurged to epidemic proportions in recent years. Small blood-sucking wingless insects, bedbugs are a public health nuisance. Their bites cause itchy rashes, though they do not transmit other diseases. While bedbug-sniffing dogs can be used to pinpoint infestations, sometimes getting rid of the pests is extremely difficult. Bedbugs lay a lot of eggs, can hide in small cracks, and can go a year without feeding. Once controlled by DDT, a pesticide no longer in use in the United States, bedbugs are finding the modern world to their liking. They are spread by an increase in domestic and international travel and can be found in hotels ranging from cheap to expensive. Many are resistant to the pesticides we do use. Some bedbugs in New York City are 264 times more resistant to a common pesticide than similar bedbugs in Florida. Once established, bedbugs are terribly difficult to eradicate.

ONE PARTICULARLY BAD OUTBREAK. In 2010, a particularly bad summer outbreak of bedbugs prompted five states to ask the Department of Defense for resources to get rid of the pests. The state of Ohio petitioned the Environmental Protection Agency (EPA) for permission to use a pesticide, Propoxur, for in-home use. The petition was denied because of concerns about the effect of the chemical

on children, leaving Ohio in a quandry. Bedbug research found that commonly used foggers, or pesticide bombs, used in other types of domestic infestation, are ineffective against the insects. In 2012, the Centers for Disease Control and Prevention and the EPA

LEARNING OBJECTIVES

1. **The Need for Pest Control:** Define the major groups of pests and the different methods we use to control them.

2. **Chemical Treatment, Promises and Problems:** Explain the serious problems that accompany overuse of new and more effective chemical pesticides, such as DDT.

3. **Alternative Pest Control Methods:** Explain the major types of alternatives to using pesticides to control pests.

4. **Integrated Pest Management (IPM):** Explain the main principles and give examples of the integrated pest management approach to reducing the use of pesticides.

5. **Pests and Policy:** List and describe the federal and international policies for controlling pests and those for control of pesticides.

gave a joint statement about bedbugs as a public health concern. Among the non-chemical means of controlling bedbugs, the CDC recommends vacuuming regularly, sealing cracks, treating with heat, removing clutter, and monitoring. A dry soil called diatomaceous earth, which is non-toxic but scratches the waxy coating found on bedbugs, can also be distributed.

A GREAT ILLUSTRATION. Bedbugs are a great illustration of the issues surrounding controlling pests: the costs of pests, the difficulties with chemical controls, and the need for a combination of treatments used in concert. The heyday of chemical pesticides as a silver bullet is over due to the reality of their environmental effects and pesticide resistance. Now we have to use new and innovative approaches to pests, whether they are found in agricultural settings, in structures, in the flour on your pantry shelf, or in beds. In this chapter, we will connect the issues of pests in our daily lives to environmental science.

13.1 The Need for Pest Control

Since earliest times, humans have suffered frustration and losses brought on by destructive pests. To this day, farmers, herdsmen, and homeowners wage a constant battle against the insects, plant pathogens, and unwanted plants that compete with them for the biological use of crops, animals, and homes.

One dictionary defines a **pest** as "an insect or other small animal that harms or destroys garden plants, trees, etc."[1] This definition includes a broad variety of organisms that interfere with humans or with our social or economic endeavors such as pathogens, nuisance wild animals, annoying insects, and molds. Other chapters consider nuisance wild animals, invasive species in natural ecosystems, and pathogens (Chapters 6, 7, and 17). The emphasis in this chapter is on those pests that interfere with agriculture, households, stored products, wood products, and structures. Pests are often defined by the types of activity they affect or what they live on. Sometimes we divide them by their taxonomic group.

Types of Pests and the Importance of Pest Control

Agricultural pests are organisms that feed on agricultural crops, ornamental plants, or animals. The most notorious of these organisms are various insects, but certain fungi, viruses, worms, snails, slugs, rats, mice, and birds also fit into the category (**Fig. 13–1a–c**). **Weeds** are plants that compete with agricultural crops, forests, and forage grasses for light and nutrients (Fig. 13–1d–e). Some unwanted plants poison cattle or have other serious effects; others simply detract from the appearance of lawns and gardens. Insects, plant pathogens, and weeds destroy an estimated 37% (before and after harvest) of potential agricultural production in the United States, at an estimated yearly loss of $122 billion to consumers and producers. Elsewhere, pests also destroy important crops. In the late 1990s, for example, Russia suffered a 70% yield loss from a potato fungus, *Phythophtora infestans*, considered to be global agriculture's worst crop disease. Another fungal pest, soybean rust, *Phakopsora pachyrhizi*,

has spread throughout South America, causing $2 billion in losses in Brazil alone in 2004. When it was blown into the United States by winds from Hurricane Irene and began to spread northward, people had to act quickly. Now there is a soybean rust alert system to treat it quickly when it is detected. A 2011 plague of mice in Australia caused more than $300 million in crop losses.

Some agricultural pests are veterinary pests (Fig. 13–1f), those affecting domestic animals. These include the screwworm fly, for example, or fleas, ticks, and mites on animals. Often such pests are disease **vectors**. They may be similar to pests attacking humans, the medical pests. Houseflies are significant pests because they can carry diseases onto food, especially from manure sources. Darkling beetles are significant pests of poultry production worldwide; they carry human and animal diseases and destroy insulation. Mosquitoes (Fig. 13–1g) carry dengue and yellow fevers and malaria; snails carry schistosomiasis, and chagas bugs carry a protozoan parasite. Other medical pests, such as the bedbug mentioned earlier, carry no diseases. The malaria-carrying mosquito, one of the most important medical pests, will be discussed at some length later (see Stewardship, "DDT for Malaria Control") though many medical pests are discussed elsewhere (Chapter 17).

Forest pests destroy living trees and wood products. The mountain pine beetle was already discussed as devastating to forest ecosystems (Chapter 7). Other wood pests include beetles such as the emerald ash borer (Fig. 13–1h). Originally from Russia, China, Japan and Korea, the beautiful green beetle was first discovered in the United States in 2002. Since then it has spread to at least 14 states and parts of Canada. Its spread is monitored with purple traps visible along highways (**Fig. 13–2**). It is estimated to have killed 50 million to 100 million ash trees as it destroys the layers directly under the bark by burrowing. The borer threatens the nearly 7.5 billion trees (about 37.5 million in urban and suburban settings) of several ash species in North America. As they die, the trees lose limbs, require removal, and are dangerous. A 2012 report estimated that the emerald ash borer and another forest pest, the Asian longhorned beetle, cost $1.7 billion in local government expenditures annually and an additional $830 million in lost property values.

---

[1]Random House Webster's Unabridged Dictionary, 2005.

Figure 13–1 Important pests. (a) Soybean rust. (b) Cotton boll weevil. (c) Male medfly. (d) Giant hogweed.
(e) Striga (witchweed). (f) Dog tick. (g) *Anopheles* mosquito. (h) Emerald ash borer.

Figure 13–2 Monitoring. Traps for emerald ash borers are found along roads. They are important parts of monitoring.

Figure 13–3 Stored product pests. This Khapra beetle, hiding in a bag of rice, is one of the worst stored product pests worldwide. It is small, resistant to pesticides, can live with little water, and can go long periods without food.

Stored product pests are those which live in items such as cereals and processed foods. Rice weevils and flour moths may be familiar pests in kitchen pantries. The Khapra beetle (*Trogoderma granarium*) **(Fig. 13–3)** is one you probably have not seen because of the efforts of federal agents to keep it out of the country. It is tiny; can eat many types of food, including those low in moisture; is resistant to many pesticides; and can even survive without food for long periods. In 2011, it was intercepted 100 times in various products at U.S. borders. Biofouling organisms are organisms that settle on surfaces in aquatic environments. Some make shipping more expensive and less fuel efficient; others clog intake pipes and screens for industries, cover piers, and settle bouys. Anything in the water can be covered. Biofouling is described later (see Sound Science, "Marine Fouling Organisms: Keeping One Step Ahead of the Barnacles"). Fabric pests include those which live on cloth, leather, feathers, and the like. They include dermestid beetles, carpet beetles, and others. There are other types of pests as well, specializing on paper and other products.

Pests and Climate Change. It is hard predict the exact effects climate change will have on pests, but scientists are trying to figure out trends as the world changes. Many insects including several of medical or agricultural importance are already increasing their ranges, and pathogens will respond to changes in precipitation, temperature, and weather events. Researchers anticipate some increase in crop pests such as nematodes and weevils that attack the roots of the plantain, a crop in the tropics, as temperatures increase. In other places, drought may lower mosquito populations. Overall, scientists expect an increase in pests, but a great deal of research still needs to be done.

The Pesticide Expense. Part of the credit for modern human prosperity can be attributed to pest control. Throughout the chapter, we will see many examples of these different types of pests and efforts to control them. Chemical pesticides have become one of the most frequently used tools in our effort to control pests, an effort that is quite costly. The EPA estimated that efforts to control pests in the United States in 2007 (the latest available survey) involved the use of 1.133 billion pounds (514,000 metric tons) of **herbicides** (chemicals that kill plants) and other **pesticides** (chemicals that kill animals and insects considered to be pests) annually **(Fig. 13–4)**, at a direct cost of $12.5 billion. According to the EPA, 5.21 billion pounds (2.4 million metric tons) of pesticides were used worldwide in 2007, at a cost of $39 billion.[2] Many of the changes in agricultural technology, such as monoculture and the widespread use of genetically identical crops, have boosted yields and also brought an increase in the proportion of crops lost to pests—from 31% in the 1950s to 37% today. During the past half-century, the use of herbicides and other pesticides multiplied manyfold, leading to a disturbing and unsustainable dependency on them.

Different Philosophies of Pest Control

Medical practice employs several basic means of treating infectious diseases. One approach is to give the patient a massive dose of antibiotics or other medicines, hoping to either eliminate the pathogen or stop it before it can get established. Another approach is to stimulate the patient's

[2]EPA Pesticide Sales and Usage 2006–2007 Market Estimates Report. (2011). http://www.epa.gov/opp00001/pestsales/07pestsales/table_of_contents2007.htm

PESTICIDE USE IN THE UNITED STATES

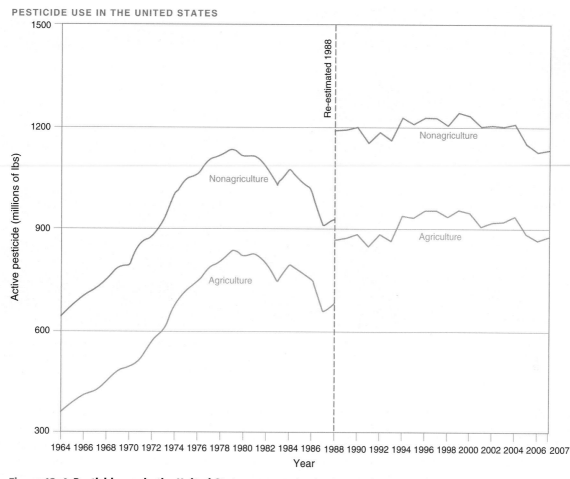

Figure 13–4 Pesticide use in the United States. Nonagricultural and agricultural uses of pesticides (active ingredients) are shown for the period 1964–2007. (*Source:* Data from the Environmental Protection Agency, Office of Pesticide Programs.)

immune system with a vaccine to produce long-lasting protection against any future invasion. Yet another is to prevent a patient from becoming ill in the first place by maintaining a healthy lifestyle and cutting down on activities that transmit disease (for example, by washing hands or avoiding sneezing on others). In practice, all of these means are often used to keep a particular pathogen under control.

The same basic philosophies govern the control of pests. Sometimes we use **chemical treatment**. Like the use of antibiotics and other medicines, chemical treatment seeks to eradicate or greatly lessen the numbers of the pest organism. Although it has had much success, this approach gives only short-term protection. Furthermore, the chemical often has side effects that are highly damaging to other organisms.

A different philosophy could be called **ecological pest control**. Like stimulating the body's immune system, or preventing pests by maintaining ecosystem health, this approach seeks to give long-lasting protection by developing control agents on the basis of knowledge of the pest's life cycle and of ecological relationships. Such agents, which may be other organisms or chemicals, work in one of two ways: Either they are highly specific for the pest species being fought, or they manipulate one or more aspects of the ecosystem.

Ecological control emphasizes the protection of people and domestic plants and animals from damage from pests rather than eradication of the pest organism. Thus, the benefits of pest control can be obtained while maintaining the integrity of the ecosystem.

These two philosophies are actually combined in the approach called **integrated pest management (IPM)**. IPM is an approach to controlling pest populations by using all suitable methods—chemical and ecological—in a way that brings about long-term management of pest populations and also has minimal environmental impact. This approach is increasing in usage, especially where pesticides are seen as undesirable because of health risks and in developing countries where the cost of pesticides is often prohibitive. We will discuss IPM further in Section 13.4.

13.2 Chemical Treatment: Promises and Problems

Pesticides are categorized according to the group of organisms they kill. There are insecticides (for insects), rodenticides (for mice and rats), fungicides (for fungi), herbicides (for plants),

STEWARDSHIP

The use of DDT for controlling malaria is one of those situations where stewardship decisions involve conflicting views of the common good. Malaria exacts a terrible toll in death and disease: 150 million to 275 million bouts of sickness and 660,000 deaths in the year 2011. The good news is that malaria deaths have dropped 25% since 2000, but the bad news is that we thought malaria would have been wiped out long ago. Fifty years ago, malaria seemed to be on the way out as new synthetic drugs killed the protozoan parasites that caused the disease and DDT wiped out mosquitoes everywhere. But resistance emerged in both parasites and mosquitoes, leaving us with fewer control options. DDT is one of a group of 12 persistent organic pollutants targeted for phasing out by the United Nations Environment Program, but it is still used against malarial mosquitoes in parts of the world because it is effective and inexpensive.

About 20 species of *Anopheles* mosquito carry malaria, which itself is caused by four different but related parasite species. Members of the *Anopheles* genus are night hunters, biting people as they rest or sleep. Many people in countries with malaria cannot afford bed nets to physically keep mosquitoes away. In about 11 countries, DDT is still the main line of defense against the mosquito, in part because it is inexpensive. The typical application involves spraying inside houses and eaves at a concentration of 2 g/m², once or twice a year. Even where

mosquitoes are resistant to DDT, it acts as a repellent or irritant that keeps the mosquitoes from staying around. In 2006, the WHO reversed a 30-year policy and officially put DDT on a par with bed nets and antimalarial drugs as major tools for controlling malaria.

One major downside of this approach is the difficulty of keeping DDT out of agricultural fields, where it can enter food chains and poison top predators. Another downside is the high exposure of some individuals when DDT is sprayed indoors. Alternative pesticides, such as synthetic pyrethroids, are already available, but resistance to these is appearing, too. An integrated public-health approach is needed

wherein mosquito breeding places are eliminated, bed nets are employed, and malaria cases are treated promptly to intercept the life cycle of the parasite. However, these steps are seldom all present in many developing-country locales.

There is no magic bullet. The story of DDT and malaria illustrates the history of problems with chemical pesticides: development of resistance by pests, resurgences and secondary-pest outbreaks, and adverse environmental and human health effects. These problems can be extremely widespread and persistent and have led to an increased need for alternative methods of pest control.

and so on. None of these chemicals, however, is entirely specific to the organisms it was designed to control; they can all pose hazards to other organisms, including humans.

Development of Chemical Pesticides and Their Successes

Finding effective materials to combat pests is an ongoing endeavor. The early substances (frequently referred to as **first-generation pesticides**) used to control agricultural pests included toxic heavy metals such as lead, arsenic, and mercury. Scientists now recognize that these substances may accumulate in soils, inhibit plant growth, and poison animals and humans. In addition, toxic heavy metals lose their effectiveness as pests become increasingly resistant to them. For example, in the early 1900s, citrus growers were able to kill 90% of injurious **scale insects** (minute insects that suck the juices from plant cells) by placing a tent over

an infested tree and piping in deadly cyanide gas for a short time. By 1930, this same technique killed as little as 3% of these pests.

The next step had its origins in the science of organic chemistry in the early 1800s. During the 19th century, chemists synthesized thousands of organic compounds, but for the most part, these compounds sat on shelves because uses for them had not yet been found. By the 1930s, however, with agriculture expanding to meet the needs of a rapidly increasing population and with first-generation pesticides failing, farmers needed something new. In time, second-generation pesticides, as they came to be called, were developed as a result of synthetic organic chemistry. One of the most famous is the pesticide DDT, which will illustrate many of the major ideas in the chapter.

The DDT Story. In the 1930s, a Swiss chemist named Paul Müller began systematically testing some organic chemicals

SOUND SCIENCE

Marine Fouling Organisms: Keeping One Step Ahead of the Barnacles

Finding a home is a rite of passage for marine plankton that settle from a free swimming stage to a sessile (attached, sedentary) form. Barnacles are a good example; they have early life stages moving about in currents and eventually change to a form that has to land on some type of structure and attach in order to continue developing. This means that solid surface area is often the limiting factor for the growth of marine communities. Every ship that plies the oceans is not only a container for human use, but a moving habitat of usable real estate for small marine organisms.

These "biofouling" organisms settle on the underwater parts of ships, piers, intake pipes for power plants, and often ruin them. On the hulls of ships, biofouling causes increased drag, which drastically increases the use of fuel by as much as 40%, raising both the economic cost and environmental impacts of shipping. Biofouling is estimated to cost the shipping industry $60 billion a year. Governments and industry spend billions to prevent it. Solutions aren't obvious.

Antifouling measures (the prevention or removal of fouling organisms) often involve biocides: some type of toxin embedded on the surface such as in a paint. Unfortunately, tributylyin (TBT), considered the most toxic pollutant ever deliberately released into the ocean, is also the most commonly used in antifouling. You might expect that something that kills barnacles and algae would be harmful to oysters and mussels, and it is. TBT is used on more than 70% of the world's vessels because it is effective, but a ban on TBT coatings on new ships makes finding

new solutions urgent. New research focuses on non-toxic coatings. Some are polymers, which create a low-friction environment that is hard to cling to. Others make it energetically difficult for organisms to attach. They are made of chemicals that have alternating positive and negative charges that repel organisms trying to attach. Other research focuses on mimicking biology. Whales often have barnacles on them, but sharks don't. It looks like the tiny points that cover sharks may keep fouling organisms off. So a biomimetic (biology mimicking) film with tiny points similar to those in sharkskin is in the works. This is a type of nanotechnology that looks very promising.

Fouling organisms cost a great deal and waste fuel, increasing climate change. But it isn't obvious how to solve the problem without toxins; at least, it hasn't been until new research has provided alternatives. Finding these new solutions can be the fun part of the job for some scientists.

Sources: Vietti, P. (2009, June 4). "New Hull Coatings Cut Fuel Use, Protect Environment." Office of Naval Research, http://www.eurekalert.org/pub_releases/2009-06/oonr-nhc060409.php Retrieved June 2012.

Evans, S. M., Leksono, T., & McKinnell, P. D. (1995, January), "Tributyltin pollution: A diminishing problem following legislation limiting the use of TBT-based anti-fouling paints." *Marine Pollution Bulletin, 30*(1), 14–21.

for their effect on insects. In 1938, he hit on the chemical dichlorodiphenyltrichloroethane (DDT), a chlorinated hydrocarbon that had first been synthesized some 50 years earlier. Just traces of DDT killed flies in Müller's laboratory.

DDT appeared to be nothing less than the long-sought "magic bullet," a chemical that was extremely toxic to insects and yet seemed nontoxic to humans and other mammals. It was inexpensive to produce, broad spectrum (effective against a multitude of insect pests), and persistent (it did not break down readily in the environment) and, hence, provided lasting protection. This last attribute lowered cost by eliminating both the material and the labor expense of repeated treatments.

In War . . . During World War II, DDT quickly became indispensable **(Fig. 13–5)**. The military used it to control body lice that spread typhus fever. As a result, World War II was one of the first wars in which fewer people died of typhus

Figure 13–5 DDT in war. During and after World War II, DDT was sprayed to kill vectors of typhus, malaria, and other diseases with great success. This image from 1945 shows a US soldier spraying DDT directly on a fellow soldier.

than of battle wounds. On the island of Saipan, DDT helped in the fight against dengue fever. The World Health Organization (WHO) of the United Nations used DDT throughout the tropical world to control mosquitoes and greatly reduced the number of deaths caused by malaria. There is little question that DDT saved millions of lives. In fact, the virtues of DDT were so outstanding that Müller was awarded the Nobel Prize for medicine in 1948 for his discovery.

And in Peace . . . Postwar uses of DDT expanded dramatically. The chemical was sprayed on forests to control spruce budworm, on salt marshes to kill nuisance mosquitoes, on suburbs to control the beetles that spread Dutch elm disease. DDT even proved highly effective in controlling agricultural insect pests. Indeed, DDT was so effective that many crop yields increased dramatically. Growers could ignore other, more painstaking methods of pest control such as crop rotation and the destruction of old crop residues. They could grow more productive but less resistant varieties. They could grow certain crops in a broader range of conditions. In short, DDT gave growers more options for growing the most economically productive crop.

The success of DDT led to the development of a great variety of synthetic organic pesticides. Currently, the U.S. Environmental Protection Agency (EPA) Pesticide Program regulates more than 18,000 pesticide products, most of which are synthetic organic pesticides. Examples and some characteristics of these insecticides and herbicides are listed in Table. 13–1.

Problems Stemming from Chemical Pesticide Use

Overuse of chemical pesticides has led to a wide range of environmental and human health problems. Human health effects are among the most concerning, especially when pesticides get into water bodies and ground water. People who work directly with pesticides, especially with inadequate safety gear, and children, whose developing bodies are heavily affected by toxins, are among the most vulnerable.

Adverse Human Health Effects. As with all toxic substances, pesticides can be responsible for both acute and chronic health effects.

Acute Effects. According to the American Association of Poison Control Centers, more than 86,419 persons called poison control centers after having been exposed to pesticides in the United States during 2010. More than 14,000 were treated in a health care facility, but only 121 died, a fact in part attributable to the system of poison control call centers throughout the United States, which collected this data. Most were farm workers or employees of pesticide companies who came in direct contact with the chemicals. Different pesticides affect different human systems: glyphosate poisoning, for example, can lead to nausea, abdominal pain, shock, and respiratory failure; pyrethrin poisoning leads to allergic reactions, seizures, pneumonia, and coma.

There is no global documentation of pesticide poisoning; however, the WHO has recently proposed standards for diagnosis and identification of acute pesticide poisoning cases. While the WHO has estimated 3 million serious poisoning episodes per year leading to hospitalization or death, numbers are far higher when detailed household surveys are taken, leading to estimates 10 times as high for all acute pesticide poisoning cases. Most of these cases occur in the developing countries, where regulations and training are often lax and information on pesticides may be poorly understood. The use of pesticides by untrained persons is considered to be the major cause of these poisonings **(Fig. 13–6)**, but in many cases, children and families come in contact with the

TABLE 13–1 Characteristics of Some Pesticides

Insecticide Class	Examples	Toxicity to Mammals	Persistence
Organophosphates	Parathion, malathion, phorate, chloropyrifos	High	Moderate (weeks)
Carbamates	Carbaryl, methomyl, aldicarb, aminocarb, carbofuran	High to moderate	Low (days)
Chlorinated hydrocarbons	DDT, toxaphene, dieldrin, chlordane, lindane	Relatively low	High (years)
Pyrethroids	Permethrin, bifenthrin, esfenvalerate, decamethrin	Low	Low (days)
Herbicide Class	**Examples**	**Effects on Plants**	
Triazines	Atrazine, simazine, cyanizine	Interfere with photosynthesis, especially in broadleaf plants	
Phenoxy	2-, 4-D; 2-, 4-, 5-T; methylchlorophenoxybutyrate (MCPB)	Cause hormonelike effects in actively growing tissue	
Acidamine	Alachlor, propachlor	Inhibit germination and early seedling growth	
Dinitroaniline	Trifluralin, oryzalin	Inhibit cells in roots and shoots; prevent germination	
Thiocarbamate	Ethylpropylthiolcarbamate (EPTC), cycloate, butylate	Inhibit germination, especially in grasses	
Phosponate	Glyphosate	Inhibit amino acid synthesis; kill a wide variety of plants	

Source: Private Pesticide Applicator Training Manual, 1993, U.S. Environmental Protection Agency, Region VIII.

Figure 13–6 Pesticides used unsafely. This farmer in Thailand is spraying pesticide on soybeans, without the benefit of protective gear.

pesticides through aerial spraying, the dumping of pesticide wastes, incorrect storage of pesticides in the home, or the use of pesticide containers to store drinking water.

Chronic Effects. Pesticides are applied to fields and orchards to reduce pest damage to crops. They are also used to protect harvested food so that it is brought undamaged to market. Because of the wide-ranging use of pesticides by most farmers, consumers are inevitably exposed to pesticide residues on their food and farmers are exposed to low levels even when they avoid acute exposure. Symptoms of pesticide poisoning (headaches, blurred vision, fainting, and nausea) are often similar to other illnesses, and exposure may not be diagnosed. The public-health concern is that pesticides might have chronic effects including the potential for causing cancer, as indicated by animal testing. In fact, evidence of carcinogenicity has been observed for many pesticides. Epidemiological evidence has implicated organochlorine pesticides in various cancers, including lymphoma and breast cancer.

Other chronic effects include dermatitis, neurological disorders including permanent brain and nerve damage, birth defects, and infertility. For example, male sterility has been clearly linked to dibromochloropropane (DBCP), once used to control nematodes and now banned. Two disorders recently associated with pesticide use are suppression of the immune system and disruption of the endocrine system. New epidemiological evidence suggests that long-term exposure to pesticides can trigger Parkinson's disease in some people. Workers in India and Russia exposed to pesticides experienced abnormally low white blood cell counts. (White blood cells are crucial elements of the immune system.)

Endocrine Disruption. Laboratory tests have shown that a number of pesticides, including atrazine and alachlor

(herbicides) and DDT, endosulfan, diazinon, and methoxychlor (insecticides), interfere with reproductive hormones. These **endocrine disruptors** are described at length elsewhere (Chapter 22). A rise in the incidence of breast cancer among humans and reports of abnormal sexual development in alligators, fish, and other animals have suggested the possibility that very low levels of a number of chemicals are able to mimic or disrupt the effects of estrogenic hormones (Chapter 10). Estrogneic hormones are sexual hormones that are highly potent at low concentrations. The Food Quality Protection Act (FQPA) of 1996 directed the EPA to develop procedures for testing chemicals for endocrine disruption activity, which it has done through the Endocrine Disruptor Screening Program (EDSP).

Adverse Environmental Effects. The story of DDT is a good illustration of the environmental impacts of pesticides as well.

There Go the Birds. In the 1950s and 1960s, ornithologists (people who study birds) observed drastic declines in populations of many species of birds that fed at the tops of food chains. Fish-eating birds such as the bald eagle and osprey (Fig. 13–7) were close to extinction. Investigators at the U.S. Fish and Wildlife National Research Center near Baltimore, Maryland, showed that the problem was reproductive failure. The eggs were breaking in the nest before hatching. Furthermore, the eggs contained high concentrations of dichlorodiphenyldichloroethylene (DDE), a product of the partial breakdown of DDT by the animal's body. DDE interferes with calcium metabolism, causing birds to lay thin-shelled eggs.

Because of accumulation, small, seemingly harmless amounts received over a long period of time may reach toxic levels in one organism. This phenomenon—referred to as **bioaccumulation**—can be understood as follows: Many synthetic organic chemicals are highly soluble in lipids (fats or fatty compounds), but less soluble in water. In the body, they become stored in lipids rather than being excreted by

Figure 13–7 Osprey. Populations of fish-eating birds such as the osprey (shown here feeding its young), brown pelican, and bald eagle were decimated in the 1950s and 1960s by the effects of widespread spraying with DDT. With the banning of the pesticide, these populations have greatly recovered. The bald eagle was taken off the endangered species list in 1994.

the kidneys. Thus, synthetic organics like pesticides and their breakdown products that are absorbed with food or water are trapped and held by the body's lipids, while the water and water-soluble wastes are passed in the urine. The body cannot fully metabolize them and has no mechanism to excrete them so they gradually accumulate in the body and may produce toxic effects.

Up the Chain. Bioaccumulation, which occurs in the individual organism, may be compounded at higher trophic levels (Chapter 5). Each organism accumulates the contamination from its food, so the concentration of contaminant in its body is many times higher than that in its food. This effect concentrates contaminants as they go up the food chain. Eventually, this multiplying effect, **biomagnification**, causes extremely high levels of contaminants in the top levels of the biomass pyramid (Chapter 5). **Figure 13–8** shows how DDT and its metabolic products worked their way up the food chain to the top predators.

Silent Spring. In the 1950s, Rachel Carson, a U.S. Fish and Wildlife Service biologist and an accomplished science writer, began reading the disturbing scientific accounts of the effects of DDT and other pesticides on wildlife. By 1962, Carson finished a book, *Silent Spring*, that documented the effects of the almost uncontrolled use of insecticides across the United States. Its basic message was that, if insecticide use continues as usual, there might someday come a spring with no birds—and with ominous consequences for humans as well.

Silent Spring became an instant best seller and triggered a debate that has not ended. Representatives of the agricultural and chemical industries claimed that the book was an unreasonable and unscientific account that, if taken seriously, would halt human progress. At the same time, however, the book was hailed as an unparalleled break-

through in environmental understanding. Eventually, concerns about environmental and long-term health effects led to the banning of DDT in the United States and most other industrialized countries in the early 1970s. Numerous other chlorinated hydrocarbon pesticides (for example, chlordane, dieldrin, endrin, and heptachlor) were also banned because of their propensity for bioaccumulation in the environment and suspected potential for causing cancer. In the years since the banning of DDT, the bird species that were adversely affected have recovered. Rachel Carson is credited with stimulating the start of the modern environmental movement and the creation of the EPA. *Silent Spring* has become a classic, and the regulation of insecticides and other toxic chemicals is in a sense a monument to Rachel Carson, who died of cancer only two years after her book was published.

Development of Resistance by Pests. One fundamental problem is that chemical pesticides gradually lose their effectiveness. Over the years, it becomes necessary to use larger and larger quantities, to try new and more potent pesticides, or to do both to obtain the same degree of control. Synthetic organic pesticides fared no better than first-generation pesticides in this respect. For example, in 1946, 2.2 pounds (1 kg) of pesticides provided enough protection to produce about 60,000 bushels of corn. By 1971, it took 141 pounds (64 kg) to produce the same amount, and losses due to pests actually *increased* during the intervening years. Resistance is not limited to agricultural pests. Stored products pests are resistant to a fumigant used in shipments, and many medical and veterinary pests have developed pesticide resistance as well. In 2011, poison resistant "supermice" were discovered in Spain and Germany. They appeared to

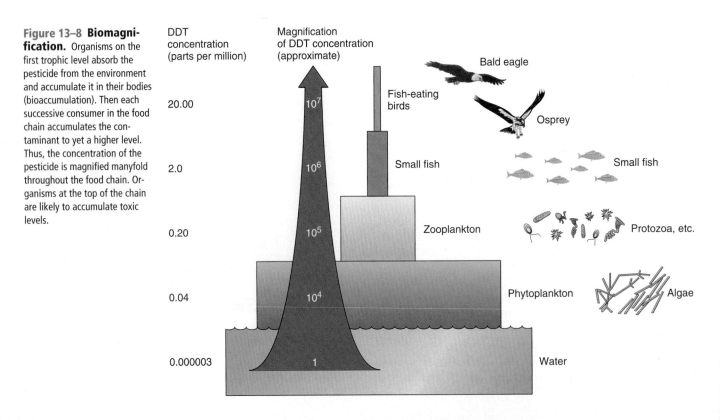

Figure 13–8 Biomagnification. Organisms on the first trophic level absorb the pesticide from the environment and accumulate it in their bodies (bioaccumulation). Then each successive consumer in the food chain accumulates the contaminant to yet a higher level. Thus, the concentration of the pesticide is magnified manyfold throughout the food chain. Organisms at the top of the chain are likely to accumulate toxic levels.

DDT concentration (parts per million)	Magnification of DDT concentration (approximate)	
20.00	10^7	Bald eagle / Fish-eating birds / Osprey
2.0	10^6	Small fish
0.20	10^5	Zooplankton / Protozoa, etc.
0.04	10^4	Phytoplankton / Algae
0.000003	1	Water

have resulted from the breeding of mice from Algeria and continental Europe.

Evolution at Work. Resistance builds up because pesticides destroy the sensitive individuals of a pest population, leaving behind only those few that already have some resistance to the pesticide. Resistance develops most rapidly in in *r*-strategists with high reproductive capacity (Chapter 4). A single pair of houseflies, for example, can produce several hundred offspring that may mature and reproduce themselves only two weeks later. Consequently, repeated pesticide applications result in the unwitting selection and breeding of genetic lines that are highly, if not totally, resistant to the chemicals that were designed to eliminate them. The Colorado potato beetle, for example, developed resistance to 52 compounds across all chemical insecticide classes, including cyanide, in 50 years. This is the same type of process that has resulted in drug-resistant diseases. In short, pesticide resistance simply confirms that natural selection is a powerful evolutionary force. Many major pest species are resistant to all of the principal pesticides.

Resurgences and Secondary-Pest Outbreaks. The second problem with the use of synthetic organic pesticides is that, after a pest has been virtually eliminated with a pesticide, the pest population not only recovers, but also explodes to higher and more severe levels. This phenomenon is known as **resurgence**. To make matters worse, small populations of insects that were previously of no concern because of their low numbers may suddenly start to explode, creating new problems. This phenomenon is called a secondary-pest outbreak, such as mirid bugs on cotton (Chapter 12). Unfortunately, the species appearing in secondary-pest outbreaks quickly became resistant to pesticides, thus compounding the problem.

Some scientists use the term **pesticide treadmill** to describe attempts to eradicate pests with synthetic organic chemicals. Overuse of chemicals increases resistance and secondary-pest outbreaks, which lead to the use of new and larger quantities of chemicals, which in turn lead to more resistance and more secondary-pest outbreaks. The same can occur on weeds or stored product pests. The use of glyphosate on GM crops and the resistance of weeds to this powerful herbicide is a good example (Chapter 12). The chemical approach breaks down because it ignores basic ecological principles. It assumes that the ecosystem is a static entity in which one species, the pest, can simply be eliminated. In reality, the ecosystem is a dynamic system of interactions, and a chemical assault on one species perturbs the system and produces other, undesirable effects. Populations of plant-eating insects are frequently held in check by other insects that parasitize or prey on them **(Fig. 13–9)**. Pesticide treatments

Insect food chains

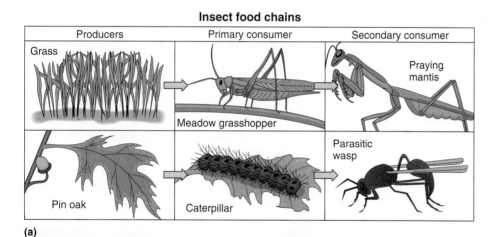

(a)

Figure 13–9 Insect food chains.
(a) Food chains exist among insects, just as they do among higher animals. (b) Ant lion (on right) capturing an ant.

(b)

often have a greater impact on these natural enemies than on the plant-eating insects they are meant to control. Consequently, with their natural enemies suppressed, the populations of both the original target pest and other plant-eating insects explode.

Nonpersistent Pesticides: Are They the Answer? A key characteristic of chlorinated hydrocarbons is their persistence; that is, they take a long time to break down. Lingering in the environment for years, they can contaminate many organisms and become biomagnified. DDT, for example, has a half-life on the order of 20 years. Because the persistent pesticides have been banned, the agrochemical industry has substituted nonpersistent pesticides for the banned compounds. The synthetic organic phosphates (for example, malathion, parathion, and chlorpyrifos), as well as carbamates such as aldicarb and carbaryl, have been used extensively in place of chlorinated hydrocarbons (see Table 13–1). These compounds are potent inhibitors of the enzyme cholinesterase, which is essential for proper functioning of the nervous system in all animals. They break down into simple nontoxic products within a few weeks after their application. Thus, there is no danger of their migrating long distances through the environment and affecting wildlife or humans long after being applied.

Toxicity. For several reasons, however, nonpersistent pesticides are not as environmentally sound as they might appear. First of all, they are persistent enough to harm nontarget organisms. The total environmental impact of a pesticide is a function not only of its persistence, but also of its toxicity, its dosage, and the location where it is applied. Many of the nonpersistent pesticides are far more toxic than DDT. They present a significant hazard to agricultural workers and others exposed to these pesticides. For instance, the organophosphates are responsible for an estimated 70% of all pesticide poisonings. The law requires the EPA to develop new health-based standards that address the risk of children's exposure to such pesticides, which has led to the banning of any use of several chemicals from a range of home garden and agricultural uses.

Bird Kill. Nonpersistent pesticides may still have far-reaching environmental impacts. For example, carbofuran is a pesticide widely used on alfalfa, rice, and soybeans. It is spread on the ground and absorbed into plants through their roots. Unfortunately, it is highly toxic to birds and other vertebrates. In the '80s, The EPA estimated that 1 million to 2 million birds were killed annually by carbofuran. In one incident, 800 to 1,200 ducks and geese died after treatment of an alfalfa field with carbofuran. Because of its toxicity to humans and wildlife, the EPA announced in 2008 that it would no longer allow residues of carbofuran on any food, essentially revoking all uses. The EPA decision was based primarily on the grounds that the pesticide posed an unacceptable risk to toddlers.

Another hazard of nonpersistent pesticides is that desirable insects may be just as sensitive to them as pest insects are. Bees, for example, which play an essential role in pol-

lination, are highly sensitive to them. Thus, the use of these compounds creates an economic problem for beekeepers and jeopardizes pollination. Also, regular spraying of neighborhoods with malathion to control mosquitoes leads inevitably to great declines in butterflies and fireflies. Many pesticides also harm predatory insects and spiders, which would otherwise keep pests down.

Finally, nonpersistent chemicals are just as likely to cause resurgences and secondary-pest outbreaks as persistent pesticides are, and pests become resistant to nonpersistent pesticides just as readily.

Synergistic Effects. As scientists try to figure out what is happening in ecosystems, the answer may be complicated by multiple factors that work together to create an unexpected outcome, something called a **synergistic effect**. A possible example is the mystery of what is killing pollinators. Pollinators are in decline worldwide. Some of the deaths are a phenomenon known as honey bee colony collapse disorder (CCD), which came into the news in 2006–2007 when beekeepers discovered their hives dying and their worker bees mysteriously gone **(Fig. 13–10)**. Meanwhile, other pollinators such as bumblebees and various types of solitary bees are also in decline. Scientists know CCD is not caused by a simple case of acute pesticide poisoning because in the case of acute poisonings, worker bees are found dead at the hive. Instead, in CCD cases, the worker bees are nowhere to be found. However, the answer might involve a mixture of pests, pathogens, and chemicals that weaken or confuse bees.

Some hypotheses are that several parasitic and pathogenic factors may be involved, including the invasive varroa mite, viruses, a gut parasite, poor nutrition, and stress. Pesticides may interact with these other factors. Several recent pesticide studies on pollinators support the involvement of chemicals in their decline. In one study on honeybees, researchers found that the bees exposed to nectar with a type of pesticide called a neonicotiniod had more

Figure 13–10 Pollinators in decline. Scientists are working to understand why pollinators are in decline. Colony collapse disorder may have different causes than other pollinator die-offs. These bumble bees died of unknown causes.

difficulty finding their way back to the hive than those exposed to a control. In another study, bumblebees treated with the same type of pesticide had lower growth rates and produced fewer queens than unexposed bees. In yet another study, honeybees were found to eat less food and perform fewer communication dances when exposed to pesticides. In answering the mystery of colony collapse disorder and other problems with pollinators, scientists are finding that chemicals may contribute to many other problems, causing a perfect storm of variables.

13.3 Alternative Pest Control Methods

Numerous ecological and biological factors affect the relationship between a pest and its host. *Ecological control* seeks to manipulate one or more of these natural factors so that crops and people are protected without jeopardizing environmental and human health. This natural approach, unlike the chemical treatment approach, depends on an understanding of the pest and its relationship with its host and with its ecosystem. The more we know about the organisms involved, the greater are our opportunities for ecological control.

The life cycle typical of moths and butterflies is shown in **Figure 13–11**. Many groups of insects have a similarly complex life cycle. The development of each stage may be influenced by numerous abiotic factors, and at each stage, the insect may be vulnerable to attack by a parasite or predator. The proper completion of each stage depends on internal chemical signals provided by insect hormones. Locating mates, finding food, and other behaviors depend on external chemical signals. All these findings suggest ways in which pest populations may be controlled without resorting to synthetic chemical pesticides.

Four general categories of ecological pest control are (1) **cultural control**, (2) control by **natural enemies**, (3) **genetic control**, and (4) **natural chemical control**. We will consider each of these in turn.

Cultural Control

A cultural control is a nonchemical alteration of one or more environmental factors in such a way that the pest finds the environment unsuitable or is unable to gain access to its target. This type of control can be practiced on all scales—by a homeowner, a farmer, or even a government.

Control of Pests Affecting Crops, Gardens, and Lawns.
Weed problems in lawns are frequently a result of cutting the grass too short. If grass is left at least 3 inches (8 cm) high, it will usually maintain a dense enough cover to keep out most crabgrass and other noxious weeds. Thus, monitoring the

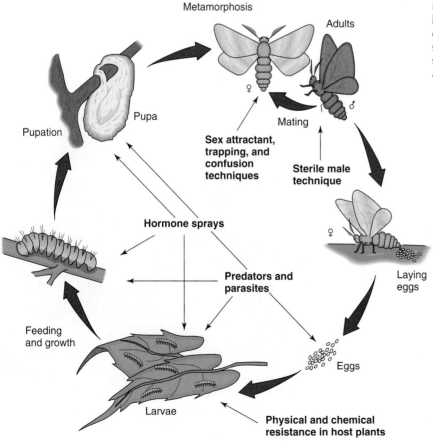

Figure 13–11 Complex life cycle of insects. Like the moth shown here, most insects have a complex life cycle that includes a larval stage and an adult stage. Biological control methods recognize the different stages and attack the insect, using knowledge of its needs and life cycle.

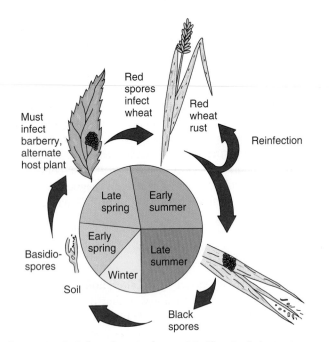

Figure 13–12 Cultural control. Part of the life cycle of wheat rust, a parasitic fungus that is a serious pest on wheat, requires that the rust infest barberry, an alternate host plant. The elimination of barberry in wheat-growing regions has been an important cultural control.

height of grass is a form of cultural weed control. A gardener may also control pests by avoiding plants that act as attractants or by planting ones that act as repellants for certain pests. Roses are attractants for common pests, while marigolds and chrysanthemums are justly famous for repelling insects.

Some parasites require an alternative host, so they can be controlled by eliminating that host **(Fig. 13–12)**. Also, hedgerows, fencerows, and shelterbelts can provide refuges where natural enemies of pests (birds, amphibians, praying mantises, and so on) can be maintained.

Crops. Managing any crop residues that are not harvested is important. Spores of plant disease organisms and insects may overwinter or complete part of their life cycle in the dead leaves, stems, or other plant residues that remain in the fields after harvesting. Plowing under or burning the material may be quite effective in keeping pest populations to a minimum. In gardens, a clean mulch of material such as grass clippings or hay will keep down the growth of weeds and protect the soil from drying and erosion. Crop rotation—the practice of changing crops from one year to the next—may provide control because pests of the first crop cannot feed on the second crop **(Fig. 13–13)** and vice versa. Similarly, breaking up monocultures, and planting mixed plantings, lowers the crop loss to any one pest. Finally, refuge plants (untreated plants near treated crops) allow natural enemies to survive. Natural enemies of pests are maintained in the uncultivated strips. In fact, this approach is used with insect-resistant GM crops as well to keep all of the insects from being exposed to the GM crop and becoming more resistant to the crop (Chapter 12).

Stopping Imports. Most of the pests that are hardest to control were unwittingly imported from other parts of the

world. Therefore, it is important to keep would-be pests out of the country. This is a major function of U.S. Customs and Border Protection and of the agriculture departments of some states. Biological materials that may carry pest insects or pathogens are either prohibited from crossing the border or subjected to quarantines, fumigation, or other treatments to ensure that they are free of pests. A recent example is the Asian longhorned beetle, which arrived in packing crates from China. The beetle infests healthy maple, birch, poplar, and ash trees and eventually kills them. As a result of this pest, new regulations require that wooden crates and pallets from China be kiln-dried or chemically treated. The Animal and Plant Health Inspections Service (APHIS) of the USDA has the job of tracking and stopping the flow of pests into the United States and between states. APHIS is described further in Section 13.5.

Cultural Control of Pests Affecting Humans. We routinely practice many forms of cultural control against diseases and parasitic organisms. For instance, disposing properly of sewage and avoiding drinking water from unsafe sources are cultural practices that protect against waterborne disease-causing organisms. Brushing hair, bathing, laundering clothes and linens, and cleaning rugs and furniture are all cultural practices that eliminate head and body lice, fleas, bedbugs, and other parasites. Special covers for mattresses can cut down on bedbugs and mites. Garbage disposal, housekeeping, and sealing cracks in buildings help keep down populations of roaches, mice, rats, flies, mosquitoes, and other pests. Sanitation in handling food as well as refrigeration, freezing, canning, and drying of foods are cultural controls that inhibit the growth of organisms that cause rotting, spoilage, and food poisoning.

If these practices are compromised, as they usually are in any major disaster, there is the very real danger of additional

Figure 13–13 Tomato wilt. This tomato plant has been devastated by tomato wilt. Growing tomatoes in the same plot maintains wilt spores in the soil, guaranteeing that wilt will strike the plants every year.

(*Source:* Author's garden.)

widespread mortality resulting from outbreaks of parasites and diseases, elements of environmental health discussed elsewhere (Chapter 17).

Control by Natural Enemies

There are four well-defined types of natural enemies: predators, parasitoids, pathogens, and herbivores. This type of control is used all over the world. In the eastern United States, tiny predator beetles have been imported in an attempt to control the (also imported) hemlock wooly adelgid, which is killing off hemlock trees. In Africa, mealybugs are controlled by parasitic wasps; these types of wasps are also effective in controlling caterpillar populations **(Fig. 13–14)**. In sub-Saharan Africa, the swarming desert locust was controlled by spraying "Green Muscle," a mix of dry spores of the fungus *Metarhizium anisopliae* with oil, on infested fields. Water hyacinth, which has blanketed many African lakes, is coming under control following the introduction of Brazilian weevils. More than 30 weed species

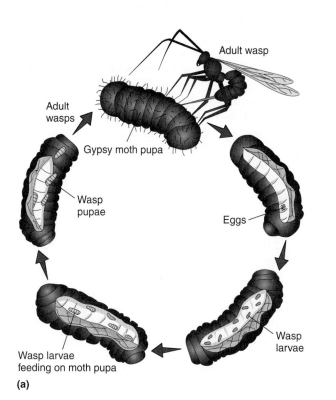

(a)

(b)

(c)

Figure 13–14 Parasitic wasps. (a) The life cycle of the parasitic wasp that uses the gypsy moth as its host. (b) A wasp depositing eggs in a gypsy moth larva. (c) Another insect parasite, a braconid wasp, lays its eggs on the pest known as the tomato hornworm, which is the larva of the sphinx moth. The wasp larvae feed on the caterpillar, and shortly before the caterpillar dies, they emerge and form the cocoons seen here.

worldwide are now limited by insects introduced into their habitats.

Protect the Natives. The problem with using natural enemies lies in finding organisms that provide control of the target species without attacking other, desirable species. Entomologists estimate that of the 50,000 known species of plant-eating insects that have the potential for being serious pests, only about 1% actually are. The populations of the other 99% are held in check by one or more natural enemies. Therefore, the first step in using natural enemies for control should be *conservation*—protecting the natural enemies that already exist. This means avoiding the use of broad-spectrum chemical pesticides, which may affect natural enemies of the target species even more than they do the target pests themselves. Restricting the use of broad-spectrum chemical pesticides will often allow natural enemies to reestablish themselves and control *secondary pests*—those that become serious problems only after the use of pesticides.

Import Aliens as a Last Resort. However, effective natural enemies are not always readily available. In many cases, the absence of a pest's natural enemies is the result of accidentally importing the pest without also importing its natural enemies. Quite often, effective natural enemies have been found by systematically combing the home region of an introduced pest and finding its various predators or parasites. Identifying the enemy species is vital, and the species must be carefully tested before intentionally releasing it, something that has not always been done well. The cane toad **(Fig. 13–15)**, imported into Australia in 1935 to control beetles in sugar cane fields, overran Queensland, eating, outcompeting, and poisoning many native species without ever controlling the beetles that prompted its original release. On the other hand, the cassava mealybug, *Phaenococcus manihoti*, ran rampant across Africa, destroying as much as 50% of a crop that was a staple for 200 million people, until it was successfully controlled by the introduced parasitoid wasp

Epidinocarsis lopezi, Modern releases are much more likely to succeed than earlier releases because of better science and more rigorous controls.

Plant Breeding

Most plant-eating insects and plant pathogens attack only one species or a few closely related species. This specificity implies a genetic incompatibility between the pest and any species that are not attacked. Most genetic-control strategies are designed to develop genetic traits in the host species that provide the same incompatibility—that is, resistance to attack by the pest. The technique has been used extensively in connection with plant diseases caused by fungal, viral, and bacterial parasites. For example, in the years 1845–1847, the potato crop in Ireland was devastated by late blight, a fungal parasite. Nearly a million people starved, and another million people emigrated to escape the same fate. The potato blight is still considered one of the worst pests of all time. Nowadays, protection against such disasters is provided in large part by growing varieties of potato that are resistant to the blight. Unfortunately, the battle between agricultural scientists and plant diseases never ends; before the 1950s, wheat stem rust used to devastate wheat fields until breeders developed wheat varieties resistant to the rust. However, a new strain of the fungus appeared in Africa in the early 2000s and is spreading eastward into Asia, infecting many varieties of wheat in the developing world.

Other plants are bred to produce substances that are lethal or at least repulsive to the would-be pest. One example **(Fig. 13–16)** is the resistance of new varieties of wheat to the Hessian fly. The fly lays its eggs on wheat leaves, and the larvae move into the main stem, weakening it. Arriving in the straw bedding of Hessian soldiers during the Revolutionary War and eventually spreading throughout much of the Midwest, the fly caused widespread devastation until scientists at the University of Kansas developed a variety of wheat that produces a chemical that is toxic to the insect; this reduced

Figure 13–15 Cane toad. Introduced into Australia to control beetles in sugar cane fields, the cane toad (*Bufo marinus*) quickly became a pest itself.

Figure 13–16 Hessian fly. This pest of wheat is controlled by maintaining wheat plants that produce a chemical that is toxic to the fly.

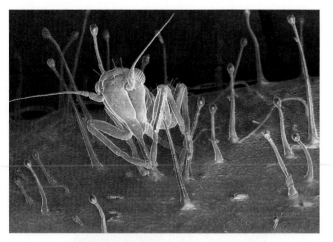

Figure 13–17 Alfalfa glandular hairs. This scanning electron micrograph shows an immature potato leafhopper trapped by sticky glandular hairs on the stem of a resistant alfalfa strain.

Figure 13–18 Bioengineered potatoes. a researcher holds two plants of a potato cultivar Fortuna, in Limburgerhof, Germany. The left plant is not genetically modified, while the plant on the right is resistant to *Phytophora infestans*, a plant pest.

losses to this pest to less than 1% where resistant varieties have been planted for several years.

Plants can be bred to have physical traits that provide physical barriers and structural traits that impede the attack of a pest. For example, hooked hairs on the leaf surfaces of some plants tend to trap and hold insects and, glandular hairs that exude a sticky substance fatally entrap others (**Fig. 13–17**).

Biotechnology. As we have seen elsewhere, biotechnology techniques can be used to incorporate pest control into plants (Chapter 12). One promising strategy is to incorporate the protein coat of a plant virus into the plant itself. When the plant "expresses" (that is, manufactures) the virus's protein coat, the plant becomes resistant to infection by the real virus. In this way, crop plants have been made resistant to more than a dozen plant viruses. Another strategy is to craft a chemical that interferes with an insect's normal cell function, silencing a gene that controls molting and sidetracking that process—thus, a gene-silencing pesticide.

Bt. Endowing plants with resistance to insects has been engineered by incorporating a potent protein produced by a bacterium, *Bacillus thuringiensis* (Bt). This protein kills the larvae of a number of plant-eating insects and is harmless to mammals, birds, and most other insects. Scientists have engineered the gene into a number of plants, including cotton, potatoes, and corn (**Fig. 13–18**). Unfortunately, resistance to Bt by some pests and outbreaks of secondary pests are dampening the success of Bt cotton (Chapter 12). Anticipating the development of resistance because of the heavy exposure of pests to the Bt toxin, the EPA has required seed companies to verify that farmers plant 20% of their fields with non-Bt versions of the crop. The non-Bt fields provide a refuge for pests that are susceptible to the Bt toxin, allowing for these insects to interbreed with potentially resistant colleagues from the Bt acreage and, hopefully, dilute resistance genes in the insect populations. Bt toxin is also used directly as a natural chemical control, described later.

Genetic Control of Pests

Another genetic-control strategy involves flooding a natural population with sterile males that have been reared in laboratories. Combating the screwworm fly provides a prime illustration. This fly, which is closely related to the housefly and looks much like it, lays its eggs in open wounds of cattle and other animals, causing pain, infection, and often death. Early in the 20th century, the problem became so severe that cattle ranching from Texas to Florida and northward was becoming economically impossible.

Screwworm flies have two essential features that make control possible: their populations are never very large, and female flies mate only once. If the female mates with a sterile male, no offspring are produced. Today sterile males are routinely used to control this pest. After huge numbers of screwworm larvae are grown in laboratories, the resulting pupae are subjected to just enough high-energy radiation to render them sterile. These sterilized pupae are then air-dropped into the infested area. Ideally, 100 sterile males are dropped for every normal female in the natural population, giving a 99% probability that wild females will mate with one of the sterile males. The technique proved so successful that it eliminated the screwworm fly from Florida in 1958–1959 and was subsequently employed to eradicate the screwworm fly in Mexico and most of Central America. The sterile-insect technique has saved billions of dollars for the cattle industry and has been used successfully on a number of agriculturally important pests.

The sterile male technique, along with spraying, has been used to eradicate the Mediterranean fruitfly, a citrus pest that has repeatedly invaded the United States and been repelled from all but Hawaii. In 1981, a serious infestation in California got out of hand, destroying millions of dollars of crops. The infestation was finally squelched with aggressive aerial spraying, the quarantining of fruits and vegetables by National Guard checkpoints, and sterile male releases. More recently, the technique was used to eradicate the tsetse fly from the island of Zanzibar, thus eliminating the disease trypano-somiasis (sleeping sickness), which the tsetse fly carries.

Natural Chemical Control

Natural chemicals can be compounds that are used directly as biocides (such as Bt toxin), pesticides made from plants, chemicals used to mimic communication chemicals, or chemicals that boost immunity.

Some natural chemicals mimic those used to communicate internally or between members of a species. **Hormones** are chemicals produced in organisms that provide "signals" that control developmental processes and metabolic functions. Some organisms also communicate via **pheromones**—chemicals secreted by one individual that influence the behavior of another individual of the same species. Hormones or hormone mimics can be used to disrupt the life cycle of a pest. Two advantages of such natural chemicals are that they are nontoxic and that they are highly specific to the pest in question. (They do not affect the natural enemies of the pest to any appreciable extent.) If the affected pest is eaten by another organism, it is simply digested. For example, caterpillar pupation is triggered by a decrease in the level of the chemical called juvenile hormone. If this chemical is sprayed on caterpillars, pupation does not occur. The larvae simply continue to feed and grow, become grossly oversized, and eventually die. One insecticide, called Mimic®, is a synthetic variation of ecdysone, the molting hormone of insects. Mimic® triggers the molting process to begin in insect larvae but not to continue, which eventually kills the larva. Mimic is specific to moths and butterflies and is viewed as a potent agent in the control of the spruce budworm, the gypsy moth, the beet armyworm, and the codling moth—all highly devastating pests.

Pheromones, produced by insects to attract a mate, can be used in pest control as well. Once identified and synthesized, pheromones may be used in either the trapping technique or the confusion technique. In the trapping technique, the pheromone is used to lure males or females into traps or into eating poisoned bait **(Fig. 13–19)**. In the confusion technique, the pheromone is dispersed over the field in such quantities that males become confused, cannot find females, and thus fail to mate. Research has identified a large number of pheromones, and these natural chemicals are now key tools employed in IPM on many insects, such as the boll weevil and the pink bollworm for cotton, the codling moth for pears and apples, the tomato pinworm for tomatoes, and several bark beetles that infest forest trees, among others.

Natural chemicals can also be used on crops to boost plant defenses. When plants (including tomato, maize, sweet pepper, and wheat) are grown from seeds dipped in jasmonic acid, a naturally occurring plant compound, aphid attacks and caterpillar damage decline. These early results may lead to promising new directions. The spraying of Bt toxin into water or on crops rather than incorporation into genetically modified plants is another example of natural chemical control.

Figure 13–19 Trapping technique. Adult Japanese beetles are lured into a trap by scented bait they find attractive. Once inside the trap, baffles prevent the beetles from escaping. These traps are available in hardware stores and are quite effective, though some users find that they are too effective, luring beetles to their yards and gardens without actually catching them. (Japanese beetles are notorious lawn grubs as larvae; as adults, they are voracious consumers of garden flowers and vegetables.)

13.4 Making a Coherent plan

The increasing availability of alternative pest control methods and the accumulating evidence of the failures of chemical treatment have led to a growing movement toward avoiding the use of pesticides, particularly on foods. This movement must contend with a strong tendency on the part of farmers to employ pesticides.

Pressures to Use Pesticides

A species becomes a pest only when its population multiplies to the point of causing significant damage. Natural controls are generally aimed at keeping pest populations below damaging levels, not at total eradication of the populations. By keeping pest populations down, natural controls avert significant damage while preserving the integrity of the ecosystem.

Therefore, the question to be asked when facing any pest species is: Is the species causing significant damage? Damage should be deemed significant only when the economic losses due to the damage considerably outweigh the cost of applying a pesticide. This point is called the **economic threshold (Fig. 13–20)**. If significant damage is not occurring, natural controls are already operating, and the situation is probably best left as is. Spraying with synthetic chemicals at this stage is more than likely to upset the natural balance and make the situation worse through resurgences. It is also not cost effective. If, however, significant damage *is* occurring, a pesticide treatment may be in order. A grower who believes that his or her plantings are at risk is likely to resort to **insurance spraying** (the use of pesticides to prevent losses to pests) such as occurs on European apple crops, where the most important threats are mildew and scab—airborne diseases that,

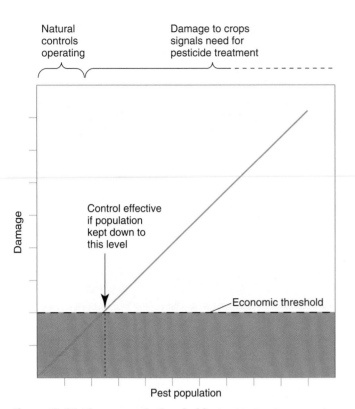

Natural controls operating

Damage to crops signals need for pesticide treatment

Damage

Control effective if population kept down to this level

Economic threshold

Pest population

Figure 13–20 The economic threshold. The objective of pest control should not be to eradicate a pest totally. All that is needed is to keep population levels below the economic threshold.

once established, cause significant economic damage. Farmers typically employ insurance spraying to control these diseases.

Consumers also put indirect pressure on growers to use pesticides because they select the best-looking fruits and vegetables, leaving the remainder to be sold at lower prices or trashed. Growers know that blemished produce means less profit, so they indulge in **cosmetic spraying**—the use of pesticides to control pests that harm only the item's outward appearance. Cosmetic spraying accounts for a significant fraction of pesticide use, does nothing to enhance crop yield or nutritional value, and results in an increase in pesticide residues remaining on the produce.

Integrated Pest Management

IPM aims to minimize the use of synthetic organic pesticides without jeopardizing crops. This is made possible by addressing all the interacting sociological, economic, and ecological factors involved in protecting crops. With IPM, the crop and its pests are seen as part of a dynamic ecosystem. The goal is not eradicating pests, but maintaining crop damage below the economic threshold. The EPA uses a four-tiered approach in describing IPM:

1. **Set action thresholds.** With the economic threshold in mind, IPM identifies a point at which pest populations or environmental conditions indicate that some control action is needed.

2. **Monitor and identify pests.** Pest populations are monitored, often by persons employed by local agricultural

extension services or farm cooperatives or by persons acting as independent consultants. These **field scouts** are trained in identifying and monitoring pest populations (traps baited with pheromones are used for this purpose) and in determining whether the population exceeds the economic threshold. Monitoring may be done by government agencies across large areas as well.

3. **Prevent pests.** Cultural and biological control practices form the core of IPM techniques. Practices such as crop rotation, polyculture (instead of monoculture), the destruction of crop residues, the maintenance of predator populations, refuge plants, and carefully timed planting and fertilizing are basic to IPM. "Trap crops" are often used. That is, early strips of a crop such as cotton are planted to lure existing pests and then the pests are destroyed by hand or by the limited use of pesticides before they can reproduce. For aquatic disease vectors such as mosquitoes, getting rid of breeding sites was once a common cultural technique that fell into disuse when pesticides became prevalent. But removing old tires, containers, water-filled ruts, and other breeding grounds is important. For rodents, making sure food and garbage are not available is part of prevention.

4. **Control pests.** If the preceding steps indicate that pest control is needed in spite of the preventive methods, measures are taken to decrease the pest population. Here selected brands of pesticides may be used in quantities that do the least damage to the natural enemies of the pest.

Making the economic benefits of natural controls known to growers is an important aspect of IPM. Because pesticides increased yields and profits when they were first used, many farmers still cling to them, believing that they offer the only way to bring in a profitable crop. The rising costs of pesticides, along with their tendency to aggravate pest problems, have eliminated their economic advantage. Pest-loss insurance, which pays the farmer in the event of loss due to pests, has enabled insured growers in developed countries to refrain from unnecessary and costly "insurance spraying."

Malarial mosquitoes provide a good example of the value of research in IPM. Control now consists of a combination of cultural controls such as pesticide impregnated nets and removal of mosquito breeding sites like puddles or water-filled containers along with chemical sprays, Bt, and encouragement of natural enemies like birds and predatory insects. There is hope that current research will yield even more solutions. In the case of pesticide-resistant mosquitoes, scientists have found that a pathogenic fungus can kill mosquito larvae. Other research has shown that infection with a bacterium that is found in many insects doesn't kill malarial mosquitoes but keeps them from passing along the malarial parasite as effectively. Recently, scientists have been able to rear large numbers of mosquitoes in a lab environment cheaply in order to try the sterile male technique. In other research, scientists have found a virus that kills older mosquitos. Only older mosquitoes can pass on malaria. Consequently, killing them as adults but before they are ready to transmit the disease, rather than constantly killing the

Figure 13–21 **Brown plant hoppers.** Immature brown plant hoppers (*Nilaparvata lugens*), shown on the stem of a rice plant.

youngest stages, will cut malarial transmission and at the same time slow the development of pesticide resistance.

Example: Indonesia. Particularly in the developing world, governmental agricultural policy often determines the extent to which IPM is adopted. Governments and aid agencies usually subsidize the purchase of pesticides, thereby strongly encouraging growers to step onto the pesticide treadmill. By contrast, Indonesia has provided a viable IPM model for other rice-growing countries. Faced with declining success in controlling the brown plant hopper **(Fig. 13–21)**, Indonesian rice growers have switched from heavy pesticide spraying to a light spraying regime that preserves the natural enemies of the insect. The success of the program can be traced to close cooperation between the Indonesian government and the FAO **(Fig. 13–22)**. FAO

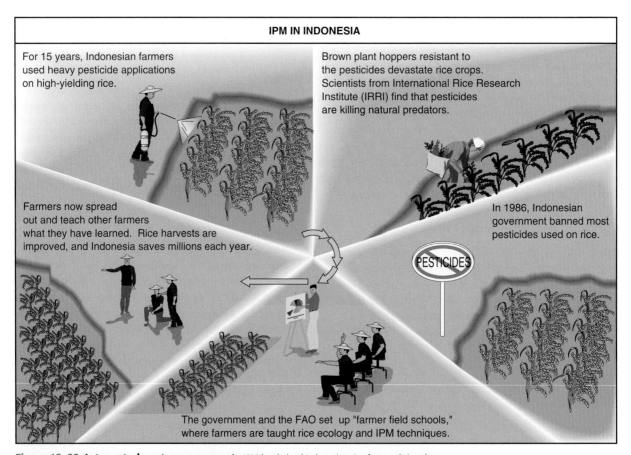

IPM IN INDONESIA

For 15 years, Indonesian farmers used heavy pesticide applications on high-yielding rice.

Brown plant hoppers resistant to the pesticides devastate rice crops. Scientists from International Rice Research Institute (IRRI) find that pesticides are killing natural predators.

Farmers now spread out and teach other farmers what they have learned. Rice harvests are improved, and Indonesia saves millions each year.

In 1986, Indonesian government banned most pesticides used on rice.

PESTICIDES

The government and the FAO set up "farmer field schools," where farmers are taught rice ecology and IPM techniques.

Figure 13–22 **Integrated pest management.** IPM has helped Indonesian rice farmers bring the brown plant hopper under control after years of frustration on the pesticide treadmill.

workers conducted training sessions for farmers that lasted throughout the growing season. Eventually, more than 200,000 farmers were taught about the rice agro-ecosystem and IPM techniques, and these farmers formed a corps that could in turn teach others.

The economic and environmental benefits of the Indonesian IPM program have been remarkable. The government has saved millions of dollars annually by not purchasing pesticides, farmers have not had to invest in pesticides and spraying equipment, the environment has been spared the application of thousands of tons of pesticide, fish are once again thriving in the rice paddies, and the health benefits of reduced pesticide use have been spread from applicators to consumers and wildlife. The FAO is sponsoring similar IPM training programs in many other developing countries.

Organically Grown Food

For consumers concerned about pesticides in their food, **organic food** is an attractive option. Walk into any supermarket and you will find counters of fresh vegetables, meats, prepared foods, and Fair Trade foods, all bearing the label of *organic*. Many farmers are turning away from the use of pesticides, chemical fertilizers, antibiotics, and hormones as they raise grains, vegetables, and livestock for the organic food trade. Typically, organic farms are small, employ traditional farming methods with diverse crops, and are tied to local economies. For the farmers, the organically raised crops often represent lower crop yields, but also lower expenses. Sales of organics have increased rapidly over the past decade, and organically grown food is now a $26 billion enterprise in the United States (up from $1 billion in 1990) and more than $55 billion worldwide. Organic foods are more costly, but recent tests have shown that these foods are far less likely to have any pesticide residues, and those that do have residues contain much lower amounts than conventionally grown foods. Most residues in organic foods came from older, persistent pesticides, such as DDT, lingering in the soil.

USDA Organic. The certification of organic foods has been a contentious process, and eventually, the industry requested help from the government. In response, in 1990 Congress passed the **Organic Foods Protection Act**, which established the National Organic Standards Board (NOSB), under USDA auspices. After many years of work, proposed guidelines for certification from the NOSB were aired in early 1998. Possibly in response to lobbying from the food industry, the board allowed genetically engineered foods, irradiated foods (foods subjected to radiation to kill bacteria), and crops fertilized with sewage sludge to be included as organic. A firestorm of protest from the organic-food community ensued. The NOSB went back to work, and in late 2002, the USDA published its final standards for certifying organic foods. As a result, no product involving genetically engineered or irradiated foods or foods fertilized with sewage sludge can be certified as 100% organic. The standards

Figure 13–23 Organic foods. This symbol on food signals that the food or crop has been produced on a certified organic farm.

also prohibit the use of conventional pesticides, antibiotics, growth hormones, and chemical fertilizers. To market certified organic foods, farmers now must have their operations scrutinized by USDA-approved inspectors. Only then may they use the USDA seal **(Fig. 13–23)** on their products. If the food is at least 95% organic, it may also display the seal, but may not claim to be 100% organic. Because it is expensive and relatively difficult to be certified organic, many farmers are not certified who use significantly less pesticides and artificial fertilizers. In addition, some organic farms are extremely large and distant, so the environmental costs of transportation, storage, and packaging of foods remain. For this reason, buying local and supporting less toxic farming benefit the environment even if organic certification is not available.

13.5 Pests, Pesticides, and Policy

When you go to the grocery store and purchase food, most likely you assume it will have minimal pesticide residues. You might even think, "Well, they wouldn't sell it if it wasn't safe." Likewise, when you go to a plant nursery and buy a plant, you probably assume that they wouldn't sell it if it was a noxious weed pest that was likely to invade your state and cost farmers millions of dollars. Both of these assumptions are at least partially true. While the fact that something is sold doesn't guarantee its safety, the United States has a wide range of guidelines followed by state and federal agencies to monitor, eradicate, and prevent pests and to protect food safety. There are also international treaties designed to lower the impact of pests and prevent the misuse of dangerous pesticides. Let's look at regulation of pests and then pesticides.

Regulation of Pests

In the United States, Animal and Plant Health Inspection Service (APHIS), a part of the USDA, was established in 1972 to monitor and control pests that could harm plants and animals. The jobs of the agency have changed over time. It has consolidated jobs formerly done by multiple agencies, including the Bureau of Animal Husbandry and various plant bureaus. It includes regulation of animals used in research and exhibition and by dealers, a function created after the Animal Welfare Act of 1966. One of the jobs of APHIS is to monitor potential pests in order to act quickly and prevent infestations that would require more pesticide use later. One example is its intensive monitoring of soybean rust when it first entered the United States. The agency has a continual Web-based tracking system with records of all counties where the pest was discovered. APHIS organized governmental agencies, universities, and farmers to track the pest, using "sentinel plots": small fields planted early that were rigorously surveyed to track fungal spread. This and other actions were effective, and the environment was different enough from its native Brazil that the fungus has never reached the epidemic proportions here that it has there.

Plant Protection Act. Plants have been regulated in the United States since the 1912 Plant Quarantine Act, the 1957 Federal Plant Pest Act, and the Federal Noxious Weed Act of 1974. These functions were consolidated into the the Plant Protection Act of 2000, which, among other things, allowed APHIS to crack down on illegal smuggling of plants and agricultural products that could harbor pests.

One program APHIS runs is the noxious weed program. Plants designated as noxious weeds cannot be moved across state lines, imported, or exported. States often have their own noxious weed lists and may have additional regulations. Plants not on the noxious weed list may also be pests. APHIS also tracks animal diseases and plant pests and regulates imports and exports to prevent movement. In 2012, APHIS was carefully regulating the movement of ash tree products and firewood to allow interstate commerce but halt the spread of the emerald ash borer, a juggling act that makes shipping difficult.

Diseases, Bioterrorism, and GM Organisms. Over the years, APHIS' role has expanded to include monitoring for pests that might attack wildlife and the monitoring of bioengineered species. It regulates the welfare of domestic animals and protects U.S. plant and animal agriculture from incoming disease. One APHIS program tracks the efforts to eradicate diseases such as cattle brucellosis, bovine tuberculosis, and pseudorabies in swine. APHIS also tracks plant diseases. In 2003, APHIS personnel were able to detect a bacterial disease of geraniums in greenhouses in four midwestern states. A devastating disease that could ruin potato crops as well as tomatoes and other plants, this particular bacterium had not been detected in the United States since 1999. APHIS agents were able to determine that the disease had been imported in geranium cuttings from Kenya.

Geranium imports from Kenya were halted until the outbreak could be contained.

APHIS also oversees the Agricultural Bioterrorism Protection Act of 2002, which regulates laboratories and vaccine companies that use, posess, or transport toxins and agents that could be harmful to humans, plants, or animals. The act aims to prevent the release of devastating diseases from a research lab or medical facility. APHIS also has the job of assessing the safety of bioengineered organisms. In 2012, it completed asessments of the possible effects of glyphosphate-resistant sugar beets. The assessments are held open to public review and then APHIS makes a regulatory determination.

U.S. federal agenies collaborate with other countries to prevent the spread of pest organisms. The International Plant Protection Convention (IPPC) ratified by the United States in 1972 set up an international system to quarantine products to prevent pest spread. One provision of the 1995 United Nations Convention on the Law of the Sea (UNCLOS) was to prevent the introduction of marine species to new environments. Other international agreements such as the Agreement concerning Cooperation in the Quarantine of Plants and their Protection against Pests and Diseases of 1959 and the Agreement on the Application of Sanitary and Phytosanitary Measures (SPS Agreement) of 1994 also are attempts by the international community to prevent the spread of plant diseases and other pests.

Regulation of Pesticides

Regulation of pesticides focuses on three main areas: safety and testing of pesticides, protection of pesticide workers, and protection of consumers of products pesticides may be on. As is the case with regulation of potential pests, there are numerous historic federal regulations about pesticides, which have been pulled together into modern laws, and there are international treaties.

Federal Regulations. The Federal Insecticide, Fungicide, and Rodenticide Act (FIFRA), established in 1947 and amended in 1972, is administered by the EPA, which has total jurisdiction over the manufacture, sale, use, and testing of pesticides. FIFRA requires manufacturers to register pesticides with the EPA. As part of registration, it must test for toxicity to animals (and, by extrapolation, to humans) and set usage standards. For example, chlordane, which was used for 40 years as a pesticide on crops, lawns, and gardens, was tested and shown to be acutely toxic to animals in high doses. Chronic exposure led to damaging effects on most vital systems and caused liver cancer in animals. Acute and chronic exposure of humans to chlordane showed similar results, except for the development of cancer. On the basis of these results, the EPA canceled all uses of chlordane for food crops in 1978, but allowed continued use against termites until 1988, when all uses of the chemical in the United States were banned. The law also stipulates that the EPA must determine whether a product "generally causes unreasonable adverse effects on the environment," and if it so determines, the product may be restricted or banned. This was the basis for the banning of DDT.

The EPA's Pesticide Program places high priority on registering "reduced-risk" pesticides, especially those that promise to replace more toxic chemicals. Many are **biopesticides**—naturally occurring substances that control pests. These include **microbial** pesticides like Bt, **plant-incorporated protectants** (in GM crops) such as corn bioengineered to manufacture Bt in its tissues, or biochemical substances that are naturally occurring, like pheromones (e.g., Mimic). These pesticides often clear the registration process within a year, compared with the typical three years a conventional pesticide might take. The EPA also approves pesticide labels that must meet certain criteria and protect pesticide workers. The agency issues requirements for safety training, protective equipment, emergency assistance, and permitting standards. The states must then demonstrate to the EPA that, on the basis of the federal regulation, they have adequate regulation and enforcement mechanisms.

FFDCA. Protecting consumers from exposure to pesticides on food was a contentious and confusing process for many years. Three agencies are involved: the EPA, the Food and Drug Administration (FDA), and the Food Safety and Inspection Service of the USDA. Basically, the EPA sets the allowable tolerances for pesticide residues, based on its assessment of toxicity, and the FDA monitors and enforces the tolerances on all foods except meat, poultry, and egg products, which are managed by the USDA.

The original enabling act was the Federal Food, Drug, and Cosmetic Act (FFDCA) of 1938. For many years, one clause of the FFDCA, the so-called **Delaney clause**, was highly controversial. The clause stated, "No [food] additive shall be deemed to be safe if it is found to induce cancer when ingested by man or animal." Because pesticides represent one of the largest categories of toxic chemicals to which people are exposed, this clause prohibited many pesticides from being used on foodstuffs when those pesticides had been found to cause cancer in laboratory tests with animals. In essence, the law stated that, if a given pesticide presents *any* risk of cancer, *no detectable residue* may remain on the food. Consequently, as analytical chemical techniques became more and more sensitive, more and more pesticides were banned for food use.

FQPA. The National Research Council examined this dilemma and recommended in 1987 that the anticancer clause be replaced with a "negligible risk" standard, based on risk-analysis procedures that had been developed subsequent to the Delaney clause. After years of debate, Congress passed the **Food Quality Protection Act (FQPA)** in 1996 and did away with the Delaney clause and amended both FIFRA and FFDCA. The major requirements of the act include:

Setting "a reasonable certainty of no harm" as the safety standard for substances applied to foods; giving special consideration to the exposure of young children to pesticide residues; defining a risk of more than one case of cancer per million people as the allowable upper limit over long periods of exposure to chemicals; requiring evaluation of all possible sources of exposure to a given pesticide, not just exposure

from food; reassessing older products according to the new requirements within 10 years; and applying the same standards to raw and processed foods (previously regulated separately).

Care for Kids. One of the emphases in the new legislation was protection of children. The focus is on children both because they typically consume more fruits and vegetables per unit of body weight than do adults and because studies have shown that children are frequently more susceptible to carcinogens and neurotoxins. Accordingly, the FQPA requires the EPA to add a tenfold safety factor in assessing children's risks from pesticides.

Government agencies have to work together to monitor food because the system is so complex. The USDA Pesticide Data Program (PDP) provides annual analyses of pesticides in fresh and processed foods. The EPA evaluates toxicity data on individual pesticides and then, using the USDA's PDP analyses and other information on risk assessment, sets limits on the amounts of a pesticide that can remain in or on foods—called tolerances (Chapter 17). To date, the EPA is on target with the directives of the FQPA, having published tolerance reassessments for more than 9,637 pesticide products (out of a total of 9,721). More than 4000 tolerances were rejected or had to be amended. For an example of a "tolerance," a less toxic organophosphate, malathion, is listed at 8 ppm tolerance for many crops, meaning that no more than 8 ppm of malathion may be found on or in the product. For milk, however, the tolerance is set at 0.5 ppm, an indication of milk's special place in children's diets.

International Pesticide Regulation

U.S. laws and international treaties both regulate the way the United States relates to the rest of the world regarding pesticides. The United States currently exports thousands of tons of pesticides (active ingredients) each year, some of which are banned in the United States itself. A number of factors—low literacy, weak laws, unfamiliarity with equipment—increase the potential for serious exposure to toxic pesticides in the developing countries. FIFRA requires "informed permission" prior to the shipment of any pesticides banned in the United States. This permission comes from the purchaser. The EPA must then notify the government as to the identity of the importing company. Internationally, there are treaties that regulate information about pesticides.

Treaties and Guidance. Two international treaties and one guidance document provide a framework for pesticides in the international arena. The Rotterdam Convention was adopted in 1998 and went into effect in February 2004. It is an international treaty ratified by 70 countries that promotes open exchange of information about hazardous chemicals including pesticides. Central to the convention is the process of **prior informed consent (PIC)**, whereby exporting countries inform all potential importing countries of actions they have taken to ban or restrict the use of pesticides or other toxic chemicals and label them clearly. Governments in the importing countries, UN agencies, and exporting countries communicate so everyone can make decisions with the information they need.

The *International Code of Conduct on the Distribution and Use of Pesticides* is a worldwide guidance document that addresses conditions of safe pesticide use. The most recent version was approved by the FAO in November 2002. The code makes clear the responsibilities of both private companies and countries receiving pesticides in promoting their safe use. In 2004, the **Stockholm convention**, a treaty designed to eliminate or restrict persistent organic pollutants, came into effect. By 2011, 178 countries had signed. Countries are expected to ban nine of the 12 most dangerous chemicals (the "dirty dozen"), use DDT only for malaria control, and limit the accidental production of two pesticide breakdown products, dioxins and furans.

The Future. These international efforts are leveling the safety field for pesticide use globally and pulling the worst chemicals out of the global stream, at least slowly. Pesticide use illustrates two trends we see in many environmental science issues: the problems of cumulative impacts and of unintended consequences (Chapter 1). As with many environmental issues, significant progress in both pest and pesticide control has benefited from grassroots action and pressure from public-interest groups and NGOs, such as the Consumers Union and the Pesticide Action Network. These and other groups are pressing for continued movement in the direction of safe and sustainable pest prevention and management. Care for the welfare of both fellow human beings and all other creatures is necessary to solve pest problems in a fashion that is sustainable in the long term.

REVISITING THE THEMES

Sound Science

It was sound science that uncovered the connections between DDT and the decline of many predatory birds—biomagnification was a great surprise to everyone. Some amazing applications of sound science are seen in the imaginative work on alternative pest controls, such as sterile male technique, the use of pheromones to control pests, studies that have uncovered natural enemies of pests, research into a wide range of ways to control malaria-carrying mosquitoes, and other parts of integrated pest management. Agricultural scientists are constantly challenged to keep up with new strains of plant pests by developing resistant varieties of the crops they attack.

Sustainability

Ecological control is a sustainable approach to controlling pests, while the chemical treatment approach guarantees that pesticides will need to be used indefinitely. Pest resistance, resurgences, and secondary outbreaks have been the norm. The ecological control methods provide the means to achieve control of most pests in a sustainable way, especially when they are controlled by natural enemies. IPM presents a reasonable alternative for managing pests

by limiting the use of pesticides. The organic food movement is proof that food can be grown without chemicals and, for consumers and the environment, is a great way to keep pesticides out of the natural world (and the refrigerator). Reducing or eliminating the use of pesticides and keeping food safe and healthy are two of the pillars of the sustainable-agriculture movement. (See Chapter 11.)

Stewardship

The widespread use of DDT during and just after World War II saved millions of lives. However, the continued, often indiscriminate, use of this persistent pesticide and others like it led to trouble in the environment, and the careful work of Rachel Carson in documenting this trouble in the 1960s was a landmark in environmental stewardship. Yet the continued use of DDT to control malaria-bearing mosquitoes shows that discerning the common good is not always easy. Because stewardship also includes human welfare and health, the focus of the federal and international regulations on the protection of children and agricultural workers is an example of stewardship. It can be argued that the work of the EPA in complying with federal regulation is stewardship in action. The agency is an aggressive promoter of reduced use of the more toxic pesticides.

REVIEW QUESTIONS

1. Define *pests*. Why do we control them? Describe at least four types.

2. Discuss two basic philosophies of pest control. How effective are they?

3. Explain the pros and cons of using the synthetic organic pesticide DDT against malarial mosquitoes.

4. What adverse environmental and human health effects can occur as a result of pesticide use?

5. Define *bioaccumulation* and *biomagnification*.

6. Why are nonpersistent pesticides not as environmentally sound as first thought?

7. Describe the four categories of natural, or biological, pest control. Cite examples of each and discuss their effectiveness.

8. Define the term *economic threshold* as it relates to pest control. What are cosmetic spraying and insurance spraying?

9. What are the four steps in IPM? Explain how IPM worked in Indonesia.

10. What is organic food, and how is it now certified?

11. How does APHIS help control pests in the United States? What international regulations help control the movement of pests across borders?

12. How does FIFRA attempt to control pesticides? What new perspective does the FQPA bring to the policy scene?

13. Describe how the Stockholm Convention, Code of Conduct, and Rotterdam Convention help regulate international use of pesticides.

THINKING ENVIRONMENTALLY

1. Consider the ethics of the United States' exportation of pesticides. A large portion of this exported material is banned in the United States itself. If the choice was up to you to continue this or put it to an end, what would you do? Remember that such pesticides have at least a positive immediate benefit to these countries. Are they a necessary evil?

2. Almost one-third of the chemical pesticides bought in the United States are for use in houses and on gardens and lawns. What should manufacturers and users do to ensure the limited and prudent use of pesticides?

3. Should the government give farmers economic incentives to switch from pesticide use to IPM? How might that work?

4. Investigate how bugs and weeds are controlled on your campus. Are IPM techniques being used? Organic fertilizers? If not, consider lobbying for their use.

5. Imagine you're the mayor of a mid-sized suburban town where mosquitoes invade from neighboring marshes. What kind of restrictions would you place on local pesticide use? Defend your answer.

MAKING A DIFFERENCE

1. When you plant a garden, select hardy varieties of plants that are known to flourish in your particular region. Pests are less likely to destroy these types because they grow much faster than they can be eaten, even without pesticides.

2. Try planting some herbs in your garden because they often have many pest-oriented benefits. They are very likely to attract the good, predator bugs that will naturally take out the pests. Sage, basil, chives, yarrow, cilantro, and garlic are all great choices.

3. Another way to bring in the predatory insects is to buy them. A great number of companies now breed and sell species such as praying mantises, lacewings, and ladybugs, all of which help get rid of your pest problems.

4. To help keep a pesticide-free lawn looking green, let the grass grow a little longer than usual. This way it will need less water and food. Select a mower that mulches grass and leaves; lawn clippings provide nitrogen, and fallen leaves are great for conditioning soil.

MasteringEnvironmentalScience®

Go to **www.masteringenvironmentalscience.com** for practice quizzes, Pearson eText, videos, current events, and more.

PART FOUR

Harnessing Energy for Human Societies

Power station in a major industrial area of Shanghai, China.

HOW ESSENTIAL ARE ENERGY SOURCES and services to human societies? Industrialized countries are so completely dependent on fossil fuels and nuclear energy that we cannot imagine an economy apart from these energy sources. Developing countries are following the same example, hoping to enhance the well-being of their people. They have a long way to go; more than 1.5 billion people lack access to electricity, and another billion have unreliable service. Our pursuit of energy sources and services exacts an underappreciated toll in human life and well-being. The processes of energy exploration, extraction, transportation and power generation carry risks that often result in accidental human deaths. Even greater fatalities and loss of health come from the air pollution that accompanies fossil fuel use. In terms of both accidents and pollution, fossil fuels are the most dangerous form of energy generation.

THERE IS MORE BAD NEWS. A recent analysis of the world's oil reserves suggests that oil production may peak within the next decade and then begin a decline that will drive prices up to new heights. Will we invade wildlife refuges and fragile coastal waters to withdraw every last drop of oil from the ground? Looming in the background is the specter of global climate change and the struggle to deal with this threat, which is tied so thoroughly to the use of fossil fuels. Then there are the direct costs of energy and the drag placed on economic recovery and growth by rising fuel costs. It wasn't so long ago that gasoline was selling for less than $2 a gallon. Added to this are the hidden costs of subsidies paid to fossil fuel companies, estimated at $550 billion a year globally.

THERE IS GOOD NEWS, TOO. Wind turbines are appearing all over the landscape on every continent. Solar panels adorn rooftops, and solar "farms" are harvesting the Sun's limitless energy. Biofuels like ethanol are powering millions of vehicles. Regulations are mandating major increases in vehicle fuel economy, and new standards are making energy use far more efficient. Renewable energy sources are on the move, already supplying 10% of electrical generation in the United States and 20% of the global electricity supply.

CAN WE CHART a path to a secure energy future? One approach is "business as usual"—pump oil, drive gas-guzzling vehicles, gouge mountains for coal, and lace the countryside with natural-gas wells and pipelines—but there are severe consequences to this approach. An alternative—the sustainable and more secure future—is to focus our energy policy on greater energy efficiency and renewable energy. A 2011 *USA Today*/Gallup poll showed that 83% of Americans favored energy legislation that provides incentives for using alternative energy sources; it was at the top of a list of eight actions Congress could take. The challenge is there, and the choice is ours to make. Part Four describes what is going on in this important area.

CHAPTER 14

Energy from Fossil Fuels

Explosion on the Deepwater Horizon rig. On April 20, 2010, a cement seal on the Macondo well ruptured and sent oil and gas gushing up on the rig above, followed by an explosion and fire that killed 11 workers. Fire boats are shown spraying water the next day on the burning remnants of the oil rig.

APPROXIMATELY 300 MILLION YEARS AGO, the Gulf of Mexico was formed by a sinking of the sea floor. In subsequent geological eras, organic matter accumulated in sediments that were deposited in the Gulf, where it was buried and converted into oil and gas. Today, oil and gas deposits in the Gulf are thought to contain up to 45 billion **barrels of oil equivalent (BOE)**, a way of combining oil and gas estimates, where 6,000 cubic feet of gas is equated with 1 barrel of oil. Petroleum exploration began in the Gulf in earnest after World War II, and 1947 marked the first offshore well drilled out of the sight of land, under 20 feet of water. Since then, some 15,000 oil and gas wells have been drilled in the Gulf of Mexico, in depths of up to 10,000 feet below the surface and a further 13,000 feet below the sea floor.

Oil and gas production in the Gulf has been prolific, currently accounting for 30% of U.S. yield. However, more than 45 billion BOE have already been withdrawn from the Gulf, and the industry has been forced to drill in deeper waters and deeper strata to find oil, as shallower wells have been exhausted. To get at the deep oil and gas, increasingly expensive and elaborate rigs are required, platforms that float on the surface and support all the necessary equipment for drilling in deep water, housing a crew, and providing a helicopter pad. One such rig

LEARNING OBJECTIVES

1. **Energy Sources and Uses: Explain how the three kinds of fossil fuel were developed over time and how they are coupled to major end uses today.**

2. **Exploiting Crude Oil: Evaluate the changes in crude oil sources and prices and their impact on the U.S. economy, and examine the concept of peak oil and its consequences.**

3. **Drilling for Natural Gas: Describe the versatility of natural gas, and explain the new source in shale and how it is extracted.**

4. **Mining Coal: Discuss the advantages and disadvantages of coal as a power source, and evaluate the methods used to mine coal.**

5. **Energy Policy and Security: Compare supply-side and demand-side energy policies that have been developed in recent years, and explain their different impacts on the future.**

was *Deepwater Horizon*; measuring 250 by 400 feet, the rig had a crew of 126 on April 20, 2010.

THE BLOWOUT AT MACONDO. Drilling at the BP-owned Macondo Well was initially begun in October 2009 by another rig, and subsequently BP moved the *Deepwater Horizon* to the site in February 2010. The new rig was successful in completing the drilling and on April 20 had inserted a cement barrier over the well surface preliminary to abandoning the well for the time being. The barrier is essential and must withstand high pressures because the oil and gas accessed by the boring are themselves under high pressure. Suddenly, the well began blowing seawater and drilling mud up the string of connected drill casings to the rig platform, forming a geyser 240 feet into the air. The drilling mud was followed by a slushy combination of mud, oil, and gas, which ignited and exploded on the rig. The explosion and subsequent fire on the rig killed 11 workers and injured many more before they could be evacuated. The rig continued to burn until, 36 hours later, it sank into the Gulf.

Under pressure from below, oil began to gush out of the well on the Gulf floor (Fig. 14–1) and continued to do so for 87 days until the well was capped off. Close to 5 million barrels of oil flowed up into the Gulf of Mexico, making an awful mess—forming a huge slick, contaminating some 625 miles of Gulf coastline, and killing countless sea turtles, birds, fish, and smaller marine organisms. It was the largest accidental oil spill in world history; the only spill greater than Deepwater Horizon was the intentional release of oil into the Persian Gulf by Iraqi forces during the first Gulf War in 1991. An intense cleanup began immediately; much oil on the surface was simply burned, sending tons of black soot into the atmosphere. Some oil was pumped up from the bottom in the vicinity of the well, some was siphoned by booms and boats on the surface, but the vast bulk of the spill either evaporated at the surface of the Gulf or was broken down by bacteria in the water column and along the shores. These natural forces were aided by the high volatility of the Gulf oil and the warm temperatures of Gulf water. Nevertheless, the Deepwater Horizon spill was an economic disaster for the Gulf Coast fishing and tourism industries during the spring and summer of 2010. Much of this loss is being covered by a $20 billion trust fund created by BP at the demand of President Barack Obama.

WHAT WENT WRONG? The immediate cause of the blowout was a failure of the cement at the base of the well that was designed to contain the gas and oil within the well pipes. Two official government reports documented the numerous causes of the accident, ultimately putting the blame on BP and its subcontractors (Transocean,

Figure 14–1 Oil on the Gulf floor. Oil gushed up from the ruptured Macondo oil well at a rate of 68,000 barrels per day and continued until the well was successfully capped 87 days later.

Halliburton).[1] Why the cement failed is not known, but the responses of the crew on board the drilling rig suggested that human error as well as equipment failure both played roles in the sequence of events that followed failure of the cement barrier. In the end, BP and the subcontractors were blamed for allowing a culture of poor risk management in the drilling operation, made worse by shortcuts taken to complete the well in a hurry because it was behind schedule and millions of dollars over budget. The legal battles over the blowout and oil spill will last for years and will add to the billions of dollars already committed to the cleanup. The government reports recommended a number of new regulations that have since been put in place, designed to lower the risks of such a catastrophic failure.

Ultimately, the environmental consequences of oil spills, drilling operations, and such operations as mountaintop coal mining are some of the many costs of doing business in a fossil fuel–driven economy. Americans use almost 19 million barrels of oil every day. In future chapters, you will learn some of the other costs, such as air pollution, which is affecting human and environmental health, and rising greenhouse gases, which are leading to global climate change.

The objective in this chapter is to gain an overall understanding of the sources of fossil fuel energy that support transportation, homes, and industries. The consequences of our dependence on fossil fuels are discussed, and we close the chapter with a serious look at public policy.

---

[1]Bureau of Ocean Energy Management, Regulation and Enforcement. *Report Regarding the Causes of the April 20, 2010 Macondo Well Blowout.* September 14, 2011. www.boemre.gov/pdfs/maps/dwhfinal.pdf. October 5, 2010; National Commission on the BP Deepwater Horizon Oil Spill and Offshore Drilling. *Deep Water: The Gulf Oil Disaster and the Future of Offshore Drilling.* Report to the President, January 2011. www.fdlp.gov. October 5, 2011.

14.1 Energy Sources and Uses

Throughout human history, the advance of technological civilization has been tied to the development of energy sources. In early times (and even today in less developed regions), the major energy source was muscle power. Some people lived in relative luxury by exploiting the labor of others—slaves, indentured servants, and minimally paid workers. Human labor was supplemented by the use of domestic animals for agriculture and transportation, by water or wind power for milling grain, and by the Sun.

By the early 1700s, inventors had already designed many kinds of machinery. The limiting factor was a continual power source to run them. The breakthrough that launched the Industrial Revolution was the development of the steam engine in the late 1700s. In a steam engine, water is boiled in a closed vessel to produce high-pressure steam, which pushes a piston back and forth in a cylinder. Through its connection to a crankshaft, the piston turns the drive wheel of the machinery. Steam engines rapidly became the power source for steamships, steam shovels, steam tractors, steam locomotives, and stationary engines to run sawmills, textile mills, and virtually all other industrial plants.

The Three Fuels: Coal, Oil, and Gas

At first, the major fuel for steam engines was firewood. Then, as demands for energy increased and firewood around industrial centers became scarce, coal was substituted. By the end of the 1800s, coal had become the dominant fuel, and it remained so into the 1940s. In addition to being used as fuel for steam engines, coal was widely used for heating, cooking, and industrial processes. In the 1920s, coal provided 80% of all energy used in the United States.

Although coal and steam engines powered the Industrial Revolution, which greatly improved life for most people,

there were many drawbacks. The smoke and fumes from the numerous coal fires made air pollution in cities far worse than anything seen today. Coal is also notoriously hazardous to mine and dirty to handle, and burning coal results in large quantities of ash that must be removed. As for steam engines, because of the size and bulk of the boiler, the engines were heavy and awkward to operate (**Fig. 14–2**). Frequently, the fire had to be started several hours before the engine was put into operation in order to heat the boiler sufficiently.

In the late 1800s, the simultaneous development of three technologies—the internal combustion engine, oil-well drilling, and the refinement of crude oil into gasoline and other liquid fuels—combined to provide an alternative to steam power. The replacement of coal-fired steam engines and furnaces with petroleum-fueled engines and oil furnaces was an immense step forward in convenience. Also, the air quality improved greatly as cities were gradually rid of the smoke and soot from burning coal. (It was only in the 1960s, with the tremendous proliferation of cars, that pollution from gasoline engines became a problem.) Further, the gasoline-driven internal combustion engine provided a valuable power-to-weight advantage that allowed rapid advances in technology. A 100-horsepower gasoline engine weighs only a tiny fraction of what a 100-horsepower steam engine and its boiler weigh, and jet engines have an even greater power-to-weight ratio. Automobiles and other forms of ground transportation would be cumbersome, to say the least, without this power-to-weight advantage, and airplanes would be impossible.

Economic development throughout the world since the 1940s has depended largely on technologies that consume fuels refined from crude oil. By 1951, crude oil became the dominant energy source for the nation. Since then, it has become the mainstay not only of the United States (making up 37% of total U.S. energy consumption), but also of most other countries, both developed and developing. Crude

Figure 14–2 Steam power. Steam-driven tractors like this 1915 Case marked the beginning of the industrialization of agriculture. Today, tractors half the size do 10 times the work and are much easier and safer to operate.

Figure 14–3 Global primary energy supply. The world consumed some 12,275 million tons of oil equivalent in 2011. The sources of the total primary energy supply are shown here. Note that 87% comes from fossil fuels. (*Source:* Data from BP Statistical Review of World Energy, 2012.)

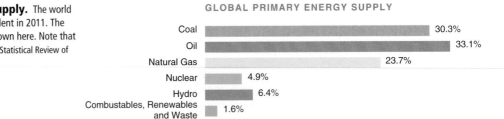

GLOBAL PRIMARY ENERGY SUPPLY

Coal	30.3%
Oil	33.1%
Natural Gas	23.7%
Nuclear	4.9%
Hydro	6.4%
Combustables, Renewables and Waste	1.6%

oil currently provides about 33% of the total annual global energy consumption (**Fig. 14–3**). In the countries of eastern Europe and in China, coal is still the dominant fuel. Coal ranks second globally, at 30%, and provides 22% of U.S. demand.

Natural gas, the third primary fossil fuel, is often found in association with oil or during the search for oil. Natural gas is largely methane, which produces only carbon dioxide and water as it combusts; thus, it burns more cleanly than coal or oil. In terms of pollution, it is the most desirable fuel of the three. Despite the obvious fuel potential of natural gas, at first there was no practical way to transport it from wells to consumers. Any gas released from oil fields was (and in many parts of the world still is) flared—that is, vented and burned in the atmosphere, a tremendous waste of valuable fuel. Gradually, the United States constructed a network of pipelines connecting wells with consumers. With the completion of these pipelines, the use of natural gas for heating, cooking, and industrial processes escalated rapidly because gas is clean, convenient (no storage bins or tanks are required on the premises), and relatively inexpensive. Currently, natural gas satisfies about 28% of U.S. energy demand and 24% of the global demand.

Thus, three fossil fuels—crude oil, coal, and natural gas—provide 86% of U.S. energy consumption and 87% of the world's consumption. The remaining percentages are furnished by nuclear power, hydropower, and renewable sources. The changing pattern of energy sources in the United States is shown in **Figure 14–4**. The picture is similar for other developed countries of the world, though the percentages differ somewhat depending on the energy resources of the country. With the help of fossil fuel energy, people and goods have become so mobile that the human community has in fact become a global community. Adding to this linkage in a powerful way is the explosion in information storage and flow, made possible by digital devices and the Internet—all powered by electricity.

Electrical Power Production

A considerable portion of the energy we gain from fossil fuels is used to generate **electrical power**, defined as the amount of work done by an electric current over a given time. The electricity itself is an **energy carrier**—it transfers energy from a **primary energy source** (coal or waterpower, for instance) to its point of use. Electricity now defines our modern technological society, powering appliances and computers in our homes, lighting up the night, and enabling the global Internet and countless other vital processes in business and industry to operate. Approximately 33% of fossil fuel production is now used to generate electricity in the United States; in 1950, the figure was only 10%.

Generators. Electric generators were invented in the 19th century. In 1831, English scientist Michael Faraday discovered that passing a coil of wire through a magnetic field causes a flow of electrons—an electrical current—in the wire. An **electric generator** is basically a coil of wire that rotates in a magnetic field or that remains stationary while a magnetic field is rotated around it. The process converts mechanical energy into electrical energy. Any time energy is converted from one form to another, some is lost through resistance and heat (in accordance with the Second Law of Thermodynamics; see Chapter 3). Also, some energy is lost in transmitting the electricity over wires from the generating plants to end users. In the end, it takes three units of primary energy

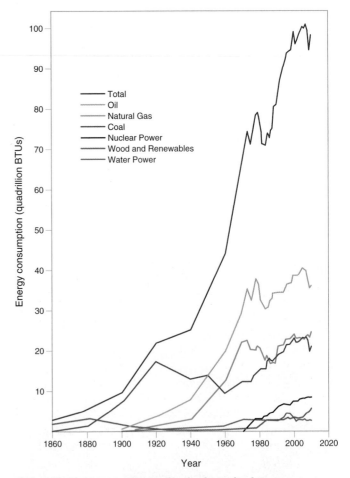

Figure 14–4 Energy consumption in the United States. Total consumption and major primary sources from 1860 to 2010 are shown here in quadrillion Btus (one Btu is the amount of heat required to raise one pound of water one degree Fahrenheit). Note how the mix of primary sources has changed over the years and how the total amount of energy consumed has continued to grow. Note also the skyrocketing increase in the use of oil after World War II (1945) as private cars and car-dependent commuting became common. (*Source:* Data from Energy Information Administration, U.S. Department of Energy, *Annual Energy Review 2010*, October 2011.)

MAJOR METHODS OF GENERATING ELECTRICITY

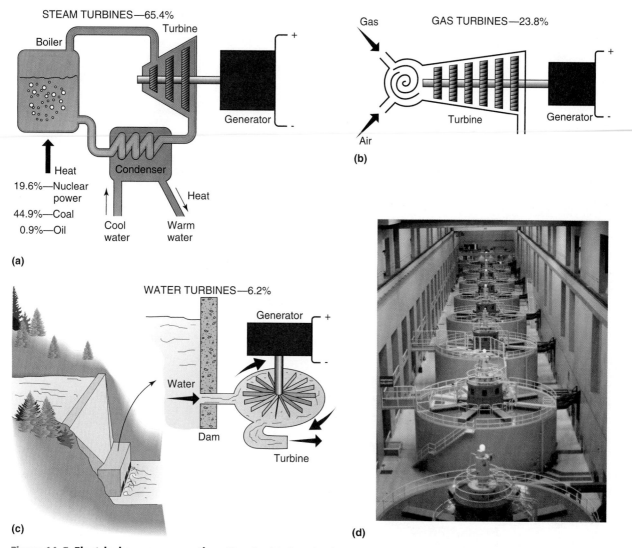

Figure 14–5 Electrical power generation. Most electricity is produced commercially by driving generators with (a) steam turbines, (b) gas turbines, and (c) water turbines. The percentage of the U.S. electricity supply derived from each source (as of 2010) is indicated. Smaller amounts of power (4.1%) are now also coming from wind energy, solar thermal energy, and photovoltaics (solar cells). (d) Each generator in this array in a hydroelectric plant is attached by a shaft to a turbine propelled by water. (*Source:* Data from Energy Information Administration, U.S. Department of Energy, *Annual Energy Review 2010,* October 2011.)

to create one unit of electrical power that can be put to use. This apparently inefficient proposition is justified because electrical power is so useful and indispensable.

In the most widely used technique for generating electrical power, a primary energy source is employed to boil water, creating high-pressure steam that drives a **turbine**—a sophisticated paddle wheel—coupled to the generator **(Fig. 14–5a)**. The combined turbine and generator are called a **turbogenerator**. Any primary energy source can be harnessed for boiling the water. Coal, oil, and nuclear energy are most commonly used at present, but burning refuse, solar energy, and geothermal energy (heat from Earth's interior) may be more widely used in the future.

In addition to steam-driven turbines, gas- and water-driven turbines are used. In a gas turbogenerator, the high-pressure gases produced by the combustion of a fuel (usually natural gas) drive the turbine directly (Fig. 14–5b).

For hydroelectric power, water under high pressure—at the base of a dam or at the bottom of a pipe beginning at the top of a waterfall—is used to drive a hydroturbogenerator (Fig. 14–5c–d). Wind turbines are also coming into use. A utility company supplying electricity to its customers will employ a mix of power sources, depending on the cost and availability of each and the demand cycle.

There are currently more than 5,500 power plants in the United States producing 4.1 billion megawatt-hrs of electricity. Over time, remarkable shifts have occurred in the sources of electricity in the United States **(Fig. 14–6)**. Although few new coal plants have been built in recent years, coal-burning power plants stay on line for decades; coal still provides more electrical power than any other source. Note, however, the exceptional rise in recent years of natural gas-powered plants and, most recently, the appearance of renewable power sources.

US ELECTRIC GENERATING CAPACITY BY YEAR IN SERVICE, 1940—2008

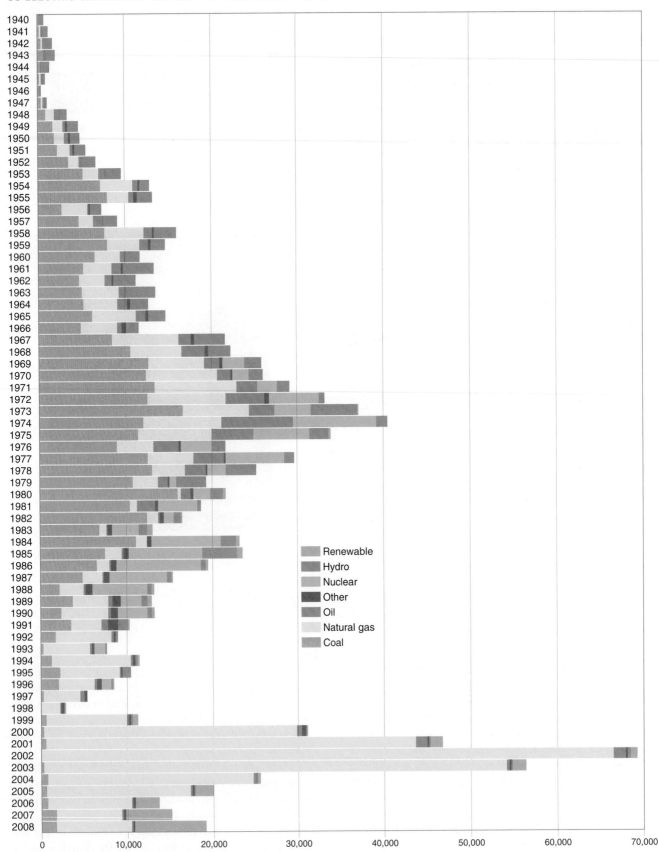

Name plate capacity (megawatts)

Figure 14–6 U.S. electric generating capacity by year in service, 1940–2008. Remarkable changes have occurred over the years in the construction of power plants, with many decades of coal plants to two decades of nuclear plants and, most recently, a huge increase in natural gas and the appearance of renewable sources. (*Source*: M. J. Bradley and Associates, *Benchmarking Air Emissions of the 100 Largest Electric Power Producers in the United States.* 2010).

TABLE 14–1 Energy Sources for Generating Electrical Power (% of total)

Source	New England	United States	World
Nuclear power	15.4	19.6	13.5
Natural gas	52.5	23.8	21.2
Coal	8.3	44.9	40.8
Oil	10.6	0.9	5.4
Hydro	10.2	6.2	16.2
Wood, refuse, renewables	3.0	4.1	2.9

Sources: ISO New England, 2011; U.S. Energy Information Administration, 2011; International Energy Agency, 2011.

Fluctuations in Demand. Where does your electricity come from? It depends on what's available. Most utility companies are linked together in what are called *pools*. In New England, for example, ISO New England ties together some 350 electric-power-generating plants through 8,000 miles of transmission lines, called the **power grid**. The utility is responsible for balancing electricity supply and demand, regardless of daily or seasonal fluctuations. The generating capacity of the ISO New England utilities is currently 33,174 megawatts (MW), produced by a variety of energy sources, as shown in Table 14–1. One MW is enough electricity to power 800 homes, on average. The greatest demands on electrical power in New England come in the summer months, when air conditioning use is at its highest. The all-time peak load (highest demand) record of 28,130 MW was set on August 2, 2006, a very hot day in New England. A power pool must accommodate daily and weekly electricity demand as well as seasonal fluctuations.

Demand Cycle. The typical pattern of daily and weekly electrical demand in the United States is shown in **Figure 14–7**. The baseload represents the constant supply of power provided by the large coal-burning and nuclear plants with capacities as high as 1,000 MW. As demand rises during the day, the utility draws on additional plants that can be turned on and off—the intermediate and peak-load power sources

that represent the utility's reserve capacity. These are the utility's gas turbines and diesel plants, or the utility may use pumped-storage hydroelectric power (water is pumped to a reservoir using off-peak power and is allowed to flow through turbines when needed). Occasionally, a deficiency in available power will prompt a brownout (a reduction in voltage) or a blackout (a total loss of power), temporarily disrupting whole regions. Such events occur most likely during times of peak power demand and may be precipitated by a sudden loss of power accompanying the shutdown of a baseload plant.

In our current economy, demand for electricity has been rising faster than supply, resulting in a decline in reserve capacity from 25% in the 1980s to the current level of 15%. Summer heat waves represent the greatest source of sudden pulses in demand, and as the climate continues to grow warmer in many parts of the United States, more and more regional utilities are being pushed to the edge of their ability to satisfy summer electricity demand. An even more serious problem is the antiquated system that controls the power transmission grid connecting power sources to users.

Blackout. Power brownouts and blackouts are a serious threat to an economy that is now highly dependent on computers and other electronic devices (brownouts are a cutback in power, blackouts are total power failures). On August 14, 2003, the largest blackout in U.S. history hit

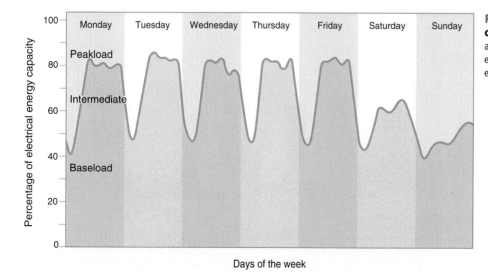

Figure 14–7 Weekly electrical demand cycle. Electrical demand fluctuates daily, weekly, and seasonally. A variety of generators must be employed to meet base, intermediate, and peak electricity needs.

eight eastern states and parts of Canada, imposing a $30 billion cost on the economies of the two countries. The blackout began when several power lines in Ohio brushed against trees and shut down. Alarm software failed, leaving controllers unaware of the problem. Transmission lines around the failure cut out, a local power plant shut down, more and more lines and plants dropped out, and within eight minutes, 50 million people across eight states and two Canadian provinces were without power. Blackouts and brownouts cost the United States an estimated $80 billion a year.

What is needed to prevent blackouts and brownouts is a self-healing **smart grid**, one that is able to monitor problems in real time, react to trouble quickly, and isolate trouble areas so that a cascading failure like the one in 2003 would be impossible. The technology for this fix is available, but funding is a huge problem, as the electric utility industry tends to run its business on very slim profit margins. However, the utilities are busy installing one component of a smart grid—*smart meters*—by the millions. The meters give instant feedback on electricity use to homeowners and the utility control center. Another component would be *smart sensors* at many points on the grid; the sensors would provide feedback on transformers and power lines, allowing instant shunting of power away from breaks and between transformers. A pilot installation of smart grid technology in Boulder, Colorado, in 2008 was highly successful. The Obama administration has already invested $4.5 billion in smart grid projects, matched by $5.5 billion from private sources.

Clean Energy? There was a time when electrical power was promoted as being the ultimate clean, nonpolluting energy source. This is true at the point of its use: Using electricity creates essentially no pollution. For a time, it was also relatively inexpensive, and electric heating systems were installed in many new homes. This is no longer true. The public knows that electricity is (1) an expensive way to heat homes and (2) generated from fossil fuels and nuclear energy. Coal-burning power plants are the major source of electric-

ity in the United States but are heavily implicated in acid deposition and global climate change (Chapters 18 and 19). Nuclear power is a much-mistrusted technology because of the potential for accidents and the problem of disposal of the nuclear waste products. (Nuclear power is discussed in Chapter 15.) Additional adverse effects of using electrical energy apply to other parts of the fuel cycle, such as mining coal and processing uranium ore. The pollution is simply transferred from one part of the environment to another. The only relatively nonpolluting energy sources are the renewables—hydroelectricity, solar, wind, and biomass—and these have their own limitations (Chapter 16).

The problem of transferring pollution from one place to another becomes even more pronounced when you consider efficiency; for example, the production of electricity from burning fossil fuels has an efficiency of only 30–35%. The energy loss is accounted for partly by some heat energy from the firebox going up the chimney and partly by a large amount of heat energy remaining in the spent steam as it comes out the end of the turbine. These unavoidable energy losses, called **conversion losses**, are a consequence of the need to maintain a high temperature difference between the incoming steam and the receiving turbine so as to maximize efficiency. Heat energy will go only toward a cooler place, so it cannot be recycled into the turbine. The most common practice is to dissipate it into the environment by means of a **condenser** (a device that turns turbine exhaust steam into water). In fact, the most conspicuous features of coal-burning and nuclear power plants are the huge cooling towers used for that purpose **(Fig. 14–8)**. Another significant energy loss occurs as the electricity is transmitted and distributed to end users by connecting wires.

An alternative to cooling towers is to pass water from a river, a lake, or an ocean over the condensing system (Fig. 14–5a). With this approach, however, the waste heat energy is transferred to the body of water, with the result that all the small planktonic organisms drawn through the con-

Figure 14–8 Cooling towers. A coal-fired power plant is equipped with cooling tower(s) to dissipate the heat from the spent steam after it has passed through the turbines.

densing system with the water are cooked and the warm water added back to the waterway may have deleterious effects on the aquatic ecosystem. Thus, waste heat energy discharged into natural waterways is referred to as **thermal pollution.**

Electricity is a form of energy that is indispensable to modern societies as they are currently structured. Worldwide, the electrical power industry produces more than 21,000 billion kilowatt-hours and revenues valued at more than $800 billion a year and, at the same time, generates a challenging array of environmental problems.

Matching Sources to Uses: Pathways of Energy Flow

In addressing whether energy resources are sufficient and where additional energy can be obtained, it is necessary to consider more than just the source of energy because some forms of energy lend themselves well to one use, but not to others. For example, the cars and trucks that travel the U.S. system of highways depend virtually 100% on liquid fuels. Almost all other machines used for moving, such as tractors, airplanes, locomotives, and ships, also depend on liquid fuels. Crude oil, together with much smaller amounts of biofuels, is

needed to support this transportation system. Nuclear power can do little to mitigate the demand for crude oil because it is suitable only for boiling water to drive turbogenerators, which will do essentially nothing for our transportation system. The same can be said for coal. Although this situation might change with more widespread use of electric vehicles, their practicality for long distances still requires a technological breakthrough in the form of a low-cost, lightweight battery that can store large amounts of power.

Primary energy use is commonly divided into four categories: (1) transportation, (2) industrial processes, (3) commercial and residential use (heating, cooling, lighting, and appliances), and (4) the generation of electrical power, which goes into categories (2) and (3) as a secondary energy use. The major pathways from the primary energy sources to the various end uses are shown in **Figure 14–9**. Transportation, which represents 28% of energy use, depends on petroleum (and some ethanol), whereas nuclear power, coal, and water power are limited to the production of electricity. Natural gas and oil are more versatile energy sources, and significant shifts have occurred in the use of these sources in recent years. For example, today only 1% of U.S. electricity is generated by oil, whereas 35 years ago it was about 15%.

ESTIMATED U.S. ENERGY USE IN 2010: ~98.0 QUADS

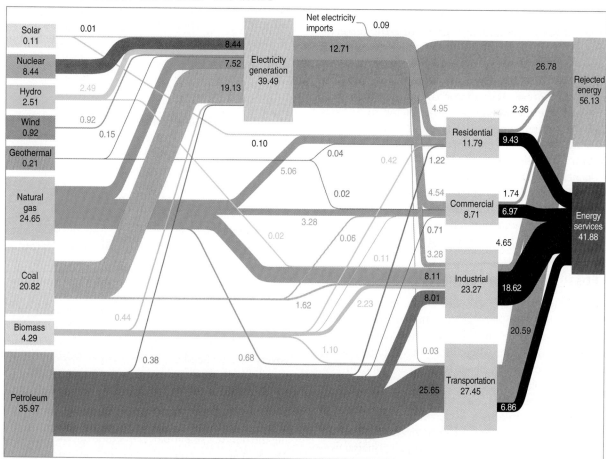

Figure 14–9 Pathways from primary energy sources to end uses in the United States. Only major pathways are shown. Note that end uses are connected to primary sources in specific ways. Note also the large percentage of energy that is wasted as it is converted to heat and lost. All energy is expressed in quads, or quadrillion Btus (10^{15} Btus). (*Source:* Lawrence Livermore National Laboratory. Data from Energy Information Administration, U.S. Department of Energy, *Annual Energy Review 2010*, October 2011.)

Figure 14–9 also indicates the high proportion of consumed energy that goes directly to waste heat ("rejected energy") rather than being used for its intended purpose—some 57%. Some waste is inevitable, as dictated by the Second Law of Thermodynamics, but the current losses are much greater than necessary. The efficiency of energy use can be at least doubled for some uses, including cars that get twice as many miles per gallon and appliances that consume half as much power. For uses like electric power generation, however, somewhat smaller increases in efficiency can be expected. In general, increasing the efficiency of energy use would have the same effect as increasing the energy supply, at far less expense. Such increases in efficiency represent a largely untapped "reserve" of energy that is saved, the **conservation reserve** discussed in more detail in Section 14.5.

In contemplating the energy future, we need to consider more than gross supplies and demands. That is, we must also examine how particular primary sources can be matched to particular end uses. Furthermore, what are the most cost-effective ways of balancing supplies and demands—by increasing supplies or by decreasing consumption through conservation, efficiency, and demand management?

14.2 Exploiting Crude Oil

The United States has abundant reserves of coal, adequate reserves of natural gas (at least for the time being), and the potential for the expansion of nuclear power. Therefore, no shortages loom in these areas now. (Note that the environmental concerns regarding coal and nuclear power are being set aside for the moment.) Petroleum to fuel our transportation system, however, is another matter. The U.S. crude-oil supplies fall far short of meeting demand; we now depend on foreign sources for 68% of our crude oil, and this dependency has been growing as U.S. sources have diminished. Foreign sources come at increasing costs, such as trade imbalances, military actions, economic disruptions, and oil spills. Let us examine the situation, starting with where crude oil comes from and how it is exploited.

How Fossil Fuels Are Formed

The reason crude oil, coal, and natural gas are called fossil fuels is that all three are derived from the remains of living organisms (**Fig. 14–10**). In the Paleozoic and Mesozoic geological eras, 100 million to 500 million years ago, freshwater swamps and shallow seas, which supported abundant vegetation and phytoplankton, covered vast areas of the globe. Anaerobic conditions in the lowest layers of these bodies of water impeded the breakdown of detritus by decomposers. As a result, massive quantities of dead organic matter accumulated. Over millions of years, this organic matter was gradually buried under layers of sediment and converted by microbial action, pressure, and heat to coal, crude oil, and natural gas. Coal is highly compressed organic matter—mostly leafy material from swamp vegetation—that decomposed relatively little. Sediments buried 7,500 to 15,000 feet

are in a temperature range that breaks down large organic molecules to smaller carbon molecules, producing crude oil. Below 15,000 feet, molecular breakdown leads to the one-carbon compound methane, or natural gas.

Crude-Oil Reserves Versus Production

The total amount of crude oil remaining represents Earth's oil **resources**. Finding and exploiting these resources is the problem. The science of geology provides information about the probable locations and extents of ancient shallow seas. On the basis of their knowledge and their field experience, geologists make educated guesses as to where oil or natural gas may be located and how much may be found. These estimates are termed **undiscovered resources**. The estimates may be far off the mark because there is no way to determine whether undiscovered resources actually exist except by exploratory drilling.

If exploratory drilling confirms the presence of oil, further drilling is conducted to determine the extent and depth of the **oil field**. From that information, a fairly accurate estimate can be made of how much oil can be economically obtained from the field. That amount then becomes **proved reserves**. Amounts are given in barrels of oil (1 barrel = 42 gallons), and the content of each reserve field is expressed in probabilities. Thus, a 50% probability, or P50, of 700 million barrels for a field means that a total of 700 million barrels is just as likely as not to come from the field. The final step, the withdrawal of oil or gas from the field, is called **production** in the oil business. Production, as used here, is a euphemism. "It is production," said ecologist Barry Commoner, "only in the sense that a boy robbing his mother's pantry can be said to be producing the jam supply."[2] In reality, this step is *extraction* from Earth, or recovery.

Production from a given field cannot proceed at a constant rate because crude oil is a viscous fluid held in pore spaces in sedimentary rock, just as water is held in a sponge. Initially, the field may be under so much pressure that the first penetration produces a gusher. Gushers, however, are generally short lived, and the oil then seeps slowly from the sedimentary rock into a well from which it is pumped out. Ordinarily, conventional pumping (**primary recovery**) can remove only 25% of the oil in an oil field. Further removal, of up to 50% of the oil in the field, is often possible, but it is more costly because it involves manipulating pressure in the oil reservoir by injecting brine or steam (**secondary recovery**), which forces the oil into the wells. In some fields, even greater recovery (**enhanced recovery**) is obtained by injecting carbon dioxide, which breaks up oil droplets and enables them to flow again.

Economics—the price of a barrel of oil—determines the extent to which reserves are exploited. No oil company will spend more money to extract oil than it expects to make selling the oil. At $13 per barrel of crude oil, the price in the late 1990s, it was economical to extract no more than 25–35% of the oil in a given field. Thus, an increase in price makes more reserves available. In the 1970s and early 1980s, higher

---

[2]Barry Commoner, "Toward a Sustainable Energy Future." *E Magazine*. May/June 1991, p. 55.

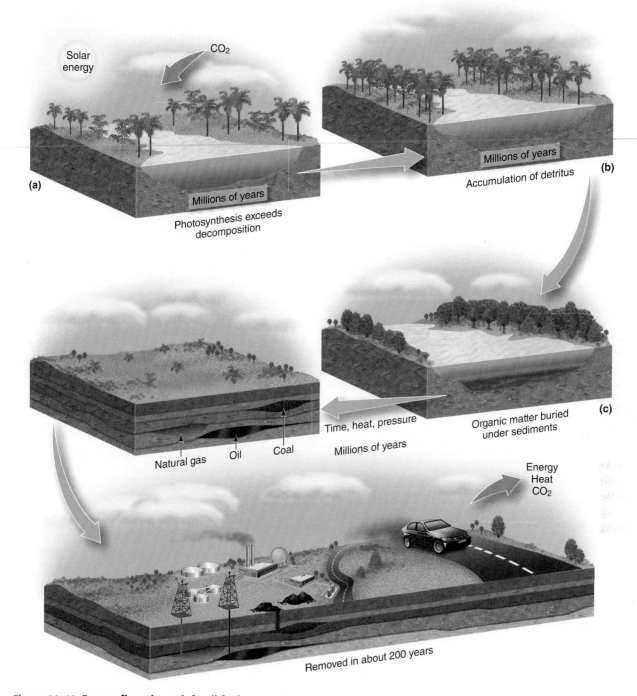

Figure 14–10 Energy flow through fossil fuels. Coal, oil, and natural gas are derived from biomass that was produced many millions of years ago. Deposits are finite, and because formative processes require millions of years, they are nonrenewable.

oil prices justified reopening old oil fields and created an economic boom in Texas and Louisiana. Then an economic recession occurred in the late 1980s, and a drop in oil prices caused the fields to be shut down again. Following that, oil prices began to rise in 1998, reached a peak of $145 a barrel in mid-2008, and then began a sharp decline as global oil demand dropped—a consequence of a severe global economic crisis. As the global economic picture began to improve, oil prices rose again, until by mid-2011 oil was once again more than $100 a barrel.

Declining U.S. Reserves and Increasing Importation

Understanding the relationships among production, price, and the exploitation of reserves can help clarify the U.S. energy dilemma. Up to 1970, the United States was fairly oil independent, producing about two-thirds of oil consumed and keeping imported oil prices low **(Fig. 14–11)**. In 1970, however, new discoveries fell short, so production decreased, while consumption continued to grow rapidly. This downturn

Figure 14–11 Oil production and consumption in the United States. Four stages can be seen: (1) Up to 1970, the discovery of new reserves allowed production to parallel increasing consumption. (2) From 1970 to 1973, a lack of new oil discoveries caused production to decline while consumption continued to climb, resulting in a vast increase in oil imports. (3) In the late 1970s to early 1980s, high oil prices promoted both lowered consumption and increased production—which included bringing the Alaskan oil field into production. (4) From 1986, when oil prices declined sharply, until the present, consumption has again increased while production has resumed its decline, making us increasingly dependent on foreign oil. (*Source:* Data from Energy Information Administration, U.S. Department of Energy, *Annual Energy Review 2010*, October 2011.)

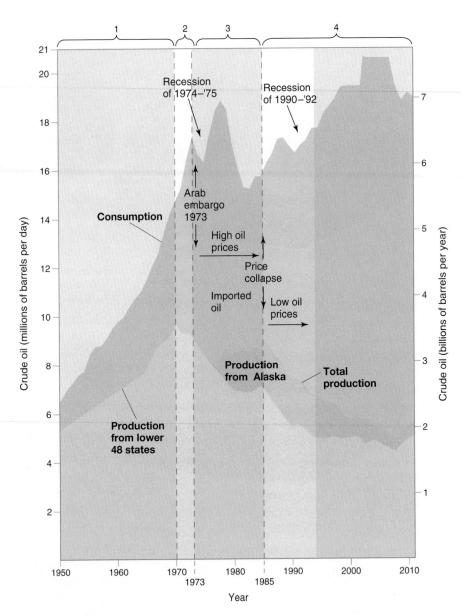

vindicated the predictions of the late geologist M. King Hubbert, who proposed that oil exploitation in a region would follow a bell-shaped curve. Hubbert observed the pattern of U.S. exploitation and predicted that U.S. production would peak (now called the *Hubbert peak*) between 1965 and 1970. At that point, about half of the available oil would have been withdrawn, and production would decline gradually as reserves were exploited. Figure 14–11 demonstrates that a Hubbert curve can be applied to U.S. oil production.

To fill the energy gap between rising consumption and falling production, the United States depended increasingly on imported oil, primarily from the Arab countries of the Middle East. European countries and Japan did likewise. Because imported oil cost only about $2.30 per barrel ($13.60 in 2012 dollars) in the early 1970s and Middle Eastern reserves were more than adequate to meet the demand, this course seemed to present few problems.

The Oil Crisis of the 1970s. It was obvious that the United States and most other highly industrialized nations were

becoming dependent on oil imports. In a classic economic move, a group of predominantly Arab countries known as the Organization of Petroleum Exporting Countries (OPEC) formed a cartel and agreed to restrain production in order to get higher prices. Imported oil began to cost more in the early 1970s, and then, in conjunction with an Arab-Israeli war in 1973, OPEC initiated an embargo of oil sales to countries that gave military and economic support to Israel (such as the United States). Spot shortages quickly occurred, which escalated into widespread panic and long lines at gas stations. We were willing to pay almost anything to have oil shipments resumed—and pay we did: OPEC resumed shipments at a price of $12 a barrel ($62 in 2012 dollars), four times the previous price.

Then, by continuing to limit oil production all through the 1970s, OPEC was able to keep supplies tight enough to force prices even higher. In the early 1980s, a barrel of oil delivered to the United States cost $36 ($91 in 2012 dollars), and the United States was paying accordingly. The high energy costs were devastating to the economy; inflation and

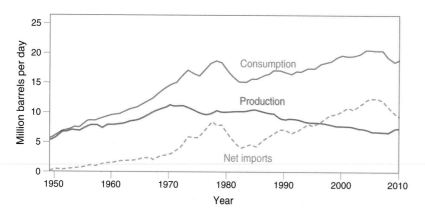

Figure 14–12 Consumption, domestic production, and imports of petroleum products. After the oil crises of the 1970s, the United States became less reliant on foreign oil. Since the mid-1980s, however, consumption of foreign oil has for the most part been rising steadily. *Note: Production figures include liquefied natural gas; net imports are imported crude oil minus exports of refined petroleum.* (*Source:* Energy Information Administration, U.S. Department of Energy, *Annual Energy Review 2010*, October 2011.)

unemployment rose and the country experienced several recessions. The purchase of foreign oil became, and remains, the single largest factor in our **balance-of-trade deficit**—where we buy more goods from other countries than we sell as exports.

Adjusting to Higher Prices. In response to the higher prices, the United States and other industrialized countries made a number of adjustments during the 1970s. The following are the most significant changes made by the United States:

- We increased domestic production by stepping up oil exploration, building the Alaska pipeline, and reopening old oil fields.

- Congress took steps to decrease consumption by setting new standards for automobile fuel efficiency at 27.5 miles per gallon or mpg (cars averaged 13 mpg in 1973), lowered speed limits to 55 miles per hour, promoted higher efficiencies for building insulation and appliances, and began to develop alternative energy sources.

- To protect against another OPEC boycott, we created a strategic oil reserve in underground caverns in Louisiana. The current stockpile is 696 million barrels of oil, equivalent to about 37 days of consumption at 19 million barrels per day.

- We encouraged oil exploration and production in a number of non-OPEC countries, especially those in the Western hemisphere.

The results of the new actions were not immediate, nor could quick returns have been expected. For example, it takes at least three to five years to design a new car model and start production. Then another six to eight years pass before enough new models are sold and old ones retired to affect the overall highway "fleet."

As we entered the 1980s, however, these efforts were definitely having an impact. The United States continued to depend substantially on foreign oil, but consumption was declining, and production, up after the Alaska pipeline was completed, was holding its own (see Figure 14–11). Elsewhere in the world, significant discoveries in Mexico, Africa, and the North Sea made the world less dependent on OPEC oil. As a result of all these factors, plus OPEC's inability to restrain its own members' production, world oil production exceeded consumption in the mid-1980s, and there was an oil "glut."

Supply exceeded demand, with predictable results. In 1986, world oil prices crashed from the high $20s to $14 per barrel. From that time until late 1999, oil prices fluctuated in the range from $10 to $20 per barrel, almost as low as they were in the early 1970s. The lower oil prices apparently benefited consumers, industry, and oil-dependent developing nations, but they hardly solved the underlying problem.

Back to Old Ways. Whereas high oil prices stimulated constructive responses, the collapse in oil prices undercut those responses:

- Oil exploration was curtailed, and production from older fields was terminated.

- Conservation efforts and incentives were abandoned. Government standards for automobile fuel efficiency were frozen at 27.5 mpg. Speed limits were raised to 65 and even 75 mph.

- Congress terminated tax incentives and other subsidies for the development or installation of alternative energy sources, destroying many new businesses engaged in solar and wind energy.

Production of crude oil in the United States continued to drop as proved reserves were drawn down. The area of declining production now includes Alaska, which reached its peak production in 1988. At the same time, consumption of crude oil rose again because of increases in the number of cars on the highways and the average number of miles per year that each car is driven. Adding to that consumption, consumers began buying vehicles that were less fuel efficient. Minivans, four-wheel-drive sport-utility vehicles (SUVs), light trucks, and other less fuel-efficient models constituted more than 50% of the new-car market. Our dependence on foreign oil has grown steadily since the mid-1980s, and now 68% of crude oil we use comes from imports (**Fig. 14–12**).

Back to the Future. The collapse in oil prices set the stage for another developing crisis: a rise in oil prices that began in late 1998. OPEC was concerned about the continuing oil glut and its low prices, so it decided to cut production just as East Asia was coming out of a serious recession. The rapid rise in demand brought on a shortfall in the oil supply, and

NET OIL IMPORTS AND PRICE OF OIL

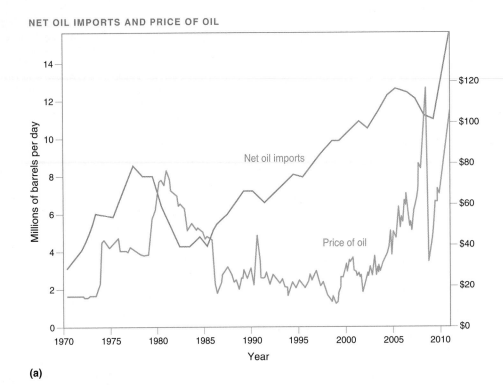

(a)

REGULAR GASOLINE PRICE IN TODAY'S DOLLARS

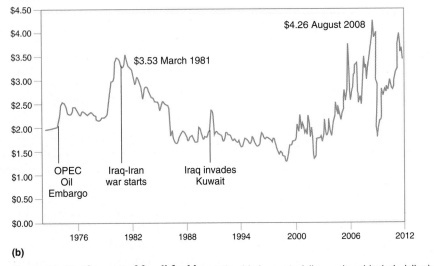

(b)

Figure 14–13 The cost of fossil fuel imports. (a) The cost in dollars per barrel (today's dollars) and net imports; (b) gasoline prices in today's dollars. (*Source:* zFacts.com, using data from U.S. Department of Energy sources.)

the price of crude oil rose precipitously **(Fig. 14–13a)**. With demand outpacing supply, oil prices kept rising, reaching a peak of $147 a barrel in mid-2008. Homeowners were strapped with high fuel-oil prices, commuters faced gasoline prices in the $4 range (Fig. 14–13b), people cut back on driving, economic growth began to slacken, and only the oil industry prospered as record profits poured in. Internationally, food prices rose to unprecedented levels, forcing many in the developing world into dependency and hunger. In the United States, shoppers purchased smaller vehicles, sales of hybrid vehicles blossomed, and the auto industry began closing factories that produced SUVs, vans, and trucks. Once again, Congress responded, raising the fuel efficiency standard, mandating the increasing use of renewable fuels, and establishing new standards for energy efficiency.

Then the bottom dropped out of the stock market in September 2008, a consequence of overpriced housing, mortgage excesses, and uncontrolled stock market speculation. A global economic meltdown followed, during which major financial institutions failed or were bought out by governments, credit suddenly became almost unavailable, and job losses accelerated. In short, the countries of the world slid into a recession that was called the worst in 75 years. The impact on oil prices was profound: Oil demand dropped, supply exceeded demand, and the price dropped back below $50 a barrel by December of the same year. Gasoline prices dropped down into the $2 a gallon range, and homeowners facing the winter could anticipate much lower fuel-oil prices. Oil-exporting countries suddenly faced a sharp decline in their revenues, and OPEC members began to consider cutting their output in order to stabilize prices. However, the low prices were short-lived. A combination of factors (heavy purchases by China, political unrest in several Arab oil-producing nations, partial recovery from the global recession) prompted oil prices to rise once again, climbing above $100 a barrel in the spring of 2011.

The Consequences of U.S. Dependency

What problems does this growing dependency on foreign oil present? As the U.S. dependency on foreign oil rises, we are faced with problems at three levels: (1) the costs of purchasing oil, (2) the risk of supply disruptions due to political instability in the Middle East, and (3) ultimate resource limitations in any case.

Costs of Purchase. The yearly cost to the United States of purchasing foreign oil dropped considerably from its high of $230 billion (corrected for inflation) in the early 1980s, due to falling prices and conservation efforts. It began a rapid rise in 2000, however, and reached $329 billion in 2010. That year, the balance-of-trade deficit was $500 billion; thus, payments for imported crude oil amounted to two-thirds of our balance-of-trade deficit. These payments have significant consequences for our economy. The price we pay at the pump is basically the same whether the oil is produced here or abroad. If the oil is produced at home, however, the money we pay for it circulates within our own economy, providing jobs and producing other goods and services. When we buy oil from a foreign country, we are draining our own economy. And as we have seen, high costs for oil can have a severe damping effect on economic growth.

Finally, there are the environmental costs of oil spills such as the *Deepwater Horizon* accident and of other forms of pollution from the drilling, refining, and consumption of fuels—although estimating total damage in dollars is sometimes difficult.

Persian Gulf Oil. The Middle East is a politically unstable region of the world. As noted earlier, it was the unexpected Arab boycott that plunged us into the first oil crisis in 1973. Maintaining access to Middle Eastern oil is also a major reason for our efforts toward negotiating peace agreements between various parties in that region. Recognizing this political instability, President Jimmy Carter in January 1980 stated that the United States would use military force if needed to ensure our access to Persian Gulf oil (the so-called **Carter Doctrine**). This doctrine was tested in the fall of 1990 when Saddam Hussein of Iraq invaded Kuwait, then producing 6 million barrels of oil per day. The U.S.-led Persian Gulf War of early 1991 threw Hussein's armies out of Kuwait and gave the United States an established military presence in the Persian Gulf region. It was this presence, however, that angered the radical Islamic terrorist Al Qaeda organization and contributed to the terrible September 11, 2001, attacks on the World Trade Center and the Pentagon.

Two More Wars. In response to 9/11, U.S. and British forces invaded Afghanistan, which had been harboring Osama bin Laden and elements of Al Qaeda. The purpose of the war was to capture bin Laden, destroy the Al Qaeda training camps, and overthrow the ruling Taliban regime. Although Afghanistan is not an oil-producing country, it can be argued that the Afghan war was a consequence of our presence in the Persian Gulf region. That war is still in progress, as the Taliban have sustained an insurgency that continues to destabilize the country, although bin Laden was finally killed in 2011 by a U.S. commando team while hiding out in Pakistan.

Then, in 2003, a U.S.–British coalition invaded Iraq in order to overthrow Saddam Hussein's rule and eliminate suspected weapons of mass destruction (**Fig. 14–14**). Some observers suggest that oil was an important part of the picture. Although it is clear that the United States and Britain did not simply commandeer the Iraqi oil, they have spent billions to restore Iraq's oil industry—and the United States and other oil-hungry Western countries are eager customers for the oil now that it is flowing again at a rate of 2.9 million barrels per day. Those countries rely on Persian Gulf oil for half of their oil imports. It is fair to say that military costs associated with maintaining access to Persian Gulf oil represent a substantial U.S. government subsidy of our oil consumption, one that has also been costly in human lives.

Other Sources. The United States has turned more and more to non–Middle Eastern oil suppliers. Canada is now the

Figure 14–14 The U.S.-Iraqi war and oil. A soldier from the U.S. 173rd Airborne stands guard at a refinery near the Baba Gurgur oil field in Kirkuk, Iraq.

number-one exporter of oil to the United States, with about 23% of the market. Mexico and Colombia are also major sources of oil. Western Hemisphere sources now provide about 40% of our imports. The relative political stability and proximity of these sources are good reasons for the United States to turn to them and away from the Middle East. Russia, a number of former Soviet republics, and several African countries (especially Nigeria) are important new sources of oil. However, most of these other sources, like the United States, are on the downside of the Hubbert curve and are seeing production declining. The conclusion? We will continue to depend on OPEC and the Persian Gulf for a substantial share of our oil imports for the present and even more so in the future.

Oil Resource Limitations and Peak Oil

Crude-oil production in the United States is decreasing because of diminishing domestic reserves. Moreover, on the basis of the "Easter-egg hypothesis," there is dim hope for major new U.S. finds. According to this hypothesis, as your searching turns up fewer and fewer eggs, you draw the logical conclusion that nearly all the eggs have been found. In terms of oil, North America is already the most intensively explored of any continental landmass. The last major find was the Alaskan oil field in 1968. Discoveries since then have been small in comparison, with increasing numbers of dry holes in between. Indeed, much of the "new" oil in the United States is coming from innovative computer mapping of geological structures and horizontal drilling technology that enable oil to be identified and withdrawn from small isolated pockets that were previously missed within old

fields. The *Deepwater Horizon* oil leak was a consequence of the need to develop wells in deeper water and deeper below the sea floor in order to keep production up. In spite of the high costs (and risks) of offshore oil, it accounts for 30% of domestic production at present. Oil resources like deep Gulf oil and potential resources (undiscovered) below the Arctic Ocean will be exploited only if prices are high enough to pay for exploration and production (see Sound Science, "Energy Returned on Energy Invested").

How much oil is still available for future generations? About 1,150 billion barrels (BBs) of oil have already been used, almost all in the 20th century. Currently, the nations of the world are using about 88 million barrels per day, or some 32 BBs a year. Rising demand by the developing countries and the continued rise in oil demand by the developed countries point to an increase of about 2% per year, so that demand could reach 40 BBs a year by 2020. How long can the existing oil reserves provide that much oil?

Hubbert's Peak. As with many issues, it depends on whom you believe. Some years ago, oil geologists Colin Campbell and Jean Laherrère provided an analysis that put the world's proved reserves at about 850 BBs.[3] This is much lower than the 1,383 BBs claimed by more recent oil industry reports (Table 14–2). On the basis of their estimates and the known amount already used, the two geologists calculated a Hubbert curve indicating that peak oil produc-

---
[3]Colin J. Campbell and Jean H. Laherrère, "The End of Cheap Oil." *Scientific American*, March 1998.

TABLE 14–2 Proved World Reserves and Annual Use of Fossil Fuel Reserves, 2011

Region	Reserves		
	Petroleum (billion barrels)*	Natural Gas (trillion ft³)*	Coal (billion metric tons)*
North America	217	382	245
South and Central America	325	268	12.5
Europe and Eurasia	141	2,779	305
Middle East	795	2,826	1.2
Africa	132	513	32
Asia Pacific	41	592	266
Total	1,653	7,361	861

	Use		
	Petroleum	Natural Gas	Coal
World use, 2011	32.1 billion barrels	113.8 trillion ft³	5.62 billion metric tons
World use, 2011 (quads)	183 quads	116 quads	99 quads

Source: BP Statistical Review of World Energy, 2012.

*The fuels may be compared by calculating their energy content in British thermal units (Btus) and then converting these figures to quadrillion Btus or quads, where 1 quad = 10^{15} Btus. One billion barrels of petroleum yield about 5.7×10^{15} Btus, or 5.7 quads. One trillion cubic feet of natural gas yield about 1.02×10^{15} Btus, or 1.02 quads. One billion metric tons of coal yield about 17.65×10^{15} Btus, or 17.65 quads.

GLOBAL OIL AND GAS PRODUCTION PROFILES

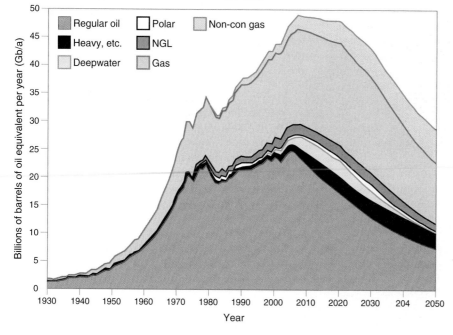

Figure 14–15 Hubbert curve of oil and gas production. The Association for the Study of Peak Oil and Gas believes that combined oil and gas production has already peaked in 2008. (*Source:* ASPO Newsletter No. 100, April 2009. http://aspoireland. files.wordpress.com/2009/12/newsletter100_200904.pdf. October 20, 2011.)

tion has already occurred. Princeton professor and former oil geologist Kenneth Deffeyes used the Hubbert technique in his book *Beyond Oil: The View from Hubbert's Peak*[4] to draw the same conclusions. Even if these scientists are wrong and the oil industry estimates are right, it would delay the peak by only a decade at most. It is a fact that some 54 countries have passed their peak in oil production; put another way, 580 of the 651 largest oil fields have passed their peak production and are declining every year. The Association for the Study of Peak Oil (ASPO), founded by Colin Campbell, believes that the peak for conventional oil is already behind us **(Fig. 14–15)**. If not, it is likely to be close, from all reports.

Downward Trajectory. At the current rate of use (which is expected to rise), the proved reserves can supply only 43 years of oil demand. New discoveries continue to add to proved reserves, but they are only a fraction of our annual use. The remaining oil will be more costly to extract, and at some point, producers will close wells rather than sustain losses on their oil production. Thus, the 21st century will become known as an era of declining oil use, just as the 20th century saw continuously rising use.

As oil becomes scarcer, the OPEC nations will once again dominate the world oil market. The Middle East possesses 48% of the world's proved reserves, and the developed world—in particular, the United States—will have to rely more and more on that region for its oil imports. In light of this fact, it would seem wise to begin to prepare for that day by reducing our dependency on foreign oil now. We can become more independent in three ways: (1) We can increase the fuel efficiency of our transportation system, which uses most of the oil; (2) we can develop alternatives to fossil fuels,

many of which are ready now; and (3) we can use the other fossil fuel resources we have to make fuel for vehicles. This leads us to some so-called non-conventional sources of oil— oil sand and oil shale—which are abundant in the United States and Canada.

Oil Sand and Oil Shale

Oil sand is a sedimentary material containing bitumen, an extremely viscous, tarlike hydrocarbon. When oil sand is heated, the bitumen can be "melted out" and refined in the same way as crude oil can. Northern Alberta, in Canada, has the world's largest oil-sand deposits (143 BBs as proved reserves), and Canada is commercially exploiting them because the cost has been competitive with oil prices. Canadian oil sands currently yield 1.3 million barrels of oil per day (almost 500 million barrels per year). The environmental damage is substantial; 118,000 acres of boreal forest and wetlands have already been heavily disturbed. The extraction process generates enormous quantities of greenhouse gases. The United States is eager to buy the oil, which represents 10% of our imported oil. It is transported via the Keystone pipelines from Alberta to southern Illinois and Oklahoma, where refineries are located. These pipelines have a limited capacity, however, and TransCanada Keystone has proposed a longer and larger pipeline, the Keystone XL, that would run 1,700 miles through seven U.S. states to Port Arthur, Texas, on the Gulf coast **(Fig. 14–16)**. Objections to this project are many: leaks in the pipeline are bound to occur; the pipeline would cross ecologically sensitive lands; and it would also cross over major recharge areas of the Oglalla aquifer, an extremely important source of irrigation water to Midwestern states. Proponents cite the large number of jobs the pipeline construction would provide, property taxes to communities it passes through, and the energy security Canadian oil provides to the United States.

[4]Kenneth Deffeyes, *Beyond Oil: The View from Hubbert's Peak* (New York: Hill and Wang, 2005).

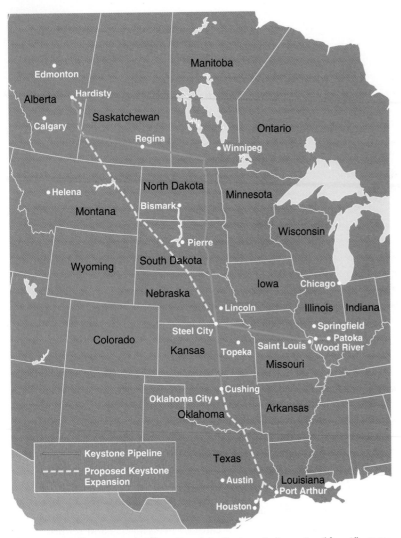

Figure 14–16 Keystone pipeline. The existing Keystone pipeline carries oil from Alberta tar sands to Oklahoma and Illinois. The proposed Keystone XL pipeline (Proposed Keystone Expansion on the map) is a larger line that can carry tar sands oil to Texas and the Gulf. (*Source*: U.S. State Department.)

stiff opposition because of its potential to damage air quality and groundwater; any oil-shale development is subject to National Environmental Protection Act analysis (environmental impact statements) and must get local and state approval.

Worldwide, the combination of oil shale and oil sand has been estimated to hold several trillion barrels of oil. Large-scale commercial development of oil sands has begun, but oil shale will remain undeveloped until oil prices rise dramatically—something that may happen as conventional oil is further along the downward slope of the global Hubbert curve. However, shale has recently begun yielding natural gas.

14.3 Drilling for Natural Gas

In the United States, natural gas is gaining popularity as a fossil fuel of choice. Almost equal quantities go to industrial, residential, and commercial use and electrical power generation (Fig. 14–9). As with oil, the cost of natural gas is subject to market fluctuations in supply and demand, and prices often rise and fall with seasonal changes. At the 2011 rate of use (24.4 trillion cubic feet), proved reserves in the United States (300 trillion cubic feet) hold only a 12-year supply of gas. This is somewhat deceiving, however, because natural gas is continually emitted from oil- and gas-bearing geological deposits and new drilling is opening up more deposits.

Shale Gas and Fracking. Until recently, most natural gas was mined from fields associated with oil and coal (**Fig. 14–17**). As a fossil fuel, gas production in the United States peaked in 1970 and

Oil shale is a fine sedimentary rock containing a mixture of solid, waxlike hydrocarbons called *kerogen*. The United States has extensive deposits of oil shale in Colorado, Utah, Texas, and Wyoming. When shale is heated to about 1,100°F (600°C), the kerogen releases hydrocarbon vapors that can be recondensed to form a black, viscous substance similar to crude oil, which can then be refined into gasoline and other petroleum products. However, about a ton of oil shale yields little more than half a barrel of oil. The mining, transportation, and disposal of wastes necessitated by an operation producing, for instance, a million barrels a day (just 5% of U.S. demand) would be a Herculean task, to say nothing of its environmental impact. Nevertheless, because the deposits contain an estimated 800 BBs of oil, the rising costs of importing oil make it inevitable that every domestic source will be considered. In November 2008, the Interior Department's Bureau of Land Management (BLM) finalized rules for how commercial oil-shale leases will be regulated, opening the way for oil companies to begin serious consideration of developing oil from the shale deposits. The process faces

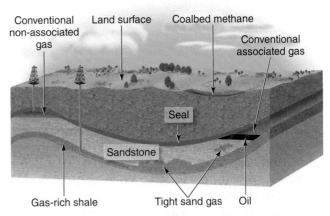

Figure 14–17 Geology of natural gas resources. Until recently, most natural gas was accessed via vertical drilling into reservoirs associated with coal and oil. Shale gas has recently been developed using horizontal drilling and "fracking," or injecting hydraulic fluid under high pressure to open up the shale and release the gas. (*Source*: U.S. Energy Information Administration.)

gradually declined. Proved reserves were not increasing as rapidly as domestic use. All this has changed, however. Using new techniques adapted from oil extraction, an abundant source of natural gas trapped in shale has dramatically changed the natural gas picture in the past few years. New shale gas resources provide an additional estimated supply of 827 trillion cubic feet, bringing the reserves to 1,111 trillion cubic feet, or a total of 45 years at today's usage rate. This is a conservative estimate. The United States is blessed with an abundance of shale gas basins, the largest being the huge Marcellus basin extending from New York to Tennessee. Some natural gas has always been found around shale rock formations, tapped by conventional vertical wells, but enormous quantities of gas remained trapped in the shale itself. The new shale mining technique involves drilling a vertical well into the shale and then turning the drill horizontally for distances up to a mile (Fig. 14–17). Then, at regular intervals along the horizontal pipe, hydraulic fluids containing sand are forced into the

pipe by huge pumps, fracturing the shale and releasing gas into spaces propped open by sand. The fluid, and then the gas, flows back up the well. The technique is called hydraulic fracturing, or **fracking**; the amount of hydraulic fluids pumped into just one well can exceed a million gallons at a single stage, and a number of stages can be employed over time at the well.

This all sounds good because of the benefits of natural gas over other fossil fuels. Burning gas in power plants releases only half as much CO_2 as coal for the same amount of electricity. The new shale sources have risen to 20% of U.S. gas production, and one consequence of this new abundance is lower prices. However, there are some serious environmental concerns with shale gas and fracking. Although the hydraulic fluids used are mostly water, other chemicals are added to the water, and much of the fluid comes back up to the surface in the process, requiring holding ponds that over time could leak into groundwater. The injected water could also contaminate underground drinking water sources at other

SOUND SCIENCE   Energy Returned on Energy Invested

Consider an oil field where all you have to do is drill a hole and oil comes gushing up. This enterprise requires little energy invested (equipment manufacture, workers, transportation of oil) compared with energy returned in the form of the barrels of oil captured. The ratio of energy returned to energy invested (**EROI**) in this case could be 100:1. That's excellent. Over time, however, the oil flow diminishes and soon has to be coaxed out with secondary or tertiary recovery methods. The EROI declines, and after a while, it drops to 1:1; there is no more point in extracting oil from that well. This is a simple illustration of an important consideration, namely, that some forms of energy are more economical to capture than others.

Oil exploitation in the United States began with very high EROI values, perhaps 100:1. Then, over time, it became increasingly expensive to coax up the oil or to import it (see illustration). Now, the EROI ratio is closer to 5:1 in the United States and 10:1 in Saudi Arabia. Tar sands oil does even worse, at 2.9:1. Other fossil fuels do better: coal at 80:1, natural gas at 15:1. We continue to use oil, however, because it is the only energy source that supports transportation. Natural gas may be favored over coal because of pipelines and of the much lower cost of gas turbines versus coal-fired power plants. So there are other considerations besides EROI that factor into the choice of energy.

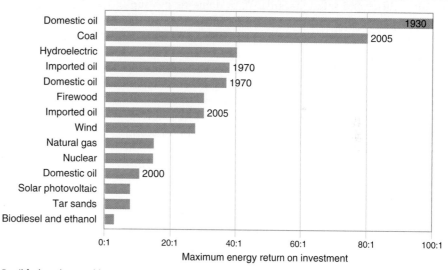

Fossil fuels and renewable energy sources are plotted as ratios of maximum energy return to investment. (*Source*: Figure 10 in Charles A. S. Hall and John W. Day Jr. "Revisiting the Limits to Growth after Peak Oil." *American Scientist 97*, May–June 2009, pp. 230–237.)

As the illustration shows, some renewable energy sources do well (hydroelectric, wind), and some do not (solar photovoltaic, biodiesel, and ethanol). If time is factored in (duration of the energy-capturing device in the case of wind and solar), these sources do better. And again, biodiesel and ethanol have a special appeal because they are liquid fuels for transportation. However, the EROI consideration helps to explain why the world isn't just rushing to renewables, as the fossil fuels are presently a more efficient energy source. However, as with

the choices of fossil fuels, there are other considerations besides EROI to factor in. For example, the fossil fuels all release the greenhouse gas CO_2 when burned to produce energy. A future where fossil fuels are consumed until they become economically impractical is an unthinkable future because of global climate change. We will need a renewable energy economy long before that happens. EROI does, however, provide some guidance for choosing renewable energy sources. Right now, wind energy looks like the best choice.

stages in the process. Unfortunately, hydraulic fracturing was exempted from regulations in the Safe Drinking Water Act in 2005.

Because of the increasing fleet of natural gas power plants (Fig. 14–6), gas is expected to surpass coal if supplies continue to grow as expected and prices stay low. And because of its extensive use in residential and commercial heating, natural gas has been replacing oil as well. If the United States gets serious about moving to a low-carbon economy, natural gas is poised to ease the way by replacing these dirtier fuels.

Global Resources. Worldwide, the estimated natural-gas resource base is even more plentiful (Table 14–2) than in the United States. Almost four times as much equivalent gas (that is, equivalent in energy) is likely to be available as oil. Much of it, however, is relatively inaccessible. Gas must either be pumped through a pipeline (a process limited by the length of the line) or be subjected to high pressures so that it remains a liquid at room temperature (liquefied natural gas, or LNG). It can then be shipped in large tankers; currently, eight LNG terminals in the United States handle these ships. Thirteen more have been approved but are not yet built. Constructing new LNG terminals has proven to be difficult, as the public views the terminals and tankers as major security and safety hazards. With domestic gas supplies now more abundant, new LNG terminals may not be needed.

Natural Gas–Run Cars. Natural gas can also meet some of the fuel needs for transportation. With the installation of a tank for compressed gas in the trunk and some modifications of the engine fuel-intake system—costing several thousand dollars—a car will run perfectly well on natural gas. Natural gas is a clean-burning fuel, producing carbon dioxide and water but virtually no hydrocarbons or sulfur oxides. Natural gas is in use in the United States by many buses and a number of private and government car fleets, but use is restricted due to the limited number of natural-gas service stations. Detroit automakers have stopped selling natural gas–powered cars, but they are abundant in Latin America and Asia-Pacific. Worldwide, some 13 million natural gas vehicles were in operation in 2010, many of them buses. It will take stronger public-policy support for this option to grow in the United States; until that happens, the natural-gas option will likely stay on the back burner.

Synthetic Oil. With the aid of the Fischer-Tropsch process, developed in the 1920s, natural gas can be chemically converted to a hydrocarbon that is liquid at room temperature and pressure—basically, a synthetic oil. Researchers are hotly pursuing ways to convert natural gas to synthetic oil more efficiently. Presently, the fuel produced is only about 10% more expensive than oil. Several major oil companies are spending billions on gearing up refineries to convert gas to liquid diesel and home-heating fuels **(Fig. 14–18)**,

Figure 14–18 Gas-to-liquid fuel plant. Sitting over a huge natural gas field, this Oryx Gas-to-Liquids plant near Doha in Qatar is one of several on the site that are expected to produce 450,000 barrels of liquid fuels per day by 2015.

and researchers are seeking an appropriate catalyst that can convert natural gas (methane) to methanol (methyl alcohol), a liquid. These options promise to make the abundant, but inaccessible, "stranded" natural gas deposits suddenly very valuable.

Applying natural gas to the transportation sector—both directly as a fuel and indirectly as a synthetic oil—may extend the oil phase of our economy, but it will not be a sustainable solution because the natural-gas reserves are also limited. What about coal, then, the most abundant fossil fuel?

14.4 Mining Coal

The world's leading producer of coal is China (2.7 billion tons in 2011), where a booming economy has stimulated energy demand. Most of this is burned to produce electricity, so much so that China has begun to import coal. Coal provides more than 70% of China's electricity. In the United States, 45% of electricity comes from coal-fired power plants. In cost per British thermal unit (Btu), coal is much less expensive than natural gas or oil. Current U.S. reserves are calculated to be 237 billion tons, and at the 2011 rate of production (1.10 billion tons/year), the supply could last 215 years. Unlike oil and natural gas, we produce more coal each year than we use, and we export about 8% of coal production annually.

Mining coal can be hazardous, and injuries and fatalities are not uncommon—in 2010, 48 fatal injuries occurred. More serious, however, is the occupational disease called black lung, or coal workers' pneumoconiosis (CWP), acquired by inhaling coal dust. Fatalities and hospitalizations from this disease have risen in recent years, prompting the

Obama administration to post new regulations designed to lower the risks posed by coal dust.

Coal can be obtained by surface mining (69%) or underground mining (31%). Both methods have substantial environmental impacts. Land subsidence and underground fires often occur in conjunction with underground mines. The town of Centralia, Pennsylvania, for example, has been abandoned and bought by the government because of a coal fire that started 40 years ago and could burn for another 100 years. Coal fires around the world produce almost as much carbon dioxide as do all the cars and trucks in the United States, contributing significantly to the world's greenhouse gas emissions. In underground mining, at least 50% of the coal must be left in place to support the roof of the mine, which is one reason why mining companies have turned to surface mining.

Surface Mining

In the more common surface, or strip, mining process, dynamite is used to break overlying layers, and then gigantic power shovels turn aside the rock above the coal seam and remove the coal. For example, **mountaintop removal mining** began in the southern Appalachians in the late 1970s as a more economical way to get at the valuable low-sulfur coal found in seams up to 1,000 feet down. The Whitesville mine **(Fig. 14–19)** is only one of many that have devastated parts of West Virginia, Kentucky, Tennessee, and Virginia. Roughly 7%—more than 470 mountains—of the 12 million acres of mountainous land in these states has been or will soon be disturbed by mountaintop removal. Already, more than 700 miles of streams have been completely buried, and another 800 miles have been degraded by the waste products that leach from the mine waste.

Lawsuits in the late 1990s over the impact of mine waste held the practice down, but a rule by the Bush administration changed all that. In 2002, mining debris was reclassified from "waste" to "fill," thus allowing companies free rein to dump the debris into stream valleys. A number of grassroots organizations, including the Ohio Valley Environmental Coalition, Mountain Justice, Appalachian Voices, and Christians for the Mountains, have been formed to combat mountaintop removal mining. They face a formidable array of politicians and coal industrialists who argue that the procedure is legal, generates thousands of jobs, and provides the country with coal it desperately needs. Indeed, West Virginia supplies about 15% of U.S. coal production.

The Obama administration has taken action to bring mountaintop removal mining under greater regulatory control. Citing the Clean Water Act, the EPA issued "final guidance" on the mining practices that strengthens and clarifies the permitting process by giving greater protection to waterways affected by mining operations. This action was prompted by a widespread failure of the mining companies to provide water discharge monitoring reports and use of a fast-track permitting process that led to violations of the Clean Water Act.

Coal Power

Virtually all of the coal the United States produces is used to generate electricity. Once it is mined, coal is transported to large power plants. A typical 1,000-MW plant burns 8,000 tons of coal a day—a mile-long train's worth. Combustion

Figure 14–19 Aerial view of mountaintop coal removal. This coal mining operation in Whitesville, West Virginia, is managed by Massey Energy. The mountain tops are stripped away and dumped in valleys to get at the coal below.

of this much coal releases 20,000 tons of carbon dioxide and 800 tons of sulfur dioxide (most of which is prevented from entering the atmosphere if the power plant is well provided with scrubbers; see Chapter 19). Then there is the waste: 800 tons of fly ash and 800 tons of boiler residue ash daily, requiring a special landfill.

Coal-burning power plants are responsible for more than generating electricity; each year pollution from them takes an estimated 13,200 lives and costs more than $100 billion in health care expenditures. The main reason for these impacts is older power plants that lack pollution controls, exempted by grandfather clauses in older Clean Air Act legislation. The EPA has recently invoked the Clean Air Act "good neighbor rule," which prohibits downwind pollution, to issue its Cross-State Air Pollution Rule. This rule requires heavy reductions in sulfur dioxide and nitrogen oxides in older power plants, effective by 2014. The EPA has also issued new disposal rules governing coal ash. The utility industry is complaining bitterly about these new rules, claiming that hundreds of thousands of jobs will be lost as they are forced to shut down many older power plants (due to the cost of retrofitting the plants with pollution controls). The utilities challenged this rule in court, and in August, 2012, a US Court of Appeals overturned the rule, stating that the EPA had overstepped its authority. The Court reaffirmed the EPA's authority to reduce downwind air pollution, but took issue with its methods.

Climate Issues

Coal combustion is the world's largest source of CO_2, which is the greenhouse gas most significantly forcing climate change (see Chapter 18). If progress is to be made in reducing CO_2 emissions, coal should be on the way out. In reality, new coal-burning power plants are coming on line as older ones are being phased out. The new plants are more efficient and are equipped with the latest pollution controls but lacking any control of CO_2 emissions. Eleven new plants were commissioned in 2010. Faced with this situation, the Sierra Club has launched its *Beyond Coal* campaign, designed to prevent the permitting and construction of new coal-fired power plants and close hundreds of existing ones. The campaign has already blocked 153 proposed coal plants. Beyond Coal recently received a huge shot in the arm from Bloomberg Philanthropies—a $50 million grant over four years for the campaign. It turns out that there are actually ways to continue to use coal and, at the same time, greatly reduce CO_2 emissions, but the technology is still being developed.

The government-and-industry-sponsored Clean Coal Power Initiative (CCPI) program is the successor to the Clean Coal Technology (CCT) program. The CCT supported a spectrum of experimental technologies applicable to existing power plants and to a new generation of coal-fired power plants. The objectives of both programs have been to remove pollutants from coal before or after burning and to achieve higher efficiencies so that greenhouse gas

emissions are reduced. Another technology developed under the program is the **integrated gasification combined cycle (IGCC) plant**, where coal is mixed with water and oxygen and heated under pressure to produce a synthetic gas (**syngas**) that is burned in a gas turbine to produce electricity. Sulfur and other contaminants are removed before burning. The CO_2 stream from the process is concentrated; one syngas plant in North Dakota pipes the CO_2 to an oil field, where it is injected into oil wells to enhance recovery.

The CCPI is providing a new round of government-industry projects designed to meet the challenges of coal use in the 21st century. Clean coal technologies would be an improvement over the current ones. But unless they include carbon capture and sequestration (CO_2 injected deep below ground) technology, they will still involve substantial greenhouse gas emissions and all of the drawbacks of mining the coal. Until a change in policy requires carbon capture from coal-burning power plants, it will not happen on a commercial scale because the technology will impose a significant cost (estimated at $30–$40 per ton).

14.5 Energy Policy

According to the Department of Energy, the period from 1973 to 1995 saw an 18% reduction in energy growth, saving the economy $150 billion a year in total energy expenditures. This was due to the steps taken in response to the oil crisis of the 1970s. Thus, public policy can bring about major improvements in energy use. Unfortunately, we also slipped back into old ways when the crisis faded into history. The state of affairs clearly demanded a new energy policy. After years of debate, Congress passed the **Energy Policy Act of 2005**, which intended to establish energy policies for the United States for years to come. It was a mixed bag. However, with the control of the 110th Congress passing to the Democratic Party, a sea change began in environmental policy, producing the **Energy Independence and Security Act of 2007** at the end of 2007. Then with the election of Barack Obama as president in 2008, a new administration had the opportunity to influence energy policy. The **American Recovery and Reinvestment Act of 2009 (Recovery Act)** directed $90 billion into clean energy; further, administrative regulations have provided some major changes in energy policy.

Supply-Side Policies

Both *supply-side* and *demand-side* policy options are available. Because energy supply and demand are such a nationwide matter, only public policy has the sweeping power to effect changes that will make a difference. The Energy Policy Act (*PA*) acted on recommendations from the Bush administration to increase the energy supply. The Energy Independence and Security Act (*EI*), however, mainly addressed the demand side. Obama administration actions and the 2009 Recovery Act (*O&R*) provide the most recent policy initiatives on energy supply and demand. The following list presents the major supply-side policies in these laws.

- **Exploring and developing domestic sources of oil and gas**

  PA: Promoted taking an inventory of offshore oil and gas resources, but backed away from opening the Arctic National Wildlife Refuge (ANWR) to drilling.

  O&R: Approved an oil pipeline into Minnesota and Wisconsin to transport oil from Alberta's tar sands. Approval of the controversial Keystone XL line from Canada to Texas is pending. Approved drilling of exploratory wells in the Arctic Ocean. Took steps to increase federal protection of an oil-rich region of ANWR.

- **Increasing the use of the vast coal reserves for energy**

  PA: Promoted energy from coal via clean coal technologies, loan guarantees, and research incentives.

  O&R: Opened up new federal land area in Wyoming's Powder River Basin to coal mining, potentially yielding 2.35 billion tons of coal.

- **Continuing subsidies to oil and nuclear industries**

  PA: Guaranteed $6.9 billion in tax incentives and billions more in loans for these industries.

  EI: Included a broad repeal of tax incentives for oil and gas in the original legislation, designed to offset the cost of many of the new elements of the law, but partisan opposition to the repeal forced the removal of this option. A limited repeal was enacted, enough to offset some estimated costs.

  O&R: Recommended cutting oil and gas subsidies in order to fund a proposed $467 billion job-creation package. This would bring in $41 billion over a decade. Partisan opposition is guaranteed, but some form of the jobs package might get approved.

- **Removing environmental and legal obstacles to energy development**

  PA: Provided tax credits for hydropower plants and streamlined permitting for drilling.

- **Providing access to remote sources of natural gas**

  CR: Recommended expediting the construction of a natural-gas pipeline to bring Alaskan natural gas to the lower 48 states.

  PA: Required periodic reporting of progress in developing a natural-gas pipeline to bring Alaskan natural gas to the lower 48 states. Construction depends on state and private monies, but established federal control over the siting and permitting of new LNG terminals.

Demand-Side Policies

In rethinking our energy strategies, we need to realize that energy has no real meaning or value by itself. Its only worth is in the work it can do. Thus, what we really want is not energy, but comfortable homes, transportation to and from

where we need to go, manufactured goods, computers that work, and so forth. As a result, many energy analysts argue that we should stop thinking in terms of where we can get additional supplies of the old fuels to keep things running as they are. Instead, we should be thinking in terms of how we can satisfy these needs with a minimum expenditure of energy and the least environmental impact.

Imagine discovering a new oil field having a production potential of at least 6 million barrels a day—three times the capacity of the Alaskan field even at its height. Furthermore, assume that this newfound field is inexhaustible and that exploiting it will not adversely affect the environment. The oil field does not exist, but saving energy through greater efficiency is the equivalent of exploiting an untapped reserve—a **conservation reserve**.

Supply-side policies dealing with fossil fuel energy are "business-as-usual" solutions to our energy security issues. They do little or nothing to reduce our vulnerability to oil price disruptions, terrorist actions, and global climate change. Demand-side policies, by contrast, have the advantage of reducing our energy needs and making it more possible to move to a future in which renewable energy can take over. At the same time, these policies will decrease our vulnerability to terrorists and to disruptions of the oil market. They will also save money and reduce pollution from burning fossil fuels. The Energy Policy Act, the Energy Independence and Security Act, and the Obama administration addressed ways to lower energy demand:

- **Increasing the mileage standards for motor vehicles**

  PA: Authorized continued "study" of the **corporate average fuel economy (CAFE)** standards after the House and Senate overwhelmingly opposed their increase. The act did, however, grant *consumer tax credits* for fuel-efficient hybrid and other advanced technology vehicles.

  EI: Raised the CAFE standards for the combined fleet of cars and light trucks gradually, so as to meet a target of 35 mpg by model year 2020. This will save about 1.1 million barrels of oil per day, half of what we import from the Persian Gulf.

  O&R: This goal has been superseded by rules from the Obama administration that mandate a CAFE target of 38 mpg for cars and 28 mpg for light trucks by 2016, New standards have also been issued for heavy trucks and buses, designed to increase fuel economy by 25%.

As noted in Section 14.2 , this *conservation reserve* has already been tapped to some extent, especially with the doubling of the average fuel efficiency of cars from 13 to 27.5 mpg. To put it bluntly, raising the CAFE requirements for all vehicles is the key to greater oil independence and energy security.

- **Increasing the energy efficiency of lighting, appliances, and buildings**

  PA: Provided tax breaks for manufacturers of energy-efficient appliances; encouraged the EPA to continue

Figure 14–20 Energy-efficient lightbulbs. Replacing standard incandescent lightbulbs with screw-in fluorescent bulbs (shown here) can cut the energy demand for lighting by 75–80%. The higher cost of the fluorescent bulbs is more than offset by the greater efficiency and much longer lifetime of the bulb.

its Energy Star program, which also addresses energy-efficient buildings; and provided tax breaks for those making energy efficiency improvements to their homes (see Chapter 16).

EI: Required a 30% efficiency increase for lightbulbs. Compact fluorescent lightbulbs (CFLs) are between 20% and 25% efficient **(Fig. 14–20)**, whereas incandescent lightbulbs are only 5% efficient. The act does not ban the incandescent bulb; however, the long-lasting CFLs already comply with the required efficiency increase and will likely become the light standard of the future. This act also set new efficiency standards for appliances: refrigerators, freezers, washers, dishwashers, dehumidifiers, and residential boilers. The act also addressed energy efficiency in buildings and industry, encouraging the development of energy-efficient "green" commercial buildings and supporting a weatherization program sponsored by the Department of Energy.

O&R: The Recovery Act extended the Energy Star program to many more facilities; provided tax credits for improving energy efficiency in buildings; developed more efficiency standards for appliances; promoted plans for a smart grid; invested in weatherizing low-income homes.

- **Encouraging industries to use combined heat and power (CHP) technologies**

PA: Took no significant positive action on CHP, but amended a previous act that required utilities to buy power from CHP plants.

EI: Defined this technology as a suitable form of energy savings, making it eligible for energy efficiency grants and research proposals.

According to CHP technology, a factory or large building installs a small power plant that produces

electricity and heats the building with the "waste" heat. Such systems can achieve an efficiency of 80%. In another energy-saving step, a new technology—the combined-cycle natural-gas unit—has been increasingly employed to generate electricity. Two turbines are employed in this unit: The first turbine burns natural gas in a conventional gas turbine (Fig. 14–6b); the second turbine is a steam turbine (Fig. 14–6a) that runs on the excess heat from the gas turbine. The use of this technology boosts the efficiency of the conversion of fuel energy to electrical energy from around 30% to as high as 50%. The cost of combined-cycle systems is half that of conventional coal-burning plants, and the pollution from such systems is much lower. A large proportion of new units being built or planned for the future consists of combined-cycle systems.

- **Promoting greater use of non-fossil fuel sources of energy (nuclear and renewable energy)**

PA: The act included a wide range of incentives to stimulate nuclear power: tax credits for generated electricity, insurance against regulatory delays, and federal loan guarantees. These incentives amount to $13 billion. The act also authorized $4.5 billion in support of renewable energy and efficiency, with tax credits for wind, solar, and biomass energy and major funding for research and development of renewable energy initiatives. The Energy Policy Act will be reviewed later (Chapters 15 and 16).

EI: Established a renewable fuel standard, encouraging the production of biofuels. However, a more extensive renewable energy portfolio standard, requiring the electric utilities to obtain 15% of their energy supply from renewable sources by 2020, was withdrawn in the end because of Republican opposition in the Senate. These and other renewable energy provisions will be covered later (Chapter 16).

O&R: The Recovery Act poured some $90 billion into clean energy, including subsidies to home renewable energy systems and businesses, electric car development, advanced biofuels, smart electric meters, factories to manufacture renewable energy components, and many others (Chapter 16). Unfortunately, many of these programs are under attack from conservatives in Congress, who favor the oil and gas industries and the electric utilities.

Final Thoughts

Conserving energy is extremely important, but keep in mind that reducing our use of fossil fuels is not eliminating that use. There are two major pathways for developing a low-carbon energy future: pursuing nuclear power and promoting renewable-energy applications. Nuclear power has been in use for 50 years, and various renewable-energy systems have also been developed. Thus, both nuclear power and renewable alternatives are at a stage where they could be expanded to provide further energy needs if suitable technological solutions and public acceptance (in the case of nuclear power) or pressure and government support (in the case of renewable energy) are provided. (Chapter 15 focuses on the nuclear option, and Chapter 16 focuses on renewable options. Broad public policy needs are summarized at the end of Chapter 16).

REVISITING THE THEMES

Sound Science

Science and technology have flourished in the age of fossil fuels, not least because they have devoted much research toward developing new and often ingenious ways of extracting the energy from those fuels. The need today, however, is for scientists and engineers to point the way toward using fossil fuels with greater efficiency. Two examples are the CHP technology that is being increasingly employed by factories and institutions and the IGCC plants developed under the Clean Coal Technology Program. A key application of IT is the work on a smart grid and its capacity for managing electricity, promoted by the Obama administration.

Sustainability

The 19th, 20th, and 21st centuries have benefited immensely from the fossil fuels buried in the Earth over past eons. These are, however, nonrenewable resources. They will not be available for very long, as human history goes. We are currently traveling on an unsustainable track, powered by fuels that are running out. Sustainable options are available, but we are not putting

our best efforts there. The most serious consequence of our current track is that we continue to pump greenhouse gases, especially CO_2, into the atmosphere, where they are already affecting the global climate and promise to do so at an accelerating rate as we burn ever more fossil fuels. Eventually, human civilization will have to make a transition to a low-carbon economy. The sooner this becomes a reality, the better off the entire planet will be.

Stewardship

The fossil fuel business is costly in human health and welfare, but we are so deeply committed to using fossil fuels that we simply chalk these concerns up to the cost of doing business. Further, what concern, if any, should we have for future generations? During the 20th century, we used half of the available oil, and our use is accelerating. A thoughtful stewardship of available resources should motivate us to refrain from pumping all the oil and natural gas to satisfy present demands and to plan for the future generations that will be facing far greater energy difficulties than we are as they struggle to complete the transition to renewable energy.

REVIEW QUESTIONS

1. What happened in the Gulf of Mexico in April 2010, and what was the cause? The consequences?

2. How have the fuels to power homes, industry, and transportation changed from the beginning of the Industrial Revolution to the present?

3. What are the three primary fossil fuels, and what percentage does each contribute to the U.S. energy supply?

4. Electricity is a secondary energy source. How is it generated, and how efficient is its generation?

5. Name the four major categories of primary energy use in an industrial society, and match the energy sources that correspond to each.

6. What is the distinction between *undiscovered resources* and *proved reserves* of crude oil, and what factors cause the amounts of each to change?

7. What did the United States do in the early 1970s to resolve the disparity between oil production and consumption? What events caused the sudden oil shortages of the mid-1970s and then the return to abundant, but more expensive, supplies?

8. What is the Carter Doctrine, and how does it relate to wars in the Persian Gulf?

9. What is meant by Hubbert's Peak, and how does it apply to the United States and the world?

10. What is the impact of Canadian oil sands on oil reserves and supply to the United States?

11. What new technology has revolutionized the natural gas scene in the United States?

12. Describe surface mining as it is practiced in the southern Appalachians.

13. What phenomenon may restrict the consumption of fossil fuels, especially coal, before resources are depleted?

14. Compare the Energy Policy Act of 2005, the Energy Independence and Security Act of 2007, and Obama administration initiatives relative to supply- and demand-side strategies.

15. To what degree can energy conservation serve to mitigate energy dependency? What are some prime examples of energy conservation?

THINKING ENVIRONMENTALLY

1. Statistics show that many developed countries use far less energy per capita than we in the United States do. Explain why this is so.

2. Predict what gasoline prices will do in the next 10 years, and rationalize your prediction.

3. Suppose your region is facing power shortages (brownouts). How would you propose solving the problem? Defend your proposed solution on both economic and environmental grounds.

4. List all the environmental impacts that occur during the "life span" of coal—that is, from mining to the electricity in your home or dorm.

5. Explore the pros and cons of the Keystone XL pipeline, fracking for shale gas, and mountaintop removal coal mining.

MAKING A DIFFERENCE

1. An easy first step for reducing your energy use is replacing your conventional lightbulbs with compact fluorescents. You've probably heard that before, but have you done it? Fluorescent bulbs do contain a small amount of mercury, so if you accidentally break one, it would be wise to open your windows; and when you clean up the glass, don't use your bare hands. The mercury levels are low enough that the last traces will vaporize in a matter of hours.

2. When you are in the market for transportation, choose a vehicle that gets high mileage per gallon in anticipation of the higher fuel costs that will certainly occur before the vehicle wears out. If you can afford it, purchase a hybrid vehicle.

3. To further cut down on power, turn your thermostat down a few degrees in the winter and up a few degrees in the summer. Wear warm or light clothes to help make up for the temperature difference. One-sixth of power use in the United States is due to air conditioning.

4. Buy a power strip. This makes it easy to turn things off without having to go to the trouble of unplugging them. Some strips can even detect idle current and shut them off without you having to even flip any switches.

MasteringEnvironmentalScience®

Go to **www.masteringenvironmentalscience.com** for practice quizzes, Pearson eText, videos, current events, and more.

CHAPTER 15

Nuclear Power

Tsunami from offshore earthquake. This photo shows a huge tsunami wave coming on shore at Miyako City, Japan, on March 11, 2011.

E ARTHQUAKES ARE COMMON EVENTS in Japan, which sits on the Pacific Rim, an area well known for its seismic activity. On March 11, 2011, at 2:46 PM local time, the Pacific tectonic plate slipped under the Okhotsk plate that lies under northern Japan, 42 miles offshore and 15 miles below the surface (see Chapter 4, Fig. 4–21, for plate tectonic map). The upthrust of 16–26 feet (5–8 meters) along a 111-mile (180-km) seabed triggered an earthquake of magnitude 9.0, the most severe in Japan's recorded history, and a tsunami that hit the northern Japanese coast with a wall of water 10–50 feet high. The earthquake was damaging, but the Japanese infrastructure was built with earthquakes in mind and the damage was not devastating. The tsunami, however, hit the coast within 10–30 minutes of the earthquake and swept over tsunami seawalls meant to protect cities and towns, pushing buildings, vehicles, and debris inland at a frightening speed. People had little time to escape the incoming waves; the death toll was close to 20,000. More than 1 million buildings were damaged or destroyed, and the economic losses from the disaster were estimated to be more than $300 billion.

Japan's 55 nuclear power plants are located on the coastline where cooling water is available, and when the March 11 earthquake struck, eleven of Japan's 55 operating nuclear power plants shut down automatically. However, several of the plants were on the east coast, and those closest to the epicenter of the earthquake suffered damage first from the earthquake and then from the tsunami. The Fukushima Daiichi power station, with six nuclear reactors, took a direct hit. Seawater swept by the reactors, cut off all electrical power, and demolished the backup generators and diesel fuel tanks. All of the reactors automatically shut down when the quake hit, but even with the nuclear fission halted, radioactive decay in the fuel rods generated heat

1. **Nuclear Energy in Perspective:** Explain how nuclear power was born in the United States, and outline the role of nuclear energy in the United States and the world.

2. **How Nuclear Power Works:** Review the basics of nuclear power, and compare the benefits and disadvantages of nuclear power compared with coal power.

3. **The Hazards and Costs of Nuclear Power Facilities:** Summarize the essentials of radioactivity and nuclear wastes; evaluate nuclear power in the light of high-level wastes, nuclear accidents, and economic considerations.

4. **More Advanced Reactors:** Describe how fast-neutron reactors are used to reprocess spent nuclear fuel, and assess the potential for fusion as a source of energy.

5. **The Future of Nuclear Power:** Examine the impact of global climate change on the nuclear energy option, and assess the future prospects for nuclear power.

and needed to be kept cool. With no cooling water available, three of the reactors suffered meltdown of the nuclear fuel in the reactor vessels, and several units were rocked by explosions set off by hydrogen gas generated by extreme temperatures in the fuel rods stripping the hydrogen from surrounding water (Fig. 15–1). All of these events released radioactive fission products into the air, estimated at about one-tenth of the total radioactivity release from the Chernobyl nuclear accident of 1986. Authorities proclaimed a mandatory evacuation zone for a 12-mile (20-km) radius from the stricken plant; 75,000 people were moved to emergency housing.

CONTAINMENT. Plant workers began the next day to inject seawater via fire extinguisher lines into the reactor vessels to cool them down. Another problem became evident when pools holding spent fuel rods also lost water and began releasing radiation. These too eventually received cooling from seawater. Within days, electrical power was restored to the facility, but conditions in the vicinity of the reactors and pools were so radioactive that workers could not get near enough to make repairs. Some 800 workers were involved in efforts to bring the reactors under control, working short periods in order to limit their exposure to the considerable radiation. The 12-mile exclusion zone received significant radioactive fallout from the stricken plant; there is the likeli-hood that this will become a "dead zone," much like that around Chernobyl. Workers erected a fabric cover over the stricken units to contain airborne radiation emanating from the reactors; meanwhile, new cooling systems for the reactors have been established, but temperature levels in the reactors and fuel pools were not brought down below the boiling point until six months after the disaster. Ultimately, the objective of this phase of containment was to achieve *cold shutdown*— where the temperature at the base of the reactor vessels (and the containment vessels below them in case meltdown penetrated the reactor vessels) is below 200°F and pressure is at 1 atmosphere. At this point, heat from radioactive decay in the fuel and vessels is under control and no more radioactive vapors are escaping. This was finally achieved at the end of 2011. Radiation releases have diminished greatly since the ini-tial meltdown in March. On a positive note, there have been no deaths from radiation exposure either to the plant workers or to citizens at large, though a number of workers at the site received significant radiation doses. This is in huge contrast to the death toll from the tsunami.

Figure 15–1 Fukushima Daiichi power plant. The power plant has six reactors, four of which were damaged by nuclear meltdowns and/or hydrogen explosions. This photo shows the wrecked building housing the No. 4 reactor.

CONSEQUENCES. Fallout from the Fukushima Daiichi nuclear disaster was not all radioactive. Shortly after the disaster, Japan's Prime Minister, Naoto Kan, announced that his government would cancel plans to build 14 more nuclear power plants, which aimed to bring Japan from its current 30% of electricity generated by nuclear power to 50%. Other countries have responded in divergent ways: China is moving full speed ahead with its plans to increase its reliance on nuclear power but, like virtually every country with nuclear power plants, is assessing safety conditions at all of its existing plants and those under construction. Germany and Switzerland announced plans to phase out nuclear energy, and Italy has put its plans to build new reactors on hold. The United States, like China, has ordered a safety review of all existing nuclear power plants. Japan's nuclear disaster is a reminder that nuclear power carries with it risks of events that have low probabilities of occurring but huge consequences if they do. Swedish Nobel physicist Hannes Alfvén said it well, in reference to nuclear power: "No acts of God can be permitted."[1]

The 2011 nuclear disaster in Japan is only one of many in the troubled history of nuclear power. In this chapter, we will delve into that history through the lenses of science and public policy as the many concerns about nuclear power are raised and discussed.

---

[1] "Energy and Environment," *Bulletin of the Atomic Scientists* (May 1972), 6.

15.1 Nuclear Energy in Perspective

Rising oil and natural gas prices, continued emissions of sulfur dioxide and nitrogen oxides from coal, and the long-term threat of climate change all suggest that an en-

ergy future based on fossil fuels is not an option. In light of these serious energy-related problems, it seems pru-dent to do everything possible to develop non-fossil-fuel sources of energy. Nuclear power is an alternative that contributes few emissions to global warming, and there is sufficient uranium to fuel nuclear reactors well into the 21st century, with the possibility of extending the nuclear fuel supply indefinitely through reprocessing technologies. Thirty-one countries now have nuclear power plants either in place or under construction, and in some of these countries, nuclear power generates more electricity than any other source. Is nuclear energy the key to the future of global energy—the best route to a sustainable relationship with our environment?

The History of Nuclear Power

Geologists have known for years that fossil fuels would not last forever. Sooner or later, other energy sources would be needed. It was anticipated that nuclear power could produce electricity in such large amounts and so cheaply that we would phase into an economy in which electricity would take over virtually all functions, including the generation of other fuels, at nominal costs. Following World War II, determined to show the world that the power of the atom could benefit humankind, the U.S. government embarked on a course to lead the world into the Nuclear Age.

The Nuclear Age. And so the government moved into the research, development, and promotion of commercial nuclear power plants (along with the continuing development of nuclear weaponry). Using this research, companies such as General Electric and Westinghouse constructed nuclear power plants that were ordered and paid for by utility companies, with assurances from the federal government, via the Price-Anderson Act of 1957, that these corporations and utilities would be covered by insurance for any legal liabilities incurred. The Atomic Energy Commission, now the Nuclear Regulatory Commission (NRC), an agency in the U.S. Department of Energy (DOE), set and enforced safety standards for the operation and maintenance of the new plants, as it does today.

In the 1960s and early 1970s, utility companies moved ahead with plans for numerous nuclear power plants (**Fig. 15-2**). By 1975, 53 plants were operating in the United States, producing about 9% of the nation's

electricity, and another 170 plants were in various stages of planning or construction. Officials estimated that by 1990 several hundred plants would be online and by the turn of the century as many as a thousand would be operating. A number of other industrialized countries got in step with their own programs, and some less developed nations were going nuclear by purchasing plants from industrialized nations.

Curtailed. After 1975, however, the picture changed dramatically. Utilities stopped ordering nuclear plants, and numerous existing orders were canceled. In some cases, construction was terminated even after billions of dollars had already been invested. Most strikingly, the Shoreham Nuclear Power Plant on Long Island, New York, after being completed and licensed at a cost of $5.5 billion, was turned over to the state in the summer of 1989, to be dismantled after generating electricity for only 32 hours. The reason was that citizens and the state deemed that there was no way to evacuate people from the area should an accident occur. Similarly, just a few weeks earlier, California citizens voted to shut down the Rancho Seco Nuclear Power Plant, located near Sacramento, after a 15-year history of troubled operation. In the United States, 28 units (separate reactors) have been shut down permanently for a variety of reasons.

Currently . . . At the end of 2011, 104 nuclear power plants were operating in the United States. The Watts Bar Unit 1 reactor in Tennessee was the last unit to come online, beginning operations in February 1996; Unit 2—partially completed, but halted in 1988—is again under construction and will be the first reactor to be completed in the 21st century in the United States (**Fig. 15-3**). The 104 existing plants (the most in any country) generate 20% of U.S. electrical power. However, in looking ahead, the U.S. Energy Information Administration projected an increase in nuclear generating capacity from the present level of 100 gigawatts (GW) to 115 GW in 2030 from new power plants expected to open as a result of the Energy Policy Act of 2005, which provides many incentives to stimulate the nuclear power industry.

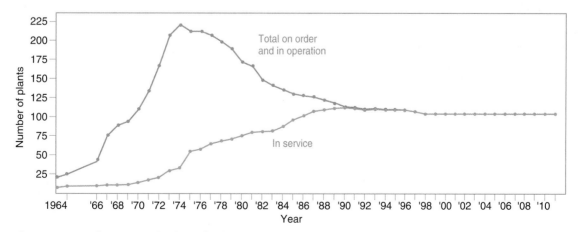

Figure 15-2 Nuclear power in the United States. Since the early 1970s, when orders for plants reached a peak, few utilities called for new plants, and many have canceled earlier orders. Nevertheless, the number of plants in service increased steadily as plants under construction were completed. The number of operating plants peaked at 112 and is holding steady at 104. (*Source:* Data from U.S. Department of Energy.)

Global Picture. After a catastrophic nuclear power plant accident at Chernobyl in the Soviet Union in April 1986, nuclear power was rethought in many countries (see Section 15.3). Yet the demand for electricity is more robust than ever, and the other means of generating electricity have their own problems. Coal-powered electricity produces more greenhouse gases and other pollutants than does any other form of energy. Oil and natural-gas supplies are more limited, and oil is vital to transportation and home heating. Hydroelectric power is already heavily developed. Wind and solar power, though increasing rapidly, is still only a small fraction of the total electricity generated.

Therefore, nuclear power plants continue to be built. Including those in the United States, the world has a total of 441 operating nuclear plants, with an additional 65 under construction. Globally, electricity generated from nuclear power has increased at a rate of 3.2% per year for the past decade. Nuclear power now generates about 13.5% of the world's electricity, but dependence on nuclear energy varies greatly among the 31 countries operating nuclear power plants (Fig. 15–4). France now produces 74% of its electricity with nuclear power and has plans to go to more than 80%. China and India, the world's population giants, also are investing heavily into nuclear energy, China with

Figure 15–3 The Watts Bar Nuclear Generating Station. Located in Spring City, Tennessee, Unit 1 of this power plant was the last reactor to come on line in the United States (in 1996). Unit 2 construction was halted in 1988 but has since resumed, with completion expected in 2013.

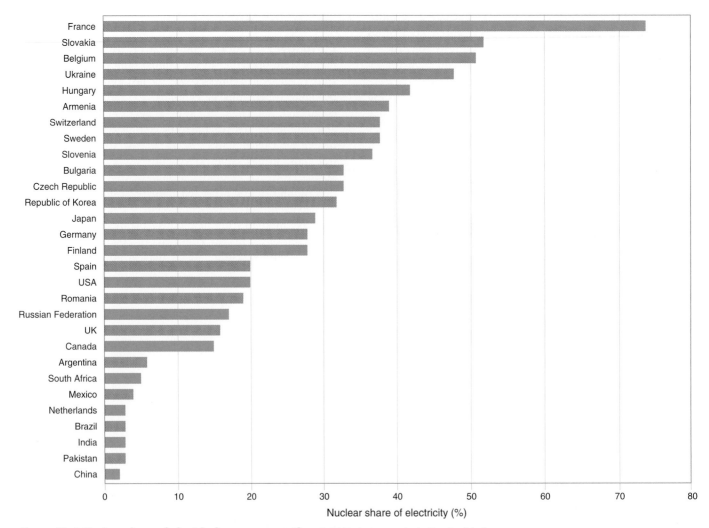

Figure 15–4 Nuclear share of electrical power generation. In 2011, those countries lacking fossil fuel reserves tended to be the most eager to use nuclear power. (*Source:* Data from International Atomic Energy Agency.)

28 and India with six new units under construction. If the safety and economic issues can be resolved, can nuclear power supply future energy needs? In order to react intelligently to this issue, we must start with a clear understanding of what nuclear power is and what its pros and cons are.

15.2 How Nuclear Power Works

The objective of nuclear power technology is to control nuclear reactions so that energy is released gradually as heat. As with plants powered by fossil fuels, the heat energy produced by a nuclear plant is used to boil water and produce steam, which then drives conventional turbogenerators. Nuclear power plants are baseload plants (see Chapter 14); they are always operating unless they are being refueled, and they are large plants, generating up to 1,400 MW.

From Mass to Energy

The release of nuclear energy is completely different from the burning of fuels or any other chemical reactions discussed so far. To begin with, materials involved in chemical reactions remain unchanged at the atomic level, though the visible forms of these materials undergo great transformation as atoms are rearranged to form different compounds. For example, the organic carbon in gasoline is still carbon after it is burned and comes out of the tailpipe as carbon dioxide. Nuclear energy, however, involves changes at the atomic level through one of two basic processes: fission and fusion. In **fission**, a large atom of one element is split to produce two smaller atoms of different elements **(Fig. 15–5a)**. In **fusion**, two small atoms combine to form a larger atom of a different element (Fig. 15–5b). In both fission and fusion, the mass of the product(s) is less than the mass of the starting material, and the lost mass is converted to energy in accordance with the law of mass-energy equivalence ($E=mc^2$), first described by Albert Einstein. Because the small loss of mass is multiplied by the speed of light squared, the amount of energy released by this mass-to-energy conversion is tremendous: The sudden fission or fusion of just one kilogram of material, for example, releases the devastatingly explosive energy of a nuclear bomb.

The Fuel for Nuclear Power Plants. All current nuclear power plants employ the fission (splitting) of uranium-235 (no technology has been devised yet to harness a fusion reaction). The element uranium, which occurs naturally in various minerals in Earth's crust, exists in two primary forms, or isotopes: uranium-238 (^{238}U) and uranium-235 (^{235}U). **Isotopes** of a given element contain different numbers of neutrons but the same number of protons and electrons.

The number 238 or 235 that accompanies the chemical name or symbol for uranium is called the **mass number** of the element. It is the sum of the number of neutrons

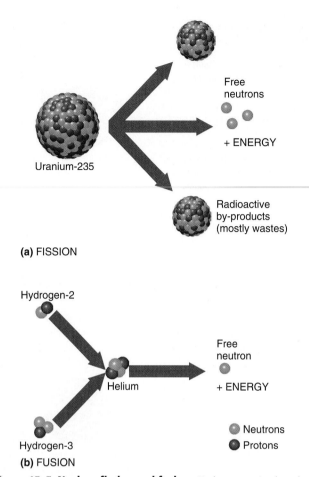

(a) FISSION

(b) FUSION

Figure 15–5 Nuclear fission and fusion. Nuclear energy is released from either (a) **fission**, the splitting of certain large atoms into smaller atoms, or (b) **fusion**, the fusing together of small atoms to form a larger atom. In both cases, some of the mass of the starting atom(s) is converted to energy.

and the number of protons in the nucleus of the atom. All atoms of any given element must contain the same number of protons, so variations in mass number represent variations in numbers of neutrons. ^{238}U contains 92 protons and 146 neutrons; ^{235}U contains 92 protons and 143 neutrons. Although all isotopes of a given element behave the same chemically, their other characteristics may differ profoundly. In the case of uranium, ^{235}U atoms will readily undergo fission, but ^{238}U atoms will not. Like uranium, many other elements exist in more than one isotopic form.

Because ^{235}U is an unstable isotope, a small but predictable number of the ^{235}U atoms in any sample of the element undergo radioactive decay and release neutrons, among other things. If one released neutron moving at just the right speed hits another ^{235}U atom, the latter becomes ^{236}U, which is highly unstable and undergoes fission immediately into lighter atoms (**fission products**). The fission reaction gives off several more neutrons and releases a great deal of energy (Fig. 15–5a). Ordinarily, these neutrons are traveling too fast to cause fission, but if they are slowed down and then strike another ^{235}U atom, they can cause fission to occur again. In this way, more neutrons and more energy are released, with the potential to repeat the

Figure 15–6 Fission reactions. (a) A simple chain reaction. When a ^{235}U atom fissions, it releases two or three high-energy neutrons, in addition to energy and split "halves." If another ^{235}U atom is struck by a high-energy neutron, it fissions, and the process is repeated, causing a chain reaction. (b) A self-amplifying chain reaction leading to a nuclear explosion. Because two or three high-energy neutrons are produced by each fission, each may cause the fission of two or three additional atoms. (c) In a sustaining chain reaction, the extra neutrons are absorbed in control rods, so that amplification does not occur.

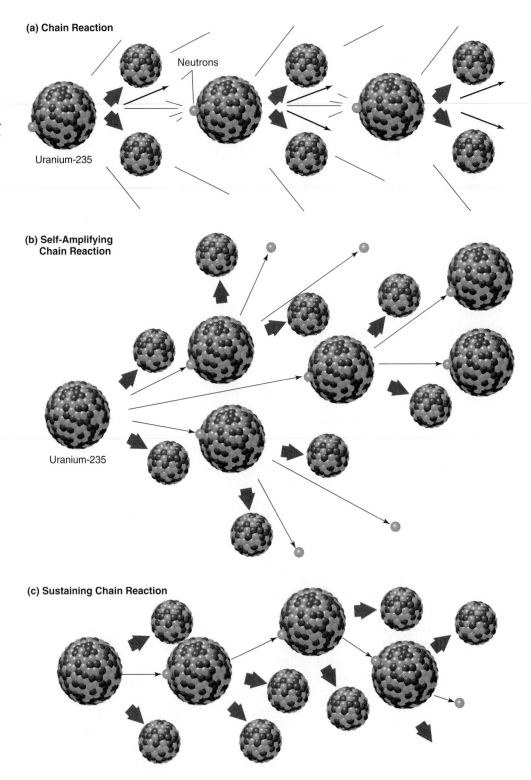

(a) Chain Reaction

Neutrons

Uranium-235

(b) Self-Amplifying Chain Reaction

Uranium-235

(c) Sustaining Chain Reaction

process. A domino effect, known as a **chain reaction**, may thus occur (**Fig. 15–6a**).

    Nuclear Fuel. To make nuclear fuel, uranium ore is mined, purified into uranium dioxide (UO_2), and enriched. Significant uranium deposits in the United States are found in Wyoming and New Mexico and are mined underground or in open pits. The ore is then subjected to **milling**, where it is crushed, treated chemically, and rendered into **yellowcake**, so called because of its yellowish color. The yellowcake, about 80% UO_2, is then delivered to a conversion facility for purification and enrichment.

Because 99.3% of all uranium found in nature is ^{238}U (thus, only 0.7% is ^{235}U), **enrichment** involves separating ^{235}U from ^{238}U to produce a material containing from 3% to 5% of ^{235}U (and the rest ^{238}U). Since these two isotopes are chemically identical, enrichment is based on their slight difference in mass. The technical difficulty of enrichment is the major hurdle that prevents less-developed countries from advancing their own nuclear capabilities.

    An Atomic Bomb. When ^{235}U is highly enriched, the spontaneous fission of an atom can trigger a self-amplifying

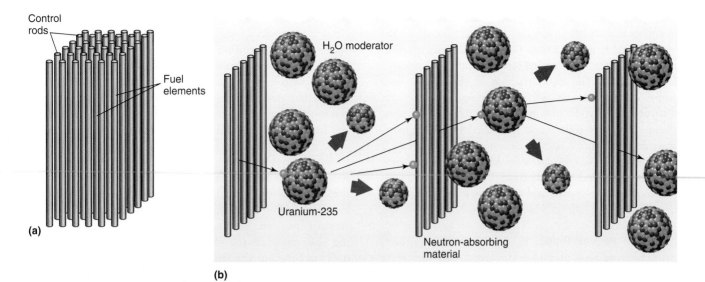

Control
rods

Fuel
elements

(a)

H₂O moderator

Uranium-235

Neutron-absorbing
material

(b)

(c)

Figure 15–7 A nuclear reactor. (a) In the core of a nuclear reactor, a large mass of uranium is created by placing uranium in adjacent tubes, called fuel elements. The rate of the chain reaction is moderated by inserting or removing rods of neutron-absorbing material (control rods) between the fuel elements. (b) The fuel and rods are surrounded by the moderator fluid, near-pure water. (c) Technicians ready the core housing to receive uranium fuel elements in this nuclear reactor.

chain reaction. In nuclear weapons, small masses of virtually pure ^{235}U or other fissionable material are forced together so that the two or three neutrons from a spontaneous fission cause two or three more atoms to undergo fission; each of these in turn triggers two or three more fissions, and so on. The whole mass undergoes fission in a fraction of a second, releasing all the energy in one huge explosion (Fig. 15–6b).

The Nuclear Reactor. A nuclear reactor for a power plant is designed to sustain a continuous chain reaction (Fig. 15–6c). Control is achieved by enriching the uranium to 3–5% ^{235}U. The modest 3–5% enrichment will not support the amplification of a chain reaction into a nuclear explosion. However, in the process of fission, some of the faster neutrons are absorbed by ^{238}U atoms, converting them into ^{239}Pu (plutonium), which then also undergoes fission when hit by another neutron. So much plutonium is produced in this way that at least one-third of the energy of a nuclear reactor comes from plutonium fission.

Moderator. A chain reaction can be sustained in a nuclear reactor only if a sufficient mass of enriched uranium is arranged in a suitable geometric pattern and is surrounded by a material called a **moderator.** The moderator slows down the neutrons that produce fission so that they are traveling at the right speed to trigger another fission. In slowing down the neutrons, the moderator gains heat. In nuclear plants in the United States, the moderator is near-pure water, and the reactors are called light-water reactors (LWRs). (The term *light water* denotes ordinary water (H_2O), as opposed to the substance known as *heavy water*, or deuterium oxide (D_2O). Deuterium is an isotope of hydrogen.)

Fuel Rods. To achieve the geometric pattern necessary for fission, the enriched uranium dioxide is made into pellets that are loaded into long metal tubes. The loaded tubes are called **fuel elements** or **fuel rods** (Fig. 15–7a). Many fuel rods are placed close together to form a reactor core inside a strong reactor vessel that holds the water, which serves as both moderator and coolant (Fig. 15–7b). Over time, fission products that also absorb neutrons accumulate in the fuel rods and slow down the rate of fission and heat production. The highly radioactive spent-fuel elements are then removed and replaced with new ones.

Control Rods. The chain reaction in the reactor core is controlled by rods of neutron-absorbing material, referred to as **control rods**, inserted between the fuel elements. The chain reaction is started and controlled by withdrawing and inserting the control rods as necessary. Consequently, a nuclear reactor is simply an assembly of fuel elements, moderator-coolant, and movable control rods (Fig. 15–7). As the control rods are removed and a chain reaction is initiated, the fuel rods and the moderator become intensely hot.

The Nuclear Power Plant. In a nuclear power plant, heat from the reactor is used to boil water to provide steam for driving conventional turbogenerators. One way to boil the water is to circulate it through the reactor, as in the **boiling-water reactor**. Two-thirds of the power plants in the United States, however, are **pressurized-water reactors (Fig. 15–8)** employing a double loop. In these, the moderator-coolant water is heated to more than 600°F by circulating it through the reactor, but it does not boil because the system is under very high pressure (155 atmospheres). The superheated water is circulated through a heat exchanger, boiling other, unpressurized water that flows past the heat exchanger tubes. This action produces the steam used to drive the turbogenerator.

Loss-of-Coolant Accident (LOCA). The double-loop design of the pressurized-water reactor isolates hazardous materials in the reactor from the rest of the

power plant. However, both reactor types have one serious drawback: If the reactor vessel should crack, the sudden loss of cooling water from around the reactor, called a **loss-of-coolant accident** (**LOCA**), could result in the core's overheating. The sudden loss of the moderator-coolant water would cause fission to cease, because the moderator would no longer be present. Nevertheless, the fuel core could still overheat, because 7% of the reactor's heat comes from radioactive decay in the newly formed fission products. In time, the uncontrolled decay would release enough heat energy to melt the materials in the core, a situation called a **meltdown**. Then the molten material falling into the remaining water could cause a steam explosion. To guard against all this, backup cooling systems keep the reactor immersed in water should leaks occur, and the entire assembly is housed in a thick concrete containment building (Fig. 15–8). In the Fukushima Daiichi nuclear plants, cooling water was no longer being circulated because the pumping mechanism had no power. A meltdown occurred in three of the units, and explosions resulted as hydrogen was separated from the superheated water (a chemical reaction where the water is split into its constituent hydrogen and oxygen).

Comparing Nuclear Power with Coal Power

Because nuclear plants are baseload plants that provide the foundation for meeting the daily and weekly electrical demand cycle, they must be replaced with other baseload plants. If nuclear power plants are phased out, one option is replacing such large plants with coal-burning plants.

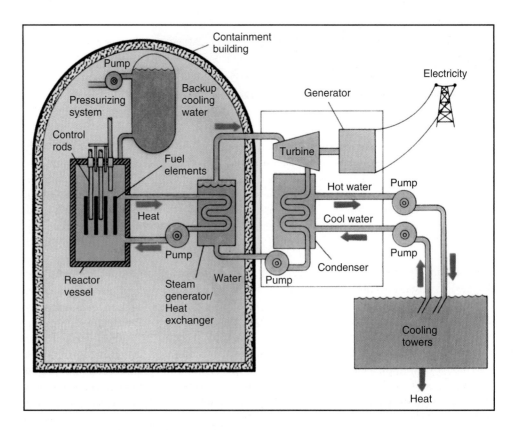

Figure 15–8 Pressurized nuclear power plant. The double-loop design isolates the pressurized water from the steam-generating loop that drives the turbogenerator. One loop is completely contained within the containment building (left). This loop, under high pressure so that the water does not boil, absorbs the heat from the nuclear reactions and transfers it to the steam generator/heat exchanger. There, water in the second loop is heated to boiling and circulated to the turbogenerator (right), where the steam turns a turbine to generate electricity. The steam is condensed by cool water, and the hot water is circulated to cooling towers.

The United States has abundant coal reserves, but is burning coal the course of action we wish to pursue? Nuclear power has some decided environmental advantages over coal-fired power (**Fig. 15–9**). Comparing a 1,000 MW nuclear plant with a coal-fired plant of the same capacity, each operating for one year, we find the following characteristics:

- **Fuel needed.** The coal plant consumes 2–3 million tons of coal. If this amount is obtained by strip mining, significant environmental destruction and acid leaching will result. If the coal comes from deep mines, there will be human costs in the form of accidental deaths and impaired health. The nuclear plant requires about 30 tons of enriched uranium, obtained from mining 75,000 tons of ore, with much less harm to humans and the environment. The fission of about 1 pound (0.5 kg) of uranium fuel releases energy equivalent to burning 50 tons of coal.

- **Carbon dioxide emissions.** The coal plant emits more than 7 million tons of carbon dioxide into the atmosphere, contributing to global climate change. The nuclear plant emits none while it is producing energy, but fossil fuel energy is used in the mining and enriching of uranium, the construction of the nuclear plant, the decommissioning of the plant after it is shut down, and the transportation and storage of nuclear waste. Because a coal plant also needs to be constructed and the coal and waste ash must be transported, much the same can be said for the coal plant.

- **Sulfur dioxide and other emissions.** The coal plant emits more than 300,000 tons of sulfur dioxide, particulates, mercury, and other pollutants, leading to acid rain and health-threatening air pollution (if it is a new plant, most of these pollutants will be captured by scrubbers). The nuclear power plant produces no acid-forming pollutants or particulates.

- **Radioactivity.** A coal plant releases 100 times more radioactivity than does a nuclear power plant because of the natural presence of radioactive compounds (uranium, thorium) in the coal. The nuclear plant releases low levels of radioactive waste gases.

- **Solid wastes.** The coal plant produces about 600,000 tons of ash that require disposal. (Much of it is reused in cement production.) The nuclear plant produces about 250 tons of highly radioactive wastes that require storage and ultimate safe disposal. (Safe disposal is an unresolved problem.)

- **Accidents.** A worst-case accident in the coal plant could result in fatalities to workers and a destructive fire, a situation common to many industries. Accidents in a nuclear plant can range from minor emissions of radioactivity to catastrophic releases that can lead to widespread radiation sickness, scores of human deaths, untold numbers of cancers, and widespread, long-lasting environmental contamination.

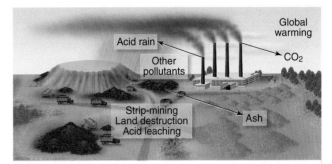

Coal

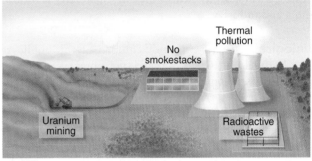

Nuclear

Figure 15–9 Nuclear power versus coal-fired power. Both methods of generating electricity have their environmental advantages and disadvantages. The nuclear power option assumes perfect containment of radioactivity and the availability of some method for waste storage and disposal.

It is the disposal of radioactive wastes and the potential for accidents that have led to public concern over nuclear power. Can these problems be overcome, or are other energy options—natural-gas-powered boilers or renewable energy—less problematic and less costly than nuclear power?

15.3 The Hazards and Costs of Nuclear Power Facilities

Assessing the hazards of nuclear power requires an understanding of radioactive substances and their danger.

Radioactive Emissions

When uranium or any other element undergoes fission, the split "halves" are atoms of lighter elements, such as iodine, cesium, strontium, cobalt, or any of some 30 other elements. These newly formed atoms—the direct products of the fission—are generally unstable isotopes of their respective elements. Unstable isotopes (called **radioisotopes**) become stable by spontaneously ejecting subatomic particles (alpha particles, beta particles, and neutrons), high-energy radiation (gamma rays and X-rays), or both. Radioactivity is measured in **curies** (after Marie Curie, the scientist who first coined the term *radioactivity*); one gram of pure radium-226 gives off one curie per second, which is approximately 37 billion spontaneous disintegrations into particles and radiation. The particles and radiation are collectively referred to as **radioactive emissions**. Materials in the reactor may also be converted to unstable isotopes and become radioactive by absorbing neutrons from the fission process (**Fig. 15–10**). These indirect products of fission, along with the direct products, are the **radioactive wastes** of nuclear power. The direct products of fission generate **high-level wastes**, so called because they are highly radioactive. The indirect products of fission, such as the reactor materials, are much less radioactive and are classified as **low-level wastes**, a category that also includes materials from hospitals and industry.

Biological Effects. A major concern of the public regarding nuclear power is that large numbers of people may be exposed to low levels of radiation, thus elevating their risk of cancer and other disorders. Is this a valid concern?

Radioactive emissions can penetrate biological tissue, resulting in radiation exposure. This exposure is measured as absorbed dose, using joules per kilogram (J/kg)—energy imparted (J) per mass of body tissue (kg). Doses are reported in Sieverts (Sv); the old unit *rem*, still in use by some agencies, is the equivalent of 0.01 Sv.

As radiation penetrates tissue, it displaces electrons from molecules, leaving behind charged particles, or ions, so the emissions are called ionizing radiation. The process of ionization may involve breaking chemical bonds or changing the structure of molecules in ways that impair their normal functions. Ordinarily, the radiation is not felt and leaves no visible mark unless the dose is high.

High Dose. In high doses, radiation may cause enough damage to prevent cell division. This is why, in medical applications, radiation can be focused on a cancerous tumor to destroy it. However, if the whole body is exposed to such levels of radiation (over 1 Sievert is considered a high dose), a generalized blockage of cell division occurs that prevents the normal replacement or repair of blood, skin, and other tissues. This result is called radiation sickness and may lead to death a few days or months after exposure. Very high levels of radiation may totally destroy cells, causing immediate death.

Low Dose. In lower doses, radiation may damage DNA, the genetic material inside cells. Cells with damaged (mutated) DNA may then begin dividing and growing out of control, forming malignant tumors or leukemia. If the damaged DNA is in an egg or a sperm cell, the result may be birth defects (mutations) in offspring. The effects of exposure to radiation may go unseen until many years after the event (10 to 40 years is typical). Other effects include the weakening of the immune system, mental retardation, and the development of cataracts.

Exposure. Health effects are directly related to the level of exposure. There is broad agreement that doses between 100 and 500 milliSieverts (1 milliSievert, or mSv = 1/1,000 of a Sievert) result in an increased risk of developing cancer. Evidence for this hypothesis comes from studies of patients with various illnesses who were exposed to high levels of X-rays in the 1930s, before the potential harm of such radiation was understood. People in these groups subsequently developed higher-than-normal rates of cancer and leukemia. It is noteworthy that the Japanese Health Ministry has raised the legal limit for workers at

Figure 15–10 Radioactive wastes and radioactive emissions. Nuclear fission results in the production of numerous unstable isotopes, designated radioactive waste. These isotopes give off potentially damaging radiation until they regain a stable structure.

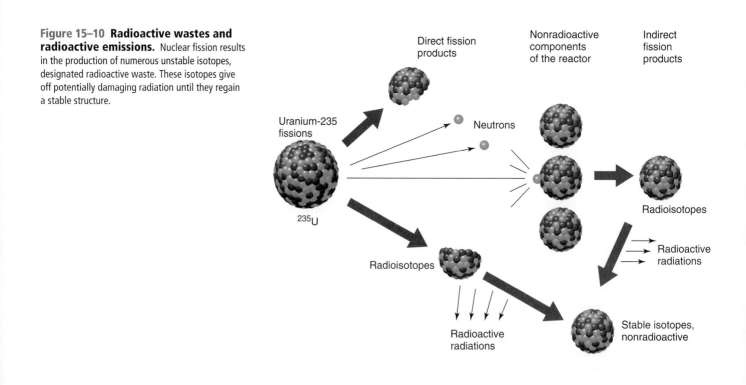

the Fukushima power plant from 100 to 250 milliSieverts, making it possible for them to spend more time working at the stricken plant, but also increasing their risk of developing cancer.

Is radiation dangerous even at doses below 100 milliSieverts? Some scientists point to the ability of living cells to repair small amounts of damage to DNA and believe that there is a threshold of radiation below which no biological effects occur. A study by the National Research Council (NRC)[2] found that there is no safe level of radiation, dismissing the idea of a threshold. Federal agencies employ the concept of no threshold and assume a direct relationship between the amount of radiation and the incidence of cancer. The NRC report predicted that 1 in 1,000 individuals would develop cancer from an exposure to 10 mSv; a tenfold increase in exposure would increase the risk to 1 in 100. Federal standards set 1.7 mSv/yr as the maximum exposure permitted for the general populace per year, except for medical X-rays.

Sources of Radiation.
Nuclear power is by no means the only source of radiation. There is also normal background radiation from radioactive materials, such as uranium and radon gas that occur naturally in the Earth's crust and cosmic rays from outer space. For most people, background radiation is the major source of radiation exposure. In addition, we deliberately expose ourselves to radiation from medical and dental X-rays and, more and more frequently, from CT scans, by far the largest source of human-induced exposure. For the average person, these deliberate exposures equal one-fifth the exposure from background sources. The average person in the United States receives a dose of about 3.6 mSv per year (Table 15–1). Thus, the argument becomes one of relative hazards. That is, does or will the radiation from nuclear power significantly raise radiation levels and elevate the risk of developing cancer?

During normal operation of a nuclear power plant, the direct fission products remain within the fuel elements and the indirect products are maintained within the containment building, which houses the reactor. No routine discharges of radioactive materials (other than permissible minor stack emissions) into the environment occur. Even very close to an operating nuclear power plant, radiation levels from the plant are lower than normal background levels. A radiation detector will pick up more radiation from the ground or the concrete of a basement floor than it will when held within 150 yards of a nuclear power plant. Careful measurements have shown that public exposure to radiation from normal operations of a power plant is less than 1% of natural background radiation.

TABLE 15–1 Relative Doses from Radiation Sources

Source	Dose
All sources (avg.)	3.6 mSv/year
Cosmic radiation at sea level*	0.26 mSv/year
Terrestrial radiation (elements in soil)	0.26 mSv/year
Radon in average house	2 mSv/year
X-rays and nuclear medicine (avg.)	0.50 mSv/year
Consumer products	0.11 mSv/year
Natural radioactivity in the body	0.39 mSv/year
Mammogram	0.30 mSv
Chest X-ray	0.10 mSv
CT scan of abdomen	10 mSv
Continued fallout from nuclear testing	0.01 mSv/year
Living near a nuclear power station (assuming perfect containment)	<0.01 mSv/year
Federal standards for the general populace	1.7 mSv/year

Source: Data from U.S. Nuclear Regulatory Commission.
*Increases 0.005 mSv for every 100 feet of elevation.

The main concern about nuclear power, therefore, should not focus on a plant's normal operation. Instead, the real problems arise from the storage and disposal of radioactive wastes and the potential for accidents.

Radioactive Wastes and Their Disposal

To understand the problems surrounding nuclear waste disposal, you must understand **radioactive decay**, a process in which, as unstable isotopes eject particles and radiation, they become stable and cease to be radioactive. As long as radioactive materials are kept isolated from humans and other organisms, the decay proceeds harmlessly.

Half-Life.
The rate of radioactive decay is such that half of the starting amount of a given isotope will decay within a certain period. In the next equal period, half of the remainder (half of a half, which equals one-fourth of the original) decays, and so on. The time for half of the amount of a radioactive isotope to decay is known as the isotope's **half-life**. The half-life of an isotope is always the same, regardless of the starting amount.

Each particular radioactive isotope has a characteristic half-life. The half-lives of various isotopes range from a fraction of a second to many thousands of years. Uranium fission results in a heterogeneous mixture of radioisotopes,

---

[2]National Research Council, *Health Risks from Exposure to Low Levels of Ionizing Radiation: Beir VII-Phase 2* (Washington, DC: National Academy of Sciences, 2005). Available at fermat.nap.edu/execsumm_pdf/11340.

the most common of which are listed in Table 15–2. Much of this material—in particular, the remaining ^{235}U and the plutonium (^{239}Pu) that has been created by the neutron bombardment of ^{238}U—can be recovered and recycled for use as nuclear fuel in an operation called **reprocessing**. Although Britain and France have a large reprocessing operation, U.S. policy prohibits the practice because of concerns over plutonium and atomic weapons (plutonium can be more easily fabricated into nuclear weapons than can uranium).

The development of nuclear power went ahead without ever fully addressing the issue of what would be done with the radioactive wastes. Proponents of nuclear power generally assumed that the long-lived high-level wastes could be solidified, placed in sealed containers, and buried in deep, stable rock formations (geologic burial) as the need for such containment became necessary. However, this has not yet happened, anywhere. Thus, the current problem of nuclear waste disposal is twofold:

*   **Short-term containment.** Storage during this period allows the radioactive decay of short-lived isotopes (Table 15–2). In 10 years, fission wastes lose more than 97% of their radioactivity. Wastes can be handled much more easily and safely after this loss occurs.

*   **Long-term containment.** The Environmental Protection Agency (EPA) had recommended a 10,000-year minimum to provide protection from the long-lived isotopes. However, some isotopes will remain radioactive for hundreds of thousands of years. Recognizing this, Congress extended the protection planning horizon to 1 million years.

Tanks and Casks. For short-term containment, spent fuel is first stored in deep swimming pool–like tanks on the sites of nuclear power plants. The water in these tanks dissipates waste heat (which is still generated to some degree) and acts as a shield against the escape of radiation. The storage pools can typically accommodate 10 to 20 years of spent fuel. However, the storage pools at U.S. nuclear plants had been filled to 50% of their capacity by 2004 and will be at 100% by 2015. After a few years of decay, the spent fuel may be placed in air-cooled dry casks for interim storage until long-term storage becomes available (Fig. 15–11). The casks are engineered to resist floods, tornadoes, and extremes of temperature. Currently, 63 facilities (mostly nuclear plants) in the United States are licensed to employ dry storage, with the prospect of many more being licensed as pool storage capacity is exhausted.

In the meantime, the wastes from the world's commercial reactors have been accumulating at a rate of 10,000 tons a year, reaching 260,000 tons at the end of 2011, almost all stored on-site at the power plants. Of these wastes, 65,000 tons are in the United States. Furthermore, because of neutron bombardment of the reactor walls, all nuclear power plants will eventually add to the stockpile of radioactive wastes.

High-Level Nuclear Waste Disposal. The United States and most other countries that use nuclear power have decided that

TABLE 15–2 Common Radioactive Isotopes Resulting from Uranium Fission and Their Half-Lives

Short-Lived Fission Products	Half-Life (days)
Strontium-89	50.5
Yttrium-91	58.5
Zirconium-95	64.0
Niobium-95	35.0
Molybdenum-99	2.8
Ruthenium-103	39.4
Iodine-131	8.1
Xenon-133	5.3
Barium-140	12.8
Cerium-141	32.5
Praseodymium-143	13.6
Neodymium-147	11.1
Long-Lived Fission Products	**Half-Life (years)**
Krypton-85	10.7
Strontium-90	28.0
Ruthenium-106	1.0
Cesium-137	30.0
Cerium-144	0.8
Promethium-147	2.6
Additional Product of Neutron Bombardment	**Half-Life (years)**
Plutonium-239	24,000

geologic burial is the safest option for disposing of the highly radioactive spent fuel, but no nation has developed plans to the point of actually carrying out the burial. Most nuclear nations have not even been able to find a site that may be suitable for receiving wastes. Where sites have been selected, questions about their safety have surfaced. The basic problem is that no rock formation can be guaranteed to remain stable and dry for tens of thousands (let alone millions) of years. Everywhere scientists look, there is evidence of volcanic or earthquake activity or groundwater leaching within the past 10,000 years or so, which is to say that it may occur again in a similar period. If such events did occur, the still-radioactive wastes could escape into the environment and contaminate water, air, or soil, with consequent effects on humans and wildlife.

Yucca Mountain. In the United States, efforts to locate a long-term containment facility have been hampered by a severe "not in my backyard" (NIMBY) syndrome. A number of states, under pressure from citizens, have passed legislation categorically prohibiting the disposal of nuclear wastes within their boundaries. In the meantime, the need to select and develop a

Figure 15–11 Dry cask storage.
These dry storage casks contain spent fuel assemblies generated at the James A. Fitzpatrick nuclear power plant in Scriba, New York. The assemblies had been stored for more than 30 years in the spent fuel pool on the plant site. Each dry cask has a 3½-inch steel liner surrounded by 21 inches of reinforced concrete and weighs 126 tons.

long-term repository has become increasingly critical, given the buildup of spent fuel in every nuclear power plant. The **Nuclear Waste Policy Act of 1982** committed the federal government to begin receiving nuclear waste from commercial power plants in 1998. At the end of 1987, Congress called a halt to the debate and selected a remote site, Yucca Mountain, in southwestern Nevada **(Fig. 15–12)**, to be the nation's civilian nuclear waste disposal site. Two other sites being considered were rejected.

Not surprisingly, Nevadans fought this selection, passing a law in 1989 that prohibits any storage of high-level radioactive waste in the state. The federal government has the power to override state prohibitions, though, and the Nevada site has undergone intensive study and exploratory construction over the past 25 years, at a cost of $10 billion. In July 2002, President George W. Bush signed a resolution (passed by Congress) voiding a move by Nevada's governor to block further development of the site. The Yucca Mountain facility was officially designated to be the nation's nuclear waste repository. The DOE submitted a license application for operation of the Yucca Mountain repository to the NRC in June 2008, and NRC staff recommended that the agency adopt DOE's environmental impact statement for the project.

Critics of the site and DOE's plans for disposal question if any degree of safety can be guaranteed, given the possibilities of earthquakes, volcanic activity, or other changes in geologic time. Indeed, studies began to suggest that water would leak through the mountain into the repository, corrode the fuel assemblies stored there, and then release radioisotopes into groundwater. When it was ruled that the site would have to be guaranteed safe for a million years, things began to look bad for the NRC's plans for long-term storage at Yucca Mountain.

Yucca Rejected. The Obama administration decided to terminate the Yucca Mountain program, citing health and safety concerns as the reason. Instead, the administration appointed a Blue Ribbon Commission to recommend a new strategy for the

nuclear power plant waste dilemma. The commission issued its report[3] early in 2012, with the following key recommendations:

- There needs to be a new approach to selecting future nuclear waste management facilities, a transparent one based on consent of the states and communities involved in the selection.

- A new federal corporation should plan and manage the waste disposal program, independent of the NRC.

- There must be a prompt effort to develop a new permanent geologic disposal facility. Even if the politics change and Yucca Mountain is given new life, the inventory of spent nuclear fuel will soon exceed the amount that could be stored at Yucca Mountain.

- Prompt efforts are needed to develop one or more consolidated interim storage facilities; after cooling in containment pools for a decade or so, the spent fuel would be transferred to dry casks at interim sites in one or several central locations.

It is interesting that one radioactive waste repository has already been constructed and is operating to receive defense-related wastes—the Waste Isolation Pilot Plant (WIPP) near Carlsbad, New Mexico. The site is above a huge salt deposit, and the wastes are conveyed to salt caves 2,150 feet underground. In this case, the city of Carlsbad was happy to have the depository, which is accompanied by significant economic benefits. The site was opened in 1999, and by 2011, it had received more than 10,000 shipments of wastes totaling 105,000 cubic yards. These wastes are largely plutonium-contaminated by-products of nuclear weapons plants and laboratories around the country. Radioactive wastes from weapons manufacture in the United States and Russia are discussed in Sound Science, "Radioactivity and the Military" (p. 383); except for the **Megatons**

Figure 15–12 Yucca Mountain, Nevada. This remote ridgeline in Nevada was chosen as the nation's sole high-level nuclear waste depository. The site is on federally protected land, approximately 100 miles northwest of Las Vegas, Nevada.

---

[3]Lee H. Hamilton and Brent Scowcroft, chairmen. *Draft Report of the Blue Ribbon Commission on America's Nuclear Future.* July 2011. www.brc.gov. Nov. 1, 2011.

to **Megawatts** program converting Russian weapons uranium to power-plant fuel, it is not a pretty picture.

Nuclear Power Accidents

Although the waste issues are a serious concern, it is the potential for accidents that generates the greatest concern over nuclear power.

Three Mile Island. On March 28, 1979, the Three Mile Island Nuclear Power Plant near Harrisburg, Pennsylvania, suffered a partial meltdown as a result of a series of human and equipment failures and a flawed design. The steam generator (Fig. 15–8) shut down automatically because of a lack of power in its feedwater pumps, and eventually, a valve on top of the generator opened in response to the gradual buildup of pressure. Unfortunately, the valve remained stuck in the open position and drained coolant water from the reactor vessel. There were no sensors to indicate that this pressure-operated relief valve was open. Operators responded poorly to the emergency, shutting down the emergency cooling system at one point and shutting down the pumps in the reactor vessel. One instrument error compounded the problem: Gauges told operators that the reactor was full of water when, actually, it needed water badly. The core was uncovered for a time and suffered a partial meltdown. Ten million curies of radioactive gas were released into the atmosphere.

The drama held the whole nation—particularly the 300,000 residents of metropolitan Harrisburg, who were poised for evacuation—in suspense for several days. The situation was eventually brought under control, and no injuries or deaths occurred, but it would have been much worse if the meltdown had been complete. The reactor was so badly damaged and so much radioactive contamination occurred inside the containment building that the continuing cleanup proved to be as costly as building a new power plant. There are no plans to restart the reactor. GPU Nuclear, operator of the plant, has since paid out $30 million to settle claims from the accident, although the company has never admitted the existence of any radiation-caused illnesses.

Chernobyl. Prior to 1986, the scenario for a worst-case nuclear power plant disaster was a matter of speculation. Then, at 1:24 AM local time on April 26, 1986, events at a nuclear power plant in Ukraine (then a part of the Soviet Union) made such speculation irrelevant (**Fig. 15–13**). Since that day, Chernobyl has served as a horrible example of nuclear energy gone awry.

Here's how it happened: While conducting a test of standby diesel generators, engineers disabled the power plant's safety systems, withdrew the control rods, shut off the flow of steam to the generators, and decreased the flow of coolant water in the reactor. However, they did not allow for the radioactive heat energy generated by the fuel core, and lacking coolant, the reactor began to heat up. The extra steam that was produced could not escape and had the effect of rapidly boosting the energy production of the reaction. In an attempt to quell the reactor, the engineers quickly inserted the carbon-tipped control rods. The carbon tips acted as moderators, slowing down the neutrons that were produced in the reaction. The neutrons, however, were still speedy enough to trigger more fission reactions, and the result

Figure 15–13 Chernobyl. An aerial view of the Chernobyl reactor disaster. This disaster was a serious nuclear power accident, directly killing at least 31 people and putting countless others in the surrounding countryside at risk for future cancer deaths.

was a split-second power surge to 100 times the maximum allowed level. Steam explosions then blew the 2,000-ton top off the reactor, the reactor melted down, and a fire was ignited in the graphite that was being used as a moderator, burning for days. At least 50 tons of dust and debris bearing 100–200 million curies of radioactivity in the form of fission products were released in a plume that rained radioactive particles over thousands of square miles, 400 times the radiation fallout from the atomic bombs dropped on Hiroshima and Nagasaki in 1945.

Consequences. As the radioactive fallout settled, 135,000 people were evacuated and relocated. The reactor was eventually sealed in a sarcophagus of concrete and steel. A barbed-wire fence now surrounds a 1,000-square-mile exclusion zone around the reactor site. The soil remains contaminated with radioactive compounds, yet for years 2,000 Ukrainian workers were bused in daily to work at the remaining reactor in the Chernobyl complex. In December 2000, the last reactor was permanently shut down.

Only two engineers were directly killed by the explosion, but 28 of the personnel brought in to contain the reactor in the aftermath of the explosion died of radiation within a few months. Over a broad area downwind of the disaster, buildings and roadways were washed down to flush away radioactive dust. Even with these precautions, many people in or near the evacuation zone were exposed to high levels of radiation, especially the short-lived radioisotope iodine-131. Predictions of hundreds of thousands of cancer deaths among those exposed followed shortly after the accident.

A report by a team of experts convened by United Nations agencies[4] has provided the most reliable assessment of the impacts of the accident on the surrounding populations.

---

[4]The Chernobyl Forum: 2003–2005, *Chernobyl's Legacy: Health, Environmental and Socio-Economic Impacts, and Recommendations to the Governments of Belarus, the Russian Federation and Ukraine*, 2nd rev. version (Vienna, Austria: International Atomic Energy Agency, 2005). Available at www.iaea.org /Publications/Booklets/Chernobyl/chernobyl.pdf.

SOUND SCIENCE

Radioactivity and the Military

Some of the worst failures in handling radioactivity have occurred at military facilities in the United States and in the former Soviet Union. A recent calculation taken from government figures indicated that the U.S. nuclear program has, over the years, sickened 36,500 Americans and killed more than 4,000, most of these in connection with the manufacture and testing of nuclear weapons (most of the sickness and death is due to cancer). Further, liquid high-level wastes stored at many U.S. facilities have leaked into the environment and contaminated wildlife, sediments, groundwater, and soil. Activities at these sites have been shrouded in secrecy; only recently have documents revealing past accidents and radioactive releases been declassified and made available to the public. Deliberate releases of uranium dust, xenon-133, iodine-131, and tritium gas have been documented at Hanford, Washington; Fernald, Ohio; Oak Ridge, Tennessee; and Savannah River, South Carolina. Cleaning up these sites is now the responsibility of the DOE, which has already spent $50 billion on this problem and estimates that the final cost could run as high as $250 billion—probably the largest environmental project the country will ever undertake.

From Russia, with Curies

USSR military weapons facilities have been even more irresponsible. The worst is a giant complex called Chelyabinsk-65, located in the Ural Mountains. For at least 20 years, nuclear wastes were discharged into the Techa River and then into Lake Karachay. At least 1,000 cases of leukemia have been traced to radioactive contamination from the Chelyabinsk facility. Even today, standing for one hour (thus receiving 6 Sv of radiation) on the shore of Lake Karachay will cause radiation poisoning and death within a week. The lake dried up in the summer of 1967, and winds blew radioactive dust with 5 million curies

An explosion at Soviet Russia's military weapons facility at Chelyabinsk so heavily contaminated the region that many local villages had to be abandoned and the residents resettled. This was the village of Metlino.

across the countryside, contaminating hundreds of thousands of people. This lake is considered to be the most polluted spot on Earth, a legacy of the Cold War and an enormous continuing source of radioactive contamination. Recognizing the dangers of spreading contamination, Russian authorities have filled the lake with hollow concrete blocks, rocks, and soil, and plants now spread their greenery over this permanently contaminated site.

Megatons to Megawatts

The end of the Cold War has brought the welcome dismantling of nuclear weapons by the United States and the nations of the former Soviet Union. Thousands of nuclear weapons are being dismantled, many as a result of earlier agreements between U.S. President Bill Clinton and Russian President Boris Yeltsin. The agreements include a joint closure of all remaining plutonium weapons production reactors, funds to aid Russia in destroying missile silos and dismantling nuclear submarines and bombs, and vast cuts in the nuclear arsenals of the two former enemies. The radioactive components of

the weapons must be handled with great care and disposed of where they are safe from illegal access and where they will pose no danger to human health for centuries to come—a daunting task.

One element of the U.S.–Russian agreements, the **Megatons to Megawatts Program**, involves a government-industry partnership wherein a private U.S. company oversees the dilution of weapons-grade uranium to the much lower enrichment of power-plant uranium, sells it to United States nuclear power plants, and pays the Russian government a market price (more than $7 billion to date). As a result, by November 2011, 17,000 nuclear warheads had been eliminated, and 433 tons of highly enriched uranium had been diluted to some 12,500 tons of power-plant fuel (see illustration). Fully half of the uranium fuel currently used by U.S. power plants comes from this program.

The closing down of much of the nuclear warfare enterprise is certainly a welcome development, but the problems of security and of disposal of the wastes will remain as a long-term legacy of the Cold War.

The main health impact has been an outbreak of thyroid cancer among children who drank milk contaminated with radioactive iodine. Thousands of cases have been diagnosed in this group, with numerous deaths. Several thousand additional deaths due to cancer are expected, but the increase will be hard to detect because of the hundreds of thousands of cancer deaths from other causes to be expected in the affected populations. However, 22 years after the accident, Russian television's *Health* program reported increases in thyroid cancer some 20 to 70 times above normal in many regions of the country and recommended that people between the ages of 20 and 40 from regions around the Ukraine seek ultrasound thyroid examinations.

The hastily constructed shelter over the damaged power plant is in a state of decay, prompting a plan for construction of a safe structure for the long-term entombment of the reactor complex. A \$2.1 billion "New Safe Confinement," funded by some 29 governments and coordinated by the European Bank for Reconstruction and Development, will consist of a steel structure large enough to house the Statue of Liberty; construction was begun in 2008 and will take six years to complete.

Could It Happen Here? Are we in the United States in danger of such an explosion occurring at a nuclear power plant? Nuclear scientists argue that the answer is no because U.S. power plants have a number of design features that should make a repeat of Chernobyl impossible. For one thing, the Chernobyl reactor used graphite as a moderator rather than the water used in LWRs. Further, LWRs are incapable of developing a power surge more than twice their normal power, well within the designed containment capacity of the reactor vessel. Finally, LWRs have more backup systems to prevent the core from overheating, and these reactors are housed in thick, concrete-walled containment buildings designed to withstand explosions such as the one that occurred in the Chernobyl reactor, which had no containment building. LWRs are not immune to accidents, however, the most serious being a core meltdown as a result of a total loss of coolant. This is what happened at three of the reactors at the Fukushima Daiichi power plant as a result of the earthquake and tsunami off the coast of Japan on March 11, 2011.

Fukushima Daiichi. The opening story of this chapter describes what can go wrong with nuclear power in the face of unexpected natural disaster. There was no anticipation of such an intense earthquake accompanied by a strong tsunami and, as a result, no way to accurately protect the power plant. Power was lost from the electrical grid and backup diesel generators, leaving only eight hours of batteries to provide power for cooling the reactors. There was no way to provide cooling to spent fuel once the water drained out of their pools. There was no way to vent dangerous gases from the containment buildings. Again, could such an accident happen in the United States? The Union of Concerned Scientists (UCS), an NGO that has spent years evaluating nuclear power plant safety, believes that it could.[5] A task force organized by the NRC disagrees.[6]

The real cost of the accidents at Three Mile Island, Chernobyl, and Fukushima must be reckoned in terms of public trust. Public confidence in nuclear energy plummeted after the first two accidents, which pointed to human error as a highly significant factor in nuclear safety—and human error is something the public understands well. Fukushima wrote a new chapter in nuclear power vulnerability to disasters, and the public also understands natural disasters. Nuclear proponents suffered a serious loss of credibility with Three Mile Island, but Chernobyl and Fukushima were their worst nightmares come true—full catastrophes, just what the antinuclear movement had predicted might someday occur. Given that there will soon be 500 nuclear power plants in operation globally, the future safety of nuclear power must be a major concern.

---

[5]Union of Concerned Scientists. "Can It Happen Here?" *Catalyst*, Summer 2011: 7–9.
[6]Nuclear Regulatory Commission. Recommendations for Enhancing Reactor Safety in the 21st Century. Near-Term Task Force. July 12, 2011. www.psr.org. Nov. 3, 2011.

Safety and Nuclear Power

As a result of Three Mile Island and other, lesser incidents in the United States, the NRC has upgraded safety standards not only in the technical design of nuclear power plants, but also in maintenance procedures and in the training of operators. Thus, proponents contend, nuclear plants were designed to be safe in the beginning, and now they are safer than ever. Some proponents of nuclear power claim that we now have the technology to build inherently safe nuclear reactors, designed in such a way that any accident would result in an automatic quenching of the chain reaction and suppression of heat from nuclear decay. In reality, however, there is no such thing as an inherently safe reactor, because that concept implies no release of radioactivity under any circumstances—an impossible expectation.

Passive Safety. Instead, nuclear scientists have designed a new generation of nuclear reactors with built-in passive safety features rather than the active safety features found in current reactors. **Active safety** relies on operator-controlled actions, external power, electrical signals, and so forth. As accidents have shown, operators may override such safety factors, and electricity, valves, and meters can fail or give false information. **Passive safety**, by contrast, involves engineering devices and structures that make it virtually impossible for the reactor to go beyond acceptable levels of power, temperature, and radioactive emissions; their operation depends only on standard physical phenomena, such as gravity and resistance to high temperatures.

New Generations of Reactors. Nuclear scientists distinguish four generations of nuclear reactors. *Generation I* reactors are the earliest, developed in the 1950s and 1960s; few are still operating. The majority of today's reactors are *Generation II* vintage, the large baseline power plants, of several designs. *Generation III* refers to newer designs with passive safety features and much simpler power plants—among them the so-called **Advanced Boiling-Water Reactors (ABWRs) (Fig. 15–14)**. General Electric (GE) has submitted the **Economic Simplified Boiling Water Reactor (ESBWR)** design for NRC approval. Compared with GE's current ABWR, this new design has many more passive safety features. In the event of a LOCA, two separate passive systems would allow water to drain by gravity to the reactor vessel. In the United States, four utility companies have applied for a federal construction and operating license based on this new reactor; of the 26 new U.S. license applications, the majority are for the Westinghouse **AP1000 Advanced Passive Reactor**, a pressurized-water reactor with new passive safety features. In 2012, the NRC awarded a license for two AP1000 reactors to be built at a site near Augusta, Georgia, the first license to be given in 34 years.

In East Asia, the nuclear reactor design of choice is now the ABWR, which produces 1,400 MW of electricity. Japan, with 55 nuclear plants, has four of these in operation; several more are under construction in Japan and Taiwan, and there are plans for many more. The new plants are designed to last for 60 years, can be constructed within five years, and are simple enough that a single operator can control them under normal conditions. They

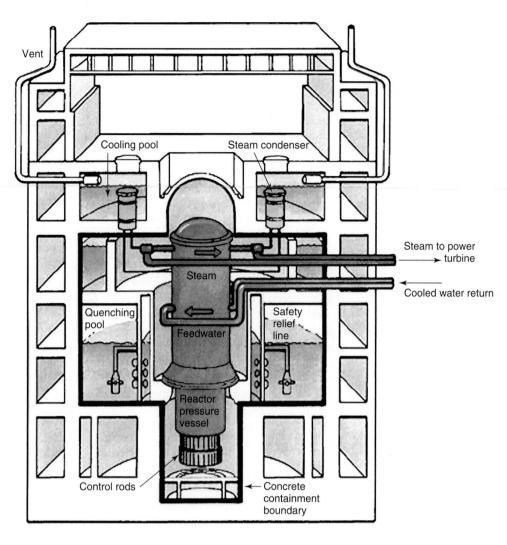

Vent

Cooling pool   Steam condenser

Steam to power
→ turbine

Steam

Cooled water return

Quenching
pool

Safety
relief
line

Feedwater

Reactor
pressure
vessel

Control rods

Concrete
containment
boundary

Figure 15–14 Advanced Boiling-Water Reactor. The core is surrounded by three concentric structures: a reactor pressure vessel, in which heat from the reactor boils water directly into steam; a concrete chamber (outlined with heavy black line) and a water pool, which together contain and quench steam vented from the reactor in an emergency; and a concrete building, which acts as a secondary containment vessel and shield. Any excessive pressure in the reactor will automatically open valves that release steam into a quenching pool, reducing the pressure. Water from the quenching pool can, if necessary, flow downward to cool the core. (*Source:* "Advanced Light-Water Reactors," by M. W. Golay and N. E. Todreas. Copyright 1990 by *Scientific American.* All rights reserved.)

generate electricity at a cost competitive with the costs of fossil fuel options.

Generation IV plants are now being designed and will likely be built within the next 20 years. One example of a Generation IV reactor is the pebble-bed modular reactor (PBMR), which will feed spherical carbon-coated uranium fuel pebbles gradually through the reactor vessel, like a gumball machine. The PBMR will be cooled with fluidized helium, an inert gas, which will also spin the turbines. These reactors will be small, will produce about 160 MW of power, and are expected to be cheap to build and inexpensive to operate. The modules can be built in a factory and shipped to the location of the power plant.

One motive behind the new designs is to restore the public's confidence in nuclear energy. Previously, nuclear proponents had emphasized the very low probabilities of accidents. As we have seen, though, improbable events can happen, and when they happen to nuclear power plants, the consequences can be dreadful. It may be that the public was on the way to a renewed appreciation of nuclear power, but this was set back on September 11, 2001.

Terrorism and Nuclear Power. What would have happened if one of the terrorist suicide groups that took over four airliners on September 11, 2001, had targeted a nuclear

power plant instead of the World Trade Center? This question has raised great concern in many venues, from the NRC, to Congress, to local hearings in municipalities that host nuclear plants. After some uncertainty following September 11, the general consensus is that a jetliner could not penetrate the very thick walls of the containment vessel protecting the reactor. After September 11, the NRC immediately beefed up the requirements for security around every power plant—adding more guards, more physical barriers, and vehicle checks at greater distances from the plant—as well as keeping the shoreline near the plants off limits. A recent National Academy of Sciences (NAS) report[7] concluded that the most vulnerable locations for terrorist attack are the spent-fuel pools. The NAS recommended policy decisions that would protect the pools from a loss-of-coolant-pool-water event brought on by terrorist activity.

New Safety Issues. The Fukushima Daiichi nuclear disaster has generated new concerns about the safety of both current and possible future nuclear power reactors. In the United States, the NRC has deployed a defense-in-depth philosophy

[7]National Research Council, *Safety and Security of Commercial Spent Nuclear Fuel Storage: Public Report* (Washington, DC: National Academies Press, 2006). Available at www.nap.edu/catalog/11263.html.

that covers everything from design and operation, to precautions against failures and mistakes, containment against the release of radioisotopes in case of accidents, and emergency plans to protect citizens. Nevertheless, the Near-Term Task Force[8] felt that the NRC's approach could be strengthened by:

- Clarifying and solidifying the regulatory framework that governs power plants. This would correct the current patchwork of regulations and voluntary industry initiatives.

- Upgrading where necessary the units that might be affected by earthquakes and flooding events and considering responses to events that are highly unlikely but would have catastrophic repercussions if they happen.

- Improving the venting and prevention against potential hydrogen gas release.

- Strengthening the on-site emergency capabilities against failures of electrical power to the plants.

The UCS analysis[9] agreed with these recommendations and added a further one:

- Require plant owners to move spent fuel after five years to dry cask storage; this lowers the risk of overheating in the pools in case of lost cooling water (as in Fukushima).

The potential for accidents and terrorist acts certainly raises questions about a more widespread promotion of nuclear power. However, as Figure 15–2 shows, utilities in the United States were already turning away from nuclear power before Three Mile Island and Chernobyl. The reasons were mainly economic.

Economics of Nuclear Power

What was behind so many canceled nuclear power plants (more than 100)? First, projected future energy demands were overly ambitious, so a slower growth rate in the demand for electricity postponed orders for all types of power plants. Second, increasing safety standards for the construction and operation of nuclear power plants caused the costs of plants to increase at least fivefold, even after inflation is considered. Current nuclear power plants cost twice as much to build per megawatt compared with a coal plant and five times as much as a natural-gas plant. Adding to the rise in costs was the withdrawal of government subsidies to the nuclear industry. Third, public protests frequently delayed the construction or start-up of a new power plant. Such delays increased costs still more because the utility was paying interest on its investment of several billion dollars even when the plant was not producing power. As these costs were passed on, consumers became yet more disillusioned with nuclear power. Finally, safety systems may protect the public, but they do not prevent an accident from financially ruining the utility. Because radioactivity prevents straightforward cleanup and repair, an accident can convert a multibillion-dollar asset into a multi-

billion-dollar liability in a matter of minutes, as Three Mile Island demonstrated. Thus, nuclear power involves a financial risk that utility executives have been hesitant to take.

Longevity? Originally, it was thought that nuclear plants would have a lifetime of about 40 years. It now appears that lifetimes will be considerably less for many of the older plants. For example, the government of Ontario decided in 2005 to shut down two 540-MW CANDU nuclear reactors more than a decade before their scheduled retirement. The reactors had been idle since 1997 because of corrosion in the pipes carrying heavy coolant water from the reactor core. The repair bill was estimated to be $1.6 billion, approximately the cost of a new reactor.

Worldwide, more than 119 nuclear plants have been shut down after an average operating lifetime of 20 years. This shorter lifetime substantially increases the cost of the power produced because the cost of the plant must be repaid over a shorter period. The lifetimes of nuclear power plants are shorter than originally expected due to embrittlement and corrosion. **Embrittlement** occurs as neutrons from fission bombard the reactor vessel and other hardware. Gradually, this neutron bombardment causes the metals to become brittle enough that they may crack under thermal stress—for example, when emergency coolant waters are introduced in the event of a LOCA. When the reactor vessel becomes too brittle to be considered safe, the plant must either be shut down or be repaired at great cost.

Corrosion is a normal consequence of steam generation. Very hot, pressurized water flows from the core into the steam generator through thousands of 3/4-inch-diameter pipes immersed in water (Fig. 15–8). The water inside and outside these pipes contains corrosive chemicals that, over time, cause cracks to develop in some of the pipes. If the main line conveying steam from the generator to the turbine were to rupture, the sudden increase in pressure in the generator could cause several cracked pipes to break at once. In that case, radioactive moderator-coolant water would be released and would overload safety systems, forcing the plant to vent radioactive gas to the outside. Cracked pipes are "repaired" basically by plugging them, cutting the overall plant power. In March 1995, using a new high-tech probe, officials discovered that up to half of the steam generator pipes in the Maine Yankee plant in Wiscasset, Maine, had developed cracks, some penetrating 80% of the pipes' thickness. As a result, the plant was shut down after only 24 years of operation. Recent problems of corrosion in reactor vessels have both surprised and alarmed NRC officials; apparently, the cooling water can be far more corrosive than originally expected.

Decommissioning. Closing down, or decommissioning, a power plant can be extremely costly. Decommissioning Maine Yankee (Fig. 15–15), now complete, cost $635 million. (By comparison, the plant cost just $231 million to build in the late 1960s.) The site now houses some 64 dry casks containing the spent-fuel assemblies, which await disposal at their final resting place, wherever that happens to be. The owners of the plant are allowed to bill former customers to recover some costs—a "stranded cost" nobody likes. Faced with these costs, some utilities are opting to repair older plants in spite of the high costs of such repairs. Other closed reactors are on sites currently

[8]Task Force, p. ix.
[9]UCS, p. 8.

Figure 15–15 Maine Yankee. The Maine Yankee Nuclear Power Plant in the process of being decommissioned, in 2003. The decommissioning is now complete, involving the removal of 200,000 tons of solid waste by rail, truck, and barge.

occupied by active units. The old, closed-down reactors are simply defueled and allowed to sit until the day when all the reactors are closed and the entire site must be decommissioned. In one closed reactor (Humboldt Bay, California), only the nuclear components were removed, and the plant was converted to a conventional natural-gas plant.

New Life? Until the Fukushima disaster, nuclear power was being viewed more favorably by policy makers, as well as by many in the environmental community, in spite of the economic issues. Most of the operating nuclear power plants in the United States were built before global climate change was recognized as a serious problem, but concerns about the greenhouse gas emissions of conventional coal-fired power plants have brought many to reconsider the advantages of nuclear power. This has come into sharp focus as the time has been drawing near for the retirement of a number of older nuclear plants, most of which will reach the end of their operating licenses within the next 10 years. As plants reach the end of their operating licenses (after 30 years or so), owners are routinely applying for license extensions. After an NRC review, these are awarded, and many plants now are licensed to operate for 60 years. With 28 license applications for new plants received and three site permits and one license issued, the NRC has a renewed set of responsibilities after years of status quo. Most of the new reactors planned are to be built on the sites of existing nuclear power plants. In the United States at least, nuclear power has a good track record, and the global warming imperative plus our growing need for electricity suggest that nuclear power will continue its revival.

15.4 More Advanced Reactors

Uranium—especially ^{235}U—is not a highly abundant mineral on Earth. At the height of optimism about nuclear energy in the 1960s, when as many as 1,000 plants were envisioned by the turn of the century, it was forecast that shortages of ^{235}U would develop. Breeder reactors, which utilize chain reactions, were seen as the solution to this problem.

Breeder (Fast-Neutron) Reactors

Recall from Section 15.2 that when a ^{235}U atom fissions, two or three neutrons are ejected. Only one of these neutrons needs to hit another ^{235}U atom to sustain a chain reaction; the others are absorbed by something else. The breeder reactor is designed so that (nonfissionable) ^{238}U absorbs the extra neutrons, which are allowed to maintain their high speed. (Such reactors are now called **fast-neutron reactors**.) When this occurs, the ^{238}U is converted to plutonium (^{239}Pu), which can then be purified and used as a nuclear fuel, just like ^{235}U. Thus, the fast-neutron reactor converts nonfissionable ^{238}U into fissionable ^{239}Pu, and because the fission of ^{235}U generally produces two neutrons in addition to the one needed to sustain the chain reaction, the fast-neutron reactor may produce more fuel than it consumes. Furthermore, 99.3% of all uranium is ^{238}U, so converting that to ^{239}Pu through fast-neutron reactors effectively increases nuclear fuel reserves more than a hundredfold.

In the United States and elsewhere, small fast-neutron reactors are operated for military purposes. France, Russia, and Japan are currently the only nations with commercial fast-neutron reactors, and these have not been in continuous use. Fast-neutron reactors present most of the problems and hazards of standard fission reactors, plus a few more. The consequences of a meltdown in a fast-neutron reactor would be much more serious than those of a meltdown in an ordinary fission reactor, due to the large amounts of ^{239}Pu, with its half-life of 24,000 years. In addition, because plutonium can be purified and fabricated into nuclear weapons more easily than can ^{235}U, the potential for the diversion of fast-neutron reactor fuel to illicit weapons production is greater. Hence, the safety and security precautions needed for fast-neutron reactors are greater. Also, fast-neutron reactors are more expensive to build and operate. However, they have the ability to extract much more energy from recycled nuclear fuel, and they produce much less high-level waste than do conventional nuclear plants.

Reprocessing

The United States has enough uranium stockpiled to satisfy current industry needs. Although the global supply of uranium is not a problem right now, a number of countries are reprocessing spent fuel through chemical processes, recovering plutonium and mixing it with ^{238}U to produce mixed-oxide (MOX) fuel that is about 5% plutonium and suitable for further use in many nuclear power plants. In this way, some of the high-level waste is reused and does not add to the stockpile of waste.

The Bush administration DOE established a program called the Advance Fuel Cycle Initiative (AFCI), moving toward establishing reprocessing in the United States as a way of reducing our growing inventory of spent fuel. However, the Obama DOE has canceled this initiative because of concerns about security and has redirected AFCI into a program that simply provides funding for academic research—on fuel separation and transmutation of plutonium to useful fuels, among other topics. One outcome of AFCI is the International Framework for Nuclear Energy Cooperation (IFNEC), a partnership of 25 countries aiming to improve the proliferation-resistance of the nuclear fuel cycle and work toward cooperative reprocessing of used fuel in fast reactors. It remains to be seen if this partnership develops into something more substantial than just consulting and research, however.

Fusion Reactors

The vast energy emitted by the Sun and other stars comes from fusion (Fig. 15–5b). The Sun, as well as other stars, is composed mostly of hydrogen. Solar energy is the result of the fusion of hydrogen nuclei (protons) into larger atoms, such as helium. Scientists have duplicated the process in the hydrogen bomb, but hydrogen bombs hardly constitute a useful release of energy. The aim of fusion technology is to carry out fusion in a controlled manner in order to provide a practical source of heat for boiling water to power steam turbogenerators.

The d–t Reaction. Because hydrogen is an abundant element on Earth (there are two atoms of it in every molecule of water) and helium is an inert, nonpolluting, nonradioactive gas, hydrogen fusion is promoted as the ultimate solution to all our energy problems; that is, fusion affords pollution-free energy from a virtually inexhaustible resource: water. However, most current designs do not use regular hydrogen (^{1}H), but rather employ the isotopes deuterium (^{2}H) and tritium (^{3}H) in what is called the **d–t reaction.** Deuterium is a naturally occurring, nonradioactive isotope that can be extracted in almost any desired amount from the hydrogen in seawater. Tritium, by contrast, is an unstable gaseous radioactive isotope that must be produced artificially. Radioactive and difficult to contain, tritium is a hazardous substance. As a result, fusion reactors could easily become a source of radioactive tritium leaking into the environment unless effective (and costly) designs are used to prevent that.

In the present state of the art, fusion power is still an energy consumer rather than an energy producer. The problem is that it takes an extremely high temperature—some 100 million degrees Celsius—and high pressure to get hydrogen atoms to fuse. In the hydrogen bomb, the temperature and pressure are achieved by using a fission bomb as an igniter. Getting to the high temperature is only one part of the problem; extracting the heat from the reacting region is also a daunting task, given the large amount of energy released from the fusion reactions.

NIF vs. ITER. Efforts are well under way to demonstrate that controlled fusion is possible. One method is to focus an array of powerful laser beams on a pellet containing hydrogen isotopes (d–t). The laser beams would crush the core of the pellet with such energy that the hydrogen atoms inside would fuse together (ignite) and release energy. This method is employed at the U.S. National Ignition Facility (NIF) in Livermore, California **(Fig. 15–16)**, a $3.5 billion facility completed in 2009 and getting close to achieving better than break-even results—releasing more energy than the energy of the laser beams. This is a demonstration project, however, and even if it works, it would only generate about 3.5 kilowatts of heat, which would only be enough to power one home. The scale-up needed to accomplish laser-driven fusion power is intimidating, but the Japanese and Europeans also have plans to work on this basic method.

A second process is the Tokamak design, in which ionized hydrogen is contained within a magnetic field while being

Figure 15–16 National Ignition Facility. The exterior of the National Ignition Facility, a 10-story building the size of three football fields, is pictured here. NIF is the world's largest and highest-energy fusion laser system and the nation's largest scientific project.

heated to the necessary temperature. Some controlled fusion has been achieved in the Tokamak devices, and in late 1994, a Tokamak facility at the Princeton Plasma Physics Laboratory in Princeton, New Jersey, achieved 10.7 MW of fusion power in 0.27 seconds of reaction. As yet, however, the break-even point has not been reached: More energy is required to run the magnets than is obtained by fusion. For example, it took 39.5 MW of energy to sustain the reaction at Princeton.

An international research team from seven countries is working on a design for an International Thermonuclear Experimental Reactor (ITER), a prototype fusion reactor of the Tokamak variety. The facility, under construction in Cadarache, France, is expected to cost more than $20 billion, and the reactor is expected to produce 10 times as much energy as it consumes. Recently, the United States rejoined the project some 10 years after withdrawing because of concerns about costs and the science itself. The most optimistic view sees the ITER coming on line by 2019. As this plan indicates, developing, building, and testing a fusion power plant will require many more years and many billions of dollars. This assumes that the engineering difficulties will be overcome, which is by no means a given. Thus, fusion is, at best, a very long-term and uncertain option. Many scientists believe that fusion power will always be the elusive pot of gold at the end of the rainbow. The standard joke about fusion power is this: It is the energy source of the future and always will be!

15.5 The Future of Nuclear Power

With fossil fuels contributing to global climate change and with alternatives such as solar and wind power in their early stages, the long-term outlook for energy is discouraging (Chapter 16). Using nuclear power seems to be an obvious choice; however, there continues to be organized opposition to nuclear energy, and it is still a costly technology, highly dependent on government patronage. After the Fukushima Daiichi power-plant disaster, most countries are examining their nuclear options, as we have seen. However, a total of 65 nuclear reactors are currently under construction **(Fig. 15–17)** and 441 are still operating, generating 13.5% of the world's electricity. Russia and other eastern European countries are holding onto their nuclear power

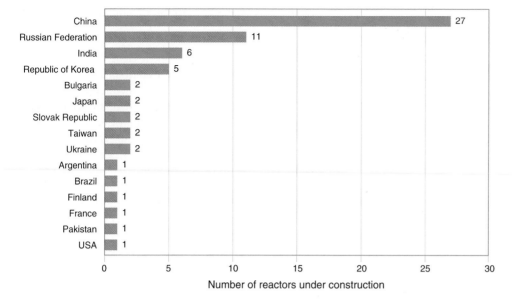

Figure 15–17 Nuclear power reactors under construction, 2010. Worldwide, 65 reactors are under active construction, with an electrical capacity of 63 Gigawatts. (*Source:* Data from International Atomic Energy Agency.)

and building more. A new plant is being built in Finland, and the United Kingdom is planning a new generation of reactors to replace its aging fleet. Asian countries are welcoming nuclear power, with China, South Korea, and India seemingly intent on using it to fulfill a major share of their energy needs. The International Atomic Energy Agency (IAEA) predicts that the Fukushima disaster will slow growth in nuclear energy but not reverse it. The agency also predicts that nuclear power will double its production of energy by 2030.

Opposition

Opposition to nuclear power is based on several premises:

- People have a general distrust of technology they do not understand, especially when that technology carries with it the potential for catastrophic accidents or the hidden, but real, capacity to induce cancer.

- Problems involving lax safety, operator failures, and cover-ups by nuclear plants and their regulatory agencies have occurred in the United States, Canada, and Japan.

- The problems of high construction costs and unexpectedly short operational lifetimes have already been mentioned. Thus, the economic argument can also be used to oppose nuclear power.

- The nuclear industry has repeatedly presented nuclear energy as extremely safe, arguing that the probabilities of accidents occurring are very low. However, when accidents do occur, probabilities become realities, and the arguments are moot.

- Nuclear power plants are viewed as prime targets for terrorist attacks. Critics argue that, therefore, it only makes sense to reduce the number of such targets, not add more of them.

- There remains the crucial problem of disposing of nuclear waste. Siting a long-term nuclear waste repository has been a difficult political as well as technological problem in country after country.

Opponents of nuclear power also cite the basic mismatch between nuclear power and the energy problem. The main energy problem for the United States is an eventual shortage of crude oil for transportation purposes, yet nuclear power produces electricity, which is not yet used much for transportation. Consequently, nuclear power simply competes with coal-fired power in meeting the demands for baseload electrical power. Given the high costs and additional financial risks of nuclear power plants, coal is cheaper, and the United States does have abundant coal reserves.

Nonetheless, there are still the environmental problems of mining and burning coal, especially global climate change. Burning coal emits more CO_2 per unit of energy produced than any other form of energy does (Chapter 14). If the long-term environmental costs were factored into the price of coal, that price would be considerably higher than it is now.

In the final analysis, nuclear power may be a unique energy option, but it is also controversial. Interestingly, public opinion about nuclear power has been shifting. For a time, polls indicated that Americans viewed nuclear power as increasingly important, in the light of surging prices for oil and gas. Some leaders in the environmental field are viewing nuclear energy more favorably, given the serious consequences for global climate change of continued use of fossil fuels. However, shortly after Fukushima, an *ABC News/Washington Post* poll indicated that 64% of Americans oppose any new nuclear construction. Opposition continues because of the perceived risks to both human health and the environment and because of the mismatch to our most critical energy problem—an impending shortage of fuel for transportation. Opponents don't believe that the benefits are worth the risks.

Rebirth of Nuclear Power

If nuclear energy is to have a brighter future, it will be because we have found the continued use of fossil fuels to be so damaging to the atmosphere that we have placed limits on their use but have not been able to develop adequate alternative energy sources. Nuclear supporters point out that U.S. nuclear power plants prevent the annual release of 164 million tons of carbon, 5.1 million tons of sulfur dioxide, and 2.4 million tons of nitrogen oxides. Also, electric cars and plug-in hybrids are now available from major car manufacturers. If this catches on, nuclear-generated electricity can indeed be substituted for oil-based fuels.

Observers agree that if the rebirth of nuclear power is to come, a number of changes will have to be made:

- The public must be convinced that every possible preventative measure has been taken to prevent terrorist attacks and sabotage.

- The industry's manufacturing philosophy will have to change to favor standard designs and factory production of the smaller reactors instead of the custom-built reactors presently in use in the United States. Recently, the DOE selected three designs for development, one of which—the Westinghouse advanced pressurized-water reactor—is a 600 MW unit similar to the one pictured in Figure 15–14. The unit, designed for modular construction, is based primarily on passive safety.

- The framework for licensing and monitoring reactors must be streamlined, but without sacrificing safety measures. This has largely been accomplished with the Energy Policy Act of 2005. The rules allow the NRC to approve a plant design prior to the selection of a specific site.

- The waste dilemma must be resolved. With the Yucca Mountain waste repository now in limbo, resolving the dilemma is becoming more difficult. The recommendations of the Blue Ribbon Commission should be adopted.

Nuclear Energy Policies. President George W. Bush made expanding nuclear energy a major component of his energy policy by signing into law the **Energy Policy Act of 2005**, which included a tax credit for the first 6,000 MW of new nuclear capacity for the first eight years of operation, insurance to protect companies building new reactors from the risk of regulatory delays, and federal loan guarantees for up to 80% of construction costs of new power plants. The act also authorized funds for the planning and construction of a nuclear plant that would produce hydrogen from water (the Next Generation Nuclear Plant), a key part of a coming fuel-cell technology for automobiles (see Chapter 16).

The newer **Energy Independence and Security Act of 2007** addressed nuclear power only tangentially; the legislation had a provision to promote American-built nuclear power plants in overseas markets.

The current Energy Secretary, Steven Chu, is a Nobel laureate atomic physicist who considers nuclear energy to be "safe and reliable." The 2012 DOE budget reflects continued funding for nuclear research and development. In his 2011 State of the Union address, President Obama included nuclear energy as one of the components of a national goal of generating 80% of U.S. electricity from "clean energy sources" by 2035. Accordingly, the 2012 budget requests included substantial increases for the loan guarantees for nuclear power plants. Although there have been some setbacks recently, it looks like nuclear power in the United States may experience a renaissance.

REVISITING THE THEMES

Sound Science

The science behind nuclear power was developed during World War II and was used to produce the two atomic bombs that ended the war with Japan in 1945. Nuclear physicists were eager to show that atomic energy could be put to beneficial uses, however, and with that in mind, the technology for nuclear power plants was developed. Sound science was, and continues to be, essential for this enterprise, especially in regard to improvements in safety and efficiency. Nevertheless, scientific uncertainties remain: For example, what is a safe level of exposure to radioactivity? How can nuclear wastes be stored safely for thousands of years to come? Can nuclear reactors be designed to be accident proof? Can fusion energy ever become a commercial reality? Answers to these questions will challenge scientists for years to come.

Sustainability

Looking toward a sustainable energy future, many believe that nuclear power has a role to play in getting us there. Because uranium must be mined and is not an abundant mineral, supplies will not last indefinitely. Reprocessing fuel can extend the life of nuclear power, but it carries risks that may be unacceptable. Still, nuclear fission may be a way of bridging the gap between fossil fuel energy and renewable energy. The most important aspect of fission is that nuclear energy does not release greenhouse gases and would be capable of providing most of our electrical power, thus enabling us to avoid relying on coal.

The current dilemma is how to deal with the buildup of the highly radioactive spent-fuel wastes. Storing them in pools and casks on the grounds of the power plants is unsustainable. Consequently, countries using nuclear power must find ways to dispose of the wastes safely, which means making sure that they are contained for thousands of years.

Stewardship

It was a major political decision to develop nuclear power for generating electricity. The technology is so expensive and fraught with risk that only government-level agencies can manage the development and ongoing regulation of nuclear power. The roles of the NRC and the EPA are crucial in protecting the U.S. public from the hazards and risks of keeping nuclear power in operation and safely handling its wastes. New reports from the Blue Ribbon Commission on America's Nuclear Future and the Near-Term Task Force provide useful recommendations for handling nuclear wastes and protecting against accidents like the Fukushima Daiichi disaster in Japan.

Government subsidies and guarantees remain in place for this enterprise, and given its potential for massive consequences if things go wrong, that may continue to be necessary. International agreements, such as those between the United States and Russia to dismantle nuclear warheads, point to the importance of maintaining close political ties among all countries with nuclear weapons capabilities or even with nuclear power plants (because these can be a source of weapons-grade nuclear material).

REVIEW QUESTIONS

1. Describe the impacts of the earthquake and tsunami that rocked Japan on March 11, 2011, on the Fukushima Daiichi nuclear power plants, and on the Japanese people.

2. Compare the outlook regarding the use of nuclear power in the United States and globally in each of the following time periods: 1960s, 1980s, and early 21st century.

3. Describe how energy is produced in a nuclear reaction. Distinguish between fission and fusion.

4. How do nuclear reactors and nuclear power plants generate electrical power?

5. What are radioactive emissions, and how are most humans exposed to them?

6. How are radioactive wastes produced, and what are the associated hazards?

7. Describe the current practices for dealing with spent fuel rods.

8. What problems are associated with the long-term containment of nuclear waste? What is the current status of the disposal situation?

9. Describe what went wrong at Three Mile Island, Chernobyl, and Fukushima.

10. What features might make nuclear power plants safer?

11. What are the four generations of nuclear reactors? Describe the advantages of Generations III and IV.

12. Discuss economic reasons that have caused many utilities to opt for coal-burning plants rather than nuclear-powered plants.

13. How do fast-neutron and fusion reactors work? Does either one offer promise for alleviating our energy shortage?

14. Explain why nuclear power does little to address our largest energy shortfall.

15. Discuss changes in nuclear power that might brighten its future.

THINKING ENVIRONMENTALLY

1. Evaluate the risks of nuclear power. Are we overly concerned or not concerned enough about nuclear accidents? Could an accident like Chernobyl or Fukushima happen in the United States?

2. Would you rather live next door to a coal-burning plant or a nuclear power plant? Defend your choice.

3. Global climate change has been cited by nuclear power proponents as one of the most important justifications for further development of the nuclear power option. Give reasons for their belief, and then cite some reasons that would counter their argument.

4. Go to the Web site www.ap1000.westinghousenuclear.com to investigate the reactor described there. Where is it being designed, and where will it likely be used in the United States? What are its features?

MAKING A DIFFERENCE

1. Pay a visit to a nuclear power plant, and explore the ways the facility (a) is protected against terrorist attack, (b) continually trains the plant's operators, (c) keeps the public informed in the event of a nuclear accident, and (d) handles its high-level wastes.

2. Visit the radiology department of your local hospital, and ask a radiologist to explain the different ways radiation is used in the hospital's medical procedures and how (or whether) the hospital tries to limit the radiation exposure of patients. Ask the radiologist what the hospital does with its low-level radioactive waste.

MasteringEnvironmentalScience®

Go to **www.masteringenvironmentalscience.com** for practice quizzes, Pearson eText, videos, current events, and more.

CHAPTER 16

Renewable Energy

Offshore wind power. The Middlegrunden wind farm, some 2 miles off Copenhagen, Denmark, consists of 20 turbines with a capacity of 40 MW. The wind farm produces about 4% of Copenhagen's power. A larger offshore wind farm for Nantucket Sound, off the southern Massachusetts coast, is soon to commence construction.

LEARNING OBJECTIVES

1. **Strategic Issues:** Explain why it is essential to replace fossil fuels with renewable energy sources, and give some details of what has to be done.

2. **Putting Solar Energy to Work:** Recall how much solar energy reaches Earth, and discuss how this energy is being turned into hot water and electricity.

3. **Indirect Solar Energy:** Summarize what is being done to employ hydropower, wind power, and biomass energy to meet energy needs.

4. **Renewable Energy for Transportation:** Discuss how renewable energy is beginning to meet the need to replace oil for transportation, and assess the technologies involved.

5. **Additional Renewable-Energy Options:** Explore the potential for geothermal energy, tidal power, and wave power to meet current and future energy needs.

6. **Policies for a Sustainable-Energy Future:** Evaluate how our national energy policies encourage both renewable energy and energy conservation.

CAPE WIND. Nantucket Sound, a shallow body of water south of Cape Cod, rarely sees a windless day. Indeed, coastal locations are well known for their constant and often cooling winds, attracting vacationers and year-round residents, who also enjoy the beaches and scenery. Summer populations in Massachusetts towns on Cape Cod swell to many times their winter numbers, putting stress on the area's energy infrastructure. So, in 2001, when Cape Wind Associates proposed an offshore wind farm in Nantucket Sound that could supply 75% of Cape Cod's electricity needs, the company expected a warm welcome. Not so. Instead, its proposal ignited a firestorm of controversy. The proposed wind farm would not be small. Some 130 turbines, each 417 feet tall, would be spread over 24 square miles of Nantucket Sound and generate 470 megawatts (MW) of power. Concerns ranged from "visual pollution" to threats to the area's tourism, navigation, fishing industry, and migrating birds. Organizations like the Alliance to Protect Nantucket Sound filed numerous lawsuits since Cape Wind first proposed the wind farm, and all were eventually dismissed. Proponents of the project, however, view these concerns as a classic case of NIMBY (not in my backyard) and point out that New England desperately needs the renewable energy of offshore winds. Opinion polls on and off Cape Cod strongly support the project. The Bureau of Ocean Energy Management (BOEM), which has authority over all offshore developments in federal waters, issued a favorable environmental impact statement, and Interior Secretary Ken Salazar signed a lease for Cape Wind to proceed. The Federal Aviation Administration recently declared the project free of interference with aeronautical operations, thus giving Cape Wind its final major permit.

During the time that Cape Wind has been trying to build the country's first offshore wind farm, nine European countries have been busy erecting 1,200 offshore turbines that have the capacity to generate some 3.2 gigawatts

Figure 16–1 Solar power bag. Konarka has developed a flexible, thin-film plastic PV system called Power Plastic that shows great promise for many solar applications.

Figure 16–2 Solar thermal power in Southern California. A solar-trough power plant. Sunlight striking the parabolic-shaped mirrors is reflected onto the central pipe, where it heats a fluid that is used in turn to boil water and drive turbogenerators.

(GW, = 1,000 megawatts) of electricity. More than 100 GW capacity is planned or under construction for offshore wind power in Europe alone. In the United States, offshore wind has gotten a boost from the Department of Interior, with its "Smart from the Start" accelerated leasing regulations and funding subsidies from the Department of Energy (DOE). This program focuses on the Atlantic Outer Continental Shelf, where 20 projects have already begun the planning and permitting process.[1] These would contribute hugely to the DOE's goal of achieving 20% of our electricity from wind energy by 2030.

Meanwhile, land-based wind turbines are sprouting up everywhere. In the hills east of San Francisco, regiments of wind turbines standing in rows up the slopes and over the crests of the hills are producing electrical power equivalent to that produced by a large coal-fired power plant. At the end of 2011, wind power capacity in the United States reached 47 GW, enough to power 13 million homes. Every year, installations of wind turbines set new records. Wind turbines are operating as well in India, Mexico, Argentina, New Zealand, and other countries around the world.

SOLAR NOW. Imagine you're in a strange environment, perhaps outdoors, out of range of any electrical outlets, and your iPad battery is running low. No problem. Just plug it in to your handy Eco Traveler Solar Power Bag and let its flexible solar panel charge up the device (Fig. 16–1). The bag works for GPS, cell phones, computers, any

number of small devices. The solar power is captured by a thin sheet of plastic polymer, called Power Plastic, made by Konarka, the company that developed the product. Power Plastic has been built into building curtain walls and the roofs of transit shelters in San Francisco and promises to appear in many more products. Power Plastic captures solar energy with an efficiency of 8.3%, which is not as efficient as silicon photovoltaic (PV) cells at 15–20%, but efficient enough considering its flexibility and adaptability.

Panels of traditional silicon-based PV cells are providing electricity in both developed and developing countries around the world. Throughout Israel and other countries in warm climates, it is now commonplace to have water heated by the Sun. Even in temperate climates, many people have discovered that proper building design, insulation, and simple solar collection devices can reduce energy bills by 75% or more. In the desert northeast of Los Angeles are "farms" with rows of trough-shaped mirrors tipped toward the Sun (Fig. 16–2). These reflectors are focusing the Sun's rays to boil water or synthetic oil and drive turbogenerators.

As these examples demonstrate, the use of energy from the Sun and wind has been making remarkable progress, to the point where renewable sources of energy are becoming cost competitive with traditional energy sources—and are far more practical in many situations. But can they replace fossil fuels as the foundational energy sources of modern societies? Given the unsustainability of our present course—due to the limited supplies of fossil fuels and their polluting impacts on global climate and the air we breathe—is it possible that renewable energy will gain the lion's share of energy in the future? Indeed, do we have any choice?

---

[1]U.S. Department of Energy. A National Offshore Wind Strategy: Creating an Offshore Wind Energy Industry in the United States. U.S. Department of Energy and U.S. Department of the Interior. February, 2011. www1.eere.energy.gov/wind/pdfs/national_offshore_wind_strategy.pdf. August 3, 2012.

16.1 Strategic Issues

The objective of this chapter is to give you a greater understanding of the potential for capturing energy from sunlight, wind, biomass, flowing water, and other sources. In 2011,

renewable energy provided 9.4% of U.S. primary energy (Fig. 16–3) and 12.6% of electrical power generation; President Barak Obama has called for 80% of the nation's electricity to be generated from clean energy (non-fossil fuel) sources by 2035. Worldwide, renewable-energy supplies 16%

PERCENT OF U.S. ENERGY USE IN 2011

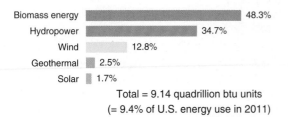

Biomass energy	48.3%
Hydropower	34.7%
Wind	12.8%
Geothermal	2.5%
Solar	1.7%

Total = 9.14 quadrillion btu units
(= 9.4% of U.S. energy use in 2011)

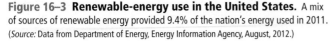

Figure 16–3 Renewable-energy use in the United States. A mix of sources of renewable energy provided 9.4% of the nation's energy used in 2011. (*Source:* Data from Department of Energy, Energy Information Agency, August, 2012.)

TABLE 16–1 Renewable Electric Power at End of 2011, GW Power Capacity

Technology	World Total	United States
Hydropower	970	79
Wind power	238	47
Biomass power	72	14
Solar PV	70	4
Geothermal power	11	3.1
Solar thermal power	1.8	0.5
Ocean (tidal) power	0.5	0
Total renewable power capacity	1,360	147

(*Source:* REN21. Global Status Report. 2012.)

of energy use and 20% of electricity supply (Table 16–1). The UN has targeted a reduction in fossil fuel energy of 80% by 2050, meaning that other sources would have to supply that much. These are ambitious and serious goals, and they raise some significant questions.

Why?

First, why should we want to replace fossil fuels with renewable sources in global energy use, in a matter of decades? The most salient reason is global climate change (see Chapter 18). It is the use of fossil fuels that has loaded the atmosphere with the greenhouse gas carbon dioxide (CO_2), raising the concentration of that one gas by more than 40% to its present concentration of 400 parts per million (ppm). As a result, global temperatures and sea level have already risen, with serious environmental consequences. If we were to continue to increase our use of fossil fuels at the present rate, atmospheric CO_2 would be over 600 ppm by midcentury and well over 1000 ppm by 2100, with global temperature and sea level increases that would be catastrophic. Most international agencies have agreed that to avoid catastrophic climate change it will be necessary to achieve a stable CO_2 level of 450 ppm by mid-century. This is why there is a call to reduce our use of fossil fuels by 80% before 2050.

There are other reasons. Oil and gas reserves will only last a matter of decades, so we are going to have to find new energy sources anyway, and in this century (Chapter 14). Also, there are 1.4 billion people who currently do not have access to electricity and many more whose access is intermittent at best. Renewable energy sources are well suited to supplying power in areas not served by a central electrical grid. And there is the matter of other pollutants, such as sulfur dioxide, mercury and nitrogen oxides, all of which are given off as we burn fossil fuels and all of which contribute significantly to poor health and fatalities.

Getting It Done

Second, is it even possible to accomplish a shift to renewables in so short a time span? Here we encounter some significant problems. Fossil fuels are great sources of concentrated energy, and we access them, distribute them, and employ them in a massive infrastructure. Our modern industrial so-

ciety "has been built around, by and for fossil fuels,"[2] as one commentator put it. It took 70 to 100 years to build the infrastructure of our fossil fuel economy, and we only have a few decades to rebuild much of this infrastructure around clean energy sources (by clean energy, Obama and others are including *nuclear power*, which does not generate CO_2). This is nothing short of a revolution, possibly on the order of the response of the United States to the onset of World War II, where factory after factory shifted its production from peacetime to wartime goods. Clearly, the manufacture of wind turbines, PV panels, transmission lines, concentrated solar power plants, electric vehicles, recharging stations, and so on would have to take place in a relatively short period of time, and this would require massive government support.

Another barrier to accomplishing this transition is the way in which government subsidies, at least in the United States, have strongly favored fossil fuels. For the years 2002–2010, government subsidies to fossil fuels were $98 billion versus $45 billion for renewables. Globally, subsidies encouraging fossil fuel consumption rose to $409 billion (versus $66 billion to renewables). This would have to change, and the fossil fuel, automotive, and utility industries pose an enormous obstacle to change, as they do everything they can to maintain their subsidies.

However, the global shift from fossil fuels to renewable or clean energy sources is starting to happen, so the question is, will it happen fast enough? There are some encouraging developments to suggest that it is possible, but much more has to be done:

• Venture capitalists are sinking a lot of money into green technologies; globally, a record $257 billion was invested in 2011 into wind farms, solar power, electric cars, and so on.

---

[2]Roberts, David. "Direct subsidies to fossil fuels are the tip of the (melting) iceberg." *Grist*, October 26, 2011.

- Thirty-nine states now have some form of a renewable portfolio standard (RPS) or alternative energy portfolio standard (AEPS), which requires electric utilities to generate a certain amount of electricity from sources other than fossil fuels; for example, California must generate 33% of its electricity from renewable sources by 2021.

- Natural gas is becoming more available and can fill a "bridge energy" role to substitute for coal (because coal generates twice as much CO_2 as natural gas per kW of electricity) and to work in tandem with renewable sources that are intermittent in their delivery, to smooth out gaps in electricity generation.

- It may be possible to continue to use coal *if*, and *only if*, the CO_2 generated in the coal power plant is captured from the waste stream and sequestered, known as carbon capture and sequestration (CCS). The technology is known, but the process adds a significant cost to the electricity generated.

- It is possible to hold greenhouse gas emissions to a limit of 450 ppm CO_2— if we stop building more fossil fuel infrastructure and allow the present infrastructure to live out normal lifespans. In other words, it is the future coal power plants and gasoline-powered automobile manufacturing plants that will commit the world to exceeding safe limits of greenhouse gas emissions.

We can now turn our attention to the different forms of renewable energy.

16.2 Putting Solar Energy to Work

Before turning our attention to the practical ways that solar energy is captured and used, let us consider some general concepts of solar energy.

Solar energy originates with thermonuclear fusion in the Sun. (Importantly, all the chemical and radioactive by-products of the reactions remain behind on the Sun.) The solar energy reaching Earth is radiant energy, entering at the top of the atmosphere at 1,366 watts per square meter, the **solar constant**. This energy ranges from ultraviolet light to visible light and infrared light (heat energy) **(Fig. 16–4)**. About half of this energy actually makes it to Earth's surface, 30% is reflected, and 20% is absorbed by the atmosphere. Thus, full sunlight can deliver about 700 watts per square meter to Earth's surface when the Sun is directly overhead. At that rate, the Sun can deliver 700 MW of power (the output of a large power plant) to an area of 390 square miles (1,000 km²). The total amount of solar energy reaching Earth is vast—almost beyond belief. Just 40 minutes of sunlight striking the land surface of the United States yields the equivalent energy of a year's expenditure of fossil fuel. The Sun delivers 10,000 times the energy used by humans.

Moreover, using some of this solar energy will not change the basic energy balance of the biosphere. Solar energy absorbed by water or land surfaces is converted to heat energy and eventually lost to outer space. Even the fraction that is absorbed by vegetation and used in photosynthesis is ultimately given off again in the form of heat energy as various consumers break down food (Chapter 3). Similarly, if humans were to capture and obtain useful work from solar energy, it would still ultimately be converted to heat and lost in accordance with the Second Law of Thermodynamics (see Chapter 3). The overall energy balance would not change.

Although solar energy is an abundant source, it is also diffuse (widely scattered), varying with the season, latitude, and atmospheric conditions. The main problem with using solar energy is one of taking a diffuse and intermittent source and concentrating it into an amount and form, such as fuel or electricity, that can be used to provide heat and run vehicles,

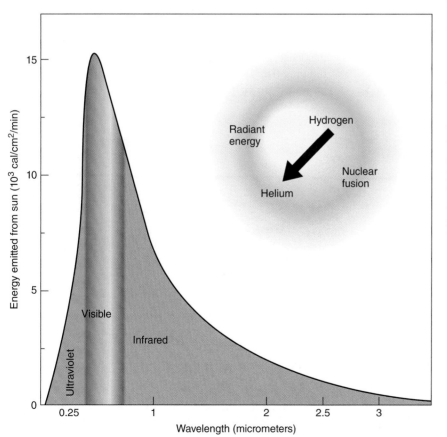

Figure 16–4 The solar-energy spectrum. Equal amounts of energy are found in the visible-light and the infrared regions of the spectrum. In the Sun, the nuclear fusion of hydrogen to form helium emits a spectrum of radiant energy that reaches Earth.

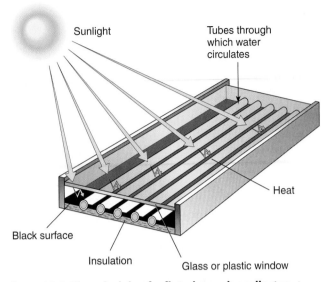

Figure 16–5 The principle of a flat-plate solar collector. As sunlight is absorbed by a black surface, it is converted to heat. A clear glass or plastic window over the surface allows the sunlight to enter the collector, which traps the heat. Air or water is heated as it passes over and through tubes embedded in the black surface.

appliances, and other machinery. These problems involve the *collection*, *conversion*, and *storage* of solar energy. Also, in the final analysis, overcoming such hurdles must be cost effective. In the sections that follow, you will see how we can overcome or, even better, sidestep these three problems in various ways to meet our energy needs in a cost-effective manner.

Solar Heating of Water

Solar hot-water heating is already popular in warm, sunny climates. A solar collector for heating water consists of a thin, broad box with a glass or clear plastic top and a black bottom in which water tubes are embedded **(Fig. 16–5)**. Such collectors are called **flat-plate collectors**. Faced toward the Sun, the black bottom gets hot as it absorbs sunlight—similar to how black pavement heats up in sunlight—and the clear cover prevents the heat from escaping, as in a greenhouse. Water circulating through the tubes is heated and conveyed to a tank, where it is stored.

In an *active system*, the heated water is moved by means of a pump. In a *passive* solar water-heating system, the system must be mounted so that the collector is lower than the tank. Thus, heated water from the collector rises by natural convection into the tank, while cooler

water from the tank descends into the collector **(Fig. 16–6)**. The tank will usually have a source of auxiliary heat (electric or gas) in order to get the temperature to a desired level or to provide heat when solar energy is insufficient.

In temperate climates, where water in the system might freeze, the system may be adapted to include a heat-exchange coil within the hot-water tank. Then antifreeze fluid is circulated between the collector and the tank. In the United States, approximately 2.5 million solar hot-water systems have been installed, but this is still only a small fraction of the total number of hot-water heaters. The reason is the initial cost, which is five to 10 times as much as gas or electric heaters. However, over time, the solar system costs less to operate than an electric or a gas system. In China, more than 18 million households are getting their hot water from solar thermal systems; China leads the world in this application of solar energy.

Solar Space Heating

The same concept for heating water with the Sun can be applied to heat spaces. Flat-plate collectors such as those used in water heating can be used for space heating. Indeed, the collectors for space heating may even be less expensive, homemade devices, because it is necessary only to have air circulate through the collector box. Again, efficiency is gained if the collectors are mounted, to allow natural convection to circulate the heated air into the space to be heated **(Fig. 16–7)**.

Building = Collector. The greatest efficiency in solar space heating, however, is gained by designing a building to act as its own collector. The basic principle is to have windows facing the Sun. In the winter, because of the Sun's angle of incidence, sunlight can come in and heat the interior of the building **(Fig. 16–8a)**. At night, insulated drapes or shades can be pulled down to trap the heat inside. The well-insulated

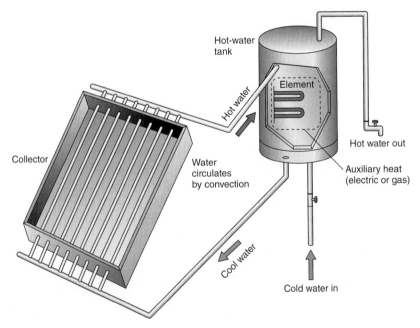

Figure 16–6 Solar water heaters. In nonfreezing climates, simple water-convection systems may suffice. In freezing climates, an antifreeze fluid is circulated. Solar heat is augmented by an auxiliary heat source in the hot-water tank.

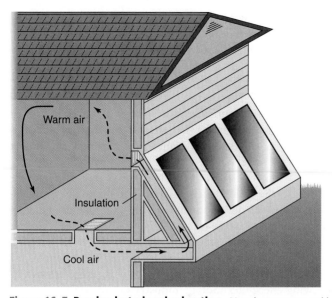

Figure 16–7 Passive hot-air solar heating. Many homeowners could save on fuel bills by adding homemade solar collectors like the one shown here. Air heated in the collector moves into the room by passive convection.

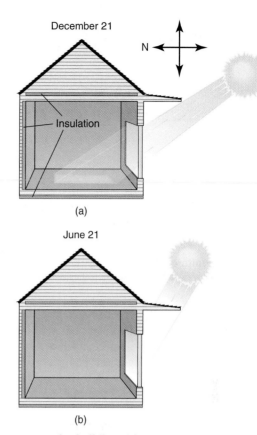

Figure 16–8 Solar building siting.
In contrast to utilizing expensive and complex active solar systems, solar heating *can* be achieved by suitable architecture and orientation of the home at little or no additional cost. (a) The fundamental feature is large, Sun-facing windows that permit sunlight to enter during the winter months. Insulating drapes or shades are drawn to hold in the heat when the Sun is not shining. (b) Suitable overhangs, awnings, and deciduous plantings prevent excessive heating in the summer.

building, with appropriately made doors and windows, acts as its own best heat-storage unit. Beyond good insulation, other systems for storing heat, such as tanks of water or masses of rocks, have not proved to be cost effective. Excessive heat load in the summer can be avoided by using an awning or overhang to shield the window from the high summer Sun (Fig. 16–8b).

Along with design, positioning, and improved insulation, appropriate landscaping can contribute to the heating and cooling efficiency of both solar and nonsolar space-heating designs, thereby conserving energy. In particular, deciduous trees or vines on the sunny side of a building can block much of the excessive summer heat while letting the desired winter sunlight pass through. An evergreen hedge on the shady side can provide protection from the cold.

Energy Stars. In almost any climate, a well-designed passive solar-energy building can reduce energy bills substantially, with an added construction cost of only 5–10%. Because about 25% of our primary energy is used for space and water heating, proper solar design, broadly adopted, would create enormous savings on oil, natural gas, and electrical power. In 2001, the Environmental Protection Agency (EPA) extended its **Energy Star Program** to buildings and began awarding the Energy Star label to public and corporate edifices. Buildings awarded the label use at least 40% less energy than others in their class and must meet a number of other criteria to qualify. By 2012, more than 17,000 buildings had earned the label, representing more than 2.5 billion square feet of space—a major contribution to **energy conservation**.

A common criticism of solar heating is that a backup heating system is still required for periods of inclement weather. Good insulation is a major part of the answer to this criticism. People with well-insulated solar homes find that they have minimal need for backup heating.

When backup is needed, a woodstove or gas heater may be enough. In any case, the criticism concerning the need for backup heating misses the point: The objective of solar heating is to reduce our dependency on conventional fuels. Even if solar heating and improved insulation reduced the demand for conventional fuels by a mere 20%, that would still represent sustainable savings of 20% of the traditional fuel and its economic and environmental costs.

Solar Production of Electricity

Solar energy can also be used to produce electrical power, providing an alternative to coal and nuclear power. There are two distinct approaches: photovoltaic systems and concentrated solar power (CSP).

Photovoltaic Cells. A solar cell—more properly called a **photovoltaic, or PV, cell**—looks like a simple wafer of material with one wire attached to the top and one to the bottom (**Fig. 16–9**). As sunlight shines on the wafer, it puts out an amount of electrical current roughly equivalent to that emitted by a flashlight battery. Thus, PV cells collect light and convert it to electrical power in one step. The cells,

Figure 16–9 Photovoltaic cell. Converting light to electrical energy, this cell provides enough energy to run the small electric motor needed to turn the fan blades.

measuring about four inches square, produce about one watt of power. Some 40 such cells can be linked to form a module that generates enough energy to light a lightbulb. Varying amounts of power can be produced by wiring modules together in panels.

How They Work. The simple appearance of PV cells belies a highly sophisticated level of science and technology. Each cell consists of two very thin layers of semiconductor material separated by a junction layer. The lower layer has atoms with single electrons in their outer orbital that are easily lost. The upper layer has atoms lacking electrons in their outer orbital; these atoms readily gain electrons. The kinetic energy of light photons striking the two-layer "sandwich" dislodges electrons from the lower layer, creating a current that can flow through a motor or some other electrical device and back to the upper side. Thus, with no moving parts, solar cells convert light energy directly to electrical power, with an efficiency of 15–20%.

Because they have no moving parts, solar cells do not wear out. However, deterioration due to exposure to the weather limits their life span to about 30 years. The major material used in PV cells is silicon, one of the most abundant elements on Earth, so there is little danger that the production of PV cells will ever suffer because of limited resources. The cost of these cells lies mainly in their sophisticated design and construction.

Uses. PV cells are already in common use in pocket calculators, watches, and numerous toys. Panels of PV cells provide power for rural homes, irrigation pumps, traffic signals, radio transmitters, lighthouses, offshore oil-drilling platforms, Earth-orbiting satellites, and other installations that are distant from power lines. It is not hard to imagine a future in which every home and building has its own source of pollution-free, sustainable electrical power from an array of PV panels on the roof. Current installations in many countries and in some 43 states in the United States employ *net metering*, where the rooftop electrical output is subtracted from the customer's use of power from the power grid.

Cost. The cost of PV power (cents per kilowatt-hour) is the cost of the installed PV systems divided by the total amount of power they may be expected to produce over their lifetime (currently as low as 12 cents per kilowatt-hour). This cost is quite comparable to that of other power alternatives (8–16 cents per kilowatt-hour for residential electricity in the United States). The first PV cells had a cost factor several hundred times that of electricity from conventional power stations transmitted through the power grid, so those cells were used mainly in areas far from the grid. PV power had its first significant application in the 1950s, in the solar panels of space satellites. This application started the development cycle rolling. As more efficient cells and less expensive production techniques evolved, costs came down dramatically. In turn, applications, sales, and potential markets expanded, creating the incentive for further development. In response to this, it is the fastest-growing energy technology industry in the world. In 2011, the industry shipped solar panels with a total capacity of 24 GW; existing solar PV panels now have the capacity to turn sunlight into more than 70 GW of electricity **(Fig. 16–10)**.

Inverters. The most complicated (and yet necessary) component of a PV system is the **inverter**. The inverter **(Fig. 16–11)** acts as an interface between the solar PV modules and the electric grid or batteries. It must change the incoming direct current (DC) from the PV modules to alternating

SOLAR PV, EXISTING WORLD CAPACITY, 1995–2011

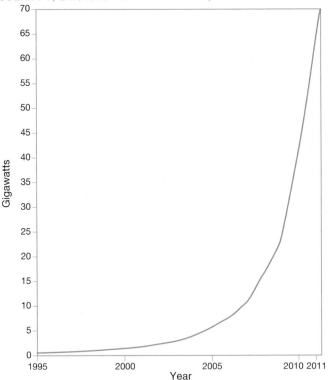

Figure 16–10 The market for PV panels. Global sales of PV panels have increased dramatically, exceeding 70,000 MW by 2011. More than three-fourths of the PV panels are connected to the electric grid. (*Source: Renewables 2012 Global Status Report.*)

Figure 16–12 PV power plant. The Copper Mountain Solar Facility in Nevada consists of 775,000 solar panels and has an installed capacity of 48 MW. It is the largest PV plant in the United States. The plant was built in 11 months at a cost of $141 million.

Figure 16–11 PV system inverter. Inverter (on right) and controllers for a PV system at Channel Islands National Park, California. The inverter is a device that connects the PV panels of a rooftop system to the electric grid or devices powered by PV electricity. It converts the direct current coming from PV panels to alternating current and also acts as a control for the system.

current (AC) compatible with the electricity coming from the grid and/or the devices that will be powered by the PV system. It must also act as a control, able to detect and respond to fluctuations in voltage or current on either DC or AC, and it must be robust enough to withstand the high temperatures of an attic (where rooftop PV systems are involved). An inverter for a rooftop system will cost several thousand dollars, but in most cases, these are also eligible for local and national subsidies. Like the cost of PV cells, the cost of inverters has also come down over time.

Utilities. The utility companies are moving toward large-scale PV installations. At the end of 2011, there were 40 plants worldwide with capacities greater than 20 MW, with 31 more planned or under construction. The largest one in the United States is the 48 MW Copper Mountain Solar Facility in Nevada (**Fig. 16–12**), providing electricity for 14,000 homes. The world's largest solar PV power plant is in Canada—the Sarnia PV Power Plant, with a 97 MW capacity. This record won't last long, as there are much

larger ones under construction. In some cases, PV is getting the nod over Concentrated Solar Power, or CSP, facilities (discussed later), because it can be added in smaller increments, allowing limited transmission lines to handle the voltages.

However, the most promising future for PV power is in the installation of PV panels on rooftops, where a huge amount of unused space is readily available. California's Edison International is busy installing 150 solar projects on warehouse and factory roofs; the company anticipates spending $875 million to provide 250 MW of electricity on commercial rooftops. For residential use, utilities in many states have established programs that provide incentives to customers to install 2- to 4-kilowatt (kW) PV systems on their roofs. A 2 kW system provides about half of the annual energy needs of a residence, and in case the homeowner is generating more than needed, the utility buys the excess electricity at the retail rate. Even though the rooftop systems represent a loss in power purchases, the utilities benefit because they avoid the need to build expensive new power plants. A federal investment tax credit of 30% of the system cost was enacted in the Emergency Economic Stabilization Act of 2008, providing incentives to homeowners to go solar. Sunny California experienced a 47% increase in rooftop solar in 2010 and passed the 1,000 MW (1 GW) mark in late 2011. The world leader in installed rooftop PV is Germany, with more than 12 GW installed capacity.

New PV Technologies. To compete with electricity from conventional sources, the cost of solar cells needs to be low. At least four new technologies may drive the cost even lower: (1) thin-film PV cells, in which cheap amorphous silicon or cadmium telluride is used instead of expensive silicon crystals, and the film can be applied as a coating on roofing tiles or glass; (2) a 3-D silicon solar cell that captures up to 25% of the incoming light energy; (3) glass coated with light-absorbing dyes that transmit energy to solar cells on the edge of the glass, while allowing much of the light to pass through to conventional solar panels; (4) the flexible plastic polymer cells illustrated in Figure 16–1. All of these techniques are either in production or close to it.

Concentrated Solar Power (CSP). If you have ever focused sunlight through a magnifying glass to burn a hole in a leaf, you have used CSP. Several technologies have been developed that convert solar energy into electricity by using reflectors (or concentrators) such as mirrors to focus concentrated sunlight onto a receiver that transfers the heat to a conventional turbogenerator. These devices work well only in regions with abundant sunlight and plenty of space, and they are producing power in the U.S. Southwest and in Spain. A major drawback to this technology is the need for a lot of water to cool the steam generated. Nevertheless, CSP is growing rapidly, now at 1.8 GW generating capacity globally, with a number of projects under construction. The two methods currently in the lead for CSP are the solar (or parabolic) trough and the power tower.

Solar Trough. One method that is proving to be cost effective is the **solar-trough** collector system, so named because the collectors are long, trough-shaped reflectors tilted toward the Sun (Fig. 16–2). The curvature of the trough is such that all of the sunlight hitting the collector is reflected onto a pipe running down the center of the system. Oil or some other heat-absorbing fluid circulating through the pipe is thus heated to very high temperatures. The heated fluid is passed through a heat exchanger to boil water and produce steam for driving a turbogenerator.

Nine solar-trough facilities in the Mojave Desert of California have been connected to the Southern California Edison utility grid since the early 1990s. Solar-trough is the most developed of CSP technologies, and most new CSP plants are solar-trough. Soon to be largest in the world, Solana, a new 250 MW solar trough plant, is under construction in the Arizona desert. One advantage of this technology is the capability of storing some of the thermal energy for release at night; the Solana plant will store energy for many hours.

Power Tower. A **power tower** is an array of Sun-tracking mirrors that focuses the sunlight falling on several acres of land onto a receiver mounted on a tower in the center of the area (Fig. 16–13). The receiver transfers the heat energy collected to a molten-salt liquid, which then flows either to a heat exchanger

to drive a conventional turbogenerator or to a tank at the bottom of the tower to store the heat for later use. Presently, the Sierra SunTower is the only power tower operating in the United States, generating 5 MW of electricity.

Spain has CSP projects under construction for some 1,300 MW, both solar-trough and power tower systems. A royal decree and government subsidies have paved the way for Spain to become a world leader in commercial CSP development (adding substantially to its rising investment in PV power plants).

The Future of Solar Energy

Solar electricity is growing at a phenomenal rate—the solar PV business is a $30 billion industry, and sales continue to grow exponentially (see Figure 16–10). The industry is hard-pressed to keep up with current demand. Solar energy does have certain disadvantages, however. First, the available technologies are still more expensive than conventional energy sources, though solar sources are getting close to parity with conventional ones. Second, solar energy works only during the day, so it requires a backup energy source, a storage battery, or thermal storage for nighttime power. Also, the climate is not sunny enough to use solar energy in the winter in many parts of the world. Against these drawbacks should be weighed the hidden costs that are not included in the cost of power from traditional sources: air pollution, strip mining, greenhouse gas emissions, and nuclear waste disposal, among others.

Matching Demand. The Sun provides power only during the day, but 70% of electrical demand occurs during daytime hours, when industries, offices, and stores are in operation. Thus, considerable savings still can be achieved by using solar panels just for daytime needs and continuing to rely on conventional sources at night. In particular, the demand for air-conditioning, which, after refrigeration, is the second largest power consumer, is well matched to energy from PV cells. In the long run, the nighttime load might be carried by forms of indirect solar energy, such as wind power and hydropower (discussed later in the chapter).

It is a fact that about 61% of our electrical power (54% worldwide) is currently generated by coal-burning and nuclear power plants (Chapter 14, Table 14–1). Therefore, the development of solar and wind electrical power can be seen as gradually reducing the need for coal and nuclear power. Unlike new nuclear or coal power plants, which take years to plan and build, solar or wind facilities can become operational within months of the decision to build them. The utility does not have to guess what power demands will be in 10 or 15 years. With solar or wind power, a utility can add capacity as it is needed, make relatively small investments at any given time, and have those investments start paying back almost immediately. Thus, this approach involves much less financial risk both for the utility and for its consumers.

One further advantage of solar and wind facilities is their relative invulnerability to terrorist attack compared with oil and gas pipelines and large power plants, especially

Figure 16–13 Sierra SunTower. Sun-tracking mirrors are used to focus a broad area of sunlight onto a molten-salt receiver mounted on the two towers. The hot salt is stored or pumped through a steam generator, which drives a conventional turbogenerator.

SUSTAINABILITY

Transfer of Energy Technology to the Developing World

The nations of the industrialized Northern Hemisphere have achieved their level of development by using energy technologies based largely on fossil fuels (and, to a lesser extent, nuclear energy). These nations' development took place during a time when fossil fuels were inexpensive. Only as those fuels became more expensive did the nations of the North begin to get serious about technologies that would make energy use more efficient. The specter of global climate change has brought a new perspective to the need to wean the industrialized North away from fossil fuel energy in the 21st century.

The same traditional fossil fuel–based technologies have been adopted by the developing nations of the Southern Hemisphere, except that these nations are lagging behind in their ability to implement the technology on a large scale. As

discussed earlier in the chapter, however, the fossil fuel–based energy pattern of the 20th century will have to be replaced largely with renewable-energy systems in the 21st century. We have only one or two generations remaining for the dominance of fossil fuel energy. If the United States and other developed nations are willing to play a significant role in helping the developing nations, it is questionable whether the same developmental path the North took should be promoted. Instead, there is the opportunity to engage in some "leapfrogging" technology transfer: The North can put its development dollars into such technologies as photovoltaics, wind turbines, and efficient public transport systems for the cities.

The solar route is especially attractive for many of the climates in the developing world, where an estimated 1.4 billion people lack electricity. For example, more than 200,000 PV systems installed in Kenya over a 20-year period now provide power to 5% of the rural Kenyan

population. The systems, with costs ranging from $300 to $1,500, are successfully marketed to people with incomes averaging less than $100 per month. Only the solar panels are imported; local companies provide the batteries and other system components. People use the energy for lighting, television, and radio. Annual sales in Kenya exceed 30,000 systems, a sign that the technology is in great demand and is fostering appropriate development in the area. Credit and financing extended to both companies and users have been key to this success. Some of the financing originates with development agencies from the North, but the majority is Kenyan. Building the infrastructure to market and service the PV systems has also been an essential part of the success story, and that can be accomplished by a relatively small number of organizations and individuals. These solar home systems are being added in many other developing countries, especially China, Bangladesh, India, Mexico, and Morocco.

---

nuclear facilities. Further, renewables produce no hazardous wastes and are geographically dispersed.

Solar power has also proved suitable for meeting the needs of electrifying villages and towns in the developing world, where, because centralized power is unavailable, rural electrification projects based on PV cells are already beginning to spread throughout the land (see Sustainability, "Transfer of Energy Technology to the Developing World"). The cost of the alternative—putting roughly 1.4 billion people who are still without electricity on the **power grid** (and thus requiring central power plants, high-voltage transmission lines, poles, wires, and transformers)—is unimaginable.

Solar electric power has a bright future. Costs of solar PV systems are declining by about 5% per year, as efficiencies of PV cells improve and manufacturers achieve economies of scale with the rising demand. Homeowners are ready to spend $10,000 or more to install rooftop PV systems, especially when subsidies are now available from the federal government and many states. The technology is now mature, and many small installation companies are competing for home solar PV installations. Finally, the solar energy resource is huge in comparison with all of the finite fossil and nuclear fuel resources, which will last only a few generations at most. We need only exploit a small fraction of solar potential to meet all of our energy needs. Undoubtedly, the future planetary energy source is going to be solar-based.

Of course, solar power is only one of a number of renewable forms of energy. Some of these other sources are derived from solar energy and have been around for a long time.

16.3 Indirect Solar Energy

Water, fire, and wind have provided energy for humans throughout history. Dams, firewood, windmills, and sails represent indirect solar energy because energy from the Sun is the driving force behind each. (Recall in Chapter 10 how solar energy powers the hydrologic cycle of evaporation and precipitation and also drives the convection currents that produce the wind). What is the potential for expanding these options from the past into major sources of sustainable energy for the future?

Hydropower

Early in technological history, it was discovered that the force of falling water could be used to turn paddle wheels, which in turn would drive machinery to grind grain, saw logs into lumber, and run simple machines. The modern culmination of this use of waterpower, or hydropower, is huge hydroelectric dams, where water under high pressure flows through channels, driving turbogenerators **(Fig. 16–14)**. The amount of power generated is proportional to both the height of the water behind the dam—that is what provides the pressure—and the volume of water that flows through (Chapter 10 describes the role played by dams in trapping and controlling rivers for flood control, irrigation, and hydropower).

About 7.9% of the electrical power generated in the United States currently comes from hydroelectric dams, most of it from about 300 large dams concentrated in the Northwest and Southeast. Worldwide, hydroelectric dams have a

Figure 16–14 Hoover Dam. About 7.9% of the electrical power used in the United States in 2011 came from large hydroelectric dams such as this one on the Colorado River in Nevada.

generating capacity of 970 GW, with more very large projects in the construction or planning stage. Hydropower generates 19% of electrical power throughout the world and is by far the most abundant form of renewable energy in use.

Trade-offs. Dams—in particular, the larger ones—are often controversial; they provide important benefits but also involve some serious consequences. The most obvious *advantages* are these:

- Hydropower eliminates the cost and environmental effects of fossil fuels and nuclear power. Hydroelectric plants have longer lives than fuel-fired ones; some plants have been in service for 100 years.

- The dams provide flood control on many rivers that have taken huge tolls in human lives, and they also provide irrigation water for agriculture. Some 40% of the world's food is generated from the irrigation facilitated by the reservoirs behind dams.

- Reservoirs behind the dams often provide recreational and tourist opportunities.

- Dams can be employed in **pumped-storage power plants**; during the night, when electricity demand is low, the dam's hydrogenerator pumps water up to an elevated reservoir through a tunnel. At times of high demand, the water is returned to the lower reservoir through the tunnel, where it spins turbines and generates electricity.

There are also some serious *disadvantages*, however:

- The reservoir created behind the dam inevitably drowns farmland or wildlife habitats and perhaps towns or

land of historical, archaeological, or cultural value. Glen Canyon Dam (on the border between Arizona and Utah), for example, drowned one of the world's most spectacular canyons.

- Dams and the large reservoirs created behind them often displace rural populations. In the past 50 years, some 40 million to 80 million people have been forced to move to accommodate the rising waters of reservoirs.

- Dams impede or prevent the migration of fish, even when fish ladders are provided. Federal surveys show that fish habitats are suffering in the majority of the nation's rivers because of damming.

- Because the flow of water is regulated according to the need for power, dams wreak havoc downstream. Water may go from near flood levels to virtual dryness and back to flood levels in a single day. Other ecological factors are also affected because sediments with nutrients settle in the reservoir, meaning that smaller-than-normal amounts reach the river's mouth.

More Dams? Even if such trade-offs were not enough, thoughts of greatly expanding waterpower in the United States are nullified by the fact that few sites conducive to large dams remain. Already, 75,000 dams six feet high or more dot U.S. rivers, with an estimated 2 million smaller structures. As a result, only 2% of the nation's rivers remain free flowing, and many of these are now protected by the Wild and Scenic Rivers Act of 1968, a law that effectively gives certain scenic rivers the status of national parks. The United States and most developed countries have already brought their hydropower to capacity, and the trend is in the direction of removing many of the dams that impede the natural flow of rivers.

Proposals for new dams outside of the United States are embroiled in controversy over whether the projected benefits justify the ecological and sociological trade-offs. For example, the 39-meter-high Nam Theun 2 Dam, a 1,075 MW project on a tributary of the Mekong River in Laos, is expected to help that country begin a process of economic development. The project received final approval in 2005 from the World Bank and Asian Development Bank for guarantees and loans, and construction was completed in 2010. The project has displaced 6,200 people and has inundated 175 square miles (450 km²) of the Nakai Plateau. In addition, it is basically being built to sell power to Thailand (only 75 MW are to be channeled to Laos at first). International Rivers (IR) is monitoring implementation of the project. Charges and countercharges have flown back and forth between the World Bank and IR concerning the impacts of the dam, especially on downstream communities; it is likely too soon to issue a verdict on the dam's overall effects.

Another huge dam, completed in 2006, is the Three Gorges Dam on the Yangtze River in China (See Chapter 10). The largest dam ever built (1.4 miles long), Three Gorges has displaced some 1.2 million people in order to generate

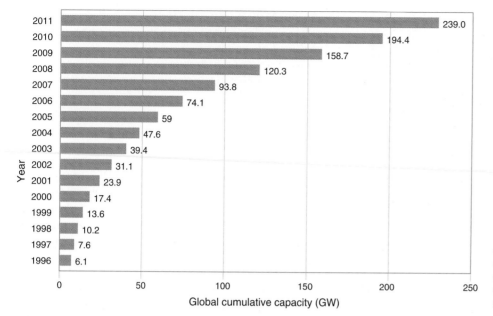

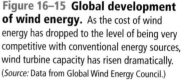

Figure 16–15 Global development of wind energy. As the cost of wind energy has dropped to the level of being very competitive with conventional energy sources, wind turbine capacity has risen dramatically. (*Source:* Data from Global Wind Energy Council.)

22,000 MW of electricity and control the flood-prone river. Offsetting the human and economic costs is the fact that it would take more than a dozen large coal-fired power plants to produce the same amount of electricity.

A special World Commission on Dams, convened to examine the impacts and controversies surrounding large dams, reported its findings in 2000. The commission began with the assertion that dams are only a means to an end, the end being "the sustainable improvement of human welfare." The report found dams to be a mixed blessing, concluding that large dams should be built only if no other options exist.[3] Many guidelines were presented for assessing costs and benefits of specific projects—in particular, for finding ways to incorporate participatory decision making, sustainability, and accountability in the planning and implementation of projects. Following up the World Commission work, the United Nations Environment Program (UNEP) established the Dams and Development Forum, which meets annually to participate in debate and decision making over dam development in various countries. Many more dams will be built, largely in the developing countries (especially China, Brazil, Turkey, and India), and these will certainly play a role in the energy future of those countries.

Wind Power

Once a standard farm fixture, windmills fell into disuse as transmission lines brought abundant lower-cost power from central generating plants. Not until the energy crisis and rising energy costs of the 1970s did wind begin to be seriously considered again as a potential source of sustainable energy. Growing rapidly, wind power now supplies around 2.5% of global electricity demand, but there are no technical, economic, or resource limitations that would prevent wind power from supplying at least 12% of the world's electricity by 2020, with a generating capacity of 1.2 thousand GW. In Denmark, which is hard pressed by a lack of fossil fuel energy, wind now supplies more than 22% of the electricity.

China is the world leader in wind energy (65 GW), with the United States close behind (47 GW capacity in 2011). In the past five years, global wind power capacity has increased at a rate of 26% per year, to 239 GW at the start of 2012 **(Fig. 16–15)**. Wind is the second-largest component of renewable electricity in the world (behind hydropower), because it has become economically competitive with conventional energy sources, and countries all over the world are encouraging the installation of wind turbines by setting targets and providing subsidies.

A new feature of the American landscape is the "wind farm," a collection of wind turbines in the same location whose power generation can occur while the land beneath the turbines can be used for traditional agricultural purposes. These are getting huge: The Horse Hollow Wind Farm near Abilene, Texas, consists of 421 turbines with a capacity of 735 MW **(Fig. 16–16)**; the Shepherd's Flat Wind Farm in eastern Oregon is under construction and when finished will occupy 30 square miles, house 338 turbines, have an installed capacity of 845 MW, and be the world's largest land-based wind farm (for a while!).

Design. Many different designs of wind machines have been proposed and tested, but the one that has proved most practical is the age-old concept of wind-driven propeller blades. The propeller shaft is geared directly to a generator. (A wind-driven generator is more properly called a wind turbine than a windmill.) As the reliability and efficiency of wind turbines have improved, the cost of wind-generated electricity has decreased. Wind farms are now producing

---

[3]*Dams and Development: A New Framework for Decision-Making. The Report of the World Commission on Dams.* Earthscan Publications. 2000.

Figure 16–16 Horse Hollow Wind Farm. Located near Abilene, Texas, this wind farm consists of 421 turbines with a capacity of 735 MW.

pollution-free, sustainable power for around five cents per kilowatt-hour, a rate that is competitive with the rates of traditional sources. Moreover, the amount of wind that can be tapped is immense. The American Wind Energy Association calculates that wind farms located throughout the Midwest could meet the electrical needs of the entire country, while the land beneath the turbines could still be used for farming. Hundreds of new turbines are sprouting from midwestern farms, and the farmers are paid royalties ($3,000 to $5,000 per turbine per year) for leasing their land to wind developers. The U.S. Department of Energy (DOE) is promoting a scenario where wind energy will provide 20% of U.S. electricity needs by 2030—an ambitious, but by no means impossible, target.

Drawbacks. Still, wind power does have some drawbacks. First, it is an intermittent source, so problems of backup or storage must be considered. Second is an aesthetic consideration: One or two wind turbines can be charming, but a landscape covered with them can be visually unappealing (to some). Third, wind turbines can be a hazard to birds and bats. Locating wind farms on migratory routes or near critical habitats for endangered species such as condors could be a problem for soaring birds in particular. Mortality from wind turbines may be upward of 40,000 per year, but this number pales in comparison to the hundreds of millions of birds killed annually by cats, automotive traffic, and glass windows.

As discussed earlier, plans for offshore wind farms are putting a new spin on wind energy. Denmark already has many offshore turbines, as do the Netherlands, the United Kingdom, and Sweden. Belgium, France, and Germany are also in the development stage. The primary lure of offshore wind is the dependability and strength of the maritime winds. Siting is perhaps easier, too, as the turbines will be far enough offshore to have much less of a visual impact than onshore sites and land does not have to be purchased. Whether offshore or onshore, the global drive to install wind power is generating electricity that is emission free; in fact,

the 43,000 MW of new wind turbines installed in 2011 displaced 125 million tons of carbon dioxide, or emissions from 65 coal-fired power plants.

Wind power and hydropower are two forms of indirect solar energy; another highly important form is biomass energy.

Biomass Energy

Burning firewood for heat is the oldest form of energy that humans have used throughout history. However, there is nothing like a new name to put life into an old concept. Thus, "burning firewood" has become "utilizing biomass energy." **Biomass energy** means energy derived from present-day photosynthesis. As Figure 16–3 indicates, biomass energy leads hydropower in renewable-energy production in the United States. (Most uses of biomass energy are for heat.)

In addition to burning wood in a stove, major means of producing biomass energy include burning municipal wastepaper and other organic waste, generating methane from the anaerobic digestion of manure and sewage sludges, running power plants on wastes from timber operations, and producing alcohol from fermenting grains and other starchy materials.

Burning Firewood. Wherever forests are ample relative to the human population, firewood—or fuelwood, as it is called—can be a sustainable energy resource; and, indeed, wood has been a main energy resource over much of human history. Fuelwood is the primary source of energy for heating and cooking for some 2.6 billion people, amounting to about 9% of total energy use from all sources.

In the United States, woodstoves have enjoyed a tremendous resurgence in recent years: About 5 million homes rely entirely on wood for heating, and another 20 million use wood for some heating. The most recent development in woodstoves is the *pellet stove*, a device that burns compressed wood pellets made from wood wastes (**Fig. 16–17**). The pellets are loaded into a hopper (controlled by computer

Figure 16–17 Pellet stove. Pellet stoves use pelletized wood waste to achieve an efficient burn that requires much less care than burns from conventional woodstoves.

chips) that feeds fuel when it is needed. The stove requires very little attention and burns efficiently, relying on one fan to draw air into it and another fan to distribute heated air to the room.

Fuelwood Crisis? Recall that two patterns of use determine how forest resources will be exploited: consumptive use and productive use (Chapter 7). Most fuelwood use in developing countries fits the *consumptive* pattern: People simply forage for their daily needs in local woodlands and forests. However, gathering wood, converting wood into charcoal, and selling the wood products in urban areas translates into important *productive* use activities for many in developing countries. All these activities have the potential to degrade local forests and woodlands. There was serious concern in the later decades of the 20th century about deforestation due to wood gathering, but though there are local shortages, it appears that global use of fuelwood peaked in the late 1990s and is now declining. As development progresses, people have more disposable income and shift up the "energy ladder" from wood to charcoal to fossil fuels. The Millennium Ecosystem Assessment judged that fuelwood demand has not become a significant cause of deforestation, though there are local areas of concern. This could change if the cost of fossil fuels continues to rise.

Burning Wastes. Facilities that generate electrical power from the burning of municipal wastes (waste-to-energy conversion) are discussed in another chapter (Chapter 21). Many sawmills and woodworking companies are now burning wood wastes, and a number of sugar refineries are burning cane wastes to supply all or most of their power. In Virginia, three 63 MW coal-fired power plants are being converted to biomass power, primarily wood waste from timber operations. Power from these sources may not meet more than a small percentage of a country's total electrical needs, but it represents a productive and relatively inexpensive way to dispose of biological wastes as well as to prevent more fossil fuel combustion.

Producing Methane. The anaerobic digestion of sewage sludge yields *biogas*, which is two-thirds methane, plus a nutrient-rich treated sludge that is a good organic fertilizer (examined further in Chapter 20). Animal manure can be digested likewise. When the disposal of manure, the production of energy, and the creation of fertilizer can be combined in an efficient cycle, great economic benefit can be achieved. In China and India, millions of small farmers maintain a simple digester in the form of an underground brick masonry fermentation chamber with a fixed dome on top for storing gas (Fig. 16–18). Agricultural wastes, such as pig manure and cattle dung, are put into the chamber and diluted 1:1 with water, and the anaerobic digestion produces the biogas, which is used for cooking, heating, and lighting. The residue slurry makes an excellent fertilizer. The concept has been expanded throughout the developing world to provide an alternative to fuelwood. Currently, some 25 million households employ these

Figure 16–18 Biogas power. Animal wastes are introduced to this unit in rural India and mixed with water. The wastes then decompose under anaerobic conditions, producing biogas.

devices. Both India and China provide subsidies for these family-scale biogas plants.

16.4 Renewable Energy for Transportation

Declining oil reserves and the enormous impact of transportation's demands for oil on our economy and on global climate suggest that the most critical need for a sustainable energy future is a new way to fuel our vehicles. Renewable energy is already answering a small but growing part of this need in the form of biofuels and may have the answer for the long term in the form of hydrogen and fuel cells.

Biofuels

Complex organic matter (primarily plants, but also animal wastes) can be processed to make fuels for vehicles. Currently, two fuels are seeing rapid expansion globally: **ethyl alcohol** (ethanol) and **biodiesel**. High oil prices, strong farmer support, government subsidies, and environmental concerns have led to a major expansion of global ethanol and biodiesel production since 2000. Although these two biofuels currently provide only 2.7% of global transportation fuels, this is just the beginning, as new technologies and government support promise to make biofuels a major competitor for fossil fuels. The current sources of biofuels are agricultural and food commodities.

Ethanol. Ethanol is produced by the fermentation of starches or sugars. The usual starting material is corn, sugarcane, sugar beets, or other grains. Fermentation is the process used in the production of alcoholic beverages. The only new part of the concept is that, instead of drinking the brew, we distill it and put it in our cars. However, production costs make ethanol more expensive than gasoline unless oil sells for more than $55 a barrel. To stimulate the industry in the United States, federal tax credits have been extended to biofuels, which make them even more competitive. **Gasohol,** 10% ethanol and 90% gasoline, is now marketed in most parts of the country.

Farm Products. In 2011, 13.9 billion gallons (334 million barrels) of ethanol for fuel, the equivalent of 221 million barrels of oil, were produced in the United States (ethanol contains about two-thirds the energy of regular gasoline, gallon for gallon). One-third of the nation's corn crop is dedicated to this use; however, an ethanol factory also produces corn oil and livestock feed. A tax credit of 45 cents a gallon helps ethanol compete with gasoline, and a **renewable fuel standard** (or RFS, which requires a minimum volume of renewable fuel in gasoline) mandated a level of 9 billion gallons of renewable fuel in 2008 (all corn-based ethanol), to rise to a limit of 15 billion gallons of corn-based ethanol by 2015. The RFS target rises to 36 billion gallons in 2022, but the increase beyond 2015 must be met by using sources other than corn.

Can we expect ethanol production from corn to make serious inroads on the oil import and greenhouse gas emissions problems? Currently, biofuels represent about 8% of U.S. fuel consumption. This could rise to 20% if the country's entire corn crop were devoted to ethanol production—an impossible assumption, given the importance of corn for animal feed and export. With some 84 million acres devoted to corn harvest in the United States, there is little suitable farmland for significant expansion. There are serious concerns that biofuel production has contributed significantly to recent rising prices of food grain, diverting corn from hungry people to ethanol production (Chapter 12).

Another issue is the energy efficiency of corn-based ethanol. For corn-based ethanol, fossil fuel energy is required to produce the corn (fertilizer, pesticides, farm machinery), to transport it, and to operate the ethanol plant. A recent analysis[4] has calculated that using ethanol instead of gasoline reduces greenhouse gas emissions, mile for mile, by only 12–18%. However, some biofuel critics argue that if new land is cleared of its forests or grasslands, the carbon emissions from the land-use change should be added to the debit from fossil fuel greenhouse gas emissions. When this is done, corn-based ethanol is a net loser in its impact on climate change. Thus, it could be reasoned that the primary strategy in using corn-based ethanol is to reduce our reliance on imported oil.

Second-Generation Biofuel. A more likely long-run technology for producing ethanol is the use of cellulosic feedstocks such as agricultural crop residues, grasses (such as switchgrass), logging residues, fast-growing trees, and fuelwoods from forests. Ethanol produced this way is called second-generation biofuel (current starch-based processes produce first-generation biofuel). One crop with exceptional potential is Miscanthus **(Fig. 16–19)**, a giant perennial grass

Figure 16–19 Miscanthus x giganteus. This hardy perennial grass grows to a height of four meters and is a prime candidate for a second-generation cellulosic biofuel.

that can be grown in poor soils and that outproduces corn and other grasses.

With the use of enzymes to break down the cellulose to sugars, demonstration facilities have shown that cellulosic feedstock technology can be a cheaper and more energy-efficient process than corn-based production. The technology is challenging, however, and the output of 6.6 million gallons in 2011 is way below the RFS target of 250 million gallons. However, several industrial-scale cellulosic plants are under construction, and production should increase dramatically once these are on line. Recall that the RFS requires that all of the increase in the target biofuel after 2015 be derived from second-generation biofuel sources. An available tax credit of $1.01 per gallon of cellulosic biofuel is moving this industry forward rapidly.

Brazil has also invested heavily in fuel ethanol; there the feedstock is sugarcane, which is much less costly to farm than corn. Brazil used to lead the world in ethanol production but has been surpassed by the United States. For years, ethanol has been added at a level of 25% of fuel in Brazil, but recently, the country mandated the production of flex-fuel vehicles that can run on any mixture of alcohol and gasoline. With the lower production costs and the capacity to shift all sugarcane acreage to ethanol production, Brazil is on the road to meeting all of its domestic fuel demand with biofuels.

Biodiesel. Imagine a truck exhaust that smells like french fries. That is not only an improvement over diesel exhaust, but also an indication that the truck is burning biodiesel, made largely from soybean oil. Sales have reached 800 million gallons of the fuel, which is very competitive with petroleum diesel fuel because of a hefty $1-per-gallon production tax credit. Still, this is only 1% of diesel fuel transportation use in the United States. Standard biodiesel is a 20% mix of soybean oil in normal diesel fuel, but practically any natural oil or fat can be mixed with an alcohol to make the fuel; one excellent source is recycled vegetable oil from frying. However, like corn-based ethanol, there are competing uses for the vegetable oils.

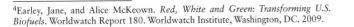

[4]Earley, Jane, and Alice McKeown. *Red, White and Green: Transforming U.S. Biofuels.* Worldwatch Report 180. Worldwatch Institute, Washington, DC. 2009.

There are some offbeat technologies surfacing to produce biodiesel, such as the factory in Missouri that takes in truckloads of turkey wastes from a Butterball turkey-processing plant and turns them into biofuels. The patented process belongs to Changing World Technologies, and its founder, Brian Appel, says that it can handle almost any organic waste imaginable and turn it into useful oil. Currently, the Missouri plant turns a profit, converting 250 tons of turkey waste per day into 20,000 gallons of oil and also selling a fertilizer product from the rendered wastes. In another technology, genetic engineering is being applied to algae to induce them to produce oils and complex alcohols. Some serious money is being invested in these enterprises, and commercial-scale production is only a few years away.

Hydrogen: Highway to the Future? Over the past decade, there has been increasing interest in powering cars with hydrogen. Conventional cars with internal combustion engines can be adapted to run on hydrogen gas in the same manner as they can be adapted to run on natural gas. (Hydrogen is not really a fuel, but an energy carrier, like electricity; it must be generated using some other energy source.) Neither carbon dioxide nor hydrocarbon pollutants are produced during the burning of hydrogen. Indeed, the only major by-product of H_2 combustion is water vapor:

$$2H_2 + O_2 \longrightarrow 2H_2O + energy$$

(Some nitrogen oxides will be produced because the burning still uses air, which is nearly four-fifths nitrogen.)

A Good Idea, But . . . If hydrogen gas is so great, why aren't we using it? The answer is that there is virtually no hydrogen gas on Earth. Any hydrogen gas in the atmosphere has long since been ignited by lightning and burned to form water. And although many bacteria in the soil produce hydrogen in fermentation reactions, other bacteria are quick to use the hydrogen because it is an excellent source of energy. Thus, there are abundant amounts of the *element* hydrogen, but it is all combined with oxygen in the form of water (H_2O) or other compounds.

Although hydrogen can be extracted from water via **electrolysis** (the reverse of the preceding reaction), electrolysis requires an *input* of energy. Hydrogen can also be derived chemically from hydrocarbon fuels such as methane, petroleum oil, and methyl alcohol, but these processes are also energy losers; that is, more energy is put into the process than is contained in the hydrogen. Moreover, because these compounds contain carbon, it would be better to use them directly than to obtain hydrogen from them and use it. Therefore, the use of hydrogen as a truly clean energy carrier must await an energy source that is itself suitably cheap, abundant, and nonpolluting—like solar energy.

Plants Do It. The "holy grail" of research and eventual commercialization is to accomplish what plant cells do: produce fuel from water, using solar energy. To date, two ways have been proposed for mimicking this process. The first is to employ a catalyst to do what the photosynthetic pathway does. Two elements are required: (1) a collector that converts photons into electrical energy (electrons), and (2) an

electrolyzer that uses the electron energy into split water to hydrogen and oxygen. Recently, researchers from the Massachusetts Institute of Technology (MIT) described a process that employs a water-oxidizing electrolyzer catalyst made from cobalt and a heavy metal alloy, all common chemicals.[5] The process uses light energy captured by a silicon solar cell and generates hydrogen and oxygen in either a wired or a wireless cell. It operates in neutral water and ambient temperature and pressure and achieves a solar-to-fuels efficiency of 4.7% (wired) and 2.5% (wireless). This may be the key step needed to get to that holy grail.

Electrolysis. The second approach is simply to pass an electrical current through water and cause the water molecules to dissociate. Hydrogen bubbles come off at the negative electrode (the cathode), while oxygen bubbles come off at the positive electrode (the anode). (You can demonstrate this process for yourself with a battery, some wire, and a dish of water. A small amount of salt placed in the water will facilitate the conduction of electricity and the release of hydrogen.) The hydrogen gas is collected, compressed, and stored in cylinders, which can then be transported or incorporated into a vehicle's chassis. One important drawback to this technology is that it is difficult to store enough hydrogen in a vehicle to allow it to operate over long distances. Also, compressing the hydrogen requires energy, thus reducing the energy efficiency involved. As an alternative, the hydrogen can be combined with metal hydrides, which can absorb the gas and release it when needed; but finding a hydride that is not too heavy has been a challenge. Another pathway that holds promise is to convert the hydrogen into liquid formic acid (HCOOH) by combining it with carbon dioxide. The hydrogen is then released by use of a catalyst.

Solar Energy to Hydrogen. A rather elaborate plan has been suggested for taking solar energy–produced hydrogen for vehicles to its logical conclusion. The plan would start by building arrays of solar-trough or photovoltaic generating facilities in the deserts of the southwestern United States, where land is cheap and sunlight plentiful. The electrical power produced by these methods would then be used to produce hydrogen gas by electrolysis. The most efficient way to move the hydrogen is through underground pipelines; hundreds of miles of such pipelines already exist where hydrogen is transported for use in chemical industries. Although this would be an expensive part of a hydrogen infrastructure, the technology for it is already well known.

Model U. The Ford Motor Company unveiled in the United States a "concept car" it has dubbed the Model U. The Model U sports a 2.3-liter internal combustion engine that runs on hydrogen and is equipped with a hybrid electric drive. The car uses four 10,000 psi tanks to store hydrogen, enough to range 300 miles before a fill-up. And there's the problem: Before such cars can become feasible, there must be a hydrogen-fueling infrastructure. Once that infrastructure is in place, cars like the Model U would likely be replaced by a completely new way of moving a vehicle: by means of fuel cells.

[5]Reece, Steven Y. et al. "Wireless Solar Water Splitting Using Silicon-Based Semiconductors and Earth-Abundant Catalysts." *Science* 334: 645–648. November 4, 2011.

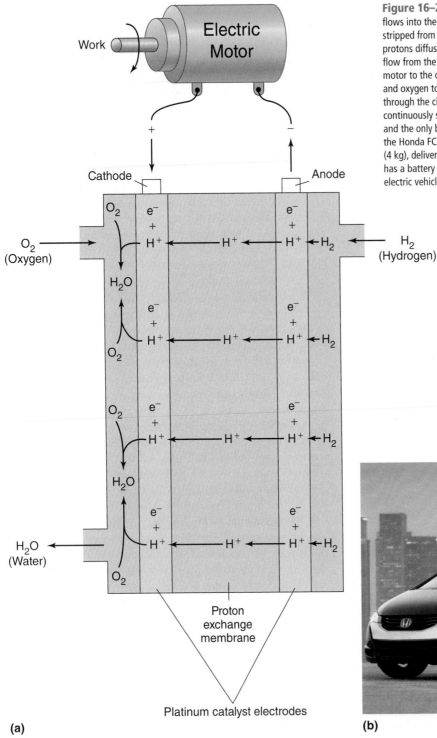

(a)

Figure 16–20 Hydrogen-oxygen fuel cell. (a) Hydrogen gas flows into the cell and meets a catalytic electrode, where electrons are stripped from the hydrogen, leaving protons (hydrogen ions) behind. The protons diffuse through a proton exchange membrane, while electrons flow from the anode to the motor. As the electrons then flow from the motor to the cathode of the fuel cell, they combine with the protons and oxygen to produce water. Thus, the chemical reaction pulls electrons through the circuit, producing electricity that runs the motor. Hydrogen is continuously supplied from an external tank, oxygen is drawn from the air, and the only by-product is water. (b) Powered by a compact fuel cell stack, the Honda FCX Clarity has a range of 270 miles on a charge of hydrogen (4 kg), delivering the equivalent of 68 miles per gallon. The vehicle also has a battery for extra assist, similar to the battery in today's hybrid gas-electric vehicles. It will be available for lease at $600 per month.

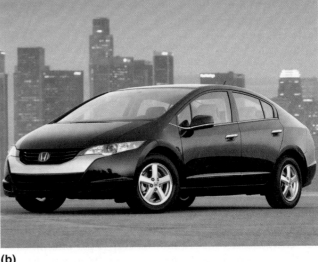

(b)

Fuel Cells. An alternative to burning hydrogen in conventional internal combustion engines uses hydrogen in **fuel cells** to produce electricity and power the vehicle with an electric motor. Fuel cells are devices in which hydrogen or some other fuel is chemically combined with oxygen in a manner that produces an electrical potential rather than initiating burning **(Fig. 16–20a).** Emissions from fuel cells consist solely of water and heat. Because fuel cells create much less waste heat than conventional engines do, energy is transferred more efficiently from the hydrogen to the vehicle—at a rate of 45–60% versus the 20% for current combustion-engine vehicles. To power a vehicle, hundreds of fuel cells are combined into a fuel-cell stack. The vehicle must also include a hydrogen storage device, a cooling system (fuel cells also generate heat), and a device to force oxygen into the fuel cells. Fuel cells are now powering 150 buses in many cities in the

United States and around the world. The fuel-cell stack develops more than 200 kW of power, from hydrogen in tanks mounted on the roof of the bus.

"Concept" vehicles operating on fuel cells have been developed by all the major car manufacturers. Honda, however, is the first manufacturer to commit to a dedicated fuel-cell vehicle assembly line. The Honda FCX Clarity (Fig. 16–20b) is rolling off the assembly lines, but at a pace of several dozen a year, it will be a long time before it makes an impact on the market. The Clarity is available for lease ($600 a month!) in California regions that are supported by hydrogen refueling stations.

If a hydrogen infrastructure is put in place and vehicles powered by fuel cells are available, we will have entered a **hydrogen economy**—one in which hydrogen becomes a major energy carrier. With solar- and wind-powered electricity also in place, we would no longer be tied to nonrenewable, polluting energy sources and would have made the transition to a sustainable-energy system. Still, the obstacles in the way of the hydrogen economy are formidable, and many believe it will be decades before it even has a chance of happening. Global climate concerns and the costs of importing oil convey an urgency to resolving this issue; reducing our use of oil in the near term is crucial, and many believe that the best pathway is greater use of hybrid vehicles and other technologies already available.

16.5 Additional Renewable-Energy Options

Aside from water, wind, biomass, and hydrogen energy, there are other renewable-energy options that have been developed or are in the concept stage.

Geothermal Energy

In various locations in the world, such as the northwestern corner of Wyoming (Yellowstone National Park), there are springs that yield hot, almost boiling, water. Natural steam vents and other thermal features are also found in such areas. They occur where the hot, molten rock of Earth's interior is close enough to the surface to heat groundwater, particularly in volcanic regions. Using such naturally heated water or steam to heat buildings or drive turbogenerators is the basis of **geothermal energy**. In 2011, geothermal energy had more than 11,200 MW of electrical power capacity (equivalent to the output of 11 large nuclear or coal-fired power stations), in countries as diverse as Nicaragua, the Philippines, Kenya, Iceland, and New Zealand. Today, the largest single facility is in the United States at a location known as The Geysers, 70 miles (110 km) north of San Francisco (**Fig. 16–21**). With more than 3,000 MW from geothermal energy, the United States is the world leader in the use of this energy source; new projects in the development stage will increase this to more than 4,500 MW. As impressive as this application of geothermal energy is, nearly double the amount

Figure 16–21 Geothermal energy. One of the 11 geothermal units operated by the Pacific Gas & Electric Company at The Geysers in Sonoma and Lake Counties, California. California geothermal energy produces 1,800 MW of power, roughly 4.3% of California's electricity needs.

is being used to directly heat homes and buildings, largely in Japan and China.

A recent study[6] by MIT suggested that the United States employ an emerging technology called enhanced geothermal systems (EGS). EGS involves drilling holes several miles deep into granite that holds temperatures of 400 degrees or more and injecting water under pressure into one hole (where it absorbs heat from the rock) and, as steam, flows up another shaft to a power plant (where it generates electricity). Such plants already exist in Australia, Europe, and Japan. This underground heat is widespread in the United States, unlike the present limited geothermal sources. The MIT team proposed a $1 billion investment to exploit the EGS potential, predicting that by 2050 some 100 gigawatts (GW) of electrical power could be developed in this way.

Heat Pumps. A less spectacular, but far more abundant, energy source than the large geothermal power plants exists anywhere pipes can be drilled into the Earth. Because the ground temperature below six feet or so remains constant, the Earth can be used as part of a heat exchange system that extracts heat in the winter and uses the ground as a heat sink in the summer. Thus, the system can be used for heating and cooling and does away with the need for separate furnace and air-conditioning systems. As **Figure 16–22** shows, a **geothermal heat pump** (GHP) system involves loops of buried pipes filled with an antifreeze solution or a refrigerant, circulated by a pump connected to the air handler (a box containing a blower and a filter). The blower moves house air through the heat pump and distributes it throughout the house via ductwork. In most cases, the up-front costs of the heat pump system are repaid in six to eight years, and then the savings really begin.

[6]Tester, Jefferson et al. *The Future of Geothermal Energy: Impact of Enhanced Geothermal Systems (EGS) on the United States in the 21st Century.* Massachusetts Institute of Technology and the Department of Energy. 2006.

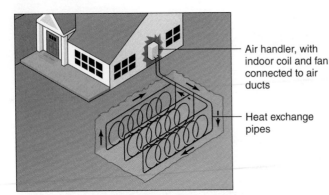

Figure 16–22 Geothermal heat pump system. The pipes buried underground facilitate heat exchange between the house and the Earth. This system can either cool or heat a house and can be installed almost anywhere, though at a higher initial cost than a conventional heating, ventilation, and air-conditioning system.

Air handler, with indoor coil and fan connected to air ducts

Heat exchange pipes

Four elementary schools in Lincoln, Nebraska, installed GHP systems for their heating and cooling. Their energy cost savings were 57% compared with the cost of conventional heating and cooling systems installed in two similar schools. Taxpayers will save an estimated $3.8 million over the next 20 years with the GHP systems. According to the EPA, GHP systems are by far the most cost-effective, energy-saving systems available. Although they are significantly more expensive to install than conventional heating and air-conditioning systems, they are trouble-free and save money over the long run. New heat pump installations in the United States are growing rapidly, with more than 1 million in operation by 2010.

Tidal and Wave Power

A phenomenal amount of energy is inherent in the twice-daily rise and fall of the ocean tides, brought about by the gravitational pull of the Moon and the Sun. Many imaginative schemes have been proposed for capturing this limitless, pollution-free source of energy. The most straightforward idea is the **tidal barrage**, in which a dam is built across the mouth of a bay and turbines are mounted in the structure. The incoming tide flowing through the turbines would generate power. As the tide shifted, the blades would be reversed so that the outflowing water would continue to generate power.

In about 30 locations in the world, the shoreline topography generates tides high enough—20 feet (6 m) or more—for this kind of use. Large tidal power plants already exist in two of those places: France and Canada. The only suitable location in North America is the Bay of Fundy, where the Annapolis Tidal Generating Station has operated since 1984, at 20 MW capacity. Thus, this application of tidal power has potential only in certain localities, and even there, it would not be without adverse environmental impacts. The barrage would trap sediments, impede the migration of fish and other organisms, prevent navigation, alter the circulation and mixing of saltwater and freshwater in estuaries, and perhaps have other, unforeseen ecological effects.

Other techniques are being explored that harness the currents that flow with the tides. For example, in 2006, Verdant Power began a project to harness the tidal energy of the East River in New York City. The project has passed its first demonstration phase, connecting six turbines to local electric customers. Verdant plans to install 30 turbines that will connect to the grid and generate 1 MW of power. San Francisco is eyeing a similar scheme to tap the energy of the 400 billion gallons of water that flow under the Golden Gate Bridge with every tide.

Standing on the shore of any ocean, an observer would have to be impressed with the energy arriving with each wave, generated by offshore winds. It might be possible to harness some of this energy, but the technological challenge is daunting. The ideal location is one that receives the wave's force before it hits the shoaling seafloor, is close enough to shore to facilitate hookup with transmission cables, and is deep enough so that the equipment will not crash on the seafloor during storm turbulence. There are a number of proposed mechanisms for capturing wave energy, and it is fair to say that this technology is in the early research-and-development stage. The U.S. DOE is supporting research on ocean energy on a 50% cost-share basis.

Ocean Thermal-Energy Conversion

Over much of the world's oceans, a thermal gradient of about 20°C (36°F) exists between surface water heated by the Sun and the colder deep water. **Ocean thermal-energy conversion (OTEC)** is the name of an experimental technology that uses this temperature difference to produce power. The technology involves using the warm surface water to heat and vaporize a low-boiling-point liquid such as ammonia. The increased pressure of the vaporized ammonia would drive turbogenerators. The ammonia vapor leaving the turbines would then be recondensed by cold water pumped up from as much as 300 feet (100 m) deep and returned to the start of the cycle.

Various studies indicate that OTEC power plants show little economic promise—unless, perhaps, they can be coupled with other, cost-effective operations. For example, in Hawaii, a shore-based OTEC plant uses the cold, nutrient-rich water pumped from the ocean bottom to cool buildings and supply nutrients for vegetables in an aquaculture operation, in addition to cooling the condensers in the power cycle. Even so, interest in duplicating such operations is minimal at present.

16.6 Policies for Renewable Energy

We have examined our planet's energy resources, requirements, and management options (Chapters 14–16). The global issues are clear: Fossil fuels, especially oil and natural gas, will not last long at the current (and increasing) rate at which they are being consumed. Even more important, every use of fossil fuels produces unhealthy pollutants and increases the atmospheric burden of carbon dioxide, accelerating the already-serious climate changes brought on by rising levels of greenhouse gases. Renewable energy means sustainable energy, and a sustainable-energy future does not include significant fossil fuel energy. Global targets, therefore, are (1) *stable atmospheric levels of greenhouse gases, especially carbon dioxide*, and (2) *a sustainable-energy development-and-consumption pattern that is based on renewable energy.* Our main focus in this chapter has been on sustainable energy *development*, but there is another energy strategy that focuses on energy *consumption*—namely, **energy conservation and efficiency.** Energy saved is energy that does not have to be developed or paid for. We laid out two policies that have energy conservation as a goal (Chapter 14). These were *increasing the mileage standards for motor vehicles* and *increasing the energy efficiency of lighting, appliances, and buildings.* Earlier in this chapter, we stressed the Energy Star program, as it encourages municipalities and businesses to achieve energy savings. We turn now to public policy in the United States and the recent and very significant changes that have encouraged renewable energy and energy conservation.

National Energy Policy

Nationally, the United States has a serious *oil-transportation* problem: We import two-thirds of our oil, much of which is consumed in transportation. This problem is an enormous drag on our economy. In fact, more than $600,000 leaves the United States *every minute* to pay for imported oil. We also have an *energy security* problem: Not only are we subject to the whims of the Organization of Petroleum Exporting Countries (OPEC) and other oil producers, but also we are vulnerable to terrorist attacks on various facets of our energy infrastructure.

These national issues, as well as other, global ones, prompted a policy response from the Bush and Obama administrations and Congress. We examined the policies pertaining to fossil fuels and nuclear power as they emerged in the **Energy Policy Act of 2005** and the **Energy Independence and Security Act of 2007** (Chapters 14 and 15). These laws have become the major expression of federal energy policy, though many other laws promoting renewable energy and conservation have been enacted. The **American Recovery and Reinvestment Act of 2009** gave a huge boost ($43 billion) to renewable energy and energy conservation, with the intent of doubling the energy from renewable sources by 2012.

For renewable energy, **supply-side policy provisions** enacted into law:

- Established a renewable fuel standard (RFS) for ethanol and biodiesel that rises from 9 billion gallons in 2008 to 15 billion gallons by 2015.

- Extended the RFS to 36 billion gallons by 2022, but all the increase over 15 billion gallons must be met by feedstocks other than corn (i.e., second-generation sources).

- Continued funding for research and development on renewable energy, including major funding increases for solar energy and biomass programs.

- Extended the production tax credit for electricity generated with wind, hydropower, geothermal, and biomass energy.

- Provided investment tax credits for geothermal heat pumps, solar water heating, biofuel, and PV systems.

- Contained provisions aimed at enabling geothermal energy to compete with fossil fuels in generating electricity.

With their focus on *energy conservation and efficiency*, **demand-side provisions** enacted into law:

- Raised the Corporate Average Fuel Economy standards for the combined fleet of cars and light trucks gradually, to meet a target of 38 miles per gallon for cars and 28 miles per gallon for light trucks by model year 2016.

- Extended the Energy Star energy-efficiency program for buildings, to include schools, retail establishments, private homes, and health care facilities.

- Provided tax credits for many types of improvements to homes and commercial buildings that enhance energy efficiency.

- Required a 30% increase in efficiency for incandescent lightbulbs, which is already met by current compact fluorescent lightbulbs (CFLs). This guarantees that the CFLs will become the light standard of the future.

- Directed EPA to inventory waste-energy potential and establish a grant program to encourage waste-energy recovery (see Chapter 21).

- Developed efficiency standards for most appliances and provided tax credits for manufacturers of energy-saving appliances.

- Continued funding for development of plug-in hybrid electric vehicles (PHEVs).

- Provided an income tax credit applicable to the purchase of new hybrid vehicles, PHEVs, or fuel cell–powered vehicles.

These recent policies did not establish a **federal renewable portfolio standard (RPS).** This would have required utilities to provide an increasing percentage of power from nonhydroelectric renewable sources (wind, solar, biomass). However, the states are setting the bar for the RPS: 37 states and the District of Columbia have implemented RPS requirements. Most states have adopted target percentages of 10–20%, with deadlines ranging from 2010 to 2025. Globally, at least 64 countries have a national target for renewable energy.

These policies also did not take any direct action on global warming, though they expanded DOE's research program for carbon capture and sequestration. Also, it can be argued that every provision that promotes greater energy efficiency and the use of renewable-energy sources is going to reduce greenhouse gas emissions in the United States.

Final Thoughts

Our look at current energy policy indicates that the United States is making a serious effort to address the need to develop renewable-energy sources and to improve the efficiency of current energy use. One development that has caught everyone's attention in recent years is the rising price of gasoline. Until recently, the United States was the only industrialized country that has seen fit to keep gasoline prices remarkably low. Fuel in all other highly developed countries is so heavily taxed that it costs consumers upward of $6 per gallon—unlike U.S. prices, which were below $1.50 for 12 of the past 20 years. When gasoline rose to $3 a gallon in 2005, consumers began to turn away from gas-guzzling sport-utility vehicles and muscle cars and have been opting for the rising number of gasoline-electric hybrid cars appearing on the lots of many car manufacturers. This supports the suggestion of a number of economists and environmental groups that we should be paying more—much more—for gasoline (and other fossil fuels). To accomplish this, we would need a policy change: a **carbon tax**—that is, a tax levied on all fuels according to the amount of carbon dioxide that they produce when consumed. Such a tax, proponents believe, would provide both incentives to use renewable sources, which would not be taxed, and disincentives to consume fossil fuels. It is hard to imagine any step that would be more effective (or more controversial!) in reducing greenhouse gas emissions in the United States. A number of European countries have already adopted such a tax.

Are the preceding developments, even the suggested carbon tax, enough to enable us to achieve a sustainable-energy system and to mitigate global climate change? Very likely, no. Yet many of them are moving us in the sustainable direction that is vital to the future of the global environment.

REVISITING THE THEMES

Sound Science

Some elegant technologies have been discussed in this chapter, developed by scientists in research laboratories and then applied to solving some very practical problems. PV devices, for example, are being intensely researched in order to bring down the costs and make solar energy competitive with conventional energy sources. Research at MIT has produced a real breakthrough in the discovery of a catalyst that can split water, imitating photosynthesis. The National Renewable Energy Laboratory in Golden, Colorado, was established by Congress in 1977 and has developed into a cutting-edge research institute that produced the solar-trough and solar-dish systems. It is now exploring hydrogen technologies and improved PV devices, as well as a number of biomass technologies and low-wind-speed turbines and other renewable-energy options. This is science at its best, with public funding put to work to develop new ideas that may revolutionize the energy scene.

Sustainability

At the risk of repetition, global civilization is on an unsustainable energy course. Our intensive use of fossil fuels has created a remarkable chapter in human history, but we are still terribly dependent on maintaining and increasing our use of oil, gas, and coal. The results have already been seen in rising atmospheric greenhouse gases and the climate changes they are producing.

Future use of fossil fuel will bring on more-intense climate changes, and long before we burn all the available oil, gas, and coal, we will have been forced to turn to other energy sources. Renewable-energy sources derived from the Sun can provide a sustainable, inexhaustible supply of energy that is virtually free from harmful environmental impacts. It is imperative that the United States, as well as other countries, embark on a course to bring about a transition to a renewable-energy economy, most likely based on hydrogen and electricity as energy carriers, to be produced from renewable sources.

Stewardship

Stewardship of energy means doing more with less and turning away from the fossil fuels that are generating the greenhouse gases that are responsible for global climate change. Government policies are rapidly moving forward in encouraging renewable energy by providing subsidies for homeowners and businesses to embrace renewable-energy options. Policies are also improving the energy efficiency and conservation picture. After all, energy saved is energy that does not have to be developed. Picture solar panels on every home, church, hospital, and school roof as well as wind turbines on college campuses, grazing lands, and offshore wind farms. Many nongovernmental organizations are involved with programs that distribute PV and biogas systems to people in poor, rural, developing-country communities. The impact is literally being felt around the globe.

REVIEW QUESTIONS

1. Why is it important for renewable energy sources to replace fossil fuels? What are the prospects for getting it done?

2. How much solar energy is available, and what happens when it is used? What are some problems with harnessing solar energy?

3. How do active and passive solar hot-water heaters work?

4. How can a building best be designed to become a passive solar collector for heat?

5. How does a PV system work, and what are some present applications of such cells?

6. What has been happening in recent years in the PV market? How are utilities encouraging PV power?

7. Describe two concentrated solar power systems, how they work, and their potential for providing power.

8. What is the potential for developing more hydroelectric power in North America versus developing countries, and what are the impacts of such development?

9. Where is wind power being harvested, and what is the future potential for wind farms?

10. What are some ways of converting biomass to useful energy, and what is the potential environmental impact of each?

11. Which biofuels are used for transportation, and what is the potential for increasing biofuels in the United States?

12. How can hydrogen gas be produced via the use of solar energy? How might hydrogen be collected and stored to meet the need for fuel for transportation in the future?

13. How do fuel cells work, and what is being done to adapt them to power vehicles?

14. What is geothermal energy, and what are two ways it is being harnessed?

15. Describe current policy for renewable energy and energy efficiency as an outcome of recent government programs.

THINKING ENVIRONMENTALLY

1. What do you believe needs to happen before solar power becomes a thoroughly mainstream form of energy production? Defend your answer.

2. What do you think could be a viable solution for backup power systems for solar heaters in the event of bad weather?

3. How much promise does hydrogen power hold? Is its eventual sustainability just a myth or something to be pursued?

4. Imagine you are charged with building a facility run entirely by renewable resources (with the possible exception of backup power). How would you evaluate which sources have the most potential?

5. Audit all of the energy needs in your daily life. How much energy do you consume? Describe how each need might be satisfied by a renewable-energy option. (Check the Internet for tools for this exercise.)

MAKING A DIFFERENCE

1. Many everyday electrical products can now be powered by photovoltaic (PV) cells. You can find solar-powered chargers for MP3 players, cell phones, digital cameras, and even laptops. While this power may not be as dependable as that from a wall charger, it certainly provides a sustainable alternative while the Sun shines (especially useful on road and camping trips, where conventional power isn't available).

2. If you live close to work, consider not driving there at all and instead taking a bicycle. The benefits are endless; great exercise, low emissions, and no money toward gas are just the beginning.

3. Take advantage of the following services from your electric company: energy conservation, home-energy audits, and subsidized compact fluorescent lighting.

4. Consult the free *Greentips* newsletter of the Union of Concerned Scientists on its Web site (www.ucsusa.org), and follow the advice given in this publication for reducing your energy footprint.

5. Adopt ways to use less fuel and electricity both at home and where you work. (Turn thermostats up in the summer and down in the winter, turn off lights when you leave a room, wear sweaters, use energy-efficient lightbulbs, and so on.)

MasteringEnvironmentalScience®

Go to **www.masteringenvironmentalscience.com** for practice quizzes, Pearson eText, videos, current events, and more.

PART FIVE

Pollution and Prevention

Paper mill at sunset, Georgetown, South Carolina.

AN INDUSTRIAL PLANT sends a steady stream of exhaust gases into the air, and the wind carries the pollutants away. If it didn't, life in the vicinity of the plant would be nasty. But where is "away"? Is it the next county, state, or country? The upper atmosphere? In reality, there is no "away" for the pollution of land, water, and air resulting from the activities of the 7.2 billion people inhabiting Earth. We are fouling our own nest, our only planet.

In his book *Red Sky at Morning*,[1] James Gustave Speth describes four long-term trends in pollution that marked the 20th century:

First, *the volumes of pollutants released increased by huge amounts*—so much so that the developed countries finally began to legislate limits on many pollutants.

Second, *the pollution evolved from gross excesses to microtoxicity*, where major releases of toxic gases and waterborne pathogens were replaced by smaller but much more toxic releases of organic chemicals and trace metals like dioxin and mercury.

Third, *industrial pollutants have shifted from the first world to the third world*—from the developed to the developing countries. China, for example, is reaping the benefits of rapid industrial growth and at the same time suffering from massive and health-damaging pollution.

Fourth, *the effects of pollution have shifted from local to global*. Air pollution, resulting from the use of fossil fuels and industrial chemicals, has brought acid rain on a regional scale and global climate change and ozone depletion on a global scale.

In this part of the book, Chapter 17 begins with a look at the links between environmental hazards (like pollution and disease) and human health. The chapters that follow examine the major forms of pollution, and you will discover that pollution is often a by-product of some useful activity (though sometimes intentionally released). We cover the range of pollutants, from global and regional air pollutants to water pollutants to municipal and industrial wastes. In every chapter, the public policies crafted to deal with each of the problems raised by pollution are discussed. Without exaggeration, the problems of pollution pose some of the most difficult and important issues facing human society.

[1]Speth, James Gustave. *Red Sky at Morning*. Yale University Press, New Haven. 2004.

414

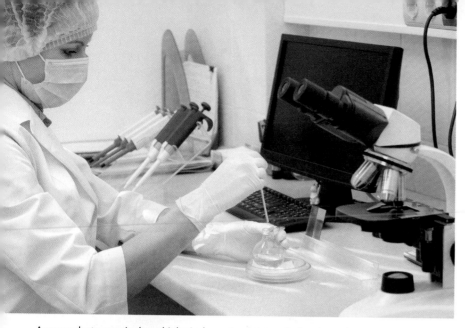

A young doctor manipulates biological agents while employing protective gear.

CHAPTER 17

Environmental Hazards and Human Health

Y OU MAY HAVE SEEN THE MOVIE. *Contagion*[2] opens with a woman developing flu-like symptoms after visiting Hong Kong. Within days, she is dead, but not before contaminating many people. Because it is so contagious, the virus quickly spreads globally—a pandemic—bringing death and chaos in its train. The film makes it clear that only the scientists can save the day, and they do so by isolating the virus, finding a cell line that will grow it, and finally getting a vaccine that can be quickly manufactured in quantity. The basic plot of the film is entirely too close to reality for comfort: in 1918, a flu pandemic killed more than 50 million people around the world. To make matters worse, the threat of a similar pandemic has recently been raised as a result of some controversial research[3] on a familiar virus, the H5N1 strain that causes avian influenza.

Avian influenza, or bird flu, appeared in 2003 in poultry and eventually spread to 63 countries, killing or leading to the culling of 400 million domestic poultry and countless wild birds and causing some $20 billion in losses. The disease also spread to humans through contact with infected birds, with 582 cases and 343 deaths. Like all influenza viruses, H5N1 contains eight genetic particles that often reshuffle when the virus reproduces in animal cells. Pigs are another source of influenza viruses. The H1N1 strain of swine influenza that initiated a pandemic in humans in 2009 was a relatively mild disease but still caused more than 16,000 deaths (Fig. 17–1); the deadly 1918 flu pandemic was

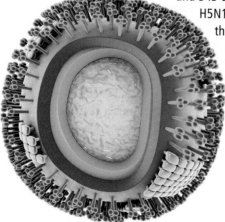

Model of the H1N1 virus particle

LEARNING OBJECTIVES

1. **Links between Human Health and the Environment:** Examine four classes of environmental hazards that bring on human misery, disease, and death: cultural, biological, physical, and chemical.

2. **Pathways of Risk:** Describe pathways whereby risk factors of poverty, tobacco use, disease transmission, toxic chemicals, and natural disasters lead to human mortality.

3. **Risk Assessment:** Evaluate the process of risk assessment and management as it is applied to human health by the EPA and the World Health Organization.

---

[2]http://www.imdb.com/title/tt1598778
[3]Berns, Kenneth I. et al. "Adaptations of Avian Flu Virus Are a Cause for Concern." *Science* 335: 660–661. February 10, 2012.

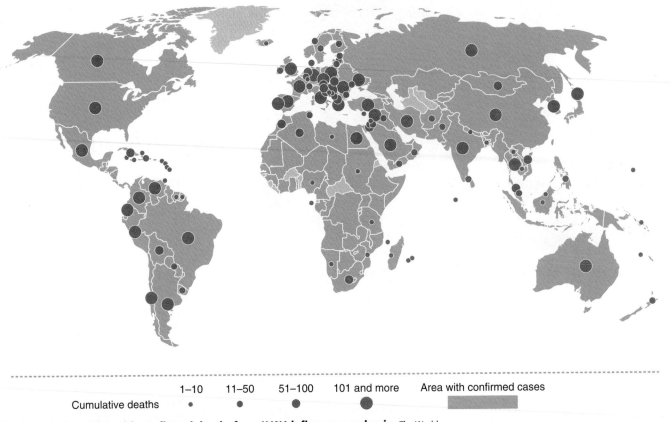

Cumulative deaths	1–10	11–50	51–100	101 and more	Area with confirmed cases

Figure 17–1 Countries with confirmed deaths from H1N1 influenza pandemic. The World Health Organization states that the disease is in the post-pandemic phase, with only scattered reports of deaths. Total worldwide deaths from the outset in 2009 to date have mounted to more than 16,000.

(*Source:* World Health Organization http://gamapserver.who.int/h1n1/cases-deaths/h1n1_casesdeaths.html. January 10, 2012.)

also caused by an H1N1 virus that likely originated in pigs. Recently, flu viruses containing genes from three sources—pigs, birds, and humans—have been found in pigs, demonstrating the ability of these viruses to mix genes.

The controversial H5N1 research[4] was carried out by two teams that set out to answer the question: Is it possible for H5N1 viruses to trigger a pandemic? The scientists genetically altered the virus to make it easily transmissible between ferrets, animals that are often used in such research because they respond to diseases much like humans do. In doing so, the scientists found that they had produced a very dangerous pathogen. The human-made virus passed from one ferret to another simply via air exchange between cages. Making matters worse, any influenza virus infecting ferrets has also been infectious to humans as well. Critics of the work pressed for it to remain unpublished, reasoning that there may be terrorists out there who could follow the procedures used and put the results to lethal use. The recombined virus could have what it takes to initiate a lethal pandemic: virulence, transmission by aerosols, and susceptibility in the human population. The research, referred to as **dual-use research** because

of its potential positive and negative applications, was reviewed by the U.S. National Science Advisory Board for Biosecurity. The board recommended that the research be published because of its public health importance, but at the same time initiated publication of a new government policy designed to give more oversight to federally-funded dual-use research. Commenting on the published research, one molecular biologist stated that these are studies that should never have been done.

The fictitious *Contagion* virus (called MEV-1) was based on a real-world disease, the Nipah virus, which was first observed in Malaysia in 1999 and has since been seen in India and Bangladesh. Recent work has shown that the Nipah virus is endemic in fruit bats and that people and pigs become infected by contact with fruit contaminated with bat urine and feces. The virus has killed more than 200 people and has shown signs of human-to-human transmission in some cases. The Nipah virus and avian and swine influenzas are examples of a number of emerging new diseases that continually remind us that we are not always able to control the hazards posed by our inevitable contacts with the natural environment and, particularly, with the animal world. A recent study found that out of 1,415 pathogens able to infect humans, 61% were **zoonotic** (transmitted between animals and humans).[5] In recent

[4]Herfts, Sander, et al. "Airborne Transmission of Influenza A/H5N1 Virus Between Ferrets." *Science* 336: 1534–1541. June 22, 2012.
Imai, M. et al. "Experimental Adaptation of an Influenza H5 HA Confers Respiratory Droplet Transmission to a Reassortant H5 HA/H1N1 Virus in Ferrets." *Nature* 486: 420–428. May 2, 2012.

[5]Taylor et al. 2001. Risk factors for human disease emergence. *Philosophical Transactions of the Royal Society B* 356: 983–989.

years, severe acute respiratory syndrome (SARS), Lassa fever, Rift Valley fever, Ebola, hantavirus, Marburg fever, and the West Nile virus have all taken human lives, and there is every reason to expect more new diseases to appear as we continue to manipulate the natural environment and change ecological relationships.

THE OLD ENEMIES. To put these diseases in perspective, however, it is not the new, emerging diseases that do the greatest damage; rather, it is the common, familiar ones that take the greatest toll of human life and health—diseases such as malaria, diarrhea, respiratory viruses, and worm infestations, which have been kept at bay in the developed countries but continue to ravage the developing countries. In the developed countries, cancer is the killer that is most closely linked to the environment, leaving

us with questions about the chemicals we are exposed to on a daily basis.

The study of the connections between hazards in the environment and human disease and death is often called **environmental health**; but it is *human* health that is the issue here, not the health of the natural environment. In this chapter, we will examine the nature of environmental hazards and the consequences of exposure to those hazards. Then we will consider some pathways whereby humans encounter the hazards, interventions that can alleviate the risks to health, and how public policy can address the need for continued surveillance and improvement in our management of environmental hazards. We begin with some basics as we examine what pollution is.

17.1 Links Between Human Health and the Environment

The U.S. Environmental Protection Agency (EPA) defines **pollution** as "the presence of a substance in the environment that, because of its chemical composition or quantity, prevents the functioning of natural processes and produces undesirable environmental and health effects." Any material that causes the pollution is called a **pollutant.**

Pollution is not usually the result of deliberate mistreatment of the environment. The pollutants are almost always the *by-products* of otherwise worthy and essential activities—producing crops, creating comfortable homes, providing energy and transportation, and manufacturing products—and of our basic biological functions. Pollution problems have become more pressing over the years because both growing population and expanding per capita use of materials and energy have increased the number and amounts of by-products that go into the environment. Also, many materials now widely used, such as aluminum cans, plastic packaging, and synthetic organic chemicals, are **nonbiodegradable.** That is, they resist attack and breakdown by detritus feeders and decomposers and consequently accumulate in the environment. An overview of the categories of pollutants that result from various activities is shown in **Figure 17–2.**

It is important to note the breadth and diversity of pollution. Any part of the environment may be affected, and almost anything may be a pollutant. The only criterion is that the addition of a pollutant results in undesirable changes. The impact of an undesirable change may be largely aesthetic—hazy air obscuring a distant view, or the unsightliness of roadside litter, for instance. The impact may be on ecosystems as a whole—the die-off of fish or forests, for example. Or the impact may be on human health—such as human wastes contaminating water supplies. The impact also may range from very local (the contamination of an individual well) to global (adding ozone-depleting chemicals to the atmosphere). We tend to think of pollution as the introduction of human-made

materials into the environment, but undesirable changes may be caused by the introduction of too much of what are otherwise natural compounds, such as fertilizer nutrients introduced into waterways and carbon dioxide introduced into the atmosphere.

Thus, the slogan "Don't pollute" is a gross oversimplification. The very nature of our existence necessitates the production of wastes. Our job in remediating present and future pollution problems is like the concept of sustainable development itself. We need to adapt the means of meeting our present needs so that by-products are managed in ways that will not jeopardize present and future generations. The general strategy in each case must be to:

1. Identify the material or materials that are causing the pollution—the undesirable change;

2. Identify the sources of the pollutants;

3. Clean up the environment already affected by pollution;

4. Develop and implement pollution-control strategies to prevent the pollutants from entering the environment; and

5. Develop and implement alternative means of meeting the need that do not produce the polluting by-product—in other words, avoid the pollution altogether.

Basically, the strategy for addressing pollution accomplishes one of the transitions to a sustainable society: "A technology transition from pollution-intensive economic production to environmentally friendly processes" (see Chapter 1).

Environmental Health

With our focus in this chapter on human health, a more precise definition of **environment** is *the whole context of human life—the physical, chemical, and biological setting of where and how people live.* Thus, the home, air, water, food, neighborhood,

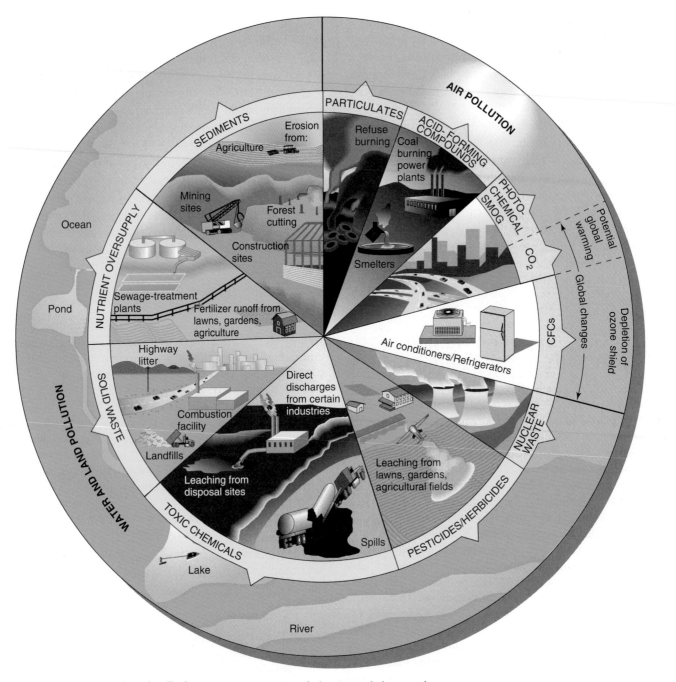

Figure 17–2 Categories of pollution. Pollution is an outcome of otherwise worthy human endeavors. Major categories of pollution and some activities that cause them are shown here.

workplace, and even climate constitute elements of the human environment. Within the human environment, there are *hazards* that can make us sick, cut our lives short, or contribute in other ways to human misfortune. In the context of environmental health, a **hazard** is anything that can cause (1) injury, disease, or death to humans; (2) damage to personal or public property; or (3) deterioration or destruction of environmental components.

The existence of a hazard does not mean that undesirable consequences inevitably follow. Instead, we speak of the connection between a hazard and something happening because of that hazard as a **risk**, defined here as the *probability* of suffering injury, disease, death, or some other loss as a result of exposure to a hazard. The notion of **vulnerability** is

also important; some people, especially the poor, are more vulnerable to certain risks than are others. Therefore, an appropriate notation of risk should be

$$\text{Risk} = \text{Hazard} \times \text{Vulnerability}$$

The presence of avian flu in poultry is a *hazard* that presents the *risk* of humans contracting the disease and dying. People working with poultry are much more *vulnerable* than are others. Because risks are expressed as probabilities, the analysis of the probability of suffering some harm from a hazard becomes a problem for scientists or other experts to address, and we will deal with the assessment of risk as it becomes the basis of policy decisions.

The Picture of Health. Health has many dimensions—physical, mental, spiritual, and emotional. The World Health Organization (WHO) defines **health** as "a state of complete physical, social, and mental well-being, and not merely the absence of disease or infirmity."[6] Unfortunately, measuring all these dimensions of health for a society is virtually impossible. Thus, to study environmental health, we will focus on *disease* and consider health to be simply *the absence of disease*. Two measures are used in studies of disease in societies: *morbidity* and *mortality*. **Morbidity** is the incidence of disease in a population and is commonly used to trace the presence of a particular type of illness, such as influenza or diarrheal disease. **Mortality** is the incidence of death in a population. Records are usually kept of the cause of death, making it possible to analyze the relative roles played by infectious diseases and factors such as cancer and heart disease. **Epidemiology** is the study of the presence, distribution, and prevention of disease in populations.

Public Health. Protecting the health of its people has become one of the most important facets of government in modern societies. In the United States, the lead federal agency for protecting the health and safety of the public is the **Centers for Disease Control and Prevention (CDC)**, an agency within the Department of Health and Human Services. The CDC's mission is "to create the expertise, information, and tools that people and communities need to protect their health—through health promotion, prevention of disease, injury and disability, and preparedness for new health threats."[7] Each state also has a health department, and most municipalities have health agents. In addition, there is a huge health care industry in the United States, with federal programs such as Medicare, Medicaid, and the Affordable Care Act; hospitals; health maintenance organizations; and local physicians and other health professionals. The delivery of health care is spread around all of these entities, but the primary responsibility for health risk management and prevention resides with the CDC and the state public-health agencies. These agencies, for example, may require public-health measures such as immunizations and quarantines, and they are responsible for monitoring certain diseases and environmental hazards and for controlling epidemics. Indeed, it was the CDC that first reported the H1N1 swine flu cases. The CDC also gathers data and provides information on health issues to state and local health care providers.

Other Countries. Virtually every country has a similar ministry of health that acts on behalf of its people to manage and minimize health risks. The health policies that are put into place, however, are subject to limitations of information and funding. Ideally, a health ministry should direct its limited resources toward strategies that will accomplish the greatest risk prevention. Fortunately, all countries have access to the programs and information of the WHO, a UN agency established in 1948 with the mission of "enabling all peoples to attain the highest possible level of health." The WHO is centered in Geneva, Switzerland, and maintains six regional offices around the world. The agency is staffed by health professionals and other experts and is governed by the UN member states through the World Health Assembly.

Life Expectancy. One universal indicator of health is human life expectancy. In 1955, average life expectancy globally was 48 years. Today it is 70 years and continuing to rise; it is expected to reach 73 years by 2025. This progress is the result of social, medical, and economic advances in the past 60 years that have increased the well-being of larger and larger segments of the human population. Recall the *epidemiologic transition*: the trend of decreasing death rates that is seen in countries as they modernize (Chapter 8). As this transition occurred in the now-developed countries, it involved a shift from high mortality (due to infectious diseases) toward the present low mortality rates (due primarily to diseases of aging, such as cancer and cardiovascular diseases).

Two Worlds. The countries of the world have undergone the epidemiologic transition to different degrees, with very different consequences. Despite the progress indicated by the general rise in life expectancy, one-sixth (almost 9 million) of the annual deaths in the world are those of children under the age of five in the developing world. In fact, a high proportion (36%) of the mortality in the low-income countries is caused by common infectious diseases (as opposed to 6% in the high-income countries).

An entirely different perspective is brought into focus as we consider some of the environmental hazards that accompany industrial growth and intensive agriculture. In addition, some of the most lethal hazards in the developed world are the outcome of purely voluntary behavior—in particular, smoking tobacco, using drugs, and engaging in risky sexual activity. We will look at four classes of environmental hazards: cultural, biological, physical, and chemical.

Environmental Hazards

There are two fundamental ways to consider hazards to human health. One is to regard the lack of access to necessary resources as a hazard. For example, lack of access to clean water and nourishing food is likely to result in harm to a person. Investigating hazards from this perspective means considering the social, economic, and political factors that prevent a person from having access to such basic needs. Although this is a fundamentally important perspective, much of it is outside the scope of this chapter (and has already been discussed in Chapters 9 and Chapter 12). Instead, the focus will be primarily on *the exposure to hazards in the environment*. What is it in the environment that brings the risk of injury, disease, or death to people?

Cultural Hazards. Many of the factors that contribute to mortality and disability are a matter of choice or at least can be influenced by choice. People engage in risky behavior and subject themselves to hazards. Thus, they may smoke cigarettes, eat too much, drive too fast, use addictive and harmful drugs, consume alcoholic beverages, sunbathe, engage in risky sexual practices, get too little exercise, or choose hazardous occupations. People generally subject themselves to these hazards because they derive some pleasure or other

[6]World Health Organization *Glossary*. Available at www.who.int/health-systemsperformance/docs/glossary.htm.

[7]Centers for Disease Control and Prevention, from www.cdc.gov/about/organization/mission.htm.

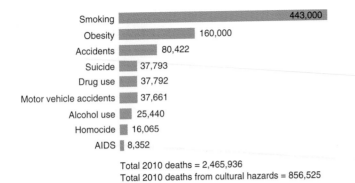

Total 2010 deaths = 2,465,936
Total 2010 deaths from cultural hazards = 856,525

Figure 17–3 Deaths from various cultural hazards in the United States, 2010. Many hazards in the environment are a matter of personal choice and lifestyle. Other hazards are due to accidents and the lethal behavior of other people.

(*Sources:* Data from the National Vital Statistics Reports, January 11, 2012; smoking deaths and obesity deaths from the Centers for Disease Control and Journal of the American Medical Association.)

benefit from them. Wanting the benefit, they are willing to take the risk that the hazard will not harm them. Factors such as living in inner cities, engaging in criminal activities, and so on are cultural sources of mortality, too. As **Figure 17–3** indicates, 35% of all deaths in the United States can be traced to cultural hazards, and in most cases, deaths from cultural hazards are preventable (if people refrain from their risky behaviors). In particular, obesity in adults has reached epidemic levels in the United States: currently, one-third of all adults are obese and another third are overweight. These conditions are estimated by calculating a person's **body mass index**: the ratio of weight in kilograms divided by the square of height in meters **(Fig. 17–4)**. Aside from the estimated 160,000 deaths a year caused by obesity, an obese person costs society some $7,000 a year in social and medical costs.

The connection between behavior and risk is especially lethal in the case of acquired immune deficiency syndrome (AIDS), caused by the human immunodeficiency virus (HIV). AIDS is one of a group of sexually transmitted diseases (STDs) that includes gonorrhea, syphilis, and chlamydia. AIDS is

taking a terrible toll in the developing world, where more than 90% of HIV-infected people live (Chapter 9). Wherever it occurs, AIDS is largely the consequence of high-risk sexual behavior. At present, there is no cure or vaccine for HIV.

Biological Hazards. Human history can be told from the perspective of the battle with pathogenic bacteria and viruses. It is a story of epidemics such as the black plague and typhus, which ravaged Europe in the Middle Ages, killing millions in every city, and of smallpox, which swept through the New World. The story takes a positive turn in the 19th century with the first vaccinations and the "golden age" of bacteriology (a span of only 30 years), when bacteriologists discovered most of the major bacterial diseases and brought bacteria into laboratory culture. The 20th century saw the advent of virology (the study of viruses and the treatment of the diseases they cause), the great discovery of antibiotics, immunizations that led to the global eradication of smallpox and the victory over polio and many other childhood diseases, and the growing influence of molecular biology as a powerful tool in the battle against disease.

The battle is not over, however, and never will be. Pathogenic bacteria, fungi, viruses, protozoans, and worms continue to plague every society and, indeed, every person. They are inevitable components of our environment. Many are there regardless of our human presence, and others are uniquely human pathogens whose access to new, susceptible hosts is mediated by the environment. Table 17–1 shows the leading global diseases and the annual number of deaths they cause.

TABLE 17–1 Global Mortality from Major Infectious Diseases, 2008

Cause of Death	Estimated Yearly Deaths*
Acute respiratory infections	3,816,000
HIV/AIDS	2,243,000
Diarrheal diseases	1,687,000
Tuberculosis	1,250,000
Malaria	655,000
Measles	328,000
Meningitis	270,000
Pertussis (whooping cough)	194,000
Hepatitis	136,000
Tetanus	128,000
Syphilis	81,000
Trypanosomiasis (sleeping sickness)	44,000
Schistosomiasis	37,000
Leishmaniasis	36,000
Dengue	13,000
Intestinal roundworms	5,000

*Total deaths from infectious diseases in 2008 are estimated at 16.0 million by the WHO.
Source: Data from World Health Organization, *The Global Burden of Disease: 2008* (January 2012).

Weight (pounds)

Height (feet and inches)	120	130	140	150	160	170	180	190	200	210	220	230	240	250
4'8"	27	29	31	34	36	38	40	43	45	47	49	52	54	56
4'10"	25	27	29	31	34	36	38	40	42	44	46	48	50	52
5'0"	25	25	27	29	31	33	35	37	39	41	43	45	47	49
5'2"	22	24	26	27	29	31	33	35	37	38	40	42	45	46
5'4"	21	22	24	26	28	29	31	33	34	36	38	40	41	43
5'6"	19	21	23	24	26	27	29	31	32	34	36	37	39	40
5'8"	18	20	21	23	24	26	27	29	30	32	34	35	37	38
5'10"	17	19	20	22	23	24	26	27	29	30	32	33	35	36
6'0"	16	18	19	20	22	23	24	26	27	29	30	31	33	34
6'2"	15	17	18	19	21	22	23	24	26	27	28	30	31	32
6'4"	15	16	17	18	20	21	22	23	24	26	27	28	29	30
6'6"	14	15	16	17	19	20	21	22	23	24	25	27	28	29

Figure 17–4 Body mass index. Body mass index is a ratio of height to weight used to measure obesity and overweight conditions. A person with a BMI of 30 and above is considered obese, while a person with a BMI of 25–29 is considered overweight.

(www.surgeongeneral.gov)

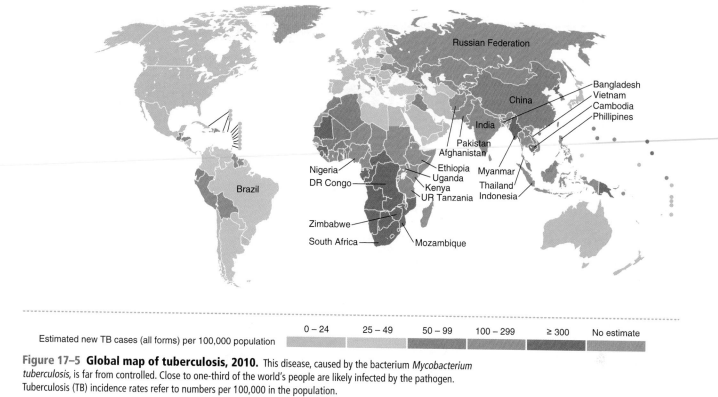

Russian Federation
Bangladesh
Vietnam
Cambodia
Phillipines
China
India
Pakistan
Afghanistan
Ethiopia
Uganda
Kenya
UR Tanzania
Myanmar
Thailand
Indonesia
Nigeria
DR Congo
Brazil
Zimbabwe
South Africa
Mozambique

| | 0 – 24 | 25 – 49 | 50 – 99 | 100 – 299 | ≥ 300 | No estimate |

Estimated new TB cases (all forms) per 100,000 population

Figure 17–5 Global map of tuberculosis, 2010. This disease, caused by the bacterium *Mycobacterium tuberculosis*, is far from controlled. Close to one-third of the world's people are likely infected by the pathogen. Tuberculosis (TB) incidence rates refer to numbers per 100,000 in the population.

(*Source:* World Health Organization, *Global Tuberculosis Control 2011.*)

Approximately one-fifth of global deaths are due to infectious and parasitic diseases. The leading causes of death in this category are the acute respiratory infections (for example, pneumonia, diphtheria, influenza, and streptococcal infections), both bacterial and viral. Pneumonia is by far the most deadly of these. Other bacterial and viral respiratory infections represent the most common reasons for visits to the pediatrician in the developed countries, but it is primarily in the developing countries that the respiratory infections lead to death, mostly in children who are already weakened by malnourishment or other diseases.

Diarrheal diseases were responsible for more than 1.2 million deaths in children under age five in 2008. One-third of these deaths are traced to a little-known virus—**rotavirus**—that infects virtually every child between the ages of three months and five years. Rotavirus is rarely diagnosed, yet it causes most cases of severe diarrhea in young children in developed countries and many childhood deaths due to diarrhea in developing countries. The difference is treatment; the infection often leads to dehydration, which is quickly life threatening, but can be reversed with rehydration therapy (the replacement of lost fluids with balanced saline solutions). The virus is as contagious as a cold virus, and fortunately, one infection usually induces lifetime immunity. Effective vaccines are now able to prevent most severe rotavirus infections; the vaccine is given to infants in three doses, starting at age 2 months. Many other cases of diarrhea are the consequence of ingesting food or water contaminated with

bacterial pathogens, such as *Salmonella*, *Campylobacter*, *Shigella*, and *E. coli*, from human wastes.

Tuberculosis. Although AIDS has overtaken tuberculosis as the disease that causes the most adult deaths worldwide, tuberculosis continues to be a major killer. Close to one-third of the world's people are infected with the microbe *Mycobacterium tuberculosis* (**Fig. 17–5**), and perhaps 10% of these will develop a life-threatening case during their lifetime. Complacency about treatment and prevention in recent years has led to a resurgence of tuberculosis, even though the bacterial pathogen that causes it has been known for more than 100 years and its genome has been sequenced. Some of the resurgence can be traced to the AIDS epidemic because HIV infection eventually leads to a compromised immune system, which allows diseases like tuberculosis to flourish. Much of the disease's comeback, however, is the result of multi-drug-resistant strains of the microbe.

Malaria. Of the infectious diseases present in the tropics, malaria is by far the most serious, accounting for an estimated 216 million cases each year and 655,000 deaths. Caused by protozoan parasites of the genus *Plasmodium*, malaria is propagated by mosquitoes of the genus *Anopheles*. The disease begins with an infected mosquito biting a person (**Fig. 17–6**). From there, the parasites invade the liver and then the red blood cells. After multiplying in the red blood cells, the parasites are released, destroying the cells. The parasites then invade other red blood cells. While in the liver and the red blood cells (most of the time), the parasites are

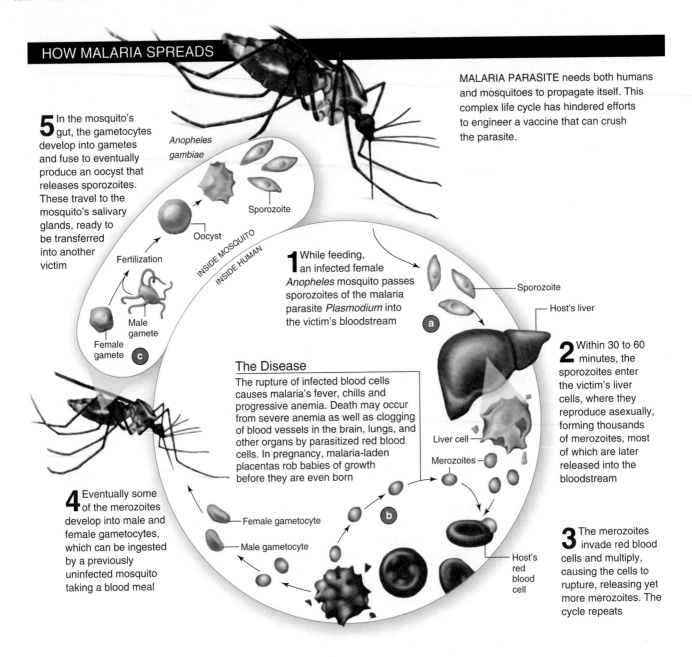

HOW MALARIA SPREADS

MALARIA PARASITE needs both humans and mosquitoes to propagate itself. This complex life cycle has hindered efforts to engineer a vaccine that can crush the parasite.

5 In the mosquito's gut, the gametocytes develop into gametes and fuse to eventually produce an oocyst that releases sporozoites. These travel to the mosquito's salivary glands, ready to be transferred into another victim

Anopheles gambiae

Sporozoite

Oocyst

Fertilization

INSIDE MOSQUITO
INSIDE HUMAN

Male gamete

Female gamete **c**

1 While feeding, an infected female *Anopheles* mosquito passes sporozoites of the malaria parasite *Plasmodium* into the victim's bloodstream

a

Sporozoite

Host's liver

2 Within 30 to 60 minutes, the sporozoites enter the victim's liver cells, where they reproduce asexually, forming thousands of merozoites, most of which are later released into the bloodstream

Liver cell

Merozoites

The Disease

The rupture of infected blood cells causes malaria's fever, chills and progressive anemia. Death may occur from severe anemia as well as clogging of blood vessels in the brain, lungs, and other organs by parasitized red blood cells. In pregnancy, malaria-laden placentas rob babies of growth before they are even born

4 Eventually some of the merozoites develop into male and female gametocytes, which can be ingested by a previously uninfected mosquito taking a blood meal

Female gametocyte

Male gametocyte

b

Host's red blood cell

3 The merozoites invade red blood cells and multiply, causing the cells to rupture, releasing yet more merozoites. The cycle repeats

Figure 17–6 Life cycle of malarial parasite. The life cycle of the protozoan parasite *Plasmodium*, which causes malaria in humans, involves the *Anopheles* mosquito and a human host.
(*Source:* Dunavan, Claire Panosian. "Tackling Malaria." *Scientific American* 293 (December 2005): 74–83. Used with permission.)

hidden from immune system cells and antibodies. This entire chain of events occurs in synchronized cycles, leading to the periodic episodes of fever, chills, and malaise that are typical of malaria. With the most virulent parasite, *P. falciparum*, 60% of the red blood cells may be destroyed in one episode, producing rapid and severe anemia. The waste products of the parasites, when they break out of the cells, produce high fever, headaches, and vomiting in the victim.

Chronic effects. Most of the diseases responsible for mortality are also leading causes of debilitation in humans of all ages. More than 4 billion episodes of diarrhea occur in the human population over the course of a year, and at any given time, more than 3.5 billion people suffer

from the chronic effects of parasitic worms such as hookworms and schistosomes.

Physical Hazards. Natural disasters—including hurricanes, tornadoes, floods, forest fires, earthquakes, landslides, and volcanic eruptions (**Fig. 17–7**)—take a toll on human life and property every year. They are the outcome of hydrological, meteorological, or geological forces. The years 2010 and 2011 were record-setting. In 2010, a new record was set for lives lost and infrastructure destroyed due to physical hazards. Heat waves, earthquakes, floods, and snowstorms led the 950 natural disasters, killing 295,000 and causing $130 billion in losses. Key events were the January 12 magnitude 7.0 earthquake in Haiti, which claimed more

(a)

(b)

(c)

(d)

Figure 17–7 Dangerous physical hazards. Natural disasters bring death and destruction in their wake.
(a) A huge offshore earthquake triggered a devastating tsunami in Japan, March 1, 2011, killing over 20,000.
(b) Heavy winter storms in the Pacific Northwest caused this tributary of the Columbia River to rampage.
(c) A magnitude 7.0 earthquake destroyed the Haitian capital, Port-au-Prince.on January 12, 2010; more than
220,000 lives were lost. (d) Joplin, Missouri, was almost leveled by a EF-5 tornado on May 22, 2011, that killed 160.

than 220,000 lives, huge floods that covered one-quarter of Pakistan, and an unprecedented heat wave in Russia that caused 55,000 deaths. The year 2011 was a record year for insurance loss ($105 billion), when 820 disasters caused a total loss of $380 billion. That year, there were major floods in Thailand and Australia, a record flood in the Mississippi River basin, and a major tornado outbreak in the Midwest and Southern states. In both years, 90% of the natural catastrophes were climate related. Table 17–2 records the mortality from major natural disasters for 2010 and 2011.

Looking at the recent record, there are *two patterns* that emerge: hazards like tsunamis, tornadoes, and earthquakes that are impossible to anticipate and hazards that are largely a consequence of choices people make about where to live.

Out of Nowhere. Each year in the United States, 780 tornadoes strike on average, spawned from the severe weather accompanying thunderstorms. Because of the tendency for cold, dry air masses from the north to mix with warm, humid air from the Gulf of Mexico, the central United States generates more tornadoes than anywhere else on Earth. Of short duration, they develop into some of the most destructive forces known in nature, with winds reaching as high as 300 miles per hour. The most intense tornadoes have killed hundreds of people. Two deadly tornadic outbreaks occurred in the spring of 2011, one in April across the Southeastern United States that killed 322 (see Fig. 17–7d) and another that almost leveled Joplin, Missouri, on May 22, killing 160. That year was a record for tornadoes in the United States—a reported 1,897.

TABLE 17–2   Mortality from Major Natural Disasters, 2010–2011

Country	Type of Disaster	Date	Deaths
Haiti	Earthquake	January 12, 2010	222,570
Russia	Heat wave	June 2010	55,736
Japan	Earthquake/tsunami	March 11, 2011	20,319
China (P. Republic)	Flood	April 14, 2010	2,968
Pakistan	Flash flood	July 28–August 10, 2010	1,985
China (P. Republic)	Landslide	August 8, 2010	1,765
China (P. Republic)	Flood	May 29–Aug. 31, 2011	1,691

Source: EM-DAT: The OFDA/CRED International Disaster Database, www.emdat.be, Université Catholique de Louvain, Brussels (Belgium).

The Japanese tsunami of March 11, 2011, and the Haitian earthquake were completely unavoidable. There is still no way to forecast earthquakes, but there are ways to mitigate their impacts. Tsunami warning systems have existed for the Pacific Ocean since 1965, but not until 2006 for the Indian Ocean. Many thousands of lives were saved by the "Large Tsunami Warning" broadcast widely on Japan's east coast immediately following the offshore earthquake. Earthquake-resistant buildings can be constructed (but were not for the city of Port-au-Prince, Haiti). This illustrates the fact that the most devastating consequences of natural hazards usually are experienced by those who are least capable of anticipating them and dealing with their effects—the poor. Most of the 20,000 fatalities brought on by the magnitude 9.0 Japanese earthquake were due to the tsunami (Japan's buildings have long been designed to withstand earthquake impact).

In Harm's Way. Much of the loss brought on by natural disasters is a consequence of poor environmental stewardship. Hillsides and mountains are deforested, leaving the soil unprotected; people build homes and towns on floodplains; villages are nestled up to volcanic mountains; cities are constructed on known geological fault lines; and coastal marshes and mangrove forests are replaced by house lots. A general human tendency is to assume that disasters happen only to other people in other places, and even if there is a risk, say, of a hurricane striking a coastal island, many people are willing to take that calculated risk in order to enjoy life on the water's edge. Trends on the U.S. "hurricane coasts" (Gulf of Mexico and southern Atlantic) are a case in point: Coastal populations continue to rise, in spite of the many hurricane strikes. One writer suggested that town zoning boards should create a *stupid zone* for such locations. Other "stupid zone" locations would be in the middle of 25-year floodplains, in highly flammable coniferous forests, on top of earthquake faults, and in valleys below dormant volcanoes.

Certainly, part of the reason for the rising human and economic costs of natural disasters is the fact that there are more people in the world to be affected. But as one climatologist put it, if natural disasters are having an increasingly greater impact, the culprit is not Mother Nature but human nature. At issue is the question of whether global climate change is behind the rising tide of natural disasters, especially intense hurricanes, tornadoes, floods, and droughts (Chapter 18). Whether it is or not, all the evidence suggests that climate change will definitely intensify future natural disasters as temperatures and sea levels rise. This is a problem we are bringing on ourselves.

Chemical Hazards. Industrialization has brought with it a host of technologies that employ chemicals such as cleaning agents, pesticides, fuels, paints, medicines, and those used directly in industrial processes (**Fig. 17–8**). The manufacture, use, and disposal of these chemicals often bring humans into contact with them. Exposure is either through the ingestion of contaminated food and drink, the breathing of contaminated air, absorption through the skin, or direct or accidental use (discussed at length in Chapter 22). **Toxicity** (the condition of being harmful, deadly, or poisonous) depends not only on *exposure*, but also on the *dose* of the toxic substance—the amount actually absorbed by a sensitive organ. Further, for most substances, there is a threshold below which no toxicity can be detected. (See Figure 22–2.)

Making things even more difficult, different people have different thresholds of toxicity for given substances. For example, children are often at greater risk than adults because children are growing rapidly and are incorporating more of their food (and any contaminants that are in it) into new tissue. Embryos are even more sensitive. Substances transmitted across the mother's placenta can have grave impacts on the embryo's development, especially in the early stages.

Many developing countries are in double jeopardy because of both their high level of infectious disease and their rising exposure to toxic chemicals through industrial

(a)

(b)

(c)

(d)

Figure 17–8 Industrial processes and hazards. Many industries produce chemicals and pollutants that contribute to the impact of toxic substances on human health. Some examples of hazardous workshops are (a) a paper mill in Port Angeles, Washington; (b) a tannery in Fez, Morocco; (c) an oil refinery; and (d) a plastics factory in China.

processes and home use. In both cases, the risks are largely preventable. In places where industrial growth is especially rapid, insufficient attention is often given to the pollutants that accompany that growth. For example, nine of the 10 cities with the worst air pollution are found in Asia, where economic growth has risen rapidly.

Carcinogens. Many chemicals, such as heavy metals, organic solvents, and pesticides, are hazardous to human health even at very low concentrations. Episodes of acute poisoning are easy to understand and are clearly preventable, but the long-term exposure to low levels of many of these substances presents the greatest challenge to our understanding. Some chemical hazards are known **carcinogens** (cancer-causing agents). Cancer takes a very large toll in the developed countries, accounting for one-fourth of mortality (see Chapter 9, Figure 9–8). Because cancer often develops over a period of 10 to 40 years, it is frequently hard to connect the cause with the effect. The twelfth Report

on Carcinogens[8] from the U.S. Department of Health and Human Services lists 51 chemicals (and three biological agents) now known to be human carcinogens and 186 more that are "reasonably anticipated to be human carcinogens" because of animal tests. Many are substances used in both developed and developing countries. Other potential effects of toxic chemicals are impairment of the immune system, brain impairment, infertility, and birth defects.

Cancer and Carcinogenesis. In the United States, 23% of all deaths in 2010—573,900—were traced to cancer, and even though it is largely a disease of older adults, cancer can strike people of all ages. How is it that a simple chemical like benzene can cause leukemia? How can aflatoxin cause liver cancer? To answer these questions, we must understand

---
[8]*Report on Carcinogens, Twelfth Edition,* June 10, 2011. U.S. Department of Health and Human Services, Public Health Service, National Toxicology Program.

cellular growth and metabolism. During normal growth, cells develop into a variety of specific types that perform unique functions in the body: blood cells, heart muscle cells, liver cells, absorptive cells that line the small intestine, and so forth. Some cells, like the stem cells that form the cells of blood, keep on dividing until death. Other cells, however, are programmed to stop dividing and simply perform their functions, often throughout life. A cancer is a cell line that has lost its normal control over growth. Recent research has revealed the presence of as many as three dozen genes that can bring about a malignant (out-of-control) cancer; these genes are like ticking time bombs that may or may not go off.

Carcinogenesis (the development of a cancer) is now known to be a process with several steps, often with long periods of time in between. In most cases, a sequence of five or more mutations must occur in order to initiate a cancer. (A mutation is a change in the informational content of the DNA that controls processes in the cell.) *Environmental* carcinogens are either chemicals that are able to bind to DNA and prevent it from functioning properly or agents, such as radiation, that can strike the DNA and disrupt it. When such an agent initiates a mutation in a cell, the cell may go for years before the next step—another mutation—occurs. Thus, it can take upward of 40 years for some cells to accumulate the sequence of mutations that leads to a malignancy. In a malignancy, the cells grow out of control and form tumors that may spread (metastasize) to other parts of the body. Early detection of cancers has improved greatly over the years, and many people survive cancer who would have succumbed to it a decade ago. The best strategy, however, is prevention, so great attention is now given to the environmental carcinogens and habits, like tobacco use, that are clearly related to cancer development.

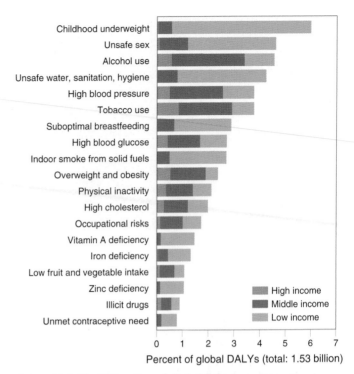

Figure 17–9 **The 19 leading global risk factors.** The top five of these 19 factors are responsible for more than one-fourth of all deaths and much of the global burden of disease. They are reported as a percentage of global DALYs (disability-adjusted life years); one DALY equals the loss of one healthy life year.

(*Source:* World Health Organization. *Global Health Risks: Mortality and Burden of Disease Attributable to Selected Major Risks.* WHO, 2009.)

17.2 Pathways of Risk

The environmental hazards identified in Section 17.1 are responsible for untold human misery and death, but if many of the consequences of exposure to the different hazards are preventable, how is it that the hazards turn into mortality statistics? In other words, what are the pathways that lead to human deaths from risks of infection, exposure to chemicals, and vulnerability to physical hazards? If prevention is the goal—and it should be—this knowledge is crucial. In 2009, the WHO, building on an earlier report on health risks, published an update: *Global Health Risks: Mortality and Burden of Disease Attributable to Selected Major Risks.*[9] The document is a thoroughly researched presentation of the major risks as factors "that raise the probability of adverse health outcomes," as the report puts it. The report shows that a remarkably small number of risk factors are responsible for a vast proportion of premature deaths and disease.

The top five factors (**Fig. 17–9**) are responsible for one-fourth of all deaths and much of the global disease burden. We will consider only a few examples of these risk factors, because a thorough examination of this topic would take up volumes.

The Risks of Being Poor

One major pathway for hazards is poverty: The WHO has stated that poverty is the world's biggest killer. Environmental risks are borne more often by the poor, not only in the developing countries, but also in the developed countries of the world. The people who are dying from infectious diseases are those who lack access to adequate health care, clean water, nutritious food, healthy air, sanitation, and shelter. Because of this, they are exposed to more environmental hazards and thus encounter greater risks. The world's number one risk factor (Fig. 17–9) is called **underweight**, a problem of malnutrition that strikes primarily children less than five years of age (see Chapter 12). This factor is strongly related to poverty: Those households living on less than $1.25 per day (one-fifth of the world's population) show the highest prevalence of underweight children. A 2008 report in *The Lancet*[10] found that more than one-third of childhood deaths under age five are caused by malnutrition.

---

[9]World Health Organization. *Global Health Risks: Mortality and Burden of Disease Attributable to Selected Major Risks.* World Health Organization, 2009. Available at www.who.int/healthinfo/global_burden_disease/GlobalHealthRisks_report_full.pdf.

[10]Robert E. Black et al. "Maternal and Child Undernutrition: Global and Regional Exposures and Health Consequences." *The Lancet* 371: 243–260. Jan. 19, 2008.

Figure 17–10 Public health clinic. This clinic on the island of Zanzibar, Tanzania, provides families with health checks to improve the health of mothers and children.

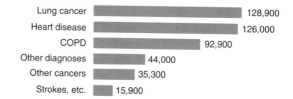

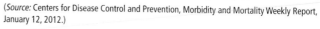

Figure 17–11 Deaths caused by smoking. Approximately 443,000 deaths in the United States are attributed to smoking each year.

(*Source:* Centers for Disease Control and Prevention, Morbidity and Mortality Weekly Report, January 12, 2012.)

Wealth. Overall, the wealthier a country becomes, the healthier is its population. However, in virtually all countries, those with wealth are able to protect themselves from many environmental hazards. People in developed countries tend to live longer, and when they die, the chief causes of death are diseases of old age—the so-called degenerative diseases, such as cancers and diseases of the circulatory system (heart disease, strokes, and so forth). By contrast, only one-third of the deaths in the low-income countries are indicative of people who are living out their life span.

There are many reasons for this discrepancy between developed and developing nations. One is education. Advances in socioeconomic development and advances in education are strongly correlated. As people—and especially women—become more educated, they act on their knowledge to secure relief from hazards to health. They may, for example, improve their hygiene, immunize their children (**Fig. 17–10**), or recognize dangerous symptoms such as dehydration and seek oral rehydration therapy. Other reasons for the discrepancy between health in the developed world and that in the developing world are the lack of safe drinking water and sanitation and the high indoor air pollution associated with the use of solid fuels such as wood, dung, and coal in simple cooking stoves.

Priorities. Education, nutrition, and the general level of wealth in a society do not tell the whole story, however. A nation may make a deliberate policy choice to put its resources into improving the health of its population rather than, say, militarization or the development of modern power sources. Countries such as Costa Rica, China, and Sri Lanka have a much longer life expectancy and lower infant mortality than their gross domestic product would predict. They have accomplished this enviable record by focusing public resources on such concerns as immunization, the upgrading of sewer and water systems, and land reform.

The Cultural Risk of Tobacco Use

Lifestyle choices such as refraining from exercise, overeating, driving fast, imbibing alcohol, climbing mountains, and

so forth carry with them a significant risk of accident and death. However, one cultural hazard, tobacco use, is the leading cause of death in the United States (**Fig. 17–11**) and ranks sixth in the global tally of risk factors (Fig. 17–9). Although tobacco use is declining in some developed countries, it is on the increase in most developing countries and remains high in the former socialist countries of Eastern Europe. Smoking often begins in childhood. Lured by the image of smoking as "cool" or by peer pressure, youngsters are very quickly addicted by the nicotine in cigarettes, an addiction that can last a (shortened) lifetime. According to the CDC, some 45 million adults (19.3%) in the United States smoke cigarettes, and half of these will die or become disabled because of their habit if they don't quit.

Marlboro Country? Tobacco is the only known product on the market that kills half of its users. Cigarette smoking has been clearly and indisputably correlated with cancer and other lung diseases. Smoking has been shown to be responsible for 29% of cancer deaths in the United States. The WHO puts the number of smokers worldwide at more than 1 billion, with 6 million deaths per year due to cigarette smoking. Eighty percent of smokers are in the developing world, where the habit is increasing at a rate of 2% per year. In China, almost 60% of men are smokers. The WHO reported that smoking is a growing epidemic in the developing countries, where its full impact is yet to come because tobacco-related disease and death don't begin until decades after the start of tobacco use.

Studies have shown that smokers living in polluted air experience a much higher incidence of lung disease than do smokers living in clean air—a synergistic effect. Certain diseases typically associated with occupational air pollution show the same synergistic relationship with smoking. For example, black lung disease is seen almost exclusively among coal miners who are also smokers, and lung disease predominates in smokers who are exposed to asbestos. Smoking is linked to half of the tuberculosis deaths in India, which leads the world in that category. Smoking also kills by impairing respiratory and cardiovascular functioning. The CDC estimates that smoking costs the United States $193 billion a year in health care costs and lost job productivity because of premature death and disability (obesity-related medical costs come in second, at $147 billion per year).

In the United States, several measures have been taken to regulate this cultural hazard. Tobacco products are highly taxed, for example, providing substantial revenues to the states ($17 billion in 2009) and the federal government ($11.5 billion in 2009). These revenues are sure to rise, as a result of the Children's Health Insurance Program Reauthorization Act of 2009. This new law raised the federal excise tax on cigarettes from $.24 to $1.01 a pack and the average state tax from $.33 to $1.20. Raising cigarette taxes has been shown to be one of the most effective ways to reduce smoking, especially in young people. Surgeons General of the United States have issued repeated warnings against smoking since 1964, and public policy has taken them seriously by requiring warning labels on smoking materials, banning cigarette advertising on television, promoting smoke-free workplaces, requiring nonsmokers' areas in restaurants, and banning smoking on all domestic airline flights. Since the warnings began, the U.S. adult smoking population has gradually dropped from 42% to 19.8%.

Secondhand Smoke. In January 1993, the EPA classified environmental tobacco smoke (ETS), meaning secondhand smoke, as a Class A (known human) carcinogen. In the words of then EPA Administrator Carol Browner, "Widespread exposure to secondhand smoke in the United States presents a serious and substantial public health risk." The EPA's action was unanimously supported by its science advisory board. The consequences of this action have been substantial: Outright bans in public areas such as shopping malls, bars, restaurants, and workplaces are now the national norm. In particular, the EPA has taken specific steps to protect children from ETS in all public places, and the Occupational Safety and Health Administration is promoting policies to protect workers from involuntary exposure to ETS in the workplace. The impact of these actions on public health should be significant.

FDA Authority. Responding to efforts by the Food and Drug Administration to curb tobacco sales to young people, tobacco companies brought a challenge to the Supreme Court. In March 2000, the U.S. Supreme Court ruled that the FDA did not have the authority to regulate tobacco. That authority, said the Court, belonged to Congress. Until 2009, the many bills filed in Congress to extend that authority to the FDA did not make it into law. However, in June 2009, the House and Senate voted overwhelmingly to give the FDA sweeping authority to regulate tobacco products, with the **Tobacco Control Act**, finally overcoming the efforts of a powerful tobacco lobby. One recent impact of the act is a requirement for new, graphic labels on cigarette packs showing some of the worst effects of smoking (**Fig. 17–12**).

FCTC and MPOWER. The battle against smoking is rapidly spreading worldwide. After several years of efforts, the WHO *Framework Convention on Tobacco Control (FCTC)* was adopted in May 2003 by the World Health Assembly. This convention is a treaty that aims to reduce the spread of smoking by requiring signatory countries to raise taxes, restrict advertising, require larger health warnings on tobacco materials, and reduce ETS (steps already taken in the United States). The treaty went into effect on February 27,

Figure 17–12 New warnings on cigarettes. The FDA has required that prominent messages be placed on packs of cigarettes to warn users of possible consequences of smoking. This photo shows one of the nine graphic warnings approved by the FDA.

(*Source:* FDA. http://www.fda.gov/TobaccoProducts/Labeling/CigaretteWarningLabels/default.htm. January 16, 2012.)

2005, and has been signed by 173 countries. Many countries have already taken the action called for in the treaty; Ireland, Spain, Norway, and Italy have banned smoking in indoor public places, and other countries have begun banning advertising and requiring printed warnings on cigarette packages.

The WHO, with funding of $500 million from Bill Gates and New York City's Mayor Michael Bloomberg, initiated a campaign in 2008 to combat the global smoking epidemic. Given the acronym MPOWER, the campaign has six components with proven strategies to reduce tobacco use:

- Monitor tobacco use and prevention policies.
- Protect people from tobacco smoke.
- Offer help to quit tobacco use.
- Warn about the dangers of tobacco.
- Enforce bans on tobacco advertising, promotion, and sponsorship.
- Raise taxes on tobacco.

The MPOWER campaign is well under way, carried out by the WHO FCTC agencies and partners. Antismoking campaigns in developed and developing countries have employed

SOUND SCIENCE
The Grisly Seven

In a world dominated by air travel, mobile phones, and the Internet, it is a stretch of the imagination to picture a billion people still infected with chronic and quite treatable tropical diseases. These are not malaria, tuberculosis, or HIV, the big three that are often equated with "tropical diseases" and which are addressed with major treatment programs and funding agencies. These diseases are, collectively, the Neglected Tropical Diseases (NTDs). Seven stand out as particularly nasty, exacting a toll of anemia, malnutrition, weakness, disfigurement, and blindness. Virtually every person experiencing grinding poverty in sub-Saharan Africa, Latin America, and Southeast Asia is infected by one or more of the "grisly seven." Not only do they accompany poverty, they also help to perpetuate it. Here are the seven, plus one more that also deserves notice.

Roundworm (*Ascariasis*). These large worms (5–14 inches) live in the small intestine and afflict some 800 million people. They are transmitted by contact with eggs in the soil and cause malnutrition, stunting, and bowel obstruction, sometimes shocking a victim by crawling out of the nose and mouth.

Whipworms (*Trichuriasis*) are one- to two-inch roundworms preferring the large intestine, spread by contaminated soil and found in 600 million people. They cause colitis and also stunted growth in children.

Hookworm. These roundworms, one half-inch long, latch on to the small intestine and, because they feed on blood, cause severe iron deficiency and anemia, also stunting child growth. They infest 600 million victims and are also transmitted by contaminated soil. Many victims are parasitized by combinations of all three of these roundworm species.

Schistosomiasis. These are half- to one-inch flukes (flatworms) that live in veins of the intestine and bladder. They are acquired by contact with freshwater; the infective agents (cercaria) are released by freshwater snails and penetrate the skin. Adult flukes lay spiny eggs that damage internal organs. These parasites infest 200 million people, causing chronic pain, anemia, stunting, and malnutrition.

Lymphatic Filariasis. Spread by mosquitoes, these 2- to 4-inch worms live in lymphatic tissues, often blocking lymph flow and causing severe leg swelling and disfigurement in severe cases, known as elephantiasis. Some 120 million are infested by these worms.

Onchocerciasis. Another roundworm disease, acquired by the bite of black flies, onchocerciasis produces worms that live in nodules under the skin, growing up to 20 inches long. Their larvae migrate to the eyes and cause blindness, and the adults cause itching and disfigurement in the skin. Most of the 30–40 million cases occur in Africa.

Trachoma. This is a bacterial infection spread by poor hygiene and houseflies. Between 60 million and 80 million people are infected by this *Chlamydia* species, different from the *Chlamydia* that causes the sexually transmitted genital infection. It is the most common form of blindness caused by infection.

Guinea Worm Disease. Two decades ago, some 3.5 million people in 21 countries were infested with this roundworm, acquired by drinking water infested with larvae. The 3-foot-long worm causes a lesion in the leg or foot and releases eggs when the victim wades in the water. The Carter Center has carried out a successful campaign to eradicate this disease, now down to some 1,000 cases in four African countries.

Every one of these neglected diseases can be treated, sometimes with a single pill, and prevented. The treatment drugs are inexpensive and effective, but perhaps because the NTDs infect people who are poor and do not often result in death, governments have not given them the attention they deserve. The WHO estimates that only 10% of people suffering from the NTDs are being reached. There are treatment programs, and treatment is focusing now on a single program of drugs that could cost as little as 50 cents. The Global Network of NTDs has been formed as a partnership between donors and public agencies and is beginning to receive funding. It is estimated that $2 billion to $3 billion would be needed for the next seven years to make a serious impact on these diseases. The Carter Center wisely selected one disease, working tirelessly to eradicate it country by country. Something like this paradigm could be applied to the grisly seven; it is believed that the NTD drugs are the most cost-effective way to improve the health and well-being of the world's poor.

Source: Peter Jay Hotez. "A Plan to Defeat Neglected Tropical Diseases." *Scientific American* 302: 90–96. January 2010.

---

the use of mass media and health warnings on tobacco product packaging, reaching a growing proportion of countries.[11] Smoke-free laws covering public places and workplaces have been newly established in 16 countries, and complete bans on tobacco advertising now cover some 6% of the world's population. Still, as the WHO report points out, governments collect some $133 billion in tobacco excise tax revenues but spend less than $1 billion on tobacco control. Given the huge health care expenditures on tobacco-related illnesses and death, there is clearly much to be gained by MPOWER's efforts to make tobacco use history.

Risk and Infectious Diseases

Epidemiology has been described as "medical ecology" because the epidemiologist traces a disease as it occurs in geographic locations as well as tracing the mode of transmission and consequences of the disease in the field. Epidemiologists point to many reasons why infectious diseases and parasites are more prevalent in the developing countries. One major pathway of risk is contamination of food and water due to inadequate hygiene and inferior sewage treatment. Some of the risk can be traced to the failure to educate people about diseases and how they can be prevented. Much, however, is simply a consequence of the general lack of resources needed to build effective public health–related infrastructure in cities, towns, and villages of the developing world. According to the UN's 2011 report on the Millennium Development Goals, an estimated 2.6 billion people are still without adequate sanitation, and 884 million lack access to safe drinking water. Many microbial diseases are transmitted in this way, and some of these—such as cholera and typhoid fever—are deadly and capable of spreading to epidemic proportions. Any diarrheal infection in small children can be fatal if it is accompanied by malnutrition or

---

[11]World Health Organization. *WHO Report on the Global Tobacco Epidemic, 2011: Warning About the Dangers of Tobacco*. Executive Summary. WHO, 2011. www.who.int/tobacco.

dehydration. By making oral rehydration therapy kits readily available throughout the country, Mexico reduced mortality rates in children by 70% in less than a decade.

Even Here? These problems are not unique to the developing world. Inadequate surveillance in developed countries sometimes leads to major outbreaks of diseases, with fatal outcomes. American consumers spend more than $1 trillion on food each year, two-thirds of which is from domestic (farm) production. At the same time, the CDC estimates that every year 48 million Americans are sickened by contaminated food, and 128,000 cases require hospitalization, 3,000 result in death. The factory-like conditions under which food processing in the developed countries takes place often give pathogens an opportunity to contaminate food on a wide scale, resulting in diarrheal outbreaks and major recalls of food products. Peanut butter, that staple food for kids of all ages, was implicated in an outbreak of *Salmonella typhimurium* in 2008–2009. (This bacterium infects the intestinal tract and brings on diarrhea, fever, and cramps.) More than 700 cases (and nine deaths) were recorded, and the source was traced to a factory in Georgia. Another *Salmonella* species brought on a nationwide recall of 500 million eggs in the summer of 2010, following the emergence of cases of infection. It was occurrences like these that brought about passage of the **FDA Food Safety Modernization Act** of 2010, expanding the FDA's authority over the food processing industry. The occurrence of such outbreaks, the resurgence of

known diseases, and the emergence of new ones like SARS, AIDS, avian flu, swine flu, and the Ebola virus are also the concern of public-health agents and epidemiologists in organizations such as the WHO and the CDC.

Tropical Diseases. The tropics, where most of the developing countries are found, have climates ideally suited for the year-round spread of disease by insects. Mosquitoes are **vectors** (organisms that transmit infectious agents) for several deadly diseases, such as yellow fever, dengue fever, Japanese encephalitis, West Nile virus, and malaria. However, there are other tropical diseases that afflict a billion people, and although they seldom kill, they are debilitating and make life miserable; these are the subject of the Sound Science essay "The Grisly Seven" (p. 429).

Malaria is by far the most serious tropical disease **(Fig. 17–13)**. Public-health attempts to control malaria have been aimed at **vector control** (e.g., attacking the *Anopheles* mosquito with the use of insecticides) and **treatment strategies** (e.g., curing people once they have been infected). Malaria was once prevalent in the southern United States, but an aggressive campaign in the 1950s to eliminate the *Anopheles* mosquito and identify and treat all human cases of malaria led to the complete eradication of the disease there.

Unfortunately, in the tropics the mosquitoes have developed resistance to all of the pesticides employed, and eradication has remained an elusive goal. DDT is still used in some developing countries to spray the walls of huts and

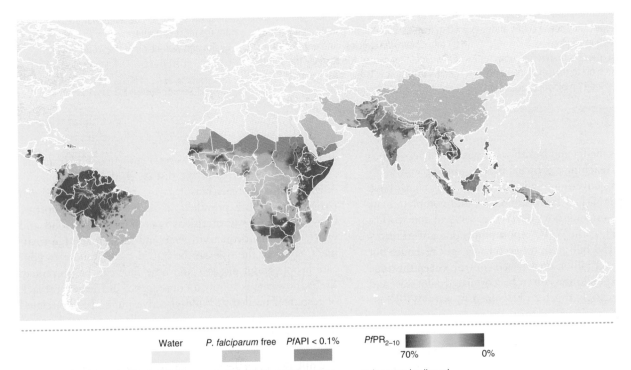

	Water	*P. falciparum* free	*PfAPI* < 0.1%	*PfPR*$_{2-10}$
				70% ——— 0%

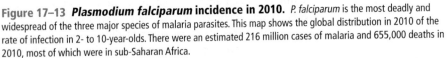

Figure 17–13 ***Plasmodium falciparum*** **incidence in 2010.** *P. falciparum* is the most deadly and widespread of the three major species of malaria parasites. This map shows the global distribution in 2010 of the rate of infection in 2- to 10-year-olds. There were an estimated 216 million cases of malaria and 655,000 deaths in 2010, most of which were in sub-Saharan Africa.

(*Source:* Malaria Atlas Project. Used with permission. http://www.map.ox.ac.uk/about-map/open-access. January 16, 2012.)

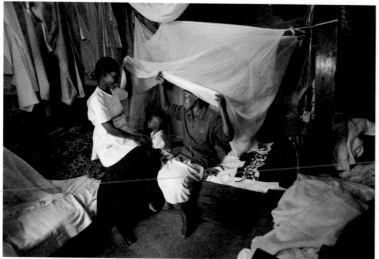

Figure 17–14 Insecticide-treated bed net. A home health worker installs a bed net in a home in Kampala, Uganda.

houses, but its use in this manner is highly controversial because of the pesticide's well-known harmful environmental and health impacts (see Stewardship, "DDT for Malaria Control," p. 322). However, this limited spraying has proven to be highly effective. Resuming their use of DDT after several years with another pesticide, South Africa and Zambia were able to dramatically reduce their malaria caseloads.

Net Results. One highly promising effort funded by the WHO's Tropical Disease Research Program found that children provided with insecticide-treated nets over their beds (bed nets) experienced a substantial reduction in mortality from all causes (Fig. 17–14). In Vietnam, a combination of bed nets and more effective drugs reduced malaria deaths by 98% in seven years. This strategy is the basis of a cost-effective, large-scale intervention throughout Africa to reduce the high mortality from malaria, and it is working. Using a combination of long-lasting insecticide-treated bed nets, indoor DDT spraying, and quick access to drug treatment, the southern Africa nation of Namibia has seen a huge decline in cases (83%), hospital admissions (92%), and deaths (96%) in just 10 years. Already, hundreds of millions of bed nets have been distributed, and huge declines in malaria incidence have been seen.

Complicating control of the disease in victims, the *Plasmodium* protozoans have developed resistance to one treatment drug after another. Until recently, chloroquine was quite effective against malaria in Africa. Now it is ineffective, and health workers have turned to artemisinin, a drug derived from the common Chinese weed *Artemisia annua*. This drug works quickly and effectively, but the WHO says it should be used only in combination with several other antimalarials, called artemisinin combination therapy (ACT), because of the danger of creating artemisinin-resistant strains of the malarial parasite when used alone. Indeed,

ACT resistance has begun to crop up in western Cambodia and is being met with a containment plan to eliminate malaria from regions where the resistance has been found.

Good News. In 2002, molecular biologists successfully sequenced the genomes of the *Anopheles* mosquito and the most lethal malaria parasite, *P. falciparum*. Armed with this information, researchers are now able to target potential weak points in both organisms in the search for new vaccines and drugs. Research continues on the development of new, more effective antimalarial drugs and on the development of an effective vaccine. Although the parasite's complicated life cycle and the fact that malaria does not readily induce long-term immunity have made a vaccine an elusive goal, a vaccine recently developed by GlaxoSmithKline—labeled RTS,S—has been field-tested and has shown modest (55% prevention) efficacy.

A number of plans have been developed over the years aimed at conquering malaria, the latest of which is the Global Malaria Action Plan (GMAP). Unlike the others, this plan has an extremely ambitious goal: reduce malaria deaths to near zero by 2015 and then gradually eliminate malaria everywhere until it is eradicated. International funding for the plan is now at $2 billion per year; one-third of that comes from the United States. Even if GMAP is not completely successful, it very likely will meet the Millennium Development Goal target to halt and begin to reverse the incidence of malaria.

Toxic Risk Pathways

How is it that people are exposed to chemical substances that can bring them harm? Airborne pollutants represent a particularly difficult set of chemical hazards to control, as they are difficult to measure and difficult to avoid. Humans breathe 30 pounds (14 kg) of air into their lungs each day. Although some of the symptoms of pollution that people suffer involve the moist surfaces of the eyes, nose, and throat, the major site of impact is the respiratory system. Three categories of impact can be distinguished:

1. **Chronic.** Pollutants cause the gradual deterioration of a variety of physiological functions over a period of years.

2. **Acute.** Pollutants bring on life-threatening reactions within a period of hours or days.

3. **Carcinogenic.** Pollutants initiate changes within cells that lead to uncontrolled growth and division (cancer), which is frequently fatal.

Indoor Air Pollution. Much attention is given to air pollutants released to the atmosphere—and rightly so (Chapter 19 discusses these pollutants). However, *indoor air pollution* can pose an even greater risk to health. The air inside the home and workplace often contains much higher

Figure 17–15 Indoor air pollution. Pollutants originate from many sources and can accumulate to unhealthy levels, leading to asthma and "sick-building syndrome" (characterized by eye, nose, and throat irritations; nausea; irritability; and fatigue) as well as other serious problems.

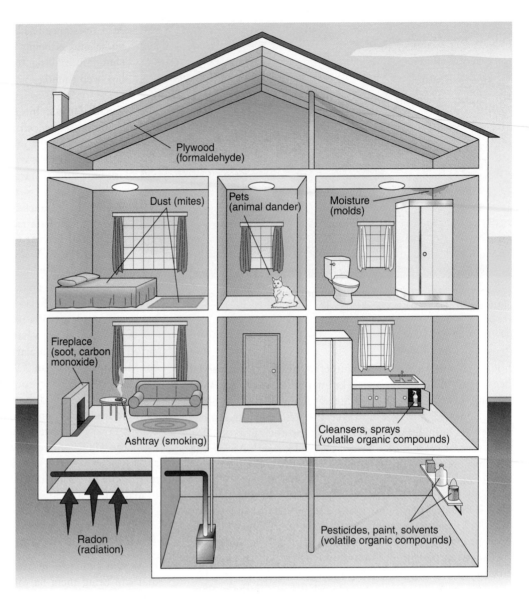

levels of hazardous pollutants than outdoor air does. In the developed countries, the overall indoor air pollution problem is threefold. First, increasing numbers and types of products and equipment used in homes and offices give off hazardous fumes. Second, buildings have become increasingly well insulated and sealed; hence, pollutants are trapped inside, where they may accumulate to dangerous levels. Third, people are exposed more to indoor pollution than to outdoor pollution. The average person spends 90% of his or her time indoors, and the people who spend the most time inside are those most vulnerable to the harmful effects of pollution—namely, small children, pregnant women, the elderly, and the chronically ill.

The sources of indoor air pollution are quite varied, as evidenced in **Figure 17–15**. One of the least excusable of these is smoking, which carries a much higher health risk than any of the other sources shown in Figure 17–15. ETS contains 40 substances that are known carcinogens and many others that are potent respiratory irritants.

Developing Countries. The most serious indoor air pollution threat is found in the developing world, where at least 3 billion people continue to rely on biofuels like wood and animal dung for cooking and heat. This threat was identified as one of the top 10 health risks by the WHO (Fig. 17–9) and ranked by the WHO as the leading *environmental* cause of death, contributing to nearly 2 million deaths annually. Fireplaces, cooking fires, and stoves are often improperly ventilated (if at all), exposing the inhabitants to very high levels of pollutants, such as carbon monoxide, nitrogen and sulfur oxides, soot, and benzene (the last two being known carcinogens). Four major problems have been associated with indoor air pollutants in the developing world: acute respiratory infections in children; chronic lung diseases, including asthma and bronchitis; lung cancer; and birth-related problems. One study in Mexico showed that women who were continually exposed to indoor smoke had 75 times the risk of developing a chronic lung disease as women who weren't exposed. Solutions to this problem include the use of ventilated stoves that burn more efficiently and the conversion from biofuels to bottled gas or liquid fuels such as kerosene.

Asthma. One common consequence of indoor air pollution is asthma, which is currently at epidemic proportions in the United States, afflicting 25 million people each year. Asthma attacks the respiratory system so that air passages tighten and constrict, causing wheezing, tightness in the chest, coughing, and shortness of breath. In acute attacks, the onset is sudden and sometimes life threatening. Substances that can trigger asthma attacks include dust, animal dander, mold, secondhand smoke, swimming pool chlorine, and various gases and particles. About half of asthma is traced to allergies, but the remainder has other, unknown causes. Asthma is expensive, too, with direct and indirect costs estimated at more than $56 billion a year in the United States, where more than 3,400 die from the condition annually. Through its Indoor Environments Division Web site, the EPA provides helpful guidelines for controlling many of the substances that trigger asthma in schools, homes, and other indoor environments.

The Hygiene Hypothesis. Oddly, asthma is much more common in developed countries, in spite of the greater exposure to indoor air pollution in developing countries. Recent research has suggested that frequent parasitic worm infections stimulate a well-regulated anti-inflammatory network in the immune systems of most people in the developing countries, protecting them against many allergic diseases like asthma. In the developed countries, where such parasitic infections are uncommon, this protection is lacking, so allergic disorders are much more common. It is interesting that children who grow up on farms in developed countries also experience much lower frequency of asthma and allergies. These observations have led to what is called the **hygiene hypothesis**: immune systems need to encounter microbes (and worms?) when we are young in order to keep inflammatory responses like asthma under control. Apparently, too much cleanliness can make you sick![12]

Toxicology. Even though the substances found in indoor air are hazardous, the link between their presence in the air and the development of a health problem is harder to establish than with infectious diseases. As with cancer, often the best that can be done is to establish a statistical correlation between exposure levels and the development of an adverse effect. The science of **toxicology** studies the impacts of toxic substances on human health and investigates the relationships between the presence of such substances in the environment and health problems like cancer (Chapter 22 discusses toxic substances and toxicology in greater detail).

Disaster Risk

Each year, more than 100 million people are affected by natural disasters. These disasters inevitably trigger emergency relief efforts and subsequent humanitarian assistance and

reconstruction—all good things. The global costs of weather-related disasters have increased dramatically in recent years, from $9 billion per year in the 1980s to $120 billion in the late 2000s. Enormous resources are devoted to these efforts, and societies need to maintain their capacity to respond swiftly and effectively to natural disasters. They receive help with this from the UN Disaster Assessment and Coordination (UNDAC) system; in 2010, the UNDAC sent teams to 12 areas.

However, much of the human death and misery brought on by natural hazards could be prevented through rigorous *disaster risk reduction* strategies. Two resources have recently become available to assist societies to move in this direction. One is the publication *Living with Risk: A Global Review of Disaster Reduction Initiatives*,[13] produced in 2004 by the Inter-Agency Secretariat of the International Strategy for Disaster Reduction. This publication, available on the Internet, applies the concepts of hazards, risks, and risk assessment to natural disasters and goes on to recommend policies and a framework for guiding and monitoring disaster risk reduction.

A second resource was created in January 2005, when 168 country delegations met in Kobe, Japan, for the UN World Conference on Disaster Reduction. At that conference, delegates adopted the *Hyogo Framework for Action* (HFA), a global blueprint for disaster risk reduction efforts for the coming decade. The framework identified five priorities for action:

1. **Make disaster risk reduction a priority:** Ensure that disaster risk reduction is a national and a local priority with a strong institutional basis for implementation.

2. **Know the risks and take action:** Identify, assess, and monitor disaster risks—and enhance early warning.

3. **Build understanding and awareness:** Use knowledge, innovation, and education to build a culture of safety and resilience at all levels.

4. **Reduce risk:** Reduce the underlying risk factors. For example, in earthquakes, it is usually houses and buildings that kill people. Building standards in earthquake-prone areas can greatly reduce the death toll of an earthquake.

5. **Be prepared and ready to act:** Strengthen disaster preparedness for effective response at all levels.

A midterm review of the HFA indicated that although significant progress has been made, it is uneven across the world; the more vulnerable segments of society still lack the resilience to hazards called for by the plan.

We now turn our attention to **risk assessment**, one of the most important tools of the toxicologist in the developed countries. Risk assessment is an approach to the problems of environmental health and is now a major element of public policy.

---

[12]Leslie, Mitch. "Gut Microbes Keep Rare Immune Cells in Line." *Science* 335: 1428. March 23, 2012.

[13]Inter-Agency Secretariat of the International Strategy for Disaster Reduction, *Living with Risk: A Global Review of Disaster Reduction Initiatives* (Geneva: United Nations International Strategy for Disaster Reduction (UN/ISDR), 2004). https://unp.un.org.

Figure 17–16 Loss of life expectancy from various risks. One way to illustrate the relative risks posed by different hazards is to rank them in terms of years of lost life associated with each risk. The data show loss of life expectancy in Denmark for men and women.

(*Source:* Knud Juel, Jan Sorensen, and Henrik Bronnum-Hansen. Risk Factors and Public Health in Denmark, Summary Report. National Institute of Public Health and University of Southern Denmark, Copenhagen, 2007)

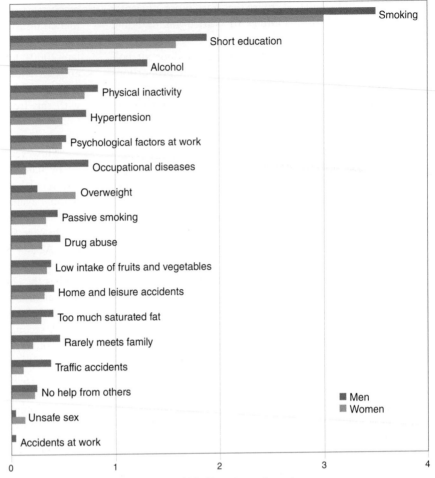

17.3 Risk Assessment

In the developed countries, our lives are freer of hazards and risks than ever before. Despite that, our society still faces hazards and the risks that they present to our health. For our own self-interest, we should know about and evaluate the risks to which we are subjected. Such knowledge enables us to make informed choices as we consider the benefits and risks of the hazards around us. In the developing countries, the heavy burden of disease so common in many of the poorer regions calls for a medical response out of humanitarian concern. But *treating the sick*, as helpful and appropriate as it is, will go on indefinitely if *prevention* is ignored. As the WHO 2009 report puts it, "Most scientific and health resources go towards treatment. However, understanding the risks to health is key to preventing disease and injuries."[14] When the risks are thoroughly understood, strategies for reducing disease and death are far easier to identify and prioritize. If governments are to act as stewards of the health of their people, they desperately need the kind of information a thorough evaluation of health risks can bring them.

Risk assessment is *the process of evaluating the risks associated with a particular hazard before taking some action in a situation in which the hazard is present.* **Figure 17–16** shows a number of risks, expressed as years of lost life expectancy, for men and women in Denmark. Alternatively, risks may be expressed as the probability of dying from a given hazard (Table 17–3). According to the table, the annual risk of dying from smoking is 1.5 per 1,000. That is, in the course of a year, 1.5 people out of every 1,000 who smoke will die as a result of their habit. Figure 17–16 suggests that this is equivalent to about three and a half years (for men) of lost life expectancy. In other words, men who never smoke will live an average of three and a half years longer than those who do.

Truthfully, not very many people actually make choices about the hazards in their lives on the basis of such risk assessments. However, risk assessment has become an important process in the development of public policy and has long been a major way of applying science to the hard problems of environmental regulation. The WHO now recommends risk assessment as an ideal way to promote the health of people everywhere. In the United States, risk assessment is now a major public-policy tool used by many agencies to address a wide variety of hazards, but it is especially employed by the Environmental Protection Agency.

---

[14]World Health Organization, *Global Health Risks*, 2009, op. cit.

TABLE 17–3 Some Common Hazards Ranked According to Their Degree of Risk

Hazard	Annual Risk*		
Cancer	1.8	per	1,000
Smoking	1.5	per	1,000
Motor vehicle accidents	1.2	per	10,000
Drug use	1.2	per	10,000
Suicide	1.1	per	10,000
Firearm injuries	1.0	per	10,000
Accidental poisoning	9.9	per	100,000
Alcohol use	8.3	per	100,000
Falls	8.1	per	100,000
HIV/AIDS	3.1	per	100,000
Work injuries	1.7	per	100,000
Drowning	1.2	per	100,000
Hang-gliding	8.6	per	1,000,000
Rock climbing	3.2	per	1,000,000
Choking by food	2.8	per	1,000,000
Air transport	2.3	per	1,000,000
Excessive heat	1.2	per	1,000,000

*Probability of dying

Source: Data from National Vital Statistics Reports 59:4, March 16, 2011, and Health and Safety Executive, UK, September 4, 2012.

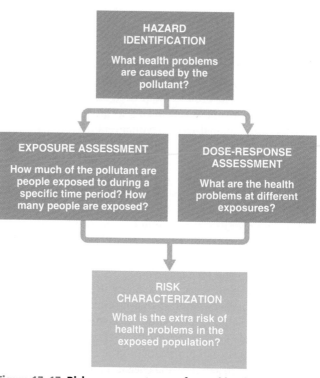

Figure 17–17 **Risk assessment as performed by the EPA.** The process of four steps in the risk assessment process is shown. (*Source:* EPA, http://epa.gov/risk/risk-characterization.htm. January 19, 2011.)

Risk Assessment by the EPA

Risk assessment began at the EPA in the mid-1970s as a way of addressing the cancer risks associated with pesticides and toxic chemicals. It has developed into a highly science-based methodology directed toward understanding how pollutants may affect human health and the environment. As currently performed at the EPA, risk assessment for human health risks has four steps: **hazard identification, dose-response assessment, exposure assessment,** and **risk characterization** (Fig. 17–17). The EPA also performs **ecological risk assessment (ERA)** in its responsibility for protecting natural organisms and ecosystems from adverse effects of human activities. The framework for ERA is conceptually quite similar to that used for human health risk assessment.

Hazard Identification: Is This Really a Hazard? Hazard identification is the process of examining evidence linking a potential hazard to its harmful effects. In the case of accidents, the link is obvious. The use of cars, for example, involves a certain number of crashes and deaths. In these cases, *historical data,* such as the annual highway death toll, are useful for calculating risks.

In other cases, the link is not so clear because there is a time delay between the first exposure and the final outcome. For example, establishing a link between exposure to certain chemicals and the development of cancer some years later is often very difficult. In cases where the linkage is not obvious, data may be needed from epidemiological studies or animal tests. An **epidemiological study** tracks how a sickness spreads through a community. Thus, to find a link between cancer and exposure to some chemical, an epidemiological study would examine all the people especially exposed to the chemical and determine whether that population has more cancer than the general population. These findings are considered to be the best data for risk assessment and have resulted in scores of chemicals being labeled as known human carcinogens.

Ask the Rats. We do not want to wait 20 years to find out that a new food additive causes cancer, so we turn to **animal testing** to find out now what *might* happen in the future. A test involving several hundred animals (usually mice or rats) takes about three years and costs more than $250,000. If a significant number of the animals develop tumors after having been fed the substance being tested, then the substance is either a possible or a probable human carcinogen, depending on the strength of the results.

Three objections to animal testing have been raised: (1) rodents and humans may have very different responses to a given chemical; (2) the doses used on the animals are often much higher than those to which humans may be exposed; and (3) some people are opposed on ethical grounds to the use of animals for such purposes. Although there are obvious differences between rodents and humans, all chemicals shown by epidemiological studies to be human carcinogens are also carcinogenic to test animals, suggesting that animal tests have some predictive value for humans.

Just Test People? In the past, chemical manufacturers that were interested in assessing the safety of pesticides (in order to have their products approved for use) paid human subjects to test the products. Such tests are far less expensive than the usual animal testing (volunteers may be paid only a few hundred dollars; rats apparently charge much higher fees). A National Academy of Sciences panel addressed this topic, stating that the use of human volunteers "to facilitate the interests of industry or of agriculture" is unjustifiable. The EPA agreed and banned any use of such tests. However, recent administrations, responding to requests from manufacturers, pressed the EPA to reevaluate its policy and allow human testing. A 2006 rule from the EPA forbids such testing on children and pregnant women but allows it for others, satisfying the pesticide manufacturers. Recent EPA actions have established new, more stringent ethical standards[15] to protect adult subjects; even now, most studies (33 per year) involve topical exposures rather than internal dosing.

Another source of information is the chemical or process itself: What are its physical and chemical properties, and what mode of action might the chemical or process take in inducing cancer or other harmful effects? The EPA is exploring new approaches to testing chemicals for toxicity to people: *ToxCast* is a chemical screening program that is now being used to screen up to 10,000 chemicals using rapid tests with human cells such as stem cells. Initial tests using chemicals already subjected to animal testing and epidemiological studies have indicated that this approach has great promise in reducing the time it takes to evaluate the risk posed by a chemical hazard.

In the end, hazard identification takes a "weight of the evidence" approach to determining the carcinogenic potential of a chemical or process. Standard descriptors are used in its conclusions, such as "likely to be carcinogenic to humans." Thus, hazard identification tells us that we *may* have a problem.

Dose-Response and Exposure Assessment: How Much for How Long?
When animal tests or human studies show a link between exposure to a chemical and an ill effect, the next step is to analyze the relationship between the concentrations of the chemical in the test (the dose) and both the incidence and the severity of the response. It is at this point that vulnerability may be introduced; for example, it is known that children are more vulnerable to some cancer-causing chemicals because they are growing. From this information, projections are made about the number of cancers that may develop in a human population exposed to different doses of the chemical. Unless there is reason to assume otherwise, a linear response is used to determine an acceptable level of exposure. This process is called dose-response assessment.

Following dose-response assessment is exposure assessment. This procedure involves identifying human groups already exposed to the chemical, learning how their exposure

came about, and calculating the doses and length of time of their exposure.

Risk Characterization: What Does the Science Say? The final step, risk characterization, is to pull together all the information gathered in the first three steps, in order to determine the risk and its accompanying uncertainties. Here is where science is able to inform the risk manager. The characterization summary should provide information about key supportive data and methods, their limitations, and, especially, the estimates of risk and the uncertainties involved. Commonly, risk is expressed as the probability of a fatal outcome due to the hazard, as in Table 17–3 for common, everyday risks.

The EPA conducts an evaluation of risk information on hazards that chemicals pose to human health. The agency employs its **Integrated Risk Information System (IRIS)** on science-based information gathered from assays and epidemiological studies and publishes IRIS assessments after they have been thoroughly reviewed in draft form. Information on more than 550 chemical substances can be found in the IRIS database (available on the Internet). One recent entry in this database is *trichloroethylene*, a widely used industrial solvent. After more than a decade of assessment, the EPA has concluded that trichloroethylene is carcinogenic to humans and also poses other health hazards to many human organ systems.

Another EPA agency, the **National Center for Environmental Assessment (NCEA)** is the EPA's program for assessment of substances and processes that are widely released in the environment, such as common air pollutants. The NCEA has published 178 risk assessments on subjects from aerosols to watersheds on its Web site. The information in IRIS and NCEA assessments is then put to use by those charged with guarding human health and the environment.

Public-Health Risk Assessment

The World Health Organization broke new ground in the area of risk assessment with its 2002 World Health Report.[16] Until then, most risk assessment took place as a function of environmental regulation. The new effort began by looking at all of the risk factors commonly known to be responsible for poor health and mortality and then asking this question: "Of all the disease burden in this population, how much could be caused by this [particular] risk?" Then certain risk factors were chosen for special attention on the basis of the following criteria:

- **Potential global impact.** The risk is likely to be among the leading causes of poor health and of mortality. For example, the risks of being overweight are now well documented in many of the developed countries, where the condition is a serious problem.

- **High likelihood of causality.** It should be possible to trace the cause-effect relationships between poor health and the factor in question. Thus, overweight persons

[15]EPA. Expanded Protections for Subjects in Human Research Studies. February 16, 2011. http://www.epa.gov/oppfead1/guidance/human-test.htm. January 19, 2012.

[16]World Health Organization. *The World Health Report 2002: Reducing Risks, Promoting Healthy Life.* Geneva: World Health Organization, 2002. www.who.int/whr/2002/en.

have a greater tendency to develop diabetes, high blood pressure, and high cholesterol; these in turn lead to greater risks of heart disease, stroke, and many forms of cancer.

- **Modifiability.** It should be possible to develop country-wide risk reduction policies that substantially lessen the impact of the factor. For instance, overeating is a matter of individual behavior that can be difficult to change, but a society can draw greater attention to the epidemic of obesity, its health risks, and the steps that can be taken to prevent overeating.

- **Availability of data.** There must be reliable data on the prevalence of the risk factor and its relationship to disease and mortality. With regard to overeating in the United States, the CDC coordinates a state-based monitoring system (the Behavioral Risk Factor Surveillance System) that collects information via telephone interviews. The CDC reports information on overeating and other, similar behavior-related health risks (participating in insufficient physical activity, failing to eat fruits and vegetables, using tobacco and alcohol) to the state and local health departments.

The more recent (2009) report[17] basically updated the 2002 effort for some 24 global risk factors.

The DALY. The data collected must be assessed with the use of a "common currency," one that facilitates a comparison of different types of risks. The WHO reports use the measurement known as the *DALY*—the *disability-adjusted life year.* One DALY represents the loss of one healthy year of a person's life. This measure, now in common usage in the global health community, assesses the burden of a disease in terms of life lost due to premature mortality and time lived with a disability (see Figure 17–9).

The WHO reports reveal some remarkable differences in risk factors in different groups of countries (Table 17–4). Each risk factor calls for a different suite of interventions. As a country evaluates its highest health risks, it will often have to prioritize its risk prevention strategies. This is especially true of the low-income developing countries, where public-health resources are severely limited. It is the opinion of the WHO that the scientific bases for the main risk factors presented in the reports are well understood. Also, the strategies for cost-effective risk prevention for the main risks are straightforward. There is broad agreement about the need for action, for example, to address the increase in tobacco consumption and the role of unsafe sex in the HIV/AIDS epidemic. The WHO has an important role to play in coordinating efforts to translate scientific information into action, and its two reports are major steps in carrying out that role.

Stewards of Health. In a very real sense, public-health ministries and agencies are the primary stewards of the health and welfare of a country's people. Public policy should clearly reflect stewardship principles that put the health of

TABLE 17–4 Ten Leading Selected Risk Factors as Percentage Causes of Disease Burden, Measured in DALYs*

Low-income countries	Percentage of total	DALYs (millions)
Underweight	9.9%	82
Unsafe water, sanitation, and hygiene	6.3%	53
Unsafe sex	6.2%	52
Suboptimal breastfeeding	4.1%	34
Indoor smoke from solid fuels	4.0%	33
Vitamin A deficiency	2.4%	20
High blood pressure	2.2%	18
Alcohol use	2.1%	18
High blood glucose	1.9%	16
Zinc deficiency	1.7%	14
Middle-income countries		
Alcohol use	7.6%	44
High blood pressure	5.4%	31
Tobacco use	5.4%	31
Overweight and obesity	3.6%	21
High blood glucose	3.4%	20
Unsafe sex	3.0%	17
Physical inactivity	2.7%	16
High cholesterol	2.5%	14
Occupational risks	2.3%	14
Unsafe water, sanitation, and hygiene	2.0%	11
High-income countries		
Tobacco use	10.7%	13
Alcohol use	6.7%	8
Overweight and obesity	6.5%	8
High blood pressure	6.1%	7
High blood glucose	4.9%	6
Physical inactivity	4.1%	5
High cholesterol	3.4%	4
Illicit drugs	2.1%	3
Occupational risks	1.5%	2
Low fruit and vegetable intake	1.3%	2

*Disability-adjusted life years: one DALY = loss of one healthy year of a person's life.
Source: World Health Report, 2009.

[17] World Health Organization, *Global Health Risks,* 2009, op. cit.

people and the environment above the economic bottom line, which often seems to drive our decisions. To quote the 2002 *World Health Report*, "Governments are the stewards of health resources. This stewardship has been defined as 'a function of a government responsible for the welfare of the population and concerned about the trust and legitimacy with which its activities are viewed by the citizenry.'" Governments can accomplish much in the arena of preventive health, but they must nurture their scarce resources carefully. It is in that sense that the WHO focuses on reducing risks in order to promote health and provides a valuable tool for accomplishing this to all countries. Beyond the government, health care providers are also stewards whose responsibility is vitally important to environmental health.

Risk Management

In human health and environmental risk assessment, the analysis of hazards and risks is a task that falls primarily to the scientific community. The task is incomplete, however, if no further consideration is given to the information gathered. Risk management naturally follows and is the responsibility of lawmakers and administrators. EPA policy separates risk assessment and risk management within the agency.

   Risk management involves (1) *a thorough review of the information available pertaining to the hazard in question and the risk characterization of that hazard* and (2) *a decision as to whether the weight of the evidence justifies a regulatory action.* Without doubt, public opinion can play a powerful role in both processes. In general, however, a regulatory decision will hinge on one or more of the following considerations:

1. **Cost-benefit analysis.** This type of analysis compares costs and benefits relative to a technology or proposed chemical product; if done fairly, the analysis can make a regulatory decision very clear cut. When the EPA issued new regulations for air-pollution particulate matter in 2006, the benefits in human health far outweighed the regulatory costs.

2. **Risk-benefit analysis.** A decision can be made with regard to the risks versus the benefits of being subjected to a particular hazard, especially when those benefits cannot be easily expressed in monetary values. The use of medical CT scans is a good example. Because they represent a radiation exposure, the scans carry a calculable risk of cancer, but the benefit derived when you scan the brain for a suspected tumor is much greater than the cancer risk involved. However, there is concern about repeated CT scans because one abdominal scan is equivalent to 500 chest X-rays.

3. **Public preferences.** As we will see, people have a much greater tolerance for risks that they feel are under their own control or are voluntarily accepted.

   Risk management has been thoroughly incorporated into the EPA's policy-making process. Its most common use

has been in the design of regulations. The process has enabled the agency to target appropriate hazards for regulation and to determine where to aim the regulations—that is, at the source, at the point of use, or at the point of disposal of the hazardous agent. As was done for toxic chemicals and cancer, a given risk level may be adopted as a standard against which policy makers can measure new risks.

Risk Perception

The U.S. public's concern about environmental problems can be traced to the fear of hazards that pose a risk to human life and health. People may perceive that their lives are more hazardous than ever before, but that is not true. In fact, our society is freer from hazards than it ever has been, as is evidenced by increased longevity. Why, then, do people protest against nuclear power plants, waste sites, and pesticide residues in food when, according to experts, these hazards pose extremely small risks (Fig. 17–16)? The answer lies in people's **risk perceptions**—their intuitive judgments about risks. In short, people's perceptions are not consistent with the results coming from a scientific analysis of risk. There are some good reasons for this discrepancy—and some lessons to be learned from it.

Hazard Versus Outrage. The reason for the inconsistency between public perception and actual risk calculations, according to Peter Sandman[18] of Rutgers University, is that the public perception of risks is often more a matter of *outrage* than one of *hazard.* Sandman holds that, while the term *hazard* expresses primarily a concern for fatalities only, the term *outrage* expresses a number of additional concerns:

1. **Lack of familiarity with a technology.** Examples include how nuclear power is produced and how toxic chemicals are handled.

2. **Extent to which the risk is voluntary.** Research has shown that people who have a choice in the matter will accept risks roughly 1,000 times as great as when they have no choice. People are much more accepting of a risk if they are in control of the elements of that risk, as in driving an automobile, swimming, and eating.

3. **Public impression of hazards.** Accidents involving many deaths (as at Bhopal) or a failure of technology (for example, at Three Mile Island) are thoroughly imprinted into public awareness by media coverage and are not quickly forgotten (Chapters 15 and 22).

4. **Overselling of safety.** The public becomes suspicious when scientists or public-relations people play up the benefits of a technology and play down its hazards.

5. **Morality.** Some risks have the appearance of being morally wrong. If it is wrong to subject humans to pesticide testing, the notion that the chemical companies clearly benefit from the testing is unacceptable. You should obey a moral imperative, regardless of the costs.

[18]Peter M. Sandman. "Risky Business." *Rutgers Magazine.* April/June 1989: 36–37.

6. **Fairness.** The benefits and the risks should be connected. If the benefits go to someone else, why should you accept any risk?[19]

Media's Role. The public perception of risk is strongly influenced by the media, which are far better at communicating the outrage elements of a risk than they are at communicating the hazard elements. Public concern over mad cow disease received extensive media coverage and had a high "outrage quotient," and beef from Canada was banned in 2003 when the disease was found in two animals. However, cigarette smoking, which causes 443,000 deaths a year in the United States alone, receives minimal media attention because it is not "news." Hence, there is no outrage factor. Indeed, it has been suggested that, if all the year's smoking fatalities occurred on one day, the media would go ballistic, and smoking would be banned the next day!

Public Concern and Public Policy. Generally speaking, *public concern*, rather than cost-benefit analysis or risk-benefit analysis conducted by scientists, often drives public policy. The EPA's funding priorities are set largely by Congress, which is supposed to be responsive to its constituency. Is this a problem?

If public concern is the primary impetus for public policy, some serious risks may get less attention than they deserve. In particular, environmental concerns are commonly

[19]V. Covello and P. Sandman, "Risk Communication, Evolution and Revolution," in *Solutions to an Environment in Peril*, Ed. A. Wolbarst (Baltimore, MD: Johns Hopkins University Press, 2001), 164–178.

perceived as much less important than they really are because of the public's preoccupation with other, more immediate issues, such as the economy, jobs, and elections. This difference underscores the importance of *risk communication*, a task that should not be left to the media. For example, the value of ecosystems and their connections to human health and welfare need far greater emphasis in the public consciousness, and this responsibility falls to the scientific and educational communities as well as to government agencies. Studies have shown that the most effective risk communication occurs by starting with what people already know, discovering what they need to know, and then tailoring the message so that it provides people with the knowledge that helps them make a more informed evaluation of risks. The process should not stop there, though: To be successful, risk communication should also involve dialogue among all the parties involved—namely, policy makers, the public, experts, and other interested groups.

The public's concern for more than the probabilities of fatalities has merit. Public concern must be heard, understood, and given a reasonable response. It may not be the best source of public policy, but it reflects certain values and concerns that could easily be omitted by an "objective," scientific assessment. The fact is that subjective judgments are going to play a role at every step in the risk-assessment process, from hazard identification to risk perception to risk management. The uncertainties involved in risk assessment should remind us that the process is only a tool—but it is the best one we have.

REVISITING THE THEMES

Sound Science

The continuing battle against pathogens and human disorders like cancer and heart disease requires the sustained attention of dedicated research laboratories around the world. This is modern science in the direct interest of humankind. Risk assessment utilizes the scientific community in the final step of the process, at which risk characterization takes place. At that point, the scientists have evaluated the risk as best they can, so it is now up to those responsible for risk management. The WHO states that the basis for the major risk factors plaguing most societies has been scientifically established. There is more work to do, of course, but the failure to prove scientific causality cannot be an excuse for inaction by those responsible for people's health.

Sustainability

A foundational strategy of sustainable development is to reduce poverty, in part because the poor are susceptible to many hazards. Numerous risk factors are a consequence of the lack of access to clean water, healthy air, sanitation, nutritious food, and health care. Infectious diseases in particular are a tremendous

impediment to the economic progress in poorer countries. It is hard to see how those countries can make much progress when HIV/AIDS, malaria, and tuberculosis take such a heavy toll on their children and young adults. The first seven Millennium Development Goals all target environmental health problems. (See Table 9–2.) If the world community takes these goals seriously, the outcome should be a revolution in public health that will move scores of countries much farther along in their quest for sustainable development.

Stewardship

Governments are the stewards of their country's health resources. Health ministries have a difficult task if they are not given the funding and personnel to carry out their mission. Fortunately, organizations such as the WHO can provide the vital information governments need to address their countries' greatest health problems. The WHO exemplifies the stewardship concept of a community of nations acting together to bring care and information to those whose health is at risk. Those who work alongside the WHO—the health ministries, health-related nongovernmental organizations, and health care providers—are the hands and hearts of the world community.

REVIEW QUESTIONS

1. Define *pollution*, *pollutant*, *nonbiodegradable*, and *environment*.

2. Differentiate between a hazard and a risk. Where does vulnerability fit in?

3. Define *morbidity*, *mortality*, and *epidemiology*.

4. Describe the public-health roles of the CDC and the WHO.

5. What are the four categories of human environmental hazards? Give examples of each.

6. List the top five risk factors that are responsible for global mortality and disease.

7. Explore the impacts of poverty and wealth on causing disease and death.

8. Document the efforts to control tobacco use both in the United States and globally.

9. What is the significance of malaria worldwide, and what are some recent developments in the battle against this disease?

10. Outline the four steps of risk assessment used by the EPA.

11. What methods are used to test chemicals for their potential to cause cancer?

12. Describe the process of risk assessment for public health as it was recently carried out by the WHO. What is a DALY?

13. Discuss the relationship between public risk perception and assessment on the one hand; and public policy, on the other.

THINKING ENVIRONMENTALLY

1. Consider Table 17–4. Discuss each group's risk factors that are primarily a consequence of human choice. Pick one of these factors, and describe how you might proceed to bring it under control in an appropriate country.

2. What global accountability is there for natural disaster relief? Consider the example of devastating weather in developing countries, triggered by climate change (which could be instigated by the more prosperous).

3. Consult the CDC Web site (www.cdc.gov), and investigate one of the emerging infectious diseases or one of the "grisly seven." What are the risks you take, and what precautions should you take, if traveling to an area where the disease has been found?

MAKING A DIFFERENCE

1. Monitor the hazards in your own life, and take steps to reduce high-risk behavior.

2. Install a carbon monoxide detector if you don't already have one—and check your home's radon levels.

3. Steer clear of phthalates, found in nearly three-quarters of today's beauty products. These are endocrine disruptors, and can obstruct hormone functions, cause infertility, miscarriage, birth defects, breast cancer, and damage organs.

4. Six percent of an average American's day is spent on the road, but that time makes for 60% of their fine particle pollution intake. Driving less is not only good for the environment, but also good for you!

5. Instead of using toxic house cleaners, consider homemade combinations of white vinegar, castile soap, baking soda, and water.

MasteringEnvironmentalScience®

Go to **www.masteringenvironmentalscience.com** for practice quizzes, Pearson eText, videos, current events, and more.

COP17/CMP7
UNITED NATIONS
DURBAN, SOUTH AFRICA

Delegates meet at Durban, South Africa, for the 17th Conference of the Parties to the Framework Convention on Climate Change, December 2011.

CHAPTER 18

Global Climate Change

"THE MALDIVES IS THE canary in the world's carbon coal mine." This was stated by Mohamed Nasheed, former president of the Maldives, an archipelago of 1190 islands in the Indian Ocean that some 320,000 people call home. Many of the islands are atolls (ring-shaped islands formed of coral), others little more than sandbars, and the capital, Malé, is a square mile of densely packed buildings surrounded by a seawall (Fig. 18–1a). The average elevation in the islands is a mere 3 feet (1 meter) above sea level, and seafront erosion has claimed many beaches and some buildings. Nasheed was working on a sovereign-wealth fund that would enable the Maldivians to purchase land somewhere (India, Sri Lanka?); the fund was to be financed by the tourism industry that is the economic engine of the islands. Over in the South Pacific, residents of numerous island nations and islands are bracing for the rising sea level that is already submerging parts of their lands. Kiribati and Tuvalu are two such island nations whose existence is threatened by environmental changes linked to global warming; as salt water encroaches on the freshwater lenses that support plant life on the islands, it is likely that the entire populations of these two countries—some 110,000—will, like the Maldivians, be forced to emigrate.

SEA LEVEL RISE. The relentless rise of sea level, now at 3.2 mm/yr, is accelerating as the oceans continue to warm and ice sheets and glaciers melt. The estimates of total sea level rise at year 2100 range from a low of 20 cm to a high of 1.8 meters,[1] and because the low estimates do not take into consideration any melting of the polar ice sheets, low-lying regions like the Maldives and Pacific

---

[1]Robert J. Nicholls and Anny Cazenave. "Sea-Level Rise and Its Impact on Coastal Zones." *Science*, 328: 1517–1520. June 18, 2010.

LEARNING OBJECTIVES

1. **Atmosphere, Weather, and Climate:** Describe the structure of the atmosphere and its gases; distinguish between weather and climate.

2. **Climate Change Science:** Explore the science that describes the greenhouse effect, summarize the evidence for recent climate change, and evaluate the various greenhouse gases and their impacts on present and future climates.

3. **Response to Climate Change:** Examine the international and U.S. measures taken to mitigate climate change and evaluate their effectiveness.

(a)

(b)

Figure 18–1 Regions under threat from rising seas. (a) The city of Malé, capital of the Maldives, in the Indian Ocean. (b) A view of Shishmaref, on the Alaskan coast.

islands are paying more attention to the high range estimates of sea level rise.

Far to the north, the island village of Shishmaref, Alaska, is falling into the sea. Built on permafrost and normally protected from ocean waves by sea ice, Shishmaref is now battered by waves as the summer sea ice is melted away, and the soil is shifting as the permafrost is melted by the unusual summer warmth. Every year, some 20 feet of the shoreline is washed away (Fig. 18–1b), sometimes taking with it the villagers' homes. The 600 inhabitants of Shishmaref are actively working toward relocating the entire community to higher ground elsewhere in Alaska. Scientists point out that the populations of these isolated islands will be merely the first wave of climate change refugees unless the global community of nations takes rapid steps to curb the fossil fuel emissions that are largely responsible for global climate change.

THE FUTURE. Fast forward to 2050. Could it be that whatever happens this far into the future depends on the actions taken by the leading nations in the world now and in the next decade to address the issue of global climate change? Ethicists claim that we are in the predicament of having to make decisions now, costly and far-reaching decisions, in order that our children and grandchildren will not be living in a world that experiences unprecedented and dangerous environmental conditions.

Can all this be true? What is happening with our climate? Indeed, what controls the climate? We will seek answers to these questions as we first investigate the atmosphere, how it is structured, and how it brings us our weather and climate. Then we will consider the scientific evidence for global climate change, examine the future projections of conditions made by elegant computer modeling, and then look at the varied responses to address global climate change.

18.1 Atmosphere, Weather, and Climate

The atmosphere is a collection of gases that gravity holds in a thin envelope around Earth. The gases within the lowest layer, the **troposphere**, are responsible for moderating the flow of energy to Earth and are involved with the biogeochemical cycling of many elements and compounds—oxygen, nitrogen, carbon, sulfur, and water, to name the most crucial ones. The troposphere ranges in thickness from 10 miles (16 km) in the tropics to 5 miles (8 km) in higher latitudes, due mainly to differences in solar energy. This layer contains practically all of the water vapor and clouds in the atmosphere; it is the site and source of our weather. Except for local temperature inversions, the troposphere gets colder with altitude **(Fig. 18–2)**. Air masses in this layer are well mixed vertically, so pollutants can reach the top within a few days. Substances entering the troposphere—including pollutants—may be changed chemically and washed back to Earth's surface by precipitation (Chapter 19). Capping the troposphere is the **tropopause**, a boundary region where air shifts from cooling with height and begins to warm.

Above the tropopause is the **stratosphere**, a layer within which temperature *increases* with altitude, up to about 40 miles above the surface of Earth. The temperature increases primarily because the stratosphere contains ozone (O_3), a form of oxygen that absorbs high-energy radiation emitted by the Sun. Because there is little vertical mixing of air masses in the stratosphere and no precipitation from it, substances that enter it can remain there for a long time. Beyond the stratosphere are two more layers, the **mesosphere** and the **thermosphere**, where the ozone concentration declines and only small amounts of oxygen and nitrogen are found. Table 18–1 summarizes the characteristics of the troposphere and stratosphere.

Weather

People sometimes make the mistake of assuming that one snowstorm or one hurricane is evidence for or against climate change. It is simply evidence that weather is happening. The

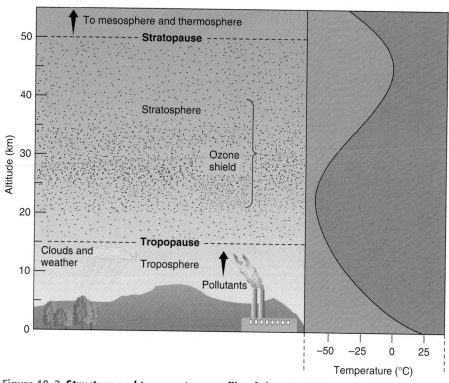

Figure 18–2 Structure and temperature profile of the atmosphere. The image on the left shows the layers of the atmosphere and the ozone shield, while the plot on the right shows the vertical temperature profile.

TABLE 18–1 Characteristics of Troposphere and Stratosphere

Troposphere	Stratosphere
Extent: Ground level to 10 miles (16 km)	Extent: 10 miles to 40 miles (16 km to 65 km)
Temperature normally decreases with altitude, down to −70°F (−59°C)	Temperature increases with altitude, up to +32°F (0°C)
Much vertical mixing, turbulent	Little vertical mixing, slow exchange of gases with troposphere, via diffusion
Substances entering may be washed back to Earth	Substances entering remain unless attacked by sunlight or other chemicals
All weather and climate take place here	Isolated from the troposphere by the tropopause

day-to-day variations in temperature, air pressure, wind, humidity, and precipitation—all mediated by the atmosphere—constitute our **weather**. **Climate** is the result of long-term weather patterns in a region, patterns that involve decades or longer. The scientific study of the atmosphere—both of weather and of climate—is **meteorology**. You can think of the atmosphere-ocean-land system as an enormous weather engine, driven by solar power and strongly affected by the rotation of Earth and its tilted axis. Solar radiation enters the atmosphere and then takes a number of possible courses **(Fig. 18–3)**. Some (29%) is reflected by clouds and Earth's surfaces, but most is absorbed by the atmosphere, oceans, and land, which are heated in the process. The land and oceans

then release energy back upward through evaporation, convection, and reradiation of infrared energy.

Flowing Air. Some of the energy that is released from Earth's surface is transferred to the atmosphere. Thus, air masses grow warmer at the surface of Earth and will tend to expand, becoming lighter. The lighter air will then rise, creating *vertical* air currents. On a large scale, this movement creates major *convection currents* (Chapter 10, Fig. 10–6). Air must flow in to replace the rising warm air, and the inflow leads to *horizontal* airflows, or wind. The ultimate source of the horizontal flow is cooler air that is sinking, and the combination produces the Hadley cell (Fig. 10–6). These

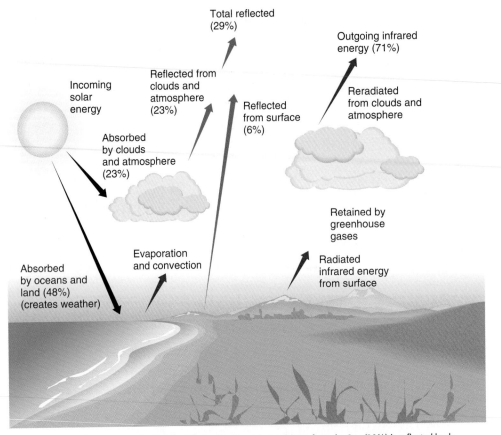

Figure 18–3 Solar-energy balance. Much of the incoming radiation from the Sun (29%) is reflected back to space, but the remainder (71%) is absorbed by the oceans, land, and atmosphere, where it creates our weather and fuels photosynthesis. Eventually, this absorbed energy is radiated back to space as infrared energy (heat).

(*Source:* Rebecca Lindsey, Climate and Earth's Energy Budget, NASA Earth Observatory, January 14, 2009.)

major flows of air create regions of high rainfall (equatorial), deserts (25° to 35° north and south of the equator), and horizontal winds (trade winds).

Convection. On a smaller scale, **convection currents** bring us the day-to-day changes in our weather as they move in a general pattern from west to east. Weather reports inform us of regions of high and low pressure, but where do these come from? Rising air (due to solar heating) creates high pressure up in the atmosphere, leaving behind a region of lower pressure close to Earth. Conversely, once the moist, high-pressure air has cooled by reradiating heat to space and losing heat through condensation (thereby generating precipitation), it flows horizontally toward regions of sinking, cool, dry air (where the pressure is lower). There the air is warmed at the surface and creates a region of higher pressure **(Fig. 18–4)**. The differences in pressure lead to airflows, which are the winds we experience. As the figure shows, the winds tend to flow from high-pressure regions toward low-pressure regions.

Jet Streams. The larger-scale air movements of Hadley cells are influenced by Earth's rotation from west to east. This creates the trade winds over the oceans and the general flow of weather from west to east. Higher in the troposphere, Earth's rotation and air-pressure gradients generate veritable rivers of air, called **jet streams**, that flow eastward at

speeds of more than 300 mph and meander considerably. Jet streams are able to steer major air masses in the lower troposphere. One example is the polar jet stream, which steers cold air masses into North America when it dips downward in latitude.

Put Together. . . Air masses of different temperatures and pressures meet at boundaries we call **fronts**, which are regions of rapid weather change. Other movements of air masses due to differences in pressure and temperature include hurricanes and typhoons and local, but very destructive, tornadoes. Finally, there are also major seasonal airflows—**monsoons**—which often represent a reversal of previous wind patterns. Monsoons are created by major differences in cooling and heating between oceans and continents. The summer monsoons of the Indian subcontinent are famous for the beneficial rains they bring and notorious for the devastating floods that can occur when the rains are heavy. Putting these movements all together—taking the general atmospheric circulation patterns and the resulting precipitation, then adding the wind and weather systems generating them, and finally mixing all this with the rotation of Earth and the tilt of the planet on its axis, which creates the seasons—yields the general patterns of weather that characterize different regions of the world. In any given region, these patterns are referred to as the region's **climate**.

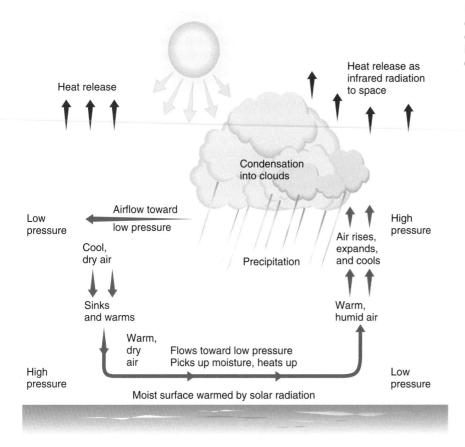

Figure 18–4 A convection cell. Driven by solar energy, convection cells produce the main components of our weather as evaporation and condensation occur in rising air and precipitation results, followed by the sinking of dry air. Horizontal winds are generated in the process.

18.2 Climate Change Science

Climate is the average temperature and precipitation expected throughout a typical year in a given region (Chapter 5). Different temperature and moisture regimes in different parts of the world have generated different types of ecosystems called biomes, reflecting the adaptations of plants, animals, and microbes to the prevailing weather patterns, or climate, of a region. The temperature and precipitation patterns themselves are actually caused by other forces—namely, the major determinants of weather previously outlined. Humans can adjust to practically any climate (short of the brutal conditions of high mountains and burning deserts), but this is not true of the other inhabitants of the particular regions we occupy. If other living organisms in a region are adapted to a particular climate, then a major change in the climate represents a serious threat to the structure and function of the existing ecosystems. The subject of climate change is such a vital issue today because we depend on these other organisms for a host of vital goods and services without which we could not survive (see Chapter 7). If the climate changes, can these ecosystems change with it in such a way that the crucial support they provide us is not interrupted? Is Earth's climate indeed changing? If so, what is causing it to change?

Synopsis of Global Climate Change

In 2007, after sifting through thousands of studies about global warming, scientists from the **Intergovernmental Panel** on **Climate Change (IPCC)** published the Fourth Assessment Report (AR4) on climate change. They concluded that *warming of the climate system is unequivocal.* Our climate is changing, the scientists said: The atmosphere is warmer **(Fig. 18–5)**, the oceans are warmer, there is more precipitation, there are more extreme weather events (like hurricanes,

GLOBAL LAND–OCEAN TEMPERATURE INDEX

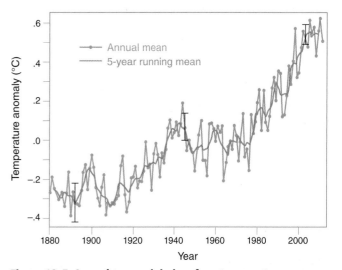

Figure 18–5 Annual mean global surface temperature anomalies. The baseline, or zero point, is the period 1951–80. The warming trend since 1975 is conspicuous.

(*Source:* NASA Goddard Institute for Space Studies, 2011.)

ATMOSPHERIC CO$_2$ AT MAUNA LOA OBSERVATORY

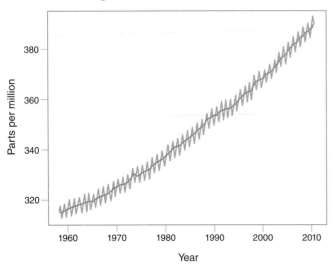

Figure 18–6 Atmospheric carbon dioxide concentrations. The concentration of CO$_2$ in the atmosphere fluctuates between winter and summer because of seasonal variation in photosynthesis. The average concentration is increasing due to human activities—in particular, burning fossil fuels and deforestation.

(*Source:* NOAA Earth System Research Laboratory, 2011.)

floods, droughts, heat waves), glaciers are melting, and the sea level is rising.

The IPCC report stated that it is *very likely* (i.e., there is 90–100% probability) that the warming is mostly caused by increased concentrations of anthropogenic (human-made) **greenhouse gases** (GHGs), atmospheric gases that trap infrared radiation. The most significant anthropogenic GHG is carbon dioxide (CO$_2$), which has risen by 40%—from 280 parts per million (ppm) to 394 ppm—since the Industrial Revolution began **(Fig. 18–6)**. The source of this added CO$_2$ is no surprise: Most of it comes from burning fossil fuels, and a smaller amount comes from burning forests (deforestation). The global use of fossil fuels is rising every year, and if nothing is done to reduce it, computer modeling tells us that the 21st century will see climate changes that are dangerous, possibly catastrophic. Sea levels will keep rising; there will be more frequent heat waves, changed weather patterns, and threats to natural ecosystems. In fact, these changes will go on for hundreds of years. Needless to say, this is not good news, and it has gotten the attention of the world. Efforts are under way to respond by way of **mitigation** (reducing GHG emissions) and **adaptation** (adjusting to coming climate changes). However, our civilization has no experience with a problem like this, and the way forward is still being debated.

IPCC and UNEP

The Fourth Assessment Report is an incredibly detailed and comprehensive study. In 1988, the UN Environment Program (UNEP) and the World Meteorological Society established the IPCC in order to provide accurate and relevant information that would lead to an understanding of human-induced

climate change. The IPCC has three working groups: one to assess the scientific issues (Working Group I), another to evaluate the impact of global climate change and the prospects for adapting to it (Working Group II), and a third to investigate ways of mitigating the effects (Working Group III). The latest IPCC assessment working groups consist of more than 2,000 experts in the appropriate fields, from 154 countries. These experts are unpaid and participate at the cost of their own research and other professional activities. Each working group produced a separate report;[2] all reports are available from the IPCC Web site (www.ipcc.ch).

The work of the IPCC has been guided by two basic questions:

> Risk assessment—Is the climate system changing, and what is the impact on society and ecosystems?

> Risk management—How can we manage the system through adaptation and mitigation?

Building on past assessments (1990, 1995, 2001) and new results of research, the three IPCC working groups released the Fourth Assessment Report in stages during 2007, the final product being *Climate Change 2007: Synthesis Report*, which was released late that year and contained key findings of the working groups.

The UNEP issued a report in 2009 (**Climate Change Science Compendium 2009**[3]) that addressed many new findings since the IPCC's assessment. Other updates have come from the American Meteorological Society[4] and the U.S. Global Change Research Program.[5] Work on the next full IPCC assessment is under way and will be released in 2013. In this chapter, we will explore the information on climate change in some detail, making full use of the IPCC and many other reports. We begin with a look into the past, which harbors "records" of climate change.

Climates in the Past

Searching the past for evidence of climate change has become a major scientific enterprise, one that becomes more difficult the further into the past we try to search. Systematic records of the factors making up weather—temperature, precipitation, storms, and so forth—have been kept for little more than 130 years. Nevertheless, these records clearly inform us that our climate is far from constant. The record of surface temperatures, from weather stations around the world and from literally millions of observations of temperatures at the

[2]See, for example, Susan Solomon et al., Eds., *Climate Change 2007: The Physical Science Basis*, Contribution of Working Group I to the Fourth Assessment Report of the Intergovernmental Panel on Climate Change (New York: Cambridge University Press, 2007).

[3]Catherine P. McMullen, Ed. *Climate Change Science Compendium 2009*. United Nations Environment Program. http://www.unep.org/compendium2009. December 13, 2011.

[4]J. Blunden et al., Eds. *State of the Climate in 2010*. Special Supplement to the Bulletin of the American Meteorological Society, Vol. 92, No. 6, June 2011. http://www1.ncdc.noaa.gov/pub/data/cmb/bams-sotc/climate-assessment-2010-lo-rez.pdf. December 13, 2011.

[5]Thomas R. Karl et al., Eds. *Global Climate Change Impacts in the United States*. Cambridge University Press. 2009. www.globalchange.gov/usimpacts. December 13, 2011.

surface of the sea, tells an interesting story (see Fig. 18–5). Since 1880, global average temperature has shown periods of cooling and warming, but since 1976, it has increased 0.6°C (1.1°F). During the 20th century, two warming trends occurred, one from 1910 to 1945 and the latest dramatic increase from 1976 until the present. The first decade of the 21st century has been the warmest decade on record.

Further Back. Observations on climatic changes can be extended much further back in time with the use of **proxies**—measurable records that can provide data on factors such as temperature, ice cover, and precipitation. For example, historical accounts suggest that the Northern Hemisphere enjoyed a warming period from 1100 to 1300 A.D. This was followed by the "Little Ice Age," between 1400 and 1850 A.D. Additional proxies include tree rings, pollen deposits, changes in landscapes, marine sediments, corals, and ice cores.

Some work done quite recently on ice cores has both provided a startling view of a global climate that oscillates according to several cycles and afforded some evidence that remarkable changes in the climate can occur within as little as a few decades. Ice cores in Greenland and the Antarctic have been analyzed for thickness, gas content (specifically, carbon dioxide (CO_2) and methane (CH_4)), two GHGs, and **isotopes**, which are alternative chemical configurations of a given compound, due to different nuclear components. Isotopes of oxygen, as well as isotopes of hydrogen, behave differently at different temperatures when condensed in clouds and incorporated into ice.

The record based on these analyses indicates that Earth's climate has oscillated between ice ages and warm periods **(Fig. 18–7a)**. During major ice ages, huge amounts of water were tied up in glaciers and ice sheets, and the sea level was lower by as much as 400 feet (120 m). Ice cores from Antarctica show eight glacial periods and eight interglacial periods,

over the past 800,000 years. During the glacial periods, temperatures and GHGs were low; during the warm interglacial periods, just the opposite was true. The GHG and temperature data match exactly. Carbon dioxide levels ranged between 150 and 280 ppm.

The most likely explanation for these major oscillations is the existence of known variations in Earth's orbit, where, in different modes of orbital configuration, the distribution of solar radiation over different continents and latitudes varies substantially. These oscillations take place according to several periodic intervals, called **Milankovitch cycles** (after the Serbian scientist who first described them), of 100,000, 41,000, and 23,000 years.

Rapid Changes. Superimposed on the major oscillations is a record of rapid climatic fluctuations during periods of glaciation and warmer times (Fig. 18–7b). One such rapid change, called the **Younger Dryas** event (*Dryas* is a genus of Arctic flower), occurred toward the end of the last ice age. Earth had been warming up for 6,000 years and then plunged again into 1,150 years of cold weather. At the end of this event, 11,700 years ago, Arctic temperatures rose 7°C in 50 years. The impact on living systems must have been enormous. It is unlikely that this warming was due to variations in solar output: There is simply no evidence that the Sun has changed over the past million years, let alone undergone major changes in just a few decades. Instead, scientists have narrowed the field of possible explanations to the link between the atmosphere and the oceans.

Ocean and Atmosphere

More than two-thirds of Earth is covered by oceans, making ours mostly a water planet. Because we all live on land, it is hard for us to imagine the dominant role the oceans play

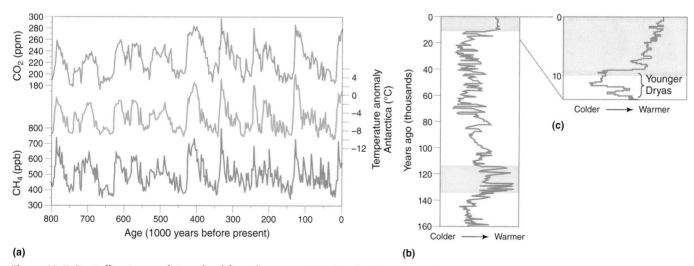

(a)

(b)

(c)

Figure 18–7 Past climates, as determined from ice cores. (a) Carbon dioxide, methane, and estimated global temperature from Antarctic ice cores, covering the past 800,000 years, compared with changes of the past century. Note how closely the temperature and GHG concentrations coincide. (b) Temperature patterns of the past 160,000 years, demonstrating climatic oscillations. (c) Higher resolution of the past 12,000 years. The Younger Dryas cold spell occurred at the start of this record. Note how rapidly the cold spell dissipated, at around 10,700 years before the present.

(*Sources:* (a) The Bohr Institute; (b) & (c) Adapted from Robert W. Christopherson, *Geosystems: An Introduction to Physical Geography,* 5th ed. (Upper Saddle River, NJ: Pearson Prentice Hall, 2005).)

in determining our climate. Nevertheless, the oceans are the major source of water for the hydrologic cycle and the main source of heat entering the atmosphere. The *evaporation of ocean water* supplies the atmosphere with water vapor, and when water vapor condenses in the atmosphere, it supplies the atmosphere with heat (latent heat of condensation, Fig. 10–4). The oceans also play a vital role in climate because of their innate *heat capacity*—the ability to absorb energy when water is heated. Indeed, the entire heat capacity of the atmosphere is equal to that of just the top 3 meters of ocean water. It is this property that makes the climate milder in coastal areas. Finally, through the movement of currents, the oceans *convey heat* throughout the globe.

Thermohaline Circulation. For many years, oceanographers have described the existence of a **thermohaline** circulation pattern that dominates oceanic currents, where *thermohaline* refers to the effects that temperature and salinity have on the density of seawater. This pattern, known formally as the **Atlantic Meridional Overturning Circulation (AMOC) (Fig. 18–8)**, has been pictured as a giant, complex conveyor belt, moving water masses from the surface to deep oceans and back again. A key area is the high-latitude North Atlantic, where salty water from the Gulf Stream moves northward on the surface and is cooled by Arctic air currents. Cooling increases the density of the water, which then sinks to depths of up to 4,000 meters—the **North Atlantic Deep Water (NADW)**. This deep-water current spreads southward through the Atlantic to the southern tip of Africa, where it is joined by cold Antarctic waters. Together, the two streams spread northward into the Indian and Pacific Oceans as deep currents. Gradually, the currents slow down and warm,

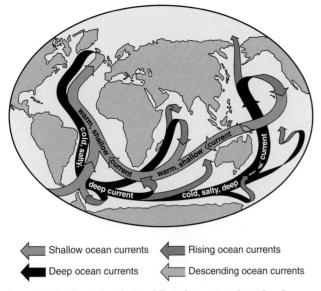

Shallow ocean currents

Deep ocean currents

Rising ocean currents

Descending ocean currents

Figure 18–8 The Atlantic Meridional Overturning Circulation. Salty water flowing to the North Atlantic is cooled and sinks, forming the North Atlantic Deep Water system. This deep flow extends southward and is joined by Antarctic water, whereupon it extends into the Indian and Pacific Oceans. Surface currents then proceed in the opposite direction, returning the water to the North Atlantic.

(*Source:* Adapted from Figure 3.15 from *Our Changing Planet,* 2nd ed., by Fred T. MacKenzie. Copyright © 1998 by Prentice Hall, Inc. Reprinted by permission of Pearson Education, Inc., Upper Saddle River, NJ 07458.)

becoming less dense and welling up to the surface, where they are further warmed and begin to move as surface waters back again toward the North Atlantic. This movement transfers enormous quantities of heat toward Europe, providing a climate that is much warmer than the high latitudes there would suggest.

Recently, the conveyor-belt paradigm has been challenged.[6] Instead of the neat flows pictured in Figure 18–8, research has revealed that eddies and oceanic winds inject and subtract water masses at many points in the overall circulation, and the actual driving force for the turnover of ocean waters is unclear. What is clear is that the turnover does occur—deep waters are formed by sinking at the poles, and warm water does return northward to affect the climate of northern Europe. This new work has called into question some of the conjecture that changes in the climate might trigger abrupt changes in the AMOC.

Abrupt Change. Recent evidence indicates that the AMOC has been interrupted in the past, changing climate abruptly. One mechanism that could have accomplished such a change is the appearance of unusually large quantities of freshwater in the North Atlantic. This would have the effect of lowering the density of the water (and, therefore, preventing much of the massive sinking that normally occurs there). With this stagnation occurring in surface Arctic waters, the northward movement of warmer, saltier water would slow down. North Atlantic marine sediments show evidence of the periodic invasion (six times in the past 75,000 years) of icebergs from the polar ice cap, which supplied huge amounts of freshwater as they melted—called Heinrich events, after the scientist who first described them. The evidence indicates that these invasions coincided with rapid cooling, as recorded in ice cores, and suggests that the AMOC shifted southward, with the NADW forming nearer to Bermuda than to Greenland. When this shift occurred, a major cooling of the climate took place within a few decades. Although it is not clear how, the normal thermohaline pattern returned and brought about another abrupt change, this time warming conditions in the North Atlantic. The Younger Dryas event likely involved just such a shift in the AMOC, brought on by the sudden release of dammed-up water from glacial Lake Agassiz into the St. Lawrence drainage. Referring to these abrupt changes in a recent article, oceanographer Wallace Broecker commented, "Earth's climate system has proven itself to be an angry beast. When nudged, it is capable of a violent response."[7]

What If . . . ? Among the likely consequences of extended global warming are increased precipitation over the North Atlantic and more melting of sea ice and ice caps. If such a pattern is sustained, it could lead to a weakening in the normal operation of the AMOC and a change in climate, especially in the northern latitudes—a possibility that Broecker has called "the Achilles' heel of our climate system." Such a

[6]M. Susan Lozier. "Deconstructing the Conveyor Belt." *Science* 328: 1507–1511. June 18, 2010.

[7]Wallace S. Broecker. "Does the Trigger for Abrupt Change Reside in the Ocean or in the Atmosphere?" *Science* 300 (June 6, 2003): 1519–1522.

change could bring major cooling to Europe. However, there are doubts about the functioning of the conveyor-belt paradigm, especially as it is presented as a smooth, simple set of ocean currents. Accordingly, most observers believe that expected increases in GHGs and the overall warming that this brings will prevent anything like a new ice age (as portrayed by the movie *The Day After Tomorrow*).

Ocean-Atmosphere Oscillations. Regional climates are notoriously erratic on time scales from a few days to decades. Studies of a number of ocean-atmosphere processes—processes that produce erratic climates on a global scale—have provided new insights into natural climate variability. There is, for example, the **North Atlantic Oscillation (NAO)**, where atmospheric pressure centers switch back and forth across the Atlantic, switching winds and storms in the process. There is also a warm-cool cycle, the **Interdecadal Pacific Oscillation (IPO)**, that swings back and forth across the Pacific over periods of several decades.

Then there is the **El Niño/La Niña Southern Oscillation (ENSO)**, where major shifts in atmospheric pressure over the central equatorial Pacific Ocean lead to a reversal of the trade winds. Incredibly, these reversals dominate the climate over much of the planet, and they can last for a year or two at a time. The 1997–2000 ENSO was the most intense of recent decades, and it spawned powerful global weather patterns that caused more than $36 billion in damages and more than 22,000 deaths. *El Niños* tend to encourage warming, while *La Niñas* do the opposite; 2010 had a strong *El Niño,* but 2011 was a strong *La Niña* year, with corresponding warmer and cooler global temperatures.

These processes are becoming better understood, and as they are, they are beginning to provide explanations for some previously puzzling developments. For example, it is clear (Fig. 18–5) that a global warming trend was halted in the mid-1940s and that it resumed 30 years later. Researchers have found evidence that this mid-century cooling was the result of shifting ocean circulation linked to the NAO. Forecasters have suggested that this and the other ocean-atmosphere oscillations have been turning greenhouse-based warming off and on. The *El Niño/La Niña* events have been happening more frequently and with greater intensity in recent decades, changes in the climate that may be linked to global warming.

The Earth as a Greenhouse

Factors that influence the climate include interactive *internal components* (oceans, the atmosphere, snow cover, sea ice) and *external factors* (solar radiation, the Earth's rotation, slow changes in our planet's orbit, and the gaseous makeup of the atmosphere). **Radiative forcing** is the influence any particular factor has on the energy balance of the atmosphere-ocean-land system. A *forcing factor* can be *positive*, leading to *warming*, or *negative*, leading to *cooling*, as it affects the energy balance. Forcing is expressed in units of watts per square meter (W/m^2); averaged over all of the planet, solar radiation entering the atmosphere is about 340 W/m^2. It is this radiation that is acted on by the various forcing factors.

Warming Processes. The interior of a car heats up when the car is sitting in sunlight with the windows closed. This heating occurs because sunlight comes in through the windows and is absorbed by the seats and other interior objects, thus converting light energy into heat energy, which is given off in the form of infrared radiation. Unlike sunlight, infrared radiation is blocked by glass and so cannot leave the car. The trapped heat energy causes the interior air temperature to rise. This is the same phenomenon that keeps a greenhouse warmer than the surrounding environment.

Greenhouse Gases. On a global scale, water vapor, CO_2, and other gases in the atmosphere play a role analogous to that of the glass in a greenhouse, hence the term *greenhouse gases (GHGs)*. Light energy comes through the atmosphere and is absorbed by Earth and converted to heat energy at the planet's surface. The infrared heat energy radiates back upward through the atmosphere and into space. The GHGs that are naturally present in the troposphere absorb some of the infrared radiation and reradiate it back toward the surface; other gases (N_2 and O_2) in the troposphere do not (Fig. 18–9). This greenhouse effect was first recognized in 1827 by French scientist Jean-Baptiste Fourier and is now firmly established. The GHGs are like a heat blanket, insulating Earth and delaying the loss of infrared energy (heat) to space. Without this insulation, average surface temperatures on Earth would be about $-19°C$ instead of $+14°C$, and life as we know it would be impossible. Therefore, our global climate depends on Earth's concentrations of GHGs. If these concentrations increase or decrease significantly, their influence as positive forcing agents will change, and our climate will change accordingly.

Ozone in the troposphere has a positive forcing effect, whereas its impact in the stratosphere is negative. However, the positive forcing predominates. Tropospheric ozone has a strong anthropogenic component linked to pollution (discussed in Chapter 19). It varies greatly in time and location and is highest in the vicinity of industrial societies.

Cooling Processes. Earth's atmosphere is also subject to negative forcing factors. For example, on average, clouds cover 50% of Earth's surface and reflect some 23% of solar radiation away to space before it ever reaches the ground. Sunlight reflected in this way is called the **planetary albedo**, and it contributes to overall cooling by preventing a certain amount of warming in the first place. The effect of the albedo is especially true for *low-lying*, fluffy clouds and can be readily appreciated if you think of how you are comforted on a hot day when a large cloud passes between you and the Sun. The radiation that was heating you and your surroundings is suddenly intercepted higher up in the atmosphere, and you feel cooler. However, *high-lying*, wispy clouds have a positive forcing effect, absorbing some of the solar radiation and emitting some infrared radiation themselves. Recent work indicates that the current net impact of clouds is a slightly positive forcing.

Snow and sea ice (the **cryosphere**) also reflect sunlight, contributing to the planetary albedo. Arctic warming has reduced this effect in recent decades, as open water and darker, unfrozen ground absorb sunlight. Cryosphere albedo is

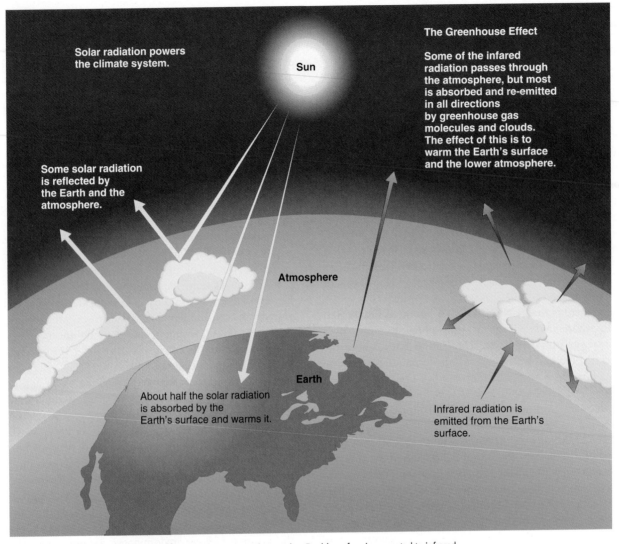

Figure 18–9 The greenhouse effect. Solar radiation that reaches Earth's surface is converted to infrared radiation and is radiated upward. Greenhouse gases in the atmosphere intercept this radiation and are warmed. The greenhouse gases in turn reradiate heat energy, and much of this is directed back toward Earth's surface.

(*Source:* S. Solomon et al., Eds. *Climate Change 2007: The Physical Science Basis,* Contribution of Working Group I to the Fourth Assessment Report of the Intergovernmental Panel on Climate Change (Cambridge, England: Cambridge University Press, 2007), FAQ 1.3, Figure 1.)

reduced even more by the widespread black carbon soot originating from anthropogenic sources (diesel engine exhaust and fires). By darkening the snow and ice, the soot promotes the absorption of radiant energy rather than its reflection.

Volcanoes. Volcanic activity can also lead to planetary cooling. When Mount Pinatubo in the Philippines erupted in 1991, some 20 million tons of particles and aerosols entered the atmosphere and contributed to a significant drop in global temperature as radiation was reflected and scattered away. This global cooling effect lasted until the volcanic debris was finally cleansed from the atmosphere by chemical change and deposition, a process that took several years. However, some sulfate aerosols from volcanic sources enter the stratosphere, where they have a longer-term cooling effect; these have increased since 2000, effectively reducing some of the global warming that otherwise might have occurred.[8]

Aerosols. Aerosols are microscopic liquid or solid particles originating from land and water surfaces (Chapter 10). Climatologists have found that industrial aerosols (from ground-level pollution) play a significant role in canceling out some of the warming from GHGs. Sulfates, nitrates, dust, and soot from industrial sources and forest fires enter the atmosphere and react with compounds there to form a high-level aerosol haze. This haze reflects and scatters some sunlight; the aerosols also contribute to the formation of clouds, which likewise increase the planetary albedo. *Sooty* aerosol (from fires) has an overall warming effect, but industrial *sulfate* aerosol creates substantial cooling, mainly through its tendency to increase global cloud cover. The mean residence time of the sulfates and soot forming the aerosol is about a week, so the aerosol does not increase over time, as the GHGs do. However, because anthropogenic sulfate is more than double that coming from natural sources, its effect is substantial and persistent. Climatologists estimate that the cooling effect of these pollutants has counteracted much

[8]S. Solomon et al. "The Persistently Variable 'Background' Stratospheric Aerosol Layer and Global Climate Change." *Science* 333: 866–870. August 12, 2011.

of the potential global warming in recent decades. Because of progress made in curbing industrial pollutants in the United States and Europe, this impact has been declining over those regions. However, industrial growth in China and India has created a major rise in aerosols over Asia.

Solar Variability. Because the Sun is the source of radiant energy that heats Earth, any variability in the Sun's radiation reaching Earth will likely influence the climate. Direct monitoring of solar irradiance goes back only a few decades, but it has confirmed that there is an 11-year cycle of changes involving solar radiation increases during times of high sunspot activity. Sunspots are known to block cosmic ray intensity, and because cosmic rays help seed clouds, the diminished cloud cover allows more solar radiation to reach Earth's surface. Indeed, variations in solar forcing are thought to be partly responsible for the early-20th-century warming trend. However, solar output began declining around 1985 and continued for at least 20 years, yet global temperatures were rising rapidly at that time and have continued to rise. The IPCC AR4 assigned a slight warming role to fluctuations in solar radiation, but found that GHGs were some 13 times more responsible for the rising global temperatures than were solar changes.

Thus . . . Global atmospheric temperatures are a balance between the positive and negative forcing from natural causes (volcanoes, clouds, natural GHGs, solar irradiance) and anthropogenic causes (sulfate aerosols, soot, ozone, increases in GHGs). The net result varies, depending on one's location. The various forcing agents bring us year-to-year fluctuations in climate and make it difficult to attribute any one event or extreme season to human causation. Nevertheless, there has clearly been enough of a shift in many features of our climate in recent decades to generate international attention.

Evidence of Climate Change

There is much natural variation in weather from year to year, and local temperatures do not necessarily follow globally averaged ones. It is a fact, however, that the first decade of the 21st century is the warmest on record (Fig. 18–5). The years 2005 and 2010 tied for a new record for global temperature—the warmest years since recordkeeping began. Since the mid-1970s, the average global temperature has risen about 0.6°C (1.1°F), almost 0.2°C per decade. The warming is happening everywhere but is greater at high latitudes in the Northern Hemisphere. Because of the continued increases in anthropogenic GHGs in the atmosphere, the observed warming is considered to be the consequence of an "enhanced greenhouse effect."

Satellites. An interesting and controversial aspect of the warming trend involves measurements of tropospheric temperatures (from Earth's surface to 8 km above) made by 13 different satellites over two decades. One early analysis of the data indicated virtually no temperature increase in the troposphere. Much was made of this apparent discrepancy with the surface temperature trends by skeptics eager to disprove global warming. This situation was addressed by the Climate Change

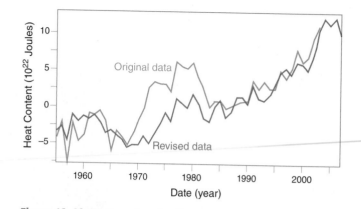

Figure 18–10 Heat capture by the oceans. The plot shows the changes in heat content of the upper 2,290 feet (700 m) of the world's oceans, as plotted by the Global Oceanographic Data Archaeology and Rescue Project (GODAR). Heat content is measured in joules (J). Two plots are seen: an original data plot and a revised data plot after corrections were made. See Sound Science, "Getting It Right."

(*Source:* Rebecca Lindsey, *Correcting Ocean Cooling,* Earth Observatory, November 5, 2008.)

Science Program (CCSP), which was created in 2002 by the Bush administration to address apparent unresolved questions about global climate change. The CCSP study, reported in May 2006, found that "the discrepancy no longer exists because errors in the satellite . . . data have been identified and corrected. New data sets have also been developed that do not show such discrepancies." Indeed, the study found a remarkable agreement between surface temperatures and tropospheric temperatures, all of which indicate a continued rise in temperature consistent with "our understanding of the effects of radiative forcing agents [showing] clear evidence of human influences on the climate system."[9] This report, the first of 21 assessments planned by the program, was approved by the Bush White House.

Ocean Heat Content. The upper 3,000 meters of the ocean have warmed measurably since 1955 (**Fig. 18–10**), a warming that dwarfs the observed warming of the atmosphere, accounting for 90% of the heat increase of Earth systems in the past several decades. There is a fascinating story about the actual measurements of ocean heat content shown in Figure 18–10 (described in Sound Science, "Getting It Right"). A key analysis of Earth's energy imbalance[10] makes the case for the ocean's continuing absorption of heat; over the past decade, the ocean has absorbed essentially all of the heat not yet seen in the atmosphere. This net absorption of heat energy has one long-term consequence and one that is more immediate.

The long-term consequence is *the impact of this stored heat* as it eventually comes into equilibrium with the atmosphere, raising temperatures over land and in the atmosphere even more. The immediate consequence is *the rise in sea level* that occurs because of thermal expansion. Sea level rose

[9]Climate Change Science Program. *Temperature Trends in the Lower Atmosphere: Steps for Understanding and Reconciling Differences,* A Report by the Climate Change Science Program and the Subcommittee on Global Change Research (Washington, DC: CCSP, 2006).

[10]James Hansen et al. "Earth's Energy Imbalance: Confirmation and Implications," *Science* 308 (June 2005): 1431–1435.

SOUND SCIENCE

Josh Willis works at NASA's Jet Propulsion Laboratory and specializes in measurements of the ocean's response to global warming. He publishes his results in peer-reviewed journals like the *Journal of Geophysical Research*. In a 2004 paper, he presented work that showed the ocean heat content increasing between 1993 and 2003, in agreement with what climatologists were seeing with the atmospheric temperature measurements. This is no trivial matter; the oceans absorb some 90% of the heat from global warming. However, Willis then did a follow-up study with colleague John Lyman, updating the time series from 2003 to 2005, showing that the oceans suddenly began to cool, and in a large way. This alarmed many climatologists and, at the same time, fed the climate change skeptics, who pounced on the data as proof that global warming wasn't happening.

One climatologist who was especially dubious about Willis's new data was Takmeng Wong, who works at NASA's Langley Research Center. Wong had been measuring the exchange of energy between the Earth and space, and his data showed that the amounts of entering and outgoing energy at the top of the atmosphere were gradually climbing out of balance. More energy was entering than leaving. This was true for the

Josh Willis, NASA Jet Propulsion Laboratory. Willis is shown here receiving the Presidential Early Career Award for Scientists and Engineers in January 2010 from John Holdren (White House Office of Science and Technology Policy Director) and Lori Garver (NASA Associate Administrator)

2003–2005 time period as well, and Wong politely asked Willis to reexamine his data because there was a clear discrepancy between the two studies. When Willis measured an even deeper ocean cooling in 2006, he began to entertain serious doubts about his data.

The source of the temperature data being employed by Willis was a flotilla of Argo floats—underwater robots that would sink to 2,000 meters, drift with the ocean currents, and then pop up to the surface, taking water temperatures as they ascended. They then transmitted the data to a satellite. Willis found

that, when tested, many new Argo floats were giving bad data—in fact, indicating cooler temperatures than other devices being employed. He also consulted an independent set of data on satellite measurements of sea level. When water heats up, it expands, and the sea level data were showing a continual rise. One final factor in the data relates to a different data set on ocean temperatures from devices called XBTs (for expandable bathythermographs, which are dropped into the ocean and relay temperatures as they sink). An independent paper had shown that the XBT data from previous years were too warm. When Willis corrected his ocean warming time series with the too-warm XBT measurements and the too-cool Argo measurements, the indications of ocean cooling went away. The warm pulses of the 1970s and 1980s disappeared, and the time series showed a gradual warming (Fig. 18–10).

The conclusion to this story is that you can never have too many independent measurements if you want to measure something as huge as a warming planet. Taken together, ocean warming, sea level rise, and energy exchange with space are now telling a consistent story, one of a warming that can be attributed only to rising GHG emissions.

Source: Rebecca Lindsey, *Correcting Ocean Cooling*, NASA Earth Observatory, November 5, 2008.

---

about 6 inches (15 cm) during the 20th century, a rate of 1.5 millimeters per year. It is now rising more rapidly, at a rate of 3.2 millimeters per year (**Fig. 18–11**). About half of this rise is due to thermal expansion, and the remainder results from the melting of glaciers and ice fields. The IPCC AR4 documents a general loss of land ice as glaciers and ice caps have thinned and melted. Significantly, historical sea level data indicate that the sea level was unchanged until 100 years ago; this modern sea level rise is a new phenomenon and is consistent with the known pattern of anthropogenic climate warming.

Other Observed Changes. The IPCC AR4 and other recent reports document a number of climate impacts, all of which are consistent with and contribute to the picture of GHG-caused climate change (note that most of these changes have potentially serious consequences for human and natural systems):

- Broad increases in warm temperature extremes and decreases in cold temperature extremes have been documented. There are also seasonal changes: Spring comes earlier and fall is later in the Northern Hemisphere,

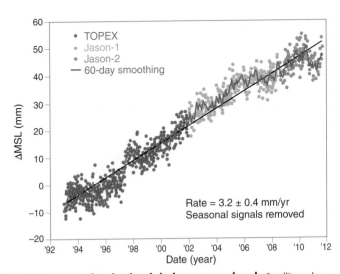

Figure 18–11 The rise in global mean sea level. Satellite radar altimeters have provided estimates of global sea level since 1993. The current rate is 3.2 mm/year. Note the 6 mm decline in 2010–2011, a consequence of the strong La Niña that produced massive rainfall events over land. Note, too, that the sea level started back up in mid-2011 as the La Niña began to subside.

(*Source:* Colorado University Sea Level Research Group, 2011. Used with permission.)

shifting ecosystems and populations out of sync. More tree deaths and forest insect damage have also been traced to the warmer temperatures. Heat waves, like those that scorched Europe in 2010 and the United States in 2011 and 2012, are increasing in intensity and frequency.

- Droughts are also increasing in frequency and intensity; the fraction of land area experiencing drought has more than doubled over the past 30 years, leading to a global decline in net primary productivity and significant reductions in corn and wheat crops. More than 60% of the United States has been under a drought that started in the late 1990s, seriously affecting agricultural crops and water resources. East Africa is experiencing its worst drought in 60 years, overwhelming refugee camps and available food aid. The increased forest fire activity in the western United States is clearly related to droughts, warmer temperatures, and earlier springs; more than 8.5 million acres burned in 2011, the third largest burn in history.

- Rising Arctic temperatures have brought about a major shrinkage of Arctic sea ice, with an accelerating decrease (**Fig. 18–12**). According to a recent assessment,[11] the Arctic Ocean is likely to have an ice-free summer by 2030.

[11]Arctic Monitoring and Assessment Program. Snow, Water, Ice and Permafrost in the Arctic, Executive Summary. 2011. www.amap.no/swipa. December 19, 2011.

While overall global temperature has increased a moderate 1.1°F, temperatures in Alaska, Siberia, and northwestern Canada have risen 5°F in summer and 10°F in winter. Spring comes two weeks earlier in the Arctic than it did at the turn of the 21st century. The polar ice cap has lost nearly 20% of its volume in the past two decades, and permafrost has been melting all over the polar regions—lifting buildings, uprooting telephone poles, and breaking up roads.

- The Greenland Ice Sheet, a huge reservoir of water, is melting and calving icebergs at unprecedented rates, undoubtedly a response to the rise in temperatures in the northern latitudes. A net loss of 530 cubic miles of water has occurred between 2003 and 2011 (**Fig. 18–13**). Record warm summers have accelerated Greenland ice melt; the largest loss (114 cubic miles) was in 2011. The two-mile-high ice sheet contains enough water to raise ocean levels 23 feet (7 m).

- Temperatures in Antarctica, which holds enough ice to raise sea levels by 185 feet (57 m), have risen by more than 0.5°C over the past 55 years; West Antarctica showed the greatest warming, at 0.85°C. The West Antarctic Ice Sheet (WAIS) is the continent's most vulnerable ice mass because it is grounded below sea level. Ocean water is constantly in contact with the edges of this ice sheet, and the ice sheet has been shrinking as the water eats away at the ice. Melting of this ice sheet would raise global sea level

SEA ICE EXTENT 09/09/2011

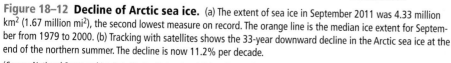

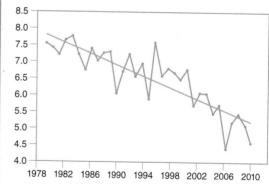

AVERAGE MONTHLY ARCTIC SEA ICE EXTENT
August 1978 to 2010

Figure 18–12 Decline of Arctic sea ice. (a) The extent of sea ice in September 2011 was 4.33 million km² (1.67 million mi²), the second lowest measure on record. The orange line is the median ice extent for September from 1979 to 2000. (b) Tracking with satellites shows the 33-year downward decline in the Arctic sea ice at the end of the northern summer. The decline is now 11.2% per decade.

(*Source:* National Snow and Ice Data Center, University of Colorado, October 2011.)

LOSS OF MASS OF THE GREENLAND ICE SHEET

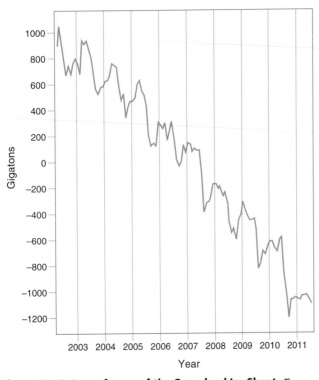

Figure 18–13 Loss of mass of the Greenland Ice Sheet. The ice sheet loss has been measured by the gravity-measuring GRACE satellites. The greatest decline occurred in 2010–2011, a total loss of 430 gigatons (114 cubic miles).

(*Source:* NOAA Arctic Report Card: Update for 2011.)

(a)

(b)

Figure 18–14 Shrinkage of the Muir glacier. The Muir glacier, a tidewater glacier in Alaska's Glacier Bay, has retreated more than 7 miles (12 km) and thinned by more than 2,600 feet (800 m) since 1941. Once covered with ice and snow (a), the land now holds trees and shrubs (b).

(*Source:* NSIDC/WDC for Glaciology, compiled 2002, updated 2006, *Online glacier photograph database*, Boulder, CO: National Snow and Ice Data Center/World Data Center for Glaciology, digital media.)

by 16 to 20 feet (5 to 6 m). The East Antarctic Ice Sheet contains eight times the mass of West Antarctic's.

- Glaciers have resumed a retreat that has been documented for more than a century, but that slowed in the 1970s and 1980s. The World Glacier Monitoring Service has followed the melting and slow retreat of mountain glaciers all over the world. The response time of glaciers to climatic trends is slow—10 to 50 years—but the data collected show an acceleration of glacial "wastage" since 1990 that is likely a response to recent rapid atmospheric warming (**Fig. 18–14**). Loss of mountain glaciers is especially serious, as these glaciers are often critical water sources for arid lands below.

- Patterns of precipitation are changing, with greater amounts from 30° N and 30° S poleward and with smaller amounts from 10° N to 30° N. Also, there have been widespread increases in heavy precipitation events, leading to more frequent flooding.

- Marine fish species populations are shifting northward as sea temperatures have warmed, impacting the commercial fisheries. Changing sea temperatures have also affected the spatial distribution of phytoplankton production over major ocean areas.

- The oceans have been soaking up some 30% of the CO_2 emissions from human sources, for which we may be thankful because this slows the rise of atmospheric CO_2 considerably. However, in doing so, the oceans are

experiencing a decrease in pH, a process known as **ocean acidification.** Although the decrease is only about 0.1 pH unit to date, there is serious concern about the future state of seawater pH in light of continued anthropogenic CO_2 production. At some level, ocean acidification will make the warm oceans inhospitable to coral reefs; the coral animals depend on an equilibrium for carbon minerals that is affected by acidic conditions. The bottom line: Human use of fossil fuels has begun to change not only the temperature, but also the chemistry of the surface oceans.

Given the accumulating evidence that the climate is changing and that the change is already causing an array of changes in biological, physical, and chemical systems on Earth, it is time to address the causes of these changes in more detail.

Rising Greenhouse Gases

According to the IPCC AR4 Synthesis Report, "Global GHG emissions due to human activities have grown since pre-industrial times, with an increase of 70% between

ESTIMATES OF GLOBAL CARBON EMISSIONS FROM FOSSIL-FUEL COMBUSTION AND CEMENT MANUFACTURE

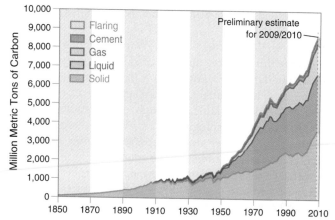

Figure 18–15 Global carbon emissions from fossil fuel combustion. Data from 1850 to 2010, from the Carbon Dioxide Information Analysis Center (CDIAC). The rise in the 2000s decade is conspicuous.

(*Source:* CDIAC, Oak Ridge National Laboratory, 2011.)

1970 and 2004."[12] They have continued to increase.[13] The most important of these anthropogenic GHGs is carbon dioxide (CO_2).

Carbon Dioxide. More than 100 years ago, Swedish scientist Svante Arrhenius reasoned that differences in CO_2 levels in the atmosphere could greatly affect Earth's energy budget. Arrhenius suggested that, in time, the burning of fossil fuels might change the atmospheric CO_2 concentration, though he believed that it would take centuries before the associated warming would be noticeable. He was not concerned about the impacts of this warming, arguing that such an increase would be beneficial. (He lived in Sweden!)

Carbon Dioxide Monitoring. In 1958, Charles Keeling began measuring CO_2 levels on Mauna Loa, in Hawaii. Measurements there have been recorded continuously, and they reveal a striking increase in atmospheric levels of the gas (Fig. 18–6). The concentrations increased exponentially until the energy crisis in the mid-1970s, rose at a rate of 1.5 ppm/year for several decades, and in the 2000s began rising at a rate of 2.0 ppm/year and higher. The data also reveal an annual oscillation of 5–7 ppm, which reflects seasonal changes of photosynthesis and respiration in terrestrial ecosystems in the Northern Hemisphere. When respiration predominates (late fall through spring), CO_2 levels rise; when photosynthesis predominates (late spring through early fall), CO_2 levels fall. Most salient, however, is the relentless rise in CO_2. As of mid-2012, atmospheric CO_2 levels were at 394 ppm, 40% higher than they were before the Industrial Revolution and higher than they have been for more than 800,000 years (Fig. 18–7a). Thus, our insulating blanket is thickening, and it is reasonable to expect that this will have a warming effect.

[12]Intergovernmental Panel on Climate Change. *Climate Change 2007: Synthesis Report*. Contribution of Working Groups I, II and III to the Fourth Assessment. Report of the Intergovernmental Panel on Climate Change (Core Writing Team, Pachauri, R. K., and Reisinger, A. (Eds.)). IPCC, Geneva, Switzerland, 2007.

[13]S. A. Montzka et al. Non-CO_2 Greenhouse Gases and Climate Change. *Nature* 476: 43–50. August 4, 2011.

TOTAL EMISSIONS: 8.3 BILLION METRIC TONS OF CARBON PER YEAR

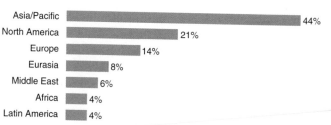

Figure 18–16 Sources of carbon dioxide emissions from fossil fuel burning. Total emissions in 2009 were approximately 8.3 GtC, or 30.3 billion metric tons of CO_2. Half of this (Europe, North America, Japan, South Korea, Taiwan, Saudi Arabia) comes from the industrialized countries.

(*Source:* Data from U.S. Energy Information Administration, December 2011.)

Sources. As Arrhenius suggested, the obvious place to look for the source of increasing CO_2 levels is our use of fossil fuels. Every kilogram of fossil carbon burned produces 3.7 kilograms of CO_2 that enters the atmosphere (each carbon atom in the fuel picks up two oxygen atoms in the course of burning and becoming CO_2). Currently, 9.1 billion metric tons (gigatons, or Gt) of fossil fuel carbon (GtC) are burned each year (**Fig. 18–15**), all added to the atmosphere as CO_2 (about 5% of this total comes from cement production and is usually included in figures reported for fossil fuel emissions). About half of this amount comes from the industrialized countries (**Fig. 18–16**). Carbon dioxide accounts for about 20% of greenhouse warming, at current levels.

Climate change skeptics sometimes claim that much of the new CO_2 comes from natural sources. This claim can be put to rest by considering the ^{14}C signature of atmospheric CO_2. ^{14}C is a radioactive isotope of carbon that is formed in the atmosphere by bombardment of nitrogen atoms by cosmic rays; it decays over time to regenerate nitrogen atoms. Fossil fuels contain no ^{14}C because of radioactive decay (the half-life of ^{14}C is 5730 years). If significant amounts of CO_2 originating from fossil fuel combustion are added to the atmosphere, the ^{14}C in the atmosphere should be declining, and it clearly is (**Fig. 18–17**).

CLEAN AIR MEASUREMENTS AT NIWOT RIDGE, COLORADO

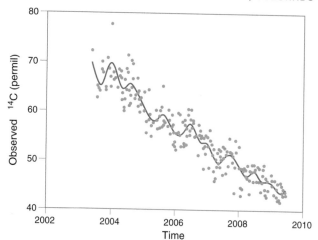

Figure 18–17 ^{14}C trend in the atmosphere. The Δ^{14}C is the ratio of ^{14}C to normal ^{12}C. The smaller the ratio, the fewer ^{14}C atoms are in the air. The decreasing trend is due to ^{14}C depleted CO_2 emitted from burning fossil fuels.

(*Source:* NOAA Earth System Research Laboratory, Issue 6: Spring 2011.)

It is estimated that land use changes such as defor-estation add another 0.9 GtC annually to the carbon al-ready coming from fossil fuel combustion. Over the past 50 years, the release of carbon from fossil fuel use and de-forestation has more than tripled. Fortunately, more than half of these carbon emissions are removed annually by so-called *sinks*.

Sinks. Careful calculations show that if all the CO_2 emitted from burning fossil fuels accumulated in the at-mosphere, the concentration would rise by 3 or more ppm per year, not the actual 2 ppm charted in Figure 18–6. Thus, there must be carbon sinks that absorb CO_2 and keep it from accumulating at a more rapid rate in the at-mosphere. There is broad agreement that the oceans serve as a sink for about 30% of the CO_2 emitted; some of this is due to the uptake of CO_2 by phytoplankton, and some is a consequence of the undersaturation of CO_2 in seawa-ter. There are limitations to the ocean's ability to absorb CO_2, however, because only the top 300 m of the ocean are in direct contact with the atmosphere. (As mentioned earlier, the deep ocean layers do mix with the upper lay-ers, but the mixing time is more than 1,000 years.) And, as the oceans acidify (precisely because they are absorb-ing CO_2), their capacity to continue this absorption will diminish.

The seasonal swings illustrated in Figure 18–6 show that biota can influence atmospheric CO_2 levels. Meas-urements indicate that terrestrial ecosystems also serve as carbon sinks. Indeed, these land ecosystems apparently have stored a net 25%, a figure that includes the losses from deforestation. As a result, terrestrial ecosystems, and especially forests, are increasingly valued for their ability to sequester carbon. A simple model of the major dynamic pools and fluxes of carbon, as presently understood, is presented in **Figure 18–18**.

Other Gases. Water vapor, methane, nitrous oxide, ozone, and chlorofluorocarbons (CFCs) also absorb infrared radia-tion and add to the insulating effect of CO_2 (Table 18–2).

Most of these gases have anthropogenic sources and are increasing in concentration, raising the concern that future warming will extend well beyond the calculated effects of CO_2 alone.

Water Vapor and Clouds. Water as vapor (a gas) or in clouds (an aerosol) absorbs infrared energy and is the most abundant GHG. Although it plays an important role in the greenhouse effect, its concentration in the tropo-sphere is quite variable. Through evaporation and pre-cipitation (the hydrologic cycle), water undergoes rapid turnover in the lower atmosphere (see Chapter 10). How-ever, recent measurements have found a general increase in water vapor in the atmosphere of 400 grams per cubic meter per decade since 1988. As temperatures over the land and oceans rise, evaporation increases, and the water vapor concentration, or humidity, rises. As the humidity rises, it traps more heat in the atmosphere, which causes even more warming. This phenomenon is called **positive feedback** and is one of the more disturbing features of future warming. The recent rise in atmospheric humid-ity is entirely consistent with the warming that has ac-companied the rise in anthropogenic GHGs. Water vapor accounts for about 50% of the greenhouse effect, and clouds an additional 25%.

Methane. Methane (CH_4), the third-most-important GHG, is 25 times more effective a GHG than is CO_2, molecule for molecule. Primarily a product of microbial fermentative reactions, methane's main natural sources are wetlands and green plants. Anthropogenic sources in-clude livestock (methane is generated in the stomachs of ruminants), landfills, coal mines, natural-gas production and transmission, rice cultivation, and manure. Although methane is gradually destroyed in reactions with other gases in the atmosphere, it is being added to the atmos-phere faster than it can be broken down. The concentra-tion of atmospheric methane has more than doubled since the Industrial Revolution, as revealed in core samples taken from glacial ice (Table 18–2). After rising at a rate of around 1.8 parts per billion (ppb) per year, methane

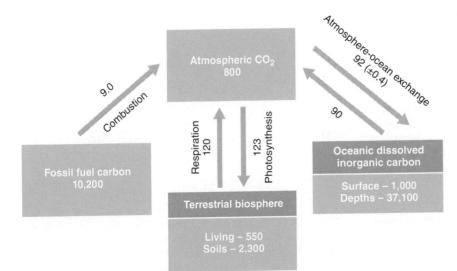

Figure 18–18 Global carbon cycle. Data are given in GtC (billions of metric tons of carbon). Pools are in the boxes, and fluxes (Gt C per year) are indicated by the arrows.

(*Source:* U.S. DOE. 2008. *Carbon Cycling and Biosequestration: Report from the March 2008 Workshop,* DOE/SC-108, U.S. Depart-ment of Energy Office of Science, pp. 2–3. http://genomicscience. energy.gov.)

TABLE 18–2 Anthropogenic Greenhouse Gases in the Atmosphere

Gas	Average Preindustrial Concentration (ppb)*	Approximate Current Concentration (ppb)	Radiative Forcing (W/m²)	Average Years in Atmosphere
Carbon dioxide (CO_2)	280,000	392,000	1.76	ca. 120
Methane (CH_4)	714	1,800	0.50	9
Nitrous oxide (N_2O)	270	322	0.17	120
Chlorofluorocarbons and halocarbons	0	1.2	0.30	50–100

Sources: IPCC AR4; recent data from CDIAC.
*Parts per billion; 1,000 ppb = 1 part per million (ppm).

leveled off from 1999–2006 but is now increasing at a rate of 6 ppb/year. Measurements indicate that two-thirds of the present methane emissions are anthropogenic. Like CO_2, methane is more abundant than it has been for at least the past 800,000 years.

Nitrous Oxide. Nitrous oxide (N_2O) levels have increased some 19% during the past 200 years and are still rising at a rate of 0.7 ppb per year. Sources of the gas include agriculture, the oceans, and biomass burning; lesser quantities come from fossil fuel burning and industry. N_2O is produced in agriculture via anaerobic denitrification processes, which occur wherever nitrogen (a major component of fertilizers) is highly available in soils (see Chapter 3). The buildup of nitrous oxide is particularly unwelcome because its long residence time (114 years) makes the gas a problem in not only the troposphere, where it contributes to warming, but also the stratosphere, where it contributes to the destruction of ozone.

Ozone. Although ozone in the troposphere is short-lived and varies at many locations, it is a potent GHG. Some ozone from the stratosphere (where it is formed) descends into the troposphere, but the greatest source is anthropogenic, through the action of sunlight on pollutants (Chapter 19). Estimates indicate that the concentration of ozone in the atmosphere has increased 36% since 1750. Major sources are automotive traffic and burning forests and agricultural wastes.

CFCs and Other Halocarbons. Emissions of halocarbons are entirely anthropogenic. Like nitrous oxide, halocarbons are long lived and contribute to both global warming in the troposphere and ozone destruction in the stratosphere. Used as refrigerants, solvents, and fire retardants, some halocarbons have 10,000 times the capacity to absorb infrared radiation that CO_2 does. The rate of production of halocarbons has declined due to adherence to the Montreal Protocol, and the concentration of halocarbons in the troposphere began declining in 1990 and is now only one-tenth of its former level.

Together the other anthropogenic GHGs are estimated to account for 5% of greenhouse warming, about one-fourth of the impact of CO_2. Although the

tropospheric concentrations of some of these gases are rising, it is hoped that they will gradually decline in importance because of steps being taken to reduce their levels in the atmosphere.

The Atmosphere's Control Knob. A recent report from the Goddard Institute for Space Studies[14] examined the role of the different greenhouse gases in maintaining temperatures on Earth. Some skeptics have claimed that because water vapor and clouds contribute the lion's share of the greenhouse effect (75%), CO_2 and the other greenhouse gases are not very important. On the contrary, Lacis and coauthors point out that, in fact, CO_2 and the other *noncondensing* gases (water readily *condenses* and can be removed from the atmosphere) are the essential components of greenhouse warming because their "forcing" role is what sustains the current levels of water vapor and clouds. They point out that without the stability of these noncondensing greenhouse gases, the water vapor and clouds could not maintain our current greenhouse conditions and Earth would be plunged into an icebound state. It is the radiative forcing of CO_2 and the other gases that controls the feedback processes involved with water vapor and clouds, even though these on average contribute 75% of the greenhouse effect. Thus, CO_2, because it is the most abundant noncondensing gas, is the key gas, the "control knob" that governs Earth's temperature.

The IPCC AR4 Synthesis Report states: "Global atmospheric concentrations of CO_2, CH_4 and N_2O have increased markedly as a result of human activities since 1750 and now far exceed pre-industrial values determined from ice cores spanning many thousands of years . . . . There is *very high confidence* (90–100%) that the global average net effect of human activities since 1750 has been one of warming, with a radiative forcing of +1.6 W/m²." **Figure 18–19** shows the understanding of the various radiative forcing. It is evident that positive anthropogenic forcing far outweighs negative.

---
[14]Andrew A. Lacis et al. "Atmospheric CO_2: Principal Control Knob Governing Earth's Temperature." *Science* 330: 356–359. October 15, 2010.

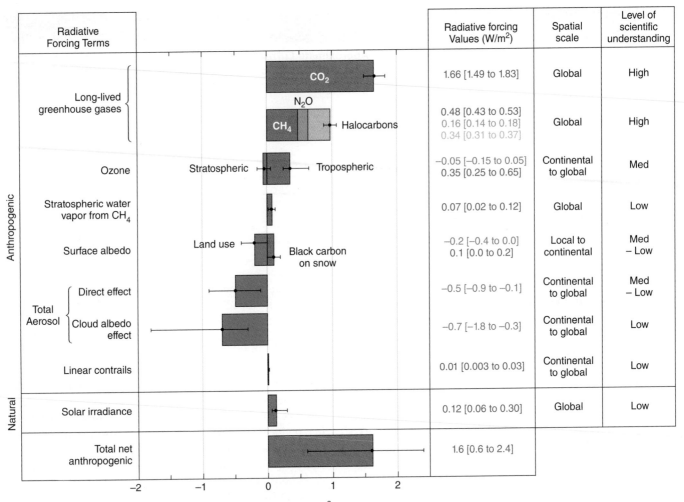

Radiative Forcing Terms			Radiative forcing Values (W/m²)	Spatial scale	Level of scientific understanding
Long-lived greenhouse gases		CO₂	1.66 [1.49 to 1.83]	Global	High
		N₂O CH₄ Halocarbons	0.48 [0.43 to 0.53] 0.16 [0.14 to 0.18] 0.34 [0.31 to 0.37]	Global	High
Ozone	Stratospheric	Tropospheric	−0.05 [−0.15 to 0.05] 0.35 [0.25 to 0.65]	Continental to global	Med
Stratospheric water vapor from CH₄			0.07 [0.02 to 0.12]	Global	Low
Surface albedo	Land use	Black carbon on snow	−0.2 [−0.4 to 0.0] 0.1 [0.0 to 0.2]	Local to continental	Med – Low
Total Aerosol	Direct effect		−0.5 [−0.9 to −0.1]	Continental to global	Med – Low
	Cloud albedo effect		−0.7 [−1.8 to −0.3]	Continental to global	Low
Linear contrails			0.01 [0.003 to 0.03]	Continental to global	Low
Solar irradiance			0.12 [0.06 to 0.30]	Global	Low
Total net anthropogenic			1.6 [0.6 to 2.4]		

Radiative Forcing (watts/meter²)

Figure 18–19 Radiative forcing. Positive and negative factors contribute to changes in energy balance within the atmosphere and at the Earth's surface. The figure presents the human factors that have led to the current sustained increase in radiative forcing (RF), in watts per square meter. The 90% confidence limits are indicated, and on the right, the relative radiative forcing and spatial scale of effects are shown, along with the level of scientific understanding.

(*Source:* S. Solomon et al., Eds. *Climate Change 2007: The Physical Science Basis,* Contribution of Working Group I to the Fourth Assessment Report of the Intergovernmental Panel on Climate Change (Cambridge, England: Cambridge University Press, 2007), Figure SPM 2.)

Future Changes in Climate

Higher temperatures, rising sea levels, heat waves, droughts, intense precipitation events, shifts in seasons, melting ice sheets, and Arctic thawing—these are happening right now, and GHG levels are continuing to rise as a result of human agency. The International Energy Agency projects a growing demand for fossil fuels as populations and economies grow. If global fossil fuel use continues on its present trajectory, emissions of GHGs will increase 35% by 2030. By 2050, emissions will increase 100%. Given the huge reserves of coal, ample supplies of natural gas, and remaining oil reserves, there is no reason to doubt this projection. However, it is clear that commitment to this future use of fossil fuels also will commit us to greater changes in climate. How much change, and when, would be a matter of guesswork if it were not for the use of climate models.

Modeling Global Climate. Weather forecasting employs powerful computers that are capable of handling large amounts of atmospheric data and applying mathematical equations to model the processes taking place in the atmosphere, oceans, and land. Forecasts of conditions for 72 hours or more have become quite accurate. Even though computing power has increased greatly in recent years, the somewhat chaotic behavior of weather parameters prevents longer-term forecasting.

Modeling climate is an essential strategy for exploring the potential future impacts of rising GHGs. Climatologists employ similarly powerful computers and have combined global atmospheric circulation patterns with ocean circulation, radiation feedback from clouds, and land surface processes to produce **atmosphere-ocean general circulation models (AOGCMs)** that are capable of simulating long-term climatic conditions. Many

ANTHROPOGENIC AND NATURAL FORCINGS

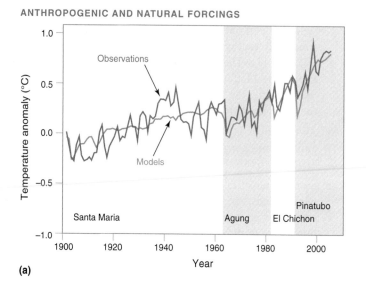

(a)

NATURAL FORCINGS ONLY

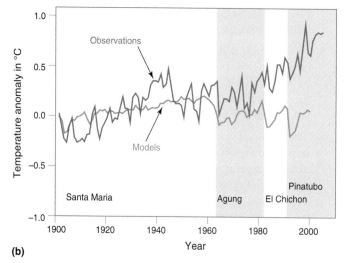

(b)

Figure 18–20 Comparison between modeled and actual data on temperature anomalies, 1900–2000. Observed values are shown by the green line, and runs from 13 models are shown in color. (a) Model results with both anthropogenic and natural forcings. (b) Model results with only natural forcings. The shaded areas mark the timing of major volcanic events.

(*Source:* S. Solomon et al., Eds. *Climate Change 2007: The Physical Science Basis,* Contribution of Working Group I to the Fourth Assessment Report of the Intergovernmental Panel on Climate Change (Cambridge, England: Cambridge University Press, 2007), Figure 23, TS.)

centers around the world are intensely engaged in exploring climate change by running models coupling the atmosphere, oceans, and land.

The best test of such models is the degree to which they can simulate present-day climate. A recent study compared some 50 climate models developed over the past two decades in order to answer this question: How well do coupled models simulate today's climate?[15] The authors concluded that the most recent models, especially those used for the IPCC AR4, have reached an unprecedented level of realism.

[15]Thomas Reichler and Junsu Kim. "How Well Do Coupled Models Simulate Today's Climate?" *Bulletin of the American Meteorological Society* 89, No. 3 (March 2008): 303–331.

Figure 18–20, from the IPCC AR4, shows how closely models track the observed temperature changes from 1900 to 2004, taking into consideration natural and anthropogenic forcing. Earlier assessment reports projected global temperature increases from 0.15 to 0.29°C per decade, based on less sophisticated models. These values compare favorably with the observed values of about 0.2°C per decade occurring since the projections.

Running the Models. The AOGCMs are employed to examine the changes that might take place over the 21st century. A number of assumptions about GHG emissions are combined with a range of climate responses to GHG concentrations (called **climate sensitivity**) to project some 35 scenarios of 21st-century climate changes. In **Figure 18–21**, we show three scenarios to illustrate a range of climate responses. None of the scenarios considers the impacts of *specific policies* to reduce GHGs:

- A2 is a high emissions scenario, reflecting a world of independent countries going their own way, with increasing population and regionally varied economic growth; technological change is slower than in other scenarios.

- A1B, a medium emissions scenario, reflects a world in which population reaches its peak in mid-century and then declines, economic growth is rapid, and new and efficient technologies are introduced; there is a balance across all energy sources, with no specific reliance on any one technology.

- B1 is a lower emissions scenario, reflecting a more globalized world with population developments as in A1B, but with effective global cooperation to reach sustainability; in particular, technologies are clean and resource efficient.

Every scenario employed in the models generates a rise in CO_2 concentrations and a corresponding rise in mean global temperature. Even if there were no increase at all in

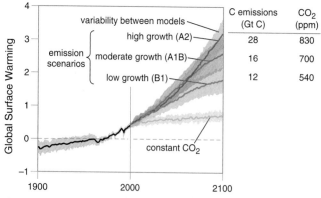

Figure 18–21 Model projections of global mean warming. Projections are for the A2, A1B, and B1 scenarios starting in 2000 and extended to 2100. The orange curve from 2000 shows the forcing "commitment" if greenhouse gas emissions were held constant in 2000. Annual carbon emissions and final CO_2 concentrations corresponding to the three scenarios are also shown.

(*Source:* S. Solomon et al., Eds. *Climate Change 2007: The Physical Science Basis,* Contribution of Working Group I to the Fourth Assessment Report of the Intergovernmental Panel on Climate Change (Cambridge, England: Cambridge University Press, 2007), Figure 32, TS.)

GHGs, current GHG concentrations represent a *commitment* to further temperature increase due to the delay in the release to the atmosphere of energy stored in the oceans. That commitment is on the order of about 0.1°C per decade for the next several decades, as shown in Figure 18–21 (the orange curve). Approximate peak carbon emissions and CO_2 concentrations corresponding with the three scenarios are also shown in Figure 18–21.

Projections. The AR4 examined the implications of the model runs and projected the potential future climate changes that could be expected, given the range of scenarios and also the current picture of climate change already under way. The major consequence of rising GHG levels during the 21st century is *rising temperatures*—the amount of rise depending on energy choices and other factors, such as population growth. Rising global temperatures are linked to two major impacts: *regional climatic changes* and *a rise in sea level*. Both of these effects show up in all of the models, though the details of regional changes from climate models are very uncertain. Some of the most significant findings are presented here:

- **Equilibrium climate sensitivity**—the sustained change in mean global temperature that occurs if atmospheric CO_2 is held at double preindustrial values, or about 550 ppm—is in the range of 2–4.5°C, with a median of 3°C. A more recent estimate of climate sensitivity,[16] based on past glacial temperature ranges, suggests a lower range: 1.7–2.6°C, median 2.3° (though there are many uncertainties in this estimate).

- The actual warming of the 21st century will be a consequence of GHG emissions, whatever they are. The model scenarios project a range of surface warming in 2100 compared with 1980–2000; for the B1, A1B, and A2 scenarios, the estimates average 1.8°C, 2.8°C, and 3.4°C, respectively. The projected rate of warming for the next two decades is 0.2°C per decade.

- The warming is expected to occur in geographical patterns similar to those of recent years: warming greatest at the higher latitudes and in continental interiors and least over the North Atlantic and the southern oceans. However, it will be warmer everywhere.

- Because of the warming climate, snow cover and sea ice will decrease, possibly fully opening up the Arctic Ocean by mid-century. Glaciers and ice caps will continue to shrink, resulting in a rise in sea level. In the Arctic, up to 90% of the upper layer of permafrost is expected to thaw, creating a positive climate feedback as methane and CO_2 are released.

- Because the oceans will be absorbing so much CO_2, they will continue to acidify, threatening coral reefs (which will already be stressed by warming tropical seas) and shellfish everywhere.

- The sea level, currently rising at a rate of 3.2 mm per year, will certainly continue to rise as a consequence of continued thermal expansion and snow and ice melt (Fig. 18–11). The range projected by the IPCC for the end of the 21st century is 7 to 23 inches (19 to 59 cm). Significantly, the projections did not include melting of the Greenland and Antarctic Ice Sheets.

- As the upper layers of the North Atlantic Ocean warm and become more fresh, the thermohaline circulation is expected to slow down, but no models show it collapsing.

- Storm intensities are expected to increase, with higher wind speeds, more extreme wave heights, and more intense precipitation; there will be fewer but more intense hurricanes in the Atlantic Basin.

- Heat waves will become more frequent and longer lasting as the climate warms. Growing seasons will lengthen, and frost days will decrease in middle and high latitudes. Outbreaks of cold air in the Northern Hemisphere will diminish by 50% or more compared with the present.

- Precipitation will decrease in already dry regions and increase in already wet regions; there will be an increase in extremes of daily precipitation. Areas of extreme drought will expand, affecting up to 30% of the world's land surface.

- The changing climatic parameters will have profound impacts on ecosystems and species, the impacts accruing as the temperature warms over time (Fig. 18–22): damage to polar ecosystems (as we have seen), bleached and dying coral reefs, major loss of Amazon rain forests, widespread extinctions of species, and loss of ecosystem resilience.

- Some impacts are positive: growing seasons would be longer in mid- to high latitudes, and temperatures would be more moderate in high-latitude countries like Canada and Russia.

A "business as usual" approach to climate change is best illustrated by the A2 scenario, which projects a temperature change between 3°C and 5°C. Note that the difference in temperature between an ice age and the warm period following it is only 5°C, so in a century, unless some serious measures are taken, Earth's climate will change dramatically. Responding to these changes will likely involve unprecedented and costly adjustments. The impact on natural ecosystems could be highly destabilizing. For the agricultural community, the greatest difficulty in coping with climatic change is *not knowing what to expect*. Already, farmers lose an average of one in five crop plantings because of unfavorable weather. As the climate shifts, the vagaries of weather will become more pronounced, and crop losses are likely to increase.

Sea Level. There is great uncertainty about the magnitude of the rise in sea level. A half-meter rise in sea level will flood many coastal areas and make them much more

---

[16]Andreas Schmittner et al. "Climate Sensitivity Estimated from Temperature Reconstructions of the Last Glacial Maximum." *Science* 334: 1385–1388. December 9, 2011.

Figure 18–22 Global warming impacts on biological systems. The impacts build up with greater temperature changes over time. The optimistic scenario is B1 from IPCC AR4; the pessimistic scenario is A2 from IPCC AR4.

(*Source:* Richard Kerr, "How Urgent Is Climate Change?" by Richard A. Kerr, from SCIENCE, November 2007, Volume 318(5854). Copyright © 2007 by AAAS. Reprinted with permission.)

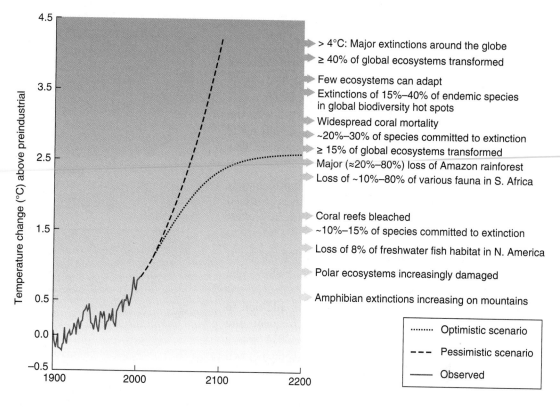

> 4°C: Major extinctions around the globe
≥ 40% of global ecosystems transformed

Few ecosystems can adapt
Extinctions of 15%–40% of endemic species in global biodiversity hot spots
Widespread coral mortality
~20%–30% of species committed to extinction
≥ 15% of global ecosystems transformed
Major (≈20%–80%) loss of Amazon rainforest
Loss of ~10%–80% of various fauna in S. Africa

Coral reefs bleached
~10%–15% of species committed to extinction

Loss of 8% of freshwater fish habitat in N. America

Polar ecosystems increasingly damaged

Amphibian extinctions increasing on mountains

.... Optimistic scenario
– – – Pessimistic scenario
——— Observed

prone to damage from storms, forcing people to abandon properties and migrate inland. A 1-meter rise would flood 17% of Bangladesh, creating tens of millions of refugees and destroying half of the country's rice lands. As we saw, for many of the small oceanic nations, a rise in sea level means obliteration, not just alteration. Models suggest that if the mean global temperature rises as much as 3°C above present levels, it will set up conditions that could cause widespread loss of the Greenland Ice Sheet. Recent studies have made it clear that the IPCC projections of sea level rise are *strikingly on the low side*. These studies indicate that 21st-century warming would lead to a sea level rise of close to 2 meters, considering the plausible rates of melting of glaciers that run into the sea (in Greenland and Antarctica). And once atmospheric GHG levels are stabilized (assuming they are), temperatures and sea levels will continue to rise for hundreds of years because of the slow response time of the oceans. The expected continued rise would cause disasters in most coastal areas, which are home to half the world's population and its business and commerce.

Climate Change Impacts in the United States

In 2002, the Bush administration launched the Climate Change Science Program (CCSP—now called the Global Change Research Program, or GCRP) to integrate the efforts of many federal agencies involved in the scientific effort to understand global climate change. One of the tasks of the CCSP has been to communicate the findings of the scientific community to policy makers and the broader public by publishing reports on climate change issues. These reports, covering a broad range of topics, are available on the Internet (www.globalchange.gov) and are very well done. The 2009 report, *Global Climate Change Impacts in the United States*,[17] draws from 19 existing CCSP reports and other peer-reviewed scientific assessments, such as those from the IPCC. It is designated as the **United Synthesis Product (USP)** and provides a detailed look at the impacts of climate change on various *sectors* (e.g., transportation, agriculture, human health) and *regions* (e.g., Northwest, Southwest, Northeast, Alaska) of the United States. Some highlights of the report are worth examining.

National Climate Change. Most of the global impacts of climate change—temperature increases, droughts, intensity of hurricanes, seasonal changes, increased intensity of heavy rain events, floods, rising sea level, large wildfires, and melting glaciers, among others—have already been observed in the United States. The projected future changes just reviewed will certainly also affect the United States. However, national and regional climates vary more than global climate, and the different regions of the United States have experienced different impacts and will do so in the

---

[17]Thomas R. Karl, Jerry M. Melillo, and Thomas C. Peterson (Eds.). *Global Climate Change Impacts in the United States*, Cambridge University Press, 2009.

future. Following are key observations from the report (and more recent updates):

- The average U.S. temperature rose more than 2°F over the previous 50 years and will rise more; heat waves could become commonplace in the next 30 years.

- Precipitation increased an average of about 5% over the previous 50 years. Shifting patterns have generally made wet areas wetter, while dry areas have become drier. This is projected to continue.

- The heaviest downpours have increased approximately 20% on average in the past century, and this is projected to continue, with the strongest increases in the wettest places.

- Southwestern North America is experiencing the worst drought since 1900; future projections suggest that droughts could worsen across the region.

- Many types of extreme weather events, in addition to heavy downpours, have become more frequent and intense during the past 40 to 50 years; 2011 was a record year for devastating weather in the United States, with 12 billion-dollar disasters and more than 1,000 deaths.[18]

- The destructive energy of Atlantic hurricanes has increased in recent decades and is projected to increase further in this century.

- Sea level rose 2 to 5 inches during the previous 50 years along many U.S. coasts; coastal cities with some 40 million people are at risk because of the likely rise in sea level linked to 21st-century GHG emissions.

- For cold-season storms outside the tropics, storm tracks are shifting northward, and the strongest storms are projected to become stronger.

- Arctic sea ice is declining rapidly, and this is projected to continue.

Some interesting graphics are used to illustrate regional climate changes. For example, for the Midwest, the summer climate in Illinois and Michigan is shown shifting southward and westward (**Fig. 18–23**) as temperatures increase and precipitation decreases. Significantly, the impacts of climate change in Alaska are greater than those in any other region of the United States. Average temperatures are projected to rise 4°F to 7°F by mid-century and could rise 8°F to 13°F by the end of the century if global emissions continue on their present course. The USP report's concluding thoughts emphasize the importance of involving scientists, policy makers, and the public in "designing, initiating, and evaluating mitigation and adaptation strategies. . . . The best decisions about these strategies will come when there is widespread understanding of the complex issue of climate change—the science and its many implications for our nation."

We turn now to our society's response to climate change.

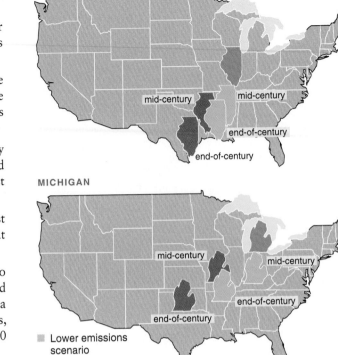

Figure 18–23 Climate on the move in the Midwest. Model projections of summer average temperature and precipitation changes in Illinois and Michigan for mid-century and end of century. The lower emissions scenario is IPPC B1, and the higher emissions scenario is IPPC A2. Note that both states are projected to get considerably warmer and have less summer precipitation.

(*Source: CSSP, Global Climate Change Impacts in the United States: Unified Synthesis Product, 2009.*)

18.3 Response to Climate Change

The world's industries and transportation networks are so locked into the use of fossil fuels that high emissions of CO_2 and other GHGs seem certain to continue for the foreseeable future, bringing on significant climate change. One possible response to this situation is **adaptation**. That is, we must anticipate some harm to natural and human systems and should plan adaptive responses to lessen the vulnerability of people, their property, and the biosphere to coming changes.

Yet it seems more sensible to reduce the rate at which emissions are added to the atmosphere and, eventually, bring about a sustainable balance—though no one thinks that doing so will be easy. This is the **mitigation** response: Take action to reduce emissions. Before we deal with these responses, we need to address an important question: Why is there still so much doubt about climate change?

Skeptics, Deniers, and Ethics

The scientific consensus on global climate change is clear: Global warming is well under way, and the consequences

---

[18]Climate Central. "Top 10 States Ravaged by Extreme Weather in 2011." *Grist*, December 30, 2011. www.grist.org. December 31, 2011.

of further climate change are potentially catastrophic. International concern about this issue has led to a number of conferences designed to address climate change. In the United States, however, public opinion polls reveal a high level of skepticism about global warming. This attitude has puzzled scientists, who have seen the evidence for global warming growing stronger all the time. Some recent work has reviewed this discrepancy,[19] pointing out that most people do not spend time examining the scientific evidence but instead form their opinions based on the views of organizations and media sources they trust. The high degree of public skepticism, then, can be traced to several factors:

- A deliberate campaign to undermine trust in climate science by organizations and corporations that stand to lose influence and profits if carbon emissions are curtailed.[20] This campaign has systematically spread disinformation in order to confuse the public.[21]

- A political divide that sees conservatives and Republicans showing much more skepticism about climate change than liberals and Democrats; this is especially evident in Congress, where the divide is virtually unanimous.

- An unfortunate release of hacked e-mails from a climate research group in the UK in late 2009 that has been called "Climategate" (more about this later) and led to an erosion of trust in climate scientists.

- The coincidence of an economic crisis in the United States that has focused the public's attention on economic issues and provided fuel for the campaigns determined to attack the public's trust in climate science and policy making.

Which Is It? True skeptics are those who seek after truth, looking for evidence that will guide them to correct conclusions. Given the substantial public skepticism about global warming, it is fair to ask: Is this really skepticism, or is it simply misguided trust?

In a very insightful book, Haydn Washington and John Cook address the issue of those who are denying climate change.[22] The authors charge that this is not a matter of skepticism; on the contrary, it is more accurate to use the label "deniers" to describe those who are challenging climate change science. In effect, they are either refusing to accept the evidence or possibly denying a truth they do not like. Of the numerous arguments presented by the deniers, one in particular has gained credence because of "Climategate." The argument is that the world's climate scientists are engaged in a conspiracy to deceive in order to continue to get lucrative grants to keep their laboratories and research empires in business.

"Climategate," as it is called, involved the illegal hacking into computers at the University of East Anglia, a center of climate research. Private e-mails between climate researchers were made public, some of which were purported to reveal conspiratorial correspondence to falsify or hide data. In fact, the correspondence was taken out of context and misinterpreted, deliberately, to suggest dishonesty on the part of the scientists. Several high-level investigations into the episode have cleared the scientists of any fraud or scientific malpractice. "Climategate," however, fed into the true conspiracy to deny that climate change was occurring, giving deniers a perfect occasion to question the scientists' integrity and thus influence public opinion against climate change science.

Who are these "deniers"? Very often, they are conservative "think tanks" that are ideologically opposed to regulation. They see the kinds of international agreements and national public policies needed to address climate change as threats to free market capitalism. They are often supported by industries that stand to lose profits if these policy changes are put in place—the fossil fuel industries (especially ExxonMobil), the utilities, and their public supporting body, the U.S. Chamber of Commerce. Washington and Cook document the work of these organizations and the range of arguments they employ, and one of the authors (Cook) supports a Web site that counters virtually every "denialist" argument (www.skepticalscience.com) with careful reasoning and data. If the stakes were not so high, one could pass off this denial phenomenon as a sociological aberration. However, given the high stakes, the denial of climate change lands us firmly in the domain of ethics.

A Perfect Moral Storm. International treaties and national policies are needed to move us in the direction of mitigation and adaptation. But why should we do it? And how should we proceed?

In a recent book,[23] philosopher Stephen Gardiner refers to our flawed attempts to address climate change as an ethical failure—in fact, a "perfect moral storm" (using the metaphor of Sebastian Junger's book *The Perfect Storm*, a true account of the loss of the fishing boat *Andrea Gail* in a rare convergence of three bad storms in the North Atlantic). Gardiner's three factors that converge are of deep moral significance: 1) climate change is a *global* "storm," one where the wealthy nations are able to shape what is done about it, possibly to the detriment of the poorer nations; 2) it is also an *intergenerational* storm, where the present generation has the power to affect future generations, who have no voice in the matter; 3) the third is a *theoretical* storm, in that the climate change threat is unprecedented and lacks adequate moral theories to give guidance for appropriate actions. The convergence of these three interacting "storms," says Gardiner, has led to our present lack of real progress in addressing the looming threat of global climate change.

[19]Sandra T. Marquart-Pyatt et al. "Understanding Public Opinion on Climate Change: A Call for Research." *Environment* 53: 38–42. July/August 2011.

[20]Naomi Oreskes and Erik M. Conway. *Merchants of Doubt*. Bloomsbury Press, New York. 2010.

[21]Steven A. Kolmes. "Climate Change: A Disinformation Campaign." *Environment* 53: 33–37. July/August, 2011.

[22]Haydn Washington and John Cook. *Climate Change Denial: Heads in the Sand*. Earthscan. Washington, DC, 2011.

[23]Stephen M. Gardiner. *A Perfect Moral Storm: The Ethical Tragedy of Climate Change*. Oxford University Press, New York. 2011.

To simplify what could become a complex philosophical discussion, here are three ethical principles that speak to the issue:

1. The first is the **precautionary principle**, as articulated in the 1992 Rio Declaration. The principle states, "In order to protect the environment, the precautionary approach shall be widely applied by States according to their capabilities. Where there are threats of serious or irreversible damage, lack of full scientific certainty shall not be used as a reason for postponing cost-effective measures to prevent environmental degradation." Employing this principle is like taking out insurance: We are taking measures to avoid a highly costly, but uncertain, situation (albeit one that is appearing quite likely). The risk of serious harmful consequences is real, the scientific consensus is strong, but the political response, as we will see, is still inadequate.

2. The second is the **polluter pays principle**. Polluters should pay for the damage their pollution causes. This principle has long been part of environmental legislation. CO_2 is where it is because the developed countries (and now many developing countries) have burned so much fossil fuel since the beginning of the industrial era. It has become clear that CO_2 is a pollutant.

3. The third is the **equity principle**. Currently, the richest 1 billion people produce 55% of CO_2 emissions, while the poorest 1 billion produce only 3%. International equity and intergenerational equity are ethical principles that should compel the rich and privileged to care about those generations that follow—especially those in the poor countries who are even now experiencing the consequences of global climate change. Climate change is going to cause harm, as we have seen. If we are able to take action that will prevent both present and future harm, it would be morally wrong not to do so.

And so, what shall we do?

Mitigation

It will be costly to take the actions needed to mitigate the causes of climate change and to adapt to changes that are unavoidable, but the costs of inaction will be far higher. If climate change is unchecked, every country will be affected, but it will be most disastrous for the world's poorest countries and people. Climatic disruptions will create hundreds of millions of refugees. Conflict and terrorism will surely follow in the wake of such deprivation, as people are radicalized by their plight. The impacts on the natural world will be equally severe, as we have seen (Fig. 18–22).

Some attempts have been made to put a price tag on inaction; these are admittedly rough estimates, but ones based on sound analyses. For the United States, one study by the Natural Resources Defense Council estimates a total economic cost of global warming at 3.6% of gross domestic product (GDP) by 2100, or almost $3.8 trillion (in today's dollars). For just four impacts—hurricane damage, energy costs, real estate losses, and water costs—the price tag comes to almost $1.9 trillion. The Stern Review,[24] an economic analysis commissioned by the British government, estimated that inaction on global climate change would result in economic loss equivalent to 5% to 20% of global GDP per year for the indefinite future. Given the economic, environmental, and social costs of the consequences of unrestrained release of GHGs, it would seem that mitigation is absolutely essential. We need to reduce emissions of GHGs in order to prevent disaster. The next question is: What should be our target?

Achieving Stabilization. The stated objective of the Framework Convention on Climate Change (FCCC) is to *stabilize the GHG content of the atmosphere at levels and on a time scale that would prevent dangerous anthropogenic interference with the climate system.* This means that GHG emissions must be brought under control to the point where they are no longer accumulating in the atmosphere, and it must be done before global temperatures rise to dangerous levels. Earth history tells us that with 3°C global warming, equilibrium sea level could be at least 80 feet (25 m) higher than today. This is clearly dangerous to human civilization and Earth systems in general and to be avoided. However, with current warming at about 0.8°C above pre-industrial levels and more warming already in the pipeline, we could get to 2°C in a couple of decades. Many feel that 2°C could well be a **tipping point**, a point beyond which irreversible—and clearly dangerous—runaway climate change occurs no matter what we do.

So we have a target: hold global temperatures to 2°C above pre-industrial levels, which corresponds to a stable concentration of CO_2 (with the other GHGs included) in the atmosphere of around 450 ppm. To achieve stabilization at 450 ppm will require major changes in our energy systems. Current carbon emissions of 33 $GtCO_2$/year will have to fall to 7 $GtCO_2$/year by 2100 and then drop even further. This is the *inconvenient truth* Al Gore referred to in his 2006 film by the same name on global warming. The IPCC AR4 Synthesis Report presented an analysis of six stabilization scenarios and their correspondence with global temperatures and sea level rise (Table 18–3). Global emissions of GHGs corresponding to the different scenarios are shown in **Figure 18–24a**, along with the time course of allowable emissions. The relationship between temperature increase above preindustrial levels and GHG stabilization-level concentration for the six scenarios is shown in Figure 18–24b. The IPCC states: "In order to stabilize the concentration of GHGs in the atmosphere, emissions would need to peak and decline thereafter. The lower the stabilization level, the more quickly this peak and decline would need to occur."[25] Clearly, the only category in Table 18–3 and Figure 18–24 that is consistent with the FCCC objective of avoiding "dangerous anthropogenic

[24]Nicholas Stern. *The Economics of Climate Change: The Stern Review* (Cambridge, England: Cambridge University Press, 2007).

[25]Intergovernmental Panel on Climate Change. *Climate Change 2007: Synthesis Report*, Contribution of Working Groups I, II and III to the Fourth Assessment. Report of the Intergovernmental Panel on Climate Change (Geneva, Switzerland: IPCC, 2007).

TABLE 18–3 Stabilization Scenarios and Corresponding Global Temperatures and Sea Level Rise

Category	CO_2-equivalent concentration at stabilization (including all GHGs) (ppm)	Peaking year for CO_2 emissions	Change in global CO_2 emissions in 2050 (% of 2000 emissions)	Global avg. temperature increase at equilibrium (°C)	Global avg. sea level rise above preindustrial at equilibrium from thermal expansion (m)
I	445–490	2000–2015	−85 to −50	2.0–2.4	0.4–1.4
II	490–535	2000–2020	−60 to −30	2.4–2.8	0.5–1.7
III	535–590	2010–2030	−30 to +5	2.8–3.2	0.6–1.9
IV	590–710	2020–2060	+10 to +60	3.2–4.0	0.6–2.4
V	710–855	2050–2080	+25 to +85	4.0–4.9	0.8–2.9
VI	855–1130	2060–2090	+90 to +140	4.9–6.1	1.0–3.7

Source: IPCC, *Climate Change 2007: Synthesis Report* (Geneva, Switzerland: IPCC, 2008).

interference" is category I. Again, from the IPCC: "Mitigation efforts over the next two to three decades will have a large impact on opportunities to achieve lower stabilization levels."

We turn now to the question of what the international community has done in response to global climate change.

International Conferences.   One of the five documents signed by heads of state at the UN Conference on Environment and Development's Earth Summit in Rio de Janeiro in 1992 was the **Framework Convention on Climate Change (FCCC)**. This convention agreed to the goal of *stabilizing GHG levels in the atmosphere*, starting by reducing GHG emissions to 1990 levels by the year 2000 in all industrialized nations. Countries were to achieve the goal by voluntary means. Five years later, it was obvious that the voluntary approach was failing. All the developed countries except those of the European Union *increased* their GHG emissions by 7% to 9% in the ensuing five years. The developing countries increased theirs by 25%.

Kyoto, 1997. Prompted by a coalition of island nations (whose very existence is threatened by global climate change), the third Conference of Parties to the FCCC met in Kyoto, Japan, in December 1997 to craft a binding agreement on reducing GHG emissions. In Kyoto, 38 industrial and former eastern bloc nations agreed to reduce emissions

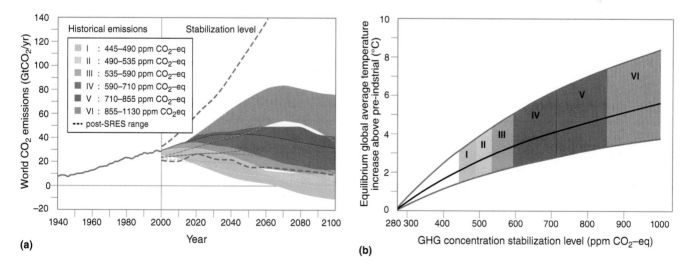

(a)

(b)

Figure 18–24 CO_2 emissions and equilibrium temperature increases for a range of stabilization levels. (a) Global CO_2 emissions for 1940 to 2000 and projected emissions ranges for six categories of stabilization scenarios from 2000 to 2100. The thickness of each band represents the uncertainty inherent in the model projections. The dashed lines (post-SRES range) show the upper and lower emissions range of more recently developed scenarios. (b) Corresponding relationship between a CO_2 stabilization target and the likely equilibrium global temperature increase above preindustrial. The black line through the middle shows the best estimate of the response of temperature to GHG concentrations (climate sensitivity); the orange and blue lines show the upper and lower bounds of climate sensitivities.

(*Source: IPCC, Climate Change 2007: Synthesis Report*, Contribution of Working Groups I, II and III to the Fourth Assessment. Report of the Intergovernmental Panel on Climate Change (Geneva, Switzerland: IPCC, 2007), Figure 5-1.)

of six GHGs to 5.2% *below* 1990 levels, *to be achieved by 2012* (the **Kyoto Protocol**). The signatories to this agreement are called the *Annex I parties*; all others (the developing countries) are *non–Annex I parties*. At the conference, the developing countries refused to agree to *any* reductions, arguing that the developed countries had created the problem and that it was only fair that the developing countries continue on their path to development as the developed countries had, energized by fossil fuels. This stance was in accord with the FCCC principle of "common but differentiated responsibilities." Under this principle, each nation must address climate change, but its priorities and efforts could differ according to national circumstances. Thus, the developing countries argued that their first priority was to alleviate poverty, whereas the developed countries had no such excuse.

The signers of the Kyoto Protocol had a great deal of flexibility in deciding how they would achieve their GHG reductions. The most obvious choices involve renewable energy, possibly nuclear energy, and a variety of energy conservation strategies. They could plant forests to act as carbon sinks or engage in emissions trading with other Annex I parties, either buying or selling assigned emissions amounts to meet targets. They could even implement projects that reduce emissions in other countries and earn credits toward their target emissions.

The targeted reductions of the Kyoto Protocol will *not* stabilize atmospheric concentrations of GHGs. The IPCC calculates that it would take immediate reductions in emissions of at least 60% worldwide to stabilize GHG concentrations at today's levels. Since this has no chance of happening, the concentration of GHGs in the atmosphere will continue to rise. The major weakness of the Kyoto Protocol is that the world's largest GHG emitters—India, China, and the United States—are not participating. The United States withdrew from the treaty in 2001, and recently, Canada, Japan, and Russia withdrew also, with the result that Kyoto now only covers 13% of global emissions. Nevertheless, Kyoto was an important first step toward binding agreements to stabilize the atmosphere.

Bali, 2007. In December 2007, a UN-sponsored climate conference held in Bali addressed the need for greater cuts in GHG emissions when the Kyoto Protocol expires in 2012. At the Bali meeting, parties to the FCCC agreed to what was essentially a "road map" to future negotiations. It was agreed that "deep cuts in global emissions will be required," to achieve the objective of avoiding dangerous climate change, and a negotiation process was established, to prepare for the next meeting.

Copenhagen, 2009. The next major climate conference was held in December 2009 in Copenhagen. There, delegates from around the world met to once again attempt to find agreement on climate change policy. On the plus side, the **Copenhagen Accord** was agreed to by 187 out of 192 countries. This accord amounted to pledges to implement actions that would limit global temperature increases to 2°C above pre-industrial levels, and a commitment by developed and developing countries to set countrywide emissions targets to be achieved by 2020. Still, nothing in the way of a binding agreement was reached there, nor was there even a timetable for reaching such agreements.

Cancún, 2010. At Cancún, negotiations by delegates appeared to get back on track after the disappointing Copenhagen meeting. Delegates approved the mitigation pledges made by all major GHG emitting countries in accordance with the Copenhagen Accord and established a Green Climate Fund of $30 billion to help the more vulnerable developing countries fund their mitigation pledges and cope with impacts of climate change. Also, mechanisms were set up to reward developing countries for maintaining their forests (REDD, for Reducing Emissions from Deforestation and Forest Degradation). Delegates postponed any decision on extending the Kyoto Protocol. There were still no binding agreements on emissions.

Durban, 2011. Once again, meeting in December as the 17th Conference of the Parties to the FCCC, delegates debated and negotiated what to do. In the end, they reached an agreement (the **Durban Platform**) to begin negotiations that would lead to a legally binding agreement by 2015, to take effect by 2020. The agreement aims at ensuring that emissions reductions will meet the goal of keeping global temperature rise below 2°C. Delegates also agreed to establish a second commitment period of the Kyoto Protocol, to begin when the first period ends (2012).

The Gap. At the Durban meeting, the United Nations Environment Program released a report[26] it was commissioned to develop. The report was based on the question of how the pledges made by each country, based on the Copenhagen Accord, help meet the 2°C target. The report showed that the pledges fall short of the target; thus, there is a *climate emissions gap* to be addressed, a gap of between 6 and 11 $GtCO_2$, depending on how well the pledges are kept **(Fig. 18–25)**. The report states that to stay within the 2°C limit, emissions would have to peak before 2020, and that won't happen unless new emissions targets are established. It is possible to

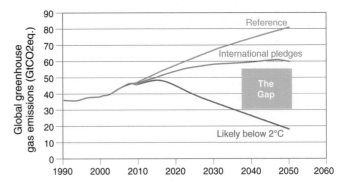

Figure 18–25 The climate emissions gap. The plot shows estimates of global GHG emissions in gigatons of CO2 equivalents (other gases equated to CO2). The reference plot (blue line) shows emissions growth in the absence of the targets and pledges, the orange line shows the pledges made by countries following the Copenhagen Accord, and the brown line shows the emissions pathway needed to keep the growth in global temperature below 2°C.

(*Source*: Climate Action Tracker, Dec. 11, 2011. Copyright © 2009 by Ecofys and Climate Analytics.)

[26]United Nations Environment Program. *Bridging the Emissions Gap: A UNEP Synthesis Report.* November 2011. www.unep.org/publications/ebooks/bridgingemissionsgap. December 28, 2011.

TABLE 18–4 Options for Mitigation of Greenhouse Gas Emissions

Cap-and-trade. Place a cap on GHG emissions; for CO_2, this can be accomplished by limiting the use of fossil fuels in industry and transportation. Rights to emit GHGs are allocated to different companies and are tradable in a market-based system. This has worked well internationally (in the EU), where different countries are allocated permits and can trade with other countries. It is anticipated that the cap will be reduced gradually until a stabilized atmosphere is achieved. This option generates potentially large revenues, which can be put to use in ways that can alleviate some of the problems created by the program.

Renewable energy. Invest in and deploy an increasing percentage of energy in the form of renewable-energy technologies: wind power, solar collectors, solar thermal energy, photovoltaics, biofuels, hydrogen-powered vehicles, and geothermal energy, among others.

Carbon capture and sequestration. Sequester CO_2 emitted from burning fossil fuels at power plants by capturing it at the site of emission and pumping it into deep, porous underground formations. This is especially vital for coal-burning power plants, as coal is the most abundant fossil fuel energy source for many major GHG emitters. This is expected to push electricity prices up about $0.04 per kilowatt-hr.

Nuclear power. Encourage the development of nuclear power, first resolving issues concerning cost-effectiveness, reliability, spent fuel, and high-level waste.

Reforestation. Stop the loss of tropical forests and encourage the planting of trees and other vegetation over vast areas now suffering from deforestation. Trees and other vegetation provide a sink for carbon that removes CO_2 from the atmosphere.

Efficiency. Make energy conservation rules much more stringent. (Tighten building codes to require more insulation, use energy-efficient lighting, and so forth.)

Mileage standards. Reduce the amount of fuels used in transportation by raising mileage standards, encouraging car pooling, and stimulating mass transit in urban areas.

Subsidies. Remove fossil fuel subsidies like depletion allowances, tax relief to consumers, support for oil and gas exploration, and the substantial costs of maintaining access to Middle Eastern oil.

Population growth. Reducing population growth would make a substantial contribution to GHG emissions reduction, especially in the economically robust developing countries.

Carbon tax. Levy a tax on all fuels according to the amount of carbon dioxide that they produce when consumed. Such a tax, proponents believe, would provide both incentives to use renewable sources, which would not be taxed, and disincentives to consume fossil fuels. This puts the polluter-pays principle into practice.

close the gap, says the report, but countries need to realize that what they are pledging just won't cut it.

Clearly, international accords are important, but *it is legislative action at the national level that brings about emissions reductions.* A research team from the UK recently analyzed national-level legislation in 16 countries that are major GHG emitters.[27] The study was encouraging: It reported on 155 existing climate change-related laws that were in effect in the countries and found that many approaches were used to accomplish the goal of reducing GHG emissions and prepare the country for the impacts of climate change. This brings us to the question: What in fact has the United States been doing about global climate change and the FCCC?

U.S. Policy. Climate change policy in the United States can be described in four venues: (1) scientific research, (2) national legislation, (3) regulatory action, and (4) state and local action. Of course, one major policy move was made by President George W. Bush in 2001 when he announced that the United States would withdraw from the Kyoto Protocol agreement.

Scientific Research. The *Global Change Research Act of 1990* established a policy of support for research on climate change; the act mandates an interagency approach, where all of the agencies and departments that might be involved in climate change research are to coordinate their activities and programs as part of the official Global Change Research Program (GCRP). Lead agencies in this effort have been NOAA, NASA, the U.S. Geological Survey, and the EPA. One example of the GCRP's effort has been the 20 reports of the Climate Change Science Program already mentioned. Budgetary support for the GCRP has been about $2 billion per year. Hundreds of U.S. scientists have been in the lead in conducting research on climate change and publishing their work, especially as they have participated in the IPCC assessments.

National Legislation. The U.S. Congress has failed to enact a comprehensive set of policies to address global climate change. It came close when a Democrat-controlled House passed the *American Clean Energy and Security Act* in 2009, which would establish an emissions reduction and trading plan called cap-and-trade (see Table 18–4), but the bill died in the Senate where Republican opposition was enough to prevent its passage. However, the *American Recovery and Reinvestment Act* has invested more than $43 billion in the

[27]Terry Townshend et al. "Legislating Climate Change on a National Level." *Environment* 53: 5–17. September/October 2011.

TABLE 18–5 Adaptation Strategies

All Countries

- **Agriculture:** Farmers shift to climate-resistant crops or crops more appropriate to the new climate regime. Irrigation is expanded to areas once fed only by rain.
- **Structural concerns:** Societies invest to reduce the costs and disruption from extreme weather events like storms, floods, droughts (e.g., seawalls, new reservoirs, revegetating coastal wetlands).
- **Emergency preparedness:** Societies invest in early-warning systems, high-quality climate information, intervention when needed (e.g., heat wave warnings, heavy precipitation events, shelters for those displaced or affected by events).
- **Reducing risks:** Societies establish a financial safety net to help their poorest members, providing insurance that accurately reflects new risks to discourage poor choices.

Developing Countries

- **Promoting development:** Investments are made in health, education, economic opportunity, and food security (to lower people's vulnerability to climate impacts and help them build adaptive capacity).
- **Controlling diseases:** Malaria, AIDS, tuberculosis, and emerging diseases are brought under control so that a changing climate does not increase their threat.
- **Enhancing economic progress:** Societies promote literacy, electronic communication, road infrastructure, and diversification of economic activity.

Source: Nicholas Stern, *The Economics of Climate Change: The Stern Review* (Cambridge, England: Cambridge University Press, 2007).

support of renewable energy and energy efficiency programs, which will certainly have an impact in lowering GHG emissions. Other legislative actions include laws that established subsidies and rebates for renewable energy sources.

Regulatory Action. The Obama administration, under the Copenhagen Accord, has committed the United States to a reduction of GHG emissions by 17% below 2005 levels, to be achieved by 2020. Major tools to accomplish this include all of the policies to promote renewable energy and energy conservation, especially the regulations that raise the corporate average fuel economy standards (Chapter 16). Most important, however, is the role of the EPA. In 2009, taking its cue from a 2007 Supreme Court ruling, the EPA determined that greenhouse gases pose a danger to public health and welfare, thus giving the agency a green light to regulate CO_2 and other GHGs. This move has infuriated many industry groups and their Republican allies, but their efforts to block the EPA from this role have so far failed. The agency has begun its regulatory efforts with the *New Source Performance Standards* for greenhouse gases from fossil-fueled power plants, announced in 2011. This permitting rule requires all new power plants to hold emissions to 1,000 pounds of CO_2 per megawatt hour, something that natural gas plants can accomplish but coal plants cannot. Limits on existing coal plants are yet to come.

State and Local Action. It is encouraging that states are taking unilateral action to address global climate change. Many states are adopting renewable portfolio standards and are mandating reporting of GHG emissions (see Chapter 16). Several regions have established GHG emissions cap-and-trade programs similar to the Clean Air Act program that was successful in bringing down sulfur dioxide emissions. Many states have set quantitative goals for reducing GHG emissions. More than 1,000 mayors have pledged to move their cities to make GHG reductions under the Climate Protection Agreement.

Mitigation Tools. Many options are available for reducing GHG emissions, many of which are already in place in the United States and globally. Table 18–5 presents a summary of most of the prominent mitigation tools. Appropriately, the majority of the options address the energy sector, where most of the anthropogenic GHGs are originating. What is needed, of course, is to "decarbonize" the energy economy, so that energy needs are met without fossil fuels. Every provision that promotes greater energy efficiency and the use of renewable energy sources is going to reduce GHG emissions in the United States (Chapter 16). Recent legislation has made progress in energy efficiency and renewable energy (as discussed in Chapters 14 and 16).

There will be economic costs and benefits to mitigation efforts. The benefits come in the form of new industries and the jobs they create, savings from efficiency measures, and the ecosystem benefits from forests that are preserved and regrown. The net costs, as calculated by the Stern Review for stabilization at 550 ppm CO_2, amount to around 2% of global GDP by 2050. These are the costs of cutting GHG emissions by reducing demand for emissions-intensive goods and services, increasing efficiencies of energy use, avoiding deforestation, and switching to lower-carbon technologies for power, heat, and transport. Recall that the Stern Review calculated the costs of inaction to be from 5% to 20% of GDP.

As a bottom line, however, climate change is already happening and will likely accelerate in the future. *Mitigation* efforts will reduce the magnitude of future climate change, but *adaptation* will also be needed.

Adaptation

Adaptation means making adjustments in anticipation of changes brought on by the rising sea levels and temperatures of the next decades. It also means reducing vulnerability to the inevitable impacts of climate change. What makes this process difficult is that the adaptations will be aiming at a moving target—and one that is uncertain at best. Adaptation is not cost free. Climate changes are bringing us into ever new situations, and some of them are developing rapidly. Climate-related natural disasters are already happening at a higher rate than ever and will only get worse.

The Stern Review detailed adaptations for developed and developing countries, many of which are presented in Table 18–5. The IPCC also weighed in with the AR4 report. In light of the risks, the IPCC stated, "The effects of climate change are expected to be greatest in developing countries in terms of loss of life and relative effects on investment and the economy." A number of international funds have been established to address these needs, the latest being the *Green Climate Fund*, launched at the Cancún international meeting. The emphasis of these funds is on the developing countries, because it is assumed that the developed countries will be far more able to afford the costs of adaptation (and can also afford to finance the programs).

Poverty and Climate Change. It is painfully evident that climate change will be much harder on the poor, as documented in a recent article by Boorse;[28] they are already struggling to meet their needs for shelter, food, employment, and health care. The article lists several telling points:

1. The poor, especially in the less developed countries, are especially vulnerable to disasters. They lack buffers to help them deal with crop failures, storms, and floods—the kinds of impacts that will increase in the changing climate.

2. The poorer countries are less able to divert resources to adaptation to the changing climate; they are already challenged to provide health care, education, and food. Failure of many to meet the Millennium Development Goals is evidence of their resource limitations (Table 9–2).

3. Rising sea levels and other climate change impacts will inevitably displace millions in the poorer countries; the flood of environmental refugees will challenge the ability (and will) of the international community to provide aid.

4. The lack of resources, especially water and arable land, will lead to greater internal (and possibly international) conflicts. The U.S. Department of Defense has identified climate change as a significant national security issue, pointing out that climate change may act as an accelerant of conflict.

A development worker[29] states, "The last thing most people living in poverty need is climate change . . . It's a very real intensifier of poverty today . . . (it) increases vulnerability to environmental shocks that are outside their control. . . ." Even now, more than 96% of disaster-related deaths take place in developing countries. It seems fair to ask, What kind of society are we if we ignore these needs and just let the climate keep on warming as we reject the changes we need to make to limit GHG emissions?

Geoengineering The threat of climate change is generating a host of futuristic schemes to help the fight against global warming.[30] Some of the schemes are clearly unworkable, while others have been floated as serious proposals. For example, there is a proposal to fertilize vast areas of the oceans with iron, thereby stimulating phytoplankton growth, which would remove CO_2 and sink to the depths. Another proposed scheme erects scrubbers, devices that would chemically absorb CO_2 in the air and then store it in deep reservoirs. Still other methods would block some of the solar radiation from reaching Earth's surface by, for example, injecting sulfate aerosols into the stratosphere or launching satellites into space that would release reflecting particles in the path of the Sun's radiation. Most climate experts dismiss these schemes on the basis of their huge costs or the likelihood of unintended consequences such as interference with weather patterns or ecosystem destruction; besides, they might not work at all.

It is fair to say that none of these schemes compares with what is already happening. Whatever the outcome, it is certain that we are conducting an enormous global experiment in geoengineering, and our children and their descendants will be living with the consequences. Global climate change is perhaps the greatest challenge facing human civilization in the 21st century.

---

[28]Dorothy Boorse. *Loving the Least of These: Addressing a Changing Environment.* National Association of Evangelicals. 2011. http://www.nae.net/images/content/Loving_the_Least_of_These.pdf. December 31, 2011.

[29]Ibid., p. 35
[30]Catherine P. McMullen, ibid., pp. 51–53.

REVISITING THE THEMES

Sound Science

Much of this chapter reflects some outstanding science, as our knowledge of climate change depends on the solid work of scientists from around the world. It is largely the scientists who are calling attention to the perils of climate change. They were the first to discover the risks involved, and they are now the chief advocates of effective action. The IPCC, with the thousands of scientists involved, has performed a masterful assessment of the scientific basis of climate change. In the United States, the CCSP/GCRP has also produced a portfolio of excellent analyses of climate change and its many dimensions. More and more of the questions and uncertainties about climate change are being resolved as scientists continue to work.

Sustainability

Earth is in the midst of an unsustainable rise in atmospheric GHG levels, the result of our intense use of fossil fuels. In short, we completely depend on a host of technologies that are threatening our future. Projections of business-as-usual fossil fuel use point to global consequences that are disastrous, but not inevitable. The United States and other developed countries will not escape these consequences, but the gravest of them affect the developing countries. A sustainable pathway is still open to us, and it involves a combination of steps we can take to mitigate the emissions and bring the atmospheric concentration of GHGs to a stable, even declining, level.

Stewardship

Stewardly care for Earth is not really something we can ignore. We must face this "perfect moral storm," and several principles have been cited as arguments for effective action to prevent dangerous climate change: the precautionary principle, the polluter-pays principle, and the equity principle. If we care about our neighbors and our descendants, we will take action both to mitigate the production of GHGs and to enable especially the poorer countries to adapt to coming unavoidable climate changes, even if taking such action is costly.

Public skepticism in the United States about climate change is a serious problem, but the public gets its cues from the media and political leaders, where denial of climate change is embedded like a destructive cancer. One can only hope that in time the public will reject the spurious arguments of the climate change deniers and seize the moral high ground in this battle for the future of our children and grandchildren.

REVIEW QUESTIONS

1. What are the important characteristics of the troposphere? Of the stratosphere?

2. What are the main components of our weather, and how is weather related to climate?

3. Provide a brief synopsis of global climate change. What are the main issues and trends?

4. What has been discovered about global climate trends in the past? How do contemporary trends relate to past records?

5. What is the Atlantic Meridional Overturning Circulation? How does it work? What are the impacts of the El Niño/ La Niña Southern Oscillation (ENSO)?

6. What is radiative forcing? Describe some warming and cooling forcing agents.

7. What evidence do global land and ocean temperatures provide for a warming Earth?

8. What additional changes contribute to the picture of human-caused warming?

9. Which of the GHGs are the most significant contributors to global warming? How do they work?

10. Describe the data for atmospheric CO_2. What are significant sources and sinks for atmospheric CO_2? Why can CO_2 be called "the atmosphere's control knob?"

11. How are computer models employed in climate change research? Describe several scenarios for 21st-century climate change.

12. What are the major IPCC projections for climate changes in the 21st century?

13. How does recent work revise the IPCC projections of future sea level rise?

14. What are some possible reasons for current skepticism about climate change in the U.S. public?

15. Give three ethical principles that can be brought to bear on "the perfect moral storm."

16. Trace the political history of the international efforts to address global climate change. What is the "climate emissions gap"?

17. Describe recent U.S. climate change policy. What is the main reason for U.S. inaction to pass climate change legislation?

18. What mitigation steps could be taken to stabilize the GHG content of the atmosphere?

19. What are some adaptation steps that should be taken to anticipate inevitable changes in climate?

THINKING ENVIRONMENTALLY

1. What steps could you take to lower your climate impact? Check the Union of Concerned Scientists Web site for suggestions.

2. Research the deniers of global climate change. What evidence do they offer in support of their views? How does it compare with the evidence from the IPCC? On balance, what do the data indicate? Check www.skepticalscience.com.

3. What recent developments most increase the urgency of action in the international arena?

4. Trace the history of the international conferences and their outcomes. What are the main areas of contention that are blocking any binding agreements?

MAKING A DIFFERENCE

Anything you can do to cut energy use will cut your CO_2 output. In general, cutting CO_2 will also cut your bills. Listed below are just a few examples.

1. Buy items with the smallest amounts of packaging. The more packaging products have, the more space they take up in the distribution trucks, and the more CO_2 emissions they create.

2. Washing your clothes in cold water instead of hot will keep 200 pounds of carbon dioxide from entering the atmosphere a year.

3. The worst form of transportation for the climate is driving alone in a car. Use public transportation when possible, bike, walk, carpool, buy a hybrid—there are many options! When deciding on a car, make fuel economy a high priority. Driving 25 fewer miles a week will earn you 1,500 fewer pounds of CO_2 per year.

4. Replace your incandescent bulbs with compact fluorescent bulbs and the new LED lights.

MasteringEnvironmentalScience®

Go to **www.masteringenvironmentalscience.com** for practice quizzes, Pearson eText, videos, current events, and more.

CHAPTER 19

Atmospheric Pollution

Smog in Mexico City. Because of its geographic location and heavy automobile traffic, Mexico City has had some of the worst air pollution anywhere.

LEARNING OBJECTIVES

1. **Air-Pollution Essentials:** Review normal atmospheric cleansing and the formation of industrial and photochemical smog.

2. **Major Air Pollutants and Their Sources:** Identify the variety and sources of the major air pollutants, and classify them as primary or secondary pollutants.

3. **Impacts of Air Pollutants:** Investigate how air pollutants impact human health and cause damage to agricultural crops and natural ecosystems.

4. **Bringing Air Pollution Under Control:** Summarize the ways that the Clean Air Act and its amendments, and regulations derived from them, have given the Environmental Protection Agency the responsibility of controlling air pollution.

5. **Destruction of the Ozone Layer:** Assess how chlorofluorocarbons and other gases have been implicated in the destruction of stratospheric ozone, and review the steps taken to bring back the ozone layer.

O N TUESDAY MORNING, October 26, 1948, the people of Donora, Pennsylvania (population 13,000), awoke to a dense fog. Donora lies in an industrialized valley along the Monongahela River, south of Pittsburgh. On the outskirts of town, a sooty sign read, "DONORA: NEXT TO YOURS, THE BEST TOWN IN THE U.S.A." The town had a large steel mill that used high-sulfur coal and a zinc-reduction plant that roasted ores laden with sulfur. At the time, most homes in the area were heated with coal. At first, the fog did not seem unusual: Most of Donora's fogs lifted by noon, as the Sun warmed the upper atmosphere and then the land. This one, however, didn't lift for five days.

SICKNESS AND DEATH.

Through Wednesday and Thursday, the air began to smell of sulfur dioxide—an acrid, penetrating odor. By Friday morning, the town's physicians began to get calls from people in trouble. At first, the calls were from elderly citizens and those with asthmatic conditions. They were having difficulty breathing. The

calls continued into Friday afternoon. People young and old were complaining of stomach pain, headaches, nausea, and choking (Fig. 19–1). Work at the mills went on, however. The first deaths occurred on Saturday morning. By 10 AM, one mortician

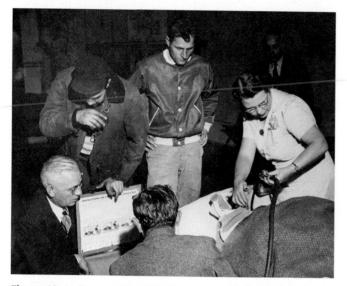

Figure 19–1 Donora Smog Victim. A nurse administers oxygen to a patient in an emergency hospital in Donora, site of a lethal smog in 1948.

had nine bodies; two other morticians had one each. On Sunday morning, the mills were shut down. Even so, the owners were certain that their plants had nothing to do with the trouble. Mercifully, the rain came on Sunday and cleared the air—but not before more than 6,000 townspeople were stricken and 20 elderly people had died. During the next month, 50 more people died.

MEMORIAL. Eventually, it was determined that the cause of the deaths was a combination of polluting gases and particles, a thermal inversion in the lower atmosphere, and a stagnant weather system that, together, brought home the deadly potential of air pollution. Today, a historical marker in the town and a small storefront museum commemorate the Donora Smog of 1948, but the lasting legacy of the smog and of those who suffered sickness and death are the state and national laws to control air pollution, culminating in the Clean Air Act of 1970 and its amendments. Donora was a landmark event.

In the previous chapter, we looked at the structure and function of the atmosphere and its links to weather and climate, and we examined the effects of greenhouse gas pollutants on climate change. In this chapter, we will consider the impacts of a host of other pollutants on human health and the environment.

19.1 Air Pollution Essentials

For a long time, we have known that the atmosphere contains numerous *gases*. The major constituents of the atmosphere are nitrogen (N_2), at a level of 78.08%; oxygen (O_2), at 20.95%; water vapor, ranging from 0% to 4%; argon (Ar), at 0.93%; and carbon dioxide (CO_2), at 0.04%. Smaller amounts of at least 40 "trace gases" are normally present as well, including ozone, helium, hydrogen, methane, nitrogen oxides, sulfur dioxide, and neon. In addition, the atmosphere contains *aerosols*: microscopic liquid and solid particles such as dust, carbon particles, pollen, sea salts, and microorganisms, originating primarily from land and water surfaces and carried up into the atmosphere. Outdoor air is a valuable and essential resource, providing gases that sustain life and maintain warmth (Chapter 18). It also shields the Earth from harmful radiation. With the advent of the Industrial Revolution, the mixture of gases and particles in our atmosphere began to change, and we learned the hard way about air pollution.

Pollutants and Atmospheric Cleansing

Air pollutants are substances in the atmosphere—certain gases and aerosols—that have harmful effects. Three factors determine the level of air pollution:

- the amount of pollutants entering the air,

- the amount of space into which the pollutants are dispersed, and

- the mechanisms that remove pollutants from the air.

The lower atmosphere, called the *troposphere* (**Fig. 19–2**), is the site and source of our weather and contains almost all of the water vapor and clouds of the atmosphere (Chapter 18). Pollutants that enter the troposphere are usually removed within hours or a few days unless they are carried up into the upper troposphere by towering cumulus clouds. If that happens, they can be carried far and persist for many days before they, too, are removed by cleansing agents.

Some pollutants are resistant to most of the cleansing mechanisms and can be carried up into the next higher layer of the atmosphere, the *stratosphere*. The most notorious of these are the ozone-depleting chemicals (chlorine and bromine compounds), which we will consider in Section 19.5.

Atmospheric Cleansing. Environmental scientists distinguish between natural and anthropogenic (human-caused) air pollutants. For millions of years, volcanoes, fires, and dust storms have sent smoke, gases, and particles into the atmosphere. Trees and other plants also emit volatile organic compounds into the air around them as they photosynthesize. However, there are mechanisms in the biosphere that remove, assimilate, and recycle these natural pollutants. First, a naturally occurring compound, the **hydroxyl radical**

Figure 19–2 Layers of the atmosphere closest to Earth's surface. The troposphere extends from the surface to roughly 6 miles above the surface, and the stratosphere is above that. The colored band shows the average temperature of the atmosphere at different altitudes. In the troposphere, temperatures generally decrease with height, while in the stratosphere temperatures increase with height.

(*Source:* CCSP 1.1. *Temperature Trends in the Lower Atmosphere: Steps for Understanding and Reconciling Differences.* Thomas R. Karl, Susan J. Hassol, Christopher D. Miller, and William L. Murray, Eds., 2006. A Report by the Climate Change Science Program and the Subcommittee on Global Change Research, Washington, DC.)

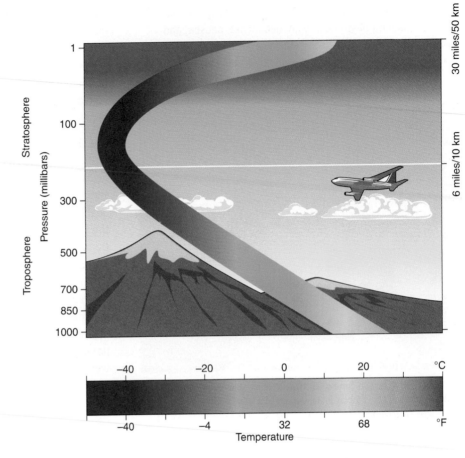

(OH), oxidizes many gaseous pollutants to products that are harmless or that can be brought down to the ground or water by precipitation **(Fig. 19–3)**. Sea salts—picked up from sea spray as air masses move over the oceans—are a second cleansing agent. These salts, now aerosols, act as excellent nuclei for the formation of raindrops. The rain then brings down many particulate pollutants (other aerosols) from the atmosphere to the ocean, cleansing pollutant-laden air. Third, sunlight breaks organic molecules apart. These three processes hold natural pollutants below toxic levels (except in the immediate area of a source, such as around an erupting volcano).

Figure 19–3 The hydroxyl radical. This is a simplified model of atmospheric cleansing by the hydroxyl radical. The first step is the photochemical destruction of ozone, which is the major process leading to ozone breakdown in the troposphere. The second step produces hydroxyl radicals, which react rapidly with many pollutants, converting them to substances that are less harmful (CO_2) or that can be returned to Earth via precipitation (HNO_3, H_2SO_4).

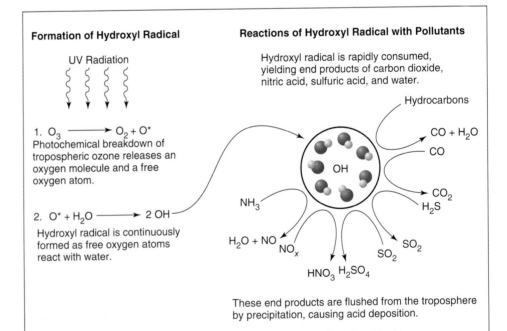

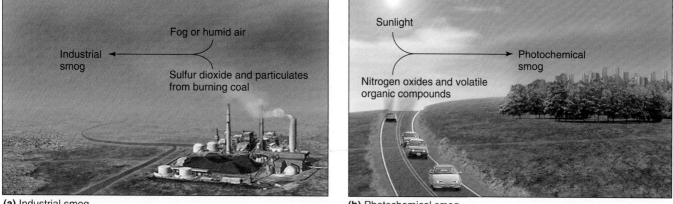

(a) Industrial smog   **(b)** Photochemical smog

Figure 19–4 Industrial and photochemical smog. (a) Industrial smog occurs when coal is burned and the atmosphere is humid. (b) Photochemical smog occurs when sunlight acts on vehicle pollutants.

Many of the pollutants oxidized by the hydroxyl radical are of concern because human activities have raised their concentrations far above normal levels. Recent studies have shown that the hydroxyl radical plays a key role in the removal of anthropogenic pollutants from the atmosphere. Highly reactive hydrocarbons are oxidized within an hour of their appearance in the atmosphere, and nitrogen oxides (NO_x) are converted to nitric acid (HNO_3) within a day. It takes months, however, for carbon monoxide (CO) to be oxidized by hydroxyl and years in the case of methane (CH_4). As Figure 19–3 shows, the photochemical breakdown of tropospheric ozone is the major source of the hydroxyl radical (not shown is a newly discovered additional source of the hydroxyl radical, from the photoexcitation of nitrogen dioxide). As we will see, higher levels of ozone result from higher concentrations of other polluting gases.

Recent research[1] indicates that the atmospheric levels of hydroxyl are quite stable, suggesting that formation and breakdown are balanced. Some hydroxyl is used up during reactions with CO and CH_4, but hydroxyl is recycled as it reacts with most other pollutants. This is good news for now, but researchers are concerned that future pollutants and climate change could affect the hydroxyl balance.

Smogs and Brown Clouds

Down through the centuries, the practice of venting the products of combustion and other fumes into the atmosphere remained the natural way to avoid their noxious effects within buildings. With the Industrial Revolution of the 1800s came crowded cities and the use of coal for heating and energy. It was then that air pollution began in earnest.

Industrial Smog. In *Hard Times*, Charles Dickens described a typical English scene: "Coketown lay shrouded in a haze of its own, which appeared impervious to the sun's rays. You only knew the town was there because there could be no such sulky blotch upon the prospect without a town." This shrouding haze became known as **industrial smog** (a combination of *sm*oke and f*og*), an irritating, grayish mixture of soot, sulfurous compounds, and water vapor **(Fig. 19–4a)**. This kind of smog continues to be found wherever industries are concentrated and where coal is the primary energy source; such smog events mostly occur in the winter. The Donora incident is a classic example. Currently, industrial smog can be found in the cities of China, India, Korea, and a number of eastern European countries.

Photochemical Smog. After World War II, the booming economy in the United States led to the creation of vast suburbs and a mushrooming use of cars for commuting to and from work. As other fossil fuels were used in place of coal for heating, industry, and transportation, industrial smog was largely replaced by another kind of smog. Increasingly, Los Angeles and other cities served by huge freeway systems began to be enshrouded daily in a brownish, irritating haze that was different from the more familiar industrial smog **(Fig. 19–5)**. Weather conditions were usually warm and sunny rather than cool and foggy, and, typically, the new haze would arise during the morning commute and begin to dissipate only by the time of the evening commute. This **photochemical smog**, as it came to be called, is produced when several pollutants from automobile exhausts—nitrogen oxides and volatile organic carbon compounds—are acted on by sunlight (Fig. 19–4b).

Inversions. Certain weather conditions intensify levels of both industrial and photochemical smogs. Under normal conditions, the daytime air temperature is highest near the ground, because sunlight strikes Earth and the absorbed heat radiates to the air near the surface. The warm air continues

[1]S. A. Montzka et al. "Small Interannual Variability of Global Atmospheric Hydroxyl." *Science* 331: 67–69. January 7, 2011.

(a)   (b)

Figure 19–5 **Los Angeles air.** (a) Los Angeles on a good day. (b) LA under a blanket of photochemical smog.

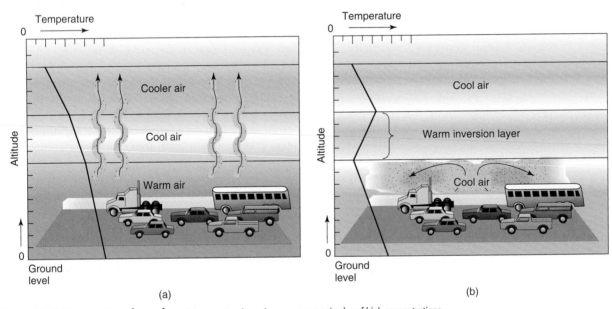

(a)   (b)

Figure 19–6 **Temperature inversion.** A temperature inversion may cause episodes of high concentrations of air pollutants. (a) Normally, air temperatures are highest at ground level and decrease at higher elevations. (b) In a temperature inversion, a layer of warmer air overlies cooler air at ground level.

to rise, carrying pollutants upward and dispersing them at higher altitudes (**Fig. 19–6a**). At times, however, a warm air layer occurs above a cooler layer. This condition of cooler air below and warmer air above is called a **temperature inversion** (Fig. 19–6b). Inversions often occur at night, when the surface air is cooled by radiative heat loss. Such inversions are usually short lived, as the next morning's sunlight begins the heating process anew and any pollutants that accumulated overnight are carried up and away. During cloudy weather, however, the Sun may not be strong enough to break up the inversion for hours or even days. A mass of high-pressure air may move in and sit above the cool surface air, trapping it. Topography can also intensify smog as cool ocean air flows into valleys and is trapped there by nearby mountain ranges. Los Angeles has both topographical features.

When such long-term temperature inversions occur, pollutants can build up to dangerous levels, prompting local health

officials to urge people with breathing problems to stay indoors. For many people, smog causes headaches, nausea, and eye and throat irritation. It may aggravate preexisting respiratory conditions, such as asthma and emphysema. In some industrial cities (such as Donora), air pollution reached lethal levels under severe temperature inversions. These cases became known as *air-pollution disasters.* London experienced repeated episodes of inversion-related disasters in the mid-1900s. One episode in 1952 resulted in 4,000 pollution-related deaths.

Atmospheric Brown Clouds. A relative newcomer on the air-pollution scene is the so-called **atmospheric brown cloud** (**ABC**), a 1- to 3-kilometer-thick blanket of pollution that frequently hovers over south and central Asia. The ABC is in fact a persistent aerosol, similar to the summer aerosol found over industrial regions in the North Temperate Zone, but dissimilar in that it persists year-round and has a different makeup (see

Chapter 18). Unique to the ABC is the high proportion of black carbon and soot derived from biomass burning (residential cooking, agricultural clearing) and fossil fuel combustion (coal and diesel exhaust). Ozone and acid rain accompany the ABC.

The impacts of the ABC are serious: major dimming, especially over large cities; reductions in rainfall due to the aerosol particles; solar heating of the atmosphere and decreased reflection by snow and ice (both of which are contributing to a shrinking of the Hindu Kush–Himalaya–Tibetan glaciers, vital water sources for the major river systems of eastern Asia); a weakening of the Indian summer monsoon; reduced crop yields due to the dimming and to ozone levels; and heightened acute and chronic health effects typical of exposure to industrial and urban air pollution.

Industrial smog from coal burning, photochemical smog from vehicular exhaust, and atmospheric brown cloud are all caused by the release of pollutants, and we now turn our attention to those pollutants.

19.2 Major Air Pollutants and Their Sources

In large measure, anthropogenic air pollutants are direct and indirect products of the *combustion* of coal, gasoline, other liquid fuels, and refuse (wastepaper, plaster, and so on). These fuels and wastes are organic compounds. With complete combustion, the products of burning organic compounds are carbon dioxide and water vapor. Unfortunately, combustion is seldom complete, and many complex chemical substances are involved. The combustion of fuels and refuse creates a host of gaseous and particulate products that in turn create unwanted problems before they are cleansed from the air. Other processes producing air pollutants are *evaporation* (of volatile substances) and *strong winds* that pick up dust and other particles.

Primary Pollutants

Table 19–1 presents the names, symbols, major sources, characteristics, and general effects of the major air pollutants. The first seven (**particulates, volatile organic compounds, carbon monoxide, nitrogen oxides, sulfur dioxide, lead,** and **air toxics**) are called **primary air pollutants** because they are the direct products of combustion and evaporation. Some primary pollutants may undergo further reactions in the atmosphere and produce additional undesirable compounds, called **secondary air pollutants** (**ozone, peroxyacetyl nitrates, sulfuric acid,** and **nitric acid**). Sources of air pollution are summarized in **Figure 19–7**. Power plants are the major source of sulfur dioxide, particulates originate from all sorts of combustion, and transportation

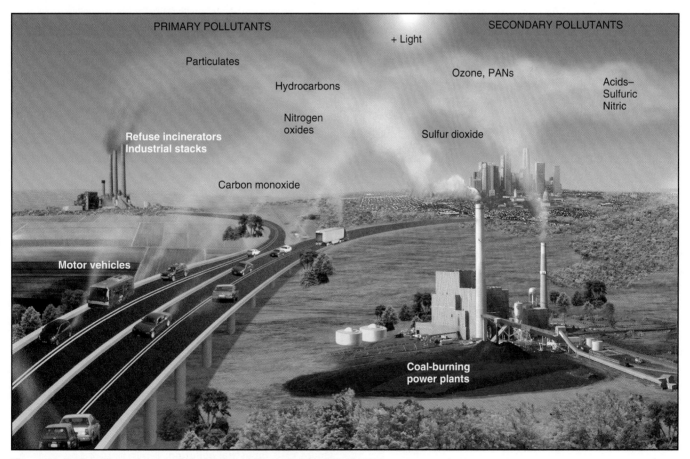

Figure 19–7 The prime sources of the major anthropogenic air pollutants. Industrial processes and transportation generate a toxic brew of primary and secondary air pollutants.

TABLE 19–1 Major Anthropogenic Air Pollutants and Their Effects

Name	Symbol	Source	Description and Effects
Primary Pollutants			
Suspended particulate matter	PM	Soot, smoke, metals, and carbon from combustion; dust, salts, metal, and dirt from wind erosion; atmospheric reactions of gases	Complex mixture of solid particles and aerosols; reduces lung function and affects respiration and cardiovascular disease
Volatile organic compounds	VOC	Incomplete combustion of fossil fuels; evaporation of solvents and gasoline; emission from plants	Mixture of compounds, some carcinogenic; major agent of ozone formation
Carbon monoxide	CO	Incomplete combustion of fuels	Invisible, odorless, tasteless gas; poisonous because of ability to bind to hemoglobin and block oxygen delivery to tissues; aggravates cardiovascular disease
Nitrogen oxides	NO_x	From nitrogen gas due to high combustion temperatures when burning fuels; also from wood burning	Reddish-brown gas and lung irritant; aggravates respiratory disease; major source of acid rain; contributes to ozone formation
Sulfur dioxide	SO_2	Combustion of sulfur-containing fuels, especially coal	Poisonous gas that impairs breathing; major source of acid rain; contributes to particle formation
Lead	Pb	Battery manufacture; lead smelters; combustion of leaded fuels and solid wastes	Toxic at low concentrations; accumulates in body and can lead to brain damage in children
Air toxics	Various	Fuel combustion in vehicles; industrial processes; building materials; solvents	Toxic chemicals, many of which are known human carcinogens (e.g., benzene, asbestos, vinyl chloride); Clean Air Act identifies 187 air toxics
Radon	Rn	Rocks and soil; natural breakdown of radium and uranium	Invisible and odorless radioactive gas can accumulate inside homes; second leading cause of lung cancer in United States
Secondary Pollutants			
Ozone	O_3	Photochemical reactions between VOCs and NO_x	Toxic to animals and plants and highly reactive in lungs; oxidizes surfaces and rubber tires
Peroxyacetyl nitrates	PAN	Photochemical reactions between VOCs and NO_x	Damages plants and forests; irritates mucous membranes of eyes and lungs
Sulfuric acid	H_2SO_4	Oxidation of SO_2 by OH radicals	Produces acid deposition, damages lakes, soils, artifacts
Nitric acid	HNO_3	Oxidation of NO_x by OH radicals	Produces acid deposition, damages lakes, soils, artifacts

accounts for the lion's share of carbon monoxide and nitrogen oxides. This diversity suggests that strategies for controlling air pollutants will be different for the different sources.

When fuels and wastes are burned, particles consisting mainly of carbon are emitted into the air; these are the particulates we see as soot and smoke (**Fig. 19–8**). In addition, various unburned fragments of fuel molecules are given off; these are the VOC emissions. Incompletely oxidized carbon is carbon monoxide (CO), in contrast to completely oxidized carbon, which is carbon dioxide (CO_2). Combustion takes place in the air, a mixture of 78% nitrogen and 21% oxygen. At high combustion temperatures, some of the atmospheric nitrogen gas is oxidized to form the gas nitric oxide (NO). In the air, nitric oxide immediately reacts with additional oxygen to form nitrogen dioxide (NO_2) and nitrogen tetroxide (N_2O_4).

Figure 19–8 Industrial pollution. Because the stack of this wood-products mill in New Richmond, Quebec Province, Canada, lacks electrostatic precipitators, the plume contains substantial amounts of suspended particulate matter that can be seen from a great distance.

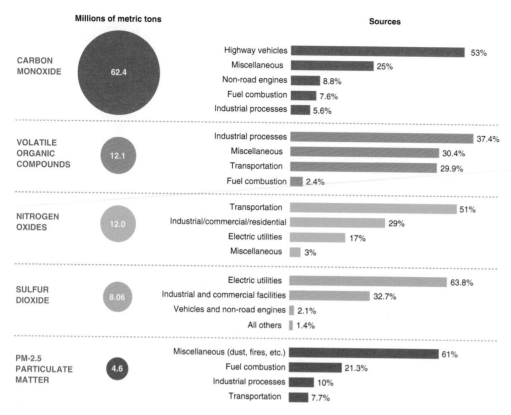

Millions of metric tons

Sources

Figure 19–9 **U.S. emissions of five primary air pollutants, by source, for 2011.** *Fuel combustion* refers to fuels burned for electrical power generation and for space heating. Note especially the different contributions by transportation and fuel combustion/electric utilities, the two major sources of air pollutants.

(*Source:* EPA National Emissions Inventory, 2012. http://www.epa.gov/ttn/chief/trends/index.html.)

These compounds are collectively referred to as the nitrogen oxides, or NO_x. Nitrogen dioxide absorbs light and is largely responsible for the brownish color of photochemical smog.

In addition to organic matter, fuels and refuse contain impurities and additives, and these substances are also emitted into the air during burning. Coal, for example, contains from 0.2% to 5.5% sulfur. In combustion, this sulfur is oxidized, giving rise to several gaseous oxides, the most common being sulfur dioxide (SO_2). Coal also contains heavy-metal impurities such as mercury, and refuse contains an endless array of "impurities."

Tracking Pollutants. The Environmental Protection Agency (EPA) operates the Clearinghouse for Inventories and Emissions Factors, which tracks trends in national emissions of the primary pollutants from all sources. The EPA also follows air quality by measuring ambient concentrations (where *ambient* means "present in the air") of the pollutants at thousands of monitoring stations across the country. According to EPA data, 2011 emissions in the United States of the first five primary pollutants amounted to 99 million tons. By comparison, in 1970, when the first Clean Air Act became law, these same five pollutants totaled 301 million tons. The relative amounts emitted in the United States in 2011, as well as their major sources, are illustrated in **Figure 19–9**. The general picture shows remarkable improvement for emissions over the past 40 years **(Fig. 19–10)**, reflecting the effectiveness of Clean Air Act regulations. This progressively improving trend over time has occurred in spite of large increases in the gross

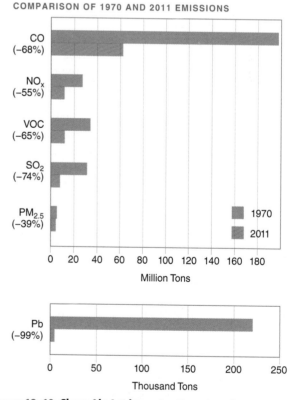

COMPARISON OF 1970 AND 2011 EMISSIONS

Figure 19–10 **Clean Air Act impacts.** Comparison of percent reductions in the emission of six criteria air pollutants between 1970 and 2011. PM$_{2.5}$ comparison values are from 1990, when data were first gathered, and 2011.

(*Source:* EPA National Emissions Inventory, 2012. http://www.epa.gov/ttn/chief/trends/index.html.)

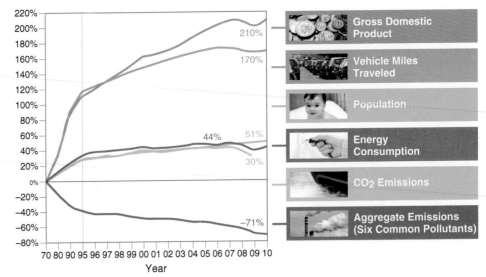

Figure 19–11 Comparison of growth versus emissions. Gross domestic product, vehicle miles traveled, energy consumption, CO$_2$ emissions, and population are compared with aggregate reductions in emissions of the six principal pollutants from 1970 to 2010.

(*Source:* EPA Air Trends, 2012. http://epa.gov/airtrends/aqtrends.html.)

domestic product, vehicular miles traveled, and population and energy consumption (**Fig. 19–11**).

Getting the Lead Out. The sixth type of primary pollutant—lead and other heavy metals—is discussed separately because the quantities emitted are far less than the levels for the first five. Before the EPA-directed phaseout in the 1980s and 1990s, lead was added to gasoline as an inexpensive way to prevent engine knock. Emitted with the exhaust from gasoline-burning vehicles, lead remained airborne and traveled great distances before settling as far away as the glaciers of Greenland. Since the phaseout, concentrations of lead in the air of cities in the United States have declined remarkably. However, lead levels are elevated around airports where piston-engine aircraft operate, the reason being that aviation gasoline still uses leaded gasoline (jet fuel has no added lead). Between 1980 and 2010, lead levels in air fell 89%. At the same time, levels of lead in children's blood have dropped 92% (some children still get lead from dust and paint chips). Lead concentrations in Greenland ice have also decreased significantly. All these declines indicate that lead restrictions in the United States and most other Northern Hemisphere nations have had a global impact. The data indicate that we have reached a steady state in lead emissions and that any new reductions must target leaded aviation gasoline, which accounts for half of the current lead air emissions. Lead smelters and battery manufacturers also add to the burden of lead.

Toxics and Radon. As with lead, the concentrations of toxic chemicals and radon in the air are small compared with those of the other primary pollutants. Some of the air's toxic compounds—benzene, for example—originate with transportation fuels. Most, however, are traceable to industries and small businesses. Radon, by contrast, is produced by the spontaneous decay of fissionable material in rocks and soils. Radon escapes naturally to the surface and seeps into buildings through cracks in foundations and basement floors, sometimes collecting in the structures.

Secondary Pollutants

Ozone and numerous reactive organic compounds are formed as a result of chemical reactions between nitrogen oxides and volatile organic carbons. Because sunlight provides the energy necessary to propel the reactions, these products are also known as **photochemical oxidants**. Note that these reactions and the ozone they produce are entirely in the troposphere; ozone in the stratosphere is "good" ozone, protecting life from damaging ultraviolet (UV) radiation. In preindustrial times, ozone concentrations ranged from 10 to 15 parts per billion (ppb), well below harmful levels. Summer concentrations of ozone in unpolluted air in North America now range from 20 to 50 ppb. Polluted air may contain ozone concentrations of 150 ppb or more, a level considered quite unhealthy if encountered for extended periods. EPA data indicate that ambient ozone levels, monitored across the United States at some 1,100 sites, declined by 20% between 1980 and 1992, leveled off during the 1990s, and then declined 11% in the first decade of the 2000s. Ozone air quality standards are the leaders in nonattainment across the country; 108 million people live in counties that still do not meet ozone standards.

Ozone Formation. A simplified view of the major reactions in the formation of ozone and other photochemical oxidants is shown in **Figure 19–12**. Nitrogen dioxide absorbs light energy and splits to form nitric oxide and atomic oxygen, which rapidly combines with oxygen gas to form ozone. If other factors are not involved, ozone and nitric oxide then react to form nitrogen dioxide and oxygen gas. A steady-state concentration of ozone results, and there is no appreciable accumulation of the gas (Fig. 19–12a). When VOCs are present, however, the nitric oxide reacts with them instead of with the ozone, causing several serious problems. First, the reaction between nitric oxide and the VOCs leads to highly reactive and damaging compounds known as peroxyacetyl nitrates, or

Major pollutants from vehicles

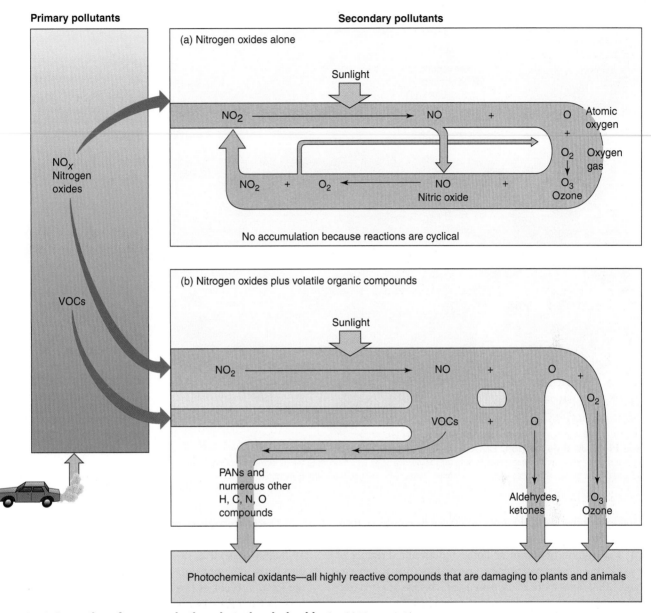

Figure 19–12 **Formation of ozone and other photochemical oxidants.** (a) Nitrogen oxides, by themselves, would not cause ozone and other oxidants to reach damaging levels, because reactions involving nitrogen oxides are cyclic. (b) When VOCs are also present, however, reactions occur that lead to the accumulation of numerous damaging compounds—most significantly, ozone, the most injurious.

PANs (Fig. 19–12b). Second, numerous aldehyde and ketone compounds are produced as the VOCs are oxidized by atomic oxygen, and these compounds are also noxious. Finally, with the nitric oxide tied up in this way, the ozone tends to accumulate. Because of the complex air chemistry involved, ozone concentrations may peak 30 to 100 miles (50 to 160 km) downwind of urban centers where the primary pollutants were generated. As a result, ozone concentrations above the air quality standards are sometimes found in rural and wilderness areas.

Sulfuric acid and nitric acid are considered secondary pollutants because they are products of sulfur dioxide and nitrogen oxides, respectively, reacting with atmospheric

moisture and oxidants such as hydroxyl. Sulfuric and nitric acids are the acids in acid rain (technically referred to as acid deposition). We now turn our attention to this important and widespread problem.

Acid Precipitation and Deposition

Acid precipitation refers to any precipitation—rain, fog, mist, or snow—that is more acidic than usual. Because dry acidic particles are also found in the atmosphere, the combination of precipitation and dry-particle fallout is called **acid deposition**. In the late 1960s, Swedish scientist Svante Odén first documented the acidification of lakes in Scandinavia and traced it to air pollutants

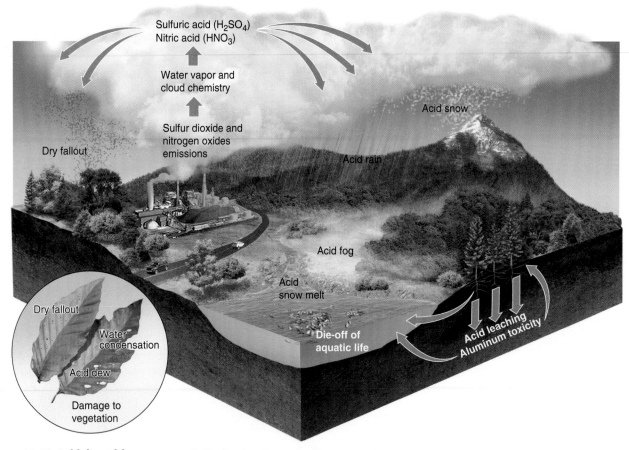

Figure 19–13 Acid deposition. Emissions of sulfur dioxide and nitrogen oxides react with hydroxyl radicals and water vapor in the atmosphere to form their respective acids, which return to the surface either as dry acid deposition or, mixed with water, as acid precipitation. Various effects of acid deposition are noted.

originating in other parts of Europe and Great Britain. Since then, careful monitoring has shown that broad areas of North America, as well as most of Europe and other industrialized regions of the world, have been experiencing precipitation that is between 10 and 1,000 times more acidic than usual. This has affected ecosystems in diverse ways, as illustrated in **Figure 19–13**.

To understand the full extent of the problem, we must first understand some principles about acids and how we measure their concentration.

Acids, Bases, and pH. Acidic properties (for example, a sour taste and corrosiveness) are due to the presence of hydrogen ions (H^+, hydrogen atoms without their electron), which are highly reactive. Therefore, an **acid** is any chemical that releases hydrogen ions when dissolved in water. The chemical formulas of a few common acids are shown in Table 19–2. Note that all of them ionize—that is, their components separate—to create hydrogen ions plus a negative ion. The higher the concentration of hydrogen ions in a solution, the more acidic the solution.

A **base** is any chemical that releases hydroxide ions (OH^-, oxygen-hydrogen groups with an extra electron) when dissolved in water. (See Table 19–2.) The bitter taste and caustic properties of alkaline, or basic, solutions are due to the presence of hydroxide ions.

The concentration of hydrogen ions is expressed as **pH**. The pH scale goes from 0 (highly acidic) through 7 (neutral) to 14 (highly basic) **(Fig. 19–14)**. As the pH scale numbers go from 0 to 7, the concentration of hydrogen ions (H^+) decreases, and the solution becomes less acidic and more neutral. As the numbers go from 7 to 14, the concentration of hydrogen ions gets lower, but the concentration of hydroxide ions (OH^-) increases, and the solution becomes more basic. The numbers on the scale (0–14) represent the negative logarithm (power of 10) of the hydrogen ion concentration, expressed in grams per liter (g/L). For example, a solution with a pH of 1 has an H^+ concentration of 10^{-1} g/L, a pH of 2 has a concentration of 10^{-2} g/L, and so on.

Because numbers on the pH scale represent powers of 10, there is a *tenfold difference* between each unit and the next. For example, pH 5 is 10 times as acidic (has 10 times as many H^+ ions) as pH 6, pH 4 is 10 times as acidic as pH 5, and so on.

Extent and Potency of Acid Precipitation. In the absence of any pollution, rainfall is normally somewhat acidic, with a pH of 5.6, because carbon dioxide in the air readily dissolves in, and combines with, water to produce carbonic acid. **Acid precipitation**, then, is any precipitation with a pH less than 5.5.

Since the middle of the 20th century, acid precipitation has been the norm over most of the Western world. The pH of rain and snowfall over a large portion of eastern North America used

TABLE 19–2 Common Acids and Bases

	Formula	Yields	H[1] Ion(s)	Plus	Negative Ion
Acid					
Hydrochloric acid	HCl	→	H^+	+	Cl^- Chloride
Sulfuric acid	H_2SO_4	→	$2H^+$	+	SO_4^{2-} Sulfate
Nitric acid	HNO_3	→	H^+	+	NO_3^- Nitrate
Phosphoric acid	H_3PO_4	→	$3H^+$	+	PO_4^{3-} Phosphate
Acetic acid	CH_3COOH	→	H^+	+	CH_3COO^- Acetate
Carbonic acid	H_2CO_3	→	H^+	+	HCO_3^- Bicarbonate

	Formula	Yields	OH[2] Ion(s)	Plus	Positive Ion
Base					
Sodium hydroxide	$NaOH$	→	OH^-	+	Na^+ Sodium
Potassium hydroxide	KOH	→	OH^-	+	K^+ Potassium
Calcium hydroxide	$Ca(OH)_2$	→	$2OH^-$	+	Ca^{2+} Calcium
Ammonium hydroxide	NH_4OH	→	OH^-	+	NH_4^+ Ammonium

to be quite acidic, reflecting the west-to-east movement of polluted air from the Midwest and industrial Canada. Many areas in this region once regularly received precipitation having a pH of 4.0 and, occasionally, as low as 3.0 (Fig. 19–15a). Acid precipitation was heavy in Europe in the latter half of the 20th century, from the British Isles to central Russia. Like North America, acid precipitation has declined greatly in these regions. It is on the rise in China and India as these countries increasingly exploit their coal resources. It has not disappeared in eastern North America, but pH values lower than 4.6 are now uncommon (Fig. 19–15b).

Sources of Acid Deposition. Chemical analysis of acid precipitation reveals the presence of two acids, sulfuric acid (H_2SO_4) and nitric acid (HNO_3). As we have seen, burning fuels produce sulfur dioxide and nitrogen oxides, so one source of the acid deposition problem is evident. These oxides enter the troposphere in large quantities from both anthropogenic and natural sources. Once in the troposphere, they are oxidized by hydroxyl radicals (Fig. 19–3) to sulfuric

and nitric acids, which dissolve readily in water or adsorb to particles and are brought down to Earth in acid deposition. This usually occurs within a week of the oxides entering the atmosphere. Another important source of acidification is traced to the use of nitrogen fertilizers, made by the Haber-Bosch process whereby atmospheric N_2 (a nonreactive form of nitrogen) is converted into reactive ammonia (NH_3) and incorporated into urea ((NH_2)$_2CO$). In turn, urea gives off NH_3 in the presence of water, which is volatile and enters the atmosphere. Whether in the atmosphere, soil, or natural waters, ammonia is eventually oxidized to nitrate (NO_3^-) by bacterial action, a process that generates acidity.

Natural sources contribute substantial quantities of pollutants to the air, including 50 million to 70 million tons per year of sulfur dioxide (from volcanoes, sea spray, and microbial processes) and 30 million to 40 million tons per year of nitrogen oxides (from lightning, the burning of biomass, and microbial processes). *Anthropogenic sources* per year are estimated at 115 million tons of sulfur dioxide, 60 million

Figure 19–14 **The pH scale.** Each unit on the scale represents a tenfold difference in hydrogen ion concentration.

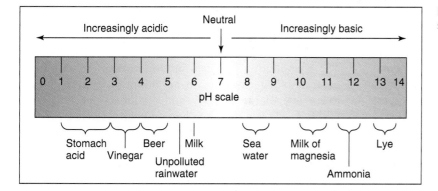

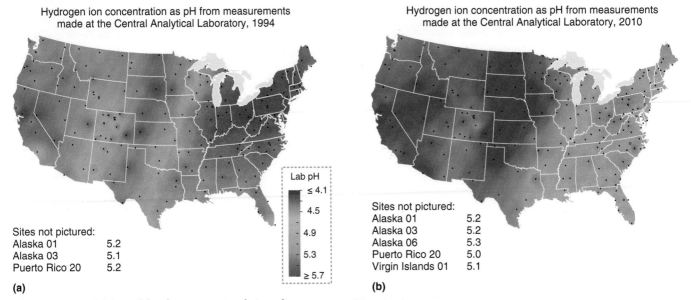

Hydrogen ion concentration as pH from measurements made at the Central Analytical Laboratory, 1994

Sites not pictured:
Alaska 01   5.2
Alaska 03   5.1
Puerto Rico 20   5.2

(a)

Lab pH
≤ 4.1
4.5
4.9
5.3
≥ 5.7

Hydrogen ion concentration as pH from measurements made at the Central Analytical Laboratory, 2010

Sites not pictured:
Alaska 01   5.2
Alaska 03   5.2
Alaska 06   5.3
Puerto Rico 20   5.0
Virgin Islands 01   5.1

(b)

Figure 19–15 Acid deposition in eastern North America. (a) 1994 rainfall pH distribution across the United States; dots show locations of measuring stations. (b) 2010 rainfall pH distribution.

(*Source:* National Atmospheric Deposition Program (NRSP-3). 2012. NADP Program Office, Illinois State Water Survey. 2204 Griffith Dr., Champaign, IL 61820).

to 70 million tons of nitrogen oxides, and 50 million tons of ammonia.[2] The vital difference between natural and anthropogenic sources is that anthropogenic sources are strongly concentrated in industrialized and agricultural regions, whereas the emissions from natural sources are spread out and are a part of the global environment. Levels of the anthropogenic oxides have increased sixfold[3] since 1900, while levels of the natural emissions have remained fairly constant.

As Figure 19–9 indicates, 8.06 million tons of sulfur dioxide were released into the air in 2011 in the United States; 92% was from fuel combustion (mostly from coal-burning power plants). Some 12.0 million tons of nitrogen oxides were released, 51% of which can be traced to transportation emissions and 41% to fuel combustion at fixed sites. In addition, 4.1 million tons of ammonia were released to the atmosphere in 2011 from fertilizer use and livestock. Nitrogen oxides and ammonia are doubly damaging because the nitrogen contributes to eutrophication in aquatic ecosystems (Chapter 20). In the eastern United States, the source of much of the acid deposition was identified as the tall stacks of 50 huge, older coal-burning power plants located primarily in the Midwest **(Fig. 19–16a)**. The tall stacks were built to alleviate local sulfur dioxide pollution at ground level. The good news is that many of these same plants are now reducing their emissions as a result of the Clean Air Act. Figure 19–16b shows the current geographic distribution of coal power plants with and without scrubbers (emissions control devices). Acid-generating emissions have decreased by 65% over the past 20 years, but there is obviously more that can be done. Of great significance, the

pattern of sources has shifted from predominantly sulfur dioxide to nitrogen oxides and ammonia.

19.3 Impacts of Air Pollutants

Air pollution is an alphabet soup of gases and particles, mixed with the normal constituents of air. The amount of each pollutant present varies greatly, depending on its proximity to the source and various conditions of wind and weather. As a result, we are exposed to a mixture that varies in makeup and concentration from day to day—even from hour to hour—and from place to place. Consequently, the effects we feel or observe are rarely, if ever, the effects of a single pollutant. Instead, they are the combined impact of the whole mixture of pollutants acting over our life span up to that point.

For example, plants may be so stressed by pollution that they become more vulnerable to other environmental factors, such as drought or attack by insects. Humans may develop lung disease due to the combined effects of ozone and NO_x. Given the complexity of this situation, it is often difficult to determine the role of any particular pollutant in causing an observed result. Nevertheless, some significant progress has been made in linking cause and effect.

Human Health

The air-pollution disasters in Donora and London demonstrated that exposure to air pollution can be deadly. Every one of the primary and secondary air pollutants (Table 19–1) is a threat to human health, particularly the health of the respiratory system **(Fig. 19–17)**. *Acute* exposure to some pollutants can be life threatening, but many effects are *chronic*, acting over a period of years to cause a gradual deterioration of physiological functions and

[2]Karen C. Rice and Janet S. Herman. "Acidification of Earth: An Assessment Across Mechanisms and Scales." *Applied Geochemistry* 27: 1–14. 2012.

[3]S. J. Smith et al. "Anthropogenic Sulfur Dioxide Emissions: 1850–2005." *Atmospheric Chemistry and Physics* 11: 1101–1116. 2011.

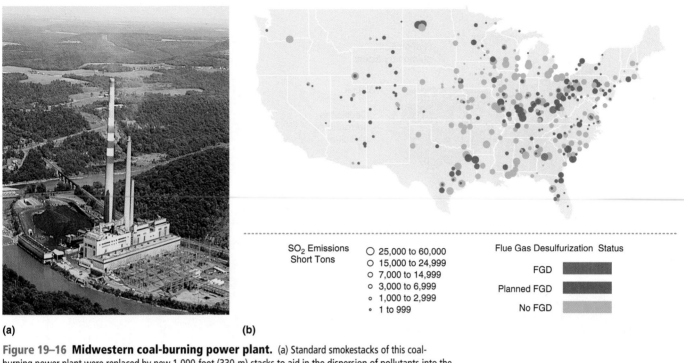

(a) **(b)**

SO$_2$ Emissions
Short Tons

○ 25,000 to 60,000
○ 15,000 to 24,999
○ 7,000 to 14,999
○ 3,000 to 6,999
○ 1,000 to 2,999
○ 1 to 999

Flue Gas Desulfurization Status

FGD
Planned FGD
No FGD

Figure 19–16 Midwestern coal-burning power plant. (a) Standard smokestacks of this coal-burning power plant were replaced by new 1,000-foot (330-m) stacks to aid in the dispersion of pollutants into the atmosphere. The taller stacks alleviated local air-pollution problems, but created a more widespread distribution of acid-generating pollutants. (b) Distribution of coal-burning power plants and their SO$_2$ emissions (2010); green indicates plants with flue gas scrubbers, blue shows plants with planned scrubbers, and gold shows plants without scrubbers to remove emissions.

(*Source:* U.S. Energy Information Administration, *Today in Energy*, Dec. 21, 2011.)

eventual premature mortality. Moreover, some pollutants are *carcinogenic*, adding significantly to the risk of lung cancer as they are breathed into the lungs.

Chronic Effects. Many people living in areas of urban air pollution suffer from chronic effects. Long-term exposure to *sulfur dioxide* can lead to bronchitis (inflammation of the bronchi). Chronic inhalation of *ozone* can cause inflammation and, ultimately, fibrosis of the lungs, a scarring that permanently impairs lung function. *Carbon monoxide* reduces the capacity of the blood to carry oxygen, and extended exposure to carbon monoxide can contribute to heart disease. Chronic exposure to *nitrogen oxides* impairs lung function and is known to affect the immune system, leaving the lungs open to attack by bacteria and viruses. Exposure to airborne *particulate matter* can bring on a broad range of health problems, including respiratory and cardiovascular pathology. Other factors—such as poor diet, lack of exercise, preexisting diseases, and individual genetic makeup—may add to the effects of pollution to bring on adverse health outcomes.

COPD. Any one or combination of these exposures can develop into **chronic obstructive pulmonary disease (COPD)**, a slowly progressive lung disease that makes it increasingly hard to breathe. COPD is estimated to affect 15 million people in the United States and is the fourth leading cause of death (128,000 deaths/year). Worldwide, COPD afflicts up to 10% of adults ages 40 and older.

The leading cause of this disease is smoking, but in many developing countries, the burning of wood and dung in indoor stoves for heating and cooking delivers much the same burden of pollutants as smoking to the women who tend the stoves and the children who are with them. COPD involves three separate disease processes: emphysema (destruction of the lung alveoli—see Fig. 19–17), bronchitis (inflammation and obstruction of the airways), and asthma.

Asthma. Those most sensitive to air pollution are small children, asthmatics, people with chronic pulmonary or heart disease, and the elderly. Asthma is an immune-system disorder characterized by impaired breathing caused by constricted air passageways. Asthma episodes are triggered by contact with allergens (dust mites, molds, pet dander) and compounds in polluted air (ozone, particulate matter, SO$_2$). According to the National Center for Health Statistics, almost 456,000 hospitalizations for asthma occur yearly, together with 1.75 million visits to hospital emergency departments. The highest prevalence is among children from birth to 4 years of age. In the past decade, the incidence of asthma in the United States doubled, affecting 25 million people (including 7 million children).

Strong Evidence. Two key studies recently analyzed by the Health Effects Institute provide strong evidence of the harmful effects of fine particles and sulfur pollution. These studies followed thousands of adult subjects living in 154

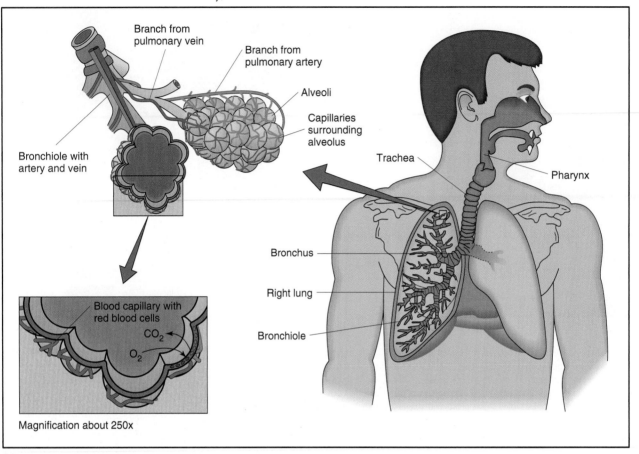

(a)

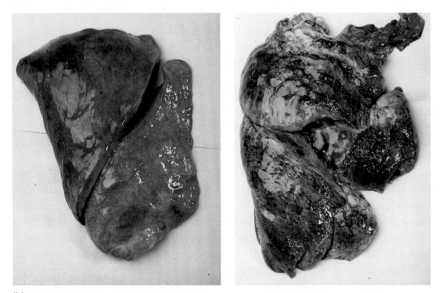

(b)

Figure 19–17 The respiratory system. (a) In the lungs, air passages branch and rebranch and finally end in millions of tiny sacs called alveoli. These sacs are surrounded by capillaries. As blood passes through the capillaries, oxygen from inhaled air diffuses into the blood from the alveoli. Carbon dioxide diffuses in the reverse direction and leaves the body in the exhaled air. (b) On the left is normal lung tissue, and on the right is lung tissue from a person who suffered from emphysema, a chronic lung disease in which some of the structure of the lungs has broken down. Cigarette smoking and heavy air pollution are associated with the development of emphysema and other chronic lung diseases.

U.S. cities for as long as 16 years. In the studies, higher concentrations of fine particles were correlated with increased mortality, especially from cardiopulmonary disease and lung cancer. The more polluted the city, the higher the mortality. The studies were used by the EPA to step up its regulatory attention to fine particles.

Fine particles were also cited as a key to heightened costs of air pollution in Southern California, along with ozone. Both pollutants exceed federal standards and were judged to cause more deaths in the region than did traffic accidents. Meeting those standards would save $28 billion annually in avoided health care costs, missed work, school absences, and premature deaths, according to a research study from California State University, Fullerton, in 2008. Health impacts of the excess pollutants include asthma, chronic bronchitis, and cardiovascular problems.

Mercury. One of the heavy-metal pollutants, mercury, deserves special attention. It is a toxic element, one that binds to key proteins in the body and can lead to many neurological disorders. In fetuses and small children, mercury impairs brain development. Power plants are the major source of mercury reaching the environment, and once present in aquatic systems, mercury is converted to a highly toxic form, methylmercury, which accumulates in fish as a consequence of bioaccumulation (Chapter 13). In recent tests, the EPA reported methylmercury concentrations that exceeded recommendations in the game fish of 49% of lakes and reservoirs. Every state has advisories warning consumers about the dangers of mercury in locally caught fish. As we will see, the EPA has recently stepped up the regulatory restrictions on mercury releases from power plants.

Acute Effects. In severe cases, air pollution reaches levels that cause death, though such deaths usually occur among people already suffering from critical respiratory or heart disease or both. The gases present in air pollution are known to be lethal in high concentrations, but such concentrations occur nowadays only in cases of accidental poisoning. Nevertheless, intense air pollution puts an additional stress on the body, and if a person is already in a weakened condition (as are, for example, the elderly or asthmatics), this additional stress may be fatal. Most of the deaths in air-pollution disasters reflect the acute effects of air pollution. However, recent research shows that even moderate air pollution can cause changes in cardiac rhythms of people with heart disease, triggering fatal heart attacks.

Carcinogenic Effects. The heavy-metal and organic constituents of air pollution include many chemicals known to be carcinogenic in high doses. According to the industrial reporting required by the EPA, 4.7 million tons of hazardous air pollutants (the air toxics—see Table 19–1) are released annually into the air in the United States. The presence of trace amounts of these chemicals in the air may be responsible for a significant portion of the cancer observed in humans. One major source of such carcinogens is diesel exhaust (Fig. 19–18). The EPA has classified diesel exhaust as a likely human carcinogen and has established the National Clean Diesel Campaign to reduce emissions from the current diesel truck fleet and establish standards for new diesel-powered engines.

In some cases, exposure to a pollutant can be linked directly to cancer and other health problems by way of epidemiological evidence. One pollutant that clearly and indisputably is correlated with cancer and other disorders is **benzene**. This organic chemical is present in motor fuels and is also used as a solvent for fatty substances and in the manufacture of detergents, explosives, and pharmaceuticals. Environmentally, benzene is found in the emissions from fossil fuel combustion that result from operating motor vehicles and burning coal and oil. Benzene

Figure 19–18 Diesel truck exhaust. Diesel truck exhaust is a major source of hazardous air pollutants as well as particulates, and the exhaust from diesel engines has been classified as a probable human carcinogen.

is also present in tobacco smoke, which accounts for half of the public's exposure to the chemical. The EPA has classified benzene as a known human carcinogen, linked to leukemia in persons encountering the chemical through occupational exposure. Chronic exposure to benzene can also lead to numerous blood disorders and damage to the immune system.

The Environment

Experiments show that plants are even more sensitive to air pollutants than are humans. Before emissions were controlled, it was common to see wide areas of totally barren land or severely damaged vegetation downwind from smelters and coal-burning power plants. The pollutant responsible was usually sulfur dioxide.

Crop Damage. Nowadays, damage to crops, orchards, and forests downwind of urban centers is caused mainly by exposure to ozone. The ozone gains access to plants through their stomata, the respiratory pores in all plant leaves. Symptoms of ozone damage include black flecking

Figure 19–19 Ambient ozone injury. This potato leaf shows the black flecking symptomatic of ozone damage.

(*Source:* U.S. Department of Agriculture Agricultural Research Service. Effects of Ozone Air Pollution on Plants. 2009. www.ars.usda.gov/Main/docs.htm?docid=12462.)

and yellowing of leaves (**Fig. 19–19**). Crops vary in their susceptibility to ozone, but damage to many important crops (such as soybeans, corn, and wheat) is observed at common ambient levels of ozone (**Fig. 19–20**). Unfortunately, ozone levels are highest in summer, when crops are growing, and rural areas often experience high ozone levels as ozone forms downwind of pollution sources. The $27 billion U.S. soybean crop suffers a loss of almost $2 billion annually due to ozone. Much of the world's grain production occurs in regions that receive enough ozone pollution to reduce crop yields. India loses $5 billion yearly to ozone crop damage, China loses $2.5 billion, and farmers in Europe lose more than $7.5 billion yearly to ozone.

Forest Damage. The negative impact of air pollution on wild plants and forest trees may be even greater than on agricultural crops.

Ozone Damage. Studies have shown that ozone damage to forest trees begins at about 40 ppb and gets more intense with higher levels. In California and southern Appalachian forests, ozone levels often reached 60 ppb and above, and seasonal losses in tree stem growth in the latter were measured at 30–50% for most species. Studies in European forests found that ozone stress affects photosynthetic carbon uptake, increases respiration of foliage, amplifies water loss, and generally disrupts carbon and nutrient metabolism. In sum, research has suggested that as temperatures climb and moisture is less predictable under conditions of expected global climate change, ozone levels will have an increasingly serious impact on forest growth.

Damage from Acid Deposition. From the Green Mountains of Vermont to the San Bernardino Mountains of California, the die-off of forest trees in the 1980s caused great concern. Red spruce forests were especially vulnerable. In New England, 1.3 million acres of high-elevation forests were devastated. Commonly, the damaged trees lost needles as acidic water drew calcium from them, rendering them more susceptible to winter freezing. Sugar maples, important forest trees in the Northeast, have shown extensive mortality, ranging from 20–80% of all trees in some forests.

Much of the damage to forests from acid precipitation is due to chemical interactions within the forest soils. Sustained acid precipitation at first adds nitrogen and sulfur to the soils, which stimulate tree growth. In time, though, these chemicals leach out large quantities of the buffering chemicals (usually calcium and magnesium salts). When these buffering salts no longer neutralize the acid rain, aluminum ions, which are toxic, are dissolved from minerals in the soil. The combination of aluminum and the increasing scarcity of calcium, which is essential to plant growth, leads to reduced tree growth. Research at the Hubbard Brook Experimental Forest in the White Mountains

Figure 19–20 Ozone impact on crop yields. Some crop species react more strongly to ozone than others.

(*Source:* U.S. Department of Agriculture Agricultural Research Service: Effects of Ozone Air Pollution on Plants. 2009. www.ars.usda.gov/Main/docs.htm?docid=12462.)

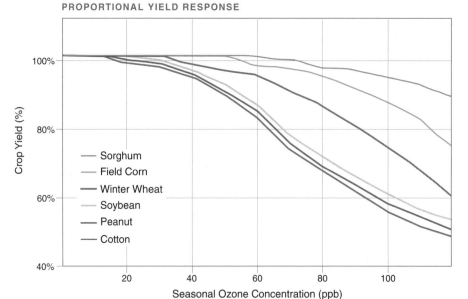

of New Hampshire has shown a marked reduction in calcium and magnesium in the forest soils from the 1960s on, which is reflected in the amount of calcium in tree rings over the same period. The net result of these changes has been a decline in forest growth. Experimental liming of soils in forests showing maple tree dieback was able to restore the trees to health, and aluminum ion concentrations were seen to decrease greatly. Although acid deposition has declined since the 1980s, few studies have actually tracked the recovery of terrestrial ecosystems.

Impacts on Aquatic Ecosystems. Some 50 years ago, anglers started noticing sharp declines in fish populations in many lakes in Sweden, Ontario, and the Adirondack Mountains of upper New York State. Since that time, as ecological damage continued to spread, studies have revealed many ways in which air pollution—especially acid deposition—alters and may destroy aquatic ecosystems.

The pH of an environment is extremely critical because it affects the function of virtually all enzymes, hormones, and other proteins in the bodies of all organisms living in that environment. Ordinarily, organisms are able to regulate their internal pH within the narrow limits necessary to function properly. A consistently low environmental pH, however, often overwhelms the regulatory mechanisms in many life forms, thus weakening or killing them. Most freshwater lakes, ponds, and streams have a natural pH in the range from 6 to 8, and organisms have adapted accordingly. The eggs, sperm, and developing young of these organisms are especially sensitive to changes in pH. Most are severely stressed, and many die, if the environmental pH shifts as little as one unit from the optimum.

As aquatic ecosystems become acidified, a shift occurs from acid-sensitive to acid-tolerant species. The acid-sensitive species die off, either because the acidified water kills them or because it keeps them from reproducing. Figure 19–13 shows that acid precipitation may leach aluminum and various heavy metals from the soil as the water percolates through it. Normally, the presence of these elements in the soil does not pose a problem because they are bound in insoluble mineral compounds and, therefore, are not absorbed by organisms. As the compounds are dissolved by low-pH water, however, the metals are freed. They may then be absorbed by and are highly toxic to both plants and animals. For example, mercury tends to accumulate in fish as lake waters become more acidic.

Acid deposition hit some regions especially hard in the latter part of the 20th century. In Ontario, Canada, approximately 1,200 lakes lost their fish life. In the Adirondacks, a favorite recreational region for New Yorkers, 346 lakes were without fish in 1990. In New England and the eastern Catskills, more than 1,000 lakes have suffered from recent acidification. The physical appearance of highly acidified lakes is deceiving. From the surface, they are clear and blue, the outward signs of a healthy condition. However, the only life found below the surface is acid-loving mosses growing on the bottom.

Neutralizing Capacity. As Figure 19–15a indicates, wide regions of eastern North America receive substantial amounts of acid precipitation, yet not all areas have acidified lakes.

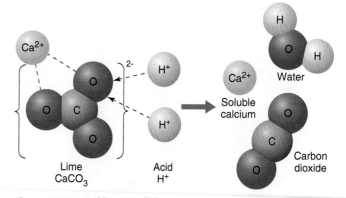

Figure 19–21 Acid neutralizing. Acids may be neutralized by certain nonbasic compounds. A chemical such as limestone (calcium carbonate) reacts with hydrogen ions as shown. Hence, the pH of a lake or river remains close to neutral despite the additional acid, when it contains adequate acid neutralizing capacity.

Apparently, many areas remain healthy, whereas others have become acidified to the point of becoming lifeless. How is this possible? The key lies in the system's **acid neutralizing capacity (ANC)**. Despite the addition of acid, a system may be protected from changes in pH by a neutralizer—a buffer that, when present in a solution, has a large capacity to absorb hydrogen ions and thus maintain the pH at a relatively constant value.

Limestone ($CaCO_3$) is a natural acid neutralizer **(Fig. 19–21)** that protects lakes from the effects of acid precipitation in many areas of the North American continent, particularly in the Midwest. Lakes and streams receiving their water from rain and melted snow that have percolated through soils derived from limestone will contain dissolved limestone. The regions that are sensitive to acid precipitation are those containing much granitic rock, which does not yield good ANC. The good news is that the sensitive regions (New England, the Adirondacks, the Northern Appalachian Plateau) are recovering their ANC, and the percentage of acidic lakes has declined in the past 18 years.[4] The decline is consistent with the decrease in acid deposition in these regions, which has been brought about by the reductions of SO_2 and NO_x emissions mandated by the Clean Air Act.

Visibility. A clear blue sky and good visibility are matters of health, but they also carry significant aesthetic value and can have a deep psychological impact on people. Can a value be put on these benefits? Many of us spend thousands of dollars and hundreds of hours commuting long distances to work so that we can live in a less polluted environment than the one in which we work. Ironically, the resulting traffic and congestion cause much of the very pollution we are trying to escape. Then we travel to national parks and wilderness areas and too often find that the visibility in *these* natural areas is impaired by what has been called "regional haze," coming from particulates and gases originating hundreds of miles away. In an encouraging move, the EPA established a

[4]Executive Office of the President. National Acid Precipitation Assessment Program Report to Congress 2011: An Integrated Assessment. USGS, 2011. http://ny.water.usgs.gov.

Figure 19–22 Effect of pollution on monuments. The corrosive effects of acids from air pollutants are dissolving away the features of many monuments and statues, as seen in the angel perched on this building buttress.

Regional Haze Rule in 1999, aimed at improving the visibility at 156 national parks and wilderness areas. The regulations call on all 50 states to establish goals for improving visibility and to develop long-term strategies for reducing the emissions (especially particles) that cause the problem.

Effects on Materials and Aesthetics. Walls, windows, and other exposed surfaces turn gray and dingy as particulates settle on them. Paints and fabrics deteriorate more rapidly, and the sidewalls of tires and other rubber products become hard and checkered with cracks because of oxidation by ozone. Metal corrosion is increased dramatically by sulfur dioxide and acids derived from sulfur and nitrogen oxides, as are weathering and the deterioration of stonework.

Limestone and marble (which is a form of limestone) are favored materials for the outsides of buildings and for monuments (collectively called artifacts). The reaction between acid deposition and limestone is causing these structures to erode at a tremendously accelerated pace. Monuments and buildings that have stood for hundreds or even thousands of years with little change are now dissolving and crumbling away (Fig. 19–22). The corrosion of buildings, monuments, and outdoor equipment by acid precipitation costs billions of dollars for replacement and repair each year in the United States. In an extreme case of corrosion, a 37-foot bronze Buddha in Kamakura, Japan, is slowly dissolving away as precipitation from Korea and China bathes the statue in highly acidic rain.

Although the decay of artifacts is a tragic loss in itself, it should also stand as a grim reminder of how we are dissolving away the buffering capacity of ecosystems. As the saying goes, "What goes up must come down." The sulfur and nitrogen oxides pumped into the troposphere in the United States gradually increased from some 11.4 million tons in 1900 to their peak of 53 million tons in 1973. Although this number has been reduced by almost two-thirds, the pollutants still come back down as acid deposition, generally to the east of their origin because of the way weather systems flow. And the soils retain much of the sulfate deposition they have received. The deposits cross national boundaries, too. As a result, Canada receives half of its acid deposition from the United States, and Scandinavia gets it mostly from Great Britain and other western European nations. Japan receives the wind-borne pollution from widespread coal burning in Korea and China. For good reason, the problem of acid deposition has been addressed at national and international levels. Accordingly, we turn next to the efforts made to curb air pollution.

19.4 Bringing Air Pollution Under Control

By the 1960s, it was obvious that pollutants produced by humans were overloading natural cleansing processes in the atmosphere. The unrestricted discharge of pollutants into the atmosphere could no longer be tolerated.

Clean Air Act

Federal legislation regarding air pollution was first introduced in 1955 as the **Air Pollution Control Act.** Several minor amendments followed until, under grassroots pressure from citizens, the U.S. Congress enacted a major bill, the **Clean Air Act Amendments of 1970 (CAA).** Together with amendments passed in 1977 and 1990, this law, administered by the EPA, represents the foundation of U.S. air-pollution control efforts. The act called for identifying the most widespread pollutants, setting **ambient standards** (levels that need to be achieved to protect environmental and human health), and establishing control methods and timetables to meet standards.

NAAQS. The CAA mandated the setting of standards for four of the primary pollutants—particulates, sulfur dioxide, carbon monoxide, and nitrogen oxides—and for the secondary pollutant ozone. At the time, these five pollutants were recognized as the most widespread and objectionable ones. Today, with the addition of lead, they are known as the **criteria pollutants** and are covered by the **National Ambient Air Quality Standards (NAAQS)** (Table 19–3). The primary standard for each pollutant is based on the presumed highest level that can be tolerated by humans without noticeable ill effects, minus a 10–50% margin of safety. For some of the pollutants, long-term and short-term levels are set. The short-term levels are designed to protect against acute effects, while the long-term standards are designed to protect against chronic effects and damage to crops, animals, vegetation, and buildings. The EPA is required to review the pollutants every five years and make adjustments if the science justifies changes. The NAAQS have far-reaching impacts on many regulations, and proposed changes always generate controversies.

TABLE 19–3 National Ambient Air Quality Standards for Criteria Pollutants (NAAQS)

Pollutant	Averaging Time*	Primary Standard
PM_{10} particulates	24 hours	150 µg/m³
$PM_{2.5}$ particulates**	1 year	15 µg/m³***
	24 hours	35 µg/m³
Sulfur dioxide	1 hour	0.075 ppm
	3 hours	0.5 ppm
Carbon monoxide	8 hours	9 ppm
	1 hour	35 ppm
Nitrogen dioxide	Annual mean	0.053 ppm
	1 hour	0.100 ppm
Ozone	8 hours	0.075 ppm
Lead	3 months	0.15 µg/m³

Source: U.S. EPA, Office of Air Quality Planning and Standards, *National Ambient Air Quality Standards.* October 2011. www.epa.gov/air/criteria. html.

*The averaging time is the period over which concentrations are measured and averaged.
**$PM_{2.5}$ is the particulate fraction having a diameter smaller than or equal to 2.5 micrometers.
***New proposed standard is 12 to 13 µg/m³.

NESHAPs. In addition, the CAA required the EPA to establish **National Emission Standards for Hazardous Air Pollutants (NESHAPs)** for a number of toxic substances (the air toxics). The Clean Air Act Amendments of 1990 greatly extended this section of the EPA's regulatory work by specifically naming 187 toxic air pollutants for the agency to track and regulate. The EPA tracks both the criteria pollutants and the hazardous air pollutants in a database, the National Emissions Inventory. This database serves as an essential source of information for the EPA in carrying out its regulatory responsibilities.

Control Strategy. The basic strategy of the 1970 Clean Air Act was to regulate the emissions of air pollutants so that the *ambient* criteria pollutants would remain below the primary standard levels. This approach is called **command-and-control** because regulations were enacted requiring industry to achieve a set limit on each pollutant through the use of specific control equipment. The assumption was that human and environmental health could be significantly improved by a reduction in the output of pollutants. If a particular region was in violation for a given pollutant, a local government agency would track down the source(s) and order reductions in emissions until the region came into compliance.

Unfortunately, this strategy proved difficult to implement. Most of the regulatory responsibility fell on the states and cities, which were often unable or unwilling to enforce control.

In spite of these difficulties, however, total air pollutants were reduced by some 25% during a time when both population and economic activity increased substantially, and even in the noncompliant regions, the severity of air-pollution episodes lessened. Accordingly, some significant changes were made in the law and in the regulations that followed.

1990 Amendments. The Clean Air Act Amendments of 1990 (CAAA) targeted specific pollutants more directly and enforced compliance more aggressively, through such means as the imposition of sanctions. As with the earlier law, the states do much of the work in carrying out the mandates of the 1990 act. Each state must develop a State Implementation Plan (SIP) that is designed to reduce emissions of every NAAQS pollutant whose control standard (Table 19–3) has not been attained. One major change was the addition of a permit application process (already in place for the release of pollutants into waterways). Polluters must apply for a permit that identifies the kinds of pollutants they release, the quantities of those pollutants, and the steps they are taking to reduce pollution. Permit fees provide funds the states can use to support their air-pollution control activities.

Under the CAAA, regions of the United States that have failed to attain the required levels must submit *attainment plans* based on **reasonably available control technology (RACT)** measures. Offending regions must convince the EPA that the standards will be reached within a certain time frame. The state SIPs and attainment performances are enforced by EPA sanctions that impose costly withholding of federal funds in case of failures.

Dealing with the Pollutants

The Clean Air Act and its amendments give the EPA broad authority to write regulations for reducing air pollutants, using a process known as rulemaking. The process involves a proposed rule stage and a final rule stage. EPA will give notice first of a proposed rule, which allows a period of one to three months for public review and comments. Next, the agency is required to consider the comments received, and when the final rule is published, responses to the comments are usually discussed, and any changes made on the basis of the comments are noted. Then the final rule is published in the Federal Register (available on the Internet), the official daily publication for a host of federal notices, regulations, executive orders, and other presidential documents. Once published, a final rule has the force of law until it is amended or repealed. We will consider the air pollution regulations by the different categories of pollutants.

Reducing Particulates. Prior to the 1970s, the major sources of particulates were industrial stacks and the open burning of refuse. The CAA mandated the phaseout of open burning of refuse and required that particulates from industrial stacks be reduced to "no visible emissions." To reduce stack emissions, industries were required to install filters, electrostatic precipitators, and other devices (though many facilities in place before the 1970 CAA were exempted). These measures markedly reduced the levels of particulates since the 1970s, but particulates continued to be released from steel mills, power plants, cement plants, smelters, construction sites, and diesel engines. Wood-burning stoves and wood and grass fires also contribute to the particulate load, making regulation even more difficult. Before 1987, the NAAQS particulate standard was "total suspended particulates," and this was replaced in 1987 by the PM_{10} standard (all particles 10 micrometers or less in diameter), based on the understanding at the time of particles that could reach the interior regions of the respiratory tract.

The EPA added a new ambient air quality standard for particulates—$PM_{2.5}$—in 1997, on the basis of information indicating that smaller particulate matter (less than 2.5 micrometers in diameter) has the greatest effect on health. The coarser material (PM_{10}) is still regulated but is not considered to play as significant a role in health problems as do the finer particles. There is overwhelming evidence that the finer particles (2.5 microns or less) go right into the lungs when breathed. Responding to this evidence, the EPA announced a revision to the NAAQS in 2006, reducing the 24-hour primary standard for $PM_{2.5}$ from 65 $\mu g/m^3$ to 35 $\mu g/m^3$. The EPA left the annual standard at 15 $\mu g/m^3$, in spite of recommendations from the EPA's Clean Air Scientific Advisory Committee that it be lowered. More recently (June, 2012), the EPA proposed strengthening the annual standard by lowering it to a range of 12 to 13 $\mu g/m^3$. This proposal came as a response to legal action by the American Lung Association. Industry groups maintain that the current standard is protective enough, but the EPA defended the change by showing benefits far outweighing the costs of implementation.

Controlling Air Toxics. By EPA estimates, the amount of air toxics emitted annually into the air in the United States is around 4.7 million tons. These come from stationary sources (factories, refineries, power plants), mobile sources (cars, trucks, buses), and indoor sources (solvents, building materials). Cancer and other adverse health effects, environmental contamination, and catastrophic chemical accidents are the major concerns associated with this category of pollutants. Under the CAAA, Congress identified 187 toxic pollutants. It then directed the EPA to identify major sources of these pollutants and to develop **maximum achievable control technology (MACT)** standards, the NESHAPs. Besides affecting control technologies, the standards include options for substituting nontoxic chemicals, giving industry some flexibility in meeting MACT goals. State and local air-pollution authorities are responsible for seeing that industrial plants achieve the goals. In response to the CAAA, the EPA has mobilized more than 300 monitoring sites to track emissions of the affected substances **(Fig. 19–23)**. To reduce the contribution from vehicles, the agency requires cleaner burning fuels (reformulated gasoline) in urban areas. Industrial sources are being addressed by setting emission standards for some 82 stationary-source categories (e.g., paper mills, oil refineries).

After years of piecemeal and delayed regulations, the EPA published final rules in December 2011 that for the first time will force the older coal- and oil-fired power plants to control their pollutants. The major focus of these new standards is mercury, a contaminant in coal and oil that is released to the air when these fuels are burned. The standards—the **Mercury and Air Toxics Standards**—require all coal- and oil-fired power plants to limit their emissions of mercury, acid gases, and other metallic toxic pollutants, preventing 90% of the mercury and 88% of acid gases from being emitted. Because of the controls needed to accomplish these reductions, emissions of SO_2 and fine particles will also be greatly reduced. The human health value of these new controls is estimated at $37 billion to $90 billion and will prevent 11,000 premature deaths per year; the estimated costs to industry to effect these controls is $9.6 billion per year. The industry has up to four years to comply with these new standards. Industry lobbies claimed that this and other new rule changes amount to a "train wreck" for the electric power industry, predicting a wave of coal-plant closures and major increases in electricity costs. A recent Congressional Research Service report[5] found these claims to be wildly overstated. The study concluded that there will be closures of old coal-fired power plants, but these are inefficient and dirty and are largely being replaced by newer natural gas–fired plants.

Limiting Pollutants from Motor Vehicles. Cars, trucks, and buses release nearly half of the pollutants that foul our air. Vehicle exhaust sends out VOCs, carbon monoxide, and nitrogen oxides that lead to ground-level ozone and PANs. Additional VOCs come from the evaporation of gasoline and oil vapors from fuel tanks and engine systems. The 1970 CAA mandated a 90% reduction in these emissions by 1975. This timing proved to be unrealistic, but enough improvements have been made over the years that today's new cars have actually achieved a 90% emissions reduction. This improvement is fortunate because driving in the United States has been increasing much more rapidly than the population has. Between

Figure 19–23 Air toxics monitoring station. A radiation monitoring device has been added to this air toxics monitoring station located on the grounds of a school in Anaheim, California.

[5]James E. McCarthy and Claudia Copeland. "EPA's Regulation of Coal-Fired Power: Is a 'Train Wreck' Coming?" Congressional Research Service R41914. August 8, 2011. www.crs.gov.

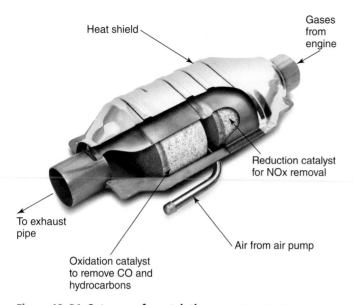

Figure 19–24 **Cutaway of a catalytic converter.** Engine gas is routed through the converter, where catalysts promote chemical reactions that change the harmful hydrocarbons, carbon monoxide, and nitrogen oxides into less harmful gases, such as carbon dioxide, water, and nitrogen.

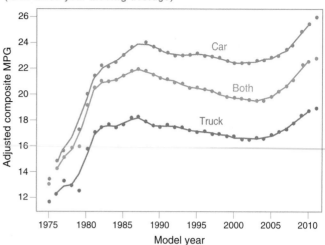

ADJUSTED FUEL ECONOMY BY MODEL YEAR
(with three-year moving average)

Figure 19–25 **Fuel economy and light-duty vehicles.** Changes in the fuel economy in model years from 1975 to 2011. "Truck" refers to pickups, family vans, and SUVs.

(*Source:* EPA, Office of Transportation and Air Quality. Light-Duty Automotive Technology, Carbon Dioxide Emissions, and Fuel Economy Trends: 1975 Through 2011. EPA-420-R-12-001a, March 2012. April 30, 2012.)

1970 and 2011, the number of vehicle miles increased from 1 trillion to 3 trillion miles per year, and between 1980 and 2010, the number of registered vehicles increased more than 56%. It is hard to imagine what the air would be like without the improvements mandated by the CAA.

Emissions. The reductions in automobile emissions have been achieved with a general reduction in the size of passenger vehicles, along with a considerable array of pollution-control devices, among which is one that affords the computerized control of fuel mixture and ignition timing, allowing more complete combustion of fuel and decreasing VOC emissions. To this day, however, the most significant control device on cars is the **catalytic converter (Fig. 19–24)**. As exhaust passes through this device, two different catalysts are at work: a reduction catalyst that converts NO_x emissions to harmless nitrogen gas and an oxidation catalyst that oxidizes unburned hydrocarbons and carbon monoxide to carbon dioxide.

Emissions standards set in 1968 were subsequently tightened because of continuing failure to meet air quality standards in many regions of the United States. The latest of these were the Tier 2 standards, phased in over the 2004 to 2009 model years. Emissions standards require attainment of specific levels or below of volatile organic carbons, carbon monoxide, nitrogen oxides, formaldehyde, and particulate matter. All vehicles used for personal transportation, up to larger SUVs and passenger vans, are subject to the Tier 2 standards. These are the emissions that are tested by state-certified inspection stations, now required in most states.

Efficiency. Although the Clean Air Act does not address fuel efficiency, it is obvious that less fuel burned means fewer pollutants emitted. Two negative factors have recently affected fuel efficiency and consumption rates. First, the elimination of federal speed limits in 1996 reduced fuel efficiencies because of the higher speeds. Second, the sale of

sport-utility vehicles (SUVs), vans, and pickup trucks surged from 25% of the combined light-duty vehicle sales in 1990 to 50% in 2004. Automobile sales finally passed those of heavier vehicles in 2008, a response to both high fuel prices and the economic recession. The EPA's "adjusted" fuel economy estimates for model year 2011 are as follows (these are real-world estimates, based on five different driving conditions): automobiles 25.9 mpg and "trucks" (includes minivans, SUVs, and pickups) 18.9 mpg, giving a fleet average of 22.8 mpg. The efficiencies have risen above earlier peak values (achieved in 1987) since model year 2009 and are getting better every year **(Fig. 19–25)**.

CAFE Standards. Under the authority of the Energy Policy and Conservation Act of 1975 and its amendments, the National Highway Traffic Safety Administration was given the authority to set *corporate average fuel economy (CAFE)* standards for motor vehicles. The intention of the law was to conserve oil and promote energy security. The standard set in 1984 for passenger cars required a fleet average of 27.5 mpg. The standard for light trucks (pickups, SUVs, minivans) was set at 20.7 mpg and eventually raised to 22.2 mpg. Note that the EPA's "adjusted" vehicle performances are lower than CAFE standards by 25% due to the different measurement methods used; the "unadjusted" EPA measurements of fuel economy are the basis for manufacturers' compliance with the CAFE standards.

The **Energy Independence and Security Act of 2007** mandated a new CAFE fuel economy standard of 35 mpg for the combined fleet of cars and light trucks, to be achieved by 2020. Congress passed this legislation as a major step toward a new energy future of reduced reliance on foreign oil and increased renewable energy, with a view toward reducing CO_2 greenhouse gas emissions (Chapter 18). In 2009, the Obama EPA set more stringent fuel-economy standards, calling for an annual fuel economy increase of 5%, beginning with the 2011 model year. By 2016, the CAFE fleet standard would reach 35.5 mpg. For the first time, regulations also were set for greenhouse gas emissions, to be lowered to 250 grams of CO_2 per mile by 2016. Then, working with automakers, unions, and environmental groups, the Obama administration announced a new round of CAFE standards: the combined fleet would have to achieve a 54.5 mpg and 163 g/mile CO_2 emissions target by 2025. Achieving these goals will certainly require some innovative technological advances in vehicle design, but at the same time, achieving the goals will cut oil imports in half, eliminate billions of tons of CO_2 emissions, and save drivers thousands of dollars per year in fuel costs.

Getting Around. Even though many Americans continue to prefer SUVs and larger automobiles, there are definite signs of change, particularly in light of the sustained high prices of gasoline in recent years. One option is the **hybrid electric vehicle**, which combines a conventional gasoline engine and a battery-powered electric motor to achieve substantial improvements in fuel economy. Such vehicles have been marketed by Honda and Toyota for almost a decade and are the most fuel-efficient cars in the United States. The Toyota Prius averages 46 mpg, while the Honda Insight gets 43 mpg, according to EPA figures.

All of the major automakers are jumping on the hybrid bandwagon, and higher prices of gasoline have boosted hybrid sales in the United States. Sales of hybrids soared in 2007, with 352,000 units sold, but with the downturn in the economy, hybrid sales dropped to 268,000 vehicles in 2011, about 2.1% of the passenger vehicle sales. Most newer hybrids are "green" versions of standard vehicles, including SUVs. For example, the Toyota Camry hybrid employs the same 2.5-liter gasoline engine as the standard four-cylinder Camry, but combines it with a large battery nestled behind the rear seat. Two electric motors work as generators that charge the battery as the car slows down. The car starts out under electric power, and with acceleration, the gasoline engine kicks in. When the car is sitting at a stoplight or coasting down a hill, the gasoline engine shuts off and starts automatically when speeding up. The EPA rates the 2012 Camry hybrid at 43 mpg, tops for all mid-sized hybrid sedans.

Prices for hybrids are several thousand dollars above those for comparable standard vehicles, but with gasoline ranging as high as $4 or more a gallon, the savings catch up with the extra cost in a few years. As more and more of the new hybrids are sold, prices are expected to drop substantially. For a time, a federal tax credit was available for hybrid buyers that offset much of the cost differential. Many states still offer tax incentives for hybrid buyers.

Another encouraging trend is the development of plug-in hybrid cars and electric cars. The plug-in hybrid has a larger battery pack than the conventional hybrid and can be driven up to 40 miles on a charge; after that, a down-sized gasoline engine either recharges the battery or serves as the source of propulsion. Electric vehicles are just that—all electric, with rechargeable batteries that give the vehicle a range up to 100 miles, depending on the speed of driving. Both in-home and commercial charging stations are becoming available to enable speedy recharging of the batteries. Both types of vehicles have now hit the mass market but, except for the Nissan Leaf, are somewhat pricey. It is certain that all of these innovative vehicles will become commonplace as the United States moves toward the 2025 fuel efficiency target.

Managing Ozone. Because ozone is a secondary pollutant, the only way to control ozone levels is to address the compounds that lead to ozone formation. For a long time, it was assumed that the best way to reduce these levels was simply to reduce emissions of VOCs. The steps pertaining to motor vehicles in the CAAA addressed the sources of about half of the VOC emissions. Point sources (industries) accounted for another 45% of such emissions, and area sources (numerous small emitters, such as dry cleaners, print shops, and users of household products) represented the remaining 5%. RACT measures have already been mandated for many point sources, and much progress has been made through EPA, state, and local regulatory efforts to reduce emissions from those sources. Since passage of the CAAA, VOC emissions have declined by 50%.

However, recent understanding of the complex chemical reactions involving NO_x, VOCs, and oxygen has thrown some uncertainty into this strategy of simply emphasizing a reduction in VOCs. The problem is that *both* NO_x and VOC concentrations are crucial to the generation of ozone (Fig. 19–12). Either one or the other can become the rate-limiting factor in the reaction that forms ozone. Thus, as the ratio of VOCs to NO_x changes, the concentration of NO_x can become the controlling chemical factor, pointing to the need to reduce NO_x emissions as well. This happens more commonly in air-pollution episodes that occur over a period of days than in the daily photochemical smog of the urban city.

New Ozone Standards? The revised ozone standard announced in 1997 (0.08 ppm instead of 0.12 ppm) met with strong opposition from industry groups, which obtained a court injunction in 1999 prohibiting the EPA from enforcing the new standard. However, the U.S. Supreme Court upheld the EPA in the dispute. This and other delays put off implementation of the revised standard to 2004. Cost-benefit estimates indicated that the anticipated health benefits far outweighed the costs of compliance. Then the EPA's Science Advisory Committee recommended in 2007 that a new ozone standard in the range of 0.06–0.07 ppm be set, based on human health concerns. EPA Administrator Stephen Johnson decided on a level of 0.075 ppm and published the new standard in early 2008—to the dismay of public-health advocates. The new ozone standard is expected to prevent a substantial number of premature deaths, but only a third of those that the lower standard would have prevented. EPA Administrator Lisa Jackson proposed lowering the standard but was asked by President Obama to withdraw the recommendation because the NAAQS would be reconsidered in 2013 anyway.

Down with NO_x. In a response to petitions from states in the Northeast that were having trouble achieving ozone and particulates ambient standards because of out-of-state emissions, the EPA has implemented regulations to reduce NO_x emissions from *mobile sources* and from *power plants and large industrial boilers and turbines*. For the mobile sources, which account for 59% of the smog-generating emissions in the eastern United States, the EPA established the Tier 2 standards, which were phased in gradually between 2004 and 2009. Under these regulations, emissions for all SUVs, pickup trucks, and passenger vans are held to the same standards as those for passenger cars. For NO_x, this means a gradual reduction to 0.07 grams per mile for the fleet of cars sold by each manufacturer, an 83% reduction over 2003 passenger car standards.

CAIR. States in the eastern half of the country continue to bear the burden of pollutants that are carried long distances by wind and weather, making it more difficult for them to attain the NAAQS levels for ozone and particulate matter. To address this, the Bush EPA announced a new rule in 2005 to accomplish further reductions of SO_2 and NO_x emissions. The **Clean Air Interstate Rule (CAIR)** set new lower caps on SO_2 and NO_x in 28 states in the Midwest and East, many of which are upwind of states experiencing nonattainment of ozone and particulate matter. CAIR would reduce SO_2 emissions 50% beginning in 2010 and 65% beginning in 2015. NO_x emissions would be reduced by similar percentages.

The new rules had a mixed reception. The industries and utilities were disappointed because the rules lacked the curb on lawsuits that a congressional act would provide, but they were pleased with the extended time frame for compliance. However, CAIR received a legal setback when a federal appeals court overturned the CAIR regulations, stating that the program had "more than several fatal flaws," including not being strict enough. However, the EPA, numerous states, and the Environmental Defense Fund successfully petitioned the court to reconsider its decision. In December 2008, the court reversed its ruling and kept CAIR in force until the EPA fixed the rule in accord with CAA requirements.

CSAPR. In 2011, the Obama EPA announced the **Cross-State Air Pollution Rule (CSAPR)** to replace CAIR. This rule addresses the same pollutants as CAIR and several additional states in the Midwest **(Fig. 19–26)**. The rule requires reductions in NO_x and SO_2 to begin in 2012 and tighten further in 2014, with the final reductions in power plant SO_2 emissions of 73% from 2005 levels and NO_x reductions of 54%. The rule involves some significant improvements in the manner in which upwind states are required to protect downwind states. In late 2012, the CSAPR rule was overturned by a US Court of Appeals, claiming that EPA overstepped its authority. CAIR is back in force.

To accomplish the emissions reductions, both CAIR and CSAPR involve a system that was first implemented with the Acid Rain Program (ARP), called **cap-and-trade**. A target cap, or budget, is set by EPA for SO_2 or NO_x, in units of millions of tons of emissions. At the start, power plants are granted emission allowances based on formulas in the regulations. The plants are free to choose how they will achieve their allowances, whether by installing emissions control systems or purchasing allowances from utilities that are under their budgeted allowance. The exchange of allowances occurs as a market, with prices fluctuating in response to supply and demand. For example, the price for SO_2 at the end of 2010 was $19 per ton, while NO_x went for $325 per ton.

The impacts of these regulations for human health are significant. The EPA estimates that the regulations would, by 2014, have prevented 13,000 to 34,000 premature deaths, 15,000 nonfatal heart attacks, and 400,000 cases of aggravated asthma, among other benefits. The monetary benefits were estimated at $120 billion to $280 billion per year, while the costs of implementation would have been approximately $2.4 billion per year.[6]

Coping with Acid Deposition

Scientists working on acid-rain issues in the 1980s calculated that a 50% reduction in acid-causing emissions in the United States would effectively prevent further acidification of the environment. This reduction was not expected to correct the already bad situations, but together with natural buffering processes, it was estimated to be capable of preventing further environmental deterioration. Because we knew that about 50% of acid-producing emissions came from coal-burning power plants, control strategies focused on these sources.

Political Developments. Although evidence of the link between power-plant emissions and acid deposition was well established by the early 1980s, no legislative action was taken until 1990. The problem was one of different regional interests. Throughout the 1980s, a coalition of politicians from the Midwestern states, representatives of high-sulfur

■ States controlled for both fine particles (annual SO_2 and NO_X) and ozone (ozone season NO_X) (20 States)

■ States controlled for fine particles only (annual SO_2 and NO_X) (3 States)

■ States controlled for ozone only (ozone season NO_X) (5 States)

States not covered by the Cross-State Air Pollution Rule

Figure 19–26 Map of states covered by the Cross-State Air Pollution Rule (CSAPR). The CSAPR is the 2011 EPA rule that replaces the Clean Air Interstate Rule (CAIR), designed to prevent downwind states from NO_x and SO_2 pollutants.

(*Source:* EPA CSAPR Fact Sheet. www.epa.gov/airtransport/statesmap.html. May 2, 2011.)

---

[6]EPA. Cross-State Air Pollution Rule (CSAPR). www.epa.gov/airtransport/. April 20, 2012.

coal producers, and representatives from the electric power industry effectively blocked all attempts at passing legislation that would take action on acid deposition.

On the other side of the issue were New York and the New England states, as well as most of the environmental and scientific communities, which argued that it was both possible and necessary to address acid deposition and that the best way to do so was to control emissions from power plants. Also, since 70% of Canada's acid-deposition problem came from the United States, diplomatic pressure toward a resolution was applied.

Action. With the passage of Title IV of the CAAA, the outcome of the controversy became history. However, two decades of action were lost because of political delays. In the wake of the passage of the 1990 act, Canada and the United States signed a treaty stipulating that Canada would cut its power plant SO_2 emissions by half and cap them at 3.5 million tons by the year 2000. Canada has done well; its emissions are now about 1.8 million tons and are expected to stay down. So has the United States.

Title IV of the Clean Air Act Amendments of 1990. Title IV of the CAAA addressed the acid-deposition problem by mandating reductions in both sulfur dioxide and nitrogen oxide levels. The major provisions of the Title IV ARP are as follows:

1. By 2010, total power plant SO_2 emissions were to be reduced 10 million tons below 1980 levels. This is the 50% reduction called for by scientists, and it involved the setting of a permanent cap of 8.9 million tons on such emissions.

2. In a major departure from the command-and-control approach, Title IV promoted a free-market approach to regulation of SO_2. This was the original cap-and-trade program that the CAIR and CSAPR have been grafted on.

3. New utilities do not receive allowances. Instead, they have to buy into the system by purchasing existing allowances. Thus, there are a finite number of allowances.

4. Nitrogen oxide emissions from power plants were to be reduced by 2 million tons by 2001. This was to be accomplished by regulating the boilers used by the utilities and by mandating the continuous monitoring of emissions.

Accomplishments of Title IV. The utilities industry responded to the new law with three actions:

1. Many utilities have switched to low-sulfur coal, available in Appalachia and the western United States.

2. Many older power plants have added scrubbers—"liquid filters" that put exhaust fumes through a spray of water containing lime. SO_2 reacts with the lime and is precipitated as calcium sulfate ($CaSO_4$). Because the technology for high-efficiency scrubbing is well established, more and more power plants are installing scrubbers, which have been required for all coal-burning power plants built after 1977.

3. Many utilities are trading their emission allowances. The purchase of allowances often represents a less costly way to achieve compliance than by purchasing low-sulfur coal or adding a scrubber. A typical transaction might involve a trade in the rights to emit, say, 10,000 tons of sulfur dioxide at a cost of $19 a ton. The combination of approaches has created so many efficiencies that it has cut compliance costs to 10% of what was expected.

To carry out Title IV of the CAAA, the EPA initiated a two-phase approach. Phase I aimed at a reduction of 3.5 million tons of SO_2 by 2000. Remarkably, the goal was met and even exceeded—at a cost far below the gloomy industry predictions (**Fig. 19–27**). Phase II began in 2000 and targeted the remaining sources of SO_2 in order to reach the 8.9-million-ton cap by 2010. This goal was met for the first time in 2007, when power plant SO_2 emissions totaled 8.9 million tons of SO_2. The emissions continue to drop, reaching 5.17 million tons in 2010, as CAIR reductions have been implemented and as more and more power plants have switched from coal to natural gas. This is a decrease of 67% from 1990 levels (Fig. 19–27).

The 2001 goal for a 2-million-ton reduction in NO_x emissions by fixed sources was met in 2000 and every year since; in fact, emissions reductions of NO_x are more than double the Title IV reduction objective. The Acid Rain Reduction Program was responsible for a large portion of these reductions, but the CAIR has also played a significant role in recent years, as we have seen. Current (2010) fixed source emissions of NO_x are 2.1 million tons, a decrease of 67% from 1990 levels. However, these sources account for less than half of all NO_x emissions, and the trend in total NO_x emissions has been a slower decline than that for SO_2.

Field News. The news from the field reflects the trends. Under the Acid Rain Program, air quality has improved, with significant benefits to human health, reductions in acid deposition, some recovery of freshwater lakes and streams, and improved forest conditions. Concentrations of sulfate in rain and deposition on the land have shown a significant decline (41–69%) over a large part of the eastern United States in the past 18 years. The acidified streams and lakes of the Adirondacks and New England have seen some recovery, in the form of pH increases and increases in acid neutralizing capacity.

One reason for the slow recovery is the continued impact of nitrogen deposition. Apparently, nitrogen plays a much larger role in acid deposition than was once believed. In passing the CAAA, Congress did not set curbs on nitrogen emissions, but simply opted to reduce the emissions from fixed sources. The other reason that slow progress has been made in recovery is the long-term buildup of sulfur deposits in soils. Residual sulfur takes a long time to flush from natural ecosystems. Accordingly, most scientists believe that the recovery of the affected ecosystems will require further reductions in both sulfur and nitrogen emissions. Acid rain still falls on eastern forests, as Figure 19–15b indicates.

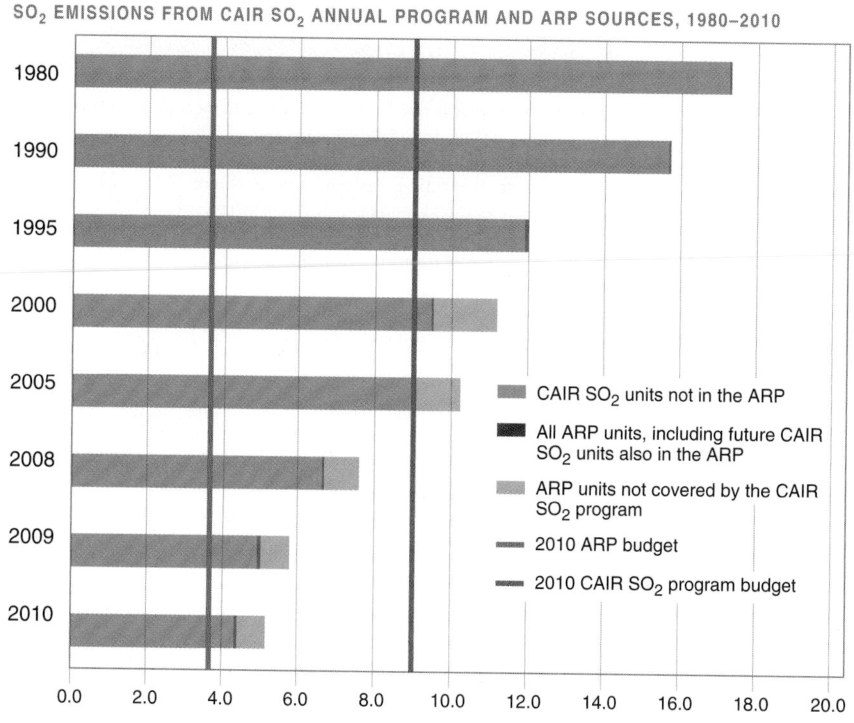

SO₂ EMISSIONS FROM CAIR SO₂ ANNUAL PROGRAM AND ARP SOURCES, 1980–2010

Figure 19–27 Power plant SO₂ emissions covered under the Acid Rain and CAIR programs.
Emissions from all power plants were under ARP allowances for the first time in 2006 and even more so in 2007.

(*Source:* Executive Office of the President. National Acid Precipitation Assessment Program Report to Congress 2011: An Integrated Assessment. USGS, 2011. http://ny.water.usgs.gov.)

The 2011 *National Acid Precipitation Assessment Program Report to Congress*[7] concluded that further reductions in SO_2 and NO_x will be needed in order to achieve substantial recovery of acidic lakes and streams and improvement of forest soils. These reductions are in the pipeline, as we have seen.

Costs and Benefits

We have air pollution because we want the goods and services that inevitably generate it. We want to be warm or cool, depending on the weather; we use electricity to power our homes, institutions, and manufacturing facilities; we crave freedom of movement, so we have our motor vehicles, trains, boats, and planes; and we want the array of goods produced by industries, from food to DVDs to computers to toys—the list is endless. We embrace all of these because we derive some benefit from them. The benefits may be essential or trivial, but we are accustomed to having them, as long as we can pay for them. Satisfying all of these needs and wants is what drives the American economy—and it produces millions of tons of pollutants in the process.

In recent decades, we have been learning that air pollution does not have to get worse—that in fact it can be remedied enough to keep most of us from getting sick and the environment from being degraded. But the remedies come with a price. Without question, measures taken to reduce air pollution carry an economic cost. Some critics have charged that air-pollution controls are not cost effective; that is, the benefits are not nearly as great as the costs. They see lost

opportunities for economic growth and tend to disregard the *costs avoided* (from improved health).

The Acid Rain Program has recently been examined from a benefit-cost perspective.[8] Benefits from improvements in human health have received the greatest attention from economic and scientific analyses, but some progress has been made in evaluating the aesthetic and environmental benefits as well. Health benefits extend well beyond the acid deposition issues, because the reductions in NO_x and SO_2 reduce the levels of $PM_{2.5}$ particulates and ozone. The estimated value of the program's health benefits in the year 2010 ranges from $174 billion to $427 billion, and the program prevented some 160,000 premature deaths. Two environmental benefit assessments have been made: Benefits from improved visibility were $40 billion in 2010, while environmental benefits to New York State alone were estimated at $1.1 billion per year. The estimated 2010 costs of administering and implementing the ARP were $3 billion, far below early estimates. The benefit-cost ratio speaks for itself. Acid rain is just one of many problems brought on by our involvement in major technologies, and although decades of inaction passed before it was addressed by public policy, the CAAA has made real improvements in the troposphere, ecosystems suffering from acidification, and human health.

We turn now to yet another problem of unintended consequences, where the use of some apparently harmless chemicals led to destruction of the stratospheric ozone that protects Earth from harmful ultraviolet radiation.

[7]Executive Office of the President, op cit.

[8]U.S. Environmental Protection Agency, Office of Air and Radiation. The Benefits and Costs of the Clean Air Act from 1990 to 2020. Summary Report, March 2011. www.epa.gov/air/sect812/aug10/summaryreport.pdf.

TABLE 19–4 Key Findings of the 2010 Scientific Assessment of Ozone Depletion

1. The Montreal Protocol is working as intended: There is clear evidence of a decrease in the atmospheric burden of ozone-depleting substances (ODS) and some early signs of stratospheric ozone recovery.

2. Springtime polar ozone depletion continues to be severe in cold stratospheric winters. Large Antarctic holes continue, and large Arctic ozone depletion occurs during colder Arctic winters. Antarctic ozone levels are projected to return to normal levels around 2060–2075.

3. Ozone abundances over regions outside of the poles no longer continue to decline, in spite of contributions of polar ozone depletion to extrapolar regions. It is likely that global ozone will return to normal levels around the middle of the 21st century.

4. Our basic understanding that anthropogenic ODS have been the principle cause of the ozone depletion over the past decades has been strengthened.

5. The abundances of anthropogenic ODS in the troposphere and the stratosphere show a downward trend from their peak values in the 1990s.

6. There are significant connections between ozone depletion efforts and climate change: Some of the replacements for ODS are potent greenhouse gases, and some greenhouse gases promote cooling in the stratosphere, which affects ozone depletion reactions.

Source: World Meteorological Organization, *Scientific Assessment of Ozone Depletion: 2010*, Global Ozone Research and Monitoring Project Report No. 52 (Geneva, Switzerland: World Meteorological Organization), 20.

19.5 Destruction of the Ozone Layer

The stratospheric ozone layer protects Earth from harmful ultraviolet radiation. The depletion of this layer is another major atmospheric challenge that is the result of human technology. Environmental scientists have traced the problem to a widely used group of chemicals: the **chlorofluorocarbons, or CFCs.** As with global warming, there were skeptics who were unconvinced that the problem was serious. However, scientists have achieved a strong consensus on the ozone depletion problem, reflected in the work of the Scientific Assessment Panel of the Montreal Protocol. Table 19–4 summarizes some of the major findings of the panel's 2010 assessment.[9]

Radiation and Importance of the Shield

Solar radiation emits electromagnetic waves with a wide range of energies and wavelengths **(Fig. 19–28).** Visible light is that part of the electromagnetic spectrum that can be detected by the eye's photoreceptors. Ultraviolet (UV) wavelengths are slightly shorter than the wavelengths of violet light, which are the shortest

Figure 19–28 The electromagnetic spectrum. Ultraviolet radiation, visible light, infrared radiation, and many other forms of radiation are at different wavelengths on the electromagnetic spectrum.

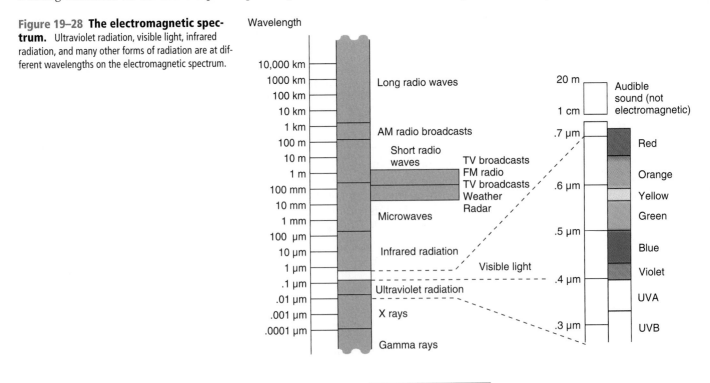

[9]World Meteorological Organization. Global Ozone Research and Monitoring Project—Report No. 52. Scientific Assessment of Ozone Depletion: 2010. http://ozone.unep.org/Assessment_Panels/SAP/Scientific_Assessment_2010/index.shtml.

Ozone in the Atmosphere

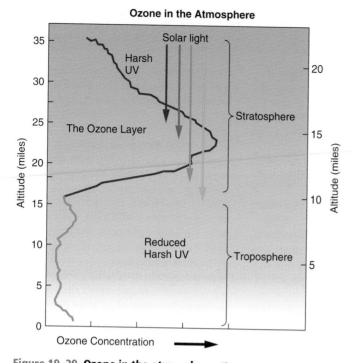

Figure 19–29 Ozone in the atmosphere. The vertical distribution of ozone in the troposphere (green) and stratosphere (red), showing the location of the stratospheric ozone layer that absorbs 99% of UV radiation.

(*Source:* Synthesis and Assessment Product 2.4. Report by the U.S. Climate Change Science Program and the Subcommittee on Global Change Research. November 2008.)

wavelengths visible to the human eye. UVB radiation consists of wavelengths that range from 280 to 315 nanometers (0.28 μm to 0.32 μm), whereas UVA radiation is from 315 to 400 nanometers (0.32 μm to 0.40 μm). Because energy is inversely related to wavelength, UVB is more energetic and, therefore, more dangerous, but UVA can also cause damage.

Upon penetrating the atmosphere and being absorbed by biological tissues, UV radiation damages protein and DNA molecules at the surfaces of all living things. (This damage is what occurs when you get a sunburn.) If the full amount of UV radiation falling on the stratosphere reached Earth's surface, it is doubtful that any life could survive. We are spared the more damaging effects from UV rays because most UV radiation (more than 99%) is absorbed by ozone in the stratosphere. For that reason, stratospheric ozone is commonly referred to as the **ozone shield (Fig. 19–29)**.

However, even the small amount (less than 1%) of UVB radiation that does reach us is responsible for sunburns and more than 700,000 cases of skin cancer and precancerous ailments per year in North America as well as for damage to plant crops and other life-forms. (See Sound Science, "Coping with UV Radiation.")

Formation and Breakdown of the Shield

Ozone is formed in the stratosphere when UV radiation acts on oxygen (O_2) molecules. The high-energy UV radiation first causes some molecular oxygen (O_2) to split apart into free

oxygen (O) atoms, and these atoms then combine with molecular oxygen to form ozone via the following reactions:

$$O_2 + UVB \rightarrow O + O \tag{1}$$

$$O + O_2 \rightarrow O_3 \tag{2}$$

Not all of the molecular oxygen is converted to ozone, however, because free oxygen atoms may also combine with ozone molecules to form two oxygen molecules in the following reaction:

$$O + O_3 \rightarrow O_2 + O_2 \tag{3}$$

Finally, when ozone absorbs UVB, it is converted back to free oxygen and molecular oxygen:

$$O_3 + UVB \rightarrow O + O_2 \tag{4}$$

Thus, the amount of ozone in the stratosphere is dynamic. There is an equilibrium due to the continual cycle of reactions of formation (Eqs. 1 and 2) and reactions of destruction (Eqs. 3 and 4). Because of seasonal changes in solar radiation, ozone concentration in the Northern Hemisphere is highest in summer and lowest in winter. Also, in general, ozone concentrations are highest at the equator and diminish as latitude increases—again, a function of higher overall amounts of solar radiation. However, the presence of other chemicals in the stratosphere can upset the normal ozone equilibrium and promote undesirable reactions there.

Halogens in the Atmosphere. Chlorofluorocarbons (CFCs) are a type of halogenated hydrocarbon (see Chapter 22). CFCs are nonreactive, nonflammable, nontoxic organic molecules in which both chlorine and fluorine atoms have replaced some hydrogen atoms. At room temperature, CFCs are gases under normal (atmospheric) pressure, but they liquefy under modest pressure, giving off heat in the process and becoming cold. When they revaporize, they reabsorb the heat and become hot. These attributes led to the widespread use of CFCs (more than 1 million tons per year in the 1980s) for the following applications:

- In refrigerators, air conditioners, and heat pumps as the heat-transfer fluid;
- In the production of plastic foams;
- By the electronics industry for cleaning computer parts, which must be meticulously purified; and
- As the pressurizing agent in aerosol cans.

Rowland and Molina. All of the preceding uses led to the release of CFCs into the atmosphere, where they mixed with the normal atmospheric gases and eventually reached the stratosphere. In 1974, chemists Sherwood Rowland and Mario Molina published a classic paper[10] (for which they were awarded the Nobel Prize in Chemistry in 1995) concluding that CFCs could damage the stratospheric ozone layer through the release of chlorine atoms and, as a result, UV radiation would increase and cause more skin cancer (their story is described in Chapter 2

---

[10]M. J. Molina and F. S. Rowland, "Stratospheric Sink for Chlorofluoro-Methanes: Chlorine-Atom Catalyzed Distribution of Ozone," *Nature* 249 (1974): 810–812.

SOUND SCIENCE

Coping with UV Radiation

People living in Chile, Australia, and New Zealand are no strangers to the effects of the thinner ozone shield. UV alerts are given in those parts of the Southern Hemisphere during spring months as lobes of ozone-depleted stratospheric air move outward from the Antarctic. At times, the ozone above those countries can be less than half its normal concentration, which means that greater intensities of UV radiation reach Earth's surface. It is predicted that one out of every three Australians will develop serious and perhaps fatal skin cancer in his or her lifetime.

Even without a thinner ozone layer, UV radiation is a serious problem. Concerned over the rising incidence of skin cancer and cataract surgery in the United States, the EPA, together with the National Oceanic and Atmospheric Administration and the Centers for Disease Control and Prevention, has initiated a new *UV Index* (see the table at the end of the essay). The index takes the form of daily forecasts of UV exposure, issued by the Weather Service for 58 cities. Satellite measurements of stratospheric ozone are combined with other weather patterns. The goal of the index is to remind people of the dangers of UV radiation and to prompt them to take appropriate action to avoid cancer, premature aging of the skin, eye damage, cataracts, and blindness. For at least the next two decades, the dangers will be quite noticeable, ranging from a greater incidence of sunburn to a likely epidemic of malignant melanoma.

Less obvious, but no less important, are the chronic effects of exposure to the Sun. The normal aging of skin—coarse wrinkling, yellowing,

and the development of irregular patches of heavily pigmented and unpigmented skin—is now known to be largely the result of damage from the Sun. This can happen even in the absence of episodes of sunburn. Further, a large proportion of the million cataract operations performed annually in the United States can be traced to UV exposure, another of the chronic effects of this kind of radiation.

The most serious impact of chronic exposure to the Sun is skin cancer; more than 1 million new cases occur each year in the United States, according to the National Cancer Institute. Three types of skin cancer are traced to UV exposure: basal cell carcinoma (BCC), squamous cell carcinoma (SCC), and melanoma. Most of the skin cancers (75%) are BCCs, slow growing and therefore treatable. SCC accounts for about 20% of skin cancers and can also be cured if treated early. However, it metastasizes (spreads away from the source) more readily than BCC and thus is potentially fatal. Melanoma (6%) is the

most deadly form of skin cancer, as it metastasizes easily. Melanoma is often traced to occasional sunburns during childhood or adolescence or to sunburns in people who normally stay out of the sun.

What is an appropriate response to this disturbing information? First, know when UV intensity is greatest (three hours on either side of midday), and take precautions accordingly. Protect your eyes with sunglasses and a hat, and apply sunscreen with a sun protection factor (SPF) rating of 15 or higher during such times. Think twice before considering sunbathing or using tanning beds, and never proceed without sunscreen that is strong enough to prevent sunburn. Always protect children with sunscreen. Their years of potential exposure are many, and their skin is easily burned. And if you detect a patch of skin or a mole that is changing in size or color or that is red and fails to heal, consult a dermatologist for treatment.

UV Index (EPA)

Exposure Category	Index Value	Minutes to Burn for "Never Tans" (Most Susceptible)	Minutes to Burn for "Rarely Burns" (Least Susceptible)
Minimal	0–2	35	>120
Low	3–4	20	>120
Moderate	5–6	12	90
High	8–9	8	55
Very high	10–12	6	40

to illustrate the public-policy life cycle). Rowland and Molina reasoned that, although CFCs would be stable in the troposphere (where they have been found to last 70 to 110 years), in the stratosphere they would be subjected to intense UV radiation, which would break them apart, releasing free chlorine atoms via the following reaction:

$$CFCl_3 + UV \rightarrow Cl + CFCl_2 \qquad (5)$$

Ultimately, all of the chlorine of a CFC molecule would be released as a result of further photochemical breakdown. The free chlorine atoms would then attack stratospheric ozone to form chlorine monoxide (ClO) and molecular oxygen:

$$Cl + O_3 \rightarrow ClO + O_2 \qquad (6)$$

Furthermore, two molecules of chlorine monoxide may react to release more chlorine and an oxygen molecule:

$$ClO + ClO \rightarrow 2Cl + O_2 \qquad (7)$$

Reactions 6 and 7 are called the **chlorine catalytic cycle** because chlorine is continuously regenerated as it reacts with ozone. Thus, chlorine acts as a **catalyst**, a chemical that promotes a chemical reaction without itself being used up in the reaction. Because every chlorine atom in the stratosphere can last from 40 to 100 years, it has the potential to break down 100,000 molecules of ozone. Thus, CFCs are judged to be damaging because they act as transport agents that continuously move chlorine atoms into the stratosphere. The damage persists because the chlorine atoms are removed from the stratosphere only very slowly. **Figure 19–30** shows the basic processes of ozone formation and destruction, including recent refinements to our knowledge of those processes that will be explained shortly.

EPA Action. After studying the evidence, the EPA became convinced that CFCs were a threat and, in 1978, banned their

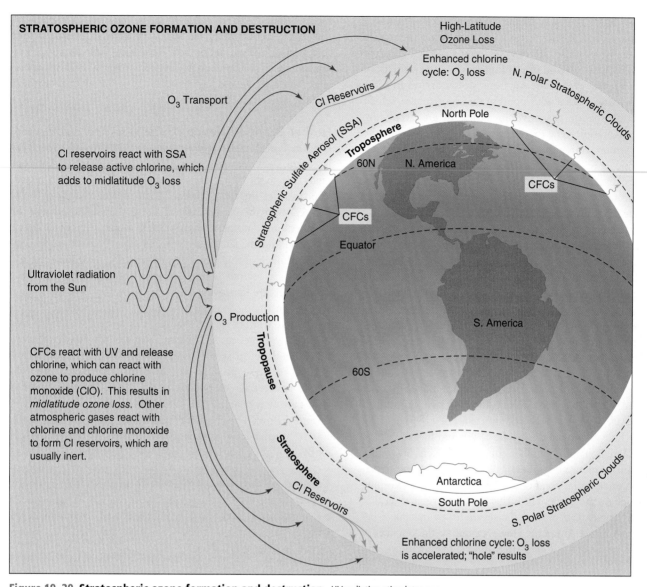

STRATOSPHERIC OZONE FORMATION AND DESTRUCTION

High-Latitude Ozone Loss

Enhanced chlorine cycle: O_3 loss

N. Polar Stratospheric Clouds

O_3 Transport

Cl Reservoirs

North Pole

Stratospheric Sulfate Aerosol (SSA)

Troposphere

60N · N. America

CFCs

CFCs

Cl reservoirs react with SSA to release active chlorine, which adds to midlatitude O_3 loss

Equator

Ultraviolet radiation from the Sun

O_3 Production

S. America

Tropopause

CFCs react with UV and release chlorine, which can react with ozone to produce chlorine monoxide (ClO). This results in *midlatitude ozone loss*. Other atmospheric gases react with chlorine and chlorine monoxide to form Cl reservoirs, which are usually inert.

60S

Stratosphere

Cl Reservoirs

Antarctica

South Pole

S. Polar Stratospheric Clouds

Enhanced chlorine cycle: O_3 loss is accelerated; "hole" results

Figure 19–30 Stratospheric ozone formation and destruction. UV radiation stimulates ozone production at the lower latitudes, and ozone-rich air migrates to high latitudes. At the same time, CFCs and other compounds carry halogens into the stratosphere, where they are broken down by UV radiation, to release chlorine and bromine. Ozone is subject to high-latitude loss during winter, as the chlorine cycle is enhanced by the polar stratospheric clouds. Midlatitude losses occur as chlorine reservoirs are stimulated to release chlorine by reacting with stratospheric sulfate aerosol.

use in aerosol cans in the United States. Manufacturers quickly switched to nondamaging substitutes, such as butane, and things were quiet for several years. CFCs continued to be used in applications other than aerosols, however, and skeptics demanded more convincing evidence of their harmfulness.

Atmospheric scientists reasoned that any substance carrying reactive halogens to the stratosphere had the potential to deplete ozone. These substances include halons, methyl chloroform, carbon tetrafluoride, and methyl bromide. Chemically similar to chlorine, bromine also attacks ozone and forms a monoxide (BrO) in a catalytic cycle. Because of its extensive use as a soil fumigant and pesticide, bromine plays a major role in ozone depletion. (Bromine is 16 times as potent as chlorine in ozone destruction.)

The Ozone "Hole." In the fall of 1985, British atmospheric scientists working in Antarctica reported a gaping

"hole" (actually, a serious thinning) in the stratospheric ozone layer over the South Pole. There, in an area the size of the United States, ozone levels were 50% lower than normal. The hole would have been discovered earlier by NASA satellites monitoring ozone levels except that computers were programmed to reject data showing a drop as large as 30% as being due to instrument anomalies. Scientists had assumed that the loss of ozone, if it occurred, would be slow, gradual, and uniform over the whole planet. The ozone hole came as a surprise, and if it had occurred anywhere but over the South Pole, the UV damage would have been extensive. As it is, the limited time and area of ozone depletion there have not apparently brought on any catastrophic ecological events so far.

News of the ozone hole stimulated an enormous scientific research effort. A unique set of conditions was found to be responsible for the hole. In the summer, gases such as nitrogen dioxide and methane react with chlorine monoxide and chlorine to trap the chlorine, forming so-called chlorine reservoirs (Fig. 19–30) and preventing much ozone depletion.

Polar Vortex. When the Antarctic winter arrives in June, it creates a vortex (like a whirlpool) in the stratosphere, which confines stratospheric gases within a ring of air circulating around the Antarctic. The extremely cold temperatures of the Antarctic winter cause the small amounts of moisture and other chemicals present in the stratosphere to form the south polar stratospheric clouds. During winter, the cloud particles provide surfaces on which chemical reactions release molecular chlorine (Cl_2) from the chlorine reservoirs.

When sunlight returns to the Antarctic in the spring, the Sun's warmth breaks up the clouds. UV light then attacks the molecular chlorine, releasing free chlorine and initiating the chlorine cycle, which rapidly destroys ozone. By November, the beginning of the Antarctic summer, the vortex breaks down, and ozone-rich air returns to the area. However, by that time, ozone-poor air has spread around the Southern Hemisphere. Shifting patches of ozone-depleted air have caused UV radiation to increase to 20% above normal in Australia. Television stations there now report daily UV readings and warnings for Australians to stay out of the sun. On the basis of current data, estimates indicate that in Queensland, where the ozone shield is thinnest, two out of three Australians are expected to develop skin cancer by the age of 70. The ozone hole intensified during the 1990s and has leveled off between 22 million and 27 million square kilometers, an area as large as North America (**Fig. 19–31**). Because severe ozone depletion occurs under polar conditions, observers have been keeping a close watch on the Arctic. The higher temperatures and weaker vortex formation there prevent the severe losses that have become routine in the Antarctic. However, one of the impacts of global climate change is lower stratospheric temperatures, and this is beginning to influence Arctic ozone levels (Chapter 18). The 2011 Arctic winter and spring brought an unusually long cold period, and an ozone hole developed that covered 2 million square kilometers and involved a loss of 80% of the ozone layer. This loss allowed high levels of UV radiation over northern Canada, Russia, and Europe in the spring.

Further Ozone Depletion. Ozone losses have not been confined to Earth's polar regions, though they are most spectacular there. A worldwide network of ozone-measuring stations sends data to the World Ozone Data Center in Toronto, Canada. Reports from the center reveal ozone depletion levels of 3.5% and 6% over the period 2006–2009 in the midlatitudes of the Northern and Southern Hemispheres, respectively. Polar ozone depletion and the subsequent mixing of air masses are responsible for much of this loss. There is virtually no ozone depletion over tropical regions. The good news from the network is that ozone levels are improving.

Is the ozone loss significant to our future? The EPA has calculated that the ozone losses of the 1980s will eventually have caused 12 million people in the United States to develop skin cancers over their lifetime and that 93,000 of these cancers will be fatal. Americans are estimated to have developed more than 900,000 new cases of skin cancer a year in the 1990s. The ozone losses of that decade allowed more UVB radiation than ever to reach Earth. Studies have confirmed increased UVB levels, especially at higher latitudes.

Coming to Grips with Ozone Depletion The dramatic growth of the hole in the ozone layer (Fig. 19–31) galvanized a response around the world. Scientists and politicians in the United States and other countries have achieved treaties designed to avert a UV disaster.

Figure 19–31 Ozone loss and extent of ozone hole. (a) The decline in ozone from 1980 to 2011 is shown in Dobson units for September 21 to October 16 each year (the average amount of ozone in the stratosphere is 300 Dobson units). (b) The size of the ozone hole is plotted for the same time frame, showing the origin of the hole and its leveling off in the last few years. No data were acquired in 1995, shown as a blue column in both plots.

(*Source:* NASA Ozone Hole Watch. http://ozonewatch.gsfc.nasa.gov/. May 15, 2012.)

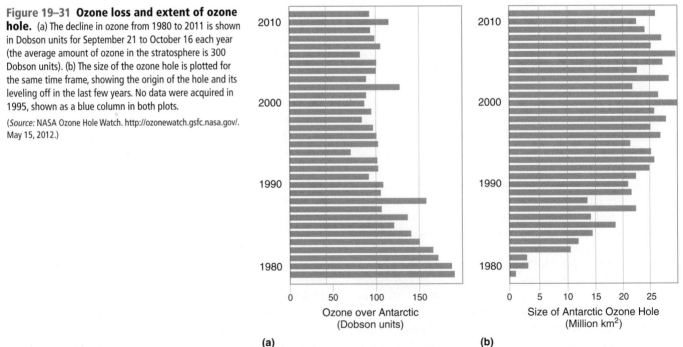

Ozone over Antarctic
(Dobson units)

(a)

Size of Antarctic Ozone Hole
(Million km²)

(b)

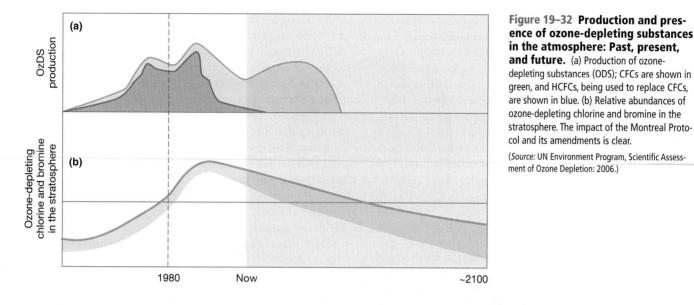

Figure 19–32 Production and presence of ozone-depleting substances in the atmosphere: Past, present, and future. (a) Production of ozone-depleting substances (ODS); CFCs are shown in green, and HCFCs, being used to replace CFCs, are shown in blue. (b) Relative abundances of ozone-depleting chlorine and bromine in the stratosphere. The impact of the Montreal Protocol and its amendments is clear.

(*Source:* UN Environment Program, Scientific Assessment of Ozone Depletion: 2006.)

Montreal Protocol. In 1987, under the auspices of its environmental program, the United Nations convened a meeting in Montreal, Canada, to address ozone depletion. Member nations reached an agreement, known as the **Montreal Protocol**, to scale back CFC production 50% by 2000. All 196 United Nations member countries have ratified the agreement.

The Montreal Protocol was written even before CFCs were so clearly implicated in driving the destruction of ozone and before the threat to Arctic and temperate-zone ozone was recognized. Because ozone losses during the late 1980s were greater than expected, an amendment to the protocol was adopted in June 1990. The amendment required participating nations to phase out the major chemicals destroying the ozone layer by 2000 in developed countries and by 2010 in developing countries. In the face of evidence that ozone depletion was accelerating even more, another amendment to the protocol was adopted in November 1992, moving the *target date for the complete phaseout of CFCs to January 1, 1996.* Timetables for phasing out all of the suspected ozone-depleting halogens were shortened at the 1992 meeting.

Quantities of CFCs were still being manufactured after 1992 to satisfy legitimate demand in the developing countries. However, according to the Montreal Protocol, all manufacturing of CFCs was to have been halted by December 31, 2005. The results of the protocol can be seen clearly in **Figure 19–32**, which shows the time course of production of ozone-depleting substances (ODS) and the abundance of those chemicals in the stratosphere. Ozone depletion continues to occur because the ODS are long-lived in the stratosphere.

Action in the United States. The United States was the leader in the production and use of CFCs and other ODS, with DuPont Chemical Company being the major producer. Following years of resistance, DuPont pledged in 1988 to phase out CFC production by 2000. In late 1991, a company spokesperson announced that, in response to new data on ozone loss, DuPont would accelerate its phaseout. Many of the large corporate users of CFCs (AT&T, IBM, and Northern Telecom, for example) phased out their CFC use by 1994.

The Clean Air Act Amendments of 1990 also address this problem—in Title VI, Protecting Stratospheric Ozone. Title VI is a comprehensive program that restricts the production, use, emissions, and disposal of an entire family of chemicals identified as ozone depleting. For example, the program calls for a phase out schedule for the hydrochlorofluorocarbons (HCFCs), a family of chemicals being used as less damaging substitutes for CFCs until nonchlorine substitutes are available. Halons—used in chemical fire extinguishers—were banned in 1994. The CAAA also regulates the servicing of refrigeration and air-conditioning units.

The Montreal Protocol had its 25th anniversary in 2012. As a result of this and subsequent agreements, CFCs are no longer being produced or used. Substitutes for CFCs are readily available and, in some cases, are even less expensive than CFCs. The most commonly used substitutes are HCFCs, which still contain some chlorine and are scheduled for a gradual phaseout (by 2030). Although they contain chlorine, HCFCs are largely destroyed in the troposphere and are much less capable of destroying stratospheric ozone than CFCs. The most *promising* substitutes are hydrofluorocarbons (HFCs), which contain no chlorine and are judged to have no ozone-depleting potential. They do, however, have significant global warming potential. Because CFCs are potent greenhouse gases, phasing out CFCs and other ozone-depleting substances is helping the efforts to slow down global climate change (Chapter 18). Ironically, HFCs are likely to have a marked influence on global climate change if their use increases in the future.

Other ozone-depleting substances are still a concern. Methyl bromide continues to be produced and released into the atmosphere; the compound is employed as a soil fumigant to control agricultural pests. Under the Montreal Protocol, it was scheduled to be completely phased out by 2005 in the developed countries and 10 years later in the developing countries. In the meantime, bromine is likely to play a greater role in ozone depletion relative to chlorine. Responding to requests from the agricultural industry, the United States has repeatedly received "critical use" exemptions to the ban for use of this ozone-depleting chemical. However, the EPA is phasing out these uses, and the current (2012) level is just 4.6% of 1991 use (1,180 tons vs. 25,500 tons).

Another ODS is nitrous oxide (N_2O). Figure 19–12a shows how nitric oxides are able to destroy ground-level ozone. This can also happen in the stratosphere, where N_2O is broken down into NO_x and the nitrogen oxides deplete ozone there, along with the chlorine and bromine compounds. Even though the current ozone depletion in the stratosphere is largely due to accumulated chlorine and bromine compounds from past releases, nitrous oxide is the predominant anthropogenic ODS currently being released and is expected to remain the largest throughout the 21st century. Oddly, this substance is not yet addressed by the Montreal Protocol and its amendments.[11]

Final Thoughts. The ozone story is a remarkable episode in human history. From the first warnings in 1974 that something might be amiss in the stratosphere because of a practically inert and highly useful industrial chemical, through the development of the Montreal Protocol and the final steps of CFC phaseout, the world has shown that it can respond collectively and effectively to a clearly perceived threat. The ozone layer is expected to recover completely by midcentury in the midlatitudes and a couple of decades later in the polar regions. The scientific community has played a crucial role in this episode, first alerting the world and then plunging into intense research programs to ascertain the validity of the threat. Scientists continue to influence the political process, which has forged a response to the threat. Although skeptics were stressing the uncertainties in our understanding of ozone loss and its consequences, the strong consensus in the scientific community convinced the world's political leaders that action was clearly needed. This is a most encouraging development as we look forward to the actions that must be taken during the 21st century to prevent catastrophic global climate change.

REVISITING THE THEMES

Sound Science

Atmospheric science has come a long way in the past half-century. Some of its triumphs include understanding causes and effects that are operative in photochemical smog, discovering the cleansing role of the hydroxyl radical, tracing the sources of lead and mercury in humans, working out causes and effects in the acid-rain story, and alerting the world to ozone shield destruction. Sound science is increasingly important in measuring ambient pollutant levels and in devising technologies to prevent more pollution. Many scientists have put their reputations and careers on the line as they have fought for policy changes to correct the human and environmental damage they have uncovered.

Sustainability

Air pollution was on an unsustainable track, with industrial smog producing health disasters, photochemical smog making city living intolerable for many, crops and natural ecosystems deteriorating seriously, and the ozone shield being destroyed. Less obvious, but just as important, chronic exposure to air pollutants was responsible for poor health and increased mortality for millions. Are we now back on a more sustainable track? The important indicators would be long-term declines in the following: lead in the environment, annual amounts of emissions of pollutants, acid and sulfate deposition in the Northeast, ambient levels of pollutants, asthma hospitalizations, and locations that are noncompliant. Most of these parameters are moving in the right direction, but much remains to be done. The ozone story represents a sustainable pathway. Levels of ozone-destroying chemicals are now declining, and it is very likely that the ozone layer will be "healthy" by midcentury.

Stewardship

Stewardship means doing the right thing when faced with difficult problems and choices. It is fair to say that our society has responded well in caring for children threatened by high lead levels in their blood and, in general, in taking the costly actions we have documented in this chapter. Further, in addressing acid deposition, we have even shown care for the natural world. The need for stewardship is still very strong, however, and thankfully, there are many in nongovernmental and other organizations, and indeed many in the EPA, that are determined to stay the course and make sure that effective action is taken. Often, though, they have to do battle with industry lobbies and special interests that are concerned with the economic bottom line and that seem unable to embrace the common good represented by effective action to curb pollution.

REVIEW QUESTIONS

1. What naturally occurring cleanser helps to remove pollutants from the atmosphere? Where does it come from? What other mechanisms also act to cleanse the atmosphere of pollutants?

2. Describe the origins of industrial smog, photochemical smog, and atmospheric brown clouds. What are the differences in the cause and appearance of each?

3. What are the seven major primary air pollutants and their sources?

4. What are the major secondary pollutants, and how are they formed?

5. What is the difference between an acid and a base? What is the pH scale?

---

[11]A. R. Ravishankara et al. "Nitrous Oxide (N_2O): The Dominant Ozone-Depleting Substance Emitted in the 21st Century." *Science* 326: 123–125. October 2, 2009.

6. What two major acids are involved in acid deposition? Where does each come from?

7. What impacts does air pollution have on human health? Give the three categories of impact, and distinguish among them.

8. Describe the negative effects of pollutants on crops, forests, and other materials. Which pollutants are mainly responsible for these effects?

9. How can a shift in environmental pH affect aquatic ecosystems? What protects these ecosystems in some regions?

10. What are the criteria pollutants and the National Ambient Air Quality Standards, and how are the standards used?

11. Discuss ways in which the Clean Air Act Amendments of 1990 address the failures of previous legislation.

12. Describe the new Mercury and Air Toxics Standards and the impacts they will have on industry and on human health.

13. What are the major steps being taken to reduce emissions from passenger vehicles? What are the CAFE standards, and how are they changing?

14. How do CAIR and CSAPR address downwind ozone pollution?

15. What has been happening in the United States to reduce acid deposition, and how successful has it been?

16. How is the protective ozone shield formed? What causes its natural breakdown?

17. How do CFCs affect the concentration of ozone in the stratosphere and contribute to the formation of the ozone hole?

18. Describe the international efforts that are currently in place to protect our ozone shield. What evidence is there that such efforts have been effective?

THINKING ENVIRONMENTALLY

1. Motor vehicles release close to half the pollutants that dirty our air. What alternatives might be introduced to encourage a decrease in our use of automobiles and other vehicles with internal combustion engines?

2. How would our pattern of life be different in the absence of the Clean Air Act and its amendments? Write a short essay describing the possibilities.

3. The wind blows across Country A and into Country B. Country A has fossil fuel–burning power plants, but a relatively low level of air pollution. Country B has power plants run by moving water (hydroelectric plants), but a higher

level of pollution. What's happening? What could the two countries do to correct the situation? Explain how this pertains to international air-pollution laws.

4. A large amount of air pollution is attributed to manufacturing. Some would argue that it is not technologically possible, or economically efficient, to eliminate this pollution. How would you respond to this argument?

5. What arguments might the auto and utility industries present to delay regulatory action on particulates and ozone? How would health officials respond to their arguments?

MAKING A DIFFERENCE

1. At all costs, avoid burning trash in your yard. It's illegal in many places and can lead to birth defects, cancers, and lung diseases.

2. Check your home heating system. Simple wood smoke contains fine particles that are associated with serious negative health effects. The tiny size of these pollutants allows them to be easily inhaled, traveling deep into your lungs where they can cause respiratory and cardiovascular problems, including premature death. Woodstoves, fireplaces, and other wood-burning devices put out hundreds of times more air pollution than other sources of heat such as natural gas, propane, oil, and electricity.

3. Discuss air-pollution issues with your employer. Many employers are realizing the benefits of considering air quality issues. Green travel plans, alternative-powered vehicles, and the provision of cycling facilities are just three ways employers can help to reduce their impacts on air quality, realize a financial saving, and maintain a healthy workforce.

4. If you live in an urban area, take advantage of public transportation. It helps ease urban congestion as well as atmospheric conditions in the area.

5. When you buy a new car, consider purchasing a hybrid-electric or an all-electric vehicle.

MasteringEnvironmentalScience®

Go to **www.masteringenvironmentalscience.com** for practice quizzes, Pearson eText, videos, current events, and more.

CHAPTER 20

Water Pollution and Its Prevention

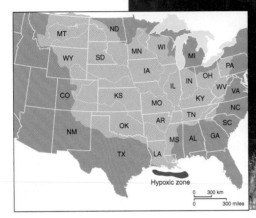

The Mississippi River watershed. The Mississippi River watershed consists of 40% of the landmass of the lower 48 states. Runoff from this huge area delivers 500 million tons of sediment and more than 1 million tons of nutrients to the Gulf of Mexico yearly. Below: A type of phytoplankton, viewed through a microscope (SEM).

(*Source:* Mississippi River/Gulf of Mexico Watershed Nutrient Task Force. Gulf Hypoxia Annual Report 2010.)

LEARNING OBJECTIVES

1. **Perspectives on Water Pollution:** Summarize the impacts that disease organisms, organic wastes, chemical pollutants, sediments, and nutrients have on human health and the environment.

2. **Wastewater Treatment and Management:** Examine the methods used by cities and towns to deal with human and domestic wastes.

3. **Eutrophication:** Describe the undesirable changes that take place in aquatic ecosystems when nutrients are introduced.

4. **Public Policy and Water Pollution:** Explain how the Clean Water Act and its amendments give the Environmental Protection Agency jurisdiction over water-pollution issues and what responsibility the states have for managing pollution and wastewater programs.

THE MISSISSIPPI RIVER WATERSHED encompasses 40% of the land area of the United States; the river collects water from this huge area and eventually delivers it to the Gulf of Mexico. Because much of the watershed is America's agricultural heartland, the Mississippi River is a reflection of what happens on the nation's farms. When U.S. farm production experienced tremendous growth during the latter half of the 20th century, more and more fertilizers were spread on agricultural fields, and more animals were raised on feedlots. At the same time, wetlands bordering river tributaries were drained and no longer intercepted agricultural runoff. The result was a tripling of soluble nitrogen delivered to the river and, eventually, to the Gulf of Mexico.

DEAD ZONE. The impact of this major flow of nitrogen to the Gulf of Mexico was first detected in 1974 by

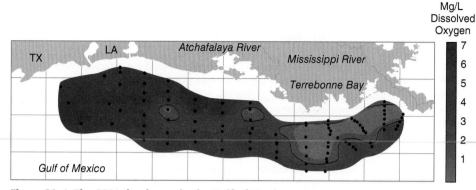

Figure 20–1 The 2011 dead zone in the Gulf of Mexico. Each year oxygen disappears from middle depths to the bottom over thousands of square miles of the Gulf of Mexico. In 2011, the dead zone was 6,765 mi², an area almost the size of New Jersey.

(*Source:* NOAA News, Aug. 4, 2011. With permission from LUMCON.)

marine scientists who found areas in the gulf where oxygen had disappeared from bottom sediments and much of the water column above them. This absence of oxygen is deadly to the bottom-dwelling animals and any oxygen-breathing creatures in the water column that are not able to escape. At first, the hypoxic (lacking oxygen) area—or **dead zone**, as it is called—was thought to be similar to those in other bodies of water that receive high levels of nutrients—a minor disturbance that would disappear seasonally. But then the hypoxic area suddenly doubled in size following the 1993 floods in the Midwest, and it continues to grow **(Fig. 20–1)**. It averaged 6,700 mi² (17,300 km²) from 2005 to 2011. The shrimpers and fishers who work these waters have learned to avoid the hypoxic area.

THE CULPRIT. To the scientists working on the problem, the connection between nitrogen and hypoxia was clear. Nitrogen is a common limiting factor in coastal marine waters. Where abundant, it promotes the dense growth of phytoplankton—photosynthetic microorganisms that live freely suspended in the water. Zooplankton—microscopic animals in the water—rapidly consume the phytoplankton and multiply. As the abundant phytoplankton and zooplankton (and their fecal pellets) die or sink toward the bottom, they are decomposed by bacteria, a process that consumes dissolved oxygen. Eventually, the oxygen is used up, creating a dead zone. In the Gulf of Mexico, the zone can extend up from the bottom of the sea to within a few meters of the surface. Typically, it stays from May through September, finally being dispersed by the mixing of the water column as colder weather sets in.

Because the gulf's fishery is a $2.8 billion dollar enterprise, the problem eventually got national attention. In 1998, Congress passed the Harmful Algal Bloom and Hypoxia Research and Control Act, establishing an interagency task force of scientists charged with assessing the dead zone in the Gulf of Mexico, as well as harmful algal blooms—another water-pollution problem discussed later in the chapter (see

Sustainability, "Harmful Algal Blooms"). The Mississippi River/Gulf of Mexico Watershed Nutrient Task Force released its first report in May 2000, confirming the connection between nitrogen and the dead zone and laying out options for reducing the nitrogen load coming downriver. Two strategies were key to mitigation of the problem: (1) reduction of fertilizer use, to greatly reduce farm runoff and lessen the nitrogen loads of streams and rivers in the Mississippi basin; and (2) the restoration and promotion of nitrogen retention and denitrification processes in the basin.

The task force proposed an action plan in 2001 with the goal of reducing the size of the hypoxic area to 1900 mi² (5000 km²) by 2015. The plan was voluntary and depended largely on conservation funding from the U.S. Department of Agriculture (USDA) to farmers upriver. Seven years into the plan, the dead zone had not shown any signs of shrinking. The latest (2008) action plan recommended a reduction target of 45% of the nitrogen and phosphorus and also acknowledged that the 1900 mi² goal by 2015 was likely unattainable. Significant new funding has come from the 2008 Farm Bill, and the USDA is directing $320 million to agricultural conservation programs in 13 participating states to aid farmers in reducing or treating runoff from farms.

The Gulf of Mexico dead zone turns out to be one more costly lesson about **eutrophication** (enrichment by nutrients): It is unreasonable to allow vital water resources to receive pollutants and then expect those resources to continue to provide us with their usual bounty of goods and services. Since the 1960s, the number of coastal dead zones has doubled every decade; worldwide, more than 415 are now known, affecting a total area of over 95,000 square miles.

In this chapter, we will focus on water *quality* (Chapter 10 dealt with water *quantity*—the global water cycle and water resources). Eutrophication and other forms of water pollution are discussed, and we take a close look at wastewater treatment.

SUSTAINABILITY

Imagine driving along Palos Verdes Drive on the California coast and having a pelican fall out of the sky and crash into your windshield. You'd probably rather not imagine that, but every spring along the coast of Southern California for the past 10 years, dead marine birds, sea otters, dolphins, and sea lions have been washing up on the beaches, and pelicans have literally dropped out of the sky—victims of an algal toxin, *domoic acid*. This toxin is acquired by birds and mammals that have ingested marine shellfish and small fish that have passed the toxin along by ingesting smaller planktonic organisms that have been feeding on a diatom occurring in dense "blooms" in the water. Domoic acid doesn't harm invertebrates or fish, but in mammals and birds, it causes brain damage that incapacitates them, often leading to death. The first reported outbreak occurred in Prince Edward Island, Canada, in 1987, when more than 100 people were sickened and three died from eating blue mussels. The spring visit of the toxin on the California coast is a recent phenomenon, and observers note that it seems to be getting worse every year.

Turning eastward, in 1988, Dr. Joanne Burkholder, a North Carolina State University plankton biologist, examined some open sores that were found on menhaden fish from a North Carolina estuary. She identified a new organism, a microscopic dinoflagellate (a type of alga) she named *Pfiesteria piscicida*. (Recently, a second toxic species, *Pfiesteria shumwayae*, has been identified.) *Pfiesteria* is a shapeshifter: it can be an amoeba, or a flagellate, or a cyst. Sometimes it secretes a toxin that stuns the fish, and then it turns into a form that proceeds to feed on the fish, which have been disoriented by the toxin. Sometimes it behaves like an ordinary alga and lives by photosynthesis.

Pfiesteria has been implicated in at least 50 major fish kills in Albemarle and Pamlico sounds, North Carolina, and has been identified in waters from Virginia, Maryland, Delaware, Florida, and even northern Europe. In the summer of 1997, fish kills struck several tributaries of the Chesapeake Bay, particularly the Pocomoke River. The fish kills and catches of fish with ulcerous lesions finally led the state to close the river to all fishing and harvesting of shellfish. When news of the fish kills in the Chesapeake Bay and the impacts on humans became public, there was a predictable panic response, and consumers all around the Chesapeake Bay stopped eating seafood. In Maryland alone, the economic loss was estimated at between $15 million and $20 million in sales during 1997.

The domoic acid and *Pfiesteria* episodes join a long list of what are called *harmful algal blooms* (*HABs*). Red tides, consisting of concentrations of different species of toxic dinoflagellates, have been occurring in coastal waters all around North America. They may be responsible for more than $1 billion in economic damage to coastal systems and fisheries. Scientists believe that nutrients—especially nitrogen and phosphorus—are the basic culprits in feeding the growth of the algae in estuaries and coastal waters. Recently, five federal agencies joined to sponsor the Ecology and Oceanography of Harmful Algal Blooms (ECOHAB), a national research program studying HABs in U.S. coastal waters. Like the dead zones, the blooms of harmful algae are increasing in frequency. These blooms are virtually always associated with nutrient enrichment, and they are found in freshwater lakes as well as coastal marine environments. They are a nasty reminder that our work in cleaning up the nation's waterways still has a long way to go.

Lesions on menhaden fish caused by *Pfiesteria*.

20.1 Perspectives on Water Pollution

In the early days of the Industrial Revolution in the United States, little attention was given to the pollution of lakes, rivers, and the coastal ocean. Industries emptied their wastes into the closest waterway, and the general means of disposing of human excrement was the outdoor privy (or behind the nearest bush). Seepage from the privy frequently contaminated drinking water and caused disease, especially in places where privies and wells were located near one another. In the late 1800s, Louis Pasteur and other scientists showed that sewage-borne bacteria were responsible for many infectious diseases. This important discovery led to intensive efforts to rid cities of human excrement as expediently as possible. Cities already had drain systems for storm water, but using these systems for human wastes had been prohibited. With the urgency of the situation, however, minds quickly changed. The flush toilet was introduced, and sewers were tapped into storm drains. Thus, Western civilization adopted the one-way flow of flushing domestic and industrial wastewater into natural waterways. Introducing sewage into a waterway was symptomatic of an approach that believed "the solution to pollution is dilution."

As a result of this practice, receiving waters that had limited capacity for dilution became open cesspools of foul odors, vermin, and filth as the overload of organic matter depleted dissolved oxygen and most aquatic life suffocated. For increasing distances around or downstream from the sewage outfall, the water became unfit for any recreational use because of the sewage contamination. Any further use of the water for human consumption required extensive treatment to remove pathogens.

Industrial wastes only added insult to injury. From the dawn of the Industrial Age to the relatively recent past, it was common practice to flush all waste liquids and contaminated water into sewer systems or directly into natural waterways. Many human health problems followed, but they either were

Figure 20–2 Cuyahoga River on fire. In 1969, the Cuyahoga River in Cleveland, Ohio, caught fire, burning seven bridges. Flammable material from industrial sources was being discharged into the river. This incident captured national attention and contributed to the public outrage over water pollution, leading to the Clean Water Act of 1972.

not recognized as being caused by the pollution or were accepted as the "price of progress."

The Federal Water Pollution Control Act of 1948 was the first federal foray into water pollution legislation, providing technical assistance to state and local governments, but otherwise leaving things up to the states and local municipalities. In the 1950s, as industrial production expanded and synthetic organics came into widespread use in the developed countries, many streams and rivers essentially became open chemical sewers as well as sewers for human waste. These waters not only were devoid of life, but also were themselves hazardous. Finally, in 1969, the Cuyahoga River, which flows through

Cleveland, Ohio, was carrying so much flammable material that it actually caught fire, which destroyed seven bridges before the fire burned itself out **(Fig. 20–2)**.

Worsening pollution (from both chemicals and sewage) and increasing recognition of the adverse health effects finally created a degree of public outrage that pushed Congress to pass the **Clean Water Act of 1972 (CWA)**. This legislation gave the newly created Environmental Protection Agency (EPA) the responsibility to carry out a very ambitious goal—namely, to restore and maintain the chemical, physical, and biological integrity of the nation's waters. The CWA was to achieve zero discharge of pollutants by 1985, a totally naïve (and impossible) task, given the nature of the pollutants involved. However, imperfect as it is, the Clean Water Act (and its amendments) is still recognized as one of the most far-reaching and effective environmental laws ever enacted.

Water Pollution: Sources and Types

Water pollutants originate from a host of human activities and reach surface water or groundwater through an equally diverse host of pathways. For purposes of regulation, it is customary to distinguish between **point sources** and **nonpoint sources** of pollutants **(Fig. 20–3)**. Point sources involve the discharge of substances from factories, sewage systems, power plants, underground coal mines, and oil wells. These sources are relatively easy to identify and, therefore, are easier to monitor and regulate than nonpoint sources, which are poorly defined and scattered over broad areas. Some of the most prominent nonpoint sources of pollution are agricultural runoff (from farm animals and croplands), storm-water drainage (from streets, parking lots, and lawns), and atmospheric deposition (from air pollutants washed to or deposited as dry particles on Earth). Pollution occurs as rainfall and snowmelt move over and through the ground, picking up pollutants as they go.

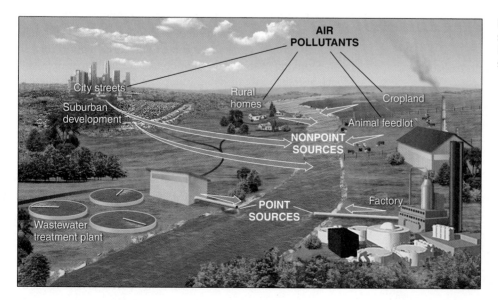

Figure 20–3 Point and nonpoint sources. Point sources are far easier to identify and correct than the diffuse nonpoint sources.

Two basic strategies are employed in attempting to bring water pollution under control: (1) reduce or remove the sources and (2) treat the water before it is released so as to remove pollutants or convert them to harmless forms. Water treatment is the best option for point sources. Source reduction can be employed for both kinds of sources and is the best option for nonpoint sources.

Pathogens. The most serious and widespread water pollutants are infectious agents that cause sickness and death (see Chapter 17). The excrement from humans and other animals infected with certain **pathogens** (disease-causing bacteria, viruses, and other parasitic organisms) contains large numbers of these organisms or their eggs. Table 20–1 lists the most common waterborne pathogens. Even after symptoms of disease disappear, an infected person or animal may still harbor low populations of the pathogen, thus continuing to act as a carrier of disease. If wastes from carriers contaminate drinking water, food, or water used for swimming or bathing, the pathogens can gain access to, and infect, other individuals **(Fig. 20–4)**.

Public Health. Before the connection between disease and sewage-carried pathogens was recognized in the mid-1800s, disastrous epidemics were common in cities. Epidemics of typhoid fever and cholera killed thousands of people prior to the 20th century. Today public-health measures that prevent this disease cycle have been adopted throughout the developed world and, to an increasing extent, in the developing world. The following measures were far more important than modern medicine in controlling waterborne diseases:

1. Purification and disinfection of public water supplies with chlorine or other agents (Chapter 10)

Figure 20–4 The Ganges River in India. In many places in the developing world, the same waterways are used simultaneously for drinking, washing, and disposal of sewage. A high incidence of disease, infant and childhood mortality, and parasites is the result.

2. Sanitary collection and treatment of human and animal wastes

3. Maintenance of sanitary standards in all facilities in which food is processed or prepared for public consumption

4. Instruction in personal and domestic hygiene practices

Standards regarding these measures are set and enforced by government public-health departments. A variety of other measures are enforced as well. For example, if bathing areas are contaminated with raw sewage, health departments close them to swimming. Implicit in all measures is monitoring for sewage contamination. (See Sound Science, "Monitoring for Sewage Contamination.") Our own personal hygiene, sanitation, and health precautions—such as not drinking water from untested sources, washing hands frequently, and making sure that foods like pork, chicken, and hamburger are always well cooked—remain the last and most important line of defense against disease.

Sanitation = Good Medicine. Many people attribute good health in a population to modern medicine, but good health is primarily a result of *the prevention of disease through public-health measures.* According to the 2012 report Progress on Drinking Water and Sanitation,[1] 780 million people still lack access to safe drinking water and 2.5 billion live in areas having poor (or no) sewage collection or treatment **(Fig. 20–5)**. Each year, more than 2 million deaths (mostly among children under five) are traced to waterborne diseases. Millennium Development Goal 7 addresses these serious public-health needs: "Halve, by 2015, the proportion of people without sustainable access to safe drinking water and basic sanitation" (Table 9–2). The 2012 progress report stated that the drinking water target was met in 2010, but the world is seriously behind in many regions in meeting the sanitation

TABLE 20–1	Pathogens Carried by Sewage
Disease	**Infectious Agent**
Typhoid fever	*Salmonella typhi* (bacterium)
Cholera	*Vibrio cholerae* (bacterium)
Salmonellosis	*Salmonella* species (bacteria)
Diarrhea	*Escherichia coli* (bacterium)
	Campylobacter species (bacteria)
	Cryptosporidium parvum (protozoan)
Infectious hepatitis	Hepatitis A virus
Poliomyelitis	Poliovirus
Dysentery	*Shigella* species (bacteria)
	Entamoeba histolytica (protozoan)
Giardiasis	*Giardia intestinalis* (protozoan)
Numerous parasitic diseases	Roundworms, flatworms

[1]Progress on Drinking Water and Sanitation: 2012 Update. UNICEF and World Health Organization. 2012. http://www.unicef.org/media/files/JMPreport2012.pdf.

SOUND SCIENCE

Monitoring for Sewage Contamination

An important aspect of public health is the detection of sewage in water supplies and other bodies of water with which humans come in contact. It is virtually impossible to test for each specific pathogen that might be present in a body of water. Therefore, an indirect method called the *fecal coliform test* has been developed. This test is based on the fact that huge populations of a bacterium called *E. coli* (*Escherichia coli*) normally inhabit the lower intestinal tract of humans and other animals, and large numbers of the bacterium are excreted with fecal material. In temperate regions at least, *E. coli* does not last long in the outside environment. Therefore, the presumption is that when *E. coli* is found in natural waters, it is an indication of recent and probably persisting contamination with sewage wastes. In most situations, *E. coli* is not a pathogen itself, but is referred to as an *indicator organism*: Its presence *indicates* that water is contaminated with fecal wastes and that sewage-borne pathogens may be present. Conversely, the absence of *E. coli* is taken to mean that water is free from such pathogens.

The fecal coliform test, one technique of which is shown in the accompanying figure, detects and counts the number of *E. coli* in a sample of water. Thus, the results indicate the relative degree of contamination of the water and the relative risk of pathogens. For example, to be safe for drinking, water may not have any fecal coliform in a 0.4-cup (100 mL) sample. The same sample with as many as 200 *E. coli* is still considered safe for swimming. Beyond that level, a river or beach may be posted as polluted, and swimming and other direct contact should be avoided. By contrast, raw sewage (99.9% water, 0.1% waste) has *E. coli* counts in the millions.

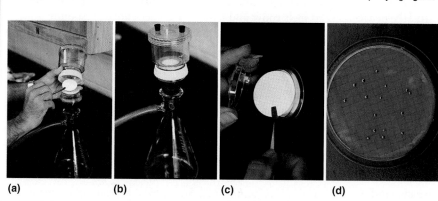

(a)   (b)   (c)   (d)

Testing water for sewage contamination by the Millipore technique. (a) A Millipore filter disk is placed in the filter apparatus. (b) A sample of the water being tested is drawn through the filter, and any bacteria present are entrapped on the disk. (c) The disk is then placed in a petri dish on a special medium that supports the growth of bacteria and that will impart a particular color to fecal *E. coli* bacteria. Next, the dish is incubated for 24 hours at 44.5°C, during which time the bacteria on the disk will multiply to form a colony visible to the naked eye. (d) *Escherichia coli* bacteria, indicating sewage contamination, are identifiable as the colonies with a metallic green sheen.

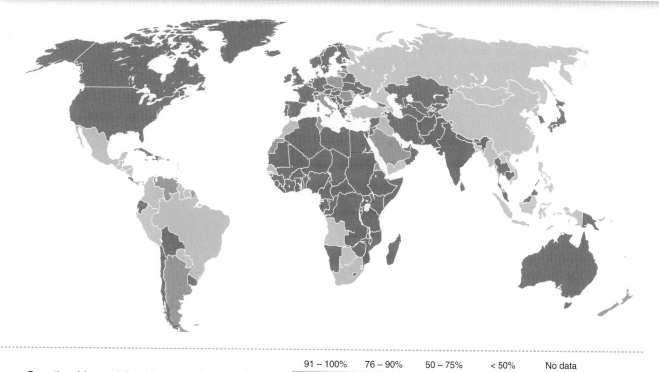

Proportion of the population using improved sanitation in 2010

91 – 100% 76 – 90% 50 – 75% < 50% No data

Figure 20–5 Worldwide distribution of improved sanitation. Less than half the population has adequate sanitation in much of the developing world. The Millennium Development Goal of reducing by half the proportion of people without improved sanitation will not be met without additional effort.

(*Source:* Progress on Drinking Water and Sanitation: 2012 Update. UNICEF and World Health Organization, 2012. www.unicef.org/media/files/JMPreport2012.pdf.)

target. Almost one-fourth of the developing world's people have no adequate place to go to the bathroom and generally relieve themselves in streams, backyards, and woods, contaminating the water and the places where children play. When it rains, the drainage ditches and alleys run with human and animal excrement.

Largely because of poor sanitation regarding water and sewage, a significant portion of the world's population is chronically infected with various pathogens. Moreover, populations in areas where there is little or no sewage treatment are extremely vulnerable to deadly epidemics of many diseases spread by way of sewage. This is especially true of cholera. Because of unsanitary conditions, an outbreak of cholera in Peru in 1990 killed almost 9,000 people as it spread through Latin America. The 2010 earthquake in Haiti created huge refugee camps, where a deadly outbreak of cholera brought 180,000 cases and 5,000 deaths. The source of the cholera there was traced to UN peacekeepers from Nepal, where the disease is endemic. Wastes from the peacekeepers' camps were simply emptied into the nearest watercourse. The World Health Organization (WHO) reports numerous cholera outbreaks annually, and cases frequently number in the thousands, with hundreds of deaths.

Organic Wastes. Along with pathogens, human and animal wastes contain organic matter that creates serious problems if it enters bodies of water untreated. Other kinds of organic matter (leaves, grass clippings, trash, and so forth) can enter bodies of water as a consequence of runoff. With the exception of plastics and some human-made chemicals, these wastes are biodegradable. As mentioned in connection with the Gulf of Mexico dead zone, when bacteria and detritus feeders decompose organic matter in water, they consume oxygen gas dissolved in the water. The amount of oxygen that water can hold in solution is severely limited. In cold water, dissolved oxygen (DO) can reach concentrations up to 10 parts per million (ppm); much less can be held in warm water. Compare this ratio with that of oxygen in air, which is 200,000 ppm (20%), and you can understand why even a moderate amount of organic matter decomposing in water can deplete the water of its DO. Bacteria keep the oxygen supply depleted as long as there is dead organic matter to support their growth and oxygen replenishment is inadequate.

BOD. Biochemical oxygen demand (BOD) is a measure of the amount of organic material in water, stated in terms of how much oxygen will be required to break it down biologically, chemically, or both. The higher the BOD measure, the greater the likelihood that dissolved oxygen will be depleted in the course of breaking down the organic material. A high BOD causes so much oxygen depletion that animal life is severely limited or precluded, as in the bottom waters of the Gulf of Mexico. Fish and shellfish are killed when the DO drops below 2 or 3 ppm, though some are less tolerant at even higher DO levels. If the system

goes anaerobic (has zero oxygen), only bacteria can survive, using their abilities to switch to fermentation or anaerobic respiration (metabolic pathways that do not require oxygen). A typical BOD value for raw sewage would be around 250 ppm. Even a moderate amount of sewage added to natural waters containing at most 10 ppm DO can deplete the water of its oxygen and produce highly undesirable consequences well beyond those caused by the introduction of pathogens (**Fig. 20–6**).

Chemical Pollutants. Because water is such an excellent solvent, it is able to hold many chemical substances in solution that have undesirable effects. Water-soluble **inorganic chemicals** constitute an important class of pollutants that include heavy metals (lead, mercury, arsenic, nickel, and so forth), acids from mine drainage (sulfuric acid) and acid precipitation (sulfuric and nitric acids), and road salts employed to melt snow and ice in the colder climates (sodium and calcium chlorides). The **organic chemicals** are another group of substances found in polluted waters. Petroleum products pollute many bodies of water, from the major oil spills in the ocean to the small streams receiving runoff from parking lots. Other organic substances with serious impacts are the pesticides that drift down from aerial spraying or that run off from land areas, as well as the various industrial chemicals, such as polychlorinated biphenyls (PCBs), cleaning solvents, and detergents (Chapter 13).

Many of these pollutants are toxic even at low concentrations (Chapter 22). Some may become concentrated as they are passed up the food chain in a process called biomagnification (Chapter 13). Even at very low concentrations, they can render water unpalatable to humans and dangerous to aquatic life. At higher concentrations, they can change the properties of bodies of water so as to prevent them from serving any useful purpose except perhaps navigation. Acid mine drainage, for example, pollutes thousands of miles of streams in coal-bearing regions of the United States (**Fig. 20–7**). Stream bottoms are coated with orange deposits of iron, and the water is so acidic that only a few hardy bacteria and algae can tolerate it.

Sediments. As natural landforms weather, especially during storms, a certain amount of sediment enters streams and rivers. However, erosion from farmlands, deforested slopes, overgrazed rangelands, construction sites, mining sites, stream banks, and roads can greatly increase the load of sediment entering waterways. Frequently, storm drains simply lead to the nearest depression or some natural streambed. Sediments (sand, silt, and clay) have direct and extreme physical impacts on streams and rivers. When erosion is slight, streams and rivers of the watershed run clear and support algae and other aquatic plants that attach to rocks or take root in the bottom. These producers, plus miscellaneous detritus from fallen leaves, support a complex food web of bacteria, protozoa, worms, insect larvae, snails, fish, crayfish, and other organisms, which keep themselves from being carried downstream by attaching to rocks

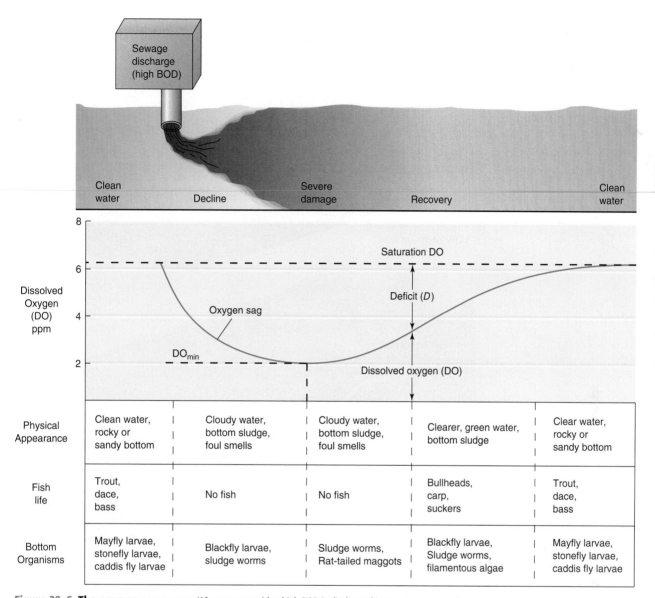

	Clean water	Decline	Severe damage	Recovery	Clean water
Physical Appearance	Clean water, rocky or sandy bottom	Cloudy water, bottom sludge, foul smells	Cloudy water, bottom sludge, foul smells	Clearer, green water, bottom sludge	Clear water, rocky or sandy bottom
Fish life	Trout, dace, bass	No fish	No fish	Bullheads, carp, suckers	Trout, dace, bass
Bottom Organisms	Mayfly larvae, stonefly larvae, caddis fly larvae	Blackfly larvae, sludge worms	Sludge worms, Rat-tailed maggots	Blackfly larvae, Sludge worms, filamentous algae	Mayfly larvae, stonefly larvae, caddis fly larvae

Figure 20–6 The oxygen sag curve. When sewage with a high BOD is discharged into a stream or small river, it creates an oxygen deficit and severely affects the stream's biology.

Figure 20–7 Acid mine drainage. Acidic water from mines contains high concentrations of sulfides and iron, which are oxidized when they come into the open, thus producing unsightly and sterile streams.

or seeking shelter behind or under rocks. Even fish that maintain their position by active swimming occasionally need such shelter to rest (**Fig. 20–8a**).

Suspended Load. Sediment entering waterways in large amounts has an array of impacts. Sand, silt, clay, and organic particles (humus) are quickly separated by the agitation of flowing water and are carried at different rates. Clay and humus are carried in suspension, making the water turbid and reducing the amount of light penetrating the water and, hence, reducing photosynthesis as well. As the material settles, it coats everything and continues to block photosynthesis. It also kills the animals by clogging their gills and feeding structures. The eggs of fish and other aquatic organisms are particularly vulnerable to being smothered by sediment.

Bed Load. Especially destructive is the **bed load** of sand and silt, which is not readily carried in suspension, but is

Figure 20–8 Impact of sediment on streams and rivers. (a) The ecosystem of a stream that is not subjected to a large sediment bed load. (b) The changes that occur with large sediment inputs. (c) Platte River at Lexington, Nebraska. The sandbars seen here constitute the bed load; they shift and move with high water, preventing the reestablishment of aquatic vegetation.

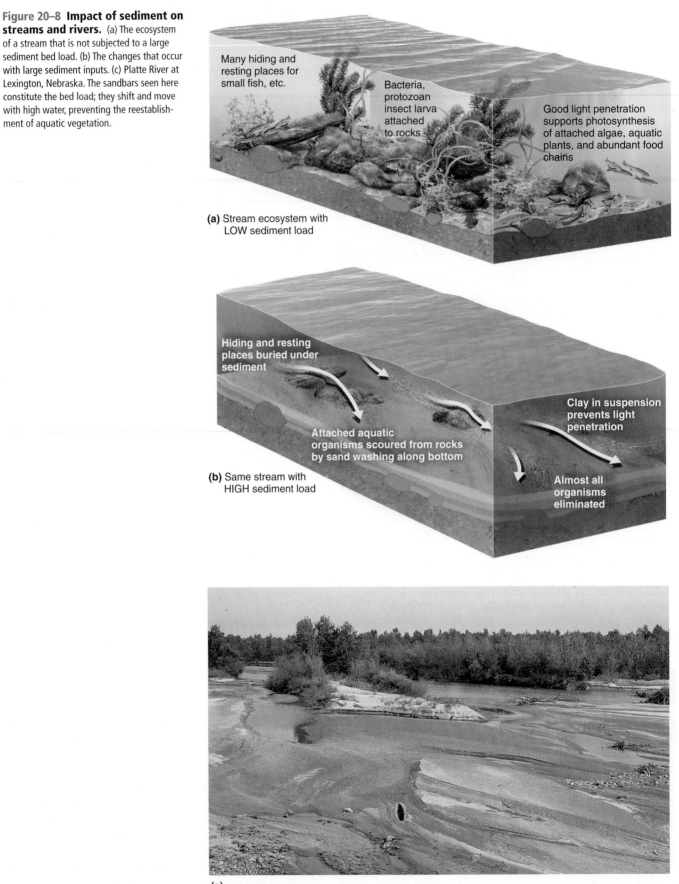

Many hiding and resting places for small fish, etc.

Bacteria, protozoan insect larva attached to rocks

Good light penetration supports photosynthesis of attached algae, aquatic plants, and abundant food chains

(a) Stream ecosystem with LOW sediment load

Hiding and resting places buried under sediment

Attached aquatic organisms scoured from rocks by sand washing along bottom

Clay in suspension prevents light penetration

Almost all organisms eliminated

(b) Same stream with HIGH sediment load

(c)

Figure 20–9 Storm-water management. Rather than letting excessive runoff from developed areas cause flooding and other environmental damage, runoff can be funneled into a retention pond, as shown here. Then it can drain away slowly, maintaining natural stream flow, or it can recharge groundwater. The retention pond may be designed to retain a certain amount of water and thus create a pocket of wildlife habitat in an otherwise urban or suburban setting.

gradually washed along the bottom. As particles roll and tumble along, they scour organisms from the rocks. They also bury and smother the bottom life and fill in the hiding and resting places of fish and crayfish. Aquatic plants and other organisms are prevented from reestablishing themselves because the bottom is a constantly shifting bed of sand (Fig. 20–8b and c). Modern storm-water management is designed to reduce the bed load, usually via storm drains that are periodically emptied of their sediment. Many newer housing developments include a storm-water retention reservoir, which is simply a pond that receives and holds runoff from the area during storms. Water may infiltrate the soil or may create a pocket of natural wetland habitat supporting wildlife (Fig. 20–9).

Sediments do not receive the attention the news media give to hazardous wastes and certain other pollution problems, but erosion is so widespread throughout the world that few streams and rivers escape the harsh impact of excessive sediment loads.

Nutrients. Some of the inorganic chemicals carried in solution in all bodies of water are classified as nutrients—essential elements required by plants. The two most important nutrient elements for aquatic plant growth are **phosphorus** and **nitrogen,** and they are often in such low concentrations in water that they are the limiting factors for phytoplankton or other aquatic plants. More nutrients mean more plant growth, so nutrients become water pollutants when they are added from point or nonpoint sources and stimulate undesirable plant growth in bodies of water. The most obvious point sources of excessive nutrients are sewage outfalls. As we will see in Section 20.2, it is costly to remove nutrients from sewage during the treatment process, so that is not always done. If the sewage is not treated at all, higher levels of nutrients will enter the water, which is what's happening in much of the developing world.

Agricultural runoff is the most notorious nonpoint source of nutrients. The nutrients are applied to agricultural crops as chemical fertilizers, manure, and dissolved in irrigation water, and are present in crop residues. When nutrients are applied in excess of the needs of plants or when runoff from agricultural fields and feedlots is heavy, the nutrients are picked up by water that eventually enters streams, rivers, lakes, and the ocean. Other nonpoint sources of nutrients include acid rain and snow, lawns and gardens, golf courses, and storm drains.

Water Quality Standards. When is water considered polluted? Many water pollutants—like pesticides, cleaning solvents, and detergents—are substances that are found in water only because of human activities. Others, such as nutrients and sediments, are always found in natural waters, and they are a problem only under certain conditions. In both cases, *pollution* means any quantity that is harmful to human health or the environment or prevents full use of the environment by humans or natural species. The mere presence of a substance in water does not necessarily pose a problem. Rather, it is the *concentration* of the pollutant that must be of primary concern.

Criteria Pollutants. But what concentration is worrisome? To provide standards for assessing water pollution, the EPA has established the National Recommended Water Quality Criteria. On the basis of the latest scientific knowledge, the EPA has listed 167 chemicals and substances as **criteria pollutants.** The majority of these are toxic chemicals, but many are also natural chemicals or conditions that describe the state of water, such as nutrients, hardness (a general measure of dissolved calcium and magnesium salts) and pH (a measure of the acidity of the water). The list identifies the pollutant and then recommends concentrations for freshwater, saltwater, and human consumption (usually of fish and shellfish). Values are given for the **criteria maximum concentration (CMC),** the highest single concentration beyond which environmental impacts may be expected, and the **criterion continuous concentration (CCC),** the highest sustained concentration beyond which undesirable impacts may be expected. The criteria are *recommendations* meant to be used by the states, which are given the primary responsibility for upholding water-pollution laws. States and Native American tribes may revise these criteria, but their revisions are subject to EPA approval.

Drinking water standards are much stricter. For these, the EPA has established the Drinking Water Standards and Health Advisories, a set of tables that are updated periodically. These standards, covering some 94 contaminants, are enforceable under the authority of the **Safe Drinking Water Act (SDWA) of 1974.** They are presented as **maximum contaminant levels (MCLs).** To see how these two sets of standards work, consider the heavy metal arsenic.

Arsenic. Arsenic is listed as a known human carcinogen by the Department of Health and Human Services; it occurs naturally in groundwater, reaching high concentrations in some parts of the world. The CMC and CCC values for arsenic are 340 µg/L and 150 µg/L (1 µg/L = 1 part per billion) for freshwater bodies and 69 µg/L and 36 µg/L for saltwater bodies. The drinking-water MCL concentration, however, is 10 µg/L, and therein lies a story.

For years, the MCL concentration for arsenic was 50 µg/L, despite warnings by the U.S. Public Health Service and the

WHO that this was much too high a level. (Both bodies recommended 10 µg/L.) The problem is the cancer risk. From standard risk assessment processes, the cancer mortality risk was assessed at 1 in 100 for people drinking water regularly at the 50 µg/L level (Chapter 17). This ratio is more than 100 times higher than the permitted risk for any other contaminant. Before President Clinton left office in early 2001, the EPA determined that the MCL for arsenic should be lowered, but the pending rule was withdrawn by the next administration, on the premise that "sound science" did not clearly support the new rule. Many states and rural areas were complaining that attaining the lower arsenic concentration would be too costly. In the end, after a National Academy of Sciences report and urgings from both the House and the Senate, then EPA administrator Christine Whitman upheld the Clinton era rule and set the arsenic MCL at 10 µg/L in October 2001.

Other Applications. Two important applications of water quality criteria are the **National Pollution Discharge Elimination System (NPDES)** and **Total Maximum Daily Load (TMDL)** programs. The NPDES program addresses point-source pollution and issues *permits* that regulate discharges from wastewater-treatment plants and industrial sources. The TMDL program, by contrast, evaluates all sources of pollutants entering a body of water, especially nonpoint sources, according to the water body's ability to assimilate the pollutant. For both of these programs, water quality criteria provide an essential means of assessing the condition of the receiving body of water as well as evaluating the levels of the various pollutant discharges. We will examine these two programs in more detail when we consider the problem of eutrophication.

The good news is that, according to the EPA, 92% of the people in the United States have access to drinking water that meets the *drinking water standards*. However, on the basis of data supplied to the EPA by the states, 41,266 rivers, lakes, and estuaries are not meeting the recommended *water quality standards*, and this excludes more than 60% of U.S. waters that have not been assessed at all. The major problems (in order) are pathogens, mercury, nutrients, other heavy metals, sediments, and oxygen depletion. As we proceed, we will address the specifics of public-policy responses to this serious national problem. We turn now to ways of dealing with the biological wastes that are a large part of our pollution problem.

20.2 Wastewater Treatment and Management

To alleviate the problem of sewage-polluted waterways, facilities were designed and constructed to treat the outflow before it entered the receiving waterway. The first wastewater-treatment plants in the United States were built around 1900. However, the combined volumes of sewage and storm water soon proved impossible to handle. During heavy rains, wastewater would overflow the treatment plant and carry raw sewage into the receiving waterway. Gradually, regulations were passed requiring municipalities to install separate systems—**storm drains** to collect and drain runoff from precipitation and **sanitary sewers** to receive all the wastewater from sinks, tubs, and toilets in homes and other buildings. (Note the distinction between the two terms; it is incorrect to speak of storm drains as sewers.)

The ideal modern system, then, is one in which all wastewater is collected separately from storm water and is fully treated to remove pollutants before the water finally is released. But progress toward this goal has been extremely uneven. Up through the 1970s, even in the United States and other developed countries, thousands of communities still discharged untreated sewage directly into waterways, and for many more, the degree of treatment was minimal. Indeed, the increasing sewage pollution of waterways and beaches was the major impetus behind the passage of the Clean Water Act.

Before addressing methods of treating wastewater, let us clarify which contaminants and pollutants we are talking about.

The Pollutants in Raw Wastewater

Raw wastewater is not only the flushings from toilets, but also the collection of wastewater from all other drains in homes and other buildings. A sewer system brings all tub, sink, and toilet drains from all homes and buildings together into larger and larger sewer pipes, just as the branches of a tree eventually come together into the trunk. This total mixture collected from all drains, which flows from the end of the "trunk" of the collection system, is called **raw sewage** or **raw wastewater**. Because we use such large amounts of water to flush away small amounts of dirt, especially as we stand under the shower or often just run the water, most of what goes down sewer drains is water. Raw wastewater is about 1,000 parts water for every 1 part waste—99.9% water to 0.1% waste.

Given our voluminous use of water, the quantity of raw wastewater output is on the order of 150–200 gallons (600–800 L) per person per day. That is, a community of 10,000 persons will produce 1.5 million to 2.0 million gallons (6–8 million L) of wastewater each day. With the addition of storm water, raw wastewater is diluted still more. Nevertheless, the pollutants are sufficient to make the water brown and foul smelling.

The pollutants in wastewater generally are divided into the following four categories, which correspond to the techniques used to remove them:

1. **Debris and grit:** rags, plastic bags, coarse sand, gravel, and other objects flushed down toilets or washed through storm drains.

2. **Particulate organic material:** fecal matter, food wastes from garbage-disposal units, toilet paper, and other matter that tends to settle in still water.

3. **Colloidal and dissolved organic material:** very fine particles of particulate organic material, bacteria, urine, soaps, and detergents and other cleaning agents.

4. **Dissolved inorganic material:** nitrogen, phosphorus, and other nutrients from excretory wastes and detergents.

In addition to the four categories of pollutants in "standard" raw wastewater, variable amounts of pesticides, heavy metals, and other toxic compounds may be found because

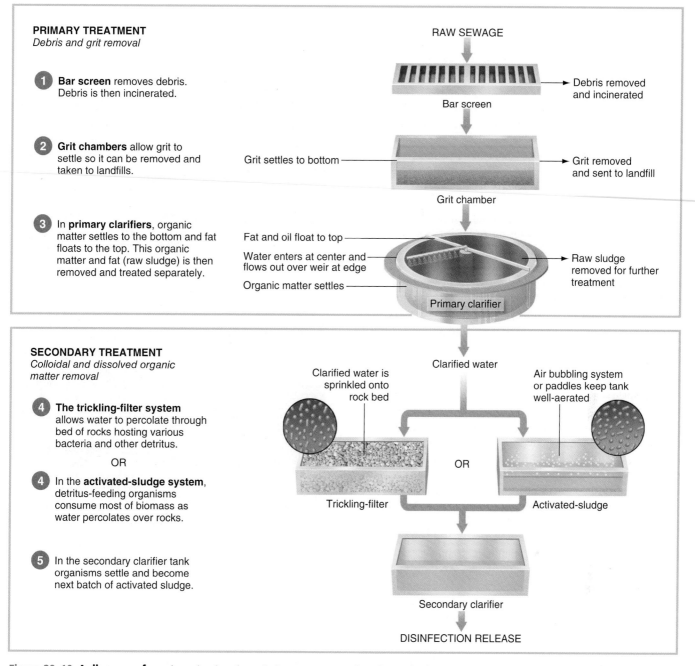

PRIMARY TREATMENT
Debris and grit removal

1 **Bar screen** removes debris. Debris is then incinerated.

2 **Grit chambers** allow grit to settle so it can be removed and taken to landfills.

3 In **primary clarifiers**, organic matter settles to the bottom and fat floats to the top. This organic matter and fat (raw sludge) is then removed and treated separately.

RAW SEWAGE

Bar screen

Debris removed and incinerated

Grit settles to bottom

Grit chamber

Grit removed and sent to landfill

Fat and oil float to top
Water enters at center and flows out over weir at edge
Organic matter settles
Primary clarifier

Raw sludge removed for further treatment

SECONDARY TREATMENT
Colloidal and dissolved organic matter removal

4 The **trickling-filter system** allows water to percolate through bed of rocks hosting various bacteria and other detritus.

OR

4 In the **activated-sludge system**, detritus-feeding organisms consume most of biomass as water percolates over rocks.

5 In the secondary clarifier tank organisms settle and become next batch of activated sludge.

Clarified water

Clarified water is sprinkled onto rock bed

Air bubbling system or paddles keep tank well-aerated

Trickling-filter

OR

Activated-sludge

Secondary clarifier

DISINFECTION RELEASE

Figure 20–10 A diagram of wastewater treatment. Raw sewage moves from the grit chamber to undergo primary treatment, in which sludge is removed. The clarified water then undergoes secondary treatment (shown here as activated-sludge or trickling-filter treatment).

people often pour unused portions of products containing such materials down sink, tub, or toilet drains. Also, industries may discharge various toxic wastes into sewers.

Removing the Pollutants from Wastewater

The challenge of treating wastewater is more than installing a technology that will do the job; it is also finding one that will do the job *at a reasonable cost*. A diagram of wastewater treatment procedures is shown in **Figure 20–10**.

Primary Treatment. Because debris and grit will damage or clog pumps and further treatment processes, removing this

material is a necessary first step. Usually, removal involves two steps: the screening out of debris and the settling of grit. Debris is removed by letting raw sewage flow through a **bar screen**, a row of bars mounted about 1 inch apart. Debris is mechanically raked from the screen and taken to an incinerator. After passing through the screen, the water flows through a **grit chamber**—a swimming pool–like tank—in which velocity is slowed just enough to permit the grit to settle. The settled grit is mechanically removed from these tanks and taken to landfills.

After debris and grit are removed, the water then flows very slowly through large tanks called **primary clarifiers**. Because its flow is slow, the water is nearly motionless for several hours. The particulate organic material, about 30–50%

of the total organic material, settles to the bottom, where it can be removed. At the same time, fatty or oily material floats to the top, where it is skimmed from the surface. All the material that is removed, both particulate organic material and fatty material, is combined into what is referred to as **raw sludge**, which must be treated separately.

Primary treatment involves nothing more complicated than putting polluted water into a tank, letting material settle, and pouring off the water. Nevertheless, such treatment removes much of the particulate organic matter at minimal cost.

Secondary Treatment. **Secondary treatment** is also called biological treatment because it uses organisms—natural decomposers and detritus feeders. Basically, an environment is created that enables these organisms to feed on the colloidal and dissolved organic material and break it down to carbon dioxide, mineral nutrients, and water via their cell respiration. The wastewater from primary treatment is a food- and water-rich medium for the decomposers and detritus feeders. The only thing that needs to be added to the water is oxygen to enhance the organisms' respiration and growth. Either of two systems may be used to add oxygen to the water: a *trickling-filter system* or an *activated-sludge system*. In the **trickling-filter system**, the water exiting from primary treatment is sprinkled onto, and allowed to percolate through, a bed of fist-sized rocks 6–8 feet (2–3 m) deep **(Fig. 20–11)**. The spaces between the rocks provide good aeration. Like a natural stream, this environment supports a complex food web of bacteria, protozoans, rotifers (organisms that consume protozoans), various small worms, and other detritus feeders attached to the rocks. The organic material in the water, including pathogenic organisms, is absorbed and digested by decomposers and detritus feeders as it trickles by.

The **activated-sludge system** (Fig. 20–10, bottom) is a more efficient, and the most common, secondary-treatment system. Water from primary treatment enters a large tank that is equipped with an air-bubbling system or a rapidly churning system of paddles. A mixture of detritus-feeding organisms, referred to as **activated sludge**, is added to the water as it enters the tank, and the water is vigorously aerated as it moves through the tank. Organisms in this well-aerated environment reduce the biomass of organic material, including pathogens, as they feed. As the organisms feed on each other, they tend to form into clumps, called *floc*, that settle readily when the water is stilled. Then, from the aeration tank, the water is passed into a **secondary clarifier** tank, where the organisms settle out, and the water—now with more than 90% of the organic material removed—flows on. The settled organisms are pumped back into the aeration tank. (They are the activated sludge that is added at the beginning of the process.) Surplus amounts of activated sludge, which occur as populations of organisms grow, are removed and added to the raw sludge. The organic material is oxidized by microbes to form carbon dioxide, water, and mineral nutrients that remain in the water solution.

Biological Nutrient Removal. The original secondary-treatment systems were designed and operated so as to simply maximize biological digestion, because the elimination

(a)

(b)

Figure 20–11 Trickling filters for secondary treatment.
(a) The water from primary clarifiers is sprinkled onto, and trickles through, a bed of rocks 6–8 feet deep. (b) Various bacteria and other detritus feeders adhering to the rocks consume and digest organic matter as it trickles by in the water, which is collected at the bottom of the filters and goes on for final clarification and disinfection.

of organic material and its resulting BOD was considered the prime objective. Ending up with a nutrient-rich discharge was not recognized as a problem. Today, with increased awareness of the problem of cultural eutrophication, secondary activated-sludge systems have been added and are being modified and operated in a manner that both removes nutrients and oxidizes detritus in a process known as **biological nutrient removal (BNR)**.

Nitrogen. Recall that, in the natural nitrogen cycle, various bacteria convert nutrient forms of nitrogen (ammonia and nitrate) back to non-nutritive nitrogen gas in the atmosphere through denitrification (Chapter 3, Fig. 3–20). For the biological removal of nitrogen, then, the activated-sludge system is partitioned into zones, and the environment in each zone is controlled in a manner that promotes the denitrifying process **(Fig. 20–12)**. The resulting inert nitrogen gas is vented into the air.

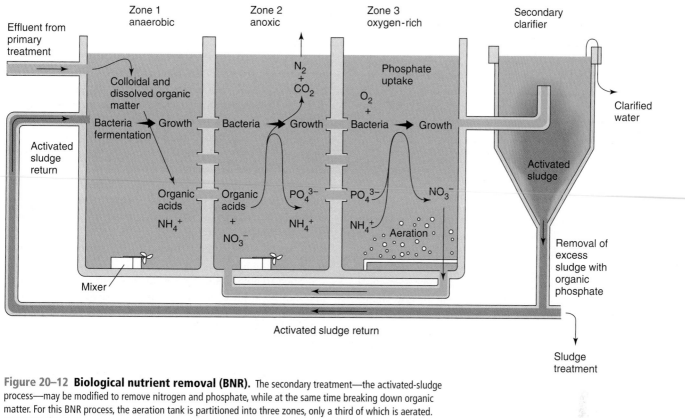

Figure 20–12 Biological nutrient removal (BNR). The secondary treatment—the activated-sludge process—may be modified to remove nitrogen and phosphate, while at the same time breaking down organic matter. For this BNR process, the aeration tank is partitioned into three zones, only a third of which is aerated. As seen in the diagram, ammonium (NH_4^+) is converted to nitrate (NO_3^-) in zone 3. Recycled to zone 2, which is without oxygen gas (anoxic), the nitrate supplies the oxygen for cell respiration. In the process, the nitrate is converted to nitrogen gas, which is released into the atmosphere. Phosphate is taken up by bacteria in zone 3 and removed with the excess sludge.

Phosphorus. In an environment that is rich in oxygen but relatively lacking in food (the environment of zone 3 in Figure 20–12), bacteria take up phosphate from the solution and store it in their bodies. Thus, phosphate is removed as the excess organisms are removed from the system. These organisms, together with the phosphate they contain, are added to, and treated with, the raw sludge, ultimately producing a more nutrient-rich, treated sludge product.

Various chemical treatments are often used as an alternative to BNR. One such process is simply to pass the effluent from standard secondary treatment through a filter of lime, which causes the phosphate to precipitate out as insoluble calcium phosphate. Another is to treat the effluent with ferric chloride, which produces insoluble ferric phosphate, or with an organic polymer, which gives rise to a floc. With these methods, most of the phosphorus is removed along with the sludge.

Final Cleansing and Disinfection. With or without BNR, the wastewater is subjected to a final clarification and disinfection (Fig. 20–10, bottom). Although few pathogens survive the combined stages of treatment, public-health rigors still demand that the water be disinfected before being discharged into natural waterways. The most widely used disinfecting agent is chlorine gas because it is both effective and relatively inexpensive. However, chlorine gas is dangerous to work with. More commonly, sodium hypochlorite

(Clorox®) provides a safer way of adding chlorine to achieve disinfection. Because chlorine reacts with organic chemicals, all chlorine treatments are known to produce a variety of toxic disinfection by-products, like chlorinated hydrocarbons. And residual chlorine introduced to waterways is itself toxic and can harm natural biota. These impacts are minimized if the effluent is well diluted by natural water.

Other disinfecting techniques are coming into use, although they are more costly than chlorine disinfection. One alternative is ozone gas, which kills microorganisms and breaks down to oxygen gas, improving water quality in the process. Ozone is unstable and, hence, explosive, so it must be generated at the point of use, a step that demands considerable capital investment and energy. Another disinfection technique is to pass the effluent through an array of ultraviolet lights mounted in the water. (Standard fluorescent lights without the white coating emit ultraviolet light.) The ultraviolet radiation kills microorganisms but does not otherwise affect the water.

Final Effluent. After all these steps, the wastewater from a modern treatment plant has a lower organic and nutrient content than many bodies of water into which it is discharged. BOD values are commonly 10 to 20 ppm, as opposed to 200 ppm in the incoming sewage. In other words, discharging the wastewater may actually contribute toward improving water quality in the receiving body. In water-short areas, there is every reason to believe that the treated water itself, with minimal additional treatment, could be recycled into the municipal water-supply system.

Bear in mind that the techniques we have described here represent the state of the art. Many cities, even in the developed world, are still operating with antiquated systems that provide lower-quality treatment. A few coastal cities still persist in discharging sewage that has received only primary treatment directly into the ocean, although in the United States this practice has almost disappeared.

Treatment of Sludge

Recall that the particulate organic matter that settles out or floats to the surface of sewage water in primary treatment forms the bulk of *raw sludge*, though the sewage also contains excesses from activated-sludge and BNR systems. Raw sludge is a gray, foul-smelling, syrupy liquid with a water content of 97–98%. Pathogens are certain to be present in raw sludge because it includes material directly from toilets. Indeed, raw sludge is considered to be a biologically hazardous material. However, as nutrient-rich organic material, it has the potential to be used as organic fertilizer if it is suitably treated to kill pathogens and if it does not contain other toxic contaminants.

Three commonly used methods for treating sludge and converting it into organic fertilizer are **anaerobic digestion**, **pasteurization**, and **composting**. Because this is a developing industry, it is unclear which method will prove most cost effective and environmentally acceptable over the long run. Also, none of these methods is capable of removing toxic substances such as heavy metals and nonbiodegradable synthetic organic compounds. The presence of such toxins can preclude the use of sludge as fertilizer.

Anaerobic Digestion. Anaerobic digestion is a process that allows bacteria to feed on the sludge *in the absence of oxygen*. The raw sludge is put into large airtight tanks called **sludge digesters (Fig. 20–13)**. In the absence of oxygen, a consortium of anaerobic bacteria breaks down the organic matter. The end products of this decomposition are carbon dioxide, methane, and water. Thus, a major by-product of anaerobic processes is **biogas**, a gaseous mixture that is about two-thirds *methane*. The other third is made up of carbon dioxide and various foul-smelling organic compounds that give sewage its characteristic odor. Because of its methane content, biogas is flammable and can be used for fuel. In fact, it is commonly collected and burned to heat the sludge digesters because the bacteria working on the sludge do best when maintained at about 104°F (38°C).

After four to six weeks, anaerobic digestion is more or less complete, and what remains is called **treated sludge** or **biosolids**, consisting of the remaining organic matter, which is now a relatively stable, nutrient-rich, humus-like material suspended in water. Pathogens have been largely, if not entirely, eliminated.

Such treated sludge makes an excellent organic fertilizer that can be applied directly to lawns and agricultural fields in the liquid state in which it comes from the digesters, providing the benefit of both the humus and the nutrient-rich water. Alternatively, the sludge may be dewatered by means of belt presses, whereby the sludge is passed between rollers that squeeze out most of the water **(Fig. 20–14a)**. This leaves the organic material as a semisolid **sludge cake** (Fig. 20–14b), which, after disinfection, is easy to stockpile, distribute, and spread on fields with traditional manure spreaders.

Pasteurization. After the raw sludge is dewatered, the resulting sludge cake may be put through ovens that operate like oversized laundry dryers. In the dryers, the sludge is pasteurized—that is, heated sufficiently to kill any pathogens (exactly the same process that makes milk safe to drink). The product is dry, odorless organic pellets. Milwaukee, Wisconsin, which has a particularly rich sludge resulting from the brewing industry, has been using this process for more than 60 years. The city bags and sells the pellets throughout the country as an organic fertilizer under the trade name Milorganite®.

Composting. Another process sometimes used to treat sewage sludge is composting. Raw sludge is mixed with wood chips or some other water-absorbing material to reduce the water content. It is then placed in **windrows**—long, narrow piles that allow air to circulate easily through the material and that can be turned with machinery. Bacteria and other decomposers break down the organic material to rich humus-like material that makes an excellent treatment for poor soil.

Alternative Treatment Systems

Despite the expansion of sewage collection systems, many homes in rural and suburban areas lie outside the reach of a municipal system. For these homes, *on-site* treatment systems are required. Currently, 25% of the U.S. population is served by such systems. The traditional and still most common on-site system is the **septic system (Fig. 20–15)**. Wastewater flows into the tank, where particulate organic material settles to the bottom. The tank acts like a primary clarifier

Figure 20–13 Anaerobic sludge digesters. In these tanks at the Deer Island Sewage Treatment Plant, Boston, the bacterial digestion of raw sludge in the absence of oxygen leads to the production of methane gas, which is tapped from the tops of the tanks, and humus-like organic matter, which can be used as a soil conditioner. The egg-like shape of the tanks facilitates mixing and digestion.

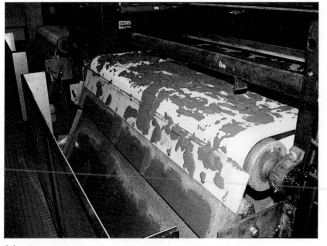

(a)

(b)

Figure 20–14 Dewatering treated sludge. (a) Sludge (98% water) may be dewatered by means of belt presses, as shown here. The liquid sludge is run between canvas belts going over and between rollers, so that much of the water is pressed out. (b) The resulting "sludge cake" is a semisolid humus-like material that can be used as an organic fertilizer.

in a municipal system. Water containing colloidal and dissolved organic material, as well as dissolved nutrients, flows into the gravel or crushed stone lining the leaching field and gradually percolates into the soil. Organic material that settles in the tank is digested by bacteria, but accumulations still must be pumped out regularly. Soil bacteria decompose the colloidal and dissolved organic material that comes through the leaching field. Prerequisites for this traditional system are suitable land area for the drain field and subsoil that allows sufficient percolation of water.

Unfortunately, on-site systems frequently fail, resulting in unpleasant sewage backup into homes and pollution of groundwater and surface waters. Homeowners are often unaware of how the systems work, and most local regulatory programs do not require homeowners to be accountable for them. Substantial nutrient and pathogen pollution of groundwater and surface waters is traced to failed on-site systems.

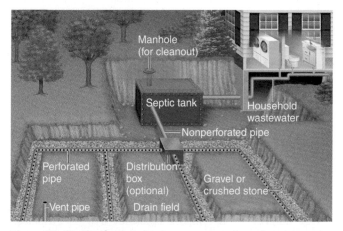

Manhole
(for cleanout)

Septic tank

Household
wastewater

Nonperforated pipe

Perforated
pipe

Distribution
box
(optional)

Gravel or
crushed stone

Vent pipe

Drain field

Figure 20–15 Septic system. Sewage treatment for a private home, using a septic tank and leaching field. Normally, the pipes and the tank are buried underground. They are shown uncovered here only for illustration.

(*Source:* http://www.catawbariverkeeper.org/issues/library-of-documents-on-sewage-issues-and-treatment/septictankdiagram.jpg/view?searchterm=septic+system)

For this reason, the EPA has made several resources available to local and state agencies responsible for maintaining water quality: voluntary national guidelines, a handbook for management, and a manual for on-site wastewater systems.

Septic System Primer. The following are some simple suggestions for successful septic system maintenance:

1. Use caution in disposing of materials down the drain. Do not dump coffee grounds, cat litter, diapers, grease, feminine hygiene products, pesticides, paint, gasoline, household chemicals (like chlorine-containing compounds), or other products that could either fill up or clog the septic tank or kill the bacteria that make the system work.

2. Maintain your system by having it inspected regularly and pumped out at least every five years—more frequently if you have a larger family. Keep records of maintenance, and have a map of the location of your septic tank and the leaching field.

3. If you have a garbage disposal unit, get rid of it; the added materials will shorten the life of the system and force it to be pumped out more frequently.

4. Keep heavy equipment and vehicles off your system and leaching field.

Composting Toilet Systems. A viable and relatively inexpensive alternative to septic systems, especially for more remote and occasional use, is the **composting toilet** system (**Fig. 20–16**). The objective of the system is to provide a sanitary means of treating human wastes that will destroy human pathogens and produce a stable humus-like end product that can be safely disposed of. There are many types of these systems available, but they have some common components:

• A bathroom toilet that connects to a composting reactor. This can have a removable bowl for easy cleaning.

Figure 20–16 Composting toilet system. (a) A toilet in the bathroom leads to (b) the composting reactor, which receives the wastes and allows them to decompose. These systems are often used in vacation cottages, parks, and roadside rest stops. They reduce human wastes to as little as 10% of their original volume.

- A composting reactor that receives solid and liquid wastes from the toilet. The reactor can be right under the toilet seat (self-contained) or located in the basement or on the ground outside.

- An exhaust system (may have a fan) to remove the odors and vapors that result from decomposition of the waste matter.

- Some means of ventilation in the reactor to promote aerobic breakdown of waste. This can be active or passive. *Active systems* involve rotating drums, automatic mixers, fans, and the like, and *passive systems* simply allow the waste to decompose in cool temperatures.

- A means of managing and draining excess liquid and leachate (liquid that has percolated through the wastes) and an access door for removing the end product.

These systems require active management if they are constantly in use, and the end product must legally be buried or removed by a licensed septage hauler. They are capable of reducing toilet wastes to 10–30% of their original volume. Their most common use is in vacation cottages, roadside rest stops, and parks, but where building codes permit, they can also be employed in urban settings. The Institute of Asian Research in Vancouver, British Columbia, is a 30,000-square-foot office complex that is not connected to the city sewer system and uses only composting toilets to handle human wastes.

Using Effluents for Irrigation. The nutrient-rich water coming from the standard secondary-treatment process is beneficial for growing plants. The problem is that we don't want to put that water into waterways, where it will stimulate the growth of undesirable algae. But why not use it for irrigating plants we do want to grow? This is a way of completing the nutrient cycle. Indeed, the concept has been put into practice in a considerable number of locations as an alternative to upgrading the treatment to remove nutrients. Again, effluents must not be contaminated with toxic materials.

To illustrate, the nutrient-rich effluent from standard secondary treatment from St. Petersburg, Florida, was causing cultural eutrophication in Tampa Bay. Now, St. Petersburg uses the effluent to irrigate 4,000 acres (1,600 hectares) of urban open space, from parks and residential lawns to a golf course. Revenues from the water sales help offset operating costs. Similarly, Bakersfield, California, receives a $30,000 annual income from a 5,000-acre (2,000 hectares) farm irrigated with its treated effluent. And Clayton County, Georgia, is irrigating 2,500 acres (1,000 hectares) of woodland with partially treated sewage. Hundreds of similar projects are under way around the country.

A number of developing countries irrigate croplands with raw (untreated) sewage effluents. The crops respond well, but parasites and disease organisms can easily be transferred to farmworkers and consumers. Therefore, it is important to emphasize that only *treated* effluents should be used for irrigation. In Matimangwe, Mozambique, sewage is treated in dry pits and turned into a rich fertilizer that dramatically increased the food harvest, using a latrine system called EcoSan. The wastes are mixed with soil and ash and left for eight months, while the lack of moisture and long incubation kill off pathogens and turn the wastes into a rich compost.

Reconstructed Wetland Systems. In treating wastewater, it is also possible to make use of the nutrient-absorbing capacity of wetlands in suitable areas and under suitable climatic conditions. The project may be part of a wetlands recovery program, or artificial wetlands may be constructed.

In the 1960s and 1970s, for example, much of the land around Orlando, Florida, which was originally wetlands, was drained and converted to cattle pasture. At the time, Orlando was discharging 20 million gallons (75 million L) per day of nutrient-rich effluent into the St. Johns River following secondary treatment. Through the Orlando Easterly Wetlands Reclamation Project, 1,200 acres (480 hectares) of pastureland were converted back to wetlands. The project involved scooping soil from pastures and building berms (mounds of earth) around them to create a chain of shallow lakes and ponds. In addition, 2 million wetland plants ranging from bulrushes and cattails to various trees were planted. The effluent entering the upper end now percolates through the wetlands for about 30 days before entering the St. Johns River virtually pure. The project has recreated a wildlife habitat that supports some threatened and endangered species like the snail kite. Wetland systems can be designed for small as well as large areas and are becoming an increasingly popular alternative for small communities. The key to success for such systems is to ensure that they are kept in balance and not provided more input than they are able to handle.

Many aquatic systems are simply not able to act like wetlands and absorb extra nutrients without major changes in ecosystem function. This response, called eutrophication, is the most widespread water-pollution problem in the United States.

20.3 Eutrophication

Recall that the term *trophic* refers to feeding (Chapter 5). Literally, *eutrophic* means "well nourished." As we will see, there are some quite undesirable consequences of high levels of "nourishment" in bodies of water. Although eutrophication can be an entirely natural process, the introduction of pollutants into bodies of water has greatly increased the scope and speed of eutrophication.

Different Kinds of Aquatic Plants

To understand eutrophication, you need to be able to distinguish between *benthic* plants and *phytoplankton*.

Benthic Plants. Benthic plants (from the Greek *benthos*, meaning "deep") are aquatic plants that grow attached to,

or are rooted in, the bottom of a body of water. All common aquarium plants and sea grasses are benthic plants. As shown in **Figure 20–17a**, benthic plants may be categorized as **submerged aquatic vegetation (SAV)**, which generally grows totally under water, or **emergent vegetation**, which grows with the lower parts in water but the upper parts emerging from the water.

To thrive, SAV requires water that is clear enough to allow sufficient light to penetrate to allow photosynthesis. As water becomes more turbid (cloudy), light is diminished. In extreme situations, it may be reduced to penetrating to just a few centimeters beneath the water's surface. Thus, increasing turbidity decreases the depth at which SAV can survive.

Another important feature of SAV is that it absorbs its required mineral nutrients from the bottom sediments through the roots, just as land plants do. SAV is not limited by water that is low in nutrients. Indeed, enrichment of the water with nutrients is counterproductive for SAV because it stimulates the growth of phytoplankton.

Phytoplankton. Phytoplankton consist of numerous species of photosynthetic algae, protists, and chlorophyll-containing bacteria (cyanobacteria, formerly referred to as blue-green algae) that grow as microscopic single cells or in small groups, or "threads," of cells. Phytoplankton live suspended in the water and are found wherever light and nutrients are available (Fig. 20–17b). In extreme situations, water may become pea-soup green (or tea colored, depending on the species involved), and a scum of phytoplankton may float on the surface and absorb essentially all the light. However, phytoplankton reach such densities only in nutrient-rich water, because, not being connected to the bottom, they must absorb nutrients from the water. A low level of nutrients in the water limits the growth of phytoplankton accordingly.

Considering the different requirements of SAV and phytoplankton, the balance between them is altered when nutrient levels in the water are changed. As long as water remains low in nutrients, populations of phytoplankton are suppressed, the water is clear, and light can penetrate to support the growth of SAV. As nutrient levels increase, phytoplankton can grow more prolifically, making the water turbid and thus shading out the SAV.

The Impacts of Nutrient Enrichment

A lake in which light penetrates deeply—one in which the bottom is visible beyond the immediate shoreline—is **oligotrophic** (low in nutrients). Such a lake is fed by a watershed that holds its nutrients well. A forested watershed, for example, holds nitrogen and phosphorus tightly and allows very little of those elements to enter the water draining into the streams and rivers. If the streams feed into a lake, it will reflect their low nutrient content.

In an oligotrophic lake, the low nutrient levels limit the growth of phytoplankton and allow enough light to penetrate to support the growth of SAV, which draws its nutrients from the bottom sediments. In turn, the benthic plants

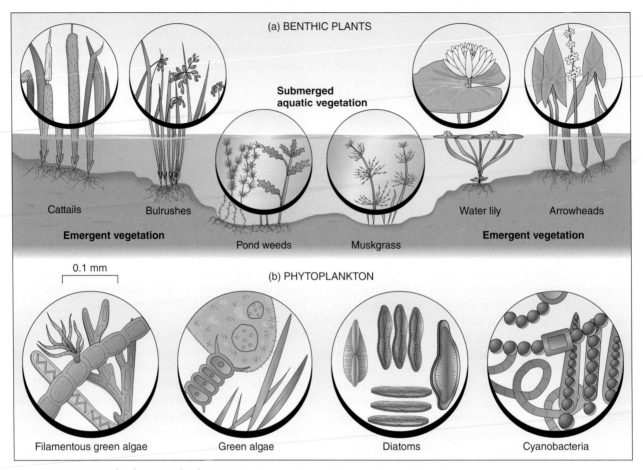

Figure 20–17 **Aquatic photosynthesizers.** (a) Benthic, or bottom-rooted, plants. These are subdivided into submerged aquatic vegetation (SAV) and emergent vegetation. (b) Phytoplankton, various photosynthetic organisms that are either single cells or small groups or filaments of cells, float freely in the water.

support the rest of a diverse aquatic ecosystem by providing food, habitats, and dissolved oxygen. The oligotrophic body of water is prized for its aesthetic and recreational qualities as well as for its production of game fish.

The Process of Eutrophication. As the water of an oligotrophic body becomes enriched with nutrients, numerous changes are set in motion. First, the nutrient enrichment allows the rapid growth and multiplication of phytoplankton, increasing the turbidity of the water. The increasing turbidity shades out the SAV that lives in the water. With the die-off of SAV, there is a loss of food, habitats, and dissolved oxygen from their photosynthesis.

Phytoplankton have remarkably high growth and reproduction rates. Under optimal conditions, phytoplankton biomass may double every 24 hours, a capacity far beyond that of benthic plants. Thus, phytoplankton soon reach a maximum population density, and continuing growth and reproduction are balanced by die-off. Dead phytoplankton settle out, resulting in heavy deposits of detritus on the lake or river bottom. In turn, the abundance of detritus supports an abundance of decomposers, mainly bacteria. The explosive growth of bacteria, consuming oxygen via respiration, creates an additional demand for dissolved oxygen. The result is the depletion of dissolved oxygen, creating hypoxic conditions, with the consequent suffocation of fish and shellfish.

A dead zone has been created; it can be as limited as a small lake basin or as extensive as the Baltic Sea, and the cause is always nutrient enrichment.

In sum, *eutrophication* refers to this whole sequence of events, starting with nutrient enrichment and proceeding to the growth and die-off of phytoplankton, the accumulation of detritus, the growth of bacteria, and, finally, the depletion of dissolved oxygen and the suffocation of higher organisms **(Fig. 20–18)**. For humans, a eutrophic body of water is unappealing for swimming, boating, and sportfishing. Also, if the lake is a source of drinking water, its value may be greatly impaired because phytoplankton rapidly clog water filters and may cause a foul taste. Even worse, some species of phytoplankton secrete various toxins into the water that may kill other aquatic life and be injurious to human health as well. (See Sustainability, "Harmful Algal Blooms.")

Shallow Lakes and Ponds. In lakes and ponds whose water depth is 6 feet (2 m) or less, eutrophication takes a somewhat different course. There, SAV may grow to a height of a meter or more, reaching the surface. Thus, with nutrient enrichment, the SAV is not shaded out but grows abundantly, sprawling over and often totally covering the water surface with dense mats of vegetation that make boating, fishing, or swimming impossible **(Fig. 20–19)**. Any vegetation beneath

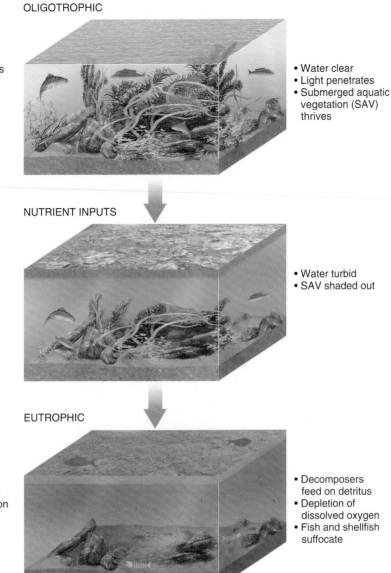

OLIGOTROPHIC

- Low in nutrients
- Phytoplankton limited

- Water clear
- Light penetrates
- Submerged aquatic vegetation (SAV) thrives

NUTRIENT INPUTS

- Nutrient-rich
- Phytoplankton thrive

- Water turbid
- SAV shaded out

EUTROPHIC

- Nutrient-rich
- Rapid turnover of phytoplankton
- Accumulation of detritus of dead algae

- Decomposers feed on detritus
- Depletion of dissolved oxygen
- Fish and shellfish suffocate

Figure 20–18 Eutrophication. As nutrients are added from sources of pollution, an oligotrophic system rapidly becomes eutrophic and undesirable.

these mats is shaded out. As the mats of vegetation die and sink to the bottom, they create a BOD that often depletes the water of dissolved oxygen, causing the death of aquatic organisms other than bacteria.

Natural Versus Cultural Eutrophication. In nature, apart from human impacts, eutrophication is part of the process of natural succession (discussed in Chapter 5). Over periods of hundreds or thousands of years, bodies of water are subject to gradual enrichment with nutrients. Thus, *natural eutrophication* is a normal process. Wherever nutrients come from sewage-treatment plants, poor farming practices, urban runoff, and certain other human activities, however, humans have inadvertently managed to vastly accelerate the process of nutrient enrichment. The accelerated eutrophication caused by humans is called **cultural eutrophication**.

Figure 20–19 Eutrophication in shallow lakes and ponds. In shallow water, sufficient light for photosynthesis continues to reach the submerged aquatic vegetation. The oversupply of nutrients stimulates growth, so that vegetation reaches the surface and forms mats. The vegetation has been raked into piles prior to removal.

Combating Eutrophication

There are two approaches to combating the problem of cultural eutrophication. One is to *attack the symptoms*—the growth of vegetation, the lack of dissolved oxygen, or both. The other is to *get at the root cause*—excessive inputs of nutrients and sediments.

Attacking the Symptoms. Attacking the symptoms is appropriate in certain situations where immediate remediation is the goal and costs are not prohibitive. Methods of attacking the symptoms of eutrophication include (1) applying chemical treatments (with herbicides), (2) aerating, (3) harvesting aquatic weeds, and (4) drawing water down.

Applying Herbicides. Herbicides are often applied to ponds and lakes to control the growth of nuisance plants. To control phytoplankton growth, copper sulfate and diquat are frequently used. For controlling SAV and emergent vegetation, fluridone, glyphosate, and 2,4-D are employed. However, at the concentrations required to bring the vegetation under control, these compounds are sometimes toxic to fish and aquatic animals. Also, fish are often killed after herbicide is applied, because the rotting vegetation depletes the water of dissolved oxygen. In sum, herbicides provide only cosmetic treatment, and as soon as they wear off, the vegetation grows back rapidly.

Aerating. Decomposers deplete the water of dissolved oxygen, thus suffocating other aquatic life, in the final and most destructive stage of eutrophication. *Artificial aeration* of the water can avert this terminal stage. An aeration technique currently gaining in popularity is to lay a network of plastic tubes with microscopic pores on the bottom of the waterway to be treated. High-pressure air pumps force microbubbles from the pores, and the bubbles dissolve directly into the water. The method is proving effective in speeding up the breakdown of accumulated detritus, improving water quality, and enabling the return of more desirable aquatic life. Despite its high cost, the technique is applicable to harbors, marinas, and some water-supply reservoirs, where the demand for better water quality justifies the cost.

Harvesting. In shallow lakes or ponds, where the problem is bottom-rooted vegetation reaching and sprawling over the surface, *harvesting* the aquatic weeds (see Figure 20–19) may be an expedient way to improve the water's recreational potential and aesthetics. Commercial mechanical harvesters are used, and nearby residents sometimes get together to remove the vegetation by hand. The harvested vegetation makes good organic fertilizer and mulch. But even harvesting has a limited effect: The vegetation soon grows back because roots are left in the nutrient-rich sediments.

Drawing Water Down. Because many recreational lakes are dammed, another option for shallow-water weed control is to draw the lake down for a period each year. This process kills most of the rooted aquatic plants along the shore, although they grow back in time.

All of these approaches are only temporary fixes for the problem and have to be repeated often, at significant cost, to keep the unwanted plant growth under control.

Getting at the Root Cause. Controlling eutrophication requires long-term strategies for correcting the problem, which ultimately means reducing the inputs of nutrients and sediments—two of the types of pollution discussed at the beginning of this chapter. The first step is to identify the major point and nonpoint sources of nutrients and sediments. Then it is a matter of developing and implementing strategies for correction. Which source or factor is most significant will depend on the human population and the land uses within the particular watershed. Therefore, each watershed must be analyzed as a separate entity, and appropriate measures must be taken to reduce the levels of nutrients and sediments exiting from that watershed. In the sections that follow, we shall discuss major strategies used to control the root causes of eutrophication.

First, consider the concept of limiting factors, in which the lack of only one nutrient can suppress growth (Chapter 3). In natural freshwater systems, phosphorus is the most common limiting factor. In marine systems (like the coastal Gulf of Mexico), the limiting nutrient is most often nitrogen. Both in the environment and in biological systems, phosphorus (P) is present as phosphate (PO_4^{3-}), and nitrogen is present in a variety of compounds, most commonly as nitrate (NO_3^-) or the ammonium ion (NH_4^+).

Ecoregional Nutrient Criteria. Beginning in 2001, the EPA began publishing water-quality nutrient criteria aimed at preventing and reducing the eutrophication that affects so many bodies of water. The agency listed its recommended criteria for *causative* factors—*nitrogen* and *phosphorus*—and for *response* factors—*chlorophyll a* as a measure of both phytoplankton density and water clarity. The EPA divided the country into ecoregions and determined criteria levels for each specific region. For example, ecoregion VIII, which is composed of the northern Great Lakes and the northeastern states, has the following criteria for lakes and reservoirs: total phosphorus 8.0 µg/L, total nitrogen 0.24 mg/L, chlorophyll *a* 2.43 µg/L, and Secchi depth (a measurement of transparency) 4.93 meters. Other ecoregions often have higher criteria levels, reflecting their differences from the pristine waters around the northern Great Lakes. Like the other water quality criteria, the nutrient criteria are provided as targets for the states as they address their water-pollution (and especially their eutrophication) problems.

National Pollution Discharge Elimination System (NPDES): Control Strategy for Point Sources. In heavily populated areas, discharges from sewage-treatment plants have been major sources of nutrients entering waterways. The levels of phosphate in effluents from these plants were elevated even more in areas where laundry detergents containing phosphate were used. Phosphate contained in detergents goes through the system and out with the discharge unless the system has a nutrient removal component. In regions where eutrophication has been identified as a problem, a key step toward prevention has been to ban the sale of phosphate-based laundry detergents. Because many states established bans on the use of these products, the major laundry detergent manufacturers voluntarily ended their manufacture in 1994. However, the bans do not usually cover dishwashing detergents, some brands of which contain as much as 8.7% phosphate.

Bans on detergents with phosphate and upgrades of sewage-treatment plants have brought about marked improvements in waterways that were heavily damaged by effluents from these plants. In a sense, however, these are the easy measures, as the target for correction (the effluent) is an obvious *point source*, and methods for correcting the problem are clear cut. The NPDES permitting process is the basic regulatory tool for reducing point-source pollutants. Under the Clean Water Act, "anyone discharging pollutants from any point source into waters of the U.S. [must] obtain an NPDES permit from EPA or an authorized state." More than 140,000 facilities are now regulated by way of NPDES permits. The permit must be drawn up in the context of the total pollutant sources (point and nonpoint) affecting a watershed or body of water.

Total Maximum Daily Load (TMDL): Control Strategy for Nonpoint Sources. Correction becomes more difficult, but no less important, when the source is diffuse, as in the case of farm and urban runoff. Remediation of such *nonpoint sources* will involve thousands—perhaps millions—of individual property owners adopting new practices regarding land management and the use of fertilizer and other chemicals on their properties. Nevertheless, that is the challenge. Section 303(d) of the Clean Water Act requires states to develop management programs to address nonpoint sources of pollution. Thus, there is a legal mandate to address the issues of agricultural and urban runoff.

The EPA has opted to develop regulations via the TMDL program. The concept is straightforward:

- Identify the pollutants responsible for degrading a body of water.

- Estimate the pollution coming from all sources (point and nonpoint).

- Estimate the ability of the body of water to assimilate the pollutants while retaining water quality.

- Determine the maximum allowable pollution load.

- Within a margin of safety, allocate the allowable level of pollution among the different sources, such that the water quality standards are achieved.

As with most attempts at solving environmental problems, the devil is in the details. The states are responsible for administering TMDL pollution abatement, but the EPA maintains oversight and approval of the results. State administrators have a number of options for managing nonpoint-source pollution, including regulatory programs, technical assistance, financial assistance, education and training, and demonstration projects. By 2011, the states had listed more than 41,000 water bodies as being impaired. More than 5,100 were on the list because of nutrient overload. The EPA has approved a total of 46,000 TMDLs submitted by the states to correct the impairments and has substantially closed the gap between the state listings and the approved TMDLs.

Best Management Practices. Reducing or eliminating pollution from nonpoint sources will involve different strategies for different sources. For example, where a significant portion of the watershed is devoted to agricultural activities, the major sources of nutrients and sediments are likely to be (1) erosion and leaching of fertilizer from croplands and (2) runoff of animal wastes from barns and feedlots. All the practices that can be used to minimize such erosion, runoff, and leaching are lumped under a single term, **best management practices**. This includes all the methods of soil conservation available (discussed in Chapter 11). Table 20–2 lists examples of these practices for a variety of nonpoint sources.

Once control measures have been put in place, the polluted body of water must be monitored to determine whether

TABLE 20–2 Best Management Practices for Reducing Pollution from Nonpoint Sources

Agriculture	Forestry
Animal waste management	Ground cover maintenance
Conservation tillage	Limiting disturbed areas
Contour farming	Log removal techniques
Strip-cropping	Pesticide and herbicide management
Cover crops	Proper management of roads
Crop rotation	Removal of debris
Fertilizer management	Riparian zone management
Integrated pest management	**Mining**
Range and pasture management	Underdrains
Terraces	Water diversion
Construction	**Multicategory**
Limiting disturbed areas	Buffer vegetated strips
Nonvegetative soil stabilization	Detention and sedimentation basins
Runoff detention and retention	Devices to encourage infiltration
Urban	Vegetated waterways
Flood storage	Interception and diversion of runoff
Porous pavements	Sediment traps
Runoff detention and retention	Streamside management zones
Street cleaning	Vegetative stabilization and mulching

Source: U.S. Environmental Protection Agency, *Guidance for Water Quality-Based Decisions: The TMDL Process*, EPA440-4-91-001 (Washington, DC: USEPA, 1991).

water quality standards are being attained. Several years of data collection may be necessary to detect genuine trends in quality. When water quality standards are reached and sustained, the body of water is removed from the list of impaired waters. To date, a total of 1,748 previously impaired bodies of water have been removed from the list (not a very impressive record). If standards are not met, the TMDL process must be revisited and new pollution allowances allocated.

Recovery. The good news is that cultural eutrophication and other forms of water pollution can be controlled and often reversed, provided that a total watershed management approach is undertaken. For a small lake fed from a modest-sized watershed, the task may be quite straightforward because major sources of nutrients can be identified and addressed. For large bodies of water, such as the Chesapeake Bay—with a watershed of 63,700 square miles (165,000 km²) in six states and the District of Columbia—the task is enormous. The Chesapeake Bay has suffered from heavy nutrient pollution, with cultural eutrophication destroying all but a tenth of the 600,000 acres of vital sea grasses that once formed the basis of the bay's rich ecosystem. A dead zone appears every year, involving up to 40% of the main bay for 10 months of the year. The Chesapeake Bay Program, a state-federal partnership begun in 1987, set the ambitious goal of achieving federal clean water standards by 2010. Unfortunately, new development in the watershed contributed nutrients and sediments as fast as the restoration efforts reduced them. Although there have been signs that the dead zone has been decreasing, other water quality indicators have shown little change. At the end of 2010, the EPA stepped in with an enforceable "road map" to put the estuary on a path to recovery.[2] The agency set a bay-wide TMDL that will limit specific nutrient inflows from each of the bay's tributaries. This will give the states clear targets for implementing the restoration activities.

Lake Washington. One remarkable success story is Lake Washington, east of Seattle **(Fig. 20–20)**. This 34-square-mile lake is at the center of a large metropolitan area. During the 1940s and 1950s, 11 sewage-treatment plants were sending state-of-the-art treated water into the lake at a rate of 20 million gallons per day. At the same time, phosphate-based detergents came into wide use. The lake responded to the massive input of nutrients by developing unpleasant "blooms" of noxious blue-green algae. The water lost its clarity, the desirable fish populations declined, and masses of dead algae accumulated on the shores of the lake.

Citizen concern led to the creation of a system that diverted the treatment-plant effluents into nearby Puget Sound, where tidal flushing would mix them with open-ocean water. The diversion was complete by 1968, and the lake responded quickly. The algal blooms diminished, the water regained its clarity, and by 1975, recovery was complete. Careful studies

Figure 20–20 Lake Washington. A testimony to the fact that bodies of water can recover from cultural eutrophication, Lake Washington is now thriving because sources of nutrients enriching the lake have been removed.

by a group of limnologists from the University of Washington showed that phosphate was the culprit. Before the diversion, the lake was receiving 220 tons of phosphate per year from all sources. Afterward, the total dropped to 40 tons per year, well within the lake's normal capacity to absorb phosphate by depositing it in deep bottom sediments. One clear lesson learned was that sewage-treatment effluent must never be allowed to enter a lake unless nutrients are removed as part of the treatment. The major lesson, however, was that eutrophication can be reversed by paying attention to nutrient inputs and addressing the various sources. Lake management, in other words, is a matter of controlling the phosphate loading into a lake from all sources: human and animal sewage, agricultural and yard fertilizers, street runoff, and failing septic systems. Many lakes are now protected by associations of concerned citizens who see themselves as stewards of their lake.

20.4 Public Policy and Water Pollution

In the United States, the responsibility for overseeing the health of the nation's waters rests with the EPA. However, the EPA can develop regulations only if Congress gives it the authority to do so. Hence, the foundation for public policy must be the laws passed by Congress. Some major legislative milestones in protecting the nation's waters are listed in Table 20–3. The landmark legislation is the Clean Water Act of 1972 (CWA), which gave the EPA jurisdiction over (and for the first time required permits for) all point-source discharges of pollutants; simply put, all discharges into U.S. waters are unlawful unless authorized by a permit. This act and subsequent amendments have provided $85 billion to help cities and towns build treatment plants to meet the federal requirement for secondary treatment of all sewage. As the table shows, the 1987 amendments to the CWA established a revolving loan fund—the Clean Water State Revolving Fund (SRF) program—to replace the direct-grants program. To build treatment facilities, local governments

[2]Juliet Eilperin. "EPA Unveils Massive Restoration Plan for Chesapeake Bay." *The Washington Post*, December 30, 2010.

TABLE 20–3	Legislative Milestones in Protecting Natural Waters and Water Supplies

1899—Rivers and Harbors Act: First federal legislation protecting the nation's waters in order to promote commerce.

1948—Water Pollution Control Act: Authorizes the federal government to provide technical assistance and funds to state and local governments to promote efforts to protect water quality.

1972—Clean Water Act: Establishes a comprehensive federal program to achieve the goal of protecting and restoring the physical, chemical, and biological integrity of the nation's waters. Requires permits for any discharge of pollution, strengthens water quality standards, and encourages the use of the best achievable pollution control technology. Provides billions of dollars for construction of sewage-treatment plants.

1972—Marine Protection, Research, and Sanctuaries Act: Prevents unacceptable dumping in oceans.

1974—Safe Drinking Water Act: Authorizes the EPA to regulate the quality and safety of public drinking-water supplies, maintain drinking-water standards for numerous contaminants, and set requirements for the chemical and physical treatment of drinking water.

1977—Clean Water Act amendments: Strengthen controls on toxic pollutants and allow states to assume responsibility for federal programs.

1987—Water Quality Act: Enacts major amendments to the Clean Water Act that (1) create a revolving loan fund to provide ongoing support for the construction of treatment plants, (2) address regional pollution with a watershed approach, and (3) address nonpoint-source pollution, with requirements for states to assess the problem and develop and implement plans for dealing with it. Makes available $400 million in grants to carry out its provisions.

1996—Safe Drinking Water Act amendments: Impose numerous requirements for managing water supplies, establish a revolving loan fund to help municipalities update their water system infrastructure, and provide flexibility for the EPA to base its selection of contaminants on health risk and costs and benefits.

2002—Great Lakes Legacy Act: Enacted amendments to CWA provisions for the Great Lakes and authorized $50 million annually for fiscal years 2004–2008 for the EPA to conduct projects designed to remediate sediment problems in the Great Lakes.

borrow at low interest rates, and as they repay the loans, the funds received are used for more loans. The fund may also be used to control nonpoint-source pollution. To date, more than $74 billion in SRF loans have been made by the states. However, the EPA estimates that an additional $298 billion is needed to meet the funding needs for Clean Water infrastructure.

Reauthorization

Reauthorization of the Clean Water Act is long overdue. Congress has been hung up on debates over whether regulations should be strengthened or weakened, whether federal regulations in the wetlands permit program intrude on private land-use rights, and how regulatory relief should be provided to industries, states, cities, and individuals required to take actions to comply with the regulations. Its approach for the past two decades has been to reauthorize the provisions of the CWA and its amendments, usually maintaining a status quo in the EPA's annual appropriations for CWA programs.

Problems and Progress

The EPA has identified nonpoint-source pollution as the nation's number-one water-pollution problem, with the construction of new wastewater-treatment facilities not far behind. Other significant issues—including storm-water discharges, combined and separate sewer overflows, wetlands protection, mountaintop mining filling of streams,

and animal feeding operations—are also receiving the EPA's attention in the form of new regulations. Recent attempts by Republicans in Congress to block the EPA from carrying out many of its mandates on these issues have failed to pass.[3]

Nevertheless, much progress has been made in the 40 years since the enactment of the Clean Water Act. This act is seen as one of the most successful of our nation's environmental laws. The number of people in the United States served by adequate sewage-treatment plants has risen from 85 million to 223 million. Soil erosion has been reduced by 1 billion tons annually, and two-thirds of the nation's waterways are safe for fishing and swimming—double the number from 1972. Many of the nation's most heavily used rivers, lakes, and bays—including the Androscoggin River in Maine, Boston Harbor, the upper Mississippi River, the South Platte River, the upper Arkansas River, Lake Erie, the Illinois River, and the Delaware River—have been cleaned up and restored. Fish now swim in rivers once so polluted that only bacteria and sludge worms could survive. Levels of toxic chemicals in the Great Lakes have been greatly reduced. A national sense of stewardship has been applied to the rivers, lakes, and bays that are our heritage from a previous generation, and public policy has been enacted and billions of dollars spent to bring our waters back from a polluted condition.

---

[3]Claudia Copeland. "Water Quality Issues in the 112th Congress: Oversight and Implementation." *CRS Report for Congress* R41594. November 7, 2011. www.crs.gov.

REVISITING THE THEMES

Sound Science

Solid scientific research and communication were essential in establishing the basic knowledge about pollutants and their impacts on human health and the environment. They continue to be highly important as the EPA carries out its research and regulatory activities for administering the Clean Water Act and the Safe Drinking Water Act. The criteria pollutant standards for natural waters and the maximum contaminant standards for drinking waters must be based on accurate risk assessments as the EPA develops and defends its regulatory activities. Sound science was particularly crucial in the investigation of dead zones and algal blooms and their causes.

Sustainability

Our natural waters represent an enormous bank of ecosystem capital, providing essential goods and services to people everywhere. The sustainability of whole ecosystems is threatened when we fail to handle our wastes in a conscientious manner. The opening story of the dead zone demonstrates how it is possible for nutrient overload to degrade an enormous area of the Gulf of Mexico. Unfortunately, a similar situation exists in the Chesapeake Bay and in many other bodies of water. These are just a couple of the more spectacular examples. The more important battlegrounds are the thousands of smaller lakes, rivers, and coastal estuaries that are also suffering from human impacts. More needs to be done, especially at the state level, to improve our national water quality; Congress can play an important role as it establishes funding for the states to carry out its mandates. Many states are behind in assessing water quality, and the need for updating sewage-treatment systems is enormous.

Stewardship

Stewardship means intentionally caring for people and the natural world. The public-health measures put in place in the past two centuries represent the stewardship of human resources. Also, as we divert public and private resources to cleaning up our waterways, we are acting as good stewards in the best sense of the concept. Lakes and watersheds often benefit from people and associations that have provided leadership in correcting the problems that lead to eutrophication. The Lake Washington story is an excellent example. Stewardship also means caring for our neighbors' needs, and nowhere are those needs more apparent than in the need for clean water and sanitation in so many developing countries. A sense of the global reach of water pollution is difficult to imagine, yet billions of people in the developing countries are held back from achieving their physical and intellectual potential because of the terrible effects of polluted water and lack of sanitation.

REVIEW QUESTIONS

1. What practices and consequences led to passage of the Clean Water Act of 1972?

2. Discuss each of the following categories of water pollutants and the problems they cause: pathogens, organic wastes, chemical pollutants, and sediments.

3. How are water quality standards determined? Distinguish between water quality criteria pollutants and maximum contaminant levels.

4. Name and describe the facility and the process used to remove debris, grit, particulate organic matter, colloidal and dissolved organic matter, and dissolved nutrients from wastewater.

5. Why is secondary treatment also called biological treatment? What is the principle involved? What are the two alternative techniques used?

6. What are the principles involved in removing biological nutrients from waste, and what is accomplished by doing so? Where do nitrogen and phosphate go in the process?

7. Name and describe three methods of treating raw sludge, and give the end product(s) that may be produced from each method.

8. How may sewage from individual homes be handled in the absence of municipal collection systems? What are some problems with these on-site systems?

9. Describe and compare submerged aquatic vegetation and phytoplankton. Where and how does each get nutrients and light?

10. Explain the difference between oligotrophic and eutrophic waters. Describe the sequential process of eutrophication.

11. Distinguish between natural and cultural eutrophication.

12. What is being done to establish nutrient criteria?

13. How does the NPDES program address point-source pollution by nutrients?

14. Describe the TMDL program. How does it address nonpoint-source pollution, and what role do water quality criteria play in the program?

15. What are some of the important public-policy issues relating to water quality?

THINKING ENVIRONMENTALLY

1. A large number of fish are suddenly found floating dead on a lake. You are called in to investigate the problem. You find an abundance of phytoplankton and no evidence of toxic dumping. Suggest a reason for the fish kill.

2. Nitrogen pollution has damaged the San Diego Creek–Newport Bay ecosystem in California by encouraging the heavy growth of intertidal algae. Report on the TMDL for this system. Consult the EPA Web site: http://water.epa.gov/lawsregs/lawsguidance/cwa/tmdl/nutrients.cfm.

3. Arrange a tour to the sewage-treatment plant that serves your community. Compare it with what is described in this chapter. Is the water being purified or handled in a way that will prevent cultural eutrophication? Are sludges being converted to and used as fertilizer? What improvements, if any, are in order? How can you help promote such improvements?

4. Suppose a new community of several thousand people is going to be built in Arizona (with a warm, dry climate). You are called in as a consultant to design a complete sewage system, including the collection, treatment, and use or disposal of by-products. Write an essay describing the system you recommend, and give a rationale for the choices involved.

MAKING A DIFFERENCE

1. Check out the lakes and ponds in your area, and determine whether cultural eutrophication is occurring. If you find that it is, follow up by consulting a local environmental group, and get involved in combating the problem.

2. Investigate sewage treatment in your community. To what extent is the sewage treated? Where does the effluent go? The sludge? Are there any problems? Are there alternatives that are more ecologically sound? Consult local environmental groups for help.

3. Bring discarded household chemicals to hazardous-waste collection centers; do not pour them down the drain. Pouring chemicals down the drain could upset your septic system or contaminate treatment-plant sludge.

4. Test your soil before using fertilizers. Excess from overfertilizing can leach into groundwater or contaminate rivers or lakes. Also, avoid using fertilizers and pesticides near surface waters.

MasteringEnvironmentalScience®

Go to **www.masteringenvironmentalscience.com** for practice quizzes, Pearson eText, videos, current events, and more.

CHAPTER 21

Municipal Solid Waste: Disposal and Recovery

Flexible solar strips cover the top of the Tessman Landfill in San Antonio, Texas. Together with a biogas collection system, the landfill will generate enough energy to power 5,500 homes.

LEARNING OBJECTIVES

1. **Solid Waste: Landfills and Combustion.** Appraise the issues involved with the two-thirds of the municipal solid waste (MSW) that is combusted or disposed of in landfills.

2. **Better Solutions: Source Reduction and Recycling.** Evaluate these two options for reducing the amount of MSW that must be managed.

3. **Public Policy and Waste Management.** Compare the roles of federal, state, and local governments in management of MSW; identify some desirable policy goals that would improve sustainability of MSW management.

T HE COMMONWEALTH OF MASSACHUSETTS has set a goal of generating 20% of its electricity from renewable energy by 2020 and has found a new way of getting there. The idea: use many of the 500 closed landfills in the state to support solar panels, converting otherwise unusable and treeless wastelands into money-making assets. Massachusetts joins a growing number of states that have discovered this new use for old landfills; similar projects are under construction or in operation in at least six other states.[1] Closed landfills are ideally suited to become solar farms: They are usually municipally owned, sited away from residential areas, treeless, close to power lines, and abundant (more than 6,000 landfills have been closed in recent decades in the United States). Through a partnership with Boston-based Southern Sky Renewable Energy, the town of Canton is erecting an array of 24,000 solar panels on 15 acres of its capped Pine Street landfill. The solar array will generate 5.6 megawatts of power, enough to supply 5,600 homes, and some $16 million in revenues for the town over the next 25 years. Easthampton, Massachusetts, is installing 10,000 solar panels on its Oliver Street landfill.

[1]David Jonathan Ross. *Construction of Solar Farms on Closed Landfills in Utah.* M.S. Thesis, Department of Civil and Environmental Engineering. Brigham Young University. 2011.

Canton's Pine Street landfill is a lesson in land use that is being learned only slowly in the United States—and, for that matter, in the rest of the world. The lesson is that, if we could picture more effectively what 50 years can bring in the way of change, we could solve one of the most contentious problems facing local communities: what to do with municipal waste. As old dumps and landfills were closed because of environmental concerns, they created a temporary problem identified as the "solid-waste crisis" in the 1970s and 1980s. In fact, closed landfills may be sustainable energy goldmines, where solar arrays and wind turbines are added to landfill gas power. Other important uses for closed landfills include playing fields, nature preserves, and golf courses (**Fig. 21–1**).

STILL A CRISIS? In some ways, the solid-waste crisis is still with us, as we will see when we examine the problem of interstate trash movement. It is commonly said that we are running out of space to put our trash and garbage. Perhaps we are, but if so, it is because we are happy to purchase the goods displayed so prominently in our malls and advertised in the media, but we are reluctant to accept the consequences of getting rid of them responsibly.

This chapter is about solid-waste issues. We examine current patterns of disposal—landfills, combustion, and

Figure 21–1 From landfills to playing fields. Thomas W. Danehy Park in Cambridge, Massachusetts, a former landfill that has been recycled into a recreational park. In addition to playing fields, the park features a half-mile "glassphalt" pathway (built with recycled glass and asphalt), shown in the foreground.

recycling—and look for sustainable solutions to our solid-waste problems. The ideal would be to imitate natural ecosystems and reuse everything. Some solutions do well in conforming to this principle, whereas others do not—and possibly cannot.

21.1 Solid Waste: Landfills and Combustion

The focus of this chapter is **municipal solid waste (MSW)**, defined as the total of all the materials (commonly called trash, refuse, or garbage) thrown away from homes and smaller commercial establishments and collected by local governments. MSW is different from hazardous waste and nonhazardous industrial waste (Chapter 22). The latter is no small matter: Industrial facilities generate and manage 7.6 billion tons of nonhazardous industrial waste annually. Included in the category are wastes from demolition and construction operations, agricultural and mining residues, combustion ash, sewage-treatment sludge, and wastes generated by industrial processes. The states oversee these wastes because Congress has not delegated any authority to the Environmental Protection Agency (EPA) to regulate them as it has for MSW and hazardous waste. The EPA Office of Solid Waste does, however, provide annual reports on municipal solid waste in the United States, and we make full use of the latest reports in this chapter.

Disposal of Municipal Solid Waste

Over the years, the amount of MSW generated in the United States has grown steadily, in part because of a growing population, but also because of changing lifestyles and the increasing use of disposable materials and excessive

packaging. In 1960, the nation generated 2.7 pounds (1.2 kg) per person per day. In 2010, we generated a total of 250 million tons (230 million metric tons) of MSW, an average of 4.4 pounds (2 kg) per person per day and enough waste to fill 95,000 garbage trucks each day. The solid-waste problem can be stated simply: *We generate huge amounts of MSW, and it is increasingly expensive to dispose of it in ways that are environmentally responsible and protective of human health.*

The refuse generated by municipalities is a mixture of materials from households and small businesses, in the proportions shown in **Figure 21–2**. However, the proportions vary greatly, depending on the generator (commercial

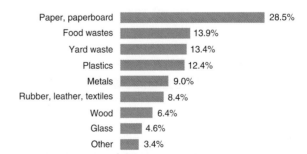

Paper, paperboard	28.5%
Food wastes	13.9%
Yard waste	13.4%
Plastics	12.4%
Metals	9.0%
Rubber, leather, textiles	8.4%
Wood	6.4%
Glass	4.6%
Other	3.4%

Figure 21–2 U.S. municipal solid-waste composition. The composition of municipal solid waste in the United States in 2010.

(*Source*: Data from EPA, Office of Solid Waste, *Municipal Solid Waste in the United States: 2010 Facts and Figures* (Washington, DC: EPA, November, 2011).)

versus residential), the neighborhood (affluent versus poor), and the time of year (during certain seasons, yard wastes, such as grass clippings and raked leaves, add greatly to the solid-waste burden). Little attention is given to what people throw away in their trash. Even if there are restrictions and prohibitions, they can be bypassed with careful packing of the trash containers. Thus, many environmentally detrimental substances—paint, used motor oil, electronic equipment, and so on—are discarded, with the feeling that they are gone forever.

Whose Job? Customarily, local governments have had the responsibility for collecting and disposing of MSW. The local jurisdiction may own trucks and employ workers, or it may contract with a private firm to provide the collection service. Traditionally, the cost of waste pickup is passed along to households via taxes. Alternatively, many municipalities have opted for a pay-as-you-throw (PAYT, as it is called) system, in which households are charged for waste collection on the basis of the amount of trash they throw away. Some municipalities put all trash collection and disposal in the private sector, with the collectors billing each home by volume and weight of trash. The MSW that is collected is then disposed of in a variety of ways, and it is at the point of disposal that state and federal regulations begin to apply.

Past Sins. Until the 1960s, most MSW was disposed of in open dumps. Older residents of smaller towns and cities can remember gathering at the dump, meeting there to look through discarded items and socialize. In most dumps, the waste was burned to reduce its volume and lengthen the life span of the dump site, but refuse does not burn well. Smoldering dumps produced clouds of smoke that could be seen from miles away, smelled bad, and created a breeding ground for flies and rats. Some cities turned to incinerators (or combustion facilities, as they are called today): huge furnaces in which high temperatures allow the waste to burn more completely than in open dumps. Without controls, however, incinerators were prime sources of air pollution. Public objections and air pollution laws forced the phaseout of open dumps and many incinerators during the 1960s and early 1970s. Open dumps were then converted to landfills.

In the United States in 2010, 54.2% of MSW was disposed of in landfills, 34.1% was recovered for recycling and composting, and the remainder (11.7%) was combusted **(Fig. 21–3)**. Over the past 10 years, the landfill and combustion components have been declining, while recycling has shown

a steady increase. The pattern is different in countries where population densities are higher and there is less open space for landfills. High-density Japan, for instance, combusts 78% and landfills 13% of its MSW. The pattern in Europe is mixed, ranging from 35% combustion and only 1% of MSW going to landfills in Germany, to 48% of nonrecycled wastes going to the landfills (with 12% combusted) in the United Kingdom, to 90% landfilled MSW in Eastern Europe. The entire EU is under strict directives to reduce wastes going to landfills, with increasingly severe penalties (fines) for nonattainment.[2]

Landfills

In a **landfill**, the waste is put on or in the ground and is covered with earth. Because there is no burning and because each day's fill is covered with at least six inches of earth, air pollution and populations of vermin are kept down. Unfortunately, aside from those concerns and the minimizing of cost, no other factors were given real consideration when the first landfills were opened. Municipal waste managers generally had little understanding of ecology, the water cycle, or what products decomposing wastes would generate, and they had no regulations to guide them. Therefore, in general, any cheap, conveniently located piece of land on the outskirts of town became the site for a landfill. Frequently, this site was a natural gully or ravine, an abandoned stone quarry, a section of wetlands, or a previous dump **(Fig. 21–4)**. Once the municipality acquired the land, dumping commenced without any precautions. After the site was full, it would be covered with earth and ignored. Only recently have abandoned landfills been seen as a valuable open-space resource.

Problems of Landfills. Landfills are subjected to biological and physical factors in the environment and will undergo change over time as a consequence of the operation of those

Figure 21–4 New Orleans dump. This burning dump, sited on wetlands, demonstrates the worst of the MSW disposal practices of the past.

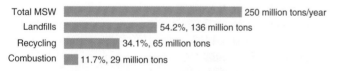

Figure 21–3 U.S. municipal solid-waste disposal. Disposal of solid waste to landfills, combustion, and recycling, 2010. An additional 20 million tons were composted.

(*Source:* Data from EPA, Office of Solid Waste, *Municipal Solid Waste in the United States: 2010 Facts and Figures* (Washington, DC: EPA, November 2011).)

[2]European Environment Agency. *Diverting Waste from Landfill: Effectiveness of Waste-Management Policies in the European Union.* EEA Report No. 7/2009. Copenhagen, 2009.

factors on the waste that is deposited. Several of the changes are undesirable because, if they are not dealt with effectively, they present the following problems:

* Leachate generation and groundwater contamination
* Methane production
* Incomplete decomposition
* Settling

Leachate Generation and Groundwater Contamination. The most serious problem by far is groundwater contamination. As water percolates through any material, various chemicals in the material may dissolve in the water and get carried along in a process called *leaching* (Chapter 10). The water with various pollutants in it is called *leachate*. As water percolates through MSW, a noxious leachate is generated that consists of residues of decomposing organic matter combined with iron, mercury, lead, zinc, and other metals from rusting cans, discarded batteries, and appliances—all generously "spiced" with paints, pesticides, cleaning fluids, newspaper inks, and other chemicals. The nature of the landfill site and the absence of precautionary measures can funnel this toxic brew directly into groundwater aquifers.

All states have some municipal landfills that currently are, or soon will be, contaminating groundwater, but Florida has some unique problems. Flat and with vast areas of wetlands, much of the state is only a few feet above sea level and rests on water-saturated limestone. No matter where Florida's landfills were located, they were either in wetlands or just a few feet above the water table. As a result, more than 145 former municipal landfill sites in Florida appear on the Superfund list. (Superfund is the federal program with the responsibility for cleaning up sites that are in imminent danger of jeopardizing human health through groundwater contamination.) All landfills in Florida are now required to have state-of-the-art landfill liners.

Methane Production. Because it is about two-thirds organic material, MSW is subject to natural decomposition. Buried wastes do not have access to oxygen, however, so their decomposition is anaerobic. A major by-product of the process is *biogas*, which is about two-thirds methane and one-third hydrogen and carbon dioxide, a highly flammable mixture (see Chapter 20). Produced deep in a landfill, biogas may seep horizontally through the soil and rock, enter basements, and even cause explosions if it accumulates and is ignited. Homes at distances up to 1,000 feet from landfills have been destroyed, and some deaths have occurred as a result of such explosions. Also, gases seeping to the surface kill vegetation by poisoning the roots. Without vegetation, erosion occurs, exposing the unsightly waste.

There is a thriving industry in 46 states that is exploiting the problem by installing "gas wells" in old and existing landfills. The wells tap the landfill gas, and the methane is purified and used as fuel. There are now more than 550 landfill gas energy projects in the United States, with 510 more in the planning stage. In 2011, commercial landfill gas in the nation produced 1,500 megawatts of electricity and 100 billion cubic feet of gas for direct-use applications, powering or

Figure 21–5 Landfill Gas Project. A well collects biogas from the Smith Creek landfill in St. Clair County, Michigan. There are now more than 550 landfill gas energy systems in operation in the U.S.

heating more than 1.7 million homes **(Fig. 21–5)**. For example, Allied Waste Services is now mining the landfill gas from the Roosevelt Regional Landfill in Washington State, a huge working landfill. The methane is fed to seven combustion turbines, with a generating capacity of 26 megawatts, enough to power 26,000 homes (average U.S. home use is 1 kW/day).

The recovery of landfill gas has significant environmental benefits. It directly reduces greenhouse gas emissions because methane is a powerful greenhouse gas and landfills are the largest anthropogenic source of methane emissions in the United States. Methane recovery also reduces the use of coal and oil, which are nonrenewable and highly polluting energy resources.

Incomplete Decomposition. The commonly used plastics in MSW resist natural decomposition because of their molecular structure. Chemically, they are polymers of petroleum-based compounds that microbes are unable to digest. Biodegradable plastic polymers have been developed from sources such as cornstarch, cellulose, lactic acid, and soybean protein as well as petroleum. As oil prices rise, plant-based biodegradable plastics are beginning to make economic inroads. Two kinds of bioplastics based on corn sugars—polyhydroxyalkanoate and polylactic acid—are in commercial production; a recent estimate from NatureWorks indicates a coming demand for up to 50 billion pounds of bioplastics a year in the next five years. Currently, bioplastics are more expensive than petroleum-based plastics and tend to be used by companies manufacturing natural and certified organic products. They are indeed biodegradable, but some of the corn-based plastics must go to a commercial composting plant to be decomposed.

A team of "garbologists" from the University of Arizona, led by William Rathje, has been carrying out research on old landfills. The research has shown that even materials formerly assumed to be biodegradable—newspapers, wood, and so on—are often degraded only slowly, if at all, in landfills. In one landfill, 30-year-old newspapers were recovered in a readable state, and layers of telephone directories, practically intact, were found marking each year. Because paper materials are 28% of MSW, this is a serious matter. The reason paper and other organic materials decompose so slowly is

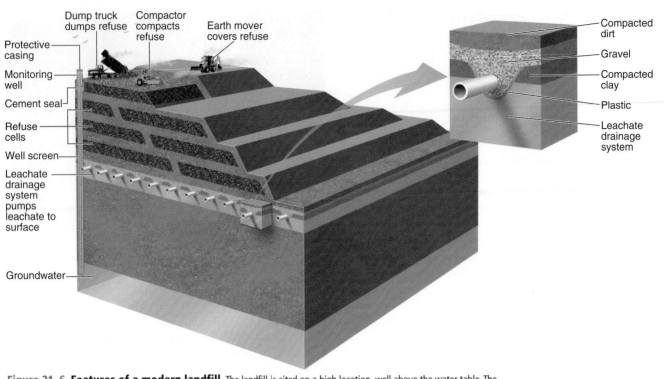

Figure 21–6 Features of a modern landfill. The landfill is sited on a high location, well above the water table. The bottom is sealed with compacted soil and a plastic liner, overlaid by a rock or gravel layer, with pipes to drain the leachate. Refuse is built up in layers as the amount generated each day is covered with soil, so the completed fill has a pyramidal shape that sheds water. The fill is provided with wells for monitoring groundwater.

the absence of suitable amounts of moisture. The more water percolating through a landfill, the better paper materials biodegrade. Unfortunately, the more percolation there is, the greater the amount of toxic leachate produced.

Settling. Finally, waste settles as it compacts and decomposes. Luckily, this eventuality was recognized from the beginning, so buildings are never put on landfills. Settling does present a problem in landfills that have been converted to playgrounds and golf courses because of the creation of shallow depressions (and sometimes deep holes) that collect and hold water. This problem can be addressed by continually monitoring the facility and using fill to restore a level surface.

Improving Landfills. Recognizing the foregoing problems, the EPA upgraded siting and construction requirements for new landfills, based on the **Resource Conservation and Recovery Act of 1976 (RCRA)** and subsequent amendments. Under current regulations,

- New landfills are sited on high ground, well above the water table, not in a geologically unstable area, and away from airports (because of bird hazards).

- The floor is contoured so that water will drain into a tile leachate-collection system. The floor and sides are covered with a plastic liner and at least two feet of compacted soil. With such a design, any leachate percolating through the fill will move into the leachate-collection system. Collected leachate can be treated as necessary.

- Layer upon layer of refuse is positioned such that the fill is built up in the shape of a pyramid. Finally, it is capped with at least 18 inches of earthen material and a layer of

topsoil and then seeded. The cap and the pyramidal shape help the landfill shed water. In this way, water infiltration into the fill is minimized, and less leachate is formed.

- The entire site is surrounded by a series of groundwater-monitoring wells that are checked periodically, and such checking must go on indefinitely.

These design features are summarized in **Figure 21–6.** Most landfills currently in operation have these improved technologies, which protect both human health and the environment.

The landfill pyramids may well last as long as the Egyptian pyramids. Fortunately, the creative siting and construction of landfills have the potential to address some highly significant future needs, because abandoned landfills can become attractive golf courses, recreational facilities, wildlife preserves, or renewable energy sites.

Siting New Landfills. Between 1988 and 2009, the number of municipal landfills declined from around 8,000 to 1,908. Because the size of landfills has increased and because recycling is on the rise, the EPA does not believe that landfill capacity is a problem. However, as the agency puts it, "regional dislocations" may be expected. These are due to the one problem associated with landfills that gets more attention than any other: *siting.* It is not that landfills take up enormous amounts of land. One particular landfill occupying 230 acres (Central Landfill) serves the entire state of Rhode Island, and Fresh Kills, once the largest landfill in the world, served much of New York City and its environs for many years and was only 2,200 acres. But as old landfills are closed, it has become increasingly difficult to find land for the new ones needed to take their place.

Figure 21–7 Fresh Kills today. This 2,200-acre landfill on Staten Island served the city of New York until 2001. It has been closed and secured and is being developed for recreation, natural areas, and renewable power facilities.

People in residential communities (where the MSW is generated) invariably reject proposals to site landfills anywhere near where they live, and those who already live close to existing landfills are anxious to close them down. Weary of the odor and heavy truck traffic, residents of Staten Island, New York, pressured the city of New York to shut down the huge Fresh Kills landfill, once the recipient of 27,000 tons of garbage per day. The city fulfilled its promise to phase out the landfill in 2001, and Staten Islanders are now anticipating the development of Fresh Kills Park,

with soccer fields, mountain bike and jogging trails, natural areas, and possibly a golf course **(Fig. 21–7)**. Gas from the huge landfill is providing the area with a new energy source. Currently, New York City exports its MSW outside of the city by truck, barge, and rail.

With spreading urbanization, few suburban areas are not already dotted with residential developments. Practically any site selected, then, is met with protests and legal suits. The problem has been repeated in so many parts of the country (and globe!) that it has given rise to some inventive acronyms: **LULU** ("locally unwanted land use"), **NIMBY** ("not in my backyard"), **NIMTOO** ("not in my term of office"), and **BANANA (build absolutely nothing anywhere near anything**)! You can well imagine how these attitudes apply to the landfill-siting problem.

Outsourcing. The siting problem has some undesirable consequences. First, it drives up the costs of waste disposal, as alternatives to local landfills are invariably more expensive. Second, it leads to the inefficient and equally objectionable practice of the long-distance transfer of trash as waste generators look for private landfills whose owners are anxious to receive trash (and cash). Often, this transfer occurs across state and even national lines, leading to resentment and opposition on the part of citizens of the recipient state or nation. Led by Pennsylvania, 11 states import more than 1 million tons of MSW a year (Table 21–1), while 13 states export more than 1 million tons per year, led by New York. Not surprisingly, the heaviest importing states are adjacent to the heaviest exporting states.

TABLE 21–1   Largest Interstate Transfers of MSW, 2005 (most recent date available)

Exporting States and Provinces	Exported MSW (millions of tons)	Importing States	Imported MSW (millions of tons)
New York	7.2	Pennsylvania	7.9
New Jersey	5.8	Virginia	5.7
Illinois	4.4	Michigan	5.4
Ontario, Canada	4.0	Indiana	2.4
Missouri	2.4	Wisconsin	2.1
Maryland	2.0	Illinois	2.1
Massachusetts	2.0	Oregon	1.8
Washington	1.7	Georgia	1.7
Minnesota	1.1	New Jersey	1.7
North Carolina	1.1	Ohio	1.7
Indiana	1.1	South Carolina	1.2
District of Columbia	1.1		
Florida	1.0		
Total, all exports	42.8	Total, all imports	42.2

Source: James E. McCarthy. *Interstate Shipment of Municipal Solid Waste: 2007 Update*, CRS Report for Congress RL34043 (Washington, DC: Congressional Research Service, June 13, 2007).

One positive impact of the siting problem is that it encourages residents to reduce their waste and recycle as much as possible. (This alternative is discussed in Section 21.2) Another impact is that the problem stimulates the use of combustion as an option for waste disposal.

Combustion: Waste to Energy

Because it has a high organic content, refuse can be burned. Currently, 86 combustion facilities are operating in the United States, burning about 29 million tons of MSW annually—11.7% of the waste stream. This process is really waste *reduction*, not waste *disposal*, because after incineration the ash must still be disposed of.

Advantages of Combustion. The combustion of MSW has real advantages:

- Combustion can reduce the weight of trash by more than 70% and the volume by 90%, thus greatly extending the life of a landfill (which is still required to receive the ash).

- Toxic or hazardous combustion products are concentrated into two streams of ash, which are easier to handle and control than the original MSW. The *fly ash* (captured from the combustion gases by air-pollution control equipment) contains most of the toxic substances and can be safely put into a landfill. The *bottom ash* (from the bottom of the boiler) can be used as fill in some construction sites and roadbeds. Some combustion facilities process the bottom ash further to recover metals and then convert the remainder into concrete blocks.

- No changes are needed in trash collection procedures or people's behavior. Trash is simply hauled to a combustion facility instead of to the landfill.

- Most combustion facilities are waste-to-energy (WTE) facilities equipped with modern emission-control technology that brings the emissions into compliance with Clean Air Act regulations.

- When burned, unsorted MSW releases about 35% as much energy as coal, pound for pound. WTE facilities produce 2,700 megawatts of electricity annually, enough to meet the power needs of 2.7 million homes. Advocates of WTE point to the wasted energy going to landfills—equivalent to 9.4 billion gallons of diesel oil a year.

- Many of these facilities add resource recovery to their waste processing, in which many materials are separated and recovered before (and sometimes after) combustion.

Drawbacks of Combustion. Combustion has some drawbacks, too:

- Air pollution and offensive odors are two problems that the public associates with combustion facilities. Large reductions in air pollutants have resulted from compliance with stringent air-pollution regulations applied to

WTE facilities, making this much less of a problem than it once was. For example, dioxin emissions (dioxin is a highly toxic substance given off when many organic substances are combusted) have been reduced by 99% and mercury emissions by 90%. Odor pollution is best controlled by isolating the plant from residential areas.

- Combustion facilities are expensive to build, and their siting has the same problem as that of landfills: No one wants to live near one. Most combustion facilities are located in industrial areas for this reason.

- Combustion ash is often loaded with metals and other hazardous substances and must be disposed of in secure landfills.

- To justify the cost of its operation, the combustion facility must have a continuing supply of MSW. For that reason, the facility enters into long-term agreements with municipalities, and these agreements can lessen the flexibility of the community's solid-waste management options.

- Even if the combustion facility generates electricity, the process wastes both energy and materials unless it is augmented with recycling and recovery.

An Operating Facility. Let us look at the operation of a typical modern WTE facility, which might serve a number of communities or a large metropolitan area. Servicing a population of 1 million or more, the plant receives about 3,000 tons of MSW per day. The waste comes in by rail and truck, and the communities pay tipping fees (the costs assessed at the disposal site) that average $68 per ton. Waste processing is efficient because, overall, about 80% of the MSW is burned for energy, 12% is recovered, and 8% is put in landfills. The process, pictured in **Figure 21–8**, is as follows:

1. Incoming waste is inspected, and obvious recyclable and bulky materials are removed.

2. Waste is then pushed onto conveyers that feed shredders capable of reducing the width of waste particles to six inches or less.

3. Strong magnets remove about two-thirds of ferrous metals for recycling before combustion.

4. The waste is then blown into boilers, where light materials burn in suspension and heavier materials burn on a moving grate.

5. Water circulated through the walls of the boilers produces steam, which drives turbines for generating electricity.

6. After the waste is burned, the bottom ash is conveyed to a processing facility, where further separation of metals may occur in a process that recovers brass, aluminum, gold, copper, and iron. In some facilities, this process nets $1,000 a day in coins alone!

7. Combustion gases are passed through a lime-based spray dryer/absorber to neutralize sulfur dioxide and other noxious gases. Then the gases go through electrostatic precipitators that remove particles. The resulting

Figure 21–8 Waste-to-energy combustion facility. Schematic flow for the separation of materials and combustion in a typical, modern WTE combustion facility. Numbers refer to steps in the process as described in the text.

air emissions are significantly lower in pollutants than those of an energy-equivalent coal-fired power plant.

8. The fly ash and bottom ash residues are put into landfills.

An appreciation for the impact of such a facility can be gained from looking at the outcome of a year's operation. In one year, 1 million tons of MSW are processed, 40,000 tons of metal are recycled, and 65 megawatts of power are generated—the equivalent of more than 60 million gallons of fuel oil and enough electricity to power 65,000 homes. All this comes from stuff that people have thrown away!

However, because of the drawbacks mentioned previously, WTE facilities are strongly opposed during the siting, permitting, and construction phases. Opponents cite public-health concerns about air pollutants like dioxins and mercury as well as traffic and property value concerns.

Costs of Municipal Solid-Waste Disposal

The costs of disposing of MSW are escalating, and not just because of the new design features of landfills. More and more, they reflect the expenses of acquiring a site and providing transportation. Landfill tipping fees average about $44 a ton (they are lower than those at WTE facilities because of the high capital costs at WTE facilities). The waste collector must recover this cost, as well as transportation costs to the site, and generate a profit (if the facility is privately operated).

With the closing of the Fresh Kills landfill, the total cost (including sanitation vehicles and workers) of disposing of New York City's MSW rose to $260 a ton. At 12,900 tons per day, this amounts to more than $1 billion a year.

Getting rid of *all* trash is becoming more expensive, and one sad consequence of this increasing expense is illegal dumping. Some towns are charging up to $5 a bag for MSW disposal, and it now costs several dollars to get rid of an automobile tire and $30 or more to dispose of a refrigerator. As a result, tires, refrigerators, yard waste, car parts, and construction waste are appearing all over the landscape. Institutions and apartments that operate with dumpsters often must put padlocks on them in order to prevent their unauthorized use by people trying to avoid disposal costs. Many states have established a corps of environmental police to track down midnight dumpers and bring them to justice.

21.2 Better Solutions: Source Reduction and Recycling

Our domestic wastes and their disposal represent a huge stream of material that flows in one direction: from our resource base to disposal sites. Just as natural ecosystems depend on recycling nutrients, we can move in the direction of sustainability only if we also learn to recycle more of our wastes. There is strong evidence that we are moving in this

Waste Management Hierarchy

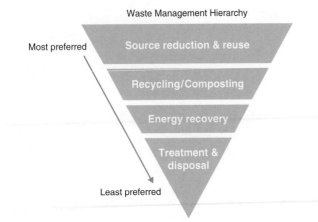

Most preferred

Source reduction & reuse

Recycling/Composting

Energy recovery

Treatment & disposal

Least preferred

Figure 21–9 EPA waste management hierarchy. This inverted triangle shows the preferable waste management options from most preferred at the top to least at the bottom.

(*Source:* EPA. Non-Hazardous Waste Managaement Heirarchy. www.epa.gov/epawaste/nonhaz/municipal/wte/nonhaz.htm. March 19, 2012.)

direction, but the best strategy of all is to reduce waste at its source—called *source reduction*. In EPA's waste management hierarchy, Source Reduction is right at the top (**Fig. 21–9**).

Source Reduction

Source reduction is defined by the EPA as "the practice of designing, manufacturing, purchasing, or using materials (such as products and packaging) in ways that reduce the amount or toxicity of trash created." Source reduction accomplishes two goals: It reduces the amount of waste that must be managed, and it conserves resources by preventing the use of virgin materials. It is noteworthy that, after rising rapidly during the latter decades of the 20th century, the amount of waste per capita in the United States peaked at 4.5 pounds in 1990 and has since leveled off. (We still lead the world in this dubious statistic, however.) The leveling off is a signal that some changes in lifestyle may be occurring that have an impact on waste generation.

Source reduction is difficult to measure because it means trying to measure something that no longer exists. The EPA measures source reduction by measuring consumer spending, which reflects the goods and products that ultimately make their way to the trash bin. For example, after 1990, consumer spending continued to grow, but the MSW stream slowed down. If the MSW had grown at the same pace as consumer spending, some 287 million tons would have been generated in 2000 instead of 232 million. Thus, some 55 million tons never made it into the waste stream in 2000, and the EPA considers this to be due to source-reduction activities.

Source reduction can involve a broad range of activities on the part of homeowners, businesses, communities, manufacturers, and institutions. Consider the following examples:

- Reducing the weight of many items has reduced the amount of materials used in manufacturing. Steel cans are 60% lighter than they used to be; disposable diapers contain 50% less paper pulp, due to absorbent-gel

technology; and aluminum cans contain only two-thirds as much aluminum as they did 20 years ago.

- The Information Age may be having an impact on the use of paper. Electronic communication, data transfer, and advertising are increasingly performed on personal computers, now present in most homes in the United States. The growth of the Internet as a medium for information transfer has been phenomenal. These developments have the potential to reduce paper waste. The U.S. Postal Service reports declining mail volume, while Internet advertising is increasing.

- Source reduction includes reuse. Many durable goods are reusable, and indeed, the United States has a long tradition of resale of furniture, appliances, and carpets. Many goods may be donated to charities and put to good use. The growing popularity of yard sales, flea markets, consignment clothing stores, and other markets for secondhand goods (**Fig. 21–10**) is an encouraging development.

- The Internet has become a major player in reuse. All sorts of used goods are offered for sale on eBay® and Craigslist, two Internet services that link private sellers and buyers. U-Exchange promotes bartering, while free goods can be advertised and acquired on the Freecycle Network™, a concept that has grown to 5,000 groups with 8.8 million members in 120 countries. There is probably a chapter in your city!

- Lengthening a product's life can keep that product out of the waste stream. If products are designed to last longer and to be easier to repair, consumers will learn that they are worth the extra cost. (You get what you pay for; cheap products usually take the shortest route to the waste bin.)

- We are all on bulk mailing lists because these lists are sold and shared widely, so we are guaranteed to receive increasing volumes of advertising. To stay off such lists, simply inform mail-order companies and other organizations involved that you do not want your name and

Figure 21–10 Waste reduction by reuse. Yard sales and flea markets are keeping many materials from becoming MSW after a single use.

address shared. Several organizations are available to help you accomplish this: Direct Marketing Association (www.dmachoice.org) helps you get rid of email and snail mail advertising; Catalog Choice™ and stopthe-junkmail.com perform a similar service.

The Recycling Solution

In addition to reuse, recycling is an obvious solution to the solid-waste problem, right after source reduction in EPA's hierarchy. More than 75% of MSW is recyclable material. There are two levels of recycling: *primary* and *secondary*. In **primary recycling**, the original waste material is made back into the same material—for example, newspapers recycled to make newsprint. In **secondary recycling**, waste materials are made into different products that may or may not be recyclable—for instance, newspapers recycled to make cardboard.

Why Recycle? Recycling is a hands-down winner in terms of energy and resource use and pollution:

- **It saves energy and resources.** One ton of recycled steel cans saves 2,500 pounds of iron ore, 1,000 pounds of coal, and more than 5,400 Btus of energy. One ton of papers recycled saves 17 trees, 6,953 gallons of water, 463 gallons of oil, and 4,000 kilowatt-hours of energy.

- **It decreases pollution.** Making recycled paper requires 64% less energy and generates 74% less air pollution and 35% less water pollution than does using wood from trees. For every ton of waste processed, a thorough recycling program will eliminate 620 pounds of carbon dioxide, 30 pounds of methane, 5 pounds of carbon monoxide, 2.5 pounds of particulate matter, and smaller amounts of other pollutants.

What Gets Recycled? The primary items from MSW that are currently being actively recycled are cans (both aluminum and steel), bottles, plastic containers, newspapers, and yard wastes. Yet there are many alternatives for reprocessing various components of refuse, and people are coming up with new ideas and techniques all the time. A few of the major established techniques, together with their current percentages of recovery by recycling, are as follows:

- Paper and paperboard (53% recovery) can be remade into pulp and reprocessed into recycled paper, cardboard, and other paper products; finely ground and sold as cellulose insulation; or shredded and composted.

- Most glass (33% recovery) that is recycled is crushed, remelted, and made into new containers; a smaller amount is used in fiberglass or "glassphalt" for highway construction.

- Some forms of plastic (13.5% recovery) can be remelted and fabricated into carpet fiber, outdoor wearing apparel, irrigation drainage tiles, building materials **(Fig. 21–11)**, and sheet plastic.

- Metals can be remelted and refabricated. Making aluminum (36% recovery) from scrap aluminum saves up to 90% of the energy required to make aluminum from virgin ore. In addition, aluminum ore is imported and represents part of the mounting U.S. trade deficit. National recycling of aluminum saves energy, creates jobs, and reduces the trade deficit.

- Yard wastes (leaves, grass, and plant trimmings—57.5% recovery) can be composted to produce a humus soil conditioner.

- Textiles (16.5% recovery) can be shredded and used to strengthen recycled paper products.

- Old tires (35% recovery) can be remelted or shredded and incorporated into highway asphalt. Not included in the 35% figure are more than a million tons of tires burned annually in special waste-to-energy facilities.

Recycling is both an environmental and an economic issue. Many people are motivated to recycle because of environmental concern, but the use of recyclable and recycled materials is also driven by economic factors. The Global Recycling Network (www.grn.com) is an Internet information exchange that promotes the trade of recyclables from MSW and the marketing of "ecofriendly" products made from recyclable materials.

Municipal Recycling

Recycling is probably the most direct and obvious way that most people can become involved in environmental issues. If you recycle, you save natural resources from being used (trees, in the case of paper), and you prevent landfills from becoming "landfulls."

Recycling's popularity is well established: Virtually every state has specific recycling goals, met with varying degrees

Figure 21–11 Deck made from Trex® decking and railing. Trex is made primarily with recycled plastic grocery bags, reclaimed pallet wrap, and waste wood.

(*Source:* www.trex.com)

Figure 21–12 MSW recycling in the United States. MSW recycled from 1960 to 2010, as total waste and percentage recycled.

(*Source:* EPA, Office of Solid Waste and Emergency Response, *Municipal Solid Waste Generation, Recycling, and Disposal in the United States: Facts and Figures for 2010* (Washington, DC: EPA, November, 2011).)

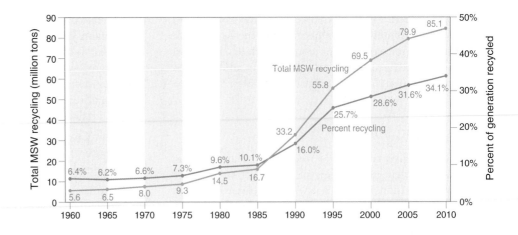

of success. EPA sources report that only 6.4% of MSW was recycled in 1960 versus 34.1% in 2010 (**Fig. 21–12**). There is a great diversity of approaches to recycling in municipalities, from recycling centers requiring miles of driving to curbside recycling with sophisticated separation processes. The most successful programs have the following characteristics:

1. There is a strong incentive to recycle where there are PAYT (pay-as-you-throw) charges for general trash and no charge for recyclable goods.

2. Recycling is not optional; mandatory regulations are in place, with warnings and sanctions for violations.

3. Residential recycling is curbside (**Fig. 21–13**), with free recycling bins distributed to households. (Curbside recycling rose rapidly and then plateaued. Currently, 71% of the U.S. populace is served via 9,066 curbside programs.)

4. The process employed is single-stream recycling, where all recyclable goods are mixed in a collection truck, allowing residents to put all materials in a single container. Separation of materials for reuse then occurs at a **materials recovery facility** (**MRF**) (see below).

Figure 21–13 Curbside recycling. Recyclable materials are picked up in a San Francisco neighborhood, where residents are issued three bins: one for garbage, one for recycling, and one for compostable materials. The city now recycles 77% of its waste.

5. Drop-off sites are provided for bulky goods like sofas, appliances, construction and demolition materials, and yard waste.

6. Recycling goals are ambitious, yet clear and feasible. Some percentage of the waste stream is targeted, and progress is followed and communicated.

Nevertheless, municipalities experience very different recycling rates. New York City, for example, has made great progress—from recycling only 5% to recycling 16%—by mandating curbside recycling for all of its 3 million households. Boston, by contrast, with only 13% recycling (with voluntary curbside pickup), has been holding Massachusetts back in its goal to achieve 56% recycling by 2010 (it was only 28% in 2008, and so the state has adopted a new goal: reduce total disposal 30% by 2020, through varying means). Worcester, farther to the west in the state, has achieved a commendable 43% recycling rate. The national prize goes to San Francisco, a city that has achieved a recycling rate of 77%.

Economics of Recycling. Recycling has its critics, who base their arguments primarily on economics. If the costs of recycling (from pickup to disposal of recyclable components) are compared with the costs of combusting waste or placing it in landfills, recycling frequently comes out second best. Markets for recyclable materials fluctuate wildly, and residents often end up subsidizing the recycling effort. Competition between landfills and combustion facilities often lowers tipping fees, creating an even greater disincentive to recycle. Thus, critics of current recycling practices argue that, unless recycling pays for itself through the sale of the materials recovered, it should not be done.

However, a true environmental assessment should compare the energy costs of recycling with the energy costs of landfill or combustion disposal, as well as comparing the energy costs of making products from recyclable goods with the energy costs of making new products from scratch—called a *life-cycle analysis*. Performed this way, the energy saved by recycling aluminum, for example, is 96%; for newsprint, 45%; for two kinds of plastic, 76–88%; and for glass, 21%.

Unfortunately, all recycling took a hit because of the recent recession. Demand for recyclable products plummeted, reached bottom in 2009, and is now on the rebound as the economy recovers. Irrespective of these economic considerations, the demonstrated support for recycling programs is strong. Experience has shown that at least two-thirds of households will recycle if presented with a curbside pickup program. The percentage goes up when recycling is combined with a PAYT program or a program that rewards participants for their participation, like RecycleBank. Dover, New Hampshire—a city of 26,000—was facing escalating costs of MSW collection and disposal after its landfill closed. The city instituted both a PAYT program and curbside recycling in 1991 and saw per household trash plummet from 6 pounds to 2.3 pounds per day in six years. At the same time, the cost per household dropped from $122 to $73 per year, and the recycling rate rose quickly to more than 50%. The city's solid-waste management costs dropped 27% in spite of a population increase.

Paper Recycling. Newspapers are by far the most important item that is recycled because of their predominance in the waste stream. It is a simple matter to tie up or bag household newspapers, and the amount recovered by recycling is increasing dramatically. Currently, far more newspaper is being recycled (72%) than discarded. Because more than 25% of the trees harvested in the United States are used to make paper, recycling paper saves trees: A 1-meter stack of newspapers equals the amount of pulp from one tree.

The market for used newspapers has fluctuated greatly over the past several decades. During the late 1980s, the market was saturated, and municipalities often had to pay to get rid of newspapers. In 1995, discarded papers were so valuable—up to $160 a ton—that thieves were stealing them off the sidewalks before the recycling trucks could pick them up. A year later, as more recycling programs came on line, the market collapsed again, and many cities once more paid as much as $25 a ton to have the newspapers hauled away—but this is still less expensive than the tipping fees at landfills. Furthermore, there is a lively international trade in used paper (Fig. 21–14). Forest-poor countries in Europe and Asia purchase wastepaper from the United States and other industrial countries in the Northern Hemisphere, where there continues to be a surplus of such paper.

After the wastepaper is incorporated into a final product, the market becomes a critical factor. Is there a demand for recycled paper? The technology for producing high-quality paper from recyclable stock has improved greatly, to the point where it is virtually impossible to distinguish the recycled product from "virgin" paper. There is often some confusion about what is meant by "recycled paper." The key is the amount of *postconsumer* recycled paper in a given product. In manufacturing processes, much paper is "wasted"; paper that is routinely recovered and rerouted back into processing is called "recycled paper." Thus, the total recycled content of a paper can be 50%, although the actual postconsumer amount recovered by recycling might be only 10% of the total amount of paper.

Glass Recycling and Bottle Laws. The glass in MSW is primarily in the form of containers, most of which held beverages. The average person drinks about a quart of liquid each day. Given that there are 312 million Americans, this daily consumption amounts to some 28 billion gallons of liquid per year for the nation as a whole. Much of this volume is packaged in single-serving containers that are used once and then thrown away. Nonreturnable glass containers constitute 3.7% of the solid-waste stream in the United States and about 50% of the nonburnable portion; they also make up a large proportion of roadside litter. Broken bottles along roads, beaches, and parklands are responsible for innumerable cuts and other injuries, not to mention flat tires. Both the mining of the materials and the process used to manufacture the beverage containers create pollution. All of these factors produce hidden costs that do not appear on the price tag of the item; however, we pay for them with taxes to clean up litter as well as with our injuries, flat tires, and environmental degradation.

In an attempt to reverse these trends, environmental and consumer groups have promoted **bottle laws** that facilitate the recycling or reuse of beverage containers. Such laws generally call for a deposit on all beverage containers, both reusable and throwaway and both glass and plastic. Retailers are required to accept the used containers and pass them along for recycling or reuse. Bottle laws have

Figure 21–14 Wastepaper exports. Bales of wastepaper ready to be loaded for shipment overseas. China imports a third of the wastepaper being sold internationally.

been proposed in virtually all state legislatures over the past decade. Nevertheless, in every case, the proposals have met with fierce opposition from the beverage and container industries and certain other special-interest groups. The reason for their opposition is economic loss to their operations. The container industry contends that bottle laws will result in the loss of jobs and higher beverage costs for the consumer.

In most cases, the industry's well-financed lobbying efforts have defeated bottle laws. However, some states—11, as of 2012—have adopted bottle laws of varying types despite industry opposition (Table 21–2). The experience of these states has proved the beverage and bottle industry's claims false. In fact, more jobs are gained than lost, costs to the consumer have not risen, a high percentage of bottles are returned, and there is a marked reduction in the can and bottle portion of litter. A final measure of the success of bottle laws is continued public approval: Despite industry efforts to repeal these laws, no state that has one has repealed it.

Repeated attempts have been made to get a national bottle law through Congress—to date, unsuccessfully. Opponents (the same ones that oppose state-level bottle bills) argue that such a law will threaten the continued success of curbside recycling, of which beverage containers represent an important source of revenue. However, as curbside recycling currently reaches only 71% of the U.S. population, a national bottle law would undoubtedly recover a greater proportion of beverage containers. (States with bottle laws report that 75% of containers are returned, a significant proportion from the curbside collections from people who don't bother to redeem the containers.) Also, a national bottle law would be labor intensive, employing tens of thousands of new workers.

In 2010, recycling and bottle laws resulted in the recovery of 33.4% of glass containers, 50% of aluminum containers, 67% of steel containers, and 29% of plastic containers, with an aggregate weight of 6.4 million tons.

TABLE 21–2   States with Bottle Laws

State	Year Passed
Connecticut	1972
Oregon	1972
Vermont	1973
Maine	1976
Michigan	1976
Iowa	1978
Massachusetts	1978
Delaware	1982
New York	1983
California	1991
Hawaii	2002

Plastics Recycling. Plastics have a bad reputation in the environmental debate for several reasons. First, plastics have many uses that involve rapid throughput—for example, packaging, bottling, the manufacture of disposable diapers, and the incorporation of plastics into a host of cheap consumer goods. Second, plastics are conspicuous in MSW and litter. Ironically, most trash is disposed of in plastic bags manufactured just for that purpose. Finally, plastics do not decompose in the environment because no microbes (or any other organisms) are able to digest them. (Imagine plastics in landfills delighting archaeologists hundreds of years from now.) Therefore, the possibility of recycling at least some of the plastic in products—containers of liquid, for example—interests the environmental community.

Bottled Water. Throughout the world, the number one "new" drink is bottled water. Health consciousness and suspicions about the safety of tap water have made the plastic water bottle the new symbol of healthy living. More than 9 billion gallons of bottled water were sold in the United States in 2011, making us the world leader in consumption of bottled water. As a consequence, a new form of litter along our highways and on our beaches is the discarded plastic water bottle. Only two states with bottle laws include bottled water and other noncarbonated beverages. People often pay more for bottled water (almost half is only bottled tap water) than for gasoline and as much as 10,000 times more than for water from the tap. Only one in six plastic water bottles is recycled, and the rest end up in litter, landfills, or combustion facilities.

For these and other reasons, opposition to bottled water is growing; more than 90 colleges and universities and several states and municipalities have either banned or restricted distribution of bottled water. The opposition considers the use of bottled water wasteful, citing the millions of barrels of oil used for manufacture and distribution of a product that can be obtained at a small fraction of the cost. Industry spokespersons counter by pointing out the role of bottled water in offering an alternative to the sugary drinks that contribute to the obesity epidemic in the United States and as an essential component of disaster relief. The market for bottled water in the United States, Canada, and Europe has declined over the past three years, but the industry has compensated by selling much more of its product in developing countries, where safe water is much harder to find.

PETE and HDPE. If you look on the bottom of a plastic container, you will see a number (or some letters) inside the little triangle of arrows that represents recyclability. The number or the letters are used to differentiate between and sort the many kinds of plastic polymers. The two recyclable plastics in most common use are high-density polyethylene (HDPE, code 2) and polyethylene terephthalate (PETE, code 1). In the recycling process, plastics must be melted down and poured into molds, and some contaminants from the original containers may carry over, making it difficult to reuse some plastics for food containers. While this somewhat restricts uses for recyclable plastic, some interesting uses are appearing, with more in the development stage. PETE, for example, is turned into carpets, jackets, film, strapping, and new PETE bottles; HDPE becomes irrigation drainage tiles, sheet plastic, and, appropriately, recycling bins.

Critics of the recycling of plastics argue that recovering plastics is more costly than starting from scratch—that is, beginning with petroleum derivatives—and that manufacturers are getting involved mainly because environmentally concerned consumers are demanding their involvement. Industry-supported critics also point out that plastics in landfills create no toxic leachate or "dangerous" biogas. Moreover, plastics in combustion facilities burn wonderfully hot and leave almost no ash. However, with the rising costs of petroleum and the many uses of recyclable plastic bottles (including new bottles), closing the loop to recover plastic bottles makes both economic and environmental sense.

Plastic Bags. Hanging from trees and bushes, blowing along highways, clogging sewer pipes, drifting on ocean currents, plastic bags are everywhere on the planet—everywhere, that is, but the trash or recycling barrel. They cause thousands of deaths of marine mammals and turtles each year. Convenient? Indeed they are, as anyone who has shopped in a supermarket knows. A hundred billion of them are distributed in the United States every year, and close to a trillion are used annually worldwide. Made of low-density polyethylene, plastic bags are difficult to recycle and, like all petroleum-based plastics, are virtually indestructible. China, once a user of 100 billion plastic bags a year, has cracked down on them, banning the production of ultrathin bags—bringing China in line with a number of other countries. In the United States, San Francisco initiated the first citywide ban on plastic grocery bags in 2007; many other cities are instituting similar bans. Instead of the thin plastic bags, customers are encouraged to carry their own cloth bags or to use paper bags or larger, thicker reusable plastic bags that hold five times as much as the standard flimsy bag. The plastics industry is fighting back with lawsuits and information campaigns touting the benefits of the plastic bag. Perhaps someday soon the "national flower of South Africa," as the bags have been called in that country, will become an endangered species.

The Composting Option. An increasingly popular way of treating yard waste and food residuals (currently 27% of MSW) is **composting**, especially as more and more states and municipalities are banning these materials from MSW collections. Composting, the natural biological decomposition (rotting) of organic matter in the presence of air, can be carried out in the backyard by homeowners, or it can be promoted by municipalities as a way of dealing with a large fraction of MSW. In addition to functioning as a way to divert wastes away from landfills and combustion facilities, composting provides a product that can be used to amend poor soils, add fertilizer to yards and gardens, and promote higher yields to agricultural crops. In 2010, 57.7% of yard trimmings were recovered from the waste stream.

Large-scale composting is carried out by recycling facilities and municipalities. Wastes brought into the facility are usually ground up into smaller pieces and then placed outdoors into rows of long piles (windrows). The windrows are turned regularly to allow oxygen to penetrate the piles,

promoting natural biological decomposition. In dry regions, water is added to maintain moisture in the windrows. After several months of this treatment, the compost may be screened for different final uses: finely screened for lawns or more coarsely for garden and landscape beds.

Small-scale composting can be done in backyards, using bins that are specially made for this use or simply piles or heaps (best to hide these from the neighbors). Instructions for managing compost are easily found on the Internet; virtually all of them will help you manage your compost to the end product of a very rich mulch that can do wonders for your garden or landscape plantings. All of them tell you to avoid putting too much food waste out, lest you attract unwanted four-legged visitors.

Regional Recycling Options

As landfills close down and MSW is transferred out of cities and towns, transfer stations are set up to receive the wastes from collection trucks and transfer them to larger vehicles for long-distance hauling to their final destination. It is an encouraging sign that a number of these transfer stations are being converted into materials recovery facilities, referred to in the trade as MRFs, or "murfs." In 2010, there were 633 MRFs operating in the United States, handling over 98,000 tons of MSW a day.

More and more, these facilities are designed to receive the collections of single-stream recycling, where all possible recyclable components are "commingled." Alternatively, some sorting takes place when waste is collected, either through curbside collection or by town recycling stations (sites to which residents can bring wastes to be recycled). In either case, the waste is trucked to the MRF, where it is moved through the facility by escalators and conveyor belts, tended by workers who inspect and sort the items collected (**Fig. 21–15**). The objective of the process is to prepare

Figure 21–15 Materials recovery facility. Single-stream recycling at a modern MRF in Baltimore. Its purpose is to sort a mixed stream of waste material (cans, bottles, newspapers, etc.) for eventual recycling.

materials for the recyclable-goods market. Glass is sorted by color, cleaned, crushed into small pebbles, and then shipped to glass companies, where it replaces the raw materials that go into glass manufacture—sand and soda ash—and saves substantially on energy costs. Cans are sorted, flattened, and sent either to detinning plants or to aluminum-processing facilities. Paper is sorted, baled, and sent to reprocessing mills or exporters. Plastics are sorted, depending on their color and type of polymer, and then sold.

The facility's advantages are its economy of scale and its ability to produce a useful end product for the recyclable-materials market. Some MRFs have gone to high technology, replacing the manual sorting with magnetic pulleys, optical sensors, and air sorters. As the movement grows, this is likely the direction the technology will take to improve the efficiency of operations and cut costs.

Less common than conventional MRFs are mixed-waste processing facilities, or so-called "dirty MRFs," which receive unsorted MSW just as if it was going to a landfill or a WTE facility. The waste is loaded on a conveyer and is sorted for recovery of recyclable materials before being landfilled or combusted. Because of extensive contamination of potentially recyclable waste, this method is less desirable and is not used as often.

21.3 Public Policy and Waste Management

The management of MSW used to be entirely under the control of local governments. In recent years, however, state and federal agencies have played an increasingly important role in waste management, partly through regulation and partly through encouragement and facilitation.

The Regulatory Perspective

At the federal level, the following legislation has been passed:

- The first attempt by Congress to address the problem was the Solid Waste Disposal Act of 1965. The legislation gave jurisdiction over solid waste to the Bureau of Solid Waste Management, but the agency's mandate was basically financial and technical rather than regulatory.

- The Resource Recovery Act of 1970 gave jurisdiction over waste management to the newly created EPA and directed attention to recycling programs and other ways of recovering resources in MSW. The act also encouraged the states to develop some kind of waste management program.

- The passage of the Resource Conservation and Recovery Act (RCRA) of 1976 saw a more regulatory (command-and-control) approach to dealing with MSW, as the EPA was given power to close local dumps and set regulations for landfills. Combustion facilities were covered by air-pollution and hazardous-waste

regulations, again under the EPA's jurisdiction. RCRA also required the states to develop comprehensive solid-waste management plans.

- The Superfund Act of 1980 (described in Chapter 22) addressed abandoned hazardous-waste sites throughout the country, 41% of which are old landfills.

- The Hazardous and Solid Waste Amendments of 1984 gave the EPA greater responsibility for setting solid-waste criteria for all hazardous-waste facilities. Because even household waste must be assumed to contain some hazardous substances, this meant that the EPA had to determine and monitor all landfill and combustion criteria more closely.

Largely in response to these federal mandates, the states have been putting pressure on local governments to develop *integrated waste management plans*, with goals for recycling, source reduction, and landfill performance.

Integrated Waste Management

It is not necessary to fasten onto just one method of handling wastes. Source reduction, waste-to-energy combustion, recycling, materials recovery facilities, landfills, and composting all have roles to play in waste management. Different combinations of these options will work in different regions of the country. A system having several processes in operation is called *integrated waste management*.

Waste Reduction. The modern United States leads the world in per capita waste production and per capita energy consumption—we really are a "throwaway society." How can we turn this situation around? True management of MSW begins in the home (or dorm room). Materialistic lifestyles, affluence, and overconsumption can pack our homes and our trash cans with stuff we simply do not need. Thousands of public storage facilities with interesting names (Stor-U-Self, for example) are packed with stuff that people can't fit into their homes but can't part with.

WasteWise. Several incentives have been put in place on the government level to encourage waste reduction. For example, the EPA sponsors the **WasteWise** program, targeting the reduction of MSW by establishing partnerships with not only local governments, schools, and organizations, but multinational corporations as well. WasteWise is a voluntary partnership that allows the partners to design their own solid-waste reduction programs. More than 2,700 organizations throughout the country have partnered with the EPA to reduce and recycle waste and, in a new focus of the program, to calculate their reductions of waste-related greenhouse gas emissions. The EPA regularly honors partner organizations that have achieved exemplary waste reduction. For example, NEC Electronics, Inc.'s Roseville, California, facility has been a WasteWise partner since 1994, winning seven WasteWise awards over the years. The facility has mastered waste diversion, diverting 82% of its solid waste from landfills in 2008. Its efforts over the years have diverted more than 4,700 tons

Figure 21–16 Pay-as-you-throw (PAYT) trash pickup. Curbside trash pickup in Hamilton, Massachusetts, features free pickup for single-stream recycling (blue crate) and compost (kitchen scraps and yard trimmings, green bin); bags for other trash are purchased at local stores (blue trash bag) for $1.75.

of waste through waste prevention and recycling, saving the company more than $7 million. The company sponsored an e-Waste Day for its community, collecting more than 54,000 pounds of **e-waste** (television sets, cell phones, monitors, batteries, computers, etc.). This is a growing problem for MSW facilities in the United States and has become an international problem because of the practice of exporting e-waste to developing countries (see Stewardship, "Between the Cracks: E-Waste," p. 548).

PAYT. The EPA also brings attention to a national trend of unit pricing, or charging households and other customers for the waste they dispose of, with the agency's PAYT program. Instead of using local taxes to pay for trash collection and disposal, which provides no incentive to reduce waste, communities levy curbside charges for all unsorted MSW—say, $1–$5 per container (**Fig. 21–16**). PAYT capitalizes on the common sentiment that people should pay their share of disposal costs based on the amount of waste they discard. This is catching on as cities and towns see the advantages of the practice; more than 7,000 cities and towns, serving 25% of the U.S. population, now employ the PAYT method. This practice is found extensively throughout Europe and Asia. A recent study of PAYT in New England communities[3] found that curbside PAYT communities generated 49% less waste than other non-PAYT communities. City managers love it because it saves money and creates jobs. As with the WasteWise program, the EPA's role is that of a facilitator, not a regulator.

EPR. Another means of bringing about waste reduction is to establish a program of *extended producer responsibility* (EPR), a concept that involves assigning some

responsibility for reducing the environmental impact of a product at each stage of its "life cycle," especially the end. EPR originated in Western Europe. Germany, the Netherlands, and Sweden have well-developed regulations that require companies to take back many items, that encourage the reuse of products, and that promote the manufacture of more durable products. In the United States, EPR is better known as *product stewardship*, a term that implies shared responsibility by producers, consumers, and municipalities. For example, Hewlett-Packard and Xerox make it easy for customers to return spent copier cartridges, and the two companies then recycle the components of the cartridges. Call2Recycle is another good example of this concept, where rechargeable batteries and cell phones are voluntarily taken back by a nonprofit corporation, Rechargeable Battery Recycling Corporation. The program collects up to 7 million pounds of batteries a year.

Again, the EPA is active in providing information to manufacturers and purchasing departments to help them design and buy more environmentally sound products.

Waste Disposal Issues. No human society can avoid generating solid waste. There will always be MSW, no matter how much we reduce, reuse, and recycle. Most experts see WTE combustion facilities holding their own, a decrease in the percentage of waste going to landfills, and a continued rise in recycling.

It will be important to break the gridlock at local and state levels on the siting of new landfills and WTE combustion facilities. Landfills will still be needed, though they should last longer than in the past. Policy makers have long known about the landfill shortage, but have opted for short-term solutions with the lowest political cost. One result of this choice is the long-distance hauling of MSW by truck and rail described earlier. If regions and municipalities are required to handle their own trash locally, as is the case in some states, they will find places to site landfills, regardless of NIMBY considerations. New landfills must use the best technologies, such as proper lining, leachate collection and treatment, groundwater monitoring, biogas collection, and final capping. If these technologies are employed, a landfill will not be a health hazard. Also, sites should be selected with a view to some future use that is attractive.

With the closing of New York City's huge Fresh Kills landfill, the city is now sending its MSW to other states. Yet most citizens in the other states object to being New York's dumping ground, and there is the deeper stewardship issue of local responsibility for local wastes as well as the inefficiency of long-distance waste hauling. It will take an act of Congress to give states or local jurisdictions the right to ban imports of waste because of the interstate commerce clause in the Constitution. The Solid Waste Interstate Transportation Act of 2009 (H.R. 274) was introduced in the 111th Congress, with bipartisan support, especially from states that are currently receiving waste from other states. The legislation would have given local and state governments the authority to limit or prohibit the transportation

---

[3]Environmental Protection Agency. "Pay-As-You-Throw Summer 2010 Bulletin." *Get Smart with Pay-As-You-Throw.* EPA 530-N-09-001. 2010.

STEWARDSHIP

We all use them—televisions, computers, DVD players, cell phones, printers, copiers, video game systems. But what do we do when they are obsolete or broken? First, we store them—an estimated seven items per household in the United States, adding up to 5 million tons. Then we get around to disposing of them. The chances are that we put them in the trash and hope they get picked up, or we take them to a transfer station and pay a small fee to turn in that old television or computer monitor. There are some better options: EcoSquid, a Web search program, helps consumers find places to take unwanted electronics, including options for resale; Best Buy, Staples, Radio Shack, and several other vendors accept electronic waste for a fee. This is no small matter, however. E-waste, as these devices are called once they are discarded, is growing far more rapidly than our ability to deal with it. The United States leads the world in e-waste, producing 3.3 million tons a year. One estimate puts the yearly global generation of e-waste at 50 million tons. To date, most of this e-waste is landfilled, with an increasing proportion recycled.

E-waste has numerous toxic components and is classified as hazardous waste. Cathode ray tubes (CRTs), which are the vacuum tubes that provide the video display on older televisions and computer monitors, contain an average of four pounds of lead. Many other heavy metals—mercury, barium, chromium, arsenic, and copper, to name a few—are found in the wiring and circuit boards of electronic devices. In a landfill, these substances can leach into groundwater. If the e-waste is incinerated, they are sent into the air. Federal law regulates the transport, treatment, and disposal of such hazardous wastes under RCRA. However, RCRA does not apply to small businesses and households, so the disposal of this e-waste is left up to the states and municipalities. To date, 23 states have implemented some kind of e-waste management, and there is broad consensus on the need for federal legislation to manage this waste (the Responsible Electronics Recycling Act has been introduced to Congress, with passage uncertain). The states that do have laws for e-waste are basically trying to make sure this waste is not landfilled or incinerated. The ideal destination for e-waste is a recycling company that will dismantle the devices, recover the useful materials, and dispose of the rest according to RCRA rules (see Chapter 22).

E-waste contains valuable materials, like indium and palladium, which are quite scarce. Precious metals like gold, silver, and copper can also be recovered, but recovery of these materials is labor intensive. "Recycling" companies may be found to take the products from transfer stations and consumers, but all too often, they are packing the e-waste into large containers and shipping it to China, India, or African countries. An estimated 80% of "recycled" e-waste takes this route, ending up in any of thousands of small shops in developing countries, where laborers pull apart the devices, burn wiring to recover copper, fry circuit boards to free the chips, and dip the chips in buckets of acid to extract gold and silver—and they are usually working without protection from the hazardous chemicals.

However, there *are* legitimate recycling companies. Recently, the nation's largest electronics recycler, Electronics Recyclers International, joined with the Electronics TakeBack Coalition and the Basel Action Network in the e-Stewards Initiative. This initiative certifies e-waste recyclers, assuring consumers and businesses that their e-waste will not be landfilled, incinerated, or exported to developing countries. With a patchwork of state and local regulations for e-waste, the e-Stewards Initiative is a welcome and much-needed measure. However, federal legislation would be better, and given the huge volume of e-waste, it can't come too soon.

of out-of-state wastes to landfills. The bill is a demand for fairness to states that are working hard to deal responsibly with their own wastes, only to see the unrestricted transport of wastes from other states. Despite having bipartisan support, this bill didn't see the light of day because of its presumed restriction of interstate commerce. Similar bills have been introduced in every Congress since the 100th, and like this bill, they have never made it out of the House Committee on Energy and Commerce.

Recycling and Reuse. Recycling is certainly well established now. It should not, however, be pursued in lieu of waste reduction and reuse. The move toward more-durable goods is an overlooked and underutilized option. In fact, waste reduction remains the most environmentally sound and least costly goal for MSW management, because wastes that are never generated do not need to be managed.

Banning the disposal of recyclable items in landfills and at combustion facilities makes very good sense. Many states have incorporated this regulation into their management programs. For example, Massachusetts phased out yard waste in 1991, metals and glass in 1992, and recyclable papers and plastic in 1994. Landfill and combustion-facility operators are supposed to conduct random truck searches and are authorized to turn back trucks with significant amounts of banned items. Massachusetts banned the disposal of cathode ray tubes (CRTs) in landfills in 2000, becoming the first state to do so. Since then, a number of states have addressed the disposal of CRTs and other forms of e-waste. (See Stewardship, "Between the Cracks: E-Waste.")

Enacting a national bottle-deposit law would be a giant stride toward the reuse and recycling of beverage containers and would also greatly reduce roadside litter in states lacking bottle laws.

Finally, closing the "recycling loop" remains a significant action to be taken by governments to encourage recycling. A number of states have opted for one or more of the following approaches: (1) minimum postconsumer levels of recycled content for newsprint and glass containers; (2) requirements stating that purchases of certain goods must include recycled products even if they are more expensive than "virgin" products; (3) requirements that all packaging be reusable or be made (at least partly) of recycled materials; (4) tax credits or incentives that encourage the use of recycled or recyclable materials in manufacturing; and (5) assistance in the development of recycling markets.

REVISITING THE THEMES

Sound Science

The role of the EPA and its scientists is crucial. The agency sets the rules for landfill operations, as mandated by Congress, and then takes a proactive role in areas where it has no regulatory authority, encouraging source reduction, recycling, and state MSW management programs. WasteWise and the PAYT program are two excellent examples of initiatives where the EPA plays a facilitating role. The greatest current public-policy needs are a national bottle law, e-waste legislation, and a law dealing with interstate waste transport.

Sustainability

Sustainable natural systems recycle materials, but we create so many products that are difficult to recycle that it becomes necessary to be intentional about source reduction and product responsibility. The greatest departure from sustainable living is the interstate transfer of trash, which wastes energy and allows cities and states to go for the cheapest immediate solution and avoid coming to grips with a responsible approach to waste management. The most positive step toward sustainability is intentional source reduction, which both conserves resources and reduces the amount of waste that must be managed.

Stewardship

Instead of allowing economic considerations to dictate MSW management choices, many municipalities and organizations are consciously choosing options that reflect (and often bear the name of) stewardship. Here is one possible stewardly waste management plan:

1. Emphasize source reduction wherever possible.

2. Employ mandatory curbside recycling and a PAYT collection program

3. If feasible, establish a MRF for efficient handling of recyclables (and possibly MSW).

4. Make sure all e-waste is delivered to a recycling company certified by e-Stewards.

5. Collect yard wastes for composting by a MRF or municipal facility.

6. Deposit residual materials in a local landfill.

7. Prohibit all interstate transfer of MSW.

This plan features local MSW management and local or regional recycling, composting, and landfilling. Every step of the plan involves decisions on the part of individuals and municipalities; these practices do not happen unless there are people who care enough to promote responsible stewardship of the "stuff" that all of us feel that we need.

REVIEW QUESTIONS

1. List the major components of MSW.

2. Trace the historical development of refuse disposal. What percentages now go to landfills, combustion, and recycling?

3. What are the major problems with placing waste in landfills? How can these problems be managed?

4. Explain the difficulties accompanying landfill siting and outsourcing. How can these processes be handled responsibly?

5. What are the advantages and disadvantages of WTE combustion?

6. What is the evidence for increasing source reduction, and what are some examples of how it is accomplished?

7. What are the environmental advantages of recycling?

8. Describe the materials that are recycled and how recycling is accomplished.

9. Discuss the attributes of successful recycling programs.

10. Distinguish between MRFs, mixed waste processing, and single-stream recycling.

11. What laws has the federal government adopted to control solid-waste disposal?

12. What is e-waste, and how is it being managed?

13. What is integrated waste management? What decisions would be part of a stewardly and sustainable solid-waste management plan?

THINKING ENVIRONMENTALLY

1. Compile a list of all the plastic items you used and threw away this week. Was it more or less than you expected? How can you reduce the number of items on the list?

2. How and where does your school dispose of solid waste? E-waste? Is a recycling program in place? How well does it work?

3. Suppose your town planned to build a combustion facility or landfill near your home. Outline your concerns, and explain your decision to be for or against the site.

4. Does your state have a bottle law? If not, what has prevented such a bill from being adopted?

5. Consider the stewardly waste management plan described in the "Stewardship" section of Revisiting the Themes. How many of these components does your city or town employ?

MAKING A DIFFERENCE

1. Investigate how municipal solid waste is handled in your school or community and what the future plans are for waste management. Promote curbside recycling if it is not happening, and investigate the possibility of adopting a PAYT system.

2. Consider your consumption patterns. Live as simply as possible so as to minimize your impact on the environment. Buy and use durable products, and minimize your use of disposables.

3. When you shop, bring your own reusable bags for your groceries. When buying small items that you can easily carry, ask that they not be put in a bag.

MasteringEnvironmentalScience®

Go to **www.masteringenvironmentalscience.com** for practice quizzes, Pearson eText, videos, current events, and more.

A customer looks over shelves of canned goods, which are likely to be lined with an expoxy resin made with bisphenol A.

CHAPTER 22

Hazardous Chemicals: Pollution and Prevention

LEARNING OBJECTIVES

1. **Toxicology and Chemical Hazards:** Explain what toxicology is and how it applies to many of the chemicals in use in our society; identify the two most toxic chemical groups in use, and assess their involvement in food chains.

2. **Hazardous Waste Disposal:** Describe the three methods employed in land disposal of hazardous wastes, and relate and give examples of what happened before land disposal came under regulation.

3. **Cleaning Up the Mess:** Explain how the Superfund program put in place by Congress deals with abandoned toxic sites, brownfields, and leaking underground storage tanks.

4. **Managing Current Toxic Chemicals and Wastes:** Review the laws put in place to (a) prevent illegal disposal of toxic wastes, (b) reduce accidents and accidental exposure, and (c) evaluate new chemicals.

5. **Broader Issues:** Discuss the issue of environmental justice, and review strategies that prevent toxic chemicals from being used.

HORMONES, OR **ENDOCRINES**, are powerful chemicals that regulate many systems and functions in vertebrates. They are produced in the body by the endocrine glands, and they work at extremely low concentrations. Most hormones are released to the blood and bind to specific receptors on target organs or cells, where they turn on genes or cause other changes in cell function. The entire hormone battery in the body is called the **endocrine system**, and in the human body, some 50 known hormones are produced by that system, controlling metabolism, growth, and reproduction. Disturbances to the endocrine system are not trivial matters—just think of diabetics, who lack the hormone insulin and would die if it were not supplied to them.

Not long ago, some very disturbing reports of abnormal sexual development in alligators and other wildlife gave credence to the possibility that very low levels of a number of synthetic chemicals released to the environment were acting to disrupt certain endocrine systems. In fact, the well-known effects of exposure to DDT by fish-eating birds are the outcome of disruption of the birds' reproductive systems, controlled by hormones (Chapter 13). And it doesn't stop there. Some chemical compounds widely used in foods and food containers have also been implicated as

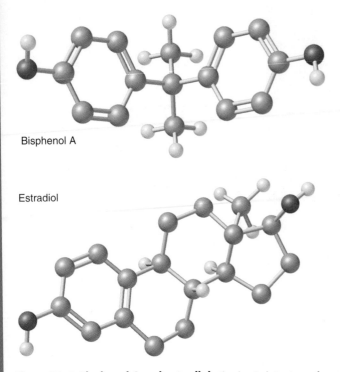

Bisphenol A

Estradiol

Figure 22–1 Bisphenol A and estradiol. The chemical structures of BPA and estradiol are shown; BPA is a known disruptor of estradiol.

endocrine disruptors, suspected of causing health problems in consumers. One such compound is bisphenol A (BPA), a component of polycarbonate plastic containers and the epoxy resin food- and beverage-can liners (**Fig. 22–1**). BPA leaches from these containers into the contents and is routinely ingested. A recent study found BPA in the urine of 93% of people in the United States Children tested showed the highest levels of exposure.

BPA = ESTROGEN? BPA is known to have estrogenic activity; that is, it simulates the effects of estradiol, one of the most important reproductive hormones in the human body. It does this by binding to the estrogen receptors found in many parts of the body. Animal tests have shown that BPA can have significant effects, such as weight gain, increased aggressiveness, altered sexual behavior, changes in the uterus and ovaries, and even disturbance of glucose metabolism that could lead to diabetes. As the results of these studies have become known, a rising tide of concern about BPA has led to its evaluation by the National Toxicology Program (NTP) Center for the Evaluation of Risks to Human Reproduction.[1] That evaluation led the Food and Drug Administration to state that there was "some concern" about the possible effects of BPA on fetal and child health. The National Institute of Health

[1]U.S. Department of Health and Human Services. Center for the Evaluation of Risks to Human Reproduction. NTP-CERHR Monograph on the Potential Human Reproductive and Developmental Effects of Bisphenol A. NIH Publication No. 08-5994. September 2008.

(NIH) has targeted $30 million that will be directed to related research on BPA and other endocrine disruptors. Consumers have demanded that BPA be removed from food and drink packaging, especially those items that are used by small children, like baby bottles and sippy cups. Canada has declared BPA a toxic chemical and is moving toward an outright ban on the use of BPA in food containers. The European Commission banned its use in baby bottles. Consumer advocacy groups are calling for application of the **precautionary principle**, which states that where there is not yet a scientific consensus, it is prudent to act as though a potentially harmful product is, in fact, harmful.

ON THE OTHER HAND . . . "Wait a minute!" say the manufacturers and chemical industry groups. They claim BPA is a very important product, in use globally at a rate of 12 billion pounds per year. Polycarbonate bottles are ubiquitous, and the epoxy can liners protect canned goods from spoilage and shelf-life deterioration. Just because the NTP report said there is some concern shouldn't hinder its usage. Similar statements, said one spokesman, could be made about anything; BPA is safe in all of its uses. Applying the precautionary principle to BPA would likely lead to a ban on a chemical that is very useful and is not harmful, according to its defenders. Despite this, in order to keep consumers happy, manufacturers have replaced polycarbonate baby bottles and sippy cups with other products. Public opinion has clearly made a difference here.

The broader concern about endocrine disruptors in food and the environment reached Congress, and the **Food Quality Protection Act (FQPA)** of 1996 directed the EPA to develop procedures for testing chemicals for endocrine disruption activity. Accordingly, the EPA established the Endocrine Disruptor Screening Program (EDSP) and has ordered manufacturers to screen an initial group of 67 pesticide chemicals, recently adding a second group of 134. In the meantime, we are all exposed to BPA whenever we consume canned goods. To lower your blood BPA level, consumer advocates suggest turning to fresh fruits and vegetables.

Stories like this, and some that can only be called disasters, have convinced the public that significant dangers are associated with the manufacture, use, and disposal of many chemicals. Very few, however, would advocate giving up the advantages of the innumerable products that derive from our modern chemical industry. Better, instead, to learn to handle and dispose of chemicals in ways that minimize the risks. Thus, over the past 35 years, regulations surrounding the production, transport, use, and disposal of chemicals have mushroomed. Where it once had almost no controls at all, the chemical industry is now heavily regulated, though some would say not very effectively.

This chapter focuses on hazardous chemicals—their nature and how they are investigated. An overview of laws and regulations designed to protect human and environmental health is presented, and the strengths and weaknesses of those measures are examined.

22.1 Toxicology and Chemical Hazards

Toxicology is the study of the harmful effects of chemicals on human health and the natural environment. A toxicologist might, for example, investigate the potential for *phenolphthalein* (a common chemical laboratory reagent that has also been used in over-the-counter laxatives) to cause health problems. The toxicologist might study the **acute** toxicity effects that could occur upon ingestion of the chemical or upon its contact with the skin, as well as the effects of **chronic** exposure over a period of years, and, finally, the **carcinogenic** (cancer-causing) potential of phenolphthalein. Indeed, such studies have shown that phenolphthalein causes tumors in mice, and because of this finding, the Food and Drug Administration (FDA) canceled all laxative uses of phenolphthalein, which is now listed as "reasonably anticipated to be a human carcinogen."

Data on toxic chemicals are made available to health practitioners and the public via a number of sources, including the National Toxicology Program (NTP) Chemical Repository and the National Institute of Environmental Health Sciences (NIEHS), the latter of which provides an annual updated *Report on Carcinogens*, with a complete list of carcinogenic agents. The Environmental Protection Agency (EPA) also makes available data on toxic chemicals in its Integrated Risk Information System (IRIS). All of these sources of information are on the Internet. Green Media Toolshed uses these and other data sources in its *Scorecard*, an exhaustive Web-based profile of 11,200 chemicals, including information on their manufacture and uses (see http://scorecard.goodguide.com/).

Dose Response and Threshold

In our discussion of risk assessment, we introduced the concepts of dose response and exposure (Chapter 17). When investigating a suspect chemical, a toxicologist would conduct tests on animals, investigate human involvement with the chemical, and present information linking the dose (the level of exposure multiplied by the length of time over which exposure occurs) with the response (some acute or chronic effect or the development of tumors). For example, studies of phenolphthalein toxicity indicate that the LDL_0 (lowest dose at which death occurred in animal testing) was 500 milligrams/kilogram. This is a low toxicity, so concern about the chemical would center on chronic or carcinogenic issues.

Human exposure to a hazard is a vital part of its **risk characterization**, and such exposure can come through the workplace, food, water, or the surrounding environment. For example, the NTP Chemical Repository gives the various uses of phenolphthalein, followed by both precautions to take in handling it and the symptoms of exposure (there are many!). Getting an accurate determination of human exposure is often the most difficult area of risk assessment.

Threshold Level. In the dose-response relationship, there is usually a *threshold*. Organisms are able to deal with

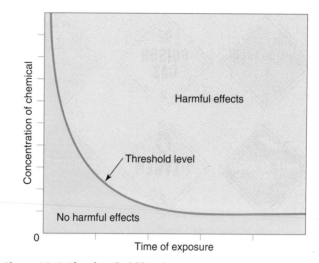

Figure 22–2 The threshold level. The threshold level for harmful effects of toxic pollutants becomes lower as the exposure time increases. It is also different for different chemicals.

certain levels of many substances without suffering ill effects. The level below which no ill effects are observed is called the **threshold level.** Above this level, the effect of a substance depends on both its concentration and the duration of exposure to it. Higher levels may be tolerated if the exposure time is short. Thus, the threshold level is high for short exposures but gets lower as the exposure time increases **(Fig. 22–2).** In other words, it is not the absolute amount, but rather the *dose*, that is important.

Where carcinogens are concerned, the EPA generally takes a zero-dose, zero-response approach. That is, there is no evidence of a discrete threshold level for any carcinogenic chemical. However, the lower the dose, the more likely it is that the response cannot be distinguished from the background level of cancers in a population. In such cases, the risk becomes very low and drops below the level for which regulatory action is needed.

These are just some basics of toxicology. The field is well established and is the most important source of sound scientific information supporting the regulatory efforts of the FDA and the EPA. The NTP was established in 1978 to provide this kind of information and has become the world's leader in assessing chemical toxicity and carcinogenicity. We turn our attention now to the chemicals themselves.

The Nature of Chemical Hazards: HAZMATs

A chemical that presents a certain hazard or risk is known as a **hazardous material (HAZMAT).** The EPA categorizes substances on the basis of the following hazardous properties:

- *Ignitability:* substances that catch fire readily (e.g., gasoline and alcohol)

- *Corrosivity:* substances that corrode storage tanks and equipment (e.g., acids)

- *Reactivity:* substances that are chemically unstable and that may explode or create toxic fumes when mixed

Figure 22–3 HAZMAT placards. These are examples of some of the placards that are mandatory on trucks and railcars carrying hazardous materials. Numbers in place of the word on the placard or on an additional orange panel identify the specific material. Placards alert workers, police, and firefighters to the kinds of hazards they face in case an accident occurs.

with water (e.g., explosives, elemental phosphorus [not phosphate], and concentrated sulfuric acid)

- *Toxicity:* substances that are injurious to health when they are ingested or inhaled (e.g., chlorine, ammonia, pesticides, and formaldehyde)

Containers in which HAZMATs are stored and vehicles that carry HAZMATs are required to display placards identifying the hazards (flammability, corrosiveness, the potential for poisonous fumes, and so on) presented by the material inside (**Fig. 22–3**).

Radioactive materials are probably the most hazardous of all and are treated as an entirely separate category (Chapter 15).

Sources of Chemicals Entering the Environment

To understand how HAZMATs enter our environment, we need to look at how people in our society live and work. First, the materials making up almost everything we use, from the shampoo and toothpaste we use in the morning to the television we watch in the evening, are products of chemical technology. Our use constitutes only one step in the **total product life cycle**, a term that encompasses all steps, from obtaining raw materials to finally disposing of the product. Implicit in our use of hair spray, for example, is that raw materials were obtained and various chemicals were produced to make both the spray and its container. Chemical wastes and by-products are inevitable in production processes. In addition, consider the risks of accidents or spills occurring in

the manufacturing process and in the transportation of raw materials, the finished product, or wastes. Finally, what are the risks of breathing hair spray? What happens to the container that still holds some of the hair spray when you throw it into the trash because the propellant is used up?

Multiply these steps by all the hundreds of thousands of products used by millions of people in homes, factories, and businesses, and you can appreciate the magnitude of the situation. More than 80,000 chemicals are registered for commercial use within the United States. At every stage—from the mining of raw materials through manufacturing, use, and final disposal—various chemical products and by-products enter the environment, with consequences for both human health and the environment (**Fig. 22–4**).

The use of many products—pesticides, fertilizers, and road salts, for example—directly introduces them into the environment. Alternatively, the intended use may leave a fraction of the material in the environment—the evaporation of solvents from paints and adhesives, for instance. Then there are the product life cycles of materials that are used tangentially to the desired item. Consider lubricants, solvents, cleaning fluids, and cooling fluids, with whatever contaminants they may contain. Likewise, there are the product life cycles of gasoline, coal, and other fuels that are consumed for energy. In addition to the unavoidable wastes and pollutants produced, in every case there is the potential for accidental releases, ranging from minor leaks from storage tanks to the release of toxic gas from a pesticide plant in Bhopal, India, in 1984 that killed more than 15,000 people.

Toxics Release Inventory. The introduction of chemicals into the environment may occur in every sector, from major industrial plants to small shops and individual homes. Whereas single events involving large amounts of one chemical may constitute a disaster and make headlines, the total amounts entering the environment from millions of homes and businesses are far greater and present much more of a health risk to society. Some idea of the quantities involved can be obtained from the **Toxics Release Inventory (TRI)**.[2] The **Emergency Planning and Community Right-to-Know Act of 1986 (EPCRA)** requires industries to report releases of toxic chemicals to the environment, and the **Pollution Prevention Act of 1990** mandates collection of data on toxic chemicals that are released to the environment, treated on-site, recycled, or combusted for energy. (The TRI does not cover small businesses, such as dry-cleaning establishments and gas stations, or household hazardous waste.) The TRI is disseminated on the Internet and provides an annual record of releases of almost 650 designated chemicals by some 21,000 industrial facilities. For 2010, the TRI reveals the following information about toxic chemicals:

- Total production-related toxic wastes: 21.8 billion pounds
- Releases to air: 860 million pounds

---

[2]U.S. Environmental Protection Agency. *2010 Toxics Release Inventory: National Analysis Overview.* January 5, 2012. http://www.epa.gov/tri.

Figure 22–4 Total product life cycle. The life cycle of a product begins when the raw materials are obtained and ends when the used-up product is discarded. At each step—or in transport between steps—wastes, by-products, or the product itself may enter the environment, causing pollution and creating various risks to human and environmental health.

- Releases to water: 230 million pounds
- Releases to land disposal sites and underground injection: 2,430 million pounds
- Total environmental releases: 3.93 billion pounds (18.0% of all production-related toxic wastes; the remainder are recycled or treated on- or off-site)

Add to the last figure the estimated 3 billion pounds of household hazardous waste released each year and an unknown quantity coming from small businesses (facilities do not have to report releases under 500 pounds), and you have some idea of the dimensions of the problem. The good news is that over the 25 years since the TRI has been in effect, the quantities of virtually all categories of toxic waste have been going down **(Fig. 22–5)**, and total releases have declined by 62%.

The Threat from Toxic Chemicals

All toxic chemicals are, by definition, hazards that pose a risk to humans. Fortunately, a large portion of the chemicals introduced into the environment are gradually broken down and assimilated by natural processes. After that, they pose no long-term human or environmental risk, even though they may be highly toxic in high-level, short-term exposures.

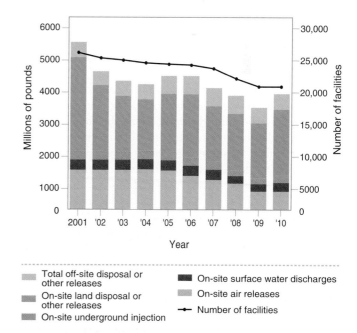

Figure 22–5 Toxics Release Inventory, 2001–2010. There has been a 30% reduction in releases of chemicals since 2001 and a total reduction since 1986 of 62.5%. The increase in 2010 is due largely to increases in the metal mining sector (*Source:* EPA, 2010 Toxics Release Inventory, January, 2012.)

Figure 22–6 Halogenated hydrocarbons. These are organic (carbon-based) compounds in which one or more hydrogen atoms have been replaced by halogen atoms (chlorine, fluorine, bromine, or iodine). Such compounds are particularly hazardous to health because they are nonbiodegradable and they tend to bioaccumulate. Shown here are tetrachloroethylene and 1,2-dibromoethane.

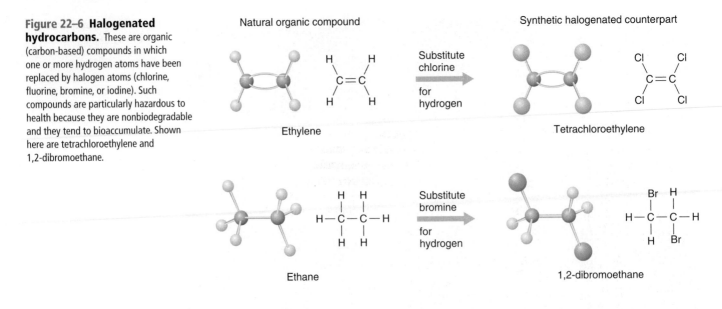

Natural organic compound — Ethylene → Substitute chlorine for hydrogen → Synthetic halogenated counterpart — Tetrachloroethylene

Ethane → Substitute bromine for hydrogen → 1,2-dibromoethane

Two major classes of chemicals do not readily degrade in the environment, however: (1) *heavy metals and their compounds* and (2) *synthetic organics*. If sufficiently diluted in air or water, most of these compounds do not pose a hazard, though there are some notable exceptions.

Heavy Metals. The most dangerous heavy metals are lead, mercury, arsenic, cadmium, tin, chromium, zinc, and copper. These metals are used widely in industry, particularly in metalworking or metal-plating shops and in such products as batteries and electronics. They are also used in certain pesticides and medicines. In addition, because compounds of heavy metals can have brilliant colors, they once were heavily used in paint pigments, glazes, inks, and dyes. Millions of children have been treated for lead poisoning because of the heavy use of lead-based paint. The paint was banned in the United States in 1978, but many older painted surfaces still hold lead and are a hazard.

Heavy metals are extremely toxic because, as ions or in certain compounds, they are soluble in water and may be readily absorbed into the body, where they tend to combine with and inhibit the functioning of some vital enzymes. Even very small amounts can have severe physiological or neurological consequences. Mercury, in particular, has contaminated rivers, lakes, and streams in the United States because of its release into the air by coal- and oil-fired power plants (mercury is present in fossil fuels in trace quantities). Using its authority under the Clean Air Act, the EPA issued in 2011 a new rule requiring a 90% reduction in the mercury released from power plants, bringing long-sought relief to the wildlife and people affected by this toxic metal pollutant.

Organic Compounds Petroleum-derived and *synthetic* organic compounds are the chemical basis for all plastics, synthetic fibers, synthetic rubber, modern paint-like coatings, solvents, pesticides, wood preservatives, and hundreds of other products. Because of their chemical structure, many synthetic organics are resistant to biodegradation. Ironically, this resistance is an important part of what makes many such compounds useful: We wouldn't want fungi and bacteria attacking and rotting the tires on our cars, and paints and wood preservatives function only insofar as they are both nonbiodegradable and toxic to decomposer organisms.

These compounds are toxic because they are often readily absorbed into the body, where they interact with particular enzymes, but their nonbiodegradability prevents them from being broken down or processed further. When a person ingests a sufficiently high dose, the effect may be acute poisoning and death. With low doses over extended periods, however, the effects are insidious and can be mutagenic (mutation causing), carcinogenic, or teratogenic (birth defect causing). Synthetic organic compounds also may cause serious liver and kidney dysfunction, sterility, and numerous other physiological and neurological problems.

Dirty Dozen. A particularly troublesome class of synthetic organics is the **halogenated hydrocarbons**, organic compounds in which one or more of the hydrogen atoms have been replaced by atoms of chlorine, bromine, fluorine, or iodine. These four elements are classed as *halogens*—hence, the name **(Fig. 22–6).** Of the halogenated hydrocarbons, the **chlorinated hydrocarbons** (also called organic chlorides) are by far the most common. Organic chlorides are widely used in plastics (polyvinyl chloride), pesticides (DDT, Kepone, and mirex), solvents (carbon tetrachlorophenol and tetrachloroethylene), electrical insulation (polychlorinated biphenyls), and many other products. Twelve **persistent organic pollutants** (called **POPs**, Table 22–1) have been banned or highly restricted globally as a result of the **Stockholm Convention on Persistent Organic Pollutants**, which was signed by 162 countries and entered into force in May 2004. Most of these are halogenated hydrocarbons. All are toxic to varying extents, and most are known animal carcinogens. Many are also suspected endocrine disruptors at very low levels. The treaty was recently expanded to add nine more halogenated hydrocarbons, most of which are no longer in use.

TABLE 22–1 The "Dirty Dozen" Persistent Organic Pollutants

Chemical Substance	Designed Use	Major Concerns
Aldrin	Pesticide to control soil insects and to protect wooden structures from termites	Toxic to humans, probable carcinogen
Chlordane	Broad-spectrum insecticide to protect crops	Biomagnification in food webs, carcinogenic
DDT	Widely used insecticide, malaria control	Biomagnification in food webs, probable carcinogen
Dieldrin	Insecticide to control termites and crop pests	Toxic, biomagnification in food webs, probable carcinogen
Endrin	Insecticide and rodenticide	Toxic, especially in aquatic systems
Heptachlor	General insecticide	Toxic, carcinogenic
Hexachlorobenzene	Fungicide	Toxic, carcinogenic
Mirex	Insecticide to control ants	High toxicity to aquatic animals, carcinogenic
Toxaphene	General insecticide	High toxicity to aquatic animals, carcinogenic
PCBs	Variety of industrial uses, especially in transformers and capacitors	Toxic, teratogenic, carcinogenic
Dioxins	No known use; by-products of incineration and paper bleaching	Toxic, carcinogenic, reproductive system effects
Furans	No known use; by-products of incineration, PCB production	Toxic, especially in aquatic systems

Source: IPCS Assessment Report: L. Ritter, et al., *Persistent Organic Pollutants: An Assessment Report on DDT, Aldrin, Dieldrin, Endrin, Chlordane, Heptachlor, Hexachlorobenzene, Mirex, Toxaphene, Polychlorinated Biphenyls, Dioxins, and Furans, International Programme on Chemical Safety,* December 1995. Available at http://www.chem.unep.ch/Pops/alts02.html.

PERC. Let us consider one of the halogenated hydro-carbon compounds, tetrachloroethylene (Fig. 22–6). This substance, also called perchloroethylene, or PERC, is colorless and nonflammable and is the major substance in dry-cleaning fluid. It is an effective solvent and is used in all kinds of industrial cleaning operations. It can also be found in shoe polish and other household products. PERC evaporates readily when exposed to air, but in the soil, it can enter groundwater easily because it does not bind to soil particles. Human exposure can occur in the workplace, especially in connection with dry-cleaning operations. It can also occur if people use products containing PERC, such as dry-cleaned garments. PERC enters the body most readily when breathed in with contaminated air. Breathing PERC for short periods can bring on dizziness, fatigue, headaches, and unconsciousness. Over longer periods, PERC can cause liver and kidney damage.

Laboratory studies also show PERC to be carcinogenic to rats and mice, and it is listed in NTP's 2011 *Report on Carcinogens*[3] as "reasonably anticipated to be a human carcinogen." Studies have shown a higher risk of cancer, as well as neurological impairment, in dry-cleaning employees. The EPA has issued rules that would phase out PERC use (by 2020) in dry-cleaning shops in residential buildings; critics wonder why it should take so long and why just in a select few shops if, as the EPA maintains, PERC is harmful to human health. The rules for PERC are currently under review by the EPA. PERC is used by 85% of U.S. dry-cleaning establishments.

Involvement with Food Chains

The trait that makes heavy metals and nonbiodegradable synthetic organics particularly hazardous is their tendency to accumulate in organisms. We have discussed the phenomena of bioaccumulation and biomagnification in connection with DDT (Chapter 13).

Minamata. A tragic episode in the early 1970s, known as Minamata disease, revealed the potential for biomagnification of mercury and other heavy metals. The disease is named for a small fishing village in Japan where the episode occurred. In the mid-1950s, cats in Minamata began to show

---

[3]U.S. Department of Health and Human Services, Public Health Service, National Toxicology Program, *Report on Carcinogens,* 12th ed. (Washington, DC: U.S. Department of Health and Human Services, 2011).

spastic movements, followed by partial paralysis, coma, and death. At first, this was thought to be a syndrome peculiar to felines, and little attention was paid to it. However, when the same symptoms began to occur in people, concern escalated quickly. Additional symptoms, such as mental impairment, insanity, and birth defects, were also observed. Scientists and health experts eventually diagnosed the cause as acute mercury poisoning.

A chemical company near the village was discharging wastes containing mercury into a river that flowed into the bay where the Minamata villagers fished. The mercury, which settled with detritus, was first absorbed by bacteria and then biomagnified as it passed up the food chain through fish to cats or to humans. Cats had suffered first and most severely because they fed almost exclusively on the remains of fish. By the time the situation was brought under control, some 50 persons had died, and 150 had suffered serious bone and nerve damage. Even now, the tragedy lives on in the crippled bodies and disabled minds of some Minamata victims.

Arctic POPs. Much to research scientists' surprise, fish, birds, and mammals all over the Arctic are showing elevated body burdens of a number of persistent organic pollutants—in particular, DDT, toxaphene, chlordane, PCBs, and dioxins. Moreover, because the Inuit people of the far north depend on these animals for food, they, too, are carrying high loads of POPs—some of the highest in the world (**Fig. 22–7**).

The real mystery is how these chemicals—all of which are toxic—get to such remote and pristine environments. No one is spraying pesticides there, and there are no industries to produce PCBs and dioxins. The key to the presence of such chemicals can be found in several factors that characterize POPs: persistence, bioaccumulation, and the potential for long-range transport. Added to these factors is the unique climate of the Arctic, where the extreme cold promotes a process called *cold condensation*. POPs are volatile chemicals carried in global air patterns to the Arctic, where they condense on the snowpack and are then washed into the water with the spring thaw. Once in the lakes and coastal waters, the chemicals are picked up by plankton and passed up the food chain through bioaccumulation and biomagnification.

Figure 22–7 An Inuit hazard. This woman is cutting the meat of a seal in Sermilik Fjord, Greenland. Seals and other animals used for food by the Inuit harbor high levels of persistent organic pollutants.

A Canadian government study found that three-fourths of Inuit women have PCB levels up to five times above those deemed safe. The potential health effects of the chemical load carried by the Arctic residents range from immune-system disorders to disruption of their hormone systems and, over time, to cancer. The latest impact has been a serious imbalance in births, where twice as many girls are born as boys in many Arctic villages. Scientists from the Arctic Monitoring and Assessment Program, an international body that advises the governments of the eight Arctic nations, measured levels of man-made chemicals in women's blood that were high enough to trigger changes in the sex of unborn children during their first few weeks of gestation. This transboundary pollution by chemicals is exactly why the UN Convention has banned them; hopefully, this action will eventually clear the Arctic air of the POPs.

22.2 Hazardous Waste Disposal

The Clean Air and Clean Water Acts, passed in the early 1970s, put an end to the widespread disposal of hazardous wastes in the atmosphere and waterways. However, their passage left an enormous loophole. If you can't vent wastes into the atmosphere or flush them into waterways, what do you do with them? Industry turned to *land disposal*, which was essentially unregulated at the time. Indiscriminate air and water disposal became indiscriminate land disposal. Thus, in retrospect, we see that the Clean Air and Clean Water Acts, for all their benefits in improving air and water quality, also succeeded in transferring pollutants from one part of the environment to another.

Methods of Land Disposal

In the early 1970s, there were three primary land disposal methods: (1) deep-well injection, (2) surface impoundments, and (3) landfills. With the conscientious implementation of safeguards, each of these methods has some merit, and each is still heavily used for hazardous-waste disposal. Without adequate regulations or enforcement, however, contamination of groundwater is virtually guaranteed.

Deep-Well Injection. Deep-well injection involves drilling a borehole thousands of feet below groundwater into a porous geological formation. A well consists of concentric pipes and casings that isolate wastes as they are injected and pipes that are sealed at the bottom to prevent wastes from backing up. Over time, the wastes often undergo reactions with naturally occurring material that make them less hazardous. This method is presently used for various volatile organic compounds, pesticides, fuels, and explosives. Currently, 143 hazardous-waste deep wells are in operation, mostly in the Gulf Coast region. The EPA's Underground Injection Control Program regulations require that the wells be limited to geologically stable areas so that the hazardous-waste liquids pumped into the well remain isolated indefinitely. The total amount of deep-well injection (230 million pounds in

2010) has declined over the years, but the technique remains useful because it has a greater potential than other methods for keeping toxic wastes from contaminating the hydrologic cycle and the food web.

Surface Impoundments. Surface impoundments are simple excavated depressions (*ponds*) into which liquid wastes are drained and held. They are the least expensive—and, hence, the most widely used—way to dispose of large amounts of water carrying relatively small amounts of chemical wastes. The impoundments may be used to carry out wastewater treatment prior to discharging the wastes. As waste is discharged into the pond, solid wastes settle and accumulate, while water evaporates **(Fig. 22–8)**. If the bottom of the pond is well sealed and if the loss of water through evaporation equals the gain from input, impoundments may receive wastes indefinitely. However, inadequate seals may allow wastes to leach into groundwater, exceptional storms may cause overflows, and volatile materials can evaporate into the atmosphere, adding to air pollution problems. To prevent these problems, hazardous waste surface impoundments are required to be constructed with a double liner system, a leachate collection and removal system, and a leak detection system. The 2010 TRI reports 833 million pounds of toxics released to on-site surface impoundments. (*On-site* refers to disposal by manufacturers or institutions on their own facilities; *off-site* means that the wastes have been transferred to a waste-treatment or disposal facility.)

Landfills. RCRA sets rigid standards for the disposal of hazardous wastes in landfills. When hazardous wastes are in a concentrated liquid or solid form, they are commonly put into drums, typically collected by a waste management company (for a hefty fee!), and then treated in accordance with their chemical and physical characteristics. *Treatment standards* for wastes are established by the EPA and characterized as **best-demonstrated available technologies (BDATs)**. These standards apply to all solid and liquid hazardous wastes and have the effect of reducing their toxicity and mobility. The BDAT technologies include stabilization and neutralization of sludges, soils, liquids, powders, and slurries; chemical oxidation of organic materials; and various specific techniques

Figure 22–9 Secure landfill. These drums of low-level nuclear waste are being buried in a secure landfill at the Hanford Nuclear Reservation near Richland, Washington.

applied to metal-bearing wastes that convert them to an insoluble solid material for landfill disposal. Some 350 million pounds of hazardous wastes were delivered to on-site landfills in 2010, and an additional 253 million pounds went to off-site landfills. Only 21 landfills in North America are licensed to receive off-site hazardous wastes, and these are owned by just a few waste management companies.

If a landfill is properly lined, supplied with a system to remove leachate, provided with monitoring wells, and appropriately capped, it may be reasonably safe and is referred to as a **secure landfill (Fig. 22–9)**. The various barriers are subject to damage and deterioration, however, requiring adequate surveillance and monitoring systems to prevent leakage.

Mismanagement of Hazardous Wastes

Because early land disposal was not regulated, in numerous instances not even the most rudimentary precautions were taken. Indeed, in many cases, deep wells were injecting wastes directly into groundwater, abandoned quarries were sometimes used as landfills with no additional precautions being taken, and surface impoundments frequently had no seals or liners whatsoever. Even worse, considerable amounts of waste failed to get to any disposal facility at all.

Midnight Dumping and Orphan Sites. The need for alternative methods to dispose of waste chemicals created an opportunity for a new enterprise: waste disposal. Many

Figure 22–8 Surface impoundment. Surface impoundments are often used for preliminary treatment of liquid wastes. Shown here is an impoundment at a paper manufacturer in Alabama.

Figure 22–10 Midnight dumping. Hazardous wastes were often left on remote or unoccupied properties by unscrupulous haulers.

reputable businesses entered the field, but—again in the absence of regulations—there were also disreputable operators. As stacks of drums filled with hazardous wastes "mysteriously" appeared in abandoned warehouses, vacant lots, or municipal landfills, it became clear that some operators simply were pocketing the disposal fee and then unloading the wastes in any available location, frequently under cover of darkness—an activity termed **midnight dumping (Fig. 22–10)**. Authorities trying to locate the individuals responsible would learn that they had gone out of business or were nowhere to be found.

Some companies or individuals simply stored wastes on their own properties and then went out of business, abandoning the property and the wastes. These locations became known as **orphan sites**—hazardous-waste sites without a responsible party to clean them up. As drums containing hazardous chemicals corroded and leaked, there was great danger of reactive chemicals combining and causing explosions and fires. One of the most infamous abandoned sites is the "Valley of the Drums" (VOD) in Kentucky **(Fig. 22–11)**.

Scope of the Mismanagement Problem. The mounting problem of unregulated land disposal of hazardous wastes was brought vividly to public attention by the events at Love Canal, near Niagara Falls, New York. The area was occupied by a school and a number of houses, all of which were perched on top of a chemical waste dump that had been covered over. The surface of the dump began to collapse, exposing tanks and barrels of chemical wastes **(Fig. 22-12)**. Fumes and chemicals began seeping into cellars. Ominously, people began reporting serious health problems, including birth defects and miscarriages. An aroused neighborhood, led by activist Lois Gibbs, began demanding that the state do something about the problem. Following confirmation of the contamination by toxic chemicals, President Jimmy Carter signed an emergency declaration in 1978 to relocate hundreds of residents. The state and federal governments closed the school and demolished a number of the homes nearest the site. In all, more than 800 families moved out of the area because of the fear of damage to their health from the toxic chemicals.

Occidental? The absence of public policy for dealing with the disposal of hazardous chemicals contributed to the situation at Love Canal. Hooker Chemical and Plastics Company had purchased an abandoned canal near Niagara Falls in 1942 and proceeded to fill it with an estimated 21,000 tons of hazardous wastes. Hooker covered the canal over with a clay cap and sold it to the school board for a small sum, reportedly after warning the board that there were chemicals buried on the property. Subsequent construction penetrated the clay cap, and rain seeped in and leached chemicals in all directions. Hooker Chemical's parent company, Occidental Petroleum, eventually spent more than $233 million on the cleanup and subsequent lawsuits.

As both government and independent researchers began surveying the extent of the problem, bad disposal practices were found to be rampant. The World Resources Institute

Figure 22–11 "Valley of the Drums," an orphan waste site. Thousands of drums of waste chemicals, many of them toxic, were unloaded near Louisville, Kentucky, around 1975 and left to rot, seriously threatening the surrounding environment, waterways, and aquifers.

Figure 22–12 Love Canal. A bulldozer uncovers one of the tanks holding toxic waste buried beneath the residential area of Love Canal.

estimated that, in the United States in the early 1980s, there existed 75,000 active industrial landfill sites, along with 180,000 surface impoundments and 200 other special facilities that were or could be sources of groundwater contamination. In most cases, the contaminated area was relatively small—200 acres (80 hectares) or less—but in total, the problem was immense and affected every state in the country. As studies and tests proceeded, thousands of individual wells and some major municipal wells were closed because they were contaminated with toxic chemicals.

Many cases of private-well contamination caused by careless disposal were discovered only after people experienced "unexplainable" illnesses over prolonged periods. While these incidents did not receive the media attention of Love Canal, they were nevertheless devastating to the people involved. Even more important than the damage already done was the recognition that the problem was ongoing.

The problems concerning toxic chemical wastes can be divided into three areas:

- Cleaning up the "messes" already created, especially where they threaten drinking water supplies

- Regulating the handling and disposal of wastes currently being produced, so as to protect public and environmental health

- Reducing the quantity of hazardous waste produced

Each of these problems is addressed in the sections that follow.

22.3 Cleaning Up the Mess

A major public health threat from toxic wastes disposed of on land is the contamination of groundwater used for drinking. In dealing with groundwater contamination, the first priority is to ensure that people have safe water. The second is to clean up or isolate the source of pollution so that further contamination does not occur.

Ensuring Safe Drinking Water

To protect the public from the risk of toxic chemicals contaminating drinking water supplies, Congress passed the **Safe Drinking Water Act (SDWA) of 1974**. Under this act, the EPA sets national standards to protect the public health, including allowable levels of specific contaminants. If any contaminants are found to exceed **maximum contaminant levels (MCLs)**, the water supply should be closed until adequate purification procedures or other alternatives are adopted. The act was amended in 1986, and the EPA now has jurisdiction over groundwater and sets MCLs for 94 contaminants. States and public water agencies are required to monitor drinking water to be sure that it meets the standards (see Chapter 20). The contaminants represent a host of chemicals that are known pollutants and are regarded as dangerous to human health (see the EPA Web site for a description of the MCLs and why they are being regulated).

Groundwater Remediation. If dumps, leaking storage tanks, or spills of toxic materials have contaminated groundwater and if such groundwater is threatening water supplies, all is not yet lost. **Groundwater remediation** is a developing and growing technology. Techniques involve drilling wells, pumping out the contaminated groundwater, purifying it, and injecting the purified water back into the ground or discharging it into surface waters. Cleaning up the source of the contamination is mandatory. If, however, the contamination is extensive, remediation may not be possible, and the groundwater must be considered unfit for use as drinking water.

Superfund for Toxic Sites

Probably the most monumental hazardous waste task is the cleanup of the tens of thousands of toxic sites resulting from the years of mismanaged disposal of toxic materials. Prompted by the public outrage over the Love Canal fiasco, Congress enacted the **Comprehensive Environmental Response, Compensation, and Liability Act of 1980 (CERCLA)**, popularly known as **Superfund**. Through a tax on chemical raw materials, this legislation provided a trust fund for the identification of abandoned chemical waste sites, protection of groundwater near the sites, remediation of groundwater if it has been contaminated, and cleanup of the sites. The Superfund program was greatly expanded by the **Superfund Amendments and Reauthorization Act of 1986 (SARA)**. Superfund, one of the EPA's largest ongoing programs, works as described in the next several subsections.

Setting Priorities. Resources are insufficient to clean up all sites at once. Therefore, a system for setting priorities has been developed:

- Wherever facilities were still operating, pressures were brought to bear on the operators to clean up the sites. Many operators who could not afford to do so, however, simply declared bankruptcy, and the sites joined those already abandoned. As sites are identified—note that many abandoned sites had long since been forgotten—their current and potential threats to groundwater supplies are initially assessed by taking samples of the waste, determining its characteristics, and testing groundwater around the site for contamination. If it is determined that no immediate threat exists, nothing more may be done.

- If a threat to human health does exist, the most expedient measures are taken immediately to protect the public. These measures may include digging a deep trench around the site, installing a concrete dike, recapping it with impervious layers of plastic and clay to prevent infiltration, and securing the area with a robust fence. Thus, the wastes are isolated, at least for the short term.

If the situation has gone past the threat stage and con-taminated groundwater is or will be reaching wells, remediation procedures are begun immediately.

- The worst sites (those presenting the most immedi-ate and severe threats) are put on the **National Priori-ties List (NPL)** and scheduled for total cleanup. A site on this list is reevaluated to determine the most cost-effective method of cleaning it up, and finally, cleanup begins with efforts to identify "potential responsible parties" (PRPs)—namely, industries in which the wastes originated. The industries are "invited" to contribute to the cleanup, and they do so either by contributing finan-cially or by participating in cleanup activities.

Cleanup Technology. If chemical wastes are contained in drums, the drummed wastes can be picked up, treated, and managed in proper fashion. For the VOD site (Fig. 22–11), the EPA cleanup started with the removal of some 4,200 drums for off-site treatment and disposal. The bigger prob-lem for many sites is the soil (often millions of tons) contami-nated by leakage. One procedure is to excavate contaminated soil and run it through an incinerator or kiln assembled on the site to burn off chemicals; synthetic organic chemicals can be broken down by incineration, and heavy metals are converted to insoluble, stable oxides. Another method is to drill a ring of injection wells around the site and a suction well in the center. Water containing a harmless detergent is injected into the injection wells and drawn into the suction well, cleansing the soil along the way. The withdrawn water is treated to remove the pollutants and is then reused for injection.

The VOD site was situated in a poorly drained shale and limestone deposit. No local use was made of the ground-water. The EPA determined that the most feasible (BDAT) treatment was simply to contain the contaminated soil and groundwater, so the agency installed a clay cap, a perime-ter drainage system, monitoring wells, and a security fence. The remedial work was completed in 1989, at a total cost of $2.5 million, and the site was deleted from the NPL in 1996. The third five-year review indicated that the cap was func-tioning well, the site was still undeveloped, and no human uses were being made of the groundwater.

Bioremediation. Another rapidly developing cleanup technology is *bioremediation.* In many cases, the soil is con-taminated with toxic organic compounds that are biodegrad-able. The problem is that they do not degrade because the soil lacks the necessary organisms, oxygen, or both to aid in biodegradation. In **bioremediation,** oxygen and organisms are injected into contaminated zones. The organisms feed on and eliminate the pollutants (as in secondary sewage treat-ment) and then die when the pollutants are gone. Consid-erable research is being done to find and develop microbes that will break down certain kinds of wastes more readily. Bioremediation, which may be used in place of detergents to decontaminate soil, is a rapidly expanding and develop-ing technology that is applicable to leaking storage tanks and spills as well as to waste-disposal sites.

Plant Food? When the soil contaminants are heavy metals and nonbiodegradable organic compounds, *phyto-remediation* has been employed with some success. **Phyto-remediation** uses plants to accomplish a number of desirable cleanup steps: stabilizing the soil, preventing further move-ment of contaminants by erosion, and extracting the con-taminants by direct uptake from the soil. After full growth, the plants are removed and treated as toxic-waste prod-ucts. However, this is a slow process and can be used only at sites where the contaminants and their concentrations are not toxic to plants. Nevertheless, scientists have found sunflowers that will capture uranium, poplar trees that soak up dry-cleaning solvents and mercury, and ferns that thrive on arsenic. Phytoremediation is now used at a level above $100 million a year.

Evaluating Superfund. Of the literally hundreds of thou-sands of various waste sites, more than 47,000 have been deemed serious enough to be given Superfund status. Over time, the EPA judged that about 33,000 of these sites did not pose a significant public-health or environmental threat. Such sites were accordingly assigned to the category "no further removal action planned" (NFRAP). Still, more than 11,300 sites remain on the master list—a figure that, it is assumed, now includes most (if not all) of the sites in the United States that pose a significant risk. Interestingly, some of the worst that have come to light are on military bases, the result of what many characterize as the totally heedless and uncon-scionable discarding of toxic materials from military opera-tions (13% of the NPL sites are "federal facilities").

Progress. As of 2012, 1,305 sites were still on the NPL (Fig. 22–13). Additions are made to the list as sites from the master list are assessed, and subtractions are made when re-mediation on a site is complete. Since 1980, when CERCLA went into effect, about 1,123 sites have received all the necessary cleanup-related construction, and 359 of these have been deleted from the list (Love Canal was removed in 2004). Cleanup of these sites takes an average of 12 years, at an average cost of $20 million. Total cleanup often takes that long because groundwater remediation is a slow proc-ess. The other NPL sites are in various stages of analysis, remediation, and construction. (An example of a Superfund site and its cleanup is presented in Stewardship, "Woburn's Civil Action.")

Who Pays? CERCLA is based on the principle that "the polluter pays." However, the history of a waste-disposal site may go back 50 years or more, and users may have included everything from schools, hospitals, and small businesses to large corporations, all of which mount legal defenses to dis-claim responsibility. When liability is difficult to track down or the responsible parties are unable to pay, the Superfund trust fund kicks in. However, more than 70% of the cleanup costs to date have been paid by the polluters (the PRPs). Funds collected for a specific site go into "special accounts" and are held there until the site cleanup is finished.

Over the years, the EPA has gained a great deal of ex-perience in dealing with Superfund sites. The technology has become quite sophisticated, and many hazardous-waste

Figure 22–13 Map of Superfund sites in the contiguous United States. Brown shows sites currently on the National Priorities List, yellow is sites proposed for the list, and green is those sites that have been cleaned up and deleted. (*Source*: Data from U.S. EPA CERCLIS database; map by skew-t, licensed under the Creative Commons Attribution-Share Alike 3.0 Unported http://creativecommons.org/licenses/by-sa/3.0/deed.en).

- ● Final National Priority List
- ● Proposed
- ● Deleted/cleaned up

STEWARDSHIP   Woburn's Civil Action

One of the earliest Superfund sites to appear on the NPL is known as the **Industri-Plex 128** site, in Woburn, Massachusetts. By the time of its official listing by the EPA in 1981, the Industri-Plex site had already become notorious. Starting in the mid–19th century, Woburn Chemical Works built a huge industrial complex of 90 buildings on 400 acres in the city of Woburn. Among the products of the complex were arsenic-based insecticides, sulfuric and acetic acids, glue from raw animal hides, chromium-tanned hides,

and organic chemicals such as phenol, benzene, and toluene. The company continued to operate until the late 1960s. When a developer attempted to excavate the site for redevelopment in the 1970s, the uncovered land yielded decaying animal hides and buried chemical waste deposits, all of which gave off vile odors.

In the 1960s, the city of Woburn had installed two wells slightly downstream of the buried wastes, and these became the subject of a civil action when a cluster of childhood leukemia cases appeared in the city. Lawyer Jan Schlichtmann[7] acted on behalf of eight families who sued W. R.

Grace & Co. and Beatrice Corporation, allegedly the parties responsible for polluting the groundwater with toxic wastes. The suit, which eventually ended in an out-of-court settlement, was portrayed in the 1998 film *A Civil Action*, in which John Travolta played the part of Schlichtmann. In spite of many inaccurate depictions in the film, Woburn residents were pleased to see their story told and the chemical companies exposed.

The EPA identified a number of "potential responsible parties," who took a financial role in the cleanup. The cleanup involved removing drums of chemical wastes, treating highly contaminated groundwater, and constructing permanent caps over some 105 acres of contaminated soils and sediments. By 1996, the EPA was entering into agreements with prospective purchasers of parcels on the site, and currently the area is fully involved in an economic renewal of the city of Woburn. The site now accommodates a regional transportation center, an interstate exchange on I-93, a retail development, an office park, and, next to the former waste site, a new mall and hotel. Woburn receives more than $20 million in tax revenue from businesses on the site. Although it is still on the NPL list, the Industri-Plex site has been transformed from a noxious waste area to a thriving commercial complex—exactly what the Superfund program is intended to accomplish.

Anderson Regional Transportation Center. Built on the Woburn Industri-Plex 128 site, this new transportation center is one of several new facilities on the site, part of an economic renewal of Woburn and responsible for some $20 million in tax revenues.

[7]See Jonathan Harr, *A Civil Action* (New York: Vintage Books, 1996).

remediation companies have become established. A great deal of progress has been made; however, the American public still rates contamination by toxic waste at the top of the list of major environmental issues.[4]

Critics. The Superfund program gets its share of criticism. Industries claim that they are unfairly blamed for pollution that reaches back to activities that were legal before the enactment of CERCLA. Many feel that overly stringent standards of cleanup are costing large sums of money without providing any additional benefit to public health. (This problem of finding a suitable balance between costs and benefits is discussed in more depth in Chapter 2.) Congress refused to renew the tax on industry at the end of 1995; the trust fund had a balance of $4 billion at the time. Since that time, the trust fund has been supported by monies in the special accounts and yearly appropriations by Congress (the special accounts had more than $1.1 billion in 2011, in 819 accounts). The Obama administration has proposed reinstating the tax on oil and chemical producers, but can do so only if Congress authorizes it. Congress did appropriate $600 million for EPA's Superfund program as a component of the **American Recovery and Reinvestment Act of 2009,** aimed at creating jobs and speeding up the cleanup of NPL sites. Some 51 sites have been addressed by these funds.

Brownfields. One highly successful recent Superfund development is the *brownfields* program. **Brownfields** are abandoned, idled, or underused industrial and commercial facilities where expansion or redevelopment is complicated by contamination by hazardous wastes. It has been estimated that environmental hazards not serious enough to be put on the Superfund NPL impair the value of $2 trillion worth of real estate in the United States. Bipartisan support led to passage of the Brownfield Act in 2002, which provides grants for site assessment and remediation work and authorized $250 million/year for the program. Previously authorized under CERCLA, the brownfields program now stands on its own. The legislation limits liability for owners and prospective purchasers of contaminated land, thus clearing the way for more cleanup of the estimated 450,000 sites that would qualify as brownfields. Since its inception, the program has leveraged more than $16 billion in cleanup and redevelopment funding and more than 75,000 jobs. States are also moving on this issue. One of the nation's most successful brownfields programs is in Massachusetts, which has provided a brownfields redevelopment fund managed by MassDevelopment, the state's economic development agency. More than 450 projects have been redeveloped or are in the development pipeline under the program.

Many brownfield sites lie in economically disadvantaged communities, and their rehabilitation contributes jobs and exchanges a functional facility for an unsightly blight on the neighborhood. For example, the city of Chicago recently acquired and cleaned up a former bus barn in west Chicago that was becoming an illegal indoor garbage dump. The Illinois EPA cleared the site for reuse, and it was acquired by Scott Peterson Meats, which erected a $5.2 million smokehouse that employs 100 workers. Frequently, the rehabilitation of brownfield sites provides industries and municipalities with centrally located, prime land for facilities that would otherwise have been carved out of suburban or *greenfield* lands (land occupied by natural ecosystems). A further advantage is that the new developments go back on the tax rolls, turning a liability into a community asset that lightens the load of residential taxpayers.

Leaking Underground Storage Tanks (LUSTs). One consequence of our automobile-based society is the millions of underground fuel-storage tanks at service stations and other facilities. Putting such tanks underground greatly diminishes the risk of explosions and fires, but it also hides leaks. Underground storage tanks were traditionally made of bare steel, so they had a life expectancy of about 20 years before they began leaking and contaminating groundwater. By 2011, more than 501,000 failures had been reported, with 6,000 new "releases" in 2011. Without monitoring, small leaks can go undetected until nearby residents begin to smell fuel-tainted water flowing from their faucets.

Underground storage tank (UST) regulations, part of RCRA, now require strict monitoring of fuel supplies, tanks, and piping so that leaks are detected early. When leaks are detected, remediation must begin within 72 hours **(Fig. 22–14)**. Rules now also require all USTs to be upgraded with interior lining and cathodic protection (to retard electrolytic corrosion of the steel), and new tanks must be provided with the same protection if they are steel. Many service stations are turning to fiberglass tanks, which do not corrode. A LUST trust fund, financed by a 0.1-cent-per-gallon tax on motor fuel, pays for federal activities involved with oversight and cleanup. The states are required to have UST programs, and they implement the federal regulations—sometimes adding more stringent requirements. By 2011, some 413,000 cleanups were completed, with a backlog of 88,000. The states have reported that LUSTs are the most common source of groundwater contamination.

22.4 Managing Current Toxic Chemicals and Wastes

The production of chemical wastes will no doubt continue as long as modern societies persist. It follows that human health and natural ecosystems can be protected only if we have management procedures to handle and dispose of wastes safely. We have already noted that the Clean Water Act and the Clean Air Act limit discharges into water and air, respectively. When problems regarding the disposal of wastes on land became evident in the mid-1970s, Congress passed RCRA in 1976 in order to control land disposal. Any

---

[4]Lydia Saad. Water Issues Worry Americans Most, Global Warming Least. March 28, 2011. http://www.gallup.com/poll/146810/Water-Issues-Worry-Americans-Global-Warming-Least.aspx.

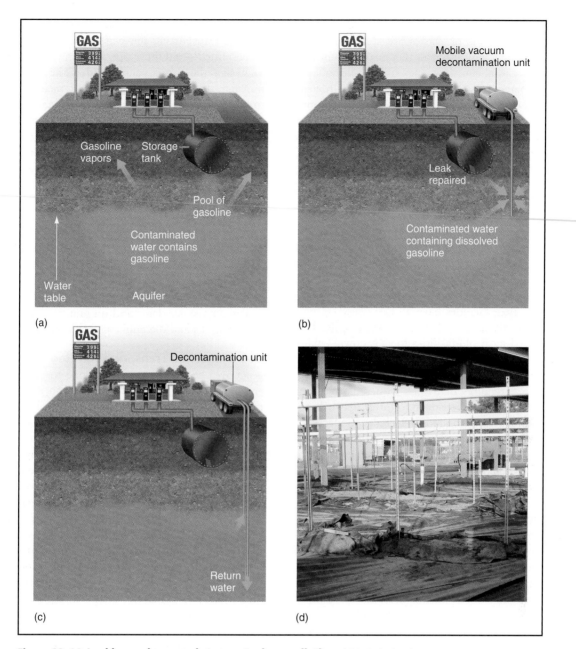

Figure 22–14 Leaking underground storage tank remediation. (a) Typical subsurface contamination from a leaking fuel tank at a gas station. (b) After the leak has been repaired, the vacuum-extraction process causes gasoline and residual hydrocarbons in the soil and on the water table to evaporate and then removes the vapors, preventing further contamination of the groundwater. (c) Contaminated groundwater is pumped out, treated, and returned to the ground. (d) Soil vapor extraction system removing volatile organic compounds in soil at an abandoned gasoline station.

company producing hazardous wastes in the United States today is regulated by these three major environmental acts.

The Clean Air and Clean Water Acts

The Clean Air Act of 1970, the Clean Water Act of 1972, and their various amendments make up the basic legislation limiting discharges into the air or water (more is said about the Clean Air Act in Chapter 19, and the Clean Water Act in Chapter 20). Under the Clean Water Act, any firm (including facilities such as sewage-treatment plants) discharging more than a certain volume into natural waterways must have a

discharge permit (specifically, a National Pollution Discharge Elimination System, or NPDES, permit). Discharge permits are a means of monitoring who is discharging what. Establishments are required to report all discharges of substances listed in the TRI. The renewal of the permits is then made contingent on reducing pollutants to meet certain standards within certain periods. Standards are being made continually stricter as technologies for pollution control improve. Some manufacturing firms discharging wastewater into municipal sewer systems are required to pretreat such water to remove any pollutant that cannot be removed by sewage-treatment plants—that is, nonbiodegradable organics and heavy metals.

Nevertheless, even this restriction does not end all water pollution. Certain amounts of wastes are still legally discharged under permits, and although they may account for a low percentage of total emissions, they still add up to large numbers, as we have seen. Moreover, small firms, homes, and farms are exempt from regulation and contribute an unknown quantity of toxics to the air and water. Further still, a great deal of pollution comes from nonpoint sources such as urban and farm runoff (Chapter 20).

The Resource Conservation and Recovery Act (RCRA)

The 1976 RCRA and its subsequent amendments are the cornerstone legislation designed to prevent unsafe or illegal disposal of all solid wastes on land. RCRA has three main features. *First*, it requires that all disposal facilities, such as landfills, be sanctioned by *permit*. The permitting process requires that these facilities have all the safety features mentioned in Section 2.2, including monitoring wells. This requirement caused most old facilities to shut down—many subsequently became Superfund sites—and new high-quality landfills with safety measures to be constructed.

Second, RCRA requires that toxic wastes destined for landfills be pretreated to convert them to forms that will not leach. Such treatment now commonly includes biodegradation or incineration in various kinds of facilities, including cement kilns **(Fig. 22–15)**. Biodegradation involves the use of systems similar to those used for secondary sewage treatment and perhaps new species of bacteria capable of breaking down synthetic organics (Chapter 20). If treatment is thorough, there may be little or nothing to put in a landfill. That is the ultimate objective.

For whatever is still going to disposal facilities, the *third* major feature of RCRA is the requirement of "cradle-to-grave" tracking of all hazardous wastes. The generator (the company that originated the wastes) must fill out a form detailing the exact kinds and amounts of waste generated. Persons transporting the waste—who are also required to be permitted—and those operating the disposal facility must each sign the form, vouching that the amounts of waste transferred are accurate. Copies of their signed forms go to the EPA. All phases are subject to unannounced EPA inspections. The generator remains responsible for any waste "lost" along the way or any inaccuracies in reporting. This provision of RCRA thus ensures that generators will deal only with responsible parties and curtails midnight dumping.

Reduction of Accidents and Accidental Exposures

A significant risk to personal and public health lies in exposures that occur as a result of leaks, accidents, and the misguided use of hazardous chemicals in the home or workplace. A considerable number of laws bear on reducing the probability of accidents and on minimizing the exposure of both workers and the public if accidents should occur.

Department of Transportation Regulations. Transport is an area that is particularly prone to accidents. As modern society uses increasing amounts and kinds of hazardous materials, the stage is set for accidents to become wide-scale disasters. To reduce this risk, **Department of Transportation regulations (DOT regs)** specify the kinds of containers and methods of packing to be used in the transport of various hazardous materials. Such regulations are intended to reduce the risk of spills, fires, and poisonous fumes that could be generated from mixing certain chemicals in case an accident occurs.

In addition, DOT regs require that every individual container and the outside of a truck or railcar carry a standard placard identifying the hazards (flammability, corrosiveness, the potential for poisonous fumes, and so on) presented by the material inside (Fig. 22–3). Such placards enable police and firefighters to identify the potential hazard and respond

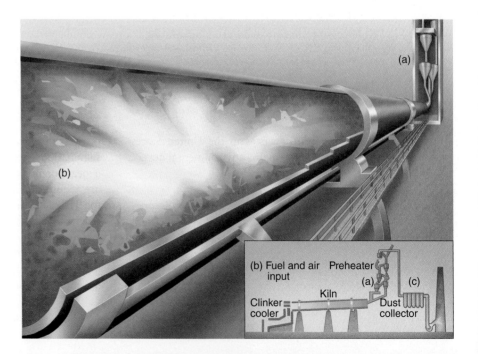

Figure 22–15 Cement kiln to destroy hazardous wastes. A cement kiln is a huge rotating "pipe," typically 15 feet in diameter and 230 feet long, mounted on an incline. (a) Solid wastes fed in with raw materials are fully incinerated and made to react with cement compounds as they gradually tumble toward the combustion chamber. (b) Flammable liquid wastes added with the fuel and air impart fuel value as they are burned. (c) Waste dust is trapped and recycled into the kiln.

appropriately in case of an accident. You may also see high-way HAZMAT signs restricting truckers with hazardous materials to using particular routes or lanes or to using a highway only during certain hours.

Worker Protection: OSHA and the "Worker's Right to Know." In the past, it was not uncommon for industries to require workers to perform jobs that entailed exposure to hazardous materials without informing the workers of the hazards involved. This situation is now addressed by amendments to the **Occupational Safety and Health Act of 1970 (OSHA)**. These amendments make up the **hazard communication standard**, or "worker's right to know." Basically, the law requires businesses, industries, and laboratories to make available both information regarding hazardous materials and suitable protective equipment. One form in which this information is presented is the **material safety data sheet (MSDS)**, which must accompany more than 600 chemicals when they are shipped, stored, and handled. These sheets contain information about the reactivity and toxicity of the chemicals, with precautions to follow when one is using the chemical. Notably, however, the responsibility to read the information and exercise proper precautions remains with the worker.

Community Protection and Emergency Preparedness: SARA, Title III. In 1984, an accident at a Union Carbide pesticide plant in Bhopal, India, caused the release of more than 40 tons of methyl isocyanate, an extremely toxic gas (**Fig. 22–16**). An estimated 600,000 people in communities surrounding the plant were exposed to the deadly fumes. Within days, more than 5,000 people died, and more than 10,000 subsequently died from gas-related diseases. At least 50,000 people suffered various degrees of visual impairment, respiratory problems, and other injuries from the exposure, and many more have experienced chronic illnesses related to the disaster. An Indian investigation of the accident found that the disaster likely happened because Union Carbide officials had scaled back safety and alarm systems at the plant in order to cut costs. Ironically, most of the deaths and injuries could have been avoided if people had known that methyl isocyanate is highly soluble in water, that a wet towel over the head would have greatly reduced one's exposure, and that showers would have alleviated aftereffects. Unfortunately, neither the people affected nor the medical authorities involved had any idea of what chemical they confronted, much less the way to protect or treat themselves.

After the Bhopal disaster, Congress passed legislation to address the problem of accidents. For expediency, the measure was added as Title III to a bill reauthorizing Superfund: the Superfund Amendments and Reauthorization Act of 1986 (SARA). Title III of SARA is better known as the **Emergency Planning and Community Right-to-Know Act (EPCRA)**, which we have already encountered in connection with the TRI.

EPCRA requires companies that handle in excess of five tons of any hazardous material to provide a "complete accounting" of storage sites, feed hoppers, and so on. The information goes to a *local emergency planning committee*, which is also required in every government jurisdiction. The committee is made up of officials representing local fire and

Figure 22–16 Toxic chemical disaster, Bhopal, India. Officials view some of the victims of a deadly gas release at a Union Carbide plant in 1984. At least 10,500 died, and many more suffered injuries.

police departments, hospitals, and any other groups that might be involved in case of an emergency, as well as the executive officers of the companies in question.

The task of the committee is to draw up scenarios for accidents involving the chemicals on-site and to have a contingency plan for every case. This means everything from having firefighters trained and properly equipped to fight particular kinds of chemical fires, to having hospitals stocked with medicines to treat exposures to the particular chemicals, to implementing procedures to follow for the different emergencies. Thus, there can be an immediate and appropriate response to any kind of accident. Information required by the act, together with the information in the TRI, enables communities to draw up a chemical profile of their local area and encourages them to initiate pollution-prevention and risk-reduction activities.

Evaluating New Chemicals

In the past, new synthetic organic compounds were introduced for a specific purpose without any testing of their potential side effects. For example, in the 1960s, a new compound known as TRIS was found to be an effective flame retardant and was widely used in children's sleepwear.

It was only later discovered that TRIS was a potent carcinogen. Treated sleepwear was immediately withdrawn from the market. It is not known how many children (if any) developed cancer from TRIS, but this and other such cases revealed a hole in the systems of protection.

Toxic Substances Control Act. Congress responded by passing the **Toxic Substances Control Act of 1976 (TSCA)**, which requires that, before manufacturing a new chemical in bulk, manufacturers submit a "pre-manufacturing notice" to the EPA, in which the potential environmental and human health impacts of the substance are assessed (including those that may derive from the ultimate disposal of the chemical). Depending on the results of the assessment, a manufacturer may be required to test the effects of the product on living things. Following the results of testing, the EPA may restrict a product's uses or keep it off the market altogether. The law, however, also tells the EPA to use the "least burdensome" approach to regulation and to compare the costs and benefits of regulation. TSCA also authorizes the EPA to develop an inventory of the chemicals already in use at the time the legislation was passed and to require testing whenever there are indications of potential risks. More than 80,000 chemical substances are currently on the inventory.

There is serious concern over the effectiveness of TSCA. A huge loophole allows chemical companies to keep their chemical recipes confidential, claiming they are trade secrets. The EPA has required testing for only 200 of the chemicals in commerce and has actually developed regulations for only five of the existing chemicals. The Government Accountability Office (the investigative arm of Congress) has labeled TSCA as a failed program, largely due to the restrictions the act places on the EPA's ability to regulate.

REACH. A new player in this field comes from the European Union (EU): **REACH (Registration, Evaluation and Authorization of Chemicals)**. Recently authorized by the EU member states, the REACH regulation has resulted in a process of registering and classifying all potentially hazardous chemical substances in use in European societies. Each chemical must be provided with a chemical safety report, which involves human health and environmental hazard assessments, as well as information on how the chemical is used once it is sold. Many chemicals not currently tested in the United States will be included. The approach, from screening to regulation, is based on the precautionary principle; the responsibility for managing the risks from chemicals and providing safety information lies with industry rather than the government (as it is in the United States). REACH has left U.S. trade representatives and many industries concerned about the $20 billion in chemical products the United States exports to Europe each year, as they are claiming that the costs of testing would be prohibitive. However, the European chemical industry is facing the same costs, which should make for a level playing field.

Laws applying to hazardous wastes are summarized graphically in **Figure 22–17**. It is noteworthy that nongovernmental consumer advocate groups have been a major force

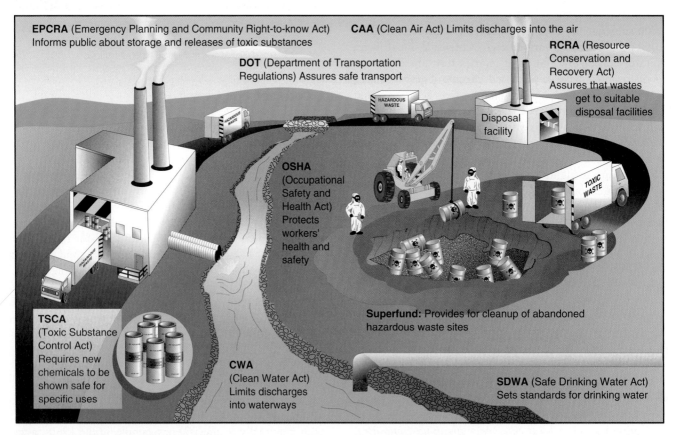

Figure 22–17 Major hazardous-waste laws. Summary of the major laws pertaining to the protection of workers, the public, and the environment from hazardous materials.

behind the passage of these laws as well as regulations requiring ingredients to be labeled on all products. Citizen action does make a difference!

22.5 Broader Issues

Before we leave the topic of hazardous wastes, we still need to address two important broader issues: the problem of environmental justice and the movement toward pollution prevention.

Environmental Justice and Hazardous Wastes

- The largest commercial hazardous-waste landfill in the United States is located in Emelle, Alabama. African Americans make up 90% of Emelle's population. This landfill receives wastes from Superfund sites and every state in the continental United States.

- Residents of Arizona's Navajo Nation reservation are exposed to uranium and radium through airborne dust and contaminated water, a consequence of 520 open uranium mines abandoned by private companies that operated the mines under government contracts.

- A recent study found that 870,000 U.S. federally subsidized housing units are within a mile of factories that have reported toxic emissions to the EPA. Most of the occupants of these apartments are minorities.

The issue is *environmental justice*, introduced in Chapter 1. The EPA defines **environmental justice** as:

> The fair treatment and meaningful involvement of all people regardless of race, color, national origin, or income with respect to the development, implementation, and enforcement of environmental laws, regulations, and policies. Fair treatment means that no group of people, including racial, ethnic, or socioeconomic group should bear a disproportionate share of the negative environmental consequences resulting from industrial, municipal, and commercial operations or the execution of federal, state, local, and tribal programs and policies.[5]

Several recent studies have shown that, all across the United States, waste sites and other hazardous facilities are more likely than not to be located in towns and neighborhoods where most of the residents are non-Caucasian. These same towns and neighborhoods are also less affluent, a further element of environmental injustice. The wastes involved are generated primarily by affluent industries and the affluent majority, but somehow the wastes tend to end up well away from where they were generated and in the backyards of people of color. It seems fair to assume that the siting of hazardous facilities is a matter of political power, and those with the power would like to have the sites well away from their own backyards.

Federal Response. The federal government has taken this problem seriously. In 1994, President Clinton issued Executive Order 12898, focusing federal agency attention on environmental justice. The EPA's response was to establish an Environmental Justice (EJ) Program, put in place early in 1998, to capture the intent of Executive Order 12898 and to further a number of strategies already established by the EPA's Office of Solid Waste and Emergency Response. For example, in connection with the EPA's Brownfields program, communities are putting abandoned properties back into productive use. Many EJ efforts are directed toward addressing justice concerns *before* they become problems. The Obama administration recently coordinated commitments by 16 federal agencies to address EJ issues and implement new strategies to combat injustices to minority communities.[6]

International EJ. One form of trade the developing countries can do without is the international trade in toxic wastes. Some countries or local jurisdictions in search of ready cash have found a source in the form of toxic-waste shipments, which invariably come from more developed countries and are often illegal. This issue appeared earlier in connection with the overseas shipments of electronic wastes (Chapter 21). Close to home, some 20% of spent vehicle batteries (20 million in 2011) are exported from the United States to Mexico, likely because environmental and safety regulations are more relaxed there. The result is worker lead exposure levels far exceeding levels allowed in the United States.

The **Basel Convention** is an international agreement that bans most international toxic-waste trade, while a more recent amendment, the Basel Ban Amendment, would prohibit all exports of hazardous waste from developed to developing countries; however, the amendment is not yet in force and the United States opposes it. The Basel Action Network, an NGO, is vigilant in publicizing and coordinating legal challenges to incidents involving toxic-waste shipments and has made it clear that this trade is a matter of international environmental justice.

Pollution Prevention for a Sustainable Society

Pollution *control* and pollution *prevention* are not the same. *Pollution control* involves adding a filter or some other device at the "end of the pipe" to prevent pollutants from entering the environment. The captured pollutants still have to be disposed of, and this entails more regulation and control. *Pollution prevention*, by contrast, involves changing the production process, the materials used, or both so that harmful pollutants won't be produced in the first place. For example, adding a catalytic converter to the exhaust pipe of your car is pollution control, whereas redesigning the engine or switching to a hybrid car so that less pollution is produced is pollution prevention.

Pollution prevention often results in better product or materials management—that is, less waste. Thus, pollution

---

[5]Environmental Protection Agency, *Final Guidance for Incorporating Environmental Justice Concerns in EPA's NEPA Compliance Analyses* (April 1998). Available at www.epa.gov.

[6]Environmental News Service. "Obama Administration Strengthens Environmental Justice Efforts." February 27, 2012. www.ens-newswire.com/ens/feb2012/2012-02-27-091.html.

prevention frequently creates a cost savings. For example, Exxon Chemical Company added simple "floating roofs" to tanks storing its most volatile chemicals, thereby reducing evaporative emissions by 90% and gaining a savings of $200,000 per year. That paid for the cost of the roofs in six months. As another example, United Musical Instruments in Nogales, Arizona, worked with the Arizona Department of Environmental Quality's Pollution Prevention Unit to install a closed-loop system that reduced water use by 500,000 gallons per year. The system also reduced hazardous waste by 58% in three years and saved the company $127,000 in its first year alone. Both are examples of the *minimization* or *elimination* of pollution.

Green Chemistry. A second angle on pollution avoidance is *substitution*—finding nonhazardous substitutes for hazardous materials. This is known as green chemistry. For example, the dry-cleaning industry, which uses large quantities of toxic organic chemicals, is exploring the use of water-based cleaning, or "wet cleaning." Preliminary results from some entrepreneurs indicate that wet cleaning is quite comparable in cost and performance to dry cleaning. There are now more than 150 wet cleaners in North America, and the number is growing.

The American Chemical Society (ACS) and the EPA have been cooperating to promote green chemistry, particularly for undergraduate and graduate chemistry courses. Case studies and laboratory modules have been developed by this cooperation, available on the ACS Web site.

Reuse. A third approach is *reuse*—cleaning up and recycling solvents and lubricants. Some military bases, for example, have been able to distill solvents and reuse them, instead of discarding these solvents in a manner that releases them into the environment. This approach has increased greatly during the past decade, as measured by the TRI data. Indeed, all three approaches have been employed increasingly by industries handling hazardous chemicals, and all have contributed to significant reductions in hazardous-waste releases into the environment. (See Fig. 22–5.)

The public disclosure of TRI data is judged to have played a crucial role in these reductions. The EPA holds that the system has been "one of the most effective environmental programs ever legislated by Congress and administered by EPA."

You, the Consumer. Finally, pollution avoidance can also be applied to the individual consumer. So far as you are able to reduce or avoid the use of products containing harmful chemicals, you are preventing those amounts of chemicals from being released into the environment. You are also reducing the by-products that result from producing those chemicals. The average American home contains as much as 100 pounds of hazardous household waste (HHW)—substances such as paints, stains, varnishes, batteries, pesticides, motor oil, oven cleaners, and more. These materials need to be stored safely, used responsibly, and disposed of properly if they are not to pose a risk of injury or death and a threat to environmental health. (Many communities hold regular HHW collection days; if yours doesn't, you could initiate one!)

Increasingly, companies are gradually beginning to produce and market **green products**, a term used for products that are more environmentally benign than are their traditional counterparts. For example, Method Products and Clorox-based Green Works are now marketing nontoxic cleaners in Target, Costco, and other large retailers. How fast and to what degree these green products replace traditional products will in large part depend on how we behave as consumers. In other words, your buying power can be an extremely potent force.

In conclusion, there are four ways to address the problems of pollution by hazardous chemicals: (1) pollution prevention, (2) recycling, (3) treatment (breaking down or converting the material to harmless products), and (4) safe disposal. The first three of these methods promote a minimum of waste, in harmony with the principles of sustainability. There is every reason to believe that we can have the benefits of modern chemical technology without destroying the sustainability of our environment by polluting it.

REVISITING THE THEMES

Sound Science

Research is under way to evaluate the impacts of bisphenol A and other potential endocrine-disrupting chemicals on humans and wildlife. It took the work of many environmental scientists to uncover the pathways by which some toxic chemicals entered ecological food webs and made their way into some of the most remote regions of Earth, the Arctic. This work has led to the banning of many persistent organic compounds.

Toxicology is a well-established and crucial branch of science that has helped us to develop a thorough inventory of hazardous chemicals and their health effects. Federal agencies and NGOs have made certain that this information is available to the public via the Internet (through the NTP, IRIS, NIEHS, and Scorecard), an excellent wedding of sound science and social needs. The "green chemistry" movement is an outcome of the efforts of chemists (and their sponsors) who are determined to design products that are nontoxic yet effective.

Sustainability

In our recent past, society was on an unsustainable course in dealing with hazardous chemicals. More and more sites were being contaminated, and we were seeing rising rates of cancer and other health problems. Although the connection between hazardous-waste sites and actual cancer cases has been difficult to prove, the rising tide of hazardous chemicals in the environment was certainly a prime suspect. The legislation and programs discussed in this chapter have turned the situation around, and the move toward pollution prevention and the recycling of hazardous chemicals indicates that we are moving in a sustainable direction in our dealings with toxic chemicals.

Stewardship

One vital task of stewardship is to protect fellow human beings from unnecessary risks. We were not doing well with this responsibility, and a few wake-up calls led to stewardly action on the part of people such as Lois Gibbs, Jan Schlichtmann, and others who put their lives and reputations on the line to bring some sanity to our permissive dealings with toxic-waste dumps. As a result, our society stopped simply accepting contamination as the "price of progress," and after a great deal of public outcry, Congress responded with legislation to address the many problems. We have made real progress in cleaning up the Superfund sites, the leaking underground storage tanks, and the brownfield sites. Industries can no longer release products without testing their potential for doing harm. The TRI, in particular, has provided the incentive for industry to clean up its act. It is not the time, however, to let down our guard. Injustice to minority groups, pressures by industries to ease up on the regulations, and the attitude that hazardous wastes no longer pose a threat indicate that this issue requires constant surveillance and repeated pressure on Congress and the EPA. New chemicals are constantly appearing, and old ones have not yet been thoroughly researched; the Superfund program is still far from completion.

REVIEW QUESTIONS

1. How do toxicologists investigate hazardous chemicals? How is this information disseminated to health practitioners and the public?

2. What four categories are used to define hazardous chemicals?

3. Define *total product life cycle*, and describe the many stages at which pollutants may enter the environment.

4. What are the two classes of chemicals that pose the most serious long-term toxic risk, and how do they affect food chains?

5. What kind of chemicals are the "dirty dozen" POPs? Why are they on a list?

6. What two laws pertaining to the disposal of hazardous wastes were passed in the early 1970s? Describe how the passage of the laws shifted pollution from one part of the environment to another.

7. Describe three methods of land disposal that were used in the 1970s. How has their use changed over time?

8. What law was passed to cope with the problem of abandoned hazardous-waste sites? What are the main features of the legislation?

9. What is being done about leaking underground storage tanks and brownfields?

10. What law was passed to ensure the safe land disposal of hazardous wastes? What are the main features of the legislation?

11. What laws exist to protect the public against exposures resulting from hazardous-chemical accidents? What are the main features of the legislation?

12. What role does the Toxic Substances Control Act play in the hazardous-waste arena?

13. Why does the EPA have an Environmental Justice Program?

14. Describe the advantages of pollution-prevention efforts.

THINKING ENVIRONMENTALLY

1. Select a chemical product or drug you suspect could be hazardous and investigate it, using the resources made available on the Internet by federal agencies and Scorecard.

2. Before the 1970s, it was not illegal to dispose of hazardous chemicals in unlined pits, and many companies did so. Should they be held responsible today for the contamination those wastes are causing, or should the government (taxpayers) pay for the cleanup? Give a rationale for your position.

3. Use the Toxics Release Inventory on the Internet to investigate the locations and amounts of toxic chemicals released in your state or region.

4. Check out the American Chemical Society Green Chemistry Web site, and find out if your college or university is on the list of schools with green chemistry programs. If not, you could ask your chemistry department to look into it.

MAKING A DIFFERENCE

1. Read labels and become informed about the potentially hazardous materials you use in the home and workplace. Whenever possible, use nontoxic substitutes. If you use toxic substances, make sure that you dispose of them properly.

2. Flammable products, such as gasoline, kerosene, propane gas, and paint thinner, should be stored in approved containers in a garage, shed, or similarly separate, well-ventilated area—never inside the house.

3. To avoid ingesting bisphenol A, eat fresh fruits and vegetables and avoid using bottles that have the recycle codes 3 or 7.

4. Use up your paint—try to buy only the amount of paint that you need, and then finish it. If possible, use only latex-based paint, which can be disposed of in household wastes. Do not put oil-based paints in the trash; save them for a household hazardous-waste collection.

MasteringEnvironmentalScience®

Go to **www.masteringenvironmentalscience.com** for practice quizzes, Pearson eText, videos, current events, and more.

PART SIX

Stewardship for a Sustainable Future

Denver, Colorado. As this view from Ferril Lake in City Park shows, Denver's setting, with the snow-capped Rockies to the west, makes it a very attractive place to live. The "Mile High City" is considered one of the most livable cities in America.

CHAPTER 23
Sustainable Communities and Lifestyles

HOW WILL WE SHAPE a sustainable future? In previous chapters, we discussed many issues that need to be addressed and resolved if we are indeed to achieve a sustainable relationship with the environment. In Chapter 23, the last one of the book, we explore one of the most difficult problems faced by every nation today: promoting livable and sustainable communities. With half of humanity now living in cities, urbanization is an irresistible trend, particularly in the developing world, where 95% of urban population growth is occurring. There, one of the most profound challenges is the slums and shantytowns that receive so many migrants from the countryside. How can the people living under such deplorable conditions be brought into the mainstream of a country's economy and achieve a sustainable well-being?

THERE ARE MANY CHALLENGES in the developed countries as well. How can these countries reverse the urban blight that plagues many older cities, dooming so many to second-class citizenship? How should both developed and developing countries deal with the urban sprawl that spreads out from cities into the countryside, consuming prime agricultural land and wildlife habitat?

IN THIS PART SIX, we explore answers to these questions. We consider examples of cities moving in a sustainable direction. We also look at some of the work being done by organizations that are grappling with the question of sustainability. Finally, we look at the vital, but difficult, task of adopting lifestyles that embrace the stewardship values capable of shaping a sustainable future. The outcome is uncertain, but we know enough now to be able to point in a sustainable and stewardly direction.

CHAPTER 23

Sustainable Communities and Lifestyles

Atlanta, Georgia

LEARNING OBJECTIVES

1. **Urban Sprawl: Explain the connections between urban sprawl, car dependency, and highway construction; examine the ways in which "smart growth" can prevent further urban sprawl.**

2. **Urban Blight: Summarize the factors leading to urban blight in older cities of the developed world; describe the reasons why cities of the developing world are also experiencing urban blight and vast slums.**

3. **Moving Toward Sustainable Communities: Examine the steps that can make cities more livable and attractive and even make them more ecologically sustainable than the suburbs.**

4. **Lifestyles and the Common Good: Evaluate some good things that can happen to move our society toward greater sustainability, and include an assessment of lifestyle options open to people who want to be responsible stewards.**

T HE WORD *BELTLINE* conjures up images of a megalane highway that runs around the outskirts of a major metropolitan area, facilitating commuting and stimulating a ring of industries, all of which guarantees that not long after such a highway is built, it will become congested. Atlanta, Georgia, has such a belt, Interstate Route 285, a notorious 10-lane highway built at first to allow through traffic to avoid the city, but now heavily used by commuters between the expanding suburbs and the corporations that have located along the highway. This is the Atlanta Beltway. The **Atlanta BeltLine** (http://beltline.org) is something else altogether, an ambitious makeover of an abandoned rail and industrial corridor that loops around the heart of Atlanta, well inside the Beltway (see photo above). One writer has called it "the country's most ambitious smart growth project."[1]

We'll talk about what smart growth means later, but this project is a remarkable visionary effort that will provide parks, trails, transit system, brownfields remediation, revitalization of distressed city neighborhoods, and much more. The BeltLine loop will be linked by

Martin Luther King Jr.'s birth house in the Old Fourth Ward, on the east side of the BeltLine.

[1]Kaid Benfield. "The Atlanta BeltLine: The Country's Most Ambitious Smart Growth Project." *Natural Resources Defense Council Switchboard.* July 26, 2011. http://switchboard.nrdc.org/blogs/kbenfield/the_countrys_most_ambitious_sm.html. February 29, 2012.

Figure 23–1 Historic Fourth Ward Park. This park, one of the new features of the Atlanta BeltLine, has replaced a drab urban landscape with a mix of green space, lakes, jogging paths, and playgrounds.

a 22-mile trail system that will include pedestrian-friendly rail-based transit segments, with spur segments that will extend into neighborhoods and existing parks and trails around the BeltLine. The east side of the BeltLine connects with the Old Fourth Ward, a neighborhood that includes the birth site of Dr. Martin Luther King, Jr., and Ebenezer Baptist Church, where King and his father served as pastors. Those buildings are now part of the Martin Luther King, Jr., National Historic Site. As part of the BeltLine project, the 17-acre Historic Fourth Ward Park has recently been completed. It has dramatically transformed a blighted neighborhood into a lovely public space connected to a small lake (**Fig. 23–1**).

HISTORY. The "Belt Line" was originally a loop of tracks that provided rail transportation to local industry and commerce in post–Civil War Atlanta. When trucks and highways came on the scene, most of the rail tracks were abandoned. The new BeltLine concept was the brainchild of Georgia Tech graduate student Ryan Gravel, who, in 1999, proposed a new transit system that would link the inner city neighborhoods. Gravel's concept quickly attracted grassroots attention, and soon the BeltLine plan included parks, trails, affordable housing, and neighborhood renewal. Public support grew as more and more civic organizations enthusiastically endorsed the BeltLine plan. Input from more than 10,000 citizens helped craft the first five-year work plan, and with funding from private and public sectors, work began in 2006. The total price tag is estimated at

$2.8 billion, but considering the potential benefits for the city of Atlanta, that seems like a bargain.

FUTURE. In concept, the Atlanta BeltLine will create:

- 2,000 acres of parks, all linked by the multi-use trail and transit loop
- 33 miles of continuous trails
- a 22-mile light rail transit system that will connect to Atlanta's regional transit network
- the cleanup and redevelopment of 1,100 acres of contaminated brownfield sites
- 5,600 units of workforce housing
- 30,000 permanent jobs and thousands of temporary construction jobs
- a $20 billion increase in the city's tax base over the next 25 years
- the connection to and renewal of some 40 city neighborhoods

Many obstacles remain before Atlanta BeltLine reaches completion. Some key right-of-way parcels of land must be acquired for the transit system, and much of the funding is uncertain. It is now in the hands of Atlanta BeltLine, Inc., a body created by the city to oversee implementation of the project; Atlanta BeltLine Partnership is a sister organization responsible for fundraising and outreach. Several parks and pieces of the trails have been built, and master plans have been completed for 10 segments of land the BeltLine comprises. There is no doubt that the Atlanta BeltLine will be years in the making, but when it is finished, it will represent a monumental improvement in urban Atlanta, making the city more livable and viable. Atlanta, with its BeltLine, joins a bevy of cities intent on transforming unused and blighted neighborhoods into showplaces of urban renewal: New York's High Line (illustrated on the cover of this text), Chicago's Millennium Reserve, Boston's Public Market, Los Angeles' River of Dreams, and Seattle's waterfront, among others.

Our objective in this chapter is first to show how urban sprawl and urban blight are related and how they can be remedied. Then we will consider the global and national efforts being undertaken to forge sustainable cities and communities. Finally, we will look at some ethical concerns and personal lifestyle issues as we consider the urgent needs of people and the environment in this new millennium.

23.1 Urban Sprawl

What is urban sprawl? Sprawl's signature manifestation is a far-flung urban-suburban network of low-density residential areas, shopping malls, industrial parks, and other facilities loosely laced together by multilane highways. The word *sprawl* is used because the perimeters of the city have simply been extended outward into the countryside,

one development after the next, with little plan as to where the expansion is going and no notion as to where it will stop. Almost everywhere we go near urban areas, we are confronted by farms and natural areas giving way to new subdivisions being built, new highways being constructed, and old roadways being upgraded and expanded (**Fig. 23–2**). In fact, cars and sprawl are codependent; they need each other!

Figure 23–2 New-highway construction. The hallmark of development over the past several decades has been the construction of new highways, which, however, have only fed the environmentally damaging car-dependent lifestyle.

Figure 23–3 The integrated city. Before the widespread use of automobiles, cities had an integrated structure. A wide variety of small stores and offices on ground floors, with residences on upper floors, placed everyday needs within walking distance. This scene is from a historic section of Philadelphia.

The Origins of Urban Sprawl

Until the end of World War II, a relatively small percentage of people owned cars. Cities had developed in ways that allowed people to meet their needs by means of the transportation that was available—mainly, walking. Every few blocks had small groceries, pharmacies, and other stores as well as professional offices integrated with residences. Often, buildings had stores at the street level and residences above (**Fig. 23–3**). Schools were scattered throughout the city, as were parks for outdoor recreation or relaxation. Walking distances were generally short, and bicycling made going somewhere even more convenient. For more specialized needs, people boarded public transportation—electric trolleys, cable cars, and buses—that took them from neighborhoods to the "downtown" area, where big department stores, specialty shops, and offices were located. Public transportation did not change the compact structure of cities, because people

still needed to walk to the transit line. At the outer ends of the transit lines, cities gave way abruptly to natural areas and farms, which provided most of the food for the city. The small towns and villages surrounding cities—the original suburbs—were compact for the same reasons and mainly served farmers in the immediate area. At the end of World War II, this pattern began to change dramatically.

Despite the many advantages of urban life, many people found cities less-than-pleasant places to live. Especially in industrial cities, poor housing, inadequate sewage systems, inadequate refuse collection, pollution from home furnaces and industry, and generally congested, noisy conditions were common. Hence, many people had the desire—the "American dream"—to live in their own house, on their own piece of land, away from the city.

Suburbs. Because of the massive production of war materiel, consumer goods were in short supply during World War II. As a result, when the war ended, there was a pent-up demand for consumer goods. When mass production of cars resumed at the end of the war, people flocked to buy them. With private cars, people were no longer restricted to living within walking distance of their workplaces or transit lines. They could move out of their cramped city flats and into homes of their own outside the city. Their cars would allow them to drive back and forth easily to their jobs, shopping, recreation, and so on.

Housing Boom. Returning veterans and the accompanying baby boom created a housing demand that rapidly exhausted the existing inventory. Developers responded quickly to the demand for private homes. They bought farmlands and natural areas outside the cities and put up houses. The government aided this trend by providing low-interest mortgages through the Veterans Administration and the Federal Housing Administration, and interest payments on mortgages and local property taxes were made tax deductible (rent was not). These financial factors meant that, for the first time in history, making monthly payments on one's own home in the suburbs was cheaper than paying rent for equivalent or less living space in the city.

Thus, the mushrooming development around cities did not proceed according to any plan; rather, it happened wherever developers could acquire land (**Fig. 23–4**). No governing body existed to devise—much less enforce—an overall plan. Cities were usually surrounded by a maze of more or less autonomous local jurisdictions (towns, villages, townships, and counties). Local governments were quickly thrown into the catch-up role of trying to provide schools, sewers, water systems, other public facilities, and, most of all, roads to accommodate the uncontrolled growth. Local zoning laws kept residential and commercial uses separate. It became more difficult to walk to a grocery store or an office.

Rust Belt. Pittsburgh, Cleveland, Youngstown, Gary—these and many other cities of the Midwest can be counted as full-fledged representatives of what has been called the **Rust Belt**. Once known as the "Foundry of the Nation," these cities and their outskirts had specialized in heavy manufacturing of industrial and consumer products up to the mid-20th

Figure 23–4 Urban sprawl. Prime land is being sacrificed for development around most U.S. cities and towns. The locations of developments are determined more by the availability of land than by any overall urban planning. Here we see sprawl surrounding the fastest-growing major city in the United States: Las Vegas, Nevada.

century. Then a combination of factors contributed to a decline in the Rust Belt: the migration of manufacturing to southern states where labor was less costly, the globalization of trade that gave foreign manufacturers with their far lower labor costs access to U.S. markets, the expansion of free trade agreements, and the outsourcing of jobs and factories because of these two developments. The result was a wholesale loss of the economic foundations of cities of the Rust Belt during the latter half of the 20th century, and this loss added to the mass migration of people from the cities to the suburbs. The term "Rust Belt" is a consequence of the all-too-common images of rusting and crumbling buildings and factories (**Fig. 23–5**).

Highways. The influx of commuters into previously rural areas soon resulted in traffic congestion, creating a need for new and larger roads. To raise money to build and expand highways, Congress passed the Highway Revenue Act of 1956, which created the **Highway Trust Fund**. This legislation placed a tax on gasoline and earmarked the revenues to be used exclusively for building new roads.

Ironically, the Highway Trust Fund perpetuates development. A new highway not only alleviates existing congestion, but also encourages development at farther locations, because *time*, not *distance*, is the limiting factor for commuters. (When people describe how far they live from work, it is almost always put in terms of minutes, not distance.) The average person is willing to spend 20 to 40 minutes each way on a daily commute. With cars and expressways, people can live many miles from work and still get there in an acceptable amount of time.

Vicious Cycle. New highways that were intended to reduce congestion actually fostered the development of open land and commuting by more drivers from more distant locations (**Fig. 23–6**). Soon traffic conditions became as congested as ever. Average commuting distance has doubled since 1960, but average commuting time has remained about the same. The increase in commuting distance, however, requires more fuel, which generates more money for the Highway Trust Fund. Thus, the whole process is repeated in a continuous

(a)

(b)

Figure 23–6 The new-highway traffic-congestion cycle.
(a) Growing traffic congestion creates the need for new and upgraded highways.
(b) However, upgraded highways encourage development in more distant locations, thus creating more traffic congestion and, hence, the need for more highways.

Figure 23–5 Rust Belt image. This abandoned factory is symbolic of many such relics of the heavy manufacturing that once took place in large midwestern cities.

TABLE 23–1 Decline in Population of American Cities, 1950–2009

City	Population (thousands)						Percent Change, 1950–2009
	1950	1970	1990	2000	2005	2009	
Baltimore, MD	950	905	736	651	640	637	–33
Boston, MA	801	641	574	589	597	645	–19
Buffalo, NY	580	463	328	293	278	270	–53
Cleveland, OH	915	751	505	478	450	431	–53
Detroit, MI	1,850	1,514	1,028	951	921	911	–51
Minneapolis, MN	522	434	368	383	375	385	–26
Philadelphia, PA	2,072	1,949	1,586	1,517	1,460	1,547	–26
St. Louis, MO	857	662	396	348	353	357	–58
Washington, DC	802	757	607	572	582	600	–25
U.S. population (millions)	151	218	249	281	296	307	+103

Source: Data from Population Division, U.S. Bureau of the Census, 2012.

cycle. As a result, government policy, as an unintended consequence, helped support and promote sprawl.

Residential developments are inevitably followed (or sometimes led) by shopping malls, industrial parks, big-box stores, and office complexes. These commercial centers are usually situated in such a way that the only possible access to them is by car. This situation, in turn, has only changed the direction of commuting: Whereas, in the early days of suburban sprawl, the major traffic flow was into and out of cities, it is now between suburban centers. Multilane highways connecting suburban centers are perpetually congested with traffic going in both directions.

Exurbs, Too. In broad perspective, then, urban sprawl is a process of **exurban migration**—that is, a relocation of residences, shopping areas, and workplaces from their traditional spots in the city to outlying areas. The populations of many eastern and midwestern U.S. cities, excluding the suburbs, have been declining over the past 60 years as a result of exurban migration and the loss of large-scale manufacturing around Rust Belt cities (Table 23–1). At the same time, the suburban population has increased from 35 million in 1950 to 150 million in 2010, half of the U.S. population. Exurban migration is continuing in a leapfrog fashion as people from older suburbs move to **exurbs** (communities farther from cities than suburbs are).

The "love affair" with cars is not just a U.S. phenomenon: Around the world, in both developed and developing countries, people aspire to own cars and adopt the car-dependent lifestyle. Consequently, urban sprawl is occurring around many developing-world cities as people become affluent enough to own cars. Love of the car-dependent lifestyle, however, is insufficient to make it sustainable. In fact, this lifestyle is at the crux of a number of the unsustainable trends described in earlier chapters.

Measuring Sprawl

A team of researchers from Rutgers University, Cornell University, and Smart Growth America (a nongovernmental

organization) worked for several years on a comprehensive effort to define, measure, and evaluate sprawl and its impacts. The research team labeled sprawl as "the process in which the spread of development across the landscape far outpaces population growth."[2] The team members described four dimensions characterizing sprawl and used them to analyze metropolitan regions in order to create a *sprawl index*:

- Population widely dispersed in low-density development; thus, measure *residential density*.

- Sharply separated homes, stores, and workplaces; thus, measure *neighborhood mix* of these three.

- A network of roads marked by poor connections and huge superblocks that concentrate traffic onto a few routes; thus, measure *accessibility of the street networks*.

- A general lack of well-defined downtowns and activity centers; thus, measure *strength of activity centers and downtowns*.

Each of the preceding four dimensions was standardized numerically so as to give 100 points as the average of the factor; higher means less sprawl, and lower means more. Scores for each factor were combined into a total *sprawl score* that also averaged 100. The team analyzed 83 metropolitan areas, representing almost half of the U.S. population. The results ranged from 14 to 178 (**Fig. 23–7**), with the Riverside–San Bernardino, California, metropolitan area the most sprawled and New York City the least. The 10 most sprawling metropolitan regions are ranked in Table 23–2. The Smart Growth America authors pointed out that, even though an area ranks higher on the sprawl score, this does not mean that sprawl is not a problem. Each of the four dimensions and the overall score were used to examine a range of travel- and transportation-related outcomes as the authors examined the impacts of sprawl on daily life.

---

[2]Reid Ewing, Rolf Pendall, and Don Chen, *Measuring Sprawl and Its Impact* (Washington, DC: Smart Growth America, 2002). Available at www.smartgrowthamerica.org/research/measuring-sprawl-and-its-impact.

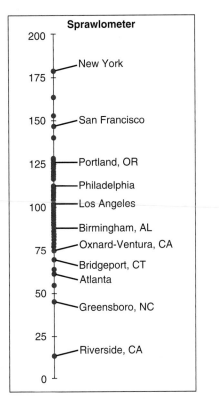

Figure 23–7 Sprawl scores for 83 metropolitan regions. The "sprawlometer" shows rankings of numerous metropolitan areas (each round dot represents a separate area). Low scores mean high sprawl, and vice versa.

(*Source: Measuring Sprawl and Its Impact, 2002,* Smart Growth America.)

TABLE 23–2 Ten Most Sprawling Metropolitan Regions

	Overall Sprawl Index Score	Rank
Riverside–San Bernardino, CA	14.2	1
Greensboro–Winston-Salem–High Point, NC	46.8	2
Raleigh–Durham, NC	54.2	3
Atlanta, GA	57.7	4
Greenville–Spartanburg, SC	58.6	5
West Palm Beach–Boca Raton–Delray Beach, FL	67.7	6
Bridgeport–Stamford–Danbury, CT	68.4	7
Knoxville, TN	68.7	8
Oxnard–Ventura, CA	75.1	9
Forth Worth–Arlington, TX	77.2	10

Source: Reid Ewing, Rolf Pendall, and Don Chen. *Measuring Sprawl and Its Impact* (Washington, DC: Smart Growth America, 2002).

Sprawl and the Recession

The expansion of the suburbs and exurbs in recent years has been remarkable. Builders and developers embarked on the creation of subdivisions of large, expensive homes (McMansions, as they are called), and new buyers were lured by the prospect of ownership of properties that would simply increase in value, as housing had done in the late 1900s and early 2000s. (Between 1997 and 2006, typical American home prices rose by 124%.) They were further encouraged by banks and mortgage companies that extended loans (called subprime mortgages) to many high-risk buyers. A high percentage of these loans were adjustable-rate mortgages (ARMs), in which the interest rates were low for the first few years but in time went up. Many loans required little or no down payment. At the same time, financial firms bought up the subprime mortgages and used these to offer securities to investors. Speculation by buyers and investors was the primary engine that fed this financial bubble.

Bubble Burst. Eventually, the building boom led to a surplus of unsold homes, and housing prices began to decline in 2006. The ARMs were soon reset to higher interest rates, and homeowners were not able to refinance at lower rates. Thousands, and eventually millions, of owners of the new properties defaulted on their loans. Property values declined, and fully one-fourth of all homes in the United States were worth less than their mortgaged amount (called underwater mortgages). Foreclosures soon followed in epidemic numbers. This resulted in the precipitous decline in value of securities backed by the bad loans, and the crisis extended to other sectors of the economy. Job layoffs and business failures followed, and unemployment numbers rose. The country, and indeed the global community, was plunged into a deep recession, with huge investment losses and financial crises that are still reverberating around the global economy.

Recovery. It is significant that areas with the most booming housing markets before the recession are now areas with the most underwater mortgages—regions in California, Florida, and the Southwest.[3] For example, the Las Vegas–Paradise metropolitan area in Nevada has 64% of mortgaged homes underwater; 35% of the homes were built since 2000, and 17% of homes there are now vacant. The unemployment rate there is 14%, and housing prices have dropped 60% from prerecession values. Although recovery from the recession is under way, the housing market is going to take a long time to recover because of the high inventory of foreclosed homes. Most of these homes are in the suburbs—and in sprawling subdivisions.

Environmental Impacts of Urban Sprawl

In addition to its contribution to the decline in home values and the recession, urban sprawl affects the quality of life and the environment in numerous ways. The *environmental impacts* of urban sprawl are many and serious, as the following discussion indicates.

Energy Resources. Shifting to a car-dependent lifestyle has created an ever-increasing demand for petroleum, with many

---

[3]Michael B. Sauter and Charles B. Stockdale. "American Cities with the Most Underwater Mortgages." *24/7 Wall Street,* December 28, 2011.

consequences (Chapter 14). From 1950 to 2010, U.S. oil consumption tripled, while population doubled—a 63% increase in per capita consumption. Transportation accounts for 28% of the U.S. greenhouse gas emissions that are causing climate change.

Air Pollution. Despite improvements in pollution control, many cities still fail to meet desired air quality standards. Vehicles are responsible for an estimated 80% of the air pollution in metropolitan regions. The higher use of gasoline produces greater amounts of carbon dioxide, the most serious anthropogenic greenhouse gas.

Water Resources. All the highways, parking lots, driveways, and other paved areas associated with urban sprawl lead to a substantial increase in runoff, resulting in increased flooding and erosion of stream banks. Also, water quality is degraded by the runoff of fertilizer, pesticides, crankcase oil (the oil that drips from engines), pet droppings, and so on. These nonpoint-source pollutants are the leading cause of water quality problems, according to the EPA.

Loss of Agricultural Land. Perhaps most serious in the long run is the loss of prime agricultural land. In the United States, the National Resources Inventory shows that cropland is declining at a rate of 1.1 million acres per year. Most of the food for cities used to be locally grown on small, diversified family farms surrounding the city. With so many of these farms turned into housing developments, it is estimated that food now travels an average of 1,000 miles (1,500 km) from where it is produced—mostly on huge commercial farms—to where it is eaten. The loss is not just the locally grown produce, but also the social interactions and ties with the farm community.

Loss of Landscapes and Wildlife. New developments are consuming land at an increasing pace. According to the National Resources Inventory, 1.7 million acres a year were developed from 1997 to 2007, compared with 1.4 million acres a year from 1982 to 1992 (**Fig. 23–8**). The result is the sacrifice of aesthetic, recreational, and wildlife values in metropolitan areas—the very places where they are most important.

The fragmentation of wildlife habitat due to urban sprawl has led to marked declines in many species, ranging from birds to amphibians. A 2005 report[4] found that the 35 fastest-growing metropolitan areas are home to nearly one-third of the nation's rarest and most endangered species of plants and animals.

Impacts of Sprawl on Quality of Life

Urban sprawl can also have major negative impacts on *quality of life* for people in the surrounding areas.

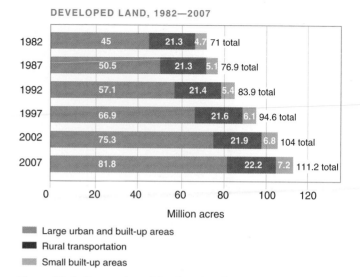

DEVELOPED LAND, 1982—2007

■ Large urban and built-up areas
■ Rural transportation
■ Small built-up areas

Figure 23–8 Conversion of land to developed uses. The buildup of developed land in the United States is shown. From 1982 to 2007, 40 million acres were converted to developed uses from natural lands and farmlands. (*Source:* U.S. Department of Agriculture. 2007 National Resources Inventory, Natural Resources Conservation Service, Washington, DC, and Center for Survey Statistics and Methodology, Iowa State University, Ames, Iowa. July 2009.)

Higher Vehicle Ownership and Driving Mileages

Cars are driven greater distances per person in high-sprawl areas. In Atlanta (sprawl score of 58), for example, daily use per person was 34 miles each day, while in Portland, Oregon (sprawl score of 126), it was only 24 miles each day. In the top 10 sprawl areas, there were 180 vehicles per 100 households, while in the lowest 10, there were 162. Chauffeuring kids to and from school or after-school activities requires more driving in sprawl areas. As these values for miles driven and cars owned aggregate over many metropolitan areas, they add up to millions more miles driven and cars on the road as well as higher gas costs for families. The higher rates of vehicle use also lead to more highway fatalities. In Riverside, California, at the top of the sprawl index, 18 of every 100,000 people die each year in highway accidents. In the eight lowest areas on the index, fewer than eight of 100,000 residents die annually on the highways.

Less Physical Activity and Greater Health Risks. The percentage of commuter trips by mass transit in the 10 highest-sprawl areas was 2% compared with 7% in the 10 lowest-sprawl areas. Similar trends are found for people walking to their workplace. People living in high-sprawl areas drive more, while those in more compact communities walk and take mass transit more. Walking and other moderate physical activities have many health benefits. The study of sprawl and its health effects found a highly significant correlation between the degree of sprawl, on the one hand, and weight, obesity, and high blood pressure, on the other.

Congestion and Higher Costs. There is no difference in commute times between high- and low-sprawl areas. This indicates that moving out to the suburbs or exurbs to get

---

[4]Reid Ewing and John Kostyack, with Don Chen, Bruce Stein, and Michelle Ernst. *Endangered by Sprawl: How Runaway Development Threatens America's Wildlife* (Washington, DC: National Wildlife Federation, Smart Growth America, and NatureServe, January 2005).

away from traffic congestion does not work. As more people move to the suburbs, the car traffic increases, and the congestion reaches out farther. Thus, metropolitan regions cannot sprawl their way out of congestion. The developments in outlying regions of metropolitan areas also all have to be serviced with schools, sewers, water, electricity, roads, and other infrastructure elements, often forcing county and town budgets (and, therefore, taxes) to escalate. More compact forms of development are less expensive to service.

The Benefits of Urban Sprawl

With all of these undesirable effects of sprawl, one might question why people put up with it. The answer is that people *perceive* that it is better to live in such areas, so they move there. There are benefits to sprawl, and it is important to examine them and compare them with the costs. This comparison was made in a study sponsored by the Federal Transit Administration.[5]

Quality-of-life issues tend to be decisive for most people, and these issues seem to be heavily weighted *in favor of sprawl*. In general, sprawl involves lower-density residential living, larger lot sizes (often 2 acres or more), larger single-family homes, better quality public schools, lower crime rates, better social services, and greater opportunity for participation in local governments. As people move farther out from an urban center, housing costs often decline, and communities tend to be more homogeneous. Given these actual or perceived benefits, it is no wonder that sprawl is the dominant form of growth occurring in major metropolitan areas.

Net Benefits? As we have seen, there is little or no advantage in commute times or traffic congestion and actually greater distances to drive for those in the outlying suburbs and exurbs. With gasoline prices approaching $4 a gallon, the commuting distance exacts a major cost to families (close to $15,000 a year for many suburban regions). Also, the traffic fatality rate is worse because speed alone allows people to drive greater distances and achieve an equivalent commuting time. Several of the perceived benefits of sprawl have serious negative side effects for society as a whole. People move to the exurbs to live in neighborhoods with lower crime rates, to have access to better schools, or to live in more homogeneous communities. In one way or another, these choices are often exclusionary in their effects. The result of thousands of people making these choices is to accentuate the concentration of poorer people in high-density, inner-city, or aging suburban communities.

Common Good? The environmental costs of sprawl are very real, but these are seldom perceived as decisive by the people moving to the sprawling suburbs. These costs are more a matter of the common good, and people tend to make choices based on *personal good* rather than the *common good*. On the whole, most observers who have studied

sprawl and its impacts believe that the costs of sprawl outweigh the benefits. They feel it is worth the effort to bring sprawl under control. The key to controlling sprawl is to provide those quality-of-life benefits that attract people, but to do it without incurring the serious environmental and social costs that are clearly associated with sprawl. This is what "smart growth" intends to do.

Reining in Urban Sprawl: Smart Growth

The way to curtail urban sprawl might appear to be simply to pass laws that bring exurban development under control. Indeed, a number of European countries and Japan have such laws. However, laws cannot be passed without the support of the public. In the United States, the strong sentiment favoring the right of a property owner to develop property as he or she sees fit has made such restrictions, with few exceptions, politically impossible to pass.

Yet there are signs that a sea change is occurring in the attitudes of Americans toward uncontrolled growth. For example, the recently emerged concept of **smart growth** is inviting communities and metropolitan areas to address sprawl and purposely choose to develop in more environmentally sustainable ways. The concept recognizes that some growth will occur and focuses on economic, environmental, and community values that together will lead to more sustainable communities. Table 23–3 presents the basic principles of smart growth. Zoning laws are being changed in many localities, and a new generation of architects and developers is beginning to focus on creating integrated communities, as opposed to disassociated facilities. By directing growth to particular places, smart growth provides protection for sensitive lands. Early indications in the marketplace—the final determining factor—are that the new communities are a great success.

TABLE 23–3   Principles of Smart Growth
Create a range of housing opportunities and choices for people of all income levels.
Create walkable neighborhoods.
Encourage the collaboration of citizens, corporations, businesses, and municipal bodies within communities.
Foster distinctive, attractive communities with a strong sense of place, setting standards for development that respond to community values.
Make development decisions predictable, fair, and cost effective.
Mix land uses to provide a more attractive and healthy place to live.
Preserve open space, farmland, natural beauty, and critical environmental areas.
Provide a variety of transportation choices.
Strengthen and direct development toward existing communities (called *infilling*).
Take advantage of compact building design instead of conventional land-hogging development.

[5]Robert W. Burchell, George Lowenstein, William R. Dolphin, and Catherine C. Galley. *The Costs of Sprawl—2000*. Washington, DC: National Academies Press, 2002.

Initiatives. Smart-growth initiatives are appearing on the ballot in many states and municipalities. These initiatives involve several key strategies:

1. **Setting boundaries on urban sprawl.** Oregon led the way in passing a land-use law in the 1970s that requires every city to set a boundary beyond which urban growth is prohibited. The boundary is a line permitting expected future growth—but controlled growth. Portland has had great success with this strategy, and today it is a highly desirable place to live and work.

2. **Saving open space.** The best way to preserve open space is to own it or the development rights to it, and states are enacting programs to acquire crucial remaining open spaces. For example, in 2004 (before the recession), voters approved 162 state and local ballot measures to provide $4.1 billion for the protection of natural areas.

3. **Developing existing urban space.** In 1997, Maryland's General Assembly passed the Smart Growth and Neighborhood Conservation initiative, which includes measures to channel new growth to community sites where key infrastructure is already in place. Abandoned and brownfield sites in metropolitan areas are already well served with roads, electricity, water, and sewerage and can be developed into communities with new homes, workplaces, and shopping areas.

4. **Creating new towns.** A key to developing sustainable communities is changing zoning laws to allow stores, light industries, professional offices, and higher-density housing to be integrated. Affordable and attractive housing can be created around industrial parks and shopping malls for the people who work and shop in those places. Safe bicycle and pedestrian access can be provided between residential areas and workplaces. Reston, Virginia; Columbia, Maryland; and Valencia, California, were established as self-contained towns along these lines in the 1960s, and they are still highly desirable as business and residential locations. Fairfax County, Virginia (next to Washington, DC), by contrast, is now almost 100% covered by sprawl-type development. It is estimated that if Fairfax had pursued smart-growth policies, the county would still be 70% open space and would be far less congested and less polluted.

Raiding the Fund. In addition to the foregoing key strategies, there have been successful moves to break the cycle created by the Highway Trust Fund, by allowing revenues to be used for purposes other than the construction of more highways. People live too far from where they work, and there are often better ways to get to work than driving automobiles. The first significant inroad was made in 1991 with the passage of the Intermodal Surface Transportation Efficiency Act (ISTEA). Under this act, a substantial amount of the money levied by the Highway Trust Fund was eligible to be used for other modes of transportation, including cycling, walking, and mass transit. The objectives of ISTEA were extended with the passage of the Transportation Equity Act for the

Twenty-First Century (TEA–21) in 1998. Then the Safe, Accountable, Flexible, Efficient Transportation Equity Act—A Legacy for Users (SAFETEA–LU) was passed in 2005. This act authorized $244 billion for 2005–2009 for the highway and transit programs, a significant increase over TEA–21 funding; mass transit has received 18% of these funds.

However, the Highway Trust Fund is not bringing in enough to sustain this level; SAFETEA-LU has expired, and the programs have been continued through short-term extensions, at lower funding levels. A new transportation act is long overdue. However, the economic stimulus package (the **American Recovery and Reinvestment Act of 2009**) contained $64 billion for roads, bridges, rail, and transit, and mass transit has been receiving much of this money, given President Obama's expressed desire to promote projects that will counteract urban sprawl and reduce our dependence on oil.

Highway Trust Fund monies are only part of the reason for the resurgence of mass transit; the high gasoline costs of 2008 drove many commuters to buses, trains, and subways, and mass transit ridership has remained high. Mass transit issues fared well in the 2008 election, with voters approving initiatives that would infuse $75 billion into transportation systems.

Fixing Suburbia. The recession and its accompanying impacts on suburbs have created a unique opportunity to redevelop suburbs in more desirable patterns. Two recent books provide some key guidelines for repairing suburban sprawl.[6] A recent survey[7] found that most Americans prefer a walkable neighborhood to a large house, especially the Baby Boomers (largely empty nesters now) and many Gen Y people (20s and 30s) who are looking for places where they can socialize. The books *Retrofitting Suburbia* by Ellen Dunham-Jones and June Williamson and *Sprawl Repair Manual* by Galina Tachieva are for planners and developers, providing specific examples of what can be done to abandoned malls and poorly designed suburban streets and roadways to convert them to more desirable neighborhoods with places to walk and village centers and greenery. For example, a 100-acre mall in Lakewood, Colorado, has been replaced with 22 blocks of green buildings and public streets, creating a real downtown for the suburban town. The authors of both books state that sprawl has many faces, and some of them are more amenable to makeover than others.

Another key strategy for reducing sprawl is to revitalize the cities. This means coping with urban blight.

23.2 Urban Blight

In the developed world, the other end of the exurban migration is the city from which people are moving. *Exurban migration* is the major factor underlying urban decay,

[6]Ellen Dunham-Jones and June Williamson. *Retrofitting Suburbia: Urban Design Solutions for Redesigning Suburbs*. Hoboken, NJ. John Wiley & Sons, Inc. 2011. Galina Tachieva. *Sprawl Repair Manual*. Washington, DC. Island Press. 2010.

[7]Belden Russonello and Stewart LLC. "The 2011 Community Preference Survey: What Americans Are Looking for When Deciding Where to Live." March 2011. www.Brspoll.com. Washington, DC.

or blight, that has occurred over the past 60 years. A completely different set of factors is responsible for urban blight in the developing world. In that case, people are moving *to* the cities at a rate that far exceeds the capacity of the cities to assimilate them. The result is urban slums that surround virtually every city in the poorer developing countries. UN–HABITAT estimates that more than 800 million people live in urban slums in the developing world, often under conditions so squalid that they defy description.

We look first at urban blight in the developed countries, with the United States as our focus.

Economic and Ethnic Segregation

Historically, U.S. cities have included people with a wide diversity of economic and ethnic backgrounds. However, moving to the suburbs required some degree of affluence—at least the ability to manage a down payment and mortgage on a home and the ability to buy and drive a car. Therefore, exurban migration often excluded the poor, the elderly, and persons with disabilities. The poor were mostly minorities because a history of racial discrimination had kept them from good education and well-paying jobs and because many had recently immigrated to the United States and were still at the bottom of the social order. Discriminatory lending practices by banks and dishonest sales practices by real estate agents (called *redlining*) kept minorities out of the suburbs even when money was not a factor. Civil rights laws passed in the 1960s made such practices illegal, but there is evidence that they still persist in various forms. Even without racial segregation, economic segregation continues to exist and has even intensified.

In short, exurban migration and urban sprawl have led to segregation of the population into groups sharing common economic, social, and cultural backgrounds. Moving into the new suburban and exurban developments (often labeled *white flight*) were the economically advantaged, whereas many areas of the cities and older suburbs were effectively abandoned to the economically depressed, who in large part were ethnic minorities, persons with disabilities, and the elderly **(Fig. 23–9)**.

The Vicious Cycle of Urban Blight

Affluent people moving to the suburbs set into motion a vicious cycle of exurban migration and urban blight that continues today. To understand how this downward cycle occurs, we must note several points about local governments. First, local (usually city or county) governments are responsible for providing public schools, maintaining local roads, furnishing police and fire protection, collecting and disposing of refuse, maintaining public water and sewers, and providing welfare services, libraries, and local parks. Second, a local government's major source of revenue to pay for these services is local property taxes, a yearly tax proportional to the market value of the property and home or other buildings on it. If property values increase or decrease, property taxes are adjusted accordingly. Third, in most cases, the central city is

(a)

(b)

Figure 23–9 Segregation by exurban migration. Exurban migration, the driving force behind urban sprawl, has also led to segregation of the population along economic and racial lines. In (a) an area of suburbia and (b) an area of inner-city Baltimore, you see the contrast.

a government jurisdiction that is separate from those of the surrounding suburbs.

Economic Dysfunction. Because of the preceding three characteristics of local government, exurban migration has the following consequences: By the economics of supply and demand, property values in the suburbs escalate with the influx of affluent newcomers. Thus, suburban jurisdictions enjoy increasing tax revenues with which they can improve and expand local services. At the same time, property values in the city decline because of decreasing demand. The lower property taxes create a powerful disincentive toward maintaining property. Many properties end up being abandoned by the owner for nonpayment of taxes. The city "inherits" such abandoned properties, but they are a liability rather than a source of revenue **(Fig. 23–10)**. The declining tax revenue resulting from falling property values is referred to as an **eroding,** or **declining, tax base,** and it has been a serious handicap for most U.S. cities since the exurban migration started in the late 1940s. Adding to the problem is the fact that many immigrants tend to settle in the inner cities, drawn there by ethnic bonds and the hope

Figure 23–10 Abandoned buildings. Many cities' ills are caused by too *little* population, not too much. When the exurban migration causes a city's population to drop below a certain level, businesses are no longer supported. The result is abandonment and urban blight. Shown here is a section of Baltimore only a few blocks from the redeveloped Inner Harbor seen in Figure 23–11.

Figure 23–11 Inner Harbor, Baltimore. High population densities help make a city, not destroy it. Redeveloping key areas of cities with heterogeneous mixtures of office buildings, shopping, restaurants, residences, and recreational facilities—all of which attract people—has helped to bring new life back into the cities.

of employment. The result is that many inner-city dwellers require public assistance of one sort or another. Thus, besides the declining tax revenues, cities bear a disproportionate burden of welfare obligations.

The eroding tax base forces city governments to cut local services, increase the tax rate, or, generally, both. Hence, the property tax on a home in a city is often two to three times greater than that on a comparably priced home in the suburbs. At the same time, schools, refuse collection, street repair, libraries, parks, and other services and facilities deteriorate from neglect. Increasing taxes and deteriorating services cause yet more people to leave the city and become part of urban sprawl, even if the car-dependent lifestyle is not their primary choice. However, it is still only the relatively more affluent individuals who can make this move; thus, the whole process of exurban migration, a declining tax base, deteriorating real estate, and worsening schools and other services and facilities is perpetuated in a vicious cycle. This downward spiral of conditions is referred to as **urban blight** or **urban decay**. It is made worse by the manufacturing losses in Rust Belt cities.

Economic Exclusion of the Inner City. The exurban migration includes more than just individuals and families: The flight of affluent people from the city removes the purchasing power necessary to support stores, professional establishments, and other enterprises. Faced with declining business, merchants and practitioners of all kinds are forced either to go out of business or to move to new locations in the suburbs. Either way, vacant storefronts are the result, and people remaining in the city lose convenient access to goods, services, and, most importantly, jobs because each business also represents employment opportunities.

Unemployment rates run at 50% or more in depressed areas of inner cities, and the jobs that do exist are mostly at the minimum wage. By contrast, the new jobs being created are in the shopping malls and industrial parks in the outlying areas, which are largely inaccessible to inner-city residents

for lack of public transportation. Therefore, not only are people who remain in the inner city poor from the outset, but also the cycle of exurban migration and urban decay has now led to their economic exclusion from the mainstream. As a result, drug dealing, crime, violence, and other forms of deviant social behavior are widespread in such areas.

To be sure, in the past two decades, many cities have redeveloped core areas. Shops, restaurants, hotels, convention and entertainment centers, office buildings, residences, and places for walking and relaxing are all combining to bring a new spirit of life and hope, as well as the practical assets of taxes and employment, back into cities **(Fig. 23–11)**. This is an encouraging new trend, so far as it goes. However, walk a few blocks in any direction from the sparkling, revitalized core of most U.S. cities, and you will find urban blight continuing unabated.

Urban Blight in Developing Countries

Two million residents of Nairobi, Kenya's capital, live in slums consisting of flimsy shacks made of scrap wood, metal, plastic sheeting, rocks, and mud **(Fig. 23–12)**. The slum neighborhoods of Nairobi surround the outskirts of the city. Burning trash and charcoal cooking fires cloud the air, and the lack of sewers adds to the smell. People get water from private companies because the city does not provide water or any other public amenities. There is no land ownership because the settlements are built on public land or on private land where landowners exact high rents. Crime and disease are endemic, and AIDS and tuberculosis spread readily in such slums.

Similar slum neighborhoods can be found around most of the developing-world cities; **UN–HABITAT** estimates that one out of every three people in the developing-world cities lives in slums. Yet in spite of the conditions, the cities are expected to continue growing well into the middle of the century. The continued rapid growth is a function of both natural population growth (which is high in many

Figure 23–12 Shantytown in Nairobi, Kenya. More than half of Nairobi's residents live in such slums, where conditions defy description.

developing countries) and migration into the city from rural areas (Chapters 8 and 9). The economies of rural areas, often based only on subsistence farming, simply do not provide the jobs needed by a growing population. So people move to the cities, where they at least have the hope of employment. The city housing supply is overwhelmed by the influx of migrants, who could not afford the rents even if housing was available. The slums actually represent a great deal of entrepreneurial energy, because the residents build the dwellings and erect their own infrastructure, usually including food stands, coffee shops, barber shops, and, in some cases, schools for their children. Indeed, the slum residents are of necessity amazing entrepreneurs, engaging in a widespread informal economy. The signature of this economy is the umbrella that covers a booth by the side of the road where goods are exchanged or sold—hence the term "umbrella people."[8]

The Needs

The vast slums, or shantytowns, surrounding the cities are a tremendous challenge to the institutional structure of developing countries. Because slum areas are unauthorized, cities rarely provide them with electricity, water, sanitation facilities, and other social amenities. Yet the slums often represent a workforce essential to the city, taking the low-paying jobs that keep the city and its inhabitants functioning. A great need in such neighborhoods is *home security*. People live in fear of bulldozers coming at any time and leveling the shantytown. In Brazil, governments are providing legal status to existing slums (*favelas*), granting lawful titles to the land. Peru has undertaken a huge titling program, giving recognition to some 1 million urban land parcels in a four-year period. Providing this level of security to the inhabitants is often the key to mobilizing further assets, such as acquiring access to credit and negotiating with the city government for utilities or additional services. People will improve their living conditions if they are given the assurance that they can stay where they are.

[8]Robert Neuwirth. "Global Bazaar: Shantytowns, Favelas and Jhopadpattis Turn Out to Be Places of Surprising Innovation." *Scientific American* 305: 56–63. September 2011.

The people in the shantytowns also need more *jobs*, and again, local governments could provide employment at low cost to accomplish many improvements in their own neighborhoods and elsewhere—collecting trash, building sewers, composting organic wastes, and growing food. Providing *cheap transportation* is another need. People need to get to where the jobs are and could often do so if they had bicycles or could afford to ride public transportation. Perhaps the greatest need of people living in the informal neighborhoods of city slums is *government representation*. Their voices are seldom heard because they are among the poorest and most powerless in a society. They are often plagued by government corruption, wherein nothing gets done unless accompanied by bribes (which these people can ill afford).

In some regions, however, community organizations are emerging to deal with the basic issues of the city slums. People are organizing and demanding their rights as citizens. Shack/Slum Dwellers International (SDI) now represents these communities to institutions like the World Bank and city governments and even held its first international conference in 2006. Change can happen, and people are learning how to make it happen. As cities devote energy and resources to providing the essential services and infrastructure to the informal slums that surround them, they will only be strengthened. The cities, and even the slums, will then become more livable.

23.3 Moving Toward Sustainable Cities and Communities

The environmental consequences of urban sprawl and the social consequences of urban blight are two sides of the same coin. A sustainable future will depend on both reining in urban sprawl and revitalizing cities. The urban blight of cities in the developing world requires deliberate policies to address the social needs of the people flocking to the shantytowns. UN–HABITAT reports that half of the world's people now live in cities, with 3 million people added weekly to developing-world cities. Looking to the future, the only possible way to sustain the global population is by having viable, resource-efficient cities, leaving the countryside for agriculture and natural ecosystems. The key word is *viable*, which means *livable*. No one wants to, or should be required to, live in the conditions that have come to typify urban blight and urban slums.

What Makes Cities Livable?

Livability is a general concept based on people's response to the question "Do you like living here, or would you rather live somewhere else?" Crime, pollution, and recreational, cultural, and professional opportunities, as well as many other social and environmental factors, are summed up in the subjective answer to that question. Although many people assume that the social ills of the city are an outcome of high population densities, crime rates and other problems in U.S. cities have climbed while populations have dwindled as a result of the exurban migration.

Figure 23–13 Livable cities. The key to having a livable city is to provide a heterogeneity of residences, businesses, and stores and to keep layouts on a human dimension so that people can meet or conduct business incidentally over coffee in a sidewalk café or stroll on a promenade through an open area. In short, the space is designed for, and devoted to, people. Shown here is Quincy Market, Boston.

Figure 23–14 Car-centered cities. In contrast to what is required for livability, city development of the past several decades has focused on moving and parking cars and creating homogeneity. Note the sterility of parking lots surrounding buildings in the foreground of this photograph of Los Angeles.

Looking at livable cities around the world, we find that the common denominators are

1. Maintaining a high population density;

2. Preserving a heterogeneity of residences, businesses, stores, and shops; and

3. Keeping layouts on a human dimension so that people can meet, visit, or conduct business informally over coffee at a sidewalk café or stroll on a promenade through an open area.

In short, the space is designed for, and devoted to, people **(Fig. 23–13)**. In contrast, city development of the past 60 years has focused on accommodating automobiles and traffic. Two-thirds of the land in cities that have grown up in the era of the automobile is devoted to moving, parking, or servicing cars, and such space is essentially alien to the human psyche **(Fig. 23–14)**. William Whyte, a well-known city planner, remarked, "It is difficult to design space that will not attract people. What is remarkable is how often this has been accomplished."[9]

A Matter of Design. The world's most livable cities are not those with "perfect" auto access between all points. Instead, they are cities that have taken measures to reduce outward sprawl, diminish automobile traffic, and improve access by foot and bicycle in conjunction with mass transit. For example, Geneva, Switzerland, prohibits automobile parking at workplaces in the city's center, forcing commuters to use the excellent public transportation system. Copenhagen bans all on-street parking in the downtown core. Paris has removed 200,000 parking places in the downtown area. Curitiba, Brazil, with a population of 1.6 million, is cited as the most livable city in all of Latin America. The achievement of Curitiba is due almost entirely to the efforts of Jaime

Lerner, who, while serving as mayor for many decades, guided development with an emphasis on mass transit rather than cars. The space saved by not building highways and parking lots has been put into parks and shady walkways, causing the amount of green area per inhabitant to increase from 4.5 square feet in 1970 to 450 square feet today.

Charinkos. In Tokyo, millions of people ride *charinkos*, or bicycles **(Fig. 23–15)**, either all the way to work or to subway stations from which they catch fast, efficient trains to their destination. Tokyo residents own more than 9 million bicycles. By sharply restricting development outside certain city limits, Japan has maintained population densities within cities and along metropolitan corridors that ensure the viability of commuter trains. Japan's cities have maintained a heterogeneous urban structure that mixes small shops, professional offices, and residences in such a way that a large portion of the population meets its needs without cars. In maintaining an economically active city, it is probably no coincidence that street crime, vagrancy, and begging are

Figure 23–15 Bicycles in Tokyo. Most people in Tokyo do not own automobiles and instead ride bicycles or walk to stations from which they take fast, efficient, inexpensive subways to reach their destinations.

[9]William H. Whyte. "The Design of Spaces," in *The City Reader*, 3rd ed., Ed. Richard T. LeGates (New York: Routledge, 2003), 432.

virtually unknown in the vast expanse of Tokyo, despite the seeming congestion.

Bicycles are a growing feature of many cities. In exchange for making special lanes, safe bikeways, and secure parking for bicycles, cities gain a great deal: decreased pollution and congestion and better health for their people because of increased physical activity. Because of its pro-biking policies, the city of Amsterdam in the Netherlands holds the record for large cities in the developed countries; fully 38% of trips within the city are made on bicycles. To be sure, cyclists are at risk of accidents with motor vehicles, but the data show that cities with the highest cycling rates have the best safety records.[10]

Portland. Portland, Oregon, is a pioneering U.S. city that has taken giant steps to curtail automobile use. The first step was to encircle the city with an urban growth boundary, a line outside of which new development was prohibited. Thus, compact growth, rather than sprawl, was ensured. Second, an efficient light-rail and bus system was built, which now carries 45% of all commuters to downtown jobs. (In most U.S. cities, only 10% to 25% of commuters ride public transit systems.) By reducing traffic, Portland was able to convert a former expressway and a huge parking lot into the now-renowned Tom McCall Waterfront Park. Portland is now ranked among the world's most livable cities and has been recognized by SustainLane (see below) as the most sustainable city in the United States.[11]

The Big Dig. Other U.S. cities are discovering hidden resources in their inner core areas. Expressways, which were built with federal highway funds to speed vehicles through the cities, frequently dissected neighborhoods, separating people from waterfronts and consuming existing open space. Now, with federal help, many cities are removing or burying the intruding expressways. Boston, for example, has recently put the finishing touches on the "Big Dig," a $15 billion project that has moved its central freeway below ground, added parks and open space, and reconnected the city to its waterfront (**Fig. 23–16**). The waterfronts themselves are often the neglected relics of bygone transportation by water and rail. In recent years, cities such as Cleveland, Chicago, San Francisco, and Baltimore have redeveloped their waterfronts, turning rotting piers and abandoned freight yards into workplaces, high-rise residences, and public and entertainment space.

Figure 23–16 Boston, Massachusetts. The Rose Fitzgerald Kennedy Greenway, on land once covered by an elevated expressway, is a series of four parks made up of gardens, plazas, and tree-lined promenades. The "Big Dig" project has revitalized the core city of Boston by replacing the central freeway with an underground tunnel system that allows traffic to flow to and through the city without disrupting city life above.

(*Source: Rose Fitzgerald Kennedy Greenway Conservancy.*)

Livable Equals Sustainable. Even though cities are concentrations of energy use, resource use, and pollution, there are certain environmental benefits to urbanization. Decreased auto traffic and a greater reliance on foot and public transportation reduce energy consumption and pollution. If the cities effectively provide the basic infrastructure needs—clean water, solid-waste and sewage disposal, electricity—the economies of scale greatly reduce per capita resource consumption in comparison with suburbs and rural areas. For example, urban populations use half as much electricity per capita as those in the suburbs. A number of factors—economic opportunities for women, access to health services, and less need for children's household labor—work to reduce fertility rates in urban populations versus rural populations.

Beyond these basics, housing can be retrofitted with passive solar space heating and heating for hot water. Landscaping can provide cooling (described in Chapter 16). A number of cities are developing vacant or cleared areas into garden plots (**Fig. 23–17**), and rooftop hydroponic gardens are becoming popular. Such gardens do not make cities agriculturally self-sufficient, but they add to urban livability, provide an avenue for recycling compost and nutrients removed from sewage, give a source of fresh vegetables, and

[10]Gary Gardner. "Power to the Pedals." *World Watch* 23: 6–11. July/August 2010.
[11]www.sustainlane.com/us-city-rankings.

Figure 23–17 Urban gardens. Urban garden plots on formerly vacant lots in Philadelphia. Urban gardening is becoming recognized as having sociological, economic, environmental, and aesthetic benefits.

have the potential to generate income for many unskilled workers. All of these factors have the important added advantage of reducing the per capita use of fuel and energy, important in the battle against climate change.

SustainLane. In the United States, the 50 most populous cities have been ranked for sustainability by the organization SustainLane. The organization's Web site is a great source of advice, personal accounts, local business and product reviews, and news about sustainability lifestyles. Its most prominent product, however, is a ranking of cities based on 16 criteria, including transit ridership, street congestion, air and water quality, green building, local food and agriculture, energy and climate change, and green economy. At the top of the ranking was Portland, Oregon, followed by San Francisco, Seattle, Chicago, New York, Boston, Minneapolis, Philadelphia, Oakland, and Baltimore. At the bottom were Las Vegas (#47), Tulsa, Oklahoma City, and Mesa. The goal of the process has been "not to shame, but to inspire, to facilitate the sharing of best practices, and to see all cities continually raise the bar on their efforts," in the words of SustainLane CEO James Elsen.[12]

To the Point. The preceding discussion of urban issues may seem to have strayed far from nature and environmental science. However, there is a close connection between the two: The decay of our cities is hastening the degradation of our larger environment. As the growing human population spreads outward from the old cities in developed countries, we are co-opting natural lands and prime farmlands and ratcheting up the rate of air pollution and greenhouse gases. Similarly, as increasing numbers in the developing world stream *into* the cities, they are degrading the human and social resources that are vital to all environmental issues. Thus, without the creation of sustainable human communities, there is little chance for the sustainability of the rest of the biosphere.

By making urban areas more appealing and economically stable, we not only improve the lives of those who choose to remain or must remain in cities, but also spare surrounding areas. Our parks, wildernesses, and farms will not be replaced by exurbs and shopping malls, but rather will be saved for future generations. In the past decade, some remarkable developments have begun to accelerate the movement toward building sustainable communities. We turn to these developments now.

Sustainable Communities

Recall that the 1992 UN Conference on Environment and Development (UNCED) created the Commission on Sustainable Development (Chapter 1). This new body was given the responsibility of monitoring and reporting on the implementation of the agreements that were made at UNCED—in particular, those related to sustainability. Following Agenda 21, one of the accords signed at UNCED, countries agreed to work on sustainable-development strategies and action plans that would put them on a track toward sustainability. Some 106 countries have now developed national sustainable-development strategies that lay out public-policy priorities. Most of these countries are actively engaged in transforming the policies into action plans, and we will investigate one of the outcomes of these activities: the *sustainable communities movement*.

Revitalizing urban economies and rehabilitating cities requires coordinated efforts on the part of all sectors of society. In fact, such efforts are beginning to occur. What is termed a *sustainable communities movement* is taking root in cities around the United States and elsewhere. In the United States, the Department of Housing and Urban Development sponsors the **Sustainable Communities and Development** program, providing grants to foster the development of more-livable urban neighborhoods. The Internet resources Sustainable Communities Online and Smart Communities Network provide a cornucopia of information, help, case studies, and linkages, demonstrating that the movement is gaining ground throughout the nation. Chattanooga, Tennessee, is now regarded as a prototype of what can occur with such a movement.

Chattanooga. Forty years ago, Chattanooga, which straddles the Tennessee River, was a decaying industrial city with high levels of pollution. Employment was falling and residents were fleeing to the suburbs, leaving abandoned properties and increasing crime. In 1969, the EPA presented the city with a special award for "the dirtiest city in America." Then Chattanooga Venture, a nonprofit organization founded by community leaders, launched *Vision 2000*, the first step of which was to bring people from all walks of life together to build a consensus about what the city *could* be like. Literally thousands of ideas were gradually distilled into 223 specific projects. Then, with the cooperation of all sectors—including government, businesses, lending institutions, and average persons—and an investment of more than $800 million, work on the projects began, providing employment in construction

[12]"Green-City Ranking Group SustainLane Explains Its Methodology," *Grist*, May 12, 2008.

for more than 7,300 people and permanent employment for 1,380. Among the projects were the following:

- With the support of the Clean Air and Clean Water Acts, local government clamped down hard on industries to control pollution.

- The Chattanooga Neighborhood Enterprise fostered the building or renovation of more than 4,600 units of low- and moderate-income housing.

- A new industry to build pollution-control equipment was spawned, as was another industry to build electric buses, which now serve the city without noise or pollution.

- A recycling center employing adults with mental disabilities to separate materials was built.

- An urban greenway demonstration farm, which schoolchildren may visit, was created.

- A zero-emissions industrial park utilizing pollution-avoidance principles was built.

- The river was cleaned up, and a 22-mile Riverwalk reclaimed the waterfront, with parks, playgrounds, and walkways built along the riverbanks.

- Theaters and museums were renovated, and freshwater and saltwater aquariums were built.

- All facilities and a renovated business district were made pedestrian friendly and accessible.

Revision. With these and numerous other projects, many of them still ongoing, Chattanooga has changed its reputation from one of the worst to one of the best places to live (Fig. 23–18). Chattanooga Venture developed a step-by-step guide to assist community groups in similar efforts to build sustainable communities, and the city's experience is being modeled in other cities throughout the United States as well as internationally. Recently, Chattanooga sponsored *Revision 2000*, with more than 2,600 participants taking up where

Figure 23–18 Chattanooga, Tennessee. With numerous projects to reduce pollution and traffic and to provide attractions and amenities for people, Chattanooga, Tennessee, has changed its reputation from one of the worst to one of the best places to live in the United States.

Vision 2000 left off. Revision 2000 identified an additional 27 goals and more than 120 recommendations for further improving Chattanooga. The city's experience demonstrates that visioning and change must occur and continue in order to successfully create and maintain sustainable community development. Although Chattanooga Venture is no longer in operation, its past work has built an expectation throughout the city that public projects will involve the public and that the process will produce results. The latest manifestation of this expectation was *Recreate 2008*, in which hundreds of people engaged in a visioning process that created a plan for revitalizing the city's park system.

23.4 Lifestyles and the Common Good

Sound Science, Stewardship, and *Sustainability*—these are the three strategic themes that have kept our eyes on the basics of how we must live on our planet. Sound science is the basis for our understanding of how the world works and how human systems interact with it, stewardship is the actions and programs that manage natural resources and human well-being for the common good, and sustainability is the practical goal that our interactions with the natural world should be working toward (Chapter 1). Are we making progress in incorporating these concepts into our society?

Our Dilemma

Our intention in this book has been to avoid dwelling on the bad news and instead to raise the hope—indeed, the certainty—that all environmental problems can be addressed successfully. Throughout the text, we have pointed to policies and possibilities that can help move human affairs in a sustainable direction. There is much at stake. When 2015 arrives, will we be able to say that we have met all the Millennium Development Goals, alleviating much of the developing world's grinding poverty and its consequences? Will we reduce our use of fossil fuels to halt global warming and the rising sea level? Can we feed the billions who will join us by midcentury and simultaneously reduce the malnutrition and hunger that still plague more than 900 million people?

Human Decisions. As the late Nobel Laureate Henry Kendall said, "Environmental problems at root are human, not scientific or technical."[13] Even though our scientific understanding is incomplete and our grasp of sustainable development is still tentative, we know enough to be able to act decisively in most circumstances. Therefore, it is human decisions, at both the personal and the societal levels, that can bring about change, and it is these decisions that define our stewardship relationship with Earth.

[13]Union of Concerned Scientists. *Keeping the Earth: Scientific and Religious Perspectives on the Environment*, videotape transcript (Pittsburgh, PA: New Wrinkle, 1996), p. 4.

However, making stewardly decisions is not a simple task. Decision making is affected by our personal values and needs, and competing values and needs exist at every turn. One group wins, and another loses. If we stop all destruction of the rain forests, the peasant who needs land to farm will not eat. If we force a halt to the harvesting of shrimp because of concerns about the bycatch, thousands of shrimpers will be out of work. Making decisions means considering competing needs and values and reaching the best conclusion in the face of the numbing complexity of demands and circumstances. It is the nature of our many dilemmas that business as usual resolves only a very small fraction of them. Even though public policies at the national and international levels are absolutely essential elements of our success in turning things around, these in the end are the outcome of decisions by very human people; and even if thoughtful decisions are made by those in power, they will fail unless they are supported by the people who are affected by them.

Here are a few examples of significant changes at the level of governments, municipalities, and industries:

- The EPA has established the Green Power Partnership, a program that encourages organizations to buy green power. The number one purchaser is Intel Corp., buying more than 2.5 billion kilowatt-hours of green power annually.

- Transition Towns (connected by the Transition Network) are more than 400 communities that are working together to respond to the challenges of peak oil and climate change.

- The American Recovery and Reinvestment Act of 2009 (the stimulus package) has invested more than $43 billion in green energy initiatives.

- The World Business Council for Sustainable Development is a network of more than 200 global companies promoting greater resource productivity and sustainable product manufacture.

- Mayors from more than 1,000 cities in the United States have signed the U.S. Mayors' Climate Protection Agreement, which pledges the cities to reduce greenhouse gas emissions to 1990 levels or below (in the absence of a federal commitment).

- Australia led the way in 2007 to eliminate incandescent lightbulbs, an action that has been followed by the European Union nations and other countries; U.S. action to remove them was put on hold by congressional conservatives, but the industry is moving ahead to switch to compact fluorescents and LEDs.

- California established the Million Solar Roofs initiative and hit the 250,000 mark (more than 1 gigawatt) at the beginning of 2012. The state's target now is 12 gigawatts of local green energy by 2020.

The Common Good. We defined the common good in the context of public policy: *to improve human welfare and to protect the natural world* (Chapter 1). What compels people to act to promote the common good? We can identify a number of values that can be the basis of stewardship and help us in our decisions involving both public policies and personal lifestyles:

- *Compassion for those less well off*, or those suffering from extreme poverty or other forms of deprivation—making compassion a major element of our foreign policy, which would energize our country's efforts in promoting sustainable development in the countries that are most in need. Compassion can motivate young people to spend two years of their life in the Peace Corps or a hunger relief agency, for example.

- *Concern for justice*—working to make just policies the norm in international economic relations. A concern for justice can move people to protest the placement of hazardous facilities in underprivileged communities.

- *Honesty*, or a concern for the truth—keeping the laws of the land and openly examining issues from different perspectives.

- *Sufficiency*—simplifying lifestyles rather than ever increasing our consumption; using no more than is necessary of Earth's resources.

- *Humility*—instead of demanding our rights, being willing to share and even defer to others.

- *Neighborliness*—being concerned for other members of our communities (even the global community), such that we do not engage in activities that are harmful to them, but instead make positive contributions to their welfare.

To achieve sustainability, we will have to couple values like these with the knowledge that can come from our understanding of how the natural world works and what is happening to it as a result of human activities. For this to work, people must be willing to grapple with scientific evidence and basic concepts and to develop a respect for the consensus that scientists have reached in areas of environmental concern. It has been our objective to convey that consensus to you in this text. Recall also that sustainable solutions have to be economically feasible, socially desirable, and ecologically viable (see Figure 1–9). This applies directly to the political decisions that are shaping our future in so many of our environmental concerns.

The good news is that many people are engaging environmental problems—working with governments and industries, bringing relief to people in need, and making exactly the kinds of decisions that are needed for effective stewardship. (See Stewardship, "The Tangier Island Covenant.") Community groups are organizing around sustainable principles, providing models of larger-scale changes that will be essential if we are to achieve a sustainable world. Traditional environmental organizations, as well as people from other sectors of society, are working together on the many issues that we have highlighted in this text, bringing about policy changes as well as changes in personal lifestyles.

STEWARDSHIP

The Chesapeake Bay, the largest estuary in the United States, is famous for its blue crabs and the unique "watermen" who fish for the crabs and other marine animals. Watermen are independent fishermen who have lived on the bay for generations and who defend their way of life fiercely. In recent years, the watermen of Tangier, a small island in the Virginia part of the eastern shore of the bay, have been in frequent battles with the Chesapeake Bay Foundation and the state regulators over crab and oyster laws that restrict when and where they can harvest. The watermen skirted restrictions by keeping undersized crabs and dredging clams and oysters out of season. They were also accustomed to throwing their trash overboard, causing the shores of the island to be littered with debris.

In 1997, the island community of Tangier became part of an unusual, but significant, process called *values-based stewardship*, a form of conflict resolution that requires parties to work within the value system of the community to bring about sustained changes in behavior toward the environment. Susan Drake, a doctoral candidate in environmental science at the University of Wisconsin, came to the island to carry out her thesis research on how conflicts are dealt with in homogeneous communities such as Tangier. Originally attracted to the project because of the highly religious community present on the island—the two churches are the center of life and authority there—Drake knew that most of the watermen considered themselves the beneficiaries of a rich bounty in an environment created by God. After working with watermen on their boats and spending time in their churches and with their families, Drake became concerned about the apparent lack of a connection between what the God-fearing watermen believed and what they actually did in their profession.

Convinced that she might be able to intervene in a positive way, Drake requested permission from her thesis committee to engage in "participatory action research," wherein the researcher may enter the conflict and attempt

to bring about a resolution. Working within the value system of the community—in this case, a value system based on faith—Drake addressed a joint meeting of the two churches on the island one Sunday morning. Referring to an image of Jesus standing behind a young fisherman at the wheel of a boat in rough seas that almost all Tangier watermen displayed in the cabins of their vessels, Drake addressed the issues of dumping trash overboard and skirting the laws. She showed the picture to the congregation and then put paper blinders over the eyes of Jesus as she listed the ways in which the watermen were disobeying the law. The day of her sermon coincided with a very high tide that gathered the debris from around the shores of the island and washed it up into the streets and people's yards.

The meeting ended with a call by Drake to the watermen to adopt a covenant she had drafted earlier with the help of some of the island's women. Fifty-six watermen went to the altar and promised to abide by the following covenant:

> The Watermen's Stewardship Covenant is a covenant among all watermen, regardless of their profession of religious faith. As watermen, we agree to (1) be good stewards of God's creation by setting a high standard of obedience to civil laws (fishery, boat, and pollution laws) and (2) commit to brotherly accountability. If any person who has committed to this covenant is overtaken in any trespass against this covenant, we agree to spiritually restore such a one in a spirit of gentleness. We also agree to fly a red ribbon on the antennas of our boats to signify that we are part of the covenant.

The next day, scores of watermen began flying red ribbons, taking trash bags out on their boats, and fishing according to the laws. Trucks began to pick up the trash on the island early in the morning. The two churches proceeded to work with the entire island community to draft a plan for the future sustainability of the island and its way of life. They formed the Tangier

Watermen's Stewardship for the Chesapeake in order to implement the plan, which is far reaching and involves the women of the island lobbying the state legislature and opening up lines of communication with scientists and government regulators involved in overseeing the bay's health. In recent years, Tangier Islanders have implemented cleanup, waste reduction, and recycling campaigns; environmental education programs; and a Coastal America Project for shoreline and wetland restoration. Island stewardship leaders also contacted a faith-based farming community in Pennsylvania, located on a tributary of the Chesapeake Bay watershed. After meeting with the watermen, the farmers decided to do all they could to prevent runoff from their farmlands that could make its way into the bay. The full story of the original covenant and later contact with the farm community is documented in a recent film.[14]

Values-based stewardship is a call to life-changing stewardship, put within the value system of a community. In this case, putting the call within the context of the faith of the watermen resulted in the establishment of structures within the community that show great promise in helping to bring the entire island toward a sustainable future.

The Tangier Island Covenant.
Susan Drake Emmerich.

[14]Center for Law and Culture. *When Heaven Meets Earth.* 2012. www.whenheavenmeetsearth.org.

Lifestyle Changes

What are some of the lifestyle changes that are needed, and are they in fact occurring? We should be encouraged and inspired by the literally millions of people in all walks of life who are acutely aware of the problems and who are making outstanding efforts to bring about solutions.

Every pathway toward solutions that we have mentioned represents the work of thousands of dedicated professionals and volunteers, ranging from scientists and engineers to businesspeople, lawyers, and public servants. Indeed, we are all involved, whether we recognize it or not. Simply by our existence on the planet, everything we do—the cars we drive, the products we use, the wastes we throw away, virtually every choice we make and action we take—has a certain environmental impact and a certain consequence for the future. Therefore, it is not a matter of choosing to have an effect, but a matter of what and how great that effect will be. It is a matter of each of us asking ourselves: Will I be part of the problem or part of the solution? The outcome will depend on how each of us responds to the challenges ahead.

There are a number of levels on which we can participate to work toward a sustainable future:

- Individual lifestyle choices
- Political involvement
- Membership and participation in nongovernmental environmental organizations
- Volunteer work
- Career choices

Lifestyle choices may involve such things as switching to a more fuel-efficient car or walking or using a bicycle for short errands; recycling paper, cans, and bottles; retrofitting your home with solar energy; starting a backyard garden and composting and recycling food and garden wastes into your soil; putting in time at a soup kitchen to help the homeless; choosing low-impact recreation, such as canoeing rather than jet-skiing; and living closer to your workplace.

Political involvement ranges from supporting and voting for particular candidates to expressing your support for particular legislation through letters or phone calls. In effect, you are exercising your citizenship on behalf of the common good.

Membership in nongovernmental environmental organizations can enhance both lifestyle changes and political involvement. As a member of an environmental organization, you will receive, and can help disseminate, information, making you and others more aware of specific environmental problems and things you can do to help. Also, your membership and contribution serve to support the lobbying efforts of the organization. A lobbyist representing a million-member organization that can follow up with thousands of phone calls and letters (and, ultimately, votes) can have a powerful impact.

In cases where enforcement of the existing law has been the weak link, some organizations, such as the Public Interest Research Groups, the Natural Resources Defense Council, and the Environmental Defense Fund, have been highly influential in bringing polluters or the government to court to see that the law is upheld. Again, this can be done only with the support of members.

Another form of involvement is *joining a volunteer organization*. Many effective actions that care for people and the environment are carried out by groups that depend on volunteer labor. Political organizations and virtually all NGOs depend highly on volunteers. Many helping organizations, such as those dedicated to alleviating hunger and to building homes for needy families **(Fig. 23–19)**, are adept at mobilizing volunteers to accomplish some vital tasks that often fill in the gaps left behind by inadequate public policies.

Finally, you may choose to devote your *career* to implementing solutions to environmental problems. Environmental careers go far beyond the traditional occupations of wildlife

Figure 23–19 Habitat for Humanity volunteers. By using charitable contributions and volunteer labor, Habitat for Humanity creates quality homes—more than 500,000 built throughout the world by 2012—that needy families can afford to buy with no-interest mortgages, enabling the money to be recycled to build more homes. Prospective home buyers participate in the construction and may gain valuable skills in the process.

ranger and park manager. Many lawyers, journalists, teachers, research scientists, engineers, medical personnel, agricultural extension workers, and others are focusing their talents and training on environmental issues or hazards. Business and job opportunities exist in pollution control, recycling, waste management, ecological restoration, city planning, environmental monitoring and analysis, nonchemical pest control, the production and marketing of organically grown produce, the manufacture of solar and wind energy components, and so on. Some developers concentrate on rehabilitation and the reversal of urban blight, as opposed to contributing more to urban sprawl. Some engineers are working on the development of pollution-free vehicles to help solve the photochemical smog dilemma of our cities. Indeed, it is difficult to think of a vocation that cannot be directed toward promoting solutions to environmental problems.

Human society is well into a new millennium, and it promises to be an era of rapid changes unprecedented in human history. Will we just survive, or will we thrive and become a sustainable global society? We are engaged in an environmental revolution—a major shift in our worldview and practice from seeing the natural world as resources to be exploited to seeing it as life's supporting structure that needs our stewardship. You are living in the early stages of this revolution, and we invite you to be one of the many who will make it happen. We close with this quote:

No one could make a greater mistake than he who did nothing because he could do only a little.

Edmund Burke, Irish orator, philosopher, and politician (1729–1797)

REVISITING THE THEMES

Sound Science

City planners and highway engineers have a new set of challenges as our society makes the changes that will move us toward greater sustainability. Creating sustainable cities and towns requires the dedicated work of social scientists, natural scientists, engineers, and many average people as they meet to provide a vision of what they would like their future to look like and then carry out that vision. Many scientific careers are great avenues for work on green solutions.

Sustainability

The story of urban sprawl is a story of poor or nonexistent public policy. Land use is traditionally under local control (if it is under any control at all), and absent any public policy dealing with land use, sprawl happens. In fact, public policy fed sprawl, as the Highway Trust Fund taxed gasoline sales in order to build roads and highways. Crafting effective policy to deal with sprawl means recognizing why people choose to live on the suburban fringes in low-density houses and changing the way we organize our communities. It requires changing public policy to do this, but smart-growth initiatives are taking root in state after state. Reining in sprawl and redesigning suburbia are essential moves toward greater sustainability of our urban surroundings.

Urban blight, too, was the outcome of poor or nonexistent public policy. Cities stood by helplessly and watched the flight of wealthier people and businesses out to the suburbs and even out of the country. In a similar way, developing-world cities are simply watching the poor flock in. It often takes people with compassion and a vision to bring about

change in these situations. Frequently, change can be brought about at the grassroots level, but it helps immensely to have leaders who care about the poor in their midst. Whether in developed or developing countries, addressing urban blight will make cities more livable and, therefore, more sustainable.

Sustainable solutions are becoming more mainstream, in the United States and around the world. Cities and towns are making an effort to respond to global climate change and limited resources as they engage in community efforts to promote sustainability and encourage sustainable lifestyle changes in their people.

Stewardship

The stewardship ethic includes a concern for justice. "White flight" into the suburbs has left minorities trapped in the inner cities, held down by racial and economic discrimination. Equally important is the migration of rural people into the cities of the developing world, where they are forced to live in slums and are unjustly denied ordinary social and environmental services. Both of these developments have energized many individuals and numerous NGOs to make common cause with those who are suffering the injustice and to help them mobilize their human and social resources to bring about change. A similar motivation to bring about change led Ryan Gravel to propose the Atlanta BeltLine, and Chattanooga Venture to begin the visioning process that eventually turned that city around. This is exactly what stewards should be doing. It is a fact that every one of us is given many opportunities to work together to create a truly sustainable society, and stewardship is the ethic that is there to guide our actions.

REVIEW QUESTIONS

1. What is the Atlanta BeltLine, and how does it propose to change urban Atlanta?

2. How did the structure of cities begin to change after World War II? What factors were responsible for the change?

3. Why are the terms *car dependent* and *urban sprawl* used to describe our current suburban lifestyle and urban layout?

4. What federal laws and policies tend to support urban sprawl?

5. Compare the costs and benefits of urban sprawl, looking at the environmental impacts and the impacts on quality of life.

6. What is *smart growth*? What are four smart-growth strategies that address urban sprawl?

7. What services do local governments provide, and what is their prime source of revenue for these services?

8. What is meant by *erosion of a city's tax base*? Why does it follow from exurban migration? What are the results?

9. How do exurban migration, urban sprawl, and urban decay become a vicious cycle?

10. What is meant by *economic exclusion*? How is it related to the problems of crime and poverty in cities?

11. What are the reasons for urban blight in the cities of many developing countries? What do the people in the shantytowns need most?

12. What are some characteristics of livable cities?

13. What is the Sustainable Communities Program? How does Chattanooga illustrate the program?

14. What are some stewardship virtues, and what role could they play in fostering sustainability?

15. State five levels at which people can participate in working toward a sustainable future.

THINKING ENVIRONMENTALLY

1. Interview an older person (say, 75 years of age or older) about what his or her city or community was like before 1950 in terms of meeting people's needs for shopping, participating in recreation, getting to school or work, and so on. Was it necessary to have a car?

2. Do a study of your region. What aspects of urban sprawl and urban blight are evident? What environmental and social problems are evident? Are they still going on? Are efforts being made to correct them?

3. Identify nongovernmental organizations in your region that are working toward reining in urban sprawl or toward preserving or bettering the city. What specific projects are under way in each of these areas? What roles are local governments playing in the process? How can you become involved?

MAKING A DIFFERENCE

1. Analyze your environmental impacts as a consumer. What modes of transportation do you use, where do you shop, what foods do you consume or throw away, what appliances do you own, how do you heat or cool your home, and how much water do you use? Consider especially your impact on the four leading consumption-related environmental problems: air pollution, global climate change, changes in natural habitats, and water pollution.

2. Investigate land-use planning and policies in your region. Is urban sprawl or urban blight a problem? Become involved in zoning issues, and support zoning and policies that will promote, rather than hinder, ecologically sound land use (*smart growth*).

3. Contact local conservation organizations or land trusts, and offer your help in protecting and preserving open land in your region.

4. Find out if there is a Habitat for Humanity project or group in your area, and if there is, volunteer some time to support its work.

MasteringEnvironmentalScience®

Go to **www.masteringenvironmentalscience.com** for practice quizzes, Pearson eText, videos, current events, and more.

APPENDIX A

Environmental Organizations

This appendix presents a list of selected nonprofit, nongovernmental organizations that are active in environmental matters. Listed are national organizations as well as some smaller, specialized ones. These organizations offer a variety of fact sheets, brochures, newsletters, publications, educational materials, and annual reports, most of which are available on the Internet. Some organizations have internship positions available for those wishing to do work for an environmental group. Web addresses are provided in the list. Most of the Web sites also provide connections via Facebook, Twitter, RSS feeds, Flickr, and YouTube. Also, a useful directory of many more environmental organizations can be accessed at **http://environmentdirectorywildlifeconservation.com**

AMERICAN FARMLAND TRUST. Focuses on the preservation of American farmland. 1200 18th St. NW, Suite 800, Washington, DC 20036. **www.farmland.org**

AMERICAN LUNG ASSOCIATION. Research, education, legislation, lobbying, advocacy: indoor and outdoor air pollution effects, smoking. 1301 Pennsylvania Ave. NW, Suite 800, Washington, DC 20004. **www.lung.org**

AMERICAN RIVERS, INC. Mission is "to preserve and restore America's rivers' systems and to foster a river stewardship ethic." Education, litigation, lobbying: wild and scenic rivers, hydropower relicensing. Information on specific rivers available. 1101 14th St. NW, Suite 1400, Washington, DC 20005. **www.amrivers.org**

BREAD FOR THE WORLD. National advocacy group that lobbies for legislation dealing with hunger. 425 3rd St. SW, Suite 1200, Washington, DC 20024. **www.bread.org**

CENTER FOR CLIMATE AND ENERGY SOLUTIONS. The successor to the Pew Center on Global Climate Change, this organization is an independent, nonpartisan, nonprofit organization working to advance strong policy and action to address the twin challenges of energy and climate change. 2101 Wilson Blvd., Suite 550, Arlington, VA 22201. **www.c2es.org**

CENTER FOR SCIENCE IN THE PUBLIC INTEREST. Research and education: alcohol policies, food safety, health, nutrition, organic agriculture. 1220 L St. NW, Suite 300, Washington, DC 20005. **www.cspinet.org**

CHESAPEAKE BAY FOUNDATION. Research, education, and litigation: environmental defense and management of the Chesapeake Bay and surrounding area. Philip Merrill Environmental Center, 6 Herndon Ave., Annapolis, MD 21403. **www.cbf.org**

CLEAN WATER ACTION PROJECT. Lobbying, education, research: water quality. 1010 Vermont Ave. NW, Suite 400 Washington, DC 20005. **www.cleanwateraction.org**

COMMON CAUSE. Lobbying: government reform, energy reorganization, clean air. 1133 19th Street NW, Washington, DC 20036. **www.commoncause.org**

CONGRESS WATCH. Lobbying: consumer health and safety, pesticides (a function of Public Citizen). 215 Pennsylvania Ave. SE, Washington, DC 20003. **www.citizen.org/congress**

CONSERVATION INTERNATIONAL. Education and research to preserve and promote awareness about the world's most endangered biodiversity. 2011 Crystal Dr., Suite 500, Arlington, VA 22202. **www.conservation.org**

CORAL REEF ALLIANCE. Works with the diving community and others to promote coral reef conservation around the world. 351 California St., Suite 650, San Francisco, CA 94104. **www.coral.org**

CLIMATE AND ENERGY. Research, education, and advocacy: affordable, clean and sustainable energy (a function of Public Citizen). 215 Pennsylvania Ave. SE, Washington, DC 20003. **www.citizen.org/cmep**

DEFENDERS OF WILDLIFE. Research, education, and lobbying: all native animals and plants, especially endangered species. 1130 17th St. NW, Washington, DC 20036. **www.defenders.org**

DUCKS UNLIMITED, INC. Hunters working to fulfill the annual life-cycle needs of North American waterfowl by protecting, enhancing, restoring, and managing important wetlands and associated uplands. One Waterfowl Way, Memphis, TN 38120. **www.ducks.org**

EARTHWATCH INSTITUTE. Environmental research is encouraged by organizing teams to make expeditions to various locations all over the world. Team members contribute to the expenses and spend from several weeks to several months on the site. 114 Western Ave., Boston, MA 02134. **www.earthwatch.org**

ENVIRONMENTAL DEFENSE FUND. Research, litigation, and lobbying: cosmetics safety, drinking water, energy, transportation, pesticides, wildlife, air pollution, cancer prevention, radiation. 257 Park Ave. South, New York, NY 10010. **www.edf.org**

ENVIRONMENTAL LAW INSTITUTE. Training, educational workshops, and seminars for environmental professionals, lawyers, and judges on institutional and legal issues affecting the environment. 200 L St. NW, Suite 620, Washington, DC 20036. **www.eli.org**

FOREST STEWARDSHIP COUNCIL. Promotes sustainable forest management and a system of certification of sustainable forest products. 212 Third Ave. North, Suite 504, Minneapolis, MN 55401. **www.fscus.org**

FREEDOM FROM HUNGER. Develops programs for the elimination of hunger worldwide. 1644 DaVinci Court, Davis, CA 95618. **www.freedomfromhunger.org**

FRIENDS OF THE EARTH. Research, lobbying: all aspects of energy development, preservation, restoration, and rational use of the Earth. 1100 15th St. NW, 11th Floor, Washington, DC 20005. **www.foe.org**

GREENBELT MOVEMENT. Environmetal organization that empowers communities, especially women, to improve livelihoods by conserving the environment, especially by planting trees in Africa. Adams Arcade, Kilimani Lane off Elgeyo Marakwet Rd. PO BOX 67545-00200, Nairobi, Kenya. **www.greenbeltmovement.org**

GREENPEACE, USA, INC. An international environmental organization dedicated to protecting the planet through nonviolent, direct action providing public education, scientific research, and legislative lobbying. 702 H St. NW, Washington, DC 20001. **www.greenpeace.org**

GROWING POWER, an urban farm organization begun in Milwaukee, WI, that grows food and provides youth employment and education in sustainable urban farming. 5500 W.

Silver Spring Drive, Milwaukee, Wisconsin 53218. www
.growingpower.org/about_us.htm

HABITAT FOR HUMANITY INTERNATIONAL. Fosters homebuilding and ownership for the poor. Education on housing issues. 121 Habitat St., Americus, GA 31709-3498. www.habitat.org

HEIFER PROJECT INTERNATIONAL. Works throughout the developing world to bring domestic animals to poor families. 1 World Ave., Little Rock, AR 72203. www.heifer.org

INSTITUTE FOR LOCAL SELF-RELIANCE. Research and education: appropriate technology for community development. 2001 S St. NW, Suite 570, Washington, DC 20009. www.ilsr.org

INTERNATIONAL FOOD POLICY RESEARCH INSTITUTE. Mission is to identify and analyze policies for sustainably ending hunger and poverty. 2033 K St. NW, Washington, DC 20006. www .ifpri.org

IZAAK WALTON LEAGUE OF AMERICA, INC. Research, education, endowment grants: conservation, air and water quality, streams. 707 Conservation Lane, Gaithersburg, MD 20878. www.iwla.org

KUPENDA FOR THE CHILDREN. Bringing help to the handicapped children of the Malindi region of the Kenyan coast, Kupenda organizes sponsorships to educate and house the children. PO Box 473, Hampton, NH 03843. www.kupenda.org

LAND TRUST ALLIANCE. Works with more than 1,700 local and regional land trusts to enhance their work. 1660 L St. NW, Suite 1100, Washington, DC 20036. www.landtrustalliance.org

LEAGUE OF CONSERVATION VOTERS. Political arm of the environmental community. Works to elect candidates to the U.S. House and Senate who will vote to protect the nation's environment and holds them accountable by publishing the National Environmental Scorecard each year, available from its Web site. 1920 L St. NW, Suite 800, Washington, DC 20036. www.lcv.org

LEAGUE OF WOMEN VOTERS OF THE UNITED STATES. Education and lobbying, general environmental issues. Publications on groundwater, agriculture, and farm policy, with additional topics available. 1730 M St. NW, Suite 1000, Washington, DC 20036. www.lwv.org

MANZANAR PROJECT. Provides low-tech solutions to hunger and poverty. Funding for women to plant mangroves along the Eritrean coast, which improves fisheries, food for domestic animals, and family income. PO Box 98, Gloucester, MA 01931. www.themanzanarproject.com

MILLENNIUM GOALS. The United Nations site providing information on and monitoring progress toward achieving the Millennium Development Goals. www.un.org/millenniumgoals

MILLENNIUM ECOSYSTEM ASSESSMENT (MA). An international work program designed to meet the needs of decision makers and the public for scientific information concerning the consequences of ecosystem change for human well-being and options for responding to those changes. The MA completed its numerous reports in 2005, many of which are referenced in this edition of the text. www.millenniumassessment.org

NATIONAL AUDUBON SOCIETY. Research, lobbying, education, litigation, and citizen action: broad-based environmental issues. 225 Varick St., New York, NY 10014. www.audubon.org

NATIONAL COUNCIL FOR SCIENCE AND THE ENVIRONMENT. Mission is to improve the scientific basis for making decisions about environmental issues. One important service provides access to the Congressional Research Service (briefings on environmental issues and the legislation addressing them). 1101 17th Street NW, Suite 250, Washington, DC 20036. ncseonline.org

NATIONAL PARK FOUNDATION. Partners with the National Park Service to encourage education, land acquisition, management of endowments, grant making. 1201 Eye St. NW, Suite 550B, Washington, DC 20005. www.nationalparks.org

NATIONAL PARKS CONSERVATION ASSOCIATION. A private organization with a mission to protect and enhance America's National Park System for present and future generations. 777 6th St. NW, Suite 700 Washington, DC 20001-3723. www .npca.org

NATIONAL RELIGIOUS PARTNERSHIP FOR THE ENVIRONMENT. An association of independent faith groups promoting stewardship of Creation in the context of different faiths. 110 Maryland Ave. NE, Suite 108, Washington, DC 20002. www.nrpe.org

NATIONAL WILDLIFE FEDERATION. Research, education, lobbying: general environmental quality, wilderness, and wildlife. 11100 Wildlife Center Drive, Reston, VA 20190. www.nwf.org

NATURAL RESOURCES DEFENSE COUNCIL. Research and litigation: water and air quality, land use, energy, pesticides, toxic waste. 40 W. 20th St., New York, NY 10011. www.nrdc.org

NATURE CONSERVANCY. A land conservation agency working around the world to protect ecologically important lands and waters for nature and people. 4525 N. Fairfax Dr., Suite 100, Arlington, VA 22203. www.nature.org

OXFAM AMERICA. Funding agency for projects to benefit the "poorest of the poor" in South America, Africa, India, Central America, the Caribbean, and the Philippines. Provides whatever resources are needed. 226 Causeway St., 5th Floor, Boston, MA 02114. www.oxfamamerica.org

PLANNED PARENTHOOD FEDERATION OF AMERICA. Education, reproductive health care services, and research: fertility control, family planning. 1110 Vermont Ave. NW, Suite 300, Washington, DC, 20005. www.plannedparenthood.org

POPULATION CONNECTION. Public education, lobbying, and research toward stable populations. 2120 L St. NW, Suite 500, Washington, DC 20037 www.populationconnection.org

THE POPULATION INSTITUTE. Education, research, and speaking engagements: population control. 107 2nd St. NE, Washington, DC 20002. www.populationinstitute.org

POPULATION REFERENCE BUREAU, INC. Organization engaged in collection and dissemination of objective population information. Excellent publications. 1875 Connecticut Ave. NW, Suite 520, Washington, DC 20009. www.prb.org

PUBLIC EMPLOYEES FOR ENVIRONMENTAL RESPONSIBILITY. A service organization that helps public employees work as "anonymous activists," enabling them to inform the public and their agencies about environmental abuses. 2000 P St. NW, Suite 240, Washington, DC 20036. www.peer.org

RACHEL CARSON COUNCIL, INC. Publication and distribution of information on pesticides and toxic substances, educational conferences, and seminars. PO Box 10779, Silver Spring, MD 20914. www.rachelcarsoncouncil.org

RAIN FOREST ACTION NETWORK. Information and educational resources: world's rain forests. 221 Pine St., 5th Floor, San Francisco, CA 94104. www.ran.org

RAINFOREST ALLIANCE. Education, medicinal plants project, timber project to certify "smart wood," and news bureau. 665 Broadway, Suite 500, New York, NY 10012. www.rainforest -alliance.org

RESOURCES FOR THE FUTURE. Research and education think tank: connects economics and conservation of natural resources, environmental quality. Publishes *Resources*. 1616 P St. NW, Washington, DC 20036. www.rff.org

SHACK/SLUM DWELLERS INTERNATIONAL. Promoting awareness of and support for the urban poor in developing countries; 14 federations and 34 country affiliates, head office in 1st Floor Campground Centre, Cnr Raapenberg and Surrey Rd. Mowbray, Cape Town 7700, S.A. **www.sdinet.org**

SIERRA CLUB. Education and lobbying: broad-based environmental issues. 85 Second St., 2nd Floor, San Francisco, CA 94105. **www.sierraclub.org**

TRUST FOR PUBLIC LAND. Works with citizen groups and government agencies to acquire and preserve open space. 101 Montgomery St., Suite 900, San Francisco, CA 94104. **www.tpl.org**

UNION OF CONCERNED SCIENTISTS. Connects citizens and scientists; works through programs and lobbying to inform the public and policy makers on science-related issues. 2 Brattle Square, Cambridge, MA 02238. **www.ucsusa.org**

U.S. PUBLIC INTEREST RESEARCH GROUP (PIRG). A federation of state PIRGs focusing on research, education, and lobbying: alternative energy, consumer protection, utilities regulation, public interest. 44 Winter St., 4th Floor, Boston, MA 02108. **www.pirg.org**

WATER ENVIRONMENT FEDERATION. Research, education, and lobbying to preserve the global water environment; provides technical education and training for water quality professionals. 601 Wythe St., Alexandria, VA 22314. **www.wef.org**

WILDERNESS SOCIETY. Research, education, and lobbying: wilderness, public lands; mission is to protect wilderness and inspire Americans to care for our wild places. 1615 M St. NW, Washington, DC 20036. **wilderness.org**

WORLD RESOURCES INSTITUTE. Global think tank that conducts research and publishes reports for educators, policy makers, and organizations on environmental issues. 10 G St. NE, Suite 800, Washington, DC 20002. **www.wri.org**

WORLD VISION. Christian humanitarian organization dedicated to working with children, families, and their communities worldwide to help them reach their full potential by tackling the causes of poverty and injustice. 34834 Weyerhaeuser Way S, Federal Way, WA 98001. **www.worldvision.org**

WORLDWATCH INSTITUTE. Research and education: energy, food, population, health, women's issues, technology, the environment. Publications: *State of the World, Vital Signs,* Worldwatch papers and reports. 1776 Massachusetts Ave. NW, Washington, DC 20036. **www.worldwatch.org**

WORLD WILDLIFE FUND. Preservation of wildlife habitats and protection of endangered species. 1250 24th St. NW, PO Box 97180, Washington, DC 20037. **www.worldwildlife.org**

APPENDIX B

Units of Measure

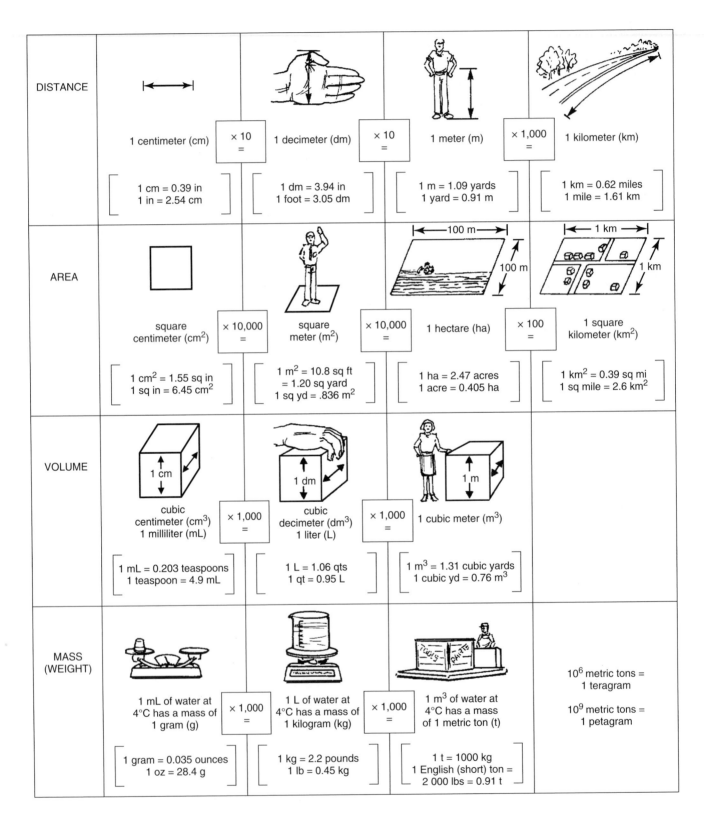

DISTANCE	1 centimeter (cm) ×10 =	1 decimeter (dm) ×10 =	1 meter (m) ×1,000 =	1 kilometer (km)
	1 cm = 0.39 in 1 in = 2.54 cm	1 dm = 3.94 in 1 foot = 3.05 dm	1 m = 1.09 yards 1 yard = 0.91 m	1 km = 0.62 miles 1 mile = 1.61 km
AREA	square centimeter (cm²) ×10,000 =	square meter (m²) ×10,000 =	1 hectare (ha) ×100 =	1 square kilometer (km²)
	1 cm² = 1.55 sq in 1 sq in = 6.45 cm²	1 m² = 10.8 sq ft = 1.20 sq yard 1 sq yd = .836 m²	1 ha = 2.47 acres 1 acre = 0.405 ha	1 km² = 0.39 sq mi 1 sq mile = 2.6 km²
VOLUME	cubic centimeter (cm³) 1 milliliter (mL) ×1,000 =	cubic decimeter (dm³) 1 liter (L) ×1,000 =	1 cubic meter (m³)	
	1 mL = 0.203 teaspoons 1 teaspoon = 4.9 mL	1 L = 1.06 qts 1 qt = 0.95 L	1 m³ = 1.31 cubic yards 1 cubic yd = 0.76 m³	
MASS (WEIGHT)	1 mL of water at 4°C has a mass of 1 gram (g) ×1,000 =	1 L of water at 4°C has a mass of 1 kilogram (kg) ×1,000 =	1 m³ of water at 4°C has a mass of 1 metric ton (t)	10⁶ metric tons = 1 teragram 10⁹ metric tons = 1 petagram
	1 gram = 0.035 ounces 1 oz = 28.4 g	1 kg = 2.2 pounds 1 lb = 0.45 kg	1 t = 1000 kg 1 English (short) ton = 2 000 lbs = 0.91 t	

Energy Units and Equivalents

1 Calorie, food calorie, or kilocalorie – The amount of heat required to raise the temperature of one kilogram of water one degree Celsius (1.8°F)

1 BTU (British Thermal Unit) – The amount of heat required to raise the temperature of one pound of water one degree Fahrenheit.

1 joule – a force of one newton applied over a distance of one meter (a newton is the force needed to produce an acceleration of 1 m per sec to a mass of one kg)

1 Calorie = 3.968 BTUs = 4,186 joules

1 BTU = 0.252 Calories = 1,055 joules

1 Therm = 100,000 BTUs

1 Quad = 1 quadrillion BTUs

1 watt = standard of electrical power

1 watt = 1 joule per second

1 watt-hour (wh) = 1 watt for 1 hr. = 3.413 BTUs

1 kilowatt (kw) = 1000 watts

1 kilowatt-hour (kwh) = 1 kilowatt for 1 hr = 3,413 BTUs

1 megawatt (Mw) = 1,000,000 watts

1 megawatt-hour (Mwh) = 1 Mw for 1 hr. = 34.13 therms

1 gigawatt (Gw) = 1,000,000,000 watts or 1,000 megawatts

1 gigawatt-hour (Gwh) = 1 Gw for 1 hr = 34,130 therms

1 horsepower = .7457 kilowatts

1 horsepower-hour = 2545 BTUs

1 cubic foot of natural gas (methane) at atmospheric pressure = 1031 BTUs

1 gallon gasoline = 125,000 BTUs

1 gallon No. 2 fuel oil = 140,000 BTUs

1 short ton coal = 25,000,000 BTUs

1 barrel = 42 gallons

APPENDIX C

Some Basic Chemical Concepts

Atoms, Elements, and Compounds

All matter, whether gas, liquid, or solid; living or nonliving; or organic or inorganic, is composed of fundamental units called **atoms**. Atoms are extremely tiny. If all the world's people, 7 billion of us, were reduced to the size of atoms, there would be room for all of us to dance on the head of a pin. In fact, we would only occupy a tiny fraction (about 1/10,000) of the pin's head. Given the incredibly tiny size of atoms, even the smallest particle that can be seen with the naked eye consists of billions of atoms.

The atoms making up a substance may be all of one kind, or they may be of two or more kinds. If the atoms are all of one kind, the substance is called an **element**. If the atoms are of two or more kinds bonded together, the substance is called a **compound**.

Through countless experiments, chemists have ascertained that there are only 94 distinct kinds of atoms that occur in nature. (An additional 21 have been synthesized.) These natural, and most of the synthetic, elements are listed in Table C–1, together with their chemical symbols. By scanning the table, you can see that a number of familiar substances—such as aluminum, chlorine, carbon, oxygen, and iron—are elements; that is, they are a single, distinct kind of atom. However, most of the substances with which we interact in everyday life—such as water, stone, wood, protein, and sugar—are not on the list. Their absence from the list is indicative that they are not elements; rather, they are compounds, which means that they are actually composed of two or more different kinds of atoms bonded together.

TABLE C–1 The Elements

Element	Symbol	Atomic Number	Atomic Weight	Element	Symbol	Atomic Number	Atomic Weight
Actinium	Ac	89	(227)	Chlorine	Cl	17	35.5
Aluminum	Al	13	27.0	Chromium	Cr	24	52.0
*Americium	Am	95	(243)	Cobalt	Co	27	58.9
Antimony	Sb	51	121.8	Copper	Cu	29	63.5
Argon	Ar	18	39.9	*Curium	Cm	96	(245)
Arsenic	As	33	74.9	*Dubnium	Db	105	(262)
Astatine	At	85	(210)	Dysprosium	Dy	66	162.5
Barium	Ba	56	137.3	*Einsteinium	Es	99	(254)
*Berkelium	Bk	97	(245)	Erbium	Er	68	167.3
Beryllium	Be	4	9.01	Europium	Eu	63	152.0
Bismuth	Bi	83	209.0	*Fermium	Fm	100	(254)
*Bohrium	Bh	107	(262)	Fluorine	F	9	19.0
Boron	B	5	10.8	Francium	Fr	87	(223)
Bromine	Br	35	79.9	Gadolinium	Gd	64	157.3
Cadmium	Cd	48	112.4	Gallium	Ga	31	69.7
Calcium	Ca	20	40.1	Germanium	Ge	32	72.6
*Californium	Cf	98	(251)	Gold	Au	79	197.0
Carbon	C	6	12.0	Hafnium	Hf	72	178.5
Cerium	Ce	58	140.1	*Hassium	Hs	108	(265)
Cesium	Cs	55	132.9	Helium	He	2	4.00

TABLE C–1 The Elements (*continued*)

Element	Symbol	Atomic Number	Atomic Weight	Element	Symbol	Atomic Number	Atomic Weight
Holmium	Ho	67	164.9	Promethium	Pm	61	(145)
Hydrogen	H	1	1.01	Protoactinium	Pa	91	231.0
Indium	In	19	114.8	Radium	Ra	88	226.0
Iodine	I	53	126.9	Radon	Rn	86	(222)
Iridium	Ir	77	192.2	Rhenium	Re	75	186.2
Iron	Fe	26	55.8	Rhodium	Rh	45	102.9
Krypton	Kr	36	83.8	Ruthenium	Ru	44	101.1
Lanthanum	La	57	138.9	*Rutherfordium	Rf	104	(261)
*Lawrencium	Lr	103	(257)	Samarium	Sm	62	150.4
Lead	Pb	82	207.2	Scandium	Sc	21	45.0
Lithium	Li	3	6.94	*Seaborgium	Sg	106	(263)
Lutetium	Lu	71	175.0	Selenium	Se	34	79.0
Magnesium	Mg	12	24.3	Silicon	Si	14	28.1
Manganese	Mn	25	54.9	Silver	Ag	47	107.9
*Meitnerium	Mt	109	(265)	Sodium	Na	11	23.0
*Mendelevium	Md	101	(256)	Strontium	Sr	38	87.6
Mercury	Hg	80	200.6	Sulfur	S	16	32.1
Molybdenum	Mo	42	95.9	Tantalum	Ta	73	180.9
Neodymium	Nd	60	44.2	Technetium	Tc	43	98.9
Neon	Ne	10	20.2	Tellurium	Te	52	127.6
*Neptunium	Np	93	237.0	Terbium	Tb	65	158.9
Nickel	Ni	28	58.7	Thallium	Tl	81	204.4
Niobium	Nb	41	92.9	Thorium	Th	90	232.0
Nitrogen	N	7	14.0	Thulium	Tm	69	168.9
*Nobelium	No	102	(254)	Tin	Sn	50	118.7
Osmium	Os	76	190.2	Titanium	Ti	22	47.9
Oxygen	O	8	16.0	Tungsten	W	74	183.8
Palladium	Pd	46	106.4	Uranium	U	92	238.0
Phosphorus	P	15	31.0	Vanadium	V	23	50.9
Platinum	Pt	78	195.1	Xenon	Xe	54	131.3
Plutonium	Pu	94	(242)	Ytterbium	Yb	70	173.0
Polonium	Po	84	(210)	Yttrium	Y	39	88.9
Potassium	K	19	39.1	Zinc	Zn	30	65.4
Praseodymium	Pr	59	140.9	Zirconium	Zr	40	91.2

*Elements that do not occur in nature.
Parentheses indicate the most stable isotope of the element.

Atoms, Bonds, and Chemical Reactions

In chemical reactions, atoms are neither created nor destroyed, nor is one kind of atom changed into another. What occurs in chemical reactions, whether mild or explosive, is simply a rearrangement of the ways in which the atoms involved are bonded together. An oxygen atom, for example, may be combined and recombined with different atoms to form any number of different compounds, but a given oxygen atom always has been, and always will be, an oxygen atom. The same can be said for all the other kinds of atoms. To understand how atoms may bond and undergo rearrangement to form different compounds, it is necessary to examine several concepts concerning the structure of atoms.

Structure of Atoms

Every atom consists of a central core called the **nucleus** (not to be confused with a cell nucleus). The nucleus of an atom contains one or more **protons** and, except for hydrogen, one or more **neutrons**. Surrounding the nucleus are particles called **electrons**. Each proton has a positive (+) electric charge, and each electron has an equal and opposite negative (–) electric charge. Thus, in any atom, the charge of the protons may be balanced by an equal number of electrons, making the whole atom neutral. Neutrons have no charge.

Atoms of all elements have this same basic structure, consisting of protons, electrons, and neutrons. The distinction between atoms of different elements is in the number of protons. The atoms of each element have a characteristic number of protons that is known as the **atomic number** of the element. (See Table C–1.) The number of electrons characteristic of the atoms of each element also differs, in a manner corresponding to the number of protons. The combined total of protons and neutrons in the nucleus of an element is the **mass number**, or **atomic weight**. The general structure of the atoms of several elements is shown in **Figure C–1**.

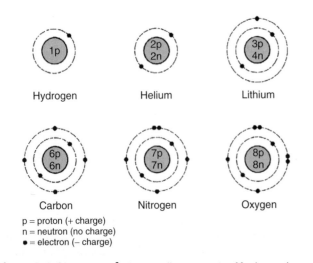

p = proton (+ charge)
n = neutron (no charge)
● = electron (– charge)

Figure C–1 Structure of atoms. All atoms consist of fundamental particles: protons (p), which have a positive electric charge; neutrons (n), which have no charge; and electrons, which have a negative charge.

The number of protons and electrons in an atom of an element (i.e., the atomic number of the element) determines the chemical properties of the element. The number of neutrons may vary. For example, most carbon atoms have six neutrons in addition to the six protons, as indicated in Figure C–1. But some carbon atoms have eight neutrons. Atoms of the same element that have different numbers of neutrons are known as **isotopes** of the element. The total number of protons plus neutrons is used to define different isotopes. For example, the usual isotope of carbon is referred to as carbon-12, while the isotope with eight neutrons is referred to as carbon-14. The chemical reactivities of different isotopes of the same element are identical. (However, certain other properties may differ.) Many isotopes of various elements prove to be radioactive, such as carbon-14. The atomic weights in Table C–1 are expressed as the average of the isotopes of an element as they occur in nature. The weights listed in parentheses are those of the most stable known isotope.

Bonding of Atoms

The chemical properties of an element are defined by the ways in which the element's atoms will react and form bonds with other atoms. By examining how atoms form bonds, we shall see how the number of electrons and protons determines these properties. There are two basic kinds of bonding: (1) **covalent bonding** and (2) **ionic bonding**.

In both kinds of bonding, it is important to recognize that electrons are not randomly distributed around the atom's nucleus. Rather, there are, in effect, specific spaces in a series of layers, or **orbitals**, around the nucleus. If an orbital is occupied by one or more electrons but is not filled, the atom is unstable; it will then tend to react and form bonds with other atoms to achieve greater stability. A stable state is achieved by having all the spaces in the orbital filled with electrons. It is important, however, to keep the charge neutral (i.e., the total number of electrons should be equal to that of the protons).

Covalent Bonding

These two requirements—filling all the spaces and keeping the charge neutral—may be satisfied by adjacent atoms sharing one or more pairs of electrons, as shown in **Figure C–2**. The sharing of a pair of electrons holds the atoms together in what it is called a **covalent bond**.

Covalent bonding, by satisfying the charge-orbital requirements, leads to discrete units of two or more atoms bonded together. Units of two or more covalently bonded atoms are called **molecules**. A few simple, but important, examples are shown in Figure C–2.

A chemical formula is simply a shorthand description of the number of each kind of atom in a given molecule. The element is given by the chemical symbol, and a subscript following the symbol gives the number of atoms of that element present in the molecule; if there is no subscript, there is only one atom of that chemical. A molecule with two or more different kinds of atoms may be called a compound, but a molecule composed of a single kind of atom—oxygen (O_2), for example—is still defined as an element.

Only a small number of elements—carbon, hydrogen, oxygen, nitrogen, phosphorus, and sulfur—have configurations of electrons that lend themselves readily to the formation of

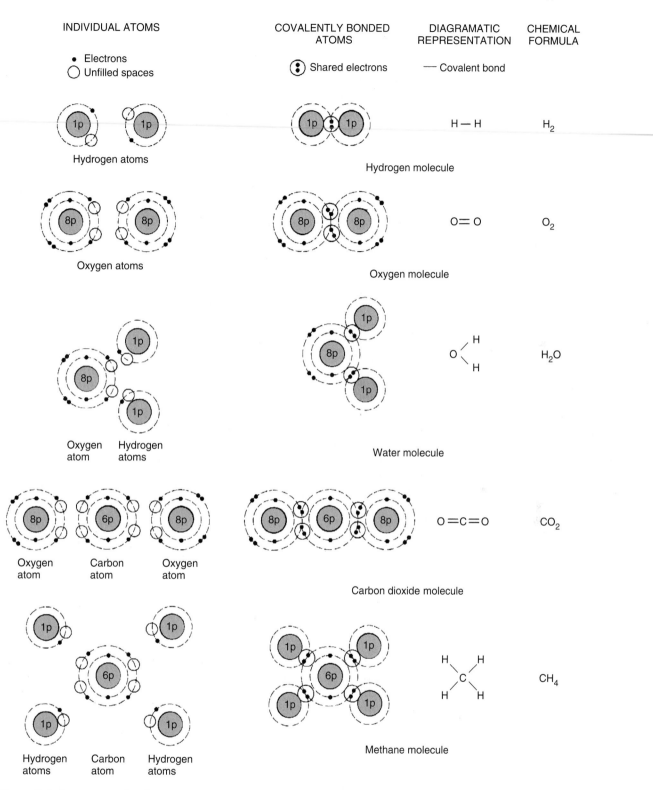

Figure C–2 Atoms to molecules. Molecules are composed of atoms covalently bonded to each other in stable configurations in which electrons are shared.

covalent bonds. Carbon in particular, with its ability to form four covalent bonds, can produce long, straight, or branched chains or rings **(Fig. C–3)**. Thus, an infinite array of molecules can be formed by using covalently bonded carbon atoms as a "backbone" and filling in the sides with atoms of hydrogen or other elements. Accordingly, it is covalent bonding among atoms of carbon and these few other elements that produces all natural organic molecules, those molecules that make up all the tissues of living things, and also synthetic organic compounds such as plastics.

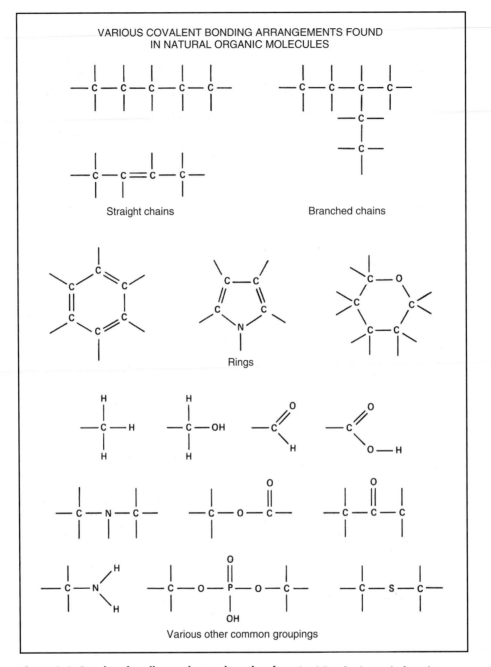

VARIOUS COVALENT BONDING ARRANGEMENTS FOUND
IN NATURAL ORGANIC MOLECULES

Straight chains   Branched chains

Rings

Various other common groupings

Figure C–3 Covalent bonding and organic molecules. The ability of carbon and a few other elements to readily form covalent bonds leads to an infinite array of complex organic molecules, which constitute all living things. A few major kinds of groupings are shown here. Note that each element forms a characteristic number of bonds: carbon, 4; nitrogen, 3; oxygen, 2; hydrogen, 1; sulfur, 2; and phosphorus, 5. Bonds left hanging (lines without atoms attached) indicate attachments to other atoms or groups of atoms.

Ionic Bonding

Another way in which atoms may achieve a stable electron configuration is to gain additional electrons to complete the filling of an orbital or to lose excess electrons in an incompleted orbital. In general, the maximum number of electrons that can be gained or lost by an atom is three. Therefore, an element's atomic number determines whether one or more electrons will be lost or gained. If an atom's outer orbital is one to three electrons short of being filled, it will always tend to gain additional electrons. If an atom has one to three electrons in excess of its last complete orbital, it will always tend to lose those electrons.

Of course, gaining or losing electrons results in the number of electrons in an atom being greater or less than the number of protons; the atom consequently has an electric charge. The charge will be one negative unit for each electron gained and one positive unit for each electron lost **(Fig. C–4)**. A covalently bonded group of atoms may acquire an electric charge in the same way. An atom or group of atoms that has acquired an electric charge in this way is called an **ion**, positive or negative. Ions are designated by a superscript following the chemical symbol, which denotes the number of positive or negative charges. The absence of superscripts indicates that the atom or molecule is neutral. Some important ions are listed in Table C–2.

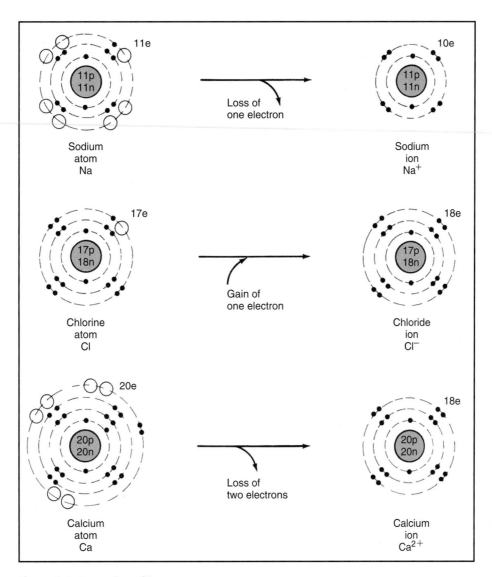

Figure C–4 Formation of ions. Many atoms will tend to gain or lose one or more electrons in order to achieve a state of complete (electron-filled) orbitals. In doing so, these atoms become positively or negatively charged ions, as indicated.

TABLE C–2 Ions of Particular Importance to Biological Systems			
Negative (–) Ions		**Positive (+) Ions**	
Phosphate	PO_4^{3-}	Potassium	K^+
Sulfate	SO_4^{2-}	Calcium	Ca^{2+}
Nitrate	NO_3^-	Magnesium	Mg^{2+}
Hydroxyl	OH^-	Iron	Fe^{2+}, Fe^{3+}
Chloride	Cl^-	Hydrogen	H^+
Bicarbonate	HCO_3^-	Ammonium	NH_4^+
Carbonate	CO_3^{2-}	Sodium	Na^+

Figure C–5 Crystals. Positive and negative ions bond together by their mutual attraction and form crystals.

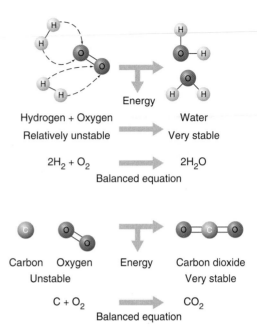

Hydrogen + Oxygen   Energy   Water
Relatively unstable   Very stable

$$2H_2 + O_2 \longrightarrow 2H_2O$$
Balanced equation

Carbon   Oxygen   Energy   Carbon dioxide
Unstable   Very stable

$$C + O_2 \longrightarrow CO_2$$
Balanced equation

Figure C–6 Stable bonding arrangements. Some bonding arrangements are more stable than others. Chemical reactions will go spontaneously toward more stable arrangements, releasing energy in the process. However, reactions may be driven in the opposite direction with suitable energy inputs.

Because unlike charges attract, positive and negative ions tend to join and pack together in dense clusters in such a way as to neutralize an atom's overall electric charge. This joining together of ions through the attraction of their opposite charges is called **ionic bonding**. The result is the formation of hard, brittle, more or less crystalline substances of which all rocks and minerals are examples (**Fig. C–5**).

It is significant to note that whereas covalent bonding leads to discrete molecules, ionic bonding does not. Any number and combination of positive and negative ions may enter into an ionic bond to produce crystals of almost any size. The only restriction is that the overall charge of positive ions be balanced by that of negative ions. Thus, substances bonded in an ionic fashion are properly called compounds but not molecules. When chemical formulas are used to describe such compounds, they define the ratio of various elements involved, not specific molecules.

Chemical Reactions

While atoms themselves do not change, the bonds between atoms may be broken and re-formed, with different atoms producing different compounds or molecules. This is essentially what occurs in all chemical reactions. What determines whether a given chemical reaction will occur? Earlier, we noted that atoms form bonds because they achieve a greater stability by doing so. However, some bonding arrangements may provide greater overall stability than others. Consequently, substances with relatively unstable bonding arrangements will tend to react to form one or more different compounds that have more stable bonding arrangements. Common examples are the reaction between hydrogen and oxygen to produce water and the reaction between carbon and oxygen to produce carbon dioxide (**Fig. C–6**).

Energy is always released when the constituents of a reaction gain greater overall stability, as indicated in the figure. Thus, when energy is released in a chemical reaction, the atoms achieve more stable bonding arrangements. Therefore, it may be said that chemical reactions always tend to go in a direction that releases energy as well as giving greater stability.

However, chemical reactions can be made to go in a reverse direction. With suitable energy inputs and under suitable conditions, stable bonding arrangements may be broken and less stable arrangements formed. As described in Chapter 3 of the text, this is the basis of the photosynthesis that occurs in green plants. Light energy causes the highly stable hydrogen-oxygen bonds of water to split and form less stable carbon-hydrogen bonds, thus creating high-energy organic compounds.

Obviously, much more could be said about the involvement of chemistry in environmental science. We would highly recommend a course (or several courses) in chemistry to the student who wants to gain a deeper understanding of how the subject is involved in environmental concepts and issues.

GLOSSARY

abiotic Pertaining to factors or things that are separate and independent from living things; nonliving.

abortion The termination of a pregnancy by some form of surgical or medicinal intervention.

acid Any compound that releases hydrogen ions when dissolved in water. Also, a water solution that contains a surplus of hydrogen ions.

acid deposition Any form of acid precipitation and also fallout of dry acid particles. (See **acid precipitation**.)

acid neutralizing capacity (ANC) In a water body, the capacity to neutralize acids due to the presence of **buffer** chemicals in solution.

acid precipitation Includes acid rain, acid fog, acid snow, and any other form of precipitation that is more acidic than normal (i.e., less than pH 5.6). Excess acidity is derived from certain air pollutants—namely, sulfur dioxide and oxides of nitrogen.

acquired immunodeficiency syndrome (AIDS) A fatal disease caused by the human immunodeficiency virus (HIV) and transmitted by sexual contact or the use of nonsterile needles (as in drug addiction).

activated-sludge system A system for removing organic wastes from water. The system uses microorganisms and active aeration to decompose such wastes. The system is used most as a means of secondary sewage treatment following the primary settling of materials.

active safety features Those safety features of nuclear reactors that rely on operator-controlled reactions, external power sources, and other features that are capable of failing. (See **passive safety features**.)

acute A reaction to a disease or chemical pollutant that brings on life-threatening reactions within hours or days.

adaptation An ecological or evolutionary change in structure or function that enables an organism to adjust better to its environment and hence enhances the organism's ability to survive and reproduce.

adiabatic cooling The cooling that occurs when warm air rises and encounters lower atmospheric pressure. **Adiabatic warming** is the opposite process, whereby cool air descends and encounters higher pressure.

Advanced Boiling-Water Reactor (ABWR) A third generation nuclear reactor design for power plants that incorporates many passive safety features. It is the design of choice for most new and planned reactors.

aeration *Soil*: The exchange within the soil of oxygen and carbon dioxide necessary for the respiration of roots. *Water*: The bubbling of air or oxygen through water to increase the amount of oxygen dissolved in the water.

aerosols Microscopic liquid and solid particles originating from land and water surfaces and carried up into the atmosphere.

Agenda 21 A definitive program to promote sustainable development that was produced by the 1992 Earth Summit in Rio de Janeiro, Brazil, and adopted by the UN General Assembly.

age structure Within a population, the different proportions of people who are old, middle aged, young adults, and children.

air pollutants Gases or aerosols in the atmosphere that have harmful effects on people or the environment.

Air Pollution Control Act Federal legislation introduced in 1955 that was the first attempt to bring air pollution under control.

air pollution disaster A short-term situation in industrial cities in which intense industrial smog brings about a significant increase in human mortality.

air toxics A category of air pollutants that includes radioactive materials and other toxic chemicals that are present at low concentrations but are of concern because they often are carcinogenic.

alga, pl. algae Any of numerous kinds of photosynthetic plants that live and reproduce while entirely immersed in water. Many species (the planktonic forms) exist as single cells or small groups of cells that float freely in the water. Other species (the seaweeds) may be large and attached.

algal bloom A relatively sudden development of a heavy growth of algae, especially planktonic forms. Algal blooms generally result from additions of nutrients, whose scarcity is normally limiting.

alkaline, alkalinity A *basic* substance; chemically, a substance that absorbs hydrogen ions or releases hydroxyl ions; in reference to natural water, a measure of the base content of the water.

altitude The vertical distance from sea level.

ambient standards Air-quality standards (set by the EPA) determining certain levels of pollutants that should not be exceeded in order to maintain environmental and human health.

amensalism An interaction between species whereby one species is harmed, while the other is unaffected.

American Recovery and Reinvestment Act of 2009 (Recovery Act) Legislation that provided money for a variety of programs, intended to stimulate the economy during the recent recession. Some $90 billion was directed into clean energy programs.

anaerobic Lacking oxygen.

anaerobic digestion The breakdown of organic material by microorganisms in the absence of oxygen. The process results in the release of methane gas as a waste product.

animal testing Procedures used to assess the toxicity of chemical substances, using rats, mice, and guinea pigs as surrogates for humans who might be exposed to the substances.

anthropogenic Referring to pollutants and other forms of impact on natural environments that can be traced to human activities.

AP1000 Advanced Passive Reactor A third generation nuclear power reactor manufactured by Westinghouse, the choice of most of the recent license applications in the United States.

aquaculture A propagation or rearing of any aquatic (water) organism in a more or less artificial system.

aquifer An underground layer of porous rock, sand, or other material that allows the movement of water between layers of nonporous rock or clay. Aquifers are frequently tapped for wells.

Arctic Climate Impact Assessment (ACIA) An assessment of the serious impacts of global warming on the Arctic region, published in 2004.

artificial selection Plant and animal breeders' practice of selecting individuals with the greatest expression of desired traits to be the parents of the next generation.

asthma A chronic disease of the respiratory system in which the air passages tighten and constrict, causing breathing difficulties; acute attacks may be life threatening.

Atlanta BeltLine A redevelopment initiative that employs an old railroad loop around the heart of Atlanta, providing parks, trails and rail transportation that will improve urban Atlanta.

Atlantic Meridional Overturning Circulation (AMOC) The pattern of ocean currents that involves global seawater movements between the surface and great depths. See **conveyor system**.

atmosphere The thin layer of gases surrounding Earth. Nitrogen, oxygen, water vapor, and carbon dioxide are major gases, while many minor gases are also present in trace amounts.

atmosphere-ocean general circulation model (AOGCM) A computer-based model for simulating long-term climatic conditions that combines global atmospheric circulation patterns with ocean circulation and cloud-radiation feedback.

atmospheric brown cloud (ABC) A 1- to 3-km-thick blanket of pollution that frequently sits over south and central Asia, made up of black carbon and soot from biomass and fossil fuel burning.

atom The fundamental unit of all elements.

atomic theory Theory that all matter is made up of particles called atoms, represented by more than 100 different elements in nature.

autotroph Any organism that can synthesize all its organic substances from inorganic nutrients, using light or certain inorganic chemicals as a source of energy. Green plants are the principal autotrophs.

Bacillus thuringensis (BT) A type of bacterium that produces a toxin that can kill the larvae of plant-eating insects; genes from this bacterium have been incorporated into plants using genomic techniques.

background radiation Radiation that comes from natural sources apart from any human activity. We are all exposed to such radiation.

bacterium, pl. bacteria Any of numerous kinds of prokaryotic microscopic organisms that exist as simple single cells that multiply by simple division. Along with fungi, bacteria constitute the decomposer component of ecosystems. A few species cause disease.

balance-of-trade deficit A deficit in money flow resulting from purchasing more from other countries than is sold to other countries.

BANANA ("build absolutely nothing anywhere near anything") An acronym expressing a common attitude about siting various facilities.

barrels of oil equivalent (BOE) A measurement that uses the energy content of a barrel of oil to compare different fossil fuel energy amounts.

Bar screen In a primary sewage treatment plant, a row of bars about one inch apart that provides initial screening of debris found in incoming sewage.

base Any compound that releases hydroxyl (OH–) ions when dissolved in water. A solution that contains a surplus of hydroxyl ions.

Basel Convention An international agreement that bans most international toxic-waste trade.

bed load The load of coarse sediment, mostly coarse silt and sand, that is gradually moved along the bottom of a riverbed by flowing water rather than being carried in suspension.

benefit-cost analysis A comparison of the benefits of any particular action or project to its costs. (See **cost-benefit ratio**.)

benthic plants Plants that grow under water and that are attached to or rooted in the bottom of the body of water. For photosynthesis, these plants depend on light penetrating the water.

benzene An organic chemical present in crude and refined oil products and tobacco smoke, a known human carcinogen.

best-demonstrated available technologies (BDATs) Treatment standards for hazardous wastes that have the effect of reducing their toxicity and mobility.

best management practice Farm management practice that serves best to reduce soil and nutrient runoff and subsequent pollution.

bioaccumulation The accumulation of higher and higher concentrations of potentially toxic chemicals in organisms. Bioaccumulation occurs in the case of substances that are ingested but that cannot be excreted or broken down (nonbiodegradable substances).

biochemical oxygen demand (BOD) The amount of oxygen that will be absorbed or "demanded" as wastes are being digested or oxidized in both biological and chemical processes. Potential impacts of wastes are commonly measured in terms of the BOD.

biodegradable Able to be consumed and broken down to natural substances, such as carbon dioxide and water, by biological organisms, particularly decomposers. Opposite: **nonbiodegradable**.

biodiesel Diesel fuel made from a mixture of vegetable oil (largely soybean oil) and regular diesel oil.

biodiversity The diversity of living things found in the natural world. The concept usually refers to the different species but also includes ecosystems and the genetic diversity within a given species.

biodiversity hot spots Thirty-four locations around the world that are home to some 60% of the world's known biodiversity.

biofuel Any fuel, but usually liquids, derived from agricultural crops (e.g., ethanol and plant oils).

biogas The mixture of gases—about two-thirds methane, one-third carbon dioxide, and small portions of foul-smelling compounds—resulting from the anaerobic digestion of organic matter. The methane content enables biogas to be used as a fuel.

biogeochemical cycle The repeated pathway of particular nutrients or elements from the environment through one or more organisms and back to the environment. The cycles include the carbon cycle, the nitrogen cycle, the phosphorus cycle, and so on.

biological evolution The theory, attributed to Charles Darwin and Alfred Russell Wallace, that all species now on Earth descended from ancestral species through a process of gradual change brought about by natural selection.

biological nutrient removal (BNR) A process employed in sewage treatment to remove nitrogen and phosphorus from the effluent coming from secondary treatment.

biological wealth The life-sustaining combination of commercial, scientific, and aesthetic values imparted to a region by its biota.

biomagnification Bioaccumulation occurring through several levels of a food chain.

biomass The mass of biological material. Usually the total mass of a particular group or category (e.g., biomass of producers).

biomass energy Energy produced by burning plant-related materials such as firewood.

biomass fuels Fuels in the form of plant-related material (e.g., firewood, grasses, organic farm wastes).

biomass pyramid The structure that is obtained when the respective biomasses of producers, herbivores, and carnivores in an ecosystem are compared. Producers have the largest biomass, followed by herbivores and then carnivores.

biome A group of ecosystems that are related by having a similar type of vegetation governed by similar climatic conditions. Examples include prairies, deciduous forests, arctic tundra, deserts, and tropical rain forests.

biopesticide Naturally occurring substances that are designed to control pests; such substances usually clear the EPA registration process quickly. See **Bacillus thuringensis**.

bioremediation The use of microorganisms for the decontamination of soil or groundwater. Usually involves injecting organisms or oxygen into contaminated zones.

biosolid In sewage treatment, biosolids are the relatively stable organic matter that remains after anaerobic digestion has taken place in sludge digesters.

biosphere The overall ecosystem of Earth. The sum total of all the biomes and smaller ecosystems, which ultimately are all interconnected and interdependent through global processes such as the water cycle and the atmospheric cycle.

biota The sum total of all living organisms. The term usually is applied to the setting of natural ecosystems.

biotechnology The use of genetic engineering techniques, enabling researchers to introduce new genes into food crops and domestic animals and thereby produce valuable products and more nutritious crops.

biotic community All the living organisms (plants, animals, and microorganisms) that live in a particular area.

biotic potential Reproductive capacity. The potential of a species for increasing its population and/or distribution. The biotic potential of every species is such that, given optimum conditions, its population will increase. (Contrast **environmental resistance.**)

biotic structure The organization of living organisms in an ecosystem into groups such as producers, consumers, detritus feeders, and decomposers.

birth control Any means, natural or artificial, that may be used to reduce the number of live births.

bisphenol A (BPA) A starter material for many plastics that is suspected of contributing to obesity, diabetes, altered sexual behavior, and cancer because of its estrogenic activity. See **endocrine disruptor.**

Blue Revolution A proposed radical change in managing water supplies intended to make more water available for human use.

blue water In the hydrologic cycle, precipitation and its further movement in infiltration, runoff, surface water, and groundwater; liquid water wherever it occurs.

BOD See **biochemical oxygen demand.**

body mass index A measurement used to calculate overweight and obesity; the ratio of weight in kilograms divided by the square of height in meters.

boiling-water reactor A nuclear reactor that employs boiling water to create steam, which is circulated to turbines that generate electricity; the water is cooled and returned to the reactor tank.

bottle law, bottle bill A law that provides for the recycling or reuse of beverage containers, usually by requiring a returnable deposit upon purchase of the item.

bottom-up regulation Basic control of a natural population that occurs as a result of the scarcity of some resource.

breeder reactor A nuclear reactor that, in the course of producing energy, also converts nonfissionable uranium-238 into fissionable plutonium-239, which can be used as fuel. Hence, a reactor that produces at least as much nuclear fuel as it consumes. Also called a **fast-neutron reactor.**

broad-spectrum pesticides Chemical pesticides that kill a wide range of pests. These pesticides also kill a wide range of nonpest and beneficial species; therefore, they may lead to environmental disturbances and resurgences. The opposite of narrow-spectrum pesticides.

brownfields Abandoned, idled, or underused industrial and commercial facilities whose further development is inhibited because of real or perceived chemical contamination.

BTU (British thermal unit) A fundamental unit of energy in the English system. The amount of heat required to raise the temperature of 1 pound of water 1 degree Fahrenheit.

buffer A substance that will maintain a constant pH in a solution by absorbing added hydrogen or hydroxyl ions. An example is the bicarbonate ion, the most important buffer in blood.

Bureau of Land Management (BLM) A federal agency that manages public lands other than national forests and national parks. The agency leases land to ranchers for grazing, among other activities.

bush meat Wild game that is harvested for food, especially in developing countries; often involves poaching and overexploitation.

calorie A fundamental unit of energy. The amount of heat required to raise the temperature of 1 gram of water 1 degree Celsius. All forms of energy can be converted to heat and measured in calories. Calories used in connection with food are kilocalories, or "big" calories, the amount of heat required to raise the temperature of 1 liter of water 1 degree Celsius.

cancer Any of a number of cellular changes that result in uncontrolled cellular growth.

cap-and-trade system A form of market-based environmental policy that sets a maximum level of pollutant (the cap), distributes pollution permits, and allows industries to trade permits to achieve their allowable pollution.

capillary water Water that clings in small pores, cracks, and spaces against the pull of gravity (e.g., water held in a sponge).

capital In classical economic theory, one of the three factors of production (alongside land and labor); more broadly, any form of wealth. See **intangible capital, ecosystem capital** and **natural capital.**

carbon dioxide sinks The oceans and terrestrial environments serve to absorb some of the carbon dioxide added to the atmosphere by human agency; these are called sinks.

carbon monoxide A highly poisonous gas, the molecules of which consist of a carbon atom with one oxygen attached. Not to be confused with carbon dioxide, a natural gas in the atmosphere.

carbon tax A tax levied on all fossil fuels in proportion to the amount of carbon dioxide that is released as they burn.

carcinogen, carcinogenic Having the property of causing cancer, at least in animals and, by implication, in humans.

carnivore An animal that feeds more or less exclusively on other animals.

carrying capacity The maximum population of a given species that an ecosystem can support without being degraded or destroyed in the long run. Represented symbolically by K.

Cartagena Protocol on Biosafety An international agreement governing trade in genetically modified organisms. The Cartagena Protocol was signed in January 2000.

Carter Doctrine As stated by President Jimmy Carter in 1980, the assertion that the United States would use military force to ensure our access to Persian Gulf oil.

catalyst A substance that promotes a given chemical reaction without itself being consumed or changed by the reaction. Enzymes are catalysts for biological reactions. Also, catalysts are used in some pollution control devices (e.g., the catalytic converter).

catalytic converter The device used by vehicle manufacturers to reduce the amounts of carbon monoxide, hydrocarbons, and nitrogen oxides in a vehicle's exhaust. The converter contains catalysts that changes these compounds to less harmful gases as the exhaust passes through.

catch shares See **individual quota system.**

cell The basic unit of life; the smallest unit that still maintains all the attributes of life. Many microscopic organisms consist of a single cell. Large organisms consist of trillions of specialized cells functioning together.

cell respiration The chemical process that occurs in all living cells whereby organic compounds are broken down to release energy required for life processes. For respiration, higher plants and animals require oxygen and release carbon dioxide and water as waste products, but certain microorganisms do not require oxygen.

cellulose The organic macromolecule that is the prime constituent of plant cell walls and hence the major molecule in wood, wood products, and cotton. Cellulose is composed of glucose molecules, but because it cannot be digested by humans, its dietary value is only as fiber, bulk, or roughage.

center-pivot irrigation An irrigation system consisting of a spray arm several hundred meters long, supported by wheels, pivoting around a central well from which water is pumped.

Centers for Disease Control and Prevention (CDC) An agency within the United States Department of Health and Human Services that is the lead federal agency for protecting the health and safety of the public.

centrally planned economy An economic system in which a ruling class makes most of the basic decisions about how the economy will be structured; typical of communist countries.

chain reaction A nuclear reaction wherein each atom that fissions (splits) causes one or more additional atoms to fission.

character displacement An effect of interspecific competition, whereby a physical change occurs in one or both competing species that lessens competition when the two species co-occur.

chemical energy The potential energy that is contained in certain chemicals; most importantly, the energy that is contained in organic compounds such as food and fuels and that may be released through respiration or burning.

chemical treatment The use of pesticides and herbicides to control or eradicate agricultural pests.

chemosynthesis The process whereby some microorganisms utilize the chemical energy contained in certain reduced inorganic chemicals (e.g., hydrogen sulfide) to produce organic matter.

chlorinated hydrocarbons Synthetic organic molecules in which one or more hydrogen atoms have been replaced by chlorine atoms. Chlorinated hydrocarbons are hazardous compounds because they tend to be nonbiodegradable and therefore to bioaccumulate; many have been shown to be carcinogenic. Also called **organochlorides**.

chlorine catalytic cycle In the stratosphere, a cyclical chemical process in which chlorine monoxide breaks down ozone.

chlorofluorocarbons (CFCs) Synthetic organic molecules that contain one or more of both chlorine and fluorine atoms and that are known to cause ozone destruction.

chlorophyll The green pigment in plants responsible for absorbing the light energy required for photosynthesis.

chromosome In a living organism, one of a number of structures on which genes are arranged in a linear fashion.

chronic A reaction to infection or pollutants that leads to a gradual deterioration of health over a period of years.

chronic obstructive pulmonary disease (COPD) A progressive lung disease that makes it hard to breathe, involving three syndromes: asthma, bronchitis, and emphysema.

CITES (Convention on International Trade in Endangered Species) An international treaty conveying some protection to endangered and threatened species by restricting trade in those species or their products.

clay Any particles in soil that are less than 0.004 millimeters in size.

Clean Air Act of 1970 Amended in 1977 and 1990, the act is the foundation of U.S. air pollution control efforts.

Clean Air Interstate Rule (CAIR) EPA rule published in 2005 establishing cap-and-trade programs for sulfur dioxide and nitrogen oxides in 28 eastern states.

Clean Water Act of 1972 (CWA) The cornerstone federal legislation addressing water pollution.

clear-cutting In harvesting timber, the practice of removing an entire stand of trees, leaving an ugly site that takes years to recover.

climate A general description of the average temperature and rainfall conditions of a region over the course of a year.

Climate Change Science Compendium 2009 A report issued by the United Nations Environment Program that addresses many new findings in climate change science since the IPCC's Fourth Assessment.

Climate Change Science Program (CCSP) See Global Change Research Program (GCRP).

climate sensitivity The response of climate to greenhouse gas concentrations, usually measured as a rise in temperature as a consequence of a rise in greenhouse gas concentration.

climax ecosystem The last stage in ecological succession. An ecosystem in which populations of all organisms are in balance with each other and with existing abiotic factors.

cogeneration Also known as **combined heat and power (CHP)**, the joint production of useful heat and electricity. For example, furnaces may be replaced with gas turbogenerators that produce electricity, while the hot exhaust still serves as a heat source. An important avenue of conservation, cogeneration effectively avoids the waste of heat that normally occurs at centralized power plants.

combined heat and power (CHP) See **cogeneration**.

combustion The practice of disposing of wastes by incineration in a special facility designed to handle large amounts of waste; modern combustion facilities also capture some of the energy by generating electricity on the site.

command-and-control strategy The basic strategy behind most public policy having to do with air and water pollution. The strategy involves setting limits on pollutant levels and specifying control technologies that must be used to achieve those limits.

commensalism A relation between two species in which one is benefited and the other is not affected.

common good A recognition of the need for protection of the natural world and improvement of human welfare, especially as applied to public policy.

commons, common-pool resources Resources (usually natural ones) owned by many people in common or, as in the case of the air or the open oceans, owned by no one but open to exploitation.

community In ecological contexts, a community is defined as the group of populations of different species of plants, animals, and microbes living together in one area.

community supported agriculture An arrangement between farms and local citizens where the citizens are shareholders in the farm, supporting the farmers and gaining access to fresh farm produce.

compaction Packing down. *Soil*: Packing and pressing out air spaces present in the soil. Compaction reduces soil aeration and infiltration and thus diminishes the capacity of the soil to support plants. *Trash*: Packing down trash to reduce the space that it requires.

competition In ecology, the interaction between members of the same species or between different species for essential resources or access to those resources.

competitive exclusion principle The concept that when two species compete very directly for resources, one eventually excludes the other from the area.

composting compost The process of letting organic wastes decompose in the presence of air. A nutrient-rich humus, or compost, is the resulting product.

composting toilet A toilet that does not flush wastes away with water but deposits them in a chamber where they will compost. (See **composting**.)

compound Any substance (gas, liquid, or solid) that is made up of two or more different kinds of atoms bonded together. (Contrast **element**.)

Comprehensive Environmental Response, Compensation, and Liability Act of 1980 (CERCLA) See **Superfund**.

Comprehensive Everglades Restoration Plan (CERP) Multi-billion-dollar plan to restore the Florida Everglades by addressing water flow and storage.

concepts Valid explanations of data from the natural world that often allow predictions but do not reach the level of validity of natural laws.

condensation The collecting of molecules from the vapor state to form the liquid state, as, for example, when water vapor condenses on a cold surface and forms droplets. Opposite: **evaporation**.

condenser A device that turns turbine exhaust steam into water in a power plant.

conditions Abiotic factors (e.g., temperature) that vary in time and space but are not used up by organisms.

confined animal feeding operation (CAFO) Animal farming where the animals are kept in crowded pens, producing wastes that are hard to manage and creating other health threats.

coniferous forest biome A biome occupying northern regions or higher mountainous altitudes dominated by coniferous trees.

conservation The management of a resource in such a way as to ensure that it will continue to provide maximum benefit to humans over the long run. *Energy*: Saving energy by cutting back on the use of heating, air conditioning, lighting, transportation, and so on and also by increasing the efficiency of energy use.

conservation biology The branch of science focused on the protection of populations and species.

conservation reserve An imaginary source of energy that results from policies promoting greater efficiency of energy use, resulting in a reduced energy requirement.

Conservation Reserve Program (CRP) A federal program whereby farmers are paid to place highly erodible cropland in a reserve. A similar program, the **Conservation Stewardship Program**, encourages farmers to conserve soil and water resources on their lands.

consumers In an ecosystem, those organisms that derive their energy from feeding on other organisms or their products.

consumption In ecological use, the process of feeding on organic matter as a source of energy; all heterotrophs are consumers.

consumptive use The harvesting of natural resources in order to provide for people's immediate needs for food, shelter, fuel, and clothing.

consumptive water use The use of water for such things as irrigation, wherein the water does not remain available for potential purification and reuse.

containment building The reinforced concrete building housing a nuclear reactor. Designed to contain an explosion, should one occur.

continental drift See **plate tectonics**.

contour farming The practice of cultivating land along the contours across, rather than up and down, slopes. In combination with strip cropping, contour farming reduces water erosion.

contraceptive A device or method employed by couples during sexual intercourse to prevent the conception of a child.

control rods A part of the core of a nuclear reactor; the rods of neutron-absorbing material that are inserted or removed as necessary to control the rate of nuclear fissioning.

covalent bond A chemical bond between two atoms, formed by sharing a pair of electrons between the atoms. Atoms of all organic compounds are joined by covalent bonds.

convection/convection currents The vertical movement of air due to atmospheric heating and cooling, moving generally in a west-to-east pattern.

Convention on Biological Diversity The biodiversity treaty signed by 158 nations at the Earth Summit in Rio de Janeiro in 1992 calling for various actions and cooperative steps between nations to protect the world's biodiversity.

conversion losses Unavoidable energy losses in energy production, usually due to the loss of heat energy.

conveyor system The giant pattern of oceanic currents that moves water masses from the surface to the depths and back again, producing major effects on the climate.

cooling tower A massive tower designed to dissipate waste heat from a power plant (or other industrial facility) into the atmosphere by condensing turbine exhaust steam into water.

Copenhagen Accord The agreement signed at the 2009 Copenhagen meeting of delegates of the Framework Convention on Climate Change, pledging actions to limit global temperature increases to 2°C above preindustrial levels.

coral bleaching A condition, usually brought on by excessively high temperatures, in hard corals where the coral animals expel their symbiotic algae and become white in appearance.

corporate average fuel economy (CAFE) Fuel economy standards for vehicles set by the National Highway Traffic Safety Administration.

corrosion In a nuclear power plant, the deterioration of pipes receiving hot, pressurized water in a circulation system that conveys heat from one part of the unit to another.

cosmetic spraying The spraying of pesticides to control pests that damage only the surface appearance of fruits and vegetables.

cost-benefit analysis See **benefit-cost analysis**.

cost-benefit ratio, benefit-cost ratio The value of the benefits to be gained from a project, divided by the costs of the project. If the ratio is greater than unity, the project is economically justified; if the ratio is less than unity, the project is not economically justified.

cost-effectiveness analysis An alternative to benefit-cost analysis, this approach accepts the merits of a goal and attempts to achieve the goal with the least costly alternative.

credit associations Groups of poor people who individually lack the collateral to assure loans but who collectively have the wherewithal to assure each other's loans. Associated with microlending.

criteria maximum concentration (CMC) A water-quality standard; the highest single concentration of a water pollutant beyond which environmental impacts may be expected.

criteria pollutants Certain pollutants whose levels are used as a gauge for the determination of air (or water) quality.

criterion continuous concentration (CCC) A water-quality standard; the highest sustained concentration of a water pollutant beyond which undesirable impacts may be expected.

critical habitat Under the Endangered Species Act, an area provided for a listed species where it can be found or could likely spread as it recovers.

critical number The minimum number of individuals of a given species that is required to maintain a healthy, viable population of the species. If a population falls below its critical number, it will likely become extinct.

Cross State Air Pollution Rule The ruling by the EPA that replaces the Clean Air Interstate Rule, requiring reductions in NO_x and SO_2 to aid states in the East to achieve ozone and particulate matter reductions.

crude birthrate (CBR) The number of births per 1,000 individuals per year.

crude death rate (CDR) The number of deaths per 1,000 individuals per year.

cryptogamic crust Made up of primitive plants, called cryptogams, the crust develops on hardened soils and can stabilize soil but also prevents water infiltration and seed germination.

cryosphere Snow and sea ice in polar regions; it is significant in its contribution to the planetary albedo as it reflects sunlight.

crystal A solid form of matter made up of regularly repeating units of atoms, ionic compounds, or molecules.

cultural control A change in the practice of growing, harvesting, storing, handling, or disposing of wastes that reduces their susceptibility or exposure to pests. For example, spraying the house with insecticides to kill flies is a chemical control; putting screens on the windows to keep flies out is a cultural control.

cultural eutrophication The process of natural eutrophication accelerated by human activities. (See **eutrophication**.)

curie The unit of measurement of radioactivity; one gram of pure radium 226 gives off one curie per second, or 37 billion spontaneous disintegrations into particles and radiation.

d–t reaction A nuclear fusion reaction employing the hydrogen isotopes deuterium and tritium; the reaction takes extremely high temperatures and is very difficult to contain.

DALY (disability-adjusted life year) In assessing the disease burden of different health risks, 1 DALY equals the loss of one healthy year of a person's life (whether through disability or loss of life).

DDT (dichlorodiphenyltrichloroethane) The first and most widely used of the synthetic organic pesticides belonging to the chlorinated hydrocarbon class.

dead zone A region of an aquatic ecosystem (usually a marine coastal area) where oxygen is low or absent from the bottom up well into the water column. Also called a **hypoxic area**.

debt crisis Brought about by the great debt incurred by many less-developed nations, this crisis involves countries building up debts that can never be paid off.

debt relief Steps, including cancellation of debts and other means of reducing poverty, taken by the World Bank and donor nations to alleviate the heavy debt of many poor nations.

declining tax base The loss of tax revenues that occurs when affluent taxpayers and businesses leave an area and property values subsequently decline. Also referred to as an **eroding tax base**.

decommissioning The retirement of nuclear power plants from service after 25 or more years because the effects of radiation will gradually make them inoperable.

decomposers Organisms whose feeding action results in decay or rotting of organic material. The primary decomposers are fungi and bacteria.

deep-well injection   A technique used for the disposal of liquid chemical wastes that involves putting them into deep dry wells where they permeate dry strata.

deforestation   The process of removing trees and other vegetation covering the soil and converting the forest to another land use, often leading to erosion and loss of soil fertility.

Delaney clause   A stipulation in the Federal Food, Drug, and Cosmetic Act of 1938 that prohibited the use of any food additive that is found to induce cancer in humans or animals; this has been replaced by the **negligible risk** standard.

demand-side policies   Energy policies that emphasize conserving energy and using more renewable and nuclear energy in the future.

demographic dividend   As a country goes through the demographic transition, there is a stage when the dependency ratio is low, giving the country the opportunity to spend more on poverty alleviation and economic growth.

demographic transition   The transition of a human population from a condition of a high birthrate and a high death rate to a condition of a low birthrate and a low death rate. A demographic transition may result from economic or social development.

demography/demographer   The study of population trends (growth, movement, development, and so on). People who perform such studies and make projections from them.

denitrification   The process of reducing oxidized nitrogen compounds present in soil or water to nitrogen gas to the atmosphere, conducted by certain bacteria and now utilized in the treatment of sewage effluents.

density-dependent   An attribute of population-balancing factors, such as predation, that increase and decrease in intensity in proportion to population density. May lead to population stability.

density-independent   Mortality to a population that occurs regardless of the population density. Tends to be destabilizing to populations.

Department of Transportation regulations (DOT regs)   Regulations intended to reduce the risk of spills, fires, and poisonous fumes by specifying the kinds of containers and methods of packing to be used in transporting hazardous materials.

dependency ratio   In a human population, the ratio of the non-working-age population (under 15 and over 65) to the working-age population.

desalination   A process that purifies seawater into high-quality drinking water via distillation or microfiltration.

desert biome   The pattern of living forms and habitat where rainfall is less than 25 centimeters (10 inches) per year.

desertification   The formation and expansion of degraded areas of soil and vegetation cover in arid, semiarid, and seasonally dry areas, caused by climatic variations and human activities. Desertified land is most commonly traced to erosion due to human mismanagement.

desert pavement   A covering of stones and coarse sand protecting desert soils from further wind erosion. The covering results from the differential erosion of finer material.

detritivores   Organisms (fungi, soil insects, bacteria) that feed on dead or decaying organic matter, important in completing the breakdown of organic matter to inorganic constituents.

detritus   The dead organic matter, such as fallen leaves, twigs, and other plant and animal wastes, that exists in any ecosystem.

detritus feeders   See **detritivores**.

deuterium (^{2}H)   A stable, naturally occurring isotope of hydrogen that contains one neutron in addition to the single proton normally in the nucleus.

developed countries   The high-income industrialized countries—the United States, Canada, western European nations, Japan, Korea, Australia, and New Zealand, as well as many middle-income countries such as those in Latin America, China, eastern Europe, and many Arab states.

developing countries   Countries in which the gross domestic product is less than $936 per capita. The category includes nations of Africa, India, other countries of southern Asia, and some former Soviet republics.

development   A term referring to the continued improvement of human well-being, usually in the developing countries.

Diclofenac   An anti-inflammatory drug often administered to animals but toxic to birds. Implicated in the die-off of vultures in the Indian subcontinent.

dioxin   A synthetic organic chemical of the chlorinated hydrocarbon class. Dioxin is highly toxic and a widespread environmental pollutant given off in the combustion of many organic compounds.

discharge permit   (technically called an **NPDES permit**) A permit that allows a company to legally discharge certain amounts or levels of pollutants into air or water.

discount rate   In economics, a rate applied to some future benefit or cost in order to calculate its present real value.

dish-engine system   A solar power system consisting of a set of parabolic concentrator dishes that focus sunlight onto a receiver, where the hot fluid is transferred to an engine that generates electricity.

disinfection   The killing (as opposed to removal) of microorganisms in water or other media where they might otherwise pose a health threat. For example, chlorine is commonly used to disinfect water supplies.

dissolved oxygen (DO)   Oxygen gas molecules (O_2) dissolved in water. Fish and other aquatic organisms depend on dissolved oxygen for respiration. Therefore, the concentration of dissolved oxygen is a measure of water quality.

distillation   A process of purifying water or some other liquid by boiling the liquid and recondensing the vapor. Contaminants remain behind in the boiler.

disturbance   A natural or human-induced event or process (e.g., a forest fire) that interrupts ecological succession and creates new conditions on a site.

DNA (deoxyribonucleic acid)   The natural organic macromolecule that carries the genetic or hereditary information for virtually all organisms.

Doha Development Round   Negotiations for trade begun in 2001 with the objective of lowering trade barriers around the world.

donor fatigue   The tendency of donor organizations or countries to withhold their aid because of long-term misuse of aid funds in the recipient countries.

dose   The mathematical product of the concentration of a hazardous material and the length of exposure to it. The effects of any given material or radiation correspond to the dose received.

dose-response assessment   Part of risk assessment, establishing the number of cancers that might develop due to exposure to different doses of a chemical.

drip irrigation   A method of supplying irrigation water through tubes that literally drip water onto the soil at the base of each plant.

drought   A local or regional lack of precipitation such that the ability to raise crops and water animals is seriously impaired.

drylands   Ecosystems characterized by low precipitation (25 to 75 cm per year) and often subject to droughts; some 41% of Earth's land is drylands.

Durban Platform   An agreement forged at Durban in 2011 at the 17th Conference of the Parties to the FCCC where delegates agreed to begin negotiations leading to a binding agreement to curb greenhouse gas emissions by 2015.

easement   In reference to land protection, an arrangement whereby a landowner gives up development rights into the future but retains ownership of the land.

ecdysone   The hormone that promotes molting in insects.

ecological economist   An economist who thoroughly integrates ecological and economic concerns; part of a new breed of economist who disagrees with classical economic theory.

ecological footprint   A concept for measuring the demand placed on Earth's resources by individuals from different parts of the world, involving calculations of the natural area required to satisfy human needs.

ecological pest control Natural control of pest populations through understanding various ecological factors and, so far as possible, utilizing those factors, as opposed to using synthetic chemicals.

ecological risk assessment (ERA) Risk assessment by the EPA designed to protect natural organisms and ecosystems from adverse effects of human activities.

ecological succession The gradual or sometimes rapid change in the species that occupy a given area, with some species invading and becoming more numerous, while others decline in population and disappear. Succession is caused by a change in one or more abiotic or biotic factors that benefit some species at the expense of others.

ecologists Scientists who study ecology.

ecology The study of all processes influencing the distribution and abundance of organisms and the interactions between living things and their environment.

economic exclusion The cutting off of access of certain ethnic or economic groups to jobs, a good education, and other opportunities and thus preventing them from entering the economic mainstream of society—a condition that prevails in poor areas of cities.

economics The social science that deals with the production, distribution, and consumption of goods and services.

economic production As seen by ecological economists, economic production is the process of converting the natural world to the manufactured world.

Economic Simplified Boiling Water Reactor (ESBWR) A Generation III nuclear power reactor proposed by General Electric; it has many more passive safety feataures than current Advanced Boiling Water Reactors.

economic systems The social and legal arrangements people in societies construct in order to satisfy their needs and wants and improve their well-being.

economic threshold The level of pest damage that, to be reduced further, would require an application of pesticides that is more costly than the economic damage caused by the pests.

economy See **economic systems.**

ecosystem An interactive complex of communities and the abiotic environment affecting them within a particular area. Ecosystems have characteristic forms, such as deserts, grasslands, tundra, deciduous forests, and tropical rain forests.

ecosystem capital The sum of goods and services provided by natural and managed ecosystems, both free of charge and essential to human life and well-being.

ecosystem management The management paradigm, adopted by all federal agencies managing public lands, that involves a long-term stewardship approach to maintaining the lands in their natural state.

ecotone A transitional region between two adjacent ecosystems that contains some of the species and characteristics of each one and also certain species of its own.

ecotourism The enterprises involved in promoting tourism of unusual or interesting ecological sites.

edges Breaks between habitats that may expose sensitive species to predators.

electrical power The amount of work done by an electric current over a given time.

electric generator A device that converts mechanical energy into electrical energy by way of wire coils that are rotated in a magnetic field.

electrolysis The use of electrical energy to split water molecules into their constituent hydrogen and oxygen atoms. Hydrogen gas and oxygen gas result.

electrolyzer In a solar receptor that is designed to produce fuel from water using solar energy, a catalyst that uses the electron energy to split water into hydrogen and oxygen.

electrons Fundamental atomic particles that have a negative electrical charge and a very small mass, approximately 9.1×10^{-31} kg. Electrons surround the nuclei of atoms and thus balance the positive charge of protons in the nucleus. A flow of electrons in a wire is identical to an electrical current.

element A substance that is made up of one and only one distinct kind of atom. (Contrast **compound.**)

El Niño/La Niña Southern Oscillation (ENSO) A climatic phenomenon characterized by major shifts in atmospheric pressure over the central equatorial Pacific Ocean. An **El Niño** involves the movement of unusually warm surface water into the eastern equatorial Pacific Ocean, and a **La Niña** reverses the pattern. These phenomena result in extensive disruption of weather around the world.

eluviation In soils, the process of leaching many minerals due to the downward movement of water; the E in E horizon of soil profiles stands for eluviation.

embrittlement Becoming brittle. Pertains especially to the reactor vessel of nuclear power plants gradually becoming prone to breakage or snapping as a result of continuous bombardment by radiation.

Emergency Planning and Community Right-to-Know Act of 1986 (EPCRA) See **SARA.**

emergent vegetation Aquatic plants whose lower parts are underwater but whose upper parts emerge from the water.

endangered species A species whose total population is declining to relatively low levels such that, if the trend continues, the species will likely become extinct.

Endangered Species Act The federal legislation that mandates the protection of species and their habitats that are determined to be in danger of extinction.

endemic species A species that is found only in a particular geographic area or habitat.

endocrine See **hormone.**

endocrine disrupters Any of a class of organic compounds, often pesticides, that are suspected of having the capacity to interfere with hormonal activities in animals.

endocrine system The collection of all of the hormones and glands that produce them in animal bodies; they control metabolism, growth, and reproduction.

energy The capacity to do work. Common forms of energy are light, heat, electricity, motion, and the chemical bond energy inherent in compounds such as sugar, gasoline, and other fuels.

energy carrier Something (e.g., electricity) that transfers energy from a primary energy source to its point of use.

energy conservation and efficiency All of those steps that are taken to avoid or reduce the use of energy, such as raising the Corporate Average Fuel Economy standards for cars and trucks.

energy flow The movement of energy through ecosystems, starting with the capture of solar energy by primary producers and ending with the loss of heat energy.

Energy Independence and Security Act of 2007 Legislation establishing energy policy by emphasizing demand-side policies such as conserving energy and using renewable energy sources.

Energy Policy Act of 2005 Legislation establishing U.S. energy policy for years to come, encouraging greater use of fossil fuels and nuclear power, and providing some support for energy conservation and renewable energy.

Energy Star Program The EPA's program focusing on energy conservation in buildings and in appliances. An Energy Star tag signals that the device or building incorporates energy-saving technology.

enhanced recovery In the removal of oil from an oil field, a process that increases recovery beyond secondary recovery by injecting carbon dioxide or hydraulic fluids into wells.

enrichment With reference to nuclear power, the separation and concentration of uranium-235 so that, in suitable quantities, it will sustain a chain reaction.

entomologist A scientist who studies insects and their life cycles, physiology, behavior, and economic significance.

entropy A degree of disorder; increasing entropy means increasing disorder.

environment The combination of all things and factors external to the individual or population of organisms in question.

environmental accounting A process of keeping national accounts of economic activity that includes gains and losses of environmental assets.

environmental health The study of the connections between hazards in the environment and human disease and death.

environmental impact The effect on the natural environment caused by human actions. Includes indirect effects, for example, through pollution, as well as direct effects such as cutting down trees.

environmental impact statement A study of the probable environmental impacts of a development project. The National Environmental Policy Act of 1968 (NEPA) requires such studies prior to proceeding with any project receiving federal funding.

environmentalism An attitude or movement involving concern about pollution, resource depletion, population pressures, loss of biodiversity, and other environmental issues. Usually also implies action to address those concerns.

environmentalist Any person who is concerned about the degradation of the natural world and is willing to act on that concern.

environmental justice The fair treatment and meaningful involvement of all people, regardless of race, color, national origin, or income, with respect to the development, implementation, and enforcement of environmental laws and regulations.

environmental movement The upwelling of public awareness and citizen action that began during the 1960s regarding environmental issues.

environmental public policy All of a society's laws and regulations that deal with that society's interactions with the environment. Its purpose is to promote the common good.

environmental racism Discrimination against people of color whereby hazardous industries are sited in nonwhite neighborhoods and towns.

environmental resistance The totality of factors such as adverse weather conditions, shortages of food or water, predators, and diseases that tend to cut back populations and keep them from growing or spreading. (Contrast **biotic potential**.)

Environmental Revolution In the view of some, a coming change in the adaptation of humans to the rising deterioration of the environment. The Environmental Revolution should bring about sustainable interactions with the environment.

environmental science The multidisciplinary branch of science concerned with environmental issues.

environmental tobacco smoke (ETS) See **secondhand smoke**.

enzyme A protein that promotes the synthesis or breaking of chemical bonds.

EPA (U.S. Environmental Protection Agency) The federal agency responsible for the control of all forms of pollution and other kinds of environmental degradation.

epidemiologic transition In human populations, the pattern of change in mortality from high death rates to low death rates; contributes to the demographic transition.

epidemiology, epidemiological study The study of the causes of disease (e.g., lung cancer) through an examination and comparison of large populations of people living in different locations or following different lifestyles or habits (e.g., smoking versus nonsmoking).

equilibrial species See *K*-strategists

equilibrium Referring to populations, the condition where births plus immigration are more or less equal to deaths plus emigration over time.

equilibrium climate sensitivity The sustained change in mean global temperature that occurs if atmospheric carbon dioxide is held at double preindustrial values, or 550 parts per million.

equity An ethical principle where people's needs are met in an impartial and just manner.

equity principle An ethical principle claiming that those who produce the most pollution should compensate those who produce

less, especially because the less fortunate often suffer the most from, say, global climate change.

erosion The process of soil particles being carried away by wind or water. Erosion moves the smaller particles first and hence degrades the soil to a coarser, sandier, stonier texture.

estimated reserves See **reserves**.

estuary A bay or river system open to the ocean at one end and receiving freshwater at the other. In the estuary, freshwater and saltwater mix, producing brackish water.

ethnobotany The study of the relationships between plants and people.

ethyl alcohol The main product of fermentation of sugars or starches, also called ethanol. It is a 2-carbon compound that can be burned in combustion engines as a substitute for gasoline.

euphotic zone In aquatic systems, the layer or depth of water through which an adequate amount of light penetrates to support photosynthesis.

eutrophic Characterized by nutrient-rich water supporting an abundant growth of algae or other aquatic plants at the surface.

eutrophication The process of becoming eutrophic.

evaporation The process whereby molecules leave the liquid state and enter the vapor or gaseous state, as, for example, when water evaporates to form water vapor. Opposite: **condensation**.

evaporative water loss The loss of moisture from soil by evaporation; it can be lessened by covering the soil with detritus or plant cover.

evapotranspiration The combination of evaporation and transpiration that restores water to the atmosphere.

evolution See **biological evolution**.

e-waste Any of a broad array of electronic devices that are discarded and often are hazardous to handle and difficult to recycle.

executive branch That facet of the government in the United States consisting of the president, his cabinet, and all administrative agencies.

exotic species A species introduced into a geographical area to which it is not native; also, species that are rare and subject to illegal trafficking.

experimentation In the practice of science, the testing of hypotheses by setting up situations where cause and effect are investigated, using careful measurements of conditions and responses.

exponential increase The growth produced when a base population increases by a given percentage (as opposed to a given amount) each year. An exponential increase is characterized by doubling again and again, each doubling occurring in the same period of time. It produces a J-shaped curve.

exposure assessment A part of risk assessment, identifying human groups exposed to a chemical and calculating the doses and length of time of their exposure.

extended product responsibility Voluntary or government programs that assign some responsibility for reducing the environmental impact of a product at various stages of its life cycle—especially at disposal.

external cost, externality Any effect of a business process not included in the usual calculations of profit and loss. Pollution of air or water is an example of a *negative* externality—one that imposes a cost on society that is not paid for by the business itself.

extinction The death of all individuals of a particular species. When this occurs, all the genes of that particular line are lost forever.

extreme poverty Poverty that is defined as living on less than $1.25 a day. In 2012, some 1.4 billion were in this condition.

exurban migration The pronounced trend since World War II of relocating homes and businesses from the central city and older suburbs to more outlying suburbs.

exurbs New developments beyond the traditional suburbs, from which most residents still commute to the associated city for work.

facilitation A mechanism of succession where earlier species create conditions that are more favorable to newer occupants and are replaced by them.

family planning Making contraceptives and other health and reproductive services available to couples in order to enable them to achieve a desired family size.

famine A severe shortage of food accompanied by a significant increase in the local or regional death rate.

FDA Food Safety Modernization Act A bill passed in 2010 that expanded the authority of the Food and Drug Administration over the food processing industry, in response to repeated outbreaks of food poisoning and pathogens in food.

fecal coliform test A test for the presence of *Escherichia coli*, the bacterium that normally inhabits the gut of humans and other mammals. A positive test result indicates sewage contamination and the potential presence of disease-causing microorganisms.

Federal Agricultural Improvement and Reform (FAIR) Act of 1996 Major legislation removing many subsidies and controls from farming.

Federal Food and Drug Administration (FDA) The federal agency with jurisdiction over foods, drugs, and all products that come in contact with the skin or are ingested.

feedback loops In ecological or environmental systems, feedback loops are processes that can amplify (positive loop) or dampen (negative loop) the behavior of the system. See **positive feedback**.

fermentation A form of respiration carried on by yeast cells and many bacteria in the absence of oxygen. Fermentation involves a partial breakdown of glucose (sugar) that yields energy for the yeast and the release of alcohol as a by-product.

fertility rate See **total fertility rate**.

fertility transition The pattern of change in birthrates in a human society from high rates to low; a major component of the demographic transition.

fertilizer Material applied to plants or soil to supply plant nutrients, most commonly nitrogen, phosphorus, and potassium but possibly others. *Organic* fertilizer is natural organic material, such as manure, that releases nutrients as it breaks down. *Inorganic* fertilizer, also called chemical fertilizer, is a mixture of one or more necessary nutrients in inorganic chemical form.

field scouts Persons trained to survey crop fields and determine whether applications of pesticides or other pest management procedures are actually necessary to avert significant economic loss.

FIFRA (Federal Insecticide, Fungicide, and Rodenticide Act) The key U.S. legislation passed to control pesticides.

fire climax ecosystems Ecosystems that depend on the recurrence of fire to maintain the existing balance.

first-generation pesticides Toxic inorganic chemicals that were the first used to control insects, plant diseases, and other pests. These pesticides included mostly compounds of arsenic and cyanide and various heavy metals, such as mercury and copper.

First Law of Thermodynamics The empirical observation, confirmed innumerable times, that energy is never created or destroyed but may be converted from one form to another (e.g., electricity to light). Also called the **Law of Conservation of Energy**. (See also **Second Law of Thermodynamics**.)

fishery Fish species being exploited, or a limited marine area containing commercially valuable fish.

fish ladder A stepwise series of pools on the side of a dam where the water flows in small falls that fish can negotiate.

fission The splitting of a large atom into two atoms of lighter elements. When large atoms such as uranium or plutonium fission, tremendous amounts of energy are released.

fission products Any and all atoms and subatomic particles resulting from splitting atoms in nuclear reactors or nuclear explosions. Practically all such products are highly radioactive.

fitness The state of an organism whereby it is adapted to survive and reproduce in the environment it inhabits.

flat-plate collector A solar collector that consists of a stationary, flat, black surface oriented perpendicular to the average angle of the sun. Heat absorbed by the surface is removed and transported by air or water (or some other liquid) flowing over or through the surface.

flood irrigation A technique of irrigation in which water is diverted from rivers by means of canals and is flooded through furrows in fields.

food aid Food of various forms that is donated or sold below cost to needy people for humanitarian reasons.

Food and Agriculture Organization of the United Nations (FAO) The UN agency responsible for coordinating international efforts to defeat hunger; also gathers and disseminates information about forests, fisheries, and farming.

food chain The transfer of energy and material through a series of organisms as each one is fed upon by the next.

Food Conservation and Energy Act of 2008 The latest farm bill, maintaining the existing high levels of support and subsidies to farms.

Food Quality Protection Act (FQPA) Passed in 1996, this act amended existing laws governing pesticide use and residues on foods, including the **Delaney clause**.

food security For families, the ability to meet the food needs of everyone in the family, providing freedom from hunger and malnutrition.

Forest Stewardship Council An alliance of organizations directed toward the certification of sustainable wood products.

food web The combination of all the feeding relationships that exist in an ecosystem.

fossil fuels Energy sources—mainly crude oil, coal, and natural gas—that are derived from the prehistoric photosynthetic production of organic matter on Earth.

FQPA (Food Quality Protection Act of 1996) Legislation that removed the Delaney clause and replaced many provisions of FIFRA.

fracking Refers to hydraulic fracturing, a technique for extracting natural gas and oil involving pumping fluids into a well to open up spaces that release gas or oil to be pumped back up the well.

fragile states Countries that are poor and conflict-prone, often lacking a stable government; an example is present-day Somalia.

fragmentation The division of a landscape into patches of habitat by road construction, agricultural lands, or residential areas.

Framework Convention on Climate Change (FCCC) A result of the 1992 Earth Summit, this international treaty was a start in negotiating agreements on steps to prevent future catastrophic climate change.

Framework Convention on Tobacco Control Adopted in 2003 and put into effect in 2005, this is a global treaty that aims to reduce the spread of smoking with the use of various strategies.

frankenfood A derogatory term used by opponents to genetic modification, applying to any food plant that has been developed with the use of genetic engineering.

free-market economy In its purest form, an economy in which the market itself determines what and how goods will be exchanged. The system is wholly in private hands.

freshwater Water that has a salt content of less than 0.1% (1,000 parts per million).

front The boundary where different air masses meet.

fuel cell A device that produces electrical energy by combining hydrogen and oxygen, used to power vehicles and other electrical machines.

fuel elements (fuel rods) The pellets of uranium or other fissionable material that are placed in tubes, or fuel rods, and that, together with the control rods, form the core of the nuclear reactor.

fuel load Dead trees and dense understories in many forestlands that tend to burn readily if ignited in a forest fire.

fuelwood The use of wood for cooking and heating; the most common energy source for people in developing countries.

fungus, pl. fungi Any of numerous species of molds, mushrooms, bracket fungi, and other forms of nonphotosynthetic plants. Fungi derive energy and nutrients by consuming other organic material. Along with bacteria, they form the decomposer component of ecosystems.

fusion   The joining together of two atoms to form a single atom of a heavier element. When light atoms such as hydrogen are fused, tremendous amounts of energy are released.

gasohol   A blend of 90% gasoline and 10% ethanol that can be substituted for straight gasoline. Gasohol serves to stretch gasoline supplies and enable gasoline to burn more cleanly.

gene   A segment of DNA that codes for one protein, which in turn determines a particular physical, physiological, or behavioral trait.

gene pool   The sum total of all the genes that exist among all the individuals of a species.

gene revolution   The widespread development of genetically modified crops.

genetically modified organism (GMO)   Any organism that has received genes from a different species through the techniques of genetic engineering; also called a transgenic organism.

genetic bank   The concept that natural ecosystems with all their species serve as a repository of genes that may be drawn upon to improve domestic plants and animals and to develop new medicines, among other uses.

genetic control   The selective breeding of a desired plant or animal to make it resistant to attack by pests. Also, attempting to introduce harmful genes—for example, those that cause sterility—into the pest populations.

genetic engineering   The artificial transfer of specific genes from one organism to another.

Genuine Progress Indicator (GPI)   An alternative measure of economic progress, the GPI calculates positive and negative economic activities, to reach a more realistic view of sustainable economic activity. It is an alternative to the **gross domestic product.**

geoengineering   Large-scale schemes, proposed to reduce global climate change, involving major intrusions into climatic or ecological systems.

Geographic Informations Systems   A digital technology enabling the location and imaging of landscapes by use of satellite signals. It employs hardware, software, and data to display many forms of geographic information.

geothermal   Refers to the naturally hot interior of Earth, where heat is maintained by naturally occurring nuclear reactions.

geothermal energy   Useful energy derived from water heated by the naturally hot interior of Earth.

geothermal heat pump   An energy system whereby heat can be obtained or deposited via pipes buried in the ground and an exchange process that allows heated or cooled air to be circulated via ductwork.

GLASOD   A map produced by the Global Assessment of Soil Degradation, prepared in the late 20th century and often cited but based on very suspect data.

Global Assessment of Land Degradation and Improvement (GLADA)   A new assessment of land degradation based on satellite remote-sensing techniques.

Global Change Research Program (GCRP)   The U.S. government-sponsored program that addresses a number of issues in climate change with research and reports.

global climate change   The cumulative effects of rising levels of greenhouse gases on Earth's climate. These effects include global warming, weather changes, and a rising sea level.

Global Climate Change Initiative   The Bush administration's voluntary response to climate change, involving a reduction in **emissions intensity.**

Global Forest Resources Assessment (FRA)   A periodic assessment of forest resources conducted by the UN Food and Agricultural Organization (FAO) and made the basis of regular FAO *State of the World's Forest* reports.

Global Gag Rule   The U.S. policy initiated by President Reagan and continued by other Republican presidents, whereby U.S. government aid is denied to any agency that provides abortions or abortion counseling.

globalization   The accelerating interconnectedness of human activities, ideas, and cultures, especially evident in economic and information exchange.

glucose   A simple sugar, the major product of photosynthesis. Glucose serves as the basic building block for cellulose and starches and as the major "fuel" for the release of energy through cell respiration in both plants and animals.

glyphosate (Roundup®)   A herbicide that kills most varieties of plants, used to control weeds with crops that have been bioengineered with resistance to the glyphosate.

golden rice   New strains of rice produced by genetic engineering that contain carotene (which the body uses to synthesize vitamin A) and iron, intended to compensate for nutritional deficiencies.

goods   Products, such as wood and food, that are extracted from natural ecosystems to satisfy human needs.

Grameen Bank   A banking network serving the poor with microloans, created by Muhammad Yunus of Bangladesh.

grassland biome   Geographical regions that have seasonal rainfall of 10 to 60 inches/yr and are dominated by grass species and large grazing animals.

grassroots   Environmental efforts that begin at the level of the populace, often in response to a perceived problem not being addressed by public policy.

gravitational water   Water that is not held by capillary action in soil but that percolates downward by the force of gravity.

graying   The increasing average age in populations in developed countries and in many developing countries that is occurring because of decreasing birthrates and increasing longevity.

gray water   Wastewater, as from sinks and tubs, that does not contain human excrement. Such water can be reused without purification for some purposes.

greenhouse effect   An increase in the atmospheric temperature caused by increasing amounts of carbon dioxide and certain other gases that absorb and trap heat, which normally radiates away from Earth.

greenhouse gases   Gases in the atmosphere that absorb infrared energy and contribute to the air temperature. These gases are like a heat blanket and are important in insulating Earth's surface. Among the greenhouse gases are carbon dioxide, water vapor, methane, nitrous oxide, chlorofluorocarbons, and other halocarbons.

green products   Manufactured products that are more environmentally benign than their traditional counterparts.

Green Revolution   The development and introduction of new varieties of (mainly) wheat and rice that have increased yields per acre dramatically in many countries since the 1960s.

green water   In the hydrologic cycle, water that is evaporated or transpired and returned as water vapor to the atmosphere; water in vapor form.

grit chamber   A part of preliminary treatment in wastewater-treatment plants; a swimming pool–like tank in which the velocity of the water is slowed enough to let sand and other gritty material settle.

gross domestic product   The total value of all goods and services exchanged in a year within a country.

groundwater   Water that has accumulated in the ground, completely filling and saturating all pores and spaces in rock or soil. Free to move more or less readily, it is the reservoir for springs and wells and is replenished by infiltration of surface water.

groundwater remediation   The repurification of contaminated groundwater by any of a number of techniques.

gully erosion   Soil erosion produced by running water and resulting in the formation of gullies.

habitat   The specific environment (woods, desert, swamp) in which an organism lives.

habitat destruction   Known to be responsible for 36% of known extinctions, the destruction of natural habitats is one of the greatest threats to wild species and ecosystems.

Hadley cell A system of vertical and horizontal air circulation predominating in tropical and subtropical regions and creating major weather patterns.

half-life The length of time it takes for half of an unstable isotope to decay. The length of time is the same regardless of the starting amount. Also refers to the amount of time it takes compounds to break down in the environment.

halogenated hydrocarbon A synthetic organic compound containing one or more atoms of the halogen group, which includes chlorine, fluorine, and bromine.

hard water Water that contains relatively large amounts of calcium or certain other minerals that cause soap to precipitate. (Contrast **soft water**.)

hazard Anything that can cause (1) injury, disease, or death to humans; (2) damage to property; or (3) degradation of the environment. *Cultural* hazards include factors that are often a matter of choice, such as smoking or sunbathing. *Biological* hazards are pathogens and parasites that infect humans. *Physical* hazards are natural disasters like earthquakes and tornadoes. *Chemical* hazards refer to the chemicals in use in different technologies and household products.

hazard communication standard The outcome of amendments to the Occupational Safety and Health Act of 1970, this law guarantees the workers' right to know about hazardous materials they encounter.

hazard identification The process of examining evidence linking a potential hazard to its possible harmful effects; a major element of risk assessment by the EPA.

hazardous material (HAZMAT) Any material having one or more of the following attributes: ignitability, corrosivity, reactivity, and toxicity.

health According to the World Health Organization, "a state of complete physical, social, and mental well-being and not merely the absence of disease or infirmity."

heavy metal Any of the high-atomic-weight metals, such as lead, mercury, cadmium, and zinc. All may be serious pollutants in water or soil because they are toxic in relatively low concentrations and they tend to bioaccumulate.

heavy water Deuterium oxide (D_2O), a form of water incorporating a heavy isotope of hydrogen.

herbicide A chemical used to kill or inhibit the growth of undesired plants.

herbivore An organism such as a rabbit or deer that feeds primarily on green plants or plant products such as seeds or nuts. Such an organism is said to be herbivorous. (Synonym: **primary consumer**.)

herbivory The feeding on plants that occurs in an ecosystem. The total feeding of all plant-eating organisms.

heterotroph Any organism that consumes organic matter as a source of energy. Such an organism is said to be heterotrophic.

high-level wastes In a nuclear reactor, the direct products of fission produce highly radioactive isotopes that make the management of reactor wastes difficult and hazardous.

Highway Trust Fund The monies collected from the gasoline tax and designated for the construction of new highways.

HIPPO An acronym for the major threats to biodiversity: Habitat destruction, Invasive species, Pollution, Population, and Overexploitation.

hormones Natural chemical substances that control the development, physiology, and behavior of an organism. Hormones are produced internally and affect only the individual organism. Hormones are coming into use in pest control. (See also **pheromones**.)

host In feeding relationships, particularly parasitism, refers to the organism that is being fed upon (i.e., the organism that is supporting the feeder).

host-parasite relationship The relationship between a parasite and the organism upon which it feeds.

hot spots See **biodiversity hot spots**.

household hazardous wastes An important component of the hazardous-waste problem in the United States.

Hubbert Peak The concept proposed by M. King Hubbert that peak oil production will follow a bell-shaped curve when half of available oil would have been withdrawn.

human capital See **intangible capital**.

Human Poverty Index, Human Development Index Based on life expectancy, literacy, and living standards, these indices are used by the United Nations Development Program to measure progress in alleviating poverty.

human system The entire system that humans have created for their own support, consisting of agriculture, industry, transportation, communication networks, and so on.

humidity The amount of water vapor in the air. (See also **relative humidity**.)

humus A dark brown or black, soft, spongy residue of organic matter that remains after the bulk of dead leaves, wood, or other organic matter has decomposed. Humus does oxidize, but relatively slowly. It is extremely valuable in enhancing the physical and chemical properties of soil.

hunger A condition wherein the basic food required for meeting nutritional and energy needs is lacking and the individual is unable to lead a normal, healthy life.

hunter-gatherers Humans surviving by hunting wild game and gathering seeds, nuts, berries, and other edible things from the natural environment.

hybrid A plant or animal resulting from a cross between two closely related species that do not normally cross.

hybrid electric vehicle An automobile combining a gasoline motor and a battery-powered electric motor that produces less pollution and gets higher gasoline mileage than do conventional gasoline-powered vehicles.

hybridization Cross-mating between two more or less closely related species.

hydrocarbons *Chemistry:* Natural or synthetic organic substances that are composed mainly of carbon and hydrogen. Crude oil, fuels from crude oil, coal, animal fats, and vegetable oils are examples. *Pollution:* A wide variety of relatively small carbon-hydrogen molecules resulting from incomplete burning of fuel that are emitted into the atmosphere. (See **volatile organic compounds**.)

hydroelectric dam A dam and an associated reservoir used to produce electrical power by letting the high-pressure water behind the dam flow through and drive a turbogenerator.

hydroelectric power Electric power that is produced from hydroelectric dams or, in some cases, natural waterfalls.

hydrogen bonding A weak attractive force that occurs between a hydrogen atom of one molecule and, usually, an oxygen atom of another molecule. Hydrogen bonding is responsible for holding water molecules together to produce the liquid and solid states.

hydrogen economy A future economy where renewable energy via hydrogen as a major energy carrier prevails.

hydrogen ions Hydrogen atoms that have lost their electrons (chemical symbol, H+).

hydrologic cycle The movement of water from points of evaporation, through the atmosphere, through precipitation, and through or over the ground, returning to points of evaporation.

hydrosphere The water on Earth, in all of its liquid and solid compartments: oceans, rivers, lakes, ice, and groundwater.

hydroxyl radical The hydroxyl group (OH), missing the electron. The hydroxyl radical is a natural cleansing agent of the atmosphere. It is highly reactive, readily oxidizes many pollutants upon contact, and thus contributes to their removal from the air.

hygiene hypothesis The theory that human immune systems need to encounter microbes and even worms at early ages in order to stimulate a regulated anti-inflammatory network; this reduces incidences of allergies and asthma.

hypothesis A tentative guess concerning the cause of an observed phenomenon that is then subjected to experiment to test its logical or empirical consequences.

hypoxia See **dead zone.**

ICPD (International Conference on Population and Development) A UN-sponsored conference held in 1994 in Cairo, Egypt, that produced a program of action that was drafted and later adopted by the United Nations.

immigration The movement of people from countries of their birth to other countries, usually with the intent to reside there permanently.

ImPACT A refinement of the IPAT formula that separates the effects of Technology (T in the equation) into two components that incorporate the different effects of consumption of resources.

incremental value In the measurement of the value of ecosystem services, the incremental value of a service may be calculated when changes in that service occur.

indicator organism An organism, the presence or absence of which indicates certain conditions. For example, the presence of *Escherichia coli* indicates that water is contaminated with fecal wastes and that pathogens may be present.

individual quota system A system of fishery management wherein a quota is set and individual fishers are given or sold the right to harvest some proportion of the quota. Also called **catch shares.**

industrialized agriculture The use of fertilizer, irrigation, pesticides, and energy from fossil fuels to produce large quantities of crops and large numbers of livestock with minimal labor for domestic and foreign sale.

Industrial Revolution During the 19th century, the development of manufacturing processes using fossil fuels and based on applications of scientific knowledge.

industrial smog The grayish mixture of moisture, soot, and sulfurous compounds that occurs in local areas in which industries are concentrated and coal is the primary energy source.

infant mortality The number of babies that die before one year of age, per 1,000 babies born.

infiltration The process in which water soaks into soil as opposed to running off the surface of the soil.

infiltration-runoff ratio The ratio of the amount of water soaking into the soil to that running off the surface. The ratio is obtained by dividing the first amount by the second.

infrared radiation Radiation of somewhat longer wavelengths than red light, which comprises the longest wavelengths of the visible spectrum. Such radiation manifests itself as heat.

infrastructure The sewer and water systems, roadways, bridges, and other facilities that underlie the functioning of a city or a society.

inherently safe reactor In theory, a nuclear reactor that is designed in such a way that any accident would be automatically corrected, with no radioactivity released.

inorganic chemical In classifying chemical pollutants, some are inorganic, such as the heavy metals (lead, mercury and so forth) and salts.

inorganic compounds/molecules *Classical definition:* All things such as air, water, minerals, and metals that are neither living organisms nor products uniquely produced by living things. *Chemical definition:* All chemical compounds or molecules that do not contain carbon atoms as an integral part of their molecular structure. (Contrast **organic compounds/molecules.**)

inorganic fertilizer See **fertilizer.**

instrumental value The value that living organisms or species have by virtue of their benefit to people; the degree to which they benefit humans. (Contrast **intrinsic value.**)

insurance spraying The spraying of pesticides when it is not really needed, in the belief that it will ensure against loss due to pests.

intangible capital One component of the wealth of nations. Comprises the *human capital,* or the population and its attributes; the *social capital,* or the social and political environment that people have created in a society; and the *knowledge assets,* or the human fund of knowledge.

integrated gasification combined cycle (IGCC) plant A coal-fired technology where coal is mixed with water and oxygen and heated under pressure to produce a synthetic gas.

integrated pest management (IPM) A program consisting of two or more methods of pest control carefully integrated and designed to avoid economic loss from pests. The objective of IPM is to minimize the use of environmentally hazardous synthetic chemicals.

Integrated Risk Information System (IRIS) A compendium of risk information published by the EPA providing data on chemical hazards; available on the Internet.

integrated waste management The approach to municipal solid waste that provides for several options for dealing with wastes, including recycling, composting, waste reduction, and landfilling, as well as incineration where unavoidable.

Interdecadal Pacific Oscillation (IPO) One of several ocean-atmosphere cycles, a warm-cool cycle that swings back and forth across the Pacific over periods of several decades.

intergenerational equity Meeting the needs of the present without jeopardizing the ability of future generations to meet their needs in an equitable way.

Intergovernmental Panel on Climate Change (IPCC) The UN-sponsored organization charged with continually assessing the science of global climate change, its potential impacts, and the means of responding to the threat.

International Whaling Commission (IWC) The international body that regulates the harvesting of whales; the IWC placed a ban on all whaling in 1986.

interspecific competition Competition for resources between members of two or more species.

intragenerational equity The attempt to achieve an equitable distribution of assets within a given generation or across existing people groups.

intraspecific competition Competition for resources between members of the same species.

intrinsic value The value that living organisms or species have in their own right; in other words, organisms and species do not have to be useful to have value. (Contrast **instrumental value.**)

invasive species An introduced species that spreads out and often has harmful ecological effects on other species or ecosystems.

inversion See **temperature inversion.**

inverter In a photovoltaic (PV) system, the device that acts as an interface between the solar PV modules and the electric grid or batteries.

ion An atom (or a group of atoms) that has lost or gained one or more electrons and, consequently, has acquired a positive or negative charge. Ions are designated by "+" or "−" superscripts following the chemical symbol for the element(s) involved.

ion-exchange capacity See **nutrient-holding capacity.**

ionic bond The bond formed by the attraction between a positive and a negative ion.

ionizing radiation Radiation that displaces ions in tissues as it penetrates them, leaving behind charged particles (ions).

IPAT formula A conceptual formula relating environmental impact (I) to population (P), affluence (A), and technology (T).

isotope A form of an element in which the atoms have more (or less) than the usual number of neutrons. Isotopes of a given element have identical chemical properties but differ in mass (weight) as a result of the superfluity (or deficiency) of neutrons. Many isotopes are unstable and radioactive. (See **radioactive decay, radioactive emissions,** and **radioactive materials.**)

ISTEA (Intermodal Surface Transportation Efficiency Act) Legislation that provides for the funding of alternative transportation (e.g., mass transit or bicycle paths) using money from the Highway Trust Fund. Succeeded by the Transportation Equity Act for the Twenty-first Century (TEA–21) and then the Safe, Accountable, Flexible, Efficient Transportation Equity Act (SAFETEA).

IUCN (International Union for Conservations of Nature) Formerly the **World Conservation Union.** An international organization that maintains a Red List of threatened species worldwide.

jet stream A river of air, generated by Earth's rotation and air-pressure gradients, high in the troposphere that flows eastward at high speeds but meanders considerably.

judicial branch   One of the three branches of government set up by the U.S. constitution, involving a system of judges at various levels on up to the Supreme Court.

junk science   Information presented as valid science but unsupported by peer-reviewed research. Often, politically motivated and biased results are selected to promote a particular point of view.

juvenile hormone   The insect hormone that, at sufficient levels, preserves the larval state. Pupation requires diminished levels of juvenile hormone; hence, artificial applications of the hormone may block pupal development.

keystone species   A species whose role is essential for the survival of many other species in an ecosystem.

kinetic energy   The energy inherent in motion or movement, including molecular movement (heat) and the movement of waves (hence, radiation and therefore light).

knowledge assets   See **intangible capital**.

K-strategist   A reproductive strategy for a species whereby there is a low reproductive rate but good survival of young due to care and protection by adults. Such species often have populations close to the carrying capacity.

Kyoto Protocol   An international agreement among the developed nations to curb greenhouse gas emissions. The Kyoto Protocol was forged in December 1997 and became binding in 2004.

labor   In classical economic theory, one of the three elements that constitute the factors of production: the work done by people as they produce goods or services.

Lacey Act   Passed in 1900, the first national act that gave protection to wildlife by forbidding interstate commerce in illegally killed animals.

land   In classical economic theory, one of the three elements that constitute the factors of production: the natural resources used in the production of goods and services.

land degradation   The processes where land is rendered less capable of supporting plant growth, often through soil erosion.

land ethic   First described by Aldo Leopold, this is an ethic that promotes protection of natural areas based on their integrity, stability, and beauty.

landfill   A site where municipal, industrial, or chemical wastes are disposed of by burying them in the ground or placing them on the ground and covering them with earth.

landscape   A group of interacting ecosystems occupying adjacent geographical areas.

land subsidence   The gradual sinking of land. The condition may result from the removal of groundwater or oil, which is frequently instrumental in supporting the overlying rock and soil.

land trust   Land that is purchased and held by various private organizations specifically for the purpose of protecting the region's natural environment and the biota that inhabit it.

La Niña   See **El Niño/La Niña Southern Oscillation (ENSO)**.

larva, pl. larvae   A free-living immature form that occurs in the life cycle of many organisms and that is structurally distinct from the adult. For example, caterpillars are the larval stage of moths and butterflies.

latitude   In biomes, latitude is the distance from the equator, which determines the average temperature for a region.

Law of Conservation of Energy   See **First Law of Thermodynamics**.

Law of Conservation of Matter   The law stating that, in chemical reactions, atoms are neither created nor changed nor destroyed; they are only rearranged.

law of limiting factors   The law stating that a system may be limited by the absence or minimum amount (in terms of that needed) of any required factor. Also known as **Liebig's law of minimums**. (See **limiting factor**.)

leachate   The mixture of water and materials that are leaching.

leaching   The process in which materials in or on the soil gradually dissolve and are carried by water seeping through the soil. Leaching may eventually remove valuable nutrients from the soil, or it may carry buried wastes into groundwater, thereby contaminating it.

lead   One of the heavy metal pollutants, lead was once added to gasoline to improve combustion. It is toxic in very small quantities and has been removed from all but aviation gasoline.

legislative branch   One of the three branches of government set up by the U.S. constitution, the legislative branch consists of the House of Representatives and the Senate, all members of which are voted into office by the people of different states.

legumes   The group of pod-bearing land plants that is virtually alone in its ability to fix nitrogen; legumes include such common plants as peas, beans, clovers, alfalfa, and locust trees but no major cereal grains. (See **nitrogen fixation**.)

Liebig's law of minimums   See **law of limiting factors**.

life expectancy   The number of years a newborn can expect to live under mortality rates current for its country of residence.

life history   The progression of changes an organism undergoes in its life from birth to death.

light-water reactor (LWR)   A reactor that uses extra-pure ordinary water as a moderator. Nuclear plants in the United States are all LWRs.

limiting factor   A factor primarily responsible for determining the growth or reproduction of an organism or a population. The limiting factor in a given environment may be a physical factor such as temperature or light, a chemical factor such as a particular nutrient, or a biological factor such as a competing species.

limits of tolerance   The extremes of any factor (e.g., temperature) that an organism or a population can tolerate and still survive and reproduce.

lipids   A class of natural organic molecules that includes animal fats, vegetable oils, and phospholipids, the last being an integral part of cellular membranes.

lithosphere   The Earth's crust, made up of rocks and minerals.

litter   In an ecosystem, the natural cover of dead leaves, twigs, and other dead plant material. This natural litter is subject to rapid decomposition and recycling in the ecosystem, whereas human litter—such as bottles, cans, and plastics—is not.

livability   A subjective index of how enjoyable a city is to live in.

loam   A soil texture consisting of a mixture of about 40% sand, 40% silt, and 20% clay; loam is usually good soil.

locavore   People who try to purchase or grow most of their food within their own region; eating locally.

logistic growth   A pattern of growth of a population that results in an S-shaped curve plotted over time, such that the population levels off at the carrying capacity (K).

longevity   The maximum life span of individuals of a given species. The known record for humans is 122 years.

loss-of-coolant accident (LOCA)   A situation in a nuclear reactor where an accident causes the loss of the water from around the reactor, resulting in overheating and a possible meltdown of the nuclear core.

low-level wastes   Radioactive wastes from nuclear power plants and other facilities that employ radioisotopes; they are easier to manage than the high-level wastes.

low-till farming   Similar to no-till farming, except that instead of using herbicides, farmers plow just once over residues of a previous crop and plant their new crop.

LULU ("locally unwanted land use")   An acronym expressing the difficulty in siting a facility that is necessary but that no one wants in his or her immediate locality.

macromolecules   Very large organic molecules, such as proteins and nucleic acids, that constitute the structural and functional parts of cells.

Magnuson Act   An act passed in 1976 that extended the limits of jurisdiction over coastal waters and fisheries of the United States to 200 miles offshore. Reauthorized in 2006 as the **Magnuson-Stevens Fishery Conservation and Management Reauthorization Act**.

malnutrition   The lack of essential nutrients such as vitamins, minerals, and amino acids. Malnutrition ranges from mild to severe and life threatening.

marine protected area   An area of the coast or open ocean that is closed to commercial fishing and mineral mining; marine reserves.

marker-assisted breeding   The science of locating genes with desirable traits using DNA sequencing and then using cross-breeding with standard crop lines, thus avoiding the use of transgenics.

market-based policy   An approach to environmental regulation that allows the use of markets to achieve a desired outcome (e.g., a **cap-and-trade system** for reducing emissions of a pollutant).

mass number   The number that accompanies the chemical name or symbol of an element or isotope. The mass number represents the number of neutrons and protons in the nucleus of the atom.

material safety data sheets (MSDS)   Documents that contain information on the reactivity and toxicity of more than 600 chemicals and that must accompany the shipping, storage, and handling of these chemicals.

materials recovery facility (MRF)   A processing plant in which regionalized recycling is carried out. Recyclable municipal solid waste, usually presorted, is prepared in bulk for the recycling market.

matter   Any gas, liquid, or solid that occupies space and has mass. (Contrast **energy**.)

maximum achievable control technology (MACT)   The best technology available for reducing the output of especially toxic industrial pollutants.

maximum contaminant level (MCL)   A drinking water standard; the highest allowable concentration of a pollutant in a drinking water source.

maximum sustainable yield (MSY)   The maximum amount of a renewable resource that can be taken year after year without depleting the resource. The maximum sustainable yield is the maximum rate of use or harvest that will be balanced by the regenerative capacity of the system.

MDG Support Team   The successor to the Millennium Project, this team works with developing countries to achieve the targets of the Millennium Development Goals (MDGs).

Medical Revolution   Medical advances and public sanitation led to spectacular reductions in mortality, beginning in the late 1800s and extending to the present.

Megatons to Megawatts Program   An exchange program between the United States and Russia whereby weapons-grade uranium from Russia is diluted to produce power-plant uranium that is used in U.S. power plants.

meltdown   The event of a nuclear reactor's getting out of control or losing its cooling water so that it melts from its own production of heat. The melted reactor would continue to produce heat and could melt its way out of its containment vessel and eventually down into groundwater, where it would cause a violent eruption of steam that could spread radioactive materials over a wide area. (See **loss-of-coolant accident (LOCA)**.)

Mercury and Air Toxics Standards   Rules published by the EPA in 2011 that require all coal- and oil-fired power plants to limit their emissions of mercury and other toxic pollutants; industries have four years to comply with the rules.

mesosphere   A layer of atmosphere beyond the stratosphere where only very small quantities of gas are found.

metabolism   The sum of all the chemical reactions that occur in an organism.

meteorology   The scientific study of the atmosphere, involving both weather and climate.

methane   A gas, CH_4. Methane is the primary constituent of natural gas and is also a product of fermentation by microbes. Methane is one of the greenhouse gases.

methyl bromide   A soil fumigant to control agricultural pests; its use releases bromine to the atmosphere, and it destroys ozone when it gets up into the stratosphere.

microbe   Any microscopic organism, although primarily a bacterium, a virus, or a protozoan.

microbial   Of or pertaining to microbes.

microclimate   The actual conditions experienced by an organism in its particular location. Owing to numerous factors—such as shading, drainage, and sheltering—the microclimate may be quite distinct from the overall climate.

microfiltration   A technology for desalination, based on forcing seawater through very small pores in a membrane filter that removes the salt.

microlending   The process of providing very small loans (usually $50–$500) to poor people to facilitate their starting a small enterprise and becoming economically self-sufficient.

microorganism   See **microbe**.

midnight dumping   The illicit dumping of materials, particularly hazardous wastes, frequently under the cover of darkness.

Milankovitch cycle   A cycle of major oscillations in the Earth's orbit, taking place over frequencies of thousands of years and known to influence the distribution of solar radiation and therefore global weather patterns.

Millennium Development Goals (MDGs)   A comprehensive set of goals aimed at addressing the most important needs of people in the developing countries to improve their well-being, adopted by the United Nations at the UN Millennium Summit in 2000.

Millennium Ecosystem Assessment (MA)   A four-year effort by over 1,360 scientists to produce reports on the state of Earth's ecosystems. The reports are aimed at policy makers and the public and are all available on the Internet.

Millennium Project   A UN-commissioned project given a mandate to develop an action plan to coordinate the efforts of various agencies and organizations to achieve the Millennium Development Goals. Succeeded by the **MDG Support Team**.

Minamata disease   A disease named for a fishing village in Japan where an "epidemic" was first observed. Symptoms—which included spastic movements, mental retardation, coma, death, and crippling birth defects in the next generation—were found to be the result of mercury poisoning.

mineral   Any hard, brittle, stone-like material that occurs naturally in Earth's crust. All minerals consist of various combinations of positive and negative ions held together by ionic bonds. A pure mineral, or crystal, is one specific combination of elements. Common rocks are composed of mixtures of two or more minerals.

mineralization   The process of gradual oxidation of the organic matter (humus) present in soil that leaves just the gritty mineral component of the soil.

mitigation   Reducing greenhouse gas emissions by a variety of strategies and policies.

mixture   A combination of elements in which there is no chemical bonding between the molecules. For example, air contains (is a mixture of) oxygen, nitrogen, and carbon dioxide.

model   In the scientific method, observations may be put together to form a larger picture of how a system works, called a model.

moderator   In a nuclear reactor, any material that slows down neutrons from fission reactions so that they are traveling at the right speed to trigger another fission. Water and graphite are two types of moderators.

molecule   The smallest unit of two or more atoms forming a compound. A molecule has all the characteristics of the compound of which it is a unit.

monsoon   The seasonal airflow created by major differences in cooling and heating between oceans and continents, usually bringing extensive rain.

Montreal Protocol   An agreement made in 1987 by a large group of nations to cut back the production of chlorofluorocarbons in order to protect the ozone shield. A 1990 amendment called for the complete phaseout of these chemicals by 2000 in developed nations and by 2010 in less-developed nations.

morbidity   The incidence of disease in a population.

mortality   The incidence of death in a population.

mountaintop removal mining   A form of **strip mining** for coal that takes off whole tops of mountains to get at the seams of coal that are buried as deeply as 1,000 feet below the surface.

municipal solid waste (MSW)   The entirety of refuse or trash generated by a residential and business community. Distinct from agricultural and industrial wastes, municipal solid waste is the refuse that a municipality is responsible for collecting and disposing of.

mutagenic   Causing mutations.

mutation   A random change in one or more genes of an organism. Mutations may occur spontaneously in nature, but their number and degree are vastly increased by exposure to radiation or certain chemicals. Mutations generally result in a physical deformity or metabolic malfunction.

mutualism   A close relationship between two organisms from which both derive a benefit.

mycorrhiza, pl. mycorrhizae   The mycelia of certain fungi that grow symbiotically with the roots of some plants and provide for additional nutrient uptake.

National Ambient Air Quality Standards (NAAQS)   The allowable levels of ambient criteria air pollutants set by EPA regulation.

National Center for Environmental Assessment (NCEA)   An agency within the EPA that performs and publishes risk assessments of substances and processes that are widely released in the environment.

National Emission Standards for Hazardous Air Pollutants (NESHAPs)   The standards for allowable emissions of certain toxic substances.

National Energy Policy Report   The 2001 report by a task force chaired by Vice President Cheney that emphasized dependence on **supply-side policies** to generate the energy needed by the United States into the future.

national forests   Public forests and woodlands administered by the National Forest Service for multiple uses, such as logging, mineral exploitation, livestock grazing, and recreation.

national parks   Lands and coastal areas of great scenic, ecological, or historical importance administered by the National Park Service, with the dual goals of protecting them and providing public access.

National Pollution Discharge Elimination System (NPDES)   An EPA-administered program that addresses point-source water pollution through the issuance of permits that regulate pollution discharge.

National Priorities List (NPL)   A list of the chemical waste sites presenting the most immediate and severe threats. Such sites are scheduled for cleanup ahead of other sites.

national wildlife refuges   Administered by the U.S. Fish and Wildlife Service, these lands are maintained for the protection and enhancement of wildlife and for the provision of public access.

natural capital   The natural assets and the services they perform. One form of the wealth of a nation is its complement of natural capital.

natural chemical control   The use of one or more natural chemicals, such as hormones or pheromones, to control a pest.

natural enemies   All the predators and parasites that may feed on a given organism. Organisms used to control a specific pest through predation or parasitism.

natural goods   The food, fuel, wood, fibers, oils, alcohols, and the like derived from the natural world, on which the world economy and human well-being depend.

natural increase   The number of births minus the number of deaths in a given population. The natural increase does not take into account immigration and emigration and is the percent of growth (or decline) of a given population during a year. It is found by subtracting the crude death rate from the crude birthrate and changing the result to a percent.

natural laws   Generalizations derived from our observations of matter, energy, and other phenomena. Though not absolute, natural laws have been empirically confirmed to a high degree and are often derivable from higher-level theory.

natural organic compounds   The organic compounds that make up living organisms, to be distinguished from synthetic organic compounds, like plastics.

Natural Resources Conservation Service (NRCS)   Formerly the Soil Conservation Service, a nationwide network of offices that provides extension services to farmers and encourage soil and water conservation throughout the United States.

natural resources   Features of natural ecosystems and species that are of economic value and that may be exploited. Also, features of particular segments of ecosystems, such as air, water, soil, and minerals.

natural sciences   Those scientific disciplines that study the natural world, from subatomic particles to the cosmos.

natural selection   The evolutionary process whereby the natural factors of environmental resistance tend to eliminate those members of a population that are least well adapted to cope with their environment and thus, in effect, tend to select those best adapted for survival and reproduction.

natural services   Functions performed free of charge by natural ecosystems, such as control of runoff and erosion, absorption of nutrients, and assimilation of air pollutants.

negligible risk standard   A new safety standard based on risk analysis, established by the **FQPA** of 1996, that replaces the **Delaney clause** in dealing with pesticide residues on food products.

Neolithic Revolution   The development of agriculture begun by human societies around 12,000 years ago, leading to more permanent settlement and population increases.

net metering   The practice of allowing individuals on the power grid to subtract the cost of the energy they get from photovoltaics or wind turbines from their electric bills.

neutron   A fundamental atomic particle found in the nuclei of atoms (except hydrogen) and having one unit of atomic mass but no electrical charge.

new forestry   Now part of the U.S. Forest Service's management practice, a forestry management strategy that places priority on protecting the ecological health and diversity of forests rather than maximizing the harvest of logs.

niche (ecological)   The total of all the relationships that bear on how an organism copes with the biotic and abiotic factors it faces.

NIMBY ("not in my backyard")   A common attitude regarding undesirable facilities such as incinerators, nuclear facilities, and hazardous-waste treatment plants whereby people do everything possible to prevent such facilities from being located near their residences.

NIMTOO ("not in my term of office")   An attitude expressing the reluctance of officeholders to make unpopular decisions on matters such as siting waste facilities.

nitric acid   A secondary air pollutant and a major component of acid deposition, nitric acid is produced when NO_x in the air reacts with the hydroxyl radical and water.

nitrogen cascade   The detrimental complex of ecological effects brought on by reactive nitrogen (Nr) that has been added to natural systems by the burning of fossil fuels and the fertilization of agricultural crops.

nitrogen fixation   The process of chemically converting nitrogen gas (N_2) from the air into compounds such as nitrates (NO_3) or ammonia (NH_3) that can be used by plants in building amino acids and other nitrogen-containing organic molecules.

nitrogen oxides (NO_X)   A group of nitrogen-oxygen compounds formed when some of the nitrogen gas in the air combines with oxygen during high-temperature combustion. Nitrogen oxides are a major category of air pollutants and, along with hydrocarbons, are a primary factor in the production of ozone and other photochemical oxidants that are the most harmful components of photochemical smog. Nitrogen oxides also contribute to acid precipitation.

nitrous oxide   A gas (N_2O). Nitrous oxide, which comes from biomass burning, fossil fuel burning, and the use of chemical fertilizers, is of concern because it is a greenhouse gas in the troposphere and it contributes to ozone destruction in the stratosphere.

nonbiodegradable   Not able to be consumed or broken down by biological organisms. Nonbiodegradable substances include plastics, aluminum, and many chemicals used in industry and agriculture. Particularly dangerous are synthetic chemicals that are also toxic and tend to accumulate in organisms. (See **biodegradable** and **bioaccumulation**.)

nonconsumptive water use The use of water for such purposes as washing and rinsing, wherein the water, albeit polluted, remains available for further uses. With suitable purification, such water may be recycled indefinitely.

nongovernmental organization (NGO) Any of a number of private organizations involved in studying or advocating action on different environmental issues.

nonpersistent Said of chemicals that break down readily to harmless compounds, as, for example, natural organic compounds that break down to carbon dioxide and water; applied to pesticides, any pesticide that breaks down within days or a few weeks into nontoxic products.

nonpoint sources Sources of pollution such as the general runoff of sediments, fertilizers, pesticides, and other materials from farms and urban areas, as opposed to specific points of discharge such as factories. Also called **diffuse sources**. (Contrast **point sources**.)

nonreactive nitrogen The main reservoir of nitrogen as nitrogen gas in the atmosphere.

nonrenewable resources Resources, such as ores of various metals, oil, and coal that exist as finite deposits in Earth's crust and that are not replenished by natural processes as they are mined. (Contrast **renewable resources**.)

North Atlantic Deep Water (NADW) The cold, saline water that sinks in the high-latitude North Atlantic and initiates a deep current that spreads the waters throughout the oceans. See **Atlantic Meridional Overturning Circulation**.

North Atlantic Oscillation (NAO) One of several ocean-atmosphere oscillations where atmospheric pressure centers switch back and forth across the Atlantic, with many impacts on winds and storms.

no-till agriculture The farming practice in which weeds are killed with chemicals (or other means) and seeds are planted and grown without resorting to plowing or cultivation. The practice is highly effective in reducing soil erosion.

NRCS (Natural Resources Conservation Service) Formerly the U.S. Soil Conservation Service). The NRCS provides information to landowners regarding soil and water conservation practices.

nuclear power Electrical power generated by using a nuclear reactor to boil water and produce steam that in turn drives a turbogenerator.

Nuclear Regulatory Commission (NRC) The agency within the U.S. Department of Energy that sets and enforces safety standards for the operation and maintenance of nuclear power plants.

Nuclear Waste Policy Act of 1982 A law that committed the federal government to begin receiving nuclear wastes from commercial power plants by 1998. This has not happened yet, as the nation debates where to take the wastes.

nucleic acids The class of natural organic macromolecules that function in the storage and transfer of genetic information.

nucleus *Biology*: The large body residing in most living cells that contains the genes or hereditary material (DNA). *Physics*: The central core of atoms, which is made up of neutrons and protons. Electrons surround the nucleus.

nutrient *Animal*: Material such as protein, vitamins, and minerals required for growth, maintenance, and repair of the body and material such as carbohydrates required for energy. *Plant*: An essential element (C, P, N, Ca) in a particular ion or molecule that can be absorbed and used by the plant.

nutrient-holding capacity The capacity of a soil to bind and hold nutrients (fertilizer) against their tendency to be leached from the soil.

observations Things or phenomena that are perceived through one or more of the basic five senses in their normal state. In addition, to be accepted as factual, observations must be verifiable by others.

Occupational Safety and Health Act of 1970 (OSHA) The act protecting the rights of workers to know about hazards in their workplace.

ocean acidification An outcome of the rise in atmospheric carbon dioxide; as the oceans take up more and more of the CO_2, the carbonate ion concentration is reduced, making it more difficult for coral animals to build their calcium carbonate skeletons.

ocean thermal-energy conversion (OTEC) The concept of harnessing the difference in temperature between surface water heated by the Sun and colder deep water to produce power.

Official Development Assistance (ODA) Financial assistance provided to developing countries from the developed countries and coordinated by United Nations agencies.

oil field The underground area in which exploitable oil is found.

oil sand Sedimentary material containing bitumen, a tar-like hydrocarbon that is subject to exploitation under favorable economic conditions.

oil shale A natural sedimentary rock that contains kerogen, a material that can be extracted and refined into oil and oil products.

oligotrophic Nutrient poor and hence unable to support much phytoplankton (said of a body of water).

omnivore An animal that feeds on both plant material and other animals.

OPEC (Organization of Petroleum Exporting Countries) A cartel of oil-producing nations that attempts to control the market price of oil.

optimal population The population of a harvested biological resource that yields the greatest harvest for exploitation; according to maximum-sustained-yield equations, the optimal population is half the carrying capacity.

optimal range With respect to any particular factor or combination of factors, the maximum variation that still supports optimal or near-optimal growth of the species in question.

optimum The condition or amount of any factor or combination of factors that will produce the best result. For example, the amount of heat, light, moisture, nutrients, and so on that will produce the best plant growth.

organically grown Grown without the use of hard chemical pesticides or inorganic fertilizer. Any produce grown according to USDA standards for organic foods is organically grown.

organic chemical In classifying pollutants, some are organic, synthetic chemicals that served some purpose but have been released into the environment and pose some serious problems.

organic compounds/molecules *Classical definition*: All living things and products that are uniquely produced by living things, such as wood, leather, and sugar. *Chemical definition*: All chemical compounds or molecules, natural or synthetic, that contain carbon atoms as an integral part of their molecular structure. The structure of organic compounds is based on bonded carbon atoms with hydrogen atoms attached. (Contrast **inorganic compounds**.)

organic fertilizer See **fertilizer**.

organic food/gardening/farming Gardening or farming without the use of inorganic fertilizers, synthetic pesticides, or other human-made materials; food raised with organic farming techniques.

Organic Foods Protection Act A law passed in 1990 that established the National Organic Standards Board, the agency that certifies organic foods.

organic phosphate Phosphate (PO_4-3) bonded to an organic molecule.

orphan site The location of a hazardous waste site abandoned by former owners.

OSHA (Occupational Safety and Health Administration) The federal agency that promulgates regulations to protect workers.

osmosis The phenomenon whereby water diffuses through a semipermeable membrane toward an area where there is more material in solution (i.e., where there is a relatively lower concentration of water). Osmosis has particular application in the salinization of those soils in which plants are unable to grow because of osmotic water loss.

outbreak   A population explosion of a particular pest. Often caused by an application of pesticides that destroys the pest's natural enemies.

overexploitation   The overharvesting of a species or ecosystem that leads to its decline.

overgrazing   The phenomenon of grazing animals in greater numbers than the land can support in the long term.

overnourishment   The condition of obesity and overweight that results from overeating.

oxidation   A chemical reaction that generally involves a breakdown of some substance through its combining with oxygen. Burning and cellular respiration are examples of oxidation. In both cases, organic matter is combined with oxygen and broken down to carbon dioxide and water.

ozone   A gas, O_3, that is a secondary air pollutant in the lower atmosphere but that is necessary to screen out ultraviolet radiation in the stratosphere. May also be used for disinfecting water.

ozone hole   First discovered over the Antarctic, a region of stratospheric air that is severely depleted of its normal levels of ozone during the Antarctic spring because of CFCs from anthropogenic (human-made) sources.

ozone shield   The layer of ozone gas (O_3) in the stratosphere that screens out harmful ultraviolet radiation from the Sun.

parasites   Organisms (plant, animal, or microbial) that attach themselves to another organism, the host, and feed on it over a period of time without killing it immediately, although usually doing harm to it. Parasites are commonly divided into ectoparasites—those that attach to the outside of their hosts—and endoparasites—those that live inside their hosts.

parasitoid   Any parasitic insect, usually a wasp, that attacks other insects and can be responsible for bringing pest insects under control.

parent material   Rock material whose weathering and gradual breakdown are the source of the mineral portion of soil.

particulates   See **PM$_{2.5}$** and **PM$_{10}$** and **suspended particulate matter.**

parts per million (ppm)   A frequently used expression of concentration. The number of units of one substance present in a million units of another. Equivalent to milligrams per liter.

passive safety features   Those safety features of nuclear facilities that involve processes that are not vulnerable to operator intrusion or electrical power failures. Passive safety features enhance the degree of safety of nuclear reactors. (See **active safety features.**)

pasteurization   The process of applying enough heat to milk or some other substance to kill pathogens and sufficient other bacteria to extend the shelf life of the product.

pastoralist   One involved in animal husbandry, usually in subsistence agriculture.

pathogen   An organism, usually a microbe, that is capable of causing disease. Such an organism is said to be pathogenic.

pathogen pollution   A situation where human wastes spread diseases to wild species.

Payments for Ecosystem Services (PES)   A market-based system that pays owners who hold properties that provide valuable services.

PAYT (pay as you throw)   A system for collecting municipal wastes that charges residents by the container for nonrecycled wastes.

peat   Compacted plant material that accumulates in wet environments where oxygen is lacking; it may be used for fuel or extracted for use in gardening.

percolation   The process of water seeping downward through cracks and pores in soil or rock.

permafrost   The ground of arctic regions that remains permanently frozen. Defines tundra because only small herbaceous plants can be sustained on the thin layer of soil that thaws each summer.

permanent polyculture   A sustainable agriculture venture where a group of perennial plants of different species are grown together to produce food and fuel.

peroxyacetyl nitrates (PANs)   A group of compounds present in photochemical smog that are extremely toxic to plants and irritating to the eyes, nose, and throat membranes of humans.

persistent organic pollutants (POPs)   Any members of a class of organic pollutants that are resistant to biodegradation and that are often toxic; for example, DDT, PCBs, and dioxin are persistent organic pollutants. Such chemicals may remain present in the environment for periods of years.

pest   Any organism that is noxious, destructive, or troublesome; usually an agricultural pest.

pesticide   A chemical used to kill pests. Pesticides are further categorized according to the pests they are designed to kill—for example, herbicides kill plants, insecticides kill insects, fungicides kill fungi, and so on.

pesticide treadmill   The idea that use of chemical pesticides simply creates a vicious cycle of "needing more pesticides" to overcome developing resistance and secondary outbreaks caused by the pesticide applications.

pH   The scale used to designate the acidity or basicity (alkalinity) of solutions or soil, expressed as the logarithm of the concentration of hydrogen ions (H^+). pH 7 is neutral; values decreasing from 7 indicate increasing acidity, and values increasing from 7 indicate increasing basicity. Each unit from 7 indicates a tenfold increase over the preceding unit.

pharma crops   Crop plants that produce pharmaceutical products with the use of genomics.

pheromones   Chemical substances secreted externally by certain members of a species that affect the behavior of other members of the same species. The most common examples are sex attractants, which female insects secrete to attract males. Pheromones are coming into use in pest control. (See also **hormones.**)

phosphorus, phosphate   An ion composed of a phosphorus atom with four oxygen atoms attached. Denoted PO_4^{-3}, phosphate is an important plant nutrient. In natural waters, it is frequently the limiting factor; therefore, additions of phosphate to natural water are often responsible for algal blooms.

photochemical oxidants   A major category of secondary air pollutants, including ozone, that are highly toxic and damaging, especially to plants and forests. Formed as a result of interactions between nitrogen oxides and hydrocarbons driven by sunlight.

photochemical smog   The brownish haze that frequently forms on otherwise clear, sunny days over large cities with significant amounts of automobile traffic. Photochemical smog results largely from sunlight-driven chemical reactions among nitrogen oxides and hydrocarbons, both of which come primarily from auto exhausts.

photosynthesis   The chemical process carried on by green plants through which light energy is used to produce glucose from carbon dioxide and water. Oxygen is released as a by-product.

photovoltaic cell (PV cell)   A device that converts light energy into an electrical current.

phthalates   A group of chemicals used to soften plastic toys and suspected of being hormone disrupters.

phytoplankton   Any of the many species of photosynthetic microorganisms that consist of single cells or small groups of cells that live and grow freely suspended near the surface in bodies of water.

phytoremediation   The use of plants to accomplish the cleanup of some hazardous chemical wastes.

planetary albedo   The reflection of solar radiation back into space due to cloud cover, snow and ice, contributing to the cooling of the atmosphere.

planetary boundaries   Safe limits on a variety of conditions on Earth that, if crossed, could become tipping points and lead to very undesirable changes that would be difficult to reverse. See **tipping point.**

plankton   Any and all living things that are found freely suspended in the water and that are carried by currents, as opposed to being able to swim against currents. Plankton includes both plant (phytoplankton) and animal (zooplankton) forms.

plant community   The array of plant species, including their numbers, ages, and distribution, that occupy a given area.

plant-incorporated protectants   In genetically modified crops, substances that have been bioengineered to incorporate substances that would repel or kill pests.

plate tectonics   The theory, proposed by Alfred Wegener, that postulates the movement of sections of the Earth's crust, creating earthquakes, volcanic activity, and the buildup of continental masses over long periods of time, also known as continental drift.

$PM_{2.5}$ and PM_{10}   Standard criterion pollutants for suspended particulate matter. PM_{10} refers to particles smaller than 10 micrometers in diameter; $PM_{2.5}$ refers to particles smaller than 2.5 micrometers in diameter. The smaller particles are readily inhaled directly into the lungs.

point sources   Specific points of origin of pollutants, such as factory drains or outlets from sewage-treatment plants. (Contrast **nonpoint sources**.)

policy life cycle   The typical course of events occurring over time in the recognition of, formulation of an approach to, implementation of a solution to, and control of an environmental problem requiring public policy action.

pollutant   A substance that contaminates air, water, or soil.

polluter-pays principle   An ethical principle whereby those who pollute the environment should pay for the damage their pollution causes.

pollution   According to the EPA, "the presence of a substance in the environment that, because of its chemical composition or quantity, prevents the functioning of natural processes and produces undesirable environmental and health effects."

Pollution Prevention Act of 1990   A law that mandates collection of data on toxic chemicals that are released to the environment, treated on site, recycled, or combusted. See **Toxics Release Inventory**.

polychlorinated biphenyls (PCBs)   A group of widely used industrial chemicals of the chlorinated hydrocarbon class. PCBs have become serious and widespread pollutants because they are resistant to breakdown and are subject to bioaccumulation. They are also carcinogenic.

population   A group within a single species whose individuals can and do freely interbreed.

population density   The number of individuals per unit of area.

population equilibrium   A state of balance between births and deaths in a population.

population explosion   The exponential increase observed to occur in a population when conditions are such that a large percentage of the offspring are able to survive and reproduce in turn. A population explosion frequently leads to overexploitation and eventual collapse of the ecosystem.

population growth rate   The change in population numbers divided by the time over which that change occurs.

population momentum   A property whereby a rapidly growing human population may be expected to grow for 50–60 years after replacement fertility (2.1 live births per female) is reached. Momentum is sustained because of increasing numbers entering reproductive age.

population profile   A bar graph that shows the number of individuals at each age or in each five-year age group, starting with the youngest ages at the bottom of the profile.

population structure   See **age structure**.

positive feedback   A phenomenon where some process leads to even more intensification of that process; an example is the evaporative rise in water vapor with global warming, adding water, a greenhouse gas, to the atmosphere.

potential energy   The ability to do work that is stored in some chemical or physical state. For example, gasoline is a form of potential energy because the ability to do work is stored in the chemical state and is released as the fuel is burned in an engine.

poverty trap   In a developing country, the situation where people are so poor that they are unable to earn or work their way out of their poverty and their condition deteriorates.

power grid   The combination of central power plants, high-voltage transmission lines, poles, wires, and transformers that makes up an electrical power system.

power tower   A solar energy converter that collects solar energy with an array of Sun-tracking mirrors and focuses the energy onto a receiver on a tower in the center of the array, where the heat energy drives a conventional turbogenerator.

prairie biome   See **grassland biome**.

precautionary principle   The principle that says that where there are threats of serious or irreversible damage, the absence of scientific certainty shall not be used as a reason for postponing cost-effective measures to prevent environmental degradation.

precipitation   Any form of moisture condensing in the air and depositing on the ground.

predation   A conspicuous process in all ecosystems, any situation where one organism feeds on another living organism.

predator   An animal that feeds on another living organism, either plant or animal.

predator-prey relationship   A feeding relationship existing between two kinds of organisms. The predator is the animal feeding on the prey. Such relationships are frequently instrumental in controlling populations of herbivores.

preservation   In protecting natural areas, the objective of preservation is to ensure the continuity of species and ecosystems, regardless of their potential utility.

pressurized-water reactor   A nuclear power plant where two loops are employed; one where water is heated in the reactor but does not boil because it is under high pressure, and a second where the pressurized water is circulated through a heat exchanger, where it boils other unpressurized water that then produces steam to drive the turbogenerator.

prey   An organism that is fed on by a predator.

primary air pollutants   Pollutants released directly into the atmosphere, mainly as a result of burning fuels and wastes, as opposed to **secondary air pollutants**.

primary clarifier   In a sewage-treatment system, a large tank that allows particulate matter to settle to the bottom before the water moves on to **secondary treatment**.

primary consumer   An organism, such as a rabbit or deer, that feeds more or less exclusively on green plants or their products, such as seeds and nuts. (Synonym: **herbivore**.)

primary energy sources   Fossil fuels, radioactive material, and solar, wind, water, and other energy sources that exist as natural resources subject to exploitation.

primary producers/primary production   Photosynthetic organisms. The activities of these organisms in creating new organic matter in ecosystems.

primary recovery   In an oil well or oil field, the oil that can be removed by conventional pumping. (See **secondary recovery**.)

primary recycling   A form of recycling where the original waste material is made back into the same material.

primary succession   The gradual establishment, through a series of stages, of a climax ecosystem in an area that has not been occupied before (e.g., a rock face).

primary treatment   The process that consists of passing the wastewater very slowly through a large tank so that the particulate organic material can settle out. The settled material is **raw sludge**, which is treated separately.

prior informed consent (PIC)   A UN-approved procedure whereby countries exporting pesticides must inform importing countries of actions the exporting countries have taken to restrict the use of the pesticides.

private land trust   See **land trust**.

produced capital   The stock of buildings, machinery, vehicles, and other elements of a country's infrastructure that are essential to the production of economic goods and services. One component of the wealth of nations.

producers   In an ecosystem, those organisms (mostly green plants) that use light energy to construct their organic constituents from inorganic compounds.

productive use   The exploitation of ecosystem resources for economic gain.

productivity   See **primary producers/primary production**.

property taxes   Taxes that the local government levies on privately owned properties, proportional to the value of the property. Property taxes are the major source of revenue for local governments.

protein   Organic macromolecules made up of amino acids; the major structural component of all animal tissues as well as enzymes in both plants and animals.

proton   A fundamental atomic particle with a positive charge, found in the nuclei of atoms. The number of protons present equals the atomic number and is distinct for each element.

protozoon, pl. protozoa   Any of a large group of eukaryotic microscopic organisms that consist of a single, relatively large complex cell or, in some cases, a small groups of cells. All have some means of movement. Amoebae and paramecia are examples.

proved reserves   See **reserves**.

proxies   These are measurable records, like tree rings, ice cores, and marine sediments, that can provide extensions back in time of climatic factors including temperatures and greenhouse gases.

public good   A category of ecosystem services that is not used up when people use it and cannot be marketed, like the air we breathe.

Public Trust Doctrine   The legal arrangement whereby wildlife resources are considered to be public resources, held in trust for all the people and protected or managed by state or federal agencies.

pumped-storage power plant   A hydroelectric plant component that pumps water up to an elevated reservoir when power demand is low and allows it to return to a lower reservoir and generate electricity when demand is higher.

purification   For water, purification occurs whenever water is separated from the solutes and particles it contains.

radiation sickness   The outcome of a high dose of radiation, often leading to death.

radiative forcing   Refers to the internal or external factors of the climate system that can influence the climate, usually to promote warming or cooling of the atmosphere as they affect the Earth's energy balance.

radioactive decay   The reduction in radioactivity that occurs as an unstable isotope (radioactive substance) gives off radiation, ultimately becoming stable.

radioactive emissions   Any of various forms of radiation or particles that may be given off by unstable isotopes. Many such emissions have very high energy and can destroy biological tissues or cause mutations leading to cancer or birth defects.

radioactive materials   Substances that are or that contain unstable isotopes and that consequently give off radioactive emissions. (See **isotope** and **radioactive emissions**.)

radioactive wastes   Waste materials that contain, are contaminated with, or are radioactive substances. Many materials used in the nuclear industry become radioactive wastes because of such contamination.

radioisotope   An isotope of an element that is unstable and that gives off radioactive emissions. (See **isotope** and **radioactive decay**.)

radon   A radioactive gas produced by natural processes in Earth that is known to seep into buildings. Radon can be a major hazard within homes and is a known carcinogen.

rain shadow   The low-rainfall region that exists on the leeward (downwind) side of a mountain range. This rain shadow is the result of the mountain range causing precipitation on the windward side.

range of tolerance   The range of conditions within which an organism or population can survive and reproduce—for example, the range from the highest to the lowest temperature that can be tolerated. Within the range of tolerance is the optimum, or best, condition.

raw sewage (raw wastewater)   The water from a sewer system as it reaches a sewage-treatment plant.

raw sludge   The untreated organic matter that is removed from sewage water by letting it settle. Raw sludge consists of organic particles from feces, garbage, paper, and bacteria.

REACH (Registration, Evaluation and Authorization of Chemicals)   The European Union's new, rigorous approach to screening hazardous chemicals for toxicity.

reactive nitrogen   All forms of nitrogen in ecosystems that are usable by organisms, as opposed to the nonreactive nitrogen in the form of nitrogen gas.

reactor vessel   A steel-walled vessel that contains a nuclear reactor.

reasonably available control technology (RACT)   Applied to the goals of the Clean Air Act, EPA-approved forms of technology that will reduce the output of industrial air pollutants. (See **maximum achievable control technology**.)

recharge area   The area over which groundwater will infiltrate and resupply an aquifer.

recruitment   The maturation and successful entry of young into an adult breeding population.

recycling   The recovery of materials that would otherwise be buried in landfills or be combusted. Primary recycling involves remaking the same material from the waste; in secondary recycling, waste materials are made into different products.

redlining   The practice of discriminatory lending and real estate showing designed to keep minorities out of white suburbs.

Red List   A list of threatened and endangered species maintained by the International Union for Conservation of Nature (IUCN) and accessible on the Internet.

relative humidity   The percentage of moisture in the air compared with how much the air can hold at the given temperature.

rem   An older unit of measurement of the ability of radioactive emissions to penetrate biological tissue. (See **sievert**.)

remittances   Money sent back to the country of birth by guest workers and immigrants to help family and friends meet their financial needs.

remote sensing   A system of imaging the Earth from satellites, or of accessing data from devices that are distant from the scientist.

renewable fuel standard (RFS)   A policy that requires a minimum volume of renewable fuel in gasoline.

renewable portfolio standard   A policy that requires utilities to provide a percentage of power from renewable energy sources.

renewable resources   Biological resources, such as trees, that may be renewed by reproduction and regrowth. Conservation to prevent overcutting and to protect the environment is still required, however. (Contrast **nonrenewable resources**.)

replacement fertility rate/level   The fertility rate (level) that will just sustain a stable population. It is 2.1 for developed countries.

reprocessing   The recovery of radioactive materials from nuclear reactors and the conversion of those materials into useful nuclear fuel.

reproductive health   Health care directed primarily toward the needs of women and infants, strongly emphasized in the 1994 Cairo population conference.

reproductive isolation   One of the processes of speciation, involving anything that keeps individuals or subpopulations from interbreeding.

reproductive strategy   A life history property of a species that involves a balance between reproduction and death. (See **r-strategist** and **K-strategist**.)

reserves   The amount of a mineral resource (including oil, coal, and natural gas) remaining in Earth that can be exploited using current technologies and at current prices. Usually given as **proved reserves** (those that have been positively identified) and **undiscovered reserves** (reserves from fields that have not yet been found).

resilience   The tendency of ecosystems to recover from disturbances through a number of processes known as resilience mechanisms (e.g., succession after a fire).

resistance   The development of more hardy pests through the effects of the repeated use of pesticides in selecting against sensitive individuals in a pest population.

resource partitioning   The outcome of competition by a group of species where natural selection favors the division of a resource in time or space by specialization of the different species.

resources   Biotic and abiotic factors that are consumed by organisms.

Resource Conservation and Recovery Act of 1976 (RCRA)   The cornerstone legislation to control indiscriminate land disposal of hazardous wastes.

respiration   See **cell respiration**.

restoration ecology   The branch of ecology devoted to restoring degraded and altered ecosystems to their natural state.

resurgence   The rapid comeback of a population (especially of pests after a severe die-off, usually caused by pesticides) and the return to even higher levels than before the treatment.

reverse osmosis   See **microfiltration**.

risk   The probability of suffering injury, disease, death, or some other loss as a result of exposure to a hazard.

risk assessment/analysis   The process of evaluating the risks associated with a particular hazard before taking some action.

risk characterization   The process of determining the level of a risk and its accompanying uncertainties after hazard assessment, dose-response assessment, and exposure assessment have been accomplished.

risk management   The task of regulators whose job is to review risk data and make regulatory decisions based on the data. The process often is influenced by considerations of costs and benefits as well as by public perception.

risk perception   Nonexperts' intuitive judgments about risks, which often are not in agreement with the level of risk as judged by experts.

roadless rule   A moratorium placed on building new roads in the national forests, put in place by the Clinton administration and contested by subsequent administrations.

rotavirus   A little-known virus that causes many childhood deaths, responsible for most cases of severe diarrhea.

Roundup®   See **glyphosate**.

*r***-strategist**   A reproductive strategy of a species that involves producing large numbers of young (a high reproductive rate) and survival of smaller numbers over time. Also called **opportunistic species**.

Rust Belt   Cities and their outskirts, often in the Midwest, that once specialized in heavy manufacturing and have since suffered a decline, characterized by abandoned factories and crumbling buildings.

Safe, Accountable, Flexible, Efficient Transportation Equity Act of 2005 (SAFETEA)   Successor to TEA–21 and ISTEA, this act authorizes funding for highway and transit programs.

Safe Drinking Water Act (SDWA) of 1974   Legislation to protect the public from the risk that toxic chemicals will contaminate drinking water supplies. The act mandates regular testing of municipal water supplies.

safety net   In a society, the availability of food and other necessities extended to people who are unable to meet their own needs for a variety of reasons.

salinization   The process whereby soil becomes saltier and saltier until, finally, the salt prevents the growth of plants. Salinization is caused by irrigation because salts brought in with the water remain in the soil as the water evaporates.

saltwater intrusion, saltwater encroachment   The phenomenon of seawater moving back into aquifers or estuaries. Such intrusion occurs when the normal outflow of freshwater is diverted or removed for use.

salvage logging   Established by the Bush administration, this refers to harvesting timber from recent burns in national forestlands; it is controversial because it is suspected of preventing natural regeneration of trees on the burned land.

sand   Mineral particles 0.063 to 2.0 mm in diameter.

sanitary sewer   A separate drainage system used to receive all the wastewater from sinks, tubs, and toilets.

SARA (Title III)   A section of the Superfund Amendments and Reauthorization Act that addresses hazardous waste accidents and promulgates community right-to-know requirements. Also known as the **Emergency Planning and Community Right-to-Know Act of 1986 (EPCRA)**. (See **Toxics Release Inventory (TRI)**.)

scale insect   Any of a group of minute insect pests that suck the juices from plant cells.

scientific method   The process of making observations and logically integrating those observations into a model of how the world works. Often involves forming hypotheses, experimenting, and conducting further testing for confirmation.

secondary air pollutants   Air pollutants resulting from reactions of primary air pollutants resident in the atmosphere. Secondary air pollutants include ozone, other reactive organic compounds, and sulfuric and nitric acids. (See **ozone, peroxyacetyl nitrates**, and **photochemical oxidants**.)

secondary clarifier   In activated sludge sewage treatment, the tank where organisms and other particles settle out after the aeration tank.

secondary consumer   An organism such as a fox or coyote that feeds more or less exclusively on other animals that feed on plants.

secondary pest outbreak   The phenomenon of a small and therefore harmless population of a plant-eating insect suddenly exploding to become a serious pest problem. Often caused by the elimination of competitors through the use of pesticides.

secondary recovery   In an oil well or oil field, the oil that can be removed by manipulating pressure in the oil reservoir by injecting brine or other substances; more costly than **primary recovery**.

secondary recycling   Recycling processes wherein waste materials are made into different products than the starting materials, for example, wastepaper to cardboard.

secondary succession   The reestablishment, through a series of stages, of a climax ecosystem in an area from which it was previously cleared.

secondary treatment   Also called **biological treatment**. A sewage-treatment process that follows primary treatment. Any of a variety of systems that remove most of the remaining organic matter by enabling organisms to feed on it and oxidize it through their respiration. Trickling filters and activated-sludge systems are the most commonly used secondary-treatment methods.

second-generation pesticides   Synthetic organic compounds used to kill insects and other pests, replacing the first-generation pesticides. Started with the use of DDT in the 1940s.

Secondhand smoke   Smoke experienced by those breathing air contaminated with tobacco smoke from others; classified as a known human carcinogen.

Second Law of Thermodynamics   The empirical observation, confirmed innumerable times, that in every energy conversion (e.g., from electricity to light), some of the energy is converted to heat and some heat always escapes from the system because it always moves toward a cooler place. Therefore, in every energy conversion, a portion of energy is lost, and since, by the **First Law of Thermodynamics**, energy cannot be created, the functioning of any system requires an energy input.

secure landfill   A landfill with suitable barriers, leachate drainage, and monitoring systems such that it is deemed secure against contaminating groundwater with hazardous wastes.

sediment Soil particles—namely, sand, silt, and clay—carried by flowing water. The same material after it has been deposited. Because of different rates of settling, deposits generally are pure sand, pure silt, or pure clay.

sedimentation The filling in of lakes, reservoirs, stream channels, and so on with soil particles, mainly sand and silt. The soil particles come from erosion. Also called **siltation**.

seep An area where groundwater seeps from the ground. Contrast with a **spring**, which is a single point from which groundwater exits.

selective cutting In forestry, the practice of cutting some mature trees from a stand but leaving enough to maintain normal ecosystem functions and a diverse biota.

selective pressure An environmental factor that causes individuals with certain traits that are not the norm for the population to survive and reproduce more than the rest of the population. The result is a shift in the genetic makeup of the population. Selective pressure is a fundamental mechanism of evolution.

septic system An on-site method of treating sewage that is widely used in suburban and rural areas. Requires a suitable amount of land and porous soil.

services Ecosystem functions that are essential to human life and economic well-being, such as waste breakdown, climate regulation, and erosion control. These can be further categorized as regulating, supporting, and provisioning services.

sex attractant A natural chemical substance (see **pheromones**) secreted by the female of many insect species that serves to attract males for the function of mating. Sex attractants may be used by humans in traps or to otherwise confuse insect pests in order to control them.

shadow pricing In cost-benefit analysis, a technique used to estimate benefits when normal economic analysis is ineffective. For example, people could be asked how much they might be willing to pay monthly to achieve some improvement in their environment.

sheet erosion The loss of a more or less even layer of soil from the land surface due to the impact of rain and runoff from a rainstorm.

shelterbelts Rows of trees around cultivated fields for the purpose of reducing wind erosion.

shelter-wood cutting In forestry practice, the strategy of cutting the mature trees in groups over a period of years, leaving enough trees to provide seeds and give shelter to growing seedlings.

sievert A unit of measurement of the ability of radioactive emissions to penetrate biological tissues. 1 sievert = 100 rem.

silt Soil particles between the size of sand particles and that of clay particles—namely, particles 0.004 to 0.063 mm in diameter.

siltation See **sedimentation**.

silviculture A term used to denote the practice of forest management.

simplification The human use of habitats that removes natural objects, such as maintaining a forest to produce one kind of tree or removing fallen logs.

sinkhole A large hole in the ground resulting from the collapse of an underground cavern.

slash-and-burn agriculture The practice, commonly found in tropical regions, of cutting and burning vegetation to make room for agriculture. The process destroys soil humus and may lead to rapid degradation of the soil.

sludge cake Treated sewage sludge that has been dewatered to make a moist solid.

sludge digesters Large tanks in which raw sludge (removed from sewage) is treated through anaerobic digestion by bacteria.

slum A concentrated habitation where squalor and disease reign, usually consisting of flimsy shacks and characterized by an almost complete lack of public amenities.

smart grid Control systems for power plants and electricity transmission lines with the ability to monitor problems, react to trouble quickly, and isolate trouble areas.

Smart Growth A movement that addresses urban sprawl by protecting sensitive lands and directing growth to limited areas, creating attractive, sustainable communities.

smog See **industrial smog** and **photochemical smog**.

social capital See **intangible capital**.

social modernization A process of sustainable development that leads developing countries through the demographic transition by promoting education, family planning, and health improvements rather than simply economic growth.

soft water Water with little or no calcium, magnesium, or other ions in solution that will cause soap to precipitate. (The soap would otherwise form a curd that makes a "ring" around the bathtub.) (Contrast **hard water**.)

soil A dynamic terrestrial system involving three components: mineral particles, detritus, and soil organisms feeding on the detritus.

soil classes Taxonomically arranged major groupings of soil that define the properties of the soils of different regions.

soil erosion The loss of soil caused by particles being carried away by wind or water.

soil fertility A soil's ability to support plant growth; often refers specifically to the presence of proper amounts of nutrients. The soil's ability to fulfill all the other needs of plants is also involved.

soil horizons Distinct layers within a soil that convey different properties to the soil and that derive from natural processes of soil formation.

soil order The highest group in the taxonomy of soils; there are 12 major orders of soil worldwide.

soil profile A description of the different naturally formed layers, called horizons, within a soil.

soil structure The composition of soil in terms of particles (sand, silt, and clay) stuck together to form clumps and aggregates, generally with considerable air spaces in between. Soil structure affects infiltration and aeration and develops as organisms feed on organic matter in and on the soil.

soil texture The relative size of the mineral particles that make up the soil. Generally defined in terms of the soil's sand, silt, and clay content.

solar cells See **photovoltaic cells**.

solar constant The radiant energy reaching the top of Earth's atmosphere from the Sun, measured at 1,366 watts per square meter.

solar energy Energy derived from the Sun; includes direct solar energy (the use of sunlight directly for heating or the production of electricity) and indirect solar energy (the use of wind, which results from the solar heating of the atmosphere, and biological materials such as wood, which result from photosynthesis).

solar-trough collectors Reflectors, in the shape of a parabolic trough, that reflect sunlight onto a tube of oil at the focal point of the trough. Thus heated, the oil is used to boil water to drive a steam turbine.

solar variability One proposed mechanism for global warming involving increases in solar output; this explanation has been discredited because of lack of convincing data.

solid waste The total of materials discarded as "trash" and handled as solids, as opposed to those that are flushed down sewers and handled as liquids.

solution A mixture of molecules (or ions) of one material in another; most commonly, molecules of air or ions of various minerals in water. For example, seawater contains salt in solution.

sound science *The basis for our understanding of how the world works and how human systems interact with it. It stems from scientific work based on peer-reviewed research and is one of the unifying themes of this text.*

source reduction According to the EPA, "the practice of designing, manufacturing, purchasing, or using materials in ways that reduce the amount or toxicity of trash created."

special-interest group In politics, people organized around a particular issue who lobby for their cause and exert their influence on the political process to bring about changes they favor.

specialization In evolution, the phenomenon whereby species become increasingly adapted to exploit one particular niche but thereby are less able to exploit other niches.

speciation The evolutionary process whereby populations of a single species separate and, through being exposed to different forces of natural selection, gradually develop into distinct species.

species All the organisms (plant, animal, or microbe) of a single kind. The "single kind" is determined by similarity of appearance or by the fact that members do or can mate and produce fertile offspring. Physical, chemical, or behavioral differences block breeding between species.

splash erosion The compaction of soil that results when rainfall hits bare soil.

springs Natural exits of groundwater to the surface.

stakeholders Groups of people who have an interest in the management of a given environmental resource, from those dependent on the resource to those managing or studying the resource.

standing crop biomass The biomass of primary producers in an ecosystem at any given time.

state capitalism A hybrid economic system with elements of both central planning and free-market economic systems, seen in China, Russia, and many Gulf oil states.

Statement on Forest Principles One of the treaties signed at the 1992 Earth Summit wherein the signatory countries agreed to a number of nonbinding principles stressing sustainable management.

steward/stewardship A steward is one to whom a trust has been given. As one of the unifying themes of this text, stewardship is *the actions and programs that manage natural resources and human well-being for the common good.*

Stockholm Convention on Persistent Organic Pollutants An international treaty that works toward the banning of most uses of the persistent organic pollutants.

storm drains Separate drainage systems used for collecting and draining runoff from precipitation in urban areas.

stratosphere The layer of Earth's atmosphere between 10 and 40 miles above the surface that contains the ozone shield. This layer mixes only slowly; pollutants that enter it may remain for long periods of time. (See **troposphere**.)

strip mining A mining procedure in which all the earth covering a desired material, such as coal, is stripped away with huge power shovels in order to facilitate removal of the material.

submerged aquatic vegetation (SAV) Aquatic plants rooted in bottom sediments and growing underwater. Submerged aquatic vegetation depends on the penetration of light through the water for photosynthesis.

subsidy A payment or other support of produced goods that helps the producers to maintain their production and earn income above what the market would normally provide.

subsistence farming Farming that meets the food needs of farmers and their families but little more. Subsistence farming involves hand labor and is practiced extensively in the developing world.

subsoil In a natural situation, the soil beneath topsoil. In contrast to topsoil, subsoil is compacted and has little or no humus or other organic material, living or dead. In many areas, topsoil has been lost or destroyed as a result of erosion or development, and subsoil is at the surface.

succession See **ecological succession**.

sulfur dioxide (SO_2) A major air pollutant and toxic gas formed as a result of burning sulfur. The major sources are burning coal (in coal-burning power plants) that contains some sulfur and refining metal ores (in smelters) that contain sulfur.

sulfuric acid (H_2SO_4) The major constituent of acid precipitation. Formed when sulfur dioxide emissions react with water vapor in the atmosphere. (See **sulfur dioxide**.)

Superfund The popular name for the **Comprehensive Environmental Response, Compensation, and Liability Act of 1980 (CERCLA)**. This act is the cornerstone legislation that provides the mechanism and funding for the cleanup of potentially dangerous hazardous-waste sites and the protection of groundwater.

Superfund Amendments and Reauthorization Act of 1986 (SARA) Amendments that strengthened the Superfund program for identifying and cleaning up hazardous waste sites.

supply-side policies Those energy policies that emphasize providing greater amounts of energy through more development of oil, coal, and gas resources.

surface impoundments Closed ponds used to collect and hold liquid chemical wastes.

surface water All bodies of water, lakes, rivers, ponds, and so on that are on Earth's surface. Contrast with **groundwater**, which lies below the surface.

suspended particulate matter (SPM) A category of major air pollutants consisting of solid and liquid particles suspended in the air. (See $PM_{2.5}$ and PM_{10}.)

sustainability A property whereby a process can be continued indefinitely without depleting the energy or material resources on which it depends. As one of the unifying themes of the text, *sustainability is the practical goal toward which our interactions with the natural world should be working.*

sustainable agriculture Agriculture that maintains the integrity of soil and water resources such that it can be continued indefinitely.

Sustainable Communities and Development Program in the United States that provides grants to encourage the development of more livable urban neighborhoods.

sustainable development Development that provides people with a better life without sacrificing or depleting resources or causing environmental impacts that will undercut the ability of future generations to meet their needs.

sustainable forest management The management of forests as ecosystems wherein the primary objective is to maintain the biodiversity and function of the ecosystem.

sustainable society A society that functions in a way so as not to deplete the energy or material resources on which it depends. Such a society interacts with the natural world in ways that maintain existing species and ecosystems.

sustainable yield The taking of a biological resource (e.g., fish or forests) in a manner that does not exceed the capacity of the resource to reproduce and replace itself.

sustained yield In forestry, the objective of managing a forest to harvest wood continuously over time without destroying the forest.

symbiosis The intimate living together or association of two kinds of organisms.

synergism The phenomenon whereby two factors acting together have a greater effect than would be indicated by the sum of their effects separately—as, for example, the sometimes fatal mixture of modest doses of certain drugs in combination with modest doses of alcohol.

synergistic effects Where two or more factors interact to affect organisms in a way that makes the impact greater than from the two acting separately.

syngas A combustible gas produced in an integrated gasification combined cycle plant by burning coal; the syngas is then burned in a gas turbine to produce electricity.

synthetic organics Any of a large group of organic compounds that may be synthesized in chemical laboratories and are known by the difficulty with which they degrade in the environment.

tariff Any tax placed on imported goods that raises the price of those goods in the recipient country.

taxonomy The science of identifying and classifying organisms according to their presumed natural relationships.

tectonic plates The huge slabs of rock in the lithosphere that have moved around and produced major changes in the surface of Earth; they are still moving, creating earthquakes and tsunamis when they do. See **plate tectonics**.

temperate deciduous forest A biome in the temperate zone characterized by deciduous trees, found where annual precipitation is 30–80 inches per year.

temperature inversion The weather phenomenon in which a layer of warm air overlies cooler air near the ground and prevents the rising and dispersion of air pollutants.

tenure Property rights over land and water resources that involve different patterns of use of those resources.

teratogenic Causing birth defects.

terracing The practice of grading sloping farmland into a series of steps and cultivating only the level portions in order to reduce erosion.

territoriality The behavioral characteristic exhibited by many animal species, especially birds and mammalian carnivores, to mark and defend a given territory against other members of the same species. A form of intraspecific competition.

territory The area defended by an animal in territorial behavior.

TESRA (Threatened and Endangered Species Recovery Act) One of a number of acts proposed in Congress in an attempt to weaken the Endangered Species Act.

theory A conceptual formulation that provides a rational explanation or framework for numerous related observations.

thermal pollution The addition of abnormal and undesirable amounts of heat to air or water. Thermal pollution is most significant with respect to the discharging of waste heat from electric generating plants—especially nuclear power plants—into bodies of water.

thermohaline Refers to temperature and salinity, specifically the cold temperature and high salinity of waters that sink in the North Atlantic and initiate the Atlantic Meridional Overturning Circulation.

thermosphere The highest layer in the atmosphere, between the mesosphere and outer space.

third world A dated term for **developing countries.**

threatened species A species whose population is declining precipitously because of direct or indirect human impacts.

threshold level The maximum degree of exposure to a pollutant, drug, or some other factor that can be tolerated with no ill effect. The threshold level varies, depending on the species, the sensitivity of the individual, the length of exposure, and the presence of other factors that may produce synergistic effects.

tidal barrage A form of renewable energy where barriers are placed across the mouth of a coastal bay and turbines are able to convert tidal water flow to power.

tilth In farming, the ability of a soil to support crop growth.

tipping point A situation in a human-impacted ecosystem where a small action or reaction is the catalyst for a major change in the state of the system.

Tobacco Control Act A law passed in 2009 giving the Food and Drug Administration greater authority to regulate tobacco products.

top-down regulation Basic control of a population (or species) occurs as a result of predation.

topsoil The surface layer of soil, which is rich in humus and other organic material, both living and dead. As a result of the activity of organisms living in the topsoil, it generally has a loose, crumbly structure, as opposed to being a compact mass.

total allowable catch (TAC) In fisheries management, a yearly quota set for the harvest of a species by managers of fisheries.

total fertility rate The average number of children that will be born alive to each woman during her total reproductive years if her fertility is average at each age.

Total Maximum Daily Load (TMDL) program An EPA-administered program to address nonpoint-source water pollution that sets pollution limits according to the ability of a body of water to assimilate different pollutants.

total product life cycle The sum of all steps in the manufacture of a product, from obtaining the raw materials through producing, using, and, finally, disposing of the product. By-products and the pollution resulting from each step are taken into account in a consideration of the total product life cycle.

toxicity The condition of being harmful, deadly, or poisonous.

toxicology The study of the impacts of toxic substances on human health and the pathways by which such substances reach humans.

Toxics Release Inventory (TRI) An annual record of releases of toxic chemicals to the environment and the locations and quantities of toxic chemicals stored at all U.S. sites. Required by the **Emergency Planning and Community Right-to-Know Act of 1986 (EPCRA).** See **SARA.**

Toxic Substances Control Act of 1976 (TSCA) A law requiring the assessment of the potential hazards of a chemical before the chemical is put on the market.

trace elements Those essential elements, like copper or iron, that are needed in only very small amounts.

trade winds The more or less constant winds blowing in horizontal directions over the surface as part of Hadley cells.

traditional agriculture Farming methods as they were practiced before the advent of modern industrialized agriculture.

TRAFFIC network An acronym for the Trade Records Analysis for Flora and Fauna in International Commerce network, monitoring trade in wildlife and wildlife parts.

tragedy of the commons The overuse or overharvesting and consequent depletion or destruction of a renewable resource that tends to occur when the resource is treated as a commons—that is, when it is open to be used or harvested by any and all with the means to do so.

trait Any physical or behavioral characteristic or talent that an individual is born with.

transgenic organism Any organism with genes introduced from other species via biotechnology to convey new characteristics.

transpiration The loss of water vapor from plants. Water evaporates from cells within the leaves and exits through the stomata.

trapping technique The use of sex attractants to lure male insects into traps.

treadle pump A pump worked like a step exercise machine that pulls water up from groundwater for irrigation, commonly used in developing countries.

treated sludge Solid organic material that has been removed from sewage and treated so that it is nonhazardous.

trickling-filter system A system in which wastewater trickles over rocks or a framework coated with actively feeding microorganisms. The feeding action of the organisms in a well-aerated environment results in the decomposition of organic matter. Used in secondary or biological treatment of sewage.

tritium (^{3}H) An unstable isotope of hydrogen that contains two neutrons in addition to the usual single proton in the nucleus. Tritium does not occur in significant amounts in nature.

trophic level A feeding level defined with respect to the primary source of energy. Green plants are at the first trophic level, primary consumers at the second, secondary consumers at the third, and so on.

trophic structure The major feeding relationships between organisms within ecosystems, organized into trophic levels.

tropical rain forest Biome in the tropic zone where rainfall averages over 95 inches/yr; locations of great biodiversity.

troposphere/tropopause The layer of Earth's atmosphere from the surface to about 10 miles in altitude. The tropopause is the boundary between the troposphere and the stratosphere above. The troposphere is well mixed and is the site and source of our weather as well as the primary recipient of air pollutants. (See also **stratosphere.**)

tsunami A series of ocean waves (often called tidal waves) originating with underwater or coastal earthquakes and moving rapidly across ocean expanses, often resulting in major inundations of coastal areas.

tundra biome In cold latitudes and high altitudes, the pattern of plants and animals able to live where permafrost persists; dominated by low-growing sedges, shrubs, lichens, mosses, and grasses.

turbid Cloudy due to particles present. (Said of water and water purity.)

turbine A rotary engine driven at a very high speed by steam, water, or exhaust gases from combustion, employed in generating electrical energy.

turbogenerator A turbine coupled to and driving an electric generator. Virtually all commercial electricity is produced by such devices.

ultraviolet radiation Radiation similar to light but with slightly shorter wavelengths and with more energy than violet light. Its greater energy causes ultraviolet light to severely burn and otherwise damage biological tissues.

underground storage tanks (USTs) Tanks used to store petroleum products at service stations, now subject to mandated protection against leakage. (See **UST legislation.**)

undernourishment A form of hunger in which the individual lacks adequate food energy, as measured in calories. Starvation is the most severe form of undernourishment.

underweight The world's number one health risk factor, this refers to the effects of undernourishment on children that prevent their normal growth.

undiscovered resources Estimates of oil and natural gas based on geological science but not explored to confirm their presence; the estimates may be way off the mark.

United Nations Convention to Combat Desertification (UNCCD) A UN agency with the mission of helping drylands countries take steps to reduce soil degradation.

United Nations Development Program (UNDP) A UN agency with offices in 131 countries, with the mission of helping developing countries to attract and use development aid effectively and, in particular, to reach the Millennium Development Goals.

United Nations Environment Program (UNEP) A UN agency with the mission of providing leadership and encouraging partnership in caring for the environment by inspiring, informing, and enabling nations and peoples to improve their quality of life without compromising that of future generations.

United Synthesis Product (USP) The 2009 report, *Global Change Impacts in the United States*, drawing from 19 existing Climate Change Science Program reports, providing a detailed look at the impacts of climate change on various sectors and regions in the U.S.

unmet need A situation where women who would like to postpone or prevent childbearing are not currently using contraception, often because it is not available to them.

upwelling In oceanic systems, the upward vertical movement of water masses, caused by diverging currents and offshore winds, bringing nutrient-rich water to the surface.

urban blight/decay The general deterioration of structures and facilities such as buildings and roadways, in addition to the decline in quality of services such as education, that has occurred in inner-city areas.

urban food forest A new design for urban landscapes where perennial plants and trees are grown with the intention of allowing people to forage them for food.

urban foragers Those urban dwellers who forage for fruits, berries, mushrooms, and other foods in urban landscapes.

urban sprawl The widespread expansion of metropolitan areas through building houses and shopping centers farther and farther from urban centers and lacing them together with more and more major highways, usually without planning.

USDA Organic A food that is certified by the U.S. Department of Agriculture to have been produced without the use of chemical fertilizers, pesticides, or antibiotics.

UST legislation Amendments to the Resources Conservation and Recovery Act of 1976 passed in 1984 to address the mounting problem of leaking underground storage tanks (USTs).

value In ethical considerations, a property attributed to objects, species, or individuals that implies a moral duty to the one assigning the value; intrinsic value and instrumental value are two types of value.

value added The increase in value from raw materials to final production.

vector/vector control An agent, such as an insect or tick, that carries a parasite from one host to another; methods of controlling a parasite or disease by attacking the vector.

virtual water In place of shipping water to water-stressed regions, food may be exported to those regions, thus preventing the direct need for water and replacing it with "virtual" water.

vitamin A specific organic molecule that is required by the body in small amounts but that cannot be made by the body and therefore must be present in the diet.

volatile organic compounds (VOCs) A category of major air pollutants present in the air in the vapor state. The category includes fragments of hydrocarbon fuels from incomplete combustion and evaporated organic compounds such as paint solvents, gasoline, and cleaning solutions. Volatile organic compounds are major factors in the formation of photochemical smog.

vulnerability A measure of the likelihood of exposure to a hazard.

Waste Isolation Pilot Plant (WIPP) A facility built by the U.S. Department of Energy in New Mexico to receive defense-related nuclear wastes.

waste-to-energy combustion A process of combustion of solid wastes that also generates electrical energy.

WasteWise An EPA-sponsored program that targets the reduction of municipal solid waste by a number of partnerships.

water cycle See **hydrologic cycle.**

water-holding capacity The ability of a soil to hold water so that it will be available to plants.

waterlogging The total saturation of soil with water. Waterlogging results in plant roots not being able to get air and dying as a result.

watershed The total land area that drains directly or indirectly into a particular stream or river. The watershed is generally named from the stream or river into which it drains.

water table The upper surface of groundwater, rising and falling with the amount of groundwater.

water vapor Water molecules in the gaseous state.

weather The day-to-day variations in temperature, air pressure, wind, humidity, and precipitation mediated by the atmosphere in a given region.

weathering The gradual breakdown of rock into smaller and smaller particles, caused by natural chemical, physical, and biological factors.

weed Any plant that competes with agricultural crops, forests, lawns, or forage grasses for light and nutrients.

wet cleaning A water-based alternative to dry cleaning that avoids the use of hazardous chemicals.

wetlands Areas that are constantly wet and are flooded at more or less regular intervals—especially marshy areas along coasts that are regularly flooded by tides.

wetland system A biological aquatic system (usually a restored wetland) used to remove nutrients from treated sewage wastewater and return it virtually pure to a river or stream. Wetland systems are sometimes used when the application of treated wastewater for irrigation is not feasible.

Wild and Scenic Rivers Act of 1968 Legislation that protects rivers designated as "wild and scenic," preventing them from being dammed or otherwise harmfully affected.

wilderness Land that is undeveloped and wild; in the U.S., land that is protected by the Wilderness Act.

Wilderness Act of 1964 Federal legislation that provides for the permanent protection of undeveloped and unexploited areas so that natural ecological processes can operate freely in them.

Wildlife Services An agency within the U.S. Department of Agriculture that removes "nuisance" animals for landowners; the agency dispatches more than 2 million animals and birds yearly.

wind farms Arrays of numerous, modestly sized wind turbines for the purpose of producing electrical power.

windrows Piles of organic material extended into long rows to facilitate turning and aeration in order to enhance composting.

wind turbines   Large "windmills" designed for the purpose of producing electrical power.

workability   The relative ease with which a soil can be cultivated.

World Bank   A branch of the United Nations that acts as a conduit for handling loans to developing countries.

World Food Program (WFP)   A UN administration that receives donations of money and food from donor countries and distributes food aid to needy regions and countries.

World Trade Organization (WTO)   A body that sets the rules for international trade agreements, made up of representatives from many countries and often dominated by the rich nations.

xeriscaping   Landscaping with drought-resistant plants that need no watering.

yard wastes   Grass clippings and other organic wastes from lawn and garden maintenance; major target for composting.

yellowcake   So called because of its yellowish color, uranium ore that has been partially purified and is ready for further purification and enrichment.

Younger Dryas event   A rapid change in global climate between 10,000 and 12,000 years ago that brought on 1,500 years of colder weather.

Yucca Mountain   A remote location in Nevada chosen to be the nation's commercial nuclear waste repository; although well along in development, its future is uncertain.

zones of stress   Regions where a species finds conditions tolerable but suboptimal. The species survives but is under stress.

zoonotic   Diseases that may be transmitted between animals and humans.

zooplankton   Any of a number of animal groups, including protozoa, crustaceans, worms, mollusks, and cnidarians, that live suspended in the water column and that feed on phytoplankton and other zooplankton.

zooxanthellae   Photosynthetic algae that live within the tissues of coral species and enable them to grow productively in warm, shallow coastal areas.

CREDITS

Photo Credits

Frontmatter
Iii: Photodisc/Getty Images/Houghton Mifflin Harcourt; vi: KIM STALLKNECHT/ AFP/Newscom; vii: Altmarkfoto/iStockphoto; viii: David Grossman / Alamy; x: AP Photo/Anonymous; xi: Mircea Bezerghean/Shutterstock; xii: altrendo travel/Getty Images; xix: AFP/Getty Images

Chapter 1
Page 1 (top left): Silia Photo/Shutterstock; (bottom right): Photodisc/Getty Images/ Houghton Mifflin Harcourt; Page 2 (top right): Joseph Sohm/Visions of America/ Glow Images, Inc.; (bottom right): Steve Byland / Fotolia; Page 3 (top left): Rachel Carson Counci/Newscom; Page 3 (top right): brt COMM / Alamy; Page 8 (top right) John Sylvester / Alamy; (top left): Casper Wilkens/iStockphoto; Page 11 (bottom left): AP Photo/Las Cruces Sun-News, Norm Dettlaff; (bottom center): Science Images/UIG/Getty Images; (top center): Joseph Sohm/Visions of America/Glow Images, Inc.; Page 15 (top left): Michael Amaranthus/Mycorrhizal Applications, Inc.; Page 18 (top left): Goldman Environmental Prize; (top right): AFP/Getty Images; Page 19 (top left): Sveta/iStockphoto; Page 21 (top left): globestock/ iStockphoto

Chapter 2
Page 23 (top): Lou Linwei / Alamy; (bottom left): Martha Bayona/iStockphoto; Page 24: PETER PARKS/AFP/Getty Images/Newscom; Page 27: KIM STALLKNECHT/ AFP/Newscom; Page 30 (left): Toklo/iStockphoto; (center): Noah Poritz/MacroWorld/ Photo Researchers, Inc.; (right): David Weintraub/Photo Researchers, Inc.; Page 35 (left): fstockfoto/Shutterstock; (center): Frontpage/Shutterstock; (top right): VisualField/iStockphoto; Page 37 (top right): AP Photo; Page 39 (top left): John Craig/Bureau of Land Management; (top right): Dainis Derics/Shutterstock

Chapter 3
Page 49 (top left): Altmarkfoto/iStockphoto; (bottom right): GlobalP/iStockphoto; Page 50 (top right): Mifippo/iStockphoto; (bottom right): Alex Staroseltsev/ Shutterstock; Page 52 (center left): Olga Khoroshunova/Fotolia; (center): Pete Oxford/Nature Picture Library; (bottom right): LYNN M. STONE/Nature Picture Library; Page 53 (top left): Jerry and Marcy Monkman/EcoPhotography.com/ Alamy; (bottom right): Richard T. Wright; Page 54 (bottom right): louise murray / Alamy; Page 57: Derek Croucher / Alamy; Page 59 (bottom left): Pascal Goetgheluck / Photo Researchers, Inc.; Page 63 (top): khunaspix/Fotolia; Page 64 (top right): Olga Khoroshunova/Fotolia

Chapter 4
Page 75 (top left): Justin Black/Shutterstock; (bottom left): Joel Sartore/Getty Images; Page 78 (top right): Kokhanchikov/Shutterstock; Page 81 (top left): The Natural History Museum / The Image Works; (center left): Nigel Hicks/Dorling Kindersley; Page 82 (top left): galam/Fotolia; (top right): Phanie / SuperStock; (bottom left): James L. Amos/National Geographic/Getty Images; (bottom right): John Watney / Photo Researchers, Inc.; Page 84 (bottom left): Images of Africa Photobank / Alamy; (bottom right): Michael S. Nolan / age fotostock / SuperStock; Page 85 (bottom left): William Leaman / Alamy; (top right): Charles Stirling (Diving) / Alamy; Page 87 (center): Stubblefield Photography/Shutterstock; (left): David Hosking / Alamy; Page 88 (top left): Stephen P. Parker/Photo Resesearchers, Inc.; (top right): Bernard J. Nebel; (bottom left): Pj Malsbury/iStockphoto; Page 89 (bottom left): Susanna Sez El Pais Photos/Newscom; Page 90 (top left): Bernard J. Nebel; (top right): Natural Visions / Alamy; Page 94 (bottom left): Jeff Lepore/Photo Researchers, Inc.; (bottom center): Craig K. Lorenz/Photo Researchers, Inc.; Page 96 (center right): Andreas Von Einsiedel/Dorlng Kindersley; (bottom right): Krys Bailey / Alamy; Page 97 (top left): LightScribe/iStockphoto; (bottom left): Doug Smith/MCT/Newscom; (bottom right): NHPA /SuperStock; Page 98 (right): John W. Bova/Photo Researchers, Inc.

Chapter 5
Page 101 (top left): Martin Rietze/Stocktrek Images/Alamy Limited; (bottom left): iStockphoto; Page 102 (center left): Kai Horstmann / vario images GmbH & Co.KG/ Alamy; (top left): ROBERT T. WELLS/UNEP / Still Pictures / The Image Works Page 107 (top left): Reproduced from Transmission electron micrograph by Breznak & Pankratz (1977) with permission of the American Society for Microbiology.; Page 111: Oksana.perkins/Shutterstock; Steve Geer/iStockphoto.com; Dmitry Savinov/Shutterstock; Galyna Andrushko/Shutterstock; Yuriy Kulyk/Shutterstock; © TTphoto/Shutterstock; Page 114: Mikedabell/iStockphoto; Johnny Lye/iStockphoto; Angelo Garibaldi/Shutterstock; Cedric Schmidt/Fotolia; iStockphoto; Mr. Eckhart/ IStockphoto; Page 116 (top left) Bernard J. Nebel; (bottom right): Bernard J. Nebel; (top left): Bernard J. Nebel; (bottom right): Piotr Gatlik/Shutterstock; Page 118 (bottom left): Anna Galejeva/Shutterstock: Page 119 (top left): szefei/Shutterstock; (top right): Tupungato/Shutterstock; Page 120: Everett Collection Inc / Alamy

Chapter 6
Page 129 (bottom left): Varin-Visage/Photo Researchers, Inc.; (top left): Jacana/ Photo Researchers, Inc.; Page 130 (bottom right): STEPHANIE STONE/AFP/Getty Images/Newscom; Page 131 (bottom right): Ryan M. Bolton/Shutterstock; Page 132 (bottom right): AP Photo/George Osodi; Page 133 (bottom left): WILDLIFE GmbH / Alamy; Page 134 (center left): iStockphoto; (center): Jeremy Woodhouse/ Blend Images/Getty Images; (center right): blickwinkel/Rose/Alamy; Page 139 (top): James P. Blair/NATIONAL GEOGRAPHIC IMAGE COLLECTION/Getty Images; (bottom left): Nagy Melinda/Shutterstock; Page 141 (top left): James H Robinson/Photo Researchers/Getty Images; (top right): mkimages/Alamy; Page 142 (top left): blickwinkel / Alamy; (bottom right): AP Photo/The Daily News-Record, Pete Marovich; Page 143 (bottom left): Piotr Gatlik/Shutterstock; (bottom left): Premaphotos/Alamy; Page 144 (center right): AP Photo/John Kuntz/The Plain Dealer; Page 145 (top right): Roberta Olenick/All Canada Photos/SuperStock; Page 146 (bottom right): AndreyKlepikov/Shutterstock; Page 148 (top): AP Photo/ The El Paso Times, Rudy Gutierrez; Page 149 (top left): Marvin Dembinsky Photo Associates/Alamy; (top center): kuma/Shutterstock; (top right): David Norris/ Photo Researchers, Inc.; (center left): A.H. Rider/Photo Researchers, Inc.; (center): Jeff Lepore/Photo Researchers, Inc.; (center right): Canon Bob/iStockphoto.com; (bottom right): Tom McHugh/Steinhart Aquarium/Photo Researchers, Inc.; (bottom center): Gorshkov25/Shutterstock; (bottom right): Nicolas Larento/Fotolia
Page 152 (top right): Richard Langs/Shutterstock
Page 153 (top left): Tom Lynn/MCT/Newscom

Chapter 7
Page 158 (top): Rich Carey / Shutterstock; (bottom): Reinhard Dirscherl/Robert Harding World Imagery; Page 161 (top left): think4photop/Shutterstock; Page 162: Andre Seale/Alamy; Page 163 (top left): Daniel Heuclin/NHPA/Photoshot; (bottom right): Jim West / Alamy; Page 164: Norma Jean Gargasz/Alamy; Page 167: Image Quest Marine; Page 169 (bottom right): jeremy sutton-hibbert/Alamy; Page 171 (bottom left): Tim Hall/Robert Harding World Imagery; Page 173 (top right): Robert Brook/Photo Researchers, Inc.; Page 176 (top left): AP Photo; Page 177 (top left): Kurt Lackovic / Alamy; (bottom right): Georgette Douwma / Photo Researchers, Inc.

Chapter 8
Page 186 (top right): David Grossman / Alamy; (bottom left) TED ALJIBE/AFP/ Getty Images/Newscom; Page 187 (left): L Ancheles/Getty Images; Ron Giling/ Lineair/Still Pictures/Specialist Stock; (right): James Baigrie/FoodPix/Getty Images; Page 190 (bottom right): Peter Johnson/CORBIS/Glow Images, Inc.; Page 191 (bottom left): Library of Congress Prints and Photographs Division [LC-DIG-nclc-01537]; Page 192 (top left): The National Archives/SSPL/Getty Images; (bottom right): NARINDER NANU/AFP/Getty Images; Page 201 (top right): Evelyn Hockstein/MCT/Newscom; Page 202 (top left): ASTER Science Team/ Nasa Images; (bottom right): J.m._Vidal European Press Agency/Newscom; Page 203 (bottom right): imagebroker.net / SuperStock

Chapter 9
Page 216 (top right): Thor Jorgen Udvang/Shutterstock; (bottom right): foto76/ Fotolia; Page 219 (center right): Amit Bhargava/Bloomberg via Getty Images; Page 221 (bottom left): F1online/Getty Images; (bottom right): Images & Stories / Alamy; Page 225 (top right): Alex Segre/Alamy; Page 227 Gallo Images / Alamy; Page 228 (top left): Courtesy of The World Bank; AP Photo/Julie Plasencia; Page 229: Caro / Alamy; Page 232: Fabienne Fossez/Alamy: Page 233: Xinhua /Landov; Page 235 (top left): Peter D. O'Neill TW : EEC/Alamy; (bottom right): STR/AFP/ Getty Images/Newscom

Chapter 10
Page 239 (top left): FOREST LUGH/EPA/Newscom; (bottom left): Stringer Shanghai/REUTERS; Page 241 (top left): Fotolia; (bottom right): Andrew Woodley / Alamy; Page 247: Dead Shot Keen / Alamy; Page 248: Caro / Alamy; Page 252: Alison Jones / DanitaDelimont.com Danita Delimont Photography/Newscom; Page 254: PacifiCorp/AP Images; Page 255 (top right): Stephen Mcsweeny/Shutterstock; (center right): Francois Gohier/Photo Researchers, Inc.; Page 256 (top): AP Photo/ Dale Wilson; Page 258 Eitan Simanor / Alamy; Page 259 (bottom left): Lowell Georgia/Photo Researchers, Inc.; (bottom right): Jake Lyell / Alamy; Page 260 (top right): berna namoglu/Shutterstock; Page 262 AP Photo/Gary Kazanjian

Chapter 11
Page 266 (bottom right): fotolE/iStockphoto; (top right): National Archives; Page 267: Ted Foxx / Alamy; Page 274 (bottom left): Bernard J. Nebel; (bottom right): Richard T. Wright; Page 275: Eye of Science / Photo Researchers, Inc.; Page 276: Philippe Psaila / Photo Researchers, Inc.; Page 278 (top): Charles R. Belinky/Photo Researchers, Inc.; (center right): Jon Arnold Images Ltd/Alamy; Page 279: Edwin Remsberg/Alamy; Page 280 (top left): U.S. Department of Agriculture; (center left):

David Wall/Alamy; Page 281: Leng Yuehan - CNImaging/Newscom; Page 282 (top left): Betty Derig/Photo Researchers; (top right): DAVID GRAY/REUTERS/Newscom; Page 284 (top right): Richard Lord/PhotoEdit; Page 285 (bottom left): Issouf Sanogo/AFP/Getty Images

Chapter 12

Page 289 (top left): AP Photo/Carrie Antlfinger; (bottom left): Basel101658/Shutterstock; Page 290 (bottom right): iStockphoto; Page 291 (top left): Pete Oxford/Danita Delimont, Agent/Alamy; Page 292 (top left): Garry D. McMichael/Photo Researchers, Inc.; (top right): Fotokostic/Shutterstock; Page 293 (top right): FDR Library; Page 294 (top): Kip Ross/National Geographic/Getty Images; Page 295 (top right): Jim West / Alamy; Page 296 (bottom left): Wayne Hutchinson/Alamy; (top right): James Steidl/Shutterstock; Page 298 (top left): Bill Barksdale/Age Fotostock; Page 299 (top left): Courtesy Danforth Plant Science Center; (bottom right): kurt_G/Shutterstock; Page 300 (bottom right): MICHEL GANGNE/AFP/Getty Images/Newscom; Page 301 (top right): Golden Rice Humanitarian Board. **www.goldenrice.org**; Page 304 (top right): Dorthea Lange/[LC-DIG-fsa-8b29516]/Library of Congress Prints and Photographs Division; Page 305 (bottom left): Susan Kuklin / Photo Researchers, Inc.; (bottom center): Todd Bannor/Alamy; (bottom right): NASA Images; Page 307 (top center): John Takai/Fotolia; Page 308 (top left): David Cole/Alamy; Page 314 (top left): John Birdsall/Age Fotostock; (top right): **www.NASA Images.gov**; (center left): Bloomberg/Getty Images; (center right): AP Photo/Orlin Wagner

Chapter 13

Page 317 (top left): Katja Heinemann/Aurora Photos/Alamy; (bottom left): Oxford Scientific/Getty Images; Page 319 (top left): Nigel Cattlin/Alamy; (top right): Richard T. Wright; (center left): Richard T. Wright; (center right): Arco Images GmbH/Alamy; (left): Nick Greaves/Alamy; (right): Allik Camazine/Alamy; (bottom left): Roger Eritja/Alamy; (bottom right): Photo Researchers, Inc.; Page 320 (top left): Drake Fleege/Alamy; (top right): U.S. Customs and Border Protection; Page 322 (top right): Ian Berry/Magnum Photos; Page 323 (bottom right): courtesy/everett collection inc/AGE Fotostock; (top right): John de la Bastide/Shutterstock; Page 325 (top left): Nigel Cattlin / Photo Researchers, Inc.; (bottom right): Ronnie Howard/Shutterstock; Page 327 (bottom left): James H. Robinson / Photo Researchers, Inc.; Page 328 (bottom right): R Koenig/Blickwinkel/AGE Fotostock; Page 330 (bottom right): Richard T. Wright; Page 331 (center right): USDA/APHIS Animal And Plant Health Inspection Service; (bottom left): Harry Rogers/Photo Researchers, Inc.; Page 332 (bottom left): juulijs/Fotolia; (bottom right): Louis Quitt/Photo Researchers, Inc.; Page 333 (top left): Christopher M. Ranger, Penn State University; (top right): A3386 Uli Deck Deutsche Presse-Agentur/Newscom; Page 334 (top right): Sonda Dawes/The Image Works; Page 336 (top right): Nigel Cattlin/Holt Studios International/Photo Researchers, Inc.

Chapter 14

Page 342 (top right): Stock Connection/SuperStock; (bottom left): Elzbieta Sekowska/Shutterstock; Page 343 (top left): AP Photo/Anonymous; (bottom left): Jason Verschoor/iStockphoto; Page 344: Handout/Reuters; Page 345: Douglas Lander/Alamy; Page 347 (center right): Mark Jensen/iStockphoto; Page 350 Martin Múransky/Shutterstock; Page 357 AP Photo/Kevin Frayer; Page 362 AP Photo/Str; Page 363 Jim West/PhotoEdit; Page 366 iStockphoto

Chapter 15

Page 369 (top left): Aflo/Mainichi Newspaper/epa/Newscom; (bottom left): Dalibor Zivotic/iStockphoto; Page 370: Kyodo/Newscom; Page 372: AP Photo/Tennessee Valley Authority; Page 375 (center right): Erich Hartmann/Magnum Photos; Page 381 (top right): AP Photo/Holtec International; (bottom right): Everett Collection Historical / Alamy; Page 382: AP Photo; Page 383: (top right): ITAR-TASS Photo Agency / Alamy; Page 387: AP Photo/Robert F. Bukaty, File; Page 388: The National Ignition Facility

Chapter 16

Page 392 (top right): Meyerbroeker/Caro/Alamy; (bottom right): Sebastien Burel/Shutterstock; Page 393 (top left): Armen Gharibyan/Alamy; (top right): MaxFX/Shutterstock; Page 398 (top left): Bernard J. Nebel; Page 399 (top right): National Renewable Energy Laboratory; (top right): Ethan Miller/Getty Images; Page 400: (bottom left): Robyn Beck/Afp/Getty Images/Newscom; Page 402: Iofoto/Shutterstock; Page 404 (top left): Planetpix/Alamy; Page 404 (bottom right): Creative Energy Corporation; Page 405 (top right): Joerg Boethling / Alamy; Page 406: Ideeone/iStockphoto; Page 408 (bottom right): Transtock Inc. / Alamy; Page 409: Courtesy of Pacific Gas and Electric Company

Chapter 17

Page 414: (top right): Mircea Bezerghean/Shutterstock; (bottom left): PhilAugustavo/iStockphoto; Page 415 (top left): Capifrutta/Shutterstock,com; (bottom left): Pasieka/Science Photo Library/Alamy; Page 423 (top left): Aflo Foto Agency/Alamy; (top right): Pammy1140/Fotolia; (center left): Mark Pearson/Alamy; (center right): Ryan McGinnis/Alamy; Page 425 (top left): Calvin Larsen/Photo Researchers, Inc.; (top right): Robert Harding/Glow Images, Inc.; (center right): Tomas Sereda/Fotolia; (center right): Bloomberg/Getty Images; Page 427: Ulrich Doering/Alamy; Page 428: AP Photo; Page 431: Jake Lyell / Alamy

Chapter 18

Page 441 (top left): Kyodo /Landov; (bottom left): Jack F/Fotolia; Page 442 (top left): Shahee Ilyas/Alamy; (center left): GABRIEL BOUYS/AFP/Getty Images; (top

center): Office of Science and Technology Policy Executive Office of the President/NASA Images; Page 454 (top right): National Snow and Ice Data Center; (top right): National Snow and Ice Data Center

Chapter 19

Page 472 (top right): Ronaldo Schemidt/AFP/Getty Image; (bottom right): Tomas Sereda/Fotolia; Page 473: Bettmann/Corbis; Page 476 (top left): Ekash/iStockphoto; (top right): Daniel Stein/iStockphoto: Page 478: Richard T. Wright; Page 480 (top right): Les Cunliffe/Fotolia; (center right): Alexandra Gl/Fotolia; (center): hartphotography/Fotolia; (right): Elenathewise/Fotolia; (bottom right): Stefan Redel/Fotolia; (bottom): martin33/Fotolia; Page 485 (top left): Grapes - Michaud/Photo Researchers, Inc.; Page 486 (center): Matt Meadows/Peter Arnold/Getty Images; (center left): Matt Meadows/Peter Arnold/Getty Images; Page 487: Seth Joel/Photographer's Choice/Getty Images; Page 488: AS Heagle/United States Department of Agriculture; Page 490: Balliolman/Fotolia; Page 492: AP Photo

Chapter 20

Page 506 (top left): Courtesy of NASA Images Images; (bottom right): Science Photo Library / Alamy; Page 508 (center right): richard ellis/Alamy; Page 509 (top left): AP Photo; Page 510: Robert Harding Picture Library / SuperStock; Page 511 (left): Bob Hudson/George Waclawin/Bernard J. Nebel; (center left): Bob Hudson/George Waclawin/Bernard J. Nebel; (center): Bob Hudson/George Waclawin/Bernard J. Nebel; (center): Bob Hudson/George Waclawin/Bernard J. Nebel; Page 513: Thomas R. Fletcher / Alamy; Page 514 (bottom): Charles R. Belinky/Photo Researchers, Inc.; Page 515: Bernard J. Nebel; Page 518 (top right): Bernard J. Nebel; (center right): Bernard J. Nebel; Page 520: eric fowke/alamy; Page 521 (top left): Bernard J. Nebel; (top right): Bernard J. Nebel; Page 522 (top left): Scott Smith/Sancor Industries Ltd; (top right): Scott Smith/Sancor Industries Ltd; Page 525 (bottom right): G. Avila / Alamy; Page 528: Jennifer Weinberg / Alamy

Chapter 21

Page 532 (top right): Jack Plunkett/Bloomberg/Getty Images; (bottom right): imagestock/iStockphoto; Page 533: Photo © Cymie Payne, courtesy of Cambridge Arts Council; Page 534: Van D. Bucher/Photo Researchers, Inc.; Page 535: Jim West / Alamy; Page 537: NYC Parks Department; Page 540: D. Hurst/Alamy; Page 541: Trex, Inc.; Page 542: Justin Sullivan/Getty Images; Page 543: John Griffin/The Image Works; Page 545: Jim R. Bounds/Bloomberg/Getty Images; Page 547: Richard T. Wright

Chapter 22

Page 551 (top left): M.Sobreira/Alamy; (bottom): Gresei/Fotolia; Page 558: imagebroker / Alamy; Page 559 (top right): AP Photo/Jackie Johnston; (bottom right): U.S. Environmental Protection Agency Headquarters; Page 560 (bottom right): Joe Traver/Liaison/Getty Images; (top left): Nancy J. Pierce/Photo Researchers, Inc.; (bottom left): Van Bucher/Photo Researchers, Inc.; Page 563 (bottom right): Richard T. Wright; Page 565 (center right): Tetra Tech EM Inc; Page 567: BEDI/AFP/Getty Images/Newscom

Chapter 23

Page 573 (top left): Curtis W. Richter/Getty Images; (bottom right): Reinhold Foeger/Shutterstock; Page 574 (top right): JRC, Inc./Alamy; (bottom): Age Fotostock/SuperStock; Page 575: ERIK S. LESSER/EPA/Landov Media; Page 576 (top left): Richard Thornton/Shutterstock; (center left): Esbin-Anderson/The Image Works; Page 577 (top right): MaxFX/Shutterstock; (bottom left): Johnny Stockshooter/Alamy; (center right): Shaun Cunningham / Alamy; (bottom right): asterix0597/iStockphoto; Page 583 (top left): Cameron Davidson /Getty Images; (center right): Bernard J. Nebel; Page 584 (top left): Bernard J. Nebel; (top right): altredo travel/Getty Images; Page 585: Marcia Chambers /dbimages/Alamy; Page 586 (top left): Medioimages/Photodisc/Getty Images; (top right): steinphoto/iStockphoto; (bottom right): Warwick Kent/Photolibrary/Getty Images; Page 588: Andrew Murphy/Alamy; Page 589 (bottom left): Chattanooga Area Convention and Visitors Bureau; Page 591 (center right): Jeffrey Pohorski; Page 592: Ilene MacDonald / Alamy

Text Credits

Chapter 1

Pages 3–4, 6: Source: *Silent Spring* by Rachel Carson. Published 1962 by Houghton Mifflin; Pages 49–50, Table 1-1: Source: Based on "Millennium Ecosystem Assessment, Ecosystems and Human Well-Being: Synthesis" from Millennium Ecosystem Assessment website; Figure 1.4: "Worldwide trends in the Human Development index," from *2010 Human Development Report*. Copyright © 2010 by United Nations Publications. Reprinted with permission. All rights reserved.; Figure 1.5: "Atmospheric CO2 at Mauna Loa Observatory (Fig. 1.5)," Scripps CO2 Program, Scripps Institution of Oceanography, La Jolla, CA. Reprinted with permission.

Chapter 2

Page 53, Table 2.1: "Total Wealth and Per Capita Wealth By Type of Wealth, 2005 (Table 2.1)," from *World Bank: World Development Report*; Page 25, Figure 2.2: Source: Based on World Development Report 1992; Page 33, Figure 2.7: Source: Based on "The Genuine Progress Indicator 2006: A Tool for Sustainable

Chapter 15

Table 15.1: Source: based on data from NRC; Page 375, Figure 15.2: Source: Data from U.S. Department of Energy; Figure 15.4: Source: Data from International Atomic Energy Agency; Figure 15.17: Source: Data from International Atomic Energy Agency.

Chapter 16

Table 16.1: Source: REN21. Global Status Report 2011; Page 403, Figure 16.10: *Renewables Global Status Report, 2009 Update, REn 21,* Secretariat, United Nations Environment Programme; Figure 16.11: National Renewable Energy Laboratory; Figure 16.15: Data from GWEC.

Chapter 17

Page 414: Source: *Red Sky at Morning: America and the Crisis of the Global Environment,* by James Gustave Speth. Published by Yale University Press, 2004; Page 419: Source: Mission Statement of the Centers for Disease Control and Prevention; Page 45–6: Source: "Risk communication: Evolution and Revolution" by Vincent Covello and Peter M. Sandman, from *Solutions to an Environment in Peril,* edited by Anthony Wolbarst. Published by John Hopkins University Press, 2001; Table 17.1: Source: based on data from The Global Burden of Disease: 2008; Table 17.2: Source: based on data from the OFDA/CRED International Disaster Database; Table 17.3: Source: based on "Risk-Benefit Analysis" by R. Wilson and E.A. Crouch; Table 17.4: "Ten Leading Selected Risk Factors as Percentage Causes of Disease Burden, Measured in DALYs," from *World Health Report 2009.* Reprinted with permission; Figure 17.1: "Timeline: Pandemic (H1N1) 2009 laboratory confirmed cases and number of deaths as reported to WHO. Status as of: 15 August 2010," from World Health Organization website. Copyright © 2010 by World Health Organization. Reprinted with permission; Page 427, Figure 17.3: Sources: Data from: "Deaths: Preliminary Data for 2009" by Kenneth D. Kochanek, et al., from *National Vital Statistics Reports,* March 16, 2011, Volume 59(4); Centers for Disease Control; and American Medical Association; Figure 17.4: Source: U.S. Department of Health & Human Services; Figure 17.5: Table from World Health Organization report: Global Tuberculosis Control 2011. Copyright © 2011 by World Health Organization. Reprinted with permission; Figure 434, Figure 17.11: Source: Centers for Disease Control and Prevention; Figure 17.16: "Loss of life expectancy of the Danes related to various risk factors. Loss of life expectancy (years) for men and women," by Krud Juel, Jan Sørensen, and Henrick Brønnum-Hansen, from *Risk Factors and Public Health in Denmark—Summary Report,* National Institute of Public Health, 2007. Copyright © 2007 by National Institute of Public Health, Denmark. Reprinted with permission; Figure 17.17: Source: Environmental Protection Agency.

Chapter 18

Table 18.3: "Climate Change 2007: Synthesis Report. Contribution of Working Groups I, II and III to the Fourth Assessment Report of the Intergovernmental Panel on Climate Change, Table SPM.6." IPCC, Geneva, Switzerland; Figure 18.14a: National Snow and Ice Data Center; Figure 18.14b: National Snow and Ice Data Center; Figure 18.5: The Stern Review Report © Crown Copyright 2006 The Economics of Climate Change: The Stern Review © Cambridge University Press 2007. Source: "Global Land-Ocean Temperature Index" graph from NASA website. Based on Fig 1A from "Global Temperature Change," by James Hansen, from *Proceedings of the National Academy of Sciences,* July 2006, Volume 103. Copyright © 2006 by National Academy of Sciences. Reprinted with permission; Figure 18.6: Source: Data from "Atmospheric carbon dioxide concentrations," NOAA; Figure 18.7a: Adapted from The Niels Bohr Institute Archives; Figure 18.7b & c: *Elemental Geosystems,* 5th Edition, by Robert W. Christopherson. Copyright © 2007 by Pearson Education. Reprinted with permission; Page 456, Figure 18.8: *Our Changing Planet: An Introduction to Earth System Science and Global Environmental Change,* 2nd Edition, by Fred T. Mackenzie. Copyright © 1998 by Pearson Education. Reprinted with permission; Page 460, Figure 18.10: Source: Graph by Robert Simmon from "Correcting Ocean Cooling" by Rebecca Lindesy, from Earth Observatory website, February 2007. National Aeronautics and Space Administration; Figure 18.11: Adapted from "The Rise In Global Mean Sea Level (2011_re14)," from *CU Sea Level Research Group* website. Copyright © 2011 by University of Colorado. Reprinted with permission; Figure 18.12: Source: National Snow and Ice Data Center. Source: National Snow and Ice Data Center; Figure 18.13: Source: NOAA Arctic Report Card: Update for 2011; Figure 18.15: Source: Carbon Dioxide Information Analysis Center; Page 465, Figure 18.16: Source: U.S. Energy Information Administration; Figure 18.17: Source: "The Cooperative Global Air Sampling Network Newsletter" from NOAA Earth System Research Laboratory, Spring 2011; Page 465, Figure 18.18: Source: Carbon Cycling and Biosequestration: Report from the March 2008 Workshop, DOE/SC-108, U.S. Department of Energy Office of Science; Page 467, Figure 18.19: "Climate Change 2007: The Physical Science basis. Working Group I Contribution to the Fourth Assessment Report of the Intergovernmental Panel on Climate Change, Figure SPM.2." Cambridge University Press. Reprinted with permission; Page 467, Figure 18.19: "Climate Change 2007: The Physical Science Basis, Contribution of Working Group I to the Fourth Assessment Report of the Intergovernmental Panel on Climate Change, Figure TS.32." Cambridge University Press. Reprinted with permission; Page 467, Figure 18.19: "Climate Change 2007: The Physical Science Basis, Contribution of Working Group I to the Fourth Assessment Report of the Intergovernmental Panel on Climate Change, Figure TS.32." Cambridge University Press. Reprinted with permission; Page 469, Figure 18.22: "How Urgent Is Climate Change?" by Richard A. Kerr, from *Science,* November 2007, Volume 318(5854). Copyright © 2007 by AAAS. Reprinted with permission; Page 471, Figure 18.23: Adapted from "Water Poverty Index," by C. A. Sullivan, et al., *Report to DFID, 2000,* from Copyright © by Centre for Ecology & Hydrology. Reprinted with permission; Page 473, Figure 18.24: "Climate Change 2007: Synthesis Report. Contribution of Working Groups I, II and III to the Fourth Assessment Report of the Intergovernmental Panel on Climate Change, Figure 5.1," Cambridge University Press. Reprinted with permission; Figure 18.25: "Global Pathways" from Climate Action Tracker website. Copyright © 2011 by Climate Action Tracker Partners. Reprinted with permission.

Chapter 19

Figure 19.3: Source: "National Ambient Air Quality Standards" from Environmental Protection Agency, October 2011; Table 19.4: Source: "Scientific Assessment of Ozone Depletion: 2010," from *Global Ozone Research and Monitoring Project Report,* No. 52, Geneva, Switzerland: World Meteorological Organization, 20; Page 489, Figure 19.9: Source: "National Emissions Inventory (NEI) Air Pollutant Emissions Trends Data" from U.S. Environmental Protection Agency website; Page 490, Figure 19.10: Source: "National Emissions Inventory (NEI) Air Pollutant Emissions Trends Data" from U.S. Environmental Protection Agency website; Page 490, Figure 19.11: Source: "Air Quality Trends" from U.S. Environmental Protection Agency website; Figure 19.15: Source: "NADP Data and Information Use Conditions" from National Atmospheric Deposition Program website; Figure 19.16: Source: "Coal plants without scrubbers account for a majority of U.S. SO2 emissions" from *Today in Energy,* from U.S. Energy Information Agency website; Page 498, Figure 19.20: Source: "Effects of Ozone Air Pollution on Plants" from U.S. Department of Agriculture website; Figure 19.25: Source: Environmental Protection Agency; Figure 19.26: Source: Environmental Protection Agency; Figure 19.27: Source: "United States Geological Survey (USGS): Water Resources of New York" from the USGS website; Page 510, Figure 19.29: Source: Report by the U.S. Climate Change Science Program and the Subcommittee on Global Change Research; Page 513, Figure 19.31: Source: "Ozone Hole Watch" from the NASA website; Page 514, Figure 19.32: Source: "Production and presence of ozone-depleting substances in the atmosphere: past, present, and future," from *Scientific Assessment of Ozone Depletion.* Copyright © 2006 by United Nations. Reprinted with permission. All rights reserved.

Chapter 20

Page 506: NASA. Table 20.2: Source: "Guidance for Water Quality-Based Decisions: The TMDL Process" from Environmental Protection Agency; Figure 20.1: Source: "The Mississippi River/Gulf of Mexico Watershed Nutrient Task Force" from the Environmental Protection Agency website; Figure 20.5: *Progress on Drinking Water and Sanitation, 2012 Update.* Copyright © 2012 by World Health Organization. Reprinted with permission.

Chapter 21

Table 21.1: Source: "Interstate Shipment of Municipal Solid Waste: 2007 Update" by James E. McCarthy, from *CRS Report for Congress,* June 13, 2007; Page 547, Figure 21.2: Source: data from Environmental Protection Agency, Office of Solid Waste, Municipal Solid Waste in the United States: 2010 Facts and Figures; Page 457, Figure 21.3: Source: data from Environmental Protection Agency, Office of Solid Waste, Municipal Solid Waste in the United States: 2010 Facts and Figures; Figure 21.9: Data from EPA, Office of Solid Waste, Municipal Solid Waste in the United States: Non-Hazardous Waste Management Hierarchy; Figure 21.12: Source: data from the Environmental Protection Agency, Municipal Solid Waste Generation, Recycling, and Disposal in the United States: Facts and Figures for 2010.

Chapter 22

Page 569: Source: data from *Environmental Protection Agency, Final Guidance for Incorporating Environmental Justice Concerns in EPA's NEPA Compliance Analyses*; Table 22.1: "The "Dirty Dozen" Persistent Organic Pollutants," from Report on DDT, Aldrin, Dieldrin, Endrin, Chlordane, Heptachlor, Hexachlorobenzene, Mirex, Toxaphene, Polychlorinated Biphenyls, Dioxins, and Furans, International Programme on Chemical Safety. Copyright © 1995 by United Nations. Reprinted with permission. All rights reserved; Figure 22.5: Source: "2010 Toxics Release Inventory" from the Environmental Protection Agency, January 2012; Figure 22.08: "U.S. Environmental Protection Agency Headquarters."

Chapter 23

Page 574: "Map of Atlanta BeltLine," from *Atlanta Beltline 2010 Annual Report.* Copyright © 2010 by Atlanta BeltLine. Reprinted with permission. All rights reserved; Page 591: *The Waterman's Stewardship Covenant* by Susan Drake Emmerich. Copyright © by Dr. Susan D. Emmerich. Reprinted with permission. Table 23.1: Source: U.S. Census Bureau; Table 23.2: "Ten Most Sprawling Metropolitan Regions," by Reid Ewing, Rolf Pendall, and Don Chen, adapted from *Measuring Sprawl and Its Impact.* Copyright © 2012 by Smart Growth America. Reprinted with permission; Page 593: Source: Edmund Burke, Irish orator, philosopher, and politician (1729–1797); Figure 23.7: "Sprawl scores for 83 metropolitan regions," adapted from *Measuring Sprawl and Its Impact.* Copyright © 2012 by Smart Growth America. Reprinted with permission; Figure 23.8: Source: U.S. Department of Agriculture. 2007 National Resources Inventory, Natural Resources Conservation Service, Washington, D.C. and Center for Survey Statistics and Methodology, Iowa State University, Ames, Iowa. July 2009; Page 605, Figure 23.16: "Boston, Massachusetts Map," from Rose Fitzgerald Kennedy Greenway Conservancy. Reprinted with permission.

INDEX